Proterozoic East Gondwana:
Supercontinent Assembly and Breakup

Geological Society Special Publications

Society Book Editors

R. J. PANKHURST (CHIEF EDITOR)

P. DOYLE

F. J. GREGORY

J. S. GRIFFITHS

A. J. HARTLEY

R. E. HOLDSWORTH

A. C. MORTON

N. S. ROBINS

M. S. STOKER

J. P. TURNER

Special Publication reviewing procedures

The Society makes every effort to ensure that the scientific and production quality of its books matches that of its journals. Since 1997, all book proposals have been refereed by specialist reviewers as well as by the Society's Books Editorial Committee. If the referees identify weaknesses in the proposal, these must be addressed before the proposal is accepted.

Once the book is accepted, the Society has a team of Book Editors (listed above) who ensure that the volume editors follow strict guidelines on refereeing and quality control. We insist that individual papers can only be accepted after satisfactory review by two independent referees. The questions on the review forms are similar to those for *Journal of the Geological Society*. The referees' forms and comments must be available to the Society's Book Editors on request.

Although many of the books result from meetings, the editors are expected to commission papers that were not presented at the meeting to ensure that the book provides a balanced coverage of the subject. Being accepted for presentation at the meeting does not guarantee inclusion in the book.

Geological Society Special Publications are included in the ISI Science Citation Index, but they do not have an impact factor, the latter being applicable only to journals.

More information about submitting a proposal and producing a Special Publication can be found on the Society's web site: www.geolsoc.org.uk.

It is recommended that reference to all or part of this book should be made in one of the following ways:

YOSHIDA, M., WINDLEY, B. F. & DASGUPTA, S. (eds) 2003. *Proterozoic East Gondwana: Supercontinent Assembly and Breakup*. Geological Society, London, Special Publications, **206**.

BRAUN, I.. & KRIEGSMAN, L. M. 2003. Proterozoic crustal evolution of southernmost India and Sri Lanka *In*: YOSHIDA, M., WINDLEY, B. F. & DASGUPTA, S. (eds) *Proterozoic East Gondwana: Supercontinent Assembly and Breakup*. Geological Society, London, Special Publications, **206**, 169–202.

GEOLOGICAL SOCIETY SPECIAL PUBLICATION NO. 206

Proterozoic East Gondwana: Supercontinent Assembly and Breakup

EDITED BY

M. YOSHIDA

Gondwana Institute for Geology and Environment, Japan
Institute of Fundamental Studies, Sri Lanka

B. F. WINDLEY

University of Leicester, UK

S. DASGUPTA

Jadavpur University, India

2003

Published by

The Geological Society

London

THE GEOLOGICAL SOCIETY

The Geological Society of London (GSL) was founded in 1807. It is the oldest national geological society in the world and the largest in Europe. It was incorporated under Royal Charter in 1825 and is Registered Charity 210161.

The Society is the UK national learned and professional society for geology with a worldwide Fellowship (FGS) of 9000. The Society has the power to confer Chartered status on suitably qualified Fellows, and about 2000 of the Fellowship carry the title (CGeol). Chartered Geologists may also obtain the equivalent European title, European Geologist (EurGeol). One fifth of the Society's fellowship resides outside the UK. To find out more about the Society, log on to www.geolsoc.org.uk.

The Geological Society Publishing House (Bath, UK) produces the Society's international journals and books, and acts as European distributor for selected publications of the American Association of Petroleum Geologists (AAPG), the American Geological Institute (AGI), the Indonesian Petroleum Association (IPA), the Geological Society of America (GSA), the Society for Sedimentary Geology (SEPM) and the Geologists' Association (GA). Joint marketing agreements ensure that GSL Fellows may purchase these societies' publications at a discount. The Society's online bookshop (accessible from www.geolsoc.org.uk) offers secure book purchasing with your credit or debit card.

To find out about joining the Society and benefiting from substantial discounts on publications of GSL and other societies worldwide, consult www.geolsoc.org.uk, or contact the Fellowship Department at: The Geological Society, Burlington House, Piccadilly, London W1J 0BG: Tel. +44 (0)20 7434 9944; Fax +44 (0)20 7439 8975; Email: enquiries@geolsoc.org.uk.

For information about the Society's meetings, consult *Events* on www.geolsoc.org.uk. To find out more about the Society's Corporate Affiliates Scheme, write to enquiries@geolsoc.org.uk.

Published by The Geological Society from:
The Geological Society Publishing House
Unit 7, Brassmill Enterprise Centre
Brassmill Lane
Bath BA1 3JN, UK
(*Orders*: Tel. +44 (0)1225 445046
Fax +44 (0)1225 442836)

Online bookshop: http://bookshop.geolsoc.org.uk

The publishers make no representation, express or implied, with regard to the accuracy of the information contained in this book and cannot accept any legal responsibility for any errors or omissions that may be made.

© The Geological Society of London 2002. All rights reserved. No reproduction, copy or transmission of this publication may be made without written permission. No paragraph of this publication may be reproduced, copied or transmitted save with the provisions of the Copyright Licensing Agency, 90 Tottenham Court Road, London W1P 9HE. Users registered with the Copyright Clearance Center, 27 Congress Street, Salem, MA 01970, USA: the item-fee code for this publication is 0305–8719/00/$15.00.

British Library Cataloguing in Publication Data
A catalogue record for this book is available from the British Library.

ISBN 1-86239-125-4

Typeset by Servis Filmsetting Limited, UK

Printed by Alden Press, Oxford, UK.

Distributors

USA
AAPG Bookstore
PO Box 979
Tulsa
OK 74101–0979
USA
Orders: Tel. +1 918 584–2555
Fax +1 918 560–2652
E-mail bookstore@aapg.org

India
Affiliated East-West Press PVT Ltd
G-1/16 Ansari Road, Daryaganj,
New Delhi 110 002
India
Orders: Tel. +91 11 327–9113
Fax +91 11 326–0538
E-mail affiliat@nda.vsnl.net.in

Japan
Kanda Book Trading Co.
Cityhouse Tama 204
Tsurumaki 1–3–10
Tama-shi
Tokyo 206–0034
Japan
Orders: Tel. +81 (0)423 57–7650
Fax +81 (0)423 57–7651
E-mail geokanda@ma.kcom.ne.jp

Contents

Preface   vii

Tectonic of Rodinia and Gondwana: continental growth, supercontinent assembly and breakup

CONDIE, K. C. Supercontinents, superplumes and continental growth: the Neoproterozoic record   1

WINDLEY, B. F. Continental growth in the Proterozoic: a global perspective   23

PISAREVSKY, S. A., WINGATE, M. T. D., POWELL, C. MCA., JOHNSON, S. & EVANS, D. A. D. Models of Rodinia assembly and fragmentation   35

YOSHIDA, M., JACOBS, J., SANTOSH, M. & RAJESH, H. M. Role of Pan-African events in the Circum-East Antarctic Orogen of East Gondwana: a critical overview   57

Australia and Gondwanaland

WINGATE, M. T. D. & EVANS, D. A. D. Palaeomagnetic constraints on the Proterozoic tectonic evolution of Australia   77

FITZSIMONS, I. C. W. Proterozoic basement provinces of southern and southwestern Australia, and their correlation with Antarctica   93

South Asia within the Gondwanaland ensemble

DASGUPTA, S. & SENGUPTA, P. Indo-Antarctic Correlation: a perspective from the Eastern Ghats Granulite Belt, India   131

DOBMEIER, C. J. & RAITH, M. M. Crustal architecture and evolution of the Eastern Ghats Belt and adjacent regions of India   145

BRAUN, I. & KRIEGSMAN, L. M. Proterozoic crustal evolution of southernmost India and Sri Lanka   169

Antarctica and its role in the Gonwanaland assembly

HARLEY, S. L. Archaean–Cambrian crustal development of East Antarctica: metamorphic characteristcs and tectonic implications   203

ZHAO, Y,. LIU, X. H., LIU, X. C. & SONG, B. Pan-African events in Prydz Bay, East Antarctica and their implications for East Gondwana tectonics   231

BAUER, W., THOMAS, R. J. & JACOBS, J. Proterozoic–Cambrian history of Dronning Maud Land in the context of Gondwana assembly   247

JACOBS, J., KLEMD, R., FANNING, C. M., BAUER, W. & COLOMBO, F. Extensional collapse of the late Neoproterozoic–Early Paleozoic East African–Antarctic Orogen in central Dronning Maud Land, East Antarctica   271

The East African Oragen

JOHNSON, P. R. & WOLDEHAIMANOT, B. Development of the Arabian-Nubian Shield: perspectives on accretion and deformation in the northern East African Orogen and the assembly of Gondwana   289

KUSKY, T. M. & MATSAH, M. I. Neoproterozoic dextral faulting on the Najd Fault System, Saudi Arabia, preceded sinistral faulting and escape tectonics related to closure of the Mozambique Ocean   327

COLLINS, A. S., JOHNSON, S. FITZSIMONS, I. C. W., POWELL, C. MCA., HULSCHER, B., ABELLO, J. & RAZAKAMANANA, T. Neoproterozoic deformation in central Madagascar: a structural section through part of the East African Orogen   363

FERNANDEZ, A. & SCHREURS, G. Tectonic evolution of the Proterozoic Itremo Group metasediments in central Madagascar   381

GRANTHAM, G. H., MABOKO, M. & EGLINTON, B. M. A review of the evolution of the Mozambique Belt and implications for the amalgamation and dispersal of Rodinia and Gondwana   401

HANSON, R. E.: Proterozoic geochronology and tectonic evolution of southern Africa   427

Preface

Supercontinent assembly and breakup has been an important topic since Wagener's discovery of Pangaea in the early twentieth century and has been recognized as an important process of the Wilson Cycle since the late 1960s. The separate proposals of a Mesoproterozoic Rodinia supercontinent by Dalziel (1991) and Hofmann (1991) concentrated the attention of many geoscientists on this topic.

The UNESCO–IUGS–IGCP-288 project 'Assembly of Gondwanaland' (1990–1996) focused on the subject of Rodinia to Gondwanaland, and was followed by IGCP-368 'Proterozoic Events in East Gondwana' (1995–2001) and by IGCP-440 'Rodinia Assembly and Breakup' (1999–2003). At the 31st IGC in Rio de Janeiro in October 2000, IGCP-368 organized a general session entitled 'Proterozoic Events in East Gondwana' to which many contributions came from both IGCP-368 and IGCP-440. This session provided the impetus in collating the present volume.

East Gondwana was traditionally thought to have formed as a major part of Rodinia during the Mesoproterozoic Grenvillian–Circum-East Antarctic Orogeny (Yoshida 1995; Unrug 1997) and survived until the Middle–Late Mesozoic when Pangaea broke up. An ice-covered Antarctica is the key component of this long-lived subsupercontinent. West Gondwana, on the other hand, assembled during the Neoproterozoic and collided with pre-existing East Gondwana at this time.

However, the accumulation of Pan-African zircon ages, mostly from Antarctica since the early 1990s, coupled with an increase in the number of reliable palaeomagnetic data from various parts of the globe, has resulted in a re-evaluation of the above classical model, creating a radical new model. According to this new model, East Gondwana did not exist during the Neoproterozoic – along with West Gondwana it was assembled during the Pan-African Orogeny and, accordingly, the whole of Gondwanaland was amalgamated at this time (e.g. Meert & Powell 2001; Powell et al. 2001). Both models still command strong support and further data are required to constrain their future viability, although the new model is becoming increasingly popular.

The present volume assembles papers on Grenvillian–Circum-East Antarctic and Pan-African events in various parts of East Gondwana (Fig.1), and presents a comprehensive review of related areas and topics. Although all papers give balanced reviews related to their topics, some stand more or less on the classical model, while others support the new model. This reflects the present debate on this subject.

The volume deals with five topics – one general and four regional. The general papers address global issues on crustal–mantle processes in Proterozoic to Early Palaeozoic times. The regional papers include comprehensive time–space–event diagrams to help provide a general overview of Late Proterozoic–Early Palaeozoic geology of the regions concerned.

Tectonics of Rodinia and Gondwana: continental growth, supercontinent assembly and breakup

Among four papers under this topic, two papers are concerned with continental growth and mantle plume activity. **Condie** discusses the role of superplumes in relation to the rate of crust formation, and concludes that a superplume event was absent during the assembly and breakup of supercontinents during Neoproterozoic time. The author suggests a possibility that the absence of the superplume activity might reflect incomplete breakup of a Palaeoproterozoic supercontinent, which was followed by the formation of Rodinia. **Windley** presents three tectonic environments for the growth of continents, namely accretionary orogens, in which juvenile material is added to pre-existing continental blocks, collisional orogens, and rifts in supercontinents. In addition to plate tectonics, mantle plume tectonics is emphasized as a major contributor to the growth of the continents.

In the following two papers, one stresses the new model for Gondwanaland assembly, whereas the second points out some shortcomings. **Pisarevsky et al.** synthesized recent palaeomagnetic and geological data, and produced a thoroughly new model of Rodinia assembly and breakup, which they propose as a working hypothesis for future work. One of their important proposals divides East Gondwana into three dispersed blocks. Their model was encouraged by the radical proposal of Fitzsimons (2000) that the Late Mesoproterozoic terrains fringing East Antarctica could be composed of three different blocks separated by two Pan-African orogenic belts. **Yoshida et al.** present a balanced overview of the Grenvillian–Circum-East Antarctic Orogen and of the Pan-African Orogen surrounding East Antarctica. They conclude that present data are insufficient to replace the classical model with the new one, and that both models require further examination and constraints. They specifically point out the importance of careful examination of geochronological data in both geological and palaeomagnetic studies, special attention being given to the lower temperature resetting of U–Pb zircon as well as Sm–Nd garnet ages.

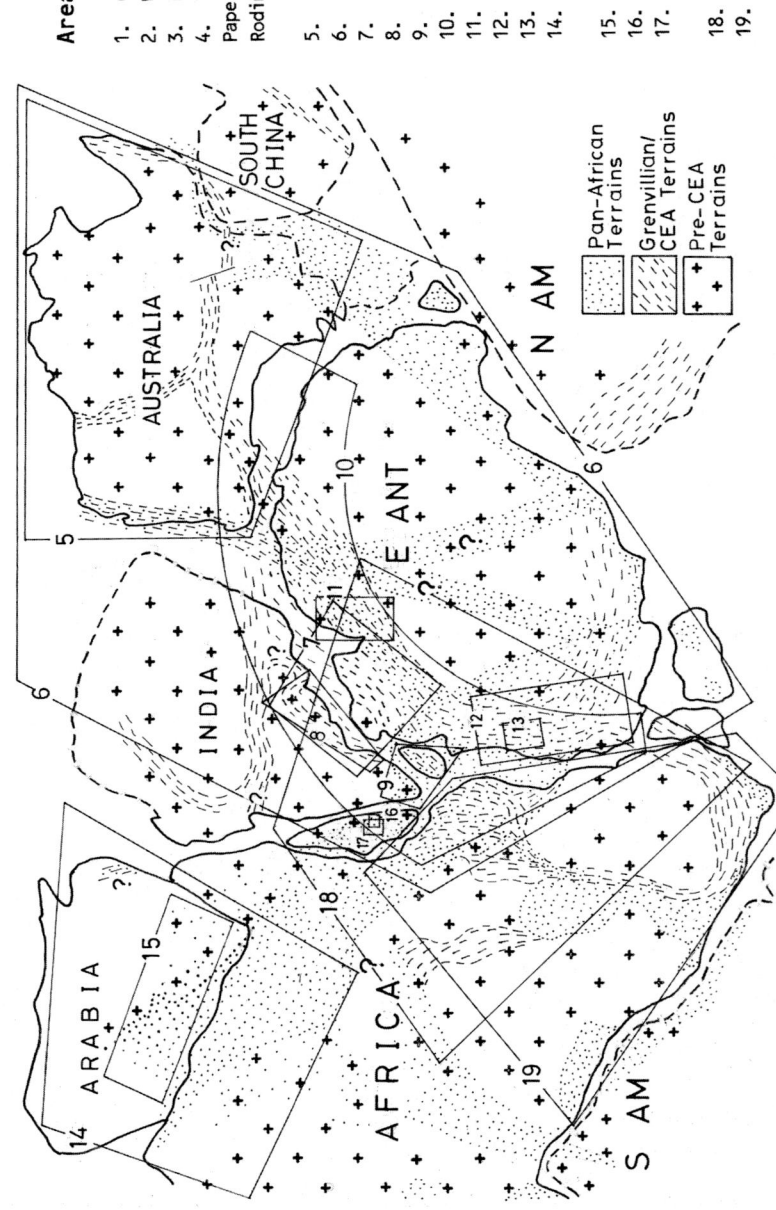

Fig. 1. Pan-African and Circum-East Antarctic (Grenvillian) terrains in East Gondwana, during ca. 1000/Ma–500/Ma. Summarized from papers in the present volume by Bauer et al. Fitzsimons et al., Kusky & Matsah and Yoshida et al. Yoshida (1995) and Unrug (1997) were also used as a base map. Broken outline of crustal blocks indicates uncertainty. Areas of study covered by papers in the present volume are also indicated. CEA: Circum-East Antarctic, N AM: North America, S AM: South America.

Australia and Gondwanaland

This section includes two papers, one with a palaeomagnetic theme and one with a geotectonic theme. **Wingate & Evans** have assembled reliable palaeomagnetic data that overlap palaeopoles from 1.7 to 1.8 Ga and 1.5 to 1.6 Ga. They conclude that the North and West Australian Cratons have occupied their present relative positions since at least c. 1.7 Ga, and that they have been joined to the South Australian Craton since at least c. 1.5 Ga, although further data are required to examine the width of the oceans between the continental blocks. **Fitzsimons** gives a comprehensive review of Proterozoic southern and western Australia and of their correlations with Antarctica. The Australia–Antarctic sector in East Gondwana is divided into the Archaean–Palaeoproterozoic Mawson Craton, the Mesoproterozoic Albany–Fraser Orogen, and the Pan-African Pinjarra Orogen. The last orogen includes Late Mesoproterozoic allochthonous blocks, which divide East Gondwana into Australo-Antarctic and Indo-Antarctic domains. The Pinjarra Orogen extends further south into ice-covered, inland Antarctica. This paper will surely encourage geoscientists to improve the new model of Gondwanaland assembly during Pan-African time.

South Asia within the Gondwanaland ensemble

All three papers in this section provide geochronological and petrological constraints on high-grade rocks occurring in India and Sri Lanka and explore different models of correlation within the framework of Rodinia. **Dasgupta & Sengupta** review the tectonothermal history of the Eastern Ghats Belt in the light of recently published isotopic data that established different geochronologic provinces in this belt, and discuss its significance in the context of Indo-Antarctic correlation. **Dobmeier & Raith** present a new provocative idea of subdivision of the crustal architecture of the Eastern Ghats Belt in eastern and southern India, and consider low-grade schist belts bordering the high-grade mobile belt as integral parts of evolution of this terrane. **Braun & Kriegsman** present an updated review of the high-grade terranes of southern India and Sri Lanka, which provides stronger support for correlation of these terranes with the Lutzow-Holm Bay area of East Antarctica.

Antarctica and its role in the Gondwanaland assembly

This section includes four papers that highlight the importance of the Pan-African orogeny in the assembly of Gondwanaland. **Harley** presents an exhaustive, updated review of history of different crustal provinces within the East Antarctic Shield. He concludes that these provinces have records of different isotopic events and were amalgamated during the Pan-African Orogeny characterized by high- to very-high-grade metamorphism. This questions the concept of an older model of a continuous Grenvillian province (e.g. Yoshida 1995) that supported the SWEAT hypothesis (Moores 1991). **Zhao et al.** promote an accretionary model and amalgamation of different blocks during the Pan-African Orogeny to explain the evolution of the Prydz Bay region of East Antarctica, discarding the traditional model of polyorogenic history (Harley & Fitzsimons 1995). This paper has major implications for the new models of configuration of Rodinia and East Gondwana, and strongly supports Fitzsimons' (2000) suggestion mentioned above. **Bauer et al.** deal with the Proterozoic–Cambrian history of both the central and western Dronning Maud Land, and present evidence of Mesoproterozoic accretionary history of the orogens to the Archaean craton. They also emphasize the strong Pan-African metamorphism and tectonism in these sectors, leading to the development of the East Antarctic Orogen (Jacobs et al. 1999) as a continuation of the East African Orogen (Stern 1994), resulting in the assembly of Gondwanaland.

Jacobs et al. characterize a 530–510 Ma Late Pan-African extensional event in the central Dronning Maud Land through interpretation of structural, petrological and isotopic data, and compare this with similar events in Madagascar and the Arabian–Nubian Shield, indicating that this event is a reflection of Pan-African collisional tectonics.

The East African Orogen

The papers in this section include two on the Arabian–Nubian Shield, two on Madagascar and two on eastern and southern Africa.

Johnson and Woldehaimanot produce the most detailed synthesis yet of the Arabian–Nubian Shield, which forms the suture between East and West Gondwana at the northern end of the East African Orogen. Subduction started at 870 Ma in the Mozambique Ocean, with arc–arc convergence and terrane suturing at 780 Ma marking the start of ocean closure and Gondwana assembly. Terrane amalgamation continued until 600 Ma, resulting in juxtaposition of East and West Gondwana, with final assembly of Gondwana being achieved by 550 Ma.

Kusky & Matsah report that dextral offset up to 10 km on one of the major faults belonging to the Najd Fault System has a maximum age of 625 ± 4.2 Ma, which provides the earliest age for the collision of East and West Gondwana. These dextral movements later switched to sinistral, when accreted terranes caught between the two continents were transported towards an oceanic margin to the north. These results provide important constraints on the terminal history of the Mozambique Ocean.

The two papers on Madagascar are concerned with the central part of that island. **Collins et al.** give a detailed structural section across the upper crustal metasedimentary Itremo Group and eastwards through the underlying high-grade gneissic mid-crustal Antananarivo Block. They consider that Gondwana accretion in this part of the East African Orogen occurred between 720 and 570 Ma. After contractional deformation the orogen collapsed, producing an extensional shear zone between the Itremo Group and the underlying gneissic block. **Fernandez & Schreurs** present a structural-based study of the tectonic evolution of the metasedimentary Itremo Group. This paper is highly controversial in comparison with the results of several other research groups studying this part of central Madagascar.

Grantham et al. summarize and review the evolution of the Mozambique orogenic belt and its extensions in Antarctica, Sri Lanka, India and Mozambique. They conclude that amalgamation of East and West Gondwana between 600 and 460 Ma occurred in a continent-scale transpressional setting during closure of the Mozambique Ocean.

Hanson presents a detailed synthesis of the Proterozoic orogenic belts on the present eastern, western and northern margins of southern Africa south of the equator. He concludes that Rodinia was assembled at 1.0 Ga and broke-up at 920–700 Ma, with rifting and within-plate magmatism into crustal blocks that amalgamated into Gondwana at 570–510 Ma, along with formation of the collisional Mozambique and Kaoko–Gariep–Saldania orogenic belts. The Damara–Lufilian–Zambezi Orogen developed largely by closure of linked, narrow ocean basins.

In closing this introduction, we recall with heartfelt gratitude the late Raphael Unrug, co-leader of IGCP-288 and 440, and Chris McA Powell, secretary of IGCP-368 and co-leader of IGCP-440. The successful activities of IGCP-368 and IGCP-440, which are reflected in the present volume, owe a great deal to their valuable collaboration and encouragement. Powell initially joined the co-editors of the present volume, and contributed much in formulating ideas on its make-up. The six years of fruitful activity of IGCP-368 were funded by UNESCO and IUGS, as well as by several Grants-in-Aid for Scientific Research of the Japan Ministry of Education, Science, Sports and Culture, to which we express our thanks.

The following reviewers kindly shared their valuable time in reviewing manuscripts submitted to the volume, and in so doing immensely improved its value: A. Bhattacharya, W. Bauer, P. Betts, J. Chiarenzelli, A. S. Collins, K. C. Condie, R. Cox, P. Dirks, I. C. W. Fitzsimons, G. Grantham, S. L. Harley, B. Hensen, J. Jacobs, B-M. Jahn, S. Johnson, N. Kelly, A. Kröner, T.M. Kusky, M. W. McElhinny, S. Maruyama, A. Nédélec, Y. Ohta, W. Preiss, M. Raith, T. D. Raub, U. Ring, M. Santosh, P. Sengupta, S. Sengupta, K. Shiraishi, R. J. Stern, B. C. Storey, T. H. Torsvik, P. Treloar and M. Whitehouse.

References

DALZIEL, I. W. D. 1991. Pacific margins of Laurentia and East Antarctica–Australia as a conjugate rift pair: evidence and implications for an Eocambrian supercontinent. *Geology*, **19**, 598–601.

FITZSIMONS, I. C. W. 2000. Grenville age basement provinces in East Antarctica: evidence for three separate collisional orogens. *Geology*, **28**, 879–882.

HARLEY, S. L. & FITZSIMONS, I. C. W. 1995. High-grade metamorphism and deformation in the Prydz Bay region, East Antarctica: terrains, events and regional correlations. *In*: YOSHIDA, M. & SANTOSH, M. (eds) *Indiana and Antarctica during the Precambrian*. Geological Society of India, Memoir, **34**, 73–100.

HOFFMAN, P. 1991. Did the breakup out of Laurentia turn Gondwana inside out? *Science*, **252**, 1409–1412.

JACOBS, J., HANSEN, B. T., HENJES-KUNST, F., ET AL. 1999. New age constraints on the Proterozoic/Lower Palaeozoic evolution of Heimefrontfjella, East Antarctica, and its bearing on Rodinia/Gondwana correlation. *Terra Antartica*, **6**, 377–389.

MEERT, J. G. & POWELL, C.MCA. 2001. Assembly and break-up of Rodinia: introduction to the special volume. *In*: POWELL, C.MCA. & MEERT, J. (eds) *Assembly and Breakup of Rodinia*. *Precambrian Research*, Special Issue, **110**, 1–8.

MOORES, E. M. 1991. Southwest US–East Antarctica (SEWEAT) connection: a hypothesis. *Geology*, **19**, 425–428.

POWELL, C.MCA., PISAREVSKY, S. & WINGATE, M. T. D. 2001. New shape for Rodinia. *Gondwana Research*, **4**, 736–737.

STERN, R. J. 1994. Arc assembly and continental collision in the Neoproterozoic east African Orogen: implications for the consolidation of Gondwanaland. *Annual Review of Earth and Planetary Science*, **22**, 319–351.

UNRUG, R. 1997. Rodinia to Gondwana; the geodynamic map of Gondwana supercontinent assembly. *GSA Today*, **7**, 1–6.

YOSHIDA, M. 1995. Assembly of East Gondwana during the Mesoproterozoic and its rejuvenation during the Pan-African period. *In*: YOSHIDA, M. & SANTOSH, M. (eds) *India and Antarctica during the Precambrian*. Geological Society of India, Memoir, **34**, 25–45.

Masaru Yoshida
Brian F. Windley
Somnath Dasgupta

Supercontinents, superplumes and continental growth: the Neoproterozoic record

KENT C. CONDIE

Department of Earth and Environmental Science, New Mexico Institute of Mining & Technology, Socorro, NM 87801, USA

Abstract: Between 1300 and 500 Ma the Neoproterozoic supercontinent Rodinia aggregated (1300–950 Ma), broke up (850–600 Ma) and a new supercontinent, Pannotia–Gondwana, formed (680–550 Ma). Only $c.$ 11% of the preserved continental crust was produced during this 800 Ma time interval and most of this crust formed as arcs, chiefly continental margin arcs. At least 50% of juvenile continental crust produced between 750 and 550 Ma is in the Arabian–Nubian Shield and in other terranes that formed along the northern border of Amazonia and West Africa. An additional 20% occurs in Pan-African orogens within Amazonia, and $c.$ 16% in the Adamastor and West African orogens. The growth rate of continental crust between 1350 and 500 Ma was similar or less than the average rate of continental growth during the Phanerozoic of 1 km^3/a, and this low rate characterizes both formation and breakup stages of the supercontinents.

The low rates of continental growth during the Neoproterozoic may be due to the absence of a superplume event associated with either Rodinia or Pannotia–Gondwana. If supercontinent breakup is required to produce a superplume event, perhaps by initiating catastrophic collapse of lithospheric slabs at the 660 km seismic discontinuity, the absence of a Meso-proterozoic–Neoproterozoic superplume event may mean that a Palaeoproterozoic supercontinent did not fully breakup prior to aggregation of Rodinia.

Although it is now recognized that continental crust has grown rapidly at several periods in the geologic past, the relationship of production rate of continental crust to the supercontinent cycle is not well understood (Condie 1998, 2000). Is there a change in the growth rate of continental crust during the supercontinent cycle? For instance, do continents grow more rapidly during the formation stage of a supercontinent than they do during the breakup stage? During the aggregation of supercontinents, juvenile crust, such as arcs, oceanic plateaus and ophiolites, can be trapped in collisional orogens, thus adding mass to the continents. However, during the breakup stage of supercontinents, oceanic crust forms more rapidly in response to new ocean ridges, and increased mantle plume activity may also produce more oceanic plateaus and flood basalts. This leads to the question of where and in what tectonic setting does most continental crust originate. Although an unknown but probably small amount of continental crust is added by underplating of mafic magmas from plume sources, it would appear that the bulk of continental growth occurs when juvenile crust is trapped in collisional orogens between colliding cratons, or in peripheral orogens, as oceanic terranes collide with continental margins. Are periods of rapid continental growth associated with superplume events in the last 1 Ga as suggested by Condie (1998, 2000) for earlier periods, and if so, what, if any, relationship exists between superplume events and supercontinents?

Using the Neoproterozoic–Early Phanerozoic supercontinents Rodinia and Gondwana as examples, in this chapter questions related to the growth of continental crust and the relationship, if any, between crustal growth and the supercontinent cycle will be addressed. The relationship of continental growth to the early rifting and breakup stages of Rodinia (1100–750 Ma) and to the later collisional events that led to formation of the younger supercontinents, Gondwana and Pannotia (750–550 Ma), will also be examined. First of all, it is important to review the configuration of plates in Rodinia, discuss uncertainties in plate location and summarize the tectonic history of Neoproterozoic supercontinents. The Sr isotopic record of seawater during the Neoproterozoic as a proxy for supercontinent evolution will also be reviewed.

Rodinia configuration

Although the configuration of plates in Rodinia is becoming better constrained with more precise isotopic ages, the connection between East Gondwana (Antarctica, Australia, India) and Laurentia is still uncertain, as is the position of many of the smaller plates such as North China and Malaysia. The SW US–East Antarctic (SWEAT) reconstruction of East Gondwana proposed by Moores (1991) and Dalziel (1991) places Antarctica adjacent to southwestern Laurentia and Australia adjacent to western Canada. Karlstrom *et al.* (1999, 2001) and Burrett & Berry (2000) suggested that Australia satisfies more

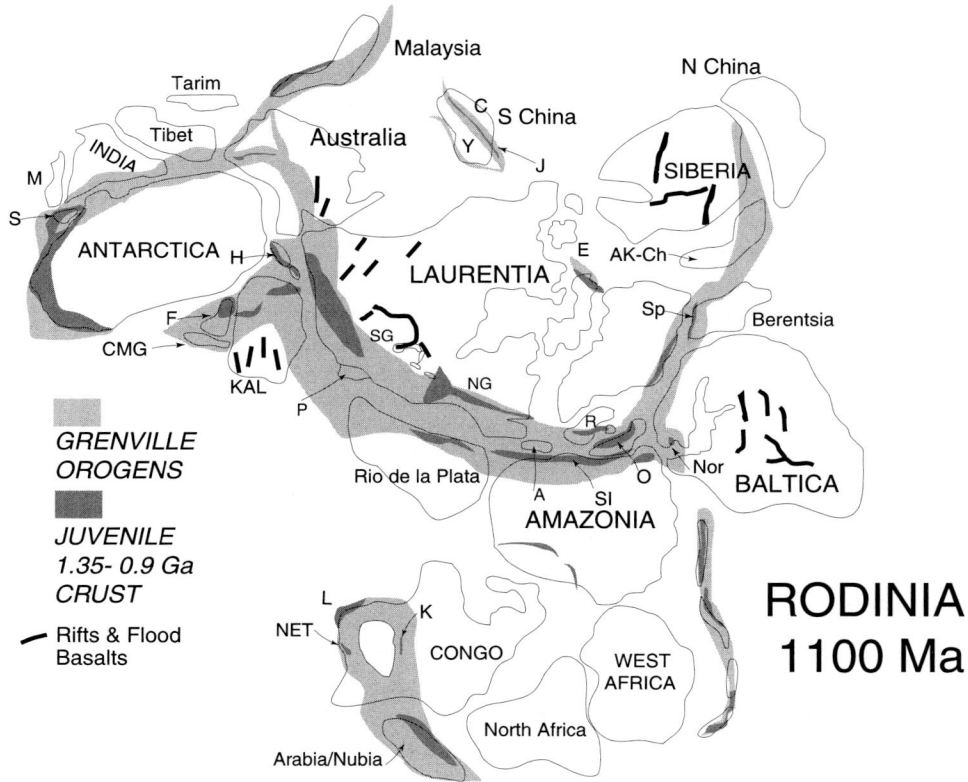

Fig. 1. Diagrammatic reconstruction of the Neoproterozoic supercontinent Rodinia showing the distribution of Grenvillian collisional orogens and 1350–900 Ma juvenile crust. Reconstruction after Hoffman (1991), Li & Powell (1995), Karlstrom *et al.* (1999) and Condie (2001*a*). Symbols: A, Arequipa; P, Precordillera; AK–Ch, Arctic Alaska–Chukotka; CMG, Coats Land–Maudheim–Grunehogna (East Antarctica); KAL, Kalahari Craton; M, Madagascar; S, Sri Lanka; Y, Yangtze Block; C, Cathaysia Block; Fk, Falkland Plateau; H, Haag Nunataks and parts of the Transantarctic Mountains; SG and NG, southern and northern Grenville Province, respectively; E, northern Ellesmere Island; R, Rockall; O, Oaxaquia; SI, San Ignacio Orogen; L, Lurio Belt; Sp, Svalbard; J, Jinning Orogen; K, Katangan Orogen; Nor, Norway; NET, NE Tanzania. Malaysia block includes Cambodia, Thailand and Vietnam.

tectonic constraints and palaeopole positions if it placed near southwestern Laurentia (AUSWUS), a fit which is tentatively adopted and illustrated in Figure 1 (all directions are relative to the present continental positions). AUSWUS matches the Grenvillian Orogen in southern Laurentia with the Arunta–Fraser–Albany Orogen of similar age in central and southwestern Australia. It also places the Yavapai–Mazatzal Palaeoproterozoic Provinces in southern Laurentia adjacent to the Broken Hill and Mount Isa terranes in eastern Australia, which have similar isotopic ages. AUSWUS also provides a somewhat better fit of palaeomagnetic data than SWEAT (Karlstrom *et al.* 2001). Palaeomagnetic data suggest that the South China block was located near Australia, with possible positions both NW and NE of Australia (Evans *et al.* 2000): the latter position is adopted in Figure 1.

Although most investigators agree that Siberia was connected to northern Laurentia during the Neoproterozoic, there is little agreement on the geometry of the fit. At least three configurations have been proposed using various piercing points, matching of alleged conjugate basins and palaeomagnetic results (Hoffman 1987; Condie & Rosen 1994; Pelechaty 1996; Frost *et al.* 1998). Still another interpretation places Siberia adjacent to the west coast of Laurentia (Sears & Price 2000). Although the problem of where Siberia and Laurentia were connected is far from being solved, the piercing points presented by Frost *et al.* (1998) are convincing, and their configuration is tentatively adopted in reconstructions given in this chapter. Contributing to the problem of the Siberia–Laurentia connection is the uncertainty of just when Siberia was rifted from Laurentia. Evidence has been cited to support both

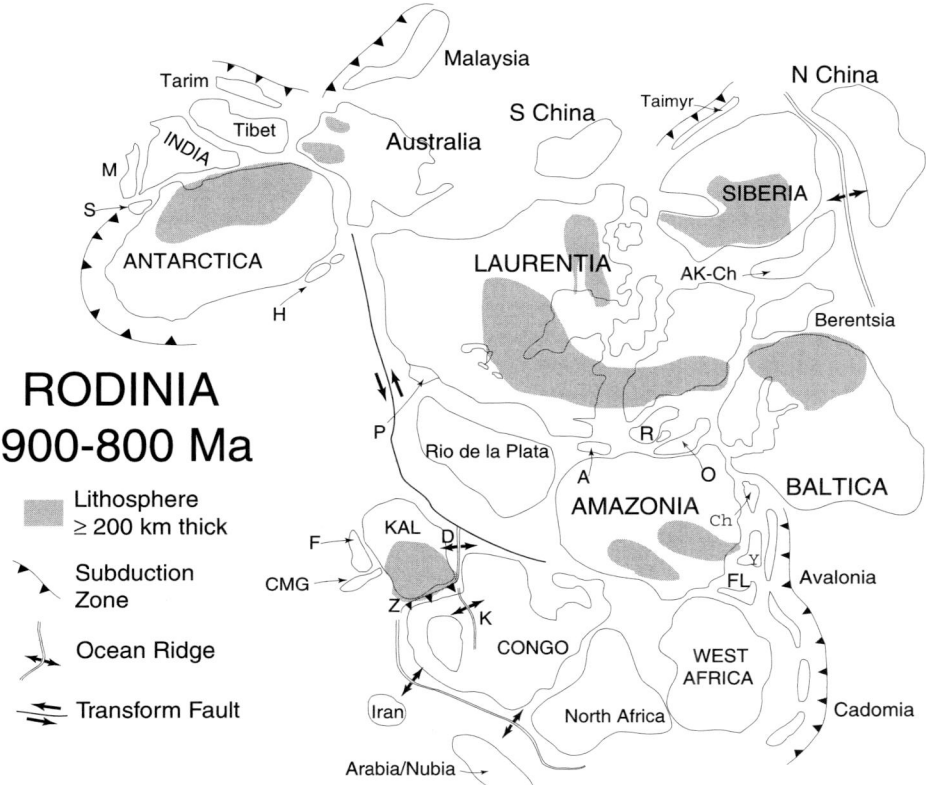

Fig. 2. Diagrammatic reconstruction of Rodinia at 900–800 Ma showing possible distribution of arc systems, rifts and flood basalts, ocean ridges, major transform faults, and thick Archean lithosphere. D, Damaran Orogen; Z, Zambezi Orogen; Ch, Chortis; Y, Yucatan. Other symbols defined in Figure 1.

Neoproterozoic and Early Cambrian rifting, but only Neoproterozoic rifting is acceptable in terms of palaeomagnetic data, the latter of which show that Laurentia and Siberia were not connected in the Early Cambrian (Kirschvink et al. 1997). If the Franklin Dyke swarms in northern Canada are related to similar aged dykes in northern Siberia, rifting of Siberia from Laurentia may have occurred at 720–700 Ma (Condie & Rosen 1994). Plate reconstructions in the Arctic region based on U–Pb zircon ages suggest that northern Alaska and Chukotka in eastern Siberia were part of the same plate during the Neoproterozoic. A striking similarity of pre-Caledonian magmatism in eastern Laurentia, Baltica and Svalbard to that in Arctic Alaska–Chukotka suggests that the latter plate was located between Siberia and Berentsia (Fig. 1; Patrick & McClellan 1995), although the exact location is unknown. Palaeomagnetic data suggest that Taimyr was located near Siberia in the Neoproterozoic; Neoproterozoic rock assemblages suggest arc affinities (Fig. 2; Vernikovskiy et al. 1998). The location of North China in Rodinia continues to be problematic. Similarities of Changcheng and Jixian Proterozoic sediments in North China to Riphean 1 and 2 sediments in Siberia are consistent with a close connection of Siberia and North China (Li & Powell 1995). Also, correlation of Late Neoproterozoic sediments and palaeomagnetic data support this interpretation (Figs 1 & 2; Halls et al. 2000).

Most investigators agree that Baltica, Amazonia and Rio de la Plata surrounded the northeastern and eastern margins of Laurentia, with bits and pieces of small plates (such as Rockall and Arequipa) tucked in between (Fig. 1). Nd and Pb isotope data suggest that Oaxaquia, today comprising much of the basement beneath Mexico, was located in NW Amazonia until at least 1100 Ma (Lopez et al. 2001). Palaeomagnetic, palaeontologic, and U–Pb zircon ages indicate that numerous terranes now found in the Appalachian–Caledonian Orogen in eastern Laurentia (such as Avalonia, Florida and Carolina), and in the Variscan and Alpine Orogens in central and southern Europe (such as Cadomia, Amorica and Iberia), began life along the northern margins of Amazonia and West Africa (Liegeois et al. 1996; Erdtmann 1998; Murphy et al. 1999). Recent palaeomagnetic data suggest that the Kalahari Plate, rather than being connected to East Antarctica as preferred by some investigators (Hoffman 1987), was located south of southern Laurentia in the Neoproterozoic. Dalziel et al. (2000) have recently proposed that it collided with southern Laurentia at c. 1100 Ma, producing the Grenvillian deformation in this area.

Although the precise configurations are unknown, cratons in Congo, North Africa (Nile Craton) and West Africa appear to have been located east of Amazonia in the Neoproterozoic (Fig. 1). As with the India–Malaysia side of East Gondwana, subduction zones surrounded part of West Gondwana, and in particular were active in what is today the Arabian–Nubian Shield. By 800 Ma, subduction had also developed adjacent to West Africa and Amazonia, producing arc systems (Fig. 2), remnants of which are the terranes that became part of eastern Laurentia and southern Europe. The positioning of terranes such as Avalonia, Cadomia and Florida adjacent to Amazonia and West Africa is supported both by palaeomagnetic results and by U–Pb zircon ages (Opdyke et al. 1987; Nance et al. 1991; Keppie & Ramos 1999; Murphy et al. 1999).

Tectonic history of Rodinia

Early Rodinian rifting (1100–900 Ma)

The first rifting in Rodinia began some 300 Ma before the supercontinent actually began to fragment. This is recorded by widespread extensional deformation beginning at c. 1100 Ma, with examples such as the Mid-continent Rift System (1108–1086 Ma) in central Laurentia (Cannon 1994; Davis & Green 1997; Timmons et al. 2001) and the Umkondo rifts in Kalahari (1100 Ma) (Fig. 1; Hanson et al. 1998). Similar, but not as well-dated, rifting occurred in Siberia and Baltica. Timmons et al. (2001) have recently recognized two rifting events in the Neoproterozoic of southwestern Laurentia, the older of which at 1100 Ma has a northwesterly trend reflecting NE–SW extension, perhaps related to the Grenville collisions (Fig. 1). These structures may correlate with similar aged structures in central Australia, lending support to the AUSWUS Rodinia reconstruction. The 1100 Ma rifting was accompanied by eruptions of flood basalts, probably associated with mantle plumes beneath the growing supercontinent. An unknown but relatively small amount of mafic juvenile crust may have underplated the continents at this time. It is noteworthy that this early rifting followed Grenvillian collisional events, which occurred between 1250 and 1190 Ma along the margin of eastern Laurentia (Rivers 1997), at 1150–1120 Ma (continuing to 980 Ma) in southern Laurentia and in the Namaqua Orogen in Kalahari (Figs 1 & 2; Dalziel et al. 2000; Condie 2001a; Knoper et al. 2001). Also recording extension are dyke swarms intruded at 1100 Ma and again at 800–750 Ma in southwestern Laurentia and in Kalahari at 1100 Ma. The 800–750 Ma dykes are also recognized in Australia (Park et al. 1995). The 20–100 Ma time interval between the 1240–1120 Ma Rodinia collisions and the first rifting may have been the time needed for a large mantle upwelling to develop beneath the new supercontinent.

Renewed collision along the Grenvillian orogens occurred between eastern Laurentia and Amazonia–Rio de la Plata at 1080–1020 Ma, and between Kalahari and southern Laurentia at 1060–980 Ma (Rivers 1997; Dalziel et al. 2000). During these collisions, the Mid-continent and related Laurentian rifts were partly closed by thrust faulting related to the collisions (Cannon 1994). Following the Grenvillian collisions in Rodinia, intracratonic rifting continued between 900 and 750 Ma, when rift basins in southwestern Laurentia (Uinta, Chuar, Pahrump) and in Siberia opened. Unlike the earlier rifting event in southwestern Laurentia, the rifts in this event (800–740 Ma) have a dominantly northern trend reflecting E–W extension, consistent with the rifting of East Gondwana from the west coast of Laurentia at this time (Timmons et al. 2001). The Uinta Rift in northern Utah, with an E–W trend, is a notable exception to the N–S trends. Aborted rifting also occurred in southern Australia between 830 and 800 Ma as various rifts in the Adelaide Basin opened (Preiss 2000). In what is now NE Africa and Arabia, between 870 and 700 Ma oceanic arcs and back-arc basins collided to form superterranes, and these later collided to form the Arabian–Nubian Shield (Abdelsalam & Stern 1996). Studies of orogenic terranes in East Antarctica suggest that an arc system existed in this area from c. 1150 to 1070 Ma, perhaps to 900 Ma (Jacobs et al. 1998) (Fig. 2).

It is important to note that none of the intracratonic rifting before c. 900 Ma actually fragmented the supercontinent. Dalziel et al. (2000) suggest that the pre-900 Ma rifts in Laurentia and Kalahari may have been impactogens related to the Grenville collisions, and as such they would not be expected to fragment the supercontinent. Arguing against an impactogen origin, however, is the fact that most of the rifting (at 1100 Ma) did not coincide with Grenvillian collisions, which occurred before 1120 Ma or after 1100 Ma. Even more important is the large amount of basaltic magma associated with the Mid-continent rift, as well as other 1100 Ma rifts,

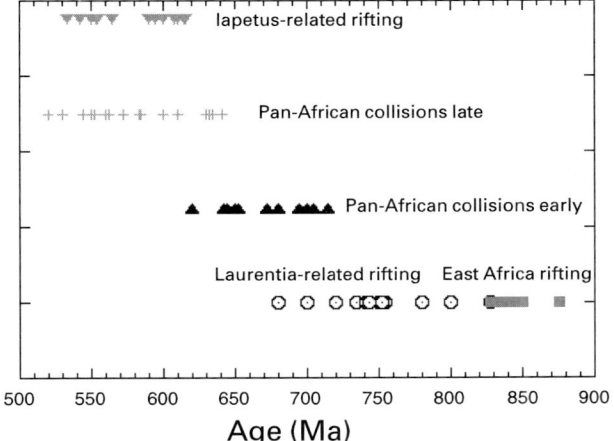

Fig. 3. Summary of the timing of rifting and collisional events in Rodinia between 900 and 500 Ma; references given in Table 1.

which favours a mantle-plume origin rather than an impactogen origin for these rifts.

A puzzling question is why none of the 1100 Ma rifts fragmented the supercontinent Rodinia, but rather they became inactive prior to major supercontinent breakup at 850–750 Ma. As described below, the major fragmentation of Rodinia occurred along peripheral rift systems around to the margins. Perhaps the thick lithosphere beneath Archean cratons in the central part of Rodinia gave additional strength to the lithosphere (Fig. 2). Thus, the rifts that eventually broke up the supercontinent were those that developed in thinner lithosphere closer to the margins of the supercontinent. Also, as pointed out by Krabbendam & Barr (2000), in large orogens such as the Grenville, collision may result in partial delamination and removal of the mantle lithosphere, thus weakening the lithosphere and making it more susceptible to later rifting.

Breakup of Rodinia (850–600 Ma)

The earliest fragmentation of Rodinia began in the East African and Katangan Orogens 900–850 Ma (Figs 2 & 3) (Stern 1994). If Kalahari was part of Laurentia at 1000 Ma (Dalziel *et al.* 2000), rifting of Kalahari from Laurentia must have begun at *c.* 900 Ma and involved largely transform faulting (Fig. 2). As recorded in the Zambezi Orogen, Kalahari collided with the Congo Craton at *c.* 820 Ma, which can be considered as the first major collision in the formation of Gondwana (Hanson *et al.* 1994). At 842 Ma, part of the Arabian–Nubian Shield was rifted from the North African Craton, perhaps in response to a mantle plume that produced an oceanic plateau, a remnant of which is preserved as the Baish Group in the western Arabian Shield (Kroner *et al.* 1992; Stein & Goldstein 1996). Rodinia continued to fragment between 850 and 750 Ma as East Gondwana and South China were rifted away from western Laurentia (Figs 3 & 4). Very little if any juvenile crust survives that accompanied the breakup along the west coast of Laurentia. Although continental blocks appear to have been rifted from the eastern margin of the Congo Craton between 900 and 850 Ma (Stern 1994), the location of these blocks today is unknown; perhaps one of the blocks is the Iran Plate (Fig. 2). Because arc-related igneous rocks are not found in the Katangan Orogen, the ocean basin that opened must have been small (Red Sea size), such that the basin margins collided before subducted slabs reached melting depths (*c.* 70 km) (Kampunzu *et al.* 1991). Extensive flood basalts in the Katangan Orogen at *c.* 850 Ma probably record a mantle plume centered near the triple junction of the Katangan, Damaran and Zambezi Orogens (Fig. 2; Kampunzu *et al.* 2000). Although the Katangan Rift continued to open between 850 and 775 Ma, it appears that only a small volume of oceanic crust was formed before it rapidly closed at 750 Ma. The Katangan Collision was later than the Zambezi Closure at 820 Ma but considerably earlier than the Damaran Closure at 600–550 Ma. Rifting continued in and south of the Arabian–Nubian Shield during the interval of 870–840 Ma, as recorded by zircon ages from the Namaqua, Gariep, Damara, Zambezi and Katangan Orogens (Table 1; Frimmel *et al.* 2001). In the East African Orogen, this rifting led to the development of a passive continental margin,

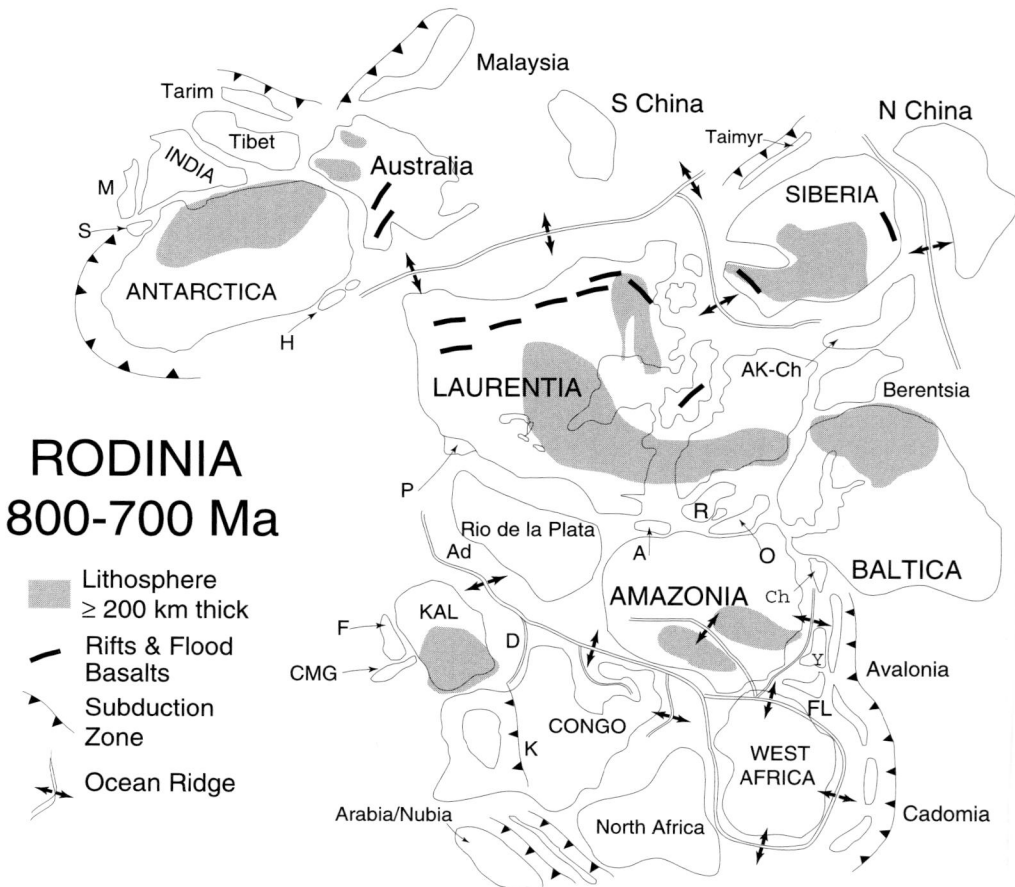

Fig. 4. Diagrammatic reconstruction of Rodinia at 800–700 Ma showing possible distribution of arc systems, rifts and flood basalts, ocean ridges, and thick Archean lithosphere. Ad, Adamastor Ocean; other symbols defined in Figures 1 & 2.

now recorded by remnants of sediments in Kenya (Stern 1994).

Most of the rifting in Avalonia occurred between 800 and 750 Ma (Table 1). Rifting at c. 760 Ma is recorded in Newfoundland, in the Maritimes in eastern Canada, in New England and in the Carolina Province in North Carolina (Harris & Glover 1988; Nance et al. 1991; Murphy et al. 1999). Arc magmatism occurred in Avalonia between c. 670 and 580 Ma, followed by collisions and accretion to Laurentia between 640 and 560 Ma (Fig. 4). Oblique shear zones with associated rift basins and post-tectonic intrusions formed after accretion at c. 540 Ma. Similar ages are reported in terranes that comprise Cadomia, many of which are between 620 and 600 Ma (Samson & D'Lemos 1998). Final accretion of Cadomian and Avalonian Terranes to Baltica or Laurentia occurred chiefly between 580 and 540 Ma

(Table 1). Although the positions of Tarim and Malaysia are not well known, rock assemblages suggest that arc systems also existed in these areas during the Neoproterozoic (Condie 2001a).

Pan-African orogenic system (750–550 Ma)

A Pan-African orogenic belt appears to have completely surrounded the West African Craton during the Neoproterozoic (Fig. 5). This orogen is exposed in NW Africa as the Anti Atlas, Ougarta, Pharusian-Tuareg, Gourma and Dahomeyan Belts on the east, and the Mauretanian, Bassaride and Rokelide Belts on the west (Caby 1987; Dostal et al. 1994; Attoh et al. 1997; Hefferan et al. 2000). All of these deformation belts developed during Pan-African collisions, chiefly between 650 and 550 Ma (Table 1; Fig. 5),

Table 1. *Distribution of Neoproterozoic juvenile crust in Rodinia*

Location	Crustal age (Ma)	Rifting age (Ma)	Accretion age (Ma)	Juvenile crust (%)	Scaled area ($\times 10^6$ km^2)*	Arc (%)	Ocean ridge (%)	WPB+ (%)	References
Arabian–Nubian Shield	900–850	900–850	700–650	90	1.02 [0.92]	50	5	45	Duyvermna et al. 1982; Reischmann et al. 1984; Kroner et al. 1987; Stern 1994; Stern & Kroner 1993; Stern & Goldstein 1996
	700–650		760–650	95	1.02 [0.92]	94	5	1	
Brazil	900–760	800–750	650–550	85	0.98 [0.83]	90	10	0	Pimentel & Fuck 1992; Babinski et al. 1996; Machado et al. 1996; Pimentel et al. 1996; de Almeida et al. 2000
Africa Adamastor					0.97 [0.32]				Porada 1989; Kukla & Stanistreet 1991; Frimmel et al. 1996a, b
Damara	750–650	750	600–550	20		20	70	10	
Kaoko–W Congo	750–650	800–750	600–570	10		0	100	0	
Gariep–Saldanian	740–720	750	600–580	50		10	60	30	
Kibaran–Lufilian	750–650	900–760	570–530	50	0.25 [0.12]	50	0	50	Kampunzu et al. 2000
West African Craton					0.642 [0.32]				Black et al. 1979; Chikaoui et al. 1980; Caby 1987; Villeneuve & Dallmeyer 1987; Dallmeyer & Villeneuve 1987; Saquaque et al. 1989; Dostal et al. 1994; Hefferan et al. 2000
Pharusian	750–600	850–800	645–560	70		90	10	0	
Mauritanide–Bassaride	700–650	850–700	660–640	40		100	0	0	
Dahomeyide	700–600	900–800	640–560	15		80	20	0	
Anti Atlas	780–600	800–780	615–565	80		95	5	0	
Rokelide	750–600	850–700	550	50		80	20	0	
East Africa	900–850	900–850	600–550	20	0.3 [0.06]	80	20	0	Maboko et al. 1985; Maboko 1995; Kroner et al. 1999; Moller et al. 2000

Table 1. (cont.)

Location	Crustal age (Ma)	Rifting age (Ma)	Accretion age (Ma)	Juvenile crust (%)	Scaled area (× 10^6 km^2)*	Arc (%)	Ocean ridge (%)	WPB+ (%)	References
Cadomia–Dalradian					0.45 [0.36]				Cabanis et al. 1987; Pharaoh et al. 1987; Winchester et al. 1987; Murphy et al. 1991; Nance et al. 1991; Liegeois et al. 1996; Hefferan et al. 2000
Carpathians	800–780			100		100	0	0	
British Isles	700–600	800–750	580–540	80		75	15	10	
America–Normandy	700–550	800–750	580–540	80		80	0	20	
Dalradian	680–550			50		50	0	50	
Avalonia–Florida					0.43 [0.39]				Feiss 1982; Opdyke et al. 1987; Harris & Glover 1988; Nance et al. 1991; Nance & Murphy 1994; Barr & White 1996; Murphy et al. 1999; Hefferan et al. 2000
Maritimes	700–600	800–750	580–570	80		75	15	10	
Carolina	700–600	800–750	600–540	80		80	10	10	
New England	675–550	800–750		100		80	10	10	
Florida	650–550	800–750		100		100	0	0	
Siberia					0.05 [0.04]				Rosen et al. 1994; Vernikovskiy et al. 1998
Taimyr	800–750		600	80		80	20	0	
Yenisey	800–600		600	80		75	15	10	

* Values in [ ] corrected for reworked component with Nd isotopes using the values given for the percentages of juvenile crust.
+ WPB, Within-plate basalts, chiefly oceanic plateau and flood basalts.

and all record remarkably similar tectonic histories, with initial rifting occurring between 850 and 700 Ma (Hefferan et al. 2000). Although pre-collisional subduction appears to have been directed chiefly outward from the West African Craton, little agreement exists as to the size of the ocean basins that opened. Indeed, if closure occurred during a 100 Ma interval on all sides of the West African Craton, it is unlikely that the intervening ocean basins were as large as originally suggested by Caby (1987). Most of the Pan-African juvenile crust trapped in the circum-West African orogens formed as continental margin arcs (Chikhaoui et al. 1980; Caby 1987; Saquaque et al. 1989; Dostal et al. 1994).

Rifting contemporary with that in West Africa occurred in Amazonia and probably also in Rio de la Plata (Castaing et al. 1994; de Almeida et al. 2000). Numerous blocks of Archean and Palaeoproterozoic crust were rifted apart in eastern Amazonia (Brazil) as small ocean basins formed between blocks. Continental margin arcs formed around the margins of many of these blocks and are the principal type of Neoproterozoic juvenile crust (Table 1). Small volumes of ophiolite were also trapped in the sutures. Closure of these basins occurred chiefly between 650 and 550 Ma, similar to basin closures around the West Africa Craton (Fig. 5). During collisions of continental blocks, which caused the Braziliano Orogeny, large volumes of pre-1 Ga continental crust were reworked and partially melted (de Almeida et al. 2000). In most cases, the final stages of collision are characterized by transcurrent shear zones.

Numerous continental margin arcs were accreted to the Arabian–Nubian Shield during the Pan-African collisions (Klemenic et al. 1985; Furnes et al. 1996; Alene et al. 2000). The sutures between oceanic terranes often contain well-preserved ophiolites (Zimmer et al. 1995). The closure of the Mozambique Ocean, in what is now Saudi Arabia, between 750 and 650 Ma resulted in collision of rifted blocks (such as the Afif Terrane) and juvenile arcs located between East and West Gondwana along north-trending sutures (Fig. 5). Continued convergence between 650 and 540 Ma led to the formation of NW-trending sinistral and NE-trending dextral transcurrent fault systems.

Mafic dyke swarms probably associated with mantle plumes indicate diachronous opening of the Iapetus Ocean, beginning at 620–600 Ma between Baltica and Greenland and spreading southward between the Appalachian Orogen and Amazonia–Rio de la Plata by 600–550 Ma (Figs 3 & 4; Bingen et al. 1998). As indicated by tectonic subsidence models (Bond & Kominz 1984), renewed rifting occurred in western Laurentia (Canadian Cordillera) at 600–550 Ma. Geologic evidence for rifting in the Death Valley area of California also supports renewed rifting at this time (Prave 1999). Although terranes were undoubtedly rifted from what is now western Canada, it is not clear where these terranes are located at present. The fragmenting pieces associated with Iapetus Ocean opening eventually collided, becoming part of Gondwana, with major collisions occurring dominantly in two stages (Fig. 5). The oldest collision occurred when part of East Gondwana (India?) collided with West Gondwana, producing deformation in the Mozambique Orogen at 725–650 Ma. Although some juvenile crust was trapped in the central and southern parts of the Mozambique Orogen, most of the crust in this part of the orogen is reworked Palaeoproterozoic and Archean crust (Maboko 1995). This early Mozambiquian collision was followed by the widespread Pan-African event at 650–550 Ma (Fig. 3). At this time the interconnecting array of rifted small ocean basins in western Gondwana closed, producing the Pan-African and Braziliano Orogenies. Most of the Neoproterozoic juvenile crust preserved today was 'captured' between colliding continental blocks during one of these collisional events.

An important question is why none of the Pan-African ocean basins in West Gondwana developed into large oceans before convergence began. Grunow et al. (1996) have suggested that it was the synchronous opening of the Iapetus Ocean and the Pan-African collisions at 650–550 Ma that was responsible for premature closure of the Pan-African basins (Figs 3 & 4). In effect, West Gondwana was caught between compressive forces on both sides, which closed the small ocean basins leading to the widespread Pan-African and Braziliano Deformation.

Rifting of Rio de la Plata from the Congo and Kalahari Plates occurred chiefly between 800 and 750 Ma as the Adamastor and Khomas Oceans opened (Fig. 4; Porada 1989; Kukla & Stanistreet 1991; Frimmel et al. 1996a, b). By 717 Ma, subduction began in the Gariep Belt in South Africa and the Adamastor Ocean began to close, with a continent–continent collision occurring at c. 545 Ma. Closure of the Adamastor Ocean and related ocean basins, and accretion of juvenile Neoproterozoic crust, occurred chiefly between 600 and 550 Ma (Table 1). Sedimentological studies in the Damara Orogen in Namibia indicate that the Congo–Rio de la Plata suturing predated the Congo–Kalahari suturing (Prave 1996).

Sr isotopic record of Rodinia

As shown by Veizer (1989), the Sr isotope record in marine carbonates can be useful in tracking the relative inputs of continental and mantle Sr into seawater. When land areas are extensive and elevations relatively high, weathering and erosion transport

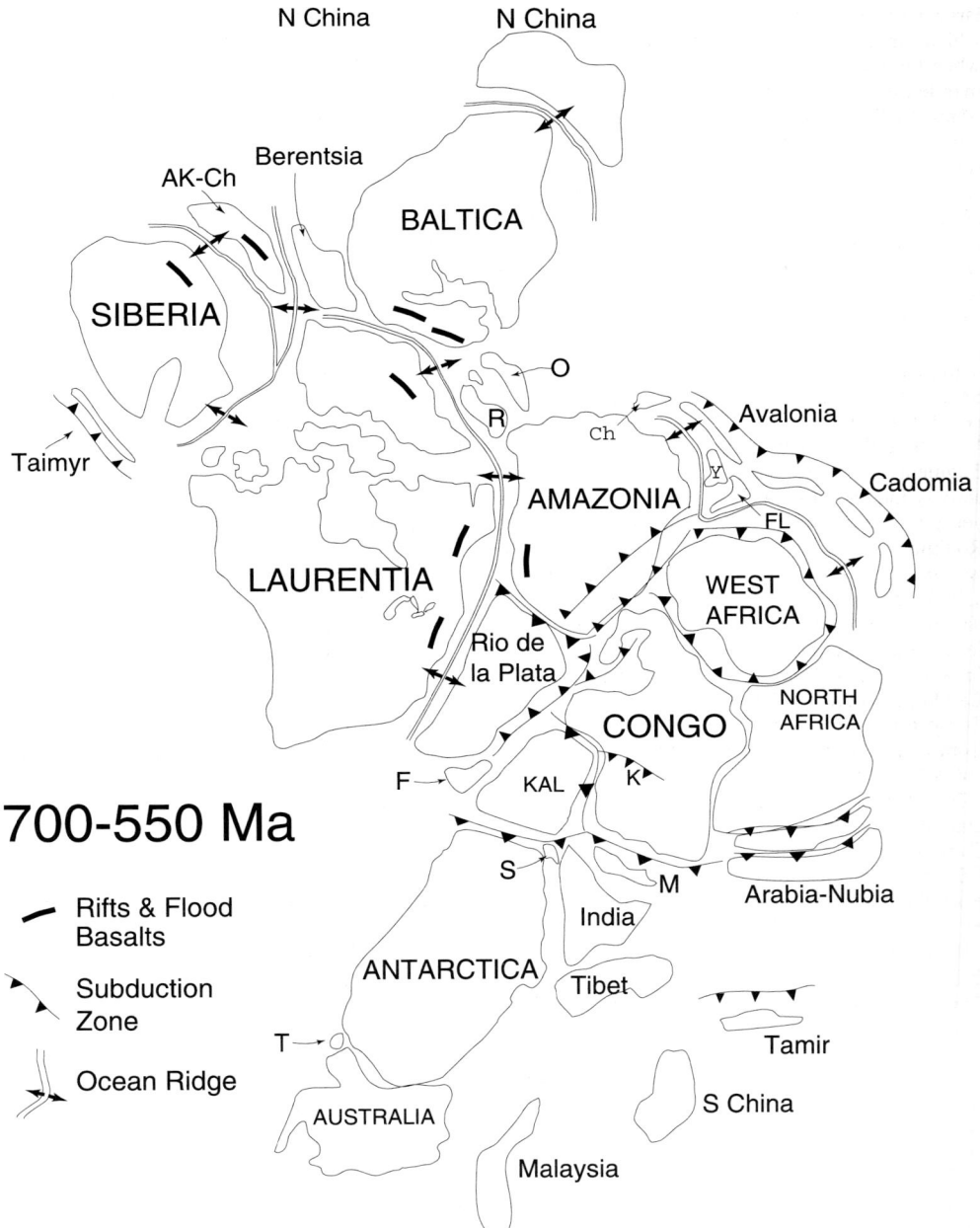

Fig. 5. Diagrammatic reconstruction of Rodinia at 700–550 Ma showing possible distribution of arc systems, rifts and flood basalts, and ocean ridges. Symbols defined in Figures 1 & 2.

large amounts of continental Sr into the oceans, which has relatively high $^{87}Sr/^{86}Sr$ ratios. In contrast, when the sea level is high and ocean ridges or/and mantle plumes are widespread and active, the input of mantle Sr into seawater is enhanced. Many of the long-term changes in the Sr isotopic composition of seawater may be related to the supercontinent cycle and perhaps also to superplume events (Condie et al. 2000). Although the Sr isotope record of seawater for the Phanerozoic is relatively well known (Veizer et al. 1999), except for the Neoproterozoic, the record for Precambrian seawater is poorly known (Asmerom et al. 1991; Jacobsen & Kaufman 1999).

In principle, it should be possible to track the formation and destruction of a supercontinent with the Sr isotopic record in seawater. During supercontinent formation, when the land area is increasing, the $^{87}Sr/^{86}Sr$ ratio of seawater should increase; during fragmentation, when ocean ridges and mantle plume activity increase, the ratio should decrease. It is not yet possible to completely track the history of Rodinia with Sr isotopes, since the seawater Sr isotope record is only well known for the last c. 900 Ma onwards (Fig. 6). There are few Sr isotope data from well-dated sediments corresponding to the formation of Rodinia between c. 1300 and 900 Ma. Sr isotope ratios from well-dated limestones of the Belt Supergroup at 1470 Ma are c. 0.7062 (Veizer & Compston 1976; Evans et al. 2000b). Carbonates from China, with less well-constrained ages in the 1000–1300 Ma range, have low Sr isotope ratios of 0.704–0.705 (Jahn & Cuvellier 1994). Grenvillian marbles from the Bancroft Terrane in eastern Canada, deposited at 1250 Ma (Rivers 1997), have a $^{87}Sr/^{86}Sr$ ratio of c. 0.7055 and marine carbonates from the 1200 Ma Society Cliffs Formation in the Arctic region (Baffin Island) have a minimum ratio of c. 0.706 (Veizer & Compston 1976; Kah et al. 2001). Although not as well dated, recent results from shallow-marine successions in Siberia and the Urals suggest that seawater Sr isotope ratios were in the range of 0.7060–0.7065 at c. 1300 Ma and then fell to c. 0.7050 by 1030 Ma (Bartley et al. 2001). If the Adrar carbonates in Mauritania are representative of seawater at 900 Ma (c. 0.707; Veizer et al. 1983), then an increase in the Sr isotopic ratio of marine carbonates between 1030 and 900 Ma may record the last stages in the formation of Rodinia. Why the earlier stages (1300–1000 Ma) in the formation of this supercontinent are not reflected in the Sr isotopes is not clear – perhaps pre-1000 Ma collisions were relatively minor.

The $^{87}Sr/^{86}Sr$ ratio decreases in seawater from c. 0.7074 at 900 Ma to a minimum of 0.706 at 850–775 Ma (Fig. 6; Jacobsen & Kaufman 1999; Walter et al. 2000). This dramatic decrease probably records the initial breakup of Rodinia (Fig. 3), with increased input of mantle Sr accompanying the breakup. This broad minimum is followed by a small but sharp increase in radiogenic Sr, levelling off between c. 700 and 600 Ma. This increase in the $^{87}Sr/^{86}Sr$ ratio may reflect some of the early plate collisions in the Arabian–Nubian Shield and elsewhere. The most significant change in the Sr isotopic ratio of Neoproterozoic seawater occurs between 600 and 500 Ma, when the $^{87}Sr/^{86}Sr$ ratio rises to nearly 0.7095 in only 100 Ma (Fig. 6). Although there is scatter in the data at this time, any fit of a curve through the data results in a rapid increase in the $^{87}Sr/^{86}Sr$ ratio. This rapid increase corresponds to the Pan-African collisions leading to the formation of the Early Palaeozoic supercontinent Gondwana. As collisions occurred, land areas were elevated and a greater proportion of continental Sr was transported into the oceans. The levelling or slight decrease in seawater Sr isotopic composition between 700 and 600 Ma is problematic, but could reflect enhanced mantle Sr input from the rifting of Siberia and Alaska–Chukotka from Laurentia (Figs 3 & 5; Kirschvink et al. 1997). Neither of the Neoprotero-

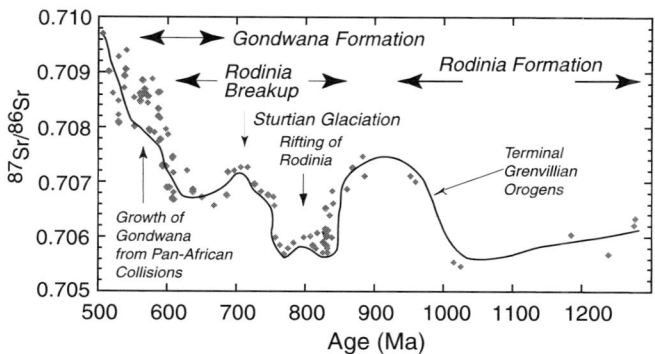

Fig. 6. Distribution of the $^{87}Sr/^{86}Sr$ ratio in seawater from 1000 to 400 Ma. Points represent published data from the least diagenetically altered marine limestones; the curve is a visual fit of the data. Principal references: Veizer & Compston (1976), Jacobsen & Kaufmann (1999) and Walter et al. (2000); other references given in text.

zoic glaciations are well represented in the Sr isotope record, a feature that may be due inadequate resolution of ages and lack of data for these glaciations.

A well-established minimum in the $^{87}Sr/^{86}Sr$ ratio at c. 470 Ma (Veizer et al. 1999) may reflect a superplume event at this time. Numerous other features, such as a period of no reversals in the Earth's magnetic field, a peak in black shale abundance and a maximum in sea level at this time, are also consistent with such a superplume event (Condie 2001b).

Juvenile continental crust (1350–500 Ma)

Condie (2001a) has shown that only 7–13% of the present continental crust formed during the 1350–900 Ma time window. As shown in Figure 1, most of this crust occurs in the long collisional Grenvillian Orogen, extending from Siberia along the eastern and southern margins of Laurentia into Australia and Antarctica. The largest volume, which is not well constrained in age or geographic distribution, is that in East Antarctica. A small but significant volume of juvenile crust also probably occurs in the triple-junction area of the Kalahari–Australia–Laurentia Plates. Almost all of the juvenile crust in the Grenville collisional belt appears to have formed between 1350 and 1200 Ma. The San Ignacio Orogen in South America, including small occurrences in Oaxaquia, Rio de la Plata and Rockall, contains up to c. 2% of the total juvenile crust, formed between 1350 and 1100 Ma. Only very small amounts of Grenvillian crust are preserved in other areas such as East Africa, the Arctic region, Malaysia, South China, Amazonia and Australia. Most of this crust also formed between 1300 and 1100 Ma, and only in the Arctic region (East Greenland, Svalbard, Ellesmere Island) is 1100–1000 Ma juvenile crust possibly present. In addition, a small volume of largely arc crust produced between 900 and 850 Ma occurs in the Arabian–Nubian Shield (Table 1).

A summary of juvenile crust produced in orogens between c. 800 and 500 Ma is included in Table 1 and shown on a Pannotia–Gondwana reconstruction in Figure 7. Estimates of the volume of juvenile crust are based on published U-Pb zircon ages and Nd isotopic data, as described in Condie (2001a). The amount of plume-related magma that may underplate the continental crust is not included in these estimates. However, because the Grenvillian and Pan-African Orogens expose varying crustal levels, from a few kilometres to nearly 30 km deep in a few cases, the estimates are thought to represent juvenile components in the upper and middle part of the continental crust in Neoproterozoic Orogens. At least 50% of the known juvenile crust of this age occurs in the Arabian–Nubian Shield, and in Avalonia–Cadomia and related terranes that formed along the border of Amazonia and West Africa (Fig. 7; Abdelsalam & Stern 1996). Another 20% is found in Pan-African orogens in Brazil, and 8% each in the Adamastor Orogen (bordering the South Atlantic today) and the circum-West Africa Orogens. The remaining 25% occurs as tectonic slices in such Neoproterozoic orogens as Taimyr, western Siberia and East Africa, and in Phanerozoic collisional orogens.

Tectonic settings of juvenile crust can be constrained using lithologic assemblages and geochemical characteristics of basalts (Condie 1997). From available data, as summarized in Table 1, it would appear that in most instances arc assemblages greatly dominate the exposed Neoproterozoic juvenile crust. The large volume of granitoids associated with most of these assemblages further suggests that continental margin arcs are most important. Data from Grenvillian orogens (1350–900 Ma) result in a similar conclusion (Condie 2001a). Although ophiolites are generally minor in Neoproterozoic orogens, a notable exception is the Adamastor Orogen and associated orogens (Table 1, row 3) in southern Africa, where mid-ocean ridge basalt (MORB) may comprise a significant proportion of the juvenile crustal assemblage. These orogens appear to have formed as continental rifts, which opened just enough to form small ocean basins before closing, trapping some of the oceanic crust.

Within-plate basalts (WPB) are also generally a minor component in most Neoproterozoic crust, averaging a few per cent of the total juvenile component (Table 1). The Katanga–Lufilian, the Arabian–Nubian Shield (900–850 Ma additions only) and Dalradian orogens are unique in that they contain c. 50% of WPB, chiefly oceanic plateau or continental flood basalts (Klemenic et al. 1985; Berhe 1990; Abdelsalam & Stern 1996; Alene et al. 2000). In the Arabian–Nubian Shield, an oceanic plateau, now represented by remnants as the Baish Group (c. 850 Ma) in Saudi Arabia, may have been the starting point for continental crust in this area (Stein & Goldstein 1996). The only ocean basin closures that captured significant volumes of oceanic crust are those of the Adamastor and Khomas Basins, now preserved in the Kaoko, West Congo, Damara and Gariep–Saldanian Orogens in southwestern Africa (Table 1). Geochemical data from basalts in the Gariep Orogen in Namibia indicate a within-plate tectonic setting for all of the mafic igneous units (Frimmel et al. 1996b): this includes bimodal volcanism during continental rifting, and seamounts and volcanic islands associated with a mantle plume in the Adamastor Ocean.

Cadomia, Avalonia and related terranes scattered along the eastern margin of Laurentia and through-

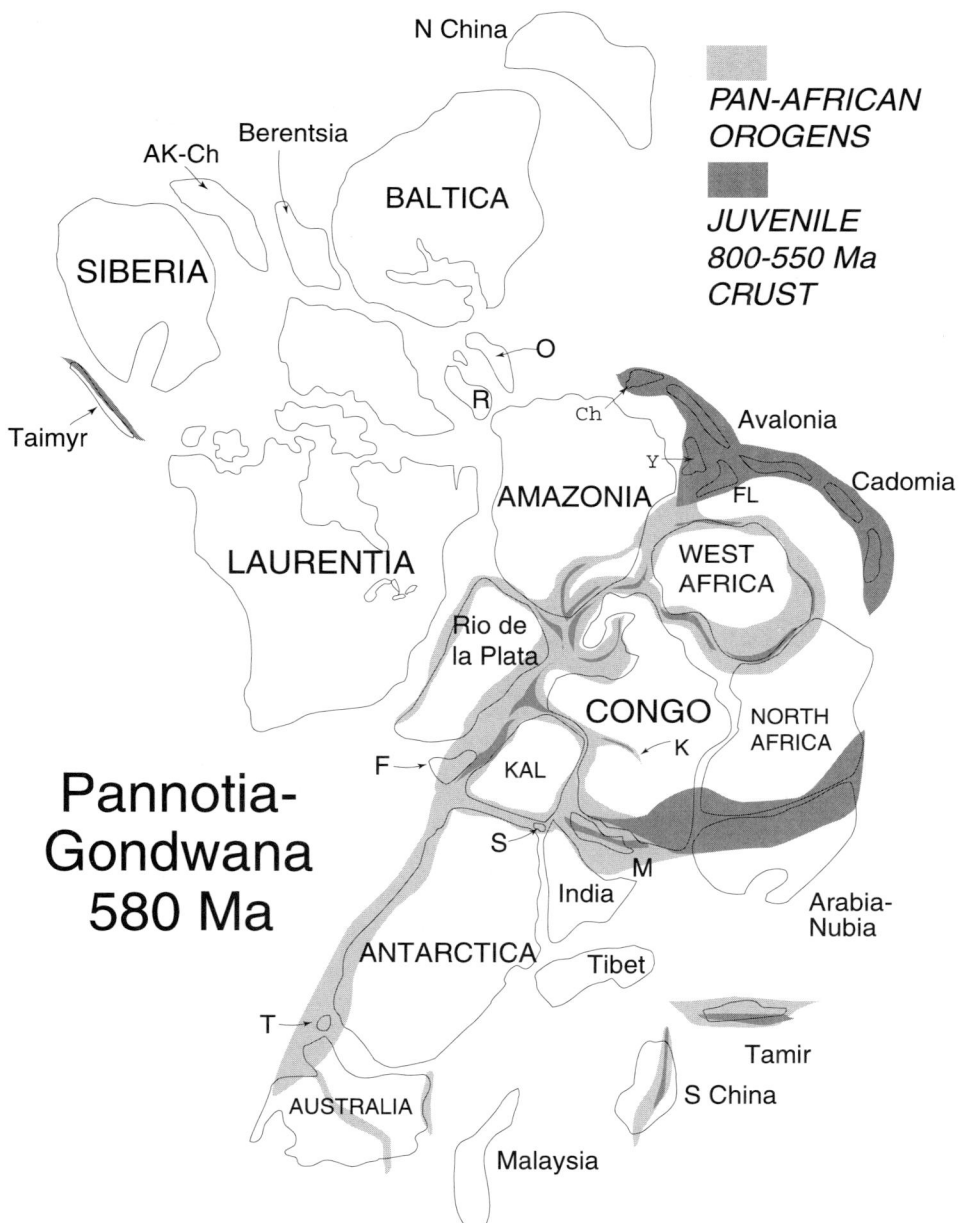

Fig. 7. Diagrammatic reconstruction of Pannotia–Gondwana at 580 Ma showing possible distribution of known juvenile crust formed between 800 and 550 Ma in relation to Pan-African orogens. T, Tasmania; other symbols defined in Figures 1 & 2.

out central Europe are comprised chiefly of juvenile continental margin arcs, and contain up to 10–20% of WPB, chiefly continental flood basalts (Dostal et al. 1990; Nance et al. 1991; Barr & White 1996; Liegeois et al. 1996; Murphy et al. 1999). Because of possible crustal contamination, the WPB compo-

nent in these terranes may not represent entirely juvenile additions to the crust. Three main phases are recognized in the evolution of Avalonia (Murphy et al. 1999): (1) an early arc phase (780–660 Ma) during which continental margin arcs developed; (2) the main arc phase (630–590 Ma) when large volumes of juvenile arc crust formed; and (3) a transition stage during which compressive deformation changed to transcurrent deformation as large transform faults developed. Most of the juvenile crust preserved in Cadomia and Avalonia formed between 700 and 600 Ma (Table 1).

Rodinia and crustal growth rate

The minimal areal distribution of juvenile continental crust produced during the lifetime of Rodinia is shown in Figure 8, on an equal-area projection of the continents. Estimates of the area of this crust were scaled from maps in published papers and corrected for reworked components using Nd isotopic data (Table 2, column 4). Results are extrapolated to geographic regions where data are not available. The estimated percentages of juvenile crust relative to the present volume of continental crust are given in column 6 of Table 2. In terms of understanding crustal growth during supercontinent evolution, it is useful to divide juvenile crust into two types, based on the type of orogen in which it is preserved. Peripheral (or accretionary) orogens, which are largely comprised of accreted arcs, ophiolites and oceanic plateaus, occur around the edges of supercontinents; examples for Rodinia are the Antarctica segment of the Grenville Orogen at 1100–900 Ma and the Arabian–Nubian and Cadomia–Avalonia Orogens at 650–550 Ma (Figs 1 & 7). Internal orogens, which form between cratons that have collided, comprise chiefly of reworked older crust (Windley 1992); most of the Grenville and Pan-African orogens fall into this category. If the juvenile crust is divided into two age groups, terranes with ages of 1350–800 Ma occupy $c.\ 10.1 \times 10^6 \text{km}^2$ (75%) and terranes with ages of 800–550 Ma comprise $c.\ 4.2 \times 10^6 \text{km}^2$ (25%) of the total Neoproterozoic crust (Table 2, column 4). Areal distributions grouped according to orogen type suggest that, in both age categories, $c.$ 50% of the juvenile crust occurs in peripheral orogens and 50% in internal orogens (Table 2, columns 2 and 3).

Because tectonic slices of Neoproterozoic rocks in younger orogens and underplated juvenile crust are not included in the estimates of juvenile crust, 15% has been added to each of the area estimates in Table 2 (column 5). Based on geologic maps of Phanerozoic orogens in Laurentia and Europe, this value is probably an upper limit for the amount of Neoproterozoic juvenile continental crust trapped in younger orogens. Results suggest that Neoproterozoic juvenile crust comprises $c.$ 11% of the total continental crust: 8% in the 1350–800 Ma category and 3% in the 800–550 Ma category (Table 2, column 6). The earlier figure of 7–13% for the 1350–800 Ma category (Condie 2001a) has been refined based on more data. For an average crustal thickness of 35 km, these data indicate a growth rate of 0.70–0.74 km^3/a (Table 2, column 8), or, if an additional 10 km is restored due to erosion, 0.90–0.95 km^3/a. These values are near or somewhat below the average growth rate of continents during the Phanerozoic of $c.$ 1 km^3/a (Reymer & Schubert 1984). It is noteworthy that continental growth rate during aggregation (1350–800 Ma) and breakup (800–550 Ma) phases of Rodinia are similar.

Assuming that the Arabian–Nubian Shield extends to the Zagros Suture in Iran and comprises Neoproterozoic juvenile crust (35 km thick), Reymer & Schubert (1986) calculated a crustal growth rate of 0.78 km^3/a, anomalously high for such a small volume of crust; similar rates were estimated by Stein & Goldstein (1996). Using only the exposed area of the Arabian Shield (0.6×10^6 km^2), Pallister et al. (1990) estimate the rate to be only 0.07 km^3/a. Our rate, based on an area of 0.92×10^6 km^2 and a crust thickness of 41 km [to be comparable to the results of Pallister et al. (1990)] is 0.13 km^3/a, about twice the rate estimated by Pallister et al. (1990) but considerably less than that estimated by Reymer & Schubert (1986). If the Palaeoproterozoic and Late Archean crust exposed in southern end of the Arabian Peninsula (Windley et al. 1996; Whitehouse et al. 1998) underlies much of the Arabian Plate, as seems probable, Reymer & Schubert(1986) and Stein & Goldstein (1996) overestimated the volume of Neoproterozoic crust, thus accounting for their anomalously high growth rates.

Rodinia and superplume events

Unlike supercontinents in the Late Archean and Palaeoproterozoic, which were associated with the production and preservation of large volumes of new continental crust (Condie 1998, 2000), the results of this study suggest that Rodinia was not accompanied by the production of significant volumes of juvenile continental crust. In fact, the levels of crustal production are over an order of magnitude lower than those of two older supercontinents, and perhaps lower than Phanerozoic crustal production levels of $c.$ 1 km^3/a. Why should Rodinia be different from older supercontinents? Perhaps the answer is the absence of a superplume event. Superplume events, which are short-lived, Earth-wide events (<100 Ma duration), during which large mantle plumes rapidly bombard the base of the lithosphere, may be responsible for the production of large volumes of juvenile crust. Both of the earlier supercontinents appear to have been associated with major superplume events at 2.7 and 1.9 Ga, and perhaps minor events at 2.5 and 1.7 Ga (Condie 1998; Isley & Abbott 1999).

In the case of a Late Archean supercontinent, it

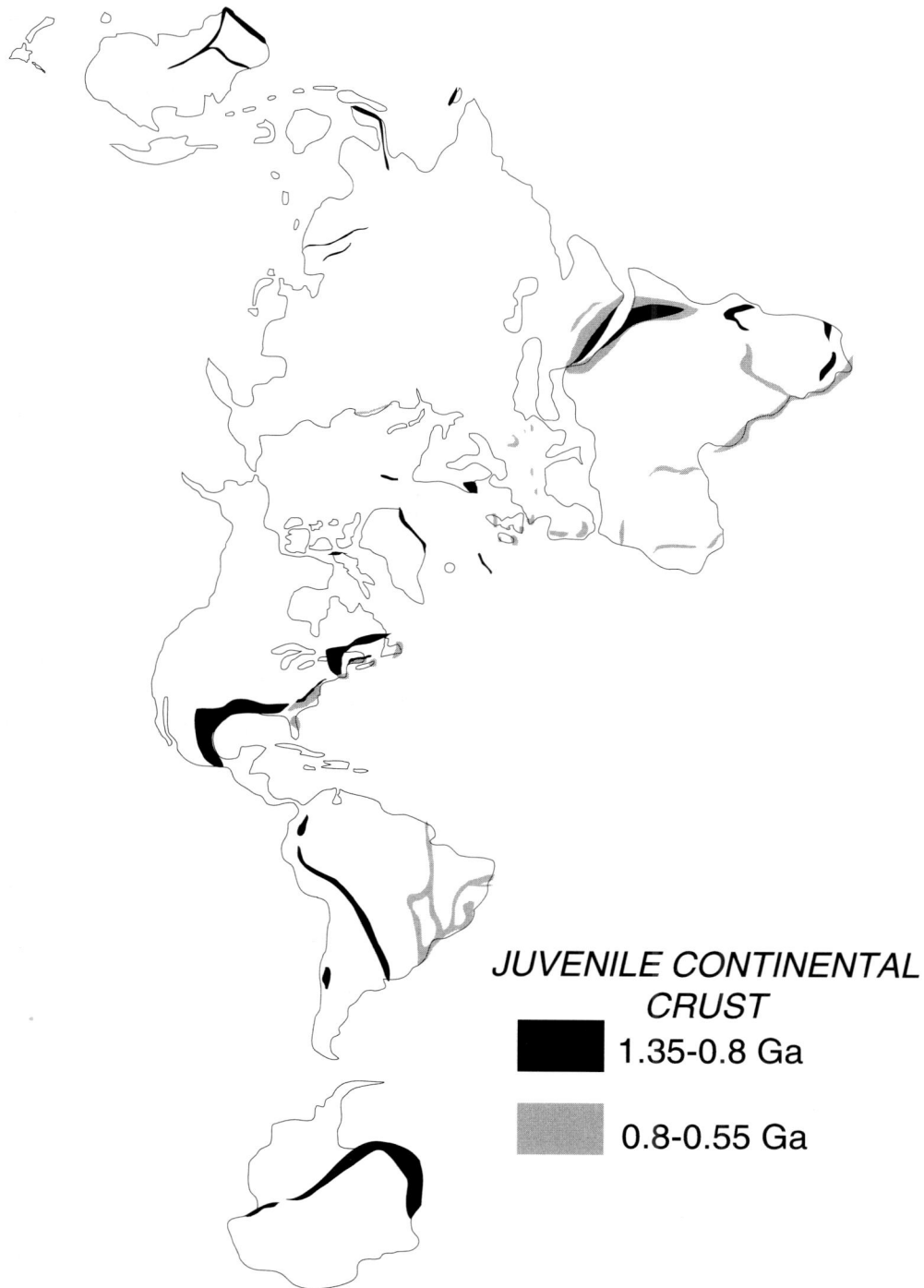

Fig. 8. Distribution of Late Mesoproterozoic and Neoproterozoic juvenile crust shown on an equal-area projection of the continents. Modified after Condie (1998).

Table 2. *Growth rate of juvenile crust in Rodinia*

Age (Ma)	Peripheral Orogen ($\times 10^6 \text{km}^2$)*	Internal Orogen ($\times 10^6 \text{km}^2$)*	Total ($\times 10^6 \text{km}^2$)	Total +15%†	Percentage‡	Growth Rate (km^2/a)§	Growth Rate (km^3/a)‖
1350–800	5.17	4.94	10.1	11.6	7.9	0.021	0.74 (0.95)
800–550	2.20	2.02	4.22	4.9	3.3	0.020	0.70 (0.90)

* Areas scaled from geologic maps and corrected for reworked crust with Nd isotope data.
† Fifteen per cent added for tectonic slices in younger orogens.
‡ Percentage of the total continental crust above sea level ($147 \times 10^6 \text{km}^2$).
§ Based on 550 Ma for the 1350–800 Ma crust and 250 Ma for the 800–550 Ma crust.
‖ Assumed crustal thickness of 35 km (45 km thickness given in parenthesis).

would appear that breakup of the supercontinent at 2.2–2.1 Ga may have triggered a 1.9 Ga superplume event (Larson 1991; Maruyama 1994; Condie 1998). If supercontinent breakup is required to produce a superplume event, perhaps by initiating catastrophic collapse of lithospheric slabs at the 660 km seismic discontinuity, the absence of a Mesoproterozoic–Neoproterozoic superplume event may mean that a Palaeoproterozoic supercontinent did not fully fragment before Rodinia began to form (Condie 2001c). Rogers (1996) suggested that two supercontinents formed at 1.9 Ga: the first, which he called Atlantica, comprises Amazonia and other Archean Cratons in South America, and the Congo, West Africa and North Africa Cratons in Africa; Nena, the second supercontinent, includes Laurentia, Baltica, Siberia and Antarctica. These two supercontinents may have survived breakup of a Palaeoproterozoic supercontinent at 1.6–1.4 Ga, and they were later incorporated intact into Rodinia (Condie 2001b). Why two supercontinents survived a Mesoproterozoic breakup is unknown. An alternative interpretation is that is that two supercontinents formed at 1.9 Ga, and that neither was large enough to provide adequate lithospheric shielding for the production of mantle upwellings large enough to break the continental lithosphere (Lowman & Jarvis 1999).

Because superplume events may have been triggered by the breakup of a Late Archean supercontinent, when Rodinia broke up at 850–600 Ma, it may also have triggered a superplume event. Is there any evidence for such an event? The recent recognition of a superchron in the Ordovician centred at c. 480 Ma presents the possibility of a superplume event at this time (Johnson *et al.* 1995). Consistent with such a superplume event is a peak in eustatic sea level and in a calculated production rate of oceanic crust (Condie 2001b). Black shales are also widespread in the Ordovician. A steep fall in the $^{87}\text{Sr}/^{86}\text{Sr}$ ratio of seawater at this time could be a response to enhanced input of mantle Sr accompanying such a superplume event.

Unlike Rodinia and Pangaea, which had lifetimes of c. 100–150 Ma (Rodinia from 900 to 800 Ma and Pangaea from 320 to 180 Ma), Gondwana appears to have survived for 370 Ma, from c. 550 to 180 Ma (Fig. 9). The Late Archean and Palaeoproterozoic supercontinents may have survived for even longer time periods. Why did Rodinia and Pangaea begin to fragment so soon after their formation? Perhaps it had to do with their relatively large sizes. As shown by the numerical models of Lowman & Jarvis (1999) and Lowman & Gable (1999), a plate carrying a supercontinent must be very large to effectively shield the underlying mantle from cooling by subduction. Gondwana, which contained only about half of the continental crust at 500 Ma, may not have been large enough to provide this shielding, and hence it survived until it became part of Pangaea. Likewise, the pre-Rodinian supercontinents contained only 40–80% of the present continental crust and thus may have required a longer period of time to effectively shield the underlying mantle to produce a mantle upwelling.

Conclusions

(1) Thick Archean lithosphere in Laurentia, Siberia and Baltica may have resulted in the Neoproterozoic supercontinent Rodinia splitting around the margins rather than through the centre during the 850–600 Ma breakup.

(2) Opening of the Iapetus Ocean at 620–550 Ma occurred at the same time as collisions between East and West Gondwana. These

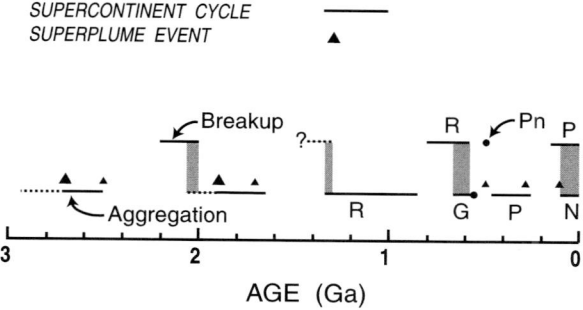

Fig. 9. The supercontinent cycle shown in relation to possible superplume events. R, Rodinia; G, Gondwana; Pn, Pannotia; P, Pangaea; N, possible new supercontinent. Size of triangles proportional to size of superplume events. Vertical grey bars are areas of assembly/breakup overlap. Modified after Condie (1998).

simultaneous events may have been responsible for closing small ocean basins in West Gondwana, producing the Pan-African and Braziliano orogenies.

(3) The Sr isotope record of seawater records the early fragmentation of Rodinia at 850–750 Ma and the formation of Pannotia–Gondwana at 600–500 Ma; the record is less clear between 750 and 600 Ma, when both fragmentation and collision events occurred simultaneously.

(4) At least 50% of the known juvenile continental crust produced between 750 and 550 Ma occurs in the Arabian–Nubian Shield and in other terranes that formed along the northern border of Amazonia and West Africa. An additional 20% is found in the Pan-African orogens of Amazonia, and c. 16% in the Adamastor and West African Orogens.

(5) Most juvenile continental crust produced during the 1350–500 Ma time interval formed in arcs, chiefly continental margin arcs. Only c. 11% of the total preserved continental crust formed during this period.

(6) About 50% of the Neoproterozoic juvenile crust occurs in peripheral orogens and 50% in internal orogens. In addition, an unknown but probably small amount of juvenile crust may have been added from mantle plumes by crustal underplating.

(7) Continental growth rates during the time period of 1350–500 Ma were similar or less than the average rate of continental growth during the Phanerozoic of 1 km^3/a. These small growth rates apply to both formation and breakup stages of supercontinents, and thus continental crustal growth is not dependent upon the stage of the supercontinent cycle.

(8) The low rates of continental growth during the Neoproterozoic may be due to the absence of superplume events associated with the formation of Rodinia or Gondwana. The breakup of Rodinia between 800 and 600 Ma, however, may have triggered a superplume event at c. 480 Ma.

(9) If supercontinent breakup is required to produce a superplume event, perhaps by initiating catastrophic collapse of lithospheric slabs at the 660 km seismic discontinuity, the absence of a Mesoproterozoic–Neoproterozoic superplume event may mean that a Palaeoproterozoic supercontinent did not fully fragment.

This paper was improved substantially from detailed reviews by Bor-ming Jahn, Jeff Chiarenzelli and Shigenori Maruyama.

References

ABDELSAMLAM, M. G. & STERN, R. J. 1996. Sutures and shear zones in the Arabian–Nubian shield. *Journal of African Earth Sciences*, **23**, 289–310.

ALENE, M., RUFFINI, R. & SACCHI, R. 2000. Geochemistry and geotectonic setting of Neoproterozoic rocks from northern Ethiopia, Arabian–Nubian shield. *Gondwana Research*, **3**, 333–347.

ASMEROM, Y., JACOBSEN, S. B., KNOLL, A. H., BUTTERFIELD, N. J. & SWETT, K. 1991. Strontium isotopic variations of Neoproterozoic seawater: implications for crustal evolution. *Geochimica et Cosmochimica Acta*, **55**, 2883–2894.

ATTOH, K., DALLMEYER, R. D. & AFFATON, P. 1997. Chronology of nappe assembly in the Pan-African Dahomeyide orogen, West Africa: evidence from $^{40}Ar/^{39}Ar$ mineral ages. *Precambrian Research*, **82**, 153–171.

BABINSKI, M., CHEMALE JR, F., HATMANN, L. A., VAN SCHMUS, W. R. & DA SILVA, L. C. 1996. Juvenile accretion at 750–700 Ma in southern Brazil. *Geology*, **24**, 439–442.

BARR, S. M. & WHITE, C. E. 1996. Tectonic setting of Avalonian volcanic and plutonic rocks in the

Caledonian highlands, southern New Brunswick, Canada. *Canadian Journal of Earth Sciences*, **33**, 156–168.

BARTLEY, J. K. ET AL. 2001. Global events across the Mesoproterozoic–Neoproterozoic boundary: Ca and Sr isotopic evidence from Siberia. *Precambrian Research*, **111**, 165–202.

BERHE, S. M. 1990. Ophiolites in NE and East Africa: implications for Proterozoic crustal growth. *Journal of the Geological Society*, London, **147**, 41–57.

BINGEN, B., DEMAIFFE, D. & VAN BREEMEN, O. 1998. The 616 Ma Egersund basaltic dike swarm, SW Norway, and Late Neoproterozoic opening of the Iapetus Ocean. *Journal of Geology*, **106**, 565–574.

BLACK, R., CABY, R., MOUSSINE-POUCHKINE, A. ET AL. 1979. Evidence for Late Precambrian plate tectonics in West Africa. *Nature*, **278**, 223–227.

BOND, G. C. & Kominz, M. A. 1984. Construction of tectonic subsidence curves for the early Paleozoic miogeocline, southern Canadian Rocky Mountains: implications for subsidence mechanisms, age of breakup, and crustal thinning. *Geological Society of America Bulletin*, **95**, 155–173.

BURRETT, C. & BERRY, R. 2000. Proterozoic Australia–Western United States (AUSWUS) fit between Laurentia and Australia. *Geology*, **28**, 103–106.

CABANIS, B., CHANTRAINE, J. & RABU, D. 1987. Geochemical study of the Brioverian volcanic rocks in the northern Armorican massif, France. In: *Implications for Geodynamic Evolution during the Cadomian*. Geological Society, London, Special Publication, **33**, 525–540.

CABY, R. 1987. The Pan-African belt of West Africa from the Sahara desert to the Gulf of Guinea. In: SCHAER, J. P. & RODGERS, J. (eds) *Anatomy of Mountain Ranges*. Princeton University Press, Princeton, NJ, 129–170.

CANNON, W. F. 1994. Closing of the Midcontinent rift – a far-field effect of Grenvillian compression. *Geology*, **22**, 155–158.

CASTAING, C., FEYBESSE, J. L., THIEBLEMONT, D., TRIBOULET, C. & CHEVREMONT, P. 1994. Paleogeographical reconstructions of the Pan-African/Brasiliano orogen: closure of an oceanic domain or intracontinental convergence between major block? *Precambrian Research*, **69**, 327–344.

CHIKHAOUI, M., DUPUY, C. & DOSTAL, J. 1980. Geochemistry and petrogenesis of Late Proterozoic volcanic rocks from NW Africa. *Contributions to Mineralogy and Petrology*, **73**, 375–388.

CONDIE, K. C. 1997. *Plate Tectonics and Crustal Evolution*. (4th edition). Butterworth-Heinemann, Oxford.

CONDIE, K. C. 2000. Episodic continental growth models: afterthoughts and extensions. *Tectonophysics*, **322**, 153–162.

CONDIE, K. C. 2001*a*. Continental growth during formation of Rodinia at 1.35–0.9 Ga. *Gondwana Research*, **4**, 5–16.

CONDIE, K. C. 2001*b*. *Mantle Plumes and their Record in Earth History*. Cambridge University Press, Cambridge.

CONDIE, K. C., & ROSEN, O. M. 1994. Laurentia–Siberia connection revisited. *Geology*, **22**, 168–170.

CONDIE, K. C., DES MARAIS, D. J. & ABBOTT, D. 2000. Geologic evidence for a mantle superplume event at 1.9 Ga. Geochemistry Geophysics and Geosystems, 1, paper no. 2000GC000095.

DALLMEYER, R. D. & Villeneuve, M. 1987. $^{40}Ar/^{39}Ar$ mineral age record of polyphase tectonothermal evolution in the southern Mauritanide orogen, SE Senegal. *Geological Society of America Bulletin*, **98**, 602–611.

DALZIEL, I. W. D. 1991. Pacific margins of Laurentia and East Antarctica–Australia as a conjugate rift pair: evidence and implications for a Eocambrian supercontinent. *Geology*, **19**, 598–601.

DALZIEL, I. W. D., MOSHER, S. & GAHAGAN, L. M. 2000. Laurentia–Kalahari collision and the assembly of Rodinia. *Journal of Geology*, **108**, 499–513.

DAVIS, D. W. & GREEN, J. C. 1997. Geochronology of the North American Midcontinent rift in western Lake Superior and implications for its geodynamic evolution. *Canadian Journal of Earth Sciences*, **34**, 476–488.

DE ALMEIDA, G. G. M., DE BRITO NEVES, B. B. & CARNEIRO, C. D. R. 2000. The origin and evolution of the South American platform. *Earth Science Reviews*, **50**, 77–111.

DOSTAL, J., DUPUY, C. & CABY, R. 1994. Geochemistry of the Neoproterozoic Tilemsi belt of Iforas: a crustal section of an oceanic island arc. *Precambrian Research*, **65**, 55–69.

DOSTAL, J., KEPPIE, J. D. & MURPHY, J. B. 1990. Geochemistry of Late Proterozoic basaltic rocks from SE Cape Breton Island, Nova Scotia. *Canadian Journal of Earch Sciences*, **27**, 619–631.

DUYVERMAN, H. J., HARRIS, N. B. W. & HAWKESWORTH, C. J. 1982. Crustal accretion in the Pan-African: Nd and Sr isotope evidence from the Arabian shield. *Earth and Planetary Science Letters*, **59**, 315–326.

ERDTMANN, B. D. 1998. Neoproterozoic to Ordovician/Silurian Baltica and Laurentia interaction with (Proto-) Gondwana: critical review of macro- and microplate transfer models. *Geologica*, **42**, 409–413.

EVANS, D. A. D., Li, Z. X., KIRSCHVINIK, J. L. & WINGATE, M. T. D. 2000*a*. A high-quality mid-Neoproterozoic paleomagnetic pole from South China, with implications for ice ages and the breakup configuration of Rodinia. *Precambrian Research*, **100**, 313–334.

EVANS, K. V., ALEINIKOFF, J. N., OBRADOVICH, J. D. & FANNING, C. M. 2000*b*. SHRIMP U–Pb geochronology of volcanic rocks, Belt Supergroup, western Montana: evidence for rapid deposition of sedimentary strata. *Canadian Journal of Earch Sciences*, **37**, 1287–1300.

FEISS, P. G. 1982. Geochemistry and tectonic setting of the volcanics of the Carolina slate belt. *Economic Geology*, **77**, 273–293.

FRIMMEL, H. E., HARNADY, C. J. H. & KOLLER, F. 1996*a*. Geochemistry and tectonic setting of magmatic units in the Pan-African Gariep belt, Namibia. *Chemical Geology*, **130**, 101–121.

FRIMMEL, H. E., KLOTZLI, U. S. & SIEGFRIED, P. R. 1996*b*. New Pb–Pb single zircon age constraints on the timing of Neoproterozoic glaciation and continental breakup in Namibia. *Journal of Geology*, **104**, 459–469.

FRIMMEL, H. E., ZARTMAN, R. E. & SPATH, A. 2001. The Richtersveld igneous complex, South Africa: U–Pb zircon and geochemical evidence for the beginning of Neoproterozoic continental breakup. *Journal of Geology*, **109**, 493–508.

FROST, B. R., AVCHENKO, O. V., CHAMBERLAIN, K. R. & FROST, C. D. 1998. Evidence for extensive Proterozoic remobilization of the Aldan shield and implications for Proterozoic plate tectonic reconstructions of Siberia and Laurentia. *Precambrian Research*, **89**, 1–23.

FURNES, H., EL-SAYED, M. M., KHALIL, S. O. & HASSANEN, M. A. 1996. Pan-African magmatism in the Wadi El-Imra district, central Eastern Desert, Egypt: geochemistry and tectonic environment. *Journal of the Geological Society*, London, **153**, 705–718.

GRUNOW, A., HANSON, R. & WILSON, T. 1996. Were aspects of Pan-African deformation linked to Iapetus opening? *Geology*, **24**, 1063–1066.

HALLS, H. C., LI, J., DAVIS, D., HOU, G., ZHANG, B. & QIAN, X. 2000. A precisely dated Proterozoic paleomagnetic pole from the North China craton, and its relevance to paleocontinental reconstruction. *Geophysical Journal International*, **143**, 185–203.

HANSON, R. E., WILSON, T. J. & MUNYANHIWA, H. 1994. Geologic evolution of the Neoproterozoic Zambezi orogenic belt in Zambia. *Journal of African Earth Sciences*, **18**, 135–150.

HANSON, R. E., MARTIN, M. W., BOWRING, S. A. & MUNYANYIWA, H. 1998. U–Pb zircon age for the Umkondo dolerites, eastern Zimbabwe: 1.1 Ga large igneous province in southern Africa–East Antarctica and possible Rodinia correlations. *Geology*, **26**, 1143–1146.

HARRIS, C. W. & GLOVER III, L. 1988. The regional extent of the 600 Ma Virgilina deformation: implication for stratigraphic correlation in the Carolina terrane. *Geological Society of America Bulletin*, **100**, 200–217.

HEFFERAN, K. P., ADMOU, H., KARSON, J. A. & SAQUAQUE, A. 2000. Anti-Atlas role in Neoproterozoic western Gondwana reconstruction. *Precambrian Research*, **103**, 89–96.

HOFFMAN, P. F. 1987. Continental transform tectonics: Great Slave Lake shear zone, NW Canada. *Geology*, **15**, 785–788.

HOFFMAN, P. F. 1991. Did the breakout of Laurentia turn Gondwanaland inside-out? *Science*, **252**, 1409–1412.

ISLEY, A. E. & ABBOTT, D. H., 1999. Plume-related mafic volcanism and the deposition of banded iron formation. *Journal of Geophysical Research*, **104**, 15461–15477.

JACOBS, J., FANNING, C. M., HENJES-KUNST, F., OLESCH, M., & PAECH, H. J. 1998. Continuation of the Mozambique belt into East Antarctica: Grenville-age metamorphism and polyphase Pan-African high-grade event in central Dronning Maud Land. *Journal of Geology*, **106**, 385–406.

JACOBSEN, S. B. & KAUFMAN, A. J. 1999. The Sr, C and O isotopic evolution of Neoproterozoic seawater. *Chemical Geology*, **161**, 37–57.

JAHN, B. M. & CUVELLIER, H. 1994. Pb–Pb and U–Pb geochronology of carbonate rocks: an assessment. *Chemical Geology*, **115**, 125–151.

JOHNSON, H. P., VAN PATTEN, D., TIVEY, M. & SAGER, W. 1995. Geomagnetic polarity reversal rate for the Phanerozoic. *Geophysical Research Letters*, **22**, 231–234.

KAH, L. C., LYONS, T. W. & CHESLEY, J. T. 2001. Geochemistry of a 1.2-Ga carbonate–evaporite succession, northern Baffin and Bylot Islands: implications for Mesoproterozoic marine evolution. *Precambrian Research*, **111**, 203–234.

KAMPUNZU, A. B., KAPENDA, D. & MANTEKA, B. 1991. Basic magmatism and geotectonic evolution of the Pan-African belt in central Africa: evidence from the Katangan and West Congolian segments. *Tectonophysics*, **190**, 363–371.

KAMPUNZU, A. B., TEMBO, F., MATHEIS, G., KAPENDA, D. & HUNTSMAN-MAPILA, P. 2000. Geochemistry and tectonic setting of mafic igneous units in the Neoproterozoic Katangan basin, central Africa: implication for Rodinia break-up. *Gondwana Research*, **3**, 125–153.

KAMPUNZU, A. B., AHALL, K. I., HARLAN, S. S., WILLIAMS, M. L., MCLELLAND, J. & GEISSMAN, J. W. 2001. Long-lived convergent orogen in southern Laurentia, its extensions to Australia and Baltica, and implications for refining Rodinia. *Precambrian Research*, **111**, 5–30.

KAMPUNZU, A. B., WILLIAMS, M. L., MCLELLAND, J., GEISSMAN, J. W. & AHALL, K. 1999. Refining Rodinia: geologic evidence for the Australia–Western U. S. connection in the Proterozoic. *GSA Today*, **9**, 1–7.

KEPPIE, J. D. & RAMOS, V. A. 1999. Odyssey of terranes in the Iaspetus and Rheic oceans during the Paleozoic. *Geological Society of America, Special Paper*, **336**, 267–276.

KIRSCHVINK, J. L., RIPPERDAN, R. L. & EVANS, D. A. 1997. Evidence for large-scale reorganization of early Cambrian continental masses by inertial interchange time polar wander. *Science*, **277**, 541–545.

KLEMENIC, P. M., POOLE, S. & ALI, S. E. M. 1985. The geo-chemistry of late upper Proterozoic volcanic groups in the Red Sea Hills of NE Sudan–evolution of a late Proterozoic volcanic arc system. *Journal of the Geological Society, London*, **142**, 1221–1233.

KNOPER, M., ANDREOLI, M. A. G., ARMSTRONG, R., TUCKER, R. D. & ASHWAL, L. D. 2001. Dextral transpression and lateral extrusion of Namaqualand during intracontinental convergence, from 1150 to 1120 Ma. *Proceedings of the IGCP 418 Meeting, Natal–Namaqua Belt*, Durban, South Africa, July 13–14.

KRABBENDAM, M. & BARR, T. D. 2000. Proterozoic orogens and the breakup of Gondwana: why did some orogens not rift? *Journal of African Earth Sciences*, **31**, 35–49.

KRONER, A., PALLISTER, J. S. & FLECK, R. J. 1992. Age of initial oceanic magmatism in the Late Proterozoic Arabian shield. *Geology*, **20**, 803–806.

KRONER, A., WINDLEY, B. F., JAECKEL, P., BREWER, T. S. & RAZAKAMANANA, T. 1999. New zircon ages and regional significance for the evolution of the Pan-African orogen in Madagascar. *Journal of the Geological Society, London*, **156**, 1125–1135.

KRONER, A. *ET AL.* 1987. Pan-African crustal evolution in the Nubian segment of NE Africa. *American Geophysical Union, Geodynamic Series*, **17**, 235–258.

KUKLA, P. A. & STANISTREET, I. G. 1991. Record of the

Damaran Khoma Hochland accretionary prism in central Namibia: refutation of an ensialic origin of a Late Proterozoic orogenic belt. *Geology*, **19**, 473–476.

LARSON, R. L. 1991. Latest pulse of Earth: evidence for a mid-Cretaceous superplume. *Geology*, **19**, 547–550.

LI, X. H. 1999. U–Pb zircon ages of granites from the southern margin of the Yangtze block: timing of Neoproterozoic Jinning orogeny in SE China and implications for Rodinia assembly. *Precambrian Research*, **97**, 43–57.

LI, Z. X. & POWELL, C. 1995. South China in Rodinia: part of the missing link between Australia – East Antarctica and Laurentia? *Geology*, **23**, 407–410.

LIEGEOIS, J. P., BERZA, T., TATU, M. & DUCHESNE, J. C. 1996. The Neoproterozoic Pan-African basement from the Alpine Lower Danubian nappe system, S. Carpathians, Romania. *Precambrian Research*, **80**, 281–301.

LOPEZ, R., CAMERON, K. L. & JONES, N. W. 2001. Evidence for Paleoproterozoic, Grenvillian, and Pan-African age Gondwanan crust beneath northeastern Mexico. *Precambrian Research*, **107**, 195–214.

LOWMAN, J. P. & GABLE, C. W. 1999. Thermal evolution of the mantle following continental aggregation in 3D convection models. *Geophysical Research Letters*, **26**, 2649–2652.

LOWMAN, J. P. & JARVIS, G. T. 1999. Effects of mantle heat source distribution on supercontinent stability. *Journal of Geophysical Research*, **104**, 12733–12746.

MABOKO, M. A. H. 1995. Neodymium isotopic constraints on the protolith ages of rocks involved in Pan-African tectonism in the Mozambique belt of Tanzania. *Journal Geological Society, London*, **152**, 911–916.

MABOKO, M. A. H., BOELRIJK, N. A. I. M., PRIEM, H. N. A. & VERDURINEN, E. A. T. 1985. Zircon U–Pb and biotite Rb–Sr dating of the Wami River granulites, eastern granulites, Tanzania: evidence for approximately 715 Ma granulite-facies metamorphism and final Pan-African cooling approximately 475 Ma. *Precambrian Research*, **30**, 361–378.

MACHADO, N., VALLADARES, C., HEILBRON, M. & VALERIANO, C. 1996. U–Pb geochronology of the central Ribeira belt and implications for the evolution of the Brazilian orogeny. *Precambrian Research*, **79**, 347–361.

MARUYAMA, S. 1994. Plume tectonics. *Journal of the Geological Society of Japan*, **100**, 24–49.

MOLLER, A., MEZGER, K. & SCHENK, V. 2000. U–Pb dating of metamorphic minerals: Pan-African metamorphism and prolonged slow cooling of high pressure granulites in Tanzania, East Africa. *Precambrian Research*, **104**, 123–146.

MOORES, E. M. 1991. Southwest U.S.–East Antarctic (SWEAT) connection: a hypothesis. *Geology*, **19**, 425–428.

MURPHY, J. B., KEPPIE, J. D. & HYNES, A. J. 1991. *Geology of the Antigonish Highlands*. Geological Survey of Canada Paper 89–10.

MURPHY, J. B., KEPPIE, J. D., DOSTAL, J. & NANCE, R. D. 1999. Neoproterozoic–Early Paleozoic evolution of Avalonia. *Geological Society of America, Special Paper*, **336**, 1–14.

NANCE, R. D. & MURPHY, J. B., 1994. Contrasting basement isotopic signatures and the palinspastic restoration of peripheral orogens: example from the Neoproterozoic Avalonian–Cadomian belt. *Geology*, **22**, 617–620.

NANCE, R. D., MURPHY, J. B., STRACHAN, R. A., D'LEMOS, R. S. & TAYLOR, G. K. 1991. Late Proterozoic tectonostratigraphic evolution of the Avalonian and Cadomian terranes. *Precambrian Research*, **53**, 41–78.

OPDYKE, N. D., JONES, D. S., MACFADDEN, B. J., SMITH, D. L., MUELLER, P. A. & SHUSTER, R. D. 1987. Florida as an exotic terrane: paleomagnetic and geochronologic investigation of lower Paleozoic rocks from the subsurface of Florida. *Geology*, **15**, 900–903.

PALLISTER, J. S., COLE, J. C., STOESER, D. B. & QUICK, J. E. 1990. Use and abuse of crustal accretion calculations. *Geology*, **18**, 35–39.

PARK, J. K., BUCHAN, K. L. & HARLAN, S. S. 1995. A proposed giant radiating dyke swarm fragmented by the separation of Laurentia and Australia based on paleomagnetism of 780 Ma mafic intrusions in western North America. *Earth & Planetary Science Letters*, **132**, 129–139.

PATRICK, B. E. & MCCLELLAND, W. C. 1995. Late Proterozoic granitic magmatism on Seward peninsula and a Barentian origin for Arctic, Alaska–Chukotka. *Geology*, **23**, 81–84.

PELECHATY, S. M. 1996. Stratigraphic evidence for the Siberia–Laurentia connection and Early Cambrian rifting. *Geology*, **24**, 719–722.

PHARAOH, T. C., WEBB, P. C., THORPE, R. S. & BECKINSALE, R. D. 1987. Geochemical evidence for the tectonic setting of late Proterozoic volcanic suites in central England. Geological Society of London, Special Publication, **33**, 541–552.

PIMENTEL, M. M. & FUCK, R. A. 1992. Neoproterozoic crustal accretion in central Brazil. *Geology*, **20**, 375–379.

PIMENTEL, M. M., FUCK, R. A. & DE ALVARENGA, C. 1996. Post-Braziliano high-K granitic magmatism in central Brazil: the role of Late Precambrian–Early Paleozoic extension. *Precambrian Research*, **80**, 217–238.

PORADA, H. 1989. Pan-African rifting and orogenesis in southern to equatorial Africa and eastern Brazil. *Precambrian Research*, **44**, 103–136.

PRAVE, A. R. 1996. Tale of three cratons: tectonostratigraphic anatomy of the Damara orogen in NW Namibia and the assembly of Gondwana. *Geology*, **24**, 1115–1118.

PRAVE, A. R. 1999. Two diamictites, two cap carbonates, two $d^{13}C$ excursions, two rifts: the Neoproterozoic Kinston Peak Formation, Death Valley, California. *Geology*, **27**, 339–342.

PREISS, W. V. 2000. The Adelaide geosyncline of South Australia and its significance in Neoproterozoic continental reconstruction. *Precambrian Research*, **100**, 21–63.

REISCHMANN, T., KRONER, A. & BASAHEL, A. 1984. Petrography, geochemistry and tectonic setting of metavolcanic sequences from the Al Lith area, SW Arabian shield. In: *Pan-African Crustal Evolution in the Arabian–Nubian Shield*. Faculty of Earth Sciences, King Abdulaziz University, Jeddah, Saudi Arabia, 366–378 (IGCP No. 164).

REYMER, A. & SCHUBERT, G. 1984. Phanerozoic addition

rates to the continental crust and crustal growth. *Tectonics*, **3**, 63–77.

REYMER, A. & SCHUBERT, G. 1986. Rapid growth of some major segments of continental crust. *Geology*, **14**, 299–302.

RIVERS, T. 1997. Lithotectonic elements of the Grenville province: review and tectonic implications. *Precambrian Research*, **86**, 117–154.

ROGERS, J. J. W. 1996. A history of continents in the past three billion years. *Journal of Geology*, **104**, 91–107.

ROSEN, O. M., CONDIE, K. C., NATAPOV, L. M. & NOZHKIN, A. D., 1994. Archean and Early Proterozoic evolution of the Siberian craton: a preliminary assessment. *In*: CONDIE, K. C. (ed.), *Archean Crustal Evolution*. Elsevier Scientific Publishers, Amsterdam 411–459.

SAQUAQUE, A., ADMOU, H., KARSON, J. & HEFFERAN, K. 1989. Precambrian accretionary tectonics in the Bou Azzer–El Graara region, Anti-Atlas, Morocco. *Geology*, **17**, 1107–1110.

SAMSON, R. D. & D'LEMOS, R. S. 1998. U–Pb geochronology and Sm–Nd isotopic composition of Proterozoic gneisses, Channel Islands, UK. *Journal Geological Society, London*, **155**, 609–618.

SEARS, J. W. & PRICE, R. A. 2000. New look at the Siberian connection: no SWEAT. *Geology*, **28**, 423–426.

STEIN, M. & GOLDSTEIN, S. L. 1996. From plume head to continental lithosphere in the Arabian–Nubian shield. *Nature*, **382**, 773–778.

STERN, R. J. 1994. Arc assembly and continental collision in the Neoproterozoic East African orogen: implications for the consolidation of Gondwanaland. *Annual Review of Earth & Planetary Sciences, 1994*, **22**, 319–351.

STERN, R. J. & KRONER, A. 1993. Late Precambrian crustal evolution in NE Sudan: isotopic and geochronologic constraints. *Journal of Geology*, **101**, 555–574.

TIMMONS, J. M., KARLSTROM, K. E., DEHLER, C. M., GEISMAN, J. W. & HEIZLER, M. T. 2001. Proterozoic multistage (1.1 and 0.8 Ga) extension in the Grand Canyon Supergroup and establishment of NW and N–S tectonic grains in the SW United States. *Geological Society of America Bulletin*, **113**, 163–180.

VEIZER, J. 1989. Strontium isotopes in seawater through time. *Annual Review of Earth & Planetary Sciences*, **17**, 141–187.

VEIZER, J. & COMPSTON, W. 1976. $^{87}Sr/^{86}Sr$ in Precambrian carbonates as an index of crustal eveolution. *Geochimica Cosmochimica Acta*, **40**, 905–914.

VEIZER, J., COMPSTON, W., CLAUER, N. & SCHIDLOWSKI, M. 1983. $^{87}Sr/^{86}Sr$ in Late Proterozoic carbonates: evidence for a mantle event at 900 Ma. *Geochimica et Cosmochimica Acta*, **47**, 295–302.

VEIZER, J. ET AL. 1999. $^{87}Sr/^{86}Sr$, ^{13}C and ^{18}O evolution of Phanerozoic seawater. *Chemical Geology*, **161**, 59–88.

VERNIKOVSKIY, V. A., VERNIKOVSKAYA, A. E. & CHERNYKH, A. I. 1998. Neoproterozoic Taymyr ophiolitic blets and opening of the PaleoPacific Ocean. *International Geology Reviews*, **40**, 528–538.

VILLENEUVE, M. & DALLMEYER, R. D. 1987. Geodynamic evolution of the Mauritanide, Bassaride, and Rokelide orogens, West Africa. *Precambrian Research*, **37**, 19–28.

WALTER, M. R., VEEVERS, J. J., CLAVER, C. R., GORJAN, P. & HILL, A. C. 2000. Dating the 840–544 Ma Neoproterozoic interval by isotopes of Sr, C, and S in seawater, and some interpretative models. *Precambrian Research*, **100**, 371–433.

WHITEHOUSE, M. J., WINDLEY, B. F., BA-BTTAT, M. A. O., FANNING, C. M. & REX, D. C., 1998. Crustal evolution and terrane correlation in the eastern Arabian shield, Yemen: geochronological constraints. *Journal of the Geological Society, London*, **155**, 281–295.

WINCHESTER, J. A., MAX, M. D. & LONG, D. B. 1987. Trace element geochemical correlation in the reworked Proterozoic Dalradian metavolcanic suites of the western Ox Mountains and NW May Inliers, Ireland. *Geological Society of London, Special Publication*, **33**, 489–502.

WINDLEY, B. 1992. Proterozoic collisional and accretionary orogens. *In*: CONDIE, K. C. (ed.), *Proterozoic Crustal Evolution*. Elsevier, Amsterdam, 419–446.

WINDLEY, B. F., WHITEHOUSE, M. J. & BA-BTTAT, M. A. O. 1996. Early Precambrian gneiss terranes and Pan-African island arcs in Yemen: crustal accretion of the eastern Arabian shield. *Geology*, **24**, 131–134.

ZIMMER, M., KRONER, A., JOCHUM, K. P., REISCHMANN, T. & TODT, W. 1995. The Gabal Gerf complex: a Precambrian N-MORB ophiolite in the Nubian shield, NE Africa. *Chemical Geology*, **123**, 29–51.

Continental growth in the Proterozoic: a global perspective

BRIAN F. WINDLEY

Department of Geology, University of Leicester, Leicester LE1 7RH, UK
(e-mail: bfw@le.ac.uk)

Abstract: During Proterozoic time, growth of the continents took place by the addition of mantle-derived, juvenile material to pre-existing continental blocks. This accretion took place largely within three tectonic environments: (1) most importantly, in accretionary orogens such as the Birimian, the Baltic Shield, the Arabian–Nubian Shield and the early Altaids in Central Asia – these orogens grew largely by the accretion of island arcs, oceanic plateaus, accretionary prisms and ophiolites; (2) in the juvenile parts of collisional orogens as in the Trans-Hudson and Grenville; (3) within supercontinents that underwent rifting and breakup, giving rise to continental flood basalts and mafic dyke swarms. In addition to plate tectonics, the role of plume tectonics is increasingly emphasized as a fundamental process in Earth evolution. A mantle superplume may increase the oceanic spreading rate, the subduction rate and thus the island-arc production rate. It may also be responsible for the formation of a supercontinent, thus preserving the juvenile parts of collisional orogens, and it may be instrumental in the fragmentation of a supercontinent, giving rise to juvenile continental flood basalts. The balance between these processes is still poorly understood, as are calculated growth rates of Proterozoic crust.

Throughout Earth history the development of the continents has been influenced by the formation of two major types of orogens; continent–continent collisional and accretionary (Windley 1995). Modern analogues of the former are the Himalayas and the European Alps, and of the latter the circum-Pacific belts extending from the Japanese islands via Taiwan and the Philippines to Indonesia, and from Alaska via the Cordillera of Canada and the USA to Mexico. These two types are just the ends of a spectrum of orogens because some are intermediate in type, e.g. the western Himalayas contains a major juvenile island arc in Kohistan. The formation of collisional orogens involved little or no crustal growth, but much reactivation of older continental crust. In contrast, accretionary orogens formed by the accretion of oceanic crust, accretionary prisms, subduction-generated island arcs, plume-generated oceanic plateaus, seamounts and oceanic islands, and thus are characterized by much crustal growth of juvenile material derived directly or indirectly from the mantle. Therefore, when considering Proterozoic orogens, we will be largely concerned with accretionary, rather than collisional, types (Windley 1992).

Most juvenile Proterozoic continental crust formed in three tectonic environments.

(1) Accretionary orogens: western and central Baltic Shield (2.50–1.75 Ga), Birimian, West Africa (2.1 Ga), SW USA, Yavapai (1.8–1.6 Ga), Arabian–Nubian Shield (1.0–0.5 Ga), Cadomian in NW Europe (0.6–0.5 Ga), the early (0.75–0.54 Ga) Altaids in Central Asia. Most accretionary orogens, present and past, are defined largely by the geological, structural and geochemical characteristics of particular rock suites that are diagnostic of, for example, island arcs, accretionary prisms and oceanic plateaus.

(2) The juvenile, minor parts of collisional orogens that involved the incorporation of mid-ocean ridge basalt (MORB)-type basalts, island arcs, oceanic plateaus and ocean island basalts, e.g. the 1.92–1.84 Ga Flin Flon Belt in the Trans-Hudson Orogen (Lucas *et al.* 1996), and within the Grenville Orogen (Condie 2001).

(3) In supercontinents or large continental blocks that were undergoing breakup as a result of mantle upwellings at 2.4–2.1, 1.5–1.3 and 1.0–0.6 Ga. Examples of products are: (a) continental flood basalts (Circum-Superior Province, 1.96 Ga; Coppermine River, northern Canada, 1.27 Ga; the Mid-Continent Rift, USA, 1.1 Ga; Kola Peninsula, 0.6 Ga); (b) giant mafic dyke swarms (worldwide at 2.4–2.0 Ga; Gardar, South Greenland, 1.2 Ga; Sudbury–Mackenzie dykes, 1.2 Ga; Grenville dykes 590 Ma).

The aim of this paper is to present the key features of the main components (e.g. island arcs, oceanic plateaus, continental flood basalts) of the Proterozoic continents that provide evidence of crustal growth. Also, some important criteria, such as Nd isotopes and seismic data, are singled out as useful constraints on juvenile growth, and, in the final section, crustal growth rates are reviewed.

Neodymium (Nd) isotopic mapping

Patchett & Arndt (1986) used Nd isotopes to demonstrate that >80% of the 1.96–1.6 Ga belt that extends from Colorado and Arizona through Michigan and South Greenland to the Baltic Shield, consists of newly differentiated, juvenile material. This belt consists of three provinces that decrease in age from NW–SE (present coordinates): (1) 1.9–1.8 Ga: the Penokean in the Lake Superior region, the Makkovik of Labrador, the Ketilidian of South Greenland and the Svecofennian of the Baltic Shield; (2) 1.8–1.7 Ga: the Yavapai of Arizona and Colorado, the Central Plains Orogen and the Killarney Belt near Lake Huron; (3) 1.7–1.0 Ga: the Matzatzal Belt of Arizona and New Mexico, the Labradorian Orogen, and the Trans-Scandinavian Batholith Belt of Sweden. The juvenile origin of the crust in Texas and Mexico is indicated by εNd values of $+3$–$+5$ (Patchett & Ruiz 1989; Ortega-Gutierrez et al. 1995).

These conclusions were confirmed by Bennett & DePaolo (1987), who concluded that crustal formation in the northern provinces contained only c. 20% of pre-existing crust, and that the southernmost province, most distal from the Archaean nucleus to the north, was derived almost entirely from Proterozoic mantle. In the formation of these provinces an aggregate area of new crust up to 1500 km wide and 5000 km long was accreted in c. 300 Ma (Hoffman 1988).

Seismic reflection studies

Over the last decade, seismic reflection profiling by COCORP-type on-land vibrators or BIRPS-type seaborne surveys have provided new insights to the crust–mantle structure of many Precambrian orogens, demonstrating that they are, in general, geometrically similar to modern orogens. Seismic profiling across the Svecofennian, Lewisian (Britain and Ireland) and Trans-Hudson Orogens have revealed that, in all cases, juvenile Palaeo-Proterozoic arc, oceanic and composite terranes have been detached from their lower crust and mantle during accretion to Archaean cratons. Thus, the juvenile terranes seen at the surface are only crustal flakes, imbricated during collisional and post-collisional events (Snyder et al. 1996). Korja & Heikkinen (1995) demonstrated that listric shear zones in the Svecofennian Orogen flatten at major detachments at depths of 35–40 and 48 km. This conclusion has major implications for interpretations of crustal thickening by juvenile magmatism, which cannot necessarily be extrapolated to Moho depths (e.g. Luosto 1997). Because Reymer & Schubert (1986) were not able to take account of this possibility in the Arabian–Nubian Shield at that time, they assumed that the juvenile greenstone belts at the surface continued all the way down to the Moho, and therefore they arrived at an anomalously high crustal growth rate. In the eastern Arabian Shield of Yemen, Windley et al. (1996) and Whitehouse et al. (1998b) demonstrated that westerly dipping, Early Precambrian gneiss terranes are imbricated with Pan-African island arcs, and this raises the likely possibility that the eastern shield under Saudi Arabia has a more complicated structure than has so far been realized.

Island arcs

Island arcs are an important component of modern accretionary orogens of the western Pacific, and of equivalent Proterozoic orogens: there are several well-documented Proterozoic examples. The Amisk Collage of the Flin Flon Belt and the Reindeer Zone in Canada is a tectonically dismembered collection of accreted juvenile terranes (Stern et al. 1995; Lucas et al. 1996; Lewry & Stauffer 1997; Leybourne et al. 1997). Between 1.904 and 1.890 Ga, tholeiitic and related calc-alkaline basalt–basaltic andesite and rare high-Ca boninites were dominant. The tholeiitic rocks are similar to modern island-arc tholeiites, having low high field strengths (HFSE) and rare earth element (REE) abundances, and chondrite-normalized light REE depletion to slight enrichment. The boninites have even lower HFSE and REE abundances. Between 1.89 and 1.864 Ga, calc-alkaline andesite–rhyolite and rare shoshonite and trachyandesite erupted with strong arc trace element signatures [e.g. high Th/Nb, La/Nb, strong negative Nb anomalies and large-ion lithophile element (LILE) enrichment], and initial εNd values ($+2.3$–$+4.6$) indicate depleted mantle contributions (Stern et al. 1995).

In the SW Baltic Shield at 1.69–1.65 Ga, felsic and basaltic lavas have markedly primitive trace element signatures and depleted Nd isotopic compositions, all consistent with derivation in an oceanic island-arc setting, possibly similar to those in the Philippine Sea (Brewer et al. 1998). Elsewhere in the Swedish Svecofennian at 1.758 Ga, meta-andesites are chemically similar to primitive modern oceanic island arcs and an εNd value of $+4.3$ indicates a depleted mantle source (Åhäll & Daly 1989).

The Birimian Orogen of West Africa underwent major crustal growth at c. 2.1 Ga, which lasted <50 Ma (Boher et al. 1992). Arc-derived andesites, dacites and rhyolites are an important component, and calc-alkaline granites make up almost half the Birimian terranes. However, it is strange that the pillowed tholeiitic basalts are overlain by pelagic cherts, shales and clastic turbidites, because that type of oceanic plate stratigraphy in the modern accretionary orogen of Japan is characteristic not of island arcs but of off-scraped oceanic floor, the sedimentary sequence indicating transport from a ridge to a trench. The detailed study by Béziat et al. (2000) demonstrated that calc-alkaline basalts and rhyolites, which are associated with pyroclastics, show classic features of arc magmatic suites, namely LILE (Large Ion Lithophile Element) and Pb enrichment, depleted HFSE patterns and high Ce/Nb and Th/Nb ratios; associated wehrlites and gabbros represent the roots of the island arc.

The Early Proterozoic Ungava Orogen in Canada is well known for the 2.00 Ga Portuniq Ophiolite. An island arc was built on the oceanic crust between 1.90 and 1.86 Ga, and a younger subduction zone gave rise to 1.85–1.83 Ga quartz diorite to granite plutons that intruded the older oceanic crust and arc (Lucas et al. 1992).

In French Guiana early continental crust accretion is characterized by the formation of volcanic centres with calc-alkaline tuff lavas and pyroclastites associated with plutonic–volcanic complexes composed largely of calc-alkaline tonalite–diorite batholiths dated at 2144±6 and 2115±7 Ma (Vanderhaeghe et al. 1998). Also in French Guiana the 2.11±0.09 Ga Inini Greenstone is dominated by calc-alkaline andesite to rhyolite, intruded by plutons of tonalite and trondhjemite, all belonging to an island arc; diamond-bearing, mantle-derived, ultramafic komatiites form part of the volcanic sequence (Capdevila et al. 1999).

Further examples of Proterozoic island arcs include: the 1.87 Ga South Harris igneous complex, Outer Hebrides, NW Scotland (Whitehouse & Bridgwater 2001), a 1.92 Ga arc in the Nagssugtoqidian Orogen of West Greenland (Whitehouse et al. 1998a); a 1.742 Ga arc in the Grand Canyon in SW USA (Ilg et al. 1996); the Kaourera Island Arc in the Kibaran Belt of central Africa that is tectonically interleaved with the 1.393 Ga Chewore Ophiolite (Johnson & Oliver, 2000), and a variety of arcs formed during the 1.950–1.700 Ga assembly of the West, North and South Australian Cratons (Myers et al. 1996). The Cadomian–Avalonian accretionary orogen of NW Europe was constructed from oceanic crust and island arcs in the period of 800–640 Ma, creating the arc-dominated terrane of Avalonia [e.g. in NW Iberia (Fernández-Suárez et al. 2000) and in Wales (Bevins et al. 1995)].

Neoproterozoic island arcs are common in the Arabian–Nubian Shield, where they were mostly formed in the period of 900–700 Ma. The oldest arcs (900–850 Ma) in Saudi Arabia consist of tholeiitic andesites and are thought to represent young immature island arcs. Thickening and melting of the immature tholeiitic crust caused the formation of more mature island arcs made up of calc-alkaline low- to high-K tonalites, trondhjemites and andesites in the period of 825–730 Ma (Blasband et al. 2000). In their plume-oriented model for the Arabian–Nubian Shield, Stein & Goldstein (1996) suggested that an oceanic plateau resisted subduction and that this enabled arcs to develop by subduction on its margins.

A recent development has been the recognition of juvenile island-arc material within deep crustal rocks, such as the enderbite–charnockite suite of the Umba Complex in the Kola Peninsula (Glebovitsky et al. 2001).

Oceanic plateaus

Many oceanic plateaus are prominent in the present Pacific Ocean and increasing evidence is accruing that similar plateaus accreted during the Mesozoic–Cenozoic to parts of the circum-Pacific accretionary orogens. An oceanic plateau implies the presence of a mantle plume not related to ridge-subduction tectonics.

Few detailed studies have yet been made of Proterozoic oceanic plateaus, but information is beginning to appear of their presence. Abouchami et al. (1990) suggested that many Birimian basalts represent oceanic plateaus, and Boher et al. (1992) argued that the island arcs they discovered (referred to above) formed on top of the assumed oceanic plateaus. Stein & Goldstein (1996) interpreted isotopic and geochemical data from the Arabian–Nubian Shield to indicate that a plume head generated an oceanic plateau, which later resisted subduction during convergence and was overprinted with continent-like characteristics. However, a worrying aspect of this work is that neither the authors or subsequent workers have described lithologies, structure and petrology, which would confirm or constrain the isotopic–geochemical model.

In the c. 1.9 Ga Flin Flon Belt of Canada, the Sandy Bay Assemblage is a c. 3 km thick, monotonous sequence of basalt flows, synvolcanic diabase and gabbro sills. Lucas et al. (1996) found that the basalts are geochemically distinct from those of arc and ocean-floor basalts. Their trace element characteristics include strong enrichment in HFSE (Nb, Zr, Ti), light REE enrichment, high Ti/V and low Zr/Nb, juvenile Nd isotopes, and, most importantly, fractionated heavy REE, suggesting the involvement of residual garnet during melting. Lucas et al. (1996) followed Stern et al. (1995) in proposing an oceanic plateau or oceanic island origin.

The Loch Maree Group in the Lewisian of NW Scotland contains basalts associated with abyssal sediments and ferruginous hydrothermal deposits (Park et al. 2001). Geochemistry of the basalts suggests derivation from an oceanic plateau or primitive arc, and the associated Ard Gneiss, which has primitive geochemical patterns, possibly formed by melting of an underplated oceanic plateau.

Continental flood basalts

Continental flood basalts represent major additions to the crust of plume-generated, mantle-derived magmas. The breakup of a 2.5 Ga supercontinent (Windley 1995) was expressed by global mafic magmatism at 2.45 Ga, the remnants of which are so abundant and extensive that they reasonably constitute a large igneous province (Heaman 1997). They include continental flood basalts, mafic dyke swarms, layered mafic–ultramafic intrusions and rift-related alkaline igneous rocks. The volume of this magmatic material certainly rivals that of Mesozoic large igneous provinces. It is known that the fragmentation of the supercontinent of Pangaea was diachronous and lasted for at least 250 Ma. Similarly, the huge number of mafic dykes and associated igneous rocks worldwide that intruded in the period of 2.45–c. 2.1 Ga probably resulted from episodic, semi-continuous attempts at further continental breakup. We are only witnessing the failed attempts at such plume-generated continental breakup preserved on exposed continental margins; many other areas are no doubt under sedimentary basins or were destroyed by collisional orogenesis.

Continental flood basalts formed at two periods in the Paleoproterozoic in the northern Baltic Shield. Numerous well-dated, large, layered mafic–ultramafic intrusions were emplaced in the period of 2.4–2.5 Ga (Amelin et al. 1995). They vary in composition from lherzolite and olivine gabbronorite to norite, anorthosite and hypersthene diorite, and represent two types of flood basalt series, whose parental magmas were generated in a mantle plume (Amelin & Semenov 1996). Interestingly, Kempton et al. (2001) found that granulite-facies xenoliths hosted in Devonian lamprophyres within the same area in the Kola Peninsula represent high-grade equivalents of the continental flood basalts. Also, at 2.4 Ga in the southeastern Baltic Shield, plume-generated komatiitic basalts were erupted in an abortive continental rift (Puchtel et al. 1997). In the Kola Peninsula the 1000 km long Pechenga Greenstone Belt has a total tectonic thickness of c. 16 km (probably thickened by imbrication) and formed for an unknown duration during the period of 2.50–1.80 Ga in a vast continental rift according to Melezhik & Sturt (1994). The rift contains many plume-derived layered gabbro–norite complexes enriched in PGE (Platinum Group Element) and chromite mineralization. Ferropicrites from c. 1.98 Ga are associated with major Ni–Cu deposits, most likely derived from Fe-rich, ^{187}Os-enriched mantle plumes (Walker et al. 1997).

A second vast continental flood basalt province in the northern Baltic Shield was reported by Puchtel et al. (1998). Submarine basalts, reaching up to 4.5 km in thickness, occur within several epicontinental basins that are remnants of a once-continuous cover over earlier basement. These plume-generated rocks occur in an area of c. 600 000 sq. km^2 of the northern Baltic Shield and represent a major contribution of juvenile material to the existing continental crust. Sm–Nd mineral and Pb–Pb whole-rock isochron ages of 1975 ± 24 and 1980 ± 57 Ma, respectively, from the upper part, and a SHRIMP U–Pb zircon age of 1976 ± 9 Ma from the lower part of the basalt pile imply a short time span of formation. Uppermost lavas have high $(Nb/Th)_N (= 1.4–2.4)$ and $(Nb/La)_N (= 1.1–1.3)$ values, an εNd (T) value of $+3.2$, and an unradiogenic Pb-isotope composition ($\mu_1 = 8.57$), all comparable with those of modern oceanic plume-derived magmas (oceanic flood basalt and oceanic island basalt). Puchtel et al. (1998) also concluded that the estimated Nb/U ratios of 53 ± 4 in the uncontaminated lavas are similar to those found in the modern mantle (c. 47), suggesting that by 2.0 Ga a volume of continental crust similar to the present-day value already existed.

Many, now isolated, segments of continental basalts that now rim the Superior Province of northern Quebec probably once belonged to an extensive continental flood basalt province, the best continuous sequence of which is exposed in the Belcher Islands (Legault et al. 1994); ages range from 1798 ± 38 (whole-rock Rb–Sr isochron) to 1960 ± 80 Ma (Pb–Pb isochron).

The 1.27 Ga Coppermine River flood basalts in the Northwest Territories of Canada were emplaced during the Mackenzie igneous event, that included the coeval, layered, mafic–ultramafic Muskox Intrusion and the vast Mackenzie Dyke Swarm that radiates from the Coppermine River basalts (Dupuy et al. 1992; Griselin et al. 1997). The basalts that overlie continental basement comprise c. 150 flows, each 4–100 m thick, many of which can be traced laterally for several tens of kilometres (maximum total thickness is 4.7 km). They were emplaced in a short time period of <5 Ma. Petrochemistry is interpreted to indicate that the lowermost lavas were produced by melting in the garnet stability field at a depth >90 km, and probably in a mantle plume beneath the continental lithosphere. Upper lavas were partly contaminated with crustal rocks as the magmas passed through the lower and upper crust.

Mafic dykes

The numbers of mafic dykes and dyke swarms intruded worldwide, especially in the Palaeoproterozoic and Late Neoproterozoic, are too numerous to enumerate; most are recorded in Ernst & Buchan (2001a). Giant, radiating dyke swarms that may be linked with the evolution of mantle plumes were reviewed by Ernst & Buchan (2001a). Many dyking events between 2.4 and 2.0 Ga can be correlated across continents, e.g. between the Canadian and Fennoscandian Shields (Park 1994; Vogel et al. 1998). From a study of the Th/Ta and La/Yb ratios of Proterozoic mafic dykes, Condie (1997a) concluded that there is an overall shift in composition of dykes from high Th/Ta ratios in the Palaeoproterozoic to low ratios in the Neoproterozoic. This reflects a decrease in the importance of Archaean subcontinental lithospheric sources and an increase in importance of plumes containing enriched mantle components such as recycled sediments and oceanic lithosphere.

The breakup of Baltica from Laurentia is marked by huge doleritic sill complexes in Fennoscandia at 1.27 Ga (Elming & Mattsson 2001). Park et al. (1995) proposed that during the breakup of Rodinia a giant, radiating, plume-generated mafic dyke swarm was emplaced at c. 780 Ma, and is now fragmented by the separation of western North America and eastern Australia. However, Wingate et al. (1998) found that the Gairdiner Dyke Swarm in Australia has a U–Th–Pb baddeleyite–zircon age of 827 ± 6 Ma, c. 40 Ma too early.

Finally, the opening of the Iapetus Ocean was marked by the emplacement of the 616 Ma Egersund Dyke Swarm in SW Norway (Bingen et al. 1998), and of the 700 km long, plume-generated, radiating 590 Ma Grenville Dyke Swarm in Canada (Seymour & Kumarapeli 1995).

Ophiolites and the ocean floor

Proterozoic ophiolites are not common, but neither are ophiolites common in the Himalayas (collisional orogen) or Japan (accretionary orogen). However, the fact that some examples with a full ophiolite stratigraphy do occur (Anon 1972), indicates that oceanic ridge and subduction processes were in operation back in time to at least the Palaeo-Proterozoic. Although volumetrically small, these ophiolites do provide key information on the role of plate tectonic processes.

The 1.998 Ga Portuniq Ophiolite in the Cape Smith Belt of Canada is situated on the continental margin of an Archaean craton. It consists of two magmatic suites (Scott et al. 1992; St-Onge et al. 1997). The lower one has a complete ophiolite stratigraphy and its composition is similar to rocks formed at present-day mid-oceanic ridges. The younger suite has sheeted mafic dykes and mafic to ultramafic cumulate rocks that are geochemically similar to tholeiites found in modern plume-generated oceanic islands, such as Hawaii.

The 1.96 Ga Jormua Ophiolite in Finland is situated on the margin of an Archaean craton (Kontinen 1987). The gabbros are similar to those in high-Ti ophiolites dredged from present-day mid-oceanic ridges. The chemistry of the basalts indicates they are comparable to Red-Sea-type basalts (Peltonen et al. 1996). The 1.901 Ga Elbow–Athapapuskow Ophiolite in the Flin Flon Belt of Canada has pillow basalts intruded by diabase sills, gabbros and mafic–ultramafic cumulates (Lucas et al. 1996).

The 1.73 Ga Payson Ophiolite in Arizona, within the juvenile Yavapai–Mazatzal Orogen, has submarine basalts, sheeted dykes (1–2 km thick) and gabbro, and was erupted upon a magmatic arc (Dann & Bowring 1997). The sheeted dykes are tholeiitic basalts with an island-arc affinity.

There are not many mid-Proterozoic orogenic belts or ophiolites. However, the Zambezi Belt between the Congo and Zimbabwe Cratons contains the remains of the 1.4 Ga (zircon age of a plagiogranite dyke) Chewore Ophiolite, which occurs as discontinuous pods associated with yoderite-bearing whiteschists within a subduction complex. Serpentinites, gabbros, sheeted dolerite dykes and basalts can be recognized at a greenschist grade (Oliver et al. 1998). One group of meta-basalts is similar to modern N-MORB, and another resembles modern island-arc basalts. This is the oldest dated remnant of Proterozoic oceanic crust in Africa.

Yakubchuk et al. (1994) pointed out that there was an ophiolite pulse in the Neoproterozoic spread over the full age range from 1000 to 570 Ma, with a pronounced concentration at 750 Ma and a lesser pulse at 600 Ma. Ophiolites of 750 Ma and older are common in Arabia, Africa and South America, whereas 600 Ma ophiolites are abundant in Central Asia. A prominent, semi-continuous belt of Neoproterozoic ophiolites extends around the margin of the Siberian Craton from the Taimyr Peninsula to the Yenesei Range, and Eastern Sayan to the Baikal Uplands and Transbaikalia (Khain et al. 1997).

In the Arabian–Nubian Shield there are 15 ophiolites with the full Penrose-type stratigraphy (but highly imbricated); eight in Saudi Arabia, five in Sudan and two in Egypt (P. R. Johnson, pers. comm.). They range in age from c. 870 to 730 Ma and most are allochthonous sheets situated in arc–arc sutures (Berhe 1997). For example, the 740 Ma Gabal Gerf Ophiolite on the Egypt–Sudan border has pillow lavas and sheeted dykes whose major- and trace-element data, including REE, are indistinguishable from modern high-Ti N-MORB. In fact, this is the only Precambrian ophiolite with N-MORB chemistry (Zimmer et al. 1995). Together with the island arcs mentioned above, these ophiolites in the Arabian–Nubian Shield document a major phase of juvenile crustal growth in the Neoproterozoic.

The early development of the Central Asian Mobile Belt or Altaids (Sengör et al. 1993) during the Neoproterozoic was dominated by oceanic plate accretion and subduction, and the emplacement of ophiolites. The largest ophiolite in Central Asia (with complete Penrose stratigraphy) is the 300 km long, 20 km wide, 569±21 Ma Bayankhongor Ophiolite in Mongolia, which probably occupies a suture zone between two continental blocks (Buchan et al. 2001). Several other ophiolites, e.g. Dariv, Khantaishir and Tuva (southern Russia), have similar ages.

Crustal growth rates

It is important, if possible, to make quantitative estimates of the growth rate of Proterozoic continental crust in specific orogens or parts of the world. Note that the rates are discussed in terms of volumes of continental crust. According to Reymer & Schubert (1986), the average production rate of continental crust during the Phanerozoic has been 1.1 ± 0.5 km^3/a. They went on to calculate and suggest that the crust formation rates during some geological periods and places, e.g. the Arabian–Nubian Shield and the Superior Province, exceeded by far the Phanerozoic rate. However, they assumed, reasonably at that time, that the juvenile greenstone belts on the surface continue all the way to the Moho; this was a necessary assumption in order to arrive at a volume calculation. However, it is now known from gravity data and especially from seismic profiling projects such as LITHOPROBE in Canada, that greenstone belts only continue to a few kilometres depth. For example, the greenstone belts of the Superior Province, which have a thickness of only c. 5 km, are situated on a major shear zone below which is reflective deeper crust (Clowes et al. 1996). The nearby Kapuskasing Uplift, which provides a complete crustal section, shows that the deeper crust consists of granulite facies mafic, and felsic, gneisses and anorthosite (Percival & West 1994). Gravity anomaly data suggest that the Barberton Greenstone Belt in South Africa does not extend beyond 6 km in depth, and 3 km is most likely (Darracott 1975). Although these examples are from Archaean juvenile greenstone belts, Snyder et al. (1996) concluded from their overall survey of seismic reflection profiles in the Early Proterozoic Svecofennian, Lewisian and Trans-Hudson Orogens that these juvenile terranes were detached from their lower crust during accretion to Archaean margins, and that the juvenile terranes are only crustal flakes, imbricated and internally deformed. This idea is supported by the discovery that Early Precambrian gneiss terranes in the eastern Arabian Shield of Yemen are imbricated with Neoproterozoic island arcs (Windley et al. 1996; Whitehouse et al. 1998b). Therefore, it can no longer be assumed that Proterozoic juvenile material in arc-rich greenstone belts continues to the Moho, in which case the very high crustal production rates quoted above are readily explicable, i.e. the actual rates were much less than those calculated.

Patchett & Arndt (1986) estimated that the overall crust production rates in central Laurentia and Baltica in the period of 1.9–1.7 Ga was c. 1.2 km^3/a, which is slightly greater than the total Phanerozoic island-arc accretion rate. Because this rate was derived only from the Yavapai–Penokean–Svecofennian Accretionary Orogen, they concluded that the global Early Proterozoic rate was around double the present rate. Condie (2001) estimated, from the aerial extent of crust on geological and tectonic maps, that crustal production rates during the formation of Rodinia in the period of 1.35–0.9 Ga fell within the 1.1 km^3/a average Phanerozoic production rate. But most parts of the Grenvillian Orogen lack sufficient geophysical data to make accurate estimates of volume. Also, the average rate for the Phanerozoic, quoted above, was calculated before the enormous extent of the largely Palaeozoic, juvenile, Altaid Orogen in Central Asia was known, and therefore must be suspect for purposes of comparison with Proterozoic rates.

The above estimates were of crustal production rates. However, the crustal net growth rate, which is equal to the mantle extraction rate minus the recycling rate, would provide a minimum estimate of crustal growth rates in the Proterozoic. The recycling rate includes all forms of recycling, including crustal delamination, subduction of sediments, oceanic crust, oceanic plateaus and island arcs, and erosion and redeposition in sedimentary basins. However, quantitative estimates of these Proterozoic processes are not known.

There seems little point in using the present-day thickness of Proterozoic continental crust anywhere in the world for these calculations, because many of the orogens concerned probably had a much greater crustal thickness at the time of their formation, and these thicknesses have been subsequently modified by erosion and/or extensional collapse. Abbott et al. (2000) calculated that the average thickness of Early Proterozoic crust was between 48 ± 9 and 60 ± 7 km (depending on the methods of calculation), but hardly anywhere do Early Proterozoic orogens today have this thickness. Only the Svecofennian Orogen has (up to) 60 km thick crust today and the original crust was even thicker, because the surface rocks are in the amphibolite or granulite facies. So, because most Proterozoic orogens today probably have a crustal thickness which is less than when originally formed, it is difficult to use present-day thicknesses in any calculations of production rate or net growth rate, and in any case the juvenile orogens at the present surface most likely do not continue to the Moho. Therefore, it seems that most estimates of the rate of growth of Proterozoic crust are premature and unreliable.

Discussion

The above brief summary of published magmatic rocks worldwide illustrates the current state of knowledge of the contribution of juvenile material to continental growth in the Proterozoic and provides a useful databank or background for discussion.

The magmatic rocks that contributed to crustal growth in the Proterozoic were generated in part by plate tectonic processes, such as oceanic plate accretion and subduction giving rise to ophiolites and island arcs, and in part by plume tectonic processes, giving rise to oceanic islands and plateaus, continental flood basalts and radiating dyke swarms. Supercontinents were important in this story because, on the one hand, they trapped fragments derived from the oceans and, on the other, because they acted as the framework for plume-generated breakup.

In the last few decades a variety of very different, theoretically based models have been produced to explain the secular development of continental crust with time. In recent years compilations of Nd isotopic data and U–Pb zircon ages suggest the following:

(1) At least 27–30% (Abbott et al. 2000, fig. 5) or 29–45% (Abbott et al. 2000, abstract) of the continental crust was extracted from the mantle by the start of the Proterozoic at 2.5 Ga.
(2) Between 50 and 52% (Abbott et al. 2000, fig. 5) or between 51 and 79% (Abbott et al. 2000, abstract) of the present volume of the continents existed by 1.8–2.0 Ga.
(3) Between 7 and 13% of the continental crust was formed between 1.35 and 0.9 Ga (Condie 2001)

In a detailed survey, Condie (1997b) concluded that of 96 post-Archaean greenstones, only c. 10% have oceanic plateau–MORB affinities, the bulk of the greenstones having arc signatures. From a study of the La/Nb ratio and Ni contents of mafic lower crustal xenoliths, Condie (1997b, 1999) discovered that about one third of the post-Archaean lower continental crust is composed of mafic rocks with mantle-plume signatures (oceanic plateaus) and the remainder are chiefly arc material. He concluded that the value of one third is a minimum for the plume component in the lower crust.

There are two fundamentally different models to explain the primary mechanisms that controlled the formation of continental crust. The first model is the **subduction model**, according to which new continents are formed at subduction zones with the result that >90% of the continental crust can be accounted for by convergent margin magmatism (Kay & Kay 1986; Rudnick et al. 1998). This model is based not only on field relations but also on trace element geochemical data using modern arc analogues. The results indicate that some Precambrian arcs formed through wet melting of upwelling asthenosphere at the initiation of oceanic subduction, others were generated from the mantle wedge during more mature stages of arc evolution, and yet others were generated through melting of young and/or warm slabs (Tarney & Jones 1994).

The second model is the **mantle plume model**, which states that new continents are formed by high degrees of partial melting within mantle plumes, giving rise to oceanic plateaus which are too thick and buoyant to subduct, and so accrete to and become part of a continent (Stein & Hofmann 1994; Abbott et al. 2000). The plume model is based on the calculation that oceanic plateaus thicker than 17 km are unsubductable (Cloos 1993). However, a variety of evidence suggests that parts or the whole of thick oceanic plateaus are able to subduct. For example, the Mesozoic Sanbagawa Belt of southern Japan contains a 2 km thick slab of an oceanic plateau originally 30 km thick, the Iratsu Body, which has been subducted to 90 km depth at 3 GPa and then raised by wedge extrusion as an eclogite into the accretionary orogen (Ota et al. 2002). Also, the lower ultramafic unit of the 1000 km long Sorachi Oceanic Plateau in northern Japan (Kimura et al. 1994), which attained a tectonized thickness of 30 km, is missing, presumed to have been subducted. Although a 4 km thick section of the 30 km thick Ontong Java Plateau is being obducted onto the Solomon Islands, three-dimensional tomographic inversion shows that a low-velocity root of the plateau has been subducted to 300 km depth (Klosko et al. 2001). Two-dimensional finite element modelling by van Hunen et al. (2002) shows that an 18 km thick oceanic plateau can subduct, causing development of a shallow-dipping or flat slab. Finally, Saunders et al. (1996) concluded that after 100 Ma a thick oceanic plateau may become potentially negatively buoyant and deeper zones will transform to eclogite, and so will be able to spontaneously subduct. With the calculation that oceanic plateaus <17 km in thickness are inherently subductable (Cloos 1993), and with the evidence of partial or complete subduction of thicker oceanic plateaus, it is not surprising that only five examples of obducted plateaus have been identified in the Mesozoic and Cenozoic geological record (Coffin & Eldholm 2001). Accordingly, the fact that modern oceanic plateaus can subduct, weakens the plume model for the growth of the continental crust.

The episodic growth of continents has been known for several decades from the episodic peaks of isotopic ages. In a recent re-evaluation, Condie (2000) concluded that the continents grew episodically, with major periods of growth (superevents) at 2.7 and 1.9 Ga, each superevent lasting only c. 800 Ma. The superevents are episodes of enhanced exchange between the lower and upper mantle, which replenish the upper mantle with juvenile trace elements, leading to rapid growth of the continents. Accordingly, Stein & Hofmann (1994) related these periods of major addition to mantle plumes that would produce oceanic plateaus. One result of a plume or superplume event would be increased sea-floor spreading rates, as envisaged by Larson (1991), which would lead to increased subduction and a higher rate of island-arc production. Another result would be increased production of oceanic plateaus which, if not subducted, would accrete and increase the volume of continental crust (Abbott *et al.* 2000).

All the above data and models take on a new perspective when tomographic data are taken into account. They are interpreted to indicate that cold lithospheric slabs are subducted down to the D″ core–mantle boundary, and some of that material is later incorporated into plumes or superplumes giving rise to oceanic plateaus, continental flood basalts and mafic dyke swarms. Condie (1998) related the superevents in the mantle at 2.7 and 1.9 Ga to catastrophic slab avalanching at the 660 km boundary.

Conclusions

Consideration of the data, ideas and models reviewed in this paper leads to the following conclusions.

(1) Both plate tectonics and plume tectonics played important roles in controlling growth of the continents during the Proterozoic. The former gave rise to oceanic plate accretion and subduction, expressed as island arcs, oceanic crust and ophiolites. The latter gave rise to oceanic islands and plateaus, continental flood basalts and mafic dyke swarms.
(2) The oceanic parts of the above scenario are mostly preserved in accretionary orogens, the modern equivalents of which are in the western Pacific (Japan to Indonesia).
(3) The two main competing models to explain the formation of the continental crust [subduction model (island arcs) v. plume model (oceanic plateaus)] have strong advocates. Problems associated with these models in relation to the Proterozoic include the following: (a) examination of the geological record shows that island arcs are the principal component of Proterozoic accretionary orogens; (b) the plume/oceanic plateau model relies heavily on the calculation that thick oceanic plateaus are too buoyant to subduct, and therefore must accrete and contribute substantially to crustal growth. However, only a few oceanic plateaus have been found in Proterozoic (and Archaean) orogens, and only five in the Mesozoic–Cenozoic geological record, and in the modern accretionary orogens of the western Pacific substantial parts of oceanic plateaus have been subducted. What is being seen today is only the remnants that have accreted to the trenches and the accretionary orogens.
(4) Calculation of crustal production rates and crustal net growth rates of Proterozoic crust are premature, because it is not known how much of the continental crust has been eroded and recycled or subducted, what parts of and how much of the components of the oceanic crust have been subducted, and so whether or not the preserved parts played an important or minor role in the accretion and growth of the continents. Finally there are still insufficient seismic surveys to confirm preliminary data which suggests that the crust in accretionary and collisional orogens has an imbricated structure, which implies that juvenile rocks at the present surface may not continue down to the Moho.

This paper was written during a 1-year sabbatical visit to the Tokyo Institute of Technology, Japan, and I wish to thank Shigenori Maruyama for support in a stimulating research environment. I am grateful to Kent Condie and Bryan Storey for incisive, helpful reviews.

References

ABBOTT, D., SPARKS, D., HERZBERG, C., MOONEY, W., NIKISHIN, A. & ZHANG, Y. S. 2000. Quantifying Precambrian crustal extraction: the root is the answer. *Tectonophysics*, **322**, 163–190.

ABOUCHAMI, W., BOHER, M., MICHARD, A. & ALBAREDE, F. 1990. A major 2.1 Ga event of mafic magmatism in West Africa: an early stage of crustal accretion. *Journal of Geophysical Research*, **95**, 17605–17629.

ÅHÄLL, K. I. & DALY, J. S. 1989. Age, tectonic setting and provenance of Östfold–Marstrand belt supracrustals: westward crustal growth of the Baltic Shield at 1760 Ma. *Precambrian Research*, **45**, 45–61.

AMELIN, Y. V. & SEMENOV, V. S. 1996. Nd and Sr isotopic geochemistry of mafic layered intrusions in the eastern Baltic shield: implications for the evolution of Paleoproterozoic continental mafic magmas. *Contributions to Mineralogy and Petrology*, **124**, 255–272.

AMELIN, Y. V., HEAMAN, L. M. & SEMENOV, V. S. 1995. U–Pb geochronology of layered mafic intrusions in the eastern Baltic Shield: implications for the timing and duration of Paleoproterozoic continental rifting. *Precambrian Research*, **75**, 31–46.

ANONYMOUS. 1972. Penrose Field Conference Report on Ophiolites. *Geotimes*, **17**, 24–25.

BENNETT, V. C. & DEPAOLO, D. J. 1987. Proterozoic crustal history of the western United States as determined by neodymium isotopic mapping. *Geological Society of America Bulletin*, **99**, 674–685.

BERHE, S. M. 1997. The Arabian–Nubian Shield. *In:* DE WIT, M. & ASHWAL, L. D. (eds) *Greenstone Belts*. Clarendon Press, Oxford, 761–771.

BEVINS, R. E., PHARAOH, T. C., COPE, J. C. W. & BREWER, T. S. 1995. Geochemical character of Neoproterozoic volcanic rocks in southwest Wales. *Geological Magazine*, **132**, 339–349.

BÉZIAT, D., BOURGES, F., DEBAT, P., LOMPO, M., MARTIN, F. & TOLLON, F. 2000. A Paleoproterozoic ultramafic–mafic assemblage and associated volcanic rocks of the Boromo greenstone belt: fractionates originating from island-arc volcanic activity in the West African craton. *Precambrian Research*, **101**, 25–47.

BINGEN, B., DEMAIFFE, D. & VAN BREEMEN, O. 1998. The 616 Ma-old Egersund basaltic dike swarm, SW Norway, and Late Neoproterozoic opening of the Iapetus Ocean. *Journal of Geology*, **106**, 565–574.

BLASLUND, B., WHITE, S., BROOIJMANS, P., DE BROODER, H. & VISSER, W. 2000. Late Proterozoic extensional collapse in the Arabian–Nubian Shield. *Journal of the Geological Society, London*, **157**, 615–628.

BOHER, M., ABOUCHAMI, A., MICHARD, F., ALBAREDE, F. & ARNDT, N. T. 1992. Crustal growth in West Africa at 2.1 Ga, *Journal of Geophysical Research*, **97**, 347–369.

BREWER, T. S., DALY, J. S. & ÅHÄLL, K. I. 1998. Contrasting magmatic arcs in the Palaeoproterozoic of the south-western Baltic Shield. *Precambrian Research*, **92**, 297–315.

BUCHAN, C., CUNNINGHAM, D., WINDLEY, B. F. & TOMURHUU, D. 2001. Structural and lithological characteristics of the Bayankhongor ophiolite zone, Central Mongolia. *Journal of the Geological Society, London*, **158**, 445–460.

CAPDEVILA, R., ARNDT, N., LETENDRE, J. & SAUVAGE, J-F. 1999. Diamonds in volcaniclastic komatiite from French Guiana. *Nature*, **399**, 456–458.

CLOOS, M. 1993. Lithospheric buoyancy and collisional orogenesis: subduction of oceanic plateaus, continental margins, island arcs, spreading ridges, and seamounts. *Geological Society of America Bulletin*, **105**, 715–737.

CLOWES, R. M., CLAVERT, A. J., EATON, D. W., HAJNAL, Z., HALL, J. & ROSS, G. M. 1996. LITHOPROBE reflection studies of Archean and Proterozoic crust in Canada. *Tectonophysics*, **264**, 65–88.

COFFIN, M. E. & ELDHOLM, O. 2001. Large igneous provinces: progenitors of some ophiolites? *Geological Society of America, Special Paper*, **353**, 59–70.

CONDIE, K. C. 1997a. Sources of Proterozoic mafic dyke swarms: constraints from Th/Ta and La/Yb ratios. *Precambrian Research*, **81**, 3–14.

CONDIE, K. C. 1997b. Contrasting sources for upper and lower continental crust: the greenstone connection. *Journal of Geology*, **105**, 729–736.

CONDIE, K. C. 1998. Episodic continental growth and supercontinents: a mantle avalanche connection? *Earth and Planetary Science Letters*, **163**, 97–108.

CONDIE, K. C. 1999. Mafic crustal xenoliths and the origin of the lower continental crust. *Lithos*, **46**, 95–101.

CONDIE, K. C. 2000. Episodic continental growth models: afterthoughts and extensions. *Tectonophysics*, **322**, 153–162.

CONDIE, K. C. 2001. Continental growth during formation of Rodinia at 1.35–0.9 Ga. *Gondwana Research*, **4**, 5–16.

DANN, J. C. & BOWRING, S. A. 1997. The Payson ophiolite and Yavapai–Mazatzal orogenic belt, central Arizona. *In:* DE WIT, M. & ASHWAL, L. D. (eds) *Greenstone Belts*. Clarendon Press, Oxford, 781–790.

DARRACOTT, B. W. 1975. The interpretation of the gravity anomaly over the Barberton Mountain Land, South Africa. *Transactions of the Geological Society of South Africa*, **78**, 123–128.

DUPUY, C., MICHARD, A., DOSTAL, J., DAUTEL, D. & BARAGAR, W. R. A. 1992. Proterozoic flood basalts from the Coppermine River area, Northwest Territories: isotope and trace element geochemistry. *Canadian Journal of Earth Sciences*, **29**, 1937–1943.

ELMING, S.-Å. & MATTSSON, H. 2001. Post-Jotnian basic intrusions in the Fennoscandian shield and the break-up of Baltica from Laurentia: a palaeomagnetic and AMS study. *Precambrian Research*, **108**, 215–236.

ERNST, R. E. & BUCHAN, K. L. 2001a. Large mafic magmatic events through time and links to mantle plume heads. *In:* ERNST, R. E. & BUCHAN, K. L. (eds) *Mantle Plumes: their identification through time*. Geological Society of America, Special Paper **352**, 483–575.

ERNST, R. E. & BUCHAN, K. L. 2001b. The use of mafic dike swarms in identifying and locating mantle plumes. *In:* ERNST, R. E. & BUCHAN, K. L. (eds) *Mantle Plumes: their Identification through Time*. Geological Society of America, Special Paper, **352**, 247–265.

FERNÁNDEZ-SUÁREZ, J., GUTIÉRREZ-ALONSO, G., JENNER, G. A. & TUBRETT, M. N. 2000. New ideas on the Proterozoic–Early Palaeozoic evolution of NW Iberia: insights from U–Pb detrital zircon ages. *Precambrian Research*, **102**, 185–206.

GLEBOVITSKY, V., MARKER, M., ALEXEJEV, N., BRIDGWATER, D., SEDOVA, I., SALNIKOVA, E. & BEREZHNAYA, N. 2001. Age, evolution and regional setting of the Palaeoproterozoic Umba igneous suite in the Kolvitsa–Umba zone, Kola Peninsula: constraints from new geological, geochemical and U–Pb zircon data. *Precambrian Research*, **105**, 247–267.

GRISELIN, M., ARNDT, N. & BARAGAR, W. R. A. 1997. Plume–lithosphere interaction and crustal contamination during formation of Coppermine River basalts, Northwest Territories, Canada. *Canadian Journal of Earth Sciences*, **34**, 958–975.

HEAMAN, L. M. 1997. Global mafic magmatism at 2.45 Ga: remnants of an ancient large igneous province? *Geology*, **25**, 299–302.

HOFFMAN, P. F. 1988. United plates of America, the birth of a craton: early Proterozoic assembly and growth of Laurentia. *Annual Reviews of Earth and Planetary Sciences*, **16**, 543–603.

ILG, B. R., KARLSTROM, K. E., HAWKINS, D. P. & WILLIAMS, M. L. 1996. Tectonic evolution of Paleoproterozoic rocks in the Grand Canyon: insights into middle-crustal processes. *Geological Society of America Bulletin*, **108**, 1149–1166.

JOHNSON, S. P. & OLIVER, G. J. H. 2000. Mesoproterozoic oceanic subduction, island-arc formation and the initiation of back-arc spreading in the Kibaran Belt of central, southern Africa: evidence from the Ophiolite Terrane, Chewore Inliers, northern Zimbabwe. *Precambrian Research*, **103**, 125–146.

KAY, R. W. & KAY, S. M. 1986. Petrology and geochemistry of the lower continental crust: an overview. *In*: DAWSON, J. B., CARSWELL, D. A., HALL, J. & WEDEPOHL, K. H. (eds) *The Nature of the Lower Continental Crust*: Geological Society of London, Special Publication, **27**, 147–159.

KEMPTON, P. D., DOWNES, H., NEYMARK, L. A., WARTHO, J. A., ZARTMAN, R. E. & SHARKOV, E. V. 2001. Garnet granulite xenoliths from the northern Baltic Shield – the underplated lower crust of a Palaeoproterozoic Large Igneous Province. *Journal of Petrology*, **42**, 731–763.

KHAIN, V. E., GUSEV, G. S., KHAIN, E. V., VERNIKOVSKY, V. A. & VOLOBUYEV, M. I. 1997. Circum-Siberian Neoproterozoic ophiolite belt. *Ofioliti*, **22**, 195–200.

KIMURA, G., SAKAKIBARA, M. & OKAMURA, M. 1994. Plumes in central Panthalassa? Deductions from accreted oceanic fragments in Japan. *Tectonics*, **13**, 905–916.

KLOSKO, E. R., RUSSO, R. M., OKAL, E. A. & RICHARDSON, W. P. 2001. Evidence for a rheologically strong chemical mantle root beneath the Ontong Java plateau. *Earth and Planetary Science Letters*, **186**, 347–361.

KONTINEN, A. 1987. An Early Proterozoic ophiolite – the Jormua mafic–ultramafic complex, northeastern Finland. *Precambrian Research*, **35**, 313–341.

KORJA, A. & HEIKKINEN, P. J. 1995. Proterozoic extensional tectonics of the central Fennoscandian shield: results from the Baltic and Bothnian Echoes from the Lithosphere experiment. *Tectonics*, **14**, 504–517.

LARSON, R. L. 1991. Geological consequences of superplumes. *Geology*, **19**, 963–966.

LEGAULT, F., FRANCIS, D., HYNES, A. & BUDKEWITSCH, P. 1994. Proterozoic continental volcanism in the Belcher Islands: implications for the evolution of the Circum Ungava fold belt. *Canadian Journal of Earth Sciences*, **31**, 1536–1549.

LEWRY, J. F. & STAUFFER, M. R. 1997. The Reindeer zone of the Trans-Hudson orogen, Canada. *In*: DE WIT, M. & ASHWAL, L. D. (eds) *Greenstone Belts*. Clarendon Press, Oxford, 739–745.

LEYBOURNE, M. I., VAN WAGONER, N. A. & AYRES, L. D. 1997. Chemical stratigraphy and petrogenesis of the Early Proterozoic Amisk Lake volcanic sequence, Flin Flon–Snow Lake greenstone belt, Canada. *Journal of Petrology*, **38**, 1541–1564.

LUCAS, S. B., ST-ONGE, M. R., PARRISH, R. R. & DUNPHY, J. M. 1992. Long-lived continent–ocean interaction in the Early Proterozoic Ungava orogen, northern Quebec, Canada. *Geology*, **20**, 113–116.

LUCAS, S. B., STERN, R. A., SYME, E. C., REILLY, B. A. & THOMAS, D. J. 1996. Intraoceanic tectonics and the development of continental crust: 1.92–1.84 Ga evolution of the Flin Flon belt, Canada. *Geological Society of America Bulletin*, **108**, 602–629.

LUOSTO, U. 1997. Structure of the Earth's crust in Fennoscandia as revealed from refraction and wide-angle reflection studies. *Geophysica*, **33**, 3–16.

MELEZHIK, V. A. & STURT, B. A. 1994. General geology and evolutionary history of the early Proterozoic Polmak–Pasvik–Pechenga–Imandra/Varzuga–Ust'Ponoy greenstone belt in the northeastern Baltic Shield. *Earth Science Reviews*, **36**, 205–241.

MYERS, J. S., SAHW, R. D. & TYLER, I. M. 1996. Tectonic evolution of Proterozoic Australia. *Tectonics*, **15**, 1431–1446.

OLIVER, G. J. H., JOHNSON, S. P., WILLIAMS, I. S. & HERD, D. A. 1998. Relict 1.4 Ga oceanic crust in the Zambezi Valley, northern Zimbabwe: evidence for Mesoproterozoic supercontinental fragmentation. *Geology*, **26**, 571–573.

ORTEGA-GUTIERREZ, F., RUIZ, J. & CENTENO-GARCIA, E. 1995. Oaxaguia, a Proterozoic microcontinent accreted to North America during the late Paleozoic. *Geology*, **23**, 1127–1130.

OTA, T., TERABAYASHI, M. & KATAYAMA, I. 2002. Thermobaric structure and metamorphic evolution of the Iratsu eclogite body in the Sanbagawa belt, central Shikoku, Japan. *Lithos*, in press.

PARK, J. K., BUCHAN, K. L. & HARLAN, S. S. 1995. A proposed giant radiating dyke swarm fragmented by the separation of Laurentia and Australia based on paleomagnetism of ca. 780 Ma mafic intrusions in western North Australia. *Earth and Planetary Science Letters*, **132**, 129–139.

PARK, R. G. 1994. Early Proterozoic tectonic overview of the northern British Isles and neighbouring terrains in Laurentia and Baltica. *Precambrian Research*, **68**, 65–79.

PARK, R. G., TARNEY, J. & CONNELLY, J. N. 2001. The Loch Maree Group: Palaeoproterozoic subduction–accretion complex in the Lewisian of NW Scotland. *Precambrian Research*, **105**, 205–226.

PATCHETT, P. J. & ARNDT, N. T. 1986. Nd isotopes and tectonics of 1.9–1.7 Ga crustal genesis. *Earth and Planetary Science Letters*, **78**, 329–338.

PATCHETT, P. J. & RUIZ, J. 1989. Nd isotopes and the origin of Grenville-age rocks in Texas: implication for Proterozoic evolution of the United States Midcontinent region. *Journal of Geology*, **97**, 685–695.

PELTONEN, P., KONTINEN, A. & HUHMA, H. 1996. Petrology and geochemistry of metabasalts from the 1.95 Ga Jormua ophiolite, northeastern Finland. *Journal of Petrology*, **37**, 1359–1383.

PERCIVAL, J. A. & WEST, G. F. 1994. The Kapuskasing Uplift: a geological and geophysical synthesis. *Canadian Journal of Earth Sciences*, **31**, 1256–1286.

PUCHTEL, I. S., ARNDT, N.T., HOFMANN, A. W *ET AL.* 1998. Petrology of mafic lavas within the Onega plateau, central Karelia: evidence for 2.0 Ga plume-related continental crustal growth in the Baltic Shield. *Contributions to Mineralogy and Petrology*, **130**, 134–153.

PUCHTEL, I. S., HAASE, K. M., HOFMANN, A. W., CHAUVEL, C., KULIKOV, V. S., GARBE-SCHÖNBERG, C.-D. & NEMCHIN, A. A. 1997. Petrology and geochemistry of crustally contaminated komatiitic basalts from the

Vetreny belt, southeastern Baltic Shield: evidence for an early Proterozoic mantle plume beneath rifted Archean continental lithosphere. *Geochimica et Cosmochimica Acta*, **61**, 1205–1222.

REYMER, A. & SCHUBERT, G. 1986. Rapid growth of some major segments of continental crust. *Geology*, **14**, 299–302.

RUDNICK, R. L., MCDONAUGH, W. P. & O'CONNELL, R. J. 1998. Thermal structure, thickness, and composition of continental lithosphere. *Chemical Geology*, **145**, 395–411.

SAUNDERS, A. D., TARNEY, J., KERR, A. C. & KENT, R. W. 1996. The formation and fate of large igneous provinces. *Lithos*, **37**, 81–95.

SCOTT, D. J., HELMSTAEDT, H. & BICKLE, M. J. 1992. Purtuniq ophiolite, Cape Smith belt, northern Quebec, Canada: a reconstructed section of Early Proterozoic oceanic crust. *Geology*, **20**, 173–176.

SENGÖR, A. M. C., NATAL'IN, B. A. & BURTMAN, V. S. 1993. Evolution of the Altaid tectonic collage and Palaeozoic crustal growth in Eurasia. *Nature*, **364**, 299–307.

SEYMOUR, K. ST. & KUMARAPELI, P. S. 1995. Geochemistry of the Grenville dyke swarm: role of plume-source mantle in magma genesis. *Contributions to Mineralogy and Petrology*, **120**, 29–41.

SNYDER, D. B., LUCAS, S. B. & MCBRIDE, J. H. 1996. Crustal and mantle reflectors from Paleoproterozoic orogens and their relation to arc–continent collisions. *In*: BREWER, T. S. (ed.) *Precambrian Crustal Evolution of the North Atlantic Regions*. Geological Society of London, Special Publication **112**, 1–23.

STEIN, M. & GOLDSTEIN, S. L. 1996. From plume head to continental lithosphere in the Arabian–Nubian shield. *Nature*, **382**, 773–778.

STEIN, M. & HOFMANN, A.W. 1994. Mantle plumes and episodic crustal growth. *Nature*, **372**, 63–68.

STERN, R. A., SYME, E. C. & LUCAS, S. B. 1995. Geochemistry of 1.9 Ga MORB- and OIB-like basalts from the Amisk collage, Flin Flon belt, Canada: evidence for an intra-oceanic origin. *Geochimica et Cosmochimica Acta*, **59**, 3131–3154.

ST-ONGE, M. R., LUCAS, S. B. & SCOTT, D. J. 1997. The Ungava orogen and the Cape Smith thrust belt. *In:* DE WIT, M. & ASHWAL, L. D. (eds) *Greenstone Belts*. Clarendon Press, Oxford, 772–780.

TARNEY, J. & JONES, C. E. 1994. Trace element geochemistry of orogenic igneous rocks and crustal growth models. *Journal of Geological Society, London*, **151**, 855–868.

VAN HUNEN, J., VAN DER BERG, A. & VLAAR, N. J. 2002. On the role of subducting oceanic plateaus in the development of shallow flat subduction. *Proceedings of an International Workshop: Role of Superplumes in the Earth System.* Tokyo Institute of Technology, Tokyo, Japan, Abstract volume, 120–123.

VANDERHAEGHE, O., LEDRU, P., THIÉBLEMONT, D., EGAL, E., COCHERIE, A., TEGYEY, M. & MILÈSI, J-P. 1998. Contrasting mechanism of crustal growth; geodynamic evolution of the Paleoproterozoic granite–greenstone belts of French Guiana. *Precambrian Research*, **92**, 165–193.

VOGEL, D. C., VUOLLO, J. I., ALAPIETI, T. T. & JAMES, R. S. 1998. Tectonic, stratigraphic and geochemical comparisons between ca. 2500–2440 Ma mafic igneous events in the Canadian and Fennoscandian Shields. *Precambrian Research*, **92**, 89–116.

WALKER, R. J., MOGAN, J. W., HANSKI, E .J. & SMOLKIN, V. F. 1997. Re–Os systematics of Early Proterozoic ferropicrites, Pechenga Complex, northwestern Russia: evidence for ancient ^{187}Os-enriched plumes. *Geochimica et Cosmochimica Acta*, **61**, 3145–3160.

WHITEHOUSE, M. J. & BRIDGWATER, D. 2001. Geochronological constraints on Paleoproterozoic crustal evolution and regional correlations of the northern Outer Hebridean Lewisian complex, Scotland. *Precambrian Research*, **105**, 227–245.

WHITEHOUSE, M. J., KALSBEEK, F. & NUTMAN, A. P. 1998a. Crustal growth and crustal recycling in the Nagssugtoqidian orogen of West Greenland: constraints from radiogenic isotope systematics and U–Pb zircon geochronology. *Precambrian Research*, **91**, 365–381.

WHITEHOUSE, M. J., WINDLEY, B. F., BA-BTTAT, F. A. O., FANNING, C. M. & REX, D. C. 1998b. Crustal evolution and terrane correlation in the eastern Arabian Shield, Yemen: geochronologic constraints. *Journal of the Geological Society, London*, **155**, 281–295.

WINDLEY, B. F. 1992. Proterozoic collisional and accretionary orogens. *In*: CONDIE, K. C. (ed) *Proterozoic Crustal Evolution*. Elsevier, Amsterdam, 419–446.

WINDLEY, B. F. 1995. *The Evolving Continents* (3rd edition). Wiley, Chichester.

WINDLEY, B. F. WHITEHOUSE, M. J. & BA-BTTAT, M. A. O. 1996. Early Precambrian gneiss terranes and Pan-African island arcs in Yemen: crustal accretion of the eastern Arabian shield. *Geology*, **24**, 131–134.

WINGATE, M. T. D., CAMPBELL, I. H., COMPSTON, W. & GIBSON, G. M. 1998. Ion microprobe U–Pb ages for Neoproterozoic basaltic magmatism in south-central Australia and implications for the breakup of Rodinia. *Precambrian Research*, **87**, 1235–1259.

YAKUBCHUK, A. S., NIKISHIN, A. M. & ISHIWATARI, A. 1994. A Late Proterozoic ophiolite pulse. *Proceedings of the 29th International Geological Congress*, part D, 273–286.

ZIMMER, M., KRÖNER, A., JOCHUM, K.P., REISCHMANN, T. & TODT, W. 1995. The Gabal Gerf complex: a Precambrian N-MORB ophiolite in the Nubian Shield, NE Africa. *Chemical Geology*, **123**, 29–51.

Models of Rodinia assembly and fragmentation

SERGEI A. PISAREVSKY, MICHAEL T. D. WINGATE, CHRIS MCA. POWELL, SIMON JOHNSON & DAVID A. D. EVANS

Tectonics Special Research Centre, Department of Geology and Geophysics, The University of Western Australia, 35 Stirling Highway, Crawley, 6009, WA, Australia.
(e-mail: spisarev@tsrc.uwa.edu.au)

Abstract: Amongst existing palaeogeographic models of the Rodinia supercontinent, or portions thereof, arguments have focused upon geological relations or palaeomagnetic results, but rarely both. A new model of Rodinia is proposed, integrating the most recent palaeomagnetic data with current stratigraphic, geochronological and tectonic constraints from around the world. This new model differs from its predecessors in five major aspects: cratonic Australia is positioned in the recently proposed AUSMEX fit against Laurentia; East Gondwanaland is divided among several blocks; the Congo–São Francisco and India–Rayner Cratons are positioned independently from Rodinia; Siberia is reconstructed against northern Laurentia, although in a different position than in all previous models; and Kalahari–Dronning Maud Land is connected with Western Australia. The proposed Rodinia palaeogeography is meant to serve as a working hypothesis for future refinements.

There is general agreement that the Earth's continental crust may have been assembled to form the supercontinent, Rodinia, in the Late Mesoproterozoic and Early Neoproterozoic. Rodinia is thought to have been produced by collisional events of broadly Grenvillian (Late Mesoproterozoic) age, and to have been relatively long-lived (c. 1100–750 Ma) (McMenamin & McMenamin 1990; Hoffman 1991).

Nonetheless, there are several versions of its composition and configuration (e.g. Hoffman 1991; Dalziel 1997; Weil et al. 1998). Laurentia is thought to have formed the core of Rodinia because it is surrounded by passive margins formed during Late Neoproterozoic breakup of the supercontinent (Bond et al. 1984). Most Rodinia models propose that Australia, Antarctica and possibly South China (Li et al. 1999) may have been situated along Laurentia's western margin (unless otherwise stated, all geographic references are in present coordinates); Baltica and Amazonia, and the Rio de la Plata Craton may have lain along its eastern margin. The precise position of Siberia is disputed, but it is generally shown as lying along either the northern or the western margin of Laurentia. The position of the Congo and Kalahari Cratons is uncertain, with at least four reconstructions having been shown for Kalahari in the last few years (Powell et al. 2001). An alternative Neoproterozoic supercontinent, Palaeopangaea, was proposed by Piper (2000), based mainly on palaeomagnetic data. This model is similar to earlier reconstructions by the same author (Piper 1987 and refs cited therein), which were criticized by both Van der Voo & Meert (1991) and Li & Powell (1999). In addition, the recent publication about Palaeopangaea (Piper 2000) contains no references for the poles employed, making the model difficult to assess. For these reasons, it will not be discussed further in this paper.

Several important results have been published recently that provide new geological, geochronological and palaeomagnetic constraints on Mesoproterozoic–Early Neoproterozoic palaeogeography. Palaeomagnetic data are necessary for quantitative constraints on Precambrian reconstructions. Unfortunately, these data are distributed very non-uniformly in time and space (Meert & Powell 2001 table 1). The majority of palaeomagnetic results for the interval during which Rodinia may have existed (c. 1100–750 Ma) come from Laurentia and Baltica, and fragments of apparent polar wander paths (APWP) can be constructed for these two blocks. Data from other cratons are sparse, making it impossible to construct an APWP for each block. The palaeopositions of these blocks are based on comparisons of individual palaeopoles, hence relative palaeolongitudes are not constrained.

The objective of this paper is to create a new model of the Rodinia supercontinent. Palaeomagnetism has been used to determine permissible fits for Rodinia; geological constraints, such as continuity of tectonic belts, and the presence of passive or active continental margins have been used to refine permissible fits into plausible reconstructions. There is also the global balance of Late Neoproterozoic rifted margins that needs to be accounted for in any acceptable reconstruction.

Selection of reliable palaeomagnetic data is the key issue for many Mesoproterozoic–Neoproterozoic reconstructions (e.g. Powell *et al.*

1993; Torsvik et al. 1996; Smethurst et al. 1998; Weil et al. 1998; Piper 2000). In the present synthesis (Table 1), only palaeomagnetic results with $Q \geq 4$ are used (Van der Voo 1990). There are few exceptions where less reliable data is referred to and all such cases are explained individually. However, existing data are insufficient to provide robust reconstructions for all cratons except Laurentia and Baltica. In addition, there are no reliable palaeomagnetic data for Amazonia, West Africa and Rio de la Plata in the interval 1100–700 Ma.

In attempting to reconstruct Rodinia, available information from the majority of Precambrian continental blocks was used. Because very little is known about the Rodinian connections of North China, NE Africa and Arabia, Avalonia, Cadomia, Omolon, and other fragments of continental crust from the Russian Far East, northern Alaska and southeastern

Table 1. *Palaeomagnetic poles at 1100–700 Ma*

Object	Age (Ma)	Pole (°N)	Pole (°E)	A_{95} (°)	Q	Reference
Laurentia						
Franklin Dykes	723+4/−2	5	163	5	II–IIII 6	Heaman et al. 1992; Park 1994
Natkusiak Formation	723+4/−2	6	159	6	III–III 6	Palmer et al. 1983; Heaman et al. 1992
Tsezotene sills and dykes	779±2	2	138	5	III–I–I 5	Park et al. 1989; LeCheminant & Heaman 1994
Wyoming Dykes	782±8 785±8	13	131	4	III–I–I 5	Harlan et al. 1997
Haliburton Intrusions A	980±10	−36	143	10	III– – –I 4	Buchan & Dunlop 1976
Chequamegon Sandstone	c. 1020*	−12	178	5	–II–I–I 4	McCabe & Van der Voo 1983
Jacobsville Sandstone J (A+B)	c. 1020*	−9	183	4	–II–I–I 4	Roy & Robertson 1978
Freda Sandstone	1050±30	2	179	4	–IIII–I 5	Henry et al. 1977; Wingate et al. 2002
Nonesuch Shale	1050±30	8	178	4	–IIII–I 5	Henry et al. 1977; Wingate et al. 2002
Lake Shore Traps	1087±2	22	181	5	IIII–I 6	Diehl & Haig 1994; Davis & Paces 1990
Portage Lake Volcanics	1095±2	27	181	2	II– –I–I 4	Halls & Pesonen 1982; Davis & Paces 1990
Upper North Shore Volcanics	1097±2	32	184	5	II– –III 5	Halls & Pesonen 1982; Davis & Green 1997
Logan Sills R	1109+4/−2	49	220	4	II–IIII 6	Halls & Pesonen 1982; Davis & Sutcliffe 1985
Abitibi Dykes	1141±2	43	209	14	IIII–I– 5	Ernst & Buchan 1993
Baltica						
Hunnedalen Dykes	≥848	−41	222	10	III–I–I 5	Walderhaug et al. 1999
Egersund Anorthosite	929–932	−44	214	4	III–III 6	Stearn & Piper 1984; Torsvik & Eide 1997
Pyätteryd Amphibolite	933–945	−43	214	11	III–I–I 5	Pisarevsky & Bylund 1998; Wang et al. 1996; Wang & Lindh 1996
Känna Gneiss	948–974	−50	225	17	III–I–I 5	Pisarevsky & Bylund 1998; Wang et al. 1996; Wang & Lindh 1996
Gällared Amphibolite	956?	−46	214	19	III–I–I 5	Pisarevsky & Bylund 1998; Möller & Söderlund 1997
Gällared Granite Gneiss	980–990	−44	224	6	III–I–I 5	Pisarevsky & Bylund 1998; Möller & Söderlund 1997

Table 1. (*cont.*)

Object	Age (Ma)	Pole (°N)	Pole (°E)	A_{95} (°)	Q	Reference
Laanila Dolerite	1045±50	−2	212	15	III–I–I 5	Mertanen *et al.* 1996
India						
Mahe Dykes, Seychelles†	748–755	80	79	11	III– – –I 4	Torsvik *et al.* 2001*a*
Malani Igneous Suite	751–771	68	88	8	IIII–I 6	Torsvik *et al.* 2001*b*
Harohalli Dykes	814±34	27	79	18	III–III 6	Radhakrishna & Mathew 1996
Wajrakarur Kimberlites	1079?	45	59	11	I–I–I–I 4	Miller & Hargraves 1994
Australia						
Mundine Well Dykes	755±3	45	135	4	IIII–I 6	Wingate & Giddings 2000
Bangemall Basin Sills	1070±6	34	95	8	IIIIII 7	Wingate *et al.* 2002
Congo						
Mbozi Complex, Tanzania	755±25	46	325	9	III–III 6	Meert *et al.* 1995; Evans 2000.
Gagwe lavas, Tanzania	795±7	25	93	10	III–I–I 5	Meert *et al.* 1995; Deblond *et al.* 2001
São Francisco						
Ilheus Dykes	1011±24	30	100	4	III–I–I 5	D'Agrella-Filho *et al.* 1990; Renne *et al.* 1990
Olivenca Dykes, normal	*c.* 1035*	16	107	8	III–I–I 5	D'Agrella-Filho *et al.* 1990; Renne *et al.* 1990
Itaju de Colonia	*c.* 1055*	8	111	10	III–III 6	D'Agrella-Filho *et al.* 1990; Renne *et al.* 1990
Olivenca Dykes, reverse	1078±18	−10	100	9	III–I–I 5	D'Agrella-Filho *et al.* 1990; Renne *et al.* 1990
Kalahari‡						
Ritscherflya Supergroup (rotated to Kalahari)	1130±12	61	29	4	II– –I–I 4	Powell *et al.* 2001
Umkondo Igneous Province	1105±5	66	37	3	II– –III 5	Powell *et al.* 2001; Wingate 2001
Kalkpunt Formation	*c.* 1065?	57	3	7	II– –III 5	Briden *et al.* 1979; Powell *et al.* 2001
Central Namaqua	*c.* 1030–1000	8	330	10	III–III 6	Onstott *et al.* 1986; Robb *et al.* 1999
Siberia						
Uchur–Maya sediments	*c.* 990–1150	−25	231	3	–IIIII 6	Gallet *et al.*, 2000
Turukhansk sediments	*c.* 975–1100	−15	256	8	–II–III 5	Gallet *et al.* 2000
South China						
Liantuo Formation	748±12	4	161	13	IIIIII 7	Evans *et al.* 2000
Oaxaquia						
Oaxaca Anorthosite	*c.* 950	47	267	23	–II–III 5	Ballard *et al.* 1989

* Age based on apparent polar wander path interpolation.
† Rotated to India 28° counterclockwise around the pole of 25.8° N, 330°E (Torsvik *et al.* 2001*a*).
‡ For ages see Powell *et al.* (2001 and refs cited therein).
SA, South Australia; NT, Northern Territory; WA, Western Australia.

Configuration of Rodinia

Laurentia and Baltica

The majority of reliable Late Mesoproterozoic–Neoproterozoic palaeomagnetic data are from Laurentia (Table 1). The Laurentian APWP can be traced within the c. 1140–1020 Ma time interval, but younger poles are sparse. It is obvious that the poles between 1020 and 720 Ma circumscribe a 'Grenville Loop', although the shape and 'direction' of this loop is debated (e.g. Park & Aitken 1986; Hyodo & Dunlop 1993; Alvarez & Dunlop 1998; Weil et al. 1998; McElhinny & McFadden 2000). Alvarez & Dunlop (1998) analysed palaeopoles from the Grenville Province, generated from rocks remagnetized during post-Grenvillian exhumation between 1000 and 900 Ma. Some of these overprints are calibrated by $^{40}Ar-^{39}Ar$ ages that support a clockwise Grenville Loop of poles in the Pacific Ocean (Fig. 1). Following a similar approach, McElhinny & McFadden (2000, table 7.4) also constructed a clockwise loop. However, the data they used for the mean poles between 940 and 800 Ma are poorly dated, so this part of the loop has been simplified with an interpolation between reliable poles, as listed in Table 1.

Similarly, Early Neoproterozoic palaeomagnetic data from Baltica (not shown in Table 1) reflect a Sveconorwegian (Grenvillian) to post-Sveconorwegian overprint (e.g. Bylund 1992; Pisarevsky & Bylund 1998; Walderhaug et al. 1999 and refs cited therein). A large group of poles with ages c. 980–930 Ma is situated in a relatively small area at c. 45°S, 235°E, whereas poles with older and younger ages occupy near-equatorial positions (Fig. 1). These poles constitute the 'Sveconorwegian Loop' on the APWP for Baltica between c. 1000 and 800–850 Ma. As with the Grenville Loop, opinions about its shape and 'direction' are divided (e.g. Bylund 1985; Elming et al. 1993; Mertanen et al. 1996; Pisarevsky & Bylund 1998). Most data in Table 1 belong to the 980–930 Ma group, with the progression of ages more in favour of a clockwise loop. A clockwise loop is also supported by remanence directions obtained from a series of cross-cutting dykes in a single quarry within the northern part of the Protogine Zone, southern Sweden (Bylund & Pisarevsky 2002).

An exception to this general scenario is the interpretation of Walderhaug et al. (1999) concerning the Hunnedalen Dykes (Fig. 1). Based on ages of 848±27 (Ar–Ar on biotite) and 855±59 Ma (Sm–Nd mineral/whole-rock isochron), and on the similarity between the Hunnedalen Dyke Pole and those from 930 Ma plutonic rocks in SW Norway, these authors proposed that all of these rocks were remagnetized during a 'late unroofing' event at 850 Ma. However, because the Ar closure temperature for biotite is c. 400°C (Berger & York 1981), significantly lower than the dykes' palaeomagnetic unblocking temperatures (>520°C), and because these dykes were emplaced at moderate depth (Walderhaug et al. 1999), it is proposed here that intrusion of Hunnedalen Dykes and fixation of the stable remanence could have occurred long before 850 Ma. The uncertainty of the Sm–Nd age is also compatible with dyke emplacement at ≥900 Ma (perhaps consanguineous with the Egersund Anorthosite) followed by unroofing or mild reheating at 850 Ma. In the present model (Fig. 1), a simplified clockwise Sveconorwegian Loop, with the traditionally accepted age of c. 950 Ma at its vertex, was used.

The Grenville and Sveconorwegian APWP Loops coincide reasonably well after a 59° clockwise rotation of Baltica around an Euler pole at 75.8°N, 95.8°W. These rotation parameters were used for the Rodinia juxtaposition of Laurentia and Baltica, a suggestion implying that Baltica and Laurentia were joined as a single entity at between least 1000 and 850 Ma. The configuration proposed by Dalziel (1997), in which west Scandinavia was connected to East Greenland, is not supported by the palaeomagnetic data. One of the arguments used for such a position for Baltica was the juxtaposition of a possible Late Mesoproterozoic orogenic belt in East Greenland with the Sveconorwegian Belt in southern Scandinavia. However, a recent SHRIMP study of detrital zircons in pre-Caledonian rocks of East Greenland by Kalsbeek et al. (2000) showed that 'if present at all, a "Grenvillian" orogen in East Greenland would be of very different character than in North America and southern Scandinavia'.

The best fit between APWP loops implies a small gap between Laurentia and Baltica, sufficient to accommodate the Rockall submarine plateau. This region of submerged continental crust has yielded U–Pb and Sm–Nd crystallization ages of between 1750 (Daly et al. 1994) and 1625 Ma (Morton & Taylor 1991), comparable to those in the Trans-Scandinavian Igneous Belt (Larson & Berglund 1992) or the Ketilidian Orogen of southern Greenland (Rainbird et al. 2001). The southern part of Rockall was possibly reworked during latest Grenvillian times [Ar–Ar on granulite at 997±5 Ma (Miller et al. 1973), which was recalculated according to Dalrymple (1979)], which is comparable to ages for the Sveconorwegian Orogeny (e.g. Gorbatschev & Bogdanova 1993). The present proposed fit also includes a palinspastic reconstruction

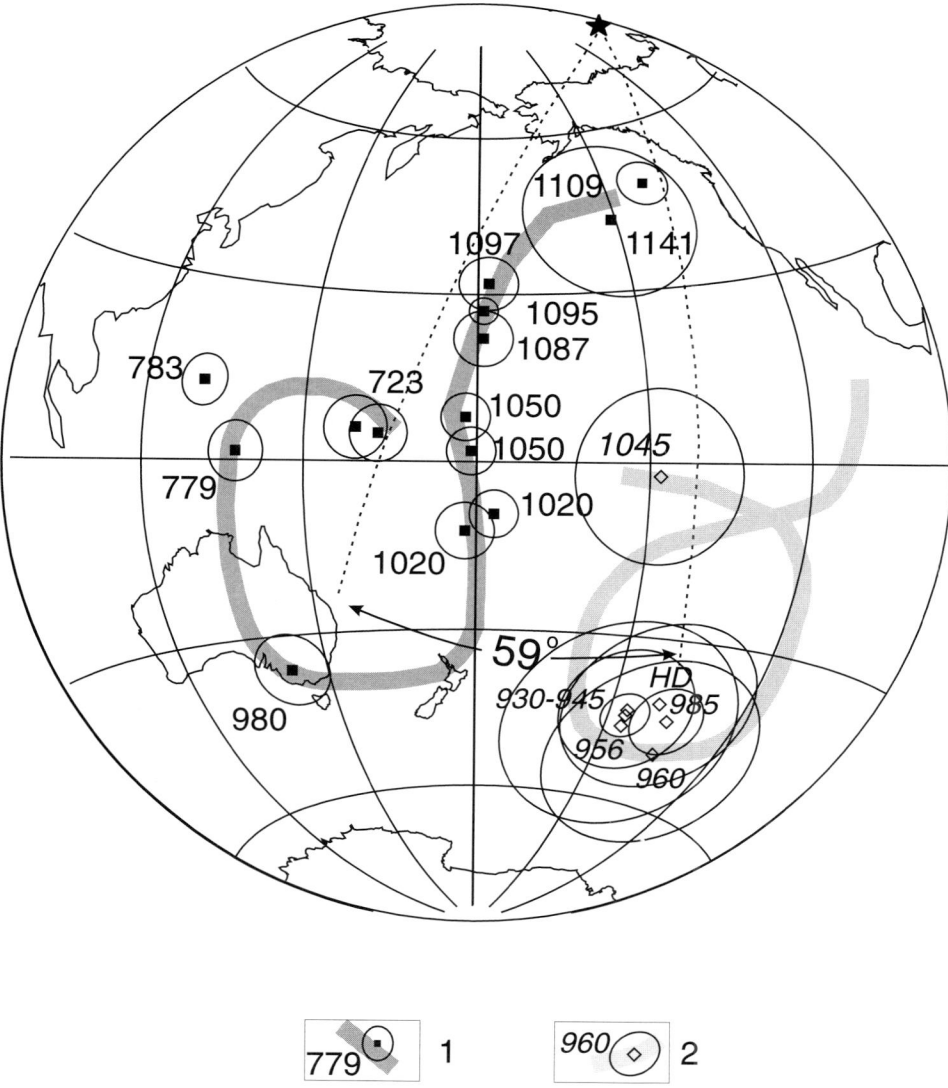

Fig. 1. Apparent polar wander paths for Laurentia and Baltica. 1, Laurentian palaeopoles and path; 2, Baltican palaeopoles and path. Star represents a Euler pole of a 59° rotation which provides a best fit for two paths. HD, Hunnedalen Dykes pole.

of northwestern Scandinavia that removes 400–500 km of Caledonian shortening (e.g. Park et al. 1994). Because c. 1050 Ma palaeopoles from the two cratons are incompatible according to the present reconstruction (Table 1; Fig. 1), it is proposed that their Rodinian fit was achieved via an oblique and substantially rotational collision between 1050 and 1000 Ma. Similar kinematics characterize the tectonic models of Park (1992) and Starmer (1996).

The proposed position of Baltica (Fig. 2) also permits juxtaposition of East Greenland and fragments of a possible Barentsia plate (East Svalbard), which may explain striking similarities between the Palaeoproterozoic and Neoproterozoic–Early Palaeozoic strata of these two areas (e.g. Gee et al. 1994; Fairchild & Hambrey 1995; Higgins et al. 2001 and refs cited therein). The results of recent studies by Higgins & Leslie (2000) and Higgins et al. (2001), which show that the pre-Caledonian margin of East Greenland restores 500–700 km to the east of the present coastline, have been incorporated into the

Fig. 2. Southern Rodinia at 990 Ma. Am, Amazonia; B, Barentsia; Ba, Baltica; Ch, Chortis; Gr, Greenland; La, Laurentia; O, Oaxaquia; P, Pampean terrane; R, Rockall; RP, Rio de la Plata; WA, West Africa. Grey regions, palinspastically restored pre-Caledonian margins of East Greenland and Baltica.

present model. The eastern (Uralian) edge of Baltica was probably a long-lived passive margin from the Late Mesoproterozoic to the Vendian (e.g. Willner et al. 2001).

Laurentia–Baltica and Amazonia

In the majority of Rodinia reconstructions, Amazonia is juxtaposed against eastern Laurentia and Baltica. According to this interpretation, the Rondonia–Sunsas Belt in southwestern Amazonia resulted from continental collision with Laurentia between 1080 and 970 Ma (e.g. Sadowski & Bettencourt 1996; Tassinari et al. 2000 and refs cited therein). There is little evidence for a Grenville-age collision in the northern part of autochthonous Amazonia, apart from the poorly understood 'Nickerian' Event at c. 1200 Ma (Gibbs & Barron 1993). Keppie & Ramos (1999) proposed that two Central American terranes – Oaxaquia (Mexico) and Chortis (Honduras and Guatemala) – were situated along the northern boundary of South America in their reconstruction for the Vendian–Cambrian boundary. Keppie & Ortega-Gutierrez (1999) suggested that these blocks originated as arcs in a Grenvillian ocean between Laurentia, Baltica and Amazonia, and were caught between the colliding cratons. Alternatively, Oaxaquia and Chortis could represent part of a continental arc formed on the present northern margin of Amazonia. In both scenarios, these blocks have experienced high-grade, collisional-style tectonometamorphism during the terminal collisions among Amazonia, Laurentia and Baltica.

In the present model, the Oaxaquia and Chortis blocks are placed along the northern margin of Amazonia, within the zone of its collision with Baltica at c. 1000 Ma (Fig. 2); the model is also constrained by palaeomagnetic data from Oaxaquia (Ballard et al. 1989) – see Table 1. Recent preliminary palaeomagnetic results of D'Agrella-Filho et al. (2001) from the Rondonia–Sunsas Province generally support a model of Amazonia–Laurentia collision after c. 1100 Ma.

Amazonia, West Africa and Rio de la Plata

Trompette (1994, 1997) proposed the existence of a single West Africa–Amazonia–Rio de la Plata megacraton in the Mesoproterozoic and Neoproterozoic. However, he did not exclude the possibility of minor relative movements between its components. For example, Onstott & Hargraves (1981) suggested that c. 1500 km of dextral shearing occurred between West Africa and Amazonia, based on comparison of Palaeoproterozoic–Mesoproterozoic palaeomagnetic data from these two blocks. The data are rather scattered and the conclusion is not very convincing, so for the purpose of this paper the Gondwanaland Amazonia–West Africa fit (Fig. 2) has been used, although it is acknowledged that such shearing was possible. Neoproterozoic palaeomagnetic data from West Africa are few and contradictory (Perrin & Prevot 1988), and therefore have not been taken into account in the present model.

Rio de la Plata is a poorly known craton, with no reliable palaeomagnetic data available for 1000–700 Ma; its Precambrian boundaries are similarly uncertain. The Rio de la Plata Block, depicted by Dalziel (1997) and Weil et al. (1998), for example, included parts of the Pampean Terrane as well as the southern extremity of the Guapore Block. In contrast, Ramos (1988) envisaged a Pampean–Rio de la Plata collision at 600–570 Ma and Trompette (1994) considered the possibility of an Amazonian affinity for the southern Guapore cratonic extension. Pimentel et al. (1999) suggested a collision between the São Francisco Craton and the Paraná Block between 790 and 750 Ma. Alkmim et al. (2001) considered the Paraná Block as part of Rio de la Plata, although supporting evidence is lacking (e.g. Trompette 1994; Cordani et al. 2000). Ramos (1988) depicted a shear zone between Rio de la Plata and the western Alto Paraguay Terrane, now covered by the Palaeozoic–Mesozoic Paraná Basin. Depending on its total displacement, this shear zone may allow the Rio de la Plata and Paraná Blocks to be considered as separate palaeogeographic entities in Early Neoproterozoic time (Ramos 1988).

In the present reconstructions, three separate blocks are proposed: (1) Rio de la Plata *sensu stricto*, which includes basement NE and SW of Buenos Aires (Cingolani & Dalla Salda 2000), and does not include the Luis Alves Block and the southern extremity of the Guapore Block; (2) the Pampean Terrane; and (3) the Paraná Block. Generally, the tectonic model of Ramos (1998) has been followed for the Rio de la Plata and Pampean Blocks, keeping them in the vicinity of Laurentia, and that of Pimentel *et al.* (1999) for the Paraná–Saõ Francisco collision.

Laurentia and Australia

The western margin of Laurentia, from northern Canada to southern USA, contains a rift–passive margin succession initiated at *c.* 750 Ma (e.g. Moores 1991; Ross *et al.* 1995 and refs cited therein). Moores (1991) proposed that Australia–East Antarctica rifted from Laurentia at that time [SW US–East Antarctic (SWEAT) hypothesis]. This configuration was used by Hoffman (1991), Dalziel (1997) and Weil *et al.* (1998) in their reconstructions of Rodinia. The SWEAT hypothesis suggested that the Grenville Belt of Laurentia continued around Antarctica, and into India and Australia. Li *et al.* (1995), using tectonostratigraphic analysis, modified the SWEAT configuration slightly by placing the South China Craton between Australia and northwestern Laurentia. Brookfield (1993), Karlstrom *et al.* (1999) and Burrett & Berry (2000) proposed an alternative Australia–Laurentia fit (AUSWUS) based on comparison of Precambrian terranes on both cratons. Until recently, Australian Late Mesoproterozoic–Neoproterozoic palaeomagnetic data were inadequate to discriminate between these hypotheses. However, a new palaeomagnetic result from the 1070 Ma Bangemall Sills, Western Australia (Wingate *et al.* 2002), supports neither of these models. According to this result, Australia at 1070 Ma was situated at lower palaeolatitudes than would be permitted by the SWEAT or AUSWUS models, placing the Cape River Province of NE Australia at a similar latitude to the southwestern end of the 1250–980 Ma Grenville Province of Laurentia (Rivers 1997; Mosher 1998). High-grade metamorphic and magmatic rocks in the Cape River Province contain 1240, 1145 and 1105 Ma zircon-age components, and may correlate to the west with Grenvillian-age rocks in the Musgrave and Albany–Fraser Orogens (Blewett *et al.* 1998). If the proposed AUSMEX reconstuction (Fig. 3) is correct, the Grenville Province may have continued through Australia. A recent palaeomagnetic study of the deep drillhole Empress 1A in the Officer Basin (Pisarevsky *et al.* 2001) indicated low palaeolatitudes for Australia between *c.* 810 and 750 Ma, also supporting this fit. While maintaining most of the geological comparisons argued in favour of AUSWUS and some of those in favour of SWEAT, the AUSMEX fit also helps to resolve some additional problems. For example, the Wyoming Craton is a more suitable source for 2.78 Ga detrital zircons in Papua New Guinea (Baldwin & Ireland, 1995) than the smaller and possibly displaced Nova Terrane proposed by Burrett & Berry (2000) in their AUSWUS fit. The AUSMEX reconstruction, however, requires conjugate passive margins for western Laurentia and eastern Australia to be identified. In the present reconstruction (Fig. 3) the latter is supposed to be in the poorly known Rio de la Plata Craton and the former in South China (see below).

Australia, Antarctica and India

A feature common to most Rodinia reconstructions (e.g. Powell *et al.* 1993, Dalziel 1997; Weil *et al.* 1998) is the assumption that East Gondwanaland (Australia, India, Madagascar and East Antarctica in their Gondwanaland configuration) has been a single tectonic entity since the end of the Mesoproterozoic. The assumption has been based mainly on the apparent continuity of a Grenville-age metamorphic belt along the India–East Antarctica Margin, and its extension into the Late Mesoproterozoic Albany–Fraser Mobile Belt of Australia, and a supposed absence of Neoproterozoic or younger sutures between the continental blocks. Late Neoproterozoic high-grade gneisses occur in the Darling Mobile Belt of Western Australia (Wilde & Murphy 1990; Harris 1994) and in the correlative Prydz–Denman Zone in East Antarctica (Fitzsimons 2000), but direct evidence for oceanic subduction is lacking and subsequent metamorphism has generally been regarded as intracontinental reactivation of

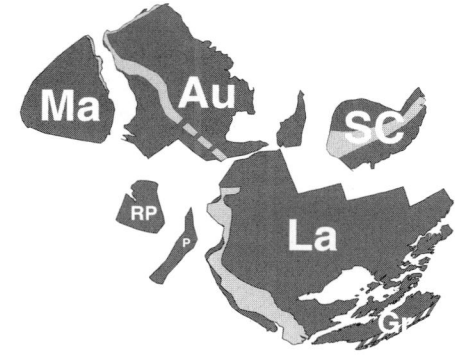

Fig. 3. Northern Rodinia at 990 Ma. Au, Australia; Ma, Mawson Craton; SC, South China. Other abbreviations as in Figure 2.

old crustal weaknesses during the Late Neoproterozoic collision of East and West Gondwanaland farther to the west.

Fitzsimons (2000) disproved the hypothesis of a single continuous Late Mesoproterozoic orogenic belt along the margin of East Antarctica. Three separate Late Mesoproterozoic–earliest Neoproterozoic orogenic belts, with different ages of metamorphism and plutonism, exist along the margins of East Antarctica. Importantly, Fitzsimons (2000) highlighted two Late Neoproterozoic orogenic zones in Antarctica, across which there are unknown amounts of displacement in Pan-African time. In a Gondwanaland reconstruction, one zone is the southern extension of the East African Orogen to the west of India, and the other is the Prydz–Denman–Darling Orogen between India and Australia–Wilkes Land (Antarctica) in East Gondwanaland.

Palaeomagnetic data for India (Table 1) also contradict the integrity of East Gondwanaland. In an East Gondwana and fit the c. 1080 Ma pole for the Wajrakarur Kimberlites in central India (Miller & Hargraves 1994) does not coincide with the 1070 Ma Bangemall Pole for Australia (Wingate et al. 2002). Additionally, India was at high palaeolatitudes at c. 810 Ma, whereas Australia was at low palaeolatitudes at that time (Pisarevsky et al. 2001).

Combining palaeomagnetic geochronological and geological data, Powell & Pisarevsky (2002) proposed a new model for the Neoproterozoic tectonic history of East Gondwanaland, in which India (together with the Rayner Block of Antarctica) was not a part of Rodinia, but collided obliquely with the rest of East Gondwanaland (West Australia and the Mawson Craton of Australia–Antarctica) between 680 and 610 Ma [or later, as discussed by Fitzsimons (2002)]. Boger et al. (2001) presented a slightly different model, but with a similar conclusion that large sections of East Antarctica and India were not parts of East Gondwanaland or Rodinia.

Kalahari and Australia

A comparison of Late Mesoproterozoic palaeopoles from the Kalahari Craton and the correlative Grunehogna fragment in East Antarctica indicates that, in Rodinia, the Kalahari–Grunehogna Craton (e.g. Jacobs et al. 1993) could have lain to the SW of Laurentia with the Namaqua–Natal orogenic belt facing outboard and away from the Laurentian Craton (Powell et al. 2001). Powell et al. (2001) also showed, by comparing palaeopoles from coeval Umkondo and Keweenawan igneous Provinces, that Kalahari could not be the 'southern continent' that indented the Grenvillian Llano Orogen between 1150 and 1100 Ma (Mosher 1998; Dalziel et al. 2000). Available geochronological data from the Namaqua Belt are similarly incompatible with those from the southwestern Grenville Belt (Powell et al. 2001). However, Kalahari could have lain off the western margin of Australia until 800–750 Ma, when breakup associated with the 755 Ma Mundine Well Mafic Dyke Swarm (Wingate & Giddings 2000) caused Kalahari to rotate anticlockwise away from the western margin of Australia. Bruguier et al. (1999) suggested that India collided with the western margin of Australia along the Darling Mobile Belt c. 1080 Ma but as discussed above, India did not achieve this position until the Late Neoproterozoic. Jacobs et al. (1998) identified c. 1080 Ma syntectonic granite sheets and plutons in Dronning Maud Land, at the same time as the Namaqua–Natal Belt was undergoing transpressional shearing. Following Fitzsimons (2002), it is proposed that here the Kalahari–Dronning Maud Land Craton joined the Australia–Mawson Craton during oblique collision between 1100 and 1000 Ma (Fig. 4a). Metamorphism in both Dronning Maud Land and the Darling Mobile Belt occurred at 1080–1050 Ma (Jacobs et al. 1998; Bruguier et al. 1999), and metasediments in both belts have indistinguishable detrital zircon populations (Fitzsimons 2002). Consistent with this model, palaeopoles from Kalahari converge with the Laurentian APWP between 1065 and 1000 Ma (Table 1; Fig. 4a).

Western Laurentia and South China

Dismissal of the SWEAT and AUSWUS hypotheses reopens the question of the conjugate to the west Laurentian Neoproterozoic passive margin. Li et al. (1995, 1999) proposed that the Cathaysia Block of South China was part of a 1.9–1.4 Ga continental strip adjoining western Laurentia prior to collision with the Yangtze Block along the Sibao Orogen by c. 1000 Ma (Li et al. 2002). The suggestion is based mainly on similarities between Neoproterozoic sedimentary successions in South China and western Canada. The Cathaysia Block could have been the source of the Mesoproterozoic detrital zircons in the Belt Supergroup in northwestern USA (Ross et al. 1992; Li et al. 1995). In the present model, South China is kept juxtaposed with western Laurentia (Fig. 3), but Australia is placed further to the south than suggested by Li et al. (1995, 1999, 2002). The Sibao Orogen may also be considered as a source of Grenville-age detrital zircons in the Mackenzie Mountains and the Amundsen Basin (Rainbird et al. 1997), and could have produced the few enigmatic 1070–1244 Ma zircons in the Buffalo Hump Formation in Washington State (Ross et al. 1992).

South China is not of sufficient size to cover the entire West Laurentian passive margin (Fig. 3),

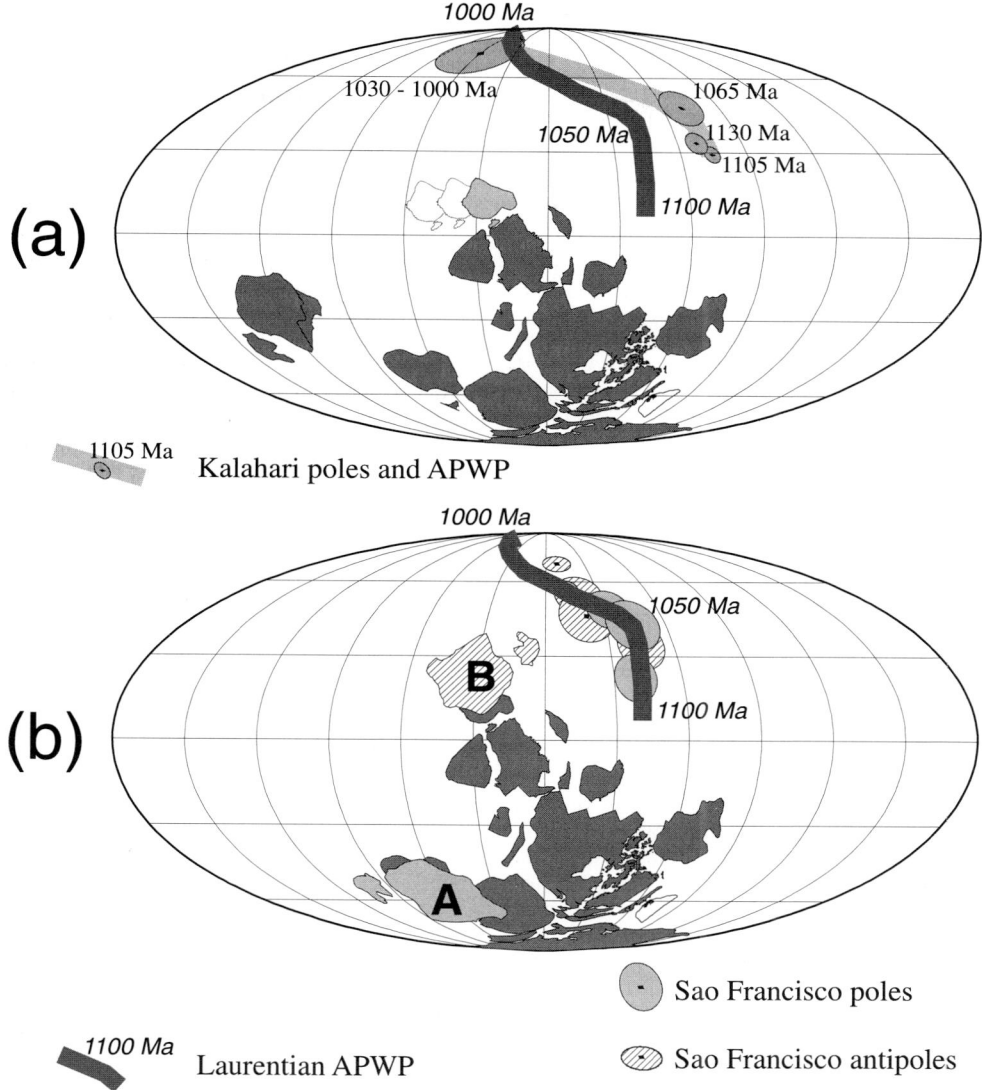

Fig. 4. (a) Proposed position of Kalahari and palaeomagnetic constraints; (b) alternative palaeopositions of Congo–São Francisco according to the two polarity options of its apparent polar wander paths (APWP) (see text).

hence other continental blocks are likely to have been attached to northwestern Laurentia in the Mesoproterozoic and Neoproterozoic – Northern Alaska, northern blocks of eastern Siberia and the Kara Plate are all possible candidates. There have been suggestions that these blocks constituted a large Precambrian craton, Arctida (Zonenshain *et al.* 1990). However, the tectonic history of these blocks is relatively well understood only back to Early Palaeozoic time (e.g. Natal'in *et al.* 1999), hence the shape, constitution and reconstruction of Arctida within Rodinia remain highly uncertain (question mark in Fig. 6). The position of the Tarim Block (Fig. 6) is in accordance with the reconstruction of Li *et al.* (1996), which is based predominantly on tectonostratigraphic comparisons.

Laurentia and Siberia

Sears & Price (1978, 2000) proposed the Siberian Craton as an alternative counterpart to western

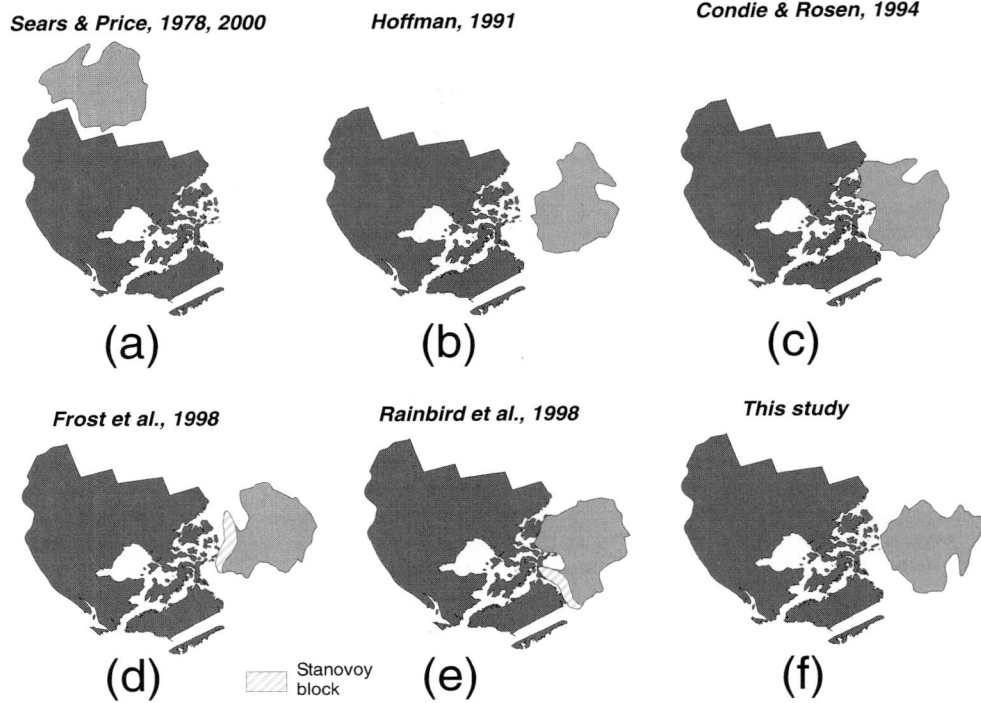

Fig. 5. Proposed palaeopositions of Siberia with respect to Laurentia.

Laurentia instead of Australia and Antarctica (Fig. 5a). Such a configuration raises several problems.

There is a mismatch of crustal age domains. Southwestern North America is dominated by juvenile, Early Proterozoic belts, whereas the Aldan Shield of Siberia is Archaean (e.g. Condie & Rosen 1994). Sears & Price (2000), citing Nd isotopic data from Ramo & Calzia (1998), argued for the presence of a substantial Archaean source component in the Death Valley area of Mojavia. However, Ramo & Calzia (1998) concluded that this Archaean component was introduced as sedimentary detritus, and was probably subducted and mixed with juvenile material at a convergent zone, either at the present western margin of the Wyoming Craton or elsewhere. Sears & Price (2000) also correlated a 1740 Ma U–Pb zircon crystallization age from the Okhotsk Massif in Siberia with similar-aged magmatic events from the Mojave, Yavapai and Mazatzal Provinces (Van Schmus & Bickford 1993). The original source of this 1740 Ma age date from the Okhotsk Massif (Kuzmin et al. 1995, table 1; see also Khudoley et al. 2001), and is the youngest in a series of 21 age determinations ranging from 3350 to 1830 Ma, all of which are systematically older than those in the Mojave, Yavapai and Mazatzal Provinces (Van Schmus & Bickford 1993).

Sears & Price (2000) juxtaposed the Palaeoproterozoic (maximum 2.4–2.5 Ga; Rosen et al. 2000), Birekte Block of Siberia [or Olenek blocks, following the determination of Condie & Rosen (1994)] against the Archaean Hearne Province/ Medicine Hat Block (Hoffman 1989). Sears & Price (2000) attempted to correlate the North Alberta Palaeoproterozoic continental and oceanic-arc terranes with the predominantly metasedimentary Hapshan Orogenic Belt which underwent granulite–facies metamorphism at 2080–1970 Ma (Rosen et al. 2000). The North Alberta arcs experienced granulite–facies metamorphsim during accretion to the Hearne Province, c. 200 Ma later, at c. 1850–1800 Ma (Ross et al. 2000).

For these reasons, the model of Sears & Price (1978, 2000) is not followed and Siberia is positioned against northern Laurentia. Within this general configuration, almost every conceivable permutation has been explored (Fig. 5b–f). These Siberia–northern Laurentia fits are based primarily on the comparison of Archaean and Palaeoproterozoic terranes that are assumed to have maintained their integrity since Late Palaeoproterozoic assembly (Hoffman 1991; Condie & Rosen 1994; Frost et al. 1998; Rainbird et al. 1998). The fit (Fig. 5b) proposed by Hoffman (1991) has additional constraints based on stratigraphic and palaeontological similarities between northern Siberian and northern Laurentian successions in the Early Cambrian (Pelechaty 1996 and refs cited therein). Frost *et al.* (1998) juxtaposed southern Siberia with northern Laurentia (Fig. 5d). Their fit is based on comparison of the Thelon Magmatic Belt with the Aldan Terrane of the Aldan Shield. The main problem with this fit is the presence of the Stanovoy Province, which contains structures that are generally perpendicular to those of the Aldan Block (highlighted in Fig. 5d and e). A chain of Early Proterozoic gabbro–anorthosite plutons along the northern margin of Stanovoy Block (Gusev & Khain 1996), and other collision-related magmatic and metamorphic events, provide evidence for collision of the Stanovoy and Aldan Blocks at 1.8–2.0 Ga (Rosen *et al.* 1994 and refs cited therein). The model of Rainbird *et al.* (1998), in which southern Siberia is juxtaposed with northern Greenland (Fig. 5e), faces the same problem. Evidence for 1000–800 Ma island arcs along the southern boundary of Siberia (Rytsk *et al.* 1999; Kuzmichev *et al.* 2001) is also not supportive of these two models. Two new palaeopoles of Gallet *et al.* (2000) are not precisely dated and cannot give unequivocal support to any of the proposed fits. They also do not exclude the possibility that Siberia was not a part of Rodinia. New Late Mesoproterozoic–Early Neoproterozoic geochronological data (Rainbird *et al.* 1998) necessitated a revision of younger Neoproterozoic Siberian palaeomagnetic data, summarized by Smethurst *et al.* (1998). If Siberia was not connected to Laurentia, the identity of the conjugate to the passive margin of northern Laurentia (Frisch & Trettin 1991) is unknown.

In the present reconstruction, a N–N fit of Siberia and Laurentia is proposed (Fig. 5f), which is not contradicted by the palaeomagnetic data of Gallet *et al.* (2000) if an age of *c.* 1000 Ma is assumed for those sediments. Subsequent minor extensions and rotations could have led to the configurations of Hoffman (1991) and Pelechaty (1996), prior to final separation in the Early Cambrian (Pelechaty 1996).

Congo–São Francisco

This craton is traditionally treated as a single entity, owing to the similarity of Archaean and Palaeoproterozoic–Mesoproterozoic rocks and bounding Late Neoproterozoic mobile belts (e.g. Teixeira *et al.* 2000; Trompette 1994). The Irumide and Mozambique Belts of east-central Africa (south and SE margins of Congo) comprise a collage of sedimentary and island-arc-related intrusive rocks with U–Pb zircon crystallization ages of between *c.* 1400 and 1000 Ma (Pinna *et al.* 1993; Kröner *et al.* 1997, 2001; Oliver *et al.* 1998; Johnson & Oliver, 2000). This region is best interpreted as facing an open ocean undergoing passive arc accretion during the Mesoproterozoic until *c.* 1000 Ma (Kröner *et al.* 1997, 2001; Johnson & Oliver 2000). The Neoproterozoic East African Orogen (east margin of Congo Craton) consists mainly of reworked Neoarchaean and Palaeoproterozoic crust, and, as yet, no Mesoproterozoic ages have been identified, so it is difficult to determine whether this margin faced an ocean or other continental blocks in Rodinia time.

The northern margin of the Congo Craton is very poorly known. A northward-deepening sedimentary succession in northern Congo and the Central African Republic (Lindian and Bangui Basins) contains putative glacial deposits, indicating a likely Neoproterozoic age (Evans 2000 and refs cited therein). This succession is deformed within the Oubanguide Belt at *c.* 620 Ma (Penaye *et al.* 1993), which may represent either a continent–continent collision (Trompette 1994) or an intracratonic Pan-African remobilization.

The *c.* 1000–910 Ma West Congolian Belt, of lower Congo and Angola, is a series of rift-related sediments and volcanics, overlain by a passive-margin succession (Tack *et al.* 2001). Owing to the palaeogeographic position of the São Francisco Craton with respect to Congo, it is possible that the West Congolian orogenic cycle, as well as its Brazilian counterpart in the Aracuaí Belt (Pedrosol-Soares *et al.* 2001), represents mainly intracratonic tectonic events. In contrast, there is evidence for the existence of ocean basins west and south of the São Francisco Craton in the early Neoproterozoic until its collision with the Paraná Block at 750–790 Ma (e.g. Pimentel *et al.* 1999).

Collectively, these data provide evidence against a long-term connection between the Congo–São Francisco Craton and any other large continental block along any of its margins, with the possible exception of the present northern and eastern margins.

Four palaeopoles from the São Francisco Craton (Table 1) generate a slightly curved APWP that is anchored at its ends by hornblende Ar–Ar dates of 1078 ± 18 and 1011 ± 24 Ma for the Olivenca and Ilheus Dykes, respectively (Renne *et al.* 1990). The similarity of this track to the Laurentian 'Keweenawan' APWP track suggests common motion of Congo–São Francisco and Laurentia during this time interval, and the possibility of restoring their relative positions by superimposing their APWP.

The low curvatures of the two APWPs' segments permit using both polarity options (Fig. 4b). If one polarity option is accepted (position A in Fig. 4b), Congo–São Francisco lies directly on top of West

Africa and Amazonia. In this case, arguments for the position of those cratons in Rodinia would need to be revised (see above). The other polarity option (position B in Fig. 4b) shows a connection (or proximity) between the Congo northern margin and Western Australia. As noted above, very little is known about this margin of Congo, so such a possibility cannot be excluded. However, this fit causes an overlap of Congo with Kalahari, and, in this case, the proposed attachment of Kalahari to Australia (Fig. 4a) would be incorrect. A third possibility is that the apparent similarity between Laurentian and São Franciscan poles is coincidental, and that these cratons moved independently. A less reliable palaeomagnetic pole from the c. 950 Ma Nyabikere Massif (Meert et al. 1994) contradicts position A in Fig 4b, but may agree with position B after minor modification. However, the palaeopole from the Gagwe Lavas (Table 1) does not permit such a juxtaposition for c. 800 Ma. Therefore, if position B of Congo at 1000 Ma is accepted, its breakup from Australia probably occurred prior to the onset of breakup in other sectors of Rodinia (see below).

There is insufficient information to prove or disprove option B, in which Congo–São Francisco is positioned close to Western Australia and correspondingly forms a part of Rodinia, so this option is left open for further investigation. However, it is recognized that the configuration preferred for the present model (Fig. 4a), with Kalahari attached to Western Australia and Congo–São Francisco as an independent plate, is also poorly constrained.

Configuration of Rodinia

The proposed reconstruction of Rodinia at c. 990 Ma is shown in Fig. 6; the corresponding rotation parameters for each block are listed in Table 2. The main differences with previous reconstructions are: (1) the position of Australia, which is juxtaposed against Laurentia in the AUSMEX fit; (2) East Antarctica was not a single block; (3) Congo–São Francisco and India–Rayner Cratons were not parts of Rodinia; (4) Siberia was situated close to northern Laurentia, but in a position different from all proposed previously; (5) Kalahari–Dronning Maud Land was attached to Western Australia.

Breakup of Rodinia

Many scientists have suggested that the breakup of Rodinia started along the western boundary of Laurentia between 820 and 720 Ma (e.g. Hoffman 1991; Powell et al. 1994; Li et al. 1999; Wingate & Giddings 2000). Rifting along the eastern margin of

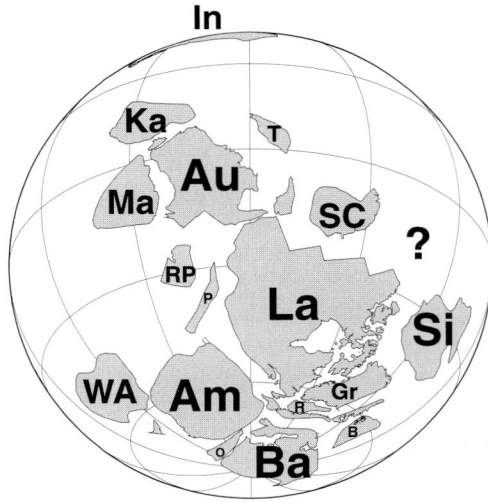

Fig. 6. The shape of Rodinia at 990 Ma. In, India; Ka, Kalahari; Si, Siberia; T, Tarim. See also notes to Figures 2 & 3. The position of Congo–São Francisco is outside the shown projection – for its proposed location see Figure 4a.

Laurentia started significantly later, c. 620–550 Ma (e.g. Bingen et al. 1998; Cawood et al. 2001). Wingate & Giddings (2000) concluded that if the SWEAT fit (Moores 1991) is correct, then breakup between Australia and western Laurentia occurred prior to 755 Ma, although a similar conclusion can be drawn for either the AUSWUS (Brookfield 1993; Karlstrom et al. 1999; Burrett & Berry 2000) or AUSMEX (Wingate et al. 2002) configurations.

Extensive mafic magmatism related to rifting is recognized in western Laurentia (Ross et al. 1995), SE Australia (Wingate et al. 1998) and South China (Li et al. 1999), although it occurred at 730–780 Ma in western Laurentia, and at 820–830 Ma in SE Australia and South China. This mismatch is difficult to explain in the context of the SWEAT or AUSWUS hypotheses. The AUSMEX configuration, however, can provide a plausible explanation. It is suggested that, initially, breakup started c. 830 Ma between South China–Laurentia–Rio de la Plata in the east, and Australia–Mawson Craton in the west (Fig. 7a). Initiation of rifting was probably related to the 827 Ma Gairdner Dyke Swarm (Wingate et al. 1998) and coeval South China dykes (Li et al. 1999). At c. 780 Ma, this spreading stopped and a new rifting event began (Fig. 7b), the evidence for which includes three 780 Ma mafic intrusive suites in western Laurentia (Park et al. 1995). A possible analogy for such a set of events is the opening of the North Atlantic Ocean (Fig. 7b, inset). Impingement of a mantle plume may have accompanied Rodinia breakup (Wingate et al. 1998; Li et al. 1999), in the

Table 2. *Euler rotation parameters*

Craton/block/terrane	Age (Ma)	Pole (°N)	Pole (°E)	Angle (°)
Laurentia to absolute framework	990	13.9	−144.1	−134.7
	800–790	31.8	−149.0	−87.0
	780–770	20.6	−148.5	−93.7
	760–750	18.0	−140.2	−95.2
Baltica to Laurentia	990–750	75.8	−95.8	−59.2
Greenland to Laurentia	990–750	67.5	−118.5	−13.8
Amazonia to Laurentia	990–750	12.0	−47.0	−110.7
West Africa to Amazonia	990–750	53.0	−35.0	−51.0
Rio de la Plata to Laurentia	990–750	9.9	−47.4	−93.7
Pampean to Rio de la Plata	990–750	70.9	−10.8	−3.8
Rockall to Laurentia	990–750	75.3	159.6	−23.5
Oaxaquia to Amazonia	990–750	12.1	81.7	53.4
Chortis to Amazonia	990–750	5.7	−78.5	139.8
Siberia to Laurentia	990	65.0	159.3	−69.6
	800–750	3.5	13.1	23.2
South China to absolute framework	990	66.4	−107.9	127.9
	800–790	65.1	176.0	143.0
	780–770	61.0	172.6	150.9
	760–750	50.4	166.9	172.7
Australia to absolute framework	990	42.6	−5.2	115.8
	800–790	56.6	51.1	72.7
	780–770	55.4	53.0	71.9
	760–750	50.8	71.3	70.8
Mawson to Australia	990–750	1.3	37.7	30.3
Kalahari to Australia	990–750	79.8	97.1	73.4
Dronning Maud Land to Kalahari	990–750	9.7	148.7	−56.3
Tarim to Australia	990–750	13.5	98.3	−153.4
India to absolute framework	990	58.6	−3.9	86.3
	800–790	32.9	6.5	59.2
	780–770	40.7	6.2	53.7
	760–750	68.7	4.3	44.9
Sri Lanka to India	990–750	9.8	82.9	−24.3
Rayner to India	990–750	4.9	−163.4	−93.2
Congo to absolute framework	990	45.6	83.8	68.3
	800–790	55.8	57.1	130.1
	780–770	46.5	65.9	129.6
	760–750	20.0	79.1	135.4
São Francisco to Congo	990–750	53.0	−35.0	51.0

same manner as Iceland in the North Atlantic (Lawer and Müller 1994).

Figure 7c shows several reliable palaeopoles with ages c. 760–750 Ma. The low-latitude position of Australia between 800 and 750 Ma is also constrained by the palaeomagnetic study of a deep drillhole in Western Australia (Pisarevsky et al. 2001). The 755 Ma Mundine Well mafic dykes in Western Australia may indicate a rifting event, possibly involving a detachment of the Kalahari Craton (Fig.

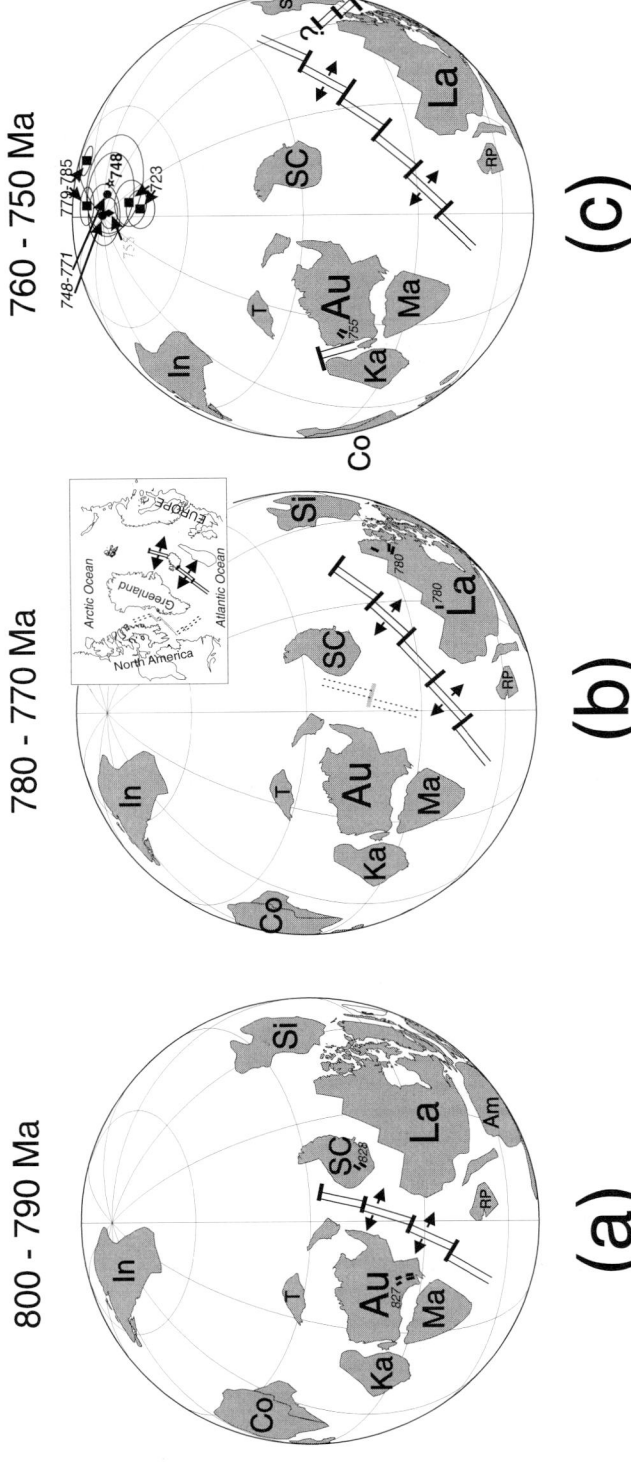

Fig. 7. Sequential breakup of Rodinia. Palaeomagnetic poles: ■, Laurentian; ●, Indian; ◆, Australian; ☆, South China. Abbreviations as in previous figures; for details, see text and Table 1.

7c). Figure 7c also shows the possible initiation of rifting between northern Laurentia and Siberia that may be represented by the 723 Ma Franklin Dykes (e.g. Heaman et al. 1992). However, this may have been an aborted rift, with final separation between Laurentia and Siberia delayed until the Early Cambrian (e.g. Pelechaty 1996).

Although India is not regarded as part of Rodinia, two sets of palaeomagnetic data can be used to position this block during the period associated with Rodinia breakup. The Harohalli Dykes of southern India, dated at 823–810 Ma (Radhakrishna & Matthew 1996), place India at high palaeolatitudes (Fig. 7a). Two high-quality poles, from the Seychelles at 755–748 Ma (Torsvik et al. 2001a), and the Malani Igneous Suite (MIS) of northwestern India at 771–751 Ma (Torsvik et al. 2001b), can be matched in a tight reconstruction at intermediate, northerly latitudes, tracing India's motion from polar to moderate latitudes (Fig. 7b and c). Traditionally, Madagascar is positioned with the Seychelles and the MIS, based on the correlation of contemporaneous, bimodal igneous intrusive rocks, interpreted as a 450 km long continental arc (Handke et al. 1999; Torsvik et al. 2001a,b; Tucker et al. 2001; Collins et al. 2002). However, identification of a significant Pan-African suture zone in eastern Madagascar (the Betsimisraka Fault; Kröner et al. 2000; Collins et al. 2002), which divides a thin eastern region of Indian or Karnataka Craton affinity from the main, central block of Neoarchaean and Palaeoproterozoic age, indicates that most of Madagascar was never connected to India prior to Neoproterozoic time. This raises two alternative palaeogeographic scenarios. The first is that the correlation between central Madagascar, the Seychelles and the MIS is incorrect (Collins et al. 2002), and that central Madagascar was not part of the Seychelles–MIS–Greater India block. The second preserves the Madagascar–Seychelles–MIS connection but extends the Betsimisraka Suture Zone through the Aravalli Belt of northwestern India, thereby separating the Seychelles and the MIS from Greater India until Neoproterozoic time (Torsvik et al. 2001b, fig. 7a). If the second scenario is correct, then there is only one reliable palaeopole for India (Harohalli Dykes; Table 1) and two poles for this separate block (Malani and Mahe; Table 1) during the 1000–550 Ma period. In this case, India would still be excluded from Rodinia and its position in Figure 7a is probably correct. Less certain would be the location of India in Figure 7b and c, but India's depicted southerly motion is consistent with eventual collision with Australia at c. 640 Ma (Powell & Pisarevsky 2002). In the present reconstructions, only the easternmost rocks of Madagascar, in their original position adjacent to India, are shown; central Madagascar is of uncertain position and is thus not shown.

The positions of Congo–São Francisco in Figure 7a and c are constrained by the Gagwe and Mbozi Poles, respectively (Table 1). The longitudes of the India and Congo–São Francisco Cratons in Figure 7 are unconstrained, as only individual poles, not APWP, are being compared. Tectonics of the Congo–São Francisco Craton during this interval of Rodinia fragmentation is intriguing: collision of the São Francisco Craton and Paraná Block had probably occurred by 790–750 Ma (Pimentel et al. 1999), and thus the onset of Gondwanaland's assembly may have overlapped in time with the final stages of Rodinia's demise.

Conclusions

A new configuration of Rodinia is proposed in this study, based on available geological data and reliable palaeopoles. Rodinia was finally assembled c. 1000 Ma, as manifest in the series of orogenic belts of 'Grenvillian' age. However, Grenvillian orogenesis was probably multistaged, as was shown for East Antarctica by Fitzsimons (2000). New palaeomagnetic data from Australia (Wingate et al. 2002) contradict the popular SWEAT and AUSWUS hypotheses, and a new AUSMEX fit is suggested. The proposed composition of Rodinia does not include India or the Congo–São Francisco Cratons; a new position of Kalahari against Western Australia is suggested.

The breakup of Rodinia probably started at c. 820–800 Ma by rifting between Australia–Mawson–Kalahari and South China–Laurentia–Rio de la Plata. At 780–770 Ma, the spreading centre jumped to a position between Laurentia and South China, and the initial branch was aborted. Kalahari detached from Australia at 760–750 Ma and the Rodinia fragments began the slow journey toward their reassembly in Gondwanaland, and, ultimately, Pangaea.

This research was supported by the Australian Research Council (ARC) through its Research Centres program. S. A. Pisarevsky acknowledges a University of Western Australia Gledden Senior Fellowship. Reconstructions were made using PLATES software from the University of Texas at Austin, and GMT software of Wessel and Smith. We are very grateful to M. W. McElhinny and T. Raub for their helpful reviews. This is Tectonics Special Research Centre publication no. 180 and a contribution to International Geological Correlation Project 440.

References

ALKMIM, F. F., MARSHAK, S. & FONSECA, M. A. 2001. Assembling West Gondwana in the Neoproterozoic: clues from the São Francisco craton region, Brazil. Geology, 29, 319–322.

ALVAREZ, V. C. & DUNLOP, D. J. 1998. A regional paleomagnetic study of lithotectonic domains in the Central Gneiss Belt, Grenville Province, Ontario. *Earth and Planetary Science Letters*, **157**, 89–103.

BALDWIN, S. & IRELAND, T. A. 1995. Tale of two eras: Pliocene–Pleistocene unroofing of Cenozoic andesite Archean zircons from active metamorphic core complexes, Solomon Sea, Papua New Guinea. *Geology*, **23**, 1023–1026.

BALLARD, M. M., VAN DER VOO, R. & URRUTIA-FUCUGAUCHI, J. 1989. Paleomagnetic results from Grenvillian-aged rocks from Oaxaca, Mexico: evidence for a displaced terrane. *Precambrian Research*, **42**, 343–352.

BERGER, G. W. & YORK, D., 1981. Geochronometry from $^{40}Ar/^{39}Ar$ dating experiments. *Geochimica Cosmochimia Acta*, **45**, 795–811.

BINGEN, B., DEMAIFFE, D. & VAN BREEMEN, O. 1998. The 616 Ma old Egersund basaltic dike swarm, SW Norway, and Late Neoproterozoic opening of the Iapetus Ocean. *Journal of Geology*, **106**, 565–574.

BLEWETT, R. S., BLACK, L. P., SUN, S-s., KNUTSON, J., HUTTON, L. J. & BAIN, J. H. C. 1998. U-Pb zircon and Sm-Nd geochronology of the Mesoproterozoic of North Queensland: implications for a Rodinian connection with the Belt supergroup of North America. *Precambrian Research*, **89**, 101–127.

BOND, G. C., NICKESON, P. A. & KOMINZ, M. A. 1984. Breakup of a supercontinent between 625 Ma and 555 Ma: new evidence and implications for continental histories. *Earth and Planetary Science Letters*, **70**, 325–345.

BOGER, S. D., WILSON, C. J. L. & FANNING, C. M. 2001. Early Paleozoic tectonism with the East Antarctic craton: the final suture between east and west Gondwana. *Geology*, **29**, 463–466.

BRIDEN, J. C., DUFF, B. A. & KRÖNER, A. 1979. Palaeomagnetism of the Koras Group, Northern Cape Province, South Africa. *Precambrian Research*, **10**, 43–57.

BROOKFIELD, M. E. 1993. Neoproterozoic Laurentia–Australia fit. *Geology*, **21**, 683–686.

BRUGUIER, O., BOSCH, D., PIDGEON, R. T., BYRNE, D. I. & HARRIS, L. B., 1999. U-Pb chronology of the Northampton Complex, Western Australia – Evidence for Grenvillian sedimentation, metamorphism and deformation and geodynamic implications. *Contributions to Mineralogy and Petrology*, **136**, 258–267.

BUCHAN, K. L. & DUNLOP, D. J. 1976. Paleomagnetism of the Haliburton Intrusions: superimposed magnetisations, metamorphism, and tectonics in the late Precambrian. *Journal of Geophysical Research*, **81**, 2951–2966.

BURRETT, C. & BERRY, R. 2000. Proterozoic Australia–Western United States (AUSWUS) fit between Laurentia and Australia. *Geology*, **28**, 103–106.

BYLUND, G. 1985. Palaeomagnetism of middle Proterozoic basic intrusives in central Sweden and the Fennoscandian apparent polar wander path. *Precambrian Research*, **28**, 283–310.

BYLUND, G. 1992. Palaeomagnetism, mafic dykes and the Protogine Zone, southern Sweden. *Tectonophysics*, **201**, 49–63.

BYLUND, G. & PISAREVSKY, S. A. 2002. Palaeomagnetism of Mesoproterozoic dykes from the Protogine Zone and the enigmatic Sveconorwegian Loop. *Geologiska Föreningen i Stockholm Förhandlingar*, **124**, 11–18.

CAWOOD, P. A., MCCAUSLAND, P. J. A. & DUNNING, G. R. 2001. Opening Iapetus: constraints from the Laurentian margin in Newfoundland. *Geological Society of America Bulletin*, **113**, 443–453.

CINGOLANI, C. & DALLA SADA, L. 2000. Buenos Aires cratonic region. *In:* CORDIANI, U. G., MILANI, E. J., THOMAZ FILHO, A. & CAMPOS, D. A. (eds) *Tectonic Evolution of South America*, 31st International Geological Congress, Rio de Janeiro. 139–150.

COLLINS, A. S., FITZSIMONS, I. C. W., H, B. & RAZAKAMANANA, T. 2002. Structure of the eastern margin of the East African Orogen in Central Madagascar. *Precambrian Research*, in press.

CONDIE, K. C. & ROSEN, O. M. 1994. Laurentia–Siberia connection revisited. *Geology*, **22**, 168–170.

CORDANI, U. G., SATO, K., TEIXEIRA, W., TASSINARI, C. G. & BASEI, M. A. S. 2000. Crustal evolution of the South American Platform. *In:* CORDIANI, U. G., MILANI, E. J., THOMAZ FILHO, A. & CAMPOS, D. A. (eds) *Tectonic Evolution of South America*, 31st International Geological Congress, Rio de Janeiro. 19–40.

D'AGRELLA-FILHO, M. S., CORDANI, U. G., DE BRITO NEVES, B. B., TRINIDADE, R. I. F. & PACCA, I. G. 2001. Evidence from South America on the formation and disruption of Rodinia. *Geological Society of Australia, Abstracts*, **65**, 27.

D'AGRELLA-FILHO, PACCA, I. G., RENNE, P. R., ONSTOTT, T. C. & TEIXEIRA, W. 1990. Paleomagnetism of Middle Proterozoic (1.01 to 1.08 Ga) mafic dykes in southeastern Bahia State–Sao Francisco Craton, Brazil. *Earth and Planetary Science Letters*, **101**, 332–348.

DALRYMPLE, G. B. 1979. Critical tables for conversion of K–Ar ages from old to new constants. *Geology*, **7**, 558–560.

DALY, J. S, FITZGERALD, R. C., BREWER, T. S., MENUGE, J. F., HEAMAN, L. M. & MORTON, A. C. 1994. Persistent geometry of Palaeoproterozoic juvenile crust on the southern margin of Laurentia–Baltica: evidence from Ireland and Rockall Bank. *Terra Abstracts*, 2, **6**, 15.

DALZIEL, I. W. D. 1997. Neoproterozoic–Paleozoic geography and tectonics: review, hypothesis, environmental speculation. *Geological Society of America Bulletin*, **109**, 16–42.

DALZIEL, I. W. D., MOSHER, S. & GAHAGAN, L. M. 2000. Laurentia–Kalahari collision and the assembly of Rodinia. *Journal of Geology*, **108**, 499–513.

DAVIS, D. W. & GREEN, J. C. 1997. Geochronology of the North American Midcontinent rift in western Lake Superior and implications for its geodynamic evolution. *Canadian Journal of Earth Sciences*, **34**, 476–488.

DAVIS, D. W. & PACES, J. B. 1990. Time resolution of geologic events on the Keweenaw Peninsula and applications for development of the Midcontinent Rift system. *Earth and Planetary Science Letters*, **97**, 54–64.

DAVIS, D. W. & SUTCLIFFE, R. H. 1985. U-Pb ages from the Nipigon plate and northern Lake Superior. *Geological Society of America Bulletin*, **96**, 1572–1579.

DEBLOND, A., PUNZALAN, L. E., BOWEN, A. & TACK, L.

2001. The Malagarazi Supergroup of SE Burundi and its correlative Bukoba Supergroup of NW Tanzania: Neo- and Mesoproterozoic chronostratigraphic constraints from Ar–Ar ages on mafic intrusive rocks. *Journal of African Earth Sciences*, **32**.

DIEHL, J. F. & HAIG, T. D. 1994. A paleomagnetic study of the lava flows within the Copper Harbour Conglomerate, Michigan: new results and implications. *Canadian Journal of Earth Sciences*, **31**, 369–380.

ELMING, S.-Å., PESONEN, L. J., LEINO, M. A. H. ET AL. 1993. The drift of the Fennoscandian and Ukrainian shields during the Precambrian: a palaeomagnetic analysis. *Tectonophysics*, **223**, 177–198.

ERNST, R. E. & BUCHAN, K. L. 1993. Paleomagnetism of the Abitibi dike swarm, southern Superior Province, and implications for the Logan Loop. *Canadian Journal of Earth Sciences*, **30**, 1886–1897.

EVANS, D. A. D. 2000. Stratigraphic, geochronological, and paleomagnetic constraints upon the Neoproterozoic climatic paradox. *American Journal of Science*, **300**, 347–433.

EVANS, D. A. D., LI, Z. X., KIRSCHVINK, J. L. & WINGATE, M. T. D. 2000. A high-quality mid-Neoproterozoic paleomagnetic pole from South China, with implications for ice ages and the breakup configuration of Rodinia. *Precambrian Research*, **100**, 313–334.

FAIRCHILD, I. J. & HAMBREY, M. J. 1995. Vendian basin evolution in East Greenland and NE Svalbard. *Precambrian Research*, **73**, 217–233.

FITZSIMONS, I. C. W. 2000. Grenville-age basement provinces in East Antarctica: evidence for three separate collisional orogens. *Geology*, **28**, 879–882.

FITZSIMONS, I. C. W. 2002. Comparison of detrital zircon ages in the Pinjarra orogen (WA) and Maud province (Antarctica): evidence for collision of western Australia with southern Africa at 1100 Ma. *Geological Society of Australia*, in press.

FRISCH, T. & TRETTIN, H. P. 1991. Precambrian successions in the northernmost part of the Canadian Shield. *In*: TRETTIN, H. P. (ed.) *Geology of the Innuitian Orogen and Arctic Platform of Canada and Greenland*. Geological Survey of Canada, Geology of Canada, **3**, 103–108.

FROST, B. R., AVCHENKO, O. V., CHAMBERLAIN, K. R. & FROST, C. D. 1998. Evidence for extensive Proterozoic remobilization of the Aldan Shield and implications for Proterozoic plate tectonic reconstructions of Siberia and Laurentia. *Precambrian Research*, **89**, 1–23.

GALLET, Y., PAVLOV, V. E., SEMIKHATOV, M. A. & PETROV, P. YU. 2000. Late Mesoproterozoic magnetostratigraphic results from Siberia: paleogeographic implications and magnetic field behaviour. *Journal of Geophysical Research*, **105**, 16481–16499.

GEE, D. G., BJÖRKLUND, L. & STØLEN, L.-K. 1994. Early Proterozoic basement in Ny Friesland – implications for the Caledonian tectonics of Svalbard. *Tectonophysics*, **231**, 171–182.

GIBBS, A. K. & BARRON, C. N. 1993. *The Geology of the Guiana Shield*. Oxford Monographs on Geology and Geophysics, **22**.

GORBATSCHEV, R. & BOGDANOVA, S. 1993. Frontiers in the Baltic Shield. *Precambrian Research*, **64**, 3–21.

GUSEV, G. S. & KHAIN, V. YE. 1996. On relations between the Baikal–Vitim, Aldan–Stanovoy, and Mongol–Okhotsk terranes (south of mid-Siberia). *Geotectonics*, **29**, 422–436.

HALLS, H. C. & PESONEN, L. J. 1982. Paleomagnetism of Keweenawan rocks. *Geological Society of America, Memoirs*, **156**, 173–201.

HANDKE, M. J., TUCKER, R. D. & ASHWAL, L. D. 1999. Neoproterozoic continental arc magmatism in west-central Madagascar. *Geology*, **27**, 351–354.

HARLAN, S. S., GEISSMAN, J. W. & SNEE, L. W. 1997. *Paleomagnetic and $^{40}Ar/^{39}Ar$ geochronologic data from late Proterozoic mafic dykes and sills, Montana and Wyoming*. USGS Professional Paper 1580.

HARRIS, L. B. 1994. Neoproterozoic sinistral displacement along the Darling Mobile Belt, Western Australia, during Gondwanaland assembly. *Journal of the Geological Society, London*, **151**, 901–904.

HEAMAN, L. M., LECHEMINANT, A. N. & RAINBIRD, R. H. 1992. Nature and timing of Franklin igneous events, Canada: implications for a Late Proterozoic mantle plume and the break-up of Laurentia. *Earth and Planetary Science Letters*, **109**, 117–131.

HENRY, S. G., MAUK, F. J. & VAN DER VOO, R. 1977. Paleomagnetism of the upper Keweenawan sediments: the Nonesuch Shale and Freda Sandstone. *Canadian Journal of Earth Sciences*, **14**, 1128–1138.

HIGGINS, A. K. & LESLIE, A. G. 2000. Restoring thrusting in the East Greenland Caledonides. *Geology*, **28**, 1019–1022.

HIGGINS, A. K., SMITH, M. P., SOPER, N. J., LESLIE, A. G., RASMUSSEN, J. A. & SØNDERHOLM, M. 2001. The Neoproterozoic Hekla Sund Basin, eastern North Greenland: a pre-Iapetian extensional sequence thrust across its rift shoulders during the Caledonian orogeny. *Journal of the Geological Society, London*, **158**, 487–499.

HOFFMAN, P. F. 1989. Precambrian geology and tectonic history of North America. *In*: BALLY, A. W. & PALMER, A. R. (eds) *The Geology of North America – An Overview*. Geological Society of America, The Geology of North America, **A**, 447–512.

HOFFMAN, P. F. 1991. Did the breakout of Laurentia turn Gondwana inside out? *Science*, **252**, 1409–1412.

HYODO, H. & DUNLOP, D. J. 1993. Effect of anisotropy on the paleomagnetic contact test for a Grenvillian dike. *Journal of Geophysical Research*, **98**, 7997–8017.

JACOBS, J., FANNING, C. M., HENJES-KUNST, F., OLESCH, M. & PAECH, H.-J. 1998. Continuation of the Mozambique Belt into East Antarctica: Grenville-age metamorphism and polyphase Pan-African high-grade events in Central Dronning Maud Land. *Journal of Geology*, **106**, 385–406.

JACOBS, J, THOMAS, R. J. & WEBER, K. 1993. Accretion and indentation tectonics at the southern edge of the Kaapvaal craton during the Kibaran (Grenville) orogeny. *Geology*, **21**, 203–206.

JOHNSON, S. P. & OLIVER, G. J. H. 2000. Mesoproterozoic oceanic subduction, island-arc formation and the initiation of back-arc spreading in the Kibaran Belt of central, southern Africa: evidence from the Ophiolite Terrane, Chewore Inliers, northern Zimbabwe. *Precambrian Research*, **103**, 125–146.

KALSBEEK, F., THRANE, K., NUTMAN, A. P. & JEPSEN, H. F. 2000. Late Mesoproterozoic to early Neoproterozoic history of the East Greenland Caledonides: evidence for Grenvillian orogenesis? *Journal of the Geological Society, London*, **157**, 1215–1225.

KARLSTROM, K. E., HARLAN, S. S., WILLIAMS, M. L., MCLELLAND, J., GEISSMAN, J. W & ÅHALL, K.-I. 1999. Refining Rodinia: geologic evidence for the Australia–Western U.S. connection in the Proterozoic. *GSA Today*, **9**, 1–7.

KEPPIE, J. D. & ORTEGA-GUTIERREZ, F. O. 1999. Middle American Precambrian basement: a missing piece of the reconstructed 1–Ga orogen. *Geological Society of America, Special Paper*, **336**, 199–210.

KEPPIE, J. D. & RAMOS, V. A. 1999. Odyssey of terranes in the Iapetus and Rheic oceans during the Paleozoic. *Geological Society of America, Special Paper*, **336**, 267–276.

KHUDOLEY, A. K., RAINBIRD, R. H., STERN, R. A. ET AL. 2001. Sedimentary evolution of the Riphean – Vendian basin of southeastern Siberia. *Precambrian Research*, **111**, 129–163.

KRÖNER, A., HEGNER, E., COLLINS, A. S., WINDLEY, B. F., BREWER, T. S., RAZAKAMANANA, T. & PIDGEON, R. T. 2000. Age and magmatic history of the Antanarivo Block, central Madagsacar, as derived from zircon geochronology and Nd isotopic systematics. *American Journal of Science*, **300**, 251–288.

KRÖNER, A., SACCHI, R., JAECKEL, P. & COSTA, M. 1997. Kibaran magmatism and Pan-African granulite metamorphism in northern Mozambique: single zircon ages and regional implications. *Journal of African Earth Sciences*, **25**, 467–484.

KRÖNER, A., WILLNER, A. P., HEGNER, E., JAECKEL, P. & NEMCHIN, A. 2001. Single zircon ages, PT evolution and Nd isotopic systematics of high-grade gneisses in southern Malawi and their bearing on the evolution of the Mozambique Belt in southeastern Africa. *Precambrian Research*, **190**, 257–291.

KUZMICHEV, A. B., BIBIKOVA, E. V. & ZHURAVLEV, D. Z. 2001. Neoproterozoic (~800Ma) orogeny in the Tuva–Mongolia Massif (Siberia): island arc-continent collision at the northeast Rodinia margin. *Precambrian Research*, **110**, 109–126.

KUZMIN, V. V., CHUCHONIN, A. P. & SHULEZHKO, I. K. 1995. Stages of metamorphic evolution of rocks of crystalline basement of the Kukhtui Uplift (Okhotsk Massif). *Reports of Russian Academy of Sciences*, **342**, 789–791 [in Russian].

LARSON, S.-Å. & BERGLUND, J. 1992. A chronological subdivision of the Transscandinavian Igneous Belt – three magmatic episodes? *Geologiska Föreningen i Stockholm Förhandlingar*, **114**, 459–461.

LAWER, L. A. & MÜLLER, R. D. 1994. Iceland hotspot track. *Geology*, **22**, 311–314.

LECHEMINANT, A. N. & HEAMAN, L. M. 1994. 779 Ma mafic magmatism in the northwestern Canadian Shield and northern Cordillera: a new regional time-marker. *Proceedings of the 8th International Conference. Geochronology, Cosmochronology and Isotope Geology, Program Abstracts*, **1107**, 197.

LI, Z. X. & POWELL, C. MCA. 1999. Palaeomagnetic study of Neoproterozoic glacial rocks of the Yangzi Block: palaeolatitude and configuration of South China in the late Proterozoic Supercontinent. *Precambrian Research*, **94**, 1–5.

LI, Z. X., LI, X. H., KINNY, P. D. & WANG, J. 1999. The breakup of Rodinia: did it start with a mantle plume beneath South China? *Earth and Planetary Science Letters*, **173**, 171–181.

LI, Z. X., LI, X. H., ZHOU, H. & KINNY, P. D. 2002. Grenville-aged continental collision in South China: new SHRIMP U–Pb zircon results and implications for Rodinia configuration. *Geology*, **30**, 163–166.

LI, Z. X., ZHANG, L. & POWELL, C. MCA. 1995. South China in Rodinia: part of the missing link between Australia–East Antarctica and Laurentia? *Geology*, **23**, 407–410.

LI, Z. X., ZHANG, L. & POWELL, C. MCA. 1996. Positions of the East Asian cratons in the Neoproterozoic supercontinent Rodinia. *Australian Journal of Earth Sciences*, **43**, 593–604.

MCCABE, C. & VAN DER VOO, R. 1983. Paleomagnetic results from the upper Keweenawan Chequamegon Sandstone: implications for red bed diagenesis and Late Precambrian apparent polar wander of North America. *Canadian Journal of Earth Sciences*, **20**, 105–112.

MCELHINNY, M. W. & MCFADDEN, P. L. 2000. *Paleomagnetism: Continents and Oceans*. Academic Press, San Diego.

MCMENAMIN, M. A. S. & MCMENAMIN, D. L. 1990. *The Emergence of Animals; The Cambrian Breakthrough*. Columbia University Press, New York.

MEERT, J. G. & POWELL, C. MCA. 2001. Assembly and break-up of Rodinia: introduction to the special volume. *Precambrian Research*, **110**, 1–8.

MEERT, J. G., HARGRAVES, R. B., VAN DER VOO, R., HALL, C. H. & HALLIDAY, A. N. 1994. Paleomagnetism and $^{40}Ar/^{39}Ar$ studies of Late Kibaran intrusives in Burundi, East Africa: implications for Late Proterozoic supercontinents. *Journal of Geology*, **102**, 621–637.

MEERT, J. G., VAN DER VOO, R. & AYUB, S. 1995. Paleomagnetic investigation of the Late Proterozoic Gagwe lavas and Mbozi complex, Tanzania and the assembly of Gondwana. *Precambrian Research*, **69**, 113–131.

MERTANEN, S., PESONEN, L. J. & HUHMA, H. 1996. Palaeomagnetism and Sm–Nd ages of the Neoproterozoic diabase dykes in Laanila and Kautokeino, northern Fennoscandia. *In:* BREWER, T. S. (ed.) *Precambrian Crustal Evolution in the North Atlantic Region*. Geological Society of London, Special Publication, **112**, 331–358.

MILLER, J. A., MATTHEWS, D. H. & ROBERTS, D. G. 1973. Rock of Grenville age from Rockall Bank. *Nature Physical Science*, **246**, 61.

MILLER, K. C. & HARGRAVES, R. B. 1994. Paleomagnetism of some Indian kimberlites and lamproites. *Precambrian Research*, **69**, 259–267.

MOORES, E. M. 1991. Southwest US–East Antarctic (SWEAT) connection: a hypothesis. *Geology*, **19**, 425–428.

MORTON, A. C. & TAYLOR, P. N. 1991. Geochemical and isotopic constraints on the nature and age of basement rocks from Rockall Bank, NE Atlantic. *Journal of the Geological Society, London*, **148**, 631–634.

MOSHER, S. 1998. Tectonic evolution of the southern Laurentian Grenville orogenic belt. *Geological Society of America Bulletin*, **110**, 1357–1375.

MÖLLER, C. & SÖDERLUND, U. 1997. Age constraints on the deformation within the Eastern Segment SW Sweden: Late Sveconorwegian granite dyke intrusion and metamorphic–deformational relations. *Geologiska Föreningen i Stockholm Förhandlingar*, **119**, 1–12.

NATAL'IN, B. A., AMATO, J. M., TORO, J. & WRIGHT, J. E. 1999. Paleozoic rocks of northern Chukotka Peninsula, Russian Far East: implications for the tectonics of the Arctic region. *Tectonics*, **18**, 977–1003.

OLIVER, G. J. H., JOHNSON, S. P., WILLIAMS, I. S. & HERD, D. A. 1998. Relict 1.4 Ga oceanic crust in the Zambezi Valley, northern Zimbabwe: evidence for Mesoproterozoic supercontinental fragmentation. *Geology*, **26**, 571–573.

ONSTOTT, T. C. & HARGRAVES, R. B. 1981. Proterozoic transcurrent tectonics: palaeomagnetic evidence from Venezuela and Africa. *Nature*, **289**, 131–136.

ONSTOTT, T. C., T. C., HARGRAVES, R. B. & JOUBERT, P. 1986. Constraints on the evolution of the Namaqua Province II: Reconnaissance palaeomagnetic and $^{40}Ar/^{39}Ar$ results from the Namaqua Province and the Kheis Belt. *Transactions of the Geological Society of South Africa*, **89**, 143–170.

PALMER, H. C., BARAGAR, W. R. A., FORTIER, M. & FOSTER, J. H. 1983. Paleomagnetism of Late Proterozoic rocks, Victoria Island, Northwest Territories, Canada. *Canadian Journal of Earth Sciences*, **20**, 1456–1469.

PARK, J. K. 1994. Palaeomagnetic constraints on the position of Laurentia from middle Neoproterozoic to Early Cambrian times. *Precambrian Research*, **69**, 95–112.

PARK, J. K., J. K. & AITKEN, J. D. 1986. Paleomagnetism of the Katherine Group in the Mackenzie Mountains: implications for post-Grenville (Hadrynian) apparent polar wander. *Canadian Journal of Earth Sciences*, **23**, 308–323.

PARK, J. K., J. K., BUCHAN, K. L. & HARLAN, S. S. 1995. A proposed giant radiating dyke swarm fragmented by separation of Laurentia and Australia based on paleomagnetism of ca. 780 Ma mafic intrusions in western North America. *Earth and Planetary Science Letters*, **132**, 129–139.

PARK, J. K., J. K., NORRIS, D. K. & LAROCHELLE, A. 1989. Paleomagnetism and the origin of the Mackenzie Arc of northwestern Canada. *Canadian Journal of Earth Sciences*, **26**, 2194–2203.

PARK, J. K., & R. G. 1992. Plate kinematic history of Baltica during the Middle to Late Proterozoic: a model. *Geology*, **20**, 725–728.

PARK, J. K., R. G., CLIFF, R. A., FETTES, D. J. & STEWARD. A. D. 1994. Precambrian rocks in northwest Scotland west of the Moine Thrust: the Lewisian Complex and the Torridonian. *In*: GIBBONS, W. & HARRIS, A. L. (eds) *A Revised Correlation of Precambrian Rocks in the British Isles*. Geological Society, London, Special Report, **22**, 6–22.

PEDROSA-SOARES, A. C., NOCE, C. M., WIEDEMANN, C. M. & PINTO, C. P. 2001. The Araçoaí–West-Congo Orogen in Brazil: an overview of a confined orogen formed during Gondwanaland assembly. *Precambrian Research*, **110**, 307–323.

PELECHATY, S. M. 1996. Stratigraphic evidence for the Siberia–Laurentia connection and Early Cambrian rifting. *Geology*, **24**, 719–722.

PENAYE, J., TOTEU, S. F., VAN SCHMUS, W. R. & NZENTI, J.-P. 1993. U–Pb and Sm–Nd preliminary geochronologic data on the Yaounde series, Cameroon: re-interpretation of the granulitic rocks as the suture of a collision in the 'Centrafrican' belt. *Comptes Rendus de l'Académie des Sciences Paris, Série II*, **317**, 789–794.

PERRIN, M. & PREVOT, M. 1988. Uncertainities about the Proterozoic and Paleozoic polar wander path of the West African craton and Gonwana: evidence for successive remagnetization events. *Earth and Planetary Science Letters*, **88**, 337–347.

PIMENTEL, M. M., FUCK, R. A. & BOTELHO, N. F. 1999. Granites and the geodynamic history of the Neoproterozoic Brazilia belt, Central Brazil: a review. *Lithos*, **46**, 463–483.

PINNA, P., JOURDE, G., CALVEZ, J. Y., MROZ, J. P. & MARQUES, J. M. 1993. The Mozambique Belt in northern Mozambique: Neoproterozoic (1100–850 Ma) crustal growth and tectogenesis, and superimposed Pan-African (800–550 Ma) tectonism. *Precambrian Research*, **62**, 1–59.

PIPER, J. D. A. 1987. *Palaeomagnetism and the Continental Crust*. Open University Press, Milton Keynes.

PIPER, J. D. A., J. D. A. 2000. The Neoproterozoic Supercontinent: Rodinia or Palaeopangaea? *Earth and Planetary Science Letters*, **176**, 131–146.

PISAREVSKY, S. A. & BYLUND, G. 1998. Palaeomagnetism of a key section of the Protogine Zone, southern Sweden. *Geophysical Journal International*, **133**, 185–200.

PISAREVSKY, S. A., S. A., LI, Z. X., GREY, K. & STEVENS, M. K. 2001. A palaeomagnetic study of Empress 1A, a stratigraphic drillhole in the Officer Basin: new evidence for the low-latitude position of Australia in the Neoproterozoic. *Precambrian Research*, **110**, 93–108.

POWELL, C.MCA. & PISAREVSKY, S.A. 2002. Late Neoproterozoic assembly of East Gondwanaland. *Geology*, **30**, 3–6.

POWELL, C. MCA., JONES, D. L, PISAREVSKY, S. A. & WINGATE, M. T .D. 2001. Paleomagnetic constraints on the position of the Kalahari craton in Rodinia. *Precambrian Research*, **110**, 33–46.

POWELL, C. MCA., LI, Z. X., MCELHINNY, M. W., MEERT, J. G. & PARK, J. K. 1993. Paleomagnetic constraints on timing of the Neoproterozoic breakup of Rodinia and the Cambrian formation of Gondwana. *Geology*, **21**, 889–892.

POWELL, C.MCA., PREISS, W. V., GATEHOUSE, C. G., KRAPEZ, B. & LI, Z. X. 1994. South Australian record of a Rodinian epicontinental basin and its mid-Neoproterozoic breakup (~700 Ma) to form the Palaeo-Pacific Ocean. *Tectonophysics*, **237**, 113–140.

RADHAKRISHNA, T. & MATHEW, J. 1996. Late Precambrian (850–800 Ma) palaeomagnetic pole for the south Indian shield from the Harohalli alkaline dykes: geotectonic implications for Gondwana reconstructions. *Precambrian Research*, **80**, 77–87.

RAINBIRD, R. H., Hamilton, M. A. & Young, G. M. 2001. Detrital zircon geochronology and provenance of the

Torridonian, NW Scotland. *Journal of the Geological Society, London*, **158**, 15–27.

RAINBIRD, R. H., MCNICOLL, V. J., THERIAULT, R. J., HEAMAN, L. M., ABBOT, J. G., LONG, J. G. & THORKELSON, D. J. 1997. Pan-continental river system draining Grenville orogen recorded by U–Pb and Sm–Nd geochronology of Neoproterozoic quartzites and mudrocks, northwestern Canada. *Journal of Geology*, **105**, 1–17.

RAINBIRD, R. H., STERN, R. A., KHUDOLEY, A. K., KROPACHEV, A. P., HEAMAN, L. M. & SUKHORUKOV, V. I. 1998. U–Pb geochronology of Riphean sandstone and gabbro from southeast Siberia and its bearing on the Laurentia–Siberia connection. *Earth and Planetary Science Letters*, **164**, 409–420.

RAMO, O. T. & CALZIA, J. P. 1998. Nd isotopic composition of cratonic rocks in the southern Death Valley region: evidence for a substantial Archean source component in Mojavia. *Geology*, **26**, 891–894.

RAMOS, V. A. 1988. Late Proterozoic–Early Paleozoic of South America – a collisional history. *Episodes*, **11**, 168–174.

RENNE, P. R., ONSTOTT, T. C., D'AGRELLA-FILHO, M. S., PACCA, I. G. & TEIXEIRA, W. 1990. ^{40}Ar/^{39}Ar dating of 1.0–1.1 Ga magnetizations from the São Francisco and Kalahari cratons: tectonic implications for Pan-African and Brasiliano mobile belts. *Earth and Planetary Science Letters*, **101**, 349–366.

RIVERS, T. 1997. Lithotectonic elements of the Grenville province: review and tectonic implications. *Precambrian Research*, **86**, 117–154.

ROBB, L. J., ARMSTRONG, R. A. & WATRES, D. J. 1999. The history of granulite-facies metamorphism and crustal growth from single zircon U–Pb geochronology: Namaqualand, South Africa. *Journal of Petrology*, **40**, 1747–1770.

ROSEN, O. M., CONDIE, K. C, NATAPOV, L. M. & NOZHKIN, A. D. 1994. Archean and Early Proterozoic evolution of the Siberian craton: a preliminary assessment. *In*: CONDIE, K. C. (ed.) *Archean Crustal Evolution.* Elsevier, Amsterdam, 411–459.

ROSEN, O. M., ZHURAVLEV, D. Z., SUKHANOV, M. K. BIBIKOVA, E. B. & ZLOBIN, V. L. 2000. Early Proterozoic terranes, collosion zones, and associated anorthosites in the northeast of the Siberian craton: isotope geochemistry and age characteristics. *Geology and Geophysics*, **41**, 163–180 [in Russian].

ROSS, G. M., BLOCH, J. D. & KROUSE, H. R. 1995. Neoproterozoic strata of the southern Canadian Cordillera and the isotopic evolution of seawater sulfate. *Precambrian Research*, **73**, 71–99.

ROSS, G. M., EATON, D. W., BOERNER, D. E. & MILES, W. 2000. Tectonic entrapment and its role in the evolution of continental lithosphere: an example from the Precambrian of western Canada. *Tectonics*, **19**, 116–134.

ROSS, G. M., PARRISH, R. R. & WINSTON, D. 1992. Provenance and U–Pb geochronology of the Mesoproterozoic Belt Supergroup (northwestern United States): implications for age of deposition and pre-Panthalassa plate reconstructions. *Earth and Planetary Science Letters*, **113**, 57–76.

ROY, J. L. & ROBERTSON, W. A. 1978. Paleomagnetism of the Jacobsville Formation and the apparent polar path for the interval ~1100 to ~670 m.y. for North America. *Journal of Geophysical Research*, **83**, 1289–1304.

RYTSK, E. YU., AMELIN, YU. V., KRYMSKY, R. SH., RIZVANOVA, N. G. & SHALAEV, V. S. 1999. Baikal–Muya belt: age evolution of crust forming. *Proceeding of the Materials of the XXXII Tectonic Meeting*. Geos Press, Moscow, **2**, 93–95 [in Russian].

SADOWSKI, G. R. & BETTENCOURT, J. S. 1996. Mesoproterozoic tectonic correlations between eastern Laurentia and the western border of the Amazon Craton. *Precambrian Research*, **76**, 213–227.

SEARS, J. W. & PRICE, R. A. 1978. The Siberian connection: a case for Precambrian separation of the North American and Siberian cratons. *Geology*, **6**, 267–270.

SEARS, J. W. & PRICE, R. A. 2000. New look at the Siberian connection: no SWEAT. *Geology*, **28**, 423–426.

SMETHURST, M. A., KHRAMOV, A. N. & TORSVIK, T. H. 1998. The Neoproterozoic and Palaeozoic palaeomagnetic data for the Siberian Platform: from Rodinia to Pangea. *Earth-Science Reviews*, **43**, 1–24.

STARMER, I. C., 1996. Accretion, rifting and collision in the North Atlantic supercontinent, 1700–950 Ma. *In*: BREWER, T. S. (ed.) *Precambrian Crustal Evolution in the North Atlantic Region*. Geological Society of London, Special Publication, **112**, 219–248.

STEARN, J. E. F. & PIPER, J. D. A. 1984. Palaeomagnetism of the Sveconorwegian mobile belt of the Fennoscandian Shield. *Precambrian Research*, **23**, 201–246.

TACK, L., WINGATE, M. T. D., LIÉGEOUS, J.–P., FERNANDEZ-ALONSO, M. & DEBLOND, A. 2001. Early Neoproterozoic magmatism (1000–910 Ma) of the Zadinian and Mayumbian Groups (Bas-Congo): onset of Rodinia rifting at the western edge of the Congo Craton. *Precambrian Research*, **110**, 277–306.

TASSINARI, C. C. G., BETTENCOURT, J. S., GERALDES, M. C., MACAMBIRA, M. J. B. & LAFON, J. M. 2000. The Amazonian craton. *In*: CORDIANI, U. G., MILANI, E. J., THOMAZ FILHO, A. & CAMPOS, D. A. (eds) *Tectonic Evolution of South America*, 31st International Geological Congress, Rio de Janeiro. 41–95.

TEIXEIRA, W., SABATE, P., BARBOSA, J., NOCE, C. M. & CARNEIRO, M. A. 2000. Archean and Paleoproterozoic tectonic evolution of the São Francisco craton, Brazil. *In*: CORDIANI, U. G., MILANI, E. J., THOMAZ FILHO, A. & CAMPOS, D. A. (eds) *Tectonic Evolution of South America*, 31st International Geological Congress, Rio de Janeiro. 101–137.

TORSVIK, T. H. & EIDE, E. 1997. Database of Norwegian Geochronology. NGU Report 98–003.

TORSVIK, T. H., ASHWAL, L. D., TUCKER, R. D. & EIDE, E. A. 2001a. Neoproterozoic geochronology and palaeogeography of the Seychelles microcontinent: the India link. *Precambrian Research*, **110**, 47–59.

TORSVIK, T. H., CARTER, L. M., ASHWAL, L. D., BHUSHAN, S. K., PANDIT, M. K. & JAMTVEIT, B. 2001b. Rodinia refined or obscured: palaeomagnetism of the Malani igneous suite (NW India). *Precambrian Research*, **108**, 319–333.

TORSVIK, T. H., SMETHURST, M. A., NEERT, J. G. ET AL. 1996. Continental break-up and collision in the Neoproterozoic and Palaeozoic – a tale of Baltica and Laurentia. *Earth-Science Reviews*, **40**, 229–258.

TROMPETTE, R. 1994. *Geology of Western Gondwana (2000–500 Ma)*. A. A. Balkema, Rotterdam.

TROMPETTE, R. 1997. Neoproterozoic (~600Ma) aggregation of Western Gondwana: a tentative scenario. *Precambrian Research*, **82**, 101–112.

TUCKER, R. D., ASHWAL, L. D., & TORSVIK, T. H., 2001. U–Pb geochronology of Seychelles granitoids: a Neoproterozoic continental arc fragment. *Earth and Planetary Science Letters*, **187**, 27–38.

VAN DER VOO, R. 1990. The reliability of paleomagnetic data. *Tectonophysics*, **184**, 1–9.

VAN DER VOO, R. & MEERT, J. G. 1991. Late Proterozoic paleomagnetism and tectonic models: a critical appraisal. *Precambrian Research*, **53**, 149–163.

VAN SCHMUS, W. R. & BICKFORD, M. E. 1993. Transcontinental Proterozoic provinces. *In*: REED, J. C., BICKFORD, M. E. & HOUSTON, R. S. ET AL. (eds) *Precambrian: Continental US*. Geological Society of America, Geology of North America, **C-2**, 171–334.

WALDERHAUG, H. J., TORSVIK, T. H., EIDE, E. A., SUNDVOLL, E. A. & BINGEN, B. 1999. Geochronology and palaeomagnetism of the Hunnedalen dykes, SW Norway: implications for the Sveconorwegian apparent polar wander loop. *Earth and Planetary Science Letters*, 169, 71–83.

WANG, X. D. & LINDH, A. 1996. Temperature–pressure investigation of the southern part of the southwest Swedish Granulite region. *European Journal of Mineralogy*, **8**, 51–67.

WANG, X. D., PAGE, L.M. & LINDH, A. 1996. $^{40}Ar/^{39}Ar$ geochronological constraints from the southeasternmost part of the eastern segment of the Sveconorwegian orogen: implications for timing of granulite facies metamorphism. *Geologiska Föreningen i Stockholm Förhandlingar*, **118**, 1–8.

WEIL, A. B., VAN DER VOO, R., MAC NIOCAILL, C & MEERT, J. G. 1998. The Proterozoic supercontinent Rodinia: paleomagnetically derived reconstruction for 1100 to 800Ma. *Earth and Planetary Science Letters*, **154**, 13–24.

WILDE, S. A., & MURPHY, D. M. K. 1990. The nature and origin of Late Proterozoic high-grade gneisses of the Leeuwin Block, Western Australia. *Precambrian Research*, **47**, 251–270.

WILLNER, A. P., ERMOLAEVA, T., STROINK, L. ET AL. 2001. Contrasting provenance signals in Riphean and Vendian sandstones in the SW Urals (Russia): constraints for a change from passive to active continental margin conditions in the Neoproterozoic. *Precambrian Research*, **110**, 215–239.

WINGATE, M. T. D. 2001. SHRIMP baddeleyite and zircon ages for an Umkondo dolerite sill, Nyanga Mountains, eastern Zimbabwe. *South African Journal of Geology*, **104**, 13–22.

WINGATE, M. T. D. & GIDDINGS, J. W. 2000. Age and palaeomagnetism of the Mundine Well dyke swarm, Western Australia: implications for an Australia–Laurentia connection at 755Ma. *Precambrian Research*, **100**, 335–357.

WINGATE, M. T. D., CAMPBELL, I. H., COMPSTON, W. & GIBSON, G. M. 1998. Ion microprobe U–Pb ages for Neoproterozoic basaltic magmatism in south-central Australia and implications for the breakup of Rodinia. *Precambrian Research*, **87**, 135–159.

WINGATE, M. T. D., PISAREVSKY, S. A. & EVANS, D. A. D. 2002. A revised Rodinia supercontinent: no SWEAT, no AUSWUS. *Terra Nova*, **14**, 121–128.

ZONENSHAIN, L. P., KUZMIN, M. I. & NATAPOV, L. M. 1990. *Geology of the USSR: A Plate-Tectonic Synthesis*. Geodynamics Series, **21**.

Role of Pan-African events in the Circum-East Antarctic Orogen of East Gondwana: a critical overview

MASARU YOSHIDA[1,2], JOACHIM JACOBS[3], M. SANTOSH[4] & H. M. RAJESH[1]

[1]*Gondwana Institute for Geology and Environment, 147–2 Hashiramoto, Hashimoto 648–0091, Japan (e-mail: gondwana@orion.ocn.ne.jp)*
[2]*Institute for Fundamental Studies, Hantana Road, Kandy, Sri Lanka*
[3]*Institute of Geosciences, University of Bremen, PF 330440, Bremen 28334, Germany*
[4]*Department of Geology, Faculty of Science, Kochi University, Akebono-cho, Kochi 780–8072, Japan*

Abstract: Recent studies of Pan-African events in East Gondwana are critically reviewed, particularly recent models of amalgamation of East Gondwana during the Pan-African period. It is pointed out that critical data are insufficient to constrain the newly proposed models and so the classical model of the Grenvillian Circum-East Antarctic Orogen cannot yet be replaced. Grenvillian tectonothermal events with a peak between 1.0 and 1.2 Ga assembled different crustal blocks of the East Antarctic Shield with different geohistories. Pan-African tectonothermal reworking took place over wide but selected areas of the orogen. Careful geochronological studies, including SHRIMP dating associated with structural and petrological investigations to correlate ages with those events, are shown to be important, since fluid-rich and/or deformational conditions are equally effective as temperature conditions for mineral recrystallization and resetting of isotopic systematics. Pan-African suture zones, one extending from the Mozambique Belt to the Shackleton Range and another connecting the Mozambique Belt to the Zambezi Belt, are equally possible, although the width of the southern Mozambique Ocean is poorly understood. The extent of Pan-African sutures in the Prydz Bay area is enigmatic, although they represent definite orogens. Palaeomagnetic studies may provide critical constraints in evaluating the sutures, provided that the age of magnetization is well established.

It has been suggested by Unrug (1992, 1997), Yoshida (1992, 1995*a*), and Yoshida *et al.* (2000) that East Gondwana formed during the late Mesoproterozoic Grenvillian Orogeny, and was variably reactivated by the Neoproterozoic–Early Palaeozoic (*c.* 850–450 Ma) Pan-African Orogeny (Fig. 1). Field studies from Enderby Land (Black *et al.* 1987; Sheraton *et al.* 1987), northern Prydz Bay (Kinny *et al.* 1993; Harley *et al.* 1995), Lützow-Holm Bay (Yoshida 1978, 1994, 1995*b*), and western and central Dronning Maud Land (e.g. Grantham *et al.* 1988, 1995; Ohta *et al.* 1990; Groenewald 1993; Groenewald *et al.* 1995; Jacobs *et al.* Wareham *et al.* 1998; Bauer *et al.* 2002) also support the above model. However, identification of extensive, strong Pan-African events in East Gondwana, especially in Antarctica (e.g. Zhao *et al.* 1992; Shiraishi *et al.* 1994; Hensen & Zhou 1995), coupled with palaeomagnetic constraints (Meert *et al.* 1995; Torsvik *et al.* 1996), led some scientists to cast doubt on the above model and thus to re-evaluate the role of Pan-African events in East Gondwana.

It was suggested by Grunow *et al.* (1996), Dalziel (1997) and Tessensohn (1997) that a Pan-African suture runs in subglacial areas inland of Dronning Maud Land (Fig. 2), and that the Kalahari Craton and a part of East Antarctica had belonged to West Gondwana and amalgamated with East Gondwana during Pan-African time. Alternatively, Meert *et al.* (1995), Hensen & Zhou (1997) and Fitzsimons (2000) proposed a radical model, such that East Gondwana itself had amalgamated during Pan-African time.

In this paper, recent models on the role of Pan-African events in East Gondwana are critically reviewed, and it is emphasized that present data are still insufficient to provide a thorough, new model. Future studies that will provide critical constraints on the tectonic models are also discussed.

Circum-East Antarctic Orogen

Yoshida & Kizaki (1983) and Tingey (1991) pointed out the extensive metamorphic complex of Grenvillian age that surrounds East Antarctica, and Yoshida (1992, 1994, 1995*a*) proposed that this Circum-East Antarctic Mobile Belt played a leading role in the assembly of East Gondwana. This proposal was supported by the occurrence of highly metamorphosed supracrustal sequence of continental margin signature and convergent tectonic characteristics with almost N–S vergence orthogonal to the

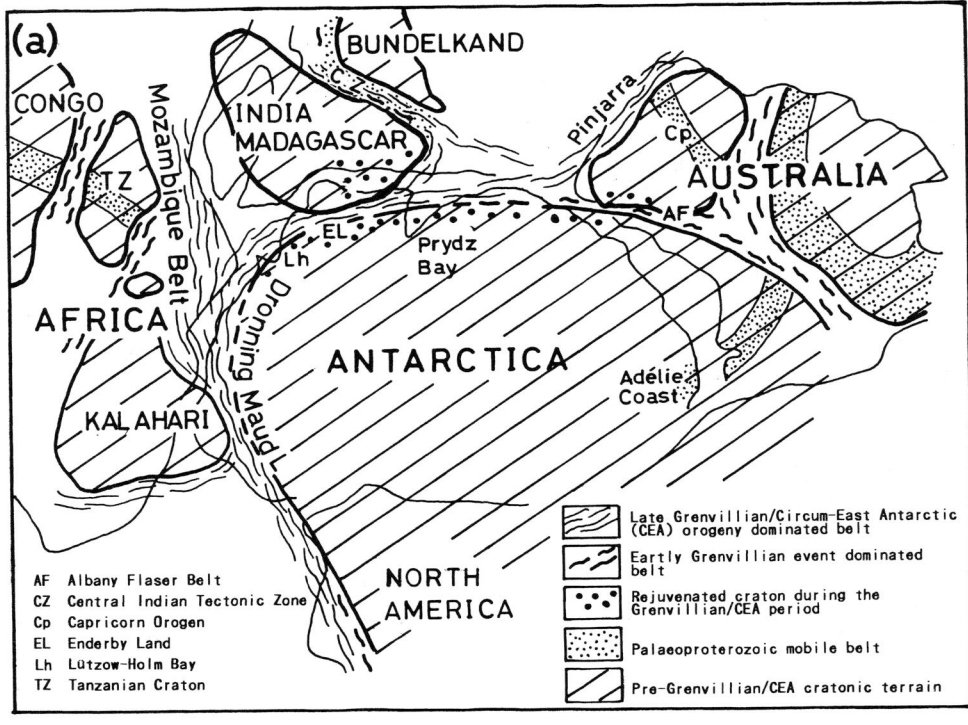

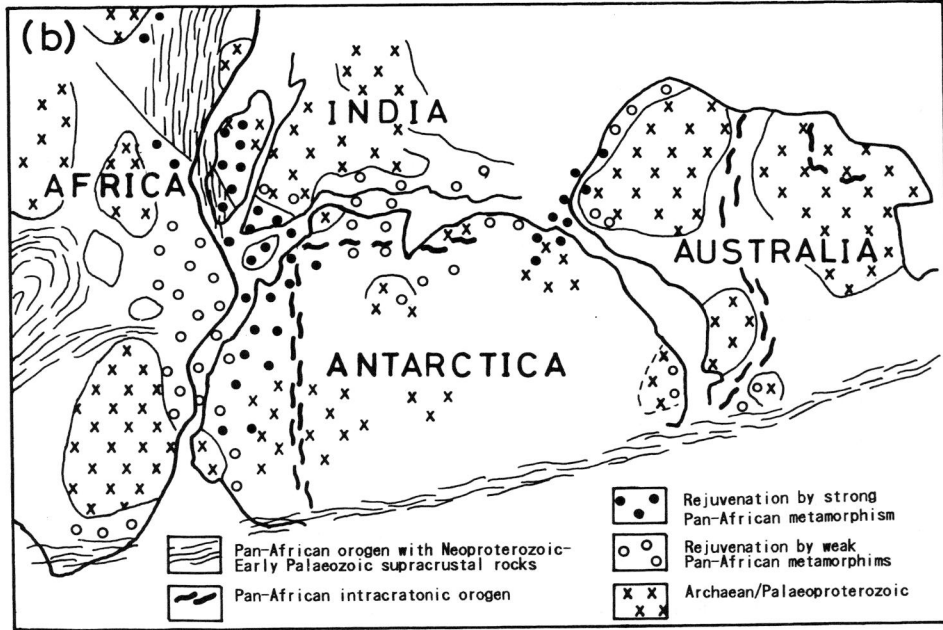

Fig. 1. (**a**) The Mesoproterozoic Grenvillian Circum-East Antarctic Orogen and its extensions in East Gondwana at c. 1000 Ma [classical model; modified after Yoshida (1996) and Yoshida *et al.* (2001)] (**b**) Pan-African events (rejuvenation/reactivation) in East Gondwana surrounding East Antarctica [classical model; modified after Yoshida (1995a, 1996)].

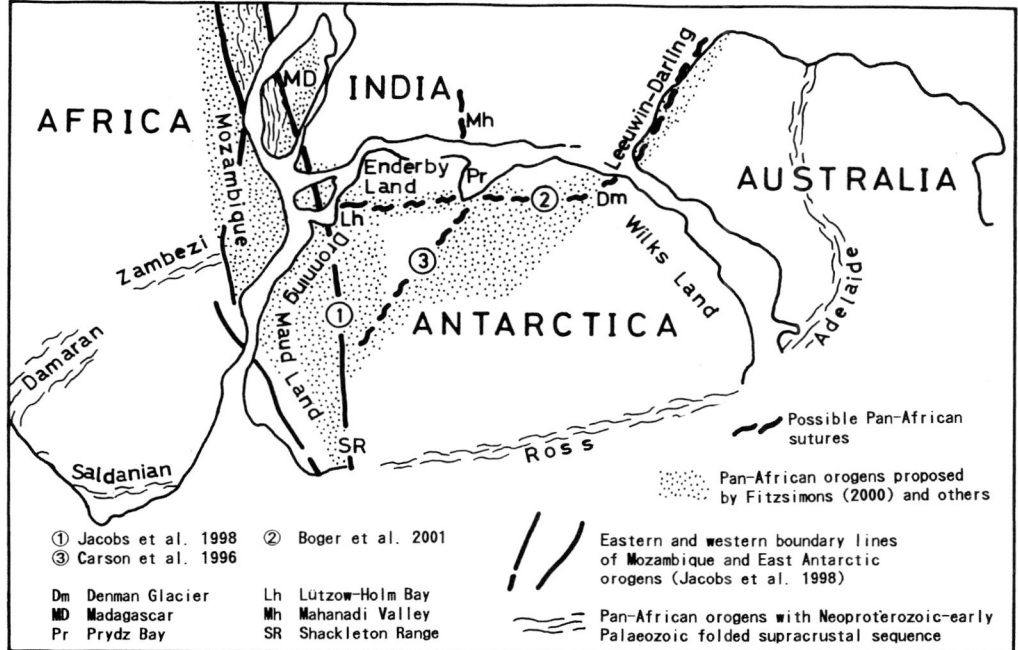

Fig. 2. Pan-African orogens and sutures surrounding East Antarctica.

Antarctic coastline. Yoshida (1996) suggested that the eastern half of the belt had a dextral transpressional and the western half an orthogonal compressional tectonic regime, and that the major part of the belt does not continue into South Australia–Wilkes Land, but turns north along the Pinjara Orogen in Western Australia (Fig. 1a). The precursor of metamorphic rocks from the Rayner Complex, Lützow-Holm Bay, Sri Lanka and South Indian Proterozoic granulite terrain were suggested to be broadly similar, judged from initial Sr isotope ratios (Yoshida 1994), Nd model ages and Sm–Nd isotopic ratios (Yoshida et al. 1999, 2000), although South India does not bear the imprints of the Circum-East Antarctic Orogeny.

There are distinct age differences within the Circum-East Antarctic Orogen. The Albany–Fraser Belt–western Wilkes Land segment contains older ages of 1.1–1.4 Ga, whereas the other segments have a 1.0–1.2 Ga age. This indicates that the collisions along the Circum-East Antarctic Orogen were not contemporaneous; they occurred within a broad Grenvillian time range (Yoshida 1995a). This was challenged by Fitzsimons (2000), who identified three different provinces along the coastal areas of East Antarctica based on the ages of high-grade tectonothermal events; from west to east (Fig. 2) they are the Maud (1090–1030 Ma), the Rayner (990–900 Ma) and the Wilkes (1330–1130 Ma) Provinces. He suggested that these terranes assembled during the Pan-African Orogeny.

Extensive Pan-African Events in East Gondwana

K–Ar ages of c. 500–700 Ma are known from almost all coastal areas fringing East Antarctica (Craddock 1970). Shackleton (1986, 1993) suggested that the collision of West and East Gondwana gave rise to the Mozambique Belt. The Pan-African ages surrounding East Antarctica reflect the above collision (Yoshida & Kizaki 1983; Yoshida 1995a).

During the 1990s, many Sm–Nd and U–Pb ages from several areas of East Gondwana confirmed the Pan-African events (e.g. Santosh et al. 1992; Shiraishi et al. 1992; Zhao et al. 1992; Hölzl et al. 1994; Muhongo & Lenoir 1994; Hensen & Zhou 1997; Jacobs et al. 1998; Kröner et al. 1997, 1999, 2000), and some were associated with compressional as well as extensional collapse tectonics, implying collisional orogeny (e.g. Kriegsman 1993; Muhongo 1994; Dirks & Wilson 1995; Groenewald et al. 1995; Appel et al. 1998; Windley et al. 1997; Collins et al. 2000; de Wit et al. 2001). These data led scientists to believe that Pan-African sutures/

orogenic belts developed within East Gondwana during the Pan-African period; details are given in following text.

Figures 1a and b show the classical view that the Pan-African rejuvenation took place extensively in East Gondwana, which had formed during the c. 1000 Ma Circum-East Antarctic Orogeny (Unrug 1992, 1997; Yoshida 1992, 1995a, 1996; Yoshida et al. 2001). Stern (1994) delineated a Pan-African (c. 750–650 Ma) suture along the Mozambique Belt from the Arabian–Nubian Shield to southern Mozambique, and considered that a Pan-African (c. 550 Ma) granulite belt including Madagascar, South India, Sri Lanka and Antarctica belonged to a parallel intracontinental mobile belt, which formed c. 100 Ma later than the major suture.

Yoshida (1995a) pointed out that the rejuvenation of old crust by the Pan-African Orogeny involved variable structural and metamorphic disturbances, which were reflected by the distribution of rejuvenation ages (c. 700 and 500 Ma) with various closure temperatures. Thus, the strongest rejuvenation indicated by whole-rock isochron ages and U–Pb zircon ages of both c. 700 and 500 Ma generally occur near the boundary areas of East and West Gondwana, whereas weak rejuvenation, as reflected by K–Ar ages occurs away from the boundary areas. One exception is the Leeuwin–Darling belt-Denman Glacier area, where strong Pan-African rejuvenation took place. The above relations reflect activity of deep crustal fractures, which were formed by the collision of the East and West Gondwanaland along weak zones within earlier mobile belts; consequently, the intensity was smaller away from the border areas. A deep fracture might also have developed along the pre-existing Pinjarra Orogen and this controlled later strong rejuvenation. This style of rejuvenation is similar to that in intraplate polygonal shears proposed by Katz (1985), where extensive collisional, extensional and horizontal shearing deformation has taken place. However, discovery of Pan-African U–Pb zircon ages, coupled with Sm–Nd internal isochron ages and related petrological evidence of Pan-African high-grade metamorphism from both the Prydz and Lützow-Holm Bay areas of East Antarctica, as well as the palaeomagnetic constraints mentioned above, led to the idea of a Pan-African suture(s) within East Gondwana.

Definition of a Pan-African Suture within East Antarctica

There are three proposed Pan-African sutures within the East Antarctic Shield (c.f. Fig. 2), although Zhao et al. (2000) proposed some modifications; (1) Shackleton Range–Lützow Holm Bay (2) Lützow-Holm Bay–southern Prydz Bay–Denman Glacier–Leeuwin–Darling Belt; (3) southern Prydz Bay–southern Prince Charles Mountains–central Dronning Maud Land.

Shackleton Range to Lützow-Holm Bay

From detailed structural and petrological studies in Sri Lanka, Kriegsman (1993) produced a kinematic interpretation of the Precambrian units of the island, which took place from c. 600 to 500 Ma. Collecting extensive structural data from East Africa, Madagascar, South India and Antarctica, he proposed a c. 300 km wide Pan-African sinistral shear zone reflecting the collisional tectonics of West and East Gondwana. Thus, Kriegsman (1993) suggested that the suture in the Mozambique Belt (Shackleton 1986, 1993) extends into East Gondwana. Grunow et al. (1996) delineated a suture within East Antarctica, connecting the Shackleton Range to the Lützow-Holm Bay area through the interior of western and central Dronning Maud Land; this was also supported by Jacobs et al. (1998) (Fig. 2. line ①). This suture generally follows the eastern boundary of Kriegsman's (1993) sinistral shear zone but passes c. 200 km to the east (African coordinate).

Fitzsimons (2000) positioned one of his major Pan-African orogens (suture zones) in this area. Dalziel (1997) further supported the suture proposed by Grunow et al. (1996) from palaeomagnetic data, and revised his previous (Dalziel 1991) Rodinia reconstruction, although Powell et al. (2001a), using palaeomagnetic data, were suspicious of the location of the Kalahari Craton given by Dalziel (1997). According to Dalziel's (1997) new reconstruction, the Kalahari Craton did not belong to East Gondwana; his idea for the removal of the Kalahari Craton from East Gondwana has generally been accepted. Most of these authors referred to a distinct suturing event in the Shackleton Range as well as extensive Pan-African events in Lützow-Holm Bay, as mentioned below.

Stern (1994) considered that the eastern Mozambique suture runs further west, to be principally composed of the c. 750–650 Ma mobile belt, including the Tanzanian Granulite Belt, and that the later Pan-African (c. 550 Ma) eastern granulite belt, including Madagascar–Sri Lanka–Lützow-Holm Bay, is intracratonic rejuvenation. However, Appel et al. (1998) suggested, from pressure–temperature (P–T) evolutional history of the Tanzanian granulites, that these rocks show an isobaric cooling $P–T$–time ($P–T–t$) path, which is characteristic of an extensional setting and not a collisional belt, and that the c. 550 Ma eastern granulite belt could be the principal suture with a collisional signature.

Extending from the Mozambique Belt of Africa, the suture is supposed to occur in Madagascar.

Collins et al. (2000) reported distinct Pan-African extensional collapse faults in Madagascar and concluded that they were the result of a Pan-African collision in the island. Handke et al. (1999) and Kröner et al. (2000) provided extensive zircon U–Pb data attesting Pan-African Orogeny from the northwestern and central part of Madagascar, including the Antananarivo Block. Kröner et al. (2000) suggested that the suture extends along the eastern boundary of the Antananarivo Block where there is a strongly sheared metasedimentary belt containing many entrained mafic and ultramafic blocks. Bauer et al. (2002) proposed two possible sutures: either the eastern one of Kröner et al. (2000) or one in the southwestern part of the island where the highly deformed and metamorphosed Neoproterozoic Vohibory Sequence contains ultramafic rocks.

Further SE (African coordinate), the suture is considered to continue into South India, either along the Palghat–Cauvery Shear Zone (*sensu lato*) (Jacobs et al. 1998; Bauer et al. 2002) or the Achankovil Shear Zone (Kriegsman 1993; Santosh & Yoshida 2001). The South Indian Proterozoic granulite terrain is composed of the northern Madurai and southern Trivandrum Blocks, both of which carry extensive Pan-African ages attesting to granulite facies metamorphism and granitoid intrusions (e.g. Rajesh et al. 1996; Bartlet et al. 1998; Bindu 1998; Bindu et al. 1998; Broun et al. 1998; Yoshida et al. 1996 and refs cited therein). The Madurai Block carries an extensive Palaeoproterozoic component, in contrast to the Trivandrum Block (Harris et al. 1994; Bartlet et al. 1998; Yoshida et al. 1996). The Palghat–Cauvery Shear Zone (*sensu lato*) is a c. 100–200 km wide zone that marks the boundary between the northern Archaean and the southern Proterozoic terrains of the South Indian Shield. Early Pan-African alkaline acid and mafic intrusions (Santosh et al. 1989; Radhakrishna et al. 1999; Miyazaki et al. 2001; Rajesh et al. 1996 and refs cited therein), as well as Pan-African rejuvenation ages (Meiβner et al. 2002), occur within this wide shear system. The Achankovil Shear Zone is a c. 20 km wide ductile shear zone with Neoproterozoic high-grade gneisses and Pan-African granitic, mafic and ultramafic masses (Santosh & Yoshida 2001). Extensive Pan-African metamorphism developed along this shear zone with gneisses derived from Late Proterozoic metasupracrustal rocks (Harris et al. 1994; Bartlett et al. 1998).

Sri Lanka is composed of three Proterozoic units – two Late Mesoproterozoic units, mostly amphibolite facies complexes on the eastern and western sides of the island (Wanni in the west and Vijayan in the east), and a Paleoproterozoic granulite facies Highland Complex in the centre. All these complexes suffered extensive Pan-African events (e.g. Hölzl et al. 1994; Kröner et al. 1994). Distinct Pan-African thrust–nappe structures developed on both east and west boundaries of the Highland Complex (Kriegsman 1993; Büchel 1994; Kleinschrodt 1994; Tani & Yoshida 1996). Serpentinite masses occur along the eastern thrust near the southern tip of the island (Geological Survey Department of Sri Lanka 1982). However, Kriegsman (1993), who reported detailed structural studies from this island, considered the island to belong to the outer margin of his principal sinistral shear zone. Grunow et al. (1996) and Jacobs et al. (1998), who delineated the suture on this small island, might have considered either of the above thrusts as the suture; they both show just a straight line crossing the island. Bauer et al. (2002) positioned the eastern boundary fault of the East Antarctic Orogen along the outer northern margin of the island.

Shiraishi et al. (1992) produced Pan-African zircon SHRIMP ages of c. 520–550 Ma on crystalline rocks from wide coastal areas of the Lützow-Holm Bay of East Antarctica, which point to a single major granulite facies metamorphism associated with prominent deformation, and suggested a Pan-African suture zone between the supposed basement of the Yamato–Belgica block to the west (Antarctic coordinate) and the Napier–Rayner block to the east. They referred to Hiroi et al. (1986), who interpreted ultramafic and mafic rocks, mostly occurring as tectonic blocks within metamorphic rocks, as a dismembered ophiolitic complex, and to Hiroi et al. (1991), who provided petrological data on paired metamorphic crystalline rocks in the Lützow-Holm Bay area (medium-pressure type) to the east and the Yamato–Belgica block (low-pressure type) to the west. Shiraishi et al. (1994) suggested this Pan-African belt continued to the Mozambique Belt, and interpreted zircon cores with c. 1.0 Ga ages as the inherited detrital grains, which provide a maximum age of sedimentation of the metasupracrustal crystalline rocks. A c. 570 Ma Rb–Sr whole-rock isochron age obtained from migmatite formed during the granulite metamorphism from Breidvågnipa, the central part of the east coast of Lützow-Holm Bay (Shimura et al. 1998), coupled with a c. 520 Ma zircon U–Pb age from syndeformational leucosome from Rundvågshetta c. 80 km SW of Breidvågnipa (Fraser et al. 1997), provided robust data for supporting the Pan-African granulite metamorphism of the Lützow-Holm Bay area proposed by Shiraishi et al. (1992, 1994). A Sm–Nd garnet–whole-rock age of c. 574 Ma from Skallen, c. 40 km south of Breidvågnipa (Yoshida et al. 1999), also supports this. Due to the above studies, the Lützow-Holm Bay area has generally been regarded as the locus of a Pan-African suture. Different ideas on the geotectonic history of the area, including the age of sedimentation of metasupracrustal rocks and the recognition of c. 1000 Ma tectonic-metamorphic events (e.g. Yoshida 1994, 1995b) are discussed elsewhere.

In central Dronning Maud Land, extensive Pan-African tectonothermal events superimposed on Grenvillian crust have been recognized (Ohta et al. 1990; Jacobs et al. 1998, 1999, 2003; Bauer et al. 2002), and Jacobs et al. (1999) termed the belt the East Antarctic Orogen. In western Dronning Maud Land, Pan-African tectonometamorphic events are relatively weak (Grantham et al. 1988, 1995; Groenewald 1993, Groenewald et al. 1995). Most of these authors considered that central and western Dronning Maud Land is a continuation of the Mozambique Belt, and Jacobs et al. (1998, 2003) and Bauer et al. (2003) showed that the possible suture (eastern boundary) of the Mozambique Belt extends inland of Dronning Maud Land to reach Lützow-Holm Bay. Jacobs et al. (1999, 2003) and Golynsky & Jacobs (2001) placed a suture in the mountain ranges of western Dronning Maud Land, where it forms the western boundary of the East African Orogen, which continues from the western boundary fault of the Mozambique Belt in southern Mozambique.

In the northern Shackleton Range, Kleinschmidt & Buggisch (1994) and Tessensohn (1997) identified a Pan-African thrust–nappe zone showing southward (continentward) vergence as the suture between the southern Antarctic and the northern Kalahari cratons. Early Cambrian sedimentary packages occur between the southern and northern basement units (Buggisch et al. 1990), supporting the Pan-African time of the suturing event. Identification of a similar thrust–nappe zone in the southern Shackleton Range, associated with possible Neoproterozoic ophiolites (c. 1.0 Ga), and extensive c. 500 Ma K–Ar ages from the tectonized zone (Talarico et al. 1999) provide additional support for the existence of a Pan-African suture in the Shackleton Range. The above reports from the Shackleton Range have been regarded as robust evidence supporting the Pan-African suturing event.

Sutures in the southern Prydz Bay area

Prydz Bay and the surrounding areas include several geologic units having distinct pre-Grenvillian or pre-Pan-African geohistories. From north to south, these include the Archaean Vestfold Block, the Archaean–Proterozoic Rauer Group, the Mesoproterozoic–Neoproterozoic gneisses of southern Prydz Bay to the Amery Ice Shelf area, Mesoproterozoic gneisses in the northern Prince Charles Mountains, and Archaean–Proterozoic rocks in the southern Prince Charles Mountains. The time of amalgamation of these terranes into the present configuration has been accepted as being c. 1000 Ma (e.g. Harley & Fitzsimons 1995), but a recent proposal has suggested a Pan-African assembly (e.g. Dirks & Wilson 1995; Fitzsimons 1997; Hensen & Zhou 1997; Harley et al. 1998). In these studies, discussions on the timing of the terrane assembly in the Prydz Bay area mostly pivot on geochronology, structural and metamorphic characteristics.

In the Larsemann Hills of the southern Prydz Bay area, Zhao et al. (1992) first identified the Pan-African high-grade metamorphism based on c. 530–550 Ma ages of a syenogranite (pink granite), using the single zircon stepwise evaporation method. They also reported Ar–Ar ages of c. 486 Ma from the same sample. The granite cross-cuts surrounding gneisses, but is partly conformable with the foliation of the gneisses; it is interpreted to have formed by melting of the host rocks. Zhao et al. (1993) obtained c. 500–540 Ma Sm–Nd internal isochron ages and 536–821 Ma Pb–Pb zircon ages from metamorphic rocks, both suggesting high-grade conditions during the Pan-African event. Furthermore, Zhao et al. (1995) proposed Neoproterozoic sedimentation of metasupracrustal rocks based on zircon Pb–Pb evaporation ages of c. 1200–766 Ma from single grains, which they considered to be of detrital origin. They considered that there was only one major tectonothermal event during the late Pan-African period, which was comparable with that in the Lützow-Holm Bay area.

The reports of Zhao et al. (1995) attracted workers to re-evaluate the Pan-African events in Prydz Bay and nearby areas (e.g. Dirks et al. 1993; Dirks & Wilson 1995; Carson et al. 1996; Fitzsimons 1997, 2000; Hensen & Zhou 1997; Boger et al. 2000, 2001; Zhao et al. 2000). However, there are different opinions on the age of geological events in the Larsemann Hills (Dirks et al. 1993; Tong et al. 1995; Zhang et al. 1996), as discussed in the next section.

Carson et al. (1996) proposed a suture which continues to the Leeuwin–Darling Belt of Western Australia, through the Denman Glacier area to the east and to the Lützow-Holm Bay area to the west (Fig. 2, line ②). Hensen & Zhou (1995) obtained c. 498 and 988–903 Ma Sm–Nd internal isochron and garnet–whole-rock ages, respectively, from gneisses from Søstrene Island, c. 20 km SW of the Larsemann Hills. From the northern Prince Charles Mountains some 450 km WSW of the Larsemann Hills, Hensen et al. (1997) reported extensive Late Pan-African (c. 630–550 Ma) overprinting to c. 825–790 Ma, events based on Sm–Nd garnet–whole-rock ages. Pan-African zircon SHRIMP ages were also reported from the Grove Mountains (Zhao et al. 2000), c. 380 km to the south of the Larsemann Hills, where metamorphic rim ages of c. 530 Ma occur with inherited core ages from c. 870 to 950 Ma. These data provide further support for the Pan-

African events surrounding the Prydz Bay area. Shiraishi et al. (1997), based on SHRIMP zircon dating, showed that the western part of the Rayner Complex of Enderby Land included many Pan-African components, thus providing a database indicating a Pan-African suture that extends from the Prydz Bay area to Lützow-Holm Bay. These data and ideas support the model of Carson et al. (1996).

Fitzsimons (2000) identified that one of the major Pan-African orogenic (suture) belts extends nearly N–S in Prydz Bay (Fig. 2), the Denman Glacier and in the Leeuwin–Darling belt. Zhao et al. (2000) considered that a Pan-African suture runs N–S at Prydz Bay and continues to the central Transantarctic Mountains through the subglacial Antarctic Shield. Boger et al. (2001) reported c. 490–550 and 1600–2800 Ma zircon SHRIMP ages from rocks of the Mawson Escarpment, some 500 km SW of the Larsemann Hills, along the eastern bank of the Lambert Glacier. They considered that older ages were inherited and detrital, and pointed to the maximum age for the deposition of the original rocks. They suggested that the Pan-African suture in the southern Prydz Bay area should extend SW to reach the central Dronning Maud Land suture passing through the Mawson Escarpment (Fig. 2. line ③). Definition of Pan-African sutures running not only on the margins but also at the centre of East Gondwana, such as the Prydz Bay area, led to the idea that Antarctica and, naturally, East Gondwana were finally amalgamated during the Pan-African period. This view is discussed below.

Pan-African amalgamation of East Gondwana

Delineation of a Pan-African orogen or a suture in Prydz Bay has resulted in a drastic change from the classical idea that East Gondwana assembled during the c. 1.0 Ga Grenvillian Circum-East Antarctic Orogeny to a radical view that East Gondwana did not exist during most of the Neoproterozoic, but was assembled during the Pan-African Orogeny contemporaneous with the formation of West Gondwana.

Following the successive reports of Pan-African Orogeny at Larsemann Hills as mentioned above, Carson et al. (1996), Hensen & Zhou (1997), Fitzsimons (2000) and Boger et al. (2001) proposed a suture(s) in the Prydz Bay area running NE–SW, E–W, or N–S, and all these authors emphasized the amalgamation of East Gondwana during the Pan-African period.

Using palaeomagnetic data, Meert et al. (1995) and Meert & Van der Voo (1997) proposed the progressive closure of sutures from west to east, i.e. first the East African Orogeny along the Mozambique Belt and secondly, the Kuunga Orogeny between the Mawson Protocontinent (Fanning et al. 1999) and the India–Madagascar–Sri Lanka–coastal East Antarctica continent (Fig. 3a). Torsvik et al. (1996) agreed with them in principle.

Fitzsimons (2000) considered the Grenville-aged coastal areas of the East Antarctic Shield to be a collage of three different segments, separated by two major Pan-African sutures – the East African Orogen, including the wide areas from Lützow-Holm Bay to central Droning Maud Land, and the Prydz–Denman–Darling Orogen. Fitzsimons (2000) was promptly followed by several papers of continental reconstruction during the Neoproterozoic period, incorporating the idea of dispersal of East Antarctica in pre-Pan-African times (e.g. Meert, 2001; Powell et al. 2001b; Torsvik et al. 2001) (Fig. 3b). However, the data supporting this new model are, as yet, insufficient, as discussed below.

Discussion

Identification of a suture

By definition, a suture is a plate boundary fault (or shear zone) with different terranes on either side; the terranes come from geographically remote localities. Therefore, to define a suture (Schackleton 1986), it is important to establish the geological signatures and past geographical locations of the terranes on either side of the fault. Thus, it is important to subdivide the Circum-East Antarctic Orogen into three terranes (Fitzsimons 2000) in order to identify sutures within it. Tectonothermal events during the suturing periods are expected to occur within the fault/shear zone. This zone may be associated with dismembered ophiolitic and supracrustal rocks from the continental margin to offshore. The age of the sediments, in the case of the Pan-African suture, is expected to be mostly Neoproterozoic, not more than some hundred years older than the collision event. The supracrustal rocks are intensely deformed under the compressional tectonic regime. Igneous rocks with arc affinities and high P/low T metamorphic rocks, which characterize subduction, are generally expected to be present. However, association of all the above signatures with the shear zone area are not a fundamental requisite for the identification of a suture. Bearing in mind these relations, the sutures within East Gondwana during the Pan-African period will now be discussed.

Sutures within East Gondwana

To understand the tectonic signature of the sutures within East Gondwana, the tectonothermal record of

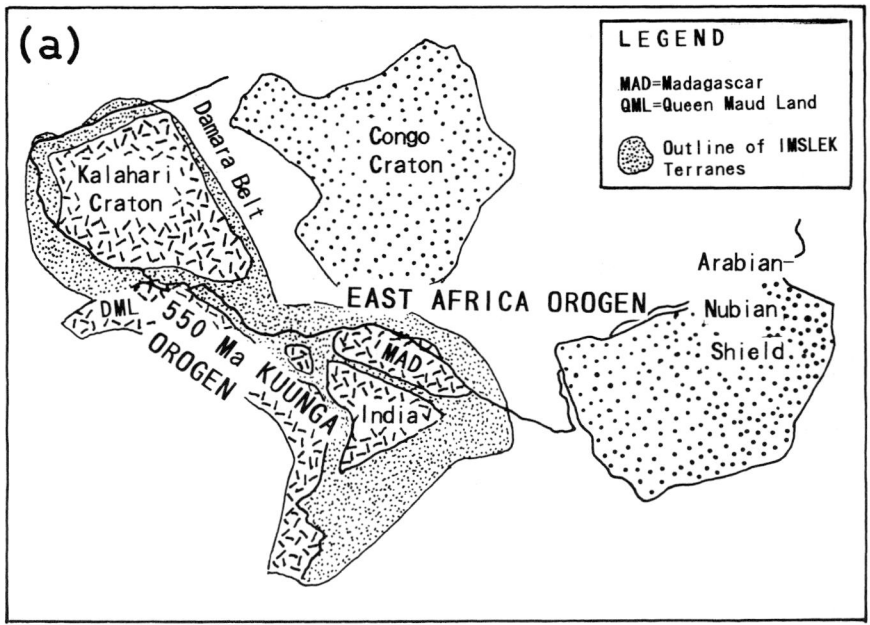

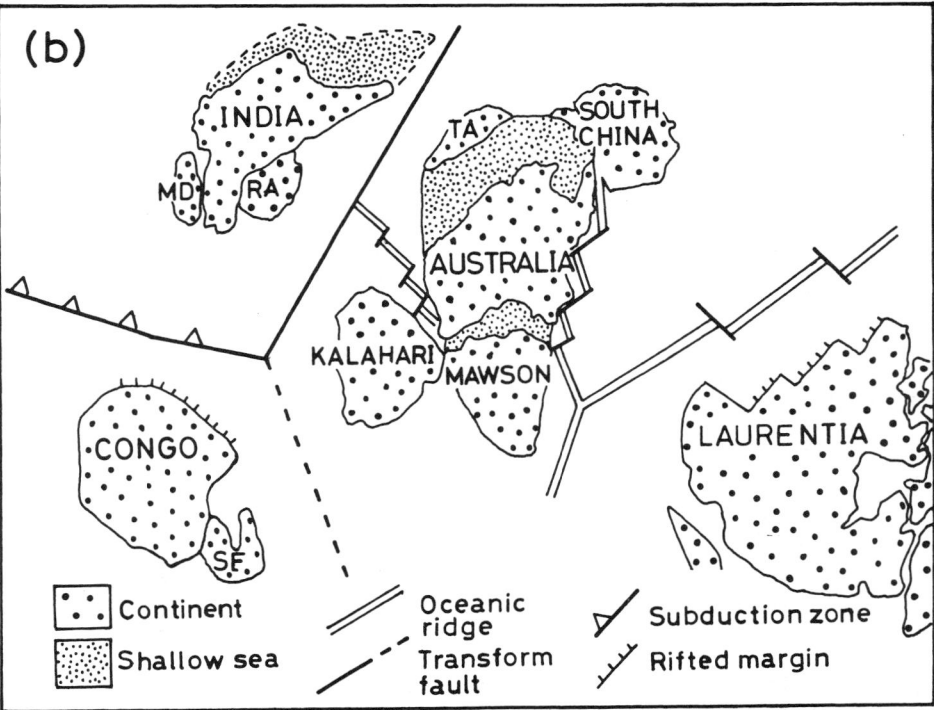

Fig. 3. Pan-African amalgamation of East Gondwana. (**a**) The double-stage collision model of Meert *et al.* (1995) during c. 800–550 Ma. DML, Dronning Maud Land; MAD, Madagascar. (**b**) Palaeogeography surrounding 'East Gondwana' at c. 750 Ma (Powell *et al.* 2001b). MD, Madagascar; RA, Rayner Complex of Enderby Land; TA, Tarim Block; SF, ???

rocks within and around the sutures have to be further characterized. However, it is not an easy task, as discussed below.

It is clear that mounting evidence, as well as global tectonic constraints, e.g. palaeomagnetic studies (Powell et al. 2001a), point to the existence of a Pan-African suture zone extending from the Mozambique Belt to the Shackleton Range through Dronning Maud Land, as mentioned above. In relation to this, some topics from the Lützow-Holm Bay area, the Zambezi Belt and the southern Prydz Bay area will now be considered.

The Lützow-Holm Bay area to the east of central Dronning Maud Land provides a key to constrain the signature and exact location of the Pan-African suture. Grunow et al. (1996) and Jacobs et al. (1998) marked a suture in this area referring to the work of Shiraishi et al. (1992, 1994), who suggested (from zircon SHRIMP studies) that there developed only one major orogeny during the Late Pan-African period. They also suggested that the protolith of high-grade metasupracrustal rocks formed in Neoproterozoic time, and that ultramafic and mafic rocks constitute a dismembered Neoproterozoic ophiolite. However, reasonable correlation of zircon growth events with metamorphism, and the relationship of zoning patterns of zircons to the mechanism and history of zircon growth, were inadequately discussed. In addition, petrochemistry and geochronology of these rocks were not given in enough detail to reasonably enable identification of ophiolite and to date its formation. Thus, high-grade Pan-African metamorphism, lack of c. 1.0 Ga event, the Neoproterozoic age of the sedimentation of protolith of paragneisses and occurrence of Neoproterozoic ophiolites were suspected (Yoshida 1994, 1995a,b). Identification or rejection of Grenvillian events have not been discussed in detail in relation to c. 700 Ma and older ages by SHRIMP (Shiraishi et al. 1994) and whole-rock isochron methods (Maegoya et al. 1968; Shirahata 1983; Shibata et al. 1986; Nakajima et al. 1988). There is compelling field evidence for superposed deformations and multiple metamorphism in this area (e.g. Yoshida 1978, 1979, 1994; Yoshida et al. 1983). Pan-African zircon SHRIMP ages could be explained by either high-grade metamorphism or lower grade fluid-rich metamorphism associated with deformation, neither of which would rule out the older events. These aspects are particularly relevant because there is clear evidence of superposition of fluid-rich medium- and low-grade metamorphic events associated with periods of deformation and granite intrusion after the major granulite facies metamorphism (Yoshida 1978, 1979, 1994; Yoshida et al. 1983).

However, subsequent geochronological and petrological evidence (Fraser et al. 1997; Shimura et al. 1998; Yoshida et al. 1999) provided a robust constraint on the high-grade metamorphism during the Pan-African period, as mentioned above. Lack of a c. 1000 Ma event and Neoproterozoic ages of the sedimentation of protolith of paragneisses are enigmatic, and if the sedimentation is Neoproterozoic, the geological relationship of the paragneisses and possible c. 1000 Ma gneisses should be examined. Thus, further detailed studies are required to fully resolve the signature of the Pan-African Orogeny in the Lützow-Holm Bay area.

If the classical idea of the existence of Neoproterozoic East Gondwana, and the principal role of the Mozambique Suture between West and East Gondwana (although West Gondwana was itself amalgamated during the Pan-African Orogeny), is considered, then either the Zambezi–Damara belts or the Shackleton Range–Dronning Maud Land zone, or both, are possible candidates for the major continuation of the Mozambique Suture. Recent geological, structural and geochronological data from the Zambezi Belt (e.g. Dirks et al. 1999; Vinyu et al. 1999) and central Dronning Maud Land (e.g. Jacobs et al. 1998, 1999, 2003; Bauer et al. 2003)–Shackleton Range (Buggisch et al. 1990; Talarico et al. 1999) support the idea that they contain the suture.

Recent identification of a Late Pan-African granulite belt in northern (Ring & Toulkerides 1995) and southern (Kröner et al. 2001) Malawi, and the occurrence of eclogite and high-P granulites with a clockwise $P-T$ evolutionary history in the former terrain, suggest that a major suture extends along the western margin of the southern Mozambique Belt. This may help delineate the principal suture from the Mozambique Belt to the Damara Belt through the Zambezi Belt as suggested by Unrug (1992, 1997). There are, however, opposing geological data for the identification of the above sutures. There is good evidence for the continuation of a Grenvillian belt on both sides of the Shackleton Range–Dronning Maud Land zone, as mentioned above, and also for the Zambezi Belt area (and also the Mwenbeshi Fault) (Hanson et al. 1988; Stern 1994). This evidence may show that the suture in the southern part of Mozambique Belt has a restricted width, or that it has a recycled signature. Thus, the extent of separation of the continental blocks, i.e. the width of the southern Mozambique Ocean, is problematic; this can only be identified by palaeomagnetic studies, which are still insufficient to be definitive (e.g. Meert 2001).

Pan-African high-grade metamorphism in the southern Prydz Bay area (Zhao et al. 1992, 1993; Dirks et al. 1993; Dirks & Wilson 1995; Hensen & Zhou 1995; Carson et al. 1996; Fitzsimons 1997; Hensen et al. 1997) is generally agreed, although similar relations in the Lützow-Holm Bay area had existed, as discussed in detail by Dirks & Wilson (1995).

Zircon Pb–Pb ages of Zhao et al. (1992) are derived from a late metamorphic discordant pink granite. Zhao et al. (1992) suggested that local, weak foliation in the granite, paralleling that in its host gneiss, is an indication of the near-contemporaneous relation of the granite with a late phase of major metamorphism. The implications of the age and metamorphism, therefore, are controlled by the evaluation of this granite in the tectonothermal history of the area. Dirks et al. (1993) showed that this granite intruded during late D4–syn-D5, still during high-grade metamorphism. However, the temperature of the granite itself was not directly constrained and, also, zircon growth under the later low–middle-grade conditions could not be ruled out (Dirks & Wilson 1995), as discussed below. The detection of Grenvillian U–Pb zircon ages (Zhao et al. 1995) and Sm–Nd garnet–whole-rock ages (Hensen & Zhou 1995) suggest not only the possible existence of Grenvillian metamorphism in this area, but might also reflect the lower grade, Pan-African metamorphism that did not extensively reset the above ages.

However, discovery of Pan-African zircon SHRIMP ages from syn-D2 Progress Granite (c. 515 Ma; Carson et al. 1996) from Larsemann Hill and from partial melting leucosome (c. 530 Ma; Fitzsimons 1996, 1997) from Brattstrand Bluffs south of Rauer Island, provide strong evidence for high-grade metamorphism during the Late Pan-African period. Zhao et al. (1995) suggested the time of sedimentation of protolith of paragneisses in the Larsemann Hills to be Neoproterozoic, younger than the c. 770 Ma detrital zircons. Their suggestion was later supported by Fitzsimons (1997), who regarded most paragneisses in the southern Prydz Bay to belong to the Neoproterozoic cover unit deposited over the older Mesoproterozoic orthogneiss unit. However, discussion on this topic is not complete. The extent and signature of c. 1000 Ma events in the southern Prydz Bay area also remain enigmatic.

As discussed by Dirks & Wilson (1995), it is important that detailed petrological–structural analysis should accompany geochronological studies, especially for the recognition of the time of emplacement of dated rocks in the tectonothermal evolution. Equally important is the extent of the Pan-African granulite metamorphism in this area, in relation to the recognition and evaluation of $P-T$ evolutionary history of each area, and evaluation of closure temperatures of U–Pb and Sm–Nd isotopic systematics in minerals. The last topic, which is common to many Pan-African metamorphic terrains, is discussed in more detail below.

Recyclicity of orogens

Fitzsimons (2000) suggested the amalgamation of three different crustal blocks with different pre-Pan-African ages along the East Antarctic coast in the Pan-African period, with two Pan-African orogens between them as mentioned above. The development of Pan-African tectonothermal events in the above areas, as well as the Leeuwin–Darling–Danman Glacier area, can be regarded as Pan-African orogens. This identification can stand regardless of the grade of the associated metamorphism, but it is important to know whether the orogenic belt is intra- or intercratonic and, if the latter, whether the separation was small or large.

There is a good correlation between the Adélie Coast of East Antarctica and southern Australia (Fig. 1a), which together form a Palaeoproterozoic continental block. As pointed out by Fanning et al. (1999), this block may form the Mawson Protocontinent, although there are insufficient data to separate it from other areas of East Gondwana during Grenvillian time. Also, both Enderby Land and western Dronning Maud Land have distinct pre-Grenvillian geohistories.

Regarding all areas of the Circum-East Antarctic Orogen, such as Wilkes Land, the Prydz Bay area, Enderby Land, Lützow-Holm Bay and Dronning Maud Land, there are insufficient data to demonstrate the existence of two or three different pre-Pan-African continental blocks. Throughout this orogenic belt, there are indications of Grenvillian events with an age range of c. 0.90–1.20 Ga (e.g. Yoshida & Kizaki 1983; Yoshida 1995a; Fitzsimons 2000). In Wilkes Land–SW Australia, there is a good correlation of the Albany–Fraser Belt with western Wilkes Land (c. 1.15–1.40 Ga, e.g. Harris 1995; Clark et al. 2000) in the eastern segment. The Pinjarra (Darling) Belt is continuous with the Obruchev Hills–Denman Glacier area in the western segment (cf. Harris 1995; Post et al. 1997), but this belt is somewhat younger (c. 1.02–1.08 Ga).

The Late Grenvillian age range of c. 1.0–1.2 Ga, which was obtained by various methods in the western segment of Australia–Antarctica as mentioned above, is consistent with ages in the Rauer Group (Kinny et al. 1993), the southern Prydz Bay area (Hensen & Zhou 1995; Zhang et al. 1996), the Rayner Complex (Black et al. 1987; White & Clarke 1993; Hensen et al. 1997), the Lützow-Holm Bay area (Shiraishi et al. 1994; Yoshida 1994) and Dronning Maud Land (Jacobs et al. 1998, 1999; Bauer et al. 2003). The Rayner Complex of Enderby Land and northern Prince Charles Mountains was affected by intrusion of late orogenic granitic/charnockitic rocks and metamorphism in the period of c. 900–1000 Ma (Black et al. 1987; Tingey 1991; Young & Ellis 1991; Hensen et al. 1997; Zhao et al. 1997; Boger et al. 2000).

As suggested by Yoshida & Kizaki (1983) and Yoshida (1994, 1995a), there is a good correlation of tectonothermal history between the Lützow-Holm Bay area (Yoshida 1978, 1994; Yoshida et al. 1983), the Rayner Complex (Black et al. 1987; Sheraton et al. 1987; Tingey 1991), rocks of the Prydz Bay area (Kinny et al. 1993; Harley et al. 1995), and western and central Dronning Maud Land (Grantham et al. 1988, 1995; Groenewald 1993; Groenewald et al. 1995; Jacobs et al. 1998; Bauer et al. 2003). Yoshida (1994) and Yoshida et al. (1999, 2000) suggested that precursors of rocks from the Rayner Complex and Lützow-Holm Bay have similar Sr and Nd isotopic signatures.

It is evident that extensive Pan-African orogenic belts transect (or follow) the Circum-East Antarctic Orogen, as demonstrated by Fitzsimons (2000). However, almost all these Pan-African belts carry clear evidence of Grenvillian precursors (e.g. Shiraishi et al. 1992, 1994; Zhao et al. 1993; Yoshida 1994; Hensen & Zhou 1995; Zhang et al. 1996; Hensen et al. 1997; Jacobs et al. 1998), strongly suggesting that the Grenvillian Belt (Circum-East Antarctic Orogen) surrounds East Antarctica from western Dronning Maud Land to western Wilkes Land, and branches into the Pinjarra Orogen to the north and into the Albany Belt to the east (Yoshida 1996). The latter may be the intracontinental belt as suggested by Harris (1995). There are also 1050–1080 Ma K–A-ages from central Wilkes Land (Craddock 1970), suggesting that an intracontinental branch of the Albany Belt extends further east along the Antarctica–Australia boundary.

The age range from c. 1.0 to 1.2 Ga dates the timing of culmination of orogenic events in the different segments of the extensive Circum-East Antarctic Orogen, as pointed out by Yoshida (1995a). The only early phase of the Circum-East Antarctic Orogeny is represented by the c. 1.3–1.4 Ga events in the Wilkes Land–Albany–Fraser Belt (Fig. 1a): two contemporaneous phases of the Grenvillian Orogeny have been reported in other Grenvillian belts (e.g. Kampunzu et al. 1998; Wasteneys et al. 1999).

During Pan-African time, considerable parts of the Circum-East Antarctic Orogen underwent extensive metamorphism and deformation, although their intensity varied in each area. The Pan-African Leeuwin Complex–Darling Belt–Denman Glacier belt is considered to be underlain by the Grenvillian Pinjarra (Darling) Orogen, which is traced from the Northampton Complex of Western Australia to the Boucher Hills east of the Denman Glacier through the basement of the Perth Basin (cf. Harris 1995). A comparison of the Palaeoproterozoic events in the Central Indian Tectonic Zone with those of the Capricorn Orogen in Western Australia, and the c. 1.0–1.2 Ga events in the former with those in the Albany Belt (Yoshida et al. 2001) appear to point to an important role of the Pinjarra Orogen in the development of East Gondwana during the Circum-East Antarctic Orogeny.

In summary, based on the considerations of all the above arguments, the classical model that the Pan-African events are mostly the result of superposition–reworking of pre-existing Grenvillian belt (e.g. Stern 1994; Yoshida 1995a; Unrug 1997) cannot yet be revised. Based on this model, Pan-African orogens delineated through the southern Prydz Bay area may be interpreted as intracratonic orogens. The Shackleton Range–Dronning Maud Land Zone is a good candidate for the principal suture, along with the Zambezi–Damala Belts, that continues from the Mozambique Belt. Even in this case, central and western Dronning Maud Land dominantly formed by the rejuvenation of the Grenvillian belt, as mentioned above. Further studies are awaited from both areas, as well as palaeomagnetic studies from related Gondwanan crustal fragments. Yoshida (1995a) considered that the above model of the intracratonic Pan-African rejuvenation was similar to that proposed by Katz (1985), which included various structural disturbances such as thrust–nappe structures with or without ophiolites, horizontal shearing and extension, principally under intracratonic conditions. Note that the zone of disturbance selectively takes place along an older mobile belt, thus giving rise to the recyclic signature of the mobile belt. Regardless of whether intra- or intercratonic, extensive areas of the Circum-East Antarctic Orogen, which suffered later Pan-African events, are more or less attributed to the Katz model, in that the younger events selectively occurred in previously existing mobile belts, thus presenting a good example of the recyclicity of orogens.

Possible approach to constrain the role of Pan-African events in East Gondwana

Regarding the idea of assembly of the major parts of East Gondwana during the Pan-African period, it must be emphasized that the present stage of knowledge in East Gondwana is still insufficient to accept the recently proposed new models. To further constrain the role of Pan-African events in East Gondwana within and around East Antarctica, detailed geological, palaeomagnetic and geophysical studies are required. Detailed structural studies associated with petrological and geochronological work, such as Kriegsman (1993), Dirks et al. (1999), Jacobs et al. (1999), Collins et al. (2000) and de Wit et al. (2001), will provide robust constraints on the role of Pan-African events in East Gondwana, although all such studies should be accompanied by careful examination of geochronological data (e.g.

Dirks et al. 1993; Harley & Fitzsimons 1995; Yoshida 1995b).

It should be taken into account that dry and less-deformed conditions in low-strain zones prevent the new growth and/or Pb loss of zircons and monazites, even under very-high-grade conditions (e.g. Kröner et al. 1994; Braun et al. 1998; Kröner et al. 2001). Thus, the lack of detection of a c. 1.0 Ga age does not necessarily indicate the absence of Grenvillian events, as pointed out by Kinny et al. (1993) in Rauer Island, Prydz Bay.

In contrast, fluid-rich and deformational conditions strongly enhance recrystallization, and drastically lower the closure temperature of Pb diffusion in minerals (e.g. Gebauer & Grünenfelder 1976; Kerrich & King 1993; Kinny et al. 1993; Yeats et al. 1996; Hartmann et al. 2000; da Silva et al. 2000). These conditions may be even more important than the temperature conditions for recrystallization/Pb loss of zircons and monazite (da Silva et al. 2000; Dobmeier & Simmat 2002). If this idea is accepted, then fluid-rich, low-grade metamorphism and deformation must be considered equally when dealing with the zircon and monazite ages, as with high-temperature conditions. Identification of the superposition of low-grade metamorphism in a high-grade terrain mainly depends on detailed structural and metamorphic studies, but not on isotopic dating. Extensive evidence of Pan-African zircon ages from several areas of East Antarctica, if it is associated with poor structural–metamorphic constraints, does not necessarily indicate extensive high-grade metamorphism in those areas, but may reflect low-grade, fluid-rich and/or high-shearing metamorphism. Kinny et al. (1993) clearly pointed out that Rauer Island, northern Prydz Bay area, could be such a case, an idea supported by Harley et al. (1995), although recent discovery of zircon SHRIMP ages from syn-peak metamorphic leucosomes from Brattstrand Bluffs (Fitzsimons 1997) may demand re-evaluation of the above idea.

There is a possibility that this problem is, to some extent, also common with Sm–Nd internal isochron ages, which are often considered to represent the high-grade metamorphic age and are used to infer the high-grade conditions of Pan-African metamorphism (Zhao et al. 1993; Hensen & Zhou 1995; Hensen et al. 1997). This is based on the very high (c. 900°C; Cohen et al. 1988) or high (c. 700–750 °C; Hensen & Zhou 1995) closure temperature of Sm–Nd systematics in garnet. However, the closure temperature of garnet for Sm–Nd isotopic systematics is estimated to be c. <600°C (Mezger et al. 1992), suggesting that the garnet–whole-rock Sm–Nd age reflects a cooling age and does not show directly the time of garnet growth during the high-grade metamorphism. Furthermore, the behaviour of Sm–Nd isotopic systematics within minerals under fluid-rich and deformational conditions is poorly understood; these conditions may even lower the above closure temperatures (e.g. Hunphries & Cliff 1982; Lahaye et al. 1995). Hensen & Zhou (1995) suggested a higher closure temperature (>700–750°C) for the Sm–Nd system in garnet, pointing out the quick cooling rate as well as dry and static conditions for the preservation of older ages. Further geochronology associated with detailed microstructural and petrological studies, both from within and outside minerals for dating analysis, associated with structural–petrological studies constraining the time of formation of the dated rocks, minerals and related reactions is awaited, and this should provide convincing conclusions.

Another important, but generally discarded, aspect when evaluating the Pan-African events in East Gondwana, is the c. 700–800 Ma age sporadically observed along the Circum-East Antarctic Orogen (Yoshida 1994, 1995a). This age is known from Enderby Land by minor granite–pegmatite activity, reflecting limited partial melting of crustal material (Black et al. 1987; Tingey 1991). This age range was also reported from the western Rayner Complex (zircon SHRIMP ages; Shiraishi et al. 1997) and the northern Prince Charles Mountains (Sm–Nd garnet–whole-rock ages; Hensen et al. 1997).

It is noteworthy that ages of extensive intrusions of granitoid rocks in central Madagascar also fall in the above age span (Handke et al. 1999; Kröner et al. 2000). Similar granitic activity with the same age span is reported from Sri Lanka (Hölzl et al. 1994), South India (Santosh et al. 1989; Rajesh et al. 1996; Bartlet et al. 1998; Braun et al. 1998) and Eastern Ghats (Shaw et al. 1997; Bindu 1998; Krause et al. 2001; Dobmeier & Simmat 2002), and could indicate a further extension of this early Pan-African event. Rb–Sr isochron ages and zircon SHRIMP ages of c. 700 Ma from the Lützow-Holm Bay area, and the Yamato and Belgica Mountains (Shibata et al. 1986; Nakajima et al. 1988; Shiraishi et al. 1994), may also be noticed from the above point of view. Yoshida (1994) correlated this age from the Lützow-Holm Bay area with major amphibolite facies metamorphism and associated tectonic events, superimposed on granulite facies metamorphism, which he interpreted to be at c. 1000 Ma.

The fact that many reports of granitic activity of this age derived from crustal melting in Enderby Land (Sheraton et al. 1987) and Madagascar (Kröner et al. 2000) appears to suggest that thermal perturbation, possibly related to mantle upwelling, played a principal role in the events of that age range. The occurrence of a c. 790 Ma anorthosite mass in the Eastern Ghats of eastern India (Krause et al. 2001), as well as several alkaline plutons in South

India (Santosh et al. 1989; Rajesh et al. 1996), supports this possibility. Referring to the geochemical signatures of similar rocks from surrounding areas, Hensen et al. (1997) suggested a continental-arc setting continuous from c. 1000 Ma events, preceding a collision–collapse event at c. 500 Ma. Further collection of data is necessary to evaluate this event, which is important in order to obtain a full tectonic picture of East Gondwana during Late Mesoproterozoic–Early Palaeozoic time.

Palaeomagnetic studies may provide valuable constraints on the configuration of East Gondwanan crustal fragments during the Grenvillian–Pan-African period, provided that such attempts are complimented with careful geological, petrological and tectonic constraints. The extent of former separation of cratons on both sides of a suture can often be ascertained by palaeomagnetic studies as mentioned above. However, palaeomagnetic studies in Precambrian terrains are made difficult by the paucity of appropriate rocks for the specific study and the difficulty in constraining the time of magnetization, especially in relation to remagnetization phenomena due to metamorphic and often even polymetamorphic overprinting in these terrains. Insufficient geological information from Precambrian Gondwanan crustal fragments further adds to the difficulties.

Buchan et al. (2000) pointed out that there are few reliable palaeomagnetic data that are useful for Precambrian–Early Palaeozoic supercontinent reconstructions, and Meert (2001) stressed that the present database from Proterozoic East Gondwana is quite insufficient to present any definite model for the palaeogeography during Proterozoic times. Torsvik et al. (2001) suggested that East Gondwana did not exist during most of Neoproterozoic time and that it assembled during the Pan-African period; this conjecture was mostly based on c. 750 Ma palaeomagnetic data from Australia, India and the Seychelles. However, data supporting their argument are scarce and some may not be conclusive on the age of magnetization. Indeed, Rathore et al. (1999) on Malani igneous rocks and Suwa et al. (1994) on Seychelles granites suggested late Pan-African rejuvenation; these data may throw open the question of the age of magnetization in these areas, critical to the above conclusion of Torsvik et al. (2001) who did not discuss these points in detail.

Geophysical data, especially geomagnetic and gravity, should provide valuable constraints on the above problems, especially because Antarctica, the key to the East Gondwana assembly, is mostly covered by ice. Recent compilation of geomagnetic and gravity data from Antarctica gives hope for their future application (Golynsky et al. 1996; von Frese et al. 1999; Golynsky & Jacobs 2001). One example, combining geomagnetic data with geologic studies from western Dronning Maud Land, clearly suggests subglacial geological structures, possibly reflecting the trace of the western boundary of the East African Orogen (Golynsky & Jacobs 2001).

Seismological studies within deep crustal transects across critical orogenic belts, such as that demonstrated in the Canadian Shield (White et al. 2000), may also provide useful criteria to constrain the crustal structure across a possible suture [e.g. proposal by Brown et al. (2001)].

Conclusions

A critical review of studies in East Gondwana leads to the contention that the classic model of the assembly of East Gondwana during the Grenvillian Circum-East Antarctic Orogeny is still valid, and that the new models invoking the assembly of East Gondwana during the Pan-African Orogeny should be further examined. Differences in age of peak metamorphism in different sectors of the orogen can be explained as a reflection of local inhomogeneities in thermal and tectonic evolution, all within a broad Grenvillian time span. Some of the salient conclusions arising from the present review are summarized below.

(1) The Circum-East Antarctic Orogen was rejuvenated during Pan-African time by extensive deformation and metamorphic events, which formed some Pan-African orogens within East Antarctica. However, the common presence of a pre-Pan-African component from the Pan-African orogens, the lack of clear Neoproterozoic offshore supracrustal sediments and ophiolites, except in the Shackleton Range, as well as poor evidence for high-P/low-T-type metamorphism suggest an intracratonic origin of these orogens. The clockwise P–T–t path, with a maximum pressure of c. 10–11 kbar detected from many Pan-African orogens, could have formed by intracratonic collision–subduction events.

(2) New models delineate either one continuous suture within East Antarctica, that extends from the Shackleton Range to Lützow-Holm Bay and further to Madagascar and the Mozambique Suture, or two or more sutures within East Antarctica passing through Lützow-Holm and Prydz Bays. The latter model involves the recent concept that East Gondwana assembled during the Pan-African Orogeny. These models are also supported by various new data and derive strength from zircon SHRIMP dating. A careful evaluation of the above new models and their basis of interpretation prompt the conclusion that the

present stage of knowledge is insufficient to replace the classical model with the new one. Both models should be examined equally in the future.

(3) A Pan-African suture extending from the Shackleton Range to Madagascar through Dronning Maud Land has received support based on evidence from geological and palaeomagnetic studies. A similar view is also suggested for the Zambezi Belt, which is equally important to constrain the role of the southern Mozambique suture(s). Detailed palaeomagnetic studies are awaited to further constrain these ideas.

(4) The present review brings out the importance of evaluating geochronological data in the light of detailed structural and petrological investigations, particularly in imaging the polymetamorphic/polyorogenic signature of orogens. More information on the closure temperatures and geochemical behaviour of the U–Pb system in zircon and monazite, the Sm–Nd system in garnet, and the behaviour of isotope systems under fluid-rich and fluid-absent conditions under various grades of deformation and metamorphism are warranted.

(5) Palaeomagnetic studies can also yield important clues. However, the present database is insufficient to obtain robust constraints on Pan-African Gondwana tectonics. Further data combining detailed petrology and geochronology to evaluate the age of magnetization are awaited.

References

APPEL, P., MOLLER, A. & SCHENK, V. 1998. High-pressure granulite facies metamorphism in the Pan-African belt of eastern Tanzania: *P–T–t* evidence against granulite formation by continental collision. *Journal of Metamorphic Geology*, **16**, 491–509.

BARTLETT, J. M., DOUGHERTY-PAGE, J. S., HARRIS, N. B. W., HAWKESWORTH, C. J. & SANTOSH, M. 1998. The application of single zircon evaporation and model Nd ages to the interpretation of polymetamorphic terrains: an example from the Proterozoic mobile belt of South India. *Contributions to Mineralogy and Petrology*, **131**, 181–195.

BAUER, W., THOMAS, R. J. & JACOBS, J. 2002. Proterozoic–Cambrian history of Dronning Maud Land in the context of Gondwana assembly. *In:* YOSHIDA, M., WINDLEY, B. F. & DASGUPTA, S. (eds) *Proterozoic East Gondwana – Supercontinent Assembly and Breakup.* Geological Society, London, Special Publication, **XX**, xx–xx.

BINDU, R. S. 1998. *Geochronological study of Precambrian terrains of South India and surrounding Gondwana areas – first attempt on electron microprobe chemical U–Pb–Th method.* DSc Thesis, Osaka City University.

BINDU, R. S., YOSHIDA, M. & SANTOSH, M. 1998. Electron microprobe dating of monazite from the Chittikara granulite, South India: evidence for polymetamorphic events. *Journal of Geosciences, Osaka City University*, **41**, 77–83.

BLACK, L. P., HARLEY, S. L., SUN, S. S. & McCULLOCH M. T. 1987. The Rayner Complex of East Antarctica: complex isotopic systematics within a Proterozoic mobile belt. *Journal of Metamorphic Geology*, **5**, 1–26.

BOGER, S. D., CARSON, C. J., FANNING, C. M. & WILSON, C. J. L. 2000. Neoproterozoic deformation in the Radok Lake region of the northern Prince Charles Mountains, east Antarctica: evidence for a single protracted event. *Precambrian Research*, **104**, 1–24.

BOGER, S. D., WILSON, C. J. L. & FANNING, C. M., 2001. Early Palaeozoic tectonism within the East Antarctic craton: the final suture between east and west Gondwana? *Geology*, **29**, 463–466.

BRAUN, I., MONTEL, J-M. & NICOLLET, C. 1998. Electron microprobe dating of monazites from high-grade gneisses and pegmatites of the Kerala Khondalite Belt, southern India. *Chemical Geology*, **146**, 65–85.

BROWN, L., KRÖNER, A., POWELL, C., WINDLEY, B. & KANAO, M. 2001. Deep seismic exploration of East Gondwana: the LEGENDS initiative. *Gondwana Research*, **4**, 846–850.

BUCHAN, K. L., MERTANEN, S., PARK, R. G., PESONEN, L. J., ELMING, S-A., ABRAHAMSEN, N. & BYLUND, G. 2000. Comparing the drift of Laurentia and Baltica in the Proterozoic: the importance of key palaeomagnetic poles. *Tectonophysics*, **319**, 167–198.

BÜCHEL, G. 1994. Gravity, magnetic and structural patterns at the deep-crustal plate boundary zone between West and East Gondwana in Sri Lanka. *In:* RAITH, M. & HOERNES, S. (eds) *Tectonic, Metamorphic and Isotopic Evolution of Deep Crustal Rocks, with Special Emphasis on Sri Lanka.* Precambrian Research, Special Volume, **66**, 77–91.

BUGGISCH, W., KLEINSCHMIDT, G., KREUZER, H. & KRUMM, S. 1990. Stratigraphy, metamorphism and nappe-tectonics in the Shackleton Range (Antarctica). *Geodatische und geophysikalische Veroffentlichungen, Reihe I, Berlin*, **15**, 64–86.

CARSON, C. J., FANNING, C. M. & WILSON, C. J. L. 1996. Timing of the Progress Granite, Larsemann Hills: evidence for Early Palaeozoic orogenesis within the East Antarctic Shield and implications for Gondwana assembly. *Australian Journal of Earth Sciences*, **43**, 539–553.

CLARK, D. J., HENSEN, B .J. & KINNY, P. D. 2000. Geochronological constraints for a two-stage history of the Albany–Fraser Orogen, Western Australia. *Precambrian Research*, **102**, 155–183.

COHEN, A. S., O'NIONS, R. K. SIEGENTHALER, R., GRIFFIN, W. L. 1988. Chronology of the pressure-temperature history recorded by a granulite terrane. *Contributions to Mineralogy and Petrology*, **98**, 303–311.

COLLINS, A., RAZAKAMANANA, T. & WINDLEY, B. 2000. Neoproterozoic extensional detachment in central Madagascar: implications for the collapse of the East African Orogen. *Geological Magazine*, **137**, 39–51.

CRADDOCK, C. 1970, Radiometric age map of Antarctica. *In*: BUSHNELL, V. C. & CRADDOCK, C. (eds) *Geologic maps of Antarctica.* Antarctic Map Folio Series, Folio 12, Plate XIX.

DA SILVA, L. C., HARTMANN, L. A., MCNAUGHTON, N. J. & FLETCHER, I. 2000. Zircon U–Pb SHRIMP dating of a Neoproterozoic overprint in Paleoproterozoic granite–gneissic terranes, southern Brazil. *American Mineralogist*, **85**, 649–667.

DALZIEL, I. W. D. 1991. Pacific margins of Laurentia and East Antarctica–Australia as a conjugate rift pair: evidence and implications for an Eocambrian supercontinent. *Geology*, **19**, 598–601.

DALZIEL, I. W. D. 1997. Neoproterozoic–Paleozoic geography and tectonics: review, hypothesis, environmental speculation. *GSA Bulletin*, **109**, 16–42.

DE WIT, M. J., BOWRING, S. A., ASHWAL, L. D., RANDRIANASOLO, L. G., MOREL, V. P. I. & RAMBELOSON, R. A. 2001. Age and tectonic evolution of Neoproterozoic ductile shear zones in southwestern Madagascar, with implications for Gondwana studies. *Tectonics*, **20**, 1–45.

DIRKS, P. G. G. M. & WILSON, C. J. L. 1995. Crustal evolution of the East Antarctic mobile belt in Prydz Bay: continental collision of 500 Ma? *Precambrian Research*, **75**, 189–207.

DIRKS, P. G. G. M., CARSON, C. J. & WILSON, C. J. L. 1993. The deformational history of the Larsemann Hills, Prydz Bay: the importance of the Pan-African (500 Ma) in East Antarctica. *Antarctic Science*, **5**, 179–102.

DIRKS, P. G. G. M., KRÖNER, A., JELSMA, H. A., SITHOLE, T. A. & VINYU, M. L. 1999. Structural relations and Pb–Pb zircon ages for the Makuti gneisses: evidence for a crustal-scale Pan-African shear zone in the Zambezi Belt, northwest Zimbabwe. *Journal of African Earth Sciences*, **28**, 427–442.

DOBMEIER, C. & SIMMAT, R. 2002. Post-Grenvillian transpression in the Chilka Lake area, Eastern Ghats Belt – implications for the geological evolution of peninsular India. *Precambrian Research*, **113**, 243–268.

FANNING, C. M., MOORE, D. H., BANNETT, V. C., DALY, S. J., MENOT, R. P., PEUCAT, J. J. & OLIVER, R. L. 1999. The 'Mawson Continent': the East Antarctic Shield and Gawler Craton, Australia. *Program and Abstracts, 8th ISAES, The Royal Society of NZ*, 103.

FITZSIMONS, I. C. W. 1996. Metapelitic migmatites from Brattstrand Bluffs, East Antarctica – metamorphism, melting and exhumation of the mid crust. *Journal of Petrology*, **37**, 395–414.

FITZSIMONS, I. C. W. 1997. The Brattstrand Paragneiss and the Søstrene Orthogneiss: a review of Pan-African metamorphism and Grenvillian relics in southern Prydz Bay. *In*: RICCI, C. A. (ed.) *The Antarctic Region: Geological Evolution and Processes.* Terra Antarctica Publication, Siena, 121–130.

FITZSIMONS, I. C. W. 2000. Grenville age basement provinces in East Antarctica: evidence for three separate collisional orogens. *Geology*, **28**, 879–882.

FRASER, G., MCDOUGALL, I., ELLIS, D. & WILLIAMS, I. S. 1997. Timing, extent, and rate of isothermal decompression following granulite-facies metamorphism at Rundvågshetta, East Antarctica. *Proceedings of the 17th Symposium on Antarctic Geosciences, Program and Abstracts*, 15–16 October 1997, National Institute of Polar Research, Tokyo, Japan, 30–31.

GEBAUER, D. & GRÜNENFELDER, M. 1976. U–Pb zircon and Rb–Sr whole-rock dating of low-grade metasediments, Example: Montagne Noire (southern France). *Contributions to Mineralogy and Petrology*, **59**, 13–32.

GEOLOGICAL SURVEY DEPARTMENT OF SRI LANKA. 1982. *Geological Map of Sri Lanka, 8 miles to one inch.* Geological Survey of Sri Lanka, Colombo.

GOLYNSKY, A. V. & JACOBS, J. 2001. Grenville-age versus Pan-African magnetic anomaly imprints in western Dronning Maud Land, East Antarctica. *Journal of Geology*, **109**, 136–142.

GOLYNSKY, A. V., MASOLOV, V. N., GOGI, Y., SHIBUYA, K., TARLOWSKY, C. & WELLMAN, P. 1996. Magnetic anomalies of Precambrian terranes of the East Antarctic shield coastal region (20°–50°E). *Proceedings of NIPR Symposium on Antarctic Geosciences*, **9**, 24–39.

GRANTHAM, G. H., GROENEWALD, P. B. & HUNTER, D. R. 1988. Geology of the northern H.U. Sverdrupfjella, western Dronning Muad Land and implications for Gondwana reconstructions. *South African Journal of Antarctic Research*, **18**, 2–10.

GRANTHAM, G. H., JACKSON, C., MOYES, A. B., GROENEWALD, P. B., HARRIS, P. D., FERRAR, G. & KRYNAUW, J. R. 1995. The tectonothermal evolution of the Kirwangeggan–H.U.Sverdrupfjella areas, Dronning Maud Land, Antarctica. *Precambrian Research*, **75**, 209–230.

GROENEWALD, P. B. 1993. Correlation of cratonic and orogenic provinces in southeastern African and Dronning Maud Land, Antarctica. *In*: FINDLAY, R. J., UNRUG, R., BANKS & WEEVERS (eds) *Gondwana Eight.* Balkema, Rotterdam, 111–123.

GROENEWALD, P. B., MOYES, A. B., GRANTHAM, G. H. & KRYNAUW, J. R. 1995. East Antarctic crustal evolution: geological constraints and modelling in western Dronning Maud Land. *Precambrian Research*, **75**, 231–151.

GRUNOW, A., HANSON, R. & WILSON, T. 1996. Were aspects of Pan-African deformation linked to Iapetus opening? *Geology*, **24**, 1063–1066.

HANDKE, M. J., TUCKER, R. D. & ASHWAL, L. D. 1999. Neoproterozoic continental arc magmatism in west-central Madagascar. *Geology*, **27**, 351–354.

HANSON, R., WILSON, T. J., BRUECKNER, H. K., ONSTOTT, T. C., WALDLAW, M. S., JOHNS, C. C. & HARDCASTLE, K. C. 1988. Reconnaissance geochronology, tectonothermal evolution, and regional significance of the middle Proterozoic Choma–Kalomo Block, southern Zambia. *Precambrian Research*, **42**, 39–61.

HARLEY, S. L. & FITZSIMONS, I. C. W. 1995. High-grade metamorphism and deformation in the Prydz Bay region, East Antarctica: terranes, events and regional correlations. *In*: YOSHIDA, M. & SANTOSH, M. (eds). *India and Antarctica during the Precambrian.* Geological Society of India, Memoir, **34**, 73–100.

HARLEY, S. L., SNAPE, I. & BLACK, L. P. 1998. The evolution of a layered metaigneous complex in the Rauer Group, East Antarctica: evidence for a distinct Archaean terrane. *Precambrian Research*, **89**, 175–205.

HARLEY, S. L., SHAPE, I. & FITZSIMONS, I. C. W. 1995. Regional correlations and terrane assembly in East Prydz Bay: evidence from the Rauer Group and Vestfold Hills. *Terra Antarctica*, **2**, 49–60.

HARRIS, L. 1995. Correlation between the Albany, Fraser and Darling Mobile Belts of Western Australia and Mirnyy to Windmill Islands in the East Antarctic Shield: implications for Proterozoic Gondwanaland reconstructions. *In*: YOSHIDA, M. & SANTOSH, M. (eds) *India and Antarctica during the Precambrian*. Geological Society of India, Memoir, **34**, 47–71.

HARRIS, N. B. W., SANTOSH, M. & TAYLOR, P. N. 1994. Crustal evolution in South India: constraints from Nd isotopes. *Journal of Geology*, **102**, 139–150.

HARTMANN, L. A., LEITE, J. A. D., SILVA, L. C. *ET AL*. 2000. Advances in SHRIMP geochronology and their impact on understanding the tectonic and metallogenic evolution of southern Brazil. *Australian Journal of Earth Sciences*, **47**, 829–844.

HENSEN, B. J. & ZHOU, BO. 1995. Retention of isotopic memory in garnets partially broken down during an overprinting granulite-facies metamorphism: implications for the Sm–Nd closure temperature. *Geology*, **23**, 225–228

HENSEN, B. J. & ZHOU, B. 1997. East Gondwana amalgamation by Pan-African collision? Evidence from Prydz Bay, East Antarctica. *In*: RICCI, C. A. (ed.) *The Antarctic Region: Geological Evolution and Processes*. Terra Antarctica Publication, Siena, 115–119.

HENSEN, B. J., ZHOU, B. & THOST, D. E. 1997. Recognition of multiple high-grade metamorphic events with garnet Sm–Nd chronology in the northern Prince Charles Mountains, Antarctica. *In*: RICCI, C. A. (ed.) *The Antarctic Region: Geological Evolution and Processes*. Terra Antarctica Publication, Siena, 97–104.

Hiroi, Y., SHIRAISHI, K. & MOTOYOSHI, Y. 1991. Late Proterozoic paired metamorphic complexes in East Antarctica, with special reference to the tectonic significance of ultramafic rocks. *In*: THOMSON, M. R. A., CRAME, J. A. & THOMSON, J. W. (eds) *Geological Evolution of Antarctica*. Cambridge University Press, Cambridge, 83–87.

Hiroi, Y., SHIRAISHI, K., MOTOYOSHI, Y., KANISAWA, S., YANAI, K. & KIZAKI, K. 1986. Mode of occurrence, bulk chemical compositions, and mineral textures of ultramafic rocks in the Lützow–Holm Complex, East Antarctica. *Memoirs of National Institute of Polar Research, Tokyo (Special Issue)*, **43**, 62–84.

HÖLZL, S., HOFMANN, A. W., TODT, W. & KOHLER, H. 1994. U/Pb geochronology of the Sri Lankan basement. *In*: RAITH, M. & HOERNES, S. (eds) *Tectonic, Metamorphic and Isotopic Evolution of Deep Crustal Rocks, with Special Emphasis on Sri Lanka*. Precambrian Research, Special Volume, **66**, 123–149.

HUNPHERIES, B. & CLIFF, R. A. 1982. Sm–Nd dating and coloning history of land. Scourian granulites, Sutherland, NW Scot. *Nature*, **295**, 515–517.

JACOBS, J., FANNING, C. M., HENJES-KUNST, F., OLESCH, M. & PAECH, H. J. 1998. Continuation of the Mozambique Belt into East Antarctica: Grenville-age metamorphism and polyphase pan-African high-grade events in Central Dronning Maud Land. *Journal of Geology*, **106**, 385–406.

JACOBS, J., HANSEN, B. T., HENJES-KUNST, F. *ET AL*. 1999. New age constraints on the Proterozoic/Lower Palaeozoic evolution of Heimefrontfjella, East Antarctica, and its bearing on Rodinia/Gondwana correlation. *Terra Antarctica*, **6**, 377–389.

JACOBS, J., KLEMD, R., FANNING, M., BAUER, W. & COLOMBO, F. 2003, Extensional collapse of the Late Neoproterozoic/Early Paleozoic East Antarctic Orogen: evidence from central Dronning Maud Land. *In:* YOSHIDA, M., WINDLEY, B. F. & DASGUPTA, S. (eds) *Proterozoic East Gondwana: Supercontinent Assembly and Break-up*, Geological Society, Special Publication **206**, 271–287.

KAMPUNZU, A. B., AKANYANG, P., MAPEO, R. B. M., MODIE, B. N. & WENDORFF, M. 1998. Geochemistry and tectonic significance of Mesoproterozoic Kgwebe metavolcanic rocks in northwest Botswana: implications for the evolution of the Kibaran Namaqua–Natal Belt. *Geological Magazine*, **133**, 669–683.

KERRICH, R. & KING, R. 1993. Hydrothermal zircon and baddeleyite in Val-dOr Archaean mesozonal gold deposits: characteristics, compositions, and fluid-inclusion properties, with implications for timing of primary gold mineralizations. *Canadian Journal of Earth Sciences*, **30**, 2334–2351.

KATZ, M. B. 1985. The tectonics of Precambrian craton-mobile belts: progressive deformation of polygonal miniplates. *Precambrian Research*, **27**, 307–319.

KINNY, P. D., BLACK, L. P. & SHERATON, J. W. 1993. Zircon ages and the distribution of Archaean and Proterozoic rocks in the Rauer Islands. *Antarctic Science*, **5**, 193–206.

KLEINSCHMIDT, G. & BUGGISCH, W. 1994. Plate tectonic implications of the structure of the Shackleton Range, Antarctica. *Polarforschung*, **63**, 57–62.

KLEINSCHRODT, R. 1994. Large-scale thrusting in the lower crustal basement of Sri Lanka. *In*: RAITH, M. & HOERNES, S. (eds) *Tectonic, Metamorphic and Isotopic Evolution of Deep Crustal Rocks, with Special Emphasis on Sri Lanka*. Precambrian Research, Special Volume, **66**, 39–57.

KRAUSE, O., DOBMEIER, C., RAITH, M. M. & MEZGER, K. 2001. Age of emplacement of massif-type anorthosites in the Eastern Ghats Belt, India: constraints from U-Pb zircon dating and structural studies. *Precambrian Research*, **109**, 25–38.

KRIEGSMAN, L. 1993. *Geodynamic evolution of the Pan-African lower crust in Sri Lanka: Structural and petrological investigations into a high-grade gneiss terrain*. Geologica Ultraiectina, Mededelingen van de Faculteit Aardwetenschappen Universiteit Utrecht No. 114.

KRÖNER, A., HEGNER, E., COLLINS, A., WINDLEY, B., BREWER, T. S., RAZAKAMANANA, T. & PIDGEON, R. 2000. Age and magmatic history of the Antananarivo Block, central Madagascar, as derived from zircon geochronology and Nd isotopic systematics. *American Journal of Science*, **300**, 251–288.

KRÖNER, A., JAECKEL, P. & WILLIAMS, I. S. 1994. Pb-loss patterns in zircons from a high-grade metamorphic terrain as revealed by different dating methods: U–Pb and Pb–Pb ages for igneous and metamorphic zircons from northern Sri Lanka. *Precambrian Research*, **66**, 151–181.

KRÖNER, A., SACCHI, R., JAECKEL, P. & COSTA, M. 1997. Kibaran magmatism and Pan-African granulite metamorphism in northern Mozambique: single zircon ages and regional implications. *Journal of African Earth Sciences*, **25**, 467–484.

KRÖNER, A., WILLNER, A. P., HEGNER, E., JAECKEL, P. & NEMCHIN, A. 2001. Single zircon ages, PT evolution and Nd isotopic systematics of high-grade gneisses in southern Malawi and their bearing on the evolution of the Mozambique belt in southeastern Africa. *Precambrian Research*, **109**, 257–291.

KRÖNER, A., WINDLEY, B. F., JAECKEL, P., BREWER, T. S. & RAZAKAMANANA, T. 1999. New zircon ages and regional significance for the evolution of the Pan-African Orogen in Madagascar. *Journal of the Geological Society, London*, **156**, 1125–1135.

LAHAYE, Y., BYERLY, A. N., CLARY, B., CATHERIN, C., FOURCADE, S. & CRUAU, G. 1995. The influence of alteration on the trace-element and Nd isotopic compotions of Komatiites. *Chemical Geology*, **126**, 43–64.

MAEGOYA, T., NOHDA, S. & HAYASE, I. 1968. Rb–Sr dating of the gneissic rocks from the east coast of Lützow-Holm Bay, Antarctica. *Memoirs of the College of Science, University of Kyoto, Series B*, **35**, 131–138.

MEERT, J. G. 2001. Growing Gondwana and rethinking Rodinia: a palaeomagnetic perspective. *Gondwana Research*, **4**, 279–288.

MEERT, J. G. & VAN DEL VOO, R. 1997. The assembly of Gondwana 800–550 Ma. *Journal of Geodynamics*, **23**, 223–235.

MEERT, J. G., VAN DER VOO, R., & AYUB, S. 1995. Palaeomagnetic investigation of the Neoproterozoic Gagwe lavas and Mbozi complex, Tanzania and the assembly of Gondwana. *Precambrian Research*, **74**, 225–244.

MEIBNER, B., DETOIS, P., SRIKANTAPPA, C. & KOEHLER, H. 2002. Geochronological evolution of the Moyar, Bhavani and Palghat shear zone of southern India: implications for east Gondwana correlations. *Precambrian Research*, **14**, 149–175.

MEZGER, K. & COSCA, M. 1999. The thermal history of the Eastern Ghats Belt (India), as revealed by U–Pb and $^{40}Ar/^{39}Ar$ dating of metamorphic and magmatic minerals; implications for the SWEAT correlation. *Precambrian Research*, **94**, 251–271.

MEZGER, K., ESSENE, E. J. & HALLIDAY, A. N. 1992. Closure temperatures of the Sm–Nd system in metamorphic garnets. *Earth and Planetary Science Letters*, **113**, 397–409.

MIYAZAKI, T., KAGAMI, H., SHUTO, K., MORIKIYO, T., RAM MOHAN, V. & RAJASEKARAN, K. C. 2001. Precise Rb–Sr age of the Yelagiri syenites and some lower-crustal effect on the Yelagiri and Senator syenites, Tamil Nadu, South India. *Gondwana Research*, **4**.

MUHONGO, S. 1994. Neoproterozoic collision tectonics in the Mozambique Belt of East Africa: evidence from the Uluguru mountains, Tanzania. *Journal of African Earth Sciences*, **19**, 153–168.

MUHONGO, S. & LENOIR, J. L. 1994. Pan-African granulite facies metamorphism in the Mozambique Belt of Tanzania: U–Pb zircon geochronology. *Journal of the Geological Society, London*, **151**, 343–347.

NAKAJIMA, T., SHIBATA, K., SHIRAISHI, K., MOTOYOSHI, Y. &

HIROI, Y. 1988. Rb–Sr whole-rock ages of metamorphic rocks from Eastern Queen Maud Land, East Antarctica (2): Tenmondai Rock and Rundvågshetta (Abstract). *Proceedings of NIPR Symposium on Antarctic Geosciences*, **2**, 172.

OHTA, Y., TØRUDBAKKEN, B. O. & SHIRAISHI, K. 1990. Geology of Gjelsvikfjella and western Mühlig-Hofmannfjella, Dronning Maud Land, East Antarctica. *Polar Research*, **8**, 99–126.

POST, N. J., HENSEN, B. J., & KINNY, P. D. 1997. Two metamorphic episodes during a 1340–1180 Ma convergent tectonic event in the Windmill Islands, East Antarctica. *In*: RICCI, C. A. (ed.) *The Antarctic Region: Geological Evolution and Processes*. Terra Antarctica Publication, Siena, 157–161.

POWELL, C. McA., Jones, D. L., Pisarevsky, S. & Wingate, M. T. D. 2001a. Palaeomagnetic constraints on the position of the Kalahari craton in Rodinia. *Precambrian Research*, **110**, 33–46.

POWELL, C. McA., PISAREVSKY, S. & WINGATE, M. T. D. 2001b. New Shape for Rodinia. *Gondwana Research*, **4**, 736–737.

RAJESH, H. M., SANTOSH, M. & YOSHIDA, M. 1996. The felsic magmatic province in East Gondwana: implications for Pan-African tectonics. *In*: YOSHIDA, M., SANTOSH, M. & ARIMA, M. (eds) *Precambrian India within East Gondwana*. Journal of Southeast Asian Earth Sciences, Special Issue, **14**, 275–291.

RATHORE, S. S., WENKATESAN, T. R. & SRIVASTAVA, R. K. 1999. Rb–Sr isotope dating of Neo-Proterozoic (Malani Group) magmatism from SW Rajasthan, India: evidence of younger Pan-African event by $^{40}Ar/^{39}Ar$ studies. *Gondwana Research*, **2**, 271–281.

RING, U. & TOULKERIDES, T. 1995. Tectonometamorphic and geochronologic constraints on Pan-African, granulite-facies metamorphic rocks in Malawi (southeastern Africa). *Centennial Geocongress (1995), Extended Abstracts*, Johannesburg, **1**, 261–264.

SANTOSH, M. & YOSHIDA, M. 2001. Pan-African extensional collapse along the Gondwana suture. *In*: DIVI, R. S. & YOSHIDA, M. (eds) *Tectonics and Mineralization in the Arabian Shield and its Extensions*. Gondwana Research, Special Issue, **4**, 188–191.

SANTOSH, M., IYER, S. S., VASCONCELLOS, M. B. A. & ENSWEILER, J. 1989. Late Precambrian alkaline plutons in southwest India: geochronologic and rare earth element constraints on Pan-African magmatism. *Lithos*, **24**, 65–79.

SANTOSH, M., KAGAMI, H., YOSHIDA, M. & NAND-KUMAR, V. 1992. Pan-African charnockite formation in East Gondwana: geochronologic (Sm–Nd and Rb–Sr) and petrogenetic constraints. *Bulletin of the Indian Geologists' Association*, **24**, 1–10.

SHACKLETON, R. M. 1986. Precambrian collision tectonics in Africa. *In*: COWARD, M. P. & RIES, A. C. (eds) *Collision Tectonics*. Geological Society, London, Special Publication, **19**, 329–349.

SHACKLETON, R. M. 1993. Tectonics of the Mozambique Belt in East Africa. *In*: PRICHARD, H. M., ALABASTER, T., HARRIS, N. B. W. & NEARY, C. R. (eds) *Magmatic Processes and Plate Tectonics*. Geological Society, London, Special Publication, **76**, 345–362.

SHAW, R. K., ARIMA, M., KAGAMI, H., FANNING, C. M., SHIRAISHI, K. & MOTOYOSHI, Y. 1997. Proterozoic events in the Eastern Ghats Granulite Belt, India: evidence from Rb–Sr, Sm–Nd systematics, and SHRIMP dating. *Journal of Geology*, **105**, 645–656.

SHERATON, J. W., TINGEY, R. J., BLACK, L. P., OFFE, L. & ELLIS, D. J. 1987. *Geology of Enderby Land and Western Kemp Land, Antarctica*. BMR Bulletin 223.

SHIBATA, K., YANAI, K. & SHIRAISHI, K. 1986. Rb–Sr whole-rock ages of metamorphic rocks from Eastern Queen Maud Land, East Antarctica. *Memoirs of National Institute of Polar Research, Special Issue*, **43**, 133–148.

SHIMURA, T., FRASER, G. L., TSUCHIYA, N. & KAGAMI, H. 1998. Genesis of the migmatites of Breidvågnipa, East Antarctica. In: MOTOYOSHI, Y. & SHIRAISHI, K. (eds) *Origin and Evolution of Continents*. Memoirs of National Institute of Polar Research, Special Issue, **53**, 109–136.

SHIRAHATA, H. 1983. Lead isotopic composition in metamorphic rocks from Skarvsnes, East Antarctica. In: OLIVER, R. L., JAMES, P. R. & JAGO, J. B. (eds) *Antarctic Earth Science*. Australian Academy of Science, Canberra, 55–58.

SHIRAISHI, K., ELLIS, D. J., FANNING, C. M., HIROI, Y., KAGAMI, H. & MOTOYOSHI, Y. 1997. Re-examination of the metamorphic and protolith ages of Rayner Complex, Antarctica: evidence for the Cambrian (Pan-African) regional metamorphic event. In: RICCI, C. A. (ed.) *The Antarctic Region. Geological Evolution and Processes*. Terra Antarctica Publication, Siena, 79–88.

SHIRAISHI, K., ELLIS, D. J., HIROI, Y., FANNING, C. M., MOTOYOSHI, Y. & NAKAI, Y. 1994. Cambrian orogenic belt in East Antarctica and Sri Lanka: implications for Gondwana assembly. *Journal of Geology*, **102**, 47–65.

SHIRAISHI, K., HIROI, Y., ELLIS, D. J., FANNING, C. M., MOTOYOSHI, Y. & NAKAI, Y. 1992. The first report of a Cambrian orogenic belt in east Antarctica – an ion microprobe study of the Lützow–Holm Complex. In: YOSHIDA, Y., KAMINUMA, K. & SHIRAISHI, K. (eds) *Recent Progress in Antarctic Earth Science*. Terra Scientific Publishing Company, Tokyo, 29–35.

SINHA, A. K. & GLOVER, III, L. 1978. U–Pb systematics during dynamic metamorphism, a study from the Brevared fault zone. *Contributions to Mineralogy and Petrology*, **66**, 305–310.

STERN, R. J. 1994. Arc assembly and continental collision in the Neoproterozoic east African Orogen: implications for the consolidation of Gondwanaland. *Annual Review of Earth and Planetary Science*, **22**, 319–351.

SUWA, K., TOKIEDA, K. & HOSHINO, M. 1994. Palaeomagnetic and petrological reconstruction of the Seychelles. *Precambrian Research*, **69**, 281–292.

TALARICO, F., KLEINSCHMIDT, G. & HENJES-KUNST, F. 1999. An ophiolite complex in the northern Shackleton Range, Antarctica. *Terra Antarctica*, **6**, 293–315.

TANI, Y. & YOSHIDA, M. 1996. The structural evolution of the Arena Gneisses and its bearing on Proterozoic tectonics of Sri Lanka. In: YOSHIDA, M., SANTOSH, M. & ARIMA, M. (eds) *Precambrian India within East Gondwana*. Journal of Southeast Asian Earth Sciences, Special Issue, **14**, 309–329.

TESSENSOHN, F. 1997. Shackleton Range, Ross Orogen and SWEAT hypothesis. In: RICCI, C. A. (ed.) *The Antarctic Region: Geological Evolution and Processes*. Terra Antarctica Publication, Siena, 5–12.

TINGEY, R. J. 1991. The Regional geology of Archaean and Proterozoic rocks in Antarctica. In: TINGEY, R. J. (ed.) *The Geology of Antarctica*. Oxford Monograph on Geology and Geophysics 27, Clarendon Press, Oxford, 1–73.

TONG, T. & LIU, X. 1997. The prograde metamorphism of the Larsemann Hills, East Antarctica: evidence for an anticlockwise P–T path. In: RICCI, C. A. (ed.) *The Antarctic Region: Geological Evolution and Processes*. Terra Publication, Siena, 105–114.

TONG, T., ZHANG, L. & CHEN, F. 1995. The U–Pb zircon chronology of mafic granulite from the Larsemann Hills, East Antarctica and its possible geological implications. *Terra Antarctica*, **2**, 123–126.

TORSVIK, T. H., CARTER, L. M., ASHWAL, L. D., BHUSHAN, S. K., PANDIT, M. K. & JAMTVEIT, B. 2001. Rodinia refined or obscured: palaeomagnetism of the Malani igneous suite (NW India). *Precambrian Research*, **108**, 319–333.

TORSVIK, T. H., SMETHURST, M. A., MEERT, J. G. ET AL. 1996. Continental break-up and collision in the Neoproterozoic and Palaeozoic – a tale of Baltica and Laurentia. *Earth Science Reviews*, **40**, 229–258.

UNRUG, R. 1992. The supercontinent cycle and Gondwana assembly: component cratons and the timing of suturing events. *Journal of Geodynamics*, **16**, 215–246.

UNRUG, R. 1997. Rodinia to Gondwana: the geodynamic map of Gondwana supercontinent assembly. *GSA Today*, **7**, 1–6.

VINYU, M. L, HANSON, R. E., MARTIN, M. W., BOWRING, S. A., JELSMA, H. A., KROL, M. A. & DIRKS, P. H. G. M. 1999. U–Pb and $^{40}Ar/^{39}Ar$ geochronological constraints on the tectonic evolution of the easternmost part of the Zambezi orogenic belt, northeast Zimbabwe. *Precambrian Research*, **98**, 67–82.

VON FRESE, R. R. B., KIM, H. R., TAN, L. ET AL. 1999. Satellite magnetic anomalies of the Antarctic crust. *Annali di Geofisica*, **42**, 309–326.

WAREHAM, C. D., PANKHURST, R. J., THOMAS, R .J., STOREY, B. C., GRANTHAM, G. H., JACOBS, J. & EGLINGTON, B. M. 1998. Pb, Nd, and Sr isotope mapping of Grenville-age crustal provinces in Rodinia. *Journal of Geology*, **106**, 647–658.

WASTENEYS, H., MCLELLAND, J. & LUMBERS, J. 1999. Precise zircon geochronology in the Adirondack Lowlands and implications for revising-plate tectonic models of the Central metasedimentary belt and Adirondack Mountains, Grenville Province, Ontario and New York. *Canadian Journal of Earth Science*, **36**, 967–984.

WHITE, D. J., ZWANZIG, H. V. & HAJNAL, Z. 2000. Crustal suture preserved in the Paleoproterozoic Trans-Hudson orogen, Canada. *Geology*, **28**, 527–530.

WHITE, R. W. & CLARKE, G. L.. 1993. Timing of Proterozoic deformation and magmatism in a tectonically reworked orogen, Rayner Complex, Colbeck Archipelago, East Antarctica. *Precambrian Research*, **63**, 1–26.

WINDLEY, B. F., BREWER, T. S., COLLINS, A., KRÖNER, A., JAECKEL, P. & RAZAKAMANANA, T. 1997. The Pan-African orogen of Madagascar. In: COX, R. M. &

ASHWAL, L. D. (eds), *Proceedings of the UNESCO-IUGS-IGCP-348/368 International Field Workshop on Proterozoic Geology of Madagascar, Antananarivo, Madagascar, 16–30 August*, Gondwana Research Group Miscellaneous Publication No. **5**, Field Science Publishers, Osaka.

WYNE, D. M. & SINHA, A. K. 1988. Physical and chemical response of zircons to deformation. *Contributions to Mineralogy and Petrology*, **98**, 109–121.

YEATS, C. J., MCNAUGHTON, N. J. & GROVES, D. I. 1996. SHRIMP U/Pb geochronological constraints on Archaean volcanic-hosted massive sulphide and lode gold mineralization at Mount Gibson, Yilgarn Craton, Western Australia. *Economic Geology*, **91**, 1354–1371.

YOSHIDA, M. 1978. Tectonics and petrology of charnockites in Lützow-Holmbukta, East Antarctica. *Journal of Geosciences, Osaka City University*, **21**, 65–125.

YOSHIDA, M. 1979. Metamorphic conditions of the polymetamorphic Lutzow–Holmbukta region, East Antarctica. *Journal of Geosciences, Osaka City University*, **22**, 97–140.

YOSHIDA, M. 1992. Late Proterozoic to early Palaeozoic events in East Gondwanan crustal fragments. *Abstracts of the 29th International Geological Congress*, Kyoto, v.2/3, 265.

YOSHIDA, M. 1994. Tectonothermal history and tectonics of Lützow–Holm Bay area, East Antarctica: a reinterpretation. *Journal of Geological Society of Sri Lanka*, **5**, 81–93.

YOSHIDA, M. 1995a. Assembly of East Gondwanaland during the Mesoproterozoic and its rejuvenation during the Pan-African period. *In*: YOSHIDA, M. & SANTOSH, M. (eds) *India and Antarctica during the Precambrian*. Geological Society of India, Memoir, **34**, 25–45.

YOSHIDA, M. 1995b. Cambrian orogenic belt in East Antarctica and Sri Lanka: implications for Gondwana assembly: a discussion. *Journal of Geology*, **103**, 467–468.

YOSHIDA, M. 1996. Southernmost Indian Peninsula and the Gondwanaland. *In*: SANTOSH, M. & YOSHIDA, M. (eds) *The Archaean and Proterozoic Terrains of Southern India within East Gondwana*, Gondwana. Research Group Memoir 3, Field Science Publishers, Osaka, 15–24.

YOSHIDA, M. & KIZAKI, K. 1983. Tectonic situation of Lutzow–Holm Bay in East Antarctica and its significance in Gondwanaland. *In*: OLIVER, R. L., JAMES, P. R. & JAGO, J. B. (eds) *Antarctic Earth Science*. Australian Academy of Science, Canberra, 36–39.

YOSHIDA, M. BINDU, R. S., KAGAMI, H., RAJESHAM, T., SANTOSH, M. & SHIRAHATA, H. 1996. Geochronologic constratins of granulite terrnes of South India and their implications for the Precambrian assembly of Gondwana. *In*: YOSHIDA, M., SANTOSH, M. & ARIMA, M. (eds) *Precambrian India within East Gondwana*. Journal of Southeast Asian Earth Sciences, Special Issue, **14**, 137–147.

YOSHIDA, M., DIVI, R. S. & SANTOSH, M. 2001. Precambrian Central India and its role in the Gondwanaland–Rodinia context. *In*: DIVI, R. S. & YOSHIDA, M. (eds) *Tectonics and Mineralization in the Arabian Shield and its Extensions*. Gondwana Research, Special Issue, **4**, 208–211.

YOSHIDA, M., KAGAMI, H. & UNNIKRISHNAN, W. 1999. Neodymium model ages from Eastern Ghats and Lützow-Holm Bay: constraints on isotopic provinces in India–Antarctic sector of East Gondwana. *In*: RAO, A. T., DIVAKARA RAO, V. & YOSHIDA, M. (eds) *Eastern Ghats Granulites*. Gondwana Research Group, Memoir 5, Field Science Publishers, Osaka, 161–172.

YOSHIDA, M., SANTOSH, M. & ARIMA, M. 2000. Pre-Pan African and Pan-African events in south India and their implications for Gondwana tectonics. *Proceedings of the International Seminar UNESCO–IUGS–IGCP-368: Precambrian Crust in Eastern and Central India*, October 29–30, 1998. Geological Survey of India, Special Publication, **57**, Geological Survey of India, Calcutta, 9–25.

YOSHIDA, M., SUZUKI, M., SHIRAHATA, H., KOJIMA, H. & KIZAKI, K. 1983. A review of the tectonic and metamorphic history of the Lützow-Holm Bay region, East Antarctica. *In*: OLIVER, R. L., JAMES, P. R. & JAGO, J. B. (eds) *Antarctic Earth Science*. Australian Academy of Science, Canberra, 44–47.

YOUNG, D. N. & ELLIS, D. J. 1991. The intrusive Mawson charnockites: evidence for a compressional plate margin setting of the Proterozoic mobile belt of East Antarctica. *In*: THOMSON, M. R. A., CRAME, J. A. & THOMSON, J. W. (eds) *Geological Evolution of Antarctica*. Cambridge University Press, Cambridge, 25–31.

ZHANG, L. S., TONG, L. X., LIU, X. H. & SCHARER, U. 1996. Conventional U–Pb age of the high-grade metamorphic rocks in the Larsemann Hills, East Antarctica. *In*: PANG, Zh. H. *et al.* (eds) *Advances in Solid Earth Science*. Science Press, Beijing, 27–35.

ZHAO, J-X., ELLIS, D. J., KILPATRICK, J. A. & MCCULLOCH, M. T. 1997. Geochemical and Sr–Nd isotopic study of charnockites and related rocks in the northern Prince Charles Mountains, East Antarctica: implications for charnockite petrogenesis and Proterozoic crustal evolution. *Precambrian Research*, **37**, 37–66.

ZHAO, Y., FANNING, C. M., LIU, X-C. & LIN, X-H. 2000. The Grove Mountains, a segment of a Pan-African Orogenic Belt in East Antarctica. *Abstracts of the 31st IGC*, San Paulo.

ZHAO, Y., LIU, X., SONG, B., ZHANG, Z., LI, J., YAO, Y. & WANG, Y. 1995. Constraints on the stratigraphic age of metasedimentary rocks from the Larsemann Hills, East Antarctica: possible implications for Neoproterozoic tectonics. *Precambrian Research*, **75**, 175–188.

ZHAO, Y., SONG, B., WANG, Y., REN, L., LI, J. & CHEN, T. 1992. Geochronology of the late granite in the Larsman Hills, East Antarctica. *In*: YOSHIDA, Y., KAMINUMA, K. & SHIRAISHI, K. (eds) *Recent Progress in Antarctic Earth Science*. Terra Scientific Publishing Company, Tokyo, 155–161.

ZHAO, Y., SONG, B., ZHANG, Z. *ET AL*. 1993. Early Paleozoic (Pan African) thermal event of the Larsemann Hills and its neighbours, Prydz Bay, East Antarctica. *Science in China (Series B)*, **38**, 74–84.

Palaeomagnetic constraints on the Proterozoic tectonic evolution of Australia

MICHAEL T. D. WINGATE & DAVID A. D. EVANS

Tectonics Special Research Centre, Department of Geology and Geophysics, The University of Western Australia, 35 Stirling Highway, Crawley, WA 6009, Australia (e-mail: mwingate@tsrc.uwa.edu.au)

Abstract: Recent plate tectonic models advocate assembly of Proterozoic Australia by tectonic processes that involved large-scale horizontal motions, whereas previous models suggested that the continent evolved as an essentially intact block of lithosphere. Geological and geochemical observations alone are insufficient to test whether the major cratonic blocks of Australia were together or widely separated during the Proterozoic; only palaeomagnetism can provide quantitative constraints on relative plate motions during the Precambrian. Despite deficiencies in the palaeomagnetic record for Proterozoic Australia, groups of overlapping palaeopoles for 1.7–1.8 and 1.5–1.6 Ga permit the North and West Australian cratonic assemblages to have occupied their present relative positions since at least *c.* 1.7 Ga, and to have been joined to the South Australian cratonic assemblage since at least *c.* 1.5 Ga. Nonetheless, additional geological, geochronological and palaeomagnetic data are required to test whether large oceans closed between any of the continental blocks.

It is generally accepted that processes similar to those of modern plate tectonics operated during the Proterozoic, if not earlier (Burke *et al.* 1976; Hoffman 1988; Windley 1995; Hamilton 1998; Collerson & Kamber 1999). The Proterozoic geology of Australia, however, has only relatively recently been interpreted in a plate tectonic context (e.g. Myers 1990; Tyler & Thorne 1990; Myers *et al.* 1996; Tyler *et al.* 1998; Krapez & Martin 1999). Plate tectonic models advocate growth of Australia by amalgamation of continental fragments, and invoke processes such as subduction, arc magmatism, terrane accretion and continent–continent collision, that occur at modern plate margins and involve large horizontal displacements.

The traditional view in Australia has been that the continent evolved as an essentially intact block of crust, by processes of mainly intracratonic rifting and vertical tectonics, although possibly with creation and destruction of small intercratonic ocean basins (e.g. Harrington *et al.* 1973; Rutland 1973; Etheridge *et al.* 1987, Wyborn 1988). The 'single-continent' model is based mainly on: (1) the prevalence of low-*P* high-*T* metamorphism; (2) the absence of rock types diagnostic of subduction, such as relicts of oceanic crust, paired metamorphic belts and calc-alkaline magmatism (e.g. Etheridge *et al.* 1987; Wyborn 1988; Mortimer *et al.* 1988*a*; Collins & Shaw 1995); (3) the observation that a single apparent polar wander path (APWP) can be constructed through all palaeopoles for Precambrian Australia, even though they are derived from different tectonic units (e.g. McElhinny & Embleton 1976; Veevers & McElhinny 1976; Idnurm & Giddings 1988; Plumb 1993; Idnurm *et al.* 1995); and (4) crustal thicknesses and seismic signatures more compatible with reworking of pre-existing crust than with terrane accretion (Drummond 1988).

Within Proterozoic orogenic belts, distinguishing between plate margin interaction and intraplate models on the basis of geological criteria can be problematic. Despite abundant evidence for compressional deformation, and for the prior truncation and subsequent juxtaposition of terranes of contrasting age, structure and metamorphic characteristics, the geology yields no information on how widely separated the terranes might have been prior to their inferred amalgamation. Neither do geochemical studies yield this information, although they can provide qualitative constraints on tectonic environments (e.g. Krapez & Martin 1999; Sheppard *et al.* 1999*a*) and have shown that subduction may have operated since 3.8 Ga (Collerson & Kamber 1999). The presence of ophiolites constitutes direct evidence for closure of an ocean basin, but does not indicate its original size, and few ophiolites have been recognized in Proterozoic orogenic belts (Moores 1986; Helmstaedt & Scott 1992; Windley 1995).

Only palaeomagnetism can reveal the past relative positions and motions of crustal fragments in Australia during these ancient times. Many new palaeomagnetic and isotopic age data have been generated in the last decade. This paper will provide a brief summary of current geological and palaeomagnetic information for Proterozoic Australia, focus on apparent agreements and conflicts between the two data sets, and attempt to reassess models for the continent's evolution.

Tectonic summary

Recent geological syntheses portray the assembly of Australia as a protracted and complex series of events (e.g. Myers et al. 1996; Tyler et al. 1998). Precambrian rocks of Australia comprise three main regions, referred to here as the North, West and South Australian cratonic assemblages, each of which contains blocks of Archaean and/or Palaeoproterozoic crust, is bounded by Palaeoproterozoic–Mesoproterozoic mobile belts, and is partly overlain by younger sedimentary basins (Fig. 1). These three regions were interpreted by Myers et al. (1996) to have formed independently in the Palaeoproterozoic, prior to their final amalgamation along Late Mesoproterozoic (1.3–1.0 Ga) orogenic belts (Figs. 2 & 3). Several geological and palaeomagnetic observations, however, favour coherence of these regions since at least the Late Palaeoproterozoic, as discussed below. The following simplified overview of the Proterozoic tectonic evolution of each cratonic assemblage focuses mainly on observations and interpretations that are suggestive of tectonic environments.

West Australian cratonic assemblage (WAC)

The WAC (Fig. 1a) contains three geologically distinct terranes inferred to have assembled by c. 1.8 Ga. The Gascoyne Complex is thought to represent a Late Archaean–Palaeoproterozoic microcontinent joined to the northern margin of the Archaean Yilgarn Craton along the 2.0–1.96 Ga Glenburgh Orogen (Sheppard & Occhipinti 2000). Sheppard et al. (1999b) interpreted the temporal and spatial distribution of granitoids to indicate north-dipping subduction and subsequent collision. Myers (1989) suggested that thrust sheets of deformed and metamorphosed mafic and ultramafic plutonic and volcanic rocks (the Trillbar Complex and Narracoota Volcanics; Fig. 1a), transported southwards onto the Yilgarn Craton, represent obducted parts of a volcanic arc.

Structural and magmatic observations were interpreted to indicate that an ocean basin was consumed by N–S oblique convergence and south-dipping subduction between the Archaean Yilgarn and Pilbara Cratons during the 1.83–1.78 Ma Capricorn Orogeny (Tyler & Thorne 1990; Tyler 1999; Hall et al. 2001). High-P kyanite-bearing mineral assemblages in the Capricorn Orogen are consistent with subduction and continental collision (Baker et al. 1987; Tyler & Thorne 1990; Fitzsimons, pers. comm.). Subsequent deformation included westward extrusion of material caught between the Pilbara and Yilgarn Cratons (Tyler & Thorne 1990) and regional extension after c. 1.62 Ga (Nelson 1998) to accommodate sediments of the Bangemall Basin (Fig. 1a).

North Australian cratonic assemblage (NAC)

Synchronous deformation, metamorphism and magmatism occurred during the 1.89–1.87 Ga Barramundi Orogeny across much of the NAC (Fig. 1b), including the Pine Creek, Tennant Creek, northern Arunta and Mount Isa Blocks, and possibly the Georgetown Block (Etheridge et al. 1987; Page 1988). Etheridge et al. (1987) argued that the Barramundi orogeny was intracontinental, based on the predominance of low-P metamorphism and the absence of features diagnostic of modern intercontinental orogeny. They proposed a model of mafic underplating and continental extension driven by small-scale mantle convection, followed by low-P/high-T compression driven by crust–mantle delamination and A-subduction. Wyborn (1988) argued that geochemical signatures of 1.88–1.84 Ga felsic igneous rocks across most of Australia are consistent with such a model.

The Kimberley Block is thought to have accreted to the NAC along the Halls Creek Orogen at c. 1.85–1.82 Ga (Bodorkos et al. 1999). Layered mafic–ultramafic bodies and associated mafic to felsic sheets of the Tickalara Complex (Fig. 1b) have geochemical signatures consistent with either an oceanic island-arc-/back-arc basin or an ensialic marginal basin setting (Sheppard et al. 1999a). On the eastern edge of the NAC, volcanics and sediments were deposited on Archaean– Palaeoproterozoic continental crust in a passive margin setting (Sheppard et al. 1999a). Sheppard et al. (1999a) argued that their observations are more consistent with processes similar to those of modern plate margin interactions than with previous ensialic models (Etheridge et al. 1987; Page & Hancock 1988).

The Strangways Orogeny includes both early (1.78–1.77 Ga) and late (1.745–1.73–Ga) events in the northern Arunta Inlier (Collins & Shaw 1995), and was inferred by Myers et al. (1996) to reflect oblique convergence and accretion of magmatic arcs along the southern margin of the NAC. Between c. 1.8 and 1.6 Ga, sedimentation and volcanism occurred across the NAC in a series of interconnected basins (Fig. 1b), remnants of which are best preserved in the McArthur Basin, and the Mount Isa and Georgetown Inliers (Plumb et al. 1990). Giles et al. (2001) suggested that these basins formed during back-arc continental extension above a long-lived, north-dipping subduction zone along the southern margin of the NAC. The 1.68–1.66 Ga Argilke event may reflect accretion of a strip of continental crust to the same margin. On geochemical grounds, Zhao & Cooper (1992) argued for subduction and consumption of

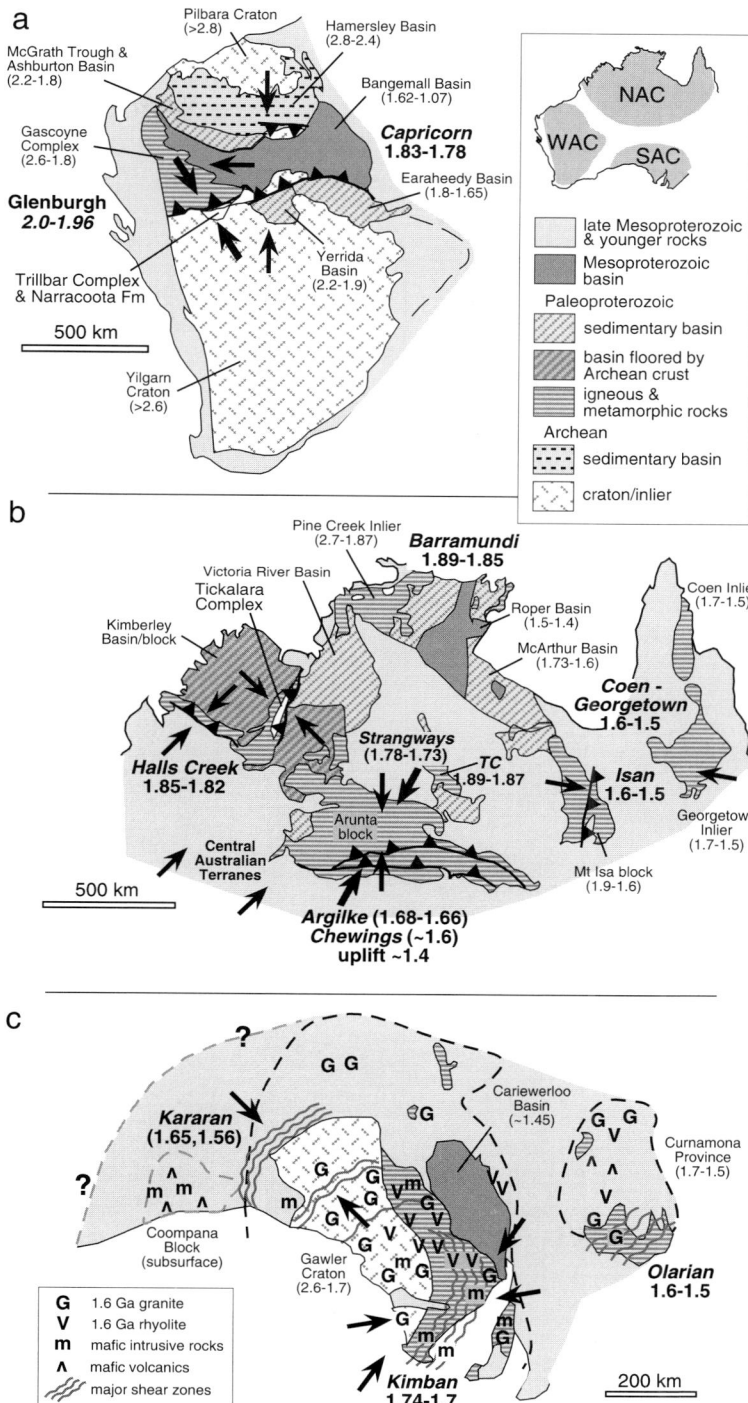

Fig. 1. (**a–c**) Simplified geology of the West, North and South Australian cratonic assemblages (WAC, NAC and SAC respectively). Orogens (italics) and associated times of tectonic activity are indicated (ages in Ga). TC, Tennant Creek Block.

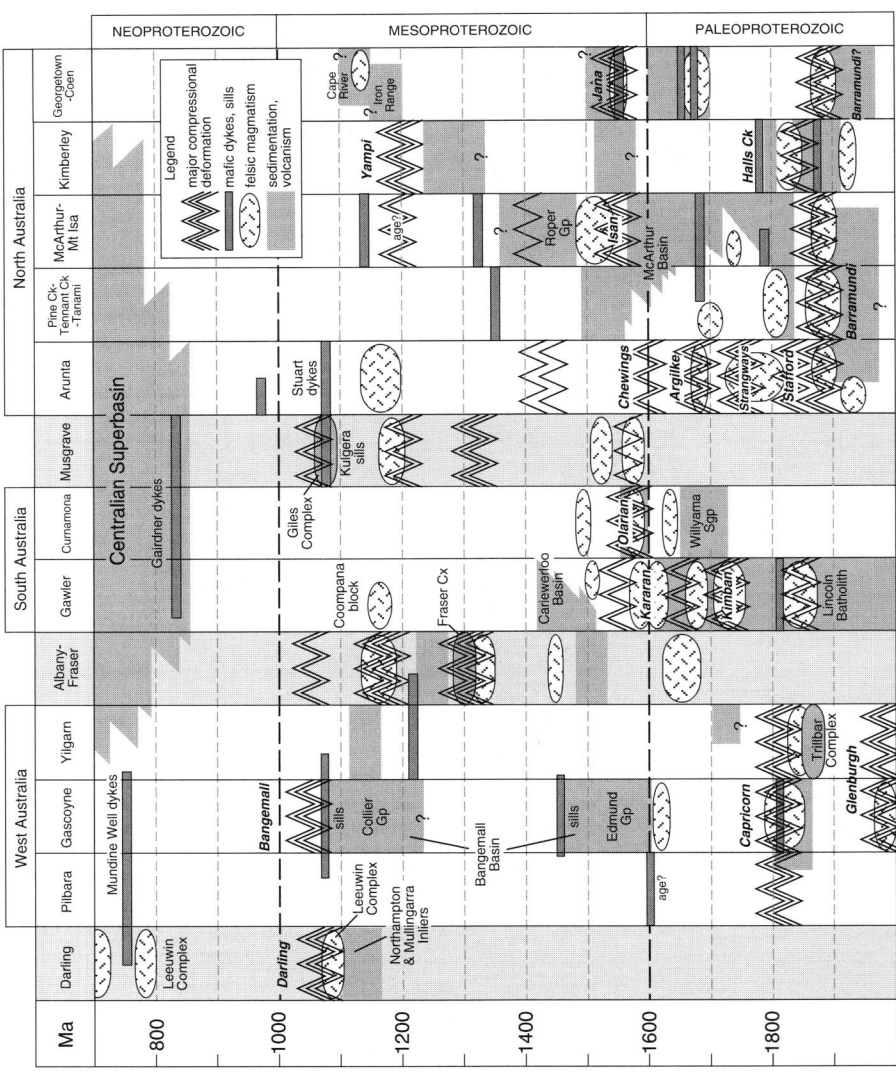

Fig. 2. Schematic time–space diagram showing major 1.8–0.7 Ga deformation, basin formation and magmatic events in the different crustal blocks of mainland Australia and in 1.3–1.1 Ga orogenic belts. The names of tectonothermal events are in italics. Compiled from literature sources cited in the text.

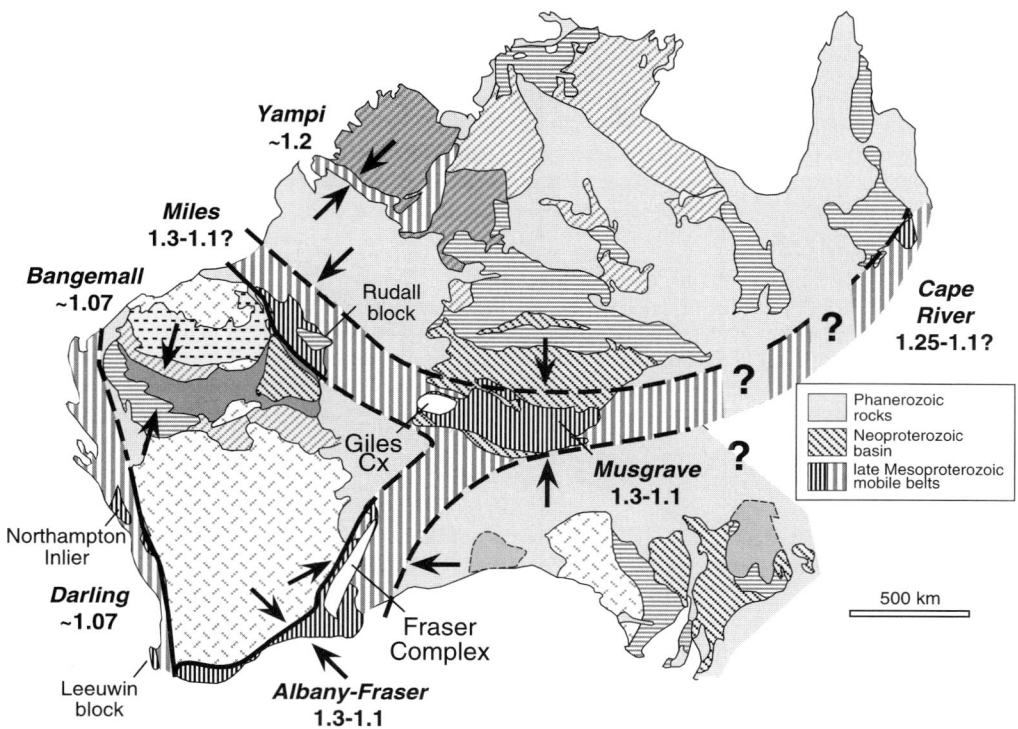

Fig. 3. Exposed Precambrian rocks of mainland Australia, illustrating 1.3–1.1 Ga orogenic belts along which the North, West, and South Australian cratonic assemblages are inferred to have amalgamated (Myers *et al.* 1996). Also shown are possible Grenville-age rocks in NE Australia and their proposed connection with the Musgrave Orogen in central Australia (Blewett *et al.* 1998). Orogens (in italics) and associated times of tectonic activity are indicated (ages in Ga). See Figure 1 for additional parts of the legend. Compiled from literature sources cited in the text.

oceanic crust during these events, although this was questioned by Collins & Shaw (1995), who cited the scarcity of diagnostic subduction-related rock types, and apparent correlations of strata across the Arunta and Tennant Creek Blocks, as evidence against accretion. Compression during the *c.* 1.6 Ga Chewings Orogeny in the southern Arunta Inlier may have been intracratonic (Myers *et al.* 1996), and was synchronous with post-tectonic pegmatite intrusion farther north, suggesting that the Arunta Inlier was intact by that time (Collins & Shaw 1995).

Broadly synchronous orogenesis and high-T metamorphism occurred at: 1.6–1.5 Ga in NE Australia; 1.58 and 1.54–1.53 Ga in the Eastern and Western Fold Belts, respectively, of the Mount Isa Inlier (Conners & Page 1995; Page & Sun 2000); 1.59–1.55 Ga in the Coen Inlier (Blewett & Black 1998); *c.* 1.55 Ga in the Georgetown Inlier (Black & Withnall 1993; Black *et al.* 1998). Myers *et al.* (1996) suggested that these events may reflect accretion of the Georgetown and Coen Blocks to the eastern margin of the NAC. However, the absence of ophiolites, paired metamorphic belts, calc-alkaline magmatism and significant vertical uplift, and the presence of low-pressure (P) high-temperature (T) metamorphism, anticlockwise $P–T$–time ($P–T–t$) paths, bimodal magmatism and broad shallow basins, suggest that the events were intraplate in character (Wyborn 1988; Blewett & Black 1998; Oliver *et al.* 1998).

South Australian cratonic assemblage (SAC)

The SAC contains two main cratonic blocks, the Archaean–Palaeoproterozoic Gawler Craton and the Palaeoproterozoic–Mesoproterozoic Curnamona Province (Fig. 1c). A detailed summary of the geological history of the Gawler Craton, including its southward extension into the Mawson Block of Antarctica, is provided by Fitzsimons (2002). Most Palaeoproterozoic and older rocks in both provinces have been deformed extensively and metamorphosed to amphibolite or granulite facies.

Voluminous felsic and mafic granitoids and minor mafic dykes of the Lincoln Batholith were emplaced at the SE margin of the Gawler Craton at c. 1.85 Ga, coincident with the final stages of deposition of Hutchison Group sediments onto Archaean Sleaford Complex basement (Mortimer et al. 1988a; Hoek & Schaefer 1998). Limited compressional deformation at this time may have accommodated successive pluton emplacement (Hoek & Schaefer 1998; Vassallo & Wilson 2002). Mortimer et al. (1988a) argued that the Lincoln Granitoids are not compatible compositionally with subduction processes and advocated a model, similar to that proposed by Etheridge et al. (1987), of magma generation by remelting of underplated mafic material in an intracratonic setting. The Lincoln Batholith was juxtaposed against the Hutchison Group along the Kalinjala Shear Zone, a major structure active during the 1.74–1.70 Ga Kimban Orogeny, which included regional deformation and metamorphism of Hutchison Group sedimentary rocks and their underlying Archaean basement, and emplacement of extensive granitoids of the 1.74–1.70 Ga Moody Suite (Daly et al. 1998; Hoek & Schaefer 1998; Vassallo & Wilson 2002). Myers et al. (1996) suggested that the Curnamona Province was amalgamated to the Gawler Craton during the Kimban Orogeny.

In the NW Gawler Craton, dip-slip structures formed during the 1.65 and 1.57–1.54 Ga Kararan Orogeny have been interpreted in terms of a 1.65 Ga collision between the Gawler Craton and a 'proto-Yilgarn' Craton (Daly et al. 1998), and may also have involved accretion of the Coompana Block (known in the subsurface only). Giles et al. (2001) and Fitzsimons (2002) suggested that a belt of 1.65–1.55 Ga synorogenic granite plutons (the Ifould Complex; Daly et al. 1998) SE of the Karari Fault Zone could represent a magmatic arc predating collision at c. 1.56 Ga. Mafic and ultramafic bodies associated with these mobile belts have been taken to imply crustal thinning (Daly et al. 1998), but have not been interpreted in terms of specific tectonic processes. The central Gawler Craton contains extensive outcrops of the 1.6–1.55 Ga subaerial felsic and minor mafic Gawler Range Volcanics and comagmatic Hiltaba Suite Granites, thought to have been generated by extensive mafic underplating and melting of Archaean and Palaeoproterozoic crust (Flint 1993; Daly et al. 1998).

The oldest rocks recognized in the Curnamona Province (Fig. 1c) are metasedimentary and bimodal meta-igneous rocks of the Willyama Supergroup that were deposited at c. 1.69 Ga (Page & Laing 1992). Inherited zircon ages of 1.78–2.7 Ga suggest the presence of Archaean and/or Early Palaeoproterozoic basement (Page & Laing 1992; Robertson et al. 1998). These rocks were subsequently deformed and metamorphosed to amphibolite and granulite facies during the c. 1.6 Ga Olarian Orogeny (Page & Laing 1992). The Curnamona Province also contains abundant late synorogenic to post-orogenic granitoid and bimodal volcanic rocks that are similar in age to, and possibly correlative with, the 1.6–1.55 Ga Gawler Range Volcanics and Hiltaba Suite Granitoids of the Gawler Craton (Daly et al. 1998; Robertson et al. 1998).

Mesoproterozoic amalgamation (?)

Both the Albany–Fraser and Musgrave Mobile Belts (Fig. 3) preserve a record of two major tectonothermal events: collision at c. 1.3 Ga, followed by intracratonic reactivation of sutures at 1.2 Ga (Clarke et al. 1995; Nelson et al. 1995; White et al. 1999; Clark et al. 2000). These observations imply that the Albany–Fraser Orogen is continuous with the Musgrave Orogen, as suggested by Myers et al. (1996). They also suggest that the NAC and WAC could have been combined prior to collision with the South Australian–Antarctic Craton, and that the Miles Orogeny, along the eastern edge of the combined Yilgarn–Pilbara Block (Fig. 3), was intracratonic.

Limited geochemical data from the Albany–Fraser and Musgrave Belts are equivocal, but suggest that c.1.3 Ga felsic orthogneisses are collision related, whereas c. 1.2 Ga granitoids have a more intraplate signature (Nelson et al. 1995; Sheraton & Sun 1995). Based on the absence of Late Mesoproterozoic plutons in the SE Yilgarn Craton, Clark et al. (2000) suggested that initial collision at 1.3 Ga involved subduction of oceanic crust beneath the South Australian–Antarctic Craton. Within the eastern Albany–Fraser Orogen, the Fraser Complex (Fig. 3) consists of imbricated slices of a layered mafic intrusion (Myers 1985) that was emplaced, metamorphosed to granulite grade, then uplifted and thrust onto the SE Yilgarn Craton margin within c. 30 Ma at 1.3 Ga (Fletcher et al. 1991). Geochemical data were interpreted by Condie & Myers (1999) to indicate that the Fraser Complex represents remnants of oceanic arcs accreted before or during collision. The presence of Archaean and/or Palaeoproterozoic crustal components within the Albany–Fraser and Musgrave Belts (e.g. Nelson et al. 1995; Camacho & Fanning 1995; Clark et al. 2000) implies that these orogens may be floored, at least in part, by continental crust. Extensive swarms of mafic dykes were emplaced subparallel to the margins of the Yilgarn Craton at c. 1.2 Ga (Evans 1999; Wingate et al. 2000; Wingate unpublished data).

Evidence for Late Mesoproterozoic activity between the NAC and WAC is provided by a U–Pb zircon age of 1310 Ma (Nelson 1996), and Rb–Sr ages of 1.3 and 1.1 Ga (Chin & De Laeter 1981), for foliated granite in the Palaeoproterozoic Rudall Block (Fig. 3). Inherited zircons ranging in age from 1.3 to 1.0 Ga in the Cape River Province and Anakie Inlier suggest the existence of a Grenville-age belt in northeastern Australia that may extend westward, beneath younger cover, to connect with the Albany–Fraser–Musgrave Orogen (Blewett et al. 1998; Hutton et al. 1998; Ferguson et al. 2001).

A late phase of deformation at c. 1060 Ma in the Musgrave Block (Sun et al. 1996; White et al. 1999) has been recognized (but not dated) in the Albany–Fraser Belt (Clark et al. 2000), and is synchronous with 1060–1090 Ma deformation and magmatism in the Darling Mobile Belt and the Bangemall Basin (Bruguier et al., 1999; Wingate et al. 2002). Fitzsimons (2001) suggested that the Mesoproterozoic blocks in the Darling Mobile Belt (Northampton and Leeuwin Blocks; Fig. 3) were accreted to the western Australian margin some time after they were deformed and metamorphosed further to the south (present coordinates), but before 'stitching' of the Northampton Block (Fig. 3) to the western Australian margin at 755 Ma by the Mundine Well Dykes (Fig. 4a; Wingate & Giddings 2000).

Neoproterozoic events

Proterozoic Australia had essentially stabilized by 1.0 Ga, and sedimentation in the continent-wide Centralian Superbasin commenced at 850–830 Ma (Walter et al. 1995). The Gairdner Dykes (Fig. 4a) reflect NE–SW extension during initial rifting in eastern and central Australia at 825 Ma (Wingate et al. 1998). The Mundine Well Dykes (Fig. 4a) are parallel to the continental margin and their emplacement may have preceded separation of an unknown continental fragment (possibly Kalahari) from the western Australian margin (Wingate & Giddings 2000; Powell & Pisarevsky 2002). Although Australia remained essentially intact during breakup events along its eastern margin at some time after c. 780 Ma (Powell et al. 1994; Preiss 2000), several intracratonic events occurred in the latest Neoproterozoic (Fig. 4b). The c. 550 Ma Petermann Ranges Orogeny involved reactivation of the Miles–Musgrave Orogen, with both north- and south-directed thrusting and considerable exhumation in the Musgrave Block (Maboko et al. 1992; Preiss & Krieg 1992; Camacho & Fanning 1995; Scrimgeour & Close 1999), as well as SW-directed thrusting at the eastern edge of the Pilbara Craton and in the King Leopold Orogen (Tyler et al. 1998). It has been suggested that an anticlockwise rotation of the NAC with respect to the rest of the continent occurred at this time (Powell et al. 1994).

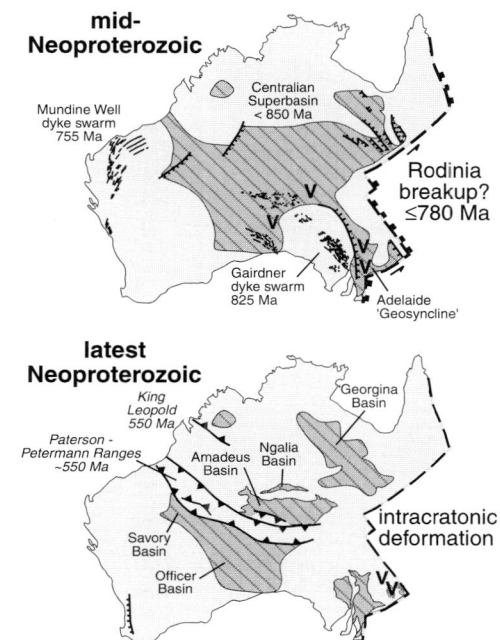

Fig. 4. (a) Sedimentation in the Centralian Superbasin started at c. 850 Ma (Walter et al. 1995). Rift-related magmatism preceding breakup along the eastern Australian margin commenced at c. 830 Ma, with emplacement of the Gairdner Dyke Swarm and associated volcanic rocks (Wingate et al. 1998). (b) Mainly intracratonic deformation occurred along latest Neoproterozoic belts, exhuming the Musgrave and Arunta Blocks, and disrupting the Centralian Superbasin.

Palaeomagnetism

Palaeomagnetism is the only method for quantitatively determining the relative positions of continental fragments during Precambrian time. Palaeomagnetic directions yield information about the latitude and orientation of a sampling locality relative to the palaeomagnetic pole at the time the magnetization was acquired. The positions of palaeopoles relative to Australia between 1.8 and 0.7 Ga (Fig. 5) show that the Australian continent underwent many changes in latitude and orientation relative to the pole during the Proterozoic.

The APWP is constructed by joining palaeopoles, backwards through time, by the shortest possible segments. Although this is the simplest approach, the result is not a unique solution. Lack of a continuous record of field reversals back to the Proterozoic leads to ambiguity in choosing a pole versus its antipole and, because of large age gaps between adjacent poles, it is inevitable that some important features of the path are overlooked. Another

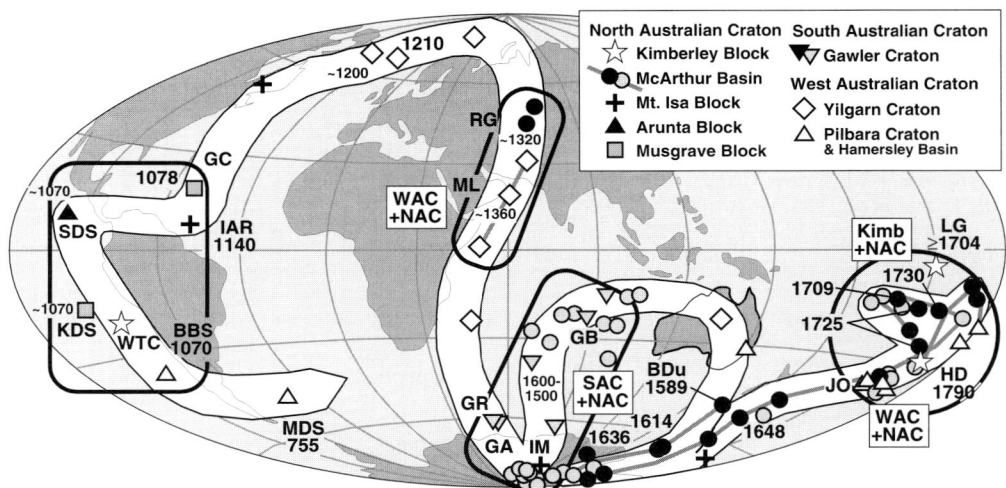

Fig. 5. Apparent polar wander path (APWP) for Australia between 1.8 and 0.7 Ga. Grey symbols for the McArthur Basin and Gawler Craton indicate overprint palaeopoles. Times at which poles from different cratonic blocks overlap are circled. Ages are in Ma (more reliable results are in larger font). Updated from Idnurm & Giddings (1988) and Idnurm *et al.* (1995), with data from Idnurm (2000), Li (2000*a, b*), Pisarevsky & Harris (2001), Wingate & Giddings (2000) and Wingate *et al.* (2002). NAC, WAC, SAC; North, West and South Australian cratonic assemblages, respectively; Kimb, Kimberley Block. For Palaeopole abbreviations refer to poles discussed in the text.

critically important uncertainty is the lack of well-dated palaeopoles. Many are dated imprecisely and/or their magnetizations cannot be demonstrated to be primary (see examples below). Nevertheless, all palaeopoles from a single crustal block will help to define the APWP for that block (provided the rocks have not been rotated by later deformation events); those reflecting secondary overprints will lie on a younger segment than primary poles for the same rocks. Most intracratonic deformation, including formation or destruction of small (≤1000 km) ocean basins, will not be detectable within typical uncertainties (A_{95}) of 5–15° in palaeopole determination. With these constraints in mind, the data can be explored in three ways.

Firstly, matching of APWP from two or more tectonic blocks over an interval of time enables, in principle, a unique reconstruction of their relative positions. Although many studies have been conducted in the last decade, the data are still insufficient to construct adequate APWP for the different blocks. The only well-defined path is that for the 1.73–1.59 Ga McArthur Basin (Fig. 5; Idnurm *et al.* 1995; Idnurm 2000). As a second example, preliminary poles reported by Idnurm & Giddings (1988) for the Morawa Lavas and enclosing sedimentary rocks appear to yield a consistently directed APWP vector for the Yilgarn Craton at *c*. 1.36 Ga (ML on Fig. 5).

Secondly, if the constituent terranes of Australia assembled during the Proterozoic via large horizontal motions, such as those that characterize Phanerozoic plate tectonic regimes, then it is unlikely that the palaeopoles from all the different blocks would fall on a single APWP. In particular, if large oceans closed between the NAC, WAC and SAC at 1.3 Ga (or earlier), the APWP should be dissimilar prior to that time, and should converge to produce a common path for times after the blocks were joined. As noted in previous palaeomagnetic analyses (e.g. Idnurm & Giddings 1988; Plumb 1993), it is possible to construct a single APWP for all Proterozoic poles, even though they are derived from different tectonic blocks (Fig. 5). All recent results also plot on the APWP defined by Idnurm & Giddings (1988). This is consistent with the different regions of Australia having evolved in essentially their present relative positions since at least 1.8 Ga, although, owing to the inadequacies in the APWP described above, significant relative movements between the crustal blocks cannot be ruled out (Plumb 1993).

Thirdly, if two or more blocks are in their correct relative positions for a particular time, the palaeopoles of that age from the different blocks should overlap. Note that when matching individual (or, in this case, average) pole positions, rather than APWP, longitude is unconstrained, hence E–W separation between blocks cannot be discerned. There are four segments of the APWP where data from different blocks overlap (Fig. 5).

Segment 1: 1.8–1.7 Ga

Palaeopoles from the McArthur Basin at c. 1.73–1.7 Ga (Idnurm et al. 1995; Idnurm 2000) are similar to those (LG and HD on Fig. 5) for the 1.79–1.7 Ga Elgee Formatton (McNaughton et al. 1999; Li 2000a) and the 1.79 Ga Hart Dolerite (McElhinny & Evans 1976) of the Kimberley Block, consistent with geological evidence for accretion of the Kimberley Block to the NAC by 1.82 Ga (Bodorkos et al. 1999; Sheppard et al. 1999a). Structural and palaeomagnetic studies indicate that a Pilbara syn-folding overprint pole (JO on Fig. 5; Schmidt & Embleton 1985) was acquired during the 1.83–1.79 Ga Capricorn Orogeny (Li 2000a). Proximity of this pole, and similar poles from Pilbara iron-ore deposits (Porath & Chamalaun 1968; Li et al. 1993; Schmidt & Clark 1994), to the McArthur Basin and Kimberley Poles implies that the NAC and WAC were in their present relative positions since at least 1.7 Ga (Li 2000a).

Segment 2: 1.6–1.5 Ga

Numerous 1.59–1.5 Ga overprint poles from the McArthur Basin (Idnurm et al. 1995; Idnurm 2000) and a 1.55–1.5 Ga post-metamorphic cooling pole (IM on Fig. 5) from Mount Isa (Tanaka & Idnurm 1994) overlap with poorly dated overprint poles from South Australian iron-ore deposits (Chamalaun & Porath 1968) and mafic dykes (GA and GB on Fig. 5; Giddings & Embleton 1976). Collectively, the data suggest that the NAC and SAC were joined by at least 1.5 Ga, although current age constraints are poor. Although the Tournefort Dykes (Parker et al. 1987), from which the GA and GB palaeopoles (Fig. 5) were obtained, intruded the 1.85 Ga Lincoln Batholith at c. 1.81 Ga (Schaefer 1998), they were metamorphosed during the Kimban Orogeny at c. 1.72 Ga (Bendall 1994) and have yielded Rb–Sr ages of 1.6–1.55 Ga (Mortimer et al. 1988b). Together with the lack of field tests to verify the stability of the GA and GB dyke magnetizations, these observations suggest that the magnetizations could be overprints, possibly with ages approximated by the Rb–Sr results.

Poles for the 1592 ± 2 Ma Gawler Range Volcanics (GR on Fig. 5; Chamalaun & Dempsey 1978) and the 1589 ± 3 Ma upper Balbirini Dolomite of the McArthur Basin (BDu on Fig. 5; Idnurm 2000) are c. 60° apart (Fig. 5), although the rocks are identical in age (Fanning et al. 1988; Page et al. 2000). If both poles are primary, this is evidence that the NAC and SAC were not in their present relative positions at 1.59 Ga. It is possible, however, that either pole might be significantly younger than the rocks themselves. It was argued by Schmidt & Clark (1992), for example, that the GR magnetization is a younger overprint, because palaeomagnetic directions they obtained from steeply dipping lower parts of the unit are similar to those obtained previously from the flat-lying upper flows; the combined data set would therefore suggest a negative fold test. Both successions, however, are essentially the same age and the upper flows were erupted at 950–1000°C (Creaser & White 1991), hence it is possible that the entire unit did not cool through its magnetic blocking temperatures until after the younger eruptions, or that the lower flows were reheated and overprinted by overlying rocks. In both cases the magnetization should have been acquired during cooling of the upper flows shortly after 1592 Ma. We thus favour a primary age for the GR pole, although the Gawler Range Volcanics should be investigated further by conducting additional field and laboratory stability tests on samples from the upper succession. Combined data from the upper and lower Balbirini Dolomite indicate a pre-folding age for the magnetization (Idnurm et al. 1995; Idnurm 2000), and we tentatively regard the BDu pole as primary.

Segment 3: 1.36–1.32 Ga

Preliminary results for the Morawa Lavas (ML on Fig. 5), and underlying and overlying sedimentary rocks, yield an APWP vector for the Yilgarn Craton at c. 1.36 Ga (Idnurm & Giddings 1988). Intrusion of dolerite at c. 1320 Ma (preliminary SHRIMP U–Pb baddeleyite age; Claoué-Long, pers. comm.) was likely responsible for overprints in the Roper Group (RG on Fig. 5) sedimentary rocks (Plumb 1993; Idnurm et al. 1995). These results are consistent with a connection between the NAC and WAC prior to 1.3 Ga.

Segment 4: 1.07 Ga

Late Mesoproterozoic palaeopoles (Fig. 5) have been obtained from the Giles Complex (GC on Fig. 5) the Stuart Dykes (SDS on Fig. 5), and the Kulgera Sills (KDS on Fig. 5) in central Australia (Facer 1971; Idnurm & Giddings 1988; Camacho et al. 1991). Although the Giles Complex is well dated at 1078 Ma (Glikson et al. 1996), the palaeopole [GC on Fig. 5; Facer 1971; recalculated by Tanaka & Idnurm (1994)] was not obtained using modern techniques and is of low reliability. The Stuart and Kulgera Intrusions yielded Sm–Nd isochron ages of 1076 ± 33 and 1090 ± 32 Ma, respectively (Zhao & McCulloch 1993). Reliability of the preliminary SDS palaeopole (Idnurm & Giddings 1988) is difficult to assess because no analytical details have been published. Reliable constraints on palaeohorizontal are not available for the Giles, Kulgera or Stuart Intrusions (no tectonic corrections were applied), and all three suites are located in crustal blocks that were deformed and probably re-oriented during the latest Neoproterozoic (Petermann Ranges) and/or Carboniferous (Alice Springs) tectonothermal events (Tanaka & Idnurm 1994; Wingate et al. 2002). The Giles Complex may also have been deformed during the later stages of the Musgrave Orogeny (White et al. 1999). A new palaeopole (BBS on Fig. 5), from dolerite sills in the Bangemall Basin, does not agree with the previous c. 1070 Ma results, but is significantly more reliable (Wingate et al. 2002). The BBS pole is inferred to be primary, is dated precisely at 1070 ± 6 Ma and structural control is well defined in adjacent sedimentary rocks. Discrepancies between the BBS pole and c. 1070 Ma poles from central Australia could be due to inadequate tectonic corrections for the central Australian poles, unrecognized overprints (GC and SDS on Fig. 5), and/or differences in age.

Palaeomagnetic data for the latest Mesoproterozoic are thus inadequate to demonstrate that Australia had amalgamated by this time, although this interpretation is supported by the crude grouping of the c. 1070 Ma poles (Fig. 5). A palaeopole (WTC on Fig. 5) for the dolomite cap of the Walsh Tillite of the Kimberley Block was proposed to indicate a Sturtian (c. 750–700 Ma) age for that unit (Li 2000b), despite other evidence indicating a correlation with the Marinoan (c. 600 Ma) glacial interval (Grey & Corkeron 1998). Compelling arguments can be made toward either interpretation and a provocative alternative is proposed here: could similarity between the WTC pole and other Australian poles for 1070 Ma indicate a Late Mesoproterozoic age for some of the glaciogenic rocks in the Kimberley Block?

Discussion

The timing of the Capricorn, Late Barramundi and Halls Creek tectonothermal events was broadly similar at c. 1.8 Ga (Fig. 2). Other similarities across the NAC include widely developed low-P/high-T metamorphic conditions and extensive magmatism with intraplate geochemical signatures (e.g. Etheridge et al. 1987; Wyborn 1988; Mortimer et al. 1988a). Basin evolution across the NAC between c. 1.8 and 1.6 Ga (Scott et al. 2000) was contemporaneous with deformation in central Australia and the Gawler Craton. The 1.74–1.7 Ga Kimban Orogeny in the Gawler Craton was contemporaneous with the Late Strangways orogenic event in the Arunta Inlier, although rocks older than c. 1.6 Ga have not been observed (or preserved) in the intervening Musgrave Block (White et al. 1999).

Several geological observations suggest amalgamation of the North, West and South Australian cratonic assemblages prior to 1.3 Ga. Orogenic events at 1.3 and 1.2 Ga in the Albany–Fraser and Musgrave Belts suggest a combined West and North Australian block by 1.3 Ga, although there was continued activity between them, in the Miles Belt (Fig. 3), between 1.3 and 1.1 Ga. Striking similarities in the ages (c. 1.8–1.55 Ga) of basin formation, magmatism, mineralization, deformation and styles of alteration among the eastern Gawler Craton, Curnamona Province and the Georgetown and eastern Mount Isa Inliers have prompted several researchers to suggest that these blocks formed part of a regionally extensive Late Palaeoproterozoic–Early Mesoproterozoic mobile belt, referred to as the Diamantina Orogen (Page & Laing 1992; Connors & Page 1995; Laing 1996; Black et al. 1998; Robertson et al. 1998). This hypothesis implies that the NAC and at least the Curnamona Province of the SAC were joined since at least 1.7 Ga. Note, however, that a Late Mesoproterozoic collisional orogen (Fig. 3) extending between the Musgrave and Cape River Blocks, as suggested by Blewett et al. (1998), would preclude the existence of a Late Palaeoproterozoic orogen linking Mount Isa and the Curnamona Province with these blocks in their present relative positions.

Based on a detailed tectonostratigraphic comparison, Giles & Betts (2000) proposed that an anticlockwise rotation of 55° and a slight eastward translation of the SAC relative to the NAC would better align what may have been continuous and linear Palaeoproterozoic–Mesoproterozoic tectonic elements. In their reconstruction, for example, mobile belts of the Arunta and Gawler–Curnamona Blocks would have formed a continuous accretionary margin between 1.88 and 1.67 Ga. Giles & Betts (2000) invoked a cycle of extension and breakup between the SAC and NAC at c. 1.45 Ga, with subsequent reamalgamation between 1.3 to 1.1 Ga to yield the present configuration of the cratons. It is intriguing that this reconstruction achieves independent support from palaeomagnetic data to first order. Accepting that both the GR and BDu poles (Fig. 5) represent their respective cratons at c. 1.59 Ga (see discussion above), rotation of the SAC craton and the GR pole according to this model (Fig. 6) better aligns the GR and BDu poles along the aggregate APWP from the McArthur Basin (Idnurm 2000). Alignment is still not exact, however, suggesting that the GR and/or BDu poles may not be primary, or that modification of the Giles & Betts (2000) model may be required.

Although it is likely that collisional orogeny did occur by interactions between the Australian crustal blocks and also that intraplate processes can occur within a plate tectonic context, there is no compel-

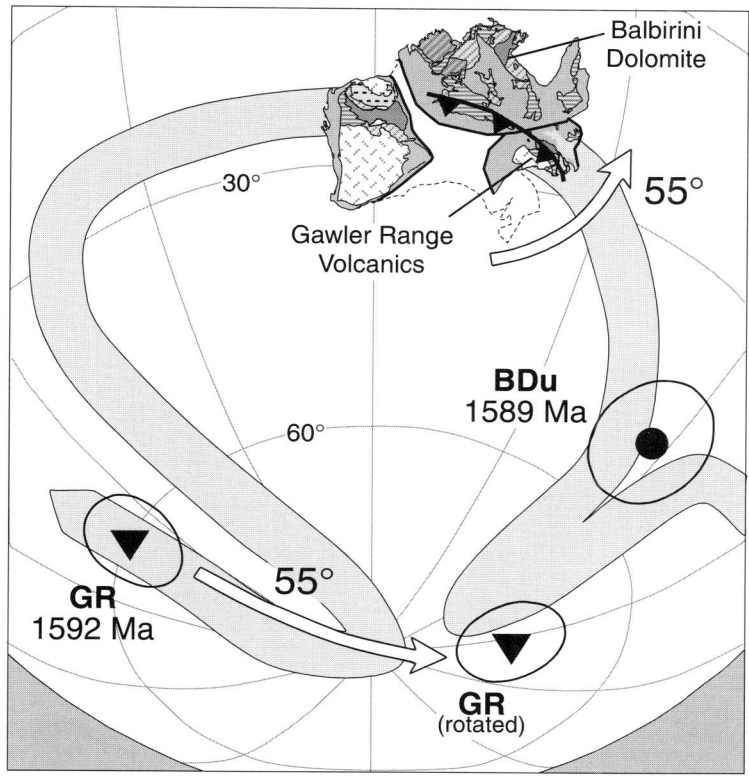

Fig. 6. Restoration of the Gawler Range Volcanics pole (GR) according to the tectonic model of Giles & Betts (2000). Thus restored, the GR pole falls on the apparent polar wander path from the McArthur Basin (Idnurm 2000), but fails to overlap precisely with the pole from the coeval upper Balbirini Dolomite (BDu; Idnurm 2000). See text for discussion.

ling evidence for large ocean basins having closed between any of the Australian blocks since at least 1.8 Ga. Geological and geochemical observations alone are insufficient to test whether the major cratonic blocks of Australia were together or were widely separated during the Proterozoic – conclusive evidence for large horizontal motions of crustal fragments in Australia during the Proterozoic must come from palaeomagnetism. Although the palaeomagnetic record is far from complete, groups of overlapping palaeopoles permit the NAC, WAC and SAC to have occupied their present relative positions since at least 1.5 Ga, and possibly the NAC and WAC to have been assembled prior to c. 1.7 Ga. These conclusions are consistent with the idea that the constituent blocks of Australia were not widely separated since at least 1.8 Ga, which is supported by several pre-1.3 Ga geological correlations, as discussed above.

Despite the recent addition of many high-quality results, the palaeomagnetic database is inadequate to rule out large horizontal motions between the Australian crustal blocks since 1.8 Ga. Ideally, APWP need to be defined for each block. Current studies, however, should focus on obtaining well-dated palaeopoles of precisely the same age from different blocks at several key points in time, from which minimum (i.e. latitudinal) separations can be determined. Reliable palaeomagnetic information will enhance, and provide key tests of, the applicability of plate tectonic interpretations in understanding Australian Proterozoic geology. If models involving large horizontal motions of the Australian fragments during the Proterozoic are correct, support is likely to be found eventually from palaeomagnetism. Until then, however, the single-continent model – perhaps incorporating minor jostlings among its constituent cratons (e.g. Giles & Betts 2000) – should not be discarded.

This research was supported by the Australian Research Council through its Research Centres Program and by an Australian Postdoctoral Fellowship to DADE. The manuscript was improved considerably by constructive comments from Peter Betts, Ian Fitzsimons, Wolfgang Preiss and Trond Torsvik. This is publication number 191 of the

Tectonics Special Research Centre, and a contribution to International Geological Correlation Program (IGCP) Project 440.

References

BAKER, J., POWELL, R., SANDIFORD, M. & MUHLING, J. R. 1987. Corona textures between kyanite, garnet, and gedrite in gneisses from the Errabiddy area, Western Australia. *Journal of Metamorphic Geology*, **5**, 357–370.

BENDALL, B. R. 1994. *Metamorphic and geochronological constraints on the Kimban orogeny, southern Eyre peninsula*. BSc (Honours) Thesis, University of Adelaide.

BLACK, L. P. & WITHNALL, I. W. 1993. The ages of Proterozoic granites in the Georgetown Inlier of northeastern Australia, and their relevance to the dating of tectonothermal events. *AGSO Journal of Australian Geology and Geophysics*, **14**, 331–341.

BLACK, L. P., GREGORY, P., WITHNALL, I. W. & BAIN, J. H. C. 1998. U–Pb zircon age for the Etheridge Group, Georgetown region, north Queensland: implications for relationship with the Broken Hill and Mt Isa sequences. *Australian Journal of Earth Sciences*, **45**, 925–935.

BLEWETT, R. S. & BLACK, L. P. 1998. Structural and temporal framework of the Coen Region, north Queensland: implications for major tectonothermal events in east and north Australia. *Australian Journal of Earth Sciences*, **45**, 597–609.

BLEWETT, R. S., BLACK, L. P., SUN, S-s., KNUTSON, J., HUTTON, L. J. & BAIN, J. H.C. 1998. U–Pb zircon and Sm–Nd geochronology of the Mesoproterozoic of North Queensland: implications for a Rodinian connection with the Belt Supergroup of North America. *Precambrian Research*, **89**, 101–127.

BODORKOS, S., OLIVER, N. H. S. & CAWOOD, P. A. 1999. Thermal evolution of the central Halls Creek Orogen, northern Australia. *Australian Journal of Earth Sciences*, **46**, 453–465.

BRUGUIER, O, BOSCH, D., PIDGEON, R. T., BYRNE, D. I. & HARRIS, L. B. 1999. U–Pb geochronology of the Northampton Complex, Western Australia – evidence for Grenvillian sedimentation, metamorphism and deformation and geodynamic implications. *Contributions to Mineralogy and Petrology*, **136**, 258–272.

BURKE, K., DEWEY, J. F. & KIDD, W. S. F. 1976. Precambrian palaeomagnetic results compatible with contemporary operation of the Wilson cycle. *Tectonophysics*, **33**, 287–299.

CAMACHO, A. & FANNING, C. M. 1995. Some isotopic constraints on the granulite and upper amphibolite facies terranes in the eastern Musgrave Block, central Australia. *Precambrian Research*, **71**, 155–181.

CAMACHO, A., SIMONS, B. & SCHMIDT, P. W. 1991. Geological and palaeomagnetic significance of the Kulgera Dyke Swarm, N.T., Australia. *Geophysical Journal International*, **107**, 37–45.

CHAMALAUN, F. H. & DEMPSEY, C. E. 1978. Palaeomagnetism of the Gawler range. Volcanics and implications for the genesis of the Middleback hematite orebodies. *Journal of the Geological Society of Australia*, **25**, 255–265.

CHAMALAUN, F. H. & PORATH, H. 1968. Palaeomagnetism of Australian haematite ore bodies – I. The Middleback Ranges of South Australia. *Geophysical Journal of the Royal Astronomical Society*, **14**, 451–462.

CHIN, R. J. & DE LAETER, J. R. 1981. The relationship of new Rb–Sr isotopic dates from the Rudall Metamorphic Complex to the geology of the Paterson Province. *Western Australia Geological Survey Annual Report*, **1980**, 80–87.

CLARK, D. J., HENSEN, B. J. & KINNY, P. D. 2000. Geochronological constraints for a two-stage history of the Albany–Fraser orogen, Western Australia. *Precambrian Research*, **102**, 155–183.

CLARKE, G. L., SUN, S.-S. & WHITE, R. W. 1995. Grenville-age belts and adjacent older terrains in Australia and Antarctica. *AGSO Journal of Australian Geology and Geophysics*, **16**, 35–39.

COLLERSON, K. D. & KAMBER, B. S. 1999. Evolution of the continents and atmosphere inferred from Th–U–Nb systematics of the depleted mantle. *Science*, **283**, 1519–1522.

COLLINS, W. J. & SHAW, R. D. 1995. Geochronological constraints on orogenic events in the Arunta Inlier: a review. *Precambrian Research*, **71**, 315–346.

CONDIE, K. C & MYERS, J. S. 1999. Mesoproterozoic Fraser Complex: geochemical evidence for multiple subduction-related sources of lower crustal rocks in the Albany–Fraser Orogen, Western Australia. *Australian Journal of Earth Sciences*, **46**, 875–882.

CONNERS, K. A. & PAGE, R. W. 1995. Relationships between magmatism, metamorphism, and deformation in the western Mount Isa Inlier, Australia. *Precambrian Research*, **71**, 131–153.

CREASER, R. A & WHITE, A. J. R. 1991. Yardea Dacite – large-volume, high-temperature felsic magmatism from the Middle Proterozoic of South Australia. *Geology*, **19**, 48–51.

DALY, S. J., FANNING, C. M. & FAIRCLOUGH, M. C. 1998. Tectonic evolution and mineral potential of the Gawler Craton, South Australia. *AGSO Journal of Australian Geology and Geophysics*, **17**, 145–168.

DRUMMOND, B. J. 1988. A review of crust/upper mantle structure in the Precambrian areas of Australia and implications for Precambrian crustal evolution. *Precambrian Research*, **40/41**, 101–116.

ETHERIDGE, M. A., RUTLAND, R. W. R. & WYBORN, L. A. I. 1987. Orogenesis and tectonic processes in the early to middle Proterozoic of northern Australia. In: KRÖNER, A. (ed.) *Precambrian Lithospheric Evolution*. American Geophysical Union Geodynamic Series, **17**, 131–147.

EVANS, T. 1999. *Extent and nature of the 1.2 Ga Wheatbelt dyke swarm, Yilgarn Craton, Western Australia*. Honours Thesis, The University of Western Australia.

FACER, R. A. 1971. Magnetic properties of rocks from the Giles Complex, central Australia. *Journal of Proceedings of the Royal Society of New South Wales*, **104**, 45–61.

FANNING, C. M., FLINT, R. B., PARKER, A. J., LUDWIG, K. R. & BLISSET, A. H. 1988. Refined Proterozoic evolution of the Gawler Craton, South Australia, through U–Pb zircon geochronology. *Precambrian Research*, **40/41**, 363–386.

FERGUSON, C. L., CARR, P. F., FANNING, C. M. & GREEN, T. J. 2001. Proterozoic–Cambrian detrital zircon and monazite ages from the Anakie Inlier, central Queensland: Grenville and pacific – Gondwana signatures. *Australian Journal of Earth Sciences*, **48**, 857–866.

FITZSIMONS, I. C. W. 2001 The Neoproterozoic evolution of Australia's western margin. *Geological Society of Australia Abstracts*, **65**, 39–42.

FITZSIMONS, I. C. W. 2002. Proterozoic basement provinces of southern and southwestern Australia, and their correlation with Antarctica. *In*: YOSHIDA, M., WINDLEY, B., DASGUPTA, S. (eds) *Proterozoic East Gondwana: Supercontinent Assembly and Breakup*. Geological Society, London, Special Publication, **xx**, xx–xx.

FLETCHER I. R., MYERS J. S. & AHMAT, A. L. 1991. Isotopic evidence on the age and origin of the Fraser Complex, Western Australia: a sample of Mid-Proterozoic lower crust. *Chemical Geology*, **87**, 197–216.

FLINT, R. B. 1993. Mesoproterozoic. *In*: DREXEL, J. F., PREISS, W. V. & PARKER, A. J. (eds) *The Geology of South Australia, Volume 1, The Precambrian*. South Australian Geological Survey Bulletin, **54**, 107–170.

GIDDINGS, J. W. & EMBLETON, B. J. J. 1976. Precambrian palaeomagnetism in Australia II: basic dykes from the Gawler Block. *Tectonophysics*, **30**, 109–118.

GILES, D. & BETTS, P. 2000. The early to middle Proterozoic configuration of Australia and its implications for Australian–US relations. *Geological Society of Australia Abstracts*, **59**, 174.

GILES, D., BETTS, P. & LISTER, G. 2001. A continental back-arc setting for the Early to Middle Proterozoic basins of north-eastern Australia. *Geological Society of Australia Abstracts*, **64**, 55.

GLIKSON, A. Y., STEWART, A. J., BALLHAUS, C. G. ET AL. 1996. Geology of the western Musgrave Block, central Australia, with particular reference to the mafic–ultramafic Giles Complex. *Australian Geological Survey Organisation Bulletin*, **239**.

GREY, K. & CORKERON, M. 1998. Late Neoproterozoic stromatolites in glaciogenic successions of the Kimberley region, Western Australia: evidence for a younger Marinoan glaciation. *Precambrian Research*, **92**, 65–87.

HALL, C. E., POWELL, C. MCA. & BRYANT, J. 2001. Basin setting and age of the Late Palaeoproterozoic Capricorn Formation, Western Australia. *Australian Journal of Earth Sciences*, **48**, 731–744.

HAMILTON, W. B. 1998. Archaean magmatism and deformation were not products of plate tectonics. *Precambrian Research*, **91**, 143–179.

HARRINGTON, H. J., BURNS, K L. & THOMPSON, B. R. 1973. Gambier–Beaconsfield and Gambier–Sorell fracture zones and the movement of plates in the Australia–Antarctica–New Zealand region. *Nature Physical Science*, **245**, 109–112.

HELMSTAEDT, H. & SCOTT, D. J. 1992. The Proterozoic ophiolite problem. *In:* CONDIE, K. C. (ed.) *Proterozoic Crustal Evolution*. Elsevier, Amsterdam, 55–95.

HOEK, J. D. & SCHAEFER, B. F. 1998. Palaeoproterozoic Kimban mobile belt, Eyre Peninsula: timing and significance of felsic and mafic magmatism and deformation. *Australian Journal of Earth Sciences*, **45**, 305–313.

HOFFMAN, P. F. 1988. United Plates of America – early Proterozoic assembly and growth of Laurentia. *Annual Reviews of Earth and Planetary Science*, **16**, 543–603.

HUTTON, L., FANNING, C. M. & WITHNALL, I. W. 1998. The Cape River area – evidence for late Mesoproterozoic and Neoproterozoic to Cambrian crust in north Queensland. *Geological Society of Australia Abstracts*, **49**, 216.

IDNURM, M. 2000. Towards a high resolution Late Palaeoproterozoic–earliest Mesoproterozoic apparent polar wander path for northern Australia. *Australian Journal of Earth Sciences*, **47**, 405–429.

IDNURM, M. & GIDDINGS, J. W. 1988. Australian Precambrian polar wander: a review. *Precambrian Research*, **40/41**, 61–88.

IDNURM, M., GIDDINGS, J. W. & PLUMB, K. A. 1995. Apparent polar wander and reversal stratigraphy of the Palaeo–Mesoproterozoic southeastern McArthur Basin, Australia. *Precambrian Research*, **72**, 1–41.

KRAPEZ, B. & MARTIN, D. McB. 1999. Sequence stratigraphy of the Palaeoproterozoic Nabberu Province of Western Australia. *Australian Journal of Earth Sciences*, **46**, 89–104.

LAING, W. P. 1996. The Diamantina orogen linking the Willyama and Cloncurry terranes, eastern Australia. *In*: PONGRATZ, J. & DAVIDSON, G. J. (eds) *New Developments in Broken Hill Type Deposits*. Centre for Ore Deposit and Exploration Studies, University of Tasmania, CODES Special Publication, **1**, 67–72.

LI, Z.-X. 2000a. Palaeomagnetic evidence for unification of the North and West Australian Cratons by ca. 1.7 Ga: new results from the Kimberley Basin of northwestern Australia. *Geophysical Journal International*, **142**, 173–180.

LI, Z.-X. 2000b. New palaeomagnetic results from the 'cap dolomite' of the Neoproterozoic Walsh Tillite, northwestern Australia. *Precambrian Research*, **100**, 359–370.

LI, Z.-X., POWELL, C. McA. & BOWMAN, R. 1993. Timing and genesis of the Hamersley iron ore deposits. *Exploration Geophysics*, **24**, 631–636.

McELHINNY, M. W. & EMBLETON, B. J. J. 1976. Precambrian and Early Palaeozoic palaeomagnetism in Australia. *Philosophical Transactions of the Royal Society of London*, **280**, 417–431.

McELHINNY, M. W. & EVANS, M. E. 1976. Palaeomagnetic results from the Hart Dolerite of the Kimberley Block, Australia. *Precambrian Research*, **3**, 231–241.

McNAUGHTON, N. J., RASMUSSEN, B. & FLETCHER, I. R. 1999. SHRIMP uranium–lead dating of diagenetic xenotime in siliciclastic sedimentary rocks. *Science*, **285**, 78–80.

MABOKO, M. A. H., McDOUGALL, I., ZEITLER, P. K. & WILLIAMS, I. S. 1992. Geochronological evidence for ~530–550 Ma juxtaposition of two Proterozoic metamorphic terranes in the Musgrave ranges, central Australia. *Australian Journal of Earth Sciences*, **39**, 457–471.

MOORES, E. M. 1986. The Proterozoic ophiolite problem, continental emergence, and the Venus connection. *Science*, **234**, 65–68.

MORTIMER, G. E., COOPER, J. A. & OLIVER, R. L. 1988a. The geochemical evolution of Proterozoic granitoids near Port Lincoln in the Gawler orogenic domain of South Australia. *Precambrian Research*, **40/41**, 387–406.

MORTIMER, G. E., COOPER, J. A. & OLIVER, R. L. 1988b. Proterozoic mafic dykes near Port Lincoln, South Australia: composition, age and origin. *Australian Journal of Earth Sciences*, **35**, 93–110.

MYERS, J. S. 1985. The Fraser Complex – a major layered intrusion in Western Australia. *Western Australia Geological Survey Report*, **14**, 57–66.

MYERS, J. S. 1989. Thrust sheets on the southern foreland of the Capricorn Orogen, Robinson Range, Western Australia. *Western Australia Geological Survey Report*, **26**, 127–130.

MYERS, J. S. 1990. Precambrian tectonic evolution of part of Gondwana, southwestern Australia. *Geology*, **18**, 537–540.

MYERS, J. S., SHAW, R. D. & TYLER, I. M. 1996. Tectonic evolution of Proterozoic Australia. *Tectonics*, **15**, 1431–1446.

NELSON, D. R. 1995. Compilation of SHRIMP U–Pb zircon data, 1994. *Western Australia Geological Survey Record*, **1995/3**.

NELSON, D. R. 1996. Compilation of SHRIMP U–Pb zircon data, 1995. *Western Australia Geological Survey Record*, **1996/5**.

NELSON, D. R. 1998. Compilation of SHRIMP U–Pb zircon geochronological data, 1997. *Western Australia Geological Survey Rec*ord, **1998/2**.

NELSON, D. R., MYERS, J. S. & NUTMAN, A. P. 1995. Chronology and evolution of the Middle Proterozoic Albany–Fraser Orogen, Western Australia. *Australian Journal of Earth Sciences*, **42**, 481–495.

OLIVER, N. H. S., RUBENACH, M. J. & VALENTA, R. K. 1998. Precambrian metamorphism, fluid flow, and metallogeny of Australia. *AGSO Journal of Australian Geology and Geophysics*, **17**, 31–53.

PAGE, R. W. 1988. Geochronology of Early to Middle Proterozoic fold belts in northern Australia: a review. *Precambrian Research*, **40/41**, 1–19.

PAGE, R. W. & HANCOCK, S. L. 1988. Geochronology of a rapid 1.85–1.86 Ga tectonic transition – Halls Creek Orogen, northern Australia. *Precambrian Research*, **40/41**, 447–467.

PAGE, R. W. & LAING, W. P. 1992. Felsic metavolcanic rocks related to the Broken Hill Pb–Zn–Ag orebody, Australia: depositional age, and timing of high-grade metamorphism. *Economic Geology*, **87**, 2138–2168.

PAGE, R. W. & SUN, S.-S. 2000. Aspects of geochronology and crustal evolution in the Eastern Fold Belt, Mt Isa Inlier. *Australian Journal of Earth Sciences*, **45**, 343–361.

PAGE, R. W., JACKSON, M. J. & KRASSAY, A. A. 2000. Constraining sequence stratigraphy in north Australian basins: SHRIMP U–Pb zircon geochronology between Mt Isa and McArthur River. *Australian Journal of Earth Sciences*, **47**, 431–459.

PARKER, A. J., BAILLIE, P. W., MCCLENAGHAN, M.P. ET AL. 1987. Mafic dyke swarms of Australia. *In*: HALLS, H. C. & FAHRIG, W. F. (eds) *Mafic Dyke Swarms*. Geological Association of Canada, Special Paper, **34**, 401–417.

PISAREVSKY, S. A. & HARRIS, L. B. 2001. Determination of magnetic anisotropy and a ca. 1.2 Ga palaeomagnetic pole from the Bremer Bay area, Albany Mobile belt, Western Australia. *Australian Journal of Earth Sciences*, **48**, 101–112.

PLUMB, K. A., AHMAD, M. & WYGRALAK, A. S. 1990. Mid-Proterozoic basins of the North Australian Craton – regional geology and mineralisation. *In*: HUGHES, F. E. (ed.) *Geology of the Mineral Deposits of Australia and Papua New Guinea*. Australasian Institute of Mining and Metallurgy, Melbourne, 881–902.

PLUMB, K. A. 1993. Proterozoic geology of Australia and palaeomagnetism. *Exploration Geophysics*, **24**, 213–218.

PORATH, H. & CHAMALAUN, F. H. 1968. Palaeomagnetism of Australian haematite ore bodies – II. Western Australia. *Geophysical Journal of the Royal Astronomical Society*, **15**, 253–264.

POWELL, C. MCA. & PISAREVSKY, S. A. 2002. Late Neoproterozoic assembly of East Gondwanaland. *Geology*, **30**, 3–6.

POWELL, C. MCA., PREISS, W. V., GATEHOUSE, C. G., KRAPEZ, B. & LI, Z. X. 1994. South Australian record of a Rodinian epicontinental basin and its mid-Neoproterozoic breakup (~700 Ma) to form the Palaeo-Pacific Ocean. *Tectonophysics*, **237**, 113–140.

PREISS, W. V. 2000. The Adelaide Geosyncline of South Australia and its significance in Neoproterozoic continental reconstruction. *Precambrian Research*, **100**, 21–63.

PREISS, W. V. & KRIEG, G. W. 1992. Stratigraphic drilling in the northeast Officer Basin: Rodda 2 Well. *Mines and Energy Review, South Australia*, **158**, 48–51.

ROBERTSON, R. S., PREISS, W. V., CROOKS, A. F., HILL, P. W. & SHEARD, M. J. 1998. Review of the Proterozoic geology and mineral potential of the Curnamona Province in South Australia. *AGSO Journal of Australian Geology and Geophysics*, **17**, 169–182.

RUTLAND, R. W. R. 1973. Tectonic evolution of the continental crust of Australia. *In*: TARLING, D. H. & RUNCORN, S. K. (eds) *Continental Drift, Seafloor Spreading and Plate Tectonics: Implications for the Earth Sciences*. Academic Press, San Diego; 1003–1025.

SCHAEFER, B. 1998. *Insights into Proterozoic tectonics from the southern Eyre Peninsula, South Australia*. PhD Thesis, The University of Adelaide.

SCHMIDT, P. W. & CLARK, D. A. 1992. Magnetic properties of Archaean and Proterozoic rocks from the Eyre Peninsula. *CSIRO Report*, **275R**.

SCHMIDT, P. W. & CLARK, D. A. 1994. Palaeomagnetism and magnetic anisotropy of Proterozoic banded-iron formations and iron ores of the Hamersley basin, Western Australia. *Precambrian Research*, **69**, 133–155.

SCHMIDT, P. W. & EMBLETON, B. J. J. 1985. Prefolding and overprint magnetic signatures in Precambrian (!2.9–2.7 Ga) igneous rocks from the Pilbara Craton and Hamersley Basin, Western Australia. *Journal of Geophysical Research*, **90**, 2967–2984.

SCOTT, D. L., RAWLINGS, D. J., PAGE, R. W., TARLOWSKI, C. Z., IDNURM, M., JACKSON, M. J. & SOUTHGATE, P. N. 2000. Basement framework and geodynamic evolution of the palaeoproterozoic superbasins of north-central Australia: an integrated review of geochemical, geochronological, and geophysical data. *Australian Journal of Earth Sciences*, **47**, 341–380.

SCRIMGEOUR, I. & CLOSE, D. 1999. Regional high-pressure metamorphism during intracratonic deformation: the Petermann orogeny, central Australia. *Journal of Metamorphic Geology*, **17**, 557–572.

SHEPPARD, S. & OCCHIPINTI, S. A. 2000. *Errabiddy and Landor 1:100000 sheets*. Western Australia Geological Survey Explanatory Notes.

SHEPPARD, S., GRIFFIN, T. J., TYLER, I. M. & TAYLOR, W. R. 1999a. Palaeoproterozoic subduction-related and passive margin basalts in the Halls Creek Orogen, northwest Australia. *Australian Journal of Earth Sciences*, **46**, 679–690.

SHEPPARD, S., OCCHIPINTI, S.A., NELSON, D. R. & TYLER, I. M. 1999b. Granites of the southern Capricorn Orogen, Western Australia. *Geological Society of Australia Abstracts*, **56**, 44–46.

SHERATON, J. W. & SUN, S.-S. 1995. Geochemistry and origin of felsic igneous rocks of the western Musgrave Block. *AGSO Journal of Australian Geology and Geophysics*, **16**, 107–126.

SUN, S.-S., SHERATON, J. W., GLIKSON, A. Y. & STEWART, A. J. 1996. A major magmatic event during 1050–1080 Ma in central Australia and an emplacement age for the Giles Complex. *AGSO Research Newsletter*, **17**, 9–10.

TANAKA, H. & IDNURM, M. 1994. Palaeomagnetism of Proterozoic mafic intrusions and host rocks of the Mt. Isa Inlier, Australia. *Precambrian Research*, **69**, 241–258.

TYLER, I. M. 1999. Palaeoproterozoic orogeny in Western Australia. *Geological Society of Australia Abstracts*, **56**, 47–49.

TYLER, I. M. & THORNE, A. M. 1990. The northern margin of the Capricorn Orogen, Western Australia: an example of an early Proterozoic collision zone. *Journal of Structural Geology*, **12**, 685–701.

TYLER, I. M., PIRAJNO, F., BAGAS, L., MYERS, J. S. & PRESTON, W. A. 1998. The geology and mineral deposits of the Proterozoic in Western Australia. *AGSO Journal of Australian Geology and Geophysics*, **17**, 223–244.

VASSALLO, J. J. & WILSON, C. J. L. 2002. Palaeoproterozoic regional-scale non-coaxial deformation: an example from eastern Eyre Peninsula, South Australia. *Journal of Structural Geology*, **24**, 1–24.

VEEVERS, J. J. & MCELHINNY, M. W. 1976. The separation of Australia from other continents. *Earth Science Reviews*, **12**, 139–159.

WALTER, M. R., VEEVERS, J. J., CALVER, C. R. & GREY, K. 1995. Neoproterozoic stratigraphy of the Centralian Superbasin, Australia. *Precambrian Research*, **73**, 173–195.

WHITE, R. W., CLARKE, G. L. & NELSON, D. R. 1999. SHRIMP U–Pb dating of Grenville–age events in the western part of the Musgrave Block, central Australia. *Journal of Metamorphic Geology*, **17**, 465–481.

WINDLEY, B. F. 1995. *The Evolving Continents* (3rd edition). John Wiley & Sons, Chichester.

WINGATE, M. T. D. & GIDDINGS, J. W. 2000. Age and palaeomagnetism of the Mundine Well dyke swarm, Western Australia: implications for an Australia–Laurentia connection at 755 Ma. *Precambrian Research*, **100**, 335–357.

WINGATE, M. T. D., CAMPBELL, I. H., COMPSTON, W. & GIBSON, G. M. 1998. Ion microprobe U–Pb ages for Neoproterozoic basaltic magmatism in south-central Australia and implications for the breakup of Rodinia. *Precambrian Research*, **87**, 135–159.

WINGATE, M. T. D., CAMPBELL, I. H. & HARRIS, L. B. 2000. SHRIMP baddeleyite age for the Fraser Dyke Swarm, southeast Yilgarn Craton, Western Australia. *Australian Journal of Earth Sciences*, **47**, 309–313.

WINGATE, M. T. D., PISAREVSKY, S. A. & EVANS, D. A. D. 2002. A revised Rodinia supercontinent: no SWEAT, no AUSWUS. *Terra Nova*, **14**, 121–128.

WYBORN, L. A. I. 1988. Petrology, geochemistry, and origin of a major Australian 1880–1840 Ma felsic volcano–plutonic suite: a model for intracontinental felsic magma generation. *Precambrian Research*, **40/41**, 37–60.

ZHAO, J-X. & COOPER, J. A. 1992. The Atnarpa Igneous Complex, southeastern Arunta Inlier, central Australia: implications for subduction at an Early Proterozoic continental margin. *Precambrian Research*, **56**, 227–253.

ZHAO, J-X. & MCCULLOCH, M. T. 1993. Sm–Nd mineral isochron ages of Late Proterozoic dyke swarms in Australia: evidence for two distinct events of mafic magmatism and crustal extension. *Chemical Geology*, **109**, 341–354.

Proterozoic basement provinces of southern and southwestern Australia, and their correlation with Antarctica

I. C. W. FITZSIMONS

Tectonics SRC, Department of Applied Geology, Curtin University of Technology, GPO Box U1987, Perth, WA 6845, Australia (e-mail: ianf@lithos.curtin.edu.au)

Abstract: Three Precambrian basement provinces extend from the southern coast of Australia into East Antarctica when reconstructed in a Gondwana configuration. These are, from east to west, the Mawson Craton, and the Albany–Fraser and Pinjarra Orogens. The Mawson Craton preserves evidence for tectonic activity from the late Archaean until the earliest Mesoproterozoic. It is exposed in the Gawler Craton of South Australia, the Terre Adélie and King George V Land coastline of East Antarctica, and the Miller Range of the central Transantarctic Mountains. It may form a significant part of the ice-covered East Antarctic Shield, although insufficient data are available to constrain its lateral extent. The Mawson Craton underwent late Palaeoproterozoic tectonism along its eastern margin (the Kimban Orogeny) and the occurrence of c. 1700 Ma eclogites in the Transantarctic Mountains implies that this was, in part, a collisional event, although elsewhere it was characterized by low P/T metamorphism. The western margin of the Mawson Craton collided with a continental fragment comprising the Nawa Domain of the Gawler Craton, the Coompana Block and the Nornalup Complex of Western Australia at c. 1560 Ma during the Kararan Orogeny. The western edge of the Nornalup Complex later collided with the Biranup and Fraser Complexes and Yilgarn Craton to form the Albany–Fraser Orogen during two stages of tectonism at c. 1350–1260 and 1210–1140 Ma. The Pinjarra Orogen truncates the western margin of the Yilgarn Craton and Albany–Fraser Orogen, and contains allochthonous 1100–1000 Ma gneissic blocks transported along the craton margin during at least two stages of Neoproterozoic transcurrent movement. It divides East Gondwana into Australo-Antarctic and Indo-Antarctic domains, which are distinct continental fragments with different Proterozoic histories that were juxtaposed by oblique collision at 550–500 Ma during the assembly of Gondwana. The path taken by the Pinjarra Orogen beneath the Antarctic ice sheet is unknown, but it is of similar width and length to the East African Orogen, and must have been a fundamental Neoproterozoic boundary of critical importance to supercontinent assembly and breakup.

East and West Gondwana were first defined in Palaeozoic reconstructions of Gondwana that noted alternative fits for the eastern fragments of India, Antarctica and Australia, both with one another and with the better constrained western fragments of Africa and South America (McElhinny & Embleton 1974; Powell *et al.* 1980). Current usage of these terms, however, focuses on the Proterozoic assembly of Gondwana and the possibility that India–Antarctica–Australia (East Gondwana) and Africa–South America (West Gondwana) represent two discrete crustal blocks with different Precambrian histories whose late Neoproterozoic collision created the Gondwana supercontinent. This distinction was first made by McWilliams (1981), who used palaeomagnetic data to argue that East and West Gondwana were independent cratonic units that collided along the Mozambique Belt of eastern Africa. This was consistent with early plate tectonic interpretations of this belt as a Tibetan-style orogen (Burke & Dewey 1972) and it is now regarded as part of the East African Orogen, a continental-scale tectonic collision zone (Shackleton 1986, 1996; Stern 1994).

Subsequent geochronological studies have progressively revised original models for the extent and internal character of East Gondwana. Pervasive late Neoproterozoic–Cambrian tectonism identified in southern India, Sri Lanka and the Dronning Maud Land sector of East Antarctica (Choudhary *et al.* 1992; Hölzl *et al.* 1994; Shiraishi *et al.* 1994; Jacobs *et al.* 1998) is an extension of the East African Orogen into Antarctica. This indicates that part of East Antarctica lies to the west of this orogen and belongs to West rather than East Gondwana (Fig. 1), as confirmed by the palaeomagnetic data of Gose *et al.* (1997). Further subdivision of East Gondwana was prompted by increasing evidence for late Neoproterozoic tectonism in the Leeuwin Complex of Western Australia, and the Denman Glacier and Prydz Bay areas of East Antarctica (Wilde & Murphy 1990; Zhao *et al.* 1992; Hensen & Zhou 1995; Sheraton *et al.* 1995; Carson *et al.* 1996; Fitzsimons *et al.* 1997). This broad orogenic zone is likely to extend beneath the Antarctic ice sheet and divide East Gondwana into Indo-Antarctic and Australo-Antarctic sectors (Fig. 1), challenging previous models for a coherent East Gondwana during the Neoproterozoic (Rogers *et al.* 1995; Yoshida 1995).

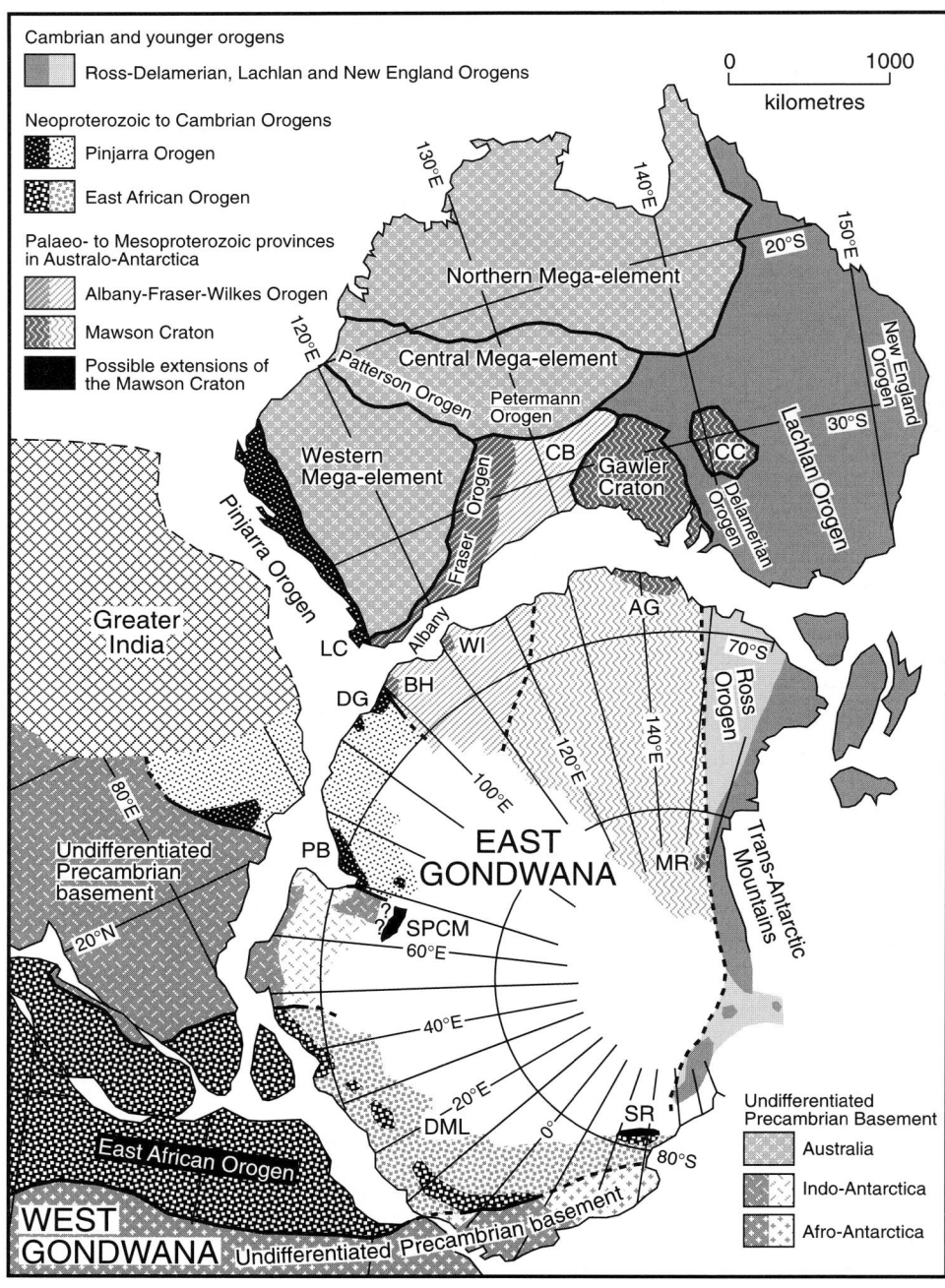

Fig. 1. Crustal orogenic provinces of easternmost Gondwana. Where two shadings are noted for one province type, the paler shading represents large areas of no outcrop where the province identity is inferred. Australian mega-elements are after Shaw *et al.* (1996). The western mega-element comprises the Archaean Yilgarn and Pilbara Cratons, and the Palaeoproterozoic Capricorn Orogen, which behaved as a single craton during the development of the Albany–Fraser and Pinjarra Orogens. The central mega-element comprises the Palaeoproterozoic–Mesoproterozoic Rudall, Arunta and Musgrave Blocks. These were reworked to varying degrees during the latest Neoproterozoic–Cambrian Petermann and Patterson Orogenies, and the Devonian–Permian Alice Springs Orogeny, which are responsible for the WNW–ESE strike of structures in this domain. The northern mega-element comprises Palaeoproterozoic basement inliers and

This paper develops these themes through a review of Proterozoic correlations between Australia and formerly contiguous rocks of East Antarctica. It focuses on three Proterozoic provinces that crop out along the southern and western coasts of Australia – the Gawler Craton, and the Albany– Fraser and Pinjarra Orogens – and their counterparts in East Antarctica (Fig. 1). Possible relationships between these crustal provinces and those of northern Australia are considered elsewhere in this volume by Wingate & Evans (2003). The Gawler Craton is part of the early Mesoproterozoic Mawson Craton that may form much of the concealed East Antarctic Shield (Fanning et al. 1995, 1996). The Albany–Fraser Orogen is part of the larger Albany–Fraser–Wilkes Orogen, the major late Mesoproterozoic Grenville-age suture in the Australo-Antarctic sector of East Gondwana. The Pinjarra Orogen is part of the proposed late Neoproterozoic orogenic zone passing through the heart of East Gondwana. The histories of these three tectonic provinces are discussed in turn, and then integrated with an emphasis on the likely lateral extent of the Mawson Craton, potential late Mesoproterozoic correlations between Australia and India, and the tectonic history of the Pinjarra Orogen and its likely path beneath the Antarctic ice cap. In contrast to many previous studies, but in keeping with recent geological and palaeomagnetic correlations (Fitzsimons 2000a,b; Boger et al. 2001; Torsvik et al. 2001; Powell & Pisarevsky 2002; Pisarevsky et al. 2003), it is argued that Greater India was only juxtaposed with southwestern Australia as a result of Neoproterozoic displacements along the Pinjarra Orogen.

Any references made here to geographical position (north, east, etc.) relate to present-day coordinates in the relevant landmass (Australia, Antarctica, etc.). To enable comparison of different Sm–Nd data sets, Nd isotope ratios have been renormalized (if necessary) to $^{146}Nd/^{144}Nd = 0.7219$, and quoted Nd model ages recalculated relative to a depleted mantle (T_{DM}) with present-day composition of $^{143}Nd/^{144}Nd = 0.51316$ and $^{147}Sm/^{144}Nd = 0.225$, assuming a decay constant of 6.54×10^{-12} and the linear ε_{Nd} growth model of Goldstein et al. (1984).

Mawson Craton

Direct comparisons between the Gawler Craton of South Australia and gneissic outcrops on the Antarctic coastline of Terre Adélie and King George V Land have been made in several studies (Moores 1991; Flöttmann & Oliver 1994), but precise U–Pb zircon geochronology in both Australia and Antarctica (Fanning et al. 1988; Fanning 1997; Peucat et al. 1999b; Goodge et al. 2001) has enabled quantitative correlation of specific tectonic events between these regions (Fig. 2), and led to the definition of the late Palaeoproterozoic Mawson Craton by Fanning et al. (1995, 1996). This section summarizes the geological history of the Gawler Craton, which is the most accessible and best-studied part of the Mawson Craton, and then considers other likely components of this extensive but poorly exposed crustal block.

Gawler Craton

Much of South Australia is underlain by the Archaean–early Mesoproterozoic Gawler Craton (Fig. 3; Webb et al. 1986; Drexel et al. 1993; Daly et al. 1998). It is best exposed in the SE, particularly along the eastern edge of the Eyre Peninsula. Elsewhere it is largely concealed by younger deposits, and its lateral extent has been defined from sparse outcrop coupled with geophysical and drillcore data. The oldest rocks in the craton are Archaean metasediments of the Sleaford and Mulgathing Complexes. These occur in the SE and NW of the craton, respectively, and are considered equivalent in age, with both recording evidence of the 2440–2400 Ma Sleafordian Orogeny.

The Sleaford Complex is exposed in the western half of the Eyre Peninsula and has not been identified east of the Kalinjala Shear Zone (Fig. 3). It has two components, the granulite facies Carnot Gneiss in the east and the Dutton Suite of granitoids emplaced into the amphibolite facies Wangary Gneiss in the west (Fig. 4a). These are interpreted as lower and upper crustal levels of the same metamorphic complex juxtaposed by a younger fault. The Carnot Gneiss is dominated by garnet–biotite–cordierite metapelite with lenses of mafic granulite, felsic augen gneiss and hypersthene–garnet gneiss equilibrated at metamorphic conditions of 850°C and 9 kbar. U–Pb zircon and monazite geochronology indicates peak metamorphism at 2440–2400 Ma, with inherited zircons yielding ages of 3150–2950 Ma (Fanning 1997). The Dutton Suite comprises granite and granodiorite intruded into the complex during the Sleafordian Orogeny, and has

---

concealed cratons. Antarctic provinces are from Fitzsimons (2000a), the margins of the East African Orogen are from Collins & Windley (2002), and the extent of Neoproterozoic reactivation in northeastern India is from Crowe et al. (2001). Greater India is that part of the subcontinent subducted beneath Tibet. AG, Terre Adélie–King George V Land; BH, Bunger Hills; CB, Coompana Block (concealed by the Officer and Eucla Basins); CC, Curnamona Craton; DG, Denman Glacier region; DML, Dronning Maud Land; LC, Leeuwin Complex; MR, Miller Range; PB, Prydz Bay; SPCM, southern Prince Charles Mountains; SR, Shackleton Range; WI, Windmill Islands.

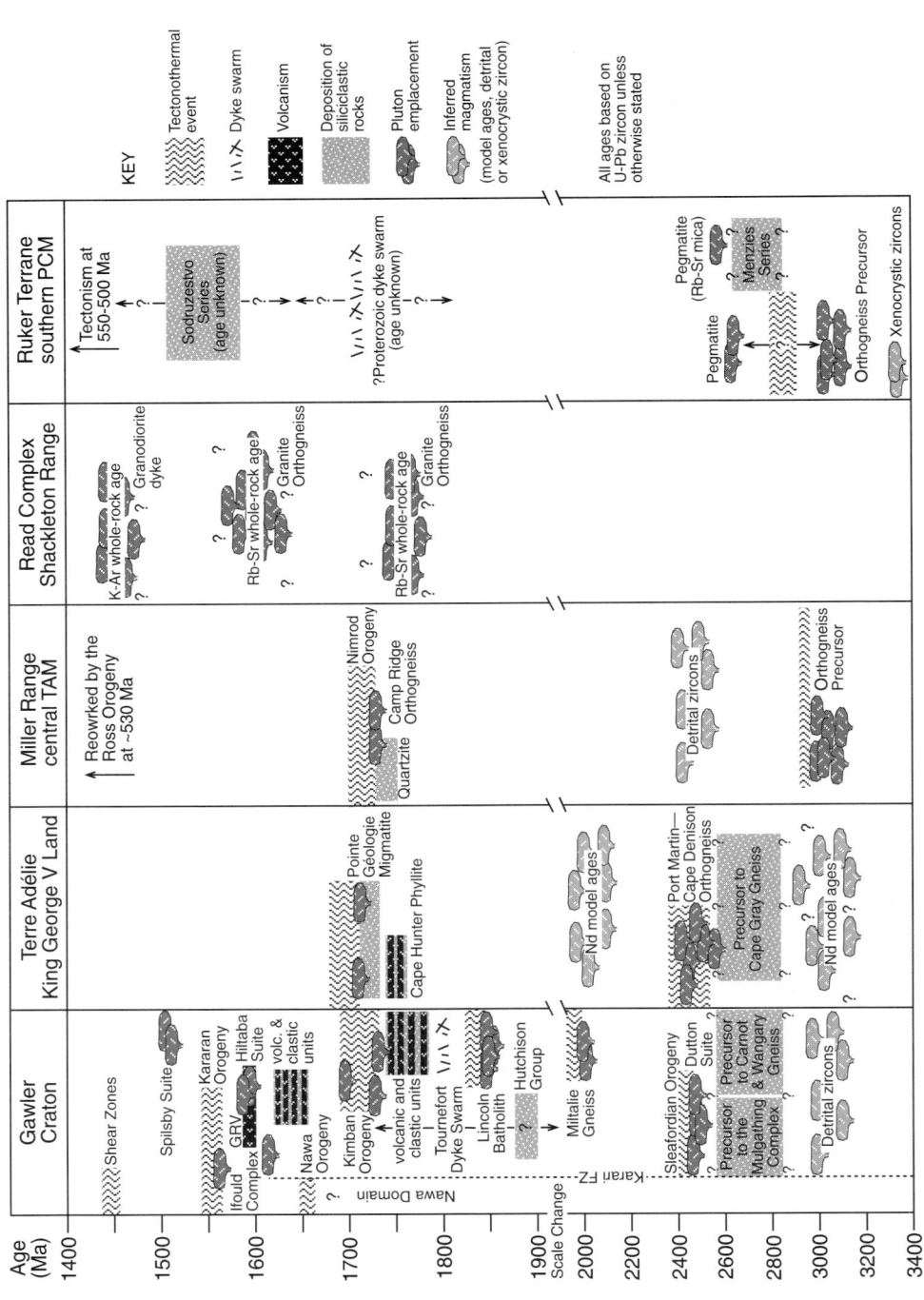

Fig. 2. Time–space diagram for potential components of the Mawson Craton. Time information is from U–Pb zircon data where possible. Data sources are summarized in the text. GRV, Gawler Range Volcanics; PCM, Prince Charles Mountains; TAM, Transantarctic Mountains.

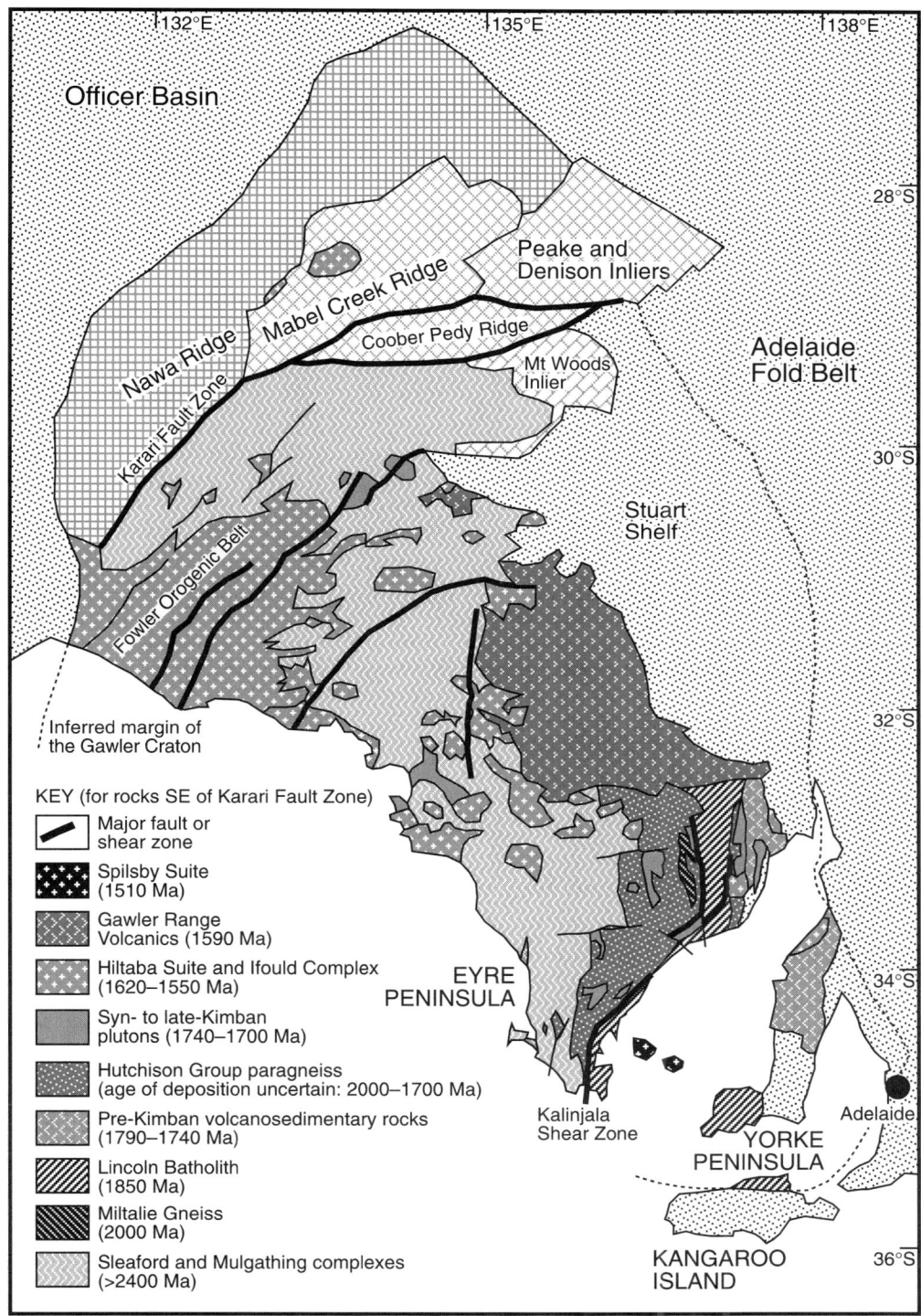

Fig. 3. Geology of the Gawler Craton, South Australia [adapted from Drexel *et al.* (1993) and Daly *et al.* (1998)]. Little is known about lithologies north of the Karari Fault Zone except for geophysical data and sparse drillcore samples.

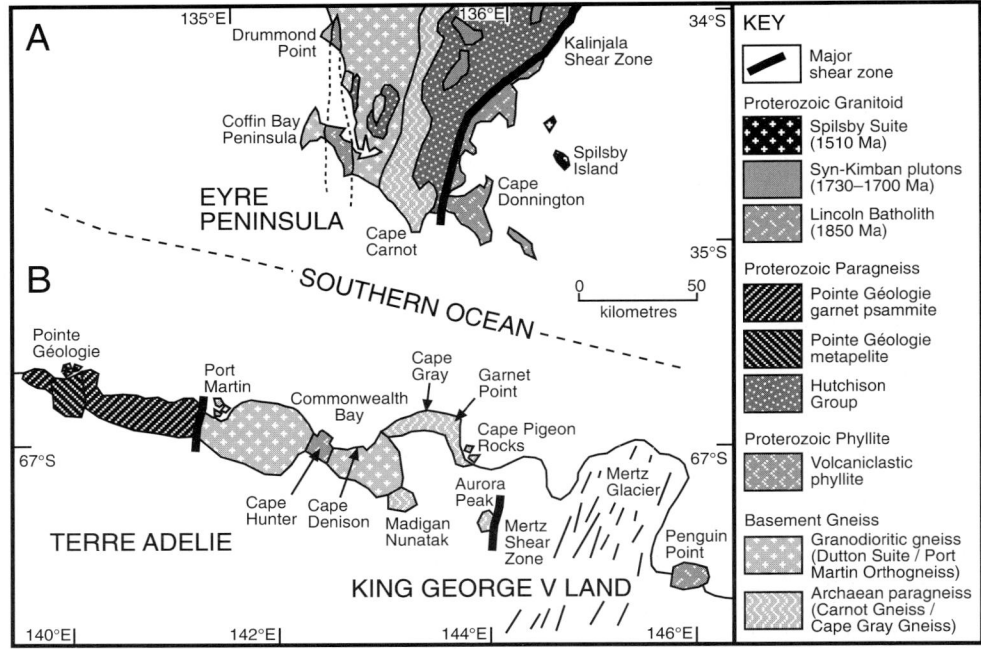

Fig. 4. Geology of the southern Eyre Peninsula (A) and Terre Adélie–King George V Land (B), adapted from Oliver & Fanning (1997) and Kleinschmidt & Talarico (2002). Relative positions of A and B based on correlation of Cape Hunter Phyllite with a similar lithology exposed on Coffin Bay Peninsula. Glaciers are denoted by closely spaced subparallel lines.

U–Pb zircon crystallization ages of 2550–2450 Ma, whereas the Wangary Gneiss has a U–Pb metamorphic zircon age of 2479±8 Ma (Fanning 1997). Although some structures in the Sleaford Complex have been ascribed to deformation during this early Palaeoproterozoic event (Drexel et al. 1993), most of the observed structures reflect reworking and reorientation during the younger Kimban Orogeny (Vassallo & Wilson 2002). The Mulgathing Complex is less well exposed and comprises a range of metasedimentary rocks, including banded iron formation, chert, carbonate, calc-silicate, quartzite and pelite, locally interleaved with komatiite and tholeiite basaltic flows, whose metamorphic grade varies from granulite to greenschist facies (Drexel et al. 1993; Daly et al. 1998). These are associated with syntectonic granitoid plutons dated at c. 2450 Ma (Fanning 1997). Complex regional structures are apparent in geophysical data from the Mulgathing Complex, but these in part reflect high-strain zones developed during the Mesoproterozoic Kararan Orogeny. Archaean crust has not been identified NW of the Karari Fault Zone, although this may simply reflect the poor outcrop in this region.

The earliest post-Sleafordian event recognized in the Gawler Craton is the emplacement of the igneous precursor to the Miltalie Gneiss at 2000±13 Ma (Fanning 1997), together with three other similar units of equivalent age. These Palaeoproterozoic granitic orthogneisses make up only a fraction of the craton (Fig. 3) and their significance is poorly understood, as is the tectonic event that produced their gneissic foliation (perhaps at c. 1964 Ma; Fanning et al. 1988). The Miltalie Gneiss is unconformably overlain by metasedimentary rocks of the Hutchison Group, which was deposited on a shallow continental shelf that deepened to the east. Multiple units of quartzite, schist, dolomite and iron formation in the Hutchison Group have been interpreted in terms of both stratigraphic and structural repetition (Parker & Lemon 1982; Vassallo & Wilson 2001). They were deformed and metamorphosed at c. 1730 Ma during the Kimban Orogeny, but their age of deposition relative to the emplacement of the 1850 Ma Lincoln Batholith is controversial.

The term Lincoln Batholith was introduced by Hoek & Schaefer (1998) to refer to granite, charnockite and gabbro-norite of the Donnington and Colbert Granitoid Suites that were emplaced at 1850±1 and 1846±14 Ma, respectively. These rocks make up much of the southeastern Eyre and the southern Yorke Peninsulas, and probably correlate with the granitic Minbrie Gneiss of the northeastern Eyre Peninsula. Originally they were regarded as the first intrusive stage of the Kimban Orogeny, and all felsic plutons emplaced between

1850 and 1700 Ma were grouped together as the Lincoln Complex (Drexel et al. 1993; Daly et al. 1998). However, precise geochronology indicates that the pulse of magmatism at 1855–1845 Ma (Fanning 1997; Daly et al. 1998) is quite distinct from the main Kimban orogenic phase at 1740–1700 Ma and is best regarded as a separate event (Hoek & Schaefer 1998; Vassallo & Wilson 2002). The Lincoln Batholith represents a significant volume of dry charnockitic magma emplaced into the mid and lower crust (Mortimer et al. 1988a), synchronous and possibly cogenetic with the Jussieu Mafic Dyke Swarm (Hoek & Schaefer 1998). Nd isotopic compositions suggest that the batholith source rocks were extracted from the mantle substantially before 1850 Ma (Turner et al. 1993), consistent with remelting of a crustal underplate. All components of the Lincoln Batholith are cut by abundant mafic tholeiites of the 1812 ± 5 Ma Tournefort Dyke Swarm (Mortimer et al. 1988b; Hoek & Schaefer 1998; Schaefer 1998), which have planar margins in Kimban low-strain zones consistent with emplacement after cooling of the batholith. Several chemical and structural groups have been distinguished in low-strain zones, with a N–S swarm being dominant (Vassallo & Wilson 1999). E–W compressional deformation of the batholith before intrusion of the Tournefort Dyke Swarm, previously ascribed to an early Kimban orogenic phase (Drexel et al. 1993; Daly et al. 1998), has been called the Lincoln Orogeny by Vassallo & Wilson (2002) and attributed to deformation during batholith emplacement.

The Hutchison Group and Lincoln Batholith are exposed to the west and east, respectively, of the Kalinjala Shear Zone, a major Kimban-age structure in the southern Eyre Peninsula (Drexel et al. 1993; Daly et al. 1998; Vassallo & Wilson 2001, 2002). Felsic gneiss with an 1850 Ma U–Pb zircon protolith age (Fanning 1997) has been correlated with the top of the Hutchison Group, suggesting that the Hutchison Group is older than the Lincoln Batholith (Daly et al. 1998) and consistent with metasedimentary enclaves in the batholith being xenoliths of Hutchison Group material (Drexel et al. 1993). Vassallo & Wilson (1999, 2001) noted, however, that the Hutchison Group in the southern Eyre Peninsula does not contain mafic dykes, whereas the nearby Sleaford Complex is cut by abundant mafic dykes that are isotopically indistinguishable from Tournefort dykes in the Lincoln Batholith (Schaefer 1998). They argued that the Hutchison Group was deposited after emplacement of the Lincoln Batholith, possibly as a result of the same E–W extension recorded by the Tournefort Dyke Swarm. Rhyolitic volcanics and clastic sedimentary rocks exposed on Eyre and Yorke Peninsulas are generally considered younger than the Hutchison Group (Drexel et al. 1993; Daly et al. 1998), although there is no conclusive evidence for this interpretation. Some of the volcanic units have been dated using U–Pb zircon techniques (Fanning 1997), including the 1790 Ma Myola Volcanics, the 1760 Ma Wardung Volcanics, and the 1740 Ma Moonta Porphyry and McGregor Volcanics.

The Kimban Orogeny produced a series of N–S-trending structures associated with complex strain patterns, high-grade metamorphism, and syn- to late tectonic granite plutons. The syntectonic Middle Camp Granite was emplaced at 1731 ± 7 Ma and the weakly deformed c. 1700 Ma Moody Granite Suite was emplaced during the final stages of the orogenic event (Fanning 1997; Daly et al. 1998), consistent with garnet–hornblende–whole-rock Sm–Nd ages of c. 1730 Ma for peak metamorphism in the Tournefort Dykes (Hand et al. 1995). Metamorphic conditions in the Hutchison Group ranged from 5 kbar and 600°C to 8 kbar and 800°C with garnet–cordierite assemblages developed locally in metapelites (Drexel et al. 1993). Ductile fabrics in the Eyre Peninsula have been attributed to two dominant deformation events (Parker & Lemon 1982; Drexel et al. 1993; Vassallo & Wilson 2001, 2002), and although there are differences in the details of proposed structural histories most workers have identified a component of dextral transpression associated with sheath folds and upright NNE–SSW-striking high-strain zones dominated by the Kalinjala Shear Zone.

There have been few attempts to interpret the Kimban Orogeny in terms of regional tectonics, although Myers et al. (1996) considered it to reflect collision of the eastern margin of the Gawler Craton with the Curnamona Craton. There is no exposed record of this collision in the Curnamona Craton, where the oldest exposed rocks are the metasedimentary Willyama Supergroup deposited at c. 1710–1640 Ma (Page et al. 2000), but the two cratons have similar Mesoproterozoic histories and must have been joined before deposition of Neoproterozoic sediments in the Adelaide Fold Belt. Hoek & Schaefer (1998) noted that the physical characteristics of the Lincoln Batholith, including an abundance of mafic enclaves and synplutonic mafic dykes, are characteristic of a magmatic arc, but Wyborn (1988) and Mortimer et al. (1988a) argued that the batholith geochemistry and lack of intermediate compositions are more suggestive of an ensialic origin than a convergent plate margin. In contrast, the Tournefort Dykes have a chemical composition consistent with their emplacement close to an active continental margin (Mortimer et al. 1988b). The western extent of the Kimban Orogeny is also rather unclear. Limited isotopic data for the Mulgathing Complex reveal some evidence for Kimban-age tectonism, including a syntectonic granite pluton dated at 1690±10Ma (Fanning 1997), but the low- to high-grade metamorphic assemblages in this complex are generally ascribed to the Sleafordian Orogeny, thus precluding a pervasive Kimban overprint. Ages of sedimentation and tectonism for high-grade paragneiss in the Peake and Denison Inliers (Fig. 3) are constrained by the 1780±12Ma Tidnamurkana Volcanics interbedded with metasedimentary rocks, and the syn-tectonic 1793±8Ma Wirriecurrie Granite. The Mount Woods Inlier (Fig. 3) contains a syntectonic granite pluton dated at 1691±25Ma and granulite facies paragneiss with 1736±14Ma metamorphic zircon (Fanning 1997), but reconnaissance geochronology of drill-core samples indicates that much (or all) of the high-grade tectonism in the NW Gawler Craton is post-Kimban in age (Daly et al. 1998).

U-Pb SHRIMP zircon geochronology of garnet-sillimanite paragneiss from the Mabel Creek Ridge implies an age of c. 1550Ma for high-grade metamorphism, whilst garnet-cordierite-sillimanite paragneiss from the Coober Pedy Ridge contains zircon with inherited cores of c. 1750 Ma and metamorphic rims of 1565±8Ma (Fanning 1997). Aeromagnetic data indicate that ductile structures in the Coober Pedy Ridge are parallel to the NE-SW-striking subvertical Karari Fault Zone, the dominant feature on geophysical images of the NW Gawler Craton, suggesting that metamorphism was associated with movement along this structure (Fig. 3; Daly et al. 1998). Southeast of the Karari Fault Zone, in the northwestern Mulgathing Complex, originally north-trending structures of assumed Archaean age have been rotated into parallelism with the fault zone along a series of anastomosing shear zones. These are associated with voluminous 1650-1550Ma synorogenic granitoid plutons, referred to as the Ifould Complex by Daly et al. (1998). This area has been named the Fowler Orogenic Belt (Daly et al. 1998) and is considered cogenetic with the Karari Fault Zone. A deformed high-grade metamafic rock from the Fowler Orogenic Belt contains metamorphic zircons dated at 1543±9Ma by SHRIMP U-Pb techniques (Fanning 1997), and structures in the drill core from the Karari Fault Zone (Rankin et al. 1989) and rare outcrop in the Fowler Orogenic Belt are consistent with sinistral transpression and east-side-up dip-slip movement.

Daly et al. (1998) attributed this tectonism to oblique collision of the Gawler Craton with another continental block along the Karari Fault Zone, and named this event the Kararan Orogeny. Pelitic rocks from drill core NW of the Karari Fault Zone, at the western end of the Nawa Ridge, have high-temperature hypersthene-sillimanite and sapphirine-quartz assemblages (8-10 kbar and 950-1000°C; Teasdale 1997; Daly et al. 1998) dated at 1653±8Ma by SHRIMP U-Pb zircon techniques (Fanning 1997). These rocks were called the Moondrah Gneiss by Daly et al. (1998), who argued that they were produced by the Kararan collision and that 1550Ma events in the Fowler Orogenic Belt were a result of post-collisional tectonism. However, fabrics in the Moondrah Gneiss are truncated by the Karari Fault Zone in geophysical images and it is possible that 1650Ma metamorphism in these rocks predated a collision at 1550Ma. This would be consistent with the lack of evidence for 1650Ma tectonism SE of the Karari Fault Zone, and suggests that the Ifould Suite could represent a pre-collision magmatic arc. Daly et al. (1998) suggested that the colliding block was the 'proto-Yilgarn Craton', but it is more likely to have been a separate block (possibly the Nawa Domain of the NW Gawler Craton and the concealed Coompana Block; Myers et al. 1996) that was later sandwiched between the Yilgarn and Gawler Cratons during the development of the Albany-Fraser Orogen at c. 1300Ma (see below).

Post-Kimban events in the central Gawler Craton include sediment deposition between 1660 and 1590Ma (Daly et al. 1998), and eruption of the 1591-1592Ma Gawler Range Volcanics, a huge dacite-rhyodacite-rhyolite province. A series of granitic plutons were intruded at this time, including the St Peter Suite at 1620Ma and the Hiltaba Suite at 1600-1585Ma, with the latter considered cogenetic with the Gawler Range Volcanics. This magmatism was synchronous with the c. 1600Ma Olarian Orogeny (Page et al. 2000), which deformed and metamorphosed rocks further east in the Curnamona Craton. Later events include emplacement of the Spilsby Suite of granitoids at 1510Ma, which are exposed east of the Eyre Peninsula, and open folding and faulting in the central Gawler Craton attributed to the Wartakan event at c. 1510Ma (Fanning et al. 1988). $^{40}Ar/^{39}Ar$ thermochronology indicates that the Gawler Craton had a protracted Mesoproterozoic exhumation history (Foster & Ehlers 1998; Fraser et al. 2002). This was followed by Neoproterozoic extension, marked by the intrusion of the 827±6Ma Gairdner Dyke Swarm (Wingate et al. 1998) and deposition of rift-related sediments in the Adelaide Fold Belt.

Terre Adélie and King George V Land

Exposure of the Mawson Craton in Terre Adélie and King George V Land is limited to sparse coastal outcrop between 138 and 146°E (Fig. 4b). The largest areas of outcrop, from west to east, are the Pointe Géologie Archipelago (site of Dumont d'Urville, the permanent French Station), Port Martin Archipelago, Capes Hunter and Denison in Commonwealth Bay (site of Mawson's Hut,

1911–1914 Australasian Antarctic Expedition), and the Way Archipelago between Cape Gray and Cape Pigeon Rocks. Rock types are dominated by amphibolite to granulite facies paragneiss and orthogneiss, with a distinctive greenschist facies phyllite at Cape Hunter (Stüwe & Oliver 1989; Tingey 1991).

The oldest rocks are exposed in the islands of eastern Commonwealth Bay and around the coast from Cape Gray to Cape Pigeon Rocks, and comprise cordierite–garnet–sillimanite metapelite interleaved with garnet–hypersthene gneiss and cut by garnet–pyroxene metamafic dykes (Fig. 4b). These granulite facies rocks equilibrated at 5–7 kbar and 750°C (Stüwe & Oliver 1989; Oliver & Fanning 1999), and the age of metamorphism at Cape Gray is $c.$ 2420 Ma (Oliver & Fanning 1999). Charnockite gneiss at Madigan Nunatak is considered part of this sequence and has U–Pb zircon ages of 2700–2350 Ma, whereas a more feldspathic variety interpreted as partial melt crystallized at 1709 ± 12 Ma, providing evidence for Palaeoproterozoic reworking of these rocks (Oliver & Fanning 1999). Further west, outcrops at Cape Denison and Port Martin are dominated by granodioritic orthogneiss with screens and enclaves of amphibolite facies metapelite, psammite and marble, cut by amphibolite dykes and post-tectonic granite (Ménot et al. 1999). Orthogneiss at Port Martin has a SHRIMP U–Pb zircon age of $c.$ 2440 Ma (Monnier et al. 1996), consistent with a conventional U–Pb zircon upper intercept age of 2366 ± 30 Ma for Cape Denison (Oliver et al. 1983). Gneissic enclaves at Port Martin contain zircon populations at 2700–2500 Ma and $c.$ 2420 Ma (Peucat et al. 1999a) and post-tectonic granite dykes contain inherited $c.$ 2400 Ma zircons with a younger generation at 1760 Ma interpreted as the crystallization age (Monnier et al. 1996). These Archaean–Palaeoproterozoic gneisses are comparable to the Sleaford Complex of the Gawler Craton, with the paragneiss of eastern Commonwealth Bay (Cape Gray Gneiss) and the Port Martin–Cape Denison orthogneiss corresponding to the Carnot Gneiss and Dutton Suite, respectively (Fig. 2; Oliver & Fanning 1999). As in the Gawler Craton, these units are regarded as different crustal levels juxtaposed by Palaeoproterozoic tectonic movement.

Greenschist facies phyllite with chlorite–muscovite–biotite $\pm$ garnet assemblages crops out at Cape Hunter, between outcrops of granodioritic orthogneiss at Cape Denison and Port Martin (Stüwe & Oliver 1989; Oliver & Fanning 1997). The low metamorphic grade of this unit led Stüwe & Oliver (1989) to suggest that it was deposited after high-grade metamorphism in the surrounding gneisses and the greenschist facies mineral assemblage was attributed to a 500 Ma metamorphic overprint identified elsewhere in Antarctica, but this rock is now considered part of a Palaeoproterozoic volcanosedimentary succession (Oliver & Fanning 1997). SHRIMP U–Pb analysis of detrital zircon in this rock has identified older populations at 2500–2400 Ma, consistent with reworking of nearby gneissic rocks, and a dominant younger population of elongate zircon crystals at 1765 ± 8 Ma, interpreted as volcanic grains and providing a maximum age for deposition (Oliver & Fanning 1997). This younger population does not have an obvious source in East Antarctica, but does match 1790–1740 Ma volcanosedimentary units of the eastern Gawler Craton, and an identical zircon population at 1766 ± 14 Ma has been identified in garnet-bearing phyllite exposed on Drummond Point and Coffin Bay Peninsula of the southwestern Eyre Peninsula. This provides a precise piercing point for Gondwana reconstructions of Terre Adélie with the Eyre Peninsula (Fig. 4; Oliver & Fanning 1997). The absence of $c.$ 1730–1690 Ma detritus is taken as evidence for pre-Kimban deposition and this unit is assumed to have been metamorphosed at a relatively shallow crustal level during the Kimban Orogeny (Oliver & Fanning 1997).

Although of much higher metamorphic grade than the Cape Hunter Phyllite, the Pointe Géologie Migmatite is of similar age and origin (Monnier et al. 1996). It comprises garnet-bearing psammite (metagreywacke), previously described as granitic gneiss (Stüwe & Oliver 1989), and pelitic gneiss with cordierite–sillimanite–garnet assemblages equilibrated at 5 kbar and 750°C (Monnier et al. 1996). Evidence of partial melting is widespread, as are syntectonic mafic dykes recrystallized to amphibolite, all cut by pink granite dykes. Metamorphism was associated with N–S-striking vertical shear zones that reorient earlier fabrics recording N–S extension (Pelletier et al. 1999), and the eastern contact of the migmatites with the Port Martin orthogneiss is most likely a ductile shear zone (Heurtebize 1952). Monazite from three samples gave a U–Pb crystallization age of 1694 ± 2 Ma, which is believed to reflect the partial melting event correlated with the Kimban Orogeny of the Eyre Peninsula, whereas SHRIMP U–Pb analysis of zircon identified detrital cores with ages of 2800, 2600 and 1740–1730 Ma, and metamorphic rims of 1710–1640 Ma with a mean age of 1688 ± 13 Ma (Peucat et al. 1999b). The protolith sediments were eroded from basement rocks like those exposed at Port Martin and Commonwealth Bay, with an additional component eroded from 1740 to 1730 Ma volcanic rocks. Metamorphic grades are equivalent to those in the Hutchison Group of the southern Eyre Peninsula, and, although the depositional age of the Hutchison Group is controversial and may be older than that for precursor sediments at Pointe Géologie (see above), these units had very similar histories during the Kimban Orogeny. T_{DM} Nd model ages for Pointe Géologie migmatite samples include a cluster at 3.2–2.8 Ga (Peucat et al. 1999b), which suggest an Archaean crustal influence consistent with 3150–2950 Ma detrital zircon in Carnot Gneiss samples from the Gawler Craton (Fanning 1997). Another population of Pointe Géologie T_{DM} Nd model ages at 2.2–1.9 Ga are similar to the age of the Miltalie Gneiss in the Gawler Craton (Fig. 2), but could reflect a mixed sedimentary provenance rather than a discrete magmatic event.

Miller Range

Amphibolite facies gneiss of the Nimrod Group is exposed over a distance of c. 100 km in the Miller Range area of the central Transantarctic Mountains (Fig. 1). It was originally regarded as Precambrian, on the basis of its high metamorphic grade, and attributed to the Proterozoic Nimrod Orogeny (Grindley et al. 1964). Outcrops comprise layered ortho- and paragneiss, including banded felsic and mafic gneiss, garnet–kyanite–sillimanite–muscovite schist, quartzite, amphibolite, migmatite and marble, with boudins of ultramafic rock and relict garnet–omphacite eclogite, cut by the locally discordant granodioritic Camp Ridge Orthogneiss. These rocks are intruded by syn- and post-tectonic Cambrian–Ordovician granite plutons of the Ross Orogen (Goodge et al. 1993b). Although metamorphic assemblages and fabrics in these rocks are Cambrian (Goodge & Dallmeyer 1992; Goodge et al. 1993a), whole-rock T_{DM} Nd model ages of 3.0–2.5 Ga (Borg et al. 1990; Borg & DePaolo 1994) and SHRIMP U–Pb zircon ages of 3150–1720 Ma (Goodge & Fanning 1999; Goodge et al. 2001) have established a prolonged Precambrian history for their protoliths, making them the only demonstrated piece of cratonic basement exposed in the Ross Orogen.

U–Pb SHRIMP zircon analysis identified magmatic zircon cores in hornblende–biotite gneiss with ages of c. 3180 and 2980 Ma, which are similar to T_{DM} Nd model ages for the Nimrod Group and are interpreted as the original crystallization age of these meta-igneous units (Goodge & Fanning 1999; Goodge et al. 2001). Three younger age populations at c. 2980–2975, 1730–1710 and 530 Ma are interpreted as metamorphic zircon growth. A partly retrogressed mafic eclogite boudin contains abundant spherical zircons with a mean upper intercept age of 1723 ± 29 Ma, interpreted as the age of high-pressure metamorphism during the Nimrod Orogeny. A poorly constrained lower intercept age of c. 510 Ma reflects the Ross Orogeny (Goodge et al. 2001). Goodge et al. (2001) argued that all of the Nimrod Group experienced eclogite-facies conditions, suggesting that the Nimrod Orogeny involved crustal thickening during plate convergence and collision, but that high-pressure assemblages were largely erased by the Ross Orogeny. The discordant Camp Ridge Orthogneiss yielded magmatic zircon dated at 1730 ± 14 Ma, interpreted as the igneous crystallization age (Goodge et al. 2001), whilst detrital zircons from a quartzite had age populations of 2555–1734 Ma (Walker & Goodge 1991), indicating that felsic plutonism and deposition occurred immediately before or during early stages of the Nimrod Orogeny.

Although 2000 km from the Gawler Craton in Gondwana reconstructions (Fig. 1), the Miller Range is broadly along-strike from high-grade rocks in the Eyre Peninsula and Terre Adélie–King George V Land, and the 1730–1700 Ma Nimrod Orogeny has been correlated with the Kimban Orogeny of South Australia (Fig. 2; Goodge et al. 2001). The 1730 Ma Camp Ridge Orthogneiss corresponds to syn-Kimban plutons (e.g. Middle Camp Granite); quartzite detrital populations indicate similar sediment sources and depositional ages to rocks in Terre Adélie and South Australia, and 3100–2900 Ma magmatism in the Miller Range matches detrital populations in the Carnot Gneiss of the Eyre Peninsula (Fanning 1997). One significant distinction, however, is the high pressure/temperature (P/T) collision inferred for the Nimrod Orogeny compared with low P/T metamorphism in the Kimban Orogeny of the Eyre Peninsula and Terre Adélie–King George V Land. One possible explanation is that the latter is dominated by a post-collision thermal pulse associated with collapse of the orogen, but more work is needed to reconcile these important differences.

Other potential fragments of the Mawson Craton

Archaean–Palaeoproterozoic basement occurs at several other locations in East Antarctica (Tingey 1991; Fitzsimons 2000a) and three of these have been suggested as possible extensions of the Mawson Craton – the Shackleton Range, the southern Prince Charles Mountains and the Bunger Hills–Denman Glacier region (Fanning et al. 1999; Goodge et al. 2001). Archaean–Palaeoproterozoic crust in the Bunger Hills–Denman Glacier region was reworked within the Albany–Fraser–Wilkes and Pinjarra Orogens and is discussed below, but the Shackleton Range and southern Charles Prince Mountains are discussed here.

The Shackleton Range is situated at the 'Atlantic' end of the Transantarctic Mountains, some 1800 km from the Miller Range (Fig. 1). Recent studies have distinguished two Proterozoic basement complexes in the Shackleton Range, separated by younger tectonic units interpreted as possible oceanic crust and an allochthonous thrust nappe (Buggisch & Kleinschmidt 1999; Talarico et al. 1999; Tessensohn et al. 1999). This thrust pile is regarded as an extension of the East African Orogen, with the southernmost and structurally lowest unit (the Read Complex) representing the East Gondwana basement that is potentially part of the Mawson Craton. The Read Complex comprises amphibolite facies paragneiss and variably deformed gabbroic to granitic plutonic rocks (Clarkson 1982; Talarico & Kroner 1999). Available geochronology, reviewed by Pankhurst et al. (1983) and Talarico & Kroner (1999), is limited to 1763 ± 21 and 1599 ± 38 Ma Rb–Sr whole-rock isochrons for samples of granitic orthogneiss, a 1454 ± 60 Ma K–Ar whole-rock age for a granodiorite dyke, and K–Ar and Rb–Sr mineral ages of 1650–1550 Ma. The older isochron is regarded as an igneous crystallization age and a low initial $^{87}Sr/^{86}Sr$ ratio for these samples indicates that they had little crustal history before this time. These late Palaeoproterozoic ages are not inconsistent with data from the Mawson Craton (Fig. 2), but there are insufficient data for a reliable comparison and there is isotopic evidence from the Transantarctic Mountains that the Read Complex and the Mawson Craton are separated by a region of younger crust (see below).

The southern Prince Charles Mountains are exposed within the Lambert Glacier Drainage System >2000 km west of Terre Adélie (Figs 1 & 5), and comprise a variety of basement gneisses and greenschist to amphibolite facies metasedimentary rocks, intruded by granite plutons and mafic dykes (Grew 1982; Tingey 1982, 1991; Kamenev et al. 1993; Mikhalsky et al. 2001b). Several terranes have been distinguished using field constraints and reconnaissance isotope data, of which the southernmost Ruker Terrane is the most likely to correspond to the Mawson Craton. The Ruker Terrane comprises amphibolite facies granitic to tonalitic orthogneiss containing 3170–3000 Ma magmatic zircon and c. 3370 Ma zircon xenocrysts, which is consistent with T_{DM} Nd model ages of 3.2–3.0 Ga (Kovach & Belyatsky 1991; Boger et al. 2001; Mikhalsky et al. 2001b). These gneisses were deformed before the emplacement of undeformed pegmatite dykes at c. 2640 Ma (Boger et al. 2001), and are overlain by quartzite and micaschist that typically have tectonic contacts with the basement, although an unconformity has been described locally (Hofmann 1982; Tingey 1982). The metasediments are subdivided into the greenschist to amphibolite facies Menzies Series cut by metamafic dykes and a younger lower greenschist facies Sodruzestvo Series that is not cut by mafic dykes. Some workers also distinguish the Ruker Series, which is cut by the dykes and includes distinctive banded iron formations. There is little consensus on absolute depositional or metamorphic ages for these rocks (Mikhalsky et al. 2001b). Isotopic data from pegmatite bodies have been used to argue that kyanite–staurolite–sillimanite schist in the Menzies Series is both Archaean (Tingey 1982) and Neoproterozoic (Grew 1982). The Sodruzestvo Series has been ascribed a Neoproterozoic age from fossil acritarchs (Iltchenko 1972) and metamorphic assemblages in these latter rocks are regarded as late Neoproterozoic–Cambrian (Tingey 1982). The limited isotopic data available for the Ruker Terrane allow tenuous correlations to be made with early events in the Gawler Craton (Fig. 2), but more age data are needed before any reliable comparisons can be made. The southern Prince Charles Mountains are discussed further below in relation to late Neoproterozoic tectonism.

Albany–Fraser–Wilkes Orogen

Late Mesoproterozoic gneiss and granite in the Albany–Fraser Orogen of Western Australia and similar rocks exposed along the Wilkes Land coast of East Antarctica have been used to discriminate between different models for the Gondwana fit of Australia and Antarctica (Lovering et al. 1981; Oliver et al. 1983; Sheraton et al. 1995). Potential links between these rocks (the Albany–Fraser–Wilkes Orogen) and either the Eastern Ghats of India (Katz 1989; Moores 1991) or the Central Indian Tectonic Zone (Harris 1993) have likewise been used as late Mesoproterozoic piecing points in Indo–Australian reconstructions, although Fitzsimons (2000b) has argued that Mesoproterozoic basement in Australia and India were juxtaposed by Neoproterozoic tectonism. This section summarizes the geology of the Albany–Fraser Orogen, and the Windmill Islands and Bunger Hills regions of the Wilkes Land coast (Fig. 6), as a prelude to later discussion of potential Gondwana correlations. Northwest- to WNW-trending dolerite dykes cut Mesoproterozoic gneiss in all of these regions, as do a number of late mylonites, but these are discussed later as part of the Pinjarra Orogen.

Albany-Fraser Orogen

The Albany–Fraser Orogen records two periods of late Mesoproterozoic tectonism at c. 1350–1260 and 1210–1140 Ma [Stages I and II of Clark et al. (2000)]. It has traditionally been divided into the E–W striking Albany Mobile Belt and NE–SW striking Fraser Mobile Belt (Fig. 7; Mathur & Shaw 1982). Harris (1995) has argued for significant differences between these two belts, although most

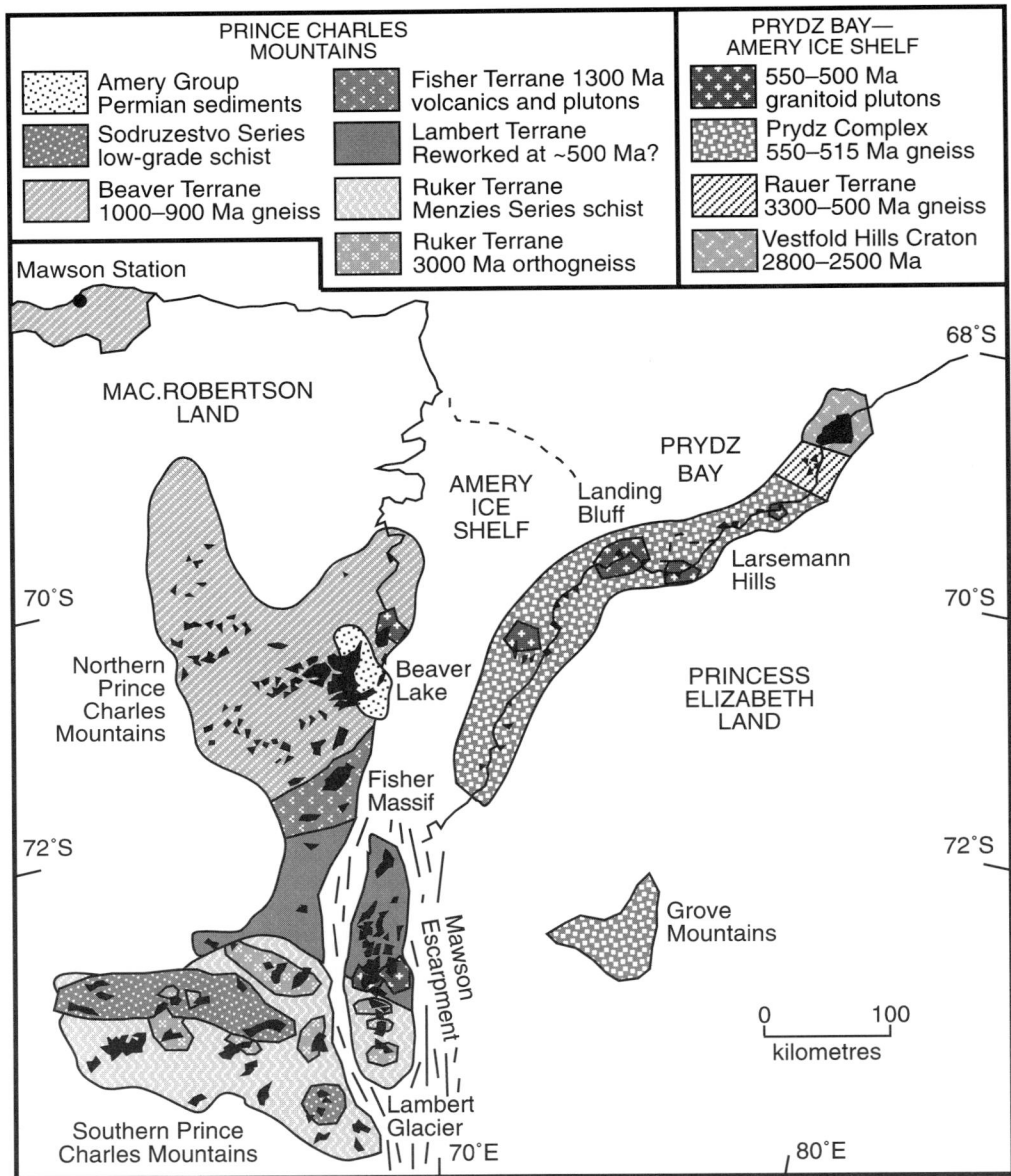

Fig. 5. Geology of the Prydz Bay–Prince Charles Mountain region [adapted from Kamenev *et al.* (1993); Fitzsimons (1997)]. Glaciers are denoted by closely spaced subparallel lines and outcrop is marked in black.

workers regard them as along-strike equivalents (Myers 1990; Clark 1999). The orogen is in faulted contact with the Archaean Yilgarn Craton to the NW, and extends eastwards under the Eucla Basin to the Coompana Block and western margin of the Gawler Craton. It is regarded as a collisional suture zone between these two cratons (Myers *et al.* 1996), and is truncated to the west and NE by the Neoproterozoic Pinjarra and Petermann Orogens, respectively (Fig. 1). To the south it was previously contiguous with outcrops on the Wilkes Land coast, as described below.

The Albany–Fraser Orogen has been divided into four lithotectonic domains, defined using differences in structural style apparent from aeromagnetic data (Beeson *et al.* 1988; Myers 1990; Whitaker

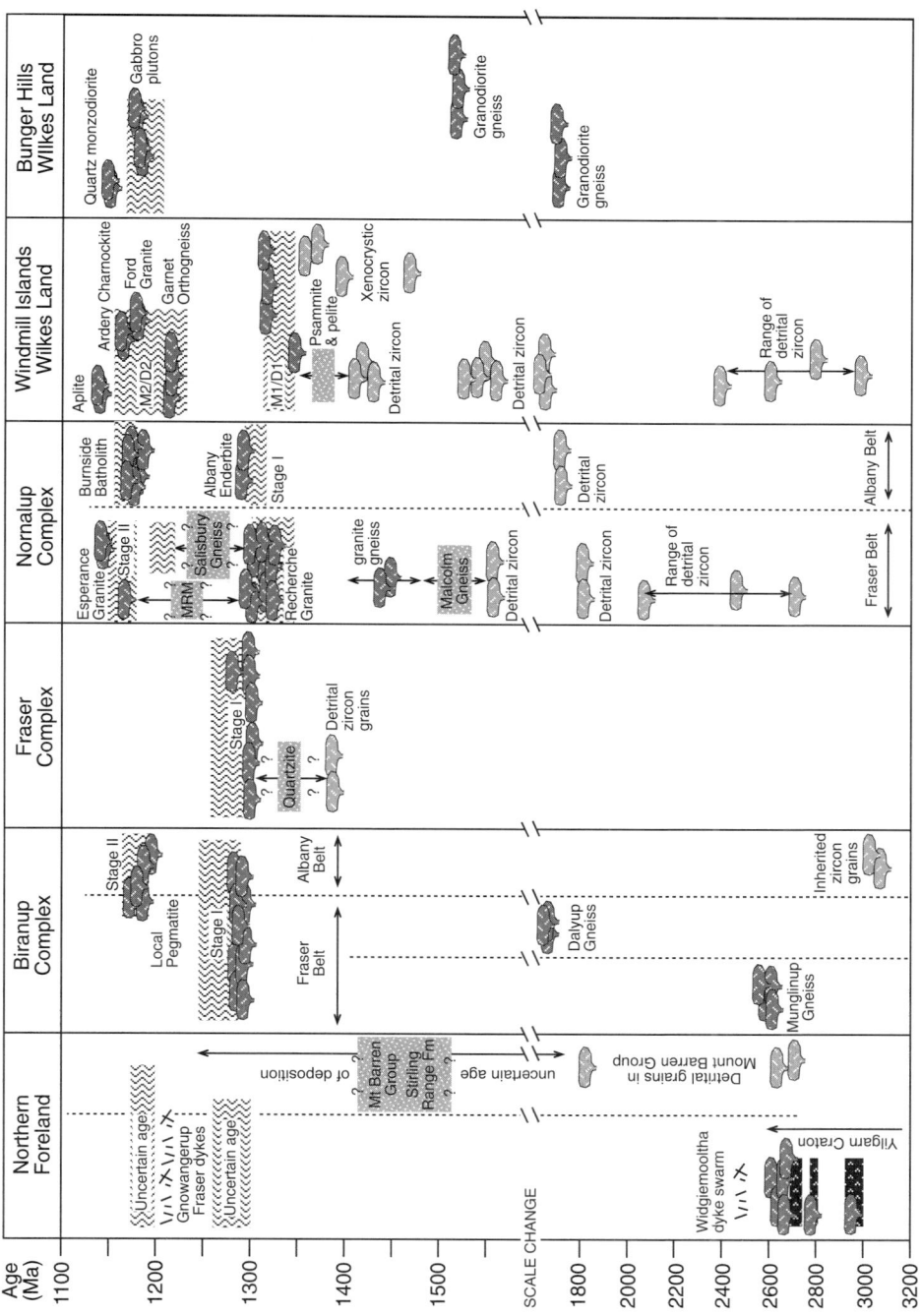

Fig. 6. Time–space diagram for components of the Albany–Fraser–Wilkes Orogen. Time information is from U–Pb zircon data. Data sources are summarized in the text and symbols are as for Figure 2. Stage I and Stage II are the two tectonic episodes distinguished in the Albany–Fraser Orogen by Clark *et al.* (2000), and M1/D1 and M2/D2 are the two tectonic episodes distinguished in the Windmill Islands by Post (2000). MRM, Mount Ragged metasedimentary rocks.

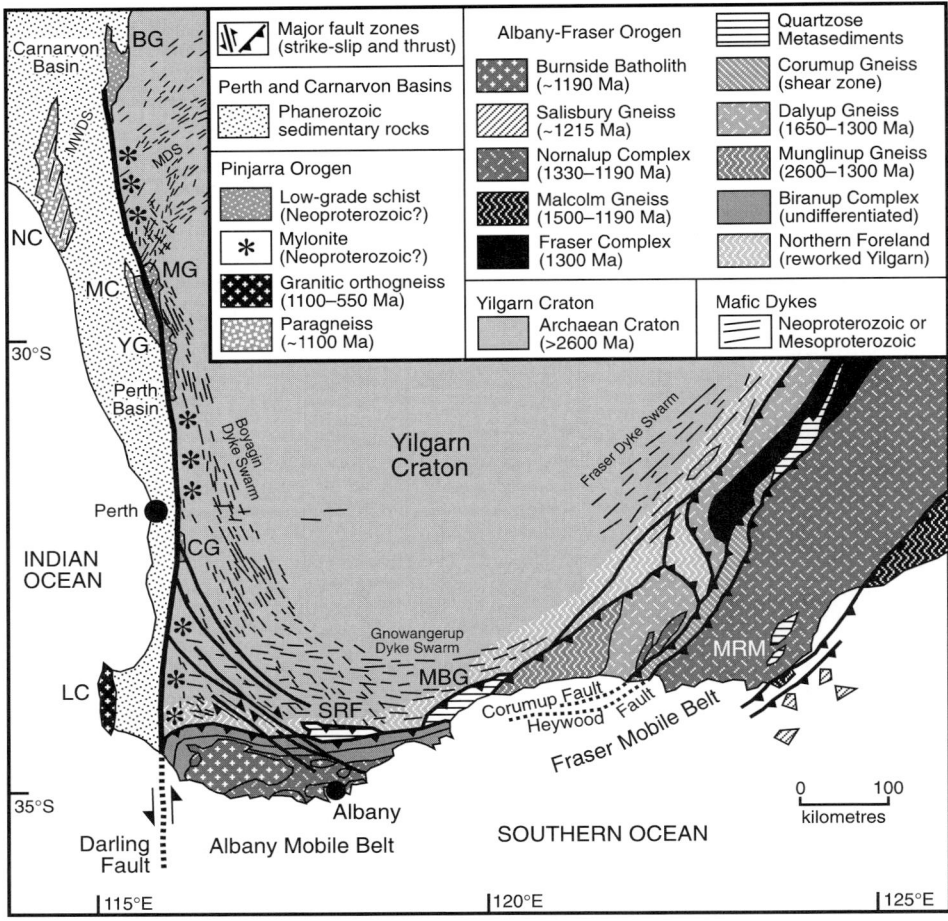

Fig. 7. Geology of southwestern Australia adapted from Myers (1990, 1995) and Clark (1999). BG, Badgeradda Group; CG, Cardup Group; LC, Leeuwin Complex; MBG, Mount Barren Group; MC, Mullingarra Complex; MDS, Muggamurra Dyke Swarm; MWDS, Mundine Well Dyke Swarm; MG, Moora Group; MRM, Mount Ragged metasedimentary rocks; NC, Northampton Complex; SRF, Stirling Range Formation; YG, Yandanooka Group.

1992, 1993) and confirmed by field and isotopic studies. Three of these domains are dominated by quartzofeldspathic lithologies and were named the Northern, Central and Southern Domains in the Albany Belt by Beeson et al. (1988). These are equivalent to the Northern Foreland, and the Biranup and Nornalup Complexes defined by Myers (1990) for both the Albany and Fraser Belts (Fig. 7), and the latter names are used here for consistency with recent geochronological studies. The fourth domain is the Fraser Complex, a gabbroic intrusive suite with a present-day strike length of 450 km.

The Northern Foreland comprises reworked 3000–2600 Ma gneiss and dolerite dykes of the southeastern Yilgarn Craton, and fault-bound quartz-rich metasedimentary rocks of the Stirling Range Formation and the Mount Barren Group, which have been thrust northwards over the craton margin (Fig. 7). Amphibolite facies recrystallization and spaced shear zones with dextral and reverse senses of movement increase in intensity southwards in the foreland to the Albany Belt (Beeson et al. 1988), and much of this tectonism occurred during Stage II of the orogeny, since it overprints the E–W trending Gnowangerup Dyke Swarm dated at c. 1210 Ma (Evans 1999). The NE–SW-trending Fraser Dyke Swarm in the poorly exposed foreland of the Fraser Belt has an identical age (Wingate et al. 2000), but its relationship to major Proterozoic faults and shear zones in this region is unclear. It is likely, however, that most of these structures formed during Stage I tectonism given the isotopic evidence for a limited Stage II overprint in the northwestern Fraser Belt. The age of the lower greenschist facies Stirling Range

Formation is controversial, with 1215 Ma U–Pb metamorphic monazite (Rasmussen et al. 2002) confirming 1150 Ma Rb–Sr whole-rock ages (Turek & Stephenson 1966) and indicating that deposition ocurred significantly earlier than the late Neoproterozoic time frame suggested by possible Ediacaran trace fossils indicate Late Neoproterozoic deposition (Cruse & Harris 1994). Maximum possible ages for deposition are provided by 2000–1900 Ma defortal zircon and monazite (Rasmussen et al. 2002). The greenschist to upper amphibolite facies Mount Barren Group has a maximum depositional age given by detrital zircons with discrete age populations at 1851 ± 30, 2634 ± 14 and 2726 ± 44 Ma (Nelson 1996). These are similar to ages in the Biranup and Nornalup Complexes (see below), but no conclusive comparisons can be made. Late Mesoproterozoic metamorphism of the Mount Barren Group is indicated by range of isotope data (Thom et al. 1981; Weatherly & McNaughton 1995), although the significance of a U–Pb SHRIMP monazite age of 1025 ± 8 Ma (Weatherly et al. 1998) is unclear, given that this is 100 Ma younger than best estimates for the end of tectonism in the Albany–Fraser Orogen.

The Biranup Complex is in thrust contact with the foreland and comprises felsic orthogneiss interlayered with minor paragneiss and metagabbro (Myers 1990). It is best studied in the Fraser Belt, where at least two different protolith components, the Munglinup and Dalyup Gneisses, have been interleaved with metagabbro and pyroxene granulite of the Fraser Complex (Fig. 7; Myers 1995). The Munglinup Gneiss was derived from Archaean monzogranite and biotite tonalite with SHRIMP U–Pb zircon crystallization ages of 2640–2575 Ma and T_{DM} Nd model ages of 2.8–2.7 Ga. It may represent reworked Yilgarn Craton (Nelson et al. 1995), although its youngest components (2595–2575 Ma) are 5–25 Ma younger than any plutons dated in adjacent parts of the craton (e.g. Hill et al. 1992; Nelson 1998), and Myers et al. (1996) interpreted it as an allochthonous Archaean block accreted to the Yilgarn Craton during the Albany–Fraser event. The Dalyup Gneiss comprises biotite–hornblende granodiorite and monzogranite gneiss with SHRIMP U–Pb zircon crystallization ages of 1695–1634 Ma and T_{DM} Nd model ages of 2.2–2.1 Ga (Nelson et al. 1995). It does not correspond to any rocks in the foreland and is exotic to the Yilgarn Craton (Nelson et al. 1995, Myers et al. 1996). Both units were intruded by 1300–1280 Ma granite and granodiorite during Stage I tectonism, which is also taken as the age of pervasive deformation in the host gneisses on the basis of c. 1250 Ma metamorphic zircon growth in one sample of Munglinup Gneiss (Nelson et al. 1995).

The Fraser Complex comprises several fault-bound slices of mafic granulite derived from at least three separate magma sources, all of which have subduction-related geochemical anomalies consistent with an oceanic-arc origin (Condie & Myers 1999). This contrasts with earlier interpretations of the complex as a layered mafic intrusion (Myers 1983). The best constraint on the timing of igneous crystallization is a Sm–Nd isochron age of 1291 ± 21 Ma for well-preserved igneous minerals in a relatively unmetamorphosed sample (Fletcher et al. 1991), but this may not be representative of the complex as a whole. Igneous minerals have been largely replaced by high-grade metamorphic assemblages developed during Stage I tectonism. Granulite facies conditions (6–7 kbar and 800°C) were attained after the complex was intruded by 1301 ± 6 Ma charnockite and overprinted during dextral transpression by higher pressure (8–10 kbar) garnet amphibolite assemblages before the emplacement of aplite dykes at 1288 ± 12 Ma (Clark et al. 1999). Subsequent exhumation associated with upper greenschist retrogression along the northern margin of the complex was complete by c. 1250 Ma, when Rb–Sr isotope systems closed (Fletcher et al. 1991). Psammitic gneiss interleaved with the Fraser Complex contains a single population of well-preserved 1388 ± 12 Ma detrital zircons (Clark et al. 1999) that does not correspond to any known basement rocks in the Albany–Fraser Orogen.

The Corumup Gneiss is a 10 km wide ductile shear zone developed between the Corumup and Heywood Faults at the southeastern margin of the Biranup Complex (Fig. 7). The protoliths of these high-strain rocks include 1700–1650 Ma Dalyup Gneiss and 1300–1280 Ma granite (Nelson et al. 1995; Clark 1999), deformed and metamorphosed during both Stage I and Stage II tectonism. Pervasive Stage I foliations developed at c. 1290 Ma (Clark 1999) and are associated with high-temperature metamorphism (6–7 kbar and 850°C) followed by a higher pressure event (9 kbar and 750°C). This history is identical to that preserved by the Fraser Complex and is assumed to reflect c. 1290 Ma thrusting of the Nornalup Complex over the Biranup Complex. Stage II tectonism produced spaced high-strain zones associated with syntectonic pegmatite at 1180–1170 Ma (Nelson et al. 1995; Clark 1999). Coaxial Stage I and Stage II structures in the Corumup Gneiss are both consistent with dextral transpression. The only isotopic evidence of c. 1200 Ma tectonism NW of the Corumup Fault is an undeformed 1187 ± 12 Ma pegmatite in the Dalyup Gneiss (Nelson et al. 1995), and the Corumup Gneiss is regarded as the northwestern limit of significant Stage II tectonism in the Fraser Belt (Clark 1999).

The Biranup Complex in the Albany Belt is poorly studied, but geophyscial data imply that it is along-strike from the Corumup Gneiss. It comprises granulite facies felsic gneiss with minor paragneiss and mafic lenses, associated with dextral transpressional shear zones developed internally and at its margin with the Northern Foreland (Beeson et al. 1988). Bimodal T_{DM} Nd model ages of 3.0–2.7 and 2.4–2.3 Ga (Fletcher et al. 1983; Black et al. 1992a) are consistent with Munglinup and, less convincingly, Dalyup Gneiss protoliths, and discordant zircons from a gneissic sample with upper intercept ages of 3100–3000 Ma (Black et al. 1992a) could reflect a Munglinup precursor. Pegmatites intruded at 1200–1180 Ma (Black et al. 1992a) cross-cut the gneissic foliation but are themselves deformed, indicating that at least some of the deformation reflects Stage II tectonism. Isotopic evidence of Stage I tectonism has not been found in the Biranup Complex of the Albany Belt, but this could reflect a lack of data rather than a significant difference between the Albany and Fraser Belts.

The Nornalup Complex is dominated in the Fraser Belt by the 1330–1288 Ma Recherche Granite (Myers 1995; Nelson et al. 1995), which has been deformed by Stage I and Stage II tectonism. This granite has T_{DM} Nd model ages of 1.9–1.8 Ga (Nelson et al. 1995) and rare c. 1440 Ma zircon xenocrysts are assumed to come from a gneissic basement to the Nornalup Complex (Clark et al. 2000). This basement is probably equivalent to the Malcolm Gneiss that crops out along the coast at the southeastern margin of the Nornalup Complex (Fig. 7). The Malcolm Gneiss comprises paragneiss units, dominated by psammite and quartzite with minor garnet–biotite–sillimanite metapelite, intruded by felsic and mafic orthogneiss, and all cut by dykes of the Recherche Granite (Clark 1999; Clark et al. 2000). A paragneiss sample yielded a T_{DM} Nd model age of 2260 Ma and contained detrital zircon populations at 2750–2030, c. 1810 and c. 1560 Ma (Nelson et al. 1995). These do not correlate conclusively with any known source terranes in the Albany–Fraser Orogen, but the youngest population provides a maximum depositional age. A poorly defined 1500–1400 Ma zircon population from a granitic orthogneiss within the Malcolm Gneiss has been interpreted as an emplacement age (Myers 1995) and provides a minimum age for deposition. An early foliation in the Malcolm Gneiss was synchronous with partial melting at peak metamorphic conditions of 4–5 kbar and 750°C, some time before the emplacement of the Recherche Granite at 1330–1288 Ma. This fabric is locally transposed by a second foliation that is also developed at the margins of the Recherche Granite, consistent with deformation during granite emplacement. A third deformation produced NW-verging folds across the Nornalup Complex, which are cut by 1313 ± 16 Ma aplite (Clark et al. 2000). These Stage I structures in the Nornalup Complex developed c. 20 Ma before Stage I structures in the Biranup Complex, suggesting that the former was deformed internally before it was thrust over the Biranup Complex at c. 1290 Ma.

Stage I events in the Nornalup Complex were followed by deposition of the Mount Ragged metasedimentary rocks, which are assumed to unconformably overlie the Recherche Granite (Fig. 7). This sequence of quartzite with minor pelite was deposited in deep-water conditions, and detrital zircon populations at c. 1780 and c. 1320 Ma are consistent with local derivation from the Dalyup Gneiss and Recherche Granite (Clark et al. 2000). Stage II tectonism produced spaced ductile shear zones in the Malcolm Gneiss and Recherche Granite, associated with amphibolite facies thrusting followed by dextral reactivation at a lower grade. Movement on these thrusts is dated at 1165 ± 5 Ma from syntectonic pegmatite in the Malcolm Gneiss, indistinguishable from the age of Stage II thrusting in the Corumup Gneiss (Clark et al. 2000). This same deformation produced folds and layer-parallel shearing in the Mount Ragged metasedimentary rocks, associated with amphibolite facies assemblages (4–5 kbar and 550°C) and rutile with a U–Pb age of 1150 ± 15 Ma (Clark et al. 2000). The late tectonic porphyritic Esperance Granite was emplaced into the Nornalup Complex and Mount Ragged metasedimentary rocks at c. 1140 Ma (Nelson et al. 1995).

Nornalup Complex rocks in the Albany Belt have a similar history to those in the Fraser Belt, but outcrops are dominated by late tectonic porphyritic Stage II granite (the Burnside Batholith) emplaced at 1190–1170 Ma (Pidgeon 1990; Black et al. 1992a) and little information is available for other units. T_{DM} Nd model ages of 2.3–2.2 Ga for paragneiss samples and 2.1–1.8 Ga for the late tectonic granite (Fletcher et al. 1983; Black et al. 1992a) are comparable to model ages for the Malcolm Gneiss and Recherche Granite, respectively. Evidence for Stage I tectonism is limited to emplacement of 1289 ± 10 Ma enderbite at Albany (Pidgeon 1990) and U–Pb zircon ages of 1304 ± 3 Ma for partial melting in nearby paragneiss (Clark 1995). This same paragneiss contained c. 1720 Ma detrital grains, consistent with erosion of a Dalyup Gneiss source, and a second generation of metamorphic zircon at 1169 ± 8 Ma synchronous with the emplacement of the Burnside Batholith and the main episode of Stage II thrusting in the Fraser Belt.

Further evidence for the nature of the Stage II event is provided by the Salisbury Gneiss, which crops out on a series of islands to the SE of the Nornalup Complex (Fig. 7) and comprises garnet–cordierite migmatite and mafic granulite with peak metamorphic conditions of 6 kbar and 800°C (Clark et al. 2000). Zircon cores of 1214 ± 8 Ma in the leucosome are interpreted as dating metamorphism and melting, which pre-dates Stage II tectonism elsewhere in the orogen and is synchronous with the emplacement of the Gnowangerup and Fraser Dyke Swarms into the southeastern Yilgarn Craton (Evans

1999; Wingate et al. 2000). Similar-aged mafic dykes occur sporadically >200km north of the Albany–Fraser Orogen (Evans 1999; Pidgeon & Nemchin 2001), indicating significant extension of the craton at this time. Zircon rims in the Salisbury Gneiss are dated at 1182 ± 13 Ma and are assumed to date high-temperature exhumation along ductile thrust zones, which are thus the same age as similar structures developed throughout the Nornalup Complex. Clark et al. (2000) proposed that thrusting of the Salisbury Gneiss over the Nornalup Complex at this time was responsible for metamorphism of the Mount Ragged metasedimentary rocks, but had little metamorphic effect on other units that had already developed Stage I high-grade assemblages. Conversely, widespread evidence for 1214 Ma partial melting in the Salisbury Gneiss implies that this unit was not previously metamorphosed at Stage I (Clark et al. 2000), suggesting a setting similar to the Mount Ragged metasedimentary rocks (i.e. a cover sequence deposited onto Nornalup basement after Stage I tectonism).

Clark (1999) and Clark et al. (2000) proposed a detailed tectonic model for the Albany–Fraser Orogen involving oblique Stage I collision between the Yilgarn and Mawson Cratons, followed by regional extension and renewed Stage II compression. Although there is no compelling evidence for a large Mesoproterozoic ocean basin between the Yilgarn and Mawson Cratons (Wingate & Evans 2003), the oceanic affinity of the Fraser Complex is consistent with an oceanic suture within the Albany–Fraser Orogen. Nelson et al. (1995) argued that the chemistry of Stage I plutons is consistent with syncollisional magmatism whereas Stage II plutons have post-collisional characteristics. Clark (1999) suggested that early Stage I high-temperature tectonism at 1330–1300 Ma could reflect accretion of exotic blocks (Dalyup Gneiss, Fraser Complex) to a continental arc along the margin of the Nornalup Complex, which he regarded as part of the Mawson Craton, although no direct evidence for that arc has been identified. This margin collided with the Yilgarn Craton causing crustal thickening and NW transport of thrust slices at c. 1290 Ma, with much of the strain focused along the Corumup Gneiss. Subsequent extension resulted in basin formation, and deposition of precursors to the Mount Ragged metasedimentary rocks and Salisbury Gneiss, followed by intrusion of the Gnowangerup and Fraser Dyke Swarms and metamorphism of the Salisbury Gneiss under low-pressure/high temperature extensional conditions. Stage II compression thrust the Salisbury Gneiss over the Nornalup Complex at c. 1180–1170 Ma, producing spaced shear zones in the Nornalup Complex and the Corumup Gneiss. Stage II tectonism was associated with emplacement of syn- to late tectonic granite at 1190–1140 Ma.

Coompana Block

Limited drill-core and regional geophysical data are the only sources of information available for the Coompana Block (Fig. 1), which is buried beneath the Eucla Basin and believed to comprise granitic gneiss intruded by undeformed mafic dykes. K–Ar hornblende and biotite ages of 1185 and 1159 Ma, respectively, are the only age data for the block (Drexel et al. 1993), indicating a thermal imprint synchronous with late Mesoproterozoic tectonism in the Albany–Fraser Orogen, although resetting of K–Ar mineral systems at this time has been recognized even further east in reactivated shear zones of the Gawler Craton (Foster & Ehlers 1998). More compelling evidence for a direct link between the Albany-Fraser Orogen, the Coompana Block and the Gawler Craton is that crust beneath the Eucla Basin has a similar geophysical signature to the eastern Nornalup Complex (Shaw et al. 1996), and that the youngest and most abundant detrital zircon population in the Malcolm Gneiss, dated at c. 1560 Ma, matches Kararan-age tectonism in the NW Gawler Craton. This implies that sedimentary precursors to the Malcolm Gneiss were eroded from the c. 1560 Ma Kararan Orogen, in which case the Gawler Craton, the Nornalup Complex and the intervening Coompana Block were a single entity by c. 1450 Ma, the minimum depositional age of the Malcolm Gneiss.

Windmill Islands

Outcrops in the Windmill Islands of East Antarctica comprise layered felsic orthogneiss and pelitic to psammitic paragneiss, associated with rare calc-silicate and banded-iron formation. These were originally believed to preserve a southwards increase in grade developed during a single tectonic event (Blight & Oliver 1977, 1982), but actually reflect a younger M2/D2 granulite-facies overprint on early M1/D1 amphibolite-facies assemblages (Paul et al. 1995; Post et al. 1997). Their tectonic history has been constrained by SHRIMP U–Pb data (Post et al. 1997; Post 2000), which are consistent with the reconnaissance U–Pb data of Oliver et al. (1983) and Williams et al. (1983), but not the discredited data of Lovering et al. (1981). No pre-Mesoproterozoic basement has yet been identified, but inherited zircons in M1 leucosome from a psammitic gneiss have age populations at 3000–2400, 1800–1700, 1600–1500 and 1450–1400 Ma, which are similar to zircon populations in the Nornalup Complex of the Fraser Belt (Fig. 6; Nelson et al. 1995; Clark et al. 2000). T_{DM} Nd model ages of 3.2–2.4 Ga for four samples of psammitic and pelitic gneiss (Post 2000) similarly imply that precursor sediments were derived from a range of Archaean–Proterozoic material. A younger 1342±21 Ma zircon population in M1 leucosome is interpreted as dating the M1 partial melting event, which occurred at upper amphibolite facies conditions (c. 4 kbar and c. 750°C) associated with biotite–sillimanite± garnet assemblages in metapelite (Post 2000). Partial melting was followed by intense deformation and syntectonic granite emplacement at 1315±6 Ma. A second stage of tectonism is associated with D2 folding and a second foliation that increases in intensity to the south. Deformation was synchronous with emplacement of syntectonic garnet-bearing orthogneiss at 1214±10 Ma, and granulite facies metamorphism (5–7 kbar and 850–900°C) produced garnet–cordierite leucosomes in metapelite at 1171±9 Ma (Post 2000). M2/D2 was also associated with the emplacement of the Ford Granite at 1173±9 Ma and Ardery Charnockite at 1163±7 Ma, followed by post-tectonic aplite intrusion at 1138±9 Ma. Syn-D2 orthogneiss and migmatite units contain 1380–1300 Ma zircon xenocrysts, interpreted as relics of the M1/D1 event and possibly a slightly earlier episode of magmatism. T_{DM} Nd model ages for syn-D1 and syn-D2 felsic intrusions are all 2.6–2.2 Ga, with no spatial trends in the data, suggesting that the Windmill Islands are a single crustal terrane (Post 2000) with a history very similar to that of the Nornalup Complex (Fig. 6).

Bunger Hills

Mesoproterozoic basement in the Bunger Hills comprises granulite facies tonalitic to granitic orthogneiss and layered psammitic to pelitic paragneiss, with minor quartzite and marble, intruded by gabbroic to granitic plutons (Fig. 8; Sheraton *et al.* 1993, 1995). T_{DM} Nd model ages of 2.2–1.9 Ga have been determined for two units of granodioritic orthogneiss, which have SHRIMP U–Pb emplacement ages of 1699±15 and 1521±29 Ma, respectively (Sheraton *et al.* 1992). The older of these correlates with the Dalyup Gneiss of the Biranup Complex, but comparisons of the younger body with the Albany–Fraser Orogen are less conclusive, with the zircon age having no obvious counterparts and the Nd model age more consistent with Nornalup Complex rocks. All these rocks were pervasively deformed and interleaved with paragneiss lithologies under granulite facies conditions (5–6 kbar and 750–800°C; Sheraton *et al.* 1993), dated at 1190±15 Ma from metamorphic zircon in one of the orthogneiss samples (Sheraton *et al.* 1992). This was followed by at least two stages of folding during NW–SE compression (Ding & James 1991; Sheraton *et al.* 1993). The later stages of folding were synchronous with intrusion of quartz monzogabbro plutons at 1171±3 and 1170±4 Ma, and a post-tectonic charnockitic quartz monzodiorite was emplaced at 1151±4 Ma (Sheraton *et al.* 1992). T_{DM} Nd model ages for the gabbro and diorite are 2.3 and 1.8 Ga, respectively. These events correlate with Stage II tectonism in the Albany–Fraser Orogen (Fig. 6; Clark *et al.* 2000), but it is unclear whether the lack of evidence for c. 1300 Ma Stage I tectonism in the Bunger Hills is geologically significant (Harris 1995) or a result of limited sampling (Clark *et al.* 2000).

Pinjarra Orogen

The current western margin of the Yilgarn Craton and the Albany–Fraser Orogen is marked by the Darling Fault (Fig. 7), which has a record of normal displacement during Gondwana breakup. This Phanerozoic structure is at least 1500 km long and reactivated an older N–S-trending tectonic boundary associated with tectonism at c. 1080 and 750–500 Ma. The precursor Proterozoic orogen has been called the Darling Mobile Belt (Glikson & Lambert 1973; Harris 1994) and the Pinjarra Orogen (Myers 1990), and the latter name is used here. It is largely concealed by the Phanerozoic Perth and Carnarvon Basins (Fig. 7), and evidence for its evolution comes from three main sources: Proterozoic gneissic inliers exposed as the Northampton, Mullingarra and Leeuwin Complexes to the west of the Darling Fault; low-grade metasedimentary schists of assumed Proterozoic age that crop out on either side of the Darling Fault; and the reworked margins of the Yilgarn Craton and Albany–Fraser Orogen immediately east of the Darling Fault, known as the Darling Fault Zone (Myers 1990).

Correlations between the Pinjarra Orogen and the Denman Glacier region of Antarctica (Fig. 9) have been used to discriminate between potential Australia–Antarctica Gondwana reconstructions (Black *et al.* 1992*b*; Harris 1995). More recently, the identification of Neoproterozoic tectonism in the Prydz Bay region of East Antarctica (Zhao *et al.* 1992; Hensen & Zhou 1995) has indicated that the Pinjarra Orogen is part of a major orogen that divides East Gondwana in two (Fig. 1). Of critical importance is whether the two tectonic episodes culminating at c. 1080 and 550 Ma reflect Mesoproterozoic collision followed by intracratonic Neoproterozoic reactivation or the incorporation of Mesoproterozoic terranes in a continent-scale Neoproterozoic suture zone. The former would need little revision of previous models for a coherent Neoproterozoic East Gondwana, but the latter requires that these models are discarded (Fitzsimons 2000*b*; Boger *et al.* 2001). Evidence for the second

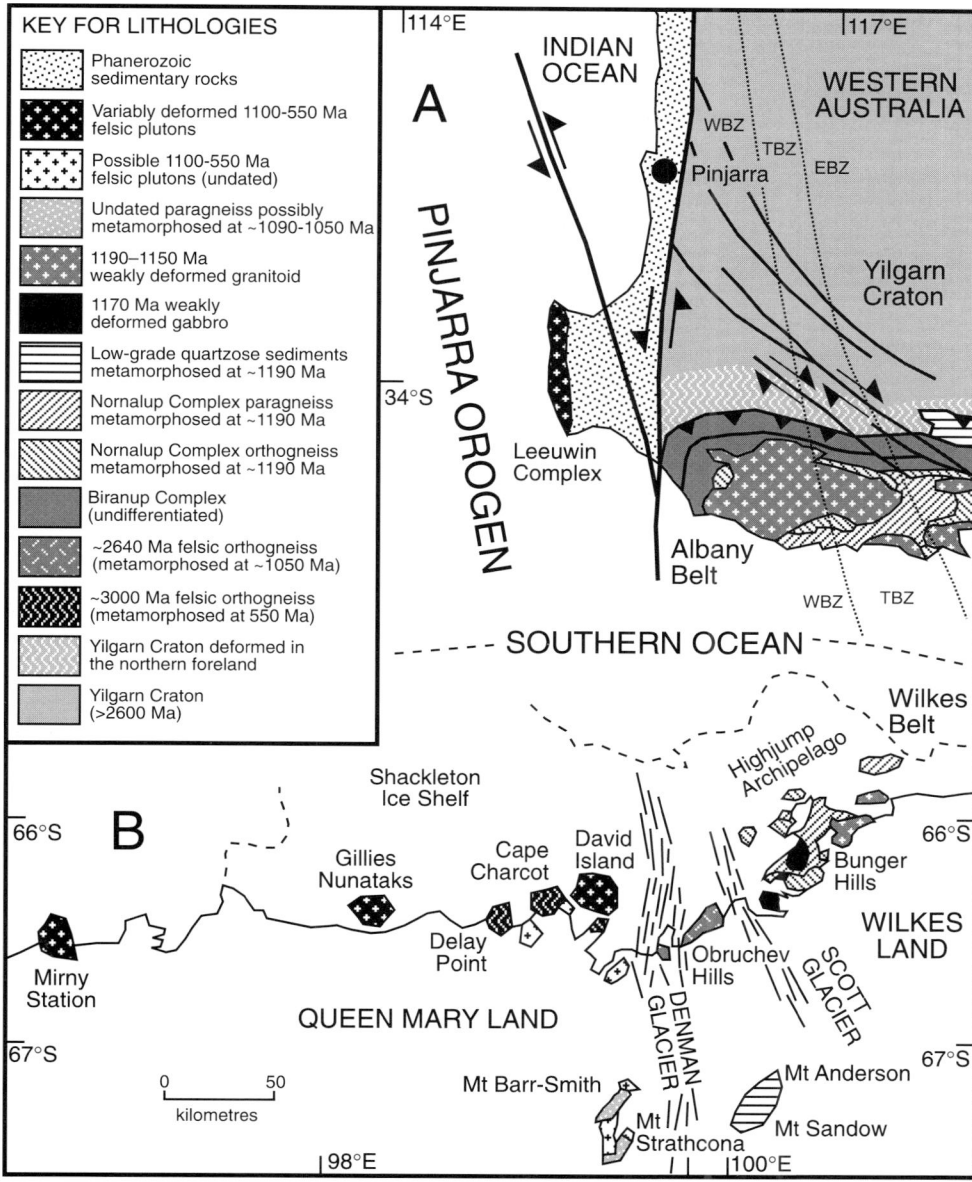

Fig. 8. Geology of southwestern Australia (A) and the Denman Glacier region (B), adapted from Myers (1990) and Sheraton *et al.* (1995). Relative positions of A and B based on a correlation of the Darling Fault with the Denman and/or Scott Glaciers (following Harris 1995). EBZ, Eastern Biotite Zone (no trace of resetting with original Yilgarn or Albany-Fraser ages preserved); TBZ, transitional biotite zone (some resetting); WBZ, western biotite zone (biotite ages reset to *c.* 500Ma) as defined by Libby & De Laeter (1998). Glaciers are denoted by closely spaced subparallel lines. Quartzose metasedimentary rocks at Mt Sandon are correlated here with the Stirling Range Formation, but isotopic data are needed to demonstrate this correlation.

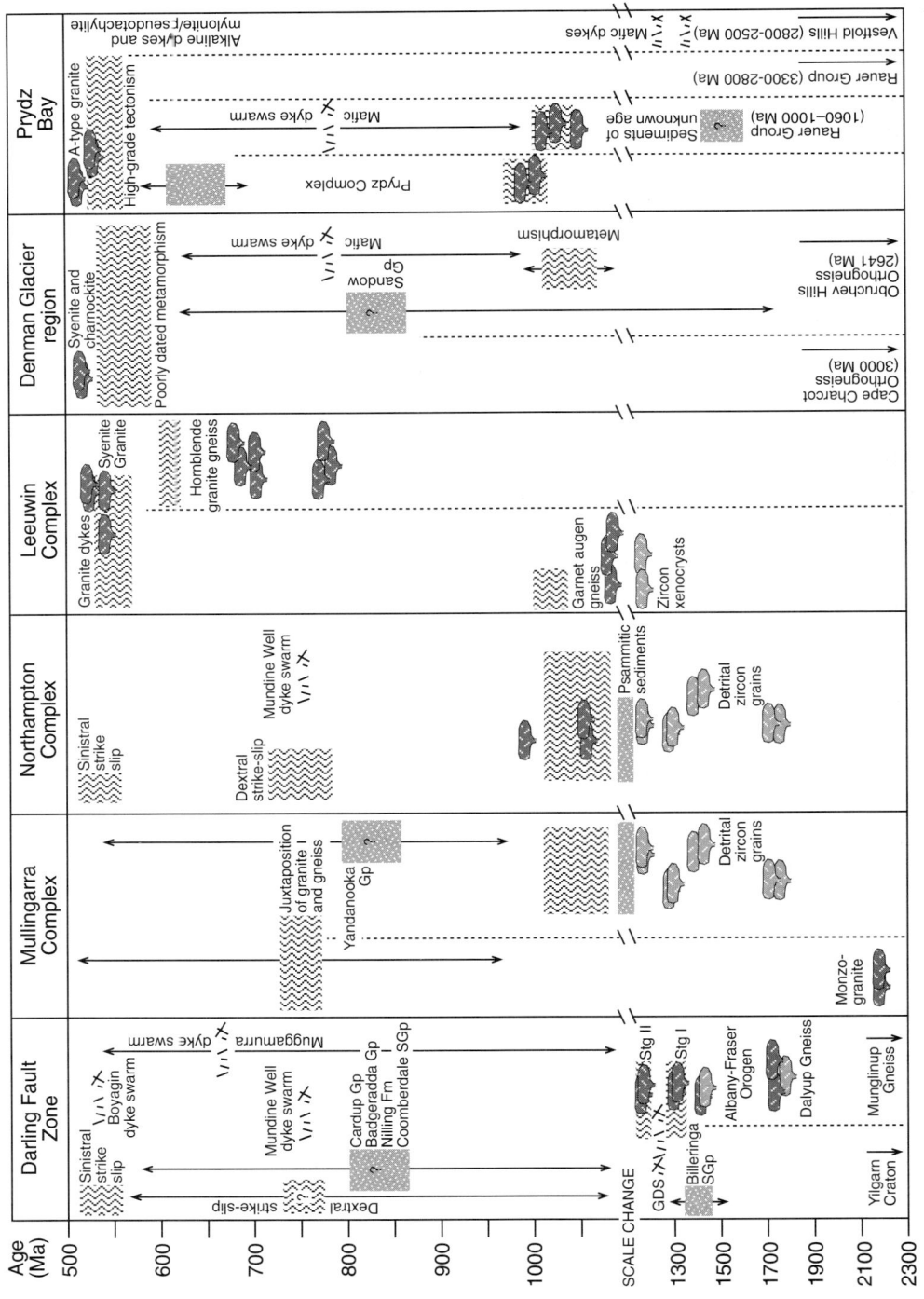

Fig. 9. Time–space diagram for components of the Pinjarra Orogen and comparable regions of Antarctica. Note the considerable uncertainty in timing of most Neoproterozoic events, apart from the emplacement of the Mundine Well Dyke Swarm and synchronous dextral deformation in the Northampton Complex, and granitoid emplacement in the Leeuwin Complex. Data sources are summarized in the text and symbols are as for Figure 2. GDS, Gnowungerup Dyke Swarm; SGp, Subgroup; Stg I and Stg II, Stage I and II tectonism of Clark *et al.* (2000).

situation is presented in this section, which reviews the geology of the Pinjarra Orogen and its Antarctic counterparts in the Denman Glacier and Prydz Bay regions.

Northampton, Mullingara, and Leeuwin Complexes

The Northampton and Mullingarra Complexes are dominated by psammitic paragneiss, with conformable lenses of pelite, quartzite and mafic gneiss (Myers 1990), whereas the Leeuwin Complex is dominated by felsic orthogneiss, ranging from granodiorite to alkali granite in composition, interleaved with anorthosite, leucogabbro and subordinate mafic lenses (Myers 1990; Wilde & Murphy 1990). The oldest rock identified in these three provinces is an unmetamorphosed monzogranite in the Mullingarra Complex with a SHRIMP U–Pb crystallization age of 2181 ± 10 Ma (Cobb et al. 2001), which is in faulted contact with metasedimentary gneiss. SHRIMP U–Pb analysis of detrital zircon from psammitic rocks in the Northampton (Bruguier et al. 1999) and Mullingarra Complexes (Cobb et al. 2001) has identified identical age distributions with major populations at 1900–1600 and 1400–1150 Ma, indicating that they belong to the same sedimentary package. These populations match the age of basement rocks in the Albany–Fraser Orogen (Fig. 9) and were most likely eroded from them (Kriegsman et al. 1999; Fitzsimons 2001). The sediments were pervasively deformed and metamorphosed to granulite facies in the Northampton Complex (6 kbar and 850°C; Kriegsman & Hensen 1998), and amphibolite facies in the Mullingarra Complex (6 kbar and 670°C; Cobb et al. 2001), accompanied by metamorphic zircon growth at 1079 ± 3 and 1058 ± 83 Ma, respectively (Bruguier et al. 1999; Cobb et al. 2001). The Northampton Complex was then intruded by a granite pluton at 1068 ± 13 Ma (Bruguier et al. 1999), followed by folding about shallow northwesterly trending axes, with similar structures identified in the Mullingarra Complex. This tectonic episode concluded with pegmatite dyke emplacement, dated in the Northampton Complex at c. 990 Ma (Bruguier et al. 1999).

Later tectonism in the Northampton Complex was marked by the emplacement of NNE-trending dolerite dykes with a K–Ar whole-rock age of c. 750 Ma (Embleton & Schmidt 1985) and a palaeomagnetic pole indistinguishable from the 755 ± 3 Ma Mundine Well Dyke Swarm of the Pilbara Craton (Wingate & Giddings 2000). The orientations of these dykes and later brittle–ductile shear zones are consistent with N–S dextral wrenching (Byrne & Harris 1993), and they are cut by sinistral, SSE-trending, greenschist to amphibolite facies shear zones that developed at c. 550 Ma according to K–Ar data from dykes adjacent to these shear zones (Embleton & Schmidt 1985). There are no reports of mafic dykes or sinistral shear zones cutting the Mullingarra Complex, although this could reflect poor outcrop rather than a significant difference in geological history.

Mesoproterozoic tectonism in the Northampton and Mullingarra Complexes has been widely regarded as evidence for a Grenville-age collisional suture between Australia and India (Myers et al. 1996; Bruguier et al. 1999), although Myers (1990, 1993) did acknowledge that these terranes could have formed elsewhere and been juxtaposed with the Yilgarn Craton after late Mesoproterozoic metamorphism. Cobb et al. (2001) and Fitzsimons (2001) have adopted this latter model, and argued that the complexes were transported to their present locations during the Neoproterozoic. There is no record of c. 1080 Ma tectonism along the adjacent edge of the Yilgarn Craton, which preserves Palaeoproterozoic Rb–Sr biotite ages (see below) even though the Mullingarra Complex is only 15 km from the craton margin. Furthermore, the undeformed Palaeoproterozoic granite body in the Mullingarra Complex must have been juxtaposed against the adjacent gneiss after 1080 Ma metamorphism, providing conclusive evidence for significant tectonic displacement. The identification of only four Archaean grains out of 132 detrital grains analysed from three samples of Northampton and Mullingarra paragneiss is also difficult to reconcile with deposition immediately adjacent to the Yilgarn Craton but up to 800 km from likely source rocks in the Albany–Fraser Orogen. Fitzsimons (2001) argued that these sediments were deposited south of their present position, closer to the Albany–Fraser Orogen, some time between 1130 and 1080 Ma. They were buried and metamorphosed at c. 1080 Ma, and later translated northwards by dextral strike-slip and accreted to the Yilgarn Craton. This translation and accretion is assumed to have taken place at c. 755 Ma, to account for dyke emplacement and brittle-ductile shearing in the Northampton Complex at this time. The magnitude of dextral displacement is poorly constrained, although Fitzsimons (2001) has suggested that it could have been as great as 750 km, since the lack of Yilgarn-age detritus in the Northampton and Mullingarra sediments might imply deposition to the south of a watershed in the Albany–Fraser Mountains. Close geological, chemical and palaeomagnetic correlations between the Northampton dykes and similar dykes in the craton suggest that the Northampton Complex has not moved significantly relative to the craton margin since 755 Ma (Wingate & Giddings 2000). This implies that dykes and brittle–ductile shears in the complex might reflect the final stages of accretion to the northwestern Yilgarn Craton.

Further constraints on this history are provided by the Leeuwin Complex, where several types of felsic gneiss have been distinguished on mineralogy and the degree of deformation (Wilde & Murphy 1990). U–Pb SHRIMP zircon data reveal three dominant periods of orthogneiss emplacement (Nelson 1996, 1999; Collins & Fitzsimons 2001). Garnet-bearing augen gneiss has crystallization ages of 1091±8 and 1091±17 Ma, and suffered lead-loss at c. 1020–1000 Ma, whereas hornblende-bearing gneiss was emplaced over an extended period from 800 to 650 Ma with some evidence for metamorphism at 630–600 Ma. All of these rocks were deformed and metamorphosed to granulite or upper amphibolite facies, and are locally highly strained and compositionally banded. Unlike the older gneisses, samples of hornblende–biotite granite and aegirine–augite syenite emplaced at 540±6 and 535±9 Ma are not compositionally banded, but are still foliated and folded. An unfoliated hornblende–biotite monzogranite dyke emplaced at 524±12 Ma provides a minimum age for this deformation. Leeuwin Complex rocks have yielded T_{DM} Nd model ages of 1.6–1.1 Ga (McCulloch 1987; Black et al. 1992b; Fletcher & Libby 1993).

This geochronological evidence for a prolonged magmatic history is difficult to reconcile with a reported lack of geochemical evidence for multiple magma sources (Wilde & Murphy 1990). However, a reassessment of published geochemical data (Wilde & Murphy 1990; Nelson 1995; Wilde 1999) by Collins & Fitzsimons (2001) found that garnet–biotite orthogneiss emplaced at 1090 Ma has trace element concentrations consistent with a syn-collisional origin, whereas 800–650 and 540–520 Ma protoliths are typical A-type granites. Fitzsimons (2001) argued that 1090 Ma magmatism in the Leeuwin Complex was probably related both spatially and genetically to 1080 Ma metamorphism of the Northampton and Mullingarra paragneiss, and, although not conclusive, geochemical evidence for a collisional setting at 1090 Ma is consistent with burial and metamorphism of juvenile Northampton and Mullingarra sediments at 1080 Ma to depths of 20 km. It is possible that the 800–650 Ma magmatism is related to transcurrent transport and accretion of the Northampton and Mullingarra Complexes to the Yilgarn Craton. Wilde & Murphy (1990) noted that the emplacement of relatively alkaline granites containing minerals such as aegirine–augite and arfvedsonite was consistent with a continental rift environment at 550–500 Ma. Another possibility was suggested by Harris (1994), who noted that this time period was characterized by sinistral transcurrent displacement both within the Leeuwin Complex and the Pinjarra Orogen as a whole, and argued that local extension in the Leeuwin Complex was generated within an extensional jog of the strike-slip system.

Low-grade schist

Low-grade metasedimentary rocks crop out at several localities adjacent to the Darling Fault (Fig. 7). The Cardup, Moora and Badgeradda Groups, and the Nilling Formation occur east of the fault and unconformably overlie the Yilgarn Craton, whereas the Yandanooka Group lies to the west and unconformably overlies the Mullingarra Complex (Myers 1990). The Cardup Group comprises folded shale, sandstone and conglomerate with a pervasive cleavage, and is cut by dolerite dykes. The Moora Group includes the Billeringa Subgroup, a gently folded series of sandstone, siltstone, basalt and tuff with a moderate cleavage, overlain by undeformed sandstone, shale and dolomite of the Coomberdale Subgroup. Both subgroups are cut by dolerite dykes and the age of the volcanics is poorly constrained by a 1390±140 Ma Rb–Sr isochron (Compston & Arriens 1968; Giddings 1976). The Badgeradda Group comprises folded sandstone, shale and conglomerate without a penetrative cleavage, whereas the adjacent Nilling Formation is composed of folded greywacke and sandstone with a spaced cleavage. The Yandanooka Group comprises undeformed but steeply dipping siltstone, conglomerate and sandstone. It is not cut by mafic dykes and is likely to be younger than the nearby Billeringa Subgroup given that the metamorphic age of the underlying Mullingarra gneiss is younger than the Rb–Sr age for the Billeringa Volcanics (Fig. 9). There are no other constraints on the depositional age of these units, which is thought to be late Mesoproterozoic–Neoproterozoic, nor on the timing of deformation. The degree of deformation and dip of the bedding commonly increase toward the Darling Fault, suggesting that deformation is related to movements along the Proterozoic precursor to this structure.

Darling Fault Zone

There is widespread evidence for reworking of the westernmost Yilgarn Craton and the Albany–Fraser Orogen during the development of the Pinjarra Orogen. The most obvious features are three swarms of subvertical dolerite dyke assumed to have intruded in the Neoproterozoic (Myers 1990). The 755 Ma Mundine Well Dyke Swarm cuts the Yilgarn Craton and Northampton Complex, as well as the Gascoyne Complex and Pilbara Craton to the north (Wingate & Giddings 2000), and its northwesterly trend is consistent with dextral movement along the

Pinjarra Orogen at this time (as discussed above). The Muggamurra Dyke Swarm cuts the northwestern Yilgarn Craton and Moora Group, and is of unknown age. It changes trend from northeasterly to east-northeasterly away from the Darling Fault. The Boyagin Dyke Swarm cuts the southwestern Yilgarn Craton, the Albany Belt, the Moora and Cardup Groups, and the Stirling Range Formation (Myers 1990; Harris & Li 1995), and changes trend from north-northwesterly to northwesterly away from the Darling Fault, subparallel to faults that cut the SW Yilgarn Craton (Fig. 7). Dyke orientation is consistent with sinistral movement along the southern Pinjarra Orogen at the time of emplacement, which is believed to be 550–500 Ma from palaeomagnetic constraints (Harris & Li 1995), and a Rb–Sr age of 590–560 Ma from the sheared margins of a dyke (Compston & Arriens 1968).

Structural reworking of the Darling Fault Zone is dominated by regional-scale sinistral rotation of structures in the western Albany Belt through 90° to bring them into parallelism with the Pinjarra Orogen (Figs 7 & 8; Beeson et al. 1995). N–S-trending amphibolite-facies mylonite zones are common within 20km of the Darling Fault along the western edge of the Albany Belt (Beeson et al. 1995) and Yilgarn Craton (Blight et al. 1981; Bretan 1986). These mylonites are best studied in the region between the Moora Group and the Albany Belt (Fig. 7), where they show evidence for at least three stages of movement (Bretan 1986). Steep dip-slip lineations probably reflect Phanerozoic reactivation of these structures during Gondwana breakup, but earlier subhorizontal lineations record a Precambrian strike-slip history, consistent with geophysical evidence that the present-day Darling Fault has reactivated an older subvertical transcurrent fault that penetrates the entire crust (Dentith et al. 1993). Kinematic data predominantly indicate sinistral strike-slip movement, consistent with the observed sense of drag on the Albany Belt, but there is also limited evidence for an earlier stage of dextral wrenching. Bretan (1986) suggested that dextral movement was late Archean on the basis of late Archaean Rb–Sr whole-rock ages reported for granite both adjacent to and within a N–S mylonite (Blight et al. 1981). However, Blight et al. (1981) acknowledged that the Rb–Sr system may not have been reset on a whole-rock scale by the mylonite, in which case deformation could be younger than Archaean. Fitzsimons (2001) noted that this record of dextral movement is consistent with structural and detrital evidence for dextral movement of the Northampton Complex, and argued that they all reflect a single deformational event at 750 Ma.

Rb–Sr biotite ages decrease systematically westwards across the southern Yilgarn Craton and Albany–Fraser Orogen (Fig. 8; Libby & De Laeter 1998; Libby et al. 1999), defining the Eastern, Transitional and Western Biotite Zones of Libby & De Laeter (1998). The latter is a broad region of isotopic resetting down to 450 Ma, attributed to the sinistral transcurrent movement that caused pervasive deformation and metamorphism in the Leeuwin Complex, and southwestern Yilgarn and Albany–Fraser margins at 550–500 Ma. It is likely that the Leeuwin Complex was translated southwards along the craton margin to its present position during this strike-slip movement, thus accounting for the sinistral drag of the Albany Belt. The Western Biotite Zone narrows northwards along the Darling Fault Zone, and there is only limited resetting of Archaean ages adjacent to the Mullingarra and Northampton Complexes where craton margin Rb–Sr biotite ages are c. 1680 Ma. Together with the lack of penetrative sinistral wrenching at 550 Ma in the Northampton and Mullingarra Complexes, this implies that late Neoproterozoic sinistral deformation was focused west of the Northampton and Mullingarra Complexes (Harris 1994).

Denman Glacier region

Both Black et al. (1992b) and Harris (1995) have emphasized similarities between the Denman Glacier region and the Pinjarra Orogen of Western Australia, with Harris noting that the Denman Glacier could be situated above a major crustal discontinuity that represents the southern extension of the Darling Fault. As discussed above, the Bunger Hills have a history of late Mesoproterozoic tectonism that corresponds closely to that of the Albany Belt, but there is no evidence for this event west of the Denman Glacier (Figs 8 & 9). This latter region is dominated by variably deformed felsic plutonic rocks, and two samples from immediately west of the glacier have been dated using SHRIMP U–Pb (Black et al. 1992b). Post-tectonic syenite from David Island has an emplacement age of 516 ± 2 Ma, consistent with a Rb–Sr whole-rock isochron age of 502 ± 24 Ma for charnockite samples collected further west near Mirny Station (McQueen et al. 1972). These both correlate with late granite and syenite in the Leeuwin Complex. Tonalitic orthogneiss from Cape Charcot has an emplacement age of 3003 ± 8 Ma, and a second zircon population at 2889 ± 9 Ma is interpreted as metamorphic. Both populations have been variably reset to lower intercept ages of 567 ± 49 and 573 ± 58 Ma, interpreted as a second metamorphic event. This rock is much older than any samples identified in the Leeuwin Complex, or elsewhere in the Pinjarra Orogen, but the absence of any Mesoproterozoic tectonism is consistent with juxtaposition of this orthogneiss with the nearby Bunger Hills sometime in the Neoproterozoic (Black et al. 1992b). Other rock types, including biotite–garnet gneissic granite, hornblende–biotite orthogneiss, and psammitic and pelitic gneiss, are not unlike rocks found in the Pinjarra Orogen, and folded low-grade quartzite, psammite and siltstone at Mounts Amundsen and Sandow are likely to correspond to the Cardup Group or the Stirling Range Formation, but more geochronology is needed before any reliable comparisons can be made.

Black et al. (1992b) reported T_{DM} Nd model ages of 3.3 Ga for the Archaean orthogneiss and ages of 2.3–1.6 Ga for five other samples from the Denman Glacier area, pointing out that these are older than T_{DM} Nd model ages for the Leeuwin Complex, so making a Pinjarra correlation unlikely. However, the Denman ages are comparable to T_{DM} Nd model ages for granitic basement to the Pinjarra Orogen intersected in drill core (2.2–2.0 Ma; Fletcher et al. 1985; Fletcher & Libby 1993). Further links with the Pinjarra Orogen are provided by Archaean orthogneiss in the Obruchev Hills, immediately east of the Denman Glacier and some 20 km west of the Bunger Hills (Fig. 8). This has a SHRIMP U–Pb emplacement age of 2641 ± 15 Ma, and a lower intercept metamorphic age of 1040 ± 53 Ma (Sheraton et al. 1992). Sheraton et al. (1990, 1993, 1995) considered this unit as part of the Bunger Hills terrane due to the Mesoproterozoic metamorphic age and the presence of mafic dykes similar to those in the Bunger Hills. This metamorphic age is, however, not equivalent to the c. 1190 Ma event in the Bunger Hills, even when the relatively large uncertainty is taken into account (Fig. 9). In contrast, it is within error of c. 1080 Ma metamorphism in the Northampton and Mullingarra Complexes, and is interpreted here as part of the Pinjarra Orogen. If this is correct, then the extension of the Darling Fault might lie beneath the Scott Glacier rather than the Denman Glacier (Fig. 8). Various generations of undated mylonite and brittle fracture overprinting late Mesoproterozoic gneiss in the Bunger Hills (Sheraton et al. 1993; Wilson 1997) might correspond to mylonite structures in the Darling Fault Zone, but late Mesoproterozoic K–Ar biotite ages from the Windmill Islands (Webb et al. 1964) indicate that the Neoproterozoic thermal overprint did not extend farther east than the Bunger Hills.

Prydz Bay region

Reviews of the Prydz Bay region have been presented by Harley & Fitzsimons (1995), Fitzsimons (1997), Harley (2003) and Zhao et al. (2003). Outcrops in eastern Prydz Bay (Fig. 5) comprise an Archaean craton in the Vestfold Hills (2800–2500 Ma; Black et al. 1991; Snape et al. 1997) and a composite terrane in the Rauer Group, with 3300–2800 and 1060–1000 Ma protoliths (Kinny et al. 1993). U–Pb zircon systematics are partially reset at 550–490 Ma in both protoliths, consistent with their juxtaposition during Cambrian tectonism (Hensen & Zhou 1997). Interleaved ortho- and paragneiss of the Prydz Complex in southern Prydz Bay was metamorphosed to granulite-facies conditions (6 kbar and 850°C; Fitzsimons 1996; Carson et al. 1997) at c. 530 Ma (Zhao et al. 1992; Hensen & Zhou 1995; Carson et al. 1996; Fitzsimons et al. 1997), and then intruded by post-tectonic A-type granites at c. 500 Ma (Tingey 1991). Zircons of 1200–700 Ma within the paragneiss are probably detrital (Zhao et al. 1995), suggesting late Neoproterozoic deposition, whereas 1000–900 Ma U–Pb zircon and garnet Sm–Nd ages in mafic units (Hensen & Zhou 1995; Zhao et al. 1995) suggest that the orthogneiss experienced c. 1000 Ma tectonism before the c. 530 Ma event. U–Pb ages of 540–500 Ma for felsic orthogneiss, syntectonic granite, and post-tectonic granodiorite and charnockite in the Grove Mountains (Zhao et al. 2000; Mikhalsky et al. 2001a) indicate that Cambrian tectonism extended inland from Prydz Bay for at least 200–300 km (Fig. 5).

High-grade tectonism in the Prydz Bay area was contemporaneous with magmatism and pervasive deformation in the southern Pinjarra Orogen at 550–500 Ma. The spatial distribution of these two regions in a Gondwana reconstruction (Fig. 1) suggests that they are likely to be parts of a single broad orogenic zone (Carson et al. 1996; Hensen & Zhou 1997; Fitzsimons 2000a), although Boger et al. (2001) and Harley (2003) have argued that they might be separate belts. Both Prydz Bay and the Denman Glacier contain c. 3000, 2600 and 1100–1000 Ma terranes juxtaposed during Neoproterozoic tectonism, and similar accretion events are identified in the Pinjarra Orogen. However, whereas Neoproterozoic sediments in the Pinjarra Orogen and the Denman Glacier region have only been mildly affected by this tectonism, those in Prydz Bay were buried to depths of 25 km during high-temperature metamorphism and deformation, followed by 10 km of exhumation after peak metamorphism (Fitzsimons 1996; Carson et al. 1997). This led Fitzsimons (1996) and Carson et al. (1997) to tentatively ascribe metamorphism in Prydz Bay to collision at a convergent plate margin followed by extensional exhumation as the orogen collapsed.

Southern Prince Charles Mountains

Boger et al. (2001) have identified Cambrian tectonism along the southeastern edge of the Lambert Terrane in the southern Prince Charles Mountains (Fig. 5). Syn- and post-tectonic granitic dykes in this area crystallized at c. 510 and 490 Ma, and garnet-bearing orthogneiss was either emplaced or deformed and metamorphosed at c. 550 Ma. The Ruker Terrane to the south has no deformation younger than 2650 Ma (Boger et al. 2001), whilst the Fisher Terrane to the north is characterized by c. 1300 Ma lower amphibolite facies metavolcanics intruded by 1020–980 Ma syenite and granite (Kinny et al. 1997). Boger et al. (2001) argued that this 550–500 Ma magmatism and deformation was evidence for the Lambert Terrane being a Cambrian suture, and the continuation of the high-grade belt exposed in Prydz Bay. Whilst available data are consistent with this model, there are some grounds for caution. The outcrops visited are the easternmost exposure in the southern Prince Charles Mountains, at the margin of the Lambert Graben. Evidence for 550–500 Ma tectonism has been noted elsewhere along this margin, in particular along the eastern shore of Beaver Lake 300 km north of Cambrian outcrops in the Lambert Terrane. Late Neoproterozoic–Cambrian tectonism at Beaver Lake included the intrusion of garnet-bearing granite dykes, amphibolite- to possibly granulite-facies metamorphism, partial melting and shear zone development (Manton et al. 1992; Scrimgeour & Hand 1997), but was interpreted as the marginal influence of tectonism focused further east rather than the gateway to an E–W suture. Evidence for a Cambrian suture in the Lambert Terrane remains inconclusive until it can be demonstrated that 550–500 Ma tectonism extends west of the Mawson Escarpment and away from the Lambert Graben that has already been established as a zone of Cambrian activity. If the continuation of the Prydz Complex does pass through the Lambert Terrane, it must change in character from a broad (>400 km wide) zone of pervasive tectonism to a narrow (c. 100 km wide) corridor between older terranes.

Discussion

The preceding review has covered a lot of ground, both in terms of geographical area and geological time, and the information presented is of relevance to several issues of critical importance to our understanding of East Gondwana. In particular, some discussion is warranted of the precise Proterozoic fit between Australia and Antarctica, the likely extent of the Mawson Craton, the path taken by the Pinjarra Orogen as it disappears under the Antarctic ice sheet, the tectonic setting of this orogen and whether it is possible to correlate Mesoproterozoic terranes across it from Australia into India.

Gondwana fit of Australia and Antarctica.

A number of Proterozoic boundaries and lithologies can be correlated between the coastlines of Australia and Antarctica, and are therefore potential piercing points that constrain the correct fit between these two continents in a Gondwana reconstruction.

(1) Small but distinctive outcrops of Palaeoproterozoic phyllite at Cape Hunter, Terre Adélie (c. 142.5°E, c. 67.0°S) and Coffin Bay Peninsula, South Australia (c. 135.5°E, c. 34.5°S) were deformed during the Kimban Orogeny, and provide a precise link between Australia and Antarctica at c. 1700 Ma (Fig. 4).
(2) The Darling Fault Zone of Western Australia (c. 115.5°E, c. 34.5°S) is a narrow structural boundary between late Mesoproterozoic gneiss of the Albany–Fraser Orogen and Neoproterozoic gneiss of the Pinjarra Orogen, which can be correlated with structures likely to lie beneath the Denman and Scott Glaciers of Queen Mary Land (c. 100.0°E, c. 66.5°S), providing a precise link between Australia and Antarctica at c. 500 Ma (Fig. 8).

The suitability of these two features as independent piercing points has been noted by Oliver & Fanning (1997) and Harris (1995), respectively, but importantly these two correlations, which are separated spatially by 1800 km and temporally by 1200 Ma, can both be satisfied by a single reconstruction between Australia and Antarctica (e.g. Fig. 1). This strongly supports the reliability of the two correlations. Many reconstructions of Gondwana are more-or-less consistent with these two correlations (e.g. Müller et al. 2000), although others (e.g. Veevers 2000) require a small but significant offset of these basement geological features. Reconstructions based on these two basement piercing points have a number of other important implications for Proterozoic correlations. Firstly, they require that the Windmill Islands are along-strike from the Nornalup Complex of the Albany–Fraser Orogen, as suggested by the geological relationships discussed above. Secondly, they require that the Transitional Biotite Zone (Libby & De Laeter 1998), marking the easternmost extent of Cambrian isotopic resetting in the Rb–Sr biotite system, should pass between the Windmill Islands and the Bunger Hills, again consistent with relationships discussed above.

Geographical extent of the Mawson Craton

Striking geological similarities between gneisses in South Australia and Terre Adélie–King George V Land indicate the existence of a Palaeoproterozoic

Australo-Antarctic continental block called the Mawson Craton (Fanning et al. 1995, 1996), and it is likely that this craton extends as far south as the Miller Range in the Transantarctic Mountains, which has a very similar Palaeoproterozoic history (Fig. 2; Goodge et al. 2001). The lateral extent of this continental block is critical to Proterozoic continental reconstructions, since it was one of the main building blocks of Rodinia, the supercontinent widely believed to have assembled at $c.1000$ Ma (Hofmann 1991; Dalziel 1991; Moores 1991). However, in the absence of suitable geophysical data for rocks buried beneath the Antarctic ice sheet, constraints on the size and shape of the Mawson Craton are limited by the sparse outcrop available.

Geochronological data from the Ruker Terrane of the southern Prince Charles Mountains and the Read Complex of the Shackleton Range do show some similarities with events in the Mawson Craton, but are far from sufficient to demonstrate an unambiguous link with it. Goodge et al. (2001) suggested that the Bunger Hills are another likely fragment of the craton, with evidence of magmatism both at $c.$ 2600 and $c.$ 1700 Ma. However, as described above, these two magmatic phases were not identified together. The 2600 Ma orthogneiss crops out in the Obruchev Hills and has no record of a Palaeoproterozoic overprint; the 1700 Ma orthogneiss occurs in the Bunger Hills, where 2300–1800 Ma model ages imply limited influence of Archaean crust. Furthermore, 1700 Ma orthogneiss in the Bunger Hills most likely correlates with the Dalyup Gneiss of the Albany–Fraser Orogen. The Dalyup Gneiss was probably accreted to the western margin of the Nornalup Complex during late Mesoproterozoic collision in the Albany–Fraser Orogen, and the Nornalup Complex is probably part of the Coompana Block that collided with the western margin of the Gawler Craton during the Kararan Orogeny at 1560 Ma. It follows that one or more Mesoproterozoic sutures are likely to lie between the Bunger Hills and the Mawson Craton. The allocation of basement outcrops to a specific craton based solely on similarities in the age of one or two events is dangerous in the absence of evidence for tectonic continuity, since tectonism can occur simultaneously in several locations on different continents. Precise age data are needed to make or break tectonic correlations, but these correlations should be based on multiple events that provide a distinctive and unique fingerprint, and also be supported by some knowledge of the regional structure, particularly when the two regions being correlated are separated by wide areas of no exposure. Conversely, different parts of a single block are unlikely to have exactly the same tectonic setting at any one time, and it should not be assumed that outcrops with different sequences of tectonic events must have belonged to different crustal fragments. These ambiguities are critical to any evaluation of likely basement correlations in East Antarctica and reinforce the need for regional geophysical data to bridge the gaps.

Southern extension of the Pinjarra Orogen

The Darling Fault Zone at the eastern margin of the Pinjarra Orogen is likely to continue southwards beneath the Scott or Denman Glaciers, but there are no features beyond the southernmost outcrop in this region (200 km from the coast) to indicate its path inland. The only constraint is that it should lie NW of all outcrops of the Mawson Craton. There are more outcrops, and hence more constraints, at the western margin of the orogen in Prydz Bay. In particular, the Prince Charles Mountains expose a 500 km N–S section through the Antarctic continent at the western edge of the Prydz Complex (Fig. 5). The Beaver Terrane of the northern Prince Charles Mountains is immediately west of the Prydz Complex and comprises early Neoproterozoic gneiss (1000–900 Ma) correlated with the Eastern Ghats of India (Fitzsimons 2000b; Dobmeier & Raith 2003). Kriegsman (1995) and Wilson et al. (1997) suggested that late Neoproterozoic tectonism in Prydz Bay continued westwards through the Beaver Terrane, but although there is local evidence for a late Neoproterozoic–Cambrian thermal and structural overprint at the eastern margin of this terrane (Manton et al. 1992; Scrimgeour & Hand 1997), elsewhere there is little or no record of late Neoproterozoic tectonism and these rocks represent the western foreland of the Prydz Complex. Fitzsimons (2000a) noted that the western margin of the Prydz Complex runs along the Lambert Graben, a Mesozoic structure beneath the Amery Ice Shelf–Lambert Glacier Ice Drainage System that is related to Gondwana breakup (Fedorov et al. 1982). He suggested that the Lambert Graben reactivated the Neoproterozoic boundary between the Prydz Complex and the Prince Charles Mountains' foreland, analogous to reactivation of the eastern margin of the Pinjarra Orogen by the Mesozoic Darling Fault. If so, then the edge of the Prydz Complex follows the Lambert Glacier to the southern Prince Charles Mountains.

Two models have been proposed for the continuation of late Neoproterozoic tectonism through the southern Prince Charles Mountains (Fig. 10). Boger et al. (2001) suggested that it runs westwards through the Lambert Terrane to intersect with the East African Orogen somewhere in Dronning Maud Land, although, as discussed above, more data are needed to demonstrate that Cambrian tectonism identified in the eastern Lambert Terrane does indeed continue westwards. If it does, then both the

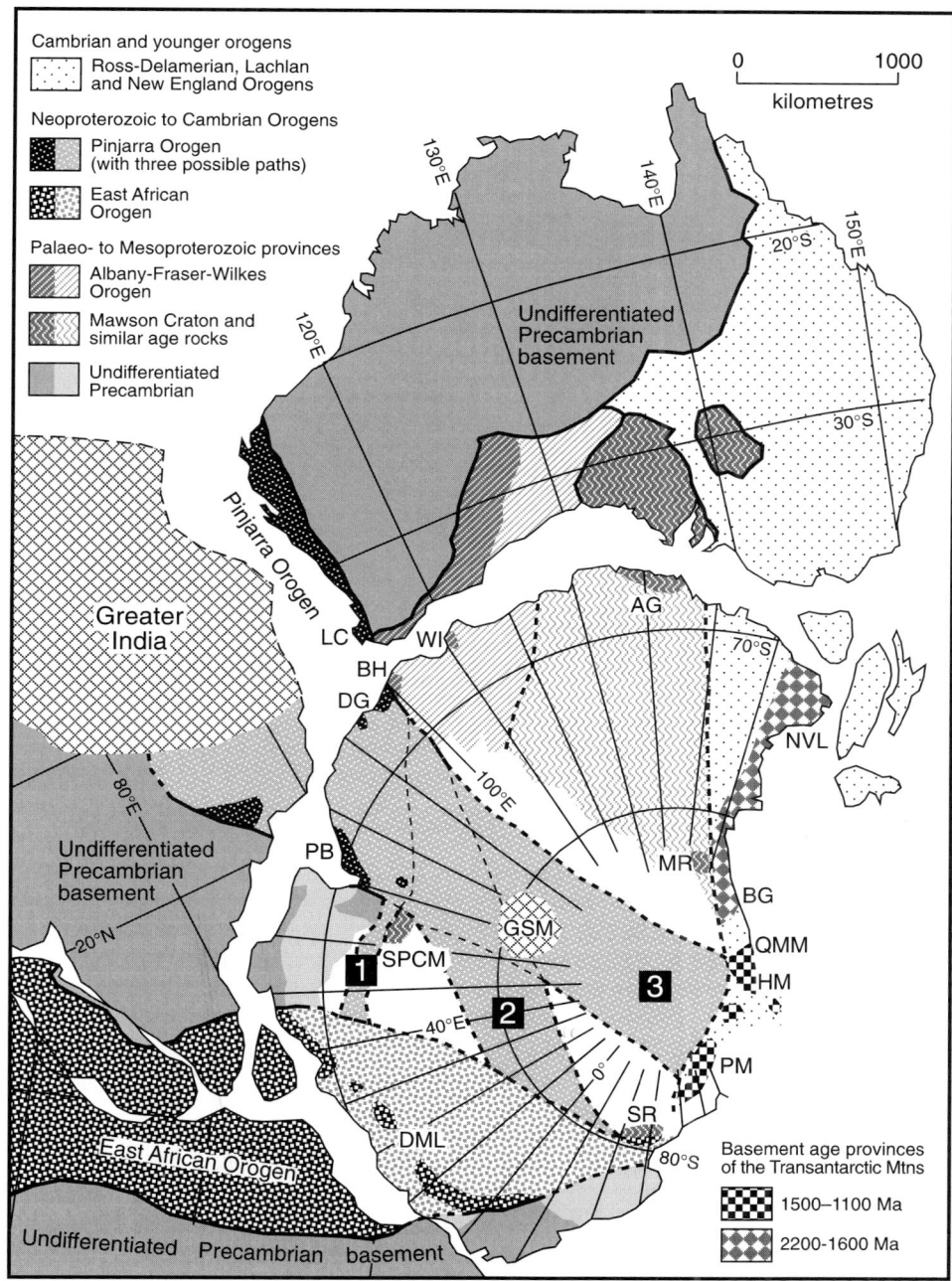

Fig. 10. Three potential pathways for the Pinjarra Orogen across East Antarctica. Major province boundaries are taken from Figure 1 and basement age provinces of the Transantarctic Mountains are after Borg & DePaolo (1994). AG, Terre Adélie–King George V Land; BG, Beardmore Glacier; BH, Bunger Hills; DG, Denman Glacier; DML, Dronning Maud Land; GSM, Gamburtsev Subglacial Mountains; HM, Horlick Mountains; LC, Leeuwin Complex; MR, Miller Range; NVL, northern Victoria Land; PB, Prydz Bay; PM, Pensacola Mountains; QMM, Queen Maud Mountains; SPCM, southern Prince Charles Mountains; SR, Shackleton Range; WI, Windmill Islands.

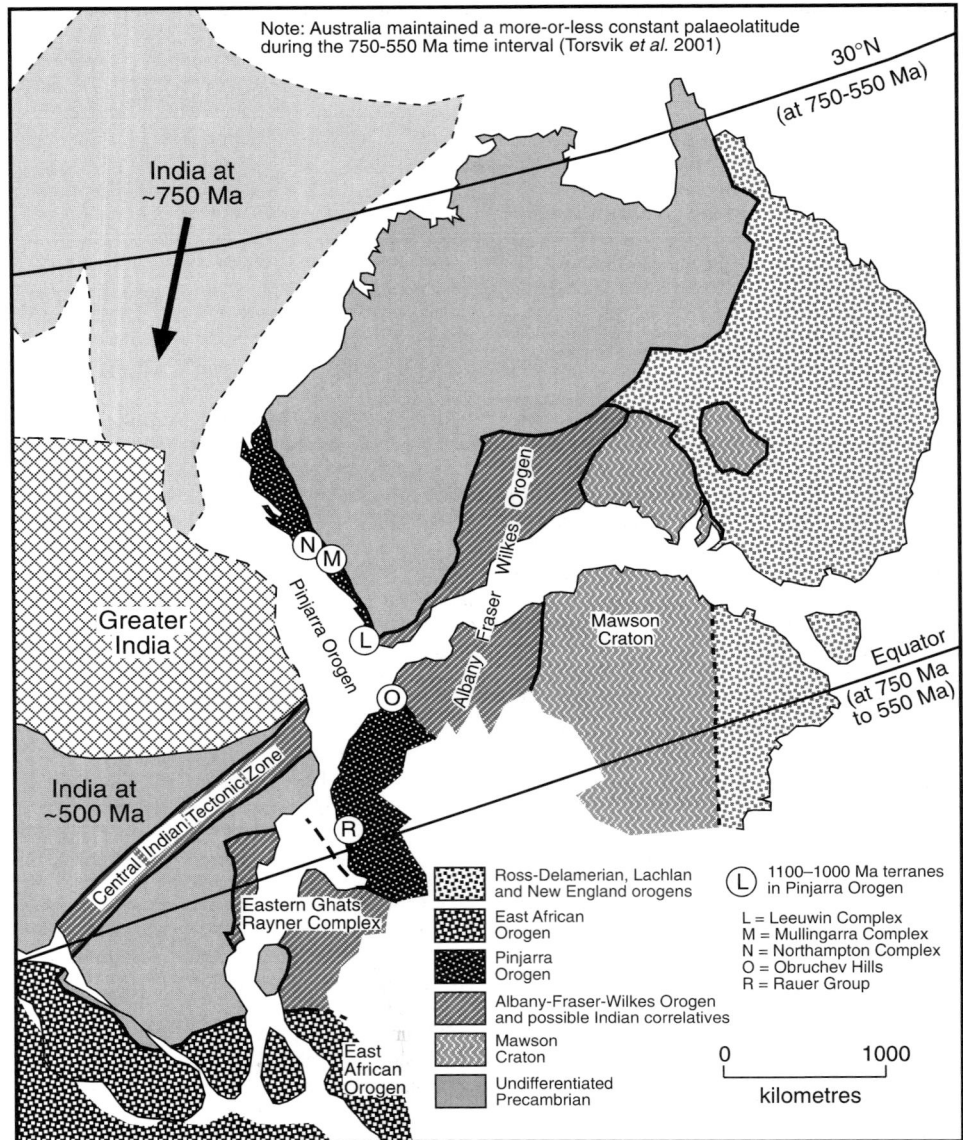

Fig. 11. Gondwana reconstruction showing late Mesoproterozoic belts in Australo-Antarctica (Albany–Fraser–Wilkes Orogen) and India (Eastern Ghats and the central Indian Tectonic Zone), adapted from Harris (1993). The Indian and Australian belts do not correlate with one another because they were juxtaposed along the Pinjarra Orogen in the latest Neoproterozoic. Likely latitude of India at c. 750 Ma is based on the palaeomagnetic data of Torsvik et al. (2001). Also shown are 1100–1000 Ma gneissic terranes in the Pinjarra Orogen, which appear to truncate the Albany–Fraser–Wilkes Orogen.

Ruker Terrane of the southern Prince Charles Mountains and the Read Complex of the Shackleton Range are likely to be part of the Mawson Craton. Alternatively, Fitzsimons (2000a) suggested that the Prydz Complex could continue a further 500 km south to the Gamburtsev Subglacial Mountains, in which case the Ruker Complex is not part of the Mawson Craton. These mountains would have elevations >3000 m after removal of the ice sheet and occur in a region where the crustal thickness is >65 km (Groushinsky & Sazhina 1982; Bentley 1991). Veevers (1994) has argued that they formed by basin inversion in response to Variscan assembly of Pangaea at c. 320 Ma, noting that they are likely to be genetically related to another area of high elevation in the Groves Mountains and to the Lambert Graben. Just as Mesozoic tectonism in the Lambert Graben and Darling Fault reactivated Proterozoic structures, it is quite possible that basin formation and inversion in the Gamburtsev Subglacial Mountains were controlled by a Neoproterozoic crustal boundary.

If the Neoproterozoic–Cambrian suture does pass through the Gamburtsev Subglacial Mountains, there are two potential continuations, one north of the Shackleton Range which would intersect the East African Orogen in western Dronning Maud Land, and the other south of the Shackleton Range, which would intersect the Ross Orogen of the Transantarctic Mountains somewhere between the Miller and Shackleton Ranges. This latter possibility is important because it can, at least in theory, be tested. Exposed rocks along the length of the Transantarctic Mountains are dominated by late Neoproterozoic–Cambrian continental margin sediments and granite plutons deformed in the Cambrian–Ordovician Ross Orogeny. Any late Neoproterozoic–Cambrian events relating to an extension of the Pinjarra Orogen in the Transantarctic Mountains will be difficult to distinguish from Ross tectonism, particularly as none of the pre-Ross basement rocks are exposed, but the chemistry of the granites should reflect the major changes in crustal basement expected across such a suture. T_{DM} Nd model ages have been determined for granites from much of the Transantarctic Mountains and are summarized by Borg & DePaolo (1994). They are typically 2.2–1.6 Ga for much of the range (Fig. 10), from northern Victoria Land to at least the Beardmore Glacier, c. 300 km south of Miller Range, although ages in the Miller Range itself are older, as discussed above. The consistency of these ages implies a continuous craton margin, consistent with the Gawler Craton and the Miller Range being part of a single early Mesoproterozoic craton. South of the Beardmore Glacier, however, T_{DM} ages are more typically 1.5–1.1 Ga, and these ages continue at least as far as the Horlick Mountains. Further evidence for distinctive crust beneath this region comes from 1100 to 1000 Ma zircon xenocrysts in c. 500 Ma granites from the Pensacola and Queen Maud Mountains, indicating that these plutons sampled Grenville-age basement (Van Schmus et al. 1997). Borg & DePaolo (1994) interpreted these younger basement ages as an allochthonous terrane accreted to the Ross margin, and infer older cratonic material to lie inboard under the ice. However, the T_{DM} ages are also identical to those of the Leeuwin Complex, and the xenocrystic zircon is of similar age to magmatism and metamorphism in the Leeuwin, Northampton and Mullingarra Complexes. An intriguing but speculative possibility is that this change in crustal character reflects a Neoproterozoic–Cambrian suture that traverses the Antarctic continent and intersects the Ross Orogen in the Horlick–Pensacola Mountains region (Fig. 10).

Tectonic setting of the Pinjarra Orogen and its Antarctic counterparts

The regional tectonic setting of late Neoproterozoic–Cambrian deformation and metamorphism in the Pinjarra–Denman and Prydz Bay regions remains uncertain, although it is clear that early models based on intracratonic magmatic underplating causing a widespread thermal overprint but limited deformation (Stüwe & Sandiford 1993) are in need of revision. The eastern margin of this tectonic zone was active throughout much of the Neoproterozoic and was characterized by N–S strike-slip movement and the juxtaposition of apparently unrelated terranes. In contrast, the western margin in Prydz Bay was dominated by events at the close of the Neoproterozoic and involved a greater degree of vertical displacement (burial and exhumation) than the eastern margin, although just like the east there is evidence that this event juxtaposed several terranes of different ages and origins (the Vestfold Hills and the Rauer Group). There is no good evidence for a Neoproterozoic magmatic arc, nor ocean closure, although this could be because this is one of the most poorly exposed orogens on Earth. The southern half is hidden by an ice cap and the northern half was a focus for rifting during Gondwana breakup, which ultimately led to a northwestern sector in Greater India being subducted beneath the Himalayas and a northeastern sector in Australia being buried by thick sedimentary sequences of the Perth and Carnarvon Basins.

An important unknown is the magnitude of the strike-slip displacement: if large, then this orogen is of global importance since it would mean that the East Gondwana configuration was not achieved until 500 Ma; if small, then this orogen has little influence on continental configurations. Geological and isotopic data summarized above imply that displacements were large and associated with accretion of allochthonous terranes, but other studies have argued for minimal offsets based on correlations between the late Mesoproterozoic Albany–Fraser Orogen and similar-aged provinces in India (Fig. 11). The most common correlation is with the Eastern Ghats (Katz 1989; Moores 1991) but precise geochronology indicates that the main phase of tectonism in the Eastern Ghats is 200 Ma younger than that in the Albany–Fraser Orogen (see review by Dobmeier & Raith 2003). As emphasized by

Fitzsimons (2000b), these two terranes are unrelated and should not be used to infer a continuous Grenville-age belt between them. A correlation has also been suggested between the Albany–Fraser Orogen and the Central Indian Tectonic Zone (Harris 1993). The evidence presented for this correlation is inconclusive, since it is based on comparison of U–Pb and Sm–Nd ages from the Albany–Fraser Orogen with Rb–Sr whole-rock data from the Central Indian Tectonic Zone. Both provinces preserve evidence for prolonged Archaean–late Neoproterozoic tectonism, but more detailed correlations are of little value without high-precision U–Pb ages from central India. Harris (1994) used this correlation to argue that any strike-slip displacement in the Pinjarra Orogen was limited, but he did not consider the evidence for $c.$ 1080 Ma tectonism along the Pinjarra Orogen. It is difficult to envisage how these late Mesoproterozoic gneissic terranes could be aligned in a distribution that truncates the older Albany–Fraser Orogen and the Central Indian Tectonic Zone without creating any offset between them. All the available evidence argues for significant Neoproterozoic displacements along the Pinjarra Orogen, and therefore for a lack of Proterozoic correlations between Australia and India.

These qualitative arguments are supported by recently published palaeomagnetic evidence for continent-scale Neoproterozoic displacements along the Pinjarra Orogen. Torsvik et al. (2001) presented palaeomagnetic data indicating that India was $c.$ 30° of latitude north of its ultimate Gondwana position relative to Australia at 750 Ma (Fig. 11). This requires southerly movement of India relative to Australia some time between 750 and 500 Ma, consistent with sinistral displacement along the Pinjarra Orogen at 550 Ma reflecting oblique collision of India with the western Australian margin during the final assembly of Gondwana (Powell & Pisarevsky 2002; Pisarevsky et al. 2003). Such a collision would explain the significant influence of 550–500 Ma tectonism on the southwestern margin of the Australian craton. Evidence for greater levels of burial and exhumation in Prydz Bay could indicate that the western margin of the orogen thickened significantly during collision, whereas the eastern margin accommodated strain largely through strike-slip displacements. The comparatively limited overprint during dextral wrenching at $c.$ 750 Ma could indicate that this earlier event involved transcurrent displacements with little component of oblique strain.

A remaining issue is a suitable name for this late Neoproterozoic–Cambrian orogen. Evidence for tectonism along Australia's western margin at both 1100–1000 and 550–500 Ma has caused some confusion in terminology. Myers et al. (1996) interpreted the earlier event in terms of a Proto-Pinjarra Orogen and used the term Pinjarra Orogen to describe the younger event. Wilde (1999) argued that the Pinjarra Orogen should refer to evidence for the 1100–1000 Ma event, but given that the terranes that preserve these events are not in situ they should not be used to define an orogen, so it is more logical to use the term Pinjarra Orogen for the Neoproterozoic belt of accreted terranes and related rocks. A name is also needed for the combined Australian–Antarctic orogen as a whole. Fitzsimons (2000b) referred to it as the Prydz–Denman–Darling Orogen and Veevers (2000) as the Prydz–Leeuwin Belt, and it would appear to correspond, at least in part, to the Kuunga Orogen of Meert & van der Voo (1997). The latter was described rather loosely as a late Neoproterozoic collision zone between India and Australia–Antarctica, but it now seems clear that this zone continues from the Pinjarra Orogen across the Antarctic continent, although its exact path remains uncertain. It is recommend here that the name Pinjarra Orogen be used for the entire belt, since it is best exposed in Australia. This term is already a collective name for various component terranes in western Australia, including the Northampton, Mullingarra and Leeuwin Complexes, and the Darling Fault Zone. There is, in fact, no basement outcrop at Pinjarra, a small town known for the Battle of Pinjarra in 1834, a bloody and one-sided clash that signified the end of Western Australian isolation from the influence of the western world. It is a fitting name for the Neoproterozoic–Cambrian suture that marks the collision of Australia with the other components of Gondwana.

Conclusions

(1) Excellent correlations can be made between Proterozoic provinces in Australia and Antarctica, which provide precise piercing points consistent with reconstructions of Gondwana immediately before breakup.

(2) The Palaeoproterozoic Mawson Craton extends from South Australia to East Antarctica, and could underlie much of the Antarctic ice sheet, although insufficient data are available at present to accurately constrain its lateral extent.

(3) The Mawson Craton underwent a major tectonic event at $c.$ 1700 Ma (the Kimban Orogeny), which can be identified along its eastern margin from the Eyre Peninsula to the central Transantarctic Mountains. The occurrence of $c.$ 1700 Ma eclogites in the Transantarctic Mountains implies that this may have been a collisional event, but low-pressure metamorphism is dominant elsewhere.

(4) A continental fragment comprising the Nawa Domain of the Gawler Craton, the buried Coompana Block and possibly the Nornalup Complex of the Albany–Fraser Orogen collided with the western margin of the Mawson Craton at $c.$ 1560 Ma.

(5) The western edge of the Nornalup Complex collided with the Biranup Complex, the Fraser Complex, and the Yilgarn Craton to form the Albany–Fraser Orogen during two stages of tectonism at $c.$ 1350–1260 and $c.$ 1210–1140 Ma. The oceanic affinity of the Fraser Complex implies that collision involved the closure of an ocean basin, although this basin need not have been large.

(6) Gneisses of 1100–1000 Ma along the Pinjarra Orogen at Australia's western margin are allochthonous blocks translated along this margin during Neoproterozoic time, and not an *in situ* Grenville-age collisional orogen between Australia and India as previously believed. At least two stages of transcurrent movement have been identified: dextral strike-slip at $c.$ 750 Ma and sinistral strike-slip at $c.$ 550–500 Ma. Proterozoic granulite belts are commonly interpreted as orthogonal collision zones, but, as shown in this case, it is also possible that they represent allochthonous blocks aligned during post-metamorphic transcurrent tectonics.

(7) Proterozoic rocks in India should not be correlated with those in Australia, and reconstructions of pre-Gondwana supercontinents should not show these two continents in their Gondwana configuration.

(8) The Pinjarra Orogen extends into Antarctica, and although its exact path is difficult to define it must divide East Gondwana into Australo-Antarctic and Indo–Antarctic domains. These two domains were juxtaposed by an oblique collision at 550–500 Ma. The East African Orogen has previously been regarded as the principal Proterozoic collisional zone in Gondwana, but the Pinjarra Orogen is of similar width and length to the East African Orogen, and was presumably just as influential in Gondwana assembly.

(9) Precise and exhaustive geochronology is crucial for tectonic reconstructions but can be inconclusive, particularly in areas of poor outcrop where similarities or differences in age are often insufficient to prove or disprove tectonic correlations without other information.

(10) Poor knowledge of Precambrian tectonic boundaries beneath the Antarctic ice sheet creates significant uncertainties in any pre-Gondwana continental reconstructions. There is a need for more information from Antarctica, particularly geophysical data but also drilling in localities like the Gamburtsev Subglacial Mountains where critical outcrops lie only a few tens of metres below the surface.

This paper is dedicated to the memories of R. L. Oliver (1921–2001) and C. McA. Powell (1943–2001). Robin showed that careful petrology, applied with characterisitc courtesy and gentle humour, is an invaluable tool for tectonic reconstructions of East Gondwana; much of his work is cited above. Chris's drive and enthusiasm were instrumental in encouraging yet another metamorphic petrologist, with few tectonic credentials, to try his hand at continent-scale correlations, and this paper would not exist without his considerable influence. A number of people have provided advice and/or access to unpublished data, including P. Cawood, M. Cobb, G. Fraser, D. Nelson, S. Pisarevsky, N. Post, S. Sheppard and M. Wingate, which together with formal reviews by W. Bauer, J. Jacobs and K. Shiraishi have greatly improved this paper. I also thank M. Yoshida for showing editorial solidarity with a very late author. This is TSRC Publication Number 184.

References

BEESON, J, DELOR, C. P. & HARRIS, L. B. 1988. A structural and metamorphic traverse across the Albany Mobile Belt, Western Australia. *Precambrian Research*, **40/41**, 117–136.

BEESON, J, HARRIS, L. B. & DELOR, C. P. 1995. Structure of the western Albany Mobile Belt (south-western Australia): evidence for overprinting by Neoproterozoic shear zones of the Darling Mobile Belt. *Precambrian Research*, **75**, 47–63.

BENTLEY, C. R. 1991. Configuration and structure of the subglacial crust. *In:* TINGEY, R. J. (ed.) *The Geology of Antarctica.* Oxford University Press, Oxford, 335–364.

BLACK, L. P., HARRIS, L. B. & DELOR, C. P. 1992*a*. Reworking of Archaean and Early Proterozoic components during a progressive, Middle Proterozoic tectonothermal event in the Albany Mobile Belt, Western Australia. *Precambrian Research*, **59**, 95–123.

BLACK, L. P., KINNY, P. D., SHERATON, J. W. & DELOR C. P. 1991. Rapid production and evolution of late Archaean felsic crust in the Vestfold Block of East Antarctica. *Precambrian Research*, **50**, 283–310.

BLACK, L. P., SHERATON, J. W., TINGEY, R. J. & MCCULLOCH, M. T. 1992*b*. New U-Pb zircon ages from the Denman Glacier area, East Antarctica, and their significance for Gondwana reconstruction. *Antarctic Science*, **4**, 447–460.

BLIGHT, D. F. & OLIVER, R. L. 1977. The metamorphic geology of the Windmill Islands, Antarctica: a preliminary account. *Journal of the Geological Society of Australia*, **24**, 239–262.

BLIGHT, D. F., OLIVER, R. L. 1982. Aspects of the geological history of the Windmill Islands, Antarctica. *In:* CRADDOCK, C. (ed.) *Antarctic Geoscience.* University of Wisconsin Press, Madison, 445–454.

BLIGHT, D. F., COMPSTON, W. & WILDE, S. A. 1981. The Logue Brook Granite: age and significance of deformation zones along the Darling Scarp. *Western*

Australia Geological Survey, Annual Report, **1980**, 72–80.
BOGER, S. D., WILSON, C. J. L. & FANNING, C. M. 2001. Early Paleozoic tectonism within the east Antarctic craton: the final suture between east and west Gondwana? *Geology*, **29**, 463–466.
BORG, S. G. & DEPAOLO, D. J. 1994. Laurentia, Australia, and Antarctica as a Late Proterozoic supercontinent: constraints from isotopic mapping. *Geology*, **22**, 307–310.
BORG, S. G., DEPAOLO, D. J. & SMITH, B. M. 1990. Isotopic structure and tectonics of the central Transantarctic Mountains. *Journal of Geophysical Research*, **95(B5)**, 6647–6667.
BRETAN, P. G. 1986. *Deformation processes within mylonite zones associated with some fundamental faults*. PhD Thesis, Imperial College, University of London, UK.
BRUGUIER, O., BOSCH, D., PIDGEON, R. T., BYRNE, D. I. & HARRIS, L. B. 1999. U–Pb chronology of the Northampton Complex, Western Australia – evidence for Grenvillian sedimentation, metamorphism and deformation and geodynamic implications. *Contributions to Mineralogy & Petrology*, **136**, 258–272.
BUGGISCH, W. & KLEINSCHMIDT, G. 1999. New evidence for nappe tectonics in the southern Shackleton Range. *Terra Antartica*, **6**, 203–210.
BURKE, K. C. & DEWEY, J. F. 1972. Orogeny in Africa. In: DESSAUVAGIE, T. F. J. & WHITEMAN, A. J. (eds) *African Geology*. University of Ibadan, Ibadan, Nigeria, 583–608.
BYRNE, D. R. & HARRIS, L. B. 1993. Structural controls on the base-metal vein deposits of the Northampton Complex. *Ore Geology Reviews*, **8**, 89–115.
CARSON, C. J., FANNING, C. M. & WILSON, C. J. L. 1996. Timing of the Progress Granite, Larsemann Hills: additional evidence for Early Palaeozoic orogenesis within the East Antarctic Shield and implications for Gondwana assembly. *Australian Journal of Earth Sciences*, **43**, 539–553.
CARSON, C. J., POWELL, R., WILSON, C. J. L. & DIRKS, P. H. G. M. 1997. Partial melting during tectonic exhumation of a granulite terrane: an example from the Larsemann Hills, East Antarctica. *Journal of Metamorphic Geology*, **15**, 105–126.
CHOUDHARY, A. K., HARRIS, N. B. W., VAN CALSTEREN, P. & HAWKESWORTH, C. J. 1992. Pan-African charnockite formation in Kerala, southern India. *Geological Magazine*, **129**, 257–264.
CLARK, W. C. 1995. *Granite petrogenesis, metamorphism and geochronology of the western Albany-Fraser Orogen, Albany, Western Australia*. BSc (Hons) Thesis, Curtin University of Technology, Australia.
CLARK, D. J. 1999. *Thermo-tectonic evolution of the Albany–Fraser Orogen, Western Australia*. PhD Thesis, University of New South Wales, Australia.
CLARK, D. J., HENSEN, B. J. & KINNY, P. D. 2000. Geochronological constraints for a two-stage history of the Albany–Fraser Orogen, Western Australia. *Precambrian Research*, **102**, 155–183.
CLARK, D. J., KINNY, P. D., POST, N. J. & HENSEN, B. J. 1999. Relationships between magmatism, metamorphism and deformation in the Fraser Complex, Western Australia: constraints from new SHRIMP U–Pb zircon geochronology. *Australian Journal of Earth Sciences*, **46**, 923–932.
CLARKSON, P. D. 1982. Geology of the Shackleton range: I. Shackleton Range Metamorphic Complex. *British Antarctic Survey Bulletin*, **51**, 257–283.
COBB, M. M., CAWOOD, P. A., KINNY, P. D. & FITZSIMONS, I. C. W. 2001. SHRIMP U–Pb zircon ages from the Mullingarra Complex, Western Australia: isotopic evidence for allochthonous blocks in the Pinjarra Orogen and implications for East Gondwana assembly. *Geological Society of Australia, Abstracts*, **64**, 21–22.
COLLINS, A. S. & FITZSIMONS, I. C. W. 2001. Structural, isotopic and geochemical constraints on the evolution of the Leeuwin Complex, southwest Australia. *Geological Society of Australia, Abstracts*, **65**, 16–19.
COLLINS, A. S. & WINDLEY, B. F. 2002. The tectonic evolution of central and northern Madagascar and its place in the final assembly of Gondwana. *Journal of Geology*, **110**, 325–340.
COMPSTON, W. & ARRIENS, P. A. 1968. The Precambrian geochronology of Australia. *Canadian Journal of Earth Sciences*, **5**, 561–583.
CONDIE, K. C. & MYERS, J. S. 1999. Mesoproterozoic Fraser Complex: geochemical evidence for multiple subduction-related sources of lower crustal rocks in the Albany–Fraser Orogen, Western Australia. *Australian Journal of Earth Sciences*, **46**, 875–882.
CROWE, W. A., COSCA, M. A. & HARRIS, L. B. 2001. $^{40}Ar/^{39}Ar$ geochronology and Neoproterozoic tectonics along the northern margin of the Eastern Ghats Belt in north Orissa, India. *Precambrian Research*, **108**, 237–266.
CRUSE, T. & HARRIS, L. B. 1994. Ediacaran fossils from the Stirling Range Formation, Western Australia. *Precambrian Research*, **67**, 1–10.
DALY, S. J., FANNING, C. M. & FAIRCLOUGH, M. C. 1998. Tectonic evolution and exploration potential of the Gawler Craton, South Australia. *AGSO Journal of Australian Geology & Geophysics*, **17**, 145–168.
DALZIEL, I. W. D. 1991. Pacific margins of Laurentia and East Antarctica–Australia as a conjugate rift pair: evidence and implications for an Eocambrian supercontinent. *Geology*, **19**, 598–601.
DENTITH, M. C., BRUNER, I., LONG, A., MIDDLETON, M. F. & SCOTT, J. 1993. Structure of the eastern margin of the Perth Basin, Western Australia. *Exploration Geophysics*, **24**, 455–462.
DING, P. & JAMES, P. R. 1991. Structural evolution of the Bunger Hills area of East Antarctica. In: THOMSON, M. R. A., CRAME, J. A. & THOMSON, J. W. (eds) *Geological Evolution of Antarctica*. Cambridge University Press, Cambridge, 13–17.
DOBMEIER, C. J. & RAITH, M. M. 2003. Crustal architecture and evolution of the Eastern Ghats Belt and adjacent regions of India. In: YOSHIDA, M., WINDLEY, B. F. & DASGUPTA, S. (eds) *Proterozoic East Gondwana: Supercontinent Assembly and Breakup*. Geological Society, London, Special Publication, **206**, 145–168.
DREXEL, J. F., PREISS, W. V. & PARKER, A. J. (eds) 1993. *The Geology of South Australia. Volume 1, The Precambrian*. South Australia Geological Survey, Bulletin, **54**.
EMBLETON, B. J. J. & SCHMIDT, P. W. 1985. Age and signifi-

cance of magnetizations in dolerite dykes from the Northampton Block, Western Australia. *Australian Journal of Earth Sciences*, **32**, 279–286.

EVANS, T. 1999. *Extent and nature of the 1200 Ma Wheatbelt dyke swarm, south-western Australia*. BSc (Hons) Thesis, University of Western Australia.

FANNING, C. M. 1997. *Geochronological synthesis of South Australia. Part II. The Gawler Craton*. South Australian Department of Mines and Energy, Open File Envelope, **8918** (unpublished).

FANNING, C. M., DALY, S. J., BENNETT, V. C., MÉNOT, R.-P., PEUCAT, J.-J., OLIVER, R. L. & MONNIER, O. 1995. The 'Mawson Block': once contiguous Archaean to Proterozoic crust in the East Antarctic Shield and Gawler Craton, Australia. *Proceedings of the VII International Symposium on Antarctic Earth Sciences*, Siena, Abstracts, 124.

FANNING, C. M., FLINT, R. B., PARKER, A. J., LUDWIG, K. R. & BLISSETT, A. H. 1988. Refined Proterozoic evolution of the Gawler Craton, South Australia, through U–Pb zircon geochronology. *Precambrian Research*, **40/41**, 363–386.

FANNING, C. M., MOORE, D. H., BENNETT, V. C. & DALY, S. J. 1996. The 'Mawson Continent': Archaean to Proterozoic crust in the East Antarctic Shield and Gawler Craton, Australia. A cornerstone in Rodinia and Gondwana. *Geological Society of Australia, Abstracts*, **41**, 135.

FANNING, C. M., MOORE, D. H., BENNETT, V. C., DALY, S. J., MÉNOT, R.-P., PEUCAT, J.-J. & OLIVER, R. L. 1999. The 'Mawson Continent': the East Antarctic Shield and Gawler Craton, Australia. *Proceedings of the 8th International Symposium on Antarctic Earth Sciences*, Wellington, Abstracts, 103.

FEDOROV, L. V., GRIKUROV, G. E., KURININ, R. G. & MASOLOV, V. N. 1982. Crustal structure of the Lambert Glacier area from geophysical data. *In*: CRADDOCK, C. (ed.) *Antarctic Geoscience*. University of Wisconsin Press, Madison, 931–936.

FITZSIMONS, I. C. W. 1996. Metapelitic migmatites from Brattstrand Bluffs, East Antarctica – metamorphism, melting and exhumation of the mid crust. *Journal of Petrology*, **37**, 395–414.

FITZSIMONS, I. C. W. 1997. The Brattstrand Paragneiss and the Søstrene Orthogneiss: a review of Pan-African metamorphism and Grenvillian relics in southern Prydz Bay. *In:* RICCI, C. A. (ed.) *The Antarctic Region: Geological Evolution and Processes*. Terra Antartica Publication, Siena, 121–130.

FITZSIMONS, I. C. W. 2000a. A review of tectonic events in the East Antarctic Shield, and their implications for Gondwana and earlier supercontinents. *Journal of African Earth Sciences*, **31**, 3–23.

FITZSIMONS, I. C. W. 2000b. Grenville-age basement provinces in East Antarctica: evidence for three separate collisional orogens. *Geology*, **28**, 879–882.

FITZSIMONS, I. C. W. 2001. The Neoproterozoic evolution of Australia's western margin. *Geological Society of Australia, Abstracts*, **65**, 39–42.

FITZSIMONS, I. C. W., KINNY, P. D. & HARLEY, S. L. 1997. Two stages of zircon and monazite growth in anatectic leucogneiss: SHRIMP constraints on the duration and intensity of Pan-African metamorphism in Prydz Bay, East Antarctica. *Terra Nova*, **9**, 47–51.

FLETCHER, I. R & LIBBY, W. G. 1993. Further isotopic evidence for the existence of two distinct terranes in the southern Pinjarra Orogen, Western Australia. *Western Australia Geological Survey, Report*, **34**, 81–83.

FLETCHER, I. R., MYERS, J. S. & AHMAT, A. L. 1991. Isotopic evidence on the age and origin of the Fraser Complex, Western Australia: a sample of Mid-Proterozoic lower crust. *Chemical Geology*, **87**, 197–216.

FLETCHER, I. R., WILDE, S. A., LIBBY, W. G. & ROSMAN, K. J. R. 1983. Sm–Nd model ages across the margins of the Archaean Yilgarn Block, Western Australia – II; southwest transect into the Proterozoic Albany–Fraser Province. *Journal of the Geological Society of Australia*, **30**, 333–340.

FLETCHER, I. R., WILDE, S. A. & ROSMAN, K. J. R. 1985. Sm–Nd model ages across the margins of the Archaean Yilgarn Block, Western Australia – III. The western margin. *Australian Journal of Earth Sciences*, **32**, 73–82.

FLÖTTMANN, T. & OLIVER, R. L. 1994. Review of Precambrian–Palaeozoic relationships at the craton margins of southeastern Australia and adjacent Antarctica. *Precambrian Research*, **69**, 293–306.

FOSTER D. A. & EHLERS, K. 1998. ^{40}Ar/^{39}Ar thermochronology of the southern Gawler Craton, Australia: implications for Mesoproterozoic and Neoproterozoic tectonics of East Gondwana and Rodinia. *Journal of Geophysical Research*, **103(B5)**, 10177–10193.

FRASER, G., LYONS, P. & DIREEN, N. G. 2002. Mesoproterozoic tectonism in the northwest Gawler Craton: ^{40}Ar/^{39}Ar geochronology, geophysical interpretations and extrapolations. *Geological Society of Australia, Abstracts*, **64**, 67.

GIDDINGS J. W. 1976. Precambrian palaeomagnetism in Australia I: basic dykes and volcanics from the Yilgarn Block. *Tectonophysics*, **30**, 91–108.

GLIKSON, A. Y. & LAMBERT, I. B. 1973. Relations in time and space between major Precambrian shield units: an interpretation of Western Australian data. *Earth & Planetary Science Letters*, **20**, 395–403.

GOLDSTEIN, S. L., O'NIONS, R. K. & HAMILTON, P. J. 1984. A Sm–Nd study of atmospheric dust and particulates from major river systems. *Earth & Planetary Science Letters*, **70**, 221–236.

GOODGE, J. W. & DALLMEYER, R. D. 1992. ^{40}Ar/^{39}Ar mineral age constraints on the Paleozoic tectonothermal evolution of high-grade basement rocks within the Ross Orogen, central Transantarctic Mountains. *Journal of Geology*, **100**, 91–106.

GOODGE, J. W. & FANNING, C. M. 1999. 2.5 b.y. of punctuated Earth history as recorded in a single rock. *Geology*, **11**, 1007–1010.

GOODGE, J. W., FANNING, C. M. & BENNETT, V. C. 2001. U–Pb evidence of ~1.7 Ga crustal tectonism during the Nimrod Orogeny in the Transantarctic Mountains, Antarctica: implications for Proterozoic plate reconstructions. *Precambrian Research*, **112**, 261–288.

GOODGE, J. W., HANSEN, V. L., PEACOCK, S. M., SMITH, B. K. & WALKER, N. W. 1993a. Kinematic evolution of the Miller Range Shear Zone, central Transantarctic Mountains, Antarctica, and implications for Neoproterozoic to early Paleozoic tectonics of the East Antarctic margin of Gondwana. *Tectonics*, **12**, 1460–1478.

GOODGE, J. W., WALKER, N. W. & HANSEN, V. L. 1993b. Neoproterozoic–Cambrian basement-involved orogenesis within the Antarctic margin of Gondwana. *Geology*, **21**, 37–40.

GOSE, W. A., HELPER, M. A., CONNELLY, J. N., HUTSON, F. E. & DALZIEL, I. W. D. 1997. Paleomagnetic data and U–Pb isotopic age determinations from Coats Land, Antarctica: implications for Late Proterozoic plate reconstructions. *Journal of Geophysical Research*, **102(B4)**, 7887–7902.

GREW, E. S. 1982. Geology of the southern Prince Charles Mountains, East Antarctica. *In:* CRADDOCK, C. (ed.) *Antarctic Geoscience.* University of Wisconsin Press, Madison, 473–478.

GRINDLEY, G. W., MCGREGOR, V. R. & WALCOTT, R. I. 1964. Outline of the geology of the Nimrod–Beardmore–Axel Heiberg Glaciers region, Ross Dependency. *In:* ADIE, R. J. (ed.) *Antarctic Geology.* North Holland Publishing Company, Amsterdam, 206–219.

GROUSHINSKY, N. P. & SAZHINA, N. B. 1982. Some features of Antarctic crustal structure. *In:* CRADDOCK, C. (ed.) *Antarctic Geoscience.* University of Wisconsin Press, Madison, 907–911.

HAND, M., BENDALL, B. R. & SANDIFORD, M. 1995. Metamorphic evidence for Palaeoproterozoic oblique convergence in the eastern Gawler Craton. *Geological Society of Australia, Abstracts*, **40**, 59.

HARLEY, S. L. 2003. Archaean–Cambrian crustal development of East Antarctica: metamorphic characteristics and tectonic implications. *In:* YOSHIDA, M., WINDLEY, B. F. & DASGUPTA, S. (eds) *Proterozoic East Gondwana: Supercontinent Assembly and Breakup.* Geological Society, London, Special Publication, **206**, 201–230.

HARLEY, S. L. & FITZSIMONS, I. C. W. 1995. High-grade metamorphism and deformation in the Prydz Bay region, East Antarctica: terrane, events and regional correlations. *In:* YOSHIDA, Y. & SANTOSH, M. (eds) *India and Antarctica during the Precambrian.* Geological Society of India, Memoir, **34**, 73–100.

HARRIS, L. B. 1993. Correlations of tectonothermal events between the central Indian tectonic Zone and the Albany Mobile Belt of Western Australia. *In:* FINDLAY, R. H., UNRUG, R., BANKS, M. R. & VEEVERS, J. J. (eds) *Gondwana Eight: Assembly, Evolution and Dispersal.* Balkema, Rotterdam, 165–180.

HARRIS, L. B. 1994. Neoproterozoic sinistral displacement along the Darling Mobile Belt, Western Australia, during Gondwanaland assembly. *Journal of the Geological Society, London*, **151**, 901–904.

HARRIS, L. B. 1995. Correlation between the Albany, Fraser, and Darling mobile belts of Western Australia and Mirnyy to Windmill Islands in the East Antarctic Shield: implications for Proterozoic Gondwanaland reconstructions. *In:* YOSHIDA, Y. & SANTOSH, M. (eds) *India and Antarctica during the Precambrian.* Geological Society of India, Memoir, **34**, 73–100.

HARRIS, L. B. & LI, Z. X. 1995. Palaeomagnetic dating and tectonic significance of dolerite intrusions in the Albany Mobile Belt, Western Australia. *Earth & Planetary Science Letters*, **131**, 143–164.

HENSEN, B. J. & ZHOU, B. 1995. A Pan-African granulite facies metamorphic episode in Prydz Bay, Antarctica: evidence from Sm–Nd garnet dating. *Australian Journal of Earth Sciences*, **42**, 249–258.

HENSEN, B. J. & ZHOU, B. 1997. East Gondwana amalgamation by Pan-African collision? Evidence from Prydz Bay, East Antarctica. *In:* RICCI, C. A. (ed.) *The Antarctic Region: Geological Evolution and Processes.* Terra Antartica Publication, Siena, 115–119.

HEURTEBIZE, G. 1952. Surs les formations géologiques de Terre Adélie. *Comptes Rendues de l'Académie des Sciences, Paris*, **234**, 2209–2211.

HILL, R. I., CHAPPELL, B. W. & CAMPBELL, I. H. 1992. Late Archaean granites of the southeastern Yilgarn Block, Western Australia: age, geochemistry, and origin. *Transactions of the Royal Society of Edinburgh: Earth Sciences*, **83**, 211–226.

HOEK, J. D. & SCHAEFER, B. F. 1998. Palaeoproterozoic Kimban mobile belt, Eyre peninsula: timing and significance of felsic and mafic magmatism and deformation. *Australian Journal of Earth Sciences*, **45**, 305–313.

HOFFMAN, P. F. 1991. Did the breakout of Laurentia turn Gondwana inside-out? *Science*, **252**, 1409–1412.

HOFMANN, J. 1982. Main tectonic features and development of the southern Prince Charles Mountains, East Antarctica. *In:* CRADDOCK, C. (ed.) *Antarctic Geoscience.* University of Wisconsin Press, Madison, 479–487.

HÖLZL, S., HOFMANN, A. W., TODT, W. & KÖHLER, H. 1994. U–Pb geochronology of the Sri Lanka basement. *Precambrian Research*, **66**, 123–149.

ILTCHENKO, L. N. 1972. Late Precambrian acritarchs of Antarctica. *In:* ADIE, R. J. (ed.) *Antarctic Geology and Geophysics.* Universitetsforlaget, Oslo, 599–602.

JACOBS, J., FANNING, C. M., HENJES-KUNST, F., OLESCH, M. & PAECH, H. J. 1998. Continuation of the Mozambique Belt into East Antarctica: Grenville-age metamorphism and polyphase Pan-African high-grade events in central Dronning Maud Land. *Journal of Geology*, **106**, 385–406.

KAMENEV, E. N., ANDRONIKOV, A. V., MIKHALSKY, E. V., KRASNIKOV, N. N. & STÜWE, K. 1993. Soviet geological maps of the Prince Charles Mountains. *Australian Journal of Earth Sciences*, **40**, 501–517.

KATZ, M. B. 1989. Sri Lanka–Indian Eastern Ghats–East Antarctica and the Australian Albany Fraser Mobile Belt: gross geometry, age relationships, and tectonics in Precambrian Gondwanaland. *Journal of Geology*, **97**, 646–648.

KINNY, P. D., BLACK, L. P. & SHERATON, J. W. 1993. Zircon ages and the distribution of Archaean and Proterozoic rocks in the Rauer Islands. *Antarctic Science*, **5**, 193–206.

KINNY, P. D., BLACK, L. P. & SHERATON, J. W. 1997. Zircon U–Pb ages and geochemistry of igneous and metamorphic rocks in the northern Prince Charles Mountains, Antarctica. *AGSO Journal of Australian Geology & Geophysics*, **16**, 637–654.

KLEINSCHMIDT, G. & TALARICO, F. 2002. The Mertz Shear Zone: new evidence of the 'Mawson Continent' from the East Antarctic Shield in George V Land. *Geological Society of Australia, Abstracts*, **67**, **237**.

KOVACH, V. P. & BELYATSKY, B. V. 1991. Geochemistry and age of granitic rocks in the Ruker granite–greenstone terrain, southern Prince Charles Mountains, East

Antarctica. *Proceedings of the 6th International Symposium on Antarctic Earth Sciences, National Institute of Polar Research,* Tokyo, Abstracts, 321–326.

KRIEGSMAN, L. M. 1995. The Pan-African event in East Antarctica: a view from Sri Lanka and the Mozambique Belt. *Precambrian Research,* **75,** 263–277.

KRIEGSMAN, L. M. & HENSEN, B. J. 1998. Back reaction between restite and melt: implications for geothermobarometry and pressure–temperature paths. *Geology,* **26,** 1111–1114.

KRIEGSMAN, L. M., MÖLLER, A. & NELSON, D. R. 1999. P–T–t path and detrital zircon geochronology of the Northampton Block, Western Australia: a Mesoproterozoic, collisional induced foreland rift. *Journal of Conference Abstracts,* **4(1),** 433.

LIBBY, W. G. & DE LAETER, J. R. 1998. Biotite Rb–Sr evidence for Early Palaeozoic tectonism along the cratonic margin in southwestern Australia. *Australian Journal of Earth Sciences,* **45,** 623–632.

LIBBY, W. G., DE LAETER, J. R. & ARMSTRONG, R. A. 1999. Proterozoic biotite Rb–Sr dates in the northwestern part of the Yilgarn Craton, Western Australia. *Australian Journal of Earth Sciences,* **46,** 851–860.

LOVERING, J. F., TRAVIS, G. A., COMAFORD, D. J. & KELLY, P. R. 1981. Evolution of the Gondwana Archaean Shield: zircon dating by ion microprobe, and relationships between Australia and Wilkes land (Antarctica). *In:* GLOVER, J. E. & GROVES, D. I. (eds) *Archaean Geology.* Geological Society of Australia, Special Publications, **7,** 193–203.

MCCULLOCH, M. T. 1987. Sm–Nd isotopic constraints on the evolution of Precambrian crust in the Australian continent. *In:* KRÖNER, A. (ed.) *Proterozoic Lithospheric Evolution.* Geodynamics Series, Volume 17. American Geophysical Union, Washington DC, 115–130.

MCELHINNY, M. W. & EMBLETON, B. J. J. 1974. Australian palaeomagnetism and the Phanerozoic plate tectonics of Eastern Gondwanaland. *Tectonophysics,* **63,** 13–29.

MCQUEEN, D. M., SCHARNBERGER, C. K., SCHARON, L. & HALPERN, M. 1972. Cambro–Ordovician paleomagnetic pole position and rubidium–strontium total rock isochron for charnockitic rocks from Mirny Station, East Antarctica. *Earth & Planetary Science Letters,* **16,** 433–540.

MCWILLIAMS, M. O. 1981. Palaeomagnetism and Precambrian tectonic evolution of Gondwana. *In:* KRÖNER, A. (ed.) *Precambrian Plate Tectonics.* Elsevier, Amsterdam, 649–687.

MANTON, W. I., GREW, E. S., HOFMANN, J. & SHERATON, J. W. 1992. Granitic rocks of the Jetty Peninsula, Amery Ice Shelf area, East Antarctica. *In:* YOSHIDA, Y., KAMINUMA, K. & SHIRAISHI, K. (eds) *Recent Progress in Antarctic Earth Science.* Terra Scientific Publishing Company, Tokyo, 179–189.

MATHUR, S. P. & SHAW, R. D. 1982. Australian orogenic belts: evidence for evolving plate tectonics? *Earth Evolution Sciences,* **4,** 281–308.

MEERT, J. G. & VAN DER VOO, R. 1997. The assembly of Gondwana 800–550 Ma. *Journal of Geodynamics,* **23,** 223–235.

MÉNOT, R.-P., PELLETIER, A., PEUCAT, J.-J., FANNING, C. M. & OLIVER, R. L. 1999. Petrological and structural constraints on the amalgamation of the Terre Adélie Craton (135–145°E), East Antarctica. *Proceedings of the 8th International Symposium on Antarctic Earth Sciences,* Wellington, Abstracts, 208.

MIKHALSKY, E. V., SHERATON, J. W. & BELIATSKY, B. V. 2001*a.* Preliminary U–Pb dating of Grove Mountains rocks: implications for the Proterozoic to Early Palaeozoic tectonic evolution of the Lambert Glacier–Prydz Bay area (East Antarctica). *Terra Antartica,* **8,** 3–10.

MIKHALSKY, E. V., SHERATON, J. W., LAIBA, A. A., TINGEY, R. J., THOST, D. E., KAMENEV, E. N. & FEDOROV, L. V. 2001*b.* Geology of the Prince Charles Mountains, Antarctica. *AGSO–Geoscience Australia, Bulletin,* **247.**

MONNIER, O., MÉNOT, R.-P., PEUCAT, J.-J., FANNING, C. M. & GIRET, A. 1996. Actualisation des données géologiques sur Terre Adélie (Antarctique Est): mise en evidence d'un collage tectonique au Proérozoïque. *Comptes Rendues de l'Académie des Sciences, Paris,* **322,** 55–62.

MOORES, E. M. 1991. Southwest U.S.–East Antarctic (SWEAT) connection: a hypothesis. *Geology,* **19,** 425–428.

MORTIMER, G. E., COOPER, J. A. & OLIVER, R. L. 1988*a.* The geochemical evolution of Proterozoic granitoids near Port Lincoln in the Gawler orogenic domain of South Australia. *Precambrian Research,* **40/41,** 387–486.

MORTIMER, G. E., COOPER, J. A. & OLIVER, R. L. 1988*b.* Proterozoic mafic dykes near Port Lincoln, South Australia: composition, age and origin. *Australian Journal of Earth Sciences,* **35,** 93–110.

MÜLLER, R. D., GAINA, C. & CLARK, S. 2000. Seafloor spreading around Australia. *In:* VEEVERS, J. J. (ed.) *Billion-year Earth History of Australia and Neighbours in Gondwanaland.* GEMOC Press, Sydney, 18–28.

MYERS, J. S. 1983. The Fraser Complex – a major layered intrusion in Western Australia. *Western Australia Geological Survey, Report,* **14,** 57–66.

MYERS, J. S. 1990. Albany–Fraser Orogen. *Western Australia Geological Survey, Memoir,* **3,** 255–263.

MYERS, J. S. 1993. Precambrian history of the West Australian Craton and adjacent orogens. *Annual Reviews of Earth & Planetary Sciences,* **21,** 453–485.

MYERS, J. S. 1995. *Geology of the Esperance 1:1 000 000 sheet.* Western Australia Geological Survey, 1:1 000 000 Explanatory Notes.

MYERS, J. S., SHAW, R. D. & TYLER, I. M. 1996. Tectonic evolution of Proterozoic Australia. *Tectonics,* **15,** 1431–1446.

NELSON, D. R. 1995. Excursion guide to the Leeuwin Complex, Western Australia. *Proceedings of the Australian Conference on Geochronology and Isotope Geoscience,* Perth 1995. Geological Survey of Western Australia.

NELSON, D. R. 1996. Compilation of SHRIMP U–Pb zircon geochronology data, 1995. *Western Australia Geological Survey, Record,* **1996/5.**

NELSON, D. R. 1998. Granite–greenstone crust formation on the Archaean Earth: a consequence of two superimposed processes. *Earth & Planetary Science Letters,* **158,** 109–119.

NELSON, D. R. 1999. Compilation of geochronology data,

1998. *Western Australia Geological Survey, Record,* **1999/2**.

NELSON, D. R., MYERS, J. S. & NUTMAN, A. P. 1995. Chronology and evolution of the middle Proterozoic Albany-Fraser orogen, Western Australia. *Australian Journal of Earth Sciences,* **42**, 481–495.

OLIVER, R. L. & FANNING, C. M. 1997. Australia and Antarctica: precise correlation of Palaeoproterozoic terrains. *In:* RICCI, C. A. (ed.) *The Antarctic Region: Geological Evolution and Processes.* Terra Antartica Publication, Siena, 163–172.

OLIVER, R. L. & FANNING, C. M. 1999. Metamorphic history of King George V Land, Antarctica, and its relationship to adjacent Adélie Land and to southern Eyre Peninsula, South Australia. *Proceedings of the 8th International Symposium on Antarctic Earth Sciences,* Wellington, Abstracts, 230.

OLIVER, R. L., COOPER, J. A. & TRUELOVE, A. J. 1983. Petrology and zircon geochronology of Herring Island and Commonwealth Bay and evidence for Gondwana reconstruction. *In:* OLIVER, R. L., JAMES, P. R. & JAGO, J. B. (eds) *Antarctic Earth Science.* Australian Academy of Science, Canberra, 64–68.

PAGE, R. W., STEVENS, B. P. J. & GIBSON, G. M. 2000. New SHRIMP zircon results from Broken Hill: towards a robust stratigraphic and event timing. *Geological Society of Australia, Abstracts,* **59**, 375.

PANKHURST, R. J., MARSH, P. D. & CLARKSON, P. D. 1983. A geochronological investigation of the Shackleton Range. *In:* OLIVER, R. L., JAMES, P. R. & JAGO, J. B. (eds) *Antarctic Earth Science.* Australian Academy of Science, Canberra, 234–237.

PARKER, A. J. & LEMON, N. M. 1982. Reconstruction of the early Proterozoic stratigraphy of the Gawler Craton, South Australia. *Journal of the Geological Society of Australia,* **29**, 221–238.

PAUL, E., STÜWE, K., TEASDALE, J. & WORLEY, B. 1995. Structural and metamorphic geology of the Windmill Islands, east Antarctica: field evidence for repeated tectonothermal activity. *Australian Journal of Earth Sciences,* **42**, 453–469.

PELLETIER, A., GAPAIS, D., MÉNOT, R.-P. & GUIRAUD, M. 1999. The 1.7 Ga tectonic and metamorphic event in the Terre Adélie Craton (East Antarctica). *Proceedings of the 8th International Symposium on Antarctic Earth Sciences,* Wellington, Abstracts, 239.

PEUCAT, J. J., MÉNOT, R. P., FANNING, C. M., PELLETIER, A. & PECORA, L. 1999a. Geochronological evidence for a late-Archaean basement in the Terre Adélie craton. *Proceedings of the 8th International Symposium on Antarctic Earth Sciences,* Wellington, Abstracts, 242.

PEUCAT, J. J., MÉNOT, R. P., MONNIER, O. & FANNING, C. M. 1999b. The Terre Adélie basement in the East-Antarctica Shield: geological and isotopic evidence for a major 1.7 Ga thermal event; comparison with the Gawler Craton in South Australia. *Precambrian Research,* **94**, 205–224.

PIDGEON, R. T. 1990. Timing of plutonism in the Proterozoic Albany Mobile belt, southwestern Australia. *Precambrian Research,* **47**, 157–167.

PIDGEON, R. T. & NEMCHIN, A. A. 2001. 1.2 Ga mafic dyke near York, south-western Yilgarn Craton, Western Australia. *Australian Journal of Earth Sciences,* **48**, 751–755.

PISAREVSKY, S. A., WINGATE, M. T. D., POWELL, C. MCA., JOHNSON, S. & EVANS, D. A. D. 2003. Models of Rodinia assembly and fragmentation. *In:* YOSHIDA, M., WINDLEY, B. F. & DASGUPTA, S. (eds) *Proterozoic East Gondwana: Supercontinent Assembly and Breakup.* Geological Society, London, Special Publication, **206**, 35–55.

POST, N. J. 2000. *Unravelling Gondwana fragments: an integrated structural, isotopic and petrographic investigation of the Windmill Islands, East Antarctica.* PhD Thesis, University of New South Wales, Australia.

POST, N. J., HENSEN, B. J. & KINNY, P. D. 1997. Two metamorphic episodes during a 1340–1180 Ma convergent tectonic event in the Windmill Islands, East Antarctica. *In:* RICCI, C. A. (ed.) *The Antarctic Region: Geological Evolution and Processes.* Terra Antartica Publication, Siena, 157–161.

POWELL, C. MCA. & PISAREVSKY, S. A. 2002. Late Neoproterozoic assembly of East Gondwana. *Geology,* **30**, 3–6.

POWELL, C. MCA., JOHNSON, B. D. & VEEVERS, J. J. 1980. A revised fit of East and West Gondwanaland. *Tectonophysics,* **63**, 13–29.

RANKIN, L. R., MARTIN, A. R. & PARKER, A. J. 1989. Early Proterozoic history of the Karari Fault Zone, northwest Gawler Craton, South Australia. *Australian Journal of Earth Sciences,* **36**, 123–133.

RASMUSSEN, B., BENGTSON, S., FLETCHER, I. R. & MCNAUGHTON, N. J. 2002. Discoidal impressions and trace-like fossils more than 1200 million years old. *Science,* **296**, 1112–1115.

ROGERS, J. J. W., UNRUG, R. & SULTAN, M. 1995. Tectonic assembly of Gondwana. *Journal of Geodynamics,* **19**, 1–34.

SCHAEFER, B. F. 1998. *Insights into Proterozoic tectonics from the southern Eyre Peninsula, South Australia.* PhD Thesis, University of Adelaide, Australia.

SCRIMGEOUR, I. & HAND, M. 1997. A metamorphic perspective on the Pan African overprint in the Amery area of Mac. Robertson Land, East Antarctica. *Antarctic Science,* **9**, 313–335.

SHACKLETON, R. M. 1986. Precambrian collision tectonics in Africa. *In:* COWARD, M. P. & RIES, A. C. (eds) *Collision Tectonics.* Geological Society, London, Special Publication, **19**, 329–349.

SHACKLETON, R. M. 1996. The final collision zone between East and West Gondwana: where is it? *Journal African Earth Sciences,* **23**, 271–287.

SHAW, R. D., WELLMAN, P., GUNN, P., WHITAKER, A. J., TARLOWSKI, C. & MORSE, M. 1996. Guide to using the Australian Crustal Elements Map. *Australian Geological Survey Organisation Record,* **1996/30**.

SHERATON, J. W., BLACK, L. P., MCCULLOCH, M. T. & OLIVER, R. L. 1990. Age and origin of a compositionally varied dyke swarm in the Bunger Hills, East Antarctica. *Chemical Geology,* **85**, 215–246.

SHERATON, J. W., BLACK, L. P. & TINDLE, A. G. 1992. Petrogenesis of plutonic rocks in a Proterozoic granulite-facies terrane – the Bunger Hills, East Antarctica. *Chemical Geology,* **97**, 163–198.

SHERATON, J. W., TINGEY, R. J., BLACK, L. P. & OLIVER, R. L. 1993. Geology of the Bunger Hills area, Antarctica: implications for Gondwana correlations. *Antarctic Science,* **5**, 85–102.

SHERATON, J. W., TINGEY, R. J., OLIVER, R. L. & BLACK, L. P. 1995. Geology of the Bunger Hills–Denman Glacier region, East Antarctica. *Australian Geological Survey Organisation Bulletin*, **244**.

SHIRAISHI, K., ELLIS, D. J., HIROI, Y., FANNING, C. M., MOTOYOSHI, Y. & NAKAI, Y. 1994. Cambrian orogenic belt in East Antarctica and Sri Lanka: implications for Gondwana assembly. *Journal of Geology*, **102**, 47–65.

SNAPE, I., BLACK, L. P. & HARLEY, S. L. 1997. Refinement of the timing of magmatism, high-grade metamorphism and deformation in the Vestfold Hills, East Antarctica, from new SHRIMP U–Pb zircon geochronology. In: RICCI, C. A. (ed.) *The Antarctic Region: Geological Evolution and Processes*. Terra Antartica Publication, Siena, 139–148.

STERN, R. J. 1994. Arc assembly and continental collision in the Neoproterozoic East African Orogen: implications for the consolidation of Gondwanaland. *Annual Reviews of Earth & Planetary Sciences*, **22**, 319–351.

STÜWE, K. & OLIVER, R. L. 1989. Geological history of Adélie Land and King George V land, Antarctica: evidence for a polycyclic metamorphic evolution. *Precambrian Research*, **43**, 317–334.

STÜWE, K. & SANDIFORD, M. 1993. A preliminary model for the 500 Ma event in the East Antarctic Shield. In: FINDLAY, R. H., UNRUG, R., BANKS, M. R. & VEEVERS, J. J. (eds) *Gondwana Eight: Assembly, Evolution and Dispersal*. Balkema, Rotterdam, 125–130.

TALARICO, F. & KRONER, U. 1999. Geology and tectono–metamorphic evolution of the Read Group, Shackleton Range: a part of the East Antarctic Craton. *Terra Antartica*, **6**, 183–202.

TALARICO, F., KLEINSCHMIDT, G. & HENJES-KUNST, F. 1999. First report of an ophiolitic complex in the northern Shackleton Range, Antarctica. *Terra Antartica*, **6**, 293–315.

TEASDALE, J. P. 1997. *Understanding poorly exposed terranes – the interpretative geology and tectonothermal evolution of the western Gawler Craton*. PhD Thesis, University of Adelaide, Australia.

TESSENSOHN, F., KLEINSCHMIDT, G., TALARICO, F., ET AL. 1999. Ross-age amalgamation of East and West Gondwana: evidence from the Shackleton Range. *Terra Antartica*, **6**, 317–325.

THOM, R., DE LAETER, J. R. & LIBBY, W. G. 1981. Rb–Sr Dating of Tectonic Events in the Proterozoic Mount Barren Group near Hopetoun. *Western Australia Geological Survey, Annual Report*, 109–112.

TINGEY, R. J. 1982. The geologic evolution of the Prince Charles Mountains – an Antarctic Archaean Cratonic Block. In: CRADDOCK, C. (ed.) *Antarctic Geoscience*. University of Wisconsin Press, Madison, 455–464.

TINGEY, R. J. 1991. The regional geology of Archaean and Proterozoic rocks in Antarctica. In: TINGEY, R. J. (ed.) *The Geology of Antarctica*. Oxford University Press, Oxford, 1–73.

TORSVIK, T. H, CARTER, L. M., ASHWAL, L. D., BHUSHAN, S. K., PANDIT, M. K. & JAMTVEIT, B. 2001. Rodinia refined or obscured: palaeomagnetism of the Malani igneous suite (NW India). *Precambrian Research*, **108**, 319–333.

TUREK, A. & STEPHENSON, N. C. N. 1966. The radiometric age of the Albany Granite and the Stirling Range Beds, southwest Australia. *Journal of the Geological Society, Australia*, **13**, 449–456.

TURNER, S. P, FODEN, J. D., SANDIFORD, M. & BRUCE, D. 1993. Sm–Nd evidence for the provenance of sediments from the Adelaide Fold Belt and south-eastern Australia with implications for episodic crustal addition. *Geochimica et Cosmochimica Acta*, **57**, 1837–1856.

VAN SCHMUS, W. R., MCKENNA, L. W. GONZALES, D. A., FETTER, A. H. & ROWELL, A. J. 1997. U–Pb geochronology of parts of the Pensacola, Thiel and Queen Maud Mountains, Antarctica. In: RICCI, C. A. (ed.) *The Antarctic Region: Geological Evolution and Processes*. Terra Antartica Publication, Siena, 187–200.

VASSALLO, J. J. & WILSON, C. J. L. 1999. Palaeoproterozoic geology of southeastern Eyre Peninsula. In: WILSON, C. J. L. (ed.) *The Great Southern Transect II: A Geological Section Incorporating the Lachlan Fold Belt, Adelaide Fold Belt and Gawler Craton, Halls Gap (Victoria) to Port Lincoln (S.A.)*. Specialist Group in Tectonics and Structural Geology, Field Guide No. 6, Geological Society of Australia.

VASSALLO, J. J. & WILSON, C. J. L. 2001. Structural repetition of the Hutchison Group metasediments, Eyre Peninsula, South Australia. *Australian Journal of Earth Sciences*, **48**, 331–345.

VASSALLO, J. J. & WILSON, C. J. L. 2002. Palaeoproterozoic regional-scale non-coaxial deformation: an example from eastern Eyre Peninsula, South Australia. *Journal of Structural Geology*, **24**, 1–24.

VEEVERS, J. J. 1994. Case for the Gamburtsev Subglacial Mountains of East Antarctica originating by mid-Carboniferous shortening of an intracratonic basin. *Geology*, **22**, 593–596.

VEEVERS, J. J. (ed.) 2000. *Billion-year Earth History of Australia and Neighbours in Gondwanaland*. GEMOC Press, Sydney.

WALKER, N. W. & GOODGE, J. W. 1991. Significance of Late Archean–Early Proterozoic U–Pb ages of individual Nimrod Group detrital zircons and Cambrian plutonism in the Miller Range, Transantarctic Mountains. *Geological Society of America, Abstracts with Programs*, **23**, 306.

WEATHERLY, S. & MCNAUGHTON, N. J. 1995. New age constraints on amphibolite facies mineral growth and deformation in metapelites: an example from the Mt Barren Group, Western Australia. *Proceedings of the Australian Conference on Geochronology and Isotope Geoscience, Workshop Programme and Abstracts*, Curtin University of Technology, Perth, 38.

WEATHERLY, S., MCNAUGHTON, N. J. & KINNY, P. D. 1998. Proterozoic tectonothermal events in the Albany-Fraser Province – how many, when and where? *Proceedings of the Tectonics Special Research Centre Inaugural Symposium*, University of Western Australia, Perth, Abstracts, 82–86.

WEBB, A. W., MCDOUGALL, I. & COOPER, J. A. 1964. Potassium–argon dates from the Vincennes Bay region and Oates Land. In: ADIE, R. J. (ed.) *Antarctic Geology*. North Holland Publishing Company, Amsterdam, 597–600.

WEBB, A. W., THOMSON, B. P., BLISSETT, A. H., DALY, S. J., FLINT, R. B. & PARKER, A. J. 1986. Geochronology of the Gawler craton, South Australia. *Australian Journal of Earth Sciences*, **33**, 119–143.

WHITAKER, A. J. 1992. *Albany Magnetic and Gravity Interpretation (1:1 000 000 scale map)*. Bureau of Mineral Resources, Geology & Geophysics, Canberra.

WHITAKER, A. J. 1993. *Esperance Magnetic and Gravity Interpretation (1:1 000 000 scale map)*. Australian Geological Survey Organisation, Canberra.

WILDE, S. A. 1999. Evolution of the western margin of Australia during the Rodinian and Gondwanan supercontinent cycles. *Gondwana Research*, **2**, 481–499.

WILDE, S. A. & MURPHY, D. M. K. 1990. The nature and origin of Late Proterozoic high-grade gneisses of the Leeuwin Block, Western Australia. *Precambrian Research*, **47**, 251–270.

WILLIAMS, I. S., COMPSTON, W., COLLERSON, K. D., ARRIENS, P. A. & LOVERING, J. F. 1983. A reassessment of the age of the Windmill Metamorphics, Casey area. *In:* OLIVER, R. L., JAMES, P. R. & JAGO, J. B. (eds) *Antarctic Earth Science*. Australian Academy of Science, Canberra, 64–68.

WILSON, C. J. L. 1997. Shear zone development and dyke emplacement in the Bunger Hills, East Antarctica. *In:* RICCI, C. A. (ed.) *The Antarctic Region: Geological Evolution and Processes*. Terra Antartica Publication, Siena, 149–156.

WILSON, T. J., GRUNOW, A. M. & HANSON, R. E. 1997. Gondwana assembly: the view from southern Africa and East Gondwana. *Journal of Geodynamics*, **23**, 263–286.

WINGATE, M. T. D & EVANS, D. A. D. 2003. Palaeomagnetic constraints on the Proterozoic tectonic evolution of Australia. *In:* YOSHIDA, M., WINDLEY, B. F. & DASGUPTA, S. (eds) *Proterozoic East Gondwana: Supercontinent Assembly and Breakup*. Geological Society, London, Special Publication, **206**, 77–91.

WINGATE, M. T. D. & GIDDINGS, J. W. 2000. Age and palaeomagnetism of the Mundine Well dyke swarm, Western Australia: implications for an Australia–Laurentia connection at 755 Ma *Precambrian Research*, **100**, 335–357

WINGATE, M. T. D., CAMPBELL, I. H., COMPSTON, W. & GIBSON, G. M. 1998. Ion microprobe U–Pb ages for Neoproterozoic basaltic magmatism in south-central Australia and implications for the break-up of Rodinia. *Precambrian Research*, **87**, 135–159.

WINGATE, M. T. D., CAMPBELL, I. H. & HARRIS, L. B. 2000. SHRIMP baddeleyite age for the Fraser dyke swarm, southeast Yilgarn Craton, Western Australia. *Australian Journal of Earth Sciences*, **47**, 309–313.

WYBORN, L. A. I. 1988. Petrology, geochemistry, and origin of a major Australian 1880–1840 Ma felsic volcano–plutonic suite: a model for intracontinental felsic magma generation. *Precambrian Research*, **40/41**, 37–60.

YOSHIDA, M. 1995. Assembly of East Gondwana during the Mesoproterozoic and its rejuvenation during the Pan-African period. *In:* YOSHIDA, Y. & SANTOSH, M. (eds) *India and Antarctica during the Precambrian*. Geological Society of India, Memoir, **34**, 22–45.

ZHAO, Y., LIU, X.-C., FANNING, C. M. & LIU, X.-H. 2000. The Grove Mountains, a segment of a Pan-African orogenic belt in East Antarctica. *Proceedings of the 31 IGC Brazil*, Abstracts.

ZHAO, Y., LIU, X-H, LIU, X-C & SONG, B. 2003. Pan-African events in Prydz Bay, East Antarctica, and its inference in the East Gondwana tectonics. *In:* YOSHIDA, M., WINDLEY, B. F. & DASGUPTA, S. (eds) *Proterozoic East Gondwana: Supercontinent Assembly and Breakup*. Geological Society, London, Special Publication, **206**, 231–245.

ZHAO, Y., LIU, X., SONG, B., ZHANG, Z., LI, J., YAO, Y. & WANG, Y. 1995. Constraints on the stratigraphic age of metasedimentary rocks from the Larsemann Hills, East Antarctica: possible implications for Neoproterozoic tectonics. *Precambrian Research*, **75**, 175–188.

ZHAO, Y., SONG, B., WANG, Y., REN, L., LI, J. & CHEN, T. 1992. Geochronology of the late granite in the Larsemann Hills, East Antarctica. *In:* YOSHIDA, Y., KAMINUMA, K. & SHIRAISHI, K. (eds) *Recent Progress in Antarctic Earth Science*. Terra Scientific Publishing, Tokyo, 155–161.

Indo-Antarctic Correlation: a perspective from the Eastern Ghats Granulite Belt, India

SOMNATH DASGUPTA & PULAK SENGUPTA

Department of Geological Sciences, Jadavpur University, Calcutta-700 032, India
(e-mail: sdg@cal3.vsnl.net.in)

Abstract: Available lithological, petrological and geochronological data on the rocks of the Eastern Ghats Belt (EGB), which has a key role to play in any model of Indo-Antarctic correlation in Precambrian and Cambrian times, have been synthesized and interpreted. Longitudinal lithological subdivisions of the belt are mostly not supported by recent geochronological data, which rather suggest a fourfold division based on protolith ages of the ortho- and paragneisses. Mineral ages show prominent tectonothermal events at Mesoproterozoic, Grenvillian and Pan-African times. However, not all of the events are preserved in all the geochronological domains. In the most well-studied Domain II, three phases of metamorphism are recorded – an early ultra high-temperature metamorphism with an anticlockwise $P–T$ path, a second granulite facies event at $c.$ 1000 Ma, with a retrograde trajectory of near-isothermal decompression, and a third of amphibolite facies. The available information, however, is not conclusive as to whether or not the entire EGB experienced polyphase UHT metamorphism. The propagation of the Grenvillian orogenic front in the EGB is identified.

Over the past decade the geoscience community has vigorously pursued a debate on the supercontinent cycle and its manifold manifestations in tectonic processes, magmatism, metamorphism and global climatic changes associated with the growth/destruction of continents and oceans. One of the major pillars in the reconstruction of the history of the Earth is the supercontinent of Rodinia (Yoshida 1995; Rogers 1996; Unrug 1996). This supercontinent is thought to have assembled at the end of the Mesoproterozoic (1100–900 Ma), largely due to suturing according to the SW US–East Antarctic (SWEAT) model (Dalziel 1991; Hoffman 1991; Moores 1991): In the SWEAT connection, the suture linking Antarctica to India is the Eastern Ghats Belt (EGB), which was supposed to have been juxtaposed against the present coastline of East Antarctica between longitude 40–70°E (Fig. 1). Notwithstanding the fact that some recent palaeomagnetic data are inconsistent with the correlation of the EGB with East Antarctica (Torsvik *et al.* 2001), this remains the most acceptable option. Rodinia is considered to have fragmented between 830 and 750 Ma (Groenewald 1993; Powell *et al.* 1994), reassembling at *c.* 500–550 Ma due to the Pan-African Orogeny into Gondwanaland (Rogers 1996). In all the models of correlation, the EGB and East Antarctica were not separated during the fragmentation of Rodinia. If such a correlation is valid, the presently exposed rocks of the EGB and East Antarctica lived together during the period *c.* 1000–500 Ma, therefore experiencing similar thermotectonic events. A convenient way to test this hypothesis is to compare the thermotectonic evolution of the erstwhile neighbours, particularly within the time frame mentioned above.

Detailed structural, petrological, geochemical and isotopic studies on East Antarctic granulites have helped to identify several distinct age provinces with characteristic tectonomagmatic–metamorphic histories [reviewed in Harley (2002)]. The scenario is not so evident in the case of the EGB, despite intensive petrological studies over the last 15 years [reviewed in Dasgupta & Sengupta (2000)]. Very recently, however, some research publications on isotope geochemistry of the EGB granulites have, to some extent, improved the situation. In this paper an updated review is provided of the metamorphic history of the EGB granulites, particularly in the light of recent isotopic results, in the context of Indo-Antarctic correlation.

Geological Framework of the EGB

The EGB, occurring along the eastern coast of India and with a strike length >1000 km (Fig. 2, inset), is dissected by two prominent NNE–SSW to NE–SW trending rifts, the Mahanadi Rift in the north and the Godavari Rift (graben) in the south (Fig. 2). A generalized geological map of the EGB is given in Figure 2.

Lithology

The geological map of the EGB (Fig. 2) shows the mode of occurrence of the different lithounits. Paragneisses are represented by khondalite (garnet – sillimanite – perthite – plagioclase – quartz gneiss), leptynite (garnet – plagioclase – perthite – quartz

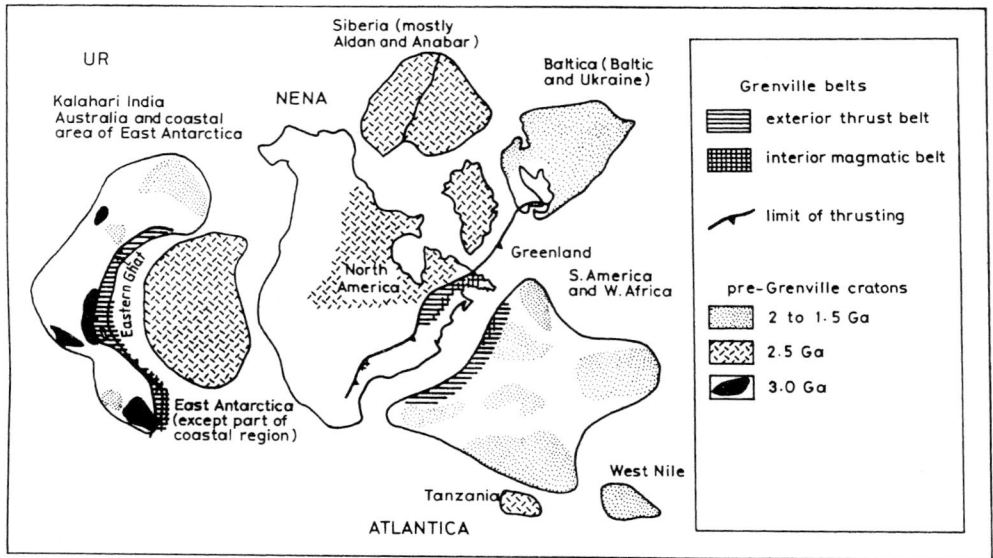

Fig. 1. Formation of the supercontinent of Rodinia according to the SWEAT connection [adopted and slightly modified after Rogers (1996) and Unrug (1996)].

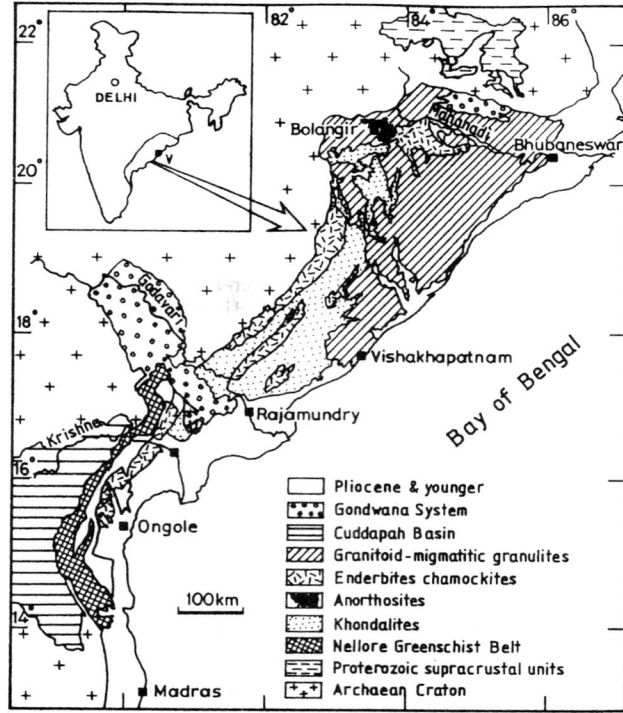

Fig. 2. A simplified geological map of the Eastern Ghats Belt (EGB). Inset: location of the EGB in the map of India.

gneiss), calc-silicate granulite (wollastonite–scapolite–grandite garnet – calcite – clinopyroxene – sphene – plagioclase) and high Mg–Al granulites (sapphirine – spinel – orthopyroxene – garnet – quartz/corundum–cordierite–sillimanite). The last two provided most of the information on P–T–fluid evolutionary history of the EGB rocks (Lal et al. 1987; Kamineni & Rao 1988; Sengupta et al. 1990, 1991, 1997, 1999; Dasgupta 1993; Dasgupta et al. 1994, 1995; Bhowmik et al. 1995; Bose et al. 2000; Rickers et al. 2001a). Mostly, the high Mg–Al granulites occur as lenses in khondalite and leptynite. Rare occurrences of this rock are noted as xenoliths in mafic granulite (e.g. Dasgupta et al. 1994; Rickers et al. 2001a). Two major members of the orthogneisses are mafic granulite (orthopyroxene– clinopyroxene–plagioclase ± garnet) and enderbite (orthopyroxene – plagioclase–quartz–perthite). True charnockite (perthite>>plagioclase) is less common and where present shows a distinct intrusive relation with the other orthogneisses (Rickers et al. 2001a). Mafic granulite and enderbite also provided excellent constraints on the P–T conditions of metamorphism (Dasgupta et al. 1991, 1992, 1993, 1994). Two generations of enderbites have been reported from some areas, where the later variety is pegmatoidal (Sengupta et al. 1999). Another significant magmatic rock in the EGB is a megacrystic granite/granitoid, which occupies large areas, particularly in the central portion. A layered mafic– ultramafic complex with chromitite occurs at Kondapalle in the southern part of the EGB (Leelanandam 1990; Sengupta et al. 1999). Massif-type anorthosites occur in the northern part of the EGB (Fig. 2) (Sarkar et al. 1981), while different varieties of alkaline rocks occur along a prominent marginal zone bordering the EGB (Leelanandam 1989).

Subdivisions of the EGB: an update

Over the past few years several attempts have been made to classify the EGB from different viewpoints. The different classification schemes proposed so far, will be discussed briefly below.

Nanda & Pati (1989) and Ramakrishnan et al. (1998) compiled a geological map of the EGB, and proposed a fourfold longitudinal subdivision on the basis of dominant lithological assemblages which, from west to east, are as follows.

(1) Western Charnockite Zone (WCZ): this is dominantly composed of charnockite and enderbite, with lenses of mafic–ultramafic rocks, and minor metapelites.
(2) Western Khondalite Zone (WKZ): this is dominated by typical metapelites (khondalite) with intercalated quartzite, calc-silicate rocks, marble and high Mg–Al granulites. These are intruded by charnockite and enderbite. Several occurrences of massif-type anorthosite are reported from this zone (Bolangir, Turkel and Jugsaipatna).
(3) Central Charnockite–Migmatite Zone (CMZ): this is dominantly composed of migmatitic supracrustal rocks, including garnet-bearing diatextite, leptynite, high Mg–Al granulite and calc-silicate rocks, which are intruded by charnockite–enderbite, porphyritic granitoids and massif-type anorthosite (Chilka Lake region).
(4) Eastern Khondalite Zone (EKZ): lithologically, this zone is similar to the WKZ, but no occurrence of anorthosite is known.

Dasgupta & Sengupta (2000) discussed some of the problems associated with such a subdivision. Lithological characteristics of all the domains show considerable overlap, and isotopic data (Rickers et al. 2001b) does not support such a classification. Further, Ramakrishnan et al. (1998) proposed a transition zone along the western margin of the EGB, which is also characterized by the presence of several alkaline intrusives. Available petrological data indicate that it is not a strictly transition zone, but that it contains a mixture of high-grade (similar to those occurring on the east) and alkaline rocks. Moreover, there is a controversy regarding the location of the so-called Transition Zone. Gupta et al. (2000) placed it further west of the area suggested by Ramakrishnan et al. (1998).

Chetty (1995) and Chetty & Murthy (1998) proposed two major subdivisions of the EGB based on interpretation of Landsat imagery data. The Nagavalli–Vamasdhara Shear Zone, trending NW–SE, was supposed to divide the EGB into two blocks: the Archaean Araku Block in the south and the Proterozoic Chilka Block in the north, and the shear zone itself, was thought to be an extension of the Napier–Rayner boundary in East Antarctica. As discussed in Sengupta et al. (1999), field geological evidence does not support this contention, neither does the isotopic data of Rickers et al. (2001b) provide support. The significance of the Sileru Shear Zone (Chetty 1995), occurring in the west, is again a hotly debated topic [see contradictions in Ramakrishnan et al. (1998) and Dobmeier & Raith (2002)].

Detailed isotopic study (Sm–Nd, Rb–Sr and Pb systematics) carried out by Rickers et al. (2001b) over the entire EGB has provided valuable clues as regards the possible distribution of different 'terranes' in the belt. Rickers et al. (2001b) have identified four crustal domains in the EGB (Fig. 3) based on Nd model ages, and supported by Rb–Sr and Pb isotopic data. Their Domain I more or less coincides with the WCZ of Ramakrishnan et al. (1998). However, the WCZ south of the Godavari Rift is a

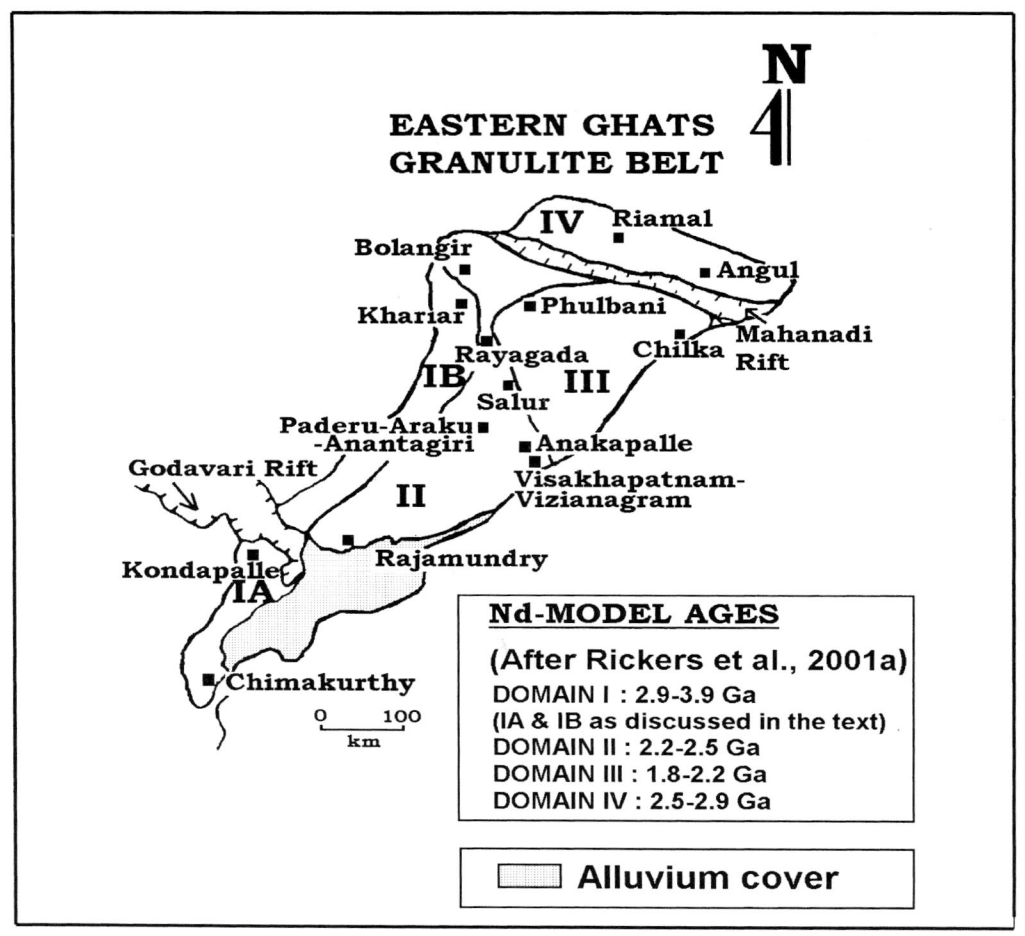

Fig. 3. Distribution of different isotopic domains [after Rickers *et al.* 2001*b*)] in the Eastern Ghats Belt with locations of areas referred in the text.

distinct Proterozoic crustal block formed from juvenile material between 2300 and 1700 Ma, and the last tectonometamorphic event in this segment occurred at 1600 Ma. However, Sengupta *et al.* also report mid-Archaean Nd model ages for the mafic granulites of the Kondapalle area, which are responsible for ultrahigh-*T* regional-scale contact metamorphism of pelitic rocks [earlier reported in Sengupta *et al.* 1999)]. It is interesting that Rickers *et al.* (2001*b*) observed notable differences between the WCZ rocks occurring on either side of the Godavari Rift: to the north they noted Archaean isotopic signatures as compared to Proterozoic signatures in the south. This reaffirms an earlier notion (Mezger *et al.* 1996; Sengupta *et al.* 1999) that the Godavari Rift is indeed a geochronological divide.

To the north of the Godavari Rift, Rickers *et al.* (2001*b*) identified three domains (Fig. 3). The domain near Chilka Lake is rather homogeneous with Nd model ages for ortho- and paragneisses converging between 1800 and 2200 Ma. Massif-type anorthosites occurring in this domain were emplaced at 790 Ma (Krause *et al.* 2001). They correlated the northern and southern boundaries of the two domains with the Mahanadi and Nagavalli–Vamasdhara Lineaments, respectively, essentially supporting Chetty & Murthy (1998). The domain sandwiched between two isotopically homogeneous domains (Fig. 3) is the most heterogeneous one. Rickers *et al.* (2001*b*) placed the domain boundaries along the Nagavalli–Vamasdhara and Sileru Lineaments on the north and west, respectively. The Nd model ages for orthogneisses show gradual variation from the areas near the Godavari Rift (3200 Ma) towards the NE (1800 Ma). Metasediments have model ages of 2100–2500 Ma. The fourth domain of

Rickers et al. (2001b), in the northern part of the EGB (Fig. 3), contains metasediments having Nd model ages of 2200–2800 Ma. Important occurrences of granulite facies rocks are noted at Angul and Riamal. The lithological assemblage is similar to that in Domain II. Additionally, massif-type anorthosite occurs at several places (e.g. Dhenkanol), which brings out similarity with the Chilka Lake Domain III. It is intriguing to note that U–Pb mineral ages from orthogneisses at Phulbani [occurring in the western part of Domain III of Rickers et al. (2001b)] are similar to those reported from similar rocks at Angul (Aftalion et al. 1988) occurring in the Domain IV. Although Phulbani and Chilka Lake areas fall in the same Domain III according to Rickers et al. (2001a), the strong Grenvillian impress in the former (Aftalion et al. 1988) is not recorded in the latter [see review by Dobmeier & Raith (2002)]. Further, it is not clear from the subdivisions of Rickers et al. (2001b) as to which domain the anorthositic rocks of Bolangir and Turkel should belong. The geological setting of these rocks is similar to that in the Angul–Dhenkanol area (Domain IV), although geochronological data on anorthosites of the latter area are not available. It is evident from the above contradictions that far more intensive structural, petrological and geochemical databases are necessary in order to understand the complexities in the distribution of the possible terranes in the EGB, their boundary relations and their thermotectonic evolution.

Despite the obvious gaps in knowledge, the present authors would prefer to discuss the metamorphic history of the EGB granulites within a reference framework of the subdivision proposed by Rickers et al. (2001b); however, for the convenience of the reader, the subdivisions proposed recently by Dobmeier & Raith (2002) will also be indicated.

Metamorphic history of the EGB: an update

Most of the petrological studies carried out by the present group of workers and others came from Domain II (Fig. 3); consequently, the petrological evolution of the EGB granulites falling in this domain are far better understood. Additionally, one detailed study is available from the southern part of Domain 1 (Sengupta et al. 1999) and one from Domain III (Sen et al. 1995).

Petrological evolution of granulites in Domain IA [Ongole Domain of the Krishna Province of Dobmeier & Raith (2002)]

From Kondapalle, south of the Godavari Rift, Sengupta et al. (1999) documented temperatures >1000°C and a deep crustal heating–cooling trajectory of evolution for high Mg–Al granulites. Dasgupta et al. (1997) documented a heating–cooling trajectory at comparable peak temperatures, but at a shallower crustal level from an area further south in Domain I. In none of the rocks studied by these workers was any evidence of late decompression preserved. Mezger et al. (1996), based on their study of U–Pb systematics of allanite, monazite and sphene, and Ar–Ar systematics of amphibole, showed that the thermal histories of the rocks occurring to the south of the Godavari Rift in Domain I (Fig. 3) are different from those in the north. In the southern portion, the last major metamorphic reconstitution took place between 1300–1600 Ma, followed only by very local resetting of the Ar–Ar geoclock in hornblende at 1100 Ma; this was further confirmed by Rickers et al. (2001b). Kovach et al. (2001) obtained an age of 1720 Ma for zircon collected from enderbitic rocks, which intruded the ultramafic rocks and ultrahigh-temperature (UHT) metapelites at Kondapalle (Sengupta et al. 1999).

Petrological evolution of granulites in Domain IB [Jeypore Province of Dobmeier & Raith (2002)]

In the northern part of Domain I, Neogi et al. (1999) reported peak P–T conditions of 9–10 kbar, >950°C, which was followed by cooling and decompression from their study on high Mg–Al granulites and mafic granulites. Gupta et al. (2000) deduced similar peak P–T conditions from geothermobarometric data on felsic and mafic granulites. Although no firm constraint can be placed on the timing of metamorphism, available isotopic data in this segment (Aftalion et al. 2000; Rickers et al. 2001b; Sarkar et al. 2000) rule out any evidence of 'Grenvillian' orogeny. However, the imprints of a Pan-African event (c. 500 Ma) were noted in places (Aftalion et al. 1998).

Petrological evolution of the EGB granulites in Domain II [Eastern Ghats Province of Dobmeier & Raith (2002)]

One of the most significant outcomes of petrological research on the EGB granulites is the recognition of ultrahigh temperature ($\cong 1000°C$) of metamorphism at lower crustal depths (30–35 km) from several areas (Kamineni & Rao 1988; Sengupta et al. 1990, 1991, 1999; Dasgupta et al. 1991, 1995; Bhowmik et al. 1995; Dasgupta & Sengupta 1995; Mohan et al. 1997; Bose et al. 2000; Rickers et al. 2001a). Dasgupta (1995) and Dasgupta & Sengupta (1995,

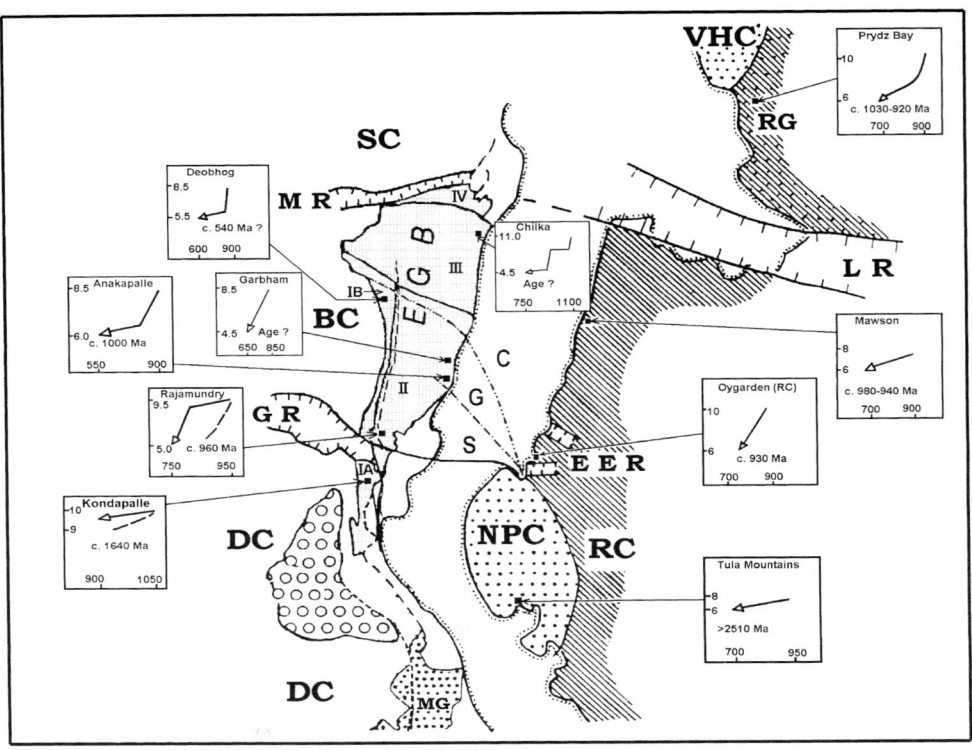

Fig. 4. Characteristic *P–T* paths of metamorphism deduced from different isotopic domains [I, II, III, IV of Rickers *et al.* (2001*b*)] of the Eastern Ghats Belt (EGB) and propagation line of the Grenvillian front in the EGB (S, present work, G; Grew *et al.* 1988; C, Chetty 1995). VHC, Vestfold Hills; RG, Rauer Group; RC, Rayner Complex; NPC, Napier Complex; MG, Madras Granulite; DC, Dharwar Craton; BC, Bastar Craton; SC, Singhbhum Craton; MR, Mahanadi Rift; GR, Godavari Rift; EER, East Antarctic Rift; LR, Lambert Rift.

2000) reviewed available information on UHT metamorphism from the EGB, and interested readers are referred to these papers for more details. Most of the information on the UHT metamorphism was obtained from sapphirine–spinel-bearing high Mg–Al granulites and calc-silicate granulites. In the former, dehydration–melting of biotite-bearing protoliths at temperatures >850°C and low pressures (*c.* 5–6 kbar) produced the high Mg–Al granulites (Sengupta *et al.* 1990; Dasgupta *et al.* 1995). Dasgupta & Sengupta (2000; Fig. 2) compiled the *P–T* trajectories of evolution of EGB granulites reported by different workers. In Figure 4, the 'calculated peak' *P–T* conditions and the *P–T* trajectories for some of the occurrences in different parts of the EGB are plotted for comparison. As the *P–T* trajectories reveal, the EGB granulites are polymetamorphic with an impress of two high-grade events. However, the imprint of all phases is not preserved uniformly in all places. Prior to discussing spatial variation in metamorphic characteristics, summarized below is the overall nature of metamorphism.

Most workers consider that the rocks evolved through an early M1 metamorphism along an anticlockwise trajectory, having a high dT/dP slope (Sengupta *et al.* 1990; Dasgupta *et al.* 1995; Mukhopadhyay & Bhattacharya 1997). In the high Mg–Al granulites, early stabilization of Zn-poor spinel + quartz testifies to a low-P/high-T condition during the prograde M1 phase. The M1 peak condition is estimated as 8–10 kbar and ≅1000°C, resulting in the stabilization of sapphirine + quartz and complex Fe–Ti–Al oxide solid solution in Mg–Al granulite, and wollastonite + scapolite in calc-silicate granulite. The rocks were metamorphosed under a fluid-absent/deficient condition at this stage. The retrograde path of M1 is one of near-isobaric cooling down to 750–800°C. The most noticeable change in the mineral assemblages during this period are the appearance of orthopyroxene + sillimanite + garnet in Mg–Al granulite, of grandite garnet in calc-silicate granulite, and coronal garnet in mafic granulite and enderbite (Sengupta *et al.* 1990, 1991, 1997, 1999; Dasgupta 1993; Dasgupta et al. 1993, 1995; Bhowmik et al. 1995; Bose et al. 2000). Similar peak P–T conditions for M1 metamorphism and subsequent near-isobaric cooling were reported by Shaw & Arima (1996) from their study on high Mg–Al granulite and calc-silicate granulite at Rayagada. Modifying their earlier conclusion, Shaw & Arima (1996) deduced peak P–T conditions of 12 kbar and 1100°C, and decompression to 9 kbar based on new textural features from high Mg–Al granulites, particularly development of sillimanite at the expense of corundum (which is intergrown with Fe–Ti oxide and spinel) and quartz. As discussed by Dasgupta & Sengupta (2000), this can be better explained by oxidation of a Fe–Ti–Al oxide, and does not require such extreme pressures.

There are no clear-cut geochronologic data on the timing of the UHT metamorphism in Domain II. Jarrick (2000), based on a U–Pb Thermal Ionization Mass Spectroscopic study of zircon and a Pb isotopic study of leached feldspars, concluded that the first high-grade metamorphism took place at 1400 Ma. Synthesizing available geochronological data, Dobmeier & Raith (2002) concluded that the M1 event took place at 1100 Ma. Clearly, two factors contributed to the uncertainties related to the estimation of the age of the M1 metamorphism: (1) UHT conditions and subsequent cooling resulting in partial resetting of mineral ages; and (2) overprinting by the later high-grade event (discussed below).

The isobarically cooled granulites were reworked by M2 metamorphism, whose peak condition is estimated to be 8–8.5 kbar and 850°C (Dasgupta et al. 1992, 1995). The retrograde path of M2 is characterized by near-isothermal decompression to 5 kbar, which is largely constrained by thermobarometric data (Dasgupta et al. 1992), and limited textural criteria. The most commonly cited textural criteria in Mg–Al granulites (Sengupta et al. 1990; Dasgupta et al. 1995; Mohan et al. 1997; Kamineni & Rao 1988) is the appearance of late cordierite at the expense of orthopyroxene + sillimanite + garnet + quartz. Although unequivocal textural evidence in favour of the late appearance of cordierite has been presented, this alone cannot testify to decompression. It is well known that the stability field of cordierite is expanded in the presence of a fluid phase. Sengupta & Dasgupta (1998) explored this possibility with the help of schematic petrogenetic grids and concluded that such a possibility can not be ruled out for the EGB granulites. Currently, the present authors are undertaking more detailed work on this aspect. The typical decompression features in quartzofeldspathic gneisses (e.g. breakdown of garnet to orthopyroxene + calcic plagioclase, Harley 1989) have been observed only locally so far. In contrast, Dasgupta et al. (1999) documented that such textures in one suite of rocks (albeit from a different domain) were produced due to Na metasomatism. In summary, decompression during the retrogression of the M2 metamorphism cannot be ruled out, but it must be emphasized that whether formation of cordierite is diagnostic of this stage or not needs to be ascertained. The age of M2 metamorphism where the peak temperature reached c. 850°C is fairly well documented as 950–1000 Ma (Grew & Manton 1986; Aftalion et al. 1988; Paul et al. 1990; Shaw et al. 1997: Mezger & Cosca 1999; Dasgupta & Sengupta 2000 and refs cited therein; Dobmeier & Raith 2002).

M3 metamorphism is represented by weak amphibolite facies overprints on the earlier assemblages, resulting in hydration/carbonation. The M3 P–T condition is estimated to be 5 kbar and 600°C. The age of M3 metamorphism in Domain II is again well characterized as Pan-African (500–550 Ma) (Mezger & Cosca 1999 and refs cited therein).

One notable exception to this general scenario in Domain II is the occurrence at Anakapalle described by Dasgupta et al. (1994), and later by Rickers et al. (2001a) (Fig. 4). Dasgupta et al. (1994) documented an early decompression to the tune of 1.5 kbar from peak metamorphic conditions, followed by cooling. This later cooling was also confirmed by Sengupta et al. (1997). The early decompression does not necessarily indicate a clockwise P–T trajectory, as mistakenly referred to in some later works (e.g. Mohan et al. 1997). It was pointed out by Dasgupta et al. (1994) that this decompression could be a part of an overall anticlockwise trajectory (see Harley 1989; Anovitz 1991). Rickers et al. (2001a) reported multistage evolution of Mg–Al granulites from the Anakapalle area, also starting with decompression from ultrahigh temperatures. The Mg–Al granulites studied by Rickers et al. (2001a) occur as xenoliths in mafic granulites, which could explain both the ultrahigh temperatures of metamorphism (baking by the enclosing magma) and decompression (emplacement as shallower crustal levels). The authors have pointed out that the evolutionary path recorded in the xenoliths does not represent the general scenario in the EGB.

Petrological evolution of granulites occurring in Domain III [Eastern Ghats Province of Dobmeier & Raith (2002)]

Sen et al. (1995) suggested a multistage evolutionary model of evolution for Mg–Al granulites with peak P–T conditions of >12 kbar and 1000°C, followed by several phases of decompression and cooling (Fig. 4). The present authors' unpublished data on the same suite of rocks reveal peak temperatures of only up to 900°C; similarly, peak pressure is estimated to be c. 8 kbar. The extremely high-P (>11 kbar) and high-T (>1000°C) conditions inferred by Sen et al. (1995) depend on two key factors: (1) presumed stability of orthopyroxene + sillimanite + quartz as the peak assemblage; and (2) use of a fluid-saturated KFMASH petrogenetic grid for interpretation of the assemblages. Non-appearance of sapphirine + quartz in magnesian bulk compositions and a low Al content in orthopyroxene both contradict the assumption (1). Similarly, presence of water-undersaturated melt during peak conditions [considered by Sen et al. (1995)] cannot be reconciled with the fluid-saturated grid. Under these conditions, the fluid-saturated grid is expected to overestimate P–T conditions (Harley 1998).

Massif-type anorthosites occurring in Domain III, earlier considered to be Mesoproterozoic (Sarkar et al. 1981), have now been dated by U–Pb systematics of zircon as 790 Ma (Chilka Lake) (Krause et al. 2001). Detailed geological investigation of the Chilka Lake area (Dobmeier & Raith 2000; unpublished data of the present authors) show that the anorthosites are deformed, and develop garnet locally at the expense of orthopyroxene and plagioclase. Recently, Dobmeier & Raith (2002) summarized available isotopic data from this domain (Dobmeier & Simmat 2001), concluding that the granulite facies overprint occurred at 690–660 Ma. From the Chilka Lake area, Dobmeier & Raith (2000) documented development of patchy charnockite on a foliation in leptynite through dehydration of biotite-bearing assemblages of a suite of felsic rocks. This foliation is also developed in the neighbouring 790 Ma massif-type anorthosite. In view of the isotopic signature, this event has been assigned an age of 690–660 Ma. The record of 690–660 Ma granulite facies metamorphism in this part of the EGB has important implications on Indo-Antarctic correlation, since such an event is not documented in the supposedly contiguous rocks occurring in Prydz Bay, East Antarctica (Harley 2002). Isotopic data show weak thermal overprint at 500 Ma in the domain (Dobmeier & Raith 2002), the petrological significance of which is not clear.

Spatial and temporal variations in P–T trajectories in the EGB, and a possible Indo-Antarctic Correlation

In the southern part of the WCZ (south of the Godavari Rift, Domain IA), Sengupta et al. (1999) documented UHT of metamorphism of pelitic granulites along a heating–cooling trajectory at lower crustal depths (Fig. 4). As argued earlier, metamorphism was induced by voluminous mafic magma emplacement in Late Archaean time. Sengupta et al. (1999) further documented extensive Mesoproterozoic (1600 Ma) high-grade reworking of the rocks in this segment along prominent shear zones. There is no evidence of the Grenvillian granulite facies metamorphism in Domain Ia. In the northern part of the WCZ (Domain Ib), Neogi et al. (1999) and Gupta et al. (2000) reported peak P–T conditions of 9–10 kbar and >950°C, which was followed by cooling and decompression. Although no firm constraint can be placed on the timing of metamorphism, available isotopic data in this segment (Aftalion et al. 2000; Sarkar et al. 2000; Rickers et al. 2001b) do not show any imprint of Grenvillian orogeny. It should be emphasized at this point that the tectonothermal history of the WCZ is important in understanding the amalgamation of the EGB with adjoining cratons, but not in understanding the Indo-Antarctic correlation.

In the Chilka Lake Domain III (Fig. 3), metamorphism of anorthosite and patchy charnockite formation in leptynite (Dobmeier & Raith 2000), as discussed earlier, possibly occurred at 690–660 Ma at distinctly lower P–T conditions. Breakdown of garnet and biotite in leptynite due to decompression in the presence of a CO_2-rich fluid was considered to be the mechanism of patchy charnockite formation (Dobmeier & Raith 2000). Based on petrological data and fabric analysis, they refuted the model of Bhattacharya et al. (1993, 1994), who considered the patches to be detached parts (boudins) of massive enderbite.

Most of the petrological studies have been carried out in Domain II of Rickers et al. (2001b) (Fig. 3). The P–T trajectories are diverse, although Dasgupta & Sengupta (2000) synthesized them to obtain an internally coherent picture. But for Mukhopadhyay & Bhattacharya (1997), all the workers agreed upon an UHT of M1 metamorphism (Lal et al. 1987; Kamineni & Rao 1988; Sengupta et al. 1990, 1991; Dasgupta et al. 1991, 1995; Bhowmik et al. 1995; Dasgupta & Sengupta 1995; Mohan et al. 1997). As argued by Dasgupta & Sengupta (2000), an anticlockwise P–T trajectory can be reconciled for M1, with a near-isobaric cooling retrograde path.

The petrographic data of Lal et al. (1987) were reinterpreted by Pal & Bose (1997) and Sengupta et al. (1997) to bring it into harmony with other results. The P–T conditions and the P–T trajectory of M1 is incidentally similar to that inferred from the Archaean Napier Complex in East Antarctica [Fig. 4; reviewed in Harley (2002)]. However, currently, there are no isotopic data indicating Mid–Late Archaean metamorphism in this part of the EGB. The limited Pb isotope data of Jarick (2000) rather shows c. 1400 Ma age for this phase. On the other hand, Dobmeier & Raith (2002), placed this early metamorphism at 1100 Ma.

There is more or less unanimity regarding a

Grenvillian (1000–950 Ma) age for M2 metamorphism in Domain II (Grew & Manton 1986; Paul et al. 1990; Mezger et al. 1996; Mezger & Cosca 1999; Rickers et al. 2001b), thereby indicating a strong similarity with the Rayner Complex in East Antarctica (Black et al. 1987; Harley, 2002). This is also true for the deduced P–T conditions and P–T trajectories (Fig. 4). Although the protolith history of the two blocks differs considerably [cf. Black et al. (1987) and Rickers et al. (2001b)], one can visualize a commonality during the Grenvillian Orogeny. Obviously, this lends credence to the model of suturing of the EGB with the Rayner Province in East Antarctica by the SWEAT Orogen as a part of Rodinia at c. 1000 Ma.

This brings us to the question of locating the Grenvillian orogenic front in the EGB and its possible continuity in East Antarctica. The boundary between the Napier and Rayner Complexes is the location of the front in East Antarctica. This boundary has been extended into the EGB in the area north of Anakapalle (Domain II) (Grew et al. 1988), along the Nagavalli–Vamasdhara Shear Zone (Chetty 1995), which is supposedly the contact between Domains II and III in the model of Rickers et al. (2001b), and finally along the Godavari Rift (Sengupta et al. 1999) (Fig.4). Recent isotopic data of Mezger & Cosca (1999), Aftalion et al. (2000), Jarick (2000), Crowe et al. (2001) and Rickers et al. (2001b), clearly identifies the regions affected by the c. 1000 Ma orogeny in the EGB. Petrologically, these areas are characterized by M2 metamorphism and its characteristic retrograde path. Combining the petrological data with the isotopic data, it appears that the Grenvillian front propagated into the EGB along the Godavari Rift and then swung northwards, parallel to the boundary of the WCZ [Fig. 4; northern part of Domain I of Rickers et al. (2001b)].

The isotopic data of Dobmeier & Simmat (2001) and the petrological interpretation of Dobmeier & Raith (2002) indicate that the effect of granulite facies overprint in parts of Domain III at 690–660 Ma cannot be fitted into any model of Indo-Antarctic correlation.

M3 metamorphic conditions never reached granulite grade in Domain II, which can be correlated with the 500–550 Ma Pan-African Orogeny from all sources of isotopic data (Mezger et al. 1996; Kovach et al. 1997; Mezger & Cosca 1999; Crowe et al. 2001). Recently, Datta et al. (2001) have described the formation of patchy charnockite on a pervasive foliation in leptynite from an area near Vizianagram (Domain II). The mechanism of formation is suggested to be similar to that inferred by Dobmeier & Raith (2000) from the Chilka Lake area (Domain III). Although no firm commitment about the age of formation of the patches can be made at this stage, available geochronological data in adjoining areas (e.g. Kovach et al. 1997) would suggest a Pan-African age. The propagation of the Pan-African orogenic front is not very evident because the effects of the M3 metamorphism were only studied locally.

Conclusions

The polyphase evolution of the EGB can be demonstrated from the extensive petrological database in combination with the available geochronological data. This information has provided new insights into the mutual positions of India and Antarctica right through the Precambrian. Four isotopic domains have been identified by Rickers et al. (2001b) in the EGB. Of those only one (Domain I ≡ WCZ) corresponds to the lithotectonic subdivisions of the EGB proposed by Ramakrishnan et al. (1998). This reiterates the contention of Dasgupta & Sengupta (2000) that the significance of the proposed lithotectonic subdivisions is poorly understood. Domain I rocks are distinctly different in their thermotectonic evolutionary history from the Rayner Complex (major granulite facies metamorphism at 1000 Ma; Black et al. 1987) within the period of 1000–500 Ma, and from the Archaean Napier Complex in East Antarctica. The geochronological framework of the EGB presented earlier makes it clear that there is little correlation in the thermotectonic histories of the EGB and all the provinces of East Antarctica before the onset of the Grenvillian Orogeny (1100–950 Ma). This has cast doubt on the proposed correlation between the Napier Complex and the southern part of the EGB. Although more isotopic constraints are necessary, the present database indicates two independent blocks (the Napier and Domain Ia) that were welded together, possibly during the Mesoproterozoic orogenic event [see discussion in Sengupta et al. (1999)]. This contention is further corroborated by the recognition of relatively widespread mid-Mesoproterozoic granulite facies reworking in the EGB and the East Antarctic Provinces (i.e. Black et al. 1987; Kelly et al. 2000 and refs cited therein). The other domains in the EGB share a common geologic history at 1000 Ma with the Rayner Complex. The 'Grenvillian' front, within the SWEAT model, sutured East Antarctica with the EGB, and propagated along the Godavari Rift, swinging northwards parallel to the WCZ. The Pan-African granulite metamorphism, so well recognized in the Rauer Group (Prydz Bay) in East Antarctica (Harley & Fitzsimons 1995; Hensen & Zhou 1995; Fitzsimons et al. 1997), may have only one counterpart in the Chilka Lake area of the EGB. Recently, Boger et al. (2001) proposed a separate cratonic block comprising India and parts of the Prince Charles Mountain in East Antarctica, which was distinct from two other cratonic assemblies, e.g. West Gondwana (South America–Africa) and East Gondwana (the rest of East Antarctica–Australia), and speculated that the orogenesis in Prydz Bay may have continued in India, somewhere north of the EGB under the alluvium of the rivers Ganges and Brahmaputra. The geological relationships (albeit with limited geochronologic data) presented in this work, would rather suggest continuation of the Prydz Bay orogenies into the Chilka Lake (Domain III) area of the EGB. It is interesting to note that the geologic setting of the Chilka Lake and Prydz Bay areas was quite dissimilar prior to Pan-African time. These two blocks appeared to have evolved independently prior to sharing a common history in the Neoproterozoic.

Our research on the EGB was supported by projects sponsored by the Council of Scientific & Industrial Research, the Department of Science & Technology, the Government of India and the German Research Foundation. Fellowships from the Alexander von Humboldt Foundation were particularly helpful. P. K. Bhattacharya was instrumental in initiating research on the EGB. Over the years we have received support from our colleagues from abroad: M. Fukuoka, M. Yoshida, H. Kagami, M. Raith, S. Hoernes and K. Mezger. Our students, past and present – S. Sanyal, S. Bhowmik, S. Karmakar, S. Chakraborti, J. Sen, U.K. Bhui, S. Pal, S. Bose, S. Bhattacharya, D. K. Mukhopadhyay, P. Sengupta, R. Dasgupta, S. Sarkar, N. Datta and A. Gangopadhyay – helped us in many ways, and provided continuous impetus. K. Rickers shared some of her unpublished data on isotopes. Our families bore with us patiently and supported us throughout. We are thankful to all the organizations and persons mentioned above. We thank two reviewers for their constructive suggestions. This is a contribution to IGCP Projects 368 and 440. We dedicate this paper to the memory of Professor Chris Powell, who was one of the prime movers in bringing out this volume.

References

AFTALION, M., BOWES, D. R., DASH, B. & DEMPSTER, T. J. 1988. Late Proterozoic charnockites in Orissa, India: a U–Pb and Rb–Sr isotopic study. *Journal of Geology*, **96**, 663–675.

AFTALION, M., BOWES, D. R., DAS, B. & FALLICK, A. E. 1998. Pan-African thermal history of the Mid-Proterozoic Khariar alkali syenite in the Eastern Ghats, Orissa, India, a U–Pb and K–Ar isotopic study. *Proceedings of the International Seminar on Precambrian Crust in Eastern and Central India. Bhubaneswar*, India. 10–12 (abstracts).

AFTALION, M., BOWES, D. R., DASH, B. & FALLICK, A. E. 2000. Late Pan-African thermal history in the Eastern Ghats terrane, India from U–Pb and K–Ar isotopic study of the Mid-Proterozoic Khariar alkali syenite, Orissa. *Geological Survey of India, Special Publication*, **57**, 26–33.

ANOVITZ, L. M. 1991. Al zoning in pyroxene and plagioclase: window on late prograde to early retrograde P–T paths in granulite terranes. *American Mineralogist*, **76**, 1328–1343.

BHATTACHARYA, S., SEN, S. K. & ACHARYYA, A. 1993. Structural evidence supporting a remnant origin of patchy charnockites in the Chilka Lake area, India. *Geological Magazine*, **130**, 363–368.

BHATTACHARYA, S. 1994. The structural setting of the Chilka Lake granulite–migmatite–anorthosite suite with emphasis on the time relation of charnockites. *Precambrian Research*, **66**, 393–409.

BHOWMIK, S. K., DASGUPTA, S., HOERNES, S. & BHATTACHARYA, P. K. 1995. Extremely high-temperature calcareous granulites from the Eastern Ghats, India: evidence for isobaric cooling, fluid buffering and terminal channelized fluid flow. *European Journal of Mineralogy*, **7**, 689–703.

BLACK, L. P., HARLEY, S. L., SUN, S. S. & MCCULLOCH, M. T. 1987. The Rayner Complex of East Antarctica: complex isotopic systematics within a Proterozoic mobile belt. *Journal of Metamorphic Geology*, **5**, 1–26.

BOGER, S. D., WILSON, C. J. L. & FANNING, C. M. 2001. Early Paleozoic tectonism within the East Antarctic craton: the final suture between east and west Gondwana? *Geology*, **29**, 463–466.

BOSE, S., FUKUOKA, M., SENGUPTA, P. & DASGUPTA, S. 2000. Evolution of high Mg–Al granulites from Sunkarametta, Eastern Ghats, India: evidence for a lower crustal heating–cooling trajectory. *Journal of Metamorphic Geology*, **18**, 223–240.

CHETTY, T. R. K. 1995. A correlation of Proterozoic shear zones between Eastern Ghats Belt, India and Enderby land, east Antarctica, based on LANDSAT imagery. *Memoirs of the Geological Society of India*, **34**, 205–220.

CHETTY, T. R. K. & MURTHY, D. S. N. 1998. Regional tectonic framework of the Eastern Ghats Mobile Belt. *Proceedings of the Workshop on Eastern Ghats Mobile Belt. A New Interpretation*. Geological Survey of India, Special Publication, **44**, 39–50.

CROWE, A. A., COSCA, M. A. & HARRIS, L. B. 2001. $^{40}Ar/^{39}Ar$ geochronology and Neoproterozoic tectonics along the northern margin of the Eastern Ghats belt, in north Orissa, India. *Precambrian Research*, **108**, 237–266.

DALZIEL, I. W. D. 1991. Pacific margins of Laurentia and East Antarctica–Australia as a conjugate rift pair: evidence and implications for an Eocambrian supercontinent. *Geology*, **19**, 598–601.

DASGUPTA, S. 1993. Contrasting mineral parageneses in high-temperature calc-silicate granulites: examples from the Eastern Ghats, India. *Journal of Metamorphic Geology*, **11**, 193–202.

DASGUPTA, S. 1995 Pressure–temperature evolutionary history of the Eastern Ghats granulite province: recent advances and some thoughts. *In:* YOSHIDA, M. & SANTOSH, M. (eds) *India and Antarctica during the Precambrian*. Memoir of the. Geological Society of India, **34**, 101–110.

DASGUPTA, S. & SENGUPTA, P. 1995. Ultrametamorphism in Precambrian granulite terrains – evidence from Mg–Al granulites and calc-granulites of the Eastern Ghats, India. *Geological Journal*, **30**, 307–318.

DASGUPTA, S. & SENGUPTA, P. 2000. Tectonothermal evolution of the Eastern Ghats Belt, India. *Geological Survey of India, Special Publication*, **55**, 259–274.

DASGUPTA, S. EHL, J., RAITH, M. M., SENGUPTA, P. & SENGUPTA, P. 1997. Mid-crustal contact metamorphism around the Chimakurthy mafic–ultramafic complex, Eastern Ghats Belt, India. *Contributions to Mineralogy and Petrology*, **129**, 182–197.

DASGUPTA, S., SANYAL, S., SENGUPTA, P. & FUKUOKA, M. 1994. Petrology of granulites from Anakapalle – evidence of Proterozoic decompression in the Eastern Ghats, India. *Journal of Petrology*, **34**, 433–459.

DASGUPTA, S., SENGUPTA, P., EHL, J., RAITH, M. & BARDHAN, S. 1995. Reaction textures in a suite of spinel granulites from Eastern Ghats Belt, India: evidence for polymetamorphism and a partial petrogenetic grid in the system KFMASH and the roles of ZnO and Fe_2O_3. *Journal of Petrology*, **36**, 435–461.

DASGUPTA, S., SENGUPTA, P., FUKUOKA, M. & BHATTACHARYA, P. K. 1991. Mafic granulites from the Eastern Ghats, India: further evidence for extremely high temperature crustal metamorphism. *Journal of Geology*, **99**, 124–133.

DASGUPTA, S., SENGUPTA, P. FUKUOKA, M. & CHAKRABORTI, S. 1992. Dehydration melting, fluid buffering and decompressional P–T path in a granulite complex from the Eastern Ghats, India. *Journal of Metamorphic Geology*, **10**, 777–788.

DASGUPTA, S., SENGUPTA, P., MONDAL, A. & FUKUOKA, M. 1993. Mineral chemistry and reaction textures in metabasites from the Eastern Ghats belt, India and their implications. *Mineralogical Magazine*, **57**, 113–120.

DASGUPTA, S., SENGUPTA, P., SENGUPTA, P. R., E. H. L., J. & RAITH, M. M. 1999. Petrology of gedrite-bearing rocks in mid-crustal ductile shear zones from the Eastern Ghats Belt, India. *Journal of Metamorphic Geology*, **17**, 765–778.

DATTA, N., GANGOPADHYAY, A., SENGUPTA, P. & SARKAR, S. 2001. Incipient charnockite formation in leptynite from Vizianagaram, A.P.: implication for the Pan-African(?) crustal reworking in the Eastern Ghats belt. *ISRGA*, Osaka, Japan (abstract).

DOBMEIER, C. & RAITH, M. M. 2000. On the origin of 'arrested' charnockitization in the Chilka Lake area, Eastern Ghats belt, India – a reappraisal. *Geological Magazine*, **137**, 27–37.

DOBMEIR, C. J. & RAITH, M. M. 2002. Crustal architecture and evolution of the Eastern Ghats Belt and adjacent regions of India. In: Yoshida, M., Windley, B. F. & Dasgupta, S. (eds) *Proterozoic East Gondwana: Supercontinent Assembly and Breakup*. Geological Society, London, Special Publication, **xx**, xx–xx.

DOBMEIER, C. & SIMMAT, R. 2001. Post-Grenvillian transpression in the Chilka Lake area, Eastern Ghats Belt–implications for the amalgamation history of peninsular India. *Precambrian Research*, **113**, 243–268.

FITZSIMONS, I. C. W., KINNEY, P. D. & HARLEY, S. L. 1997. Two stages of zircon and monazite growth in anatectic leucogneiss: SHRIMP constraints on the duration and intensity of Pan-African metamorphism in Prydz Bay, East Antarctica. *Terra Nova*, **9**, 47–51.

GREW, E. S. & MANTON, W. I. 1986. A new correlation of sapphirine granulites in the Indo-Antarctic metamorphic terrane: Late Proterozoic dates from the Eastern Ghats. *Precambrian Research*, **33**, 123–139.

GREW, E. S. & JAMES, P. R. 1988. U–Pb data on granulite facies rocks from Fold island, Kemp coast, East Antarctica. *Precambrian Research*, **42**, 63–75.

GROENEWALD, P. B. 1993. Correlation of cratonic and orogenic provinces in south-eastern Africa and Dronning Maud Land, Antarctica. *In:* FINDLEY, R. H., UNRUG, R., BANKS, M. R. & VEEVERS, J. J. (eds) eds. *Gondwana Eight: Assembly, Evolution and Dispersal*. A.A. Balkema, Rotterdam, 111–122.

GUPTA, S. BHATTACHARYA, A., RAITH, M. M. & NANDA, J. K. 2000. Contrasting pressure–temperature–deformation history across a vestigal craton – mobile belt boundary: the western margin of the Eastern Ghats belt at Deobhog, India. *Journal of Metamorphic Geology*, **18**, 683–697.

HARLEY, S. L. 1989. The origin of granulites: a metamorphic perspective. *Geological Magazine*, **126**, 215–247.

HARLEY, S. L. 1998. On the occurrence and characterisation of ultrahigh–temperature crustal metamorphism. *In:* TRELOAR, P.J. & O'BRIEN, P.J. (eds) *What Drives Metamorphism and Metamorphic Reactions ?* Geological Society London, Special Publication, **138**, 81–107.

HARLEY, S. L. 1995. High grade metamorphism and deformation in the Prydz bay region, East Antarctica: terranes, events and regional correlation. *Memoir of the Geological Society of India*, **34**, 73–100.

HARLEY, S. L. 2002. Archaen–Cambrian crustal development of East Antarctica: metamorphic characteristics and tectonic implications. In: Yoshida, M., Windley, B. F. & Dasgupta, S. (eds) *Proterozoic East Gondwana: Supercontinent Assembly and Breakup*. Geological Society, London, Special Publications, **xx**, xx–xx.

HENSEN, B. J. & ZHOU, B. 1995. A Pan-African granulite facies metamorphic episode in Prydz Bay, Antarctica: evidence from Sm–Nd garnet dating. *Australian Journal of Earth Sciences*, **42**, 249–258.

HOFFMAN, P. F. 1991. Did the breakup of Laurentia turn Gondwanaland inside-out ? *Science*, **252**, 1409–1412.

JARICK, J. 2000. *Die thermotektonomeramorphe entwicklung des Eastern Ghats Belt, Indien – ein test der SWEAT hypothese*. PhD thesis, J. W. Goethe Universitaet, Frankfurt.

KAMINENI, D. C. & RAO, A. T. 1988. Sapphirine-bearing quartzite from the Eastern Ghats granulite terrain, Vizianagaram, India. *Journal of Geology*, **96**, 209–220.

KELLEY, N. M., CLARKE, G. L., CARSON, G. L. & WHITE, R. W. 2000. Thrusting in the lower crust: evidence from the Oygarden Island, Kemp Land, East Antarctica. *Geological Magazine*, **137**, 219–234.

KOVACH, V. P., RAITH, M. M., SALNIKOVA, E. B. *ET AL.* 2001. Geochronological evidence for late PaleoProterozoic magmatic and metamorphic events in the Eastern Ghats belt, India: implications for the India–East Antarctica correlation. *EUG XI*, Strassburg, 598 (abstracts).

KOVACH, V. P., KOVACH, V. P., SALNIKOVA, E. B., KOTOV, A. B., YAKOVA, S. J. & RAO, A. T. 1997. Pan-African U–Pb zircon age from apatite–magnetite veins of

Eastern Ghats granulite Belt, India. *Journal of Geological Society of India*, **50**, 421–424.

KRAUSE, O., DOBMEIER, C., RAITH, M. M. & MEZGER, K. 2001. Age of emplacement of massif-type anorthosites in the Eastern Ghats belt, India: constraints from U–Pb zircon dating and structural studies. *Precambrian Research*, **109**, 25–38.

LAL, R. K., ACKERMAND, D. & UPADHYAY, H. 1987. *P–T–X* relationships deduced from corona textures in sapphirine–spinel–quartz assemblages from Paderu, South India. *Journal of Petrology*, **28**, 1139–1168.

LEELANANDAM, C. 1989. The Prakasam alkaline province in Andhra Pradesh, India. *Journal of the Geological Society of India*, **42**, 435–447.

LEELANANDAM, C. 1990. The anorthosite complexes and Proterozoic mobile belt of Peninsular India: a review. In: NAQUI, S. M. (ed.) *Precambrian Continental Crust and its Economic Resources*. Elsevier, Amsterdam, 409–435.

MEZGER, K. & COSCA, M. A. 1999. The thermal history of the Eastern Ghatsbelt (India), as revealed by $U-Pb$ and $^{40}Ar-^{34}Ar$ dating of metamorphic and magmatic minerals: implications for the SWEAT correlation. *Precambrian Research*, **94**, 251–271.

MEZGER, K., COSCA, M. A. & RAITH, M. 1996. Thermal history of the Eastern Ghats Belt (India) deduced from U–Pb and $^{40}Ar-^{39}Ar$ dating of metamorphic minerals. *Journal of Conference Abstracts, V.M. Goldschmidt Confernce*, **1**, 401.

MOHAN, A., TRIPATHI, P. & MOTOYOSHI, Y. 1997. Reaction history of sapphirine granulites and a decompressional *P–T* path in a granulite complex from the Eastern Ghats. *Proceedings of the Indian Academy of Sciences (Earth and Planetary Sciences)*, **106**, 115–129.

MOORES, E. M. 1991. Southwest U.S.–East Antarctic (SWEAT) connection: A hypothesis. *Geology*, **19**, 425–428.

MUKHOPADHYAY, A. K. & BHATTACHARYA, A. 1997. Tectonothermal evolution of the gneiss complex at Salur in the Eastern Ghats Granulite Belt of India. *Journal of Metamorphic Geology*, **15**, 719–734.

NANDA, J. K. & PATI, U. C. 1989. Field relations and petrochemistry of the granulites and associated rocks in the Ganjam–Koraput sector of the Eastern Ghats belt. *Indian Minerals*, **43**, 247–264.

NEOGI, S., DASGUPTA, S., SENGUPTA, P. & DAS, N. 1999. Ultrahigh temperature decompression in a suite of granulites from the Eastern Ghats Belt–Bastar craton contact and its significance. *Memoir of the Gondwana Research Group*, **5**, 115–138.

PAL, S. & BOSE, S. 1997. Mineral reactions and geothermobarometry in a suite of granulite facies rocks from Paderu, Eastern Ghats Granulite Belt: a reappraisal of the *P–T* trajectories. In: SEN, S. K. (ed.) *Eastern Ghats granulites*. Special volume for the Proceedings Earth and Planetary Science, Indian Academy of Sciences, **106**, 77–89.

PAUL, D. K., RAYBARMAN, T. K., MACNAUGHTON, M. J., FLETCHER, I. R., POTTS, P. J., RAMKRISHNAN, M. & AUGUSTINE, P. F. 1990. Archaean–Proterozoic evolution of Indian charnockites. Isotopic and geochemical evidence from granulites of the Eastern Ghats Belt. *Journal of Geology*, **98**, 253–263.

POWELL, C. MCA., LI, Z. X., PCELLHINNY, M. W., MEERT, J. G. & PARK, J. K. 1994. Paleomagnetic constraints on the timing of the Neoproterozoic breakup of Rodinia and the Cambrian formation of Gondwana. *Geology*, 22, 889–892.

RAMAKRISHNAN, M., NANDA, J. K. & AUGUSTINE, P. F. 1998. Geological evolution of the Proterozoic Eastern Ghats mobile belt. *Proceedings of a Workshop on Eastern Ghats Mobile Belt*. Geological Survey of India Special Publication, **44**, 1–21.

RICKERS, K., RAITH, M. M. & DASGUPTA, S. 2001a. Multistage reaction textures in xenolithic high Mg–Al granulites at Anakapalle, Eastern Ghats Belt, India: examples of contact polymetamorphism and infiltration-driven metasomatism. *Journal of Metamorphic Geology*, **19**, 561–580.

RICKERS, K. & MEZGER, K. 2001b. Crustal evolution of the Eastern Ghats Belt, India: evidence from radiogenic isotopes. *Precambrian Research*, **112**, 183–212.

ROGERS, J. J. W. 1996 . History of 3 billion years old continents. *Journal of Geology*, **104**, 91–107.

SARKAR, A. N., BHANUMATHI, L. & BALASUBRAHMANYAN, M. N. 1981. Petrology, geochemistry and geochronology of the Chilka Lake igneous complex, Orissa state, India. *Lithos*, **14**, 92–111.

SARKAR, A., PATI, U. C., PANDA, P. K., PATRA, P. C., KUNDU, H. K. & GHOSH, S. 2000. Late Archaean charnockitic rocks from the northern marginal zones of the Eastern Ghats Belt: a geochronological study. *Geological Survey of India Special Publication*, **57**, 171–179.

SEN, S. K., BHATTACHARYA, S. & ACHARYA, A. 1995. A multi-stage pressure–temperature record in the Chilka Lake granulites: the epitome of the metamorphic evolution of Eastern Ghats, India? *Journal of Metamorphic Geology*, **14**, 287–298.

SENGUPTA, P. & DASGUPTA, S. 1998. Stability relations of the sapphirine- and spinel-bearing assemblages in simple and complex systems and origin of 'apparent' *P–T* trajectories in granulites. *Proceedings of the International Seminar on Precambrian Crust in Eastern and Central India*, Bhubaneswar, India, 102–103 (abstracts).

SENGUPTA, P., DASGUPTA, S., BHATTACHARYA, P. K., FUKUOKA, M., CHAKRABORTI, S. & BHOWMICK, S. 1990. Petro-tectonic imprints in the sapphirine granulites from Anantagiri, Eastern Ghats Mobile Belt, India. *Journal of Petrology*, **31**, 971–996.

SENGUPTA, P., KARMAKAR, S., DASGUPTA, S. & FUKUOKA, M. 1991. Petrology of spinel granulites from Araku, Eastern Ghats, India and a petrogenetic grid for sapphirine-free rocks in the system FMAS. *Journal of Metamorphic Geology*, **9**, 451–459.

SENGUPTA, P., SANYAL, S., DASGUPTA, S., FUKUOKA, M., EHL, J. & PAL, S. 1997. Controls of mineral reactions in high-grade garnet–wollastonite–scapolite–bearing calcsilicate rocks: an example from Anakapalle, Eastern Ghats, India. *Journal of Metamorphic Geology*, **15**, 551–564.

SENGUPTA, P., SEN, J., DASGUPTA, S., RAITH, M. M., BHUI, U. K. & EHL, J. 1999. Ultrahigh temperature metamorphism of meta-pelitic granulites from Kondapalle, Eastern Ghats Belt: implications for the Indo-Antarctic correlation. *Journal of Petrology*, **40**, 1065–1087.

SHAW, R. K. & ARIMA, M. 1996. High temperature metamorphic imprint from calc-granulites of Rayagada, Eastern Ghats, India: implication of isobaric cooling-path. *Contribution to Mineralogy and Petrology*, **126**, 169–180.

SHAW, R. K., ARIMA, M., KAGAMI, H., FANNING, C. M., SHIRAISHI, K. & MOTOYOSHI, Y. 1997. Proterozoic events in the Eastern Ghats Granulite Belt, India: evidence from Rb–Sr, Sm–Nd systematics, and SHRIMP dating. *Journal of Geology*, **105**, 645–656.

SIMMAT, R. & RAITH, M. M. 1998. EPMA monazite dating of metamorphic events in the Eastern Ghats Belt of India. *Proceedings of the International Seminar on Precambrian Crust in Eastern and Central India*, Bhubaneswar, India, 126–128 (abstracts).

TORSVIK, T. S., CARTER, L. M., ASHWAL, L. D., BHUSAN, S. K., PANDIT, M. K. & JAMTVEIT, B. 2001. Rodinia refined or obscured: palaeomagnetism of the Malani igneous suite (NW India). *Precambrian Research*, **108**, 319–333.

UNRUG, R. 1996. The assembly of Gondwanaland. *Episodes*, **19**, 11–20.

YOSHIDA, M. 1995. Assembly of East Gondwanaland during the Mesoproterozoic and its rejuvenation during the Pan-African period. *Memoir Geological Society of India*, **34**, 25–46.

Crustal architecture and evolution of the Eastern Ghats Belt and adjacent regions of India

CHRISTOPH J. DOBMEIER[1], MICHAEL M. RAITH[2]

[1]*Institut für Geologische Wissenschaften, FR Geologie, Freie Universität Berlin, Malteserstraße 74–100, 12249 Berlin, Germany (e-mail: dobmeier@zedat.fu-berlin.de)* [2]*Mineralogisch-Petrologisches Institut, Universität Bonn, Poppelsdorfer Schloss, 53115 Bonn, Germany*

Abstract: Extending along the east coast of peninsular India, the Eastern Ghats expose a deep section through a composite orogenic belt that once formed part of the Proterozoic mobile belt system within East Antarctica and East India. The critical evaluation of the existing geological and isotopic data strongly suggests that this orogenic belt includes not only the granulite facies Eastern Ghats Belt but also the Nellore-Khammam Schist Belt and lower grade units at the southern margin of the Singhbhum Craton. The present authors propose its subdivision into four crustal provinces with widely different geological evolutions. The Rengali and Jeypore Provinces formed at the margin of the Bhandara Craton in the Late Archaean. In the Krishna Province, volcanosedimentary rocks equivalent to the Cuddapah Supergroup accumulated, probably on the Dharwar Craton in the Palaeoproterozoic, and the major tectonometamorphic event took place between 1.67 and 1.55 Ga, subsequent to a short-lived igneous activity. The Eastern Ghats Province, which shows considerable similarities with the Rayner Province of East Antarctica, was strongly affected by pervasive deformation, high-grade metamorphism and crustal-derived magmatism between 1.1 and 0.9 Ga, which extensively modified the crustal structure of present eastern peninsular India. Neoproterozoic and Early Phanerozoic tectonothermal activities were largely restricted to pre-existing shear zones, but the present configuration of the composite orogenic belt may have been achieved only during the Pan-African Orogeny.

Most of peninsular India exposes Precambrian crust which was formed, reworked and consolidated over several major orogenic events. Two of these events were associated with the assembly and breakup of the supercontinents Rodinia and Gondwana in Mesoproterozoic and Neoproterozoic–Early Phanerozoic times, respectively. The subcontinent's crustal evolution during these periods is well documented in two high-grade crustal provinces situated at the southern tip (Southern Granulite Terrain) and eastern coast (Eastern Ghats Belt) (Fig. 1).

Unravelling the Proterozoic history of India was long impeded by the very high grade of metamorphism and the strong deformation that most rocks had experienced, and the resulting unclear primary relations between them. In his fundamental treatise of the Precambrian of peninsular India, Fermor (1936) regarded the granulite terrains of east and south India as constituents of a coherent crustal province, the Charnockite Province, that swerves around an older non-charnockitic Province. This notion remained widely accepted, with minor modifications as summarized in Gopalakrishnan (1998), until geochronological methods allowed reliable dating of magmatic and metamorphic rocks, establishing the geological significance of the attained ages. The obtained isotope data showed that the Eastern Ghats Belt and the Southern Granulite Terrain underwent pervasive metamorphism at appreciably different times [as reviewed in Jayananda & Peucat (1996) and Mezger & Cosca (1999)].

In the case of the Eastern Ghats Belt, the considerable increase of knowledge over the past decade calls for a critical evaluation of new and old data, and permits the formulation of an up-to-date synthesis of its crustal architecture and geological evolution. This offers new prospects for the discussion on the assembly of the supercontinents Rodinia and Gondwana.

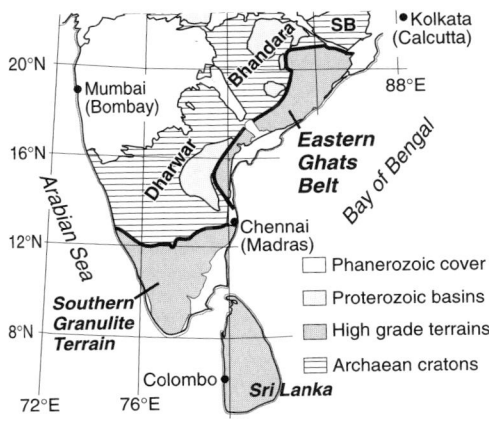

Fig. 1. Main geological units of peninsular India. SB, Singhbhum Craton.

Background information on adjacent Archaean cratonic terrains

Two high-grade crustal provinces skirt the Archaean protocontinental core of peninsular India (Fig. 1), which may be subdivided into the Dharwar, Bhandara (or Bastar), Singhbhum (or East India) and Aravalli (or Bundelkhand) Cratons (Naqvi & Rogers 1987). As the latter is not in contact with the Eastern Ghats Belt, its geological evolution will not be described below. The unified cratonic forelands of the granulite terrains are termed Proto-India.

Dharwar Craton

Intense research during the past decades has greatly advanced our knowledge of the crustal make-up and evolution of this important cratonic terrain. Excellent syntheses of existing and new data were provided by Chadwick et al. (2000) and Jayananda et al. (2000). The craton (Figs 1 & 2a) comprises two distinct parts that are separated by a crustal-scale, N–S-trending sinistral shear zone (Chitradurga-Kollegal Shear Zone). The western part is made up of a polyphase gneissic basement of tonalite–trondhjemite–granodiorite (TTG) associations, with narrow high-grade belts of greenstone-type volcano-sedimentary sequences (Sargur Group). The Sargur Group was consolidated between 3.4 and 2.9 Ga, before the deposition of a low-grade volcanosedimentary sequence (Dharwar Supergroup) in extended intracratonic basins at 2.9–2.6 Ga. The final intrusion of granite plutons occurred at c. 2.6 Ga. The eastern part of the craton consists of a series of N–S-trending arc-type plutonic complexes, the Dharwar Batholith, emplaced during a major episode of calc-alkaline magmatism at 2.55–2.51 Ga. Interdispersed are thin, low- to medium-grade greenstone-type schist belts, interpreted as intra-arc volcanosedimentary sequences. This major accretionary event led to the final stabilization of the entire craton at c. 2.5 Ga. Chadwick et al. (2000) have inferred an Andean margin setting in which the western cratonic domain forms the foreland to an accretionary arc system, represented by the batholith and schist belts. Jayananda et al. (2000) opposed the subduction scenario and argued that the geochemical and isotopic characteristics of Late Archaean magmatism in the Dharwar Batholith, and the associated structural and metamorphic features, are best explained in terms of a plume model.

Bhandara Craton

Compared to the neighbouring Dharwar and Singhbhum Cratons, the Bhandara Craton (Figs 1 & 2a) has not been the target of integrated research in recent times. Its geological evolution thus remains poorly constrained. As in the adjacent cratons, its basement consists of polyphase TTG gneisses and granitoids with minor infolded schist belts of dominantly intracratonic sedimentary sequences (Crookshank 1963; Narayanaswami et al. 1963). The limited isotopic data indicate a mid-Archaean (c. 3.5 Ga) protolith age for trondhjemitic gneisses, and widespread felsic magmatism and metamorphism at 2.65–2.41 Ga (Sarkar et al. 1993, 1994; Choudhary et al. 1996; Singh & Chabria 1999).

Singhbhum Craton

The nucleus of the Singhbhum Craton (Figs 1 & 2a) comprises a gneissic basement, voluminous granites and a low-grade volcanosedimentary cover sequence (Naqvi & Rogers 1987; Mukhopadyay 2001). Deposition of the gneiss precursors occurred after c. 3.5 Ga and was followed by the intrusion of tonalite at c. 3.44 Ga (Sengupta et al. 1996; Mishra et al. 1999). Basaltic melts were emplaced within the sedimentary sequence at c. 3.3 Ga (Bakshi et al. 1987; Sharma et al. 1994), possibly during intense deformation and metamorphism, prior to the multi-stage emplacement of the Bonai, Singhbhum and Mayurbhanj Granites, and the deposition of the low-grade volcanosedimentary Iron Ore Group between 3.4 and 3.1 Ga (Mishra et al. 1999). The age of unconformable deposition of a predominantly clastic cover sequence (Singhbhum, Dhanjori and Kolhan Groups) and the emplacement of mafic dyke swarms is not constrained (Mukhopadyay 2001).

Proterozoic intracratonic basins

The eastern margins of the Bhandara and Dharwar Cratons are covered by sedimentary sequences of Proterozoic basins (Figs 1 & 2b). The largest and best investigated, the Cuddapah Basin, unconformably overlies the granitoid basement of the eastern Dharwar Craton (Nagaraj Rao et al. 1987; Raman & Murty 1997). This very-low- to low-grade metamorphic volcanosedimentary sequence, with an estimated aggregate thickness of 9–13 km, has been subdivided into a lower Cuddapah Supergroup and an upper Kurnool Group. The Kurnool Group is restricted to the western basin and overlies the Cuddapah Supergroup with a marked unconformity. Both sequences consist of mature clastic deposits and intercalated calcareous sediments. Dolerite dykes and a few lamproites transect the entire Cuddapah Supergroup, while mafic flows and sills, ignimbrites, and ash-fall tuffs occur mostly in its lower portion (Chaudhuri et al. 2002). The Pb–Pb

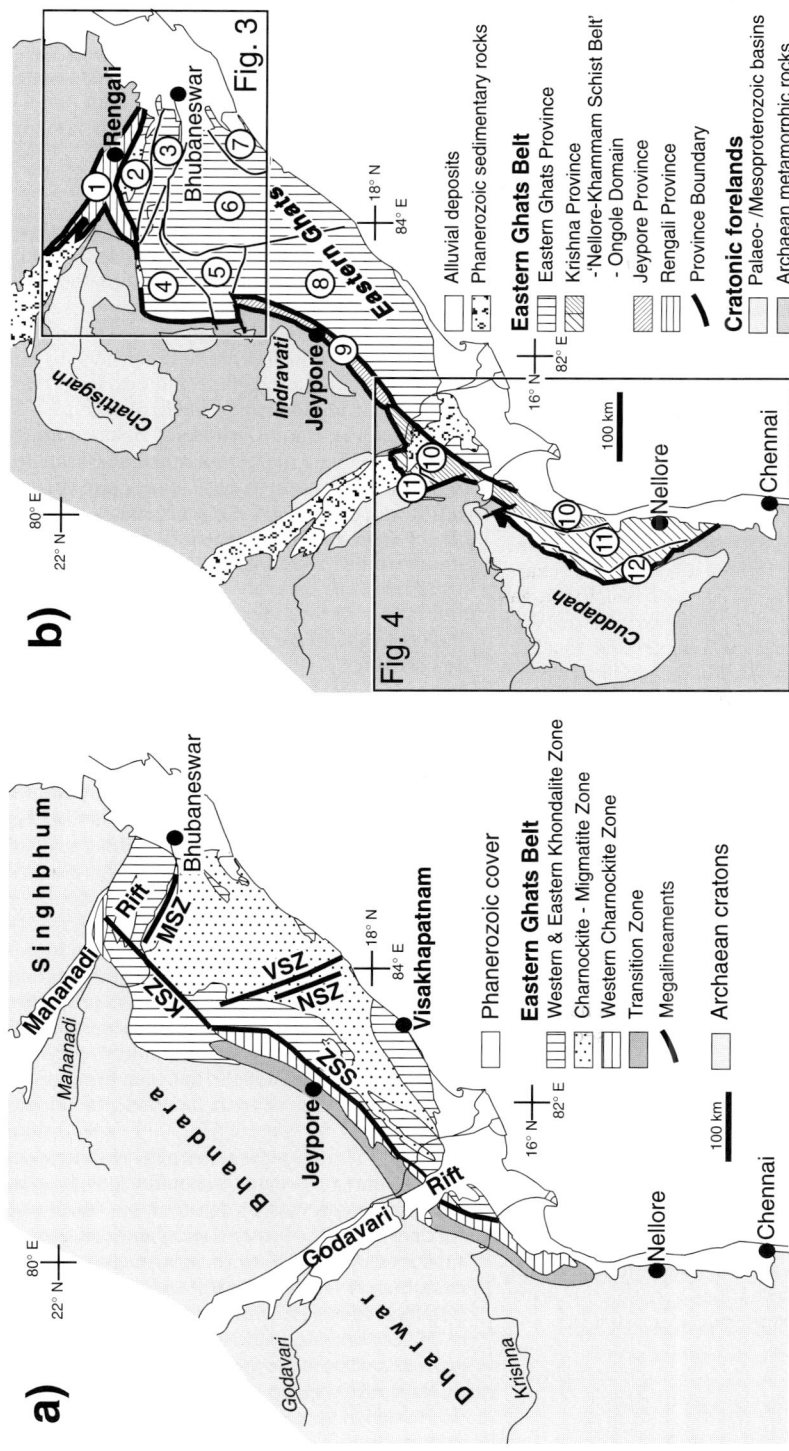

Fig. 2. Geological outline of the Eastern Ghats and adjacent regions. (**a**) Simplified geological map of the Eastern Ghats and adjacent regions. (**a**) Simplified geological map of Ramakrishnan et al. (1998) picturing the most important megalineaments [following Chetty (1995, 2001)]. Megalineaments: KSZ, Koraput–Sonapur Shear Zone; MSZ, Mahanadi Shear Zone; NSZ, Nagavalli Shear Zone; SSZ, Sileru Shear Zone; VSZ, Vamsadhara Shear Zone. (**b**) New subdivision in crustal provinces with distinct geological evolution. The provinces are further divided into domains separated by megalineaments and shear zones [following Nash et al. (1996) and Crowe et al. (2001)]. The domains, from north to south, are indicated by encircled numbers: 1, Rengali; 2, Angul; 3, Tikarpara; 4, Khariar; 5, Rampur; 6, Phulbani; 7, Chilka Lake; 8, Visakhapatnam; 9, Jeypore; 10, Ongole; 11, Vinjamuru; 12, Udayagiri. Map resources are indicated in Figs 3 & 4; boxes indicate Figs 3 & 4.

ages of c. 1.8 Ga, interpreted to date uranium mineralization in stromatolitic dolomites of the basal sequence (Zachariah et al. 1999), provide a minimum age for carbonate sedimentation and comply with a Rb–Sr whole-rock age of 1817 ± 24 Ma for a mafic sill from the intermediate part of the sequence (Bhaskar Rao et al. 1995). A K–Ar total fusion age of 1350 ± 52 Ma for micas of the Chelima Lamproite (Chalapati Rao et al. 1996) postdates sedimentation of the Cuddapah Supergroup. The age of sedimentation of the Kurnool Group is not constrained by palaeontological or radiometric data, but the presence of diamondiferous conglomerates at the base of the sequence suggests a minimum age of c. 1090 Ma, the emplacement age of kimberlite pipes in the eastern Dharwar Craton (Chalapati Rao et al. 1996), which are regarded as the most likely source rocks (Saha 2001). From west to east, the Cuddapah Basin deepens, the grade of metamorphism increases and strain becomes more intense. A marked increase in the intensity of deformation occurs at the Rudravaram Line, which separates the weakly deformed (low-amplitude open folds, tilted blocks) western sector from the Nallamallai Fold Belt that is characterized by isoclinal, overturned folds and thrusts (Meijerink et al. 1984). The Kurnool Group may have been deposited only after this deformation (Meijerink et al. 1984).

The basins of the Bhandara Craton are less well studied. The major basins are the Chattisgarh, containing the Chattisgarh Soupergroup, and the Indravati containing the Indravati Group (Chaudhuri et al. 2002 and refs cited therein). Both groups represent similar platformal sequences of unmetamorphosed conglomerates, sandstones, shales, locally stromatolitic limestones, cherts and dolomites, with an aggregate thickness >600 m (Chaudhuri et al. 2002). No reliable age data exist to constrain the time of sedimentation, but basaltic dykes with Nd model ages of between 2.1 and 2.9 Ga, and probable emplacement ages of c. 750 Ma (Fachmann 2001), transect the entire sequence in the Chattisgarh Basin.

General geological outline of the Eastern Ghats Belt – a summary of earlier ideas

Traditionally, the term Eastern Ghats Belt (or Eastern Ghats Mobile Belt or Eastern Ghats Granulite Belt) denotes a contiguous terrain of granulite facies rocks at the northeastern coast of peninsular India. To the north and west, it borders on Archaean rocks of the Singhbhum, Dharwar and Bhandara Cratons, and to the SE it disappears underneath alluvial plains and the Bay of Bengal (Fig. 1). Based on field work by the Geological Survey of India and an earlier concept of Nanda & Pati (1989), Ramakrishnan et al. (1998) compiled a 1:1 000 000 geological map of the Eastern Ghats Belt and proposed its subdivision into four major lithological units, all aligned parallel to the contact with the cratons (Fig. 2a). These units, from west to east, are:

(1) **Western Charnockite Zone:** an assemblage of charnockites, enderbites, basic granulites and minor enclaves of metasedimentary migmatites forming the westernmost unit.
(2) **Western Khondalite Zone:** a sequence of garnet–sillimanite–alkali feldspar gneisses [termed khondalite by Walker (1902)] with intercalations of quartzites, calc-silicate rocks and high-Mg–Al granulites. The metasedimentary sequence has been intruded by enderbite and charnockite plutons, and several massif-type anorthosite complexes.
(3) **Charnockite–Migmatite Zone:** an intensely migmatized association of supracrustal granulites intruded by voluminous porphyritic granitoids and massif-type anorthosite.
(4) **Eastern Khondalite Zone:** a zone that strongly resembles the Western Khondalite Zone.

The contact with the Archaean cratons is assumed to be a discrete boundary shear zone, the Eastern Ghats Boundary Fault in the north and NW, but a wide transition zone along most of the western margin (Fig. 2a) has been inferred from the presence of orthopyroxene-bearing rocks within, presumably, cratonic gneisses (Ramakrishnan et al. 1998). Alternatively, Chetty (1995, renewed 2001) presented a terrane model based on extensive satellite imagery studies. Megalineaments (Fig. 2a), interpreted as shear zones or thrusts (Chetty & Murthy 1994, 1998b; Mahalik 1994; Chetty 1995, 2001; Ramakrishnan et al. 1998; Biswal & Jena 1999; Biswal 2000), outline the contact with the cratonic forelands and separate the belt into several structural domains, postulated to represent suspect terranes. Earlier presented, but not considered further, conceptions on the litho(-tectonic) divisions of the belt may be taken from the literature (Gopalakrishnan 1998; Ramakrishnan et al. 1998).

Models for the tectonothermal evolution generally include a first event of granulite facies metamorphism and pervasive deformation in the Late Archaean–Early Proterozoic, and a protracted similar event in the Mesoproterozoic followed by the intrusion of late granitoids and alkaline rocks at c. 0.9 Ga (Chetty & Murthy 1998b; Ramakrishnan et al. 1998; Sarkar & Paul 1998). Sarkar & Paul (1998) provide a comprehensive account of geochronological data published up to 1994.

Table 1. *Correlation between the proposed subdivision of the Eastern Ghats Belt and adjoining regions in provinces and the subdivision of Ramakrishnan et al. (1998)*

Subdivision of Ramakrishnan et al. (1998)		New subdivision – province	Orogen
Eastern Ghats Belt	Nellore–Khammam Schist Belt	Krishna Province	Krishna
	Transition Zone		
	Western Charnockite Zone	Jeypore Province	?
	Western Khondalite Zone	Rengali Province	Eastern Ghats
	Charnockite–Migmatite Zone	Eastern Ghats Province	
	Eastern Khondalite Zone		

Crustal architecture of the Eastern Ghats Belt – a new subdivision

Recent research based on adequate isotope and structural data evidently shows that the internal crustal architecture and regional differentiation of the tectonothermal and magmatic events are much more intricate than previously expected. A redefinition of the Eastern Ghats Belt thus seems unavoidable, as the granulite terrain appears to consist of several crustal segments with distinct geological histories. The evaluation further shows that granulite facies grade is not a suitable criterion for the discrimination of these segments, as two of them consist of both granulite-grade and lower grade rock assemblages. In consequence, it is proposed that the term Eastern Ghats Belt and the above given synonyms are abandoned. Table 1 lists the classification of Ramakrishnan et al. (1998) and compares it to the present authors' new province-based scheme. A province is defined as a crustal segment with a distinct geological history. A province may be further divided into several domains characterized by specific features (e.g. lithology, structure, metamorphic grade) which distinguish them from neighbouring domains. Figure 2b shows the extension of the newly defined provinces and their subdivision into domains.

Recently published data and ideas to develop a model for the geological evolution of each province, and of the entire composite orogenic belt exposed in the Eastern Ghats and adjacent areas, are reviewed and discussed below. The provinces are presented in order of the age of crustal consolidation, starting with the oldest.

Jeypore Province

The newly defined Jeypore Province comprises the northern portion of the Western Charnockite Zone of Ramakrishnan et al. (1998) and extends from SW of Bhawanipatna (Fig. 3) for c. 300 km towards the SW along the western margin of the Bhandara Craton (Fig. 2b). In the absence of detailed mapping, the southern termination cannot be exactly located but field observations suggest that the province terminates north of Upper Sileru (Fig. 4). As the presence of metasedimentary rocks reported by Ramakrishnan et al. (1998) cannot be confirmed for this sector of the Western Charnockite Zone, it appears that the Jeypore Province exclusively consists of granulite facies meta-igneous rocks. The igneous character of the protoliths is established by locally abundant ultrabasic or basic cumulates and xenoliths, as well as by geochemical data (Subba Rao et al. 1998; Kovach et al. 2001; authors' unpublished data). Rocks of intermediate composition, enderbites and charnoenderbites, dominate. Basic granulites occur as disrupted bands, lenses and xenoliths within the enderbitic rocks, as well as massive, regionally extensive bodies (Nanda & Pati 1989). Basic granulites, locally garnet-bearing, form a large complex located between the Machkund (Subba Rao et al. 1998) and Balimela Dam sites, and charnockites concentrate around Jeypore (authors' unpublished observation). In places, NW–SE trending, unfoliated fine-grained mafic dykes occur, which show only a low-grade metamorphic overprint.

Subba Rao et al. (1998) concluded, from a limited geochemical data set, that the igneous protoliths of basic granulites and enderbites originated from poorly fractionated tholeiitic and intermediate

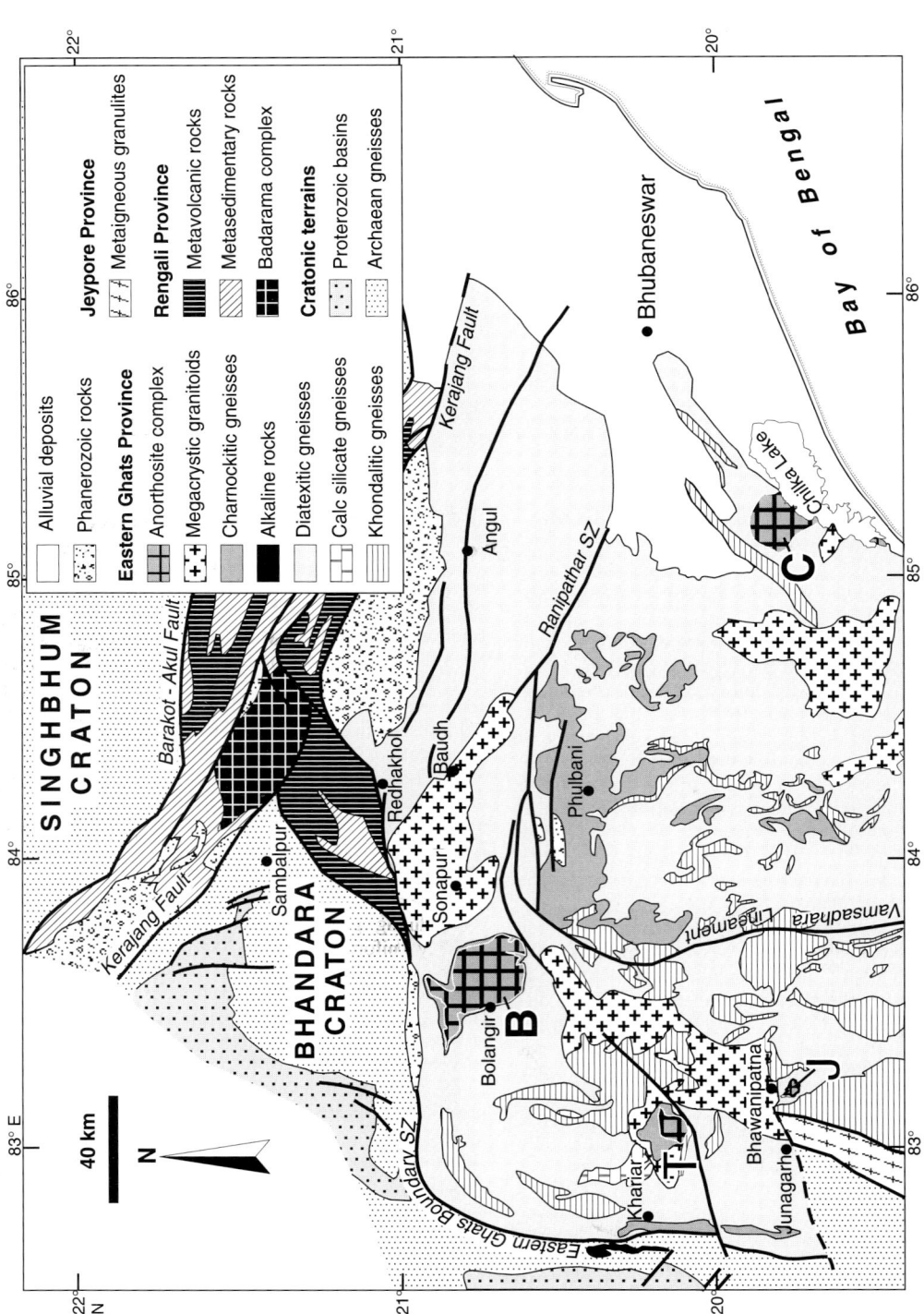

Fig. 3. Geology of the northern Eastern Ghats. Compiled from Madhavan & Khurram (1989), Nash et al. (1996), Maji et al. (1997) and Nanda & Panda (1999), Crowe et al. (2001), Fachmann (2001), Dobmeier (2002). Capitals identify Proterozoic anorthosite complexes: B, Bolangir; C, Chilka Lake; J, Jugsaypatna; T, Turkel.

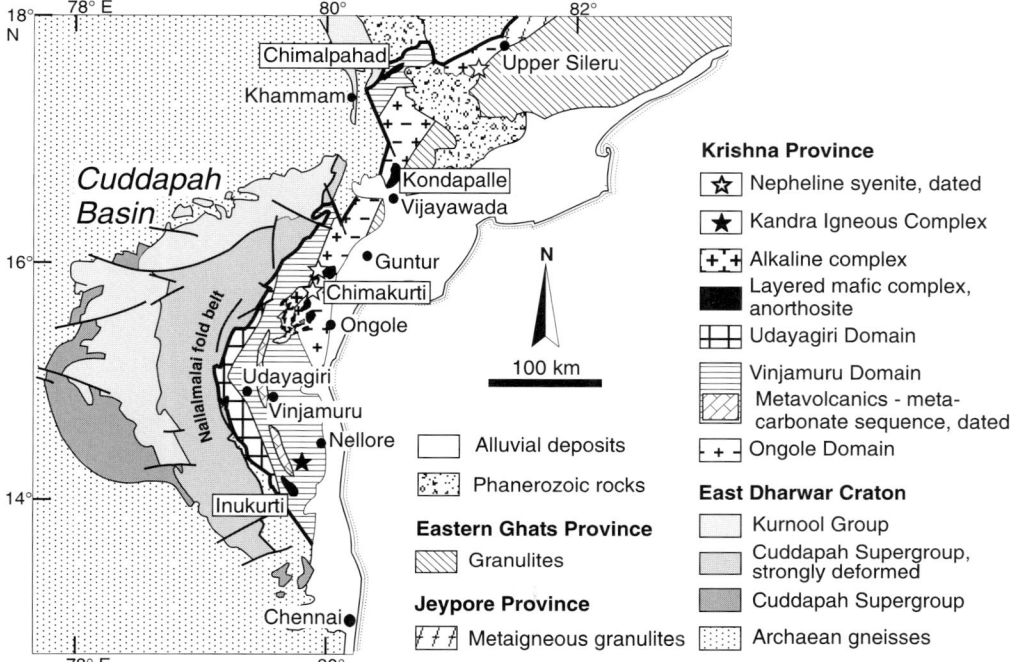

Fig. 4. Geology of the Krishna Province and adjoining areas. Compiled from Raman & Murty (1997), Nagasai Sharma & Ratnakar (2000), Okudaira et al. (2000), with information from Dasgupta *et al.* (1997), Babu (1998) and Ramakrishnan *et al.* (1998).

magmas, respectively, which had evolved through crustal assimilation and fractionation, suggesting emplacement in a rift environment. The dominance of intermediate rocks with calc-alkaline affinity [chemical analyses of Subba Rao *et al.* (1998) and authors' unpublished data], however, points to arc-related magmatism in an Andean-type active continental margin setting.

Archaean Nd model ages of 3.9–3.0 Ga (T_{DM}) for enderbites, in combination with strongly retarded Pb isotopic signatures for feldspars, which imply Pb homogenization in Archaean times, suggest a formation of the igneous protoliths before 3 Ga (Kovach *et al.* 2001; Rickers *et al.* 2001). Preliminary U–Pb data from complexly zoned zircon grains point to a high-grade metamorphic event at c. 2.8 Ga and indicate the absence of any significant later thermal overprint (Kovach *et al.* 2001). The age of metamorphism, however, is not yet constrained by independent isotopic data.

Prolific development of orthopyroxene-bearing leucosomes in enderbites and charnoenderbites through dehydration–melting involving biotite and hornblende, indicate that peak conditions of granulite facies metamorphism probably exceeded the P–T estimate of c. 5.6 kbar and 790°C obtained from thermobarometry on garnetiferous basic granulites of the Machkund area (Subba Rao *et al.* 1998). The structurally controlled segregation of leucosomes provides evidence for the contemporaneity of granulite facies metamorphism and deformation.

The deformation history is, however, poorly known, as no structural studies exist. The present authors' observations from the sector between Jeypore and Balimela indicate that ductile deformation during granulite facies metamorphism was largely uniform within the massive meta-igneous rocks, giving rise to a weak continuous schistosity and/or a compositional layering. Locally, narrow mylonitic shear zones, containing mineral assemblages indicative of a medium-grade metamorphic overprint, transect the pervasive fabric at acute angles. These shear zones trend parallel to the margins of the Jeypore Province and dip 40–50° towards the SE. A well-developed mineral lineation plunges steeply, and C/S fabrics and asymmetric clasts indicate transport of the hanging wall towards the NW.

Rengali Province

Detailed satellite imagery analyses supplemented by extensive ground-truth control have established that the northern part of the granulite terrain of the Eastern Ghats, where the regional trend of the dominant foliations changes from SW–NE to WNW–ESE, is fragmented into several fault-bounded crustal domains characterized by distinct lithological, structural and geochronological features (Mahalik 1994; Nash et al. 1996; Crowe et al. 2001; Fachmann 2001; Crowe 2003).

Mahalik (1994) distinguished two structural domains dominated by medium- to high-grade quartzofeldspathic gneisses, the Rengali and Angul assemblages, and another three domains consisting of biotite schists locally grading into amphibolites or migmatitic hornblende gneisses, massive quartzite units and pelitic schists. He assigned the high-grade gneiss assemblages to the Eastern Ghats Belt and the low- to medium-grade volcanosedimentary sequences to the Singhbhum Craton, although one of the low-grade domains separates the gneiss domains. By contrast, Nash et al. (1996) and Fachmann (2001) regarded the metavolcanosedimentary pile as cover sequence over locally migmatitic and charnockitic gneisses forming cores to regional antiforms. Based on lithological similarities between the meta-volcanosedimentary sequence and the Iron Ore Group, they assumed that the entire assemblage formed part of the Singhbhum Craton. The Rengali Domain was confined to the area between the Barakot and the Kerajang (or North Orissa Boundary) Fault Zones (Fig. 3), while southward adjoining gneisses were allocated to the Angul Domain of the Eastern Ghats (Province). Fachmann (2001) also reported NE–SW-trending mafic dykes from the Rengali Domain and neighbouring provinces (Singhbhum Craton, Angul Domain, Chattisgarh Basin). Recently, Crowe et al. (2001) reinterpreted the geological framework of the area, highlighting that an amphibolite facies rock assemblage equivalent to the Rengali Domain occurs to the south of the Kerajang Fault Zone in a SW-trending, fault-bounded, triangular domain that terminates at the Eastern Ghats Boundary Fault. Further, on the basis of isotope data, a coherent block of granulite facies quartzofeldspathic gneisses and mafic granulites, the Badarama Complex, was interpreted to be included within the Rengali Province (Fig. 3).

Identical ^{207}Pb–^{206}Pb zircon SHRIMP ages of 2802 ± 3 and 2801 ± 10 Ma for hornblende orthogneisses from the Badarama Complex and the eastern termination of the Rengali Province, respectively, are interpreted to date crystallization of the precursor granitoids, and thus point to extensive magmatism at 2.8 Ga within the Rengali Province (Crowe 2003). ^{207}Pb–^{206}Pb zircon ion microprobe ages of 2811 ± 3 (leucocratic granitic gneiss) and 2803 ± 4 Ma (hornblende orthogneiss) for xenoliths embedded in biotite granite corroborate this event (Mishra et al. 2000). The interpretation of Mishra et al. (2000) that the late Archaean ages provide a minimum age for granulite facies metamorphism in the Eastern Ghats Province cannot be accepted, as none of the analysed xenoliths contain mineral assemblages indicative of granulite facies. In addition, the samples were taken near the contact with the Singhbhum Craton and far away from the contact with the Eastern Ghats Province. A ^{207}Pb–^{206}Pb zircon ion microprobe age of c. 3.5 Ga from a biotite-bearing quartzofeldspathic gneiss xenolith (Mishra et al. 2000) strongly argues for derivation of the xenoliths from the Singhbhum Craton, where similar zircon ages are reported from paragneisses (Mishra et al. 1999). Rb–Sr whole-rock ages of 2924 ± 25 Ma for the host biotite granite (Mishra et al. 2000), and 2745 ± 103 and 2735 ± 44 Ma for two charnockite samples (Sarkar et al. 1998), have to be critically evaluated in the light of likely disturbances of the Rb–Sr isotope system during later high-grade metamorphism. However, these rocks still seem to reflect the 2.8 Ga magmatic event. Likewise, Nd isotopic signatures of late mafic dykes near Rengali ($T_{DM} = 2.1$–2.7 Ga, $\varepsilon_{Nd}(T) = -9.9 \pm 1.1$) have been interpreted in terms of varying contamination of the primary basaltic melts with Archaean crustal material (Fachmann 2001).

The age and P–T evolution of metamorphism in the Rengali Province are poorly constrained. The assemblage garnet–staurolite–kyanite in pelitic schists (Crowe 2003) requires conditions of >5 kbar and 550–700°C (see Spear & Cheney 1989), whereas the presence of arrested-type or massive charnockite (Nash et al. 1996; Crowe et al. 2001) and garnet + hornblende-bearing migmatites (Crowe 2003) indicates considerably higher peak temperatures. However, upper amphibolite to granulite facies assemblages are largely restricted to the SW-trending arm of the Rengali Province (Crowe et al. 2001). A minimum age for the amphibolite facies metamorphism is inferred from an ^{40}Ar–^{39}Ar plateau age for hornblende, indicating cooling below c. 500°C at 699 ± 8 Ma (Crowe et al. 2001), and a Sm–Nd mineral isochron age of 792 ± 85 Ma for the emplacement of a mafic dyke which experienced only a low-grade overprint (Fachmann 2001).

The pronounced WNW–ESE lithostructural trend results from synmetamorphic, multiple tight to isoclinal folding of the metavolcanosedimentary sequences with gently, to the east or west plunging fold axes (Nash et al. 1996; Crowe et al. 2001; Fachmann 2001). Pervasive secondary foliations in gneisses and metavolcanosedimentary rocks are parallel to the subvertical axial planes of the folds. At a later increment of the bulk deformation, strain was partitioned in and at ductile shear zones. This deformation was accompanied by retrogression to amphibolite and greenschist facies conditions. Steeply plunging mineral lineations on the generally steeply, to the north or south, dipping mylonitic foliation within major shear zones indicate N–W directed compression during amphibolite facies metamorphism (Fachmann 2001). As no shear sense indicators are reported, it remains undecided if compression was associated with thrusting (Fachmann 2001) or right-lateral transpression (Nash et al. 1996). ^{40}Ar–^{39}Ar plateau ages for white mica and biotite from pervasively retrogressed schists and amphibolites indicate reactivation of the Kerajang Fault Zone between 490 and 470 Ma. Truncation of Late Carboniferous–Early Triassic deposits of the Talchir Basin (Fig. 3) proves brittle reactivation of the Kerajang Fault (Mahalik 1994; Nash et al. 1996; Fachmann 2001).

Pooled apparent apatite fission track ages of between 270 ± 13 and 252 ± 9 Ma suggest that the Rengali Domain entered the partial annealing zone for apatite (120–60°C) at the latest in the Late Permian (Lisker & Fachmann 2001). Samples from major fault zones varying between 208 ± 8 and 181 ± 13 Ma further indicate a localized thermal overprint, and possible reactivation of the fault zones in the Jurassic (Lisker & Fachmann 2001).

Krishna Province

The Krishna Province comprises the granulites of the newly defined Ongole Domain [equivalent to the southern part of the Western Charnockite Zone of Ramakrishn et al. (1998)] and the low- to medium-grade Nellore–Khammam Schist Belt (Fig. 4). Merging of the two crustal terrains, so far considered as unrelated geological units, appears justified by congruent geochronological data for magmatic and metamorphic events (Fig. 5), which point to a common Palaeoproterozoic evolution.

Nellore-Khammam Schist Belt

Along-strike, the Nellore–Khammam Schist Belt has been divided into the Nellore Schist Belt in the south and the Khammam Schist Belt in the north, both being separated by a faulted block of cratonic gneisses (Fig. 4; Raman & Murty 1997; Okudaira et al. 2000). The Nellore Schist Belt has been split into a western upper unit and an eastern lower unit, which are suspected to be separated by a fault or thrust (Narayana Rao 1983; Vasudevan & Rao 1983). The Khammam Schist Belt has been correlated with the lower unit (Okudaira et al. 2000). Following the subdivision of crustal provinces in fault or shear-zone-bounded domains by Nash et al. (1996), the present authors propose provisionally renaming the upper unit the Udayagiri Domain and the lower unit the Vinjamuru Domain (Fig. 4) until the nature of the contact is clarified.

The low-grade Udayagiri Domain consists of a greenschist facies volcanosedimentary sequence of pelites, psammites and conglomerates with local intercalations of cherts, limestones and felsic volcanics. Basic meta-igneous rocks are extremely rare (Raman & Murty 1997). By contrast, basic volcanic rocks are ubiquitous in the amphibolite facies Vinjamuru Domain, which also comprises metasedimentary schists, gneisses and migmatites, and locally abundant felsic metavolcanics. Marbles and calc-silicate gneisses occur together with intermediate and felsic metavolcanic rocks, banded barite–magnetite layers, and kyanite-bearing schists to the west of Vinjamuru (Raman & Murty 1997; Vasudevan et al. 2003). The southern Vinjamuru Domain hosts several economically important muscovite pegmatites (Babu 1998). Anorthositic rocks concentrate near Inukurti in the extreme south (Kanungo & Chetty 1978; Kanungo et al. 1986) and at Chimalpahad near the southwestern margin of the Godavari Rift (Fig. 4). Anorthosites at Inukurti are embedded in voluminous amphibolites, suggesting that they form cumulate layers and domes within a layered intrusion. The Chimalpahad layered mafic complex synkinematically intruded a sequence of partly migmatitic metasedimentary gneisses and amphibolites (Raman & Murty 1997; Leelanandam & Narashima Reddy 1998). Coarse cumulate gabbros with rare interlayered ultramafic rocks, metamorphosed pillow basalts and mafic (sheeted?) dykes constitute the Kandra Igneous Complex near the southwestern margin of the Vinjamuru Domain. Nagaraj Rao et al. (1991) interpreted this assemblage as a remnant ophiolite sequence. An ophiolitic melange of chaotically intermingled blocks of ultramafic rocks, metamorphosed pillow basalts and sedimentary rocks, including finely laminated cherts, embedded in a serpentinized matrix is reported from an exposure near the contact with the Ongole Domain at Kanigiri (Leelanandam 1990). The ophiolitic nature of both occurrences is not independently confirmed.

On the basis of chemical discrimination diagrams, Mukhopadyay et al. (1994) proposed a back-arc setting for the amphibolites of the Vinjamuru Domain, whereas Hari Prasad et al. (2000) argued that the protoliths formed in two different tectonic settings comparable to recent ocean island arcs and continental margin arcs.

The timing of felsic magmatism in the Vinjamuru Domain has been constrained by ^{207}Pb–^{206}Pb single grain evaporation ages for magmatic zircons from metarhyolites (1868 ± 6 and 1771 ± 8 Ma; Vasudevan et al. 2003). ^{207}Pb–^{206}Pb evaporation ages for some of the numerous zircon xenocrysts show that the felsic rocks, which are seen as products of

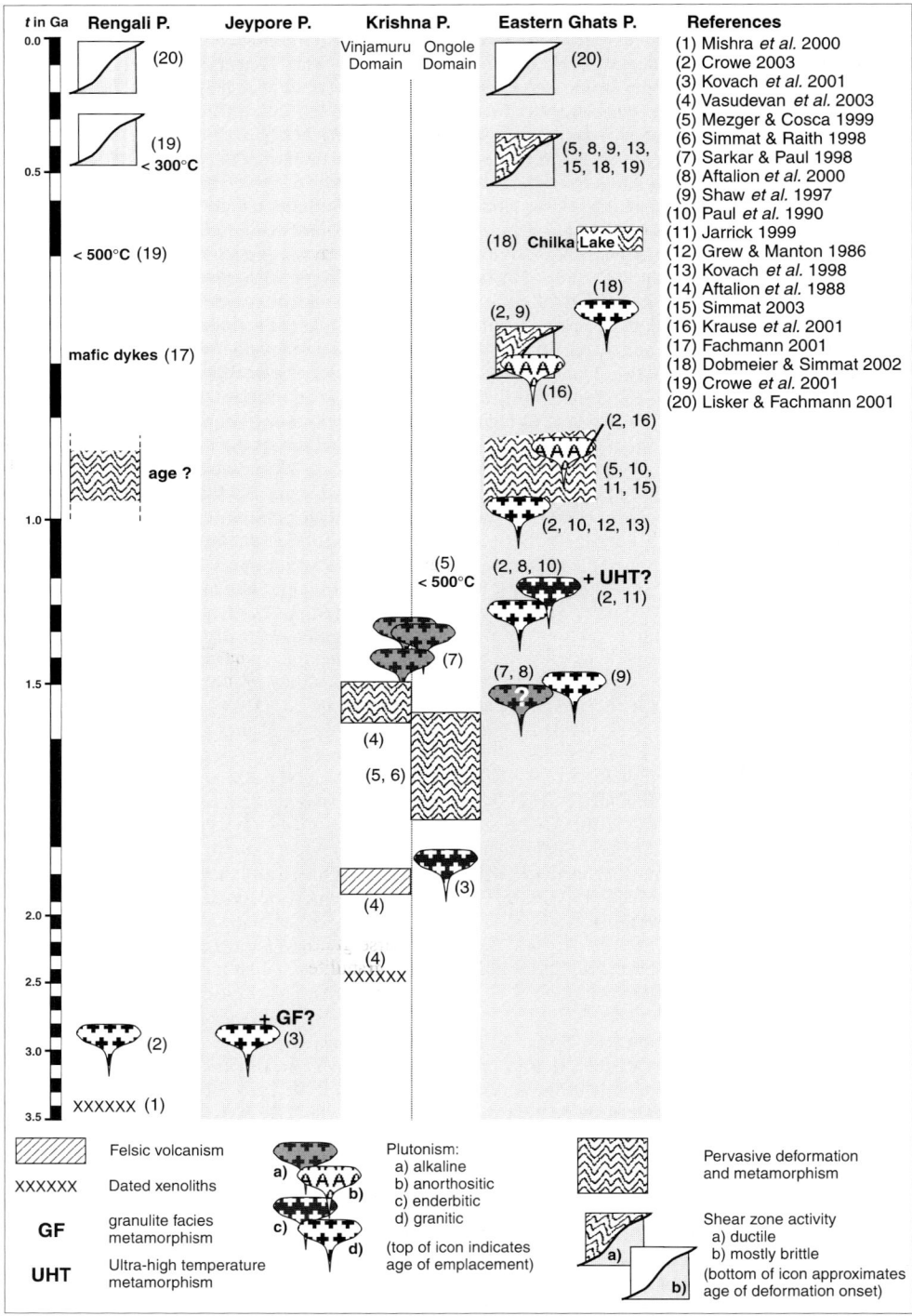

Fig. 5. Comparison of isotope data and tectonometamorphic events in the defined crustal provinces. For data sources see text.

intracrustal melting, incorporated crustal material as old as 2431 Ma. A Rb–Sr whole-rock age of 1534±14 Ma is interpreted to reflect incomplete homogenization of the Sr isotopic system, probably during the metamorphic overprint (Vasudevan et al. 2003). Several K–Ar, Rb–Sr and Pb–U–Th isotope data for muscovite and samarskite from muscovite pegmatites cluster at c. 1.6 Ga, but these data were generated in the late 1940s to the late 1960s [for a review see Babu (1998)] so their reliability is uncertain. A K–Ar whole-rock age of 989±23 Ma for a metabasalt and a K–Ar muscovite age of 806 Ma for a metapelitic schist indicate final cooling of the southern Vinjamuru Domain in Neoproterozoic times (Gosh et al. 1994).

Metamorphism in the Udayagiri Domain did not exceed greenschist facies conditions (Babu 1998). Geothermobarometric calculations for amphibolites and kyanite-bearing metapelites of the Vinjamuru Domain yielded maximum P–T conditions of 7–10 kbar and 700–750°C, with apparent pressures increasing northwards (Babu 1998; Okudaira et al. 2000). Further details of the P–T evolution are not known.

Several structural studies in both domains have established that a pervasive schistosity, S1, formed during peak metamorphic conditions and was subsequently multiply folded. S1 trends parallel to the arcuate margins of the province and dips steeply to the east. Associated tight to isoclinal and westward-facing folds, F1, with subhorizontal fold axes, were repeatedly folded into upright open folds with either near-coaxial or near-perpendicular fold axes (Babu 1998; Raman & Murty 1997; Okudaira et al. 2000). Muscovite-bearing pegmatites were emplaced along the foliations and in fold hinges (Babu 1998), but it remains unclear if their emplacement postdates deformation. The development of NW–SE trending left-lateral shear zones at greenschist facies conditions in rigid amphibolites near the Godavari Rift (Hari Prasad et al. 2000; Okudaira et al. 2000) may be associated with rift formation.

Ongole Domain

The granulite facies rocks of the Ongole Domain extend from Ongole in the south to the southwestern margin of the Godavari Rift and reappear on the other side of the rift (Fig. 4). This is exemplified by the occurrence of characteristic metasedimentary lithologies (interlayered diatexitic garnet–spinel–corundum–sillimanite gneisses, garnet–sillimanite quartzite, calc-silicate rocks) up to Upper Sileru.

The domain essentially consists of garnetiferous basic granulites and large volumes of multi-intrusive enderbitic granulites interlayered with locally porphyritic leptynites. Metasedimentary rocks mostly occur in its eastern part as rafts and partly resorbed layers with black-wall contacts in the meta-igneous rocks. Diatexitic pelitic garnet–sillimanite–spinel granulites dominate, and are interbanded with quartzitic to psammitic gneisses and rare calc-silicate granulites, suggesting a typical association of shallow-marine sediments as protoliths. A layered mafic–ultramafic complex at Kondapalle (Fig. 4) has intruded the high-grade metasedimentary package at deep crustal levels, and was dismembered and invaded by enderbite (Sengupta et al. 1999). The layered complex consists of gabbronorite grading into finely interbanded leucogabbronorite and anorthosite with clinopyroxenite- and chromite-bearing enstatite cumulate layers (Leelanandam 1997). The entire granulite assemblage is transected by allanite- and monazite-bearing pegmatites and microgranites.

The deposition age of the sediments, the oldest component in the assemblage, is not known. Nd model ages (T_{DM}) reflecting the mean crustal residence age of the provenance areas range between 2.8 and 2.6 Ga, and compare well with Nd model ages for granitoids of the eastern Dharwar Craton (Rickers et al. 2001). From this, these authors concluded that the clastic material was presumably derived from the Archaean granitoids without incorporation of juvenile material. The intrusion of the enderbite–charnockite precursors has been dated by near-concordant U–Pb ages for magmatic zircons at c. 1.72–1.70 Ga (Kovach et al. 2001). These data, together with Nd model ages (T_{DM}) of 2.5–2.3 Ga (Rickers et al. 2001), suggest a derivation from mantle-derived magma through assimilation–fractional crystallization processes involving substantial input of Archaean crustal material. The short-lived magmatism was closely followed by granulite facies metamorphism. Concordant U–Pb monazite ages of 1672±3 Ma, which date the synkinematic formation of coarse-grained leucosome networks in the enderbitic granulites (Kovach et al. 2001) and the emplacement of a synkinematic pegmatite (Mezger & Cosca 1999), approximate the age of granulite facies metamorphism. A reproducible ^{207}Pb–^{206}Pb allanite age from a late kinematic pegmatitic vein (c. 1598 Ma; Mezger & Cosca 1999) and the results of electron probe microanalysis (EPMA) monazite dating (age populations of 1.64–1.55 and 1.45–1.39 Ga; Simmat 2003), argue for a prolonged or polyphase metamorphism. The Pb isotope compositions of feldspars from metasedimentary rocks cluster tightly at the 1.5 Ga isochron (Rickers et al. 2001), consistent with last isotopic homogenization of Pb during this prolonged metamorphism. A Rb–Sr whole rock age of 1507±59 Ma for a mangeritic granulite may not date emplacement (Sarkar & Paul 1998) but rather reflect incomplete homogenization of the Sr isotopic system during metamorphism. A

^{40}Ar–^{39}Ar total degassing age for hornblende documents final cooling below c. 500°C at 1111 Ma (Mezger & Cosca 1999).

The P–T evolution of the dominant granulite facies metamorphism in the Ongole Domain has not been studied in detail (Sengupta et al. 1999; Dasgupta & Sengupta 2002). Metapelitic country rocks from the immediate contact with the Kondapalle layered complex exhibit mineral assemblages indicative of pressures of 8 kbar and temperatures >1000°C that are attributed to deep-crustal, ultrahigh-temperature contact metamorphism preceeding the regional granulite facies event (Sengupta et al. 1999). The ultrahigh-temperature event is not documented by the EPMA monazite age data of these rocks (Simmat & Raith 1998; Simmat 2003). Detailed studies of the regional deformation that accompanied granulite facies metamorphism are lacking. In general, deformed mafic dykes, potassium-feldspar-bearing leucosomes in diatexitic granulites and quartz veinlets evidence multiple isoclinal folding of the rocks. Compositional layering and axial plane foliations trend mostly parallel to the margins of the domain and dip moderately to steeply to the E/SE or NW. Orthopyroxene grains define a SE–NW plunging mineral lineation. Locally present centimetre-wide subvertical shear zones are filled with orthopyroxene-bearing leucosomes and transected at acute angles by narrow mylonite zones. Shear sense indicators uniformly suggest transport of the hanging wall towards the NW. Axes of late open folds plunge moderately to steeply to the NE. Brittle shear zones dipping moderately to the NE have been observed only near the Godavari Rift and thus may be associated with rift formation.

The emplacement of variegated alkaline to calc-alkaline plutons and layered mafic–ultramafic complexes (Prakasam Alkaline Province; Prasada Rao et al. 1988; Leelanandam 1998) into granulites of the Ongole Domain, and into migmatitic gneisses and amphibolites of the Vinjamuru Domain, blurred the boundary between both domains (Fig. 4). The mafic–ultramafic complexes consist of olivine-bearing clinopyroxenites, and gabbroic to noritic and anorthositic rocks (Dasgupta et al. 1997; Rao et al. 1998b). In the alkaline plutons, nepheline syenite clearly dominates over hornblende syenite, syenite and quartz syenite. Additionally, ferrosyenite, shonkinite and alkali granite occur as subordinate components. Whether small volumes of gabbro are genetically related is debated (Leelanandam 1998; Nagsai Sharma & Ratnakar 2000).

Sarkar & Paul (1998) listed Rb–Sr whole-rock ages for nepheline syenites of the Purimetla (1369±28 Ma), Uppalapadu (1348±41 Ma) and Elchuru (1242±33 Ma) alkaline plutons, and suggested that these ages date emplacement. Overall, however, the chronology of magmatism and the effects of regional deformation and metamorphism on the plutons are poorly constrained. Dasgupta et al. (1997) reported that the Chimakurthy layered complex (Fig. 4) and the subsequently intruded alkaline rocks were affected by progressive deformation. The metamorphic regime during this inferred regional deformation event has not been specified. By contrast, Nagasai Sharma & Ratnakar (2000) stated that the rocks of the Chanduluru composite pluton are devoid of deformation features except for weakly strained margins. The chemical characteristics of the calc-alkaline pluton point to emplacement in an Andean-type arc. Clearly, further studies are required.

At the contact with the Chimakurthy mafic–ultramafic complex, cordierite-bearing metapelitic country rocks show imprints of a mid-crustal, ultrahigh-temperature contact metamorphism along a near-isobaric heating–cooling trajectory within the temperature interval of c. 750–1000°C (Dasgupta et al. 1997). Two-pyroxene thermometry established maximum temperatures of c. 1100°C in the centre of the complex (Rao et al. 1998b).

Eastern Ghats Province

The Eastern Ghats Province consists of an intensely deformed and metamorphosed assemblage of metasedimentary, basic and enderbitic granulites, interspersed with massif-type anorthosites and probably multi-intrusive granite–charnockite complexes. The dominant metasedimentary lithologies are khondalites and diatexitic quartzofeldspathic gneisses, both frequently interlayered with sillimanite and garnet-bearing quartzites and leucocratic garnetiferous gneisses, commonly termed leptynite. Conformable bands and metre-sized xenoliths of high-Mg–Al granulites, containing sapphirine and spinel-bearing assemblages, have been found in several localities [summarized in Sengupta et al. (1999) and Dasgupta & Sengupta (2002)]. Calcareous rocks occur as rare layers or lenses of calc-silicate granulites, except for the northwestern margin where larger occurrences exist (Bhattacharya et al. 1998; Ramakrishnan et al. 1998) (Fig. 3). Marbles are known only from a small area to the north of Visakhapatnam (Bhowmik et al. 1995). Geochemical analyses confirm the sedimentary origin of the khondalites (Nanda & Pati 1989; Sen & Bhattacharya 1997; Rao et al. 1998a), but the notion that they represent metamorphosed equivalents of deeply weathered soil profiles (Dash et al. 1987) met with strong opposition (Nanda & Pati 1991). Field and geochemical evidence implies a magmatic origin for most of the leptynites (Shaw 1996; Sen & Bhattacharya 1997; Yamamoto et al. 1998), i.e. their formation through segregation and intrusion of felsic melts generated by biotite melting in the high-grade crustal lithologies (Braun et al. 1996). Especially in the Chilka Lake area, leptynites constitute sizeable intrusive bodies of garnetiferous leucogranite (Dobmeier & Raith 2000; Dobmeier & Simmat 2002). The meta-igneous rocks range from basic to felsic in composition, but fine-grained enderbites (Murthy et al. 1998), locally migmatitic charnockites (Charan et al. 1998) and, above all, garnet- and/or orthopyroxene-bearing porphyric granitoids (Mukhopadhyay & Bhattacharya 1997; Kovach et al. 1998; Krause 1998), which constitute voluminous batholitic bodies apparently associated with megalineaments (Nash et al. 1996; Krause 1998; Ramakrishnan et al. 1998), predominate. The porphyric granitoids were not affected by the early deformation. Their bulk chemical compositions and isotope data imply a derivation from crustal sources (Krause 1998). Basic rocks form centimetre- to metre-sized layers and lenses. The high-grade gneiss association in the northern part of the province hosts four massif-type anorthosite complexes of varying size [Bolangir – Bhattacharya et al. (1998) and Dobmeier (2003); Chilka Lake – Sarkar et al. (1981) and Dobmeier & Simmat (2002); Turkel – Maji et al. (1997); Jugsaypatna – Nanda & Panda (1999)] (Fig. 3). The larger anorthosite massifs are bordered by crustal-derived felsic igneous rocks varying from monzodiorite to granite. Strongly Fe-enriched dioritic rocks showing exceptionally high concentrations of rare earth elements (REE) and high-field-strength elements (HFSE) occur at the immediate contact with the anorthosite which they intrude in cross-cutting dykes and sheets (Maji et al. 1997; Raith et al. 1997; Bhattacharya et al. 1998). These rocks are interpreted as residual melts of the anorthosite (Bhattacharya et al. 1998) and, due to the high modal abundance of zircon, have enabled dating of the age of anorthosite emplacement at Chilka Lake and Bolangir (Krause et al. 2001).

The regional distribution of khondalitic lithologies led to the subdivision of the Eastern Ghats Province into a Western and an Eastern Khondalite Zone, and an intermediate Charnockite–Migmatite Zone (Nanda & Pati 1989; Ramakrishnan et al. 1998). However, the results of Nd isotope mapping (Rickers et al. 2001) oppose such a subdivision, rather suggesting that the province is divided along the Nagavalli–Vamsadhara and Mahanadi Megalineaments (Chetty 1995; Fig. 2a) into a northeastern domain (Angul and Tikarpara Domains; Fig. 2b) with Nd model ages (T_{DM}) of 2.2–2.8 Ga, a central domain with Nd model ages restricted to 1.8–2.2 Ga and a southwestern domain with a more heterogeneous distribution of Nd model ages (paragneisses 2.1–2.5 Ga, orthogneisses 1.8–3.2 Ga). As the majority of analysed granitoids exhibit pronounced S-type characteristics (Krause 1998), the data reflect regional-scale isotopic and hence compositional heterogeneity of the crustal source. Evidently, for the central domain the presence of Archaean components can be ruled out.

U–Pb ages of detrital zircons (Shaw et al. 1997; Jarick 1999) suggest that the deposition of sediments did not cease before c. 1.35 Ga. An early episode of felsic magmatism at c. 1.45 Ga in the Eastern Ghats Province seems to be indicated by Sm–Nd whole-rock isochron ages for orthogneisses from Rajagada (Shaw et al. 1997), which has to be verified by U–Pb zircon dating. Near-coeval with this crustal magmatism, an alkaline complex was emplaced near Khariar at c. 1.5 Ga (Sarkar & Paul 1998; Aftalion et al. 2000). However, it remains unclear whether the alkaline complex (Fig. 3) intruded the Bhandara craton or the Eastern Ghats Province (Madhavan & Khurram 1989; Aftalion et al. 2000).

Discordant U–Pb zircon data (Aftalion et al. 1988), a Pb–Pb isochron age (Paul et al. 1990); and a concordant ^{207}Pb–^{206}Pb SHRIMP age (Crowe 2003) for orthopyroxene-bearing meta-igneous rocks from the Angul and Phulbani Domains point to a second pulse of felsic plutonism at 1.2–1.15 Ga. Further field and isotopic work is needed to substantiate this magmatic event and its regional extent.

An early phase of ultra high-temperature metamorphism (c. 8 kbar and 950°C) preserved in sapphirine-bearing high-Mg–Al granulites (Lal et al. 1987; Kamineni & Rao 1988; Dasgupta 1995; Dasgupta & Sengupta 2002) has been dated with the common Pb method at 1099 ± 56 Ma (Jarick 1999) and thus may be linked to this event. A further time constraint is provided by a near-concordant ^{207}Pb–^{206}Pb SHRIMP age of 1105 ± 9 Ma for zircons with euhedral zoning, interpreted to date partial melting of leucogranite (leptynite) in the Phulbani area (Crowe 2003).

The emplacement of the voluminous complexes of porphyric granitoids between 985 and 955 Ma has been firmly established by near-concordant U–Pb ages (Grew & Manton 1986; Paul et al. 1990; Kovach et al. 1998) and concordant ^{207}Pb–^{206}Pb SHRIMP ages (Crowe 2003) for magmatic zircons. An accompanying pervasive granulite facies metamorphism attained peak conditions of 7–8 kbar and 750–800°C, and was followed by a steeply decompressive retrograde P–T segment to final conditions of 4–5 kbar and 650–700°C [Mukhopadhyay & Bhattacharya 1997; also reviewed in Sengupta et al. (1999) and Dasgupta & Sengupta (2002)]. U–Pb monazite ages of 973–954 Ma from metasedimentary and meta-igneous rocks (Aftalion et al. 1988; Paul et al. 1990; Mezger & Cosca 1999), Pb–Pb single zircon evaporation ages of 952–946 Ma for metamorphic zircons from high-Mg–Al granulites (Jarick 1999) and comprehensive EPMA monazite data clustering between 1.05 and 0.95 Ga (Simmat 2003) are interpreted to date this metamorphism, which was accompanied by pervasive regional deformation.

From regionally limited structural studies (Sarkar et al. 1981; Halden et al. 1982; Bhattacharya et al. 1994; Shaw 1996), it has been concluded that the entire Eastern Ghats Province experienced three discrete major episodes of regional deformation [summarized in Bhattacharya (1997) and Sarkar & Paul (1998)], the first of which may be coeval with ultra-high-grade metamorphism (Jarick 1999; Crowe 2003). Yet, most of these structural analyses have only been based on the discrimination of fold axis populations using the plunge of the axes as discriminant. Further, it was taken for granted that the entire granulite terrain has been deformed homogeneously (cf. Bhattacharya 1997). However, the results of combined structural and geochronological studies at the Bolangir and Chilka Lake anorthosite complexes (Krause et al. 2001; Dobmeier 2003; Dobmeier & Simmat 2002) and in the Angul Domain (Halden et al. 1982) have shown that most structures are the imprints of progressive regional deformation that occurred at appreciably different time in different sectors of the Eastern Ghats Province.

In the Khariar Domain (Figs 2b & 3), the time of emplacement of the massif-type anorthosite at Bolangir (Bhattacharya et al. 1998), in a thrust shear-dominated deformational regime with N–S directed shortening (Dobmeier 2003), is constrained by an upper intercept age of 933 ± 32 Ma for slightly discordant U–Pb zircon data from ferrodiorite (Krause et al. 2001) and a near-concordant ^{207}Pb–^{206}Pb zircon SHRIMP age of 918 ± 20 Ma for garnetiferous leucogranite (Crowe 2003). Pervasive secondary foliations in the anorthosite complex and the country rocks trend W–E and dip moderately to steeply to the south, with lineations and fold axes plunging in southern directions. U–Pb titanite data for calc-silicate rocks show that cooling to c. 650°C occurred at 570–530 Ma (Mezger & Cosca 1999) and thus point to a significant Pan-African thermal overprint of this domain.

In the Angul Domain (Figs 2b & 3) an Early Neoproterozoic age of granulite facies metamorphism and deformation, during which the N–S-trending, isoclinally folded, steep compositional layering and associated schistosities were formed (Halden et al. 1982), is indicated by U–Pb and EPMA monazite data which cluster at c. 960 Ma (Aftalion et al. 1988; Mezger & Cosca 1999; Simmat 2003). U–Pb titanite data (Mezger & Cosca 1999) show that the domain was cooled below c. 650°C at c. 930 Ma. Halden et al. (1982) interpreted a Rb–Sr muscovite age of 854 ± 6 Ma to date the emplacement of pegmatites in axial planes of late folds. However, a coeval ^{40}Ar–^{39}Ar age (854 ± 4 Ma) for amphibole from the same area (Lisker & Fachmann 2001) rather suggests interpreting the Rb–Sr age as a cooling age, unrelated to deformation. The geochronological data imply that the Angul Domain was tectonically and thermally decoupled from the larger part of the Eastern Ghats Province prior to the regional Pan-African tectonothermal event.

The sequence and geometry of structures in the Chilka Lake Domain strongly resemble those described in the Angul Domain, except for the orientation of the planar and linear elements that both trend predominantly WSW–ENE (Bhattacharya et al. 1994; Dobmeier & Raith 2000). However, the U–Pb zircon and EPMA monazite dates have established that the regional deformation occurred in this domain much later. The emplacement of the Chilka Lake anorthosite complex (792 ± 2 Ma; Krause et al. 2001) and leucogranites (762–743 Ma) probably mark the initial phase of a prolonged Neoproterozoic orogenic event that culminated with a pervasive transpressive deformation and granulite facies metamorphism at 690–660 Ma (Dobmeier & Simmat 2002).

Similar ages have also been reported from the northern margin of the Khariar Domain near Bolangir [U–Pb zircon SHRIMP age of 764 ± 35 Ma (Crowe 2003) and EPMA monazite data of 820–780 Ma (authors' unpublished data)] and its eastern extension, the Kantilo–Ranipathar Shear Zone [^{207}Pb–^{206}Pb zircon SHRIMP age of 789 ± 7 Ma (Crowe 2003)], and the Rampur Domain [Rb–Sr mineral isochron data ages of 833 ± 10 and 781 ± 38 Ma, and Sm–Nd mineral isochron data ages of 815 ± 9 and 808 ± 64 Ma (Shaw et al. 1997)]. The latter data are thought to relate to thermal metamorphism accompanying the emplacement of a porphyritic granite. An ^{40}Ar–^{39}Ar amphibole age of 650 ± 4 Ma for the Koraput alkaline complex [Rb–Sr whole-rock age 864 ± 25 Ma; cited in Sarkar & Paul (1998)] indicates final cooling below c. 500°C near the contact with the Jeypore Province (Mezger & Cosca 1999).

Further ^{40}Ar–^{39}Ar stepwise heating (Mezger & Cosca 1999; Aftalion et al. 2000; Crowe et al. 2001; Lisker & Fachmann 2001) and laser spot analysis data (Lisker & Fachmann 2001), Sm–Nd and Rb–Sr mineral isochrons (Shaw et al. 1997), U–Pb mineral ages (Kovach et al. 1998; Mezger & Cosca 1999; Aftalion et al. 2000), and EPMA monazite data (Dobmeier & Simmat 2002; Simmat 2003) provide evidence for a final pulse of medium- to high-grade thermal activity associated with deformation localized in shear zones at c. 515 Ma. Moreover, these data highlight a rather complex cooling history for the northern Eastern Ghats Province between 622 and 456 Ma, which necessitates differential movements at major shear zones in Late Neoproterozoic–Cambrian time.

Pooled apparent apatite fission track ages, scattering essentially between 300 and 250 Ma, document that the Angul and Tikarpara Domains entered the partial annealing zone for apatite at the latest in Late

Permian times (Lisker & Fachmann 2001). Data for samples from major fault zones varying between 152 ± 12 and 119 ± 6 Ma indicate a localized thermal overprint, and reactivation of the fault zones during the Cretaceous, probably coeval with the onset of seafloor production in the Bay of Bengal at c. 117 Ma (Lisker & Fachmann 2001).

Boundaries of crustal provinces

Reconnaissance studies have shown that all crustal provinces are bounded by sheared margins. Most prominent of all is the western margin of the Eastern Ghats Province, which coincides with a salient satellite megalineament that runs from the Godavari Rift via Sileru and Koraput to close to Bhawanipatna along the contact with the Jeypore Province (Fig. 2). Further to the north, where the Eastern Ghats Province abuts on the Bhandara Craton, the margin is less distinct in satellite imagery but better studied on the ground. The megalineament is termed either the Sileru Shear Zone (Chetty 1995, 2001; Chetty & Murthy 1998b) or the Elchuru–Kunavaram–Koraput Shear Zone (Chetty & Murthy 1998a), while its northern extension is known as the Eastern Ghats Boundary Fault (Ramakrishnan et al. 1998) or the Lakhna Thrust (Biswal 2000). The width of the moderately to the E/SE inclined Eastern Ghats Boundary Shear Zone averages 3 km, and is composed of granulites and cratonic rocks. All rocks are complexly folded but contain a late shear foliation that trends strictly parallel to the shear zone boundary and a mylonitic foliation in frequently encountered mylonitic rocks (Gupta et al. 2000). Feldspar, quartz, pyroxene, amphibole, epidote or biotite mark an associated down-dip lineation, and shear sense indicators uniformly imply transport of the hanging wall towards the W/NW. Chetty (1995) additionally reported subsequent dextral transcurrent displacement. The folded foliations contain mineral assemblages indicative of widely different metamorphic conditions, whereas the late foliation formed at similar $P-T$ conditions everywhere. Gupta et al. (2000) documented the preservation of a narrow inverted thermal gradient within the adjacent cratonic domain, with temperature increasing from c. 350 °C in the west to c. 700°C near the contact with the granulites of the Eastern Ghats Province. The thermal profile indicates aggregation of the terrains through westward thrusting of the hot granulite assemblage over a cooler cratonic foreland. The granulites show only minor retrogression. Significantly, the shear zone is the locus of emplacement of the Khariar and Koraput Alkaline Plutons (Leelanandam 1998). Considerably different emplacement ages for the alkaline complexes of c. 1500 (Aftalion et al. 2000) and c. 864 Ma (Sarkar & Paul 1998) imply either repeated alkaline magmatism or disturbance of the Rb–Sr systematics during later tectonometamorphic events. In fact, EPMA monazite data suggest that the late foliation formed only in the early Phanerozoic, i.e. during the Pan-African event at c. 530 Ma (authors' unpublished data). Further to the north, the shear zone swings sharply to the east. Essentially, the structural evolution remains the same near the moderately south-dipping shear zone, but shear sense indicators suggest north-directed transport of the hanging wall (Dobmeier 2003). U–Pb zircon SHRIMP data (Crowe 2003) and EPMA monazite data (authors' unpublished data) from the northern Bolangir Anorthosite Complex indicate repeated activity along this segment of the shear zone in Neoproterozoic and Early Phanerozoic times.

The propagation of the Eastern Ghats Boundary Shear Zone east of the Bolangir Anorthosite is unclear (Nash et al. 1996; Ramakrishnan et al. 1998; Crowe et al. 2001). The contrasting evolution of the Angul Domain suggests that the Shear Zone continues into the Kantilo Shear Zone and the western segment of the Ranipathar Shear Zone [Mahanadi Shear Zone of Chetty (1995); Fig. 3). The western segment of the Kantilo Shear Zone is blurred by the emplacement of the Sonapur Granite at 982 ± 6 Ma (Crowe 2003). Further to the east, mylonites formed at 789 ± 7 Ma in a porphyritic granite gneiss which was emplaced at 980 ± 7 Ma (Crowe 2003). Yet, the absence of a regional Pan-African imprint in the Tikarpara Domain has to be substantiated by more radiometric data.

The contact zone of the Jeypore Province with the Bhandara Craton shows marked similarities with the Eastern Ghats Boundary Shear Zone. It is characterized by pervasive retrogression of the granulites to amphibolites and hydrated gneisses, and the formation of a narrow-spaced planar fabric in all rocks (Nanda & Pati 1989). Mylonites, however, are largely confined to the cratonic rocks, which exhibit a steadily increasing metamorphic overprint reaching lower amphibolite facies conditions towards the contact zone. Planar fabrics in the high-strain zone trend strictly parallel to the contact and dip moderately to steeply towards E/ESE. An associated down-dip mineral stretching lineation (amphibole, epidote, feldspar, quartz) and unambiguous shear sense indicators in the mylonites indicate west-directed transport of the hanging wall during retrogression (authors' unpublished field observation). The age of this deformation is not constrained. It has to be noted that this zone of retrogression/prograde metamorphism is much narrower than shown by Ramakrishnan et al. (1998).

To the north of the Godavari Rift, problems in locating the margin of the Krishna Province result from poor exposure and an apparent transitional character. The sheared Kunavaram Alkaline

Complex may have been emplaced at or near this margin at 1265 ± 58 Ma (Rb–Sr whole-rock age; Sarkar & Paul 1998) and thus is broadly synchronous with the alkaline complexes of the Prakasam Alkaline Province. Alternatively, the Rb–Sr age may be afflicted by disturbance of the Sr isotopic system during later deformation and medium-grade metamorphism. To the south of the rift, the Krishna Province essentially abuts against the Cuddapah Basin. The contact is a several kilometres wide tectonic mélange, produced by thrust imbrication and intense folding, and is intruded by several felsic plutons (Meijerink et al. 1984). Further, the available data indicate that westward thrusting of the Krishna Province caused the strong deformation of the easternmost Cuddapah Basin with creation of the Nallamalai Fold Belt. However, age and evolution of the deformation remain unconstrained.

Repeated intense reactivation of the margins of the Rengali Province – the Barakot–Akul and Kerajang Faults (Mahalik 1994; Nash et al. 1996) – at semiductile and brittle conditions in Phanerozoic times (Crowe et al. 2001; Fachmann 2001; Lisker & Fachmann 2001) makes it extremely difficult to assess their Proterozoic significance. From results of their satellite imagery studies, Nash et al. (1996) and Crowe et al. (2001) inferred considerable right-lateral displacement along the Kerajang Fault and the ductile precursor shear zone in Neoproterozoic or Early Phanerozoic times, respectively. However, no combined tectonic and geochronological studies exist to confirm this assumption.

There is no evidence for a continuation of the Eastern Ghats Belt into the Southern Granulite Terrain [see the discussion of proposed continuations in Gopalakrishnan (1998)]. Notably, the Early Mesoproterozoic period of crust formation and tectonism in the Eastern Ghats Belt coincides with an episode of tectonometamorphic inactivity in southern India. The limited geochronological database implies that crust formation in the Southern Granulite Terrain was essentially completed in Early Palaeoproterozoic time [reviewed in Braun & Kriegsman (2002)]. To date, however, the internal configuration of the Southern Granulite Terrain is not yet sufficiently known to arrive at a conclusive model of its relations to the neighbouring crustal terrains.

Discussion – geological evolution of the Eastern Ghats and adjacent areas

Pre-Rodinia evolution

The present authors review and synthesis (Fig. 5) has shown that the Eastern Ghats and adjacent areas expose a deep section through a composite orogenic belt that comprises four granulite facies, crustal provinces with widely different geological evolutions.

Isotope data establish that the Jeypore Province evolved in Archaean time (Kovach et al. 2001; Rickers et al. 2001). It now constitutes a narrow high-grade igneous belt along the eastern fringe of the Bandhara Craton. In the absence of adequate geochemical and isotopic data, the palaeotectonic setting cannot be assessed, although the lithological composition and the spatial arrangement of the province in relation to the Bhandara Craton suggests formation in an active continental margin setting. Alternatively, the highly sheared margins allows understanding of the province as an exotic terrane that has been amalgamated with Proto-India, previous to or together with the Eastern Ghats Province.

At least parts of the Rengali Province also formed in Archaean time. Crowe (2002) argues that the Rengali Province earlier constituted a part of the northern Bandhara Craton and was juxtaposed with the Eastern Ghats Province by right-lateral transport along a transcurrent shear zone possibly in Late Mesoproterozoic times. This interpretation offers the interesting possibility to directly interrelate the Rengali and Jeypore Provinces, as the latter probably experienced igneous activity and/or high-grade metamorphism at approximately the same time as when voluminous granitoids intruded the Rengali Province. However, additional data are required from the poorly studied Jeypore Province before further conclusions can be drawn.

Strikingly similar isotope data for magmatism and metamorphism documented in the granulites of the Ongole Domain, and lower grade rocks of the adjacent Nellore–Khammam Schist Belt (Fig. 5), strongly argue for a common geological history of these units which together form the Krishna Province. The extrusion of conformable felsic lavas at 1.87–1.77 Ga (Vasudevan et al. 2002) proves that the volcanosedimentary sequence of the Vinjamuru Domain evolved in the Late Palaeoproterozoic and thus evidently rules out any correlation with the Archaean greenstone belts of the eastern Dharwar Craton as suggested by earlier workers (e.g. Raman & Murty 1997; Kumar et al. 1999 and refs cited therein). The igneous activity was closely followed by a tectonothermal event between c. 1.65 and 1.55 Ga. A second event, probably associated with alkaline magmatism (Sarkar & Paul 1998), may have occurred between c. 1.45 and 1.3 Ga (Mezger & Cosca 1999; Simmat 2002). Coeval kimberlites and lamproites of the eastern Dharwar Craton and the Cuddapah Basin (Chalapati Rao et al. 1996) are most likely further products of this alkaline magmatism. The extent of the mantle-derived igneous province points to extensive magmatic underplating, but the causative geodynamic processes remain unclear.

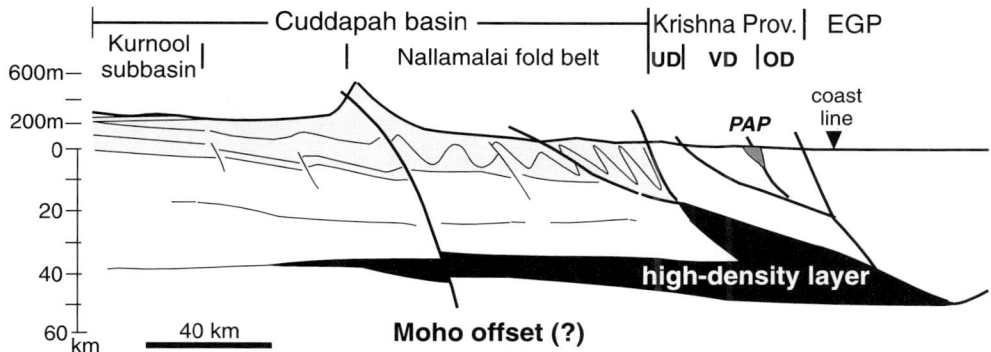

Fig. 6. Schematic cross-section through the crust of the Krishna Province and its western foreland. Based on data from Kaila *et al.* (1987). UD, Udayagiri Domain; VD, Vinjamuru Domain; OD, Ongole Domain; EGP, Eastern Ghats Province; PAP, Prakasam Alkaline Province.

The limited knowledge of the nature of its boundaries and lack of time constraints on deformation in the eastern Cuddapah Basin prohibit far-reaching interpretations of the geodynamic evolution of the Krishna Province and the adjacent craton. However, the metamorphic zonation within the Krishna Province and the Cuddapah Basin is suggestive of E–W-oriented convergence, causing an inverse metamorphic stacking of granulites (Ongole Domain), amphibolite facies rocks (Vinjamuru Domain), greenschist facies rocks (Udayagiri Domain) and very-low-grade rocks (Nallamalai Fold Belt), in the Late Palaeoproterozoic. The available deep seismic sounding data support this interpretation (Kaila *et al.* 1987; Nayak *et al.* 1998; Fig. 6).

Noticeably, the lithological successions of the Krishna Province resemble the Cuddapah Supergroup, with similar crystallization ages for igneous rocks. It thus appears likely that the Krishna Province constitutes metamorphic equivalents of the Cuddapah Supergroup. Nd model ages for paragneisses of the Ongole Domain suggest a derivation of the sedimentary material from the Dharwar Craton (Rickers *et al.* 2001), which is also considered as provenance of the Cuddapah Supergroup sediments (Crawford & Compston 1973; Meijerink *et al.* 1984).

The west-vergent stacking presumably resulted from the collision of the Napier Complex with the eastern Dharwar Craton in Late Palaeoproterozoic times. This perception is supported by recent isotope data that indicate a tectonothermal event at c. 1.6 Ga in the Napier Complex (Grew *et al.* 2001; Owada *et al.* 2001). It is further consistent with latest models for the pre-Pan-African configuration of Antarctica (Fitzsimons 2000; Boger *et al.* 2001).

The alkaline complexes exposed along the western margin of the Eastern Ghats Province and at domain boundaries within the Krishna Province may denote incipient continental rifting (Leelanandam 1998) in the Early Mesoproterozoic. Further, rifting may have initiated the deposition of the sedimentary sequences of the Eastern Ghats Province. However, this geodynamic scenario remains highly speculative since the cause and the mandatory eastern cratonic counterpart are not known.

Rodinia assembly

The presented isotope data clearly dismiss the repeatedly expressed view that the Eastern Ghats Province underwent a first granulite facies metamorphism and deformation in the Late Archaean (e.g. Yoshida *et al.* 1992; Chetty 1995; Chetty & Murthy 1998b; Ramakrishnan *et al.* 1998; Sarkar & Paul 1998). Clearly, most of the crust of the Eastern Ghats Province evolved in Proterozoic times during continent–continent collision processes related to the assembly of Rodinia (Rickers *et al.* 2001). Two models have been proposed for the crustal evolution: either the Eastern Ghats Province may have grown from SW to NE in several phases or it may consist of three distinct crustal blocks separated by the composite Vamsadhara–Nagavalli Shear Zone and the Ranipathar (or Mahanadi) Shear Zone (Fig. 2) (Rickers *et al.* 2001), both of which were the site of voluminous felsic plutonism during Early Neoproterozoic times (Fig. 3). The second model is supported by the sharp contacts between the blocks and their considerably different Nd isotope compositions. Incomplete reworking of the crust in the southwestern block, a typical feature of orogenic fronts, contrasts with the homogeneous crustal material of the central block, indicative of a position far away from the orogenic front. The central block, which corresponds to the Phulbani Domain (Fig. 2b), may represent an indenter-like protrusion of internal portions of the orogen. If so, the bounding shear zones should exhibit opposing shear sense during the same time interval, i.e. left-lateral at the Vamsadhara–Nagavalli Composite Shear Zone but right-lateral along the Ranipathar Shear Zone.

Ultrahigh-grade regional metamorphism presumably documents amalgamation of the Eastern Ghats Province with Proto-India during the global Grenvillian Orogeny at $c.$ 1.1 Ga. Imprints of this event became largely erased during subsequent extensive granitoid emplacement and coeval pervasive deformation and high-grade metamorphism between 985 and 950 Ma. It appears that crustal consolidation of most parts of the Eastern Ghats Province was achieved during this episode. Yet, the Chilka Lake Domain became pervasively reworked at high-grade metamorphic conditions between 800 (anorthosite emplacement) and 690–660 Ma (transpressional deformation). This event also led to reactivation of major shear zones, at least in the northern part of the province.

Imprints of Gondwana assembly

Isotope data establish for most of the Eastern Ghats Belt that the rocks experienced a medium-grade thermal overprint in Early Phanerozoic times, i.e. during the global Pan-African Orogeny. The combination of EPMA–monazite age data (authors' unpublished data), in addition to structural and petrological data from the Eastern Ghats Boundary Shear Zone near Junagarh (Gupta *et al.* 2000; Fig. 3), imply that final thrusting of the Eastern Ghats Province onto the Bhandara Craton occurred during this event, and that considerable parts of the Eastern Ghats Province left the high-temperature regime only at that time. The near-coeval activity of several other major shear zones leaves no doubt that the Pan-African event modified the internal organization of the Eastern Ghats Belt. It has to be pointed out, that the above proposed indentation of the Phulbani Domain has not necessarily occurred during the pervasive Late Grenvillian tectonothermal event. Unfortunately, no combined structural and geochronological data are available for the Vamsadhara–Nagavalli Lineament.

Recognition of suture zones in or at the Eastern Ghats Belt

Apparent ophiolitic rock assemblages have been reported only from the Vinjamuru Domain (Leelanandam 1990; Nagaraj Rao *et al.* 1991), but the occurrences lie well away from the domain margins and the ophiolitic nature has not been firmly established. However, the highly sheared margin of the Eastern Ghats Province separates crustal blocks of markedly different ages, and distinct geochemical and isotopical signatures, and ocean floor remnants could therefore be expected. The isotopical and geochronological data suggest that the Rengali and Jeypore Provinces constitute parts of the Bhandara Craton, and that the supracrustal rocks of the Krishna Province have accumulated on the eastern extension of the Dharwar Craton in Palaeoproterozoic times. By contrast, the geological evolution of the Eastern Ghats Province differs decidedly from the adjacent cratonic foreland. Cooling of the Ongole Domain granulites below $c.$ 500°C at $c.$ 1.1 Ga (Mezger & Cosca 1999) is in strong contrast to the protracted granulite facies metamorphism in the adjoining Eastern Ghats Province through the Grenvillian and Pan-African eras. This difference necessitates important differential movements along the Sileru Shear Zone during and following granulite facies metamorphism (Simmat & Raith 1998; Simmat 2002). Accordingly, the boundary shear zone of the Eastern Ghats Province may constitute a suture zone, even if inherent characteristics of suture zones, such as ophiolite bodies or metasedimentary rocks of indisputable oceanic affinity, are absent.

High-grade crustal terranes in the Rodinia and Gondwana context

The eastern margin of Proto-India was the locus of repeated accretionary activity during the Proterozoic era. A first orogenic cycle started with the deposition of volcanosedimentary sequences on the eastern Dharwar Craton at $c.$ 1.9–1.7 Ga. These platform sequences were segmented and transported towards the west during subsequent nappe tectonics. There is no evidence for incorporation of cratonic crust in the thrust belt. The weakly deformed and metamorphosed autochthonous western Cuddapah Basin forms the western foreland of the west-vergent thrust–nappe stack of the Krishna Province. The pile consists of three major tectonic units with markedly different degrees of metamorphism and intensities of deformation. The greenschist facies Udayagiri Domain at the base of the pile is overlain by the amphibolite facies Vinjamuru Domain and the granulite facies Ongole Domain. Collision took place between 1.7 and 1.55 Ga. Although plate reorganization during subsequent assembly and breakup of Rodinia and Gondwana makes it difficult to identify the eastern (present coordinates) continental counterpart and the possible extent of the orogen, latest isotope data suggest that the orogeny resulted from collision of the eastern Dharwar Craton with the Napier Complex of East Antarctica. The emplacement of basic–ultrabasic and alkaline intrusive complexes (Prakasam Alkaline Province) indicates post-collisional relaxation of the thickened lithosphere between 1.4 and 1.2 Ga. Uplift and cooling below c. 500°C was completed before the onset of the Grenvillian tectonothermal event.

The Eastern Ghats Province exposes the root zone of a collisional orogen predominantly consisting of intensely deformed high-grade pelitic to quartzo-feldspathic continental-derived metasedimentary sequences that were detached from their basement and intruded by voluminous synorogenic granitoids. Demonstrably, the basement of the supracrustals, which were probably deposited in Early Mesoproterozoic times, has not been incorporated into the orogen. A derivation of the sediments from Proto-India can be ruled out and an adequate provenance has not yet been identified. The welding together of this province with Proto-India probably occurred at 1.1 Ga and was associated with ultrahigh-grade metamorphism. Pervasive deformation at high-grade metamorphic conditions and extensive crustal magmatism between 985 and 950 Ma mark the climax of the orogenic activity and result from intracontinental collisional processes. The similar evolutionary history of the Rayner Province of East Antarctica (Black et al. 1987; Harley & Hensen 1990; James et al. 1991; Boger et al. 2000; Kelly et al. 2000) suggests that both provinces are segments of the same Grenvillian orogenic belt that supposedly formed during the collision of Proto-India and the Central Antarctic Craton [see discussions in Mezger & Cosca (1999), Fitzsimons (2000) and Rickers et al. (2001)]. The plate tectonic significance of an episode of strong crustal reworking in the Chilka Lake Domain dated between 800 and 660 Ma (Dobmeier & Simmat 2002) remains unclear. This age group is unknown from adjacent terrains in the Eastern Ghats Province, but provides a possible link with the Southern Granulite Terrain and Sri Lanka (see Braun & Kriegsman 2002), and the Rayner Complex (Black et al. 1987), where intrusive activity in this period has been reported. The post-Grenvillian architecture of the orogen was considerably modified during the Pan-African Orogeny. The deep crustal section was internally reorganized, uplifted and thrust over its western cratonic forelands, possibly together with the Jeypore Province, in Late Neoproterozoic–Early Phanerozoic times. This implies a long-lasting stability of the thickened Grenvillian crust. In contrast to the situation in the Lützow-Holm Bay and Prydz Bay complexes of East Antarctica (Fitzsimons 2000; Boger et al. 2001), the Southern Granulite Terrain of southernmost India, and Sri Lanka (see below), the Pan-African compressional tectonics were confined to shear zones, reflecting localized intracontinental deformation. This gives yet another similarity with the Rayner Province (Clarke 1988).

The palaeotectonic model for East Antarctica of Boger et al. (2001) lends credibility to the consideration of Torsvik et al. (2001) that NW India was not united with peninsular India by c. 750 Ma. In consequence, the short-lived magmatic event along the east coast of peninsular India at 800–780 Ma cannot be correlated with subduction processes at the Mozambique suture.

Our understanding of the architecture and evolution of the Eastern Ghats Belt has developed over many years of most enjoyable and fruitful collaborative research with A. Bhattacharya (IIT, Kharagpur), S. Dasgupta and P. Sengupta (Jadavpur University, Calcutta) – the ideas and arguments expressed in this paper would not have been possible without exchanges with these workers. The scientific support of and discussions with our colleagues are also gratefully acknowledged: S. Hoernes (Bonn), K. Mezger (Münster) and V. Kovach (Sankt Petersburg). Coworkers participating in the projects, notably J. Ehl, O. Krause, K. Rickers and R. Simmat, also made substantial contributions. Valuable reviews of the manuscript by A. Bhattacharya, S. Dasgupta, I. Fitzsimons, S. Harley, N. Kelly and L. Kriegsman are very much appreciated. Finally, S. Dasgupta and M. Yoshida are thanked both for inviting us to contribute to this special memoir and for their helpful suggestions during the writing of this paper. Our research in India was supported by the Deutsche Forschungsgemeinschaft (DFG) through grants Ra 205/16, Ra 205/20, Ra 205/23, Ra 205/25 and Do 644/1. This paper is a contribution to the IGCP projects 368 and 440.

References

AFTALION, M., BOWES, D. R., DASH, B. & DEMPSTER, T. J. 1988. Late Proterozoic charnockites in Orissa, India: U–Pb and Rb–Sr isotopic study. *Journal of Geology*, **96**, 663–676.

AFTALION, M., BOWES, D. R., DASH, B. & FALLICK, A. E. 2000. Late Pan-African thermal history in the Eastern Ghats terrane, India, from U–Pb and K–Ar isotopic study of the Mid-Proterozoic Khariar alkali syenite, Orissa. *Geological Survey of India Special Publication*, **57**, 26–33.

ASAMI, M., SUZUKI, K., GREW, E. S. & ADACHI, M. 1998. CHIME ages for granulites from the Napir Complex, East Antarctica. *Polar Geoscience*, **11**, 172–199.

BABU, V. R. R. M. 1998. The Nellore Schist Belt: an Archaean Greenstone Belt, Andhra Pradesh, India. *Gondwana Research Group Memoir*, **4**, 97–136.

BAKSHI, A. K., ARCHIBALD, D., SAHA, A. K. & SARKAR, S. N. 1987. $^{40}Ar–^{39}Ar$ incremental heating study of mineral separates from the early Archaean east India craton: implications for the thermal history of a section of the Singhbhum granite batholitic complex. *Canadian Journal of Earth Sciences*, **42**, 1985–1993.

BASHKAR RAO, Y. J., PANTULU, G. V. C., DAMODARA REDDY, V. & GOPALAN, K. 1995. Time of early sedimentation and volcanism in the Proterozoic Cuddapah basin, south India: evidence from the Rb–Sr age of the Pulivendla mafic sill. *Memoir, Geological Society of India*, **33**, 329–338.

BHATTACHARYA, A., RAITH, M., HOERNES, S. & BANERJEE, D. 1998. Geochemical evolution of the massif-type anorthosite complex at Bolangir in the Eastern Ghats Belt of India. *Journal of Petrology*, **39**, 1169–1195.

BHATTACHARYA, S. 1997. Evolution of Eastern Ghats granulite belt of India in a compressional tectonic regime and juxtaposition against Iron Ore Craton of Singhbhum by oblique collision–transpression. *Proceedings of the Indian Academy of Science (Earth and Planetary Sciences)*, **106**, 65–75.

BHATTACHARYA, S., SEN, S. K. & ACHARYYA, A. 1994. The structural setting of the Chilka Lake granulite–migmatite–anorthosite suite with emphasis on the time relation of charnockites. *Precambrian Research*, **66**, 393–409.

BHOWMIK, S. K., DASGUPTA, S. HOERNES, S. & BHATTACHARYA, P. K. 1995. Extremely high-temperature calcareous granulites from the Eastern Ghats, India: evidence for isobaric cooling, fluid buffering, and terminal channelized fluid flow. *European Journal of Mineralogy*, **7**, 689–703.

BISWAL, T. K. 2000. Fold-thrust geometry of the Eastern Ghats Mobile Belt, a structural study from its western margin, Orissa, India. *Journal of African Earth Sciences*, **31**, 25–33.

BISWAL, T. K. & JENA, S. K. 1999. Large lateral ramp in the fold-thrust belts of Mesoproterozoic Eastern Ghats Mobile Belt, Eastern India. *Gondwana Research*, **2**, 657–660.

BLACK, L. P., HARLEY, S. L., SUN, S. S. & MCCULLOCH, M. T. 1987. The Rayner Complex of East Antarctica: complex isotopic systematics within a Proterozoic mobile belt. *Journal of Metamorphic Geology*, **5**, 1–26.

BOGER, S. D., CARSON, C. J., WILSON, C. J. L. & FANNING, C. M. 2000. Neoproterozoic deformation in the Radock Lake region of the northern Prince Charles Mountains, east Antarctia; evidence for a single protracted orogenic event. *Precambrian Research*, **104**, 1–24.

BOGER, S. D., WILSON, C. J. L. & FANNING, C. M. 2001. Early Palaeozoic tectonism within the East Antarctic craton: the final suture between east and west Gondwana? *Geology*, **29**, 463–466.

BOSE, R. N. & KARTHA, T. D. G. K. 1977. Geophysical studies in parts of west coast of India. *Indian Journal of Earth Science*, **57**, 251–256.

BRAUN, I. & KRIEGSMAN, L. M. 2002. Proterozoic crustal evolution of southernmost India and Sri Lanka. *In:* YOSHIDA, M., WINDLEY, B. F. & DASGUPTA, S. (eds). *Proterozoic East Gondwana: Supercontinent Assembly and Breakup*. Geological Society, London, Special Publication.

BRAUN, I., RAITH, M. & RAVINDRA KUMAR, G. R. 1996. Dehydration–melting phenomena in leptynitic gneisses and the generation of leucogranites: a case study from the Kerala Khondalite Belt, southern India. *Journal of Petrology*, **37**, 1285–1305.

CHADWICK, B., VASUDEV, V. N. & HEGDE, G. V. 2000. The Dharwar craton, southern India, interpreted as the result of Late Archaean oblique convergence. *Precambrian Research*, **99**, 91–111.

CHALAPATI RAO, N. V., MILLER, J. A., PYLE, D. M. & MADHAVAN, V. 1996. New Proterozoic K–Ar ages for some kimberlites and lamproites from the Cuddapah Basin and the Dharwar Craton, South India: evidence for non-contemporaneous emplacement. *Precambrian Research*, **79**, 363–369.

CHARAN, S. N., SUBBA RAO, M. V. & DIVAKARA RAO, V. 1998. Prograde–retrograde mineralogy and geochemical variation trends in the charnockites of Gamparai region, Visakhapatnam district, Andhra Pradesh. *Geological Survey of India Special Publication*, **44**, 220–231.

CHAUDHURI, A. K., SAHA, D., DEB, G. K., DEB, S. P., MUKHERJEE, M. K. & GHOSH, G. 2002. The Purana basins of Southern Cratonic Province of India – a case for Mesoproterozoic fossil rifts. *Gondwana Research*, **5**, 23–33.

CHETTY, T. R. K. 1995. A correlation of Proterozoic shear zones between Eastern Ghats, India and Enderby Land, East Antarctica, based on LANDSAT imagery. *Memoir, Geological Society of India*, **34**, 205–220.

CHETTY, T. R. K. 2001. The Eastern Ghats Mobile Belt, India: a collage of juxtaposed terranes (?). *Gondwana Research*, **4**, 319–328.

CHETTY, T. R. K. & MURTHY, D. S. N. 1994. Collision tectonics in the Eastern Ghats Mobile Belt: mesoscopic to satellite scale structural observations. *Terra Nova*, **6**, 72–81.

CHETTY, T. R. K. & MURTHY, D. S. N. 1998a. Eluchuru–Kunavaram–Koraput (EKK) shear zone, Eastern Ghats Granulite Terrane, India: a possible Precambrian suture zone. *Gondwana Research Group Memoir*, **4**, 37–48.

CHETTY, T. R. K. & MURTHY, D. S. N. 1998b. Regional tectonic framework of the Eastern Ghats Mobile Belt: a new interpretation. *Geological Survey of India Special Publication*, **44**, 39–50.

CHOUDHARY, A. K., NAIK, A., MUKHOPADYAY, D. & GOPALAN, K. 1996. Rb–Sr dating of Sambalpur granodiorite, western Orissa. *Journal of the Geological Society of India*, **47**, 503–506.

CLARKE, G. L. 1988. Structural constraints on the Proterozoic reworking of Archaean crust in the Rayner Complex, MacRobertson and Kemp Land Coast, East Antarctica. *Precambrian Research*, **40**, 137–156.

CRAWFORD, A. R. & COMPSTON, W. 1973. The age of the Cuddapah and Kurnool Systems, southern India. *Journal of the Geological Society of Australia*, **19**, 453–464.

CROOKSHANK, H. 1963. Geology of southern Bastar and Jeypore from the Bailadila range to the Eastern Ghats. *Memoir, Geological Survey of India*, **87**.

CROWE, W. A. 2003. Age constraints for magmatism and deformation in the northern Eastern Ghats Belt, India: tectonic implications for the development of East Gondwanaland. *Precambrian Research*, in press.

CROWE, W. A., COSCA, M. A. & HARRIS, L. B. 2001. $^{40}Ar/^{39}Ar$ geochronology and Neoproterozoic tectonics along the northern margin of the Eastern Ghats Belt in north Orissa, India. *Precambrian Research*, **108**, 237–266.

DASGUPTA, S. 1995. Pressure–temperature evolutionary history of the Eastern Ghats granulite province; recent advances and some thoughts. *In:* YOSHIDA, M. & SANTOSH, M. (eds) *India and Antarctica during the Precambrian. Memoir, Geological Society of India*, **34**, 101–110.

DASGUPTA, S. & SENGUPTA, P. 2002. Indo–Antarctica correlation: a perspective from the Eastern Ghats Belt. *In:* YOSHIDA, M., WINDLEY, B. F. & DASGUPTA, S. (eds) *Proterozoic East Gondwana: Supercontinent Assembly and Breakup.* Geological Society, London, Special Publication, **xx**, xx–xx.

DASGUPTA, S., EHL, J., RAITH, M. M., SENGUPTA, P. & SENGUPTA, P. 1997. Mid-crustal contact metamorphism around the Chimakurthy mafic–ultramafic complex, Eastern Ghats Belt, India. *Contributions to Mineralogy and Petrology*, **129**, 182–197.

DASH, B., SAHU, K. N. & BOWES, D. R. 1987. Geochemistry and original nature of Precambrian khondalites in the Eastern Ghats, Orissa, India. *Transactions of the Royal Society of Edinburgh Earth Sciences*, **78**, 115–127.

DOBMEIER, C. 2003. Emplacement of Proterozoic massif-type anorthosites during crustal shortening: evidence from the Bolangir anorthosite complex (Eastern Ghats Belt, India). *International Journal of Earth Sciences*, in press.

DOBMEIER, C. & RAITH, M. M. 2000. On the origin of 'arrested' charnockitization in the Chilka Lake area, Eastern Ghats Belt, India: a reappraisal. *Geological Magazine*, **137**, 27–37.

DOBMEIER, C. & SIMMAT, R. 2002. Post-Grenvillian transpression in the Chilka Lake area, Eastern Ghats Belt – implications for the geological evolution of peninsular India. *Precambrian Research*, **113**, 243–268.

FACHMANN, S. 2001. *Geologische Entwicklung im Umfeld des Mahanadi-Riftes (Indien).* PhD Thesis, Technische Universität Bergakademie Freiberg, Germany.

FERMOR L. L. 1936. An attempt at the correlation of the ancient schistose formations of Peninsular India. *Memoir, Geological Survey of India*, **70**, 1–51.

FITZSIMONS, I. C. W. 2000. Grenville-age basement provinces in East Antarctica: evidence for three separate collisional orogens. *Geology*, **28**, 879–882.

GOPALAKRISHNAN, K. 1998. Extensions of Eastern Ghats Mobile Belt, India – a geological enigma. *Geological Survey of India Special Publication*, **44**, 22–38.

GOSH, D., DAS, J. N., RAO, A. K., RAY BARMAN, T., KOLLAPURI, V. K. & SARKAR, A. 1994. Fission-track and K–Ar dating of pegmatite and associated rocks of Nellore schist belt, Andhra Pradesh: evidence of Middle to Late Proterozoic events. *Indian Minerals*, **48**, 95–102.

GREW, E. S. & MANTON, W. I. 1986. A new correlation of sapphirine granulites in the Indo-Antarctic metamorphic terrain: late Proterozoic dates from the Eastern Ghats Province of India. *Precambrian Research*, **33**, 123–137.

GREW, E. S., SUZUKI, K. & ASAMI, M. 2001. CHIME ages of xenotime, monazite and zircon from beryllium pegmatites in the Napier Complex, Khmara Bay, Enderby Land, East Antarctica. *Polar Geoscience*, **14**, 99–118.

GUPTA, S., BHATTACHARYA, A., RAITH, M. & NANDA, J. K. 2000. Contrasting pressure–temperature–deformational history across a vestigial craton–mobile belt assembly: the western margin of the Eastern Ghats Belt at Deobogh, India. *Journal of Metamorphic Geology*, **18**, 683–697.

HALDEN, N. M., BOWES, D. R. & DASH, B. 1982. Structural evolution of migmatites in granulite facies terrane: Precambrian crystalline complex of Angul, Orissa, India. *Transactions of the Royal Society of Edinburgh: Earth Sciences*, **73**, 109–18.

HARI PRASAD, B., OKUDAIRA, T., HAYASAKA, Y., YOSHIDA, M. & DIVI, R. S. 2000. Petrology and geochemistry of amphibolites from the Nellore–Khammam Schist Belt, SE India. *Journal of the Geological Society of India*, **56**, 67–78.

HARLEY, S. L. & HENSEN, B. J. 1990. Archaean and Proterozoic high-grade terranes of East Antarctica (40–80°E): a case study of diversity in granulite facies metamorphism. *In:* ASHWORTH, J. R. & BROWN, M. (eds) *High-temperature Metamorphism and Crustal Anatexis.* Unwin Hyman Ltd, London, 320–370.

JAMES, P. R., DING, P. & RANKIN, L. 1991. Structural geology of the early Precambrian gneisses of northern Fold Island, Mawson Coast, East Antarctica. *In:* THOMSON, M. R. A., CRAME, J. A. & THOMSON, J. W. (eds) *Geological Evolution of Antarctica. Proceedings of the 5th International Symposium on Antarctic Earth Science.* Cambridge University Press, Cambridge, 19–23.

JARICK, J. 1999. *Die thermotektonometamorhe Entwicklung des Eastern Ghats Belt, Indien - ein Test der SWEAT-Hypothese.* PhD Thesis, Johann Wolfgang Goethe-Universität, Frankfurt am Main, Germany.

JAYANANDA, M. & PEUCAT, J.-J. 1996. Geological framework of southern India. *Gondwana Research Group Memoir*, **3**, 53–75.

JAYANANDA, M., MOYEN, J. F., MARTIN, H., PEUCAT, J. J., AUVREY, B. & MAHABALESHWAR, B. 2000. Late Archaean (2550–2520 Ma) juvenile magmatism in the eastern Dharwar craton, southern India: constraints from geochronology, Nd–Sr isotopes and whole rock geochemistry. *Precambrian Research*, **99**, 225–254.

KAILA, K. L., TEWARI, H. C., ROY CHOWDHURY, K., RAO, V. K., SRIDHAR, A. R. & MALL, D. M. 1987. Crustal structure of the northern part of the Proterozoic Cuddapah basin of India from deep seismic soundings and gravity data. *Tectonophysics*, **140**, 1–12.

KAMINENI, D. C. & RAO, A. T. 1988. Sapphirine granulites from Kakanuru area, Eastern Ghats, India. *American Mineralogist*, **73**, 692–700.

KANUNGO, D. N. & CHETTY, T. R. K. 1978. Anorthosite body in the Nellore Mica–Pegmatite Belt of Eastern India. *Journal of the Geological Society of India*, **19**, 87–90.

KANUNGO, D. N., CHETTY, T. R. K. & MALLIKARJUNA RAO, J. 1986. Anorthosites and associated rocks of Nellore–Gudur, South India. *Journal of the Geological Society of India*, **27**, 428–439.

KELLY, N. M., CLARKE, G. L., CARSON C. J. & WHITE R. W. 2000. Thrusting in the lower crust: evidence from the Oygarden Islands, Kemp Land, East Antarctica. *Geological Magazine*, **137**, 219–234.

KOVACH, V. P., BEREZHNAYA, N. G., SALNIKOVA, E. B., NARAYANA, B. L., DIVAKARA RAO, V., YOSHIDA, M. & KOTOV, A. B. 1998. U–Pb zircon age and Nd isotope systematics of megacrystic charnockites in the Eastern Ghats Granulite Belt, India, and their implication for East Gondwana reconstruction. *Journal of African Earth Sciences*, **27**, 125–127.

KOVACH, V. P., SIMMAT, R., RICKERS, K. ET AL. 2001. The Western Charnockite Zone of the Eastern Ghats Belt, India – an independent crustal province of late Archaean (2.8 Ga) and Palaeoproterozoic (1.7–1.6 Ga) terrains. *International Symposium and Field Workshop on the Assembly and Breakup of Rodinia and Gondwana, Osaka, 26–30 October 2001.* Gondwana Research, **4**, 666–667.

KRAUSE, O. 1998. *Die petrogenetische Bedeutung der porphyrischen Granitoide für die Krustenentwicklung des Eastern Ghats Belt (Indien).* PhD Thesis, Universität Bonn, Germany.

KRAUSE, O., DOBMEIER, C., RAITH, M. M. & MEZGER, K. 2001. Age of emplacement of massif-type anorthosites in the Eastern Ghats Belt, India: constraints from U–Pb zircon dating and structural studies. *Precambrian Research*, **109**, 25–38.

KUMAR, R., OKUDAIRA, T., YASUTAKA, T., HARI PRASAD, B., DIVI, R. S. & YOSHIDA, M. 1999. Structural features around the Archaean–Proterozoic terrain boundary in Khammam district, south India. *Journal of Geosciences, Osaka City University*, **42**, 237–245.

LAL, R. K., ACKERMAND, D. & UPADHYAY, H. 1987. P–T–X relationships deduced from corona textures in sapphirine–spinel–quartz assemblages from Paderu, southern India. *Journal of Petrology*, **28**, 1139–1168.

LEELANANDAM, C. 1990. The Kandra volcanics in Andhra Pradesh: a possible ophiolite. *Current Science*, **59**, 785–788.

LEELANANDAM, C. 1997. The Kondapalle layered complex, Andhra Pradesh, India: a synoptic overview. *Gondwana Research*, **1**, 95–114.

LEELANANDAM, C. 1998. Alkaline magmatism in the Eastern Ghats Belt – a critique. *Geological Survey of India Special Publication*, **44**, 170–179.

LEELANANDAM, C. & NARASHIMA REDDY, M. 1998. Precambrian anorthosites from peninsular India – problems and perspectives. *Geological Survey of India Special Publication*, **44**, 152–169.

LISKER, F. & FACHMANN, S. 2001. The Phanerozoic history of the Mahanadi region, India. *Journal of Geophysical Research, B: Solid Earth*, **106**, 22027–22050.

MADHAVAN, V. & KHURRAM, M. Z. A. K. 1989. The alkaline gneisses of Khariar, Kalahandi district, Orissa. *Memoir, Geological Society of India*, **15**, 265–289.

MAHALIK, N. K. 1994. Geology of the contact between the Eastern Ghats Belt and the North Orissa Craton, India. *Journal of the Geological Society of India*, **44**, 41–51.

MAJI, A. K., BHATTACHARYA, A. & RAITH, M. 1997. The Turkel anorthosite complex revisited. *Proceedings of the Indian Academy of Sciences: Earth and Planetary Sciences*, **106**, 313–325.

MEIJERINK, A. M. K., RAO, D. P. & RUPKE, J. 1984. Stratigraphic and structural development of the Precambrian Cuddapah basin, S. E. India. *Precambrian Research*, **26**, 57–104.

MEZGER, K. & COSCA, M. 1999. The thermal history of the Eastern Ghats Belt (India), as revealed by U–Pb and $^{40}Ar/^{39}Ar$ dating of metamorphic and magmatic minerals: implications for the SWEAT correlation. *Precambrian Research*, **94**, 251–271.

MISHRA, S., DEOMURARI, M. P., WIEDENBECK, M., GOSWAMI, J. N., RAY, S. & SAHA, A. K. 1999. $^{207}Pb/^{206}Pb$ zircon ages and the evolution of the Singhbhum Craton, eastern India: an ion microprobe study. *Precambrian Research*, **93**, 139–151.

MISHRA, S., MOITRA, S., BHATTACHARYA, S. & SIVARAMAN, T. V. 2000. Archaean granitoids at the contact of Eastern Ghats Granulite Belt and Singhbhum–Orissa Craton, in Bhubaneswar–Rengali sector, Orissa, India. *Gondwana Research*, **3**, 205–213.

MUKHOPADHYAY, A. K. & BHATTACHARYA, A. 1997. Tectonothermal evolution of the gneiss complex at Salur in the Eastern Ghats granulite belt of India. *Journal of Metamorphic Geology*, **15**, 719–734.

MUKHOPADYAY, D. 2001. The Archaean nucleus of Singhbhum: the present state of knowledge. *Gondwana Research*, **4**, 307–318.

MUKHOPADYAY, I., RAY, J. & GUHA, S. B. 1994. Amphibolitic rocks around Kotturu, Khammam district, Andhra Pradesh: structural and petrological aspects. *Indian Journal of Geochemistry*, **9**, 39–53.

MURTHY, D. S. N, CHETTY, T. R. K., BALARAM, V. & GOVIL, P. K. 1998. Palaeotectonic environment of pyroxene granulite rocks of Gokavaram–Gangavaram area of the Eastern Ghats Granulite Terrain of Andhra Pradesh. *Geological Survey of India Special Publication*, **44**, 286–296.

NAGARAJ RAO, B. K., KATTI, P. M. & ROOP KUMAR, D. 1991. Does the Kandra Igneous Complex represent an ophiolite belt? *Records of the Geological Survey of India*, **124**, 264–266.

NAGARAJ RAO, B. K., RAJUKAR, S. T., RAMALINGASWAMY, G. & RAVINDRA BABU, B. 1987. Stratigraphy, structure and evolution of the Cuddapah basin. *Memoirs, Geological Society of India*, **6**, 33–86.

NAGSAI SHARMA, V. & RATNAKAR, J. 2000. Petrology of the gabbro–diorite–syenite–granite complex of Chanduluru, Prakasam Alkaline Province, Andhra Pradesh, India. *Journal of the Geological Society of India*, **55**, 553–572.

NANDA, J. K. & PANDA, P. K. 1999. Anorthosite–leuconorite–norite complex from Jugasaipatna, Kalahandi district, Eastern Ghats Belt of Orissa sector. *Gondwana Research Group Memoir*, **5**, 89–104.

NANDA, J. K. & PATI, U. C. 1989. Field relations and petrochemistry of the granulites and associated rocks in the Ganjam–Koraput sector of the Eastern Ghats Belt. *Indian Minerals*, **43**, 247–264.

NANDA, J. K. & PATI, U. C. 1991. Geochemistry and original nature of Precambrian khondalites in the Eastern Ghats, Orissa, India: a discussion. *Transactions of the Royal Society of Edinburgh: Earth Sciences*, **82**, 87–88.

NAQVI, S. M. & ROGERS, J. J. W. 1987. Precambrian Geology of India. *Oxford Monographs on Geology and Geophysics*, **6**.

NARAYANA RAO, M. 1983. Lithostratigraphy of the Precambrian rocks of the Nellore schist belt. *Quarterly Journal of the Geological, Mining and Metallurgical Society of India*, **55**, 83–89.

NARAYANASWAMI S., CHAKRAVARTY, S. C., VEMBAN, N. A. ET AL. 1963. The geology and manganese ore deposits of the manganese belt in Madhya Pradesh and adjoining parts of Maharashtra; Part I, General introduction. *Geological Survey of India Bulletin, Series A*, **22**.

NASH, C. R., RANKIN, L. R., LEEMING, P. M. & HARRIS, L. B. 1996. Delineation of lithostructural domains in northern Orissa (India) from Landsat Thematic Mapper imagery. *Tectonophysics*, **260**, 245–257.

NAYAK, P. N., CHOUDHURY, K. & SARKAR, B. 1998. A review of geophysical studies of the Eastern Ghats Mobile Belt. *Geological Survey of India Special Publication*, **44**, 87–94.

OKUDAIRA, T., HARI PRASAD, B. & KUMAR, R. 2000. Proterozoic evolution of the Nellore–Khammam Schist Belt in the Khammam district, SE India. *Journal of Geosciences, Osaka City University*, **43**, 193–202.

OWADA, M., OSANAI, Y., TOYOSHIMA, T., TSUNOGAE, T., HOKADA, T. & KAGAMI, H. 2001. Late Archaean to Proterozoic tectonothermal events in Napier Complex, East Antarctica: correlation with East Gondwana fragments. *International Symposium and Field Workshop on the Assembly and Breakup of Rodinia and Gondwana*, Osaka, 26–30 October 2001. Gondwana Research, **4**, 724–725.

PAUL, D. K., RAY BARMAN, T., MCNAUGHTON, N. J., FLETCHER, I. R., POTTS, P. J., RAMAKRISHNAN, M. & AUGUSTINE, P. F. 1990. Archaean–Proterozoic evolution of Indian charnockites: isotopic and geochemical evidence from granulites of the Eastern Ghats Belt. *Journal of Geology*, **98**, 253–263.

PRASADA RAO, A. D., RAO, K. N. & MURTY, Y. G. K. 1988. Gabbro–anorthosite–pyroxenite complexes and alkaline rocks of Chimakurti–Elcheru area, Prakasam district, Andhra Pradesh. *Geological Survey of India Records*, **116**, 1–20.

RAITH, M., BHATTACHARYA, A. & HOERNES, S. 1997. A HFSE- and REE-enriched ferrodiorite suite from the Bolangir anorthosite complex, Eastern Ghats Belt, India. *Proceedings of the Indian Academy of Sciences: Earth and Planetary Sciences*, **106**, 299–311.

RAMAKRISHNAN, M., NANDA, J. K. & AUGUSTINE, P. F. 1998. Geological evolution of the Proterozoic Eastern Ghats Mobile Belt. *Geological Survey of India Special Publication*, **44**, 1–21.

RAMAN, P. K. & MURTY, V. N. 1997. *Geology of Andhra Pradesh*. Geological Society of India, Bangalore.

RAO, A. T., SANJEEVA RAO, P. C. & SRINIVASA RAO, K. 1998a. Geochemistry of khondalites from Visakhapatnam in the Eastern Ghats. *Geological Survey of India Special Publication*, **44**, 297–306.

RAO, A.T., SATYANARAYANA, P.V.V. & YOSHIDA, M. 1998b. Geothermometry of coexisting pyroxenes from Chimakurti Anorthosite Complex at the contact of Eastern Ghats Granulite and Nellore Granite–Greenstone Belts, Andhra Pradesh, India. *Gondwana Research Group Memoir*, **4**, 165–184.

RICKERS, K., MEZGER, K. & RAITH, M. M. 2001. Evolution of the continental crust in the Proterozoic Eastern Ghats Belt, India and new constraints for Rodinia reconstruction: implications from Sm–Nd, Rb–Sr and Pb–Pb isotopes. *Precambrian Research*, **112**, 183–212.

SAHA, D. 2001. Deformation of Neoproterozoic successions adjoining the Eastern Ghats, India – implications for tectonic convergence associated with Rodinia Gondwana assembly. *International Symposium and Field Workshop on the Assembly and Breakup of Rodinia and Gondwana*, Osaka, 26–30 October 2001. Gondwana Research, **4**, 759–761.

SARKAR, A. & PAUL, D. K. 1998. Geochronology of the Eastern Ghats Precambrian Mobile Belt – a review. *Geological Survey of India Special Publication*, **44**, 51–86.

SARKAR, A., BHANUMATHI, L. & BALASUBRAHMANYAN, M. N. 1981. Petrology, geochemistry, and geochronology of the Chilka Lake igneous complex, Orissa State, India. *Lithos*, **14**, 93–111.

SARKAR, A., PATI, U. C., PANDA, P. K., PATRA, P. C., KUNDU, H. K. & GHOSH, S. 1998. Late-Archaean charnockitic rocks from the northern marginal zones of the Eastern Ghats Belt: a geochronological study. *International Seminar on Precambrian Crust in Eastern and Central India*, October 29–30, Bhubaneswar, India. Abstract volume, 128–131.

SARKAR, G., CORFU, F., PAUL, D. K., MCNAUGHTON, N. J., GUPTA, S. N. & BISHUI, P. K. 1993. Early Archaean crust in Bastar Craton, Central India – a geochemical and isotopic study. *Precambrian Research*, **62**, 127–137.

SARKAR, G., GUPTA, S. N. & BISHUI, P. K. 1994. New Rb–Sr isotopic ages and geochemistry of granitic gneisses from southern Bastar; implications for crustal evolution. *Indian Minerals*, **48**, 7–12.

SEN, S. K. & BHATTACHARYA, S. 1997. Dehydration melting of micas in the Chilka Lake khondalites: the link between the metapelites and granitoids. *Proceedings of the Indian Academy of Science: Earth and Planetary Sciences*, **106**, 277–297.

SENGUPTA, P., SEN, J., DASGUPTA, S., RAITH, M., BHUI, U. K. & EHL, J. 1999. Ultra-high temperature metamorphism of metapelitic granulites from Kondapalle, Eastern Ghats Belt,: implications for the Indo-Antarctic correlation. *Journal of Petrology*, **40**, 1065–1087.

SENGUPTA, S., CORFU, F., MCNUTT, R. H. & PAUL, D. K. 1996. Meso-Archaean crustal history of the eastern Indian Craton: Sm–Nd and U–Pb isotopic evidence. *Precambrian Research*, **77**, 17–22.

SHARMA, M., BASU, A. R. & RAY, S. L. 1994. Sm–Nd isotopic and geochemical study of the Archaean tonalite–amphibolite association from the eastern Indian craton. *Contributions to Mineralogy and Petrology*, **117**, 45–55.

SHAW, R. K. 1996. Structural features of granulites from Rayagada, Eastern Ghats, India: some preliminary observations. *Journal of Mineralogy, Petrology and Economic Geology*, **91**, 443–454.

SHAW, R. K., ARIMA, M., KAGAMI, H., FANNING, C. M., SHIRAISHI, K. & MOTOYOSHI, Y. 1997. Proterozoic events in the Eastern Ghats Granulite Belt, India: evidence from Rb–Sr, Sm–Nd systematics, and SHRIMP dating. *Journal of Geology*, **105**, 645–656.

SIMMAT, R. 2003. *Identifizierung hochgradig metamorpher Provinzen des Eastern Ghats Belt in Indien anhand einer EMS-Studie von Monazit-Altersmustern*. PhD Thesis, Universität Bonn, Germany.

SIMMAT, R. & RAITH, M. M. 1998. EPMA monazite dating of metamorphic events in the Eastern Ghats Belt of India. *Beihefte zum European Journal of Mineralogy*, **10**, 276.

SINGH, Y. & CHABRIA, T. 1999. Late Archaean–early Proterozoic Rb–Sr isochron age of granite from Kawadgaon, Bastar district, Madhya Pradesh. *Journal of the Geological Society of India*, **54**, 405–409.

SPEAR, F. S. & CHENEY, J. T. 1989. A petrogenetic grid for pelitic schists in the system SiO_2–Al_2O_3–FeO–MgO–K_2O–H_2O. *Contributions to Mineralogy and Petrology*, **101**, 149–164.

SUBBA RAO, M. V., CHARAN, S. N. & DIVAKARA RAO, V. 1998. Geochemical signatures of the charnockite suite of rocks of the Machkund region, Orissa: implications for their petrogenesis and constraints on the evolutionary processes of the Eastern Ghats Mobile Belt. *Geological Survey of India Special Publication*, **44**, 256–267.

TORSVIK, T. H., CARTER, L. M., ASHWAL, L. D., BHUSHAN, S. K., PANDIT, M. K. & JAMTVEIT, B. 2001. Rodinia refined or obscured: palaeomagnetism of the Malani igneous suite (NW-India). *Precambrian Research*, **108**, 319–333.

VASUDEVAN, D. & RAO, T. M. 1983. The high grade schistose rocks of the Nellore schist belt, Andhra Pradesh and their geologic evolution. *Indian Minerals*, **16**, 43–47.

VASUDEVAN, D., KRÖNER, A., WENDT, I. & TOBSCHALL, H. 2003. Geochemistry, petrogenesis and age of felsic to intermediate metavolcanic rocks from the Palaeoproterozoic Nellore Schist Belt, Vinjamur, Andhra Pradesh, India. *Journal of Asian Earth Sciences*, in press.

WALKER, T. L. 1902. Geology of Kalahandi State, Central Provinces. *Memoir, Geological Survey of India*, **33**, 1–22.

YAMAMOTO, T., TANI, Y., MIYASHITA, Y., RAO, A. T. & YOSHIDA, M. 1998. Migmatite and granulites in the Patapatnam–Tekkali area, Eastern Ghats, India. *Journal of Geosciences, Osaka City University*, **41**, 123–142.

YOSHIDA, M., FUNAKI, M. & VITANAGE, P. W. 1992. Proterozoic to Mesozoic East Gondwana: the juxtaposition of India, Sri Lanka, and Antarctica. *Tectonics*, **11**, 381–391.

ZACHARIAH, J. K., BHASKAR RAO, Y. J., SRINIVASAN, R. & GOPALAN, K. 1999. Pb, Sr and Nd isotope systematics of uranium mineralised stromatolitic dolomites from the Proterozoic Cuddapah Supergroup, south India: constraints on age and provenance. *Chemical Geology*, **162**, 49–64.

Proterozoic crustal evolution of southernmost India and Sri Lanka

INGO BRAUN[1] & LEO M. KRIEGSMAN[2,3]

[1]*Mineralogisch-Petrologisches Institut, Universität Bonn, Poppelsdorfer Schloss, D-53115 Bonn, Germany (e-mail: ingo.braun@uni-bonn.de)*
[2]*National Museum of Natural History/Naturalis, PO Box 9517, NL-2300 RA Leiden, The Netherlands (e-mail: kriegsman@naturalis.nnm.nl)*
[3]*Department of Geology, University of Turku, FIN-20014 Turku, Finland*

Abstract: The Southern Granulite Terrain of India comprises four Proterozoic domains (from north to south): the Madurai Block, and the Achankovil, Ponmudi and Nagercoil Units. All but the Achankovil Unit share similarities with the Highland Complex of Sri Lanka: (1) evidence for Palaeoproterozoic crust formation (from U–Pb ages and 2.0–3.0 Ga Nd model ages); and (2) Pan-African ultrahigh-temperature metamorphism between 610 and 550 Ma. More specifically, the Ponmudi Unit and the Madurai Block share lithological and petrological characteristics with the low-P and high-P domains of the Highland Complex, respectively. The Achankovil Unit is correlated with the Wanni Complex of Sri Lanka on the basis of lithologies and Nd model ages (1.0–2.0 Ga). These domains and the Vijayan and Kadugannawa Complexes of Sri Lanka represent Late Mesoproterozoic–Early Neoproterozoic crustal domains.
Similarities in tectonic style, degree of metamorphism and Neoproterozoic U–Pb ages, suggest a common Pan-African tectonothermal evolution for the lower crustal domains of southern India, Sri Lanka and the Lützow-Holm Bay of Eastern Antarctica. Sri Lanka may have been located at the junction between the Mozambique Belt and a second belt partly involving areas that underwent a Grenvillian high-grade event during the formation of Rodinia. Madagascar shows a similar Pan-African metamorphic overprint, but is located in a different position and lacks the influence of the second belt. Alternative views are discussed.

Most of Peninsular India and Sri Lanka exposes Precambrian crust which was formed, reworked and consolidated during several major orogenic events. Two of these events were associated with the assembly and breakup of the supercontinents Rodinia and Gondwana in Mesoproterozoic, and Neoproterozoic–Early Phanerozoic times, respectively. The crustal evolution of India during these periods has mainly affected two high-grade crustal provinces situated at its southern tip [the Southern Granulite Terrain (SGT); Fig. 1)] and the eastern coast (the Eastern Ghats Belt; Dobmeier & Raith 2002).

Unravelling the geological history of India and Sri Lanka was long impeded by the lack of geochronological data other than that of Rb–Sr. A breakthrough was achieved only when geochronological methods allowed reliable dating of magmatic and metamorphic rocks, and the geological significance of the attained ages was established.

In his fundamental treatise of the Precambrian of Peninsular India, Fermor (1936) regarded the granulite facies terrains of south and east India as constituents of a coherent Archaean crustal province – the Charnockite Province – that swerves around an older non-charnockitic province. This notion remained widely accepted with minor modifications [summarized in Gopalakrishnan (1998)] until reliable isotope age data showed that the Southern Granulite Terrain and the Eastern Ghats Belt underwent pervasive metamorphism at appreciably different times during the Proterozoic (Hansen *et al.* 1985; Harris *et al.* 1994; Mezger & Cosca 1999 and refs. cited therein). Similarly, the age of high-grade metamorphism in Sri Lanka had previously been interpreted in terms of multiple events at *c.* 2000 Ma, and *c.* 1250–1100 and 900–800 Ma, based on Rb–Sr data (Crawford & Oliver 1969; Katz 1971; de Maesschalck *et al.* 1990). However, U–Pb zircon data firmly constrain the time of peak metamorphism at between *c.* 610 and 540 Ma (Hölzl *et al.* 1991; Kröner & Williams 1993; Kröner *et al.* 1994*a,b*). Other important recent results are: (1) the discovery of distinct basement provinces based on Nd model ages (Milisenda *et al.* 1988); (2) a good petrological database documenting a pressure–temperature (P–T) gradient across the granulite facies terrain (e.g. Schumacher *et al.* 1990; Faulhaber & Raith 1991); and (3) improved constraints on the tectonometamorphic evolution of the Sri Lankan basement (papers in Raith & Hoernes 1994; Raase 1998; Kriegsman & Schumacher 1999) and its position in Gondwana (Kriegsman 1995).

A considerable increase of knowledge during the last decade calls for a critical evaluation of new and old data, and permits the formulation of an up-to-date synthesis of the geological evolution of these domains which offers new prospects for discussion on the assembly of the supercontinents Rodinia and Gondwana.

Archaean domains of southern India

The crystalline basement of southern India shows a transition from amphibolite to granulite facies conditions (Fermor 1936) which was interpreted to reflect a continuous section through progressively deeper mid- to lower continental Archaean crust (Crawford 1969). Later studies have revealed intricate compositional and geochronological details which are inconsistent with such an idealized notion. The Biligirirangan Hills are a clear continuation of the old Archaean Dharwar Craton, whereas the other 'Archaean' domains (the Nilgiri Hills, and the Krishnagiri–Salem and Chennai areas) seem to contain abundant juvenile material added to the crust around the Archaean–Proterozoic boundary, coeval wih the emplacement of the Dharwar Batholith. The Southern Granulite Terrain (SGT) south of the Dharwar Craton most likely consists of a heterogeneous assembly of Late Archaean and Proterozoic lower crustal complexes. At present, the limited number of geochronological and isotope data hardly allows a proper distinction of individual terranes within the SGT, let alone identification of the limit between areas that were consolidated c. 2.5 Ga ago and terrains which were involved in Proterozoic high-grade metamorphism and deformation.

Dharwar Craton Syntheses of existing and new data on the Dharwar Craton were provided by Chadwick *et al.* (2000) and Jayananda *et al.* (2000). The craton comprises two distinct parts that are separated by a crustal-scale N–S sinistral shear zone: the Chitradurga–Kollegal Shear Zone (Fig. 1).

The western part is made up of a polyphase, 3.4–2.9 Ga old gneissic basement of tonalite–trondhjemite–granodiorite (TTG) associations with narrow high-grade belts of greenstone-type volcanosedimentary sequences (the Sargur Group).

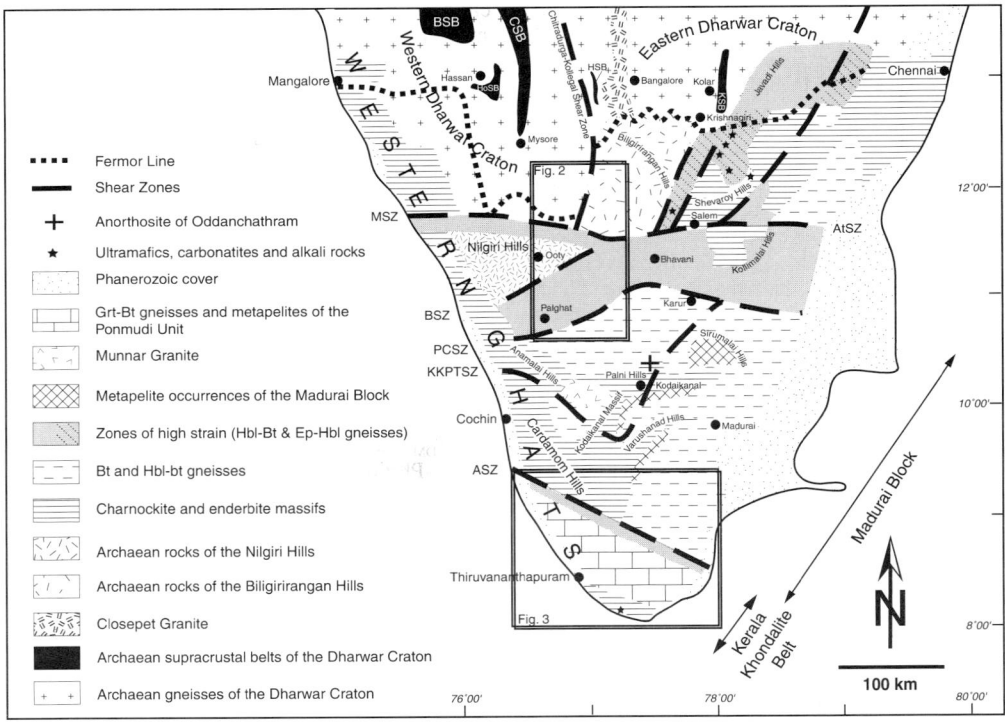

Fig. 1. Map of South India showing the main geological units and structural elements of the southern Dharwar Craton and the Southern Granulite Terrain. Information on the geology has been predominantly compiled from the geological maps of Kerala, Tamil Nadu and Karnataka (Geological Survey of India 1995*a,b*) and own data from the Kerala Khondalite Belt. The occurrence and orientation of shear zones largely follows Drury *et al.* (1984) and has been added by recent results of Ghosh (1999). MSZ, Moyar Shear Zone; BSZ, Bhavani Shear Zone; PCSZ, Palghat–Cauvery Shear Zone; KKPTSZ, Karur–Kambam–Painavu–Trichur Shear Zone; ASZ, Achankovil Shear Zone; AtSZ, Attur Shear Zone; BSB, Bababudan Schist Belt; HoSB, Holenarsipur Schist Belt; CSB, Chitradurga Schist Belt; HSB, Huliyurdurga Schist Belt; KSB, Kolar Schist Belt.

This basement was consolidated before the deposition of a low-grade volcanosedimentary sequence (the Dharwar Supergroup) in intracratonic basins at 2.9–2.6 Ga and locally intruded by granite plutons at c. 2.6 Ga.

The eastern part consists of a series of N–S-trending arc-type plutonic complexes, collectively named the Dharwar Batholith, which were emplaced during a major episode of calc-alkaline magmatism at 2.55–2.51 Ga, the most prominent of which is the N–S-trending Closepet Batholith. Interleaving low- to medium-grade greenstone-type schist belts have been interpreted as intra-arc volcanosedimentary sequences.

Nd (T_{DM}) mean crustal residence ages for TTG gneisses of the Western and Eastern Dharwar Craton range between 3.4 and 3.1 Ga (Peucat et al. 1993; Jayananda et al. 2000), whereas appropriate ages for the granitoids of the Eastern Dharwar Craton are significantly younger (2.9–2.6 Ga; Peucat et al. 1989, 1993; Jayananda et al. 1995b; Krogstad et al. 1995; Jayananda et al. 2000). In both cases, the overlap between Nd model ages and geochronological data indicate stages of crustal growth. In contrast, Nd (T_{DM}) ages of 3.4–3.2 Ga and corresponding ε_{Nd} values of −7 (at 2.54 Ga, the U–Pb age of zircon crystallization) for leucogranites from the Bangalore area point to crustal reworking during the Late Archaean (Jayananda et al. 2000). Jayananda et al. (2000) argue that the data are best explained in terms of a plume model, whereas Chadwick et al. (2000) postulate a major accretionary event in an Andean margin setting, which led to the final stabilization of the craton at c. 2.5 Ga. The latter model was recently supported by major and trace element modelling of rocks from the Closepet Batholith, according to which the granitoids were derived from magmas of two source regions: (1) enriched mantle, which prior to anatexis was metasomatized by slab-derived fluids; and (2) partially melted Dharwar TTG gneisses (Moyen et al. 2001).

Biligirirangan Hills The granulite terrain of the Biligirirangan (BR) Hills (Fig. 1) is predominantly composed of non-garnetiferous enderbites and charnockites. Along its western, eastern and southern margins it is bound by shear zones, whereas to the north a transition into the Closepet Granite and the Dharwar Craton occurs. The precursor rocks of the BR granulites most likely were derived from a tonalite–trondhjemite igneous rock suite with close similarities in chemical composition to the grey gneisses of the Dharwar Craton (Buhl 1987; Peucat et al. 1989, 1993; Raith et al. 1999). Single zircon evaporation and U–Pb zircon upper concordia intercept ages of 3.47 Ga are interpreted as the time of formation of the magmatic protoliths (Jayananda & Peucat 1996) and overlap with Nd (T_{DM}) model ages of between 3.6 and 3.2 Ga, suggesting crustal growth at c. 3.5 Ga. A U–Pb monazite age of 2.51 Ga provides evidence for Late Archaean granulite facies metamorphism (Jayananda & Peucat 1996), and Sm–Nd garnet–whole-rock and Rb–Sr biotite ages of 2473 ± 15 and 2305 Ma, respectively, reflect slow cooling during the Palaeoproterozoic (Jayananda & Peucat 1996). ε_{Nd} values of −3 – −14 (at 2.5 Ga) are comparable to appropriate data from the Dharwar Craton and further emphasize the extended crustal evolution of the granulite precursor rocks (Buhl 1987; Harris et al. 1994; Jayananda et al. 1995b). It has therefore been suggested that the BR granulites represent the exhumed deeper levels of the Dharwar Craton (Pichamuthu 1965; Ramakrishnan & Vishwanathan 1983; Raith et al. 1999).

Nilgiri Hills The Nilgiri Hills are separated from the Dharwar Craton and the Proterozoic terrains to the south by the Moyar and Bhavani Shear Zones (Fig. 1). They represent a tilted section of lower crust, indicated by a steep gradient of palaeopressures from 9 to 10 kbar along the Moyar Shear Zone to 6–7 kbar towards the Bhavani Shear Zone at fairly uniform temperatures of 730–800 °C (Raith et al. 1990; Srikantappa et al. 1992; Jayananda & Peucat 1996). The Nilgiri Hills predominantly consist of strongly deformed and migmatized garnetiferous enderbitic granulites with minor contributions of kyanite-bearing gneisses, quartzites and metabasites. In contrast to previous suggestions (Condie & Allen 1984; Peucat et al. 1989; Janardhan et al. 1994), recent geochemical investigations indicate a sedimentary origin for the precursor rocks of the Nilgiri enderbites (Raith et al. 1999). A Sm–Nd garnet–whole-rock age of 2496 ± 15 Ma is interpreted as the age of high-grade metamorphism (Jayananda & Peucat 1996). ε_{Nd} values of −2 – +4, Nd (T_{DM}) mean crustal residence ages of 2.9–2.6 Ga, and U–Pb zircon, single zircon evaporation ages and Rb–Sr and Sm–Nd whole-rock isochron ages of between 2.65 and 2.50 Ga indicate that formation and erosion of the precursor rocks, sedimentation and high-grade metamorphism must have occurred within a very short time span of 100–200 Ma (Buhl 1987; Peucat et al. 1989; Jayananda & Peucat 1996; Raith et al. 1999).

Granulites southwest of Chennai Granulites from the Chennai area involve metasediments, charnockitic and enderbitic granulites of igneous origin, and mafic granulites (Bernard-Griffiths et al. 1987 and refs cited therein). A first Sm–Nd isotope and rare earth element (REE) study of meta-igneous rocks yielded Late Archaean–Early Proterozoic Nd (T_{CHUR}) model ages of between 2.80 and 2.35 Ga, and ε_{Nd} values close to zero (0.97–1.94 at 2.55 Ga;

Bernard-Griffiths et al. 1987). In conjunction with a U–Pb zircon age of 2.6 Ga for an igneous charnockite (Vinogradov et al. 1964), these data point to a stage of Late Archaean juvenile magmatism. Further results showed that acid and mafic granulites are not cogenetic but instead were modified by metasomatic processes during granulite facies metamorphism, which has been dated at 2.55 ± 0.14 Ga on the basis of a Sm–Nd whole-rock isochron age (Bernard-Griffiths et al. 1987). However, since the distances between the sample sites were at least 100 m and the limited diffusion path lengths generally impedes homogenization of the Nd isotope systematics at this scale, it is inferred that the reported value yields a protolith age rather than the time of rehomogenization. Whether or not the formation of the magmatic rocks was immediately followed by granulite facies metamorphism, which according to petrological investigations reached P–T conditions of 6.5–7.5 kbar and 750–800°C (Bhattacharya & Sen 1986), remains to be verified by appropriate geochronological techniques.

Archaean domains within the Southern Granulite Terrain (SGT)

The SGT comprises an assembly of crustal domains of contrasting composition and tectonothermal evolution (e.g. Buhl 1987; Peucat et al. 1989; Choudhary et al. 1992; Harris et al. 1994; Raith et al. 1999), the details of which are still not well understood. Charnockite–enderbite massifs and extensive gneiss–migmatite terrains, mostly retrograde from granulite facies, are interspersed with high-grade supracrustal sequences and intruded by numerous granites [reviewed in Naqvi & Rogers (1987) and Jayananda & Peucat (1996)]. Geological maps commonly show the SGT subdivided into large crustal blocks which are transsected by crustal-scale shear zones (Drury & Holt 1980; Drury et al. 1984). A close inspection of available data, however, shows that this view is too simplistic and that the lithological architecture of wide areas of the SGT is poorly known. The Moyar and Bhavani Shear Zones represent major boundaries between the western part of the SGT and the Dharwar Craton. By contrast, the eastward continuation of the Bhavani Shear Zone and its relationship to lineaments transecting the northeastern part of the SGT is not well constrained. Farther south, the Palghat–Cauvery Shear Zone (PCSZ), which previously was regarded as the northern termination of the Proterozoic part of the SGT (Harris et al. 1994), is also not well understood. Whereas high-strain deformation and retrogression of granulites during the Proterozoic was widespread in the western part of the PCSZ, the recognition of such processes in the eastern part is not unequivocal (Ghosh 1999; Mukhopadhyay et al. 2001). Likewise, the area bounded by the Bhavani and Attur Shear Zones in the north and the PCSZ in the south comprises crustal units which were consolidated during the Archaean and Proterozoic, respectively. Based on lithological criteria, Mukhopadhyay et al. (2001) concluded that the area north of Karur predominantly consists of Archaean lower crust. This is in line with the notion of Ghosh (1999) that the Archaean basement of the Dharwar Craton extends southwards up to the so-called Karur–Kambam–Painavu–Trichur Shear Zone (KKPTSZ; Fig. 1). At present, the very limited geochronological database does not allow establishment of a detailed picture of the lithological architecture of the SGT. As a consequence, modelling the assembly and tectonometamorphic evolution of the SGT has to be done with severe limitations.

As a working model, the eastern continuation of the Bhavani Shear Zone, i.e. the northern limit of the SGT, is drawn along the eastern limit of the BR Hills, thus swinging around to a NNE trend (Fig. 1). The main contentions are: (1) that in the area east of the BR Hills, including the Shevaroy Hills, Proterozoic ages have been reported (H. Köhler pers. comm.), which are lacking from the BR and Nilgiri Hills; and (2) Late Proterozoic (c. 800 Ma) alkali and carbonatite magmatism is known from the Salem area.

An overview on the current state of knowledge on the area south of the Bhavani Shear Zone is presented below. First, the main geological features of the Archaean terrains situated between the Dharwar Craton and the PCSZ are addressed, followed by a more extensive description of the Proterozoic part of the SGT.

Krishnagiri–Salem Zone This area is regarded as part of an amphibolite–granulite transition zone extending from the Kolar Schist Belt of the Eastern Dharwar Craton down to the Attur Shear Zone in the south. The northern section of this area consists of a tonalite-granodiorite-granite suite that shows abundant incipient charnockitization. In the south, the Shevaroy Hills are mainly made up of massive charnockites and enderbites (Jayananda & Peucat 1996). The Krishnagiri–Salem Zone represents a tilted crustal section with southward increasing P–T conditions from 5 kbar and 600°C to 7–8 kbar and 800°C (Rameswar Rao et al. 1991; Hansen et al. 1995). U–Pb zircon upper concordia intercept ages and single zircon evaporation ages of between 2.55 and 2.51 Ga indicate that intrusion of the tonalitic and granitic precursor magmas occurred during the Late Archaean (Peucat et al. 1993). The overlapping Nd (T_{DM}) mean crustal residence ages of 2.9–2.7 Ga (Peucat et al. 1989) point to a short period of crustal evolution. Sm–Nd garnet–whole-rock data for mafic granulites, enderbites and paragneisses of the Shevaroy Hills (Fig. 1) yielded Palaeoproterozoic ages close to 2.35 Ga, whereas two markedly younger ages of 2082 ± 6 and 1722 ± 6 Ma either indicate incomplete resetting or another Palaeoproterozoic stage of high-grade metamorphism (H. Köhler pers. comm.).

The E–W trending Attur Shear Zone (AtSZ; Fig. 1) forms the southern limit of the Shevaroy Hills against the Kollimalai Hills, and the high-strain gneiss terrain of the PCSZ (Fig. 1). The western termination of the Krishnagiri–Salem Zone is also marked by NNE–SSW-trending shear zones, which presumably were active during the Neoproterozoic (Geological Survey of India 1995b; Jayananda & Peucat 1996). Notably, the latter must have witnessed extensive localized fluid flow that led to retrogression of the granulites and the formation of hornblende–biotite(-epidote) gneisses (Geological Survey of India 1995b). In addition, intrusion of carbonatitic, alkaline and ultramafic magmas was widespread, and occurred c. 800 Ma ago (Schleicher et al. 1998 and refs. cited therein).

Proterozoic domains of the Southern Granulite Terrain (SGT)

In agreement with previous models on the lithological make-up of the SGT, two crustal provinces are distinguished: (1) the Madurai Block in the north; and (2) the Kerala Khondalite Belt in the south. The Madurai Block covers the largest portion of the SGT and represents a composite mid- to lower crustal domain. Unlike previous suggestions (Harris et al. 1994), it is bounded to the north by the Bhavani Shear Zone (Meissner et al. 2002). Within the Kerala Khondalite Belt (KKB), three lithotectonic units have been distinguished: the central and southern parts are termed the Ponmudi and Nagercoil Units, as suggested by Srikantappa et al. (1985); the northern part of the KKB, which is roughly identical with the Achankovil Shear Zone of previous authors, displays striking differences in lithology and model ages, and is therefore termed Achankovil Unit (Fig. 1).

Lithologies

Charnockite–enderbite massifs (Fig. 1) Large parts of the Madurai Block are made up of charnockite–enderbite massifs which form the highlands of the Western Ghats (the Cardamom and Varushanad Hills, and the Anamalai and Palni Hills). To the east, in the lowlands of the Madurai Block, hornblende–biotite gneisses and quartzofeldspathic biotite gneisses, which probably represent retrogressed granulites, as well as large granite intrusives, are exposed. Locally, enclaves of metasedimentary rocks occur.

The Kodaikanal Massif predominantly comprises enderbitic two-pyroxene granulites and strongly migmatitic enderbitic gneisses of igneous origin. The emplacement of the quartz–diorite, tonalite and granodiorite gneiss precursors most likely occurred in Palaeoproterozoic times, whereas U–Pb monazite and Sm–Nd garnet–whole-rock ages of c. 550 Ma confirmed that granulite facies metamorphism, migmatization and the development of orthopyroxene-bearing assemblages took place during Pan-African Orogeny (Bartlett et al. 1998; Jayananda et al. 1995a). Synmetamorphic deformation of the gneisses led to isoclinal folding and the development of an E–W-striking and steeply north-dipping foliation (Raith et al. 1997). Numerous enclaves and rafts of intensely migmatized garnet–cordierite–sillimanite gneisses mark the contact with a metasedimentary sequence (quartzites, marbles, pelitic to psammitic granulites) in the lowlands south of the Kodaikanal Massif. Sapphirine-bearing granulites within these enclaves record an ultrahigh-grade segment of the P–T evolution (Grew 1982, 1984; Mohan et al. 1985; Mohan & Windley 1993; Raith et al. 1997; Brown & Raith 1996). In the northeastern part of the Kodaikanal Massif, the gneisses are intruded by the massif-type anorthosite of Oddanchathram (Janardhan & Wiebe 1985).

The Cardamom Hills Massif also consists of polyphase charnockites, enderbites and two-pyroxene granulites (Chacko et al. 1992). Charnockites and enderbites are amphibole- and biotite-bearing but mostly garnet-free. Their massive appearance in the field is indicative of a magmatic origin. Based on major and trace element characteristics, Chacko et al. (1992) inferred an emplacement as dry, orthopyroxene-bearing granites [C-type granitoids of Kilpatrick & Ellis (1992)], causing ultrahigh-temperature metamorphism in the supracrustal rocks of the adjacent KKB. However, field relationships and textural features are not in line with this scenario: the charnockites display a migmatitic gneissic fabric, which indicates that their emplacement preceded high-grade metamorphism and deformation. By contrast, migmatization in the garnet–biotite gneisses of the KKB mainly took place subsequent to the main phase of penetrative deformation. It therefore appears more likely that the charnockite assemblage formed during high-grade metamorphism, i.e. subsequent to magma emplacement.

In the Nagercoil Unit of the KKB, massive charnockites, enderbites and mafic granulites constitute the dominant lithologies, whereas garnet–biotite gneisses and metapelites are subordinate. Field relations and textural features reveal the predominance of magmatic rocks, and geochemical data point to

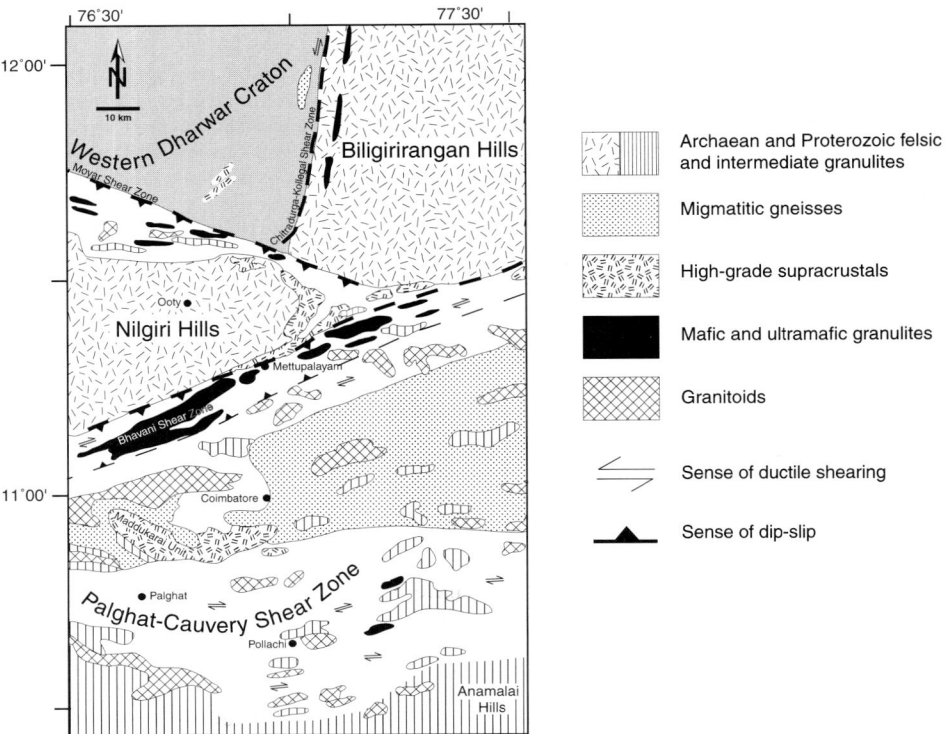

Fig. 2. Geological map of the northwestern part of the Southern Granulite Terrain along a profile from the Dharwar Craton into the Anamalai Hills (after Meissner et al. 2002). The stippled line roughly separates lithologically different domains within the Bhavani Shear Zone. North of it, mafic and ultramafic layered igneous complexes are abundant, and occur along the contact with the Nilgiri Hills. In the southern part they are almost absent and instead intrusive bodies of granitic composition are frequent.

differentiated calc-alkaline volcanic or volcanoclastic protoliths (Srikantappa et al. 1985). As for the Cardamom Hills Massif, a derivation of the charnockites from C-type granitoids has been inferred (Santosh 1996). The relative timing of magma emplacement, deformation and migmatization, however, argues against such an interpretation. Yet, the limited database prohibits firm conclusions as to the origin and evolution of the charnockite–enderbite assemblage.

Gneiss–migmatite terrains (Fig. 1) Biotite and hornblende–biotite-bearing para- and orthogneisses form the dominant rock types in the areas outside the charnockite–enderbite massifs. They host numerous occurrences of metapelitic granulites in the southern and eastern parts of the SGT, whereas the region north of Madurai is characterized by abundant granite intrusions. Between the PCSZ and the Bhavani Shear Zone, the gneiss assemblage often displays imprints of high strain and thus was mapped as 'fissile' or 'sheared gneisses' (Geological Survey of India 1995a,b). Hornblende–biotite and biotite gneisses are regarded as retrogressed equivalents of the granulites that constitute the charnockite–enderbite massifs in the Madurai Block. Geochronological studies indicate that retrogression of the granulites at the western edge of the PCSZ occurred synchronous with transcurrent deformation at c. 521 Ma (Meissner et al. 2002). The currently available limited database does not determine whether this spatial and genetic relationship holds true for the entire Madurai Block or if parts of the hornblende–biotite gneisses represent individual crustal units that originated from distinct protoliths, and thus experienced a different metamorphic evolution.

Interspersed rocks of supracrustal origin comprise garnet–biotite–sillimanite–cordierite and garnet–biotite gneisses, marbles, calc-silicate rocks and quartzites, the abundance and regional distribution of which is not well constrained. Based on LANDSAT imagery, Drury et al. (1984) inferred that a continuous supracrustal unit extends from the

Cauvery Delta towards the Achankovil Shear Zone. By contrast, the geological map of Tamil Nadu (Geological Survey of India 1995b) shows only localized occurrences of metasediments between Karur and Melur and west of Madurai. Field studies have reported further occurrences of metasediments within the Madurai Block. Notably, the slopes of the Kodaikanal Massif are known to host numerous outcrops of metasedimentary rocks (Mohan & Windley 1993). Clinohumite-bearing marbles have been reported from the southern slope of the Varushanad Hills, where they occur as enclaves in charnockitic gneisses (Janardhan et al. 2001). The similarity of these marbles with occurrences at Ambasamudram in the northeastern KKB could have significant implications for the extent of the so-called Achankovil Shear Zone, and the genetic relationship of the two crustal segments north and south of it.

In the northern part of the Palghat Gap (Figs 1 & 2), a sequence of marbles, calc-silicate rocks, pelitic schists and banded iron formations constitutes the Maddukarai Unit (Fig. 2), which previously had been interpreted as lithological equivalent of the Archaean Sargur Group of the Dharwar Craton (Gopalakrishnan et al. 1976). Since these rocks underwent only Pan-African metamorphism (Meissner et al. 2002), this correlation appears unlikely.

The Achankovil and Ponmudi Units of the KKB form the largest metasedimentary sequences of the SGT. Both units experienced polyphase ductile deformation, the details of which still need further investigation. Structural imprints of early deformation are largely erased by the pervasive deformation that accompanied Pan-African high-grade metamorphism. In pelitic and semi-pelitic garnet–biotite (–sillimanite–cordierite) gneisses, a system of NW–SE-striking open to isoclinal folds has developed. The dip of the NE-vergent axial planes systematically increases from south to north within the KKB, ending in upright folds in the Achankovil Shear Zone (authors' observation).

In the Achankovil Unit, garnet–biotite and garnet–biotite–sillimanite(–cordierite) gneisses predominate. In addition, cordierite gneisses, marbles, quartzites and pinkish granites, all of which are virtually absent in the other units of the KKB, occur in significant amounts. Strongly migmatitic cordierite–garnet–orthopyroxene gneisses are abundant in the northwestern part of the Achankovil Unit. These rocks are exceptional within the metasedimentary units of the KKB, not only because of their mineral assemblage (Sinha Roy et al. 1984; Santosh 1987; Nandakumar & Harley 2000; Cenki et al. 2002) but also due to their young Nd model ages of 1.4–1.3 Ga (Harris et al. 1994; Brandon & Meen 1995; Bartlett et al. 1998). Quartzite ridges, often in close spatial association with pinkish granites, apparently concentrate in the eastern portion of the Achankovil Unit.

The Ponmudi Unit consists predominantly of migmatitic garnet–biotite gneisses and pelitic granulites (the so-called khondalites) (Figs 1 & 2). Marbles, calc-silicate rocks and mafic granulites form minor lithological components. Among these, the humite-bearing marbles of Ambasamudram became well known and were investigated petrologically in great detail (Krishnanath 1981; Rosen et al. 1997; Satish Kumar et al. 2001). The garnet–biotite gneisses were subdivided into a sodic (K_2O/Na_2O <1) and a potassic group (K_2O/Na_2O>1) (Chacko et al. 1992). Sodic gneisses are fine to medium grained, contain alkali feldspar or plagioclase, quartz and biotite. Potassic gneisses are usually coarse grained and sometimes display a conspicuous augen texture, which is defined by alkali feldspar porphyroclasts. A characteristic feature of all garnet–biotite gneisses is the development of garnet-bearing in situ leucosomes and the generation of leucogranitic melts through biotite dehydration–melting (Braun et al. 1996). Leucogranites are occasionally extremely enriched in fluorapatite and monazite (Braun & Raith 1996; Rajesh & Santosh 1996a). Arrested charnockite formation in garnet–biotite gneisses and pelitic granulites, and localized bleaching of charnockites, are common features in the Ponmudi Unit (Ravindra Kumar & Chacko 1986; Santosh & Yoshida 1986; Raith & Srikantappa 1993 and refs cited therein). Principally, the formation of arrested charnockite domains was structurally controlled. Local coalescence of the isolated arrested charnockite domains led to the development of dense networks and even to apparently massive charnockite exposures. True massive charnockites, however, are rare. Most workers now agree that a reduction in H_2O activity through influx of CO_2 was responsible for the growth of orthopyroxene at the expense of biotite. However, it is still debated whether the CO_2 was released locally from fluid inclusions, or derived from mantle fluids or graphite-bearing lithologies [as reviewed in Raith & Srikantappa (1993)]. Recently, the idea of K-metasomatism has been introduced as an alternative explanation for the generation of arrested charnockite (Fonarev et al. 2000).

Chacko et al. (1992) concluded that the gneisses of the Ponmudi Unit are entirely of sedimentary origin, proposing that the precursor sediments of the potassic gneisses were derived from the charnockites and enderbites of the Cardamom Hills Massif. In contrast, the magmatic origin of the porphyritic potassic gneisses was substantiated from field observations and electron probe microanalysis (EPMA) monazite dating (Braun et al. 1998; Braun & Bröcker 2001).

Neoproterozoic igneous rocks Syenites and granites form small plutons of <40 km² which were emplaced into the crystalline basement of the SGT along lineaments [reviewed in Rajesh & Santosh (1996b)]. Their modal composition ranges from granite to alkali syenite and minor foid-bearing monzosyenite. Common mafic minerals are alkali-amphibole, biotite, titanite and apatite; clinopyroxene and garnet occur occasionally. The generation of these intrusives has been attributed to extensional tectonics (Rajesh & Santosh 1996b) and the available geochronological data suggest that their emplacement took place between 770 and 560 Ma (Odom 1982; Kovach et al. 1998; Ghosh 1999). In the eastern part of the Achankovil Unit, between Tirunelveli, Ambasamudram and Tenkasi (Fig. 3), several small granite intrusive bodies occur spatially associated to quartzite ridges and marbles. The fine-grained rocks are meta-aluminous (ASI (Aluminium Saturation Index) = 0.99–1.10) and contain garnet, amphibole or biotite as mafic phases (authors' observations). Field relations indicate a pre- to synkinematic magma emplacement.

The massif-type anorthosite of Oddanchathram at the northern slope of the Kodaikanal Massif is coarse grained and consists of labradoritic plagioclase with minor clino- and orthopyroxene, magnetite and ilmenite (Narasimha Rao 1964; Janardhan & Wiebe 1985; Janardhan 1996). Garnet and orthopyroxene are restricted to the contact with host enderbitic gneisses, and formed in response to granulite facies regional metamorphism. SHRIMP work on zircons from samples from the central, undeformed part of the intrusion yielded an average $^{206}Pb-^{238}U$ emplacement age of 563±9 Ma (Ghosh 1999).

Boundaries

Several crustal-scale shear zones have been postulated or identified within the SGT (Drury & Holt 1980; Drury et al. 1984; Chetty 1996; Reddi et al. 1988; Ghosh et al. 1998). However, few of these have been studied in detail and the published results are still equivocal.

The Bhavani Shear Zone (BSZ) bounds the Nilgiri and BR Hills to the south. Together with the Moyar Shear Zone, which mainly shows reverse displacement of the Nilgiri with respect to the Dharwar Craton (Raith et al. 1999), it separates the pervasively Archaean from the predominantly Proterozoic parts of southern India.

Recent investigations provide a comprehensive set of structural, petrological and geochronological data of some shear zones in the SGT (Ghosh 1999; Meissner et al. 2002). These studies revealed that the BSZ was already active during the Late Archaean and became reactivated during the Neoproterozoic. Kinematic indicators point to dextral oblique-slip movement within the shear zone and reverse faulting of the southern block against the Nilgiri Hills (Naha & Srinivasan 1996; Meissner et al. 2002). This led to

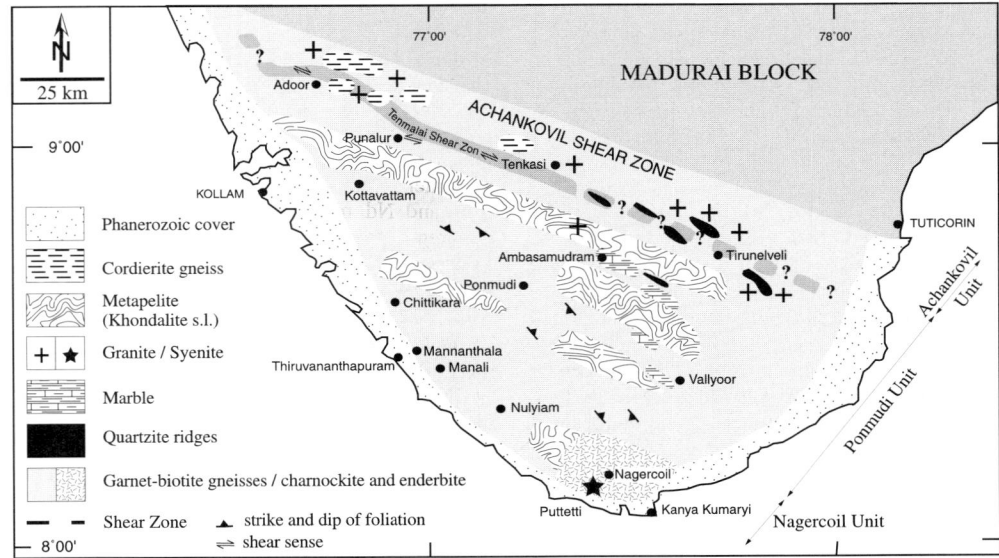

Fig. 3. Schematic geological map of the Kerala Khondalite Belt showing the subdivision into Achankovil, Ponmudi and Nagercoil Units [compiled from Sinha Roy et al. (1984), Chacko et al. (1992), Geological Survey of India (1995a,b), Sacks et al. (1997), Rajesh et al. 1998 and author's own data)]. The position and extension of the Tenmalai Shear Zone has been taken from Sacks et al. (1997).

the exhumation of high-P mafic granulite–enderbite complexes along the southern slopes of the Nilgiri Hills, which are regarded as remnants of Late Archaean oceanic crust and thus would mark a suture (Raith et al. 1999). The eastward continuation of the BSZ and its relation with other major shear zones of the SGT, notably those entering the Eastern Dharwar Craton, remain to be explored.

The E–W trending PCSZ was first identified from LANDSAT imagery and interpreted as a crustal-scale shear zone (Drury & Holt 1980). Field observations indicate a strongly heterogeneous distribution of strain within the shear zone and an overall dextral sense of shearing (Drury & Holt 1980; Radhakrishna 1989; Ramakrishnan 1993; D'Cruz et al. 2000). Other workers have questioned pervasive shearing along the whole PCSZ but did not completely discard shear movements on a local scale (Mukhopadhyay et al. 2001). Geophysical studies across the western termination of the PCSZ, known as the Palghat Gap, showed significant gradients in gravity, electric resistivity and seismic reflectivity (Bose & Kartha 1977; Subrahmanyam & Verma 1986; Mishra 1988), which coincide perfectly with a marked topographic depression. Retrogression of enderbites and charnockites to hornblende–biotite gneisses due to fluid infiltration is a widespread feature in the PCSZ. Structural studies indicate that it postdates deformation and shearing, which according to Sm–Nd and Rb–Sr mineral ages is related to the Pan-African tectonothermal event (Mukhopadhyay et al. 2001, Meissner et al. 2002).

Based on structural and geochronological data, Ghosh et al. (1998) concluded that the Archaean basement of the Dharwar Craton extends farther south into the SGT and is delimited by the KKPTSZ (Fig. 1). Accordingly, plutonic complexes such as the Oddanchathram Anorthosite were emplaced along this tectonic lineament c. 560 Ma ago. U–Pb zircon upper concordia intercept and concordant U–Pb monazite ages of 804 ± 31 and 791 ± 17 Ma from a charnockite and a khondalite sample north and NW of Dindigul led to the suggestion that the granulite-gneiss terrain south of the KKPTSZ represents the southern continuation of the Eastern Ghats Belt. Clearly, additional detailed studies are necessary to substantiate these far-reaching interpretations. Their conclusion that charnockitization of gneisses north of the KKPTSZ occurred in Late Archaean time (Ghosh et al. 1998), however, is contradicted by results of structural and geochronological studies which demonstrated a late Pan-African age for this process (Mukhopadhyay et al. 2001; Meissner et al. 2002).

The WNW–ESE trending Achankovil Lineament, first recognized on LANDSAT images, has been interpreted as a major shear zone separating two structurally different crustal blocks (Drury et al. 1984). In contrast, the aeromagnetical map of southern India (Reddi et al. 1988) distinguishes crustal domains within the Madurai Block and the KKB, whose boundaries neither match the lineament pattern of the SGT (Drury et al. 1984; Chetty 1996) nor the location of the Achankovil Shear Zone Field studies did not succeed in localizing the margins of the shear zone and provided controversial results on its kinematics (Radhakrishna et al. 1990; Sacks et al. 1997, 1998; Rajesh et al. 1998). Recently, Sacks et al. (1997) distinguished the Tenmalai Shear Zone (TSZ) within the Achankovil Shear Belt as the locus of major dextral displacement at granulite facies conditions, while Rajesh et al. (1998) reported sinistral shearing from an area south of the TSZ, which led to the suggestion of repeated movement (Sacks et al. 1998). Obviously, detailed mapping and structural work is necessary to obtain comprehensive knowledge on the structural evolution of this part of the SGT and to provide unequivocal evidence for or against the existence of a shear zone at the boundary between the KKB and the Madurai Block.

Crust-forming processes and products

Nd (T_{DM}) model ages for the SGT, except the Achankovil Unit, range between 2 and 3 Ga (Choudhary et al. 1992; Harris et al. 1994; Bartlett et al. 1995, 1998; Brandon & Meen 1995; Jayananda et al. 1995a; Meissner et al. 2002). Late Archaean crustal growth has been suggested by Ghosh (1999) on the basis of U-Pb zircon upper concordia intercept ages of 2.9–2.5 Ga for granitic gneisses of the Kollimalai Hills. Single zircon evaporation ages of 2436 ± 4, 2438 ± 12 (Bartlett et al. 1995, 1998) and 2115 ± 8 Ma (Jayananda et al. 1995a) from metagranites and a charnockite of the Kodaikanal Massif were interpreted as intrusion ages. They are in good agreement with zircon upper concordia intercept ages and Nd model ages from similar rocks of this region, and thus indicate crustal growth in Palaeoproterozoic time (Harris et al. 1994; Jayananda et al. 1995a; Ghosh 1999).

Geochronological and isotope data suggest that the Pan-African event in the SGT was essentially a period of crustal reworking. Nd isotopic systematics for metasedimentary rocks indicate derivation of the clastic material from isotopically contrasting source regions. Metapelites from the Maddukarai Unit have Nd (T_{DM}) model ages of 2.0–1.9 Ga and ε_{Nd} values of between -8.2 and -9.9 (at 550 Ma). Model ages for cordierite gneisses and calc-silicate rocks from the Achankovil Unit are younger and range from 1.4 to 1.3 Ga, with ε_{Nd} values of -3.1– -6 (Harris et al. 1994; Brandon & Meen 1995; Bartlett et al. 1998). Although Sm–Nd fractionation during partial melting may have occurred, these distinct Nd isotope systematics suggest derivation of the sediments from Palaeo-, Meso- and Neoproterozoic source regions, respectively, which are not known from the exposed crust of Peninsular India.

The emplacement of alkali granitic and carbonatitic magmas probably represents the only important episode of Neoproterozoic crustal growth in the SGT and geochronological investigations point to intrusion and cooling between 800 and 516 Ma (Odom 1982; Santosh & Drury 1988; Soman et al. 1995; compilation in Rajesh & Santosh 1996b; Miller et al. 1997; Kovach et al. 1998; Schleicher et al. 1998; Ghosh 1999). However, the only published Nd model age for an alkali granite from the Madurai Block yields a value of c. 2 Ga, indicating crustal reworking (Harris et al. 1994).

Geochronology

Although recent geochronological studies point to a polyphase magmatic and metamorphic evolution of the SGT, the available data do not yet allow discrimination between distinct episodes. Few Archaean Sm-Nd whole-rock isochron ages of c. 2.9 Ga were obtained from layered mafic intrusions within the BSZ and PCSZ by Bhaskar Rao et al. (1996), who consider them to reflect a high-grade metamorphic event that followed shortly after magma emplacement. Meissner et al. (2002), however, argue that complete resetting of the whole-rock Sm-Nd systematics during metamorphism is unlikely and that the Sm-Nd ages would rather reflect the time of magma intrusion.

Additional Archaean ages were obtained from rocks of the BSZ and along a N-S transect between Salem and Karur (Fig. 1; Buhl 1987; Ghosh 1999). SHRIMP U-Pb zircon ages of 3.0-2.9 Ga from a mafic granulite north of Karur closely resemble the values reported by Bhaskar Rao et al. (1996) for similar rocks of the BSZ and are interpreted as emplacement ages. Some of the analysed zircons have rims which yield nearly concordant ^{207}Pb-^{206}Pb ages of c. 2.5 Ga. This value is identical with ^{207}Pb-^{206}Pb zircon and monazite upper concordia intercept ages obtained from granitic and tonalitic gneisses of the BSZ and the area between Karur and Salem (Ghosh 1999). Within this group of Late Archaean ages, Ghosh (1999) further distinguished two groups with ages of 2.54-2.51 and 2.51- 2.50 Ga, which he interprets as intrusion ages of the granitic precursors and a subsequent stage of charnockitization. Field relationships show that the gneisses intruded into Archaean metasediments and mafic granulites. As has been previously shown for the Moyar Shear Zone (Raith et al. 1999), this also shows that along the BSZ and the regions further to the east, Archaean basement has been involved in deformation and metamorphism (Ghosh 1999).

Palaeoproterozoic ages have been reported from different lithologies and localities throughout the SGT, the interpretation of which is not always straightforward. Sm-Nd garnet-whole-rock data for mafic granulites, enderbites and paragneisses of the Shevaroy Hills yielded ages close to 2.35 Ga. Two markedly younger ages of 2082±6 and 1722±6 Ma could indicate either incomplete resetting or a further Palaeoproterozoic episode of high-grade metamorphism (H. Köhler pers. comm.).

Pb-Pb single zircon evaporation ages of 2438±12, 2436±4 (Bartlett et al. 1995) and 2115±8 Ma (Jayananda et al. 1995a) for orthogneisses of the Kodaikanal Massif were interpreted as intrusion ages, and apparently indicate a Palaeoproterozoic episode of crustal growth. EPMA and U-Pb TIMS (Thermal Ionization Mass Spectometry) dating of monazites and zircons from metasedimentary rocks of the KKB also yielded ages of between 2.2 and 2.0 Ga (Buhl 1987; Soman et al. 1995; Bartlett et al. 1998; Braun et al. 1998), which may either provide an upper time constraint for sedimentation or date an early episode of high-grade metamorphism.

Single zircon evaporation, Sm-Nd garnet-whole-rock and EPMA monazite ages from charnockites and metasedimentary gneisses of different localities in the SGT range from 1.9 to 1.7 Ga, and thus may point to a hitherto unknown Late Palaeoproterozoic high-grade metamorphic event (Choudhary et al. 1992; Jayananda et al. 1995a; Bartlett et al. 1998; Braun et al. 1998; Meissner et al. 2002; authors' unpublished data).

Similar geochronological results suggest a common tectonometamorphic evolution of the KKB and the Madurai Block in Pan-African times. Clear evidence for the prevalence of Pan-African metamorphism (650-550 Ma) associated with granitic magmatism (590-525 Ma) in the entire Madurai Block has been provided by Hansen et al. (1985), Jayananda et al. (1995a), Miller et al. (1997), Bartlett et al. (1998) and Ghosh (1999). From the KKB, a similar range of ages (560-516 Ma) has been obtained from U-Pb zircon and monazite, Sm-Nd mineral-whole-rock and EPMA monazite dating (Buhl 1987; Choudhary et al. 1992; Soman et al. 1995; Unnikrishnan-Warrier et al. 1995; Miller et al. 1997; Unnikrishnan-Warrier 1997; Bartlett et al. 1998; Braun et al. 1998). Pb-Pb whole-rock isochron ages of 509±25 and 523±32 Ma obtained from fluorapatite and monazite of leucogranites (Braun et al. 1996; Berger & Braun 1997) point to protracted melt generation and crystallization in the KKB, which is further supported by results of U-Pb monazite dating of granitic gneisses (Braun & Bröcker 2001).

A concordant ^{207}Pb-^{206}Pb zircon age of 548±2 Ma from the charnockitized Kalipara Granite

(NE Achankovil Unit) is interpreted as an intrusion age and thus gives a lower limit for the charnockitization process in the KKB (Ghosh 1999). A granitic dyke, which caused bleaching of the charnockitized Kalipara Granite, yielded a nearly concordant ^{207}Pb–^{206}Pb zircon age of 526 ± 3 Ma, identical to a concordant ^{207}Pb–^{206}Pb zircon age of 526 ± 1 Ma from another granite dyke in the same area. From this it follows that arrested charnockite formation must have occurred within a rather narrow time span of c. 20 Ma. Sm–Nd mineral–whole-rock isochron ages from arrested charnockites of the Ponmudi Unit (Fig. 3) range from 558 to 517 Ma and confirm that *in situ* charnockitization postdated high-grade metamorphism and migmatization of garnet–biotite gneisses (Choudhary *et al.* 1992; Unnikrishnan-Warrier *et al.* 1995). In contrast, Köhler *et al.* (1993) report significantly younger ages of c. 485 Ma from a small slab study of a gneiss–charnockite pair. These data are in good agreement with Rb–Sr mineral–whole-rock cooling ages (484–440 Ma)(Choudhary *et al.* 1992; Unnikrishnan-Warrier *et al.* 1995; Unnikrishnan-Warrier 1997), K–Ar mica ages for pegmatites (474–445 Ma; Soman *et al.* 1982), and results of EPMA monazite data from pegmatites and gneisses (520–420 Ma; Braun *et al.* 1998; authors' unpublished data).

Geochronological studies have also documented repeated activity of major shear zones in the northern segment of the SGT (Ghosh 1999; Meissner *et al.* 2002). Meissner *et al.* (2002) showed that the Moyar Shear Zone, which defines the southern boundary of the Western Dharwar Craton, was reactivated during the Neoproterozoic. Reactivation was associated with amphibolite facies retrogression and occurred at c. 614 Ma (Sm–Nd garnet–whole-rock age); this was followed by cooling to below c. 320°C between 610 and 540 Ma (Rb–Sr biotite age). In the BSZ, a Sm–Nd mineral–whole-rock age of c. 726 Ma points to resetting of the isotope system at small scales, approximately constraining the time of early Pan-African shear zone reactivation. Ghosh (1999) obtained a U–Pb zircon age of 601 ± 1 Ma for a pegmatite dyke from the BSZ (SE of Mettupalayam; Fig. 2). Sm–Nd garnet–whole-rock and Rb–Sr mica–whole-rock data for dolerites, pegmatites, and sheared granites and gneisses from the same area yielded slightly younger ages (Meissner *et al.* 2002). Growth of coronitic garnet in a metadolerite at 552 ± 13 Ma is attributed to the Pan-African thermal event; a post-kinematic garnet-bearing pegmatite emplaced at 513 ± 5 Ma provides a lower time constraint for ductile deformation. The shape of the cooling path subsequent to amphibolite facies retrogression was determined by Rb–Sr dating of micas from late pegmatites, and yielded ages of 521–472 (Bt (biotite)) and 504 ± 13 Ma (Ms(muscovite)) (Meissner *et al.* 2002). TIMS U–Pb dating of zircons from the PCSZ yielded two Pan-African ages of 600 ± 4 and 522 ± 3 Ma, interpreted as the time of granitic magmatism and subsequent resetting due to hydrothermal activity, respectively (Ghosh 1999). The older date corresponds reasonably well with Sm–Nd garnet–whole-rock ages of 610 ± 60 and 560 ± 17 Ma for metapelites from the Maddukarai Unit, whereas the younger age agrees with Sm–Nd garnet–whole-rock ages of 521 ± 8 and 491 ± 52 Ma from the same unit (Meissner *et al.* 2002). Rb–Sr biotite ages are uniform and indicate cooling below c. 300°C between 488 and 477 Ma (Meissner *et al.* 2002; R. Mallick & H. Köhler pers. comm.).

In conclusion, the data summarized above distinguish two clearly separated Palaeoproterozoic and Neoproterozoic–Early Phanerozoic age populations. The Pan-African event is well constrained and recognized in different domains of the SGT. Currently, it remains uncertain whether the Palaeoproterozoic population corresponds to an early episode of high-grade metamorphism, as ages of between 1.6 and 0.7 Ga are rare in the SGT. A poorly defined group of zircon evaporation ages of between 1100 and 900 Ma (Bartlett *et al.* 1998) and EPMA monazite data of 1.5–0.8 Ga (Braun *et al.* 1998) most likely reflect partial resetting of Palaeoproterozoic ages. By contrast, Ghosh (1999) reported a U–Pb upper concordia intercept age of 987 ± 65 Ma for a granite gneiss xenolith within the 548 ± 2 Ma Kalipara Granite of the KKB, which he interpreted as intrusion age of the gneiss precursor. A nearly concordant ^{207}Pb–^{206}Pb zircon age of 796 ± 1 Ma for a granite sample and a mean ^{206}Pb–^{238}U SHRIMP monazite age of 791 ± 17 Ma for a khondalite sample (Ghosh 1999) are in good agreement with the intrusion age of the Sevattur Carbonatite (Schleicher *et al.* 1997). These data may indicate a widespread and previously unrecognized thermal and magmatic event in the SGT. U–Pb zircon ages of c. 740 Ma for the Munnar Alkali Granite (Odom 1982) and 722 ± 13 Ma for a mafic granulite point to a short-lived bimodal magmatic pulse preceding the Pan-African reworking.

Metamorphic Evolution

Temperature estimates from charnockites, sapphirine-bearing granulites and calc-silicate rocks of the Kodaikanal Massif range from 900 to 1000°C, providing compelling evidence for ultrahigh temperature (UHT) metamorphism in these parts of the Madurai Block (Brown & Raith 1996; Raith *et al.* 1997; Satish-Kumar 2000). The breakdown of orthopyroxene–sillimanite assemblages and the replacement of kyanite by sillimanite in sapphirine-granulites indicates maximum pressures of

10–12 kbar for the Kodaikanal Massif. The $P-T$ evolution of sapphirine-bearing high-Mg–Al granulites in the Kodaikanal Massif is characterized by two stages of isothermal decompression with an intermediate stage of isobaric cooling (Raith et al. 1997). The authors further suggest that the two stages of isothermal decompression reflect two periods of crustal uplift, possibly related to Palaeo- and Neoproterozoic metamorphic events.

In the Ponmudi and Achankovil Units of the KKB, $P-T$ conditions have been assessed with Fe–Mg exchange and feldspar thermometry, Al-in-orthopyroxene thermobarometry, as well as the construction of petrogenetic grids for charnockites, cordierite gneisses, pelitic gneisses and garnet-biotite gneisses (Braun et al. 1996; Chacko et al. 1996; Satish-Kumar & Harley 1998; Nandakumar & Harley 2000; Cenki et al. 2002). The results of these studies indicate peak temperatures of c. 900°C at pressures of at least 5–6 kbar. Previously inferred high-pressure conditions for cordierite gneisses of the Achankovil Unit (Santosh 1987) were not confirmed by subsequent work (Nandakumar & Harley 2000; Cenki et al. 2002). Instead, reaction textures and phase equilibria point to maximum pressures of c. 8 kbar (Cenki et al. 2002). Chacko et al. (1987, 1996) and Nandakumar & Harley (2000) report a $P-T$ gradient with systematically increasing pressures and temperatures from 5 kbar and 750°C in the central part of the Ponmudi Unit up to 8–9 kbar and 1000°C towards the charnockite massifs of the Cardamom Hills and the Nagercoil Unit. Chacko et al. (1996) and Santosh (1996) suggested that the temperature gradient was caused by the thermal input from the emplacement of C-type magmas which, according to field relations and geochronological data, appears to be unlikely.

Santosh (1987) inferred a clockwise $P-T$ path with a first stage of near-isothermal decompression from high pressures, followed by a stage of near-isobaric cooling for cordierite gneisses of the Achankovil Unit. Results of subsequent studies on migmatitic cordierite gneisses, charnockites and calc-silicate rocks from the northern Ponmudi Unit and the Achankovil Unit, however, imply initial near-isobaric cooling to 700–800°C at 6 kbar, followed by a stage of near-isothermal decompression down to 3 kbar (Satish-Kumar & Harley 1998; Fonarev et al. 2000; Nandakumar & Harley 2000; Cenki et al. 2002).

Evidently, the units of the SGT are characterized by distinct tectonometamorphic histories. Yet, details of the $P-T$ evolution, the duration of high-grade metamorphism, and the relationship between metamorphism, magmatism and deformation are poorly constrained. Further integrated studies in other parts of the SGT may reveal the existence of other units with $P-T$ evolutions different from those discussed above for the Kodaikanal Massif and the KKB. The recognition of a final stage of isothermal decompression in both crustal domains, however, may be taken as evidence for their common metamorphic evolution during the Pan-African Orogeny.

Geological evolution of the Southern Granulite Terrain (SGT)

The charnockite–enderbite massifs, which make up the highlands and essentially consist of igneous rocks, presumably formed during an episode of Late Archaean–Early Palaeoproterozoic magmatism. Hornblende–biotite and biotite gneisses, which predominate in the eastern part of the SGT, may represent retrogressed amphibolite facies equivalents of the highland granulites. Some of these gneisses may also constitute distinct crustal units that were added to the charnockite–enderbite massifs in Proterozoic times. The supracrustal granulites of the Ponmudi Unit originated from a shallow-marine pelite–greywacke–carbonate–quartzite association. Low $\delta^{18}O$ values for garnet–biotite gneisses (8.5–10‰: authors' unpublished data) indicate rather immature sediments. Nd model ages of 2.6–1.9 Ga contrast with Nd model ages of 1.4–1.3 Ga for paragneisses of the Achankovil Unit. This may reflect derivation of the precursor sediments from distinct provenance areas or deposition at considerably different times. Dating of detrital zircons is required to further constrain the time of deposition.

The nature of the Palaeoproterozoic event remains speculative and whether it represents a magmatic or a high-grade metamorphic event, or both, requires further isotopic investigation. By contrast, the Pan-African pervasive metamorphism is well documented through petrological and geochronological data for the entire SGT. The recognition of a clockwise $P-T$ path points to collisional tectonics as the driving force of metamorphism. The emplacement of alkaline magmas and an inferred bimodal magmatic suite between 800 and 570 Ma may indicate a preceding stage of crustal extension.

The final assembly of the SGT and its amalgamation with the Archaean blocks to the north was associated with granulite facies metamorphism and subsequent upper amphibolite facies retrogression in Neoproterozoic times. The entire terrain experienced a final stage of isothermal decompression with localized fluid ingress and arrested charnockite formation. Subsequent isobaric cooling, recognized in rocks from the Palni Hills and the KKB proceeded until temperatures of c. 300°C were reached during Ordovician times.

Table 1. *Correlation of nomenclature for the Sri Lankan basement*

Vitanage (1972)	Cooray (1978)	Geological Survey of Sri Lanka (1982)	Cooray (1994), this paper	Abbreviation (model age/Ga)*
East Vijayan	East Vijayan	Vijayan	Vijayan Complex	VC (1.0–1.9)
West Vijayan	West Vijayan	,,	Wanni Complex	WC (1.0–2.0)
Highland Series	Highland Series	Highland Series	,,	,, (1.0–2.0)
,,	,,	,,	Highland Complex†	HC (2.0–3.0)
Southwest Group	Southwest Group	,,	,,	,, (2.0–3.0)
Arenas	Highland Series	,,	Kadugannawa Complex	KC (1.0–2.0)

* Nd model age range of Milisenda *et al.* (1988).
† Highland Southwest Complex of Kröner *et al.* (1991).

Sri Lanka

Units and lithologies

Since the first comprehensive review of Sri Lankan geology by Adams (1929), many subdivisions of the Sri Lankan basement rocks have been advanced (e.g. Coates 1935; Wadia 1942; Cooray 1962, 1984, 1994) and names of basement units have varied considerably. The knowledge of the basement geology of Sri Lanka increased rapidly in the period from 1987 to 1995, largely due to the efforts of Japanese–Sri Lankan and German–Sri Lankan cooperative projects (Hiroi & Motoyoshi 1990; Kröner 1991*b*; Raith & Hoernes 1994). As a result, the nomenclature of rock units and the definition of boundaries have seen important revisions (e.g. Kröner *et al.* 1991; Cooray 1994). Cooray's (1994) nomenclature (Table 1) will be followed in this paper.

Cooray (1994) distinguished four basement units: the Highland Complex (HC; Fig. 4), the Wanni Complex (WC), the Kadugannawa Complex (KC) and the Vijayan Complex (VC). The HC and WC consist mainly of interbedded metapelites, quartzites, marbles, granitic gneisses, migmatites and charnockites, whereas metabasites are abundant in the HC only (see more detailed descriptions below). The KC in central Sri Lanka, which corresponds to the Arenas of Vitanage (1972), comprises meta-igneous gneisses with chemistries ranging from gabbroic and dioritic to granitic (Stosch 1991; Voll & Kleinschrodt 1991). Metasediments are rare in this unit, comprising thin marble and calc-silicate horizons, thin quartzites and itabirites (Kriegsman 1994). The VC, exposed in eastern Sri Lanka (Fig. 4), consists of meta-igneous gneisses of tonalitic to leucogranitic composition (Milisenda 1991), with minor metaquartzites and calc-silicate rocks (Dahanayake & Jayasena 1983). The HC rocks yielded relatively old Nd model ages (2–3 Ga; Milisenda *et al.* 1988; Liew *et al.* 1991*b*), whereas rocks from the other units give significantly lower values (1–2 Ga; Milisenda *et al.* 1988; Liew *et al.* 1991*b*). Metamorphic grade reaches granulite facies throughout the HC, WC and KC, although locally retrogressed to amphibolite facies, whereas the VC rocks generally display amphibolite facies assemblages. Thermometry shows that this is not merely due to differences in fluid activity, but reflects real temperature differences (Schumacher *et al.* 1990). Late-stage, localized charnockitisation affected all units (WC – Hansen *et al.* 1987; HC – Kehelpannala 1991; VC–Kriegsman, pers. comm.).

The Southwest Group defined by earlier workers (e.g. Hapuarachchi 1983) is now considered to be part of the HC (Kröner *et al.* 1991; Cooray 1994), because lithological differences with the HC are of minor importance and petrological studies have shown a similar *P–T* evolution (e.g. Perera 1984; Faulhaber & Raith 1991). Two important differences remaining, however, are the lower abundance of metabasites and the generally lower peak pressures recorded in the Southwest Group. The following lithological descriptions are based on an average, idealized tectonostratigraphy of basement units in Sri Lanka. In view of multiple intrusion events, intense folding and strong flattening (Almond 1994; Kröner *et al.* 1994*b*), it is neither intended nor possible to give a detailed stratigraphy prior to these events. For example, the large number of marble layers in the HC may simply be due to repetition of a few original layers by thrust stacking and folding (Kröner *et al.* 1991).

Highland Complex (HC) The HC consists of *c.* 50% of metasedimentary rocks and 50% of meta-igneous gneisses (Kröner *et al.* 1991). The most obvious metasediments are marble, orthoquartzite and metapelite (Table 1). Minor constituents are calc-silicate rocks, either at the contact between marble and adjacent lithologies or as boudins within marble or orthoquartzite. Sapphirine-bearing, highly magnesian granulites have been recognized at four

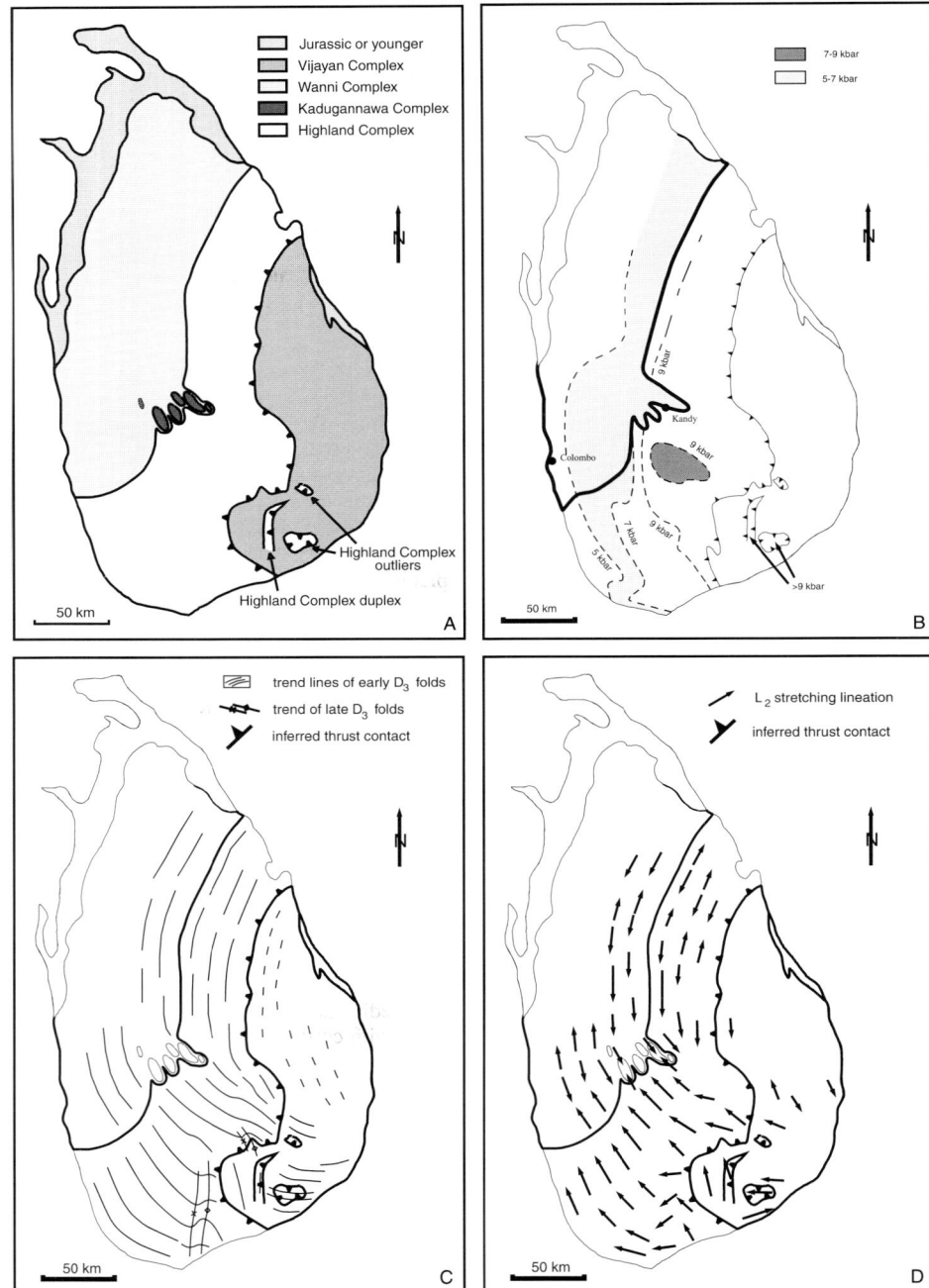

Fig. 4. Map of Sri Lanka. **A**, Units; **B**, *P–T* data after Schumacher *et al.* (1990), Faulhaber & Raith (1991) and Schumacher & Faulhaber (1994); **C**, trends of D3 folds; **D**, trends of stretching lineations (after Kriegsman 1993).

localities (Kriegsman & Schumacher 1999; Fernando et al. 2000). This supracrustal sequence also contains the metamorphic equivalents of what has been interpreted as a bimodal volcanic suite (Pohl & Emmermann 1991). These authors proposed deposition along a rifted continental margin that possibly evolved into a stable shelf region (Dissanayake & Munasinghe 1984). Meta-igneous rocks include granitic, charnockitic, and enderbitic and metabasic gneisses. Some rock types are difficult to classify as either metasedimentary or meta-igneous, notably so-called grey gneisses, which mainly consist of quartz, feldspar and biotite.

Radiometric age data from the HC show a prolonged crustal history in the time span of 3500–450 Ma (e.g. Hölzl et al. 1991, 1994) (Fig. 5). SHRIMP U–Pb ages of detrital zircons are in the range of 3.2–2.4 Ga (Kröner et al. 1987). These data and U–Pb zircon upper intercept ages of metapelites of 2.1–1.9 Ga have been interpreted as indicating deposition 1.9 Ga ago (Fig. 5; Hölzl et al. 1991, 1994). Nd model ages of 2–3 Ga and U–Pb ages of detrital zircons suggest that these sediments contained components with a prolonged previous crustal history (Milisenda et al. 1988; Liew et al. 1991a,b).

U–Pb zircon upper intercept ages of meta-igneous gneisses generally fall in the range of 1.95–1.85 Ga, which has been interpreted as the intrusion age of the precursor magmas (Baur et al. 1991; Hölzl et al. 1991, 1994; Kröner & Williams 1993). The only exceptions are a metabasite with a poorly defined upper intercept of c. 2.3 Ga (Hölzl et al. 1991, 1994) and a granitic gneiss from the Polonnaruwa area, which probably intruded c. 670 Ma ago (Baur et al. 1991). Cross-cutting relations between meta-igneous rocks and metasediments are extremely scarce. This has been attributed to either intrusion in the form of sills (Voll & Kleinschrodt 1991) or to high strain obliterating intrusive contacts (Kriegsman 1991, 1993; Kröner et al. 1991, 1994b).

Kadugannawa Complex (KC)

Rocks of the KC (Fig. 4) are exposed in the cores of six doubly plunging D3 synforms and one intervening antiform in the Kandy area (Vitanage 1972; Munasinghe & Dissanayake 1980), and structurally overlie the gneisses of the HC. They have been considered as the basal unit of the WC by some authors (e.g. Kehelpannala 1997). The main rock types are biotite–hornblende and biotite gneisses, with amphibolites and minor metasediments of quartzofeldspathic, quartzitic and pelitic composition (e.g. Vitanage 1972; Perera 1983; Cooray 1984; Kröner et al. 1991). Kleinschrodt et al. (1991) gave a detailed lithogical description of the KC and described a basic layered intrusion at the deepest levels (see also Stosch 1991). This intrusion is locally overlain by a thin metasedimentary sequence comprising marble and calc-silicate horizons, quartzites and itabirites (Kriegsman 1994), and by dioritic to tonalitic gneiss elsewhere. Kriegsman (1994) described a sequence of rocks typified by elongate amphibolitic lenses in a calc-silicate matrix, which may represent highly deformed pillow lavas. Pohl & Emmermann (1991) propose that all meta-igneous rocks of the KC constitute a single calc-alkaline suite. Zircon evaporation U–Pb data indicate intrusion between c. 1010 and 890 Ma (Kröner et al. 1991).

Directly underlying the basic intrusion in most localities is a pink, locally gneissose granite of typical S-type affinity (Hölzl et al. 1994), which has a similar Nd model age to rocks from the KC (Milisenda et al. 1988). Kröner et al. (1987) reported SHRIMP zircon U–Pb ages of c. 1100 Ma, whereas Hölzl et al. (1991), using the conventional multi-grain U–Pb technique, showed a concordant age of c. 580 Ma at a different locality. Two possible interpretations are: (1) that the ages c. 1100 Ma represent the intrusive age, consistent with the age spectrum from zircon evaporation U–Pb data (above) and 580 Ma is the time of partial remelting of these granitoids; or (2) that the older age population represents zircons inherited from the melt source of the host rocks and 580 Ma is the intrusive age. The first rock type directly underneath the pink granitic gneiss is generally an orthoquartzite. The pink granitic gneiss is taken as the lower limit of the KC (see Definition of boundaries below).

Wanni Complex (WC) The WC (Fig. 4) consists predominantly of migmatized gneisses of igneous and sedimentary parentage. The meta-igneous gneisses generally have a granitic, granodioritic and I-type tonalitic composition (Kröner et al. 1994a). Metasediments include garnet–sillimanite gneisses, often with cordierite, calc-silicate layers and, possibly, some grey gneisses containing small garnets. The alternation of meta-igneous and metasedimentary rock types is comparable to the HC, but major differences are: (1) metabasic gneisses and pure marbles are virtually absent; (2) cordierite occurs more widely, both in leucosomes and in melt pods, as in their host rocks; and (3) Nd model ages of 2.0–1.0 Ga (Milisenda et al. 1988; Liew et al. 1991b) and intrusion ages of 790–750 Ma (Baur et al. 1991; Kröner & Jaeckel 1994) to 1100–1000 Ma (Fig. 5; Burton & O'Nions 1990; Kröner et al. 1994a) are significantly younger than those in the HC. Nd model ages are comparable to the KC, but the meta-igneous gneisses of the WC are more felsic than those of the KC, and metasediments are more common. The younger age limit to sedimentation is constrained by the intrusive activity c. 790–750 Ma.

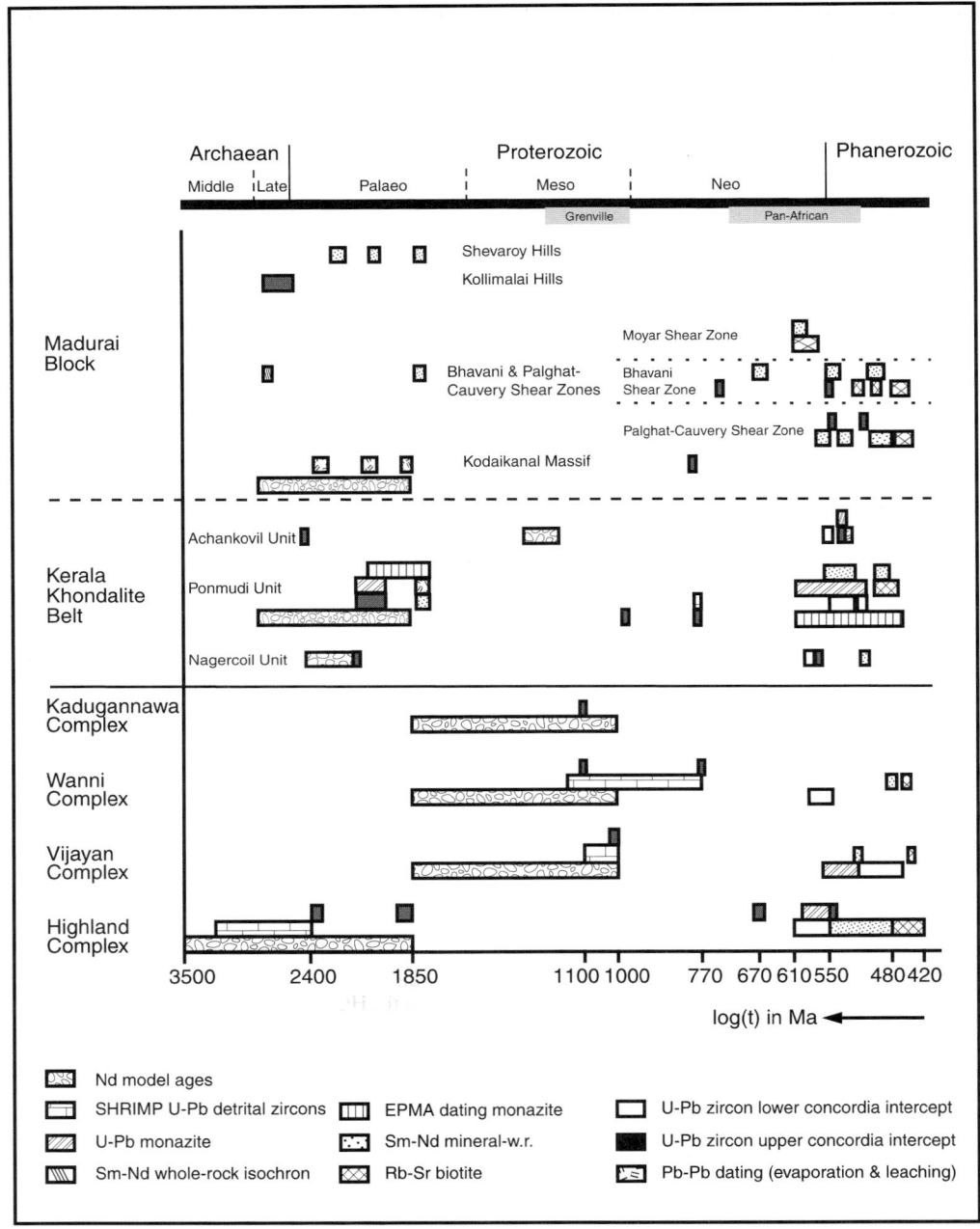

Fig. 5. Summary of geochronological data on the crystalline basement of the Southern Granulite Terrain and Sri Lanka. Timescale is logarithmic. Data compiled from the literature mentioned in the text.

Kröner *et al.* (1994*a*) report a maximum age (Pb–Pb evaporation) of 1329 ± 18 Ma for detrital zircons, which implies that at least part of the WC sediments were derived from Mesoproterozoic source rocks, in agreement with the Nd data. The western part of the WC, which was termed West Vijayan Complex in the older literature (e.g. Geological Survey of Sri Lanka 1982), consists mainly of little deformed granites, such as the late tectonic Tonigala and Galgamuwa Granites (558 ± 14 and 552 ± 8 Ma;

Hölzl et al. 1991), and rare metasediments. The Galgamuwa Granite is part of an alkaline group of magmas characterized by high Na, K, F, Zr, REE, Sr and Ba (Pohl & Emmermann 1991). Hence, the relative timing with respect to structures, the absolute timing and the composition, are all indicative of post-collisional magmatism [see papers in Liégeois (1998)]. The metamorphic grade is upper amphibolite facies, locally with charnockitic patches overgrowing the weakly developed gneissic layering, similar to those described in the rest of the WC (Hansen et al. 1987; Burton & O'Nions 1990; Milisenda et al. 1991).

Vijayan Complex (VC)
The VC (Fig. 4) consists of meta-igneous gneisses of leucogranitic to tonalitic composition (Kröner et al. 1991) with rare inclusions of metaquartzites and calcsilicate rocks (Dahanayake & Jayasena 1983). The mineral content is variable, but the most common minerals are quartz, plagioclase, biotite, amphibole and K-feldspar. The geochemical characteristics of this granitoid suite shows that they are I-type intrusions (Milisenda 1991; Pohl & Emmermann 1991). These granitoids intruded c. 1020Ma ago (Fig. 5; Hölzl et al. 1991). Nd crustal residence ages (1.9–1.0 Ga; Milisenda et al. 1988; Liew et al. 1991a,b) indicate that the VC is largely a juvenile addition to the crust, with small amounts of older crust involved (Milisenda 1991). An active continental margin setting is proposed at the time of intrusion (Milisenda 1991).

Metamorphic grade is generally amphibolite facies, but granulite-grade rocks have been reported from the east coast (e.g. Jayawardena & Carswell 1976; de Maesschalck et al. 1991). The zircon U–Pb lower fraction intercepts between 510 and 460Ma (Hölzl et al. 1991, 1994) probably reflect the age of peak metamorphism. Late-stage, patchy charnockitisation affected the VC in the Pottuvil area (Kriegsman pers. comm.).

Definition of boundaries The boundary between the HC and the VC is rather well defined, since rocks on one side (HC) consist of granulite facies rocks with abundant metasediments and old Nd model ages (Milisenda et al. 1988), whereas the other side (VC) is characterized by amphibolite facies granitoid orthogneisses virtually devoid of metasediments and with young Nd model ages (Milisenda et al. 1988). This pattern seems to be complicated, however, by the existence of a several hundred to a thousand metre thick zone, where rocks with apparently young Nd model ages seem to be interleaved with granulitic, partly metasedimentary, gneisses (Cooray 1962; Büchel 1994; Kleinschrodt 1994).

The lower limit of the KC has been drawn at the contact between the highest orthoquartzite and a continuous horizon of pink granitic gneiss, which has been studied in some detail by earlier workers (e.g. Munasinghe & Dissanayake 1980; Perera 1983; Kröner et al. 1987). This granitic gneiss has a similar Nd model age to more typical KC rocks (Milisenda et al. 1988) and is therefore included in the KC. It is not certain whether the entire sequence below this granitic gneiss forms part of the 'old' Nd model age province (HC) of Milisenda et al. (1988). The uppermost lithologies could very well belong to the 'young' WC, because lithological differences between the HC and WC are small. The structural evolution of rocks near this boundary have been described by Kriegsman (1994).

The boundary between the WC and HC is an isotopic boundary based on a large-scale sampling grid and, as such, not recognizeable in the field. Lithologies on both sides are comparable and are a poor guide to define the boundary. Kehelpannala (1991) proposed to draw it at the top of the highest marble band, which seems to crop out more or less continuously from north to south over >100km (Geological Survey of Sri Lanka 1982). This boundary is some hundreds of metres tectonostratigraphically below the lower limit of the KC and coincides with a high-strain zone called the Digana Movement Zone by Kleinschrodt et al. (1991). The WC would thus incorporate the KC, as proposed by various workers (Fernando 1970; Vitanage 1970; Kehelpannala 1997).

Large-scale structures and the scheme of structural phenomena

The most obvious structures in Sri Lanka, which have been described by most earlier workers (e.g. Adams 1929; Oliver 1957; Berger & Jayasinghe 1976; Geological Survey of Sri Lanka 1983; Silva 1985; Sandiford et al. 1988), are (1) the boundary between the HC and the VC; (2) upright folds, often doubly plunging, in the central part of the island; and (3) the large and systematic change in trend of the axial planes of these folds across the island. The presence of earlier isoclinal folds and L–S tectonite fabrics was recognized by Berger & Jayasinghe (1976), who reinterpreted some 100m scale recumbent folds described by Cooray (1962) as early isoclinal folds. Voll & Kleinschrodt (1991), Kehelpannala (1991, 1997), Kriegsman (1991, 1993) and Almond (1994) gave reviews of all deformational phenomena suggestive of a complicated polyphase structural evolution.

The contact between the HC and the VC has been interpreted as an unconformity by Adams (1929), Coates (1935), Wadia (1942), Fernando (1948) and Katz (1971) on the basis that structural trends in the VC are truncated against the HC. Alternatively,

Cooray (1962, 1978), Crawford & Oliver (1969), Berger (1973), Hapuarachchi (1972) and Newton & Hansen (1986) considered the VC as a retrogressed equivalent of the HC. However, isotopic data (Milisenda *et al.* 1988; Hölzl *et al.* 1991, 1994; Liew *et al.* 1991*a,b*) have now firmly established that the VC is younger than the HC and is neither a basement to the HC metasediments nor its retrogressed equivalent. Instead, the contact between these units represents a major thrust (Hatherton *et al.* 1975; Kröner 1986; Kriegsman 1991, 1993; Kleinschrodt 1994; Büchel 1991, 1994). The presence of serpentinized ultramafic bodies (Dissanayake & van Riel 1978), copper–iron mineralizations (Munasinghe & Dissanayake 1982) and retrogressed charnockites (Cooray 1962; Hapuarachchi 1972; Büchel 1994) testifies to the importance of this contact.

Upright folds occur throughout the island (Fig. 4). The trend of the axial planes of these folds is NNE–SSW in NE Sri Lanka, and swings to N–S and NNW–SSE in the central part of the island. More to the south the trend becomes NW–SE and W–E, locally even WSW–ENE. This means that the trend of upright folds shows a 120° bend, which is comparable to so-called syntactical bends, such as the western Alps and the Ibero-Armorican Arc (Brun & Burg 1982). Similar to those areas, some debate exists as to whether it is due to later refolding or was formed during development of the upright folds (cf. Oliver 1957; Berger & Jayasinghe 1976; Munasinghe & Dissanayake 1982; Büchel 1991, 1994).

Outcrop histories and fold styles are rather similar over large areas. Most outcrops show the following generalized history (Kriegsman 1993, 1994): (1) development of an early foliation (S1), accentuated by leucosomes due to segregation processes and partial melting; (2) isoclinal folding of S1 into recumbent F2 folds, with mineral and stretching lineations parallel to the fold axes; (3) reworking of S1 and S2 into upright or inclined F3 folds, commonly with new leucosomes parallel to the axial plane. This two-stage migmatization described in Kriegsman (1993, 1994) and Almond (1994) was recently studied in more detail by Kehelpannala & Ratnayake (2001). Variation in fold style is mostly related to variation in layer thickness. The main large-scale transitions are: (1) a gradual change in attitude of F3 axial planes over the island from upright and locally east-dipping in western Sri Lanka via upright in central Sri Lanka to west-dipping and subhorizontal in the east; (2) the above mentioned syntactical bend defined by trends of upright folds and stretching lineations; (3) a change from fold-dominated deformation in central Sri Lanka to thrust tectonics in the east and SE.

For correlation with data presented by other workers (e.g. Berger & Jayasinghe 1976; Voll & Kleinschrodt 1991; Kehelpannala 1997), it should be noted that all structures associated with isoclinal folding and the formation of a pervasive L–S tectonite fabric are classified as D2 here, all structures deforming this gneissic layering, notably thrusts and upright to inclined folds, are grouped as D3, and all structures preceding D2, commonly with uncertain kinematic constraints, are collectively called D1 (Table 2). This procedure is largely kinematic: structures that can, with some degree of confidence, be fitted into the same kinematic framework, e.g. subhorizontal N–S stretching during D2, are grouped into the same event, without violating overprinting relations. For example, Almond's (1994) D4 and D5 are interpreted as late stages of D3; D3 of Kehelpannala (1997) and Kehelpannala & Ratnayake (2001) is grouped into D2, and their D4–D6 are all grouped into D3 in this paper, because they are interpreted as manifestations of the same bulk stress field.

Metamorphic evolution and P–T path(s)

Hapuarachchi (1972), in the first comprehensive review of the metamorphism of Sri Lankan granulites, recognized several distinct metamorphic domains and postulated a general westward decrease of maximum recorded pressures. Newton & Hansen (1986) also advanced arguments for westward decreasing P and T across the Sri Lankan granulite terrain and proposed the presence of a tilted crustal section. The presence of a metamorphic field gradient was questioned by Perera (1984) but has been verified by later workers. Notably, the restriction of the high-P assemblage garnet + clinopyroxene + quartz to the eastern half of the HC (Hapuarachchi 1972) has been confirmed (Schenk *et al.* 1991).

Petrological research has concentrated on metamorphosed basaltic/gabbroic to intermediate rocks (Sandiford *et al.* 1988; Schumacher *et al.* 1990; Faulhaber & Raith 1991; Schumacher & Faulhaber 1994), felsic charnockites (Prame 1991*a*) and metamorphosed pelitic rocks (Prame 1991*b*; Raase & Schenk 1994; Hiroi *et al.* 1994). The field P–T gradient has further been substantiated by thermobarometry on metabasites and charnockites (Fig. 4; Schumacher *et al.* 1990; Faulhaber & Raith 1991), by the distribution of biotite + Fe–Ti-oxides (Schumacher *et al.* 1990) and biotite dehydration reactions (Raase & Schenk 1994), and by the trend in the Ti- content of amphiboles (Schumacher *et al.* 1990). The maximum recorded pressures decrease from *c.* 9 to 10 kbar in the central and eastern part of the HC, including the Kataragama Outlier (Fig. 4), to *c.* 5–6 kbar in western Sri Lanka (Schumacher *et al.* 1990; Faulhaber & Raith 1991; Schumacher & Faulhaber 1994). The maximum temperature

Table 2. *Deformation scheme (see text for correlation with other workers)*

Phase	Phenomena / processes*
D1	Early segregation and formation of leucosome, now parallel to lithological banding Crenulated sillimanite inclusions in garnet
D2	Several sets of isoclinal recumbent folds Formation of gneissic layering and transposition of older structures Formation of high-T stretching lineation Boudinage of metabasites and calc-silicate rocks in a quartzitic to metapelitic matrix Symmetric and asymmetric foliation boudinage Nappe tectonics WC + KC
D3	Early thrusts within HC, subsequently refolded in upright folds Upright folds in central Sri Lanka Steep zones in the limbs of some upright folds Local transposition of S2 to S3 in small-scale steep shear zones Double plunge of F3 fold axes, giving rise to elongate domes and basins Thrusting at HC/VC contact Local development of biotite lineation with E–W azimuth East-verging, asymmetric inclined folds near HC–VC contact, cut by eastward directed thrusts 10–100 km scale syntactical bending of D3 axial planes *c.* N-S trending upright folds and steep shear zones deforming earlier folds in syntactical bend Refolding of main thrust contact between HC and VC

* HC, Highland Complex; KC, Kadugannawa Complex; VC, Vijayan Complex; WC, Wanni Complex.

decreases in the same direction from *c.* 830 to 680°C (Schumacher *et al.* 1990; Faulhaber & Raith 1991; Schumacher & Faulhaber 1994). Thermometry shows that this is not merely due to differences in fluid activity, but reflects real temperature differences (Schumacher *et al.* 1990). Temperatures in the VC are comparable to those in the WC and KC, and are considerably lower than those in the HC (Schumacher *et al.* 1990). The observation that late-stage, localized charnockitization affected all units (WC – Hansen *et al.* 1987; HC – Shaw *et al.* 1987; VC and KC – Kriegsman pers. comm.) suggests that metamorphic conditions at that stage were similar for all domains.

Hapuarachchi (1972) mentioned a metamorphic evolution defined by an early high-pressure stage and the subsequent appearance of low-pressure assemblages, including andalusite. Similarly, Perera (1984) and Sandiford *et al.* (1988) described a peak P–T stage, followed by growth of typical decompression textures in metabasites. The P–T path for pelitic rocks, based on the sequence kyanite (early, inclusions in garnet), followed by sillimanite (pervasive), followed by andalusite (rare, texturally late), is clockwise (Hiroi *et al.* 1990, 1994; Raase & Schenk 1994). By contrast, reaction textures involving pyroxenes, plagioclase, garnet and quartz in some metamorphosed igneous rocks (Schumacher *et al.* 1990; Prame 1991*a*), and high temperatures (> 900°C) from pyroxene exsolution (Schenk *et al.* 1988) have been used by these authors as an indication of isobaric cooling, which is apparently not documented in the pelitic rocks and occurred prior to uplift.

Sapphirine-bearing granulites from two localities, located within the high-pressure realm of the central HC in Sri Lanka, followed a roughly clockwise P–T path that can be divided into two main stages (Kriegsman & Schumacher 1999). The early, prograde part is characterized by garnet growth and the post-peak-phase by the development of spectacular corona-type textures. In the varieties that are richer in SiO_2 (sillimanite- and quartz-bearing), cordierite is commonly formed, whereas garnet is generally consumed. Based both on the slopes of reactions that were inferred from the textures in thin section and P–T estimates, the corona/cordierite formation records near-isothermal decompression beginning at *c.* 9 kbar and 830°C, continuing to at least *c.* 7.5 kbar and 810°C. By contrast, garnet–orthopyroxene thermobarometry on the same rocks yields significantly higher temperatures of 900–950°C (Kriegsman pers. comm.).

The evidence from sapphirine-bearing granulites supports the interpretation that the sequence of kyanite followed by sillimanite followed by andalusite in metapelites formed during a single metamorphic cycle evolving along an essentially clockwise P–T path (Hiroi *et al.* 1990, 1994; Raase & Schenk 1994; see also Kriegsman 1996). In contrast, growth of garnet + quartz ± clinopyroxene from plagioclase + orthopyroxene at high temperature and subsequent garnet breakdown to a new generation of orthopyroxene + plagioclase in the metamorphosed mafic rocks has been interpreted as evidence of near-isobaric cooling at high pressures followed by decompression (Schenk et al. 1988; Schumacher et al. 1990; Schumacher & Faulhaber 1994). Support for this interpretation comes from high temperatures recorded in exsolved coexisting pyroxenes (Schenk et al. 1988).

The data for the decompression stage recorded in sapphirine-bearing granulites (Kriegsman & Schumacher 1999) and metabasites (Schumacher & Faulhaber 1994) are in good agreement. However, data for the P–T trajectory's orientation just prior to decompression from the same two rock types appears to be contradictory. This apparent contradiction may be resolved by the UHT ($>900°C$) measured by feldspar thermometry (Raase 1998) and garnet–orthopyroxene thermobarometry (Harley 1998; Kriegsman pers. comm.). Simultaneous garnet growth and garnet breakdown along the retrograde path (Faulhaber & Raith 1991) resulted from the large variation in dP/dT of reaction line slopes ($+8$–$+18$ bar/K) for the assemblage garnet + plagioclase + orthopyroxene + quartz, due to variable Ca contents of the first two phases (Kriegsman 1996). By implication, the slope of the retrograde path after isothermal decompression is $c. +15$ bar/K. Subsequent uplift is documented by the presence of andalusite near some late pegmatites (Hiroi et al. 1990; Raase & Schenk 1994).

The source of the high-temperature thermal anomaly is not well understood, but the synmetamorphic mafic intrusive bodies may have contributed to the heat budget and may also have caused a CO_2 flux into adjacent rocks (Bolder-Schrijver et al. 2000).

Age of metamorphism

Early interpretations of the timing of high-grade metamorphism in Sri Lanka were based on Rb–Sr data (Crawford & Oliver 1969; Katz 1971; de Maesschalck et al. 1990), which were interpreted in terms of metamorphic events at $c.$ 2000 Ga, and $c.$ 1250–1100 and 900–800 Ma. U–Pb and K–Ar mineral ages (Holmes 1955; Cooray 1969) suggested cooling between 600 and 450 Ma.

More recent conventional multigrain U–Pb zircon dating, however, suggests that no Pb loss occurred between $c.$ 1900 and 610 Ma (Baur et al. 1991; Hölzl et al. 1991). This, and similar, more qualitative results from the U–Th–Pb system (Liew et al. 1991a,b) can be taken as evidence for the absence of high-grade metamorphic events during this period.

The lower intercepts in concordia diagrams for chords based on discordant zircon fractions and concordant monazite ages lie in the range of 610–550 Ma in the HC, which probably reflects the time of regional high-grade metamorphism (Baur et al. 1991; Hölzl et al. 1991). Lower concordia intercept ages in the WC are in the range of 590–540 Ma (Hölzl et al. 1991, 1994; Kröner et al. 1994a, b), and in the VC between 510 and 460 Ma. These data suggest diachronous Pb loss in the Sri Lankan basement, which may reflect a diachronous culmination of peak metamorphic conditions.

Burton & O'Nions (1990) reported Sm–Nd and Rb–Sr mineral isochron ages of $c.$ 535 Ma on pods of arrested charnockite in the Kurunegala area. Since these pods formed during D3 upright folding (Kriegsman 1991, 1993), this may date the time of upright folding in that area. Since garnet Sm–Nd cooling ages vary from 560 to 480 Ma and biotite Rb–Sr cooling ages from 470 to 440 Ma, a correlation with temperature has been interpreted in terms of two-stage cooling: (1) slow cooling at a rate of 2–3°C/Ma from the peak temperature of $c.$ 830°C (560 Ma) to $c.$ 600°C (480 Ma); (2) fast cooling at a rate of 10–25°C/Ma until 450 Ma (Hölzl et al. 1991). Similar cooling ages were obtained by Burton & O'Nions (1990). Mineral cooling ages determined by ^{40}Ar–^{39}Ar laser microprobe are in the range of 493–420 Ma (Irwin et al. 1987), suggesting that the thermal anomaly may have lasted even longer.

Tectonic evolution

Gneisses of the HC in central Sri Lanka show a well-developed gneissic layering (S2) and a strong NNW–SSE-oriented mineral and stretching lineation (L2), defined by peak metamorphic assemblages in all rock types (e.g. Kriegsman 1994). Compositional banding is generally parallel to S2. Together, they have been refolded around upright folds with gently plunging fold axes (Berger & Jayasinghe 1976). Fold axes of early isoclinal folds and late upright folds are parallel to L2. The WC and KC have been interpreted as fold nappes, which were emplaced on top of the HC with a top-to-the-NNW movement sense parallel to L2 (Kriegsman 1994). This main deformational event occurred at the peak of metamorphism, as evidenced, among many other features, by extreme garnet plasticity (Kleinschrodt & McGrew 2000). Partial breakdown of peak metamorphic assemblages to retrograde symplectites of various assemblages postdates S2.

The nappe model and Nd model ages suggest that the WC and KC, and the eastern part of the HC, may represent distinct lower crustal terranes juxtaposed during a Pan-African high-grade, high-strain event (D2). The contact between these terrains is interpreted as a terrane boundary in a lower crustal environment (Kriegsman 1993, 1994). The top-to-the-NNW movement sense suggests that the KC may have been originally located south of the eastern HC. This, and the similarity in isotope signatures, indicates a possible pre-assembly link of the KC, the WC of western Sri Lanka and the VC of eastern Sri Lanka. This leads to the hypothesis that two terranes were juxtaposed during the Pan-African Orogeny in Sri Lanka: the HC terrane and the combined WC–KC–VC Terrane. Whether this implies that Sri Lanka is located at the suture between West and East Gondwana, as postulated by Kröner (1991a), or at the suture of the younger Kuunga Orogen at 550 Ma, postulated to have existed outboard of the East Africa Orogen (Meert et al. 1995), remains to be resolved.

Extensional collapse may have been triggered by delamination of the lowermost crust and upper mantle. This process may also have induced decompression melting of upwelling asthenospheric material, causing the intrusion of mafic magmas into the lower crust and providing a heat source for granulite metamorphism (Kriegsman 1993). Final unroofing at decreasing temperatures occurred during a later deformation event (D3), when granulite rocks were part of the hanging wall of a major thrust.

E–W shortening caused eastward thrusting of the granulite facies belt over the amphibolite facies of the VC and simultaneous upright folding (Kriegsman 1993, 1995). An arcuate structure, defined by the trend of upright D3 folds and L2 stretching lineations, is well developed in hanging-wall rocks of the main contact zone, but is lacking from rocks in the footwall (Kriegsman 1993, 1995). This syntactical bend can be explained by the 'corner effect' near a colliding promontory (Brun & Burg 1982), here the VC or unexposed crustal domains underlying the VC. This, and the combination of folding and thrusting, has been explained in terms of simultaneous E–W horizontal shortening and sinistral strike-slip along an E–W- to ESE–WNW-trending transcurrent shear zone south of Sri Lanka prior to Gondwana breakup (Kriegsman 1993, 1995).

The colinearity of L2 and L3, with few exceptions, and the short time span separating D2 and D3, indicate a similar kinematic framework for these deformation phases (Kriegsman 1993). It is argued that both events may be regarded as N–S extension with vertical and E–W horizontal shortening. During D2, vertical shortening exceeded E–W horizontal shortening, causing subhorizontal foliations, L–S tectonites and northward nappe emplacement. At this stage, gravitational forces probably exceeded the forces due to plate convergence. During D3, E–W horizontal shortening exceeded vertical shortening, resulting in eastward thrusting and upright folding (Kriegsman 1993).

Extensive post-collisional magmatism (high Ba–Sr) occurred at 550–560 Ma in the WC [after data in Hölzl et al. (1991) and Pohl & Emmermann (1991)], but granites of similar age and composition are also present in the HC (e.g. the Tangalla Granite in southern Sri Lanka, 550 ± 3 Ma; Hölzl et al. 1991). Similar to the situation in the Palaeoproterozoic Svecofennian Orogeny in SW Finland (Eklund et al. 1998), post-collisional magmas intruded c. 30 Ma after S-type granites (in Sri Lanka, the c. 580 Ma pink granites; Hölzl et al. 1991, 1994).

Pb loss in the VC occurred on average 80 Ma later than Pb loss in the HC and WC. Hence, peak metamorphism in the VC is really Cambrian, similar to large parts of East Antarctica. The sudden increase in the cooling rate of the HC c. 480 Ma may be related to thrusting of the HC on top of the VC, which at the same time would have heated the VC. The exact details still need some reconsideration, but this fits remarkably well with the P–T path of the HC, characterized by early isobaric cooling [from >900 to c. 800°C (Schumacher et al. 1990) rather than 830–600°C (Hölzl et al. 1991, 1994)] followed by isothermal decompression and, a little bit later, uplift accompanied by cooling. As thrusting and upright folding (D3) are coeval (Kriegsman 1993), the time of thrusting is also constrained by the time of upright folding. Syn-D3-arrested charnockization in the WC is dated at 535 Ma (Burton & O'Nions 1990), i.e. 25 Ma before the oldest Pb loss in the VC (Hölzl et al. 1991, 1994). As structures also need time to develop and may be diachronous in different areas, these age constraints should be regarded as tentative only.

In summary, a likely scenario is: (1) a back-arc(?) basin stage in the WC until 740 Ma; (2) high-temperature ($T > 900°C$ at 8–10 kbar) metamorphism at c. 610 (U–Pb zircon lower intercepts) to 580 Ma (pink granite intrusions); (3) isobaric cooling to 800°C until c. 540 Ma, overlapping with post-collisional magmatism at 550–560 Ma; (4) upright folding and thrusting of the HC on top of the VC at c. 540–530 Ma, causing uplift and cooling in the HC, but heating and partial melting in the VC; (5) Pb loss in the VC from 510 to 460 Ma, simultaneous with cooling of the entire basement area.

Discussion and conclusions

The subdivision of the Sri Lankan crystalline basement into three distinct complexes (the HC, the VC, and the joint WC–KC), which most likely were juxtaposed during the Pan-African Orogeny, is well justified by petrology, geochronology and isotope data (see above), but the correct subdivision of the SGT of India is less clear. Consequently, the linkage of the SGT with the lower crustal complexes of Sri Lanka, as well as its position within East Gondwana, necessarily retain elements of speculation.

The southeastern part of the Madurai Block and the northeastern KKB consist of a similar lithological association, comprising migmatitic metapsammites and metapelites, quartzites, marbles, calc-silicates, and granites (Cenki pers. comm.). Structural studies on the presence and nature of the Achankovil (or Tenmalai) Shear Zone along the northwestern edge of the KKB are still ambiguous (see above). In addition, similar studies in southern Tamil Nadu did not provide any evidence for the occurrence of a shear zone or a tectonic lineament in that area (Cenki pers. comm.). These findings contradict the current view that the SGT consists of two individual crustal blocks which are juxtaposed along the Achankovil Shear Zone. The lowlands east of the Western Ghats seem to form a continuous crustal section within the SGT and may represent the retrogressed equivalent of the charnockite–enderbite massifs of the Cardamom Hills and the Kodaikanal Massif.

In the following sections, the problems and arguments in favour of a relationship between the SGT and Sri Lanka, and the role of the SGT during the formation of Gondwana in the Neoproterozoic, will be discussed.

Correlation between the Southern Granulite Terrain (SGT) and Sri Lanka

The SGT of India shows abundant characteristics in common with the HC and the WC of Sri Lanka. Both crustal domains consist of meta-igneous (charnockites, enderbites) and metasedimentary rocks, the latter predominantly comprising metapelites and metapsammites of similar major and trace element, and stable isotope (O and C) composition (Prame & Pohl 1994; Braun 1997; Braun pers. comm.). Calc-silicate rocks and marbles from both units also display similarities in their $\delta^{18}O$ and $\delta^{13}C$ signatures, which indicate deposition in a shallow-marine environment and subsequent changes due to devolatilization and influx of external fluids (Hoffbauer & Spiering 1994; Satish-Kumar et al. 2001). Nd model ages of between 2 and 3 Ga, together with U–Pb zircon upper concordia intercept and single zircon evaporation ages of 2.2–1.8 Ga, indicate crustal growth during Archaean and Palaeoproterozoic times, whereas the Neoproterozoic is predominantly characterized by crustal reworking. High-temperature to UHT conditions during the Pan-African Orogeny have been deduced for both units. Temperatures of $c.$ 1000°C derived from sapphirine granulites from the Kodaikanal Massif were recently confirmed by carbon isotope thermometry on marbles from an adjacent area (Raith et al. 1997; Satish-Kumar 2000) and match the results of two-feldspar and garnet–orthopyroxene thermometry on rocks from the HC fairly well (Harley 1998; Raase 1998; Kriegsman pers. comm.). Calc-silicate rocks from the KKB and the Buttala Klippe, an outlier of the HC in SE Sri Lanka, record temperatures of $>$ 837°C and 875–900°C at pressures of $<$ 6.6 kbar and $c.$ 9 kbar, respectively (Satish-Kumar & Harley 1998; Mathavan & Fernando 2001). Reaction textures indicate isobaric cooling from peak metamorphic temperatures followed by near-isothermal decompression (Schumacher et al. 1990; Prame 1991a; Satish-Kumar & Harley 1998; Kriegsman & Schumacher 1999; Nandakumar & Harley 2000; Cenki et al. 2002). A similar pressure gradient as in the SGT is recognized in the granulites of the HC with a high-pressure ($>$9 kbar) domain in its central part and a significant drop down to mid-crustal values of 5–6 kbar towards the southwestern edge (Schumacher & Faulhaber 1994). It is thus suggested that (parts of) the Madurai Block can be correlated with the high-pressure domains of the HC, while its southwestern part may be linked with the Ponmudi and Nagercoil Units of the KKB. The latter is supported by the paucity of metabasic gneisses, and the presence of graphite in high-quality vein deposits in the KKB and the southwestern HC (Dissanayake & Chandrajith 1999).

Strongly migmatitic cordierite–garnet–(orthopyroxene) gneisses are abundant in the northwestern part of the Achankovil Unit (Fig. 3), and display similar textural and compositional features as their counterparts in the southwestern part of Sri Lanka (Prame & Pohl 1994). The occurrence of Mesoproterozoic Nd model ages (1.4–1.3 Ga) in these and other rocks from the Achankovil Unit may point to a genetic relationship with the WC (Harris et al. 1994; Brandon & Meen 1995; Bartlett et al. 1998; Cenki pers. comm.). In addition, biotite- and/or amphibole-bearing granites, which are widespread in the eastern part of the SGT, share similar textural and compositional features with late to post-tectonic, including typical post-collisional, granites from the WC (Pohl & Emmermann 1991; Hölzl et al. 1994). Likewise, the recognition of magmatic activity in the northeastern SGT $c.$ 800 Ma ago (Ghosh 1999) also points to a correlation with the WC. Yet, unless more petrological, isotope and geochronological data become available, such interpretations must remain speculative.

Rodinia connection

Zircon upper and lower concordia intercept ages of 2.30–1.85 and 0.61–0.55 Ga in rocks of the HC record Palaeoproterozoic sediment provenance ages and/or magmatism, as well as Pan-African high-grade metamorphism and associated magmatism (Hölzl et al. 1994). In contrast, a Grenvillian age population is completely absent. This age distribution is identical with that of the SGT (2.50–1.85 and 0.61–0.52 Ga age groups; Fig. 5). It indicates that both complexes formed part of the East African Orogen (see Kröner et al. 2000, fig. 14) but rules out an active role in the formation and breakup of Rodinia.

This is in obvious contrast to the VC and WC, where Grenvillian magmatism is well constrained by Nd isotope and U–Pb age data (Milisenda et al. 1988; Burton & O'Nions 1990; Hölzl et al. 1991, 1994; Liew et al. 1991a,b; Kröner et al. 1994a). With this in mind, the recognition of a Late Mesoproterozoic (Grenvillian) age in the SGT (987±65 Ma; Ghosh 1999) is of major importance for the geodynamic reconstruction of this region.

Southern Granulite Terrain (SGT) and Sri Lanka in the Gondwana context

Fitzsimons (2000a,b), reviewing the Grenvillian and Pan-African domains of Antarctica, postulated the existence of two orogenic belts of Pan-African age transecting Antarctica: the East African Orogen (= the Mozambique Belt) continuing into the Lützow-Holm Bay, central Dronning Maud Land, and the Shackleton Range; and the Prydz–Denman–Darling Orogen, which would line up with the Pinjarra Orogen in Western Australia. The evidence for the first orogen is convincing, but the orientation of the Prydz–Denman–Darling Orogen is not well constrained. Fitzsimons' main arguments for letting it run towards Antarctica's interior are: (1) a marked decrease of Cambrian reworking to the west of the Lambert Graben; (2) indications for Pan-African magmatism and cooling ages in the southern Prince Charles Mountains. As for the first argument, it is clear that Cambrian (Pan-African) reworking was not equally pervasive in all areas. In Kriegsman's (1993, 1995) model, the second Pan-African belt starting around Sri Lanka was pervasive in Lützow-Holm Bay, but only caused low- to medium-grade reworking (upright folding and shear zones) in the Rayner Complex to the east. Further east, the next domain with obvious high-grade Pan-African reworking is the southern Prydz Bay area [see review of data in Fitzsimons (2000b)], contrasting with the Lambert Graben area west of it. Hence, the degree and grade of Cambrian reworking decreases eastwards from Lützow-Holm Bay but westwards from Prydz Bay, the two situations thus being mirror images. Structural trends in Lützow-Holm Bay indicate that the Pan-African tectonic zone transects the coastline at an oblique angle. The present authors speculate that it continues eastwards, then swings around and joins up with the Prydz Bay area, where it again transects the coastline at an oblique angle. Retrograde Pan-African reworking in the Eastern Ghats Belt (Crowe et al. 2001) may be a related phenomenon.

The presence of Ediacaran trace and body fossils in sediments of the low-grade Stirling Range Formation, immediately north of the Grenvillian Albany–Fraser Belt in Western Australia, indicates that it represents a Neoproterozoic basin (Cruse et al. 1993; Cruse & Harris 1994). Strong folding is therefore attributed to a Neoproterozoic–Lower Palaeozoic (Pan-African) event (Cruse et al. 1993; Harris & Beeson 1993; Cruse & Harris 1994). Similar-looking fossils are reported from the greenschist facies Sandow Group to the south of the Grenvillian Bunger Hills, east of the Denman Glacier in Antarctica (Sheraton et al. 1995). These rocks were also strongly folded during the Pan-African event (Sheraton et al. 1995). Together, this indicates that Neoproterozoic basins both north and south of the Grenvillian Albany–Fraser–Wilkes Province (sensu Fitzsimons 2000a,b) were subjected to Pan-African tectonics. Harris & Beeson (1993) postulate the existence of a zone of Lower Palaeozoic (late Pan-African) dextral reactivation of pre-existing (Mesoproterozoic–Grenvillian) basement structures, affecting the Stirling Ranges north of the Albany–Fraser Belt and the Vindhyan Supergroup north of the Central Indian Tectonic Zone. Rb–Sr ages of 1014±18 and 939±22 Ma on <2 mm illite-rich clay fractions from the lower Vindhyan show that these sediments were indeed deposited in a Neoproterozoic basin (Bansal et al. 1999). This zone is at high angles to the Prydz–Denman–Darling 'Orogen', where Fitzsimons (2000a,b) postulated sinistral strike-slip in the same period. The two belts could be regarded as continent-scale conjugates in a setting of NW–SE shortening (present-day Australian coordinates). Importantly, this shortening direction is consistent with the shortening direction in the Sri Lanka–Lützow-Holm Bay area (Kriegsman 1995).

Clearly, what is needed to decide on the correct model is a thorough overview of structural data in East Antarctica. In addition, it could be fruitful to replace thinking in terms of more or less straight belts by geodynamic concepts taking modern-day plate tectonic complexities (e.g. SE Asia) into account.

Using U–Pb age data from magmatic zircons of reworked Maud and Rayner material, Fitzsimons (2000b) inferred that two Grenville-age collisional orogens, the Maud and Rayner Provinces, are separated by the Lützow-Holm Belt and that the Pan-African suture between both provinces lies towards the eastern termination of the Lützow-Holm Belt, adjacent to the Rayner Complex (Fig. 6). Sri Lanka

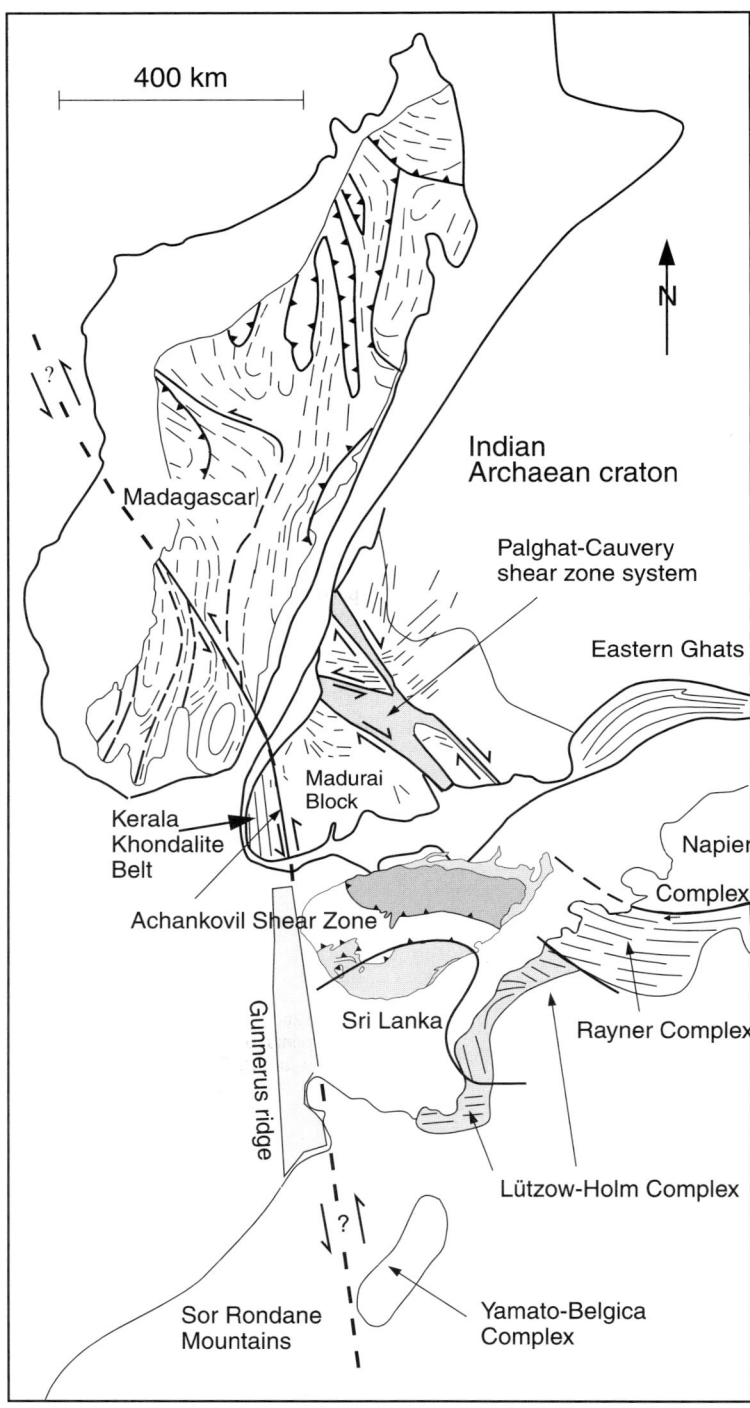

Fig. 6. Position of Peninsular India and Sri Lanka in Gondwana, modified after Kriegsman (1995) and incorporating new data on Madagascar from Kröner et al. (2000).

is commonly shown juxtaposed to Lützow-Holm Bay in a more northerly position with respect to southern India than at present (Fig. 6; e.g. Lawver & Scotese 1987; Powell et al. 1988; Kriegsman 1995; Fitzsimons 2000a,b; Chand et al. 2001; Collins & Windley 2002), or, less frequently, shifted southwards (De Wit et al. 1988; Dissanayake & Chandrajith 1999; Nédélec et al. 2000; Muhongo et al. 2001). U–Pb zircon crystallization ages of orthogneisses from the VC and the WC between 1100 and 1000 Ma (Hölzl et al. 1991, 1994; Kröner & Jäckel 1994; Kröner et al. 1994a, b) correspond with published data from the Maud Province [Jacobs et al. (1998) and compilation in Fitzsimons (2000b)], and thus would imply that these two complexes of Sri Lanka formed part of West Gondwana. However, unless further detailed geochronological and isotope studies provide additional constraints, the continuation of the suture between East and West Gondwana into Sri Lanka and southern India has to be left open (Fig. 6; Fitzsimons 2000b).

Ages between 790 and 770 Ma (Baur et al. 1991; Hölzl et al. 1994), recognized in orthogneisses of the sediment-rich WC, represent a stage of early Pan-African magmatism, possibly related to extensional tectonics. As Kröner (2001) reports subduction-related magmatism in the Mozambique Belt in this time interval, the WC could represent a coeval back-arc basin. Similar ages have recently been obtained from alkali rocks and carbonatites from the northeastern part of the SGT (c. 800 Ma; Schleicher et al. 1998; Miyazaki et al. 2000), as well as from granitic gneisses and (alkali) granites of the Madurai Block [770–570 Ma; compilation in Rajesh & Santosh 1996b) and Ghosh (1999)]. Such ages are unknown from East Antarctica, but they have been reported from the Leeuwin Complex in southwestern Australia (Nelson 1996), located opposite to the Denman Glacier of East Antarctica in Gondwana reconstructions (e.g. Fitzsimons 2000a,b).

The juxtaposition of the SGT with the crystalline basement of southern Madagascar has been inferred from the predominance of metasedimentary lithologies in both units, the occurrence of cordierite gneisses along the Ranotsara and the Achankovil Shear Zones, which were interpreted to have been connected in Gondwana, the distribution of Palaeoproterozoic and Neoproterozoic ages, and similarities in P–T evolution (Nicollet 1990; Windley et al. 1994; Yoshida et al. 1999; Nédélec et al. 2000; de Wit et al. 2001; Collins & Windley 2002). Accordingly, Gondwana reconstructions showed the terrains south of the shear zones in a neighbouring position as part of the East African Orogen (or the Mozambique Belt), whereas the areas to the north were considered unaffected by Pan-African tectonism. This view has been modified due to geochronological studies which showed a Pan-African imprint in the Madurai Block, as well as in major parts of central Madagascar (Hansen et al. 1985; Jayananda et al. 1995a, b; Bartlett et al. 1998; Paquette & Nédélec 1998; Tucker et al. 1999; Kröner et al. 2000). As a consequence, recent models correlate shear zones in central and south Madagascar with either the Achankovil Shear Zone, the KKPTSZ or the PCSZ of southern India (see Fig. 1). Kröner et al. (2000) questioned the validity of these interpretations, which in their view were not supported by lithological and geochronological data. Notably, the absence of early and middle Neoproterozoic ages (820–640 Ma) in the SGT, which are widespread in central Madagascar (Tucker et al. 1999; Kröner et al. 2000), was taken as main argument against a link between the Proterozoic parts of Madagascar and southern India. However, the few early Neoproterozoic ages recently recognized in the SGT correspond as well with data from central Madagascar as they do with those from the WC of Sri Lanka. Given the limited number of geochronological data from the SGT, in particular from the Madurai Block, further studies could very well reveal the occurrence of a widespread Early Neoproterozoic event in this terrain. In addition, stratiform granites, some of them with alkaline affinities, are widespread in the eastern Madurai Block (Cenki pers. comm; authors' own observations). Similarities in chemical and mineralogical compositions, and textural features (Braun pers. comm.) suggest that they could be equivalent to the stratoid granites of central Madagascar (Nédélec et al. 1995). Again, more geochronological and isotope data are needed for the SGT granites to test this hypothesis.

We would like to thank S. Dasgupta and M. Yoshida for giving us the opportunity to write this extended overview on the geology of the Southern Granulite Terrain and Sri Lanka and for their patience in waiting for the final version. Comments by M. Raith, C. Dobmeier, G.R. Ravindra Kumar and B. Cenki on an earlier version, as well as the critical review by A. Kröner, significantly improved the manuscript.

References

ADAMS, F. D. 1929. The geology of Ceylon. *Canadian Journal of Research*, **1**, 425–511.

ALMOND, D. C. 1994. *Solid-Rock Geology of the Kandy Area, Sri Lanka*. Institute of Fundamental Studies, Kandy, Sri Lanka, 1–74.

BANSAL, M., VIJAN, A. R., MAHAR, Y. S., GHOSH, N. KRISHNA PRASAD, S. & PRABHU, B. N. 1999. Rb-Sr dating of lower Vindhyan sediments from Son Valley, Madhya Pradesh, India. *In*: CHAVADI, V. C. & UGARKAR, A. G. (eds) *Special Volume of the First Convention & National Seminar*. Indian Mineralogist, **33**, 1–9.

BARTLETT, J. M., DOUGHERTY-PAGE, J. S., HARRIS, N. B. W., HAWKESWORTH, C. J. & SANTOSH, M. 1998. The application of single zircon evaporation and Nd model ages to the interpretation of polymetamorphic terrains: an example from the Proterozoic mobile belt of south India. *Contributions to Mineralogy and Petrology*, **131**, 181–195.

BARTLETT, J. M., HARRIS, N. B. W., HAWKESWORTH, C. J. & SANTOSH, M. 1995. New isotope constraints on the crustal evolution of South India and Pan-African granulite metamorphism. *Memoir, Geological Society of India*, **34**, 391–397.

BAUR, N., KRÖNER, A., TODT, W., LIEW, T. C. & HOFMANN, A. W. 1991. U–Pb isotopic systematics of zircons from prograde and retrograde transition zones in high-grade orthogneisses, Sri Lanka. *Journal of Geology*, **99**, 527–545.

BERGER, A. R. 1973. The Precambrian metamorphic rocks of Ceylon. A critique of a radical interpretation. With reply by Katz, M. B. *Geologische Rundschau*, **62**, 347–350.

BERGER, M. & BRAUN, I. 1997. Pb–Pb dating of apatite by a stepwise dissolution technique. *Chemical Geology*, **142**, 23–40.

BERGER, A. R. & JAYASINGHE, N. R. 1976. Precambrian structure and chronology in the Highland Series of Sri Lanka. *Precambrian Research*, **3**, 559–576.

BERNARD-GRIFFITHS, J., JAHN, B.-M. & SEN, S. K. 1987. Sm–Nd isotopes and REE geochemistry of Madras granulites, India: an introductory statement. *Precambrian Research*, **37**, 343–355.

BHASKAR RAO, Y. J., CHETTY, T. R. K., JANARDHAN, A. S. & GOPALAN, K. 1996. Sm–Nd and Rb–Sr ages and P–T history of the Archaean Sitampundi and Bhavani layered meta-anorthosite complexes in Cauvery shear zone, south India: evidence for Late Proterozoic reworking of Archaean crust. *Contributions to Mineralogy and Petrology*, **125**, 237–250.

BHATTACHARYA, A. & SEN, S. K. 1986. Granulite metamorphism, fluid buffering, and dehydration melting in the Madras charnockites and metapelites. *Journal of Petrology*, **27**, 1119–1141.

BOLDER-SCHRIJVER, L. J. A., KRIEGSMAN, L. M. & TOURET, J. L. R. 2000. Primary carbonate/CO_2 inclusions in sapphirine-bearing granulites from central Sri Lanka. *Journal of Metamorphic Geology*, **18**, 259–269.

BOSE, R. N. & KARTHA, T. D. G. 1977. Geophysical studies in parts of west coast (Kerala) of India. *Indian Journal of Earth Science*, **S. Ray Volume**, 251–266.

BRANDON, A. D. & MEEN, J. K. 1995. Nd isotopic evidence for the position of southernmost Indian terranes within East Gondwana. *Precambrian Research*, **70**, 269–280.

BRAUN, I. 1997. *Partielle Anatexis und Granitgenese unter granulitfaziellen Bedingungen: Der Kerala Khondalite Belt (Südindien) als Fallbeispiel*. PhD. Thesis, University of Bonn.

BRAUN, I. & BRÖCKER, M. 2001. Granitic magmatism in the Kerala Khondalite Belt, southern India: results of EPMA- and U-Pb dating of monazite. *Journal of Conference Abstracts*, **6**, 375.

BRAUN, I. & RAITH, M. 1996. Occurrence of fluorapatite in granitic veins from the Kerala Khondalite Belt, southern India. *Journal of the Geological Society of India*, **48**, 629–639.

BRAUN, I., MONTEL, J.-M. & NICOLLET, C. 1998. Electron microprobe dating of monazites from high-grade gneisses and pegmatites of the Kerala Khondalite Belt, southern India. *Chemical Geology*, **146**, 65–85

BRAUN, I., RAITH, M. & RAVINDRA KUMAR, G. R. 1996. Dehydration–melting phenomena in leptynitic gneisses and the generation of leucogranites: a case study from the Kerala Khondalite Belt, southern India. *Journal of Petrology*, **37**, 1285–1305.

BROWN, M. & RAITH, M. 1996. First evidence of ultrahigh temperature decompression from the granulite province of southern India. *Journal of the Geological Society, London*, **153**, 819–822.

BRUN, J.-P. & BURG, J.-P. 1982. Combined thrusting and wrenching in the Ibero-Armorican arc: a corner effect during continental collision. *Earth and Planetary Science Letters*, **61**, 319–332.

BÜCHEL, G. 1991. Gravimetric investigations across tectonic boundaries between the Highland/ Southwestern Complex and the Vijayan Complex in Sri Lanka. *In*: KRÖNER, A. (ed) The crystalline crust of Sri Lanka, Part 1, Summary of research of the German–Sri Lankan Consortium. *Geological Survey Department of Sri Lanka, Professional Paper*, **5**, 89–93.

BÜCHEL, G. 1994. Gravity, magnetic and structural patterns at the deep-crustal plate boundary zone between West- and East-Gondwana in Sri Lanka. *In*: RAITH, M. & HOERNES, S. (eds) *Tectonic, Metamorphic and Isotopic Evolution of Deep Crustal Rocks, with Special Emphasis on Sri Lanka*. Precambrian Research, **66**, 77–91.

BUHL, D. 1987. *U–Pb und Rb–Sr – Altersbestimmungen und Untersuchungen zum Strontiumisotopenaustausch an Granuliten Südindiens*. PhD Thesis, Universität Münster.

BURTON, K. W. & O'NIONS, R. K. 1990. The timescale and mechanism of granulite formation at Kurunegala, Sri Lanka. *Contributions to Mineralogy and Petrology*, **106**, 66–89.

CENKI, B., KRIEGSMAN, L. M. & BRAUN, I. 2002. Melt-producing and melt-consuming reactions in anatectic granulites: P–T evolution of the Achankovil cordierite gneisses, South India. *Journal of Metamorphic Geology*, in press.

CHACKO, T., LAMB, M. & FARQUHAR, J. 1996. Ultra-high temperature metamorphism in the Kerala Khondalite Belt. *In*: SANTOSH, M. & YOSHIDA, M. (eds) *The Archaean and Proterozoic Terrains in Southern India within East Gondwana*. Gondwana Research Group Memoir, **3**, 157–165.

CHACKO, T., RAVINDRA KUMAR, G. R., MEEN, J. K. & ROGERS, J. J. 1992. Geochemistry of high-grade supracrustal rocks from the Kerala Khondalite Belt and adjacent massif charnockites, South India. *Precambrian Research*, **55**, 469–489.

CHACKO, T., RAVINDRA KUMAR, G. R. & NEWTON, R. C. 1987. Metamorphic P–T conditions of the Kerala (South India) Khondalite Belt: a granulite facies supracrustal terrain. *Journal of Geology*, **96**, 343–358.

CHADWICK, B., VASUDEV, V. N. & HEGDE, G. V. 2000. The Dharwar craton, southern India, interpreted as the result of Late Archaean oblique convergence. *Precambrian Research*, **99**, 91–111.

CHAND, S., RADHAKRISHNA, M. & SUBRAHMANYAM, C.

2001. India-East Antarctica conjugate margins; rift-shear tectonic setting inferred from gravity and bathymetry data. *Earth and Planetary Science Letters*, **185**, 225-236.

CHETTY, T. R. K. 1996. Proterozoic shear zones in southern granulite terrain, India. *In*: SANTOSH, M. & YOSHIDA, M. (eds) *The Archaean and Proterozoic Terrains in Southern India within East Gondwana*. Gondwana Research Group Memoir, **3**, 77-89.

CHOUDHARY, A. K., HARRIS, N. B. W., VAN CALSTEREN, P. & HAWKESWORTH, C. J. 1992. Pan-African charnockite formation in Kerala, South India. *Geological Magazine*, **129**, 257-264.

COATES, J. S. 1935. The geology of Ceylon. *Ceylon Journal of Science*, **19**, 101-215.

COLLINS, A. S. & WINDLEY, B. F. 2002. The tectonic evolution of central and northern Madagascar and its place in the final assembly of Gondwana. *Journal of Geology*, **110**, 325-339.

CONDIE, K. C. & ALLEN, P. 1984. Origin of Archaean charnockites from southern India. *In*: KRÖNER, A., HANSON, G. N. & GOODWIN, A. M. (eds) *Archaean Geochemistry*. Springer, Berlin, 183-203.

COORAY, P. G. 1962. Charnockites and their associated gneisses in the Precambrian of Ceylon. Quart. *Journal of the Geological Society of London*, **118**, 239-273.

COORAY, P. G. 1969. The significance of mica ages from the crystalline rocks of Ceylon. *In*: *Age Relations in Highgrade Metamorphic Terrains*. Special Paper, Geological Association of Canada, **5**, 47-56.

COORAY, P. G. 1978. Geology of Sri Lanka. *Proceedings of the Regional Congress on Geology, Mineral and Energy Resources of Southeast Asia (GEOSEA)*, **3**, 701-710.

COORAY, P. G. 1984. *An Introduction to the Geology of Sri Lanka (Ceylon)* (2nd edition). National Museums of Sri Lanka Publication, Colombo.

COORAY, P. G. 1994. The Precambrian of Sri Lanka: a historical review. *In*: RAITH, M. & HOERNES, S. (eds) *Tectonic, Metamorphic and Isotopic Evolution of Deep Crustal Rocks, with Special Emphasis on Sri Lanka*. Precambrian Research, **66**, 3-18.

CRAWFORD, A. R. 1969. Reconnaissance Rb-Sr dating of the Precambrian rocks of southern Peninsular India. *Journal of the Geological Society of India*, **10**, 117-166.

CRAWFORD, A. R. & OLIVER, R. L. 1969. The Precambrian geochronology of Ceylon. *Geological Society, Australia, Special Publications*, **2**, 283-306.

CROWE, W. A., COSCA, M. A. & HARRIS, L. B. 2001. $^{40}Ar/^{39}Ar$ geochronology and Neoproterozoic tectonics along the northern margin of the Eastern Ghats Belt in north Orissa, India. *Precambrian Research*, **108**, 237-266.

CRUSE, T. & HARRIS, L. B. 1994. Ediacaran fossils from the Stirling Range Formation, Western Australia. *Precambrian Research*, **67**, 1-10.

CRUSE, T., HARRIS, L. B. & RASMUSSEN, B. 1993. The discovery of Ediacaran trace and body fossils in the Stirling Range Formation, Western Australia; implications for sedimentation and deformation during the 'Pan-African' orogenic cycle. *Australian Journal of Earth Sciences*, **40**, 293-296.

DAHANAYAKE, K. & JAYASENA, H. A. H. 1983. General geology and petrology of some Precambrian crystalline rocks from the Vijayan Complex of Sri Lanka. *Precambrian Research*, **19**, 301-315.

D'CRUZ, E., NAIR, P. K. R. & PRASANNAKUMAR, V. 2000. Palghat gap - a dextral shear zone from the South Indian granulite terrain. *Gondwana Research*, **3**, 21-32.

DEANS, T. & POWELL, J. L. 1968. Trace elements and strontium isotopes in carbonatites, fluorites and limestones from India and Pakistan. *Nature*, **218**, 750-752.

DE MAESSCHALCK, A. A., OEN, I. S., HEBEDA, E. H., VERSCHURE, R. H. & ARPS, C. E. S. 1990. Rubidium-Strontium whole-rock ages of Kataragama and Pottuvil charnockites and East Vijayan gneiss; indication of a 2 Ga metamorphism in the Highlands of southeast Sri Lanka. *Journal of Geology*, **98**, 772-779.

DE WIT, M. J., BOWRING, S. A., ASHWAL, L. D., RANDRIANASOLO, L. G., MOREL, V. P. I. & RAMBELOSON, R. A. 2001. Age and tectonic evolution of Neoproterozoic ductile shear zones in southwestern Madagascar, with implication for Gondwana studies. *Tectonics*, **20**, 1-45.

DE WIT, M., JEFFERY, M., BERGH, H. & NICOLAYSEN, L. 1988. *Geological map of sectors of Gondwana reconstructed from their position ca. 150 Ma*. American Association of Petroleum Geologists, Tulsa, Oklahoma, USA, 2 sheets.

DISSANAYAKE, C. B. & VAN RIEL, B. J. 1978. Petrology and geochemistry of a recently discovered nickeliferous serpentinite in Sri Lanka. *Journal of the Geological Society of India*, **19**, 464-471.

DISSANAYAKE, C. B. & MUNASINGHE, T. 1984. Reconstruction of the Precambrian sedimentary basin in the granulite belt of Sri Lanka. *Chemical Geology*, **47**, 221-247.

DISSANAYAKE, C. B. & CHANDRAJITH, R. 1999. Sri Lanka-Madagascar Gondwana Linkage: Evidence for a Pan-African mineral belt. *Journal of Geology*, **107**, 223-235.

DOBMEIER, C. J. & RAITH, M. M. 2002. Crustal architecture and evolution of the Eastern Ghats Belt and adjacent regions of India. *In*: YOSHIDA, M., WINDLEY, B. F. & DASGUPTA, S. (eds) *Proterozoic East Gondwana: Supercontinent Assembly and Breakup*. Geological Society, London, Special Publication, **xx**, xx-xx.

DRURY, S. A. & HOLT, R. W. 1980. The tectonic framework of the South Indian craton: a reconnaissance involving LANDSAT imagery. *Tectonophysics*, **65**, T1-T5.

DRURY, S. A., HARRIS, N. B. W., HOLT, R. W., REEVES-SMITH, G. W. & WIGHTMAN, R. T. 1984. Precambrian tectonics and crustal evolution in South India. *Journal of Geology*, **92**, 1-20.

EKLUND, O., KONOPELKO, D., RUTANEN, H., FRÖJDÖ, S. & SHEBANOV, A. D. 1998. 1.8 Ga Svecofennian post-collisional shoshonitic magmatism in the Fennoscandian shield. *In*: LIÉGEOIS, J.-P. (ed.) *Post-collisional Magmatism*. Lithos, **45**, 87-108.

FAULHABER, S. & RAITH, M. 1991. Geothermometry and geobarometry of high-grade rocks: a case study on garnet-pyroxene granulites in southern Sri Lanka. *Mineralogical Magazine*, **55**, 17-40.

FERMOR, L. L. 1936. An attempt at the correlation of the ancient schistose formations of Peninsular India. *Geological Survey of India, Memoir*, **70**, 1-52.

FERNANDO, L. J. D. 1948. The geology and mineral deposits of Ceylon. *Bulletin of the Imperial Institute*, **46**, 303–325.

FERNANDO, L. J. D. 1970. The geology and mineral resources of Ceylon. *Proceedings of the Second Seminar of Geochemical Prospecting Methods and Techniques*. UN Mineral Resources Development Series, **38**, 381–390.

FERNANDO, G. W. A. R., BAUMGARTNER, L. P. & HOFMEISTER, W. 2000. Metasomatic corundum–sapphirine–spinel-bearing rocks at Rupaha, Sri Lanka. *Beihefte zum European Journal of Mineralogy*, **12**, 46.

FITZSIMONS, I. C. W. 2000a. A review of tectonic events in the East Antarctic shield and their implications for Gondwana and earlier supercontinents. *Journal of African Earth Sciences*, **31**, 3–23.

FITZSIMONS, I. C. W. 2000b. Grenville-age basement provinces in East Antarctica: evidence for three separate collisional orogens. *Geology*, **28**, 879–882.

FONAREV, V. I., KONILOV, A. N. & SANTOSH, M. 2000. Multistage metamorphic evolution of the Trivandrum Granulite Block, southern India. *Gondwana Research*, **3**, 293–314.

GEOLOGICAL SURVEY OF INDIA. 1995a. *Geological and mineral map of Kerala (scale 1:500 000)*.

GEOLOGICAL SURVEY OF INDIA. 1995b. *Geological and mineral map of Tamil Nadu and Pondicherry (scale 1:500 000)*.

GEOLOGICAL SURVEY OF SRI LANKA. 1982. *Geological map of Sri Lanka (scale 8 miles to 1 inch)*. Geological Survey Department, Colombo, Sri Lanka.

GEOLOGICAL SURVEY OF SRI LANKA. 1983. *Structural map of Sri Lanka (scale 8 miles to 1 inch)*. Geological Survey Department, Colombo, Sri Lanka.

GHOSH, J. P. 1999. *U–Pb geochronology and structural geology across major shear zones of the Southern Granulite Terrain of India and organic carbon isotope stratigraphy of the Gondwana coal basins of India: their implications for Gondwana studies*. PhD Thesis, University of Cape Town, South Africa.

GHOSH, J. P., ZARTMAN, R. E. & DE WIT, M. J. 1998. Re-evaluation of tectonic framework of southernmost India: new U–Pb geochronological and structural data, and their implication for Gondwana reconstruction. *In*: ALMOND, J., ANDERSON, J., BOOTH, P. ET AL. (eds) *Gondwana 10: Event stratigraphy of Gondwana*. Journal of African Earth Sciences, **27/1A**, 86.

GOPALAKRISHNAN, K. 1998. Extensions of Eastern Ghats mobile belt, India; a geological enigma. *In*: RAMAKRISHNAN, M., PAUL, D. K. & MISHRA, R. N. (eds) *Proceedings of Workshop on the Eastern Ghats mobile belt*. Geological Survey of India, Special Publication, **44**, 22–38.

GOPALAKRISHNAN, K., SUGAVANAM, E. B. & VENKATA RAO, V. 1976. Are there rocks older than Dharwars? A reference to rocks in Tamil Nadu. *Indian Mineralogist*, **16**, 26–34.

GREW, E. S. 1982. Sapphirine, kornerupine, and sillimanite + orthopyroxene in the charnockitic region of South India. *Journal of the Geological Society of India*, **23**, 469–505.

GREW, E. S. 1984. Note on sapphirine and sillimanite + orthopyroxene from Panrimalai, Madurai District, Tamil Nadu. *Journal of the Geological Society of India*, **25**, 116–119.

HANSEN, E. C., HICKMAN, M. H., GRANT, N. K. & NEWTON, R. C. 1985. Pan-African age of Peninsular gneiss near Madurai, south India. *EOS*, **66**, 419–420.

HANSEN, E. C., JANARDHAN, A. S., NEWTON, R. C., PRAME, W. K. B. N. & RAVINDRA KUMAR, G. R. 1987. Arrested charnockite formation in southern India and Sri Lanka. *Contributions to Mineralogy and Petrology*, **96**, 225–244.

HANSEN, E. C., NEWTON, R. C., JANARDHAN, A. S. & LINDENBERG, S. 1995. Differentiation of Late Archaean crust in Eastern Dharwar Craton, Krishnagiri–Salem area, south India. *Journal of Geology*, **103**, 629–651.

HAPUARACHCHI, D. J. A. C. 1972. Evolution of the granulites and subdivision of the granulite facies in Ceylon. *Geological Magazine*, **109**, 435–443.

HAPUARACHCHI, D. J. A. C. 1983. The granulite facies in Sri Lanka. *Geological Survey Department of Sri Lanka Professional Paper*, **4**.

HARLEY, S. L. 1998. An appraisal of peak temperatures and thermal histories in ultrahigh-temperature (UHT) crustal metamorphism; the significance of aluminous orthopyroxene. *In*: MOTOYOSHI, Y. & SHIRAISHI, K. (eds) *Origin and Evolution of Continents; Proceedings of the International Symposium*. Memoirs of National Institute of Polar Research Special Issue, **53**, 49–73.

HARRIS, L. B. & BEESON, J. 1993. Gondwanaland significance of lower Palaeozoic deformation in central India and SW Western Australia. *Journal of the Geological Society, London*, **150**, 811–814.

HARRIS, N. B. W., JACKSON, D. H., MATTEY, D. P., SANTOSH, M. & BARTLETT, J. 1993. Carbon-isotope constraints on fluid advection during contrasting examples of incipient charnockite formation. *Journal of Metamorphic Geology*, **11**, 833–843.

HARRIS, N. B. W., SANTOSH, M. & TAYLOR, P. N. 1994. Crustal evolution in South India: constraints from Nd isotopes. *Journal of Geology*, **102**, 139–150.

HATHERTON, T., PATTIARCHCHI, D. B. & RANASINGHE, R. B. 1975. Gravity map of Sri Lanka (scale 1:1,000,000). Geological Survey Department of Sri Lanka Professional Paper, **3**.

HIROI, Y. & MOTOYOSHI, Y. (eds) 1990. *Study of Geologic Correlation between Sri Lanka and Antarctica (1988–1989)*. Chiba University, Japan.

HIROI, Y., ASAMI, M., COORAY, P. G. ET AL. 1990. Arrested charnockite formation in Sri Lanka: field and petrographical evidence. *In*: HIROI, Y. & MOTOYOSHI, Y. (eds) *Study of Geologic Correlation between Sri Lanka and Antarctica (1988–1989)*. Chiba University, Japan, 1–18.

HIROI, Y., OGO, Y. & NAMBA, K. 1994. Evidence for prograde metamorphic evolution of Sri Lankan pelitic granulites, and implications for the development of continental crust. *In*: RAITH, M. & HOERNES, S. (eds) *Tectonic, Metamorphic and Isotopic Evolution of Deep Crustal Rocks, with Special Emphasis on Sri Lanka*. Precambrian Research, **66**, 245–263.

HOFFBAUER, R. & SPIERING, B. 1994. Petrologic phase equilibria and stable isotope fractionations of carbonate-silicate parageneses from granulite-grade rocks of Sri Lanka. *In*: RAITH, M. & HOERNES, S. (eds) *Tectonic, Metamorphic and Isotopic Evolution of Deep Crustal Rocks, with Special Emphasis on Sri Lanka. Precambrian Research*, **66**, 325–350.

HOLMES, A. 1955. Dating the Precambrian of Peninsular India and Ceylon. *Proceedings of the Geological Society of Canada*, **7**, 81–105.

HÖLZL, S., HOFMANN, A. W., TODT, W. & KÖHLER, H. 1994. U–Pb geochronology of the Sri Lankan basement. *In*: RAITH, M. & HOERNES, S. (eds) *Tectonic, Metamorphic and Isotopic Evolution of Deep Crustal Rocks, with Special Emphasis on Sri Lanka*. Precambrian Research, **66**, 123–149.

HÖLZL, S., KÖHLER, H., KRÖNER, A., JAECKEL, P. & LIEW, T. C. 1991. Geochronology of the Sri Lankan basement. *In*: KRÖNER, A. (ed.) *The Crystalline Crust of Sri Lanka, Part 1, Summary of Research of the German–Sri Lankan Consortium*. Geological Survey Department of Sri Lanka, Professional Paper, **5**, 237–257.

IRWIN, J. J., KIRSCHBAUM, C., LIM, T. *ET AL.* 1987. Lasermicroprobe ^{39}Ar–^{40}Ar ages of high grade metamorphic rocks from Sri Lanka. *EOS*, **68**, 431.

JACOBS, J. FANNING, C. M., HENJES-KUNST, F., OLESCH, M. & PEACH, H.-J. 1998. Continuation of the Mozambique Belt into East Antarctica: Grenville-Age metamorphism and polyphase Pan-African highgrade events in central Dronning Maud Land. *Journal of Geology*, **106**, 385–406

JANARDHAN, A. S. 1996. The Oddanchathram anorthosite body, Madurai Block, southern India. *In*: SANTOSH, M. & YOSHIDA, M. (eds) *The Archaean and Proterozoic Terrains in Southern India within East Gondwana*. Gondwana Research Group Memoir, **3**, 385–390.

JANARDHAN, A. S. & WIEBE, R. A. 1985. The petrology and geochemistry of Oddanchatram anorthosite and associated basic granulites, Tamil Nadu, South India. *Journal of the Geological Society of India*, **26**, 163–176.

JANARDHAN, A. S., JAYANANDA, M. & SHANKARA, M. A. 1994. Formation and tectonic evolution of granulites from the Biligirirangan and Nilgiri Hills, S. India: geochemical and isotope constraints. *Journal of the Geological Society of India*, **44**, 27–40.

JANARDHAN, A. S., SRIRAMGURU, K., BASAVA, S. & SHANKARA, M. A. 2001. Geikielite–Mg–Al–Spinel–Titanoclinohumite association from a marble quarry near Rajapalaiyam area, part of the 550 Ma old southern granulite terrain, southern India. *Gondwana Research*, **4**, 359–366.

JAYANANDA, M. & PEUCAT, J.-J. 1996. Geochronological framework of Southern India. *In*: SANTOSH, M. & YOSHIDA, M. (eds.) *The Archaean and Proterozoic Terrains in Southern India within East Gondwana*. Gondwana Research Group Memoir, **3**, 53–75.

JAYANANDA, M., JANARDHAN, A.S., SIVASUBRAMANIAN, P. & PEUCAT, J.-J. 1995a. Geochronologic and isotopic constraints on granulite formation in the Kodaikanal area, southern India. *In*: YOSHIDA, M. & SANTOSH, M. (eds) *India and Antarctica during the Precambrian*. Memoir, Geological Society of India, **34**, 373–390.

JAYANANDA, M., MARTIN, H., PEUCAT, J.-J. &
MAHABALESHWAR, B. 1995b. Late Archaean crust–mantle interactions: geochemistry of LREE-enriched mantle derived magmas. Example of the Closepet batholith, southern India. *Contributions to Mineralogy and Petrology*, **119**, 314–329.

JAYANANDA, M., MOYEN, J. F., MARTIN, H., PEUCAT, J. J., AUVRAY, B. & MAHABALESWAR, B. 2000. Late Archaean (2550–2520 Ma) juvenile magmatism in the eastern Dharwar Craton, southern India; constraints from geochronology, Nd–Sr isotopes and whole rock geochemistry. *Precambrian Research*, **99**, 225–254.

JAYAWARDENA, D. E., DE S. & CARSWELL, D. A. 1976. The geochemistry of 'charnockites' and their constituent ferromagnesian minerals from the Precambrian of south-east Sri Lanka (Ceylon). *Mineralogical Magazine*, **40**, 541–554.

KATZ, M. B. 1971. The Precambrian metamorphic rocks of Ceylon. *Geologische Rundschau*, **60**, 1523–1549.

KEHELPANNALA, K. V. W. 1991. Structural evolution of high-grade terrains in Sri Lanka with special reference to the areas around Dodangaslanda and Kandy. *In*: KRÖNER, A. (ed.) *The Crystalline Crust of Sri Lanka, Part 1, Summary of Research of the German–Sri Lankan Consortium*. Geological Survey Department of Sri Lanka, Professional Paper, **5**, 69–88.

KEHELPANNALA, K. V. W. 1997. Deformation of a highgrade Gondwana fragment, Sri Lanka. *Gondwana Research*, **1**, 47–68.

KEHELPANNALA, K. V. W. & RATNAYAKE, R. M. J. W. K. 2001. Polyphase migmatization of layered basic rocks in the Wanni Complex of Sri Lanka. *Gondwana Research*, **4**, 174–178.

KILPATRICK, J. A. & ELLIS, D. J. 1992. C-type magmas: igneous charnockites and their extrusive equivalents. *Transactions of the Royal Society of Edinburgh: Earth Sciences*, **83**, 155–164.

KLEINSCHRODT, R. 1994. Large scale thrusting in the lower crustal basement of Sri Lanka. *In*: RAITH, M. & HOERNES, S. (eds) *Tectonic, Metamorphic and Isotopic Evolution of Deep Crustal Rocks, with Special Emphasis on Sri Lanka*. Precambrian Research, **66**, 39–57.

KLEINSCHRODT, R. & MCGREW, A. 2000. Garnet plasticity in the lower continental crust: implications for deformation mechanisms based on microstructures and SEM-electron channeling pattern analysis. *Journal of Structural Geology*, **22**, 795–809.

KLEINSCHRODT, R., VOLL, G. & KEHELPANNALA, W. 1991. A layered basic intrusion, deformed and metamorphosed in granulite facies of the Sri Lanka basement. *Geologische Rundschau*, **80**, 779–800.

KÖHLER, H., ROLLER, G., FEHR, T. & HOERNES, S. 1993. Isotopengeochemische Untersuchungen an Granuliten Südindiens. *European Journal of Mineralogy*, **5**, 129.

KOVACH, V. P., SANTOSH, M., SALNIKOVA, E. B., BEREZHNAYA, N. G., BINDU, R. S., YOSHIDA, M. & KOTOV, A. B. 1998. U–Pb zircon age of the Puttetti alkali syenite, southern India. *Gondwana Research*, **1**, 408–410.

KRIEGSMAN, L. M. 1991. Structural geology of the Sri Lankan basement – a preliminary review. *In*: KRÖNER, A. (ed.) *The Crystalline Crust of Sri Lanka, Part 1, Summary of Research of the German–Sri Lankan*

Consortium. Geological Survey Department of Sri Lanka, Professional Paper, **5**, 52–68.

KRIEGSMAN, L. M. 1993. Structural and petrological investigations into the Sri Lankan high-grade terrain–geodynamic evolution and setting of a Gondwana fragment. *Geologica Ultraiectina*, **114**.

KRIEGSMAN, L. M. 1994. Evidence for a fold nappe in the high-grade basement of central Sri Lanka: terrane assembly in Pan-African lower crust? *In*: RAITH, M. & HOERNES, S. (eds) *Tectonic, Metamorphic and Isotopic Evolution of Deep Crustal Rocks, with Special Emphasis on Sri Lanka.* Precambrian Research, **66**, 59–76.

KRIEGSMAN, L. M. 1995. The Pan-African event in East Antarctica: a view from Sri Lanka and the Mozambique Belt. *In*: DIRKS, P. H. G. M., PASSCHIER, C. W. & HOEK, J. D. (eds) *Tectonics of East Antarctica*. Precambrian Research, **75**, 263–277.

KRIEGSMAN, L. M. 1996. Divariant and trivariant reaction line slopes in FMAS and CFMAS: theory and applications. *Contributions to Mineralogy and Petrology*, **126**, 38–50.

KRIEGSMAN, L. M. & SCHUMACHER, J. C. 1999. Petrology of sapphirine-bearing granulites from Central Sri Lanka. *Journal of Petrology*, **40**, 1211–1239.

KRISHNANATH, R. 1981. Coexisting humite–chondrodite–spinel magnesium calcite assemblage from the calc-silicate rocks of Ambasamudram, Tamil Nadu, India. *Journal of the Geological Society of India*, **22**, 235–242.

KRÖNER, A. 1986. Composition, structure, and evolution of the early Precambrian lower continental crust: constraints from geological observations and age relationships. *American Geophysical Union, Geodynamics Series*, **14**, 107–119.

KRÖNER, A. 1991a. African linkage of Precambrian Sri Lanka. *Geologische Rundschau*, **80**, 429–440.

KRÖNER, A. (ed.) 1991b *The Crystalline Crust of Sri Lanka, Part 1, Summary of Research of the German–Sri Lankan Consortium*. Geological Survey Department of Sri Lanka, Professional Paper, **5**.

KRÖNER, A. & JAECKEL, P. 1994. Zircon ages from rocks of the Wanni Complex, Sri Lanka. *Journal of the Geological Society of Sri Lanka*, **5**, 41–57.

KRÖNER, A. & WILLIAMS, I. S. 1993. Age of metamorphism in the high-grade rocks of Sri Lanka. *Journal of Geology*, **101**, 513–521.

KRÖNER, A., HEGNER, E., COLLINS, A. S., WINDLEY, B. F., BREWER, T. S., RAZAKAMANANA, T. & PIDGEON, R. T. 2000. Age and magmatic history of the Antananarivo Block, Central Madagascar, as derived from zircon geochronology and Nd isotopic systematics. *American Journal of Science*, **300**, 251–288.

KRÖNER, A., WILLNER, A. P., HEGNER, E., JAECKEL, P. & NEMCHIN, A. 2001. Single zircon ages, PT evolution and Nd isotopic systematics of high-grade gneisses in southern Malawi and their bearing on the evolution of the Mozambique Belt in southeastern Africa. *Precambrian Research*, **109**, 257–291.

KRÖNER, A., COORAY, P. G. & VITANAGE, P. W. 1991. Lithotectonic subdivision of the Precambrian basement in Sri Lanka. *In*: KRÖNER, A. (ed.) *The Crystalline Crust of Sri Lanka, Part 1, Summary of Research of the German–Sri Lankan Consortium*. Geological Survey Department of Sri Lanka, Professional Paper, **5**, 5–21.

KRÖNER, A., JAECKEL, P. & WILLIAMS, I. S. 1994a. Pb-loss patterns in zircons from a high-grade metamorphic terrain as revealed by different dating methods: U–Pb and Pb–Pb ages for igneous and metamorphic zircons from northern Sri Lanka. *In*: RAITH, M. & HOERNES, S. (eds) *Tectonic, Metamorphic and Isotopic Evolution of Deep Crustal Rocks, with Special Emphasis on Sri Lanka*. Precambrian Research, **66**, 151–181.

KRÖNER, A., KEHELPANNALA, K. V. W. & KRIEGSMAN, L. M. 1994b. Origin of compositional layering and mechanism of crustal thickening in the high-grade gneiss terrain of Sri Lanka. *In*: RAITH, M. & HOERNES, S. (eds) *Tectonic, Metamorphic and Isotopic Evolution of Deep Crustal Rocks, with Special Emphasis on Sri Lanka*. Precambrian Research, **66**, 21–37.

KRÖNER, A., WILLIAMS, I. S., COMPSTON, W., BAUR, N., VITANAGE, P. W. & PERERA, L. R. K. 1987. Zircon ion microprobe dating of high-grade rocks in Sri Lanka. *Journal of Geology*, **95**, 775–791.

KROGSTAD, E. J., HANSON, G. N. & RAJAMANI, V. 1995. Sources of continental magmatism adjacent to Late Archaean Kolar suture zone, south India: distinct isotopic and elemental signatures of two Late Archaean magmatic series. *Contributions to Mineralogy and Petrology*, **122**, 159–173.

LAWVER, L. A. & SCOTESE, C. R. 1987. A revised reconstruction of Gondwanaland. *In*: MCKENZIE, G. D. (ed.) *Gondwana Six: Structure, Tectonics, and Geophysics*. American Geophysical Union, Geophysical Monograph, **40**, 17–23.

LIEGEOIS, J.-P. (ed.) 1998. Post-collisional magmatism. *Lithos*, **45**, 1–563.

LIEW, T. C., MILISENDA, C. C. & HOFMANN, A. W. 1991a. Isotopic contrasts, chronology of elemental transfers and high-grade metamorphism: the central Sri Lankan granulites and the Lewisian (Scotland) and Nuk (SW Greenland) gneisses. *Geologische Rundschau*, **80**, 279–288.

LIEW, T. C., MILISENDA, C. C., HÖLZL, S., KÖHLER, H. & HOFMANN, A. W. 1991b. Isotopic characterization of the high-grade basement rocks in Sri Lanka. *In*: KRÖNER, A. (ed.) *The Crystalline Crust of Sri Lanka, Part 1, Summary of Research of the German–Sri Lankan Consortium*. Geological Survey Department of Sri Lanka, Professional Paper, **5**, 258–267.

MATHAVAN, V. & FERNANDO, G. W. A. R. 2001. Reactions and textures in grossular-wollastonite-scapolite calc-silicate granulites from Maligawila, Sri Lanka: evidence for high-temperature isobaric cooling in the meta-sediments fo the Highland Complex. *Lithos*, **59**, 217–232

MEERT, J. G., VAN DER VOO, R. & AYUB, S. 1995. Paleomagnetic investigation of the Neoproterozoic Gagwe lavas and Mbozi Complex, Tanzania and the assembly of Gondwana. *Precambrian Research*, **74**, 225–244.

MEISSNER, B., DETERS, P., SRIKANTAPPA, C. & KÖHLER, H. 2002. Geochronological evolution of the Moyar, Bhavani and Palghat shear zones of southern India: implications for Gondwana correlations. *Precambrian Research*, **114**, 149–175.

MEZGER, K. & COSCA, M. 1999. The thermal history of the

Eastern Ghats Belt (India), as revealed by U–Pb and $^{40}Ar/^{39}Ar$ dating of metamorphic and magmatic minerals: implications for the SWEAT correlation. *Precambrian Research*, **94**, 251–271.

MILISENDA, C. C. 1991. Compositional characteristics of the Vijayan Complex. *In*: KRÖNER, A. (ed.) *The Crystalline Crust of Sri Lanka, Part 1, Summary of Research of the German–Sri Lankan Consortium.* Geological Survey Department of Sri Lanka, Professional Paper, **5**, 135–140.

MILISENDA, C. C., LIEW, T. C., HOFMANN, A. W. & KRÖNER, A. 1988. Isotopic mapping of age provinces in Precambrian high-grade terrains: Sri Lanka. *Journal of Geology*, **96**, 608–615.

MILISENDA, C. C., LIEW, T. C., HOFMANN, A. W. & KÖHLER, H. 1994. Nd isotopic mapping of the Sri Lanka basement: update, and additional constraints from Sr isotopes. *In*: RAITH, M. & HOERNES, S. (eds) *Tectonic, Metamorphic and Isotopic Evolution of Deep Crustal Rocks, with Special Emphasis on Sri Lanka.* Precambrian Research, **66**, 95–110.

MILLER, J. S., SANTOSH, M., PRESSLEY, R. A., CLEMENTS, A. S. & ROGERS, J. J. W. 1997. A Pan-African thermal event in southern India. *Journal of Southeast Asian Earth Sciences*, **14**, 127–136.

MISHRA, D. C. 1988. Geophysical evidence for a thick crust south of Palghat–Tiruchi Gap in the high grade terrain of south India. *Journal of the Geological Society of India*, **31**, 79–81.

MIYAZAKI, T., KAGAMI, H., SHUTO, K., MORIKIYO, T., RAM MOHAN, V. & RAJASEKARAN, K. C. 2000. Rb–Sr geochronology, Nd–Sr isotopes and whole-rock geochemistry of Yelagiri and Sevattur syenites, Tamil Nadu, South India. *Gondwana Research*, **3**, 39–54.

MOHAN, A. & WINDLEY, B. F. 1993. Crustal trajectory of sapphirine-bearing granulites from Ganguvarpatti, South India: evidence for an isothermal decompression path. *Journal of Metamorphic Geology*, **11**, 867–878.

MOHAN, A., LAL, R. K. & ACKERMAND, D. 1985. Granulites of Ganguvarpatti, Madurai District, Tamil Nadu. *Indian Journal of Earth Sciences*, **12**, 255–278.

MOTOYOSHI, Y., MATSUBARA, S. & MATSUEDA, H. 1989. P–T evolution of the granulite-facies rocks of the Lutzow-Holm Bay region, East Antarctica. *In*: DALY, J. S., CLIFF, R. A. & YARDLEY, B. W. D. (eds) *Evolution of Metamorphic Belts*. Geological Society, London, Special Publications, **43**, 325–329.

MOYEN, J.-F., MARTIN, H. & JAYANANDA, M. 2001. Multielement geochemical modelling of crust–mantle interactions during late-Archaean crustal growth: the Closepet granite (South India). *Precambrian Research*, **112**, 87–105.

MUHONGO, S., KRÖNER, A. & NEMCHIN, A. A. 2001. Single zircon evaporation and SHRIMP ages for granulite-facies rocks in the Mozambique Belt of Tanzania. *Journal of Geology*, **109**, 171–190.

MUKHOPADHYAY, D., SENTHIL KUMAR, P., SRINIVASAN, R., BHATTACHARYA, T. & SENGUPTA, P. 2001. Tectonics of the eastern sector of the Palghat–Cauvery lineament near Nammakal, Tamilnadu. *DCS-DST News*, 9–13.

MUNASINGHE, T. & DISSANAYAKE, C. B. 1980. Pink granites in the Highland Series of Sri Lanka – a case study. *Journal of the Geological Society of India*, **21**, 446–452.

MUNASINGHE, T. & DISSANAYAKE 1982. A plate tectonic model for the geologic evolution of Sri Lanka. *Journal of the Geological Society of India*, **23**, 369–380.

MUTHUSWAMI, T. N. & GNANASEKARAN, R. 1962. The structure and phase petrology of the metamorphic complex Devadanapatti, Madurai district. *Annamalai University Journal*, **23**, 183–196.

NAHA, K. & SRINIVASAN, R. 1996. Nature of the Moyar and Bhavani shear zones, with a note on its implication on the tectonics of the southern Indian Precambrian shield. *Proceedings of the Indian Academy of Science: Earth and Planetary Sciences*, **106**, 237–247.

NANDAKUMAR, V. & HARLEY, S. L. 2000. A reappraisal of the pressure–temperature path of granulites from the Kerala Khondalite Belt, southern India. *Journal of Geology*, **108**, 687–703.

NAQVI, M. S. & ROGERS, J. J. W. (eds) 1987. *Precambrian of South India*. Memoir, Geological Society of India, **4**.

NARASIMHA RAO, P. 1964. Anorthosites of Oddanchathram, Palni Taluk, Madras State. *Indian Mineralogist*, **5**, 99–104.

NÉDÉLEC, A., STEPHENS, E. W. & FALLICK, A. E. 1995. The Panafrican stratoid granites of Madagascar: alkaline magmatism in a post-collisional extensional setting. *Journal of Petrology*, **36**, 1367–1391.

NÉDÉLEC, A., RALISON, B., BOUCHEZ, J.-L. & GRÉGOIRE, V. 2000. Strructure and metamorphism of the granitic basement around Antananarivo: a key to the Pan-African history of central Madagascar and its Gondwana connections. *Tectonics*, **19**, 997–1020.

NELSON, D. R. 1996. *Compilation of SHRIMP U–Pb zircon geochronology data, 1995*. Western Australia Geological Survey, Record 1996/5.

NEWTON, R. C. & HANSEN, E. C. 1986. The South India–Sri Lanka high-grade terrain as a possible deep-crust section. *In*: DAWSON, D. B., CARSWELL, D. A., HALL, J. & WEDEPOHL, K. H. (eds) *The Nature of the Lower Continental Crust*. Geological Society, London, Special Publication, **24**, 297–307.

NICOLLET, C. 1985. Les gneiss rubanés à cordierite et grenat d'Ihosy: un marqueur thermo-barométrique dans la Sud de Madagascar. *Precambrian Research*, **28**, 175–185.

NICOLLET, C. 1990. Crustal evolution of the granulites of Madagascar. *In*: VIELZEUF, D. & VIDAL, Ph. (eds.) *Granulites and Crustal Evolution*. NATO ASI Series C, **311**. Kluwer Academic Publishers, Dordrecht, 291–310.

ODOM, A. L. 1982. Isotopic age determinations of rock and mineral samples from Kerala, India. U.N. New York, N.Y., Case No. 81–10084 (unpublished).

OLIVER, R. L. 1957. The geological structure of Ceylon. *Ceylon Geographer*, **11**, 9–16.

PAQUETTE, J.-L. & NÉDÉLEC, A. 1998. A new insight into Pan-African tectonics in the East–West Gondwana collision zone by U–Pb zircon dating of granites from central Madagascar. *Earth and Planetary Science Letters*, **155**, 45–56.

PERERA, L. R. K. 1983. The origin of the pink granites of Sri Lanka – another view. *Precambrian Research*, **20**, 17–37.

PERERA, L. R. K. 1984. Co-existing cordierite-almandine –

a key to the metamorphic history of Sri Lanka. *Precambrian Research*, **25**, 349–364.

PEUCAT, J. J., MAHABALESHWAR, B. & JAYANANDA, M. 1993. Age of younger tonalitic magmatism and granulite metamorphism in the south Indian transition zone (Krishnagiri area), comparison with older Peninsular gneisses from the Gorur–Hassan area. *Journal of Metamorphic Geology*, **11**, 879–888.

PEUCAT, J. J., VIDAL, P., BERNARD-GRIFFITHS, J. & CONDIE, K. C. 1989. Sr, Nd and Pb isotopic systems in the Archaean low- to high-grade transition zone of southern India: syn-accretion vs. post-accretion granulites. *Journal of Geology*, **97**, 537–550.

PICHAMUTHU, C. S. 1965. Regional metamorphism and charnockitization in Mysore state, India. *Indian Mineralogist*, **6**, 46–49.

POHL, J. R. & EMMERMANN, R. 1991. Chemical composition of the Sri Lankan Precambrian basement. *In*: KRÖNER, A. (ed) *The crystalline crust of Sri Lanka, Part 1, Summary of Research of the German–Sri Lankan Consortium*. Geological Survey Department of Sri Lanka, Professional Paper, **5**, 94–124.

POWELL, C.MCA., ROOTS, S. R. & VEEVERS, J. J. 1988. Pre-breakup continental extension in East Gondwanaland and the early opening of the eastern Indian Ocean. *Tectonophysics*, **155**, 261–283.

PRAME, W. K. B. N. 1991a. Metamorphism and nature of the granulite-facies crust in southwestern Sri Lanka.: characterization by pelitic/psammopelitic rocks and associated granulites. *In*: KRÖNER, A. (ed.) *The Crystalline Crust of Sri Lanka, Part 1, Summary of Research of the German–Sri Lankan Consortium*. Geological Survey Department of Sri Lanka, Professional Paper, **5**, 188–199.

PRAME, W. K. B. N. 1991b. Petrology of the Kataragama Complex, Sri Lanka: evidence for high P,T granulite-facies metamorphism and subsequent isobaric cooling. *In*: KRÖNER, A. (ed.), *The Crystalline Crust of Sri Lanka, Part 1, Summary of Research of the German–Sri Lankan Consortium*. Geological Survey Department of Sri Lanka, Professional Paper, **5**, 200–224.

PRAME, W. K. B. N. & POHL, J. 1994. Geochemistry of pelitic and psammopelitic Precambrian metasediments from southwestern Sri Lanka: implications for two contrasting source terrains and tectonic settings. *In*: RAITH, M. & HOERNES, S. (eds) *Tectonic, Metamorphic and Isotopic Evolution of Deep Crustal Rocks, with Special Emphasis on Sri Lanka*. Precambrian Research, **66**, 223–244.

RAASE, P. 1998. Feldspar thermometry; a valuable tool for deciphering the thermal history of granulite-facies rocks, as illustrated with metapelites from Sri Lanka. *The Canadian Mineralogist*, **36**, 67–86.

RAASE, P. & SCHENK, V. 1994. Petrology of granulite-facies metapelites for the Highland Complex, Sri Lanka: implications for the metamorphic zonation and the P–T path. *In*: RAITH, M. & HOERNES, S. (eds) *Tectonic, Metamorphic and Isotopic Evolution of Deep Crustal Rocks, with Special Emphasis on Sri Lanka*. Precambrian Research, **66**, 265–294.

RADHAKRISHNA, B. P. 1989. Suspect tectono-stratigraphic terrane elements in the Indian subcontinent. *Journal of the Geological Society of India*, **34**, 1–24.

RADHAKRISHNA, T., MALUSKI, H., MITCHELL, J. G. & JOSEPH, M. 1999. $^{40}Ar/^{39}Ar$ and K/Ar geochronology of the dykes from the south Indian granulite terrain. *Tectonophysics*, **304**, 109–129.

RADHAKRISHNA, T., MATHAI, J. & YOSHIDA, M. 1990. Geology and structure of the high-grade rocks from Punalur–Achankovil sector, south India. *Journal of the Geological Society of India*, **35**, 263–272.

RADHAKRISHNA, T., PEARSON, D. G. & MATHAI, J. 1995. Evolution of Archaean southern Indian lithospheric mantle: a geochemical study of Proterozoic Agali-Coimbatore dykes. *Contributions to Mineralogy and Petrology*, **121**, 351–363.

RAITH, M. & HOERNES, S. (eds) 1994. *Tectonic, Metamorphic and Isotopic Evolution of Deep-crustal Rocks, with Special Emphasis on Sri Lanka*. Precambrian Research, **66**.

RAITH, M. & SRIKANTAPPA, C. 1993. Arrested charnockite formation at Kottavattam, southern India. *Journal of Metamorphic Geology*, **11**, 815–832.

RAITH, M., KARMAKAR, S. & BROWN, M. 1997. Ultra-high-temperature metamorphism and multistage decompressional evolution of sapphirine granulites from the Palni Hills Ranges, southern India. *Journal of Metamorphic Geology*, **15**, 379–399.

RAITH, M., SRIKANTAPPA, C., ASHAMANJARI, K. G. & SPIERING, B. 1990. The granulite terrane of the Nilgiri Hills (southern India): characterization of high-grade metamorphism. *In*: VIELZEUF, D. & VIDAL, PH. (eds) *Granulites and Crustal Evolution*. NATO ASI Series C, 311. Kluwer Academic Publishers, Dordrecht, 339–365.

RAITH, M. M., SRIKANTAPPA, C., BUHL, D. & KÖHLER, H. 1999. The Nilgiri Enderbites, south India: nature and age constraints on protolith formation, high-grade metamorphism and cooling history. *Precambrian Research*, **98**, 129–150.

RAJESH, H. M. & SANTOSH, M. 1996a. Fluorapatite from alkaline pegmatites of the Kerala Khondalite Belt: a petrologic and fluid inclusion study. *Journal of the Geological Society of India*, **48**, 637–646.

RAJESH, H. M. & SANTOSH, M. 1996b. Alkaline magmatism in peninsular India. *In*: SANTOSH, M. & YOSHIDA, M. (eds) *The Archaean and Proterozoic Terrains in Southern India within East Gondwana*. Gondwana Research Group Memoir, **3**, 91–115.

RAJESH, H. M., SANTOSH, M. & YOSHIDA, M. 1998. Dextral Pan-African shear along the southwestern edge of the Achankovil Shear Belt, South India: constraints on Gondwana Reconstructions: a discussion. *Journal of Geology*, **106**, 105–110.

RAMAKRISHNAN, M. 1993. Tectonic evolution of the granulite terrains of southern India. *In*: RADAKRISHNA, B. P. (ed.) (1993): *Continental Crust of South India*. Memoir, Geological Society of India, **25**, 35–44.

RAMAKRISHNAN, M. & VISHWANATHAN, M. N. 1983. Crustal evolution in central Karnataka: a review of present data and models. *In*: NAQVI, M. S. & ROGERS, J. J. W. (eds) *Precambrian of South India*. Memoir, Geological Society of India, **4**, 96–109.

RAMESWAR RAO, D., CHARAN, S. N. & NATARAJAN, R. 1991. P–T conditions and geothermal gradients of gneiss–enderbitic rocks: Dharmapuri area, Tamil Nadu, India. *Journal of Petrology*, **32**, 539–554.

RATSCHBACHER, L., FRISCH, W., LINZER, G. & MERLE, O.

1991. Lateral extrusion in the eastern Alps. Part 2: structural analysis. *Tectonics*, **10**, 257–271.

RAVINDRA KUMAR, G. R. & CHACKO, T. 1986. Mechanisms of charnockite formation and breakdown in southern Kerala; implications for the origin of the southern Indian granulite terrain. *Journal of the Geological Society of India*, **28**, 277–288.

RAVINDRA KUMAR, G. R. & CHACKO, T. 1994. Geothermobarometry of mafic granulites and metapelite from the Palghat Gap, south India: petrological evidence for isothermal uplift and rapid cooling. *Journal of Metamorphic Geology*, **12**, 479–492.

REDDI, A. G. B., MATHEW, M. P., SINGH, B. & NAIDU, P. S. 1988. Aeromagnetic evidence of crustal structure in the granulite terrain of Tamil Nadu–Kerala. *Journal of the Geological Society of India*, **32**, 368–381.

ROSEN, K., GIERÉ, R. & RAITH, M. 1997. Petrology of clinohumite-, humite- and chondrodite-bearing marbles from the Kerala Khondalite Belt, southern India. *Terra Nova*, **10**, 679.

SACKS, P. E., NAMBIAR, C. G. & WALTERS, L. J. 1997. Dextral Pan-African shear along the southwestern edge of the Achankovil Shear Belt, South India: constraints on Gondwana Reconstructions. *Journal of Geology*, **105**, 275–284.

SACKS, P. E., NAMBIAR, C. G. & WALTERS, L. J. 1998. Dextral Pan-African shear along the southwestern edge of the Achankovil Shear Belt, South India: constraints on Gondwana Reconstructions: a reply. *Journal of Geology*, **106**, 110–114.

SANDIFORD, M., POWELL, R., MARTIN, S. F. & PERERA, L. R. K. 1988. Thermal and baric evolution of garnet granulites from Sri Lanka. *Journal of Metamorphic Geology*, **6**, 351–364.

SANTOSH, M. 1987. Cordierite gneisses of south Kerala, India. Petrology, fluid inclusion and implications on uplift history. *Contributions to Mineralogy and Petrology*, **97**, 343–356.

SANTOSH, M. 1996. The Trivandrum and Nagercoil granulite blocks. *In*: SANTOSH, M. & YOSHIDA, M. (eds) *The Archaean and Proterozoic Terrains in Southern India within East Gondwana.* Gondwana Research Group Memoir, **3**, 243–277.

SANTOSH, M. & DRURY, S. A. 1988. Alkali granites with Pan-African affinities from Kerala, S. India. *Journal of Geology*, **96**, 616–626.

SANTOSH, M. & Yoshida, M. 1986. Charnockite in the breaking: evidence from the Trivandrum region, Kerala. *Journal of the Geological Society of India*, **28**, 306–310.

SANTOSH, M., HARRIS, N. B. W., JACKSON, D. H. & MATTEY, D. P. 1990. Dehydration and incipient charnockite formation: a phase equilibria and fluid inclusion study from South India. *Journal of Geology*, **98**, 915–926.

SATISH-KUMAR, M. 2000. Ultrahigh–temperature metamorphism in Madurai Granulites, southern India: evidence from Carbon isotope thermometry. *Journal of Geology*, **108**, 479–486.

SATISH-KUMAR, M. & Harley, S.L. 1998. Reaction textures in scapolite–wollastonite–grossular calc-silicate rock from the Kerala Khondalite Belt, southern India: evidence for high-temperature metamorphism and initial cooling. *Lithos*, **44**, 83–99.

SATISH-KUMAR, M., WADA, H., SANTOSH, M. & YOSHIDA, M. 2001. Fluid-rock history of granulite facies humite-marbles from Ambasamudram, southern India. *Journal of Metamorphic Geology*, **19**, 395–410.

SCHENK, V., RAASE, P. & SCHUMACHER, R. 1988. Very high temperatures and isobaric cooling before tectonic uplift in the Highland Series of Sri Lanka. *Terra Cognita*, **8**, 265.

SCHENK, V., RAASE, P. & SCHUMACHER, R. 1991. Metamorphic zonation and P–T history of the Highland Complex in Sri Lanka. *In*: KRÖNER, A. (ed.) *The Crystalline Crust of Sri Lanka, Part 1, Summary of Research of the German–Sri Lankan Consortium.* Geological Survey Department of Sri Lanka, Professional Paper, **5**, 150–163.

SCHLEICHER, H., KRAMM, U., PERNICKA, E., SCHIDLOWSKI, M., SCHMIDT, F., SUBRAMANIAN, V., TODT, W. & VILADKAR, S. G. 1998. Enriched subcontinental upper mantle beneath southern India: evidence from Pb, Nd, Sr, and C–O isotopic studies on Tamil Nadu Carbonatites. *Journal of Petrology*, **39**, 1765–1785.

SCHLEICHER, H., TODT, W., VILADKAR, S. G. & SCHMIT, F. 1997. Pb/Pb age determinations on Newania and Sevathur carbonatites of India: evidence for multistage histories. *Chemical Geology*, **140**, 261–273.

SCHUMACHER, R. & FAULHABER, S. 1994. Summary and discussion of P–T estimates from garnet–pyroxene–plagioclase–quartz-bearing granulite-facies rocks from Sri Lanka. *In*: RAITH, M. & HOERNES, S. (eds) *Tectonic, Metamorphic and Isotopic Evolution of Deep Crustal Rocks, with Special Emphasis on Sri Lanka.* Precambrian Research, **66**, 295–308.

SCHUMACHER, R., SCHENK, V., RAASE, P. & VITANAGE, P. W. 1990. Granulite facies metamorphism of metabasic and intermediate rocks in the Highland Series of Sri Lanka. *In*: ASHWORTH, J. R. & BROWN, M. (eds) *High-temperature Metamorphism and Crustal Anatexis.* Unwin Hyman, London, 235–271.

SHAW, H. F., NIEMEYER, S., GLASSLEY, W., RYERSON, F. J. & ABEYSINGHE, P. B. 1987. Isotopic and trace element systematics of the amphibolite to granulite facies transition in the Highland Series of Sri Lanka. *EOS*, **68**, 464.

SHERATON, J. W., TINGEY, R. J., OLIVER, R. L. & BLACK, L. P. 1995. Geology of the Bunger Hills–Denman Glacier region, East Antarctica. *Australian Geological Survey Organization, Bulletin*, **244**, 1–124.

SINHA-ROY, S., MATHAI, J. & NARAYANASWAMY 1984. Structure and metamorphic characteristics of cordierite-bearing gneiss of South Kerala. *Journal of the Geological Society of India*, **25**, 231–244.

SILVA, K. K. M. W. 1985. Geological structure of Sri Lanka – a new perspective. *In*: DISSANAYAKE, C. B. & COORAY, P. G. (eds) *Recent Advances in the Geology of Sri Lanka.* International Center for Training and Exchanges in the Geosciences, Occasional Publication, **6**, 101–108.

SILVA, K. K. M. W 1987. Mineralization and wall-rock alteration at Bogala graphite deposit, Bulatkohupitiya, Sri Lanka. *Economic Geology*, **82**, 1710–1722.

SMITH, A. G. & HALLAM, A. 1970. The fit of the southern continents. *Nature*, **225**, 139–144.

SOMAN, K., LOBZOVA, R. V. & SIVADAS, K. M. 1986. Geology, genetic types, and origin of graphite in

South Kerala, India. *Economic Geology*, **81**, 997–1002.

SOMAN, K., NAIR, N. G. K., GOLUBYEV, V. N. & ARAKELYAN, M. M. 1982. Age data on pegmatites of south Kerala and their tectonic significance. *Journal of the Geological Society of India*, **23**, 458–462.

SOMAN, K., NARAYANASWAMY & VAN SCHMUS, W. R. 1995. Preliminary U–Pb zircon ages of high-grade rocks in southern Kerala, India. *Journal of the Geological Society of India*, **45**, 127–136.

SRIKANTAPPA, C., RAITH, M. & SPIERING, B. 1985. Progressive charnockitization of a leptynite–khondalite suite in southern Kerala, India – Evidence for formation of charnockites through decrease in fluid pressure? *Journal of the Geological Society of India*, **26**, 849–872.

SRIKANTAPPA, C., RAITH, M. & TOURET, J. L. R. 1992. Synmetamorphic high density fluid inclusions in the lower crust: evidence from Niligiri granulites, South India. *Journal of Petrology*, **33**, 733–760.

STOSCH, H.-G. 1991. Geochemistry of the mafic intrusion in the synforms of the Kandy area. *In*: KRÖNER, A. (ed.) *The Crystalline Crust of Sri Lanka, Part 1, Summary of Research of the German–Sri Lankan Consortium*. Geological Survey Department of Sri Lanka, Professional Paper, **5**, 125–134.

SUBRAHMANYAM, C. & VERMA, R. K. 1986. Gravity fields, structure and tectonics of the eastern Ghats. *Tectonophysics*, **126**, 195–212.

TAKIGAMI, Y., YOSHIDA, M. & FUNAKI, M. 1999. ^{40}Ar–^{39}Ar ages of dolerite dykes from Sri Lanka. *Polar Geoscience*, **12**, 176–182.

TORSVIK, T. H., TUCKER, R. D., ASHWAL, L. D., CARTER, L. M., JAMTVEIT, B., VIDYADHARAN, K. T. & VENKATARAMANA, P. 2000. Late cretaceous India–Madagascar fit and timing of break-up related magmatism. *Terra Nova*, **12**, 220–224.

TUCKER, R. D., ASHWAL, L. D., HANDKE, M. J., HAMILTON, M. A., LE GRANGE, M. & RAMBELOSON, R. A. 1999. U–Pb geochronology and isotope geochemistry of the Archaean and Proterozoic rocks of north-central Madagascar. *Journal of Geology*, **107**, 135–153.

UNNIKRISHNAN-WARRIER, C. 1997. Isotopic signature of Pan-African rejuvenation in the Kerala Khondalite Belt, southern India: implications for East Gondwana Reassembly. *Journal of the Geological Society of India*, **50**, 179–190.

UNNIKRISHNAN-WARRIER, C., SANTOSH, M. & YOSHIDA, M. 1995. First report of Pan-African Sm–Nd and Rb–Sr mineral isochron ages from regional charnockites of southern India. *Geological Magazine*, **132**, 253–260.

VINOGRADOV, A., TUGARINOV, A., ZHYCOV, C., STAPNIKOVA, N., BIBIKOVA, E. & KHORRE, K. 1964. Geochronology of Indian Precambrian. *XXII International Geological Congress Report*, **X**, 531–567.

VITANAGE, P. W. 1970. A study of the geomorphology and morphotectonics of Ceylon. *Proceedings of the Second Seminar of Geochemical Prospecting Methods and Techniques*. UN Mineral Resources Development Series, **38**, 391–405.

VITANAGE, P. W. 1972. Post-Precambrian uplifts and regional neotectonic movements in Ceylon. *Proceedings of the 24th International Geological Congress*, Montreal, Canada, Section 3, 642–654.

VITANAGE, P. W. 1985. Tectonics and mineralization in Sri Lanka. *Geological Society of Finland Bulletin*, **57**, 157–168.

VOLL, G. & KLEINSCHRODT, R. 1991. Sri Lanka: structural, magmatic and metamorphic development of a Gondwana fragment. *In*: KRÖNER, A. (ed.) *The Crystalline Crust of Sri Lanka, Part 1, Summary of Research of the German–Sri Lankan Consortium*. Geological Survey Department of Sri Lanka, Professional Paper, **5**, 22–51.

WADIA, D. N. 1942. Geological Survey. *Ceylon Administration Reports*, **part II**, 14–15.

WINDLEY, B. F., RAZAFINIPARANY, A., RAZAKAMANANA, T. & ACKERMAND, D. 1994. Tectonic framework of the Precambrian of Madagascar and its Gondwana connections: a review and reappraisal. *Geologische Rundschua*, **83**, 642–659.

YOSHIDA, M., FUNAKI, M. & VITANAGE, P. W. 1992. Proterozoic to Mesozoic East Gondwana: the juxtaposition of India, Sri Lanka, and Antarctica. *Tectonics*, **11**, 381–391.

YOSHIDA, M., RAJESH, H. M. & SANTOSH, M. 1999. Juxtaposition of India and Madagascar: a perspective. *Gondwana Research*, **3**, 449–462.

Archaean–Cambrian crustal development of East Antarctica: metamorphic characteristics and tectonic implications

S. L. HARLEY

Department of Geology & Geophysics, University of Edinburgh, Kings Buildings, West Mains Road, UK (e-mail: sharley@glg.ed.ac.uk)

Abstract: The East Antarctic Shield consists of a variety of Archaean and Proterozoic–Cambrian high-grade terranes that have distinct crustal histories and were amalgamated at various times in the Precambrian–Cambrian. High-grade Pan-African tectonism at 600–500 Ma is recognized from four distinct belts: the Dronning Maud Land, Lützow-Holm Bay, Prydz Bay and Denman Glacier Belts. These high-grade belts juxtapose distinct Mesoproterozoic and Neoproterozoic crustal provinces (Maud, Rayner and Wilkes), the Rauer Terrane, and have also marginally affected Archaean cratonic remnants in the Napier Complex and southern Prince Charles Mountains. The Wilkes Province experienced its principal tectonothermal events prior to 1130 Ma and was not affected by the younger events that characterize the Maud Province (1150 and 1030–990 Ma), the Rayner Province (990–920 Ma) and the Rauer Terrane (1030–990 Ma). These differences between the isotopic/event records of the basement provinces now separated by the Pan-African belts require that the older provinces were not formerly parts of a continuous 'Grenville' belt as proposed in the SW US–East Antartic model. East Antarctica was not a single unified crustal block within either East Gondwana or Rodinia until the Cambrian, which is now demonstrated to be the key phase of high-grade and ultrahigh-temperature (UHT) metamorphism associated with supercontinent assembly.

The high-grade Pan-African tectonism is characterized by extensive infracrustal melting, clockwise P–T paths, rapid post-peak exhumation along isothermal decompression paths to shallow- or mid-crustal levels by 500 Ma and the generation, at least locally, of UHT conditions. A significant flux of heat from the mantle into the deep and initially overthickened crust is required to produce these observed metamorphic effects. Whilst the thermal evolution can be explained by models that invoke the removal of most of the lithospheric mantle following crustal thickening and prior to rapid extension of the remaining crust, these one-dimensional models are inconsistent with present crustal thicknesses of 25–35 km in the Pan-African domains of the East Antarctic Shield.

The Proterozoic–Cambrian evolution of East Gondwana has become a major focus of attention with respect to the testing and refinement of models of supercontinent formation and destruction (Yoshida 1995; Rogers 1996; Unrug 1997; Fitzsimons 2000a). In particular, the concept of a Late Mesoproterozoic supercontinent, Rodinia, produced by the 1100–900 Ma amalgamation of several continental fragments, including East Antarctica, East Africa, India and Australia [SW US–East Antarctic (SWEAT): Moores 1991; Dalziel 1991], has stimulated extensive geological and isotopic re-evaluation of the basement domains that constitute much of the margins of these now separate continental areas. In the light of these studies it is apparent that the Rodinia model in its original and modified configurations is not compatible with the distinct pre-1000 and 1000–500 Ma records now established for regions previously linked in the SWEAT hypothesis (Fitzsimons 2000a, b). In contrast, several of the basement domains have been shown to record the effects of intense late Pan-African (600–500 Ma) high-grade metamorphism and deformation, strongly suggesting that East Gondwana was not finally assembled until the Cambrian (Hensen & Zhou 1997; Fitzsimons 2000a, b).

In line with these developments, a new emphasis has emerged in the analysis of basement domains or terranes that formerly were unified in East Gondwana. The key questions now of concern for each domain, whether in India, Australia or east Antarctica, include: (1) what are the relative metamorphic grades, and what is the importance of Late Mesoproterozoic–Early Neoproterozoic (1300–900 Ma) and Pan-African (600–500 Ma) events?; (2) what P–T paths and T–time records are preserved, and can these be used to constrain the tectonic settings of the metamorphic events?; (3) to what extent do the basement areas between younger tectonic belts preserve distinct crustal histories, and up to which times?; (4) are there systematic age or event differences between the belts now generally referred to as late Pan-African, and how do these impact on accretionary models for East Gondwana assembly?

The geology of the East Antarctic Shield, which preserves distinct high-grade terrains that were once adjacent to Africa, India and Australia in pre-Permian continental reconstructions (e.g. Dalziel 1991), is central to the Rodinia / East Gondwana debate and to the key questions noted above. In this contribution the general features of the major crustal domains or terranes that constitute the East Antarctic

From YOSHIDA, M., WINDLEY, B. F. & DASGUPTA, S. (eds) 2003. *Proterozoic East Gondwana: Supercontinent Assembly and Breakup*. Geological Society, London, Special Publications, **206**, 203–230. 0305–8719/03/$15 © The Geological Society of London.

Shield, building on the excellent recent reviews of Fitzsimons (2000a, b), will be discussed. The metamorphic evidence that bears on the nature of Mesoproterozoic and Pan-African tectonism in selected terranes within the East Antarctic Shield will be examined, with a view to addressing some of the questions posed above and defining the remaining gaps in our understanding of both the East Antarctic Shield and the assembly of East Gondwana.

East Antarctic Shield

The East Antarctic Shield consists of a variety of Archaean and Proterozoic–Cambrian high-grade terranes that have distinct crustal histories and were amalgamated at various times in the Precambrian–Cambrian. The key tectonothermal events or cycles recorded in the East Antarctic Shield are Late Archaean–earliest Proterozoic (i.e. <2840 but > 2480 Ma), Palaeoproterozoic (2200–1700 Ma), Mesoproterozoic–Neoproterozoic (1400–910 Ma) and latest Proterozoic–Cambrian (600–500 Ma, i.e. Pan African). Using this as a template, the basement domains or terranes may be divided into four, partially overlapping, categories according to their age, their degree of reworking, and the nature, style and ages of the reworking events (Fig. 1).

(1) Archaean (>2500 Ma)–Palaeoproterozoic (>1690 Ma) terranes with limited or negligible later overprinting effects resulting from later tectonic events, i.e. the records of the earlier tectonic events are preserved largely intact. The terranes in which the Archaean–Palaeoproterozoic geological histories are largely intact are the Napier Complex of Enderby Land, the Grunehogna Craton of western Dronning Maud Land, the Vestfold Block of Prydz Bay and the Mawson Block of Terre Adélie. The Ruker Terrane may also largely preserve its ancient (>2650 Ma: Boger et al. 2001) tectonic history without significant younger overprints over much of the southern Prince Charles Mountains.

(2) Archaean or mixed Archaean/Palaeoproterozoic areas with strongly polyphase histories in which the early record is partially to largely obscured as a result of overprinting by younger tectonic events, e.g. at 1100–930 Ma and 600–500 Ma, or even in both of these episodes. The key high-grade areas within this category are the Rauer Terrane of Prydz Bay and the eastern margins of the Napier Complex. Parts of the Ruker Terrane of the southern Prince Charles Mountains may be placed in this group, though its geological complexity is rather different in character to the high-grade areas noted above and the significance of any overprints may be spatially variable (Mikhalsky et al. 2001).

(3) Proterozoic terranes (provinces) dominated by Mesoproterozoic–Neoproterozoic tectonism (1400–910 Ma) and only affected to a minor or limited extent by overprinting at 600–500 Ma. These have been grouped into three broad provinces (Fitzsimons 2000a) – the Maud (Dronning Maud Land), Rayner (Enderby, Kemp and MacRobertson Lands: Fig. 2) and Wilkes (Wilkes Land coast) Provinces.

(4) Regions or belts that preserve extensive evidence for Pan-African high-grade metamorphism, magmatism and deformation at times within the broad interval of 600–500 Ma. Though dominated by the Pan-African tectonothermal events, these belts may also preserve evidence for earlier (Proterozoic and Archaean) metamorphic and deformation events. Terranes of category (2) above, though similar in this respect to those distinguished in this category, differ in that the Pan-African events in those complex areas do not dominate the preserved tectonothermal records. The recognition of intense 600–500 Ma Pan-African tectonism and high-grade metamorphism in the East Antarctic Shield has been one of the major breakthroughs in Antarctic geology in the past decade. This, coupled with the preservation of distinct earlier crustal histories in intervening terranes, has refuted the notion of a Circum-East Antarctic Grenville orogenic belt produced by the c. 1100–900 Ma collision of a unified East Antarctic Craton with other parts of East Gondwana (Fitzsimons 2000a).

There are four belts in which high-grade metamorphism at 600–500 Ma is associated with intense deformation, melting and the emplacement of syn- to late tectonic intrusives. The Lützow-Holm region is the first belt in which this style of Pan-African tectonism has been established (e.g. Shiraishi et al. 1994), but it has since been recognized as dominant in the Dronning Maud Land Belt of central Dronning Maud Land (cDML on Fig. 1) and the Yamato–Belgica Mountains (YB on Fig. 1), the Prydz Bay area around the Larsemann Hills and Brattstrand Bluffs coastline (PZ on Fig. 1), and the Denman Glacier region (DG on Fig. 1). Pan-African tectonic activity at 600–500 Ma is also manifested in areas of less intense overprinting, typified by localized shearing, pegmatite intrusion and the emplacement of larger crustally derived felsic igneous complexes. In such regions the earlier Archaean or Proterozoic tectonic events can be recognized but there is a thermal overprint that affects mineral

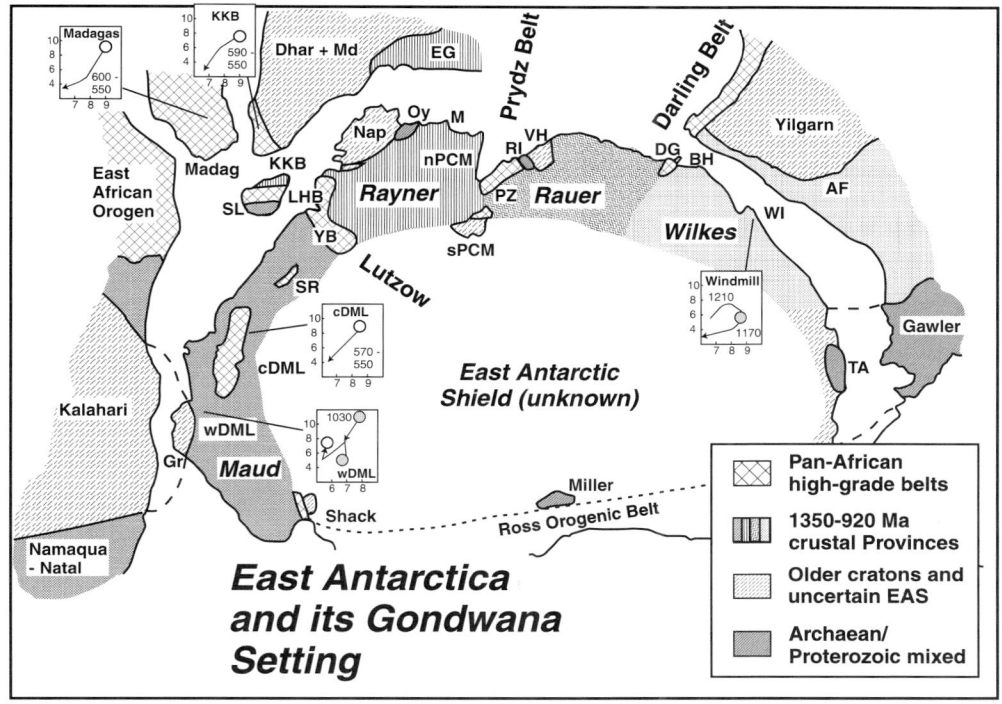

Fig. 1. Map of East Antarctica in its reconstructed Gondwana context, modified from Fitzsimons (2000a, b) and drawn with latitudes and longitudes referred to the present Antarctic coordinates. The Mesoproterozoic–Neoproterozoic provinces recognized by Fitzsimons (2000a) are labelled (Wilkes, Rayner and Maud), along with a possible fourth province (Rauer). Pan-African-age belts (Lutzow, Prydz and Darling) are also distinguished. Specific areas are as follows: AF Albany–Fraser; BH, Bunger Hills; cDML, central Dronning Maud Land; DG Denman Glacier; Dhar + Md, Dhawar and Madras; EG, Eastern Ghats; Gr, Grunehogna; KKB, Kerala Khondalite Belt; LHB, Lützow-Holm Bay; M, Mawson; Madag, Madagascar; Nap, Napier Complex; nPCM, northern Prince Charles Mountains; Oy, Oygardan Islands; PZ, Prydz Bay; RI, Rauer Islands; SL, Sri Lanka; sPCM, southern Prince Charles Mountains; Shack Shackleton Range; SR, Sør Rondane; TA, Terre Adélie; VH, Vestfold Hills; wDML, western Dronning Maud Land; WI Windmill Islands; YB, Yamato–Belgica. Small rectangular boxes are $P-T$ boxes depicting the $P-T$ histories of selected areas, with ages (in Ma) annotated. On each box the vertical axis is pressure (P) in kbar and the horizontal axis is temperature (T) in 100°C units. Shaded circles are the $P-T$ conditions inferred for Mesoproterozoic events (e.g. 1170 Ma in the case of the Windmill Islands), whereas unshaded circles indicate $P-T$ conditions associated with Pan-African tectonism.

cooling ages, including 300–500°C mica blocking-temperature ages (K–Ar, Ar–Ar, Rb–Sr isotopic systems), and often even higher temperature (650–750°C) monazite (U–Pb) and garnet (Sm–Nd mineral isochron) ages.

Whilst it is not possible to provide a full overview of each and every recognized domain or terrane in the East Antarctic Shield, it is nevertheless useful to outline the key features of each to provide a full context for evaluation of the younger tectonic zones and allow comparisons to be made with basement regions in other parts of former East Gondwana (e.g. India, Australia, Sri Lanka). Hence, in this paper the geology of key Archaean and Palaeoproterozoic terranes, as well as that of the Mesoproterozoic and Pan-African high-grade terranes, are reviewed prior to further discussion of the possible tectonic significance of the metamorphic $P-T$ records of some the Pan-African granulites. The four types of basement domains/terranes distinguished above are regrouped for this purpose within the geographic sectors that reflect their broad nearest-neighbour correlations in East Gondwana prior to the Mesozoic: Antarctica–Australia, Antarctica–India-Sri Lanka and Antarctica–Africa–Madagascar. The Antarctica–Australia and Antarctica–Africa–Madagascar sectors, which are considered in more depth by other authors in this volume, are briefly described prior to a more detailed appraisal of the terranes in the Antarctica–India–Sri Lanka sector.

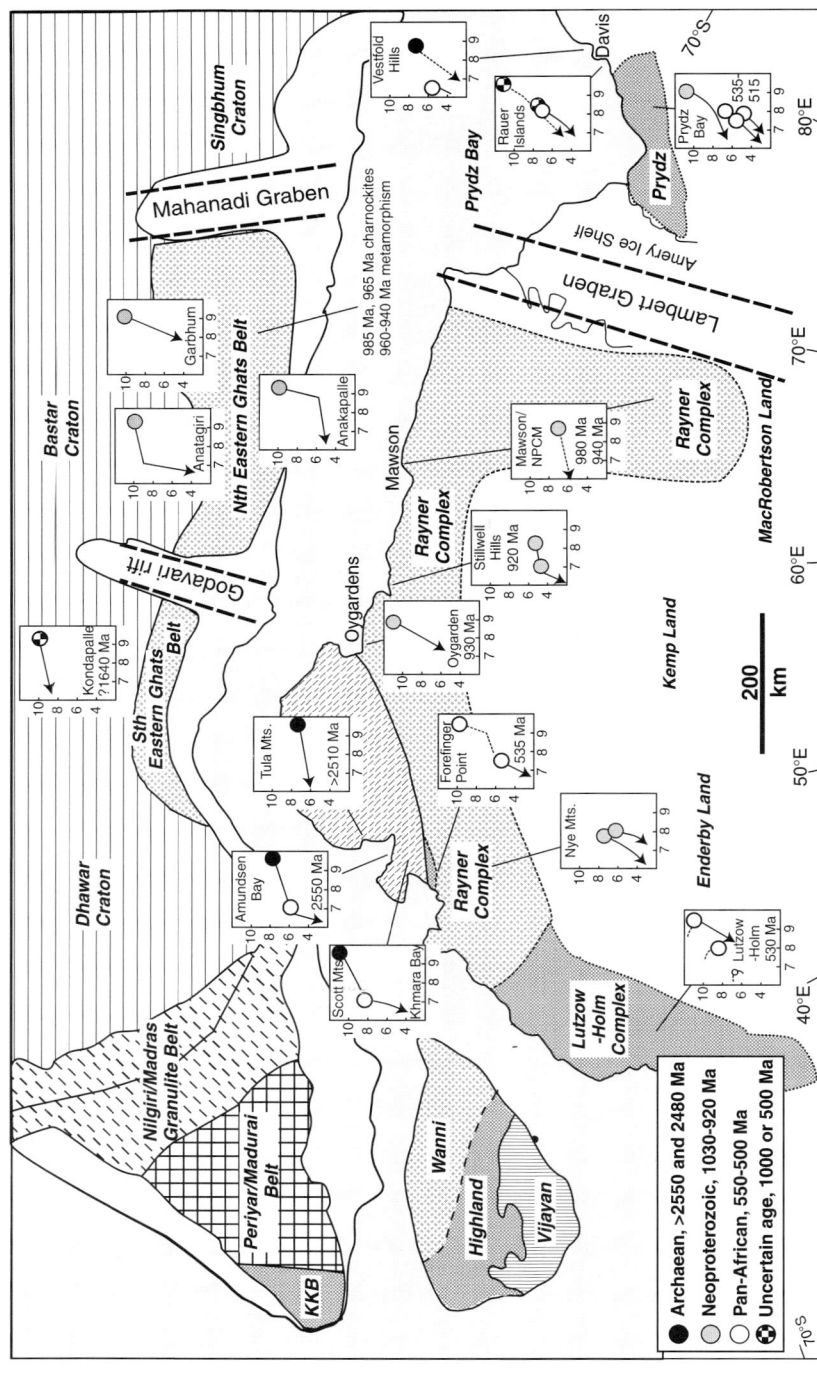

Fig. 2. *P–T* diagrams and simplified time/event information for high-grade terranes and regions in the Antarctica–Indian sector from Lützow-Holm Bay and Sri Lanka (west) to the Vestfold Hills and Singhbum Craton (east). On each small *P–T* box the vertical axis is pressure (*P*) in kbar and the horizontal axis is temperature (*T*) in units of 100 °C (multiply by 100 to arrive at *T* in °C). Note that the principal Indian shield domains, distinguished as the Dhawar Craton (3400–2500 and 2700–2500 Ma), the Nilgiri/Madras Granulite Belt (3000–2500 Ma) and the Periyar/Madurai Belt (2600–2000, 1000 and 550 Ma), are left unshaded or are shaded with unique patterns as they are not considered to directly correlate with the terranes within Antarctica. Ages of events in the Rauer Islands (Mesoproterozoic v. Pan-African) are uncertain because of polymetamorphism; similarly, for Kondapalle in the southern Eastern Ghats Belt. See Dobmeier & Raith (2002) and Dasgupta & Sengupta (2002) for details of the domain structure of the Eastern Ghats Belt, and the text for sources of data. Latitudes and longitudes are referred to present Antarctic coordinates.

Terranes of the Antarctica–Australia Sector

Archaean–Palaeoproterozoic Mawson Block in Terre Adélie

A major region of Archaean–Proterozoic complexity is the Mawson Block of Terre Adélie and other areas on the eastern fringe of the East Antarctic Shield (TA on Fig. 1; Fanning *et al.* 1999; Peucat *et al.* 1999). This block or craton includes the Gawler Craton and adjacent areas of Australia (Fig. 1), and also correlates with high-grade age-event features in the Nimrod Group of the Miller Range in the central Transantarctic Mountains (Miller on Fig. 1; Goodge *et al.* 2001). The basement geology in the Terre Adélie region includes vestiges of 3050–3150 gneisses (e.g. Carnot Gneisses, Fanning *et al.* 1999) and 2560–2450 Ma supracrustals that were metamorphosed in the earliest Proterozoic (2440 Ma). Pelites, greywackes and volcanics were deposited later, probably in a rift environment, between 1775 and 1700 Ma (Point Geologié; Fanning *et al.* 1999). These younger units were then metamorphosed at 1710–1690 Ma to high grades (5–6.5 kbar and 700–750°C) in the west of the region, and mainly to greenschist and amphibolite grades in the east. The 1710–1690 Ma Kimban tectonothermal event in the higher grade domain involved extensive melting of cordierite-bearing pelites, accompanied by early recumbent folding and shearing that reflect N–S extension related to horizontal flattening (Peucat *et al.* 1999). The resulting flatlying fabrics are variably overprinted by subvertical high-grade shears and coeval melt veins which, in some areas, dominate the final structure (Peucat *et al.* 1999). In lower grade areas this event is manifested only in minor magmatism and localized retrograde schist zones. The overall Palaeoproterozoic evolution of the Terre Adélie area is considered to reflect a rapid cycle of sedimentation and high-temperature/low-pressure metamorphism in response to lithospheric thinning without any significant prior crustal thickening (Peucat *et al.* 1999). This style of metamorphism differs from that recorded in the Nimrod Group of the Miller Range of the central Transantarctic Mountains, where Goodge *et al.* (2001) have identified eclogite facies metamorphism of comparable age (1700 Ma). The features of the Mawson Block and basis for its correlations within the East Antarctic Shield are discussed in depth by Fitzsimons (2002), and hence will not be pursued further here.

Archaean in the Denman Glacier / Obruchev Hills area

In the Denman Glacier region (DG on Fig. 1) granitic and tonalitic orthogneiss precursors with ages of between 3000 and 2640 Ma have been recorded, and a granulite facies metamorphic event proposed at 2890 Ma (Black *et al.* 1992). However, the overall geological setting and deformational history of these ancient gneisses, which are intruded by 515 Ma syenites and granites, is not yet clear. As the details of this area and its relationships to present and formerly adjoining regions in Antarctica and Australia are considered in depth by Fitzsimons (2002) they will not be elaborated on further here.

Mesoproterozoic Wilkes Province

The Wilkes Province occurs to the east of the Denman Glacier area briefly described above. The Wilkes Province lacks evidence for any high-grade tectonothermal events younger than 1130 Ma. It is dominated by the 1210–1170 Ma medium-pressure granulite facies metamorphism (6–8 kbar and 800–900°C) of supracrustal sequences and granitoids on a clockwise *P–T* path. Post-peak, high-temperature, near-isothermal decompression (ITD) to *c.* 4–5 kbar and subsequent cooling at mid-crustal depths (10–12 km) was largely complete by 1140 Ma (Post *et al.* 1997). Compositionally varied (felsic and mafic) plutonic rocks, including charnockites, were emplaced between 1170 and 1150 Ma during the post-peak exhumation phase of this major tectonic event (Sheraton *et al.* 1993). In the Windmill Islands (WI on Fig. 1) earlier magmatic and tectonic events at 1340–1315 Ma are also recognized. These involved granite intrusion into 1400–1350 Ma supracrustals and low-pressure/high-temperature metamorphism at 4–5 kbar and 750°C and during crustal thickening (Post *et al.* 1997).

This Mesoproterozoic geological history is clearly very distinct from those of the Maud and Rayner Provinces (described below). However, the geological histories preserved in the Bunger Hills and Windmill Islands areas (BH and WI, respectively, on Fig. 1) of the Wilkes Province correlate very well with those documented from the Albany– Fraser Mmobile Belt of SW Western Australia (AF on Fig. 1) (Clark *et al.* 2000). The pre-1300–1200 Ma metamorphic/deformational events recognized in the Windmill Islands are similar in age and style to collision-related events in the eastern parts of the Albany–Fraser Belt (the Nornalup Complex: Clark *et al.* 2000), whilst the dominant phase of granulite metamorphism and intrusive activity at 1210–1140 Ma is shared by all areas (Sheraton *et al.* 1993; Post *et al.* 1997; Clark *et al.* 2000; Fitzsimons 2000*b*, 2002).

Pan-African high-grade events in the Denman Glacier Belt

Cambrian high-grade tectonism is likely to have been important in the Denman Glacier area (DG on Fig. 1), to the west of the Bunger Hills. This area is correlated with the Darling Mobile Belt or Pinjarra Orogen of Western Australia on the basis of general continental reconstructions, as well as the available geochronology of both the region itself and bounding areas to the west and east (Fitzsimons 2000*a*, 2002). However, although a 520–500 Ma isotopic and rock record is present (Black *et al.* 1992), the geology of the Denman Glacier region and the relationships between these ages and the principal high-grade deformation fabrics remain to be clarified before it can be established to be part of single major Pan-African Orogen linked with the Prydz Belt (PZ on Fig. 1) some 1000 km to the west (Fitzsimons 2000*a*, 2002; Boger *et al.* 2001).

Terranes of the Antarctica–Africa–Madagascar sector

Grunehogna Craton, Dronning Maud Land

On the far western side of the East Antarctic Shield a small area of 3000 Ma granitic basement that crops out in west Dronning Maud Land is termed the Grunehogna Craton (Grantham *et al.* 1995; Gr on Fig. 1). This Archaean basement, overlain by 1100–1000 Ma shelf or platform sediments and volcanics of the Ritscherflya Supergroup, and intruded by 800 Ma mafic sills, is interpreted to be a fragment of the Kalahari Craton of southern Africa that has accidentally been left attached to East Antarctica (Jacobs *et al.* 1993).

Mesoproterozoic–Neoproterozoic Maud Province

The major Proterozoic events in the Maud Province of Dronning Maud Land are felsic plutonism at 1180–1130 and 1080 Ma, followed by regional high-grade metamorphism and deformation at 1050–1030 Ma (Grantham *et al.* 1995; Jacobs *et al.* 1996), and further thrust-related deformation that may be as young as 980 Ma. For example, in western Dronning Maud Land (wDML on Fig. 1), magmatic and supracrustal rocks that occur to the east of a major discontinuity, the Heimefront Shear Zone (Jacobs *et al.* 1997), were metamorphosed under amphibolite to granulite facies conditions at 1080–1000 Ma and contemporaneously intruded by voluminous crustally derived granites and charnockites (Grantham *et al.* 1995). The present structural configuration of these amphibolite to granulite facies gneisses is telescoped by 1010–980 Ma NW-directed thrusting and by further NW-directed thrusting in the Pan-African at 520–500 Ma (Grantham *et al.* 1995; Jacobs *et al.* 1996). In the Sverdrupfjella area of western Dronning Maud Land a western amphibolite facies zone is dominated by hornblende–biotite gneisses derived from volcanic-arc sediment protoliths, interleaved with marbles and calc-silicate gneisses derived from shallow-marine sedimentary precursors. These units are overthrust by sheets that incorporate granulite facies supracrustals (pelites, metagreywackes and meta-arenites) and mafic to ultramafic gneisses interpreted as metamorphosed tholeiitic basalts and gabbros. This assortment of rock types and stacked units is considered to represent the collided and telescoped remains of a Mesoproterozoic volcanic arc and marginal basin that was initially formed by 1600–1200 Ma subduction beneath an East Antarctic craton (Grantham *et al.* 1995).

Collision with the Kalahari Craton is considered to have produced the dominant high-grade gneissic fabrics now observed in many parts of the Maud Province, and resulted in peak P–T conditions (in the eastern Sverdrupfjella) of 11 kbar and 800°C, probably followed by high-temperature decompression to 6–7 kbar at 700°C. This ITD path could either be real or, like others in polymetamorphic terranes, an artifact produced by the overprinting of one event (e.g. 1000 Ma) by a later unrelated event (e.g. 520 Ma) proceeding at high temperatures but at mid-crustal levels. In this respect it is notable that the Grunehogna Craton to the west preserves >1000 Ma K–Ar mica ages, indicating shallow crustal depths since that time, whereas 500–480 Ma mica Ar–Ar ages prevalent in the Maud Province require final cooling and shallow crustal conditions only from the Ordovician (Jacobs *et al.* 1996). These young cooling ages most likely reflect the amphibolite–greenschist facies metamorphism and reactivation of the 1000 Ma gneisses, and their intrusion by A-type granites, in the late Pan-African. It is important to note, however, that some recent work suggests that the metamorphic grade in the latter event may have locally reached upper amphibolite–granulite facies, caused partial melting and produced the principal high-grade gneissic assemblages in some areas at 540–520 Ma (Board pers. comm.).

Pan-African high-grade tectonism in the Dronning Maud Land Belt

In western Dronning Maud Land, early (1000 Ma) thrusts with top-to-the-west and NW movement directions may have been reactivated at moderate to high grades in Pan-African tectonism at 520–500 Ma, a time during which both A-type and S-type granitoids were also intruded into the Maud Province gneisses (Grantham et al. 1995). The Urfjell Group, a sequence of immature quartzites and arkoses deposited between 620 and 510 Ma and now preserved in tectonic contact with the Maud Province gneisses, experienced its deformation in this Pan-African event.

High-grade metamorphism and deformation at 600–500 Ma is recognized from across a significant part of central Dronning Maud Land (cDML on Fig. 1). The Pan-Afican tectonism affects both juvenile and post-1000 Ma crustal precursors (e.g. 600 Ma intrusives), and older orthogneisses (e.g. 1190 Ma metagranitoids), as well as supracrustals that had already undergone high-grade metamorphism at 1090–1030 Ma, ages compatible with those obtained in the Mesoproterozoic Maud Province (Paech 1997; Jacobs et al. 1998). Nappe structures, sheath folding, thrusts that disrupt and re-orient earlier structures, and the presence of laterally variable high-strain zones all point to the Pan-African metamorphism being accompanied by horizontal shortening and extensive displacement of crustal blocks. Such features are compatible with this region being an eastern and southern extension of the East African orogenic belt (Fig. 1; Jacobs et al. 1998; Markl et al. 2000; de Wit et al. 2001). In central Dronning Maud Land the intrusion of 625–610 Ma anorthosite and 540–500 Ma syenites, charnockites and A-type granites bracket an initial Pan-African high-grade event (570–550 Ma) that progressed at 8 kbar and 830°C (Jacobs et al. 1998; Piazolo & Markl 1999). A later amphibolite facies overprint (3–4 kbar and 650°C), and accompanying localized melting and granite emplacement, occurred at 530–510 Ma; by 500–490 Ma the high-grade gneisses had cooled to <300°C. This complex series of magmatic and metamorphic events between 625 and 500 Ma has been interpreted in terms of an initial phase of mafic underplating of the crust, leading to the anorthosites, followed later by collision between parts of West Gondwana (to the NW of central Dronning Maud Land) and the contemporaneous Antarctic Craton at 570–550 Ma. This was then terminated by post-collisional orogenic collapse at 530–500 Ma (Markl et al. 2000; de Wit et al. 2001).

Further to the east, in the the Sør Rondane Mountains (SR on Fig. 1), 1180 Ma orthogneiss precursors and abundant volcanogenic metasediments were metamorphosed under amphibolite–granulite facies conditions and intruded by felsic plutons at c. 1065–1000 Ma (Shiraishi et al. 1999). These gneisses were overprinted in the northeastern Sør Rondane Mountains by a high-grade metamorphic event that progressed at 650–630 Ma in a western zone, but at 600 Ma progressed further eastwards, suggesting that two formerly distinct but similar crustal domains or sheets were subsequently juxtaposed in this part of the Sør Rondane Mountains. Late Pan-African medium-grade metamorphism, migmatization and granitic magmatism then partially overprinted these earlier events at 570–530 Ma (Shiraishi et al. 1999). Still further east in the Yamato–Belgica Complex (YB on Fig. 1) mid–late Proterozoic supracrustals and felsic intrusives were also metamorphosed to granulite facies at 660–500 Ma, and intruded by granites late in the Pan-African (Shiraishi et al. 1997).

Terranes of the Antarctica–India–Sri Lanka sector

Archaean Napier Complex, Enderby Land

The Archaean Napier Complex (Nap on Fig. 1; Fig. 2) of Enderby Land is a remarkable ultrahigh-temperature (UHT) metamorphic complex that records the highest grades of metamorphism seen in the continental crust (Harley 1998a; Harley & Motoyoshi 2000). Whilst some precursors of granitic and tonalitic gneisses locally range in age up to 3800 Ma (Harley & Black 1997), the dominant granulite facies orthogneisses were derived from the metamorphism of somewhat younger (2980–2850 Ma) granites and granodiorites (Sheraton et al. 1987; Harley & Black 1997). These, the more ancient precursors, and gneisses derived from varied supracrustals, were strongly deformed and interleaved under UHT metamorphic conditions in the Late Archaean. A strong and pervasive flat-lying gneissic fabric and early isoclinal folds are refolded and reoriented about several later phases of folds and high-strain zones, so that original relationships between the units are obscured and almost all fabrics are layer-parallel. Intense down-dip linear fabrics defined by the dimensional orientation of the highest grade minerals (e.g. sillimanite, sapphirine, orthopyroxene, quartz) are common, as is boudinage and megaboudinage of competent units during intense layer-parallel extension (e.g. Sheraton et al. 1980, 1987; Harley & Black 1987; Toyoshima et al. 1999).

The Napier Complex is rightly regarded as the best example of UHT crustal metamorphism worldwide. The lithological diversity of the layered paragneisses, which include mafic pyroxene granulites, metamorphosed ironstones and magnetite–quartz rocks, pelites and Fe- and Mg-rich metaquartzites, has enabled the formation of varied UHT mineral assemblages that each record real or apparent near-isobaric cooling (IBC) in their assemblage-specific reaction textures (Ellis et al. 1980; Ellis 1983; Ellis

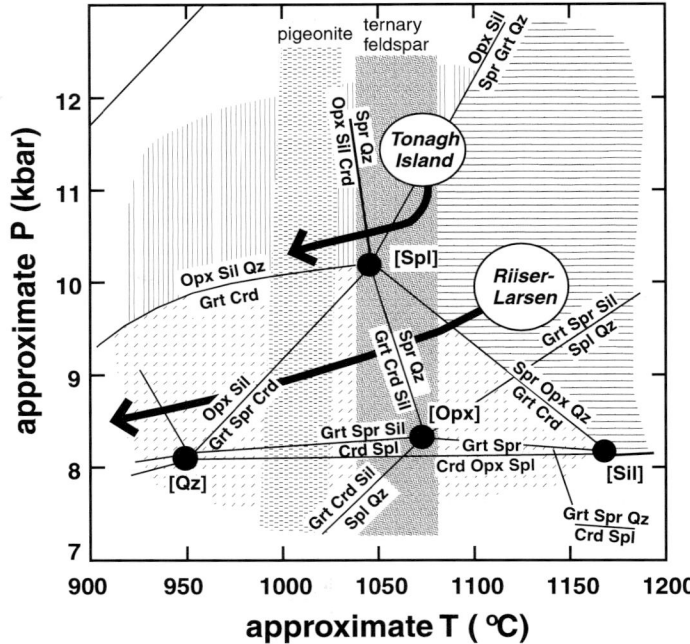

Fig. 3. P–T diagram of the model FMAS-system grid for low f_{O2} pelites, modified after Hensen (1971), Hensen & Green (1973) and Hensen & Harley (1990). Fields of key FMAS assemblages (Spr + Opz + Qz, Opx + Sil + Qz, and Grt + Crd + Sil + Qz) are distinguished by shading. P–T vectors for the Napier Complex are added based on the observations described in the text: Al_2O_3 zoning in Opx coexisting with Spr + Qz (Harley & Motoyoshi 2000) implies cooling, and formation of Opx + Spr + Sil from Grt (Hollis & Harley 2002). Also shown are the T ranges for peak metamorphism obtained from feldspar thermometry (Hokada 2001) and stability of metamorphic pigeonites and subcalcic ferro-augites (Sandiford & Powell 1986; Harley 1987a). Mineral abbreviations: Crd, cordierite; Grt, garnet; Opx, orthopyroxene; Qz, quartz; Sil, sillimanite; Spl, spinel; Spr, sapphirine.

& Green 1985; Harley 1985, 1987a; Sandiford 1985; Hokada et al. 1999). The critical peak minerals and mineral assemblages that indicate UHT metamorphism at 7–11 kbar and 1050–1120°C include: sapphirine + quartz + orthopyroxene or garnet (Ellis et al. 1980; Grew 1980; Motoyoshi & Hensen 1989; Harley & Motoyoshi 2000); aluminous orthopyroxene (9–11 wt% Al_2O_3) + sillimanite + quartz (Grew 1980; Harley 1985; Sandiford 1985; Hokada et al. 1999); osumilite and osumilite + garnet (Ellis et al. 1980; Grew 1982; Hensen & Motoyoshi 1992; Harley 1998a); ternary mesoperthitic feldspars in pelitic gneisses (Grew 1982; Harley 1985; Sheraton et al. 1987; Hokada et al. 1999; Hokada 2001); and inverted Mg-bearing pigeonite in metamorphosed ironstones (Sandiford & Powell 1986; Harley 1987a). Some of these varied constraints on the UHT metamorphism, including the key FeO–MgO–Al_2O_3–SiO_2 (FMAS) assemblages involving orthopyroxene, sapphirine, garnet, sillimanite and quartz, are illustrated in Figure 3. Harley & Motoyoshi (2000) have demonstrated that peak temperatures in this UHT metamorphism were at least 1120 ± 20°C in order to account for observed chemical zoning in orthopyroxene coexisting with sapphirine + quartz, in which orthopyroxene Al_2O_3 contents decrease from 12 wt% in cores to 9 wt% in rims and grains recrystallized with sapphirine.

Post-peak near-IBC cooling occurred at <8 kbar in the northern parts of the UHT region (Amundsen Bay and Tula Mountains) and at 9–10 kbar in areas further south (Scott Mountains: Harley & Hensen 1990; Harley 1998a). This gradient is supported by regional differences in core Al_2O_3 contents in orthopyroxenes equilibrated with garnet + sillimanite (Harley 1985), differences in reaction coronas on sapphirine + quartz (cordierite + sillimanite + garnet in the north; orthopyroxene + sillimanite + garnet in the south: Sheraton et al. 1980, 1987) and the stability of osumilite + garnet in the areas to the north of Amundsen Bay compared with orthopyroxene + sillimanite + K-feldspar + quartz southwards and eastwards (Hensen & Motoyoshi 1992; Harley 1998a). IBC is also documented from garnet and garnet + quartz coronas developed on orthopyroxene in pelites, on orthopyroxene + plagioclase in intermediate rocks, and on orthopyroxene + clinopyroxene in mafics and pyroxenites (Ellis & Green 1985; Harley 1985; Sheraton *et al.* 1987; Osanai *et al.* 1999).

Whilst supporting this general model of post-peak IBC, new textural studies (Hollis & Harley 2002) indicate that a phase of late- to post-deformational ITD through c. 1–3 kbar and $>1000°C$ preceded IBC in the UHT region. This intervening phase of ITD is suggested by the occurrence in tonalitic gneisses of recrystallized plagioclase moats and orthopyroxene collars on garnets elongated in the main UHT fabrics. In support of this, in some magnesian metapelites, garnet is initially resorbed to produce sapphirine + orthopyroxene + sillimanite lamellar intergrowths, which are then extensively replaced and overgrown by later garnet. This textural sequence suggests that the reaction garnet = sapphirine + orthopyroxene + sillimanite, an FMAS divariant reaction that progresses at temperatures lower than or near to spinel [Spl] on Fig. 3), was initially crossed from lower temperature/higher pressure to higher temperature/lower pressure (e.g. with ITD), and then crossed in the reverse direction in the major phase of IBC (Hollis & Harley 2002). Most spectacularly, garnets flattened parallel to the gneissic fabric in a mafic granulite show two-stage coronas that form a 'hub-and-spoke' texture. In this case radial orthopyroxene + plagioclase intergrowths that surround the garnet 'hub' have later been used as sites for the formation of fine rinds of a second garnet ('spokes'), in addition to clinopyroxene and quartz. This textural progression is considered to have resulted from initial garnet breakdown with ITD through the reaction garnet + clinopyroxene + quartz = orthopyroxene + plagioclase, followed by reversal of this reaction to form the second garnet 'spokes' in the IBC phase. The interpretation of the initial textures as indicating ITD prior to IBC is, of course, not unique, given that the reaction textures only indicate that reactions with positive dP/dT slopes have been crossed, and this could occur either with heating or decompression. Also, the textural sequence could result from the superposition of two or more distinct metamorphic events that are unrelated in time, rather than forming as a result of a continuous P–T history, so that the P–T path established from them is artificial or apparent rather than real (Hensen *et al.* 1995). Given these caveats, in this case geobarometry of the secondary garnet-bearing assemblages formed in coronas indicate pressures some 1–3 kbar lower than those inferred from the peak UHT assemblages and hence support the concept of a period of ITD prior to cooling, if the P–T evolution is continuous rather than punctuated.

A counter-clockwise prograde P–T path has been proposed for the UHT event in the Napier Complex by Motoyoshi & Hensen (1989), on the basis of their interpretation of sapphirine–quartz intergrowths as pseudomorphs after earlier cordierite on an up-pressure heating path. This form of P–T path is consistent with the observations that only sillimanite (never kyanite) is observed as inclusions in garnet and that spinel is often mantled by sapphirine in sillimanite-bearing pelites, and with the recent observation of early cordierite included in later garnet and sapphirine in an aluminous pelite (Hollis & Harley 2002). However, this simple interpretation of the prograde path must be regarded with some caution in the light of the multi-event histories that have been proposed to account for complexities in the geochronology of the terrane (Hensen & Motoyoshi 1992; Motoyoshi & Hensen 1989; Harley & Black 1997). The ostensibly prograde reaction textures seen in the UHT region (Motoyoshi & Hensen 1989) may be artefacts of event superposition.

The age of the UHT metamorphism is in the Napier Complex is controversial, despite a vast amount of zircon U–Pb and other isotopic data being available. Harley & Black (1997) have ascribed a c. 2840 Ma age to the UHT event, and attributed the ubiquitous zircon age clusters at 2480–2450 Ma to fabric formation and fluid access in an unrelated later high-grade event (D_3). In contrast, other workers (Grew & Manton 1979; Shiraishi *et al.* 1997; Grew 1998) interpret these younger zircon ages to approximate the age of the UHT event, implying that that all pre-2500 Ma zircons are inherited/protolith grains (orthogneisses) or detrital grains (paragneisses). Recent zircon U–Pb age data from a tonalitic orthogneiss from Tonagh Island (Carson *et al.* 2002) establish that the protolith (granodiorite) is no older than 2626 ± 28 Ma, providing an upper limit on the age of UHT. These workers intepret the abundant 2480 Ma zircon overgrowths and grains in this rock to record the UHT event. However, *in-situ* SHRIMP analyses of zircons from syn- to post-deformational leucosome and pegmatite (author's unpublished data) indicate that the UHT metamorphism can be no younger than 2550 Ma, even though most 'metamorphic' zircon rims in paragneisses also studied *in situ* give younger near-concordant ages (2479 ± 3, 2479 ± 7 and 2511 ± 18 Ma: author's unpublished data). The latter zircons contain fracture networks, partially healed on submicrometre to 10 μm scales, that may be sites of ancient Pb loss. Concordant U–Pb data from such zircons may only define the minimum age of fracture seal rather than the age of UHT metamorphism. Concordant zircon ^{207}Pb–^{206}Pb ages between 2550 and 2440 Ma, obtained from a quartzo feldspathic gneiss from Tonagh Island (Shiraishi *et al.* 1997), could also be interpreted in terms of UHT at or earlier than 2550 Ma followed by post-peak radiogenic Pb loss. At present, the only complete consensus on the age problem is that the pervasive high-grade deformation and UHT metamorphism in the Napier Complex was complete by the earliest Palaeoproterozoic (2480 Ma).

Interpretations of the contribution of synmetamorphic magmatism to the UHT event, the timescale of IBC and deep-crustal residence, and hence models for the nature of the UHT metamorphism in the Napier Complex, are hampered by this lack of agreement on its age. The precursors of strongly yttrium-(Y) and heavy rare earth element (HREE)-depleted orthogneisses, interpreted to have formed through the deep-seated ($>$12 kbar) melting of mafic precursors (Ellis 1987; Sheraton et al. 1987), generally appear to be older than 2800 Ma (Harley & Black 1997) and hence are not relevant to the tectonic setting of the UHT event if it is 2550 Ma or younger in age. Models of UHT metamorphism caused by magmatic accretion into and above the presently exposed crust of the Napier Complex, though consistent with a counter-clockwise P–T history (Motoyoshi & Hensen 1989), require that the principal magmatic additions are broadly synchronous with the UHT metamorphism. Hence, the general lack of magmatic rocks with proven ages of 2550–2450 Ma in the Napier Complex argues against infracrustal magmatic accretion as a cause of the UHT metamorphism.

Any thermal modelling of the UHT event has to be able to explain how temperatures of up to 1120 °C can be generated in crust undergoing deformation at depths of c. 25–40 km. This transient thermal gradient of over 30–40 °C/km could not be sustained down to depths of over 50–60 km without vastly exceeding any reasonable solidus for even refractory peridotite in the contemporaneous mantle. It follows from this that the lithosphere–asthenosphere transition at the time of UHT metamorphism was unlikely to be $>$10–15 km deeper than the presently exposed UHT crustal rocks. Hence, convective removal of the lithospheric thermal boundary layer (e.g. Platt & England 1994) or complete removal of the lithospheric mantle (Platt et al. 1998) following crustal thickening would provide plausible mechanisms for generating the large-scale UHT metamorphism. The phase of UHT near-ITD prior to IBC proposed by Hollis & Harley (2002) could reflect the final stages of rapid and short-lived extensional crustal thinning during and following mantle lithosphere removal, whereas the phase of IBC would postdate this thinning (e.g. Harley 1989; Harley & Hensen 1990). However, such collision–extension models are not consistent with the counter-clockwise prograde P–T history advocated by Motoyoshi & Hensen (1989) and others for the Napier Complex. As a consequence, the UHT metamorphism of the Napier Complex remains enigmatic and rather inexplicable in terms of its tectonic setting.

Mafic dyke swarms intruded the Napier Complex in the Proterozoic (1200 Ma), and steep shear zones and high-strain zones of Cambrian and probably Mesoproterozoic ages also cut and offset the Archaean gneisses (Sheraton et al. 1987). On its eastern flank, in the Oygarden Islands area, 3500–2500 Ma Archaean protoliths and gneisses correlated with those in the Napier Complex are strongly reworked under probable granulite facies conditions at 1600 Ma, and subsequently by Rayner Province deformation and metamorphism at 930–920 Ma (Kelly et al. 2000). In contrast, the main phase of renewed deformation of the Napier Complex on its western and southwestern margins appears to be in amphibolite and granulite facies high-strain zones associated with Pan-African tectonism at 540–520 Ma (Shiraishi et al. 1997; Carson et al. 2002).

Archaean Vestfold Block, Prydz Bay

The Vestfold Hills (VH on Fig. 1; Fig. 2) is a distinct orthogneiss-dominated granulite facies basement block that experienced its main magmatic accretion events and crust formation in the time interval of 2520–2480 Ma (Black et al. 1991; Snape et al. 1997). An older (2520–2000 Ma) group of tonalitic orthogneisses and pyroxene granulites, and supracrustal rocks including metavolcanics, Fe-rich semipelitic and quartzitic sediments and Mg–Al rich claystones (Oliver et al. 1982; Collerson et al. 1983; Harley 1993), underwent high-grade metamorphism and intense deformation prior to 2500 Ma. The intrusion and subsequent deformation of a varied group of granitic to gabbroic magmatic rocks, the 2500–2480 Ma Crooked Lake Gneisses, then followed (Black et al. 1991; Snape et al. 1997). Granulite facies metamorphism at 5–8 kbar and 800–900 °C, constrained by sapphirine–orthopyroxene–spinel and cordierite-bearing assemblages (Harley 1993), was broadly contemporaneous with the main magmatism and completed by the earliest Proterozoic (2470 Ma). The gneiss complex was exhumed to only 3–4 kbar by 2460 Ma (Snape et al. 1997), residing at these shallow crustal depths until at least 2240 Ma when the intrusion of a norite body caused low-pressure pyrometamorphism of sapphirine–enstatite xenoliths at 2.5 kbar and 1170 °C (Harley & Christy 1995). After 2460 Ma the gneisses were cut by several generations of mylonite, pseudotachylite and shear zones (Passchier et al. 1991; Hoek et al. 1992; Dirks et al. 1993b), and intruded by extensive dyke swarms and larger igneous complexes at various times in the Proterozoic (e.g. $>$2470, 2240, 1754, 1380 and 1241 Ma: Lanyon et al. 1993; Seitz 1994). The relatively rapid cooling and partial exhumation of the Vestfold Block in the Palaeoproterozoic is confirmed by Archaean–Early

Proterozoic monazite ages (2460 Ma: Kinny et al. unpublished data). Such old cooling ages, recorded even in gneisses that have experienced later deformation in the SW corner of the Vestfold Hills, indicate that the Archaean Vestfold Block did not experience pervasive high-grade tectonothermal overprints at 1000 Ma or at 550–500 Ma, despite its present-day proximity to the Rauer Terrane and Prydz Belt (described below).

Ruker Terrane, Southern Prince Charles Mountains

The Ruker Terrane of the southern Prince Charles Mountains (sPCM on Fig. 1) comprises an Archaean (c. 3200–2950 Ma) granitic gneiss basement overlain by and interleaved with deformed metasedimentary and metavolcanic rocks (Tingey 1991; Mikhalsky et al. 2001). This terrane is widely regarded as representative of a larger region of the 'inboard' East Antarctic Shield that collided or interacted with outboard terranes (e.g. the Archaean Napier Complex) and other shield areas (e.g. the Dharwar and Bastar Cratons of India) at 990–910 and 550–500 Ma (e.g. Fitzsimons 2000b).

The deformed metasedimentary and metavolcanic rocks of the Ruker Terrane have been divided into three tectonostratigraphic units, termed series (Mikhalsky et al. 2001). Two of these (the Menzies and Ruker Series) are cut by deformed mafic dykes and interpreted to be Archaean and Archaean-Palaeoproterozoic in depositional age. Quartzites, pelitic and calcareous metasediments and amphibolites of the Menzies Series exhibit medium-pressure Barrovian-style metamorphism (staurolite + kyanite ± garnet) and may preserve evidence for two distinct metamorphic events. Mafic to felsic metavolcanic rocks and associated metadolerite sills, metapelitic schist, slate, phyllite and banded ironstones of the Ruker Series, however, are only metamorphosed into the greenschist facies. The third supracrustal unit, the Sodruzhestvo Series, consists of calcareous schist, pelite, phyllite and slate, and minor marble, quartzite and conglomerate metamorphosed at greenschist facies. As the Sodruzhestvo Series is not cut by the metamorphosed and commonly deformed Mesoproterozoic dolerite dykes that transect the Menzies and Ruker Series it is most likely Late Mesoproterozoic or Neoproterozoic in its depositional age (Tingey 1991; Mikhalsky et al. 2001), though this age attribution is highly uncertain.

The ages of principal regional metamorphic event that have affected the older units in the Ruker Terrane are uncertain and contentious. In the case of the granitic basement, metamorphism is constrained by cross-cutting pegmatites to be older than 2650 Ma (Boger et al. 2001). However, the main regional tectonothermal event recorded in all of the tectonostratigraphic units metasediments is tentatively inferred by Mikhalsky et al. (2001) to correspond with the dominant Rayner Province tectonism at 990–910 Ma, though this interpretation is at odds with the age attributions of other workers (see Fitzsimons 2002). Greenschist facies retrogression of the Menzies Series occurred during Pan-African metamorphism and deformation that was associated with the emplacement of abundant A-type granitic rocks at 550–500 Ma (Sheraton et al. 1996; Mikhalsky et al. 2001). The greenschist and higher grade metamorphism of the younger sedimentary unit, the Sodruzhestvo Series, may have also occurred in this event. The importance of this Pan-African overprinting is confirmed by recent work on the north eastern margin of the southern Prince Charles Mountains, where the Archaean tectonism in the south is overprinted northwards by extensive deformation, medium- to high-grade metamorphism and granite emplacement in the early Palaeozoic (Boger et al. 2001).

Complexity in Archaean–Mesoproterozoic Pan African tectonism: The Rauer Terrane, Prydz Bay

The Rauer Islands on the eastern side of Prydz Bay (RI on Fig. 1) consist of both Archaean and Proterozoic crustal components. The complex geological evolution of this terrane involves polyphase reworking of diverse Archaean crustal precursors in Archaean(?) and Mesoproterozic granulite facies events, and further overprinting of all lithologies at 530–500 Ma by mylonites and amphibolite to granulite facies high-strain zones associated with Pan-African tectonism in the adjacent Prydz Belt (Sheraton et al. 1984; Harley 1987b; Harley & Fitzsimons 1991, 1995; Harley et al. 1992, 1995, 1998; Kinny et al. 1993; Sims et al. 1994).

The Archaean in the Rauer Islands is dominated by >3300 and c. 2840–2800 Ma tonalitic orthogneisses (Kinny et al. 1993). The latter orthogneisses intrude and enclose spectacular Archaean layered igneous complexes that preserve many of their original igneous features, despite being metamorphosed and folded (Harley et al. 1995, 1998). Distinctive supracrustal units with probable Archaean depositional ages include marbles and Mg-rich aluminous pelites and quartzites. The Archaean orthogneisses, extensively dyked and then deformed again in younger events (Harley et al. 1992; Sims et al. 1994), are interleaved with Mesoproterozoic supracrustals and 1030–1000 Ma felsic to mafic intrusives (Kinny et al. 1993) that have also experienced high-grade metamorphism and partial melting. The younger supracrustal units (Filla Supracrustals) are typified by Fe-rich pelites, quartzites and semipelites, calc-silicate rocks, ironstones and mafic volcanics (Harley 1987b; Harley & Fitzsimons 1991). The intensity of deformation postdating the 1030 Ma granitoids, 1000 Ma leucogranites and various generations of pegmatite is highly variable, and many high-strain zones may be polyphase, involving the reactivation of gneissic fabrics in the overprinting Pan-African deformation (Sims et al. 1994).

The general range of peak granulite facies P–T conditions that affected both Archaean and Mesoproterozoic protoliths in the Rauer Islands is 6–8.5 kbar and 840 ± 40°C (Harley 1988; Harley & Fitzsimons 1991; Harley & Buick 1992). These conditions were followed by rapid near-ITD of the deep rocks through some 7–10 km during and subsequent to intense deformation, as indicated by mineral reaction textures and thermobarometry. The key reaction textures include grossular-rich garnet replacement by wollastonite + plagioclase and grossular-poor garnet by orthopyroxene + plagioclase symplectites, orthopyroxene–plagioclase rinds on garnet + quartz, and cordierite coronas on garnet in sillimanite-bearing metapelites. Such textures are also very widespread in other Mesoproterozoic and Pan-African high-grade terranes in the East Antarctic Shield (Harley & Hensen 1990) and cannot be used to discriminate between the two major tectonothermal episodes without supporting age information (Hensen et al. 1997).

This problem of the timing of ITD and the potential for polyphase textural development in distinct metamorphic events is particularly acute in the Rauer Terrane because of its undoubted polyphase history (Harley 1987b; Kinny et al. 1993; Harley et al. 1995, 1998), and it is not yet clear which parts of the complex P–T record reflect the 1000 and 500 Ma events. Notably, 1030°C UHT metamorphism is recorded at Mather Peninsula in the Rauer Islands by the stability of Mg-rich orthopyroxene + garnet + sillimanite assemblages that are sequentially overprinted by sapphirine- and cordierite-bearing mineral reaction textures (considered in more detail in the Discussion) that indicate ITD from depths of 10–12 kbar through to 6–7 kbar at temperatures of 900–1000°C (Harley & Fitzsimons 1991; Harley 1998b). This UHT–ITD P–T history is very similar to that recorded in some 550–500 Ma granulites from other areas in the East Antarctic Shield, but has not as yet been unequivocally demonstrated to be of a similar age.

Despite the extensive isotopic evidence for medium- or high-grade metamorphism in the Rauer Islands at 550–500 Ma, the preservation of monazite U–Pb ages near 1000 Ma in the metasedimentary gneisses of the Filla Supracrustals confirm the importance of the 1030–1000 Ma event (Kinny & Harley unpublished data). It is considered likely, therefore, that Archaean and Proterozoic high-grade gneisses with pre-existing metamorphic histories have been interleaved in a short-lived Pan-African tectonic event focused along anastomosing high-strain zones, as first proposed by Hensen & Zhou (1995, 1997). From the perspective of the amalgamation history of the East Antarctic Shield, and indeed of East Gondwana, this c. 1000 Ma magmatic and metamorphic record in the Rauer Terrane is distinct in age and event sequence from that seen in the Rayner Province (described below). Hensen & Zhou (1995) argued for docking of the Rauer Terrane with the Vestfold Block at c. 1000 Ma on the basis of a weak Rb–Sr isotopic imprint in the latter area, and suggested that both areas were then affected to varying extents by the c. 530–500 Ma tectonism that formed the Prydz Belt to the SW. However, the paucity of any record of 1000 Ma magmatism, high-grade metamorphism and deformation in the Vestfold Hills also allows the possibility that the Vestfold Block and Rauer Terrane were not juxtaposed until later, perhaps only at 500 Ma (Harley et al. 1998).

Mesoproterozoic–Neoproterozoic Rayner Province

The Rayner Province (Figs 1 & 2) is dominated by the 1000–910 Ma upper amphibolite to granulite facies metamorphism of pre-1000 Ma tonalitic to granitic orthogneisses and less abundant supracrustals, and by the emplacement of voluminous granitic and charnockitic intrusives during the course of the high-grade history principally in the time interval of 990–950 Ma (Black et al. 1987; Sheraton et al. 1987; Clarke 1988; White & Clarke 1993; Kinny et al. 1997; Boger et al. 2000; Kelly et al. 2000). The pre-1000 Ma orthogneisses display Y-depleted and Sr-undepleted geochemistries consistent with their generation from melting of plagioclase-poor mafic source rocks, probably in a volcanic-arc setting (Sheraton et al. 1996). The supracrustals, which include pelitic, psammo-pelitic and rarer calcareous sediments of largely Mesoproterozoic depositional age, were interleaved and deformed with orthogneiss precursors prior to charnockite emplacement, but also have undergone post-charnockite granulite metamorphism and deformation that had terminated by 910 Ma (Boger et al. 2000; Clarke 1988; Kelly et al. 2000; Dunkley et al. 2002). A Pan-African overprint is present but appears to be largely confined to pegmatites and greenschist to amphibolite facies shear zones.

The high-temperature (1000°C) magmatic charnockites of the Mawson coastline (Young et al.

1997) and northern Prince Charles Mountains (Sheraton et al. 1996; Kinny et al. 1997) postdate initial flat-lying fabrics in the gneisses but are deformed along with the older orthogneisses and supracrustals in the later fold phases. These charnockites show a spectrum of compositions, from high-field-strength-element (HFSE)-rich varieties, which may be fractionated from mantle-derived magmas, to more siliceous granitic compositions. The latter types often exhibit depletions in Y and HREE, and hence are inferred to have been produced by deep-level (40–50 km) partial melting of overthickened crust, probably associated with emplacement of mantle-derived magma at the base of that crust (Kilpatrick & Ellis 1992; Sheraton et al. 1996). The dry magmas may then have migrated upwards to their emplacement depths (15–20 km) during extensional or transtensional phases of the orogeny.

In detail, there are significant spatial variations in the ages, intensities and styles of the collisional–transpressional events in the Rayner Province. The main high-grade metamorphism and principal deformation episodes in the west and northern parts of the Rayner Province in Enderby and Kemp Lands are 930–910 Ma in age (Black et al. 1987; Kelly et al. 2002). In the Oygarden Group of Kemp Land, initial deformation in the Rayner Structural Episode at c. 930 Ma involved deep-crustal ductile thrusting on an east-directed transport axis, leading to isoclinal recumbent folds and sheath folds in thrusted slices that structurally overly high-grade mylonites (929 ± 10 and 924 ± 17 Ma: Kelly et al. 2002). This event affected Archaean orthogneisses and paragneisses which are correlated with those of the Napier Complex (Kelly pers. comm.), as well as Proterozoic lithologies. Later strain partitioning resulted in west-trending, steeply south-dipping, high-strain zones separated by lower strain zones characterized by the upright folding of earlier structures, consistent with E–W stretching coupled with constriction in a N–S direction. The final steeply dipping high-strain zones, typified by E–W trends and steep SE-plunging mineral rodding lineations, were formed 925 Ma (Kelly et al. 2000, 2002).

The Rayner Province history further east along the Mawson coast and in the northern Prince Charles Mountains is somewhat different. Along the Mawson coastline mid-Proterozoic supracrustals and felsic orthogneisses, including Archaean (2700 Ma) precursors, were deformed by east-over-west thrusting, and intruded by leucogranites and granites at 1000–990 Ma, contemporaneous with the development of north-trending recumbent folds (Clarke 1988; White & Clarke 1993; Dunkley et al. 2002). Further felsic magmatism, including the emplacement of charnockites, preceded and accompanied upright folding that mostly occurred prior to 940 Ma, the age of a discordant felsic dyke. Later E–W-trending upright folding took place at up to 910 Ma, an age obtained both from zircons formed in anatectic leucosomes and zircons isotopically reset under high-grade conditions in other rocks (Dunkley et al. 2002). In the northern Prince Charles Mountains, granulite (and amphibolite) facies metamorphism associated with flat-lying fabrics and coaxially refolded recumbent isoclinal folds progressed at 1000–980 Ma, again prior to the emplacement of charnockites and granite sheets (Fitzsimons & Thost 1992; Boger et al. 2000). The dominant high-grade fabrics produced in this early event were then overprinted by E–W-trending steep folds and high-strain zones that formed at 940 Ma under high-grade conditions accompanied by localized partial melting (Boger et al. 2000). By c. 900 Ma the granulites that typify much of the northern Prince Charles Mountains were thrusted to the north over amphibolite facies felsic and intermediate gneisses shallow-dipping discrete mylonite zones, in response to continued N–S convergence (Boger et al. 2000). The younger ages of c. 910 Ma for metamorphic zircons at Cape Bruce (Dunkley et al. 2002) and the northern Prince Charles Mountains are similar to those obtained from a quartzofeldspathic gneiss from Mount Sibiryakov in the western Rayner Complex in Enderby Land (907 ± 14 Ma Shiraishi et al. 1997). A charnockite from Sandercock Nunatak in this region yields an age of 977 ± 11 Ma, consistent with those in the northern Prince Charles Mountains and at Mawson.

Near-ITD is a prominent feature of the 930–920 Ma metamorphism of the Rayner Province adjacent to the Napier Complex, both in Enderby Land itself (Ellis 1983; Black et al. 1987) and in the Oygarden Group (Kelly et al. 2000). In the latter area, for example, peak P–T conditions of 10 ± 2 kbar and $>800°C$ are deduced from garnet-bearing mafic granulites and unusual orthopyroxene–corundum, kornerupine- and sapphirine-bearing mineral assemblages. Post-deformational breakdown of garnet to orthopyroxene + plagioclase in mafic granulites and to sapphirine + orthopyroxene in Mg–Al gneisses are consistent with ITD to 6 kbar, whilst temperatures were still $>700°C$ in the Oygarden Group, a P–T path similar in form to those previously deduced by Ellis (1983) and Black et al. (1987) for the Rayner Complex in Kemp and Enderby Lands.

In contrast, the 1000–980 to 940–910 Ma P–T paths exhibited by granulites exposed eastward along the Mawson coastline and in the northern Prince Charles Mountains involve cooling with little or only minor decompression. Peak P–T conditions established in these areas are typically 5–7 kbar and 700–830°C (Fitzsimons & Harley 1992; White & Clarke 1993; Dunkley et al. 2002). Essentially, near-IBC is documented from garnet corona textures formed on scapolite and wollastonite in calc-silicate rocks (Fitzsimons & Harley 1992), and garnet growth in orthogneisses (Dunkley et al. 2002) and in metapelites (Hand et al. 1994). This IBC may have occurred prior to 940 Ma at Cape Bruce on the Mawson coastline (Dunkley et al. 2002), but is constrained to have ensued only during and after 940–910 Ma in the northern Prince Charles Mountains (Boger et al. 2000).

The differences between these P–T histories may reflect the progressive development of the collisional belt or the timing of mantle-induced heating and magmatism in relation to thickening. For example, the marginal zones of the Rayner Province may have experienced rapid exhumation by lateral extrusion adjacent to relatively rigid 'buttresses' (e.g. the Napier Complex) during final collision, whereas domains within the broad interior of the Rayner Province evolved at depth for longer (1000–930 Ma) and only underwent significant exhumation (perhaps along local high-strain zones) after cooling had ensued. This complexity in the ages of the main metamorphic and deformational events probably reflects the development of the Rayner Province during diachronous or oblique collision between a southern (present coordinates) cratonic block within the East Antarctic Shield (most likely the Ruker Terrane of the south Prince Charles Mountains), parts of India and the Archaean Napier Complex. Indeed, the Rayner Complex of Enderby Land and the Kemp Land coast (Figs 1 & 2) preserves some evidence for pre-Rayner events [e.g. 1488 Ma anorthositic and granitic intrusives in Enderby Land (Black et al. 1987) and 1600 Ma tectonothermal event in Kemp Land (Kelly et al. 1999)] that are not evident elsewhere in the province.

Further evidence bearing on this collisional setting is found in the central Prince Charles Mountains, where the Rayner Province includes a lower grade region, the Fisher Terrane (Sheraton et al. 1996). This consists of highly deformed epidote- to garnet-amphibolite facies metavolcanic and intrusive rocks, mainly mafic to intermediate in composition (Beliatsky et al. 1994), cut by pre- to post-tectonic granites and mafic dykes. The geochemistry of the metavolcanic rocks, as well as associated granitoids, are consistent with their formation in an active continental margin and magmatic-arc setting (Sheraton et al. 1996; Kinny et al. 1997). Although detailed relationships between the Fisher Terrane and the surrounding high-grade gneisses of the Rayner Province are uncertain, the similarites in protolith and metamorphic ages have been used as evidence that both formed in the same accretionary regime outboard of an older cratonic block, the Ruker Terrane of the southern Prince Charles Mountains (Beliatsky et al. 1994; Sheraton et al. 1996).

It is speculated that from 1000 Ma the closure of an oceanic basin that existed between the Ruker Terrane (to the south in present coordinates) and the Indian Shield (to the north in present coordinates) led, through a combination of accretion and collision, to the development of the Rayner Province in Antarctica and the Eastern Ghats Granulite Belt in India. Overall, the evidence for diverse protoliths in terms of lithologies (sediments, volcanics, intrusives) and ages (Proterozoic and Archaean), coupled with the metamorphic P–T paths terminated by extensive ITD, at least in those parts of the Rayner Province adjacent to the Napier Complex, are consistent with such a collisional tectonic setting. The source of heat to allow temperatures to reach 850°C or more during and after collision must ultimately be the mantle, unless the overlying crust present prior to denudation had an anomalously high internal heat production – a scenario for which there is no evidence in the preserved rock units.

Pan-African high-grade tectonism in the Lützow-Holm Bay Belt

The Lützow-Holm Bay belt exposes an extensive supracrustal sequence, including pelites, marbles and calc-silicate rocks, which is interleaved with mafic and felsic gneisses and boudinaged ultramafics. Some relict Archaean crustal precursors may be present, as shown by the presence of ancient zircons (2900–1500 Ma), that are interpreted as detrital, but most sedimentary precursors of the abundant supracrustal paragneisses are considered to have been deposited after 1000 Ma, incorporating material derived from crust similar to the Rayner Province (Shiraishi et al. 1994, 1997). The Lützow-Holm Bay Belt (excluding the Forefinger Point region; Fig. 2) shows an E–W metamorphic field gradient from amphibolite facies (east) to medium-pressure (6–8 kbar) granulite facies (west), and even into UHT metamorphic grades at Rundvågshetta, in the SW of the terrane (Figs 2 & 4; Motoyoshi & Ishikawa 1997; Fraser et al. 2000). Uniquely for the high-grade terranes of the East Antarctic Shield, convincing evidence for a clockwise P–T history is preserved in the form of relict kyanite and staurolite inclusions in higher grade peak assemblages such as garnet–sillimanite–orthopyroxene (e.g. Motoyoshi & Ishikawa 1997).

The main high-grade tectonothermal episode in the Lützow-Holm Bay area is constrained from U–Pb ages on metamorphic zircons to have occurred in the interval of 550–520 Ma (Shiraishi et al. 1994, 1997; Fraser et al. 2000), somewhat younger than the ages for initial Pan-African high-grade metamorphism in central Dronning Maud Land and the Sør Rondane Mountains. At least two phases of near-

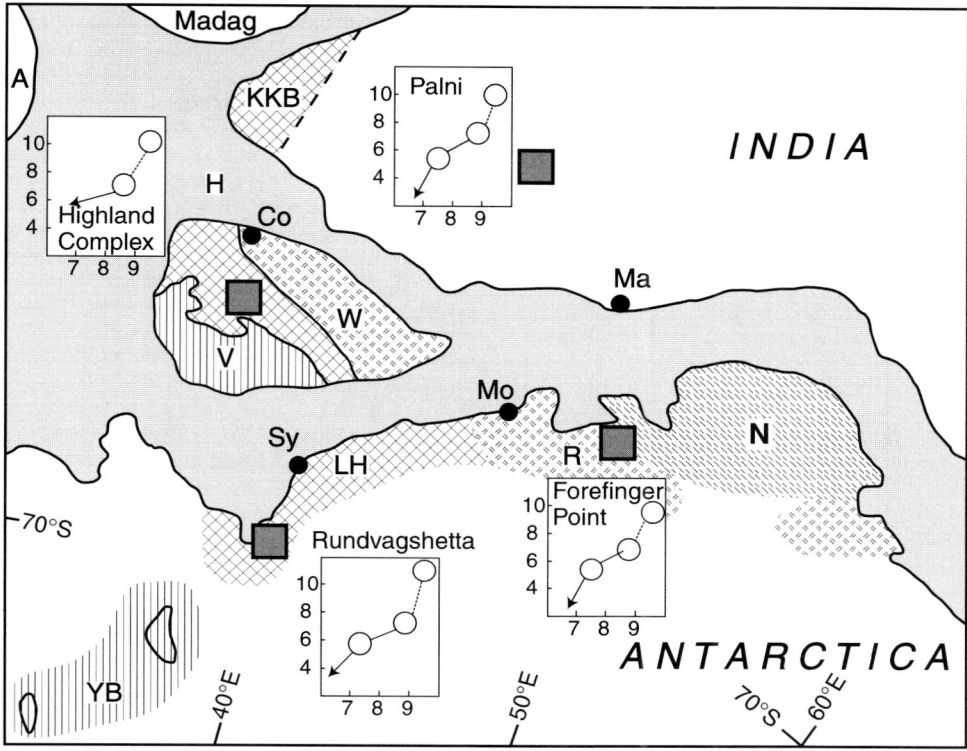

Fig. 4. The Yamato–Belgica (YB), Lützow-Holm Bay (LH), Rayner (R) and Napier (N) Complexes of East Antarctica in relation to adjacent areas in the Gondwana reconstruction, showing disposition of terrains and the occurrences of probable 550–500 Ma Pan-African ultrahigh-temperature (UHT) metamorphism (shaded squares). P–T box information for these UHT cases is as for Figure 2. Other terrains are: V, Vijayan Complex; H, Highland Complex; W, Wanni Complex. These and the Kerala Khondalite Belt (KKB) are shaded according to their probable correlative areas after Shiraishi et al. (1994) A, Africa; Colombo; M, Madagascar; Ma, Madras; Mo, Molodezhnaya; Sy, Syowa.

isoclinal folding are recognized, accompanied by boudinage of metabasites and ultramafics, and the segregation of partial melts into boudin necks and layer-parallel vein systems. These highest grade structures and fabrics were then reoriented in upright fold structures. Fraser et al. (2000) have further refined the age of the peak metamorphism at Rundvågshetta as 517 ± 9 Ma using zircons from a synboudinage leucosome, consistent with the younger end of the spectrum of Pan-African ages obtained in earlier studies both along the Lützow-Holm Bay coast (532–521 Ma: Shiraishi et al. 1994) and at Forefinger Point in Enderby Land (537–520 Ma: Shiraishi et al. 1997). This high-grade metamorphism was swiftly followed by exhumation and cooling, so that mineral cooling ages (e.g. U–Pb monazite, and mica K–Ar, Ar–Ar and Rb–Sr) are in the range of 515–500 Ma (Fig. 5). Cooling to c. 300°C was preceded by high-temperature decompression manifested in the development of ITD textures such as orthopyroxene + plagioclase symplectites and cordierite coronas on garnets in metabasites and pelites, respectively. Hence, the mineral assemblage, reaction texture and thermochronology evidence indicates that the granulites of the western part of Lützow-Holm Bay were exhumed on a regional scale through at least c. 10–20 km within 20 Ma of their peak metamorphism and deformation (Fig. 5). Furthermore, in at least two localities (Forefinger Point and Rundvågshetta; Fig. 4) within the Lützow-Holm Bay domain, Pan-African metamorphism attained UHT conditions of 10–12 kbar and at least 900–1000°C, leading to the formation of orthopyroxene–sillimanite and sapphirine-bearing assemblages (Harley et al. 1990; Motoyoshi et al. 1994; Motoyoshi & Ishikawa 1997; Fraser et al. 2000). Rapid ITD indicated by mineral reaction textures involving the breakdown of high-Mg garnet, also characteristic of these localities, is considered in more detail in the Discussion.

Metamorphic and deformational evidence is therefore consistent with both lateral and vertical

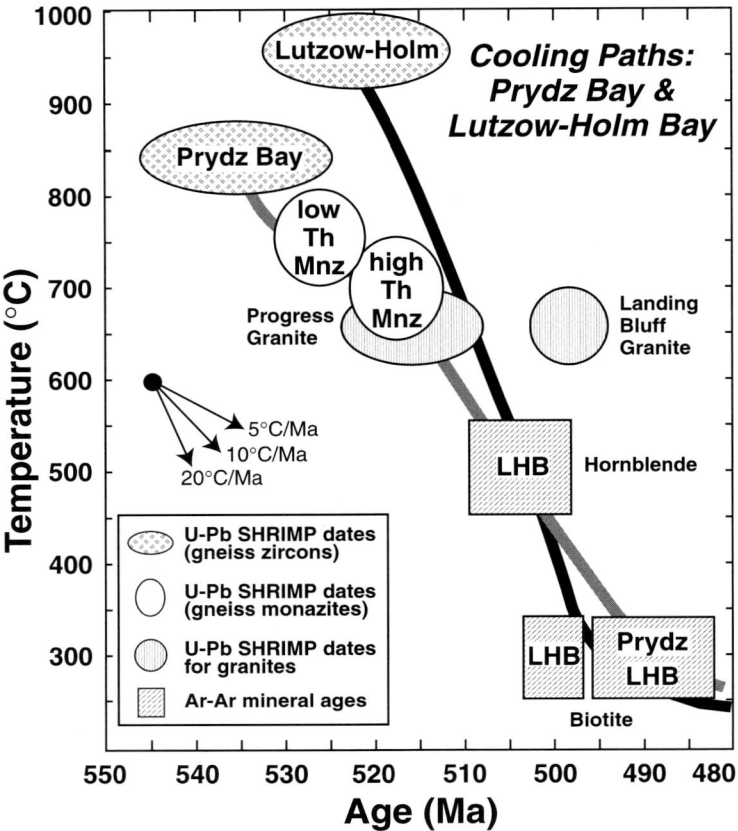

Fig. 5. Pan-African temperature–time (T–t) cooling paths for southern Prydz Bay, modified from Fitzsimons *et al.* (1997), and SW Lützow-Holm Bay at Rundvågshetta, based on the data of Fraser *et al.* (2000). Mnz, Monazite.

crustal displacements during the progress of the 550–500 Ma Pan-African tectonothermal events in the Lützow-Holm Belt. Indeed, this terrane has been interpreted to be part of a Cambrian orogenic belt that would have included the Highland Complex of Sri Lanka, and the Kerala Khondalite Belt of SW India and parts of eastern Madagascar (Shiraishi *et al.* 1994). The issue of whether oceanic crust was formed in this region prior to 550 Ma and then involved in the subsequent Pan-African tectonism has not been resolved as yet.

Pan-African high-grade tectonism in the Prydz Bay Belt

In the Prydz Bay area SW of the Rauer Islands (PZ in Fig. 1; Fig. 2), a highly migmatized granulite facies supracrustal sequence is tectonically interleaved with an orthogneiss-dominated sequence that partially preserves older histories (e.g. 960–920 Ma) and may have either been basement to the supracrustals or allochthonous (Fitzsimons & Harley 1991, 1992; Thost *et al.* 1991; Harley & Fitzsimons 1995). The orthogneisses locally preserve polyphase garnet breakdown textures involving orthopyroxene, plagioclase and spinel, which from garnet Sm–Nd dating (Hensen & Zhou 1995; Hensen *et al.* 1995) indicate an early phase of very-high-temperature (900°C) decompression prior to 905 Ma, overprinted by a later moderate-temperature (700–800°C) decompression that is ascribed to the Pan-African high-grade tectonic event principally manifested in the supracrustal sequence (Thost *et al.* 1991; Hensen & Zhou 1995).

The supracrustals are characterized by garnet- and cordierite-bearing leucogneisses, and extensively migmatized garnet–cordierite–sillimanite–quartz–K-feldspar gneisses that may also contain hercynitic spinel, graphite and ilmenite. Peak garnet–cordierite-sillimanite–quartz and garnet–spinel–cordierite–sillimanite assemblages are partially overprinted by reaction textures in which spinel +

cordierite form at the expense of garnet + sillimanite and by spinel + biotite + quartz assemblages in post-peak high-strain zones that cut the peak fabrics (Fitzsimons 1996). All mineral assemblages and fabrics in the supracrustals reflect intense Pan-African tectonism (Ren *et al.* 1992; Zhao *et al.* 1992, 1995; Dirks *et al.* 1993*a*; Hensen & Zhou 1995) characterized by pervasive ductile deformation, high degrees of partial melting and at least 10 km of post-peak exhumation as indicated by the ITD reaction textures noted above (Fitzsimons & Harley 1991, 1992; Carson *et al.* 1995, 1996; Fitzsimons *et al.* 1997). Peak metamorphism, melting and deformation under $P-T$ conditions of 5.5–7.0 kbar and 800–860°C (Stüwe & Powell 1989; Fitzsimons & Harley 1992; Fitzsimons 1996) occurred at 535–525 Ma (Fitzsimons *et al.* 1997), and was synchronous with compressional deformation. Extensional, top-to-the-south ductile deformation, focused along flat-lying high-strain zones, occurred at *c.* 520–510 Ma. This deformation was synchronous with post-peak exhumation, further granite emplacement (515 Ma Progress Granite: Carson *et al.* 1996) and leucosome generation, and with localized hydrous retrogression of peak assemblages through the release of volatiles from residual melt bodies (Carson *et al.* 1995; Fitzsimons 1996; Fitzsimons *et al.* 1997). Ages obtained from garnet–whole-rock Sm–Nd isochrons, in the range of 517 ± 4–509 ± 5 Ma for all but one sample from the region (Hensen & Zhou 1995), do not correspond with the peak of metamorphism but rather with the higher temperature segment of the post-peak exhumation history (Fig. 4).

As with the Lützow-Holm Bay region, crust initially was thickened and metamorphosed at considerable depths (even up to 40 km if UHT at Mather Peninsula, Rauer Islands, can be attributed to the Pan-African events – see below) and then underwent ITD to pressures of only 4 kbar prior to rapid cooling. Rb–Sr and Ar–Ar thermochronology of biotites suggests that by 500–490 Ma the presently exposed granulites of this area were already cooled to <300°C and probably resident at depths of only < 10 km in the crust (Fig. 5; Fitzsimons *et al.* 1997). The metamorphic record, in particular the rapid implied exhumation at an average rate of *c.* 1 mm/a or more, coupled with the deformational evidence for compression followed by extension, are compatible with collisional orogenesis followed by extensional collapse of the thickened and anomalously hot crust (as considered in the Discussion). Though not yet as well documented, 510–450 Ma metamorphism and intense deformation in parts of the Mawson Escarpment and in the Grove Mountains may represent continuations of the Pan-African tectonism seen in Prydz Bay, and provide further evidence for its effects on disparate crustal blocks (Boger *et al.* 2001).

Discussion

Some correlations and comparisons between the Rayner Province and the Eastern Ghats

The evolution of the Rayner Province described in a preceding section can be compared with that of the formerly adjacent Eastern Ghats of India (Dasgupta 1995; Dasgupta & Sengupta 2000). As discussed in depth by Dasgupta & Sengupta (2002) and by Dobmeier & Raith (2002), the Eastern Ghats belt may be divided into four domains on the basis of Nd model ages and metamorphic records. Three of these domains that are restricted to the northern Eastern Ghats Belt (domains 2–4: Dasgupta & Sengupta 2002; Dobmeier & Raith 2002) preserve strong evidence for high-grade metamorphism (8–9 kbar and 850°C) and charnockite emplacement in the time interval of 990–930 Ma, superimposed, in some cases, on earlier UHT (8–10 kbar and 1000°C) metamorphism (Fig. 2; Dasgupta & Sengupta 2000). These domains correlate well in their Neoproterozoic metamorphic and magmatic event records (Mezger & Cosca 1999) with the Rayner Province in the Mawson coast and northern Prince Charles Mountains areas, where the main events occur at 990–950 and 940–920 Ma (Kinny *et al.* 1997; Boger *et al.* 2000), though it is clear that more isotopic data are required to take this comparison further to the individual domain level.

A domain adjacent to the Dhawar and Bastar Cratons of India (domain 1: Dasgupta & Sengupta 2002), and which occupies essentially all of the southern Eastern Ghats Belt (i.e. west of the Godivari Rift, relative to Antarctic coordinates, on Fig. 2), is distinctive in containing Archaean crust extensively reworked under granulite conditions at 1670–1550 Ma but not overprinted by high-grade metamorphism and deformation at 990–920 Ma [see Dobmeier & Raith (2002) for details]. To date, no equivalent domain or zone devoid of any imprint of 990–920 Ma high-grade tectonism has been recognized in the Rayner Province. However, Kelly *et al.* (1999, 2000) have recognized a *c.* 1600 Ma high-grade tectonic event in reworked Archaean gneisses from the Oygarden Islands, extensively overprinted, as described above, by late-Rayner metamorphism and deformation at 930–920 Ma. One plausible amalgamation scenario arising from these preliminary observations is that the easternmost margin of the Napier Complex, or an Archaean craton very similar to the Napier in its age characteristics, was reworked during a 1670–1550 Ma collision with India, after which time it remained joined to it to form the central and eastern Indian Shield. Later, at 990–920 Ma, the Ruker Terrane of the southern Prince Charles Mountains, and its outboard magmatic arc and sedimentary sequences (northern Prince Charles Mountains), collided obliquely with this combined India + Napier Complex Shield to produce the Rayner Province and its characteristic structural progression, as well as the main domains of the northern Eastern Ghats Belt (Kelly pers. comm.). The resulting India–Napier–Rayner–Ruker continent (Enderbia) was then involved in collision/accretion with the East Africa/Maud Province continental block(s) on its western margin at 600–500 Ma, and with the Rauer Terrane and other crustal blocks on its eastern margin at 550–500 Ma.

Cambrian juxtaposition of distinct Proterozoic crustal provinces in East Antarctica

The 600–500 Ma, or Pan-African, high-grade belts summarized above juxtapose the distinct Mesoproterozoic and Neoproterozoic Grenvillian provinces (Maud, Rayner and Wilkes), and the more complex Rauer Terrane. The ages of the main events in these provinces across the East Antarctic Shield vary markedly from west to east. The more easterly areas (Bunger Hills and Windmill Islands of the Wilkes Province) experienced their principal tectonothermal events prior to 1130 Ma and were not affected by the younger (broadly 1080–930 Ma) events that characterize the Maud and Rayner Provinces and the Rauer Terrane further west. Furthermore, the main Grenvillian tectonothermal events in the Rayner Province culminate in the time period of 990–920 Ma, whereas those in the Rauer Terrane and Maud Province, which are distinct from each other in terms of their earlier histories, culminate at 1030–990 Ma. As pointed out by Fitzsimons (2000a, b) and others, these fundamental differences between the isotopic/event records of the basement provinces now separated by the Pan-African Dronning Maud Land, Lützow-Holm, Prydz and Denman Belts require that the older provinces were not formerly parts of a continuous Grenville belt. East Antarctica was not a single unified crustal block within either East Gondwana or Rodinia until the Cambrian, which is now demonstrated to be the key phase of high-grade and UHT crustal metamorphism associated with supercontinent assembly. This discussion will therefore conclude with a consideration of the nature of Pan-African tectonism in the light of the metamorphic records preserved in the 550–500 Ma high-grade belts in the Antarctica–India sector.

UHT Metamorphism, P–T–t paths and the nature of Pan-African tectonism in the Antarctic–Indian sector

Crustal melting, clockwise P–T paths, rapid post-peak exhumation to shallow- or mid-crustal levels (by 500–490 Ma in most cases) and the generation, at least locally, of UHT conditions are key features of the 550–500 Ma Pan-African tectonism that must be explained by any tectonic models for these events. UHT–ITD metamorphism is preserved at several widely dispersed localities in the Antarctic–Indian sector, recognized from aluminous garnet-bearing granulites from Palni and nearby areas in southern India (Raith et al. 1997), in the Highland Series of Sri Lanka (Kriegsman & Schumacher 1999) at Rundvågshetta in the Lützow-Holm Complex (Motoyoshi & Ishikawa 1997), at Forefinger Point in Enderby Land (Harley et al. 1990; Shiraishi et al. 1997), and at Mather Peninsula, Rauer Islands (Harley & Fitzsimons 1991; Harley 1998b) (Figs 2 & 4). As noted in earlier sections, zircon U–Pb geochronology and garnet Sm–Nd dating indicates that the UHT event occurred at 550–500 Ma at Forefinger Point (530 ± 8 Ma: Shiraishi et al. 1997), Rundvågshetta (517 ± 9 Ma: Fraser et al. 2000) and Sri Lanka (Holzl et al. 1994), but at Mather Peninsula, Rauer Islands, the age is still under debate. Though not definitively UHT (i.e. not $>900°C$), the extensive 860°C metamorphism in the Prydz Belt adjacent to the Rauer Islands is also considered in this discussion, as its 530–500 Ma ITD P–T–t path is well constrained (Fig. 5; Fitzsimons et al. 1997) and relevant.

The Antarctic (and India–Sri Lanka) UHT localities preserve remarkably similar mineral assemblage and reaction texture relationships. At Mather Peninsula, peak P–T estimates are 11–12 kbar and $1033 \pm 30°C$ (Harley 1998b), comparable to the peak P–T conditions of 11 kbar and 1016–1085°C indicated for UHT rocks from Rundvågshetta using retrieval calculations from garnet–orthopyroxene Fe–Mg–Al thermometry (Harley 1998c). Reaction textures in which garnet and orthopyroxene + sillimanite are replaced by lower pressure equivalents such as sapphirine + orthopyroxene + sillimanite, sapphirine + orthopyroxene + cordierite and spinel + cordierite (Fig. 6) define the UHT–ITD path (reactions A–F in Fig. 7). The sapphirine- and cordierite-forming continuous reactions interpreted from preserved textures, such as Grt = Spr + Opx + Sil, Opx + Sil + Qz = Crd, Opx + Sil = Spr + Crd and Grt = Spr + Opx + Crd, are consistent with ITD at temperatures greater than the [Qtz] invariant point of Hensen & Harley (1990). The P–T evolution deduced in each case involves initial ITD from 10–12 to 6–7 kbar at temperatures of $>850–900°C$ –

representing exhumation of at least 10–20 km at essentially UHT conditions prior to cooling down to the blocking temperatures of Rb or Ar in biotite.Though not as extreme, the ITD evolution in the Prydz Belt is similar, with exhumations of at least 10 km achieved in <20 Ma (Fig. 5; Fitzsimons et al. 1997).

Fraser et al. (2000) have modelled the UHT–ITD P–T–t history at Rundvågshetta, taking into account the constraints that: (1) initially rapid cooling to 300°C by 500 Ma occurred following significant exhumation of the UHT granulites to depths of 12–15 km (i.e. ITD preceeded cooling); (2) further cooling was significantly slower, to 250°C by c. 420 Ma. Hence, although the simple averaged cooling rate from 520 to 500 Ma is 30°C/Ma and the averaged exhumation rate is 1.15 mm/a, Fraser et al. (2000) model the P–T–t history with initially rapid exhumation at 3 mm/a and dT/dt of c. 10°C/Ma for some 7 Ma, followed by negligible exhumation (0.25 mm/a) with high but slowing rates of cooling (from c. 60 down to 8°C/Ma) at mid-crustal depths. On the basis of a similar P-T path and time span of apparent cooling from >900°C at 530 Ma to 250–300°C at 497 Ma (Black et al 1987; Shiraishi et al 1997), a similar exhumation history probably applies to Forefinger Point. In the absence of definitive age data, the other localities noted in Figures 5 & 7 do not constrain the models further. The likely initial post-peak exhumation rates of 3 mm/a are in the range of those estimated by Harley (1989) for the preservation of ITD textures in granulites, and are comparable to rates in seen today in some active collision zones undergoing later extension (e.g. Platt et al. 1998) and along large-scale transpressional fault zones.

One-dimensional thermal modelling of lithospheric thickening/thinning/extension scenarios (Platt & England 1994; Platt et al. 1998) associated with collision demonstrate that the key factors contributing to the attainment of HT/UHT conditions are a high internal heat production in the initially thickened crust, significant post-thickening residence in the deep crust prior to the onset of exhumation (i.e. prolonged deep-crustal incubation), rapid removal of most of the lithospheric mantle and rapid extensional unroofing. As shown in the models developed by Platt et al. (1998) for application to the thermal evolution of the Betics and Alboran Sea, even with optimal crustal heat production, incubation (30–60 Ma) and timescales of extension/exhumation (6–20 Ma), it is difficult to generate temperatures >900°C in crustal rocks at 8–11 kbar without having earlier replaced essentially all of the subjacent mantle lithosphere by asthenosphere.

Figure 8 illustrates three such models in which an initially 30 km thick crust and subjacent 95 km thick lithospheric mantle is doubled in thickness, allowed to incubate for selected time spans, thinned from below by removal of all of the lithospheric mantle beneath 62.5 km and then stretched by a factor of three over a range of times between 6 (fast) and 60 Ma (slow). The resultant P–T paths for rocks buried to 55 km show that UHT conditions near 900°C at 12–8 kbar, and followed by ITD to 5–6 kbar, can be attained at exhumation rates of c. 1.8 mm/a. Such rocks would be underlain by 7.5 km of crust and mantle lithosphere prior to extension, and by only 2.5 km at the end of extension in this set of models (stretching factor $\beta = 3$). The Pan-African ITD P–T paths of the HT granulites from Prydz Bay, Lützow-Holm Bay and central Dronning Maud Land (Shiraishi et al. 1994; Fitzsimons 1996; Piazolo & Markl 1999), as well as those of the Kerala Khondalite Belt (Nandakumar & Harley 2000) and Madagascar (Markl et al. 2000; De Wit et al. 2001), are generally consistent with the features of this generic thickening/lithospheric removal/extension model. It seems inescapable that a significant flux of additional heat from the mantle into the deep and initially overthickened crust is required to produce the observed Pan-African metamorphic effects even without the UHT localities. Similar conclusions as to the nature and style of Pan-African tectonics have been put forward independently by De Wit et al. (2001), who attributed the prolonged but punctuated deformational and thermal evolution of southwestern Madagascar in the time period of 650–490 Ma to collision (650–605 Ma), followed by the development of a thickened plateau that then underwent collapse through crustal extension from 520 to 490 Ma.

The Pan-African UHT P–T paths considered in this contribution (Fig. 7) lie at higher temperatures than those of most granulite facies crust, and can only be generated in such thickening/extension models if the granulites start their heating phase at the base of the thickened crust (60 km), if all the mantle lithosphere is removed and if the asthenospheric mantle temperature is >1300°C at 250 km, the condition used in the models of Platt et al. (1998). Replacement of the lithosphere by higher temperature (e.g. 1500°C at 250 km) sublithospheric mantle, e.g. that in a mantle plume, would allow temperatures of c. 1000°C to be attained in the initially deepest crustal rocks during subsequent extension and unroofing, but only provided that the rates of thinning and exhumation are of the order of 2–3 mm/a for several millions of years.

A consequence of these one-dimensional thickening/extension models is that the final extended crustal rocks preserving HT and UHT conditions would cool at 'mid-crustal' conditions (4–6 kbar) but be underlain by only a few kilometres structural thickness of crust: the replaced mantle would lie at shallow depths and be in close contact with some of

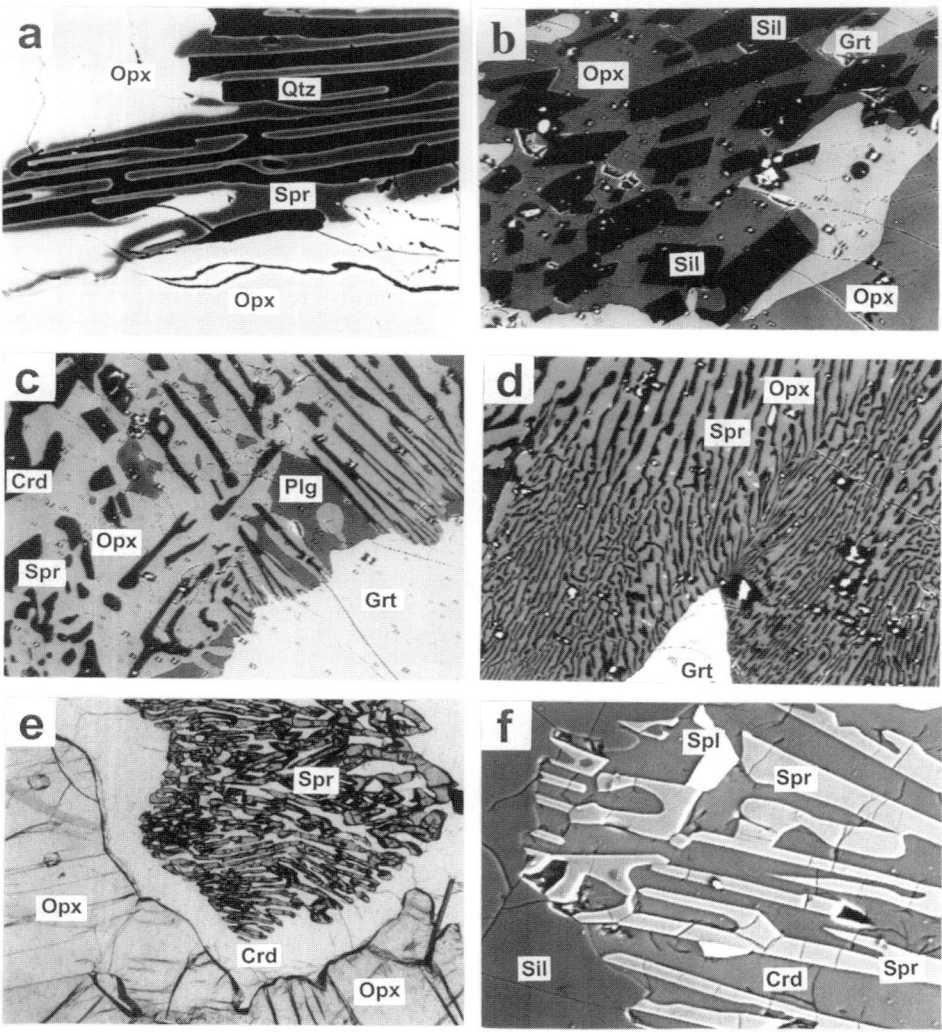

Fig. 6. Examples of reaction textures in ultrahigh-temperature (UHT) rocks from East Antarctica. (**a**) Backscattered electron image of aluminous orthopyroxene (Opx) partially replaced by a lamellar symplectite of sapphirine (Spr) and quartz (Qtz), Mount Riiser–Larsen, Napier Complex, Enderby Land. This texture is associated with a decrease in the Al_2O_3 content of the Opx (from 12 to 9 wt%) and has been interpreted as reflecting a UHT–near-isobaric cooling (IBC) (UHT–IBC) (Harley & Motoyoshi 2000) in the Late Archaean–earliest Palaeoproterozoic. Field of view, 150 μm across. (**b**) Backscattered electron image of a domain of aluminous orthopyroxene (Opx) intergrown with euhedral sillimanite (Sil) and partially replaced by garnet (Grt), Mather Peninsula, Rauer Islands. The Al_2O_3 content of the Opx coexisting with Sil (7–8.5 wt%) is less than that of Opx adjacent to Grt only (9–11 wt%) and is interpreted as consistent with prograde heating within the Opx–Sil–Grt assemblage field (Harley 1998*b*, 1998*c*). Field of view, 60 μm across. (**c**) Backscattered electron image of a domain of aluminous orthopyroxene (Opx) intergrown with sapphirine (Spr) and cordierite (Crd) in a lamellar/blocky symplectite, Mather Peninsula, Rauer Islands. Plagioclase (Plag) is also produced in this garnet (Grt)-breakdown texture, most likely from the minor grossular component in the initial garnet. The FMAS phases are considered to have formed via the continuous reaction Grt = Opx + Spr + Crd (reaction D in Fig. 7) during decompression (Harley 1998*b*). Field of view, 300 μm across. (**d**) Backscattered electron image of a domain of aluminous orthopyroxene (Opx) intergrown with sapphirine (Spr) in a lamellar symplectite that is radial on garnet (Grt), Mather Peninsula, Rauer Islands. Cordierite (Crd) also occurs in this texture, but more remote from the resorbed Grt interface. The Spr–Opx lamellae coarsen into 'leaves' when more distant from the Grt interface. The reaction is also Grt = Opx + Spr + Crd. Field of view, 40 μm across. (**e**) Photomicrograph of a layered or zonal reaction texture in which cordierite (Crd) occurs as a collar or moat between reactant orthopyroxene (Opx) and former sillimanite (Sil) that is

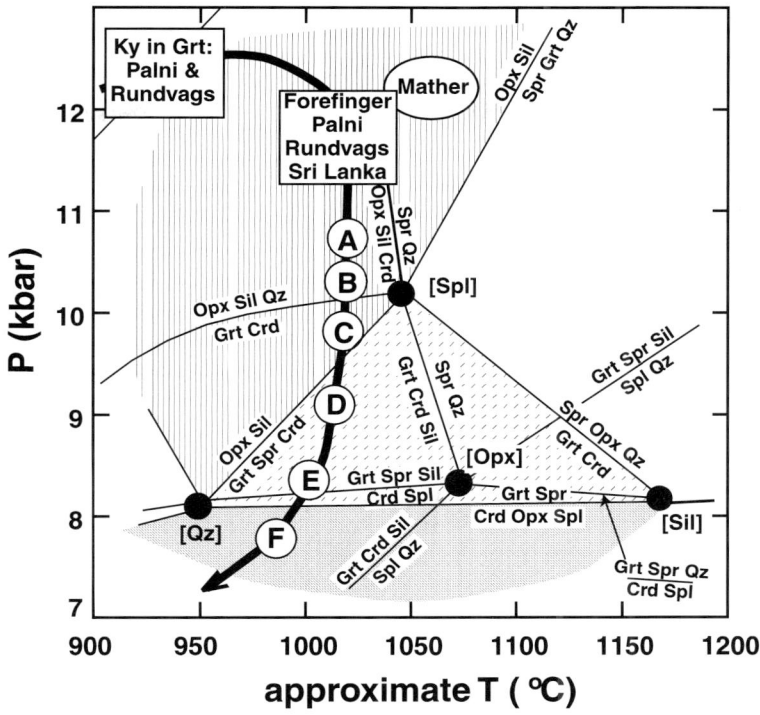

Fig. 7. *P–T* diagram illustrating *P–T* paths for the post-Archaean ultrahigh-temperature (UHT)–isothermal decompression (ITD) localities in Antarctica, southern India and Sri Lanka, referenced against the FMAS assemblage grid presented by Harley (1998a), modified from those of Hensen (1971), Hensen & Green (1973) and Hensen & Harley (1990). The ITD paths are constrained by reaction textures (denoted as A–F) and mineral compositional zoning, particularly in garnet. The continuous reactions interpreted from preserved textures are: A, Grt = Spr + Opx + Sil; B, Opx + Sil + Qz = Crd; C, Opx + Sil = Spr + Crd; D, Grt = Spr + Opx + Crd; E, Grt Sil + = Spl + Crd; F, Grt = Spl + Opx + Crd. Shaded areas indicate the *P–T* stability ranges of the key FMAS divariant assemblages seen in these UHT case studies (Grt + Opx + Spr + Sil; Opx + Sil + Spr + Crd; Grt + Opx + Spr + Crd). Mineral abbreviations as in Figure 3. *P–T* reaction path data for the various areas are: Forefinger Point, Enderby Land – Harley et al. (1990); Rundvågshetta, Lützow-Holm Bay – Motoyoshi & Ishikawa (1997), Fraser et al. (2000); Palni, southern India – Raith et al. (1997); Sri Lanka Highland Complex – Kriegsman & Schumacher (1999); Mather Peninsula, Rauer Islands – Harley (1998b).

the highest grade rocks. Whilst this could provide an explanation for the occurrence of ultramafic lenses and boudins in these terranes (e.g. Lützow-Holm Bay: Shiraishi et al. 1994), there is no evidence for extensive areas of such high-density mantle material at shallow depths in any of the Pan-African terranes considered here. For example, in the Prydz Belt and Lützow-Holm Bay there is no evidence for major additions to the crust in tectonic events since 500 Ma, and yet present crustal thicknesses are 25–35 km (Bentley 1991). Hence, the one-dimensional thermal models that best explain the UHT metamorphism are not compatible with simple observations on the nature of the present crust in these regions. In order to satisfy the constraints available from both the preserved crustal structure and thickness and the metamorphic *P–T–t* paths, it will be necessary to develop and test at least

now replaced by a symplectite of sapphirine (Spr) and Crd, Forefinger Point, Enderby Land. This texture is interpreted to relfect progress of the FMAS continuous reaction Opx + Sil = Spr + Crd (reaction C in Fig. 7) during decompression (Harley et al. 1990). Field of view, 1.5 mm across. (**f**) Backscattered electron image of a sapphirine (Spr) + cor'''dierite (Crd) symplectite replacing sillimanite (Sil) formerly adjacent to garnet (Grt) (not in the image), Mather Peninsula, Rauer Islands. Minor hercynitic spinel (Spl) is restricted to the symplectite area and is inferred to have formed with the Spr + Crd, which is interpreted to have been produced via the reaction Grt + Sil = Spr + Crd on decompression (Harley 1998b). Field of view, 60 μm across.

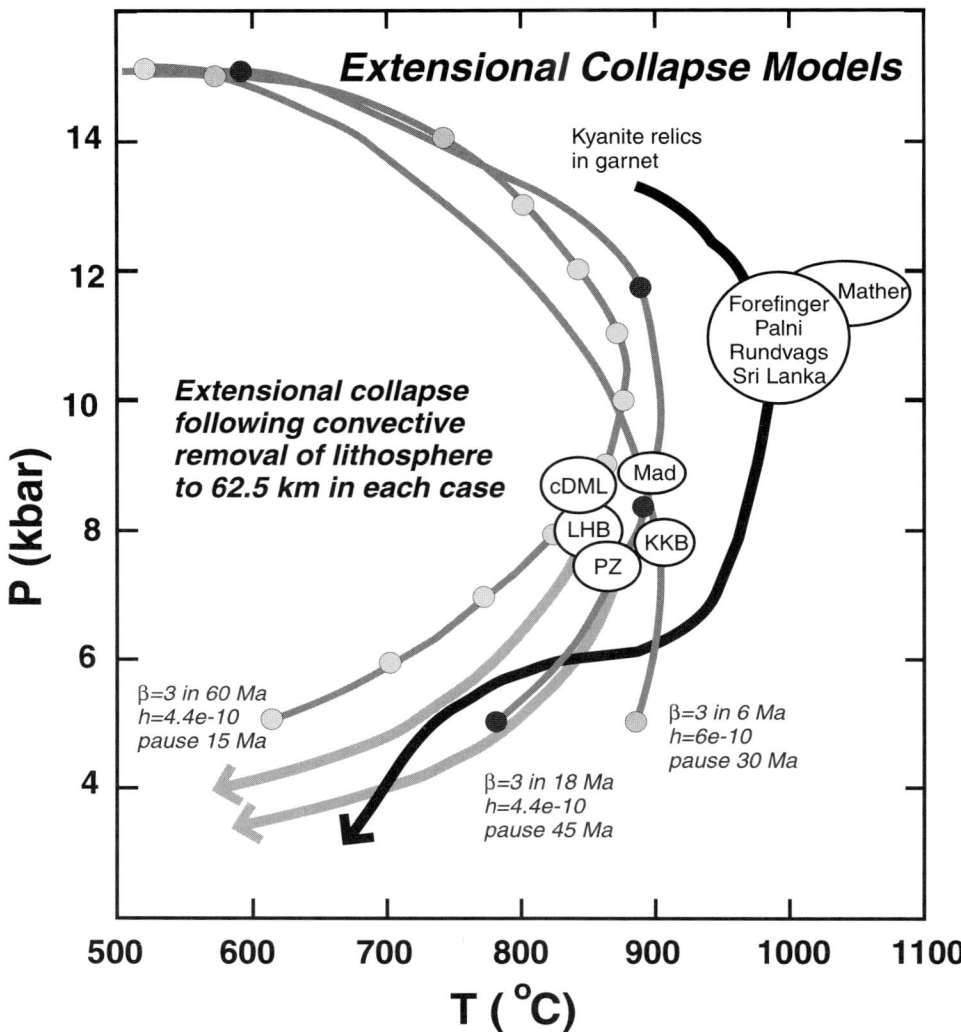

Fig. 8. Comparison of the P–T path for the ultrahigh-temperature (UHT)–isothermal decompression (ITD) localities given in Figure 7, extended down to lower pressures where cooling and then further decompression occurs (Harley et al. 1990; Motoyoshi & Ishikawa 1997; Fraser et al. 2000), with model P–T–t paths associated with extensional thinning of previously thickened crust (Platt et al. 1998). Three model paths for rocks initially buried to 55 km in overthickened crust are presented, each with filled circles that represent 6 Ma time intervals. In each case, all the mantle lithosphere below a depth of 62.5 km is removed at some time after thickening; the time lapse between lithospheric thickening and removal is the 'pause'. Crustal heat production (h) is given is units of mW kg^{-1}. Following lithosphere removal, the crust is allowed to extend (thin) by a stretching factor (β) of 3, over time intervals of 60, 18 and 6 Ma for paths a, b and c, respectively. These model paths correspond to the paths in figure 5j, 5i and 5h in Platt et al. (1998). Also shown are P–T paths for other Pan-African granulites mentioned in the text: PZ, Prydz Bay (Fitzsimons 1996); LHB, Lützow-Holm Bay (Shiraishi et al. 1994); Mad, Madagascar (Markl et al. 2000); cDML, central Dronning Maud Land (Piazolo & Markl 1999); KKB, Kerala Khondalite Belt (Nandakumar & Harley 2000).

two-dimensional thermomechanical models that incorporate the effects of lateral accretion and allow for alternating phases of transpression and transtension within the evolving Pan-African collisional belts. Such modelling is beyond the scope of this contribution, but should form the focus of future research in the light of the problems and potentials recognized in this synthesis.

Discussions over the years on various aspects of Antarctic crustal evolution with numerous Antarctic workers have influenced this synthesis, and I thank all of these and others for their contributions to Antarctic geology. B. Hensen and I. Fitzsimons are also thanked for their helpful reviews. This contribution to IGCP 368 has been supported by NERC grant GR9/2628 and by Royal Society travel grants to SLH.

References

BELIATSKY, B. V., LAIBA, A. A. & MIKHALSKY, E. V. 1994. U–Pb zircon age of metavolcanic rocks of Fisher Massif (Prince Charles Mountains, East Antarctica). *Antarctic Science*, **6**, 355–358.

BENTLEY, C. R. 1991. Configuration and structure of the subglacial crust. *In*: TINGEY, R. J. (ed.) *The Geology of Antarctica*. Oxford Monographs on Geology and Geophysics 17. Clarendon Press, Oxford.

BLACK, L. P., HARLEY, S. L., SUN, S. S. & MCCULLOCH, M. T. 1987. The Rayner Complex of East Antarctica: complex isotopic systematics within a Proterozoic mobile belt. *Journal of Metamorphic Geology*, **5**, 1–26.

BLACK, L. P., KINNY, P. D., SHERATON, J. W. & DELOR, C. P. 1991. Rapid production and evolution of late Archaean felsic crust in the Vestfold Block of East Antarctica. *Precambrian Research*, **50**, 283–310.

BLACK, L. P., SHERATON, J. W., TINGEY, R. J. & MCCULLOCH, M. T. 1992. New U–Pb zircon ages from the Denman Glacier area, East Antarctica, and their significance for Gondwana reconstruction. *Antarctic Science*, **4**, 447–460.

BOGER, S. D., CARSON, C. J., WILSON, C. J. L. & FANNING, C. M. 2000. Neoproterozoic deformation in the Radok Lake region of the northern Prince Charles Mountains, east Antarctica: evidence for a single protracted orogenic event. *Precambrian Research*, **104**, 1–24.

BOGER, S. D., WILSON, C. J. L. & FANNING, C. M. 2001. Early Palaeozoic tectonism within the east Antarctic craton: the final suture between east and west Gondwana? *Geology*, **29**, 463–466.

CARSON, C. J., AGUE, J. J. & COATH, C. D. 2002. U–Pb geochronology from Tonagh Island, east Antarctica: implications for the timing of ultra-high temperature metamorphism of the Napier Complex. *Precambrian Research*, **116**, 237–263.

CARSON, C. J., DIRKS, P. H. G. M., HAND, M., SIMS, J. P. & WILSON, C. J. L. 1995. Compressional and extensional tectonics in low–medium pressure granulites from the Larsemann Hills, East Antarctica. *Geological Magazine*, **132**, 151–170.

CARSON, C. J., FANNING, C. M. & WILSON, C. J. L. 1996. Timing of the Progress Granite, Larsemann Hills, evidence for Early Palaeozoic orogenesis within the East Antarctic Shield and implications for Gondwana assembly. *Australian Journal of Earth Sciences*, **43**, 539–553.

CLARK, D. J., HENSEN, B. J. & KINNY, P. D. 2000. Geochronological constraints for a two-stage history of the Albany–Fraser orogen, Western Australia. *Precambrian Research*, **102**, 155–183.

CLARKE, G. L. 1988. Structural constraints on the Proterozoic reworking of Archaean crust in the Rayner Complex, MacRobertson and Kemp Land Coast, East Antarctica. *Precambrian Research*, **40/41**, 137–156.

COLLERSON, K. D., REID, E., MILLAR, D. & MCCULLOCH, M. T. 1983. Lithological and Sr–Nd isotopic relationships in the Vestfold Block: implications for Archaean and Proterozoic crustal evolution in the East Antarctic Shield. *In*: OLIVER, R. L., JAMES, P. R. & JAGO, J. B. (eds) *Antarctic Earth Science*. Australian Academy of Science, Canberra, 77–84.

DALZIEL, I. W. D. 1991. Pacific margins of Laurentia and East Antarctica–Australia as a conjugate rift pair: evidence and implications for an Eocambrian supercontinent. *Geology*, **19**, 598–601.

DASGUPTA, S. 1995. Pressure–temperature evolutionary history of the Eastern Ghats granulite province: recent advances and some thoughts. *In*: YOSHIDA, M. & SANTOSH, M. (eds) *India and Antarctica During the Precambrian*. Memoir, Geological Society of India, **34**, 101–110.

DASGUPTA, S. & SENGUPTA, P. 2000. Tectonothermal evolution of the Eastern Ghats Belt, India. *Geological Survey of India Special Publication*, **55**, 259–274.

DASGUPTA, S. & SENGUPTA, P. 2002. Indo-Antarctic correlation: a perspective from the Eastern Ghats Granulite Belt, India. *In*: YOSHIDA, M., WINDLEY, B. W., & DASGUPTA, S. (eds) *Proterozoic East Gondwana: Supercontinent Assembly and Breakup*. Geological Society, London, Special Publication, **xx**, xx–xx.

DE WIT, M. J., BOWRING, S. A., ASHWAL, L. A., RANDRIANASOLO, L. G., MOREL, V. P. I. & RAMBELOSON, R. A. 2001. Age and tectonic evolution of Neoproterozoic ductile shear zones in southwestern Madagascar, with implications for Gondwana studies. *Tectonics*, **20**, 1–45.

DIRKS, P. H. G. M., CARSON, C. J. & WILSON, C. J. L. 1993a. The deformational history of the Larsemann Hills, Prydz Bay: the importance of the Pan-African (500 Ma) in East Antarctica. *Antarctic Science*, **5**, 179–192.

DIRKS, P. H. G. M., HOEK, J. D., WILSON, C. J. L. & SIMS, J. R. 1993b. The Proterozoic deformation of the Vestfold Hills basement complex, East Antarctica; implications for the tectonic evolution of adjacent granulite belts. *Precambrian Research*, **65**, 277–295.

DOBMEIER, C. J. & RAITH, M. M. 2002. Crustal architecture and evolution of the Eastern Ghats Belt and adjacent regions of India. *In*: YOSHIDA, M., WINDLEY, B. W. & DASGUPTA, S. (eds). *Proterozoic East Gondwana: Supercontinent Assembly and Breakup*. Geological Society, London, Special Publication, **xx**, xx–xx.

DUNKLEY, D. J., CLARKE, G. L. & WHITE, R. W. 2002.

Structural and metamorphic evolution of the mid to late-Proterozoic Rayner Complex, Cape Bruce, East Antarctica. *In*: GAMBLE, J. A., SKINNER, D. N. B., HENRYS, S. & LYNCH, R. (eds) *Antarctica at the Close of a Millennium*. Proceedings Volume of the 8th International Symposium on Antarctic Earth Sciences, Royal Society of New Zealand Bulletin, **35**, in press.

ELLIS, D. J. 1983. The Napier and Rayner Complexes of Enderby Land, Antarctica – contrasting styles of metamorphism and tectonism. *In*: OLIVER, R. L., JAMES, P. R. & JAGO, J. B. (eds) *Antarctic Earth Science*, Australian Academy of Science, Canberra, 20–24.

ELLIS, D. J. 1987. Origin and evolution of granulites in normal and thickened crust. *Geology*, **15**, 167–170.

ELLIS, D. J. & GREEN, D. H. 1985. Garnet-forming reactions in mafic granulites from Enderby Land, Antarctica – implications for geothermometry and geobarometry. *Journal of Petrology*, **26**, 633–662.

ELLIS, D. J., SHERATON, J. W., ENGLAND, R. N. & DALLWITZ, W. B. 1980. Osumilite–sapphirine–quartz granulites from Enderby Land, Antarctica – mineral assemblages and reactions. *Contributions to Mineralogy and Petrology*, **72**, 123–143.

FANNING, C. M., MOORE, D. H., BENNETT, V. C., DALY, S. J., MENOT, R. P., PEUCAT, J. J. & OLIVER, R. L. 1999. The 'Mawson Continent': the East Antarctic shield and Gawler Craton, Australia. *Programme & Abstracts of the 8th International Symposium Antarctic Earth Sciences*, Wellington 103.

FITZSIMONS, I. C. W. 1996. Metapelitic migmatites from Brattstrand Bluffs, East Antarctica – metamorphism, melting and exhumation of the mid crust. *Journal of Petrology*, **37**, 395–414.

FITZSIMONS, I. C. W. 2000a. Grenville-age basement provinces in East Antarctica: evidence for three separate collisional orogens. *Geology*, **28**, 879–882.

FITZSIMONS, I. C. W. 2000b. A review of tectonic events in the East Antarctic Shield and their implications for Gondwana and earlier supercontinents. *Journal of African Earth Sciences*, **31**, 3–23.

FITZSIMONS, I. C. W. 2002. Proterozoic basement provinces of southern and southwestern Australia, and their correlation with Antarctica. *In*: YOSHIDA, M., WINDLEY, B. W. & DASGUPTA, S. (eds) *Proterozoic East Gondwana: Supercontinent Assembly and Breakup*. Geological Society, London, Special Publication, **xx**, xx–xx.

FITZSIMONS, I. C. W. & HARLEY, S. L. 1991. Geological relationships in high-grade gneisses of the Brattstrand Bluffs coastline, Prydz Bay, East Antarctica. *Australian Journal of Earth Sciences*, **38**, 497–519.

FITZSIMONS, I. C. W. & HARLEY, S. L. 1992. Mineral reaction textures in high-grade gneisses: evidence for contrasting pressure-temperature paths in the Proterozoic Complex of East Antarctica. *In*: YOSHIDA, Y., KAMINUMA, K. & SHIRAISHI, K. (eds) *Recent Progress in Antarctic Earth Science*. Terra Scientific Publishing Company, Tokyo, 103–111.

FITZSIMONS, I. C. W. & THOST, D. E. 1992. Geological relationships in high-grade basement gneisses of the Northern Prince Charles Mountains, East Antarctica. *Australian Journal of Earth Sciences*, **39**, 173–193.

FITZSIMONS, I. C. W., KINNY, P. D. & HARLEY, S. L. 1997. Two stages of zircon and monazite growth in anatectic leucogneiss: SHRIMP constraints on the duration and intensity of Pan-African metamorphism in Prydz Bay, East Antarctica. *Terra Nova*, **9**, 47–51.

FRASER, G., MCDOUGALL, I., ELLIS, D. J. & WILLIAMS, I. S. 2000. Timing and rate of isothermal decompression in Pan-African granulites from Rundvagshetta, East Antarctica. *Journal of Metamorphic Geology*, **18**, 441–454.

GOODGE, J. W., FANNING, C. M. & BENNETT, V. C. 2001. U–Pb evidence of c. 1.7Ga crusted tectonism during the Nimrod Orogeny in the Transantarctic Mountains, Antartica: implications for Proterozoic plate reconstruction. *Precambrian Research*, **112**, 261–288.

GRANTHAM, G. H., JACKSON, C., MOYES, A. B., HARRIS, P. D., GROENEWALD, P. B., FERRAR, G. & KRYNAUW, J. R. 1995. P–T evolution of the H.U. Sverdrupfjella and Kirwanveggan, Dronning Maud Land, Antarctica. *Precambrian Research*, **75**, 209–229.

GREW, E. S. 1980. Sapphirine + quartz association from Archaean rocks in Enderby Land, Antarctica. *American Mineralogist*, **65**, 821–836.

GREW, E. S. 1982. Osumilite in the sapphirine–quartz terrane of Enderby Land, Antarctica: implications for osumilite petrogenesis in the granulite facies. *American Mineralogist*, **67**, 762–787.

GREW, E. S. 1998. Boron and Beryllium minerals in granulite-facies pegmatites and implications of beryllium pegmatites for the origin and evolution of the Archaean Napier Complex of East Antarctica. *In*: MOTOYOSHI, Y. & SHIRAISHI, K. (eds) *Origin and Evolution of Continents*. Memoir, National Institute of Polar Research Special Issue, **53**, 74–92.

GREW, E. S. & MANTON, W. 1979. Archaean rocks in Antarctica: 2.5 billion year uranium–lead ages of pegmatites in Enderby Land. *Science*, **206**, 443–445.

HAND, M., SCRIMGEOUR, I., POWELL, R., STÜWE, K. & WILSON, C. J. L. 1994. Metapelitic granulites from Jetty Peninsula, East Antarctica, formation during a single event or polymetamorphism? *Journal of Metamorphic Geology*, **12**, 557–573.

HARLEY, S. L. 1985. Garnet-orthopyroxene bearing granulites from Enderby Land, Antarctica: metamorphic pressure–temperature–time evolution of the Archaean Napier Complex. *Journal of Petrology*, **26**, 819–856.

HARLEY, S. L. 1987a. A pyroxene-bearing metaironstone and other pyroxene-granulites from Tonagh Island, Enderby Land, Antarctica: further evidence for very high temperature (>980°C) Archaean regional metamorphism in the Napier Complex. *Journal of Metamorphic Geology*, **5**, 341–356.

HARLEY, S. L. 1987b. Precambrian geological relationships in high-grade gneisses of the Rauer Islands, East Antarctica. A*ustralian Journal of Earth Sciences*, **34**, 175–207.

HARLEY, S. L. 1988. Proterozoic granulites from the Rauer Group, East Antarctica. I. Decompressional pressure-temperature paths deduced from mafic and felsic gneisses. *Journal of Petrology*, **29**, 1059–1095.

HARLEY, S. L. 1989. The origins of granulites: a metamorphic perspective. *Geological Magazine*, **126**, 215–247.

HARLEY, S. L. 1992. Proterozoic granulite terranes. *In*: CONDIE, K. (ed.) *Proterozoic Crustal Evolution*. Elsevier, Amsterdam, 301–359.

HARLEY, S. L. 1993. Sapphirine granulites from the Vestfold Hills, East Antarctica: geochemical and metamorphic evolution. *Antarctic Science*, **5**, 389–402.

HARLEY, S. L. 1998a. On the occurrence and characterisation of ultrahigh-temperature crustal metamorphism. *In*: TRELOAR, P. J & O'BRIEN, P. J. (eds) *What Drives Metamorphism and Metamorphic Reactions?* Geological Society, London, Special Publication, **138**, 81–107.

HARLEY, S. L. 1998b. Ultrahigh temperature granulite metamorphism (1050°C, 12 kbar) metamorphism and decompression in garnet (Mg70)–orthopyroxene-sillimanite gneisses from the Rauer Group, East Antarctica. *Journal of Metamorphic Geology*, **16**, 541–562.

HARLEY, S. L. 1998c. An appraisal of peak temperatures and thermal histories in ultrahigh-temperature (UHT) crustal metamorphism: the significance of aluminous orthopyroxene. *In*: MOTOYOSHI, Y. & SHIRAISHI, K. (eds) *Origin and Evolution of Continents*. Memoir, National Institute of Polar Research Special Issue, **53**, 49–73.

HARLEY, S. L. & BLACK, L. P. 1987. The Archaean geological evolution of Enderby Land, Antarctica. *In*: PARK, R. G. & TARNEY, J. (eds) *Evolution of the Lewisian and Comparable Precambrian High Grade Terrains*. Geological Society, London, Special Publication, **27**, 285–296.

HARLEY, S. L. & BLACK, L. P. 1997. A revised Archaean chronology for the Napier Complex, Enderby Land, from SHRIMP ion-microprobe studies. *Antarctic Science*, **9**, 74–91.

HARLEY, S. L. & BUICK, I. S. 1992. Wollastonite-scapolite assemblages as indicators of granulite pressure–temperature–fluid histories: the Rauer Group, East Antarctica. *Journal of Petrology*, **33**, 693–728.

HARLEY, S. L. & CHRISTY, A. C. 1995. Titanium-bearing sapphirine in a partially melted aluminous granulite xenolith, Vestfold Hills, Antarctica: geological and mineralogical implications. *European Journal of Mineralogy*, **7**, 637–653.

HARLEY, S. L. & FITZSIMONS, I. C. W. 1991. Pressure–temperature evolution of metapelitic granulites in a polymetamorphic terrane: the Rauer Group, East Antarctica. *Journal of Metamorphic Geology*, **9**, 231–243.

HARLEY, S. L. & FITZSIMONS, I. C. W. 1995. High-grade metamorphism and deformation in the Prydz Bay region, East Antarctica: terranes, events and regional correlations. *In*: YOSHIDA, M. & SANTOSH, M. (eds) *India and Antarctica during the Precambrian*. Memoir, Geological Society of India, **34**, 73–100.

HARLEY, S. L. & HENSEN, B. J. 1990. Archaean and Proterozoic high-grade terranes of East Antarctica (40–80°E): a case study of diversity in granulite facies metamorphism. *In*: ASHWORTH, J. R. & BROWN, M. (eds) *High-temperature Metamorphism and Crustal Anatexis*. Unwin Hyman, London, 320–370.

HARLEY, S. L. & MOTOYOSHI, Y. 2000. Alumina zoning in orthopyroxene in a sapphirine quartzite: evidence for >1120°C ultrahigh temperature metamorphism in the Napier Complex, Enderby Land, Antarctica. *Contributions to Mineralogy and Petrology*, **138**, 293–307.

HARLEY, S. L., FITZSIMONS, I. C. W., BUICK, I. S. & WATT, G. 1992. The significance of reworking, fluids and partial melting in granulite metamorphism, East Prydz Bay, Antarctica. *In*: YOSHIDA, Y., KAMINUMA, K. & SHIRAISHI, K. (eds) *Recent Progress in Antarctic Earth Science*. Terra Scientific Publishing Company, Tokyo, 119–127.

HARLEY, S. L., HENSEN, B. J. & SHERATON, J. W. 1990. Two-stage decompression in orthopyroxene–sillimanite granulites from Forefinger Point, Enderby Land, Antarctica: implications for the evolution of the Archaean Napier Complex. *Journal of Metamorphic Geology*, **8**, 591–613.

HARLEY, S. L., SNAPE, I. & BLACK, L. P. 1998. The early evolution of a layered metaigneous complex in the Rauer Group, East Antarctica: evidence for a distinct Archaean terrane. *Precambrian Research*, **89**, 175–205.

HARLEY S. L., SNAPE, I. & FITZSIMONS, I. C. W. 1995. Regional correlations and terrane assembly in East Prydz Bay: evidence from the Rauer Group and Vestfold Hills. *Terra Antartica*, **2**, 49–60.

HENSEN, B. J. 1971. Theoretical phase relations involving cordierite and garnet in the system MgO–FeO–Al_2O_3–SiO_2. *Contributions to Mineralogy and Petrology*. **33**, 191–214.

HENSEN, B. J. & GREEN, D. H. 1973. Experimental study of the stability of cordierite and garnet in pelitic compositions at high pressures and temperatures. III. Synthesis of experimental data and geological applications. *Contributions to Mineralogy and Petrology*, **38**, 151–166.

HENSEN, B. J. & HARLEY, S. L. 1990. Graphical analysis of P–T–X relations in granulite facies metapelites. *In*: ASHWORTH, J. R. & BROWN, M. (eds) *High-temperature Metamorphism and Crustal Anatexis*. Unwin Hyman, London, 19–56.

HENSEN, B. J. & MOTOYOSHI, Y. 1992. Osumilite-producing reactions in high-temperature granulites from the Napier Complex, East Antarctica: tectonic implications. *In*: YOSHIDA, Y., KAMINUMA, K. & SHIRAISHI, K. (eds) *Recent Progress in Antarctic Earth Science*. Terra Scientific Publishing Company, Tokyo, 87–92.

HENSEN, B.J. & ZHOU, B. 1995. A Pan-African granulite facies metamorphic episode in Prydz Bay, Antarctica: evidence from Sm–Nd garnet dating. *Australian Journal of Earth Sciences*, **42**, 249–258.

HENSEN, B. J. & ZHOU, B. 1997. East Gondwana amalgamation by Pan-African collision? Evidence from Prydz Bay, Eastern Antarctica. *In*: RICCI, C. A. (ed.) *The Antarctic Region: Geological Evolution and Processes*. Terra Antartica Publication, Siena, 115–119.

HENSEN, B. J., ZHOU, B. & THOST, D. E. 1995. Are reaction textures reliable guides to metamorphic histories? Timing constraints from garnet Sm–Nd chronology for 'decompression' textures in granulites from Søstrene Island, Prydz Bay, Antarctica. *Geological Journal*, **30**, 261–271.

HENSEN, B. J., ZHOU, B. & THOST, D. E. 1997. Recognition of multiple high grade metamorphic events with garnet Sm–Nd chronology in the northern Prince Charles Mountains, Antarctica. *In*: RICCI, C. A. (ed.) *The Antarctic Region: Geological Evolution and Processes*. Terra Antartica Publication, Siena, 97–104.

HOEK, J. D., DIRKS, P. G. E. M. & PASSCHIER, C. W. 1992. A late-Proterozoic extensional–compressional tectonic cycle in East Antarctica. *In*: YOSHIDA, Y., KAMINUMA, K. & SHIRAISHI, K. (eds) *Recent Progress in Antarctic Earth Science*. Terra Scientific Publishing Company, Tokyo, 137–143.

HOKADA, T. 2001. Feldspar thermometry in ultrahigh-temperature metamorphic rocks: evidence of crustal metamorphism attaining ~1100°C in the Archean Napier Complex, East Antarctica. *American Mineralogist*, **86**, 932–938.

HOKADA, T., OSANAI, Y., TOYOSHIMA, T., OWADA, M., TSUNOGAE, T. & CROWE, W. A. 1999. Petrology and metamorphism of sapphirine-bearing aluminous gneisses from Tonagh in the Napier Complex, East Antarctica. *Polar Geoscience*, **12**, 49–70.

HOLLIS, J. & HARLEY, S. L. 2002. New evidence for the peak temperatures and the near-peak pressure–temperature evolution of the Napier Complex. *In*: GAMBLE, J. A., SKINNER, D. N. B., HENRYS, S. & LYNCH, R. (eds) *Antarctica at the close of a Millennium*. Proceedings Volume of the 8th International Symposium on Antarctic Earth Sciences, Royal Society of New Zealand Bulletin, **35**, in press.

HÖLZL, S., HOFFMAN, A. W., TODT, W. & KÖHLER, H. 1994. U-Pb geochronology of the Sri Lankan basement. *Precambrian Research*, **66**, 123–149.

JACOBS, J., BAUER, W., SPAETH, G., THOMAS, R. J. & WEBER, K. 1996. Lithology and structure of the Grenville-aged (~1.1 Ga) basement of Heimefrontfjella (East Antarctica). *Geologische Rundschau*, **85**, 800–821.

JACOBS, J., FALTER, M., WEBER, K. & JESSBERGER, E. K. 1997. $^{40}Ar-^{39}Ar$ evidence for the structural evolution of the Heimefront shear zone (Western Dronning Maud Land), East Antarctica. *In*: RICCI, C. A. (ed.) *The Antarctic Region: Geological Evolution and Processes*. Terra Antartica Publication, Siena, 37–44.

JACOBS, J., FANNING, C. M., HENJES-KUNST, F., OLESCH, M. & PAECH, H.-J. 1998. Continuation of the Mozambique Belt into East Antarctica: Grenville-age metamorphism and polyphase Pan-African high-grade events in central Dronning Maud Land. *Journal of Geology*, **106**, 385–406.

JACOBS, J., THOMAS, R. J. & WEBER, K. 1993. Accretion and indentation tectonics at the southern edge of the Kaapvaal craton during the Kibaran (Grenville) orogeny. *Geology*, **21**, 203–206.

KELLY, N. M., CLARKE, G. L. & FANNING, C. M. 1999. Proterozoic reworking of Archaean crust: evidence from preliminary sensitive high resolution microbe (SHRIMP) geochronology, Oygarden Group, Rayner Complex, East Antarctica. *Programme & Abstracts of the 8th International Symposium Antarctic Earth Sciences*, Wellington, 168.

KELLY, N. M., CLARKE, G. L. & FANNING, C. M. 2002. A two-stage evolution of the Neoproterozoic Rayner Structural Episode: new U–Pb sensitive high resolution microbe (SHRIMP) constraints from the Oygarden Group, Kemp Land, East Antarctica. *Precambrian Research*, **116**, 307–330.

KELLY, N. M., CLARKE, G. L., CARSON, C. J. & WHITE, R. W. 2000. Thrusting in the lower crust: evidence from the Oygarden Islands, Kemp Land, East Antarctica. *Geological Magazine*, **137**, 219–234.

KILPATRICK, J. A. & ELLIS, D. J. 1992. C-type magmas: igneous charnockites and their extrusive equivalents. *Transactions of the Royal Society of Edinburgh: Earth Sciences*, **83**, 155–164.

KINNY, P. D., BLACK, L. P. & SHERATON, J. W. 1993. Zircon ages and the distribution of Archaean and Proterozoic rocks in the Rauer Islands. *Antarctic Science*, **5**, 193–206.

KINNY, P. D., BLACK, L. P. & SHERATON, J. W. 1997. Zircon U–Pb ages and geochemistry of igneous and metamorphic rocks in the northern Prince Charles Mountains, Antarctica. *AGSO Journal of Australian Geology and Geophysics*, **16**, 637–654.

KRIEGSMAN, L. M. & SCHUMACHER, J. 1999. Petrology of sapphirine-bearing and associated granulites from central Sri Lanka. *Journal of Petrology*, **40**, 1211–1239.

LANYON, R., BLACK, L. P. & SEITZ, H-M. 1993. U-Pb zircon dating of mafic dykes and its application to the Proterozoic geological history of the Vestfold Hills, East Antarctica. *Contributions to Mineralogy and Petrology*, **115**, 184–203.

MARKL, G., BÄUERLE, J. & GRUJIC, D. 2000. Metamorphic evolution of Pan-African granulite facies metapelites from southern Madagascar. *Precambrian Research*, **102**, 47–68.

MEZGER, K. & COSCA, M. A. 1999. The thermal history of the Eastern Ghats Belt (India) as revealed by U–Pb and $^{40}Ar/^{39}Ar$ dating of metamorphic and magmatic minerals: Implications for SWEAT correlation. *Precambrian Research*, **94**, 251–271.

MIKHALSKY, E. V., SHERATON, J. W., LAIBA, A. A., TINGEY, R. J. THOST, D. E., KAMENEV, E. N. & FEDOROV, L. V. 2001. *Geology of the Prince Charles Mountains, Antarctica*. Geoscience Australia Bulletin 247. Australian Geological Survey Organisation, Canberra.

MOORES, E. M. 1991. The southwest U.S.–East Antarctica (SWEAT) connection: a hypothesis. *Geology*, **19**, 425–428.

MOTOYOSHI, Y. & HENSEN, B. J. 1989. Sapphirine–quartz–orthopyroxene symplectites after cordierite in the Archaean Napier Complex, Antarctica: evidence for a counterclockwise P–T path. *European Journal of Mineralogy*, **1**, 467–471.

MOTOYOSHI, Y. & ISHIKAWA, M. 1997. Metamorphic and structural evolution of granulites from Rundvågshetta, Lützow-Holm Bay, East Antarctica. *In*: RICCI, C. A. (ed.) *The Antarctic Region: Geological Evolution and Processes*. Terra Antartica Publication, Siena, 65–72.

MOTOYOSHI, Y., ISHIKAWA, M. & FRASER, G. L. 1994. Reaction textures in granulites from Forefinger Point, Enderby Land, East Antarctica: an alternative interpretation on the metamorphic evolution of the Rayner Complex. *Proceedings of the NIPR Symposium on Antarctic Geosciences*, National Institute of Polar Research, Tokyo, **7**, 101–114.

NANDAKUMAR, V. & HARLEY, S. L. 2000. A reappraisal of

the pressure–temperature path of granulites from the Kerala Khondalite Belt, southern India. *Journal of Geology*, **108**, 687–703.

OLIVER, R. L., JAMES, P. R., COLLERSON, K. D. & RYAN, A. B. 1982. Precambrian geologic relationships in the Vestfold Hills, Antarctica. *In*: CRADDOCK, C. (ed.) *Antarctic Geoscience*. University of Wisconsin Press, Madison, 435–444.

OSANAI, Y, TOYOSHIMA, T., OWADA, M., TSUNOGAE, T., HOKADA, T. & CROWE, W. A. 1999. Geology of ultra-high temperature metamorphic rocks from Tonagh in the Napier Complex, East Antarctica. *Polar Geoscience*, **12**, 1–28.

PAECH, H.-J. 1997. Central Dronning Maud Land: Its history from amalgamation to fragmentation of Gondwana. *Terra Antartica*, **4**, 41–49

PASSCHIER, C. W., BEKENDAM, R. F., HOEK, J. D., DIRKS, P G. H. M. & DE BOORDER, H. 1991. Proterozoic geological evolution of the northern Vestfold Hills, Antarctica. *Geological Magazine*, **128**, 307–318.

PEUCAT, J. J., MÉNOT, R. P., MONNIER, O. & FANNING, C. M. 1999. The Terre Adélie basement in the East Antarctica Shield: geological and isotopic evidence for a major 1.7 Ga thermal event; comparison with the Gawler Craton in South Australia. *Precambrian Research*, **94**, 205–224.

PIAZOLO, S. & MARKL, G. 1999. Humite- and scapolite-bearing assemblages in marbles and calcsilicates of Dronning Maud Land, Antarctica and their implications on Gondwana reconstructions. *Journal of Metamorphic Geology*, **17**, 91–107.

PLATT, J. P. & ENGLAND, P. C. 1994. Convective removal of lithosphere beneath mountain belts: thermal and mechanical consequences. *American Journal of Science*, **293**, 307–336.

PLATT, J. P., SOTO, J.-I., WHITEHOUSE, M. J., HURFORD, A. J. & KELLEY, S. P. 1998. Thermal evolution, rate of exhumation, and tectonic significance of metamorphic rocks from the floor of the Alboran extensional basin, western Mediterranean. *Tectonics*, **17**, 671–689.

POST, N. J., HENSEN, B. J. & KINNY, P. D. 1997. Two metamorphic episodes during a 1340–1180 Ma convergent tectonic event in the Windmill Islands, East Antarctica. *In*: RICCI, C. A. (ed.) *The Antarctic Region: Geological Evolution and Processes*. Terra Antartica Publication, Siena, 157–161.

RAITH, M., KARMAKAR, S. & BROWN, M. 1997. Ultra-high temperature metamorphism and multistage decompressional evolution of sapphirine granulites from the Palni Hill Ranges, southern India. *Journal of Metamorphic Geology*, **15**, 379–400.

REN, L., ZHAO, Y. & CHEN, T. 1992. Re-examination of the metamorphic evolution of the Larsemann Hills, eastern Antarctica. *In*: YOSHIDA, Y., KAMINUMA, K. & SHIRAISHI, K. (eds) *Recent Progress in Antarctic Earth Science*. Terra Scientific Publishing Company, Tokyo, 145–153.

ROGERS, J. J. M. 1996. A history of the continents in the past three billion years. *Journal of Geology*, **104**, 91–107.

SANDIFORD, M. 1985. The metamorphic evolution of granulites at Fyfe Hills: implications for Archaean crustal thickness in Enderby Land, Antarctica. *Journal of Metamorphic Geology*, **3**, 155–178.

SANDIFORD, M. & POWELL, R. 1986. Pyroxene exsolution in granulites from Fyfe Hills, Enderby Land, Antarctica: evidence for 1000°C metamorphic temperatures in Archaean continental crust. *American Mineralogist*, **71**, 946–954.

SEITZ, H.-M. 1994. Estimation of emplacement pressure for 2350 Ma high-Mg tholeiite dykes, Vestfold Hills, Antarctica. *European Journal of Mineralogy*, **6**, 195–208.

SHERATON, J. W., BLACK, L. P. & MCCULLOCH, M. T. 1984. Regional geochemical and isotopic characteristics of high-grade metamorphics of the Prydz Bay area: the extent of Proterozoic reworking of Archaean continental crust in East Antarctica. *Precambrian Research*, **26**, 169–198.

SHERATON, J. W., OFFE, L. A., TINGEY, R. J. & ELLIS, D. J. 1980. Enderby Land, Antarctica – an unusual Precambrian high-grade metamorphic terrain. *Journal of the Geological Society of Australia*, **27**, 1–18.

SHERATON, J. W., TINDLE, A. G. & TINGEY, R. J. 1996. Geochemistry, origin and tectonic setting of granitic rocks of the Prince Charles Mountains, Antarctica. *AGSO Journal of Australian Geology and Geophysics*, **16**, 345–370.

SHERATON, J. W., TINGEY, R. J., BLACK, L. P., OFFE, L. A. & ELLIS, D. J. 1987. Geology of Enderby Land and Western Kemp Land, Antarctica. *Bulletin Australian Bureau of Mineral Resources*, **223**, 1–51.

SHERATON, J. W., TINGEY, R. J., BLACK, L. P. & OLIVER, R. L. 1993. Geology of the Bunger Hills area, Antarctica: implications for Gondwana correlations. *Antarctic Science*, **5**, 85–102.

SHIRAISHI, K., ELLIS, D. J., HIROI, Y., FANNING, C. M., MOTOYOSHI, Y. & NAKAI, Y. 1994. Cambrian orogenic belt in East Antarctica and Sri Lanka: implications for Gondwana assembly. *Journal of Geology*, **102**, 47–65.

SHIRAISHI, K., ELLIS, D. J., FANNING, C. M., HIROI, Y., KAGAMI, H. & MOTOYOSHI, Y. 1997. Re-examination of the metamorphic and protolith ages of the Rayner Complex, Antarctica: evidence for the Cambrian (Pan-African) regional metamorphic event. *In*: RICCI, C.A. (ed.) *The Antarctic Region: Geological Evolution and Processes*. Terra Antartica Publication, Siena, 79–88.

SHIRAISHI, K., FANNING, C. M., ARMSTRONG, R. & MOTOYOSHI, Y. 1999. New evidence for polymetamorphic events in the Sør Rondane Mountains, East Antarctica. *Programme & Abstracts of the 8th International Symposium of Antarctic Earth Sciences*, Wellington, 280.

SIMS, J. R., DIRKS, P. H. G. M., CARSON, C. J. & WILSON, C. J. L. 1994. The structural evolution of the Rauer Group, East Antarctica: mafic dykes as passive markers in a composite Proterozoic terrain. *Antarctic Science*, **6**, 379–394.

SNAPE, I. S., BLACK, L. P. & HARLEY, S. L. 1997. Refinement of the timing of magmatism and high-grade deformation in the Vestfold Hills, East Antarctica, from new SHRIMP U–Pb zircon geochronology. *In*: RICCI, C. A. (ed.) *The Antarctic Region: Geological Evolution and Processes*. Terra Antartica Publication, Siena, 139–148.

STÜWE, K. & POWELL, R. 1989. Low-pressure granulite facies metamorphism in the Larsemann Hills area, East Antarctica; petrology and tectonic implications for the evolution of the Prydz Bay area. *Journal of Metamorphic Geology*, **7**, 465–483.

THOST, D. E., HENSEN, B. J. & MOTOYOSHI, Y. 1991. Two-stage decompression in garnet-bearing mafic granulites from Søstrene Island, Prydz Bay, East Antarctica. *Journal of Metamorphic Geology*, **9**, 245–256.

TINGEY, R. J. 1991. The regional geology of Archaean and Proterozoic rocks in Antarctica. *In*: TINGEY R. J (ed.) *The Geology of Antarctica*. Oxford University Press, Oxford, 1–73.

TOYOSHIMA, T., OSANAI, Y., OWADA, M., TSUNOGAE, T., HOKADA, T. & CROWE, W. A. 1999. Deformation of ultrahigh-temperature metamorphic rocks from Tonagh in the Napier Complex, East Antarctica. *Polar Geoscience*, **12**, 29–48.

UNRUG, R. 1997. Rodinia to Gondwana: the geodynamic map of Gondwana supercontinent assembly. *GSA Today*, **7**, 1–6.

WHITE, R. W. & CLARKE, G. L. 1993. Timing of Proterozoic deformation and magmatism in a tectonically reworked orogen, Rayner Complex, Colbeck Archipelago, east Antarctica. *Precambrian Research*, **63**, 1–26.

YOSHIDA, M. 1995. Assembly of East Gondwanaland during the Mesoproterozoic and its rejuvenation during the Pan-African period. *In*: YOSHIDA, M. & SANTOSH, M. (eds) *India and Antarctica During the Precambrian*. Memoir, Geological Society of India, **34**, 22–45.

YOUNG, D. N., ZHAO, J.-X., ELLIS, D. J. & MCCULLOCH, M. T. 1997. Geochemical and Sr–Nd isotopic mapping of sources provinces for the Mawson charnockites, east Antarctica: implications for Proterozoic tectonics and Gondwana reconstruction. *Precambrian Research*, **86**, 1–19.

ZHAO, Y., LIU, X., SONG, B., ZHANG, Z., LI, J., YAO, Y. & WANG, Y. 1995. Constraints on the stratigraphic age of metasedimentary rocks from the Larsemann Hills, East Antarctica: possible implications for Neoproterozoic tectonics. *Precambrian Research*, **75**, 175–188.

ZHAO, Y., SONG, B., WANG, Y., REN, L., LI, J. & CHEN, T. 1992. Geochronology of the late granite in the Larsemann Hills, East Antarctica. *In*: YOSHIDA, Y., KAMINUMA, K. & SHIRAISHI, K. (eds) *Recent Progress in Antarctic Earth Science*. Terra Scientific Publishing Company, Tokyo, 155–161.

Pan-African events in Prydz Bay, East Antarctica, and their implications for East Gondwana tectonics

Y. ZHAO[1], X.H. LIU[2], X. C. LIU[1] & B. SONG[3]

[1]*Institute of Geomechanics, CAGS, Beijing, 100081, China*
[2]*Institute of Geology, CAS, Beijing, 100029, China*
[3]*Institute of Geology, CAGS, Beijing, 100037, China*

Abstract: Decompression, anatexis and the clockwise granulite-grade *P–T* evolution of high-grade rocks of Prydz Bay reflect late collisional extension that occurred *c.* 530 Ma in the Prydz Belt. Rapid cooling of the mid-crust high-grade terranes of the Prydz Belt was achieved during *c.* 517–486 Ma by tectonically driven exhumation along dextral ductile shear zones in which late tectonic partial melt bodies were emplaced. Instead of a model of polyphase metamorphism and deformation for the basement-and-cover sequences in Prydz Bay, we apply an accretionary one, i.e. an accretionary wedge with allochthonous blocks, to interpret the tectonic history in the Late Neoproterozoic of the Prydz Belt. SHRIMP U–Pb dates and Nd isotopic data available for both Prydz Bay and the Grove Mountains are used to explain amalgamation of the high-grade terranes in the Prydz Belt. This demonstrates that the assembly of the East Antarctic Craton was completed in the Pan-African event, and the East Antarctic Craton is a Pan-African-age collage rather than a keystone of East Gondwana during the Neoproterozoic.

It has been traditionally considered that the East Antarctic Craton, was the keystone of East Gondwana, formed during a Grenville-age event around the end of the Mesoproterozoic (Dalziel 1991; Hoffman 1991; Moores 1991; Tingey 1991). In this model, widespread tectonothermal activities from the latest Neoproterozoic to the Early Palaeozoic in the East Antarctica Craton, regarded as the Pan-African event, were attributed to a low- to medium-grade overprint and local granite intrusion. This notion was first challenged in the Prydz Bay region by the study of a Pan-African-age late tectonic granite (Zhao *et al.* 1991, 1992), the Progress Granite (Carson *et al.* 1996; cf. Figs 1, 2 & 3) – which is closely associated with regional anatexis and high-grade metamorphism that peaked in Cambrian times (Ren *et al.* 1992; Zhao *et al.* 1993; Carson *et al.* 1997; Fitzsimons *et al.* 1997). After a decade of debate over which event was responsible for the regional tectonothermal evolution, i.e. the Grenville-age event or Pan-African-age event, increasing evidence demonstrates that the Pan-African-age event finally led to the formation of the East Antarctica Craton (Fitzsimons 2000*a*; Zhao *et al.* 2000; Boger *et al.* 2001). This Pan-African age Prydz Belt (Fitzsimons 2000*a*) extends inland from Prydz Bay to the Grove Mountains (Zhao *et al.* 2000; Liu *et al.* 2002*a*; Fig. 1) and the southern Prince Charles Mountains (Boger *et al.* 2001). Therefore, it bisects the East Antarctic Craton for some lateral distance. In this paper the Pan-African-age event in Prydz Bay is reviewed and it is discussed whether it represents a suture assembling the Rayner and Wilkes Terranes in Cambrian times, and whether East Gondwana was similar to West Gondwana, finally juxtaposed during Gondwana amalgamation.

Regional geological background

The high-grade rocks of Prydz Bay are divided into three terranes (Fig. 2): the Archaean Vestford Hills gneisses which were formed and metamorphosed rapidly *c.* 2500 Ma (Black *et al.* 1991); the Rauer Islands gneisses, with the southwestern and northeastern parts containing Archean components, where multiple generations of deformed dykes occur (Harley *et al.* 1998) and the northwestern and central parts with dominant 1000 Ma granitic orthogneisses (Kinny *et al.* 1993) and predominant paragneisses in southern Prydz Bay with minor composite orthogneiss (Sheraton & Collerson 1983; Fitzsimons & Harley 1991; Dirks & Wilson 1995; Fitzsimons 1997).

Southern Prydz Bay composite orthogneiss, correlated with rocks of similar appearance in the Rauer Islands (Sheraton & Collerson 1983; Fitzsimons & Harley 1991; Carson *et al.* 1995; Dirks & Hand 1995), were thought to be basement rocks intruded by *c.*1000 Ma granites and metamorphosed around that time (Tingey 1991; Kinny *et al.* 1993; Carson *et al.* 1996). Dominant paragneisses represent the cover sequence (Dirks *et al.* 1993; Carson *et al.* 1995*b*), which were deposited onto and derived from erosion of this basement sometime in the Neoproterozoic, based on limited isotopic data (Zhao *et al.* 1995; Carson *et al.* 1996; Fitzsimons 1997), although no field observation can verify this

From YOSHIDA, M., WINDLEY, B. F. & DASGUPTA, S. (eds) 2003. *Proterozoic East Gondwana: Supercontinent Assembly and Breakup*. Geological Society, London, Special Publications, **206**, 231–245. 0305-8719/03/$15 © The Geological Society of London.

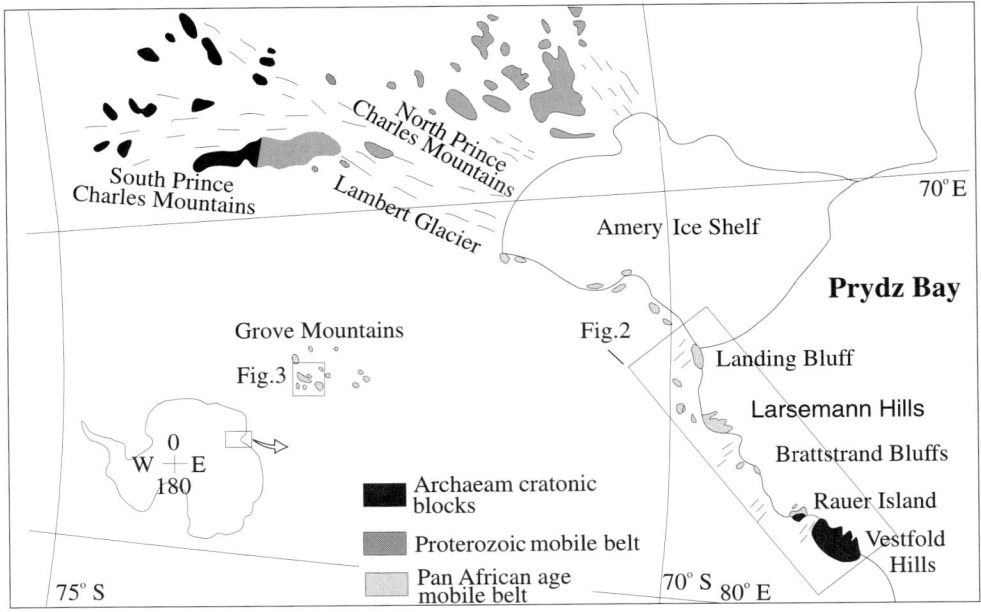

Fig. 1. Location map of the Pan-African-age Prydz Belt, East Antarctica.

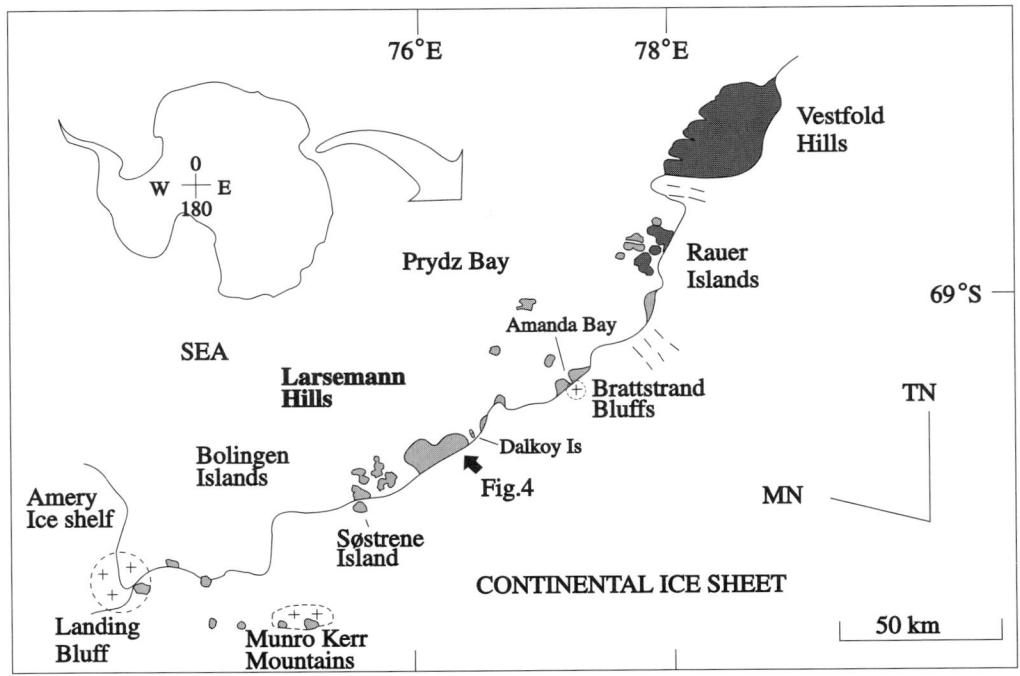

Fig. 2. Map showing syn- and post-tectonic granites and the main localities along the Prydz Bay coastline, East Antarctica: Archaean Vestfold Hills Block, southern Rauer Islands orthogneisses with Archaean components, Neoproterozoic paragneisses, 1000 Ma Søstrene Island orthogneiss.

inference and regional correlation of the high-grade terranes in the East Prydz Bay region are debatable (Harley & Fitzsimons 1995). This view is more or less from a multiple deformation and metamorphism model. If more and more data show that amalgamation of the different terranes occurred in the Cambrian (Zhao et al. 1995; Hensen & Zhou 1995; Fitzsimons et al. 1997; Fitzsimons 2000a, b), it is reasonable to treat the paragneisses as accreationary complexes commonly in orogenic belts, which were metamorphosed to the granulite facies grade in the Pan-African event.

Furthermore, isolated mafic lenses and narrow mafic layers metamorphosed at granulite grade are scattered throughout the paragneisses. Ultramafic granulite lenses can be also observed in them. A highly aluminous–ferromagnesian gneiss (high-Al–Fe–Mg gneiss), corresponding to the blue gneiss of Stüwe et al. (1989) and the cordierite-spinel gneiss of Dirks et al. (1993), is too ferromagnesian and aluminous in composition, probably representing a restite of combined metapelitic and metabasic origin. Geochemical data (Ren 1997) show that mafic gneisses could be a mafic cumulate, and volcanic rocks in origin and protoliths of paragneisses could be greywacke, subgreywacke and quartz–sandstone.

The Grove Mountains, c. 400 km south of Prydz Bay, consist of high-grade metamorphic pale and dark orthopyroxene-bearing felsic gneiss, with minor metasedimentary rocks, mafic granulite and occasionally scapolite-bearing calc-silicate rocks. They were previously regarded (on very limited data) as a Grenville terrane (Tingey 1991), but were recently documented to be part of an Early Palaeozoic orogenic belt (Zhao et al. 2000). Syntectonic sheet-like bodies and post-tectonic granitic dykes crop out at the eastern part of the Grove Mountains.

Syn- and post-tectonic granites

The post-tectonic granites of Prydz Bay, typically at the Landing Bluff, Dalkoy Island, Amanda Bay and the Polarforschung (Table 1), intruded regional high grade rocks. They were regarded as post-orogenic granites (Sheraton & Black 1988), accompanied by emplacement of planar pegmatite dykes, development of shear zones and localized amphibolite to greenschist facies overprints (Tingey 1991; Kinny et al. 1993; Harley & Fitzsimons 1995), but nothing to do with regional granulite facies metamorphism.

Recognition of a syntectonic granite – the Progress Granite (Carson et al. 1996) – in the Larsemann Hills led to recognition of the importance of the Pan-African event (Zhao et al. 1992; Dirks et al. 1993) in Prydz Bay. The Progress Granite, with a biotite–K-feldspar foliation, is thought to be syntectonic but postdates early foliation in country gneisses. It is a complex pluton, whose earlier intrusion, considered as Granitic Gneiss III by Dirks et al. (1993) or metapelite metamorphosed to granulite grade by Stüwe et al. (1989) and Carson et al. (1995b), is transitional to migmatitic paragneisses (Fig. 3). The zircon grains from the body (ZG20405) demonstrate their igneous origin. Analyses of three zircon grains with different morphologies by the Pb–Pb evaporation technique (Kober 1986, 1987) yielded consistent radiogenic $^{207}Pb/^{206}Pb$ ratios at different evaporation steps. Their apparent ages are clustered at 556 ± 7 Ma (Zhao et al. 1992), which probably represents their crystallization age.

The main pluton, the Progress Granite (Carson et al. 1996) or a K-feldspar-rich pink granite (Stüwe et al. 1989), a late tectonics syenogranite (Zhao et al. 1992), Granitic Gneiss V (Dirks et al. 1993), intrudes the earlier body, but is cut by granitic pegmatite dykes or veinlets and by a late syenogranite dyke (Zhao et al. 1992; Dirks et al. 1993; Carson et al. 1996). The Progress Granite is mainly comprised of K-feldspar, quartz, biotite and plagioclase, with subordinate garnet, magnetite, ilmenite, spinel, sillimanite, pyrite, allanite, antiperthite, monazite, apatite and zircon. It is an S-type granite (Stüwe et al. 1989), and its garnet, spinel and sillimanite were picked up in partial melting of the Larsemann Hills paragneisses. The zircon grains also contain inherited crystals; therefore, the Pb–Pb age at 547 ± 9 Ma for them is slightly older than the intrusive age due to inherited components. The Pb–Pb evaporation ages of the zircon individuals from the Progress Granite are generally similar to those of SHRIMP U–Pb ages for zircon grains of type 3 and type 2 (cf. Table 1 & Fig. 3; Zhao et al. 1992; Carson et al. 1996 table 2, Fig. 8), but they are older than those for zircon grains of type 1 due to the limits of the single zircon stepwise evaporation technique.

The Progress Granite occupies a high-strain zone with dextral movement sense. SHRIMP U–Pb zircon ages of 516 ± 7 and 481 ± 8 Ma for the Progress Granite (Carson et al. 1996) are indistinguishable from Sm–Nd garnet–whole-rock dates (517–497 Ma) for granulite-grade gneisses (Zhao et al. 1993; Hensen & Zhou 1995), an Ar–Ar hornblende age of 514 ± 2 Ma and Ar–Ar biotite age of 506–486 Ma for both the main pluton and its country gneisses (Zhao et al. 1997). Although the Progress Granite preserves D2 fabric, which is similar to that in host gneisses, its SHRIMP U–Pb age is almost the same as the age of 514 ± 7 Ma for the syenogranite dyke, which preserves the extension fabric identical with that in a D3 high-strain zone (Carson et al. 1996). The data above demonstrate that rapid

Table 1. *Isotopic data for the syn- and post-tectonic granites, pegmatites and dykes of the Prydz Belt, East Antarctica*

	Syntectonic granite			Post-tectonic granite			Dyke intrusion
Amanda Bay	Larsemann Hills	Grove Mountains	Dalkoy Island	Landing Bluff	Polar-forschung		Rauer Islands
527 ± 21Ma[1] T_{DM} = 1.9Ga[1] (granite)	547 ± 9Ma[3] 516 ± 7Ma[4] (Progress Granite) T_{DM} = 1.6Ga[1]	534 ± 5Ma[5] (Gneissic granite) T_{DM} = 1.7Ga[1]	543 ± 6Ma[1] T_{DM} = 1.7Ga[1]	493 ± 17Ma[6] (Rb–Sr) 500 ± 4Ma[6] (U–Pb)	c. 500Ma(?)[7]		500 ± 12Ma[8] (Pegmatite) 511 ± 10Ma[9] (Pegmatitic ferrodiorite)
U–Pb zircon: 535 ± 13Ma[2] 536 ± 35Ma[2]	556 ± 7Ma[3] (Earlier partial melting body) T_{DM} = 1.7Ga[1]	501 ± 7Ma[5] (Granodiorite dyke) T_{DM} = 1.9Ga[1]					548 ± 26Ma[8] (Aplite dyke D_4 deformation)
Monazite: 528 ± 4Ma[2] 527 ± 11Ma[2] 518 ± 3Ma[2] 512 ± 13Ma[2] (Brattstrand Bluffs, anatectic leucogneiss)	514 ± 7Ma[4] (Syn-D_3, Barrtangen) T_{DM} = 1.5Ga[1]						

[1]Zhao unpublished data; [2]Fitzsimons *et al.* 1997; [3]Zhao *et al.* 1992; [4]Carson *et al.* 1996; [5]Zhao *et al.* 2000 [6]Quoted by Tingey 1991; [7]Inferred by correlation; [8]Kinny *et al.* 1993; [9]Harley *et al.* 1998.

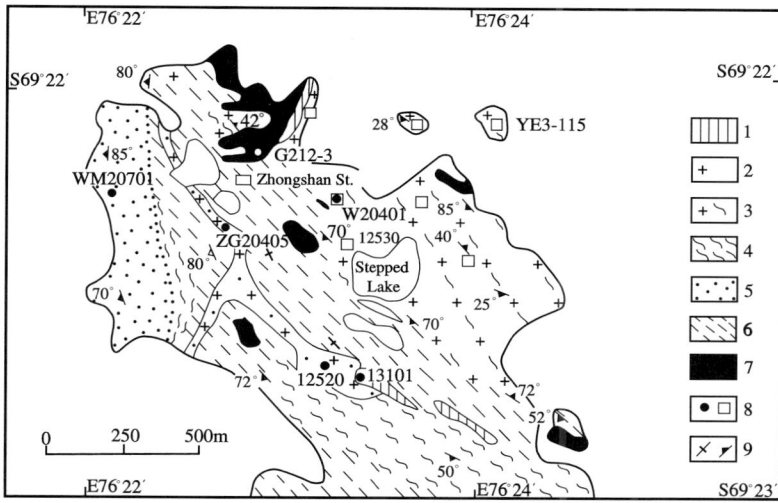

Fig. 3. The geological map of the northern Mirror Peninsular, the Larsemann Hills, East Antarctica. 1, granitic pegmatite; 2, syenogranite; 3, gneissic leucogranite; 4, high-Al-Fe-Ma gneisses; 5, migmatitic paragneiss; 6, banded migmatite; 7, mafic gneiss; 8, sampling localities for Sm–Nd, Pb–Pb and Ar–Ar samples; 9, strike and dip of the foliation.

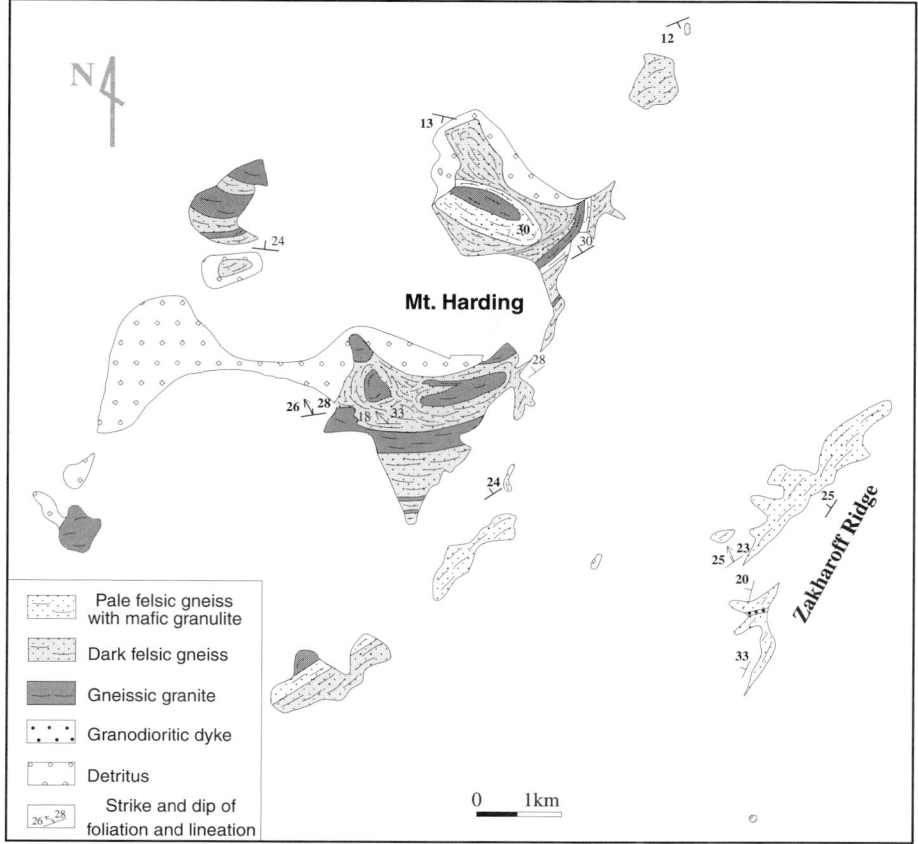

Fig. 4. The geological map of Mount Harding, the Grove Mountains, East Antarctica (After Liu *et al.* 2002*a*)

cooling from >700°C to c. 300°C happened c. 517–486 Ma, which can be attributed to the tectonically driven exhumation along the dextral ductile shear zone (Zhao et al. 1997). Retrograde fluid flow along the dextral ductile zone in a residual magma, overprints the Progress Granite and pegmatite dykes in it. When the high-grade rocks of the Larsemann Hills cooled enough to cause fractures, partial melting bodies in the lower crust could fill up the fractures to generate the granite dyke intrusions with a contact metamorphic halo.

Again in the Grove Mountains (Fig. 4), extension southward of the Prydz Belt, a syntectonic granite, a partial melting body occur at Mount Harding, the Davey Nunataks and the Escarpment (Liu et al. 2002a). This is a coarse-grained two-feldspar granite, consisting of biotite, K-feldspar (orthoclase, microcline or perthite), plagioclase, quartz and sometimes hornblende. Its layer-parallel intrusions are concordant with the host gneisses or with emplacement in them, with the orientation of K-feldspar phenocrysts parallel to the foliation of surrounding gneisses. The zircon crystals show clearly inherited internal and external overgrowths (Fig. 5; Zhao et al. 2000). SHRIMP U–Pb data of the zircon demonstrate two array ages clustered at 534 ± 5 Ma (Zhao et al. 2000). The external rims are c. 8 Ma younger than the low U internal zones. Once more, fine-grained granite dykes, 0.5–2 m wide, cross-cut the gneisses and the gneissic granite of the Grove Mountains. Contact metamorphic halos of those dykes can be observed. Two arrays of SHRIMP U–Pb ages are obtained for both inherited internal and external overgrowths of small size zircon crystals (Fig. 6; Zhao et al. 2000). The latter gives its intrusive age as 501 ± 7 Ma when the high-grade terrane was cooling to the stable geotherm, the former yields a partial melting age of 528 ± 5 Ma, inherited from granulite facies peak metamorphism and anatexis (see below).

The same case is also observed for the Brattstrand Bluff migmatitic metapelites (Fitzsimons et al. 1997). The Amanda Bay Granite, on the Brattstrand Bluff coastline, is a syn–late-tectonic pluton, which has a biotite foliation that postdates early foliation in the country gneisses (Sheraton & Collerson 1983; Fitzsimons & Harley 1991). It is a K-feldspar-rich pink granite, consisting of quartz, K-fledspar, plagioclase, biotite, and garnet, with spinel, ilmenite, apatite and zircon as common accessory phases. It is also a S-type granite and may have been derived from partial melting of the Brattstrand Bluff migmatitic paragneisses. The U–Pb zircon age of the Amanda Bay Granite is 527 ± 21 Ma (Zhao, unpublished data), overlapping (within error) with the U–Pb zircon dates at 535 ± 13 and 536 ± 35 Ma for anatectic leucog-

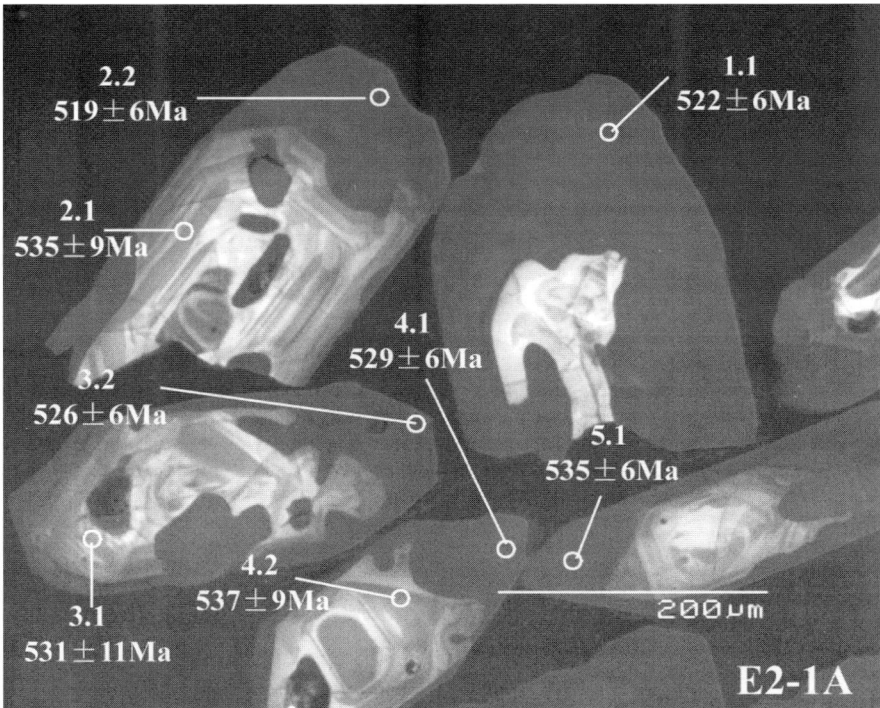

Fig. 5. Cathodoluminescence images of typical zircon grains analysed for sample E2–1A (gneissic granite).

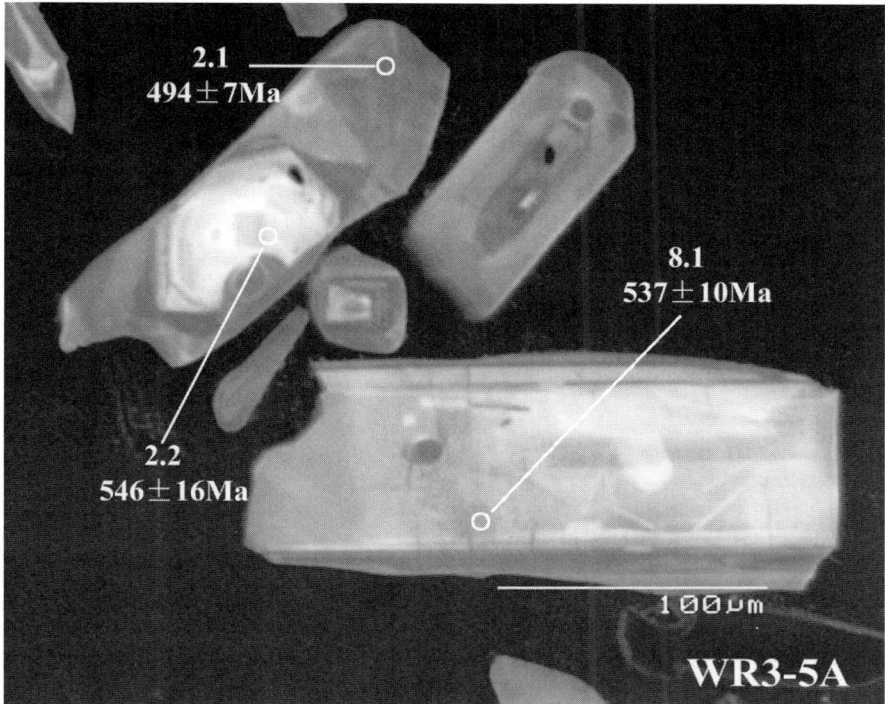

Fig. 6. Cathodoluminescence images of typical zircon grains analysed for sample WR3–5A (fine-grained granodioritic dyke).

neisses nearby (Fitzsimons et al. 1997). This age is consistent with, or close to, the ages of 541 ± 1 and 528 ± 1 Ma of peak metamorphism of the Larsemann Hills in the Prydz Belt (e.g. Zhang et al. 1996).

The two post-tectonic granites, the Landing Bluff and the Dalkoy Granites, have features of I-type granites. But all five granites discussed above have A-type chemical signatures (Sheraton et al. 1984; Sheraton & Black 1988). The Landing Bluff Granite has a Rb–Sr whole-rock age of 493 ± 17 Ma (Tingey 1991; recalculated by Sheraton et al. 1984) and a U–Pb SHRIMP zircon age of 500 ± 4 Ma (Black, unpublished data, quoted by Tingey 1991). The Dalkoy Granite has an unexpected slightly older SHRIMP U–Pb zircon age of 543 ± 6 Ma (Zhao, unpublished data). The syn- and post-tectonic granitic intrusions and granitic dykes, including pegmatite dyke, are summarized in Table 1 and will be discussed below.

Pan-African-age and Grenvillian events

Anatexis and high-grade metamorphism

Extensive partial melting during high-grade decompression produced local late syntectonic granitic intrusions and anatectic leucogneisses in the Prydz Belt. In the Larsemann Hills, small partial melt bodies occur in the Steinnes Peninsula, the southeastern Broknes Peninsula, Vikoy Island, Barrtangen and some islands to the north of Upsoy Island. They are related to rapid tectonic exhumation of the mid-crustal granulites of the Larsemann Hills during the late orogenic extension (Fitzsimons et al. 1997; Zhao et al. 1997), although extra heat is also required (Fitzsimons 1996).

This decompression melting can be attributed to a late collisional event, recognized first by Harley (1991) for the decompression P–T path of the Rauer Islands granulites. Dirks et al. (1993) and Carson et al. (1995b) divided structures of the Larsemann Hills into two groups for two separate high-grade events, corresponding to c. 1000 and 550 Ma, respectively. Fitzsimons (1996) considered those two structural and metamorphic events as a single-cycle evolution based upon SHRIMP U–Pb dates just available then (cf. Carson et al. 1996; Fitzsimons et al. 1997). D2 assemblages documenting peak conditions of c. 7 kbar and c. 780°C were followed by retrograde isothermal decompression conditions of c. 4–5 kbar and c. 750°C. Decompression melt bodies in a high-strain zone could, because of the deformation focused on them, have promoted extension and exhumation of the mid-crustal migmatites, postdating prograde thrusting and thickening. This process has been postulated for convergent orogenic belts (e.g. Carr 1992; Brown 1994; Harris & Massey 1994), and post-collisional anatexis and metapelitic migmatites were considered for the Prydz Belt granulites in relation to this issue (e.g. Fitzsimons 1996; Carson et al. 1997).

Rare, but abundant in the Larsemann Hills, boro-silicate assemblages (e.g. grandidierite–kornerupine–tourmaline; Ren et al. 1992; Carson et al. 1995a) are volatile- and fluid-rich megacrystals, especially for radial kornerupines. They were generated during decompression melting and low pressure granulite-grade metamorphism at P–T conditions of c. 6 kbar and c. 750°C, which was probably synchronous with emplacement of pegmatitic intrusions at 511 ± 10–500 ± 12 Ma (Kinny et al. 1993; Harley et al. 1998).

The Grove Mountains also appear to have experienced a single Pan-African granulite facies metamorphism, which has been dated at 529 ± 14 Ma from the zircon rims of an orthogneiss (Zhao et al. 2000). Peak metamorphic temperatures of c. 850°C at c. 6 kbar are reached from the reintegrated compositions of exsolved clinopyroxene in mafic granulites (Liu et al. 2002b). Syn- to late tectonic leucosomes often occur in the same direction as the gneissosity of the orthogneisses.

Basement and cover

It has generally been considered that the high-grade rocks cropping out along the Prydz Bay coastline comprise two main lithological associations, mafic–felsic composite orthogneisses and migmatitic paragneisses (Fitzsimons & Harley 1991; Dirks & Wilson 1995). The mafic–felsic interleaved composite orthogneisses vary in their proportions and have been proposed as basement, perhaps corresponding to the Archaean and Proterozoic orthogneisses of the southern Rauer Islands (Fig. 2), which seemingly underwent high-grade metamorphism in the Late Archaean and around the end of the Mesoproterozoic (1000–1100 Ma; Harley 1991; Kinny et al. 1993; Carson et al. 1995b; Dirks & Hand 1995; Harley & Fitzsimons 1995). The migmatitic paragneisses were assumed to represent the cover sequence derived from, and deposited on, the basement (Dirks & Wilson 1995; Hensen & Zhou 1995; Carson et al. 1996). Their deposition time was commonly assumed to be Late Mesoproterozoic (Sheraton & Collerson 1983; Dirks et al. 1993); however, no field relationship can be observed to verify this proposition.

SHRIMP U–Pb data, obtained first by Kinny et al. (1993), apparently support the assumption that the mafic–felsic composite orthogneisses were subject to a Grenville-age granulite-grade event and could serve as a basement. But the key point is that if the high-grade gneisses of the Rauer Islands experienced the pervasive Grenville-age event, the Archaean composite orthogneisses should have witnessed it and must have documented its influence. However, no c. 1000 Ma vestige can be detected from SHRIMP U–Pb zircon data for the seven samples of the Rauer Islands Archaean orthogneisses obtained both by Kinny et al. (1993) and Harley et al. (1998). Instead, Pan-African-age overprints can readily be observed for the zircon samples of both Archaean and c. 1000 Ma orthogneisses, which underwent intense deformation and pervasive granulite-grade metamorphism of Pan-African age.

The Søstrene orthogneiss was recently regarded as the basement rock (e.g. Hensen & Zhou 1995; Fitzsimons 1997), which recorded a 998 ± 12 Ma high-grade event (Hensen & Zhou 1995). However, no solid evidence shows a widespread Grenville-age high-grade overprint on the mafic–felsic composite orthogneiss along the southwestern Prydz Bay coastlines, although a 940 Ma orthogneiss was found in the northeastern Larsemann Hills (Zhao et al. 1995), so, its country gneisses must be older than 940 Ma. Recent SHRIMP U–Pb dating of zircon individuals from 'basement' rocks yields scattered age results, ranging from the latest Mesoproterozoic to the Early Neoproterozoic (Wang et al. 2001), and some of the rocks are even younger. Therefore the possibility still remains that they are metavolcanic-sedimentary composite rocks deposited in Neoproterozoic times.

For the migmatitic paragneisses, analysed inherited detrital zircon grains yielded ^{207}Pb–^{206}Pb ages ranging from 1200 to 766 Ma using thermal evaporation techniques (Zhao et al. 1995); SHRIMP U–Pb dating of the same zircon sample confirms these results (Zhao & Liu 2002). They, plus Nd isotopic data (Fig. 7; see below) demonstrate that at least some protoliths of the regional migmatitic paragneisses are Neoproterozoic in age.

Inland in the Grove Mountains (Figs 1 & 4), there is no evidence to confirm a Grenville high-grade event previously thought to exist (cf. Tingey 1991). The preservation of augite megacrysts and the appearance of equilibrium assemblages indicate that mafic granulites from the Grove Mountains were subject to a single episode of high-grade metamorphism in the Pan-African event (Liu et al. 2002b). Preliminary SHRIMP U–Pb zircon data of a felsic gneiss give scattered crystallization ages of c. 850–960 Ma (Fig. 10; Zhao et al. 2000) and a Pan-African age high-grade deformational event is indicated by SHRIMP U–Pb dating of new zircon overgrowths at 529 ± 14 Ma for the sample. The inherited zircon individuals are probably detrital and were quite possible derived mostly from the northern Prince Charles Mountains (nPCM), the 990–900 Ma

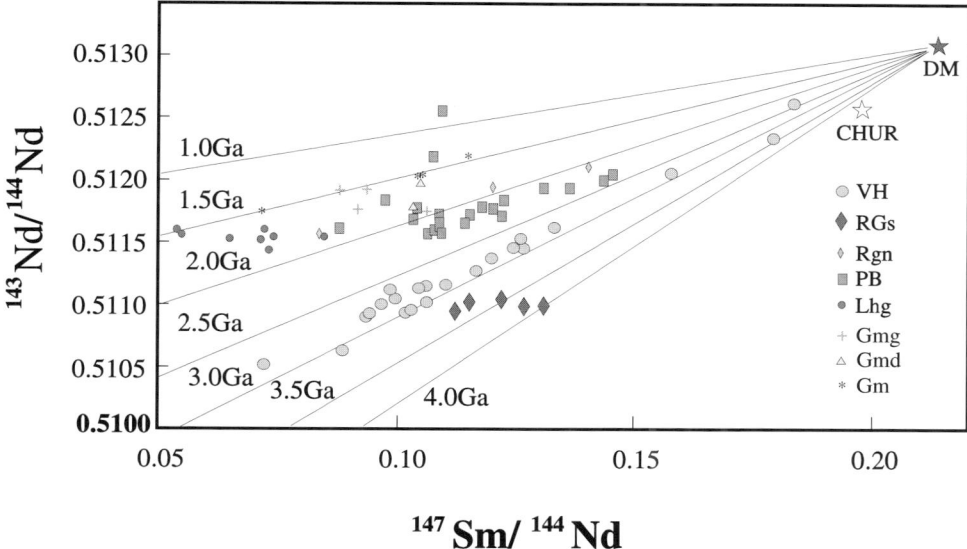

Fig. 7. Sm–Nd isotopic data for high-grade rocks of the Prydz Belt, East Antarctica. Seven reference lines correspond to crustal residence model ages of 1.0–4.0 Ga. DM, depleted mantle; CHUR, chondritic uniform reservoir; VH, Vestfold Hills, RGs, southern Rauer Islands; RGn, northern Rauer Islands; PB, Prydz Bay paragneisses; Lhg, Larsemann Hills granites; Gmg, Grove Mountains granite; Gmd, Grove Mountains granitic dyke; Gm, Grove Mountains gneisses. Data recalculated from Sheraton *et al.* (1984), Kinny *et al.* (1993), Hensen & Zhou (1995) and Zhao *et al.* (1995, unpublished data) [see Zhao *et al.* (1995) for details].

Rayner Province (Fitzsimons 2000*b*) or the Eastern Ghats–nPCM cratonic block (Boger *et al.* 2001). Nd model ages (T_{DM}) are clustered at *c.* 1700–1600 Ma (Fig. 7), which are younger than those of *c.* 1000 Ma granitic orthogneisses (Sheraton *et al.* 1984; Kinny *et al.* 1993, 1997; Zhao *et al.* 1995). Therefore this implies that crustal growth there in Neoproterozoic times could be significant.

Multiple metamorphism or terrane amalgamation

The geochronological evidence presented here strongly indicates that the Pan-African-age Prydz Belt may extend from Prydz Bay to the Grove Mountains, at least, and probably even farther to the southern Prince Charles Mountains (sPCM; Boger *et al.* 2001). There is some evidence of Grenville-age events in the Prydz Belt, but their relationships and their meaning are not yet clear.

Does the model of polyphase deformation and metamorphism work for the high-grade terranes of the Prydz Belt? If it does, it can provide a reasonable evolution for the following reasons?

(1) The southern Rauer Islands' high-grade terrane, with Archaean components, was juxtaposed with a Grenvillian orthogneiss terrane without any *c.* 1000 Ma influence (Hensen & Zhou 1995).

(2) Some 'basement orthogneisses' were not subject to a Grenvillian overprint or a Late Archaean overprint.

(3) There are many inherited zircon grains with Neoproterozoic crystallization ages (Table 2) in the Prydz Belt, and even from the 'Archaean terrane' of the southern Rauer Islands (cf. Kinny *et al.* 1993, Samples SP12 – spots 8–2, 10–1, 18–1 – and H107). They were probably deposited in Neoproterozoic times

(4) The Grove Mountains did not experience a Grenvillian episode, which is supported by at least two lines of evidence, geochronology and petrology (Zhao *et al.* 2000; Liu *et al.* 2002*b*), and Neoproterozoic crustal growth is probably significant there.

(5) Some mafic–ultramafic complexes of the Prydz Belt preserve primary igneous features, which could demonstrate a single cycle of granulite-grade evolution (Liu *et al.* 2002*b*).

(6) Quite different from the Pan-African events, which clustered at *c.* 530 Ma (Zhao *et al.* 1992), all Grenvillian relics found in the Prydz Belt cannot give a generally consistent age, which spans a long period from *c.* 1100 to *c.* 940 Ma.

Table 2. *Pb–Pb ages and U–Pb ages for detrital zircon of paragneisses from the Prydz Belt, East Antarctica*

Sample No.	Age (Ma)	Reference
WM 20701*	1200 ± 6, 982 ± 13, 789 ± 20, 872 ± 9, 766 ± 12, 950 ± 14	Zhao et al. 1995
*†	c. 850–870 (four zircon grains)	Zhao unpublished data
F22103*	343 ± 7, 914 ± 8, 878 ± 5, 904 ± 10, 918 ± 2	Zhao et al. 1995
†	c. 750–890 (seven spots)	Zhao unpublished data
LZY 1165*	821 ± 7, 716 ± 15(?)	Zhao et al. 1993
12520*	727 + 8/-9(?)	Zhao et al. 1992
SP12*†	800–900(?) (three spots 8–2, 10–1 and 18-1) (1)	Kinny et al. 1993
H107*†	c. 800–1000(?) (five spots)	Kinny et al. 1993
261*†	781, 761	Carson et al. 1996
MN1-5A†*	860–950 (17 spots)	Zhao et al. 2000

*Pb–Pb evaporation results
†SHRIMP U–Pb results.

Taking all of the above points into account, the model of polyphase metamorphism and deformation for the basement-and-cover rocks in Prydz Bay is negated. Instead, we apply an accretionary model to interpret the tectonic history of the Prydz Belt.

In this model (Fig. 8), different terranes have their different histories. They were accreted in Neoproterozoic times and then amalgamated in the Pan-African event when the Gondwana supercontinent assembled. If so, this means that oceanic crust was subducted and a palaeo-ocean existed in Neoproterozoic times. There is, at present, no direct evidence for this in observed geological record.

Implications for East Gondwana Tectonics

Decompression anatexis and clockwise granulite grade P–T evolution of high-grade rocks of Prydz Bay can be attributed to late collisional extension at c. 530 Ma (Harley 1991; Zhao et al. 1993; Carson et al. 1996; Fitzsimons et al. 1997). Rapid exhumation of the mid-crustal high-grade terranes of the Prydz Belt was achieved during c. 517–486 Ma by tectonically driven, ductile dextral shear zones where late tectonic partial melt intrusions promoted the process. Therefore, emplacement of pegmatite dykes and small granite bodies is not a minor event; they are related to the late evolution of the Pan-African event in East Antarctica and East Gondwana.

If the Pan-African event can be interpreted to involve the juxtaposition of terranes and the assembly of cratonic blocks in East Gondwana, where are the oceans that once existed between them? The paucity of evidence for ocean closure during the latest Neoproterozoic–Cambrian is the key point in solving this issue. Keeping in mind that an accretionary model is applied to interpret the tectonic history of the Prydz Belt in the late Neoproterozoic, mafic and ultramafic granulite lenses are scattered in the migmatitic paragneisses, just as in the Jurassic accretionary complex in Japan (Ichikawa 1990). The geochemical features of the lenses of the Larsemann Hills and the Grove Mountains indicate that some of them can be mid-oceanic ridge basalt (MORB) with low-Ti (TiO_2 = 1.1–1.31%), relatively low-P (P_5O_2 = 0.1–0.2%), low-rare earth elements (REE) (47–93 ppm), low REE/high REE (2.27–2.54) and $(La/Yb)_N$ (1.30–1.62), and ocean–island basalt with Ti-rich (TiO_2 = 2.68%), REE (202 ppm), LREE [$(La/Yb)_N$ = 4.8], Ti/Y = 343 and Z/Y = 3.1 (Yu et al. 2002). The boron- and magnesium-rich gneisses (Ren et al. 1992; Carson et al. 1995a) occurring in the migmatitic paragneisses of the Larsemann Hills suggest an origin associated with an accretionary complex. Further boron isotope study of them is needed to provide more detailed information.

The final suture (Fig. 9) represented by the Prydz Belt between the Grenville-age provinces or the Indo-Antarctica and East Gondwana cratonic blocks proposed by Fitzsimons (2000b), Zhao et al. (2000) and Boger et al. (2001), further bisects the East Gondwana Craton into several blocks. Some researchers (e.g. Shiraishi et al. 1994; Gose et al. 1997; Jacobs et al. 1998) have shown that the suture zone in the East African Orogen can separate Coats Land off the East Antarctic Craton. Hence, as advocated by Zhao et al. (1995, 2000) and Fitzsimons (2000a, b), the East Antarctic Craton was finally assembled during the Cambrian, i.e. it is a Pan-African age collage rather than a keystone for East Gondwana during the Neoproterozoic, which neces-

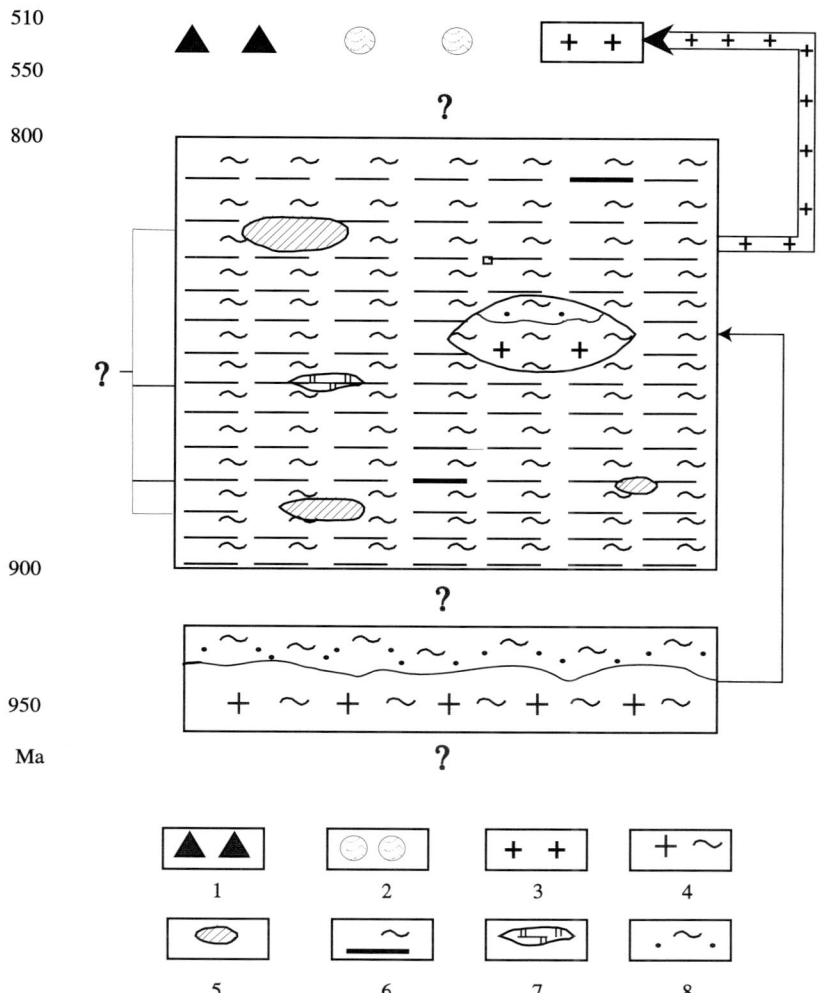

Fig. 8. Accretionary model for the tectonic history of the Prydz Belt in the Neoproterozoic. 1, Granulite facies metamorphism; 2, deformation; 3, anatexis; 4, Granvillian-age gneissic granite, e.g. 940 Ma gneissic granite; 5, mafic and/or ultramafic gneisses, similar geochemically to the metavolcanic rocks (cf. Stuwe *et al.* 1989; Ren 1997); 6, Neoproterozoic paragneisses, similar to turbidites geochemically (cf. Stuwe *et al.* 1989; Ren 1997); 7, carbonate block; 8, Mesoproterozoic paragneisses (see text for details).

sitates modification of the configuration of the Rodinia supercontinent and demonstrates that the assembly of the East Antarctic Craton was completed in the Pan-African event.

This work is a contribution to IGCP projects no. 368 on Proterozoic Events in East Gondwana and 440 on Assembly and Breakup of Rodinia. The manuscript was much improved by the careful review of S. Harley. We also thank S. Dasgupta and M. Yoshida's helpful editorial work. This work is financially supported by the National Natural Science Foundation of China (grants 49572139 and 40072028) and the National Ninth Five-Year Project for Antarctic Sciences (grant 98–927–01–06), Antarctic Administration of China (2001 DIA 50040–9). Logistic support by CHINARE is also acknowledged.

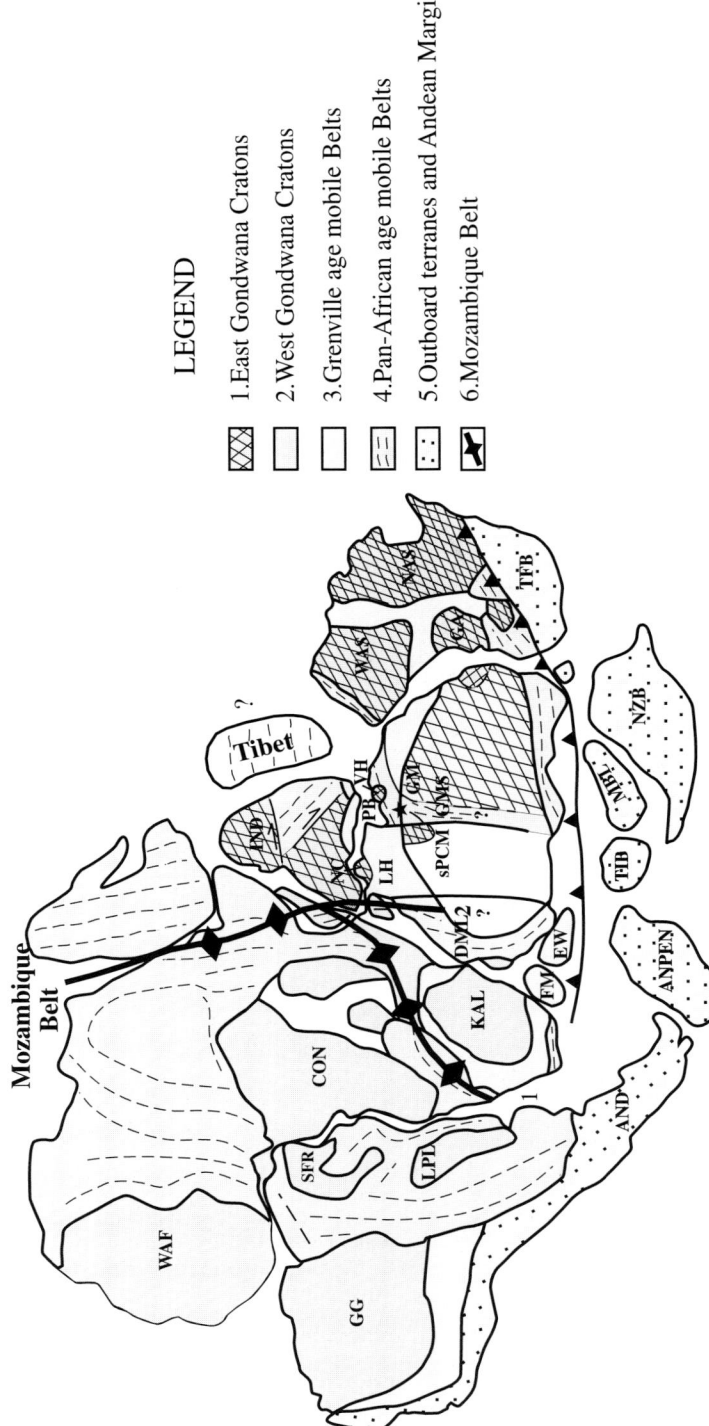

Fig. 9. Reconstruction of Gondwana (c. 500 Ma) after Yoshida (1995), Unrug (1996), Wilson *et al.* (1997) and Fitzsimons (2000*a*) [see Fitzsimons (2000*a*, fig. 5) for a detailed explanation]. AND, Andean Margin; ANPEN, Antarctic Peninsula; CON, Congo Craton; DML, Dronning Maud Land; EW, Ellsworth–Whitmore Block; FM, Falkland–Malvinas Block; GA(W), Gawler Craton; GG, Guayana & Guopore Cratons; GM,Grove Mountains; GSM, Gamburtsev Sub-glacial Mountain; IND, Indian Cratons; KAL, Kalahari Craton; LH, Lützow-Holm; LPL, La Plata Craton; MBL, Marie Byrd Land; NAS, North Australian Craton; NC, Napear Complex; NZB, New Zealand Block; PB, Prydz Bay; SFR, Saõ Francisco Craton; SPCM, south Prince Charles Mountains; TFB, Tasman Fold Belt; TIB, Thurston Island Block; VH, Vestfold Hills; WAF, West African Cratons; WAS, West Australian Craton.

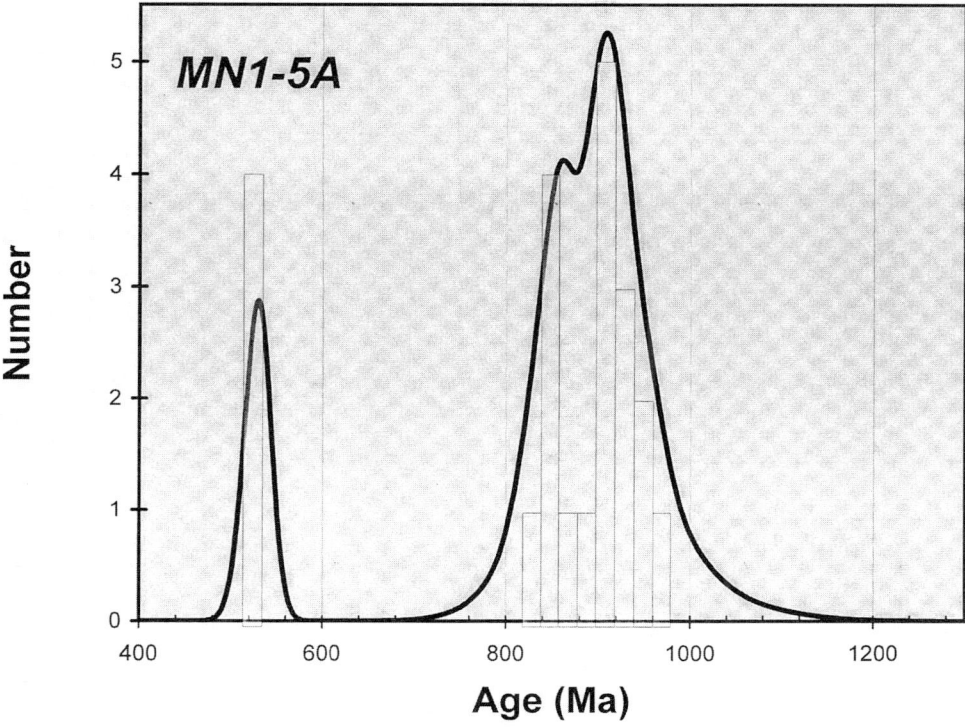

Fig. 10. U–Pb age frequency histogram of a felsic gneiss (sample MN1–5A). U–Pb ages are clustered at 529±14 Ma for regional granulite-grade metamorphism, and at 850 and 910 Ma for inherited zircon grains.

References

BLACK, L. P., KINNY, P. D., SHERATON, J. W. & DELOR, C. P. 1991. Rapid production and evolution of late Archaean felsic crust in the Vestfold Block of East Antarctica. *Precambrian Research*, **50**, 283–310.

BOGER, S. D., CARSON, C. J., WILSON, C. J. L. & FANNING, C. M. 2000. Neoproterozoic deformation in the northern Prince Charles Mountains, East Antarctica: evidence for a single protracted orogenic event. *Precambrian Research*, **104**, 1–24.

BOGER, S. D., WILSON, C. J. L. & FANNING, C. M. 2001. Early Paleozoic tectonism within the East Antarctic craton: the final suture between east and west Gondwana? *Geology*, **29**, 463–466.

BROWN, M. 1994. The generation, segregation, ascent and emplacement of granite magma: the migmatite-to-crustally derived granite connection in thickened orogens. *Earth-Science Review*, **36**, 83–130.

CARR, S. D. 1992. Tectonic setting and U–Pb geochronology of the early Tertiary Ladybird leucogranite suite, Thor-Odin Pinnacles area, southern Ominera belt, British Columbia. *Tectonics*, **11**, 258–278.

CARSON, C. J., HAND, M. & DIRKS, P. H. G. M. 1995a. Stable coexistence of grandidierite and kornerupine duirng medium pressure granulite facies metamorphism. *Mineralogical Magazine*, **59**, 327–339.

CARSON, C. J., DIRKS, P. H. G. M., HAND, M., SIMS, J. P. & WILSON, C. J. L. 1995b. Compressional and extensional tectonics in low-medium pressure granulites from the Larsemann Hills, East Antarctica. *Geological Magazine*, **132**, 151–170.

CARSON, C. J., FANNING, C. M. & WILSON, C. J. L. 1996. Timing of the Progress Granite, Larsemann Hills: additional evidence for Early Palaeozoic orogenesis within the East Antarctic Shield and implications for Gondwana assembly. *Australian Journal Earth Sciences*, **43**, 539–553.

CARSON, C. J., POWELL, R., WILSON, C. J. L. & DIRKS, P. H. G. M. 1997. Partial melting during tectonic exhumation of a granulite terrane: an example from the Larsemann Hills, East Antarctica. *Journal of Metamorphic Geology*, **15**, 105–126.

DALZIEL, I. W. D. 1991. Pacific margins of Laurentia and East Antarctica–Ausralia as a conjugate rift pair: evidence and implications for an Eocambrian supercontinent. *Geology*, **19**, 598–601.

DIRKS, P. H. G. M. & HAND, M. 1995. Clarifying temperature–pressure paths via structures in granulite from the Bolingen islands, Antarctica. *Australian Journal of Earth Sciences*, **42**, 157–172.

DIRKS, P. H. G. M. & WILSON, C. J. L. 1995. Crustal evolution of the East Antarctic mobile belt in Prydz Bay: continental collision at 500 Ma? *Precambrian Research*, **75**, 189–207.

DIRKS, P. H. G. M., CARSON, C. J. & WILSON, C. J. L. 1993.

The deformational history of the Larsemann Hills, Prydz Bay: the importance of the Pan-African (500 Ma) in East Antarctica. *Antarctic Science*, **5**, 179–192.

FITZSIMONS, I. C. W. 1996. Metapelitic migmatites from Brattstrand Bluffs, East Antarctica–metamorphism, melting and exhumation of the mid-crust. *Journal of Petrology*, **37**, 395–414.

FITZSIMONS, I. C. W. 1997. The Brattstrand Paragneiss and the Sostrene Orthogneiss: a review of Pan-African metamorphism and Grenvillian relics in southern Prydz Bay. *In*: RICCI, C. A. (ed.) *The Antarctic Region: Geological Evolution and Processes*. Terra Antartica Publication, Siena, 121–130.

FITZSIMONS, I. C. W. 2000a. A review of tectonic events in the East Antarctic Shield and their implications for Gondwana and earlier supercontinents. *Journal of African Earth Sciences*, **31**, 3–23.

FITZSIMONS, I. C. W. 2000b. Grenville-age basement provinces in East Antarctica: evidence for three separate collisional orogens. *Geology*, **28**, 879–882.

FITZSIMONS, I. C. W & HARLEY, S. L. 1991. Geological relationships in high-grade gneisses of the Brattstrand Bluffs coastline, Prydz Bay, East Antarctica. *Australian Journal of Earth Sciences*, **38**, 497–519.

FITZSIMONS, I. C. W., KINNY, P. D. & HARLEY, S. L. 1997. Two stages of zircon and monazite growth in anatectic leucogneiss: SHRIMP constraints on the duration and intensity of Pan-African metamorphism in Prydz Bay, East Antarctica. *Terra Nova*, **9**, 47–51.

GOSE, W. A., HELPER, M. A., CONNELLY, J. N., HUTSON, F. E. & DALZIEL, I. W. D. 1997. Paleomagnetic data and U-Pb isotopic age determinations from Coats Land, Antarctica: implications for Late Proterozoic plate reconstructions. *Journal of Geophysical Research*, **102**, 7887–7902.

HARLEY, S. L. 1991. Metamorphic evolution of granulites from the Rauer Group, East Antarctica: evidence for decompression following Proterozoic collision. *In*: THOMSON, M. R. A., CRAME, J. A. & THOMSON, J. W. (eds) *Geological Evolution of Antarctica*. Cambridge University Press, Cambridge, 99–105.

HARLEY, S. L. & FITZSIMONS, I. C. W. 1995. High-grade metamorphism and deformation in the Prydz Bay region, East Antarctica: terrane, events and regional correlations. *In*: YOSHIDA, Y. & SANTOSH, M. (eds) *India and Antarctica During the Precambrian*. Memoir, Geological Society of India, **34**, 73–100.

HARLEY, S. L., SNAPE, I. & BLACK, L. P. 1998. The evolution of a layered metaigneous complex in the Rauer Group, East Antarctica: evidence for a distinct Archaean terrane. *Precambrian Research*, **89**, 175–205.

HARRIS, N. B. & MASSEY, J. A. 1994. Decompression and anatexis of Himalayan metapelites. *Tectonics*, **13**, 1537–1546.

HENSEN, B. J. & ZHOU, B. 1995. A Pan-African granulite facies metamorphic episode in Prydz Bay, Antarctica: evidence from Sm–Nd garnet dating. *Australian Journal of Earth Sciences*, **42**, 249–258.

HOFFMAN, P. F. 1991. Did the breakout of Laurentia turn Gondwanaland inside out? *Science*, **252**, 1409–1412.

ICHIKAWA, K. 1990. Pre-Cretaceous terranes of Japan. *In*: ICHIKAWA, K., MIZUTANI, S., HARA, I., HADA, S. & YAO, A. (eds) *Pre-Cretaceous Terranes of Japan*. IGCP224, Osaka, 1–12.

JACOBS, J., FANNING, C. M., HENJES-KUNST, F., OLESCH, M. & PAECH, H. J. 1998. Continuation of the Mozambique Belt into East Antarctica: Grenville-age metamorphism and polyphase Pan-African high-grade events in central Dronning Maud Land. *Journal of Geology*, **106**, 385–406.

KINNY, P. D., BLACK, L. P. & SHERATON, J. W. 1993. Zircon ages and the distribution of Archaean and Proterozoic rocks in the Rauer Islands. *Antarctic Science*, **5**, 193–206.

KINNY, P. D., BLACK, L. P. & SHERATON, J. W. 1997. Zircon U-Pb ages and geochemistry of igneous and metamorphic rocks in the northern Prince Charles Mountains, Antarctica. *AGSO Journal of Australian Geology and Geophysics*, **16**, 637–654.

KOBER, B. 1986. Whole-grain evaporation for $^{207}Pb/^{206}Pb$ age investigations on single zircons using a double-filament thermal ion source. *Contributions to Mineralogy and Petrology*, **93**, 482–490.

KOBER, B. 1987. Single zircon evaporation combined with Pb emitter bedding for $^{207}Pb/^{206}Pb$ age investigations to zirconology. *Contributions to Mineralogy and Petrology*, **96**, 63–71.

LIU, X. C., ZHAO, Z., ZHAO, Y. & LIU, X. H. 2002a. Geological aspects of the Grove Mountains, East Antarctica. *Proceedings of the 8th International Symposium of Antarctic & Earth Science*, in press.

LIU, X. C., ZHAO, Z., ZHAO, Y., CHEN, J. & LIU, X. H. 2002b. Pyroxene exsolution in mafic granulites from the Grove Mountains, East Antarctica: constraints on the Pan-African metamorphic conditions. *European Journal of Mineralogy*, in press.

MOORES, E. M. 1991. Southwest U.S.–East Antarctic (SWEAT) connection: a hypothesis. *Geology*, **19**, 425–428.

REN, L. 1997. *Metamorphic geology of the high-grade area of the Larsemann Hills and its adjcent region, East Antarctica*. PhD Thesis, Chinese Academy of Geological Sciences [in Chinese].

REN, L., ZHAO Y. & CHEN T. 1992. Re-examination of the metamorphic evolution of the Larsemann Hills, eastern Antarctica. *In*: YOSHIDA, Y., KAMINUMA, K. & SHIRAISHI, K. (eds) *Recent Progress in Antarctic Earth Science*. Terra Scientific Publishing, Tokyo, 145–153.

SHERATON, J. W. & BLACK, L. P. 1988. Geochemical evolution of granitic rocks in the East Antarctica Shield, with particular reference to post-orogenic granites. *Lithos*, **21**, 37–52.

SHERATON, J. W. & COLLERSON, K. D. 1983. Archaean and Proterozoic geological relationships in the Vestfold Hills–Prydz Bay area, Antarctic. *BMR Journal of Australian Geology and Geophysics*, **8**, 119–128.

SHERATON, J. W., BLACK, L. P. & MCCULLOCH, T. 1984. Regional geochemical and isotopic characteristics of high-grade metamorphics of the Prydz Bay area: the extent of Proterozoic reworking of Archaean continental crust in East Antarctica. *Precambrian Research*, **26**, 169–198.

SHIRAISHI, K., ELLIS, D. J., HIROI, Y., FANNING, C. M., MOTOYOSHI, Y. & NAKAI, Y. 1994. Cambrian orogenic belt in East Antarctica and Sri Lanka: implications for Gondwana assembly. *Journal of Geology*, **102**, 47–65.

STÜWE, K., & POWELL, R. 1989. Low-pressure granulite facies metamorphism in the Larsemann Hills area, East Antarctica: petrology and tectonic implications for the evolution of the Prydz Bay area. *Journal of Metamorphic Geology*, **7**, 465–483.

STÜWE, K., BRAUN, H.-M. & PEER, H. 1989. Geology and structure of the Larsemann Hills area, Prydz Bay, East Antarctica. *Australian Journal of Earth Sciences*, **36**, 219–241.

TINGEY, R. J. 1991. The regional geology of Archaean and Proterozoic rocks in Antarctica. *In*: TINGEY, R. J. (ed.) *The Geology of Antarctica*. Oxford University Press, Oxford, 1–73.

UNRUG, R. 1996. The assembly of Gondwanaland. *Episodes*, **19**, 11–20.

WANG, Y., LIU, D., WILLIAMS, I. S., REN, L. & TONG, L. 2001. New zircon ages, geochemistry and their significance of the basement rocks from the Larsemann Hills, East Antarctica. *Gondwana Research*, **4**, 820.

WILSON, T. J., GRUNOW, A. M. & HANSON, R. E. 1997. Gondwana assembly: the view from southern Africa and East Gondwana. *Journal of Geodynamics*, **23**, 263–286.

YOSHIDA, M. 1995. Assembly of East Gondwana during the Mesoproterozoic and its rejuvenation during the Pan-African period. *In*: YOSHIDA, M. & SANTOSH, M. (eds) *India and Antarctica during the Precambrian*. Memoir, Geological Society of India, **34**, 22–45.

YU, L., LIU, X. H., LIU, X. C., ZHAO, Y., FANG, A. & JU, Y. 2002. Geochemical characteristics of the metabasic volcanic rocks in the Grove Mountains, East Antarctica. *Acta Petrologica Sinica*, **18**, 91–99 [in Chinese with English abstract].

ZHANG, L., TONG, L., LIU, X. H. & SCHARER, U. 1996. Conventional U–Pb age of the high-grade metamorphic rocks in the Larsemann Hills, East Antarctica. *In*: PANG, Z., ZHANG. J. & SUN, J. (eds). *Advances in Solid Earth Science*. Science Press, Beijing, 27–35.

ZHAO, Y. & LIU, X. C. 2002. *Possible Neoproterozoic accretionary tectonics of the Prydz belt, East Antarctica*. JiuWu Research Report, Ocean Press, Beijing, in press [Chinese Polar Administration edition, in Chinese].

ZHAO, Y., LIU, X. C., FANNING, C. M. & LIU, X. H. 2000. The Grove Mountains, a segment of a Pan-African orogenic belt in East Antarctica. *Abstract Volume of the 31th International Geological Congress*, Rio de Janeiro, Brazil.

ZHAO, Y., LIU, X. H., SONG, B., ZHANG, Z., LI, J., YAO, Y. & WANG, Y. 1995. Constraints on the stratigraphic age of metasedimentary rocks from the Larsemann Hills, East Antarctica: possible implications for Neoproterozoic tectonics. *Precambrian Research*, **75**, 175–188.

ZHAO, Y., LIU X. H., WANG, S. & SONG, B., 1997. Syn- and post-tectonic cooling and exhumation in the Larsemann Hills, East Antarctica. *Episodes*, **20**, 122–127.

ZHAO, Y., SONG, B., WANG, Y., REN, L., LI, J. & CHEN, T. 1991. Geochronological study of the metamorphic and igneous rocks of the Larsemann Hills, East Antarctica. *Proceedings of the 6th ISAES (Abstracts)*, Tokyo. National Institute for Polar Research, Japan, 662–663.

ZHAO, Y., SONG, B., ZHANG, Z. ET AL. 1993. Early Paleozoic ('Pan African') thermal event of the Larsemann Hills and its neighbours, Prydz Bay, East Antarctica. *Science in China (Series B)*, **23**, 1001–1008 [in Chinese: English edition: **38**, 74–95, 1995].

ZHAO, Y., SONG, B., WANG, Y., REN, L., LI, J. & CHEN, T. 1992. Geochronology of the late granite in the Larsemann Hills, East Antarctica. *In*: YOSHIDA, Y., KAMINUMA, K. & SHIRAISHI, K. (eds) *Recent Progress in Antarctic Earth Science*. Terra Scientific Publishing, Tokyo, 155–161.

Proterozoic–Cambrian history of Dronning Maud Land in the context of Gondwana assembly

W. BAUER[1], R. J. THOMAS[2] & J. JACOBS[3]

[1]*Geologisches Institut der RWTH Aachen, 52056 Aachen, Germany*
(e-mail: bauer@geol.rwth-aachen.de)
[2]*British Geological Survey, Kingsley Dunham Centre, Keyworth, Nottingham NG12 5GG, UK*
[3]*FB Geowissenschaften, Universität Bremen, PF 330440, 28334 Bremen, Germany*

Abstract: Dronning Maud Land contains a fragment of an Archaean craton covered by sedimentary and magmatic rocks of Mesoproterozoic age, surrounded by a Late Mesoproterozoic metamorphic belt. Tectonothermal events at the end of the Mesoproterozoic and in Late Neoproterozoic–Cambrian times (Pan-African) have been proved within the metamorphic belt. In western Dronning Maud Land a juvenile Mesoproterozoic basement was accreted to the craton at $c.$ 1.1 Ga. Mesoproterozoic rocks were also detected by zircon SHRIMP dating of gneisses in central Dronning Maud Land, followed by a long hiatus for which geochronological data are lacking, an amphibolite to granulite facies metamorphism and syntectonic granitoid emplacement of Pan-African age have been dated. During this orogeny older structures were completely overprinted in a sinistral tranpressive deformation regime, leading to the mainly coast-parallel tectonic structures of the East Antarctic Orogen. Putting Antarctica back in its Gondwana position, the East Antarctic Orogen continues northward in East Africa as the East African Orogen, whereas a connection to the marginal Ross Orogen at the Pacific margin of East Antarctica is suggested along the Shackleton Range. The East Antarctic–East African Orogen resulted from closure of the Mozambique Ocean and collision of West and East Gondwana, i.e. western Dronning Maud Land was part of West Gondwana. During this collision the lithospheric mantle probably delaminated, allowing the asthenosphere to underplate the continental crust and producing heat for the voluminous, typically anhydrous, Pan-African granitoids of central Dronning Maud Land.

The mountains of Dronning Maud Land (DML) cover the area from 18°W to 28°E parallel to the coastline of Antarctica (Fig. 1). The ranges occur either as elongate N–S nunatak chains, divided by minor glaciers, or as steep escarpments separating the polar plateau from the ice-covered coastal plains. The most important glaciers are the Jutulstraumen and its southwestern extension, the Penckmulde, separating the scattered nunataks of Borgmassivet and Ahlmannryggen from the escarpment-type outcrops of western DML.

The first geological information from the region was collected during the Norwegian–British–Swedish expedition of 1949–1952 (Roots 1953; Giæver 1954). Systematic geological research started when permanent stations were established in this area in the early 1960s. Today, geological maps are available for large parts of DML at scales of 1:150000 or larger (e.g. Jayaram & Bejarniya 1991; Jacobs & Weber 1993; Ohta 1993; Ohta *et al.* 1996). In addition, numerous recent papers covering the geochronology, geochemistry and structural geology of the area have been published from different working groups, allowing a concise review of the geological evolution of the entire area.

DML is one of the most important key regions for Mesoproterozoic–Cambrian plate movement reconstructions. In modern plate reconstructions of the Late Proterozoic–Cambrian Gondwana supercontinent (e.g. Martin & Hartnady 1986; Grunow *et al.* 1996; Wilson *et al.* 1997) the southward continuation of the Pan-African-aged Mozambique Belt (Holmes 1951) of East Africa has to be expected in DML. This mobile belt was formed as a result of continent–continent collision following the closure of the Mozambique Ocean (Dalziel 1991; Hoffman 1991), which led to the final amalgamation of West and East Gondwana. The precise position of the collision suture is still under debate (e.g. Shackleton 1996). Recent palaeomagnetic (Gose *et al.* 1997), geochronological (Jacobs *et al.* 1998, 1999) and aeromagnetic studies (Golynski & Jacobs 2001) show that DML can be subdivided into three distinct areas with strikingly different geological histories:

(1) an Archaean craton with an undeformed Proterozoic cover;
(2) a Late Mesoproterozoic collision orogen;
(3) a Pan-African-aged collision orogen with relicts of pre-Pan-African structures and voluminous post-tectonic intrusives.

In this paper, specific emphasis is given to the poorly known rocks of central DML, which form a critical part of the Gondwana puzzle.

From YOSHIDA, M., WINDLEY, B. F. & DASGUPTA, S. (eds) 2003. *Proterozoic East Gondwana: Supercontinent Assembly and Breakup.* Geological Society, London, Special Publications, 206, 247–269. 0305–8719/03/$15 © The Geological Society of London.

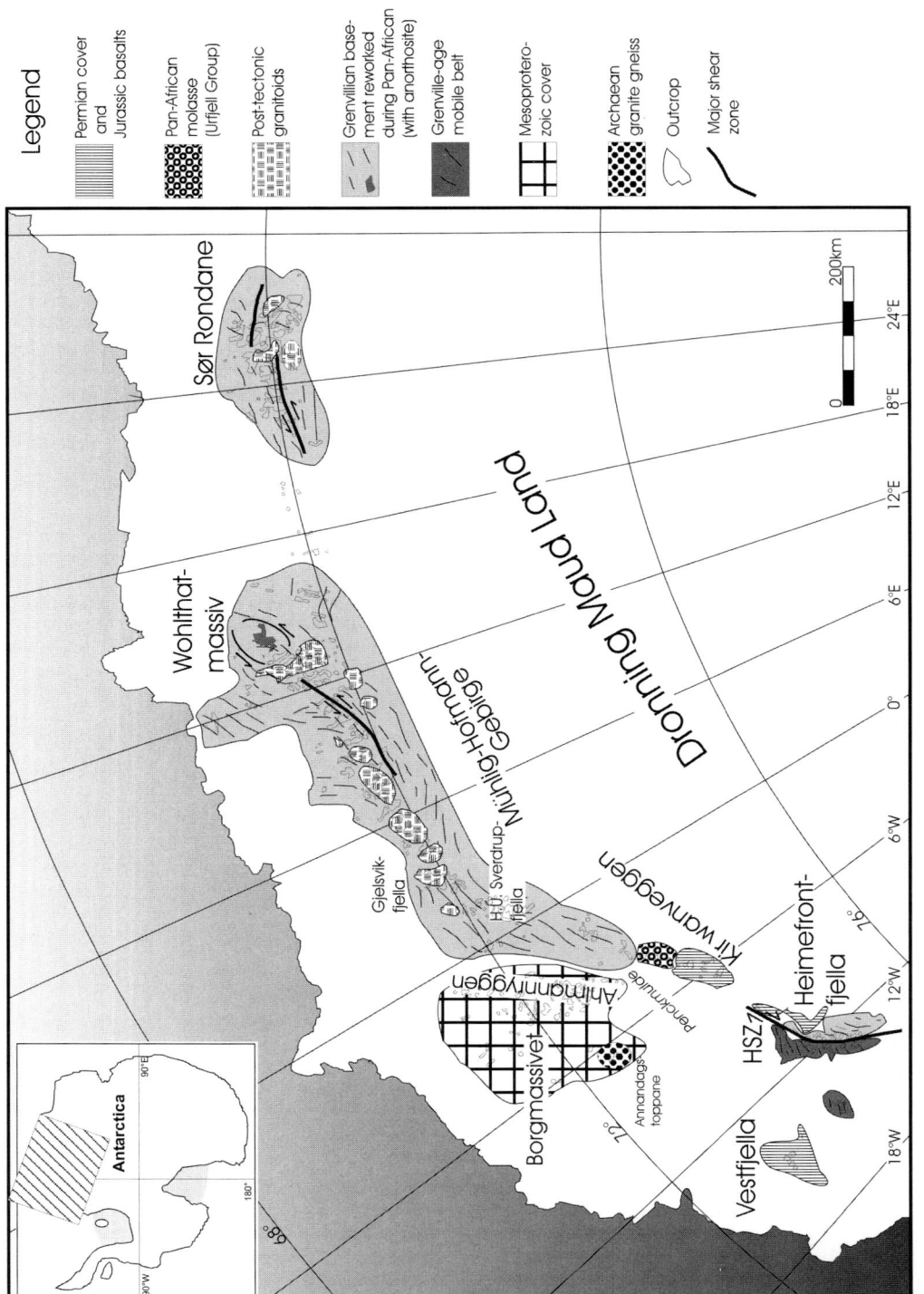

Fig. 1. Geological overview map of Dronning Maud Land.

Note: All place names in the text follow the Composite Gazetter of Antarctica (www.pnra.it/SCAR_GAZE). According to SCAR recommendations, place names should be given in their original form and not in a translated form. Due to the scientific activities of different nations, a mixture of English (Eng.), German (Ger.), Norwegian (Nor.) and Russian names exist for the area under consideration. Here the most important synonyms are given: -fjella (Nor.) = -gebirge (Ger.) = mountains (Eng.); -ryggen (Nor.) = -rücken (Ger.) = ridge (Eng.); -massivet (Nor.) = -massiv (Ger.) = massif (Eng.); -kjedane (Nor.) = -ketten (Ger.) = chains (Eng.). Due to problems with the usage of place names, several names for geological objects have been introduced in the international literature which omit the endings and thus do not follow the nomenclatural rules, e.g. Heimefront Shear Zone instead of Heimefrontfjella Shear Zone or Lützow-Holm Complex instead of Lützow-Holmbukta Complex, but are used here as they are widely accepted.

Archaean craton and its Proterozoic cover

The oldest rocks reported thus far from DML are exposed at Annandagstoppane (6°40'W, 72°35'S). Halpern (1970) published a Rb–Sr whole-rock age of c. 2960 Ma from a biotite granite. Zircons are extremely rare in this rock and the few grains that could be isolated were highly metamict (Barton et al. 1987), so precise zircon U–Pb data are not available. Barton et al. (1987) verified the Archaean Rb–Sr age with new samples, concluding that the granite was emplaced during the interval of 3115–2945 Ma. A subsequent hydrothermal event at c. 2820 Ma overprinted the rocks and homogenized the Sr isotopes on a whole-rock scale.

Geochemically, the granite is classified as a peraluminous, S-type, derived from a crustal source. Although the geochemical characteristics and provenance are different from the Archaean granites of southern Africa, the Annandagstoppane Granite is generally accepted as a dismembered fragment of the Archaean Kaapvaal Craton in East Antarctica (e.g. Groenewald et al. 1995).

In the Borgmassivet and Ahlmannryggen area east and NE of Annandagstoppane (Fig. 1), extensive volcanosedimentary rocks are exposed in numerous nunataks. Wolmarans & Kent (1982) described this succession, the Ritscherflya Supergroup, in detail. They distinguished an Ahlmannryggen Group, dominated by clastic sedimentary rocks overlain by the Jutulstraumen Group, mainly made up of mafic lava flows and tuffs. Wolmarans & Kent (1982) estimated a total thickness of c. 3000 m, although in general it is very difficult to correlate the successions between the scattered outcrops. The base is nowhere exposed, but many authors believe that these rocks must rest unconformably upon the Annandagstoppane Granite (e.g. Groenewald et al. 1995). The Ritscherflya Supergroup was metamorphosed up to the chlorite zone of the greenschist facies but does not show a penetrative deformation.

The oldest rocks of the Ahlmannryggen Group are sulphide-bearing clastic sediments, deposited in a low-energy reducing environment. In the upper part of the sequence, a change to shallow-water sediment deposition was recognized, dominated by braided and meandering river sediments, derived from the SW to west and deposited in an intracratonic basin. The Jutulstraumen Group consists of subaerial basaltic to andesitic lavas and minor sedimentary rocks, probably deposited in small ponds between the lava flows (Wolmarans & Kent 1982).

The Ritscherflya Supergroup is intruded by mafic sills and dykes (Borgmassivet Intrusions) which form up to 80% of the outcrops. The intrusions are quartz-normative tholeiites, which are geochemically very similar to the lava flows of the Jutulstraumen Group. Strong alteration of the lavas and the intrusive rocks has rendered their geotectonic setting controversial. According to Peters et al. (1991) and Krynauw et al. (1991) geochemical trace element and rare earth element (REE) patterns of lavas and intrusives, respectively, are typical of many continental tholeiites. The alteration probably masks the original isotope geochemistry of Rb and Sr that were used for dating the mafic flows and sills (Moyes & Barton 1990). Moyes et al. (1995) reported new ages obtained from Sm–Nd and Rb–Sr whole-rock data, concluding that the depositional age of the upper Ahlmannryggen Group is c. 1080 Ma, and that the Ritscherflya Supergroup was coevally intruded by the Borgmassivet Intrusives. These dates are similar to those published by Hanson et al. (1998) for the Umkondo Dolerites in eastern Zimbabwe. Barton & Moyes (1990) described similarities between the Umkondo Group and the Ritscherflya Supergroup regarding to their sedimentology and geochemistry. Therefore, it is quite reasonable that the Annandagstoppane Granite and its Proterozoic cover (the Grunehogna Craton; Fig. 2a) are an Antarctic extension of the Kaapvaal Craton and its cover sequences of the Umkondo Group.

Late Mesoproterozoic orogenic belt of western Dronning Maud Land (DML) and its westward continuation in southern Africa

The molasse-like sediments of the Ahlmannryggen Group were derived from a rising orogen during the Late Mesoproterozoic (Groenewald et al. 1991). A strongly deformed amphibolite facies mobile belt is exposed in the mountain chains and escarpments adjacent to the Grunehogna Craton. When the first c. 1 Ga U–Pb zircon dates from these rocks from H.U.

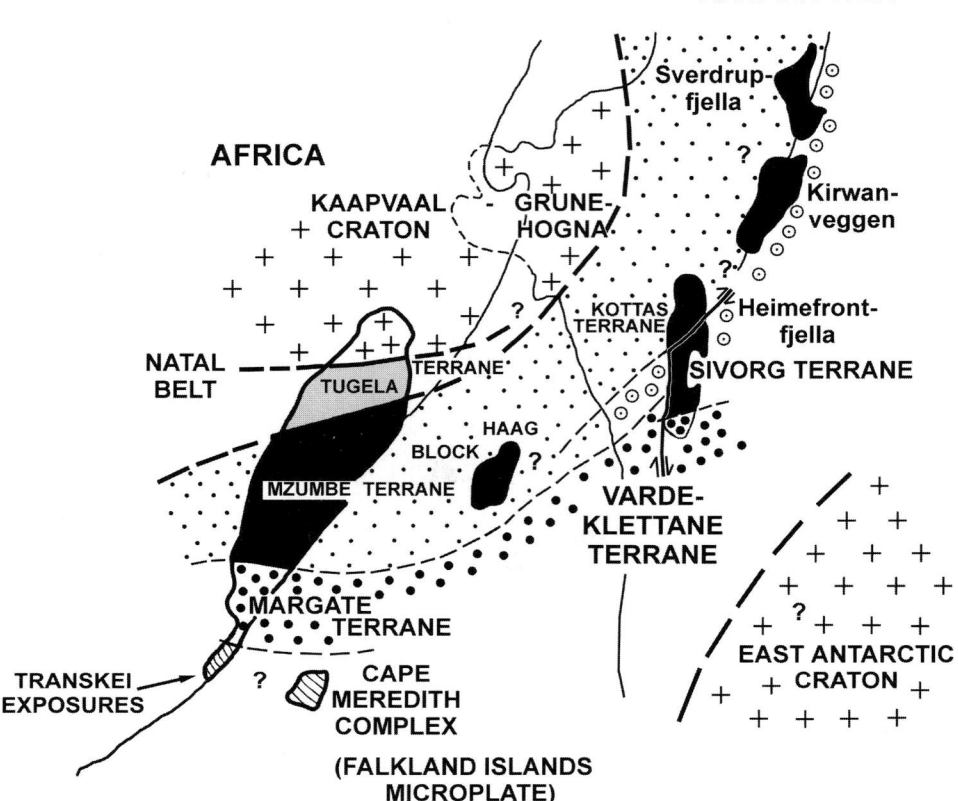

Fig. 2. Model of a Late Mesoproterozoic volcanic arc/back arc adjacent to the Kaapvaal-Grunehogna Craton according to Bauer (1995) and Groenwald et al. (1995). (a) Sketch map of SE Africa and Dronning Maud Land. Antarctic counterparts of the ophiolite nappes (Tugela Terrane) have not been found yet. Note that the boundary between the Mzumbe-Haag-Kottas Volcanic Arc and Sivorg-Kirwan-Sverdrup back arc in Kirwanveggen and H.U. Sverdrupfjella is tentative due to later overprint.

Sverdrupfjella, Kirwanveggen and Heimefrontfjella became available (Moyes & Barton 1990; Arndt et al. 1991), it was already known that the whole area had only cooled down to temperatures of <300°C in Cambro–Ordovician times (c. 500 Ma) from older K–Ar whole-rock and mica data (Ravich & Krylov 1964). All subsequent U–Pb zircon data obtained were Late Mesoproterozoic in age (with one exeption from a c. 500 Ma alkali granite at Brattskarvet in the northeastern H.U. Sverdrupfjella; Moyes et al. 1993). Together with the observed polyphase deformation patterns, it was concluded that these rocks underwent two main deformation events at c. 1.1 Ga (Namaqua–Grenvillian) and 550 Ma (Pan-African).

The first K–Ar mica ages in the range of 987–960 Ma from the westernmost nunataks of Heimefrontfjella were reported by Jacobs (1991). These data represent Early Neoproterozoic cooling without Pan-African reheating. These nunataks are separated from the main range by a several kilometre wide shear zone, known as the Heimefront Shear Zone (Jacobs et al. 1993). The shear zone trends NNE–SSW in the southern and central part of the range, swinging to a ENE–WSW direction further north. West and northwest of the shear zone the Kottas and the Vardeklettane Terranes (Jacobs et al. 1996) contain rocks only weakly affected by Pan-African events.

Whereas the granulite facies rocks of the Vardeklettane Terrane are less well studied, the Kottas Terrane is mainly made up of amphibolite facies metamorphic rocks derived from igneous protoliths. They occur as sheet-like augen gneisses, a metatrondhjemite–tonalite–diorite suite and grey Bt-Pl gneisses containing euhedral zircons of possible volcanic origin (Schmidt pers. comm.). Paragneisses and calc-silicate rocks, together with amphibolite slices, are known but are of restricted

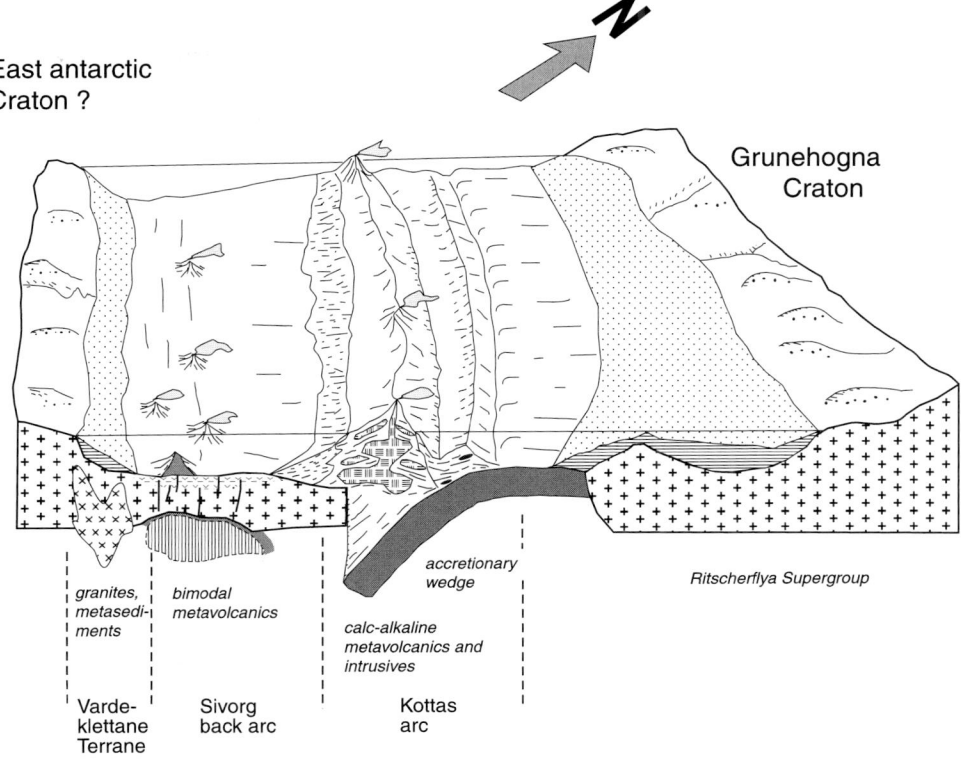

Fig. 2. (b) Block diagram showing the situation in the Heimefronfjella (orientation: Antarctic north)

extent, being exposed only in some isolated nunataks in the NE and a syncline in the North of the Kottas Terrane. The meta-igneous rocks have a typical calc-alkaline subduction-related geochemical signature, whereas the amphibolites in the supracrustal sequence have an oceanic affinity (Bauer 1995).

At least two penetrative Late Mesoproterozoic deformations can be distinguished: earlier NE vergent folds were refolded and overprinted by NNW vergent folds (Spaeth & Fielitz 1987; Bauer 1995). Undeformed, post-tectonic pegmatite and aplite veins indicate the end of the Late Mesoproterozoic deformation prior to 1060 Ma (Arndt et al. 1991).

The western extension of the Late Mesoproterozoic orogenic belt is exposed in Natal (SE Africa). Here, three distinct terranes are exposed, two of which (Mzumbe and Margate Terranes; Thomas 1989a) appear remarkably comparable to the Kottas and Vardeklettane Terranes, respectively (Fig. 2a). Much of the Mzumbe Terrane is made up of arc-related metavolcanic rocks and tonalitic gneisses (e.g. Thomas & Eglington 1990), which have a similar age and geochemistry to those of the meta-trondhjemite–tonalite–diorite suite of the Kottas Terrane. The Vardeklettane Terrane finds its counter-

part in the granulite facies of the Margate Terrane of Natal (Fig. 2a). Both terranes are composed of minor gneisses with supracrustal protoliths, intruded by anhydrous granites and charnockites. Complex magmatic, nebulous replacement and aureole charnockites are characteristic for both terranes.

East Antarctic Orogen

The term East Antarctic mobile belt was firstly used by Dirks & Wilson (1995) for an E–W-trending corridor of highly strained, mostly granulite facies, rocks of Late Mesoproterozoic–Neoproterozoic/Cambrian rocks, which extent parallel to the Antarctic coast from western DML to Prydz Bay at 78°E. In this paper the term East Antarctic Orogen is used according to the definition given by Jacobs et al. (1998), suggesting a c. 550 Ma orogen which is exposed from the eastern Heimefrontfjella, through Kirwanveggen, H.U. Sverdrupfjella, central DML and Sør Rondane (Fig. 1) up to Lützow Holmbukta and running southwestward through East Antarctica. A lithostratigraphic and structural summary is given in Table 1.

Table 1: Summary and comparison of Pan-African lithostratigraphy and tectonothermal history in Dronning Maud Land

	Eastern Heimefrontfjella (Jacobs et al. 1996)	Kirwanveggen (Grantham et al. 1995)	H.U. Sverdrupfjella (Grantham et al. 1995)	Gjelsvikfjella and Mühlig-Hofmann-Gebirge (Bauer & Jacobs 2001)	Orvinfjella (Bauer et al. 2003)	Wohlthatmassiv (Bauer et al. 2003)	Sør Rondane (Shiraishi et al. 1991)
Basement lithology	Bimodal metavolcanic sequence, major pre- and syntectonic calc-alkaline granitoids, minor metasedimentary rocks (marbles, metapelites, quartzites)	Banded gneisses, partly migmatitic, minor biotite + garnet gneisses, interlayered with amphibolites, syntectonic megacrystic orthogneisses	Quartzo-feldspathic, quartzose and pelitic paragneisses, marbles, various metabasites, syntectonic megacrystic orthogneisses	Migmatitic sequence: quartzo-feldspathic gneisses, orthogneisses, minor metasediments; post-migmatitic augen gneisses	Bimodal metavolcanics, orthogneisses	Meta-anorthosite, metasediments, minor bimodal metavolcanics, orthogneisses	Metapelites, marbles and calc-silicates in the north; mylonitized tonalitic gneisses in the south
Predominant structural trends	NNE–SSW	NE–SW	NE–SW	E–W and N–S	E–W and N–S	E–W and NNE–SSW	E–W
Major Pan-African deformation phases	D3 dextral transpressive shear zone and attributed transpressive folds	D5 discrete subhorizontal shear zones, top-to-NW and -SE; D4 regional wrapping on NE axes	D4 open upright folding on NE axes; D3 NE–SW-trending open folds	D5 locally top-to-SW shearing; D4 NW-vergent folds; D3 W-vergent folds; D2 N-vergent folds	D4 discrete extensional shearing; D3 transpressive shearing and W-vergent folds; D2 N-vergent folds	D4 discrete extensional shearing; D3 transpressive shearing and W-vergent folds; D2 N-vergent folds	D4 transposition of northern and southern part along a major shear zone; D3 open, E–W trending folds; D2 subvertical folds

	Late Mesoproterozoic deformation phases (partly tentative)	D3 SE–NW trending folds and ductile shearing D2 SE plunging isoclinal folds D1 isoclinal folding	D2 isoclinal SE-plunging folds, thrust duplexes to WNW D1 isoclinal folding	D1 mineral internal relicts?	D1 mineral internal relicts?	D1 relicts of isoclinal rootless folds	D1 isoclinal E–W folds
	D2 ENE vergent folds D1 relicts in a microscopic scale			*(Probably more than one phase, most older structures are obliterated)*			
Pan-African maximum metamorphic conditions	6.0–6.5 kbar and 580–610°C (? Late Mesoproterozoic)	12–13 kbar and 650–700°C	8–12 kbar and 750°C	8 kbar and 750°C	6.8 kbar and 830°C	8–9 kbar and 700±30°C	7.0–8.5 kbar and 750–830°C
Post-tectonic Pan-African granitoids	None	None	Minor	Voluminous	Voluminous	Voluminous	Voluminous

East Antarctic Orogen in western DML

Crossing the Heimefront Shear Zone (HSZ; Fig. 1) from west to east we step from Mesoproterozoic basement of the foreland of the Pan-African Orogen into the orogen itself. The HSZ probably originates from the indentation of the Kaapvaal–Grunehogna Craton (Fig. 2b) into the surrounding belt at the final stage of the Late Mesoproterozoic Orogeny (Jacobs et al. 1993). During the Pan-African Orogen it was reactivated as a dextral transpressive shear zone. The polyphase nature of the HSZ is illustrated by at least four types of mylonites which are recognized. Steeply inclined mylonites with a subhorizontal stretching lineation are very prominent but gently SE-dipping, discrete mylonites with a top-to-NW sense of shear also occur. In northern and central parts of Heimefrontfjella, the shear zone splits into several branches. The less deformed rafts between anastomosing shears are characterized by folding around steeply inclined axes. Amplitudes and wavelengths of these folds range from a few centimetres to several hundreds of metres.

The region east of the shear zone is mainly composed of bimodal metavolcanic rocks intruded by syntectonic granitoids during Late Mesoproterozoic times (Jacobs et al. 1996). Maximum metamorphic conditions have been calculated at 6–6.5 kbar and 600°C (Schulze 1992) but it remains unclear whether these conditions were reached during the Late Mesoproterozoic event or in Pan-African times. K–Ar mica (blocking temperature 350°C) and ^{40}Ar–^{39}Ar hornblende (blocking temperatures at least 500°C) ages indicate at least lower amphibolite facies conditions during Pan-African times (Jacobs et al. 1995, 1997).

To the NE of Heimefrontfjella, the Pan-African Orogen extends into Kirwanveggen and H.U. Sverdrupfjella, adjacent to the southern and eastern boundary of the Grunehogna Craton (Fig. 1). The metamorphic basement comprises juvenile Mesoproterozoic sequences similar to those from Heimefrontfjella. In Kirwanveggen, banded hornblende–biotite–plagioclase gneisses and garnet-bearing gneisses interlayered with amphibolites and metapelites were intruded by megacrystic orthogneisses (Grantham et al. 1995). Lenses of marbles and calc-silicates are present locally. To the NE in the H.U. Sverdrupfjella, three lithological zones parallel to the margin of the Grunehogna Craton, have been identified (Grantham & Hunter 1991). Close to the Grunehogna Craton, grey gneisses of tonalitic composition are interpreted as being derived from calc-alkaline metavolcanic rocks. The distribution of the geochemical data suggests a volcanic arc that developed on or near a continental margin as a probable source (Groenewald et al. 1995).

Further east, a sequence comprising metacarbonates and quartzofeldspathic metasediments was overthrusted by a metapelite sequence containing mafic lenses and boudins. The latter are composed of metanorite, garnet peridotite, olivine gabbronorite and garnet pyroxenite. The relict assemblages of garnet–augite–rutile–quartz and garnet–olivine–ferrosilite–spinel–hornblende indicate peak metamorphic conditions of >12 kbar and 700°–790°C (Groenewald et al. 1995). The geochemical signature of the metamafic rocks points to a mid-ocean ridge basalt (MORB) origin, whereas the associated metasedimentary sequence probably represents the transition from a stable shelf facies in a pelagic back-arc basin (Grantham & Hunter 1991).

Grantham et al. (1995) distinguished at least three deformation phases during the Late Mesoproterozoic and two Pan-African deformation phases, separated by geochronologically and structurally well-constrained intrusions. Ductile thrusts and multiple duplex structures are widely distributed in H.U. Sverdrupfjella. Along SE- and east-dipping thrust planes, both the volcanic-arc and the back-arc basin were stacked onto the Kaapvaal–Grunehogna Craton during Late Mesoproterozoic times (Groenewald et al. 1995). A phase of decompression, thermal relaxation and exhumation is connected with the intrusion of leucocratic granites at 1050 Ma (Groenewald et al. 1995).

Unfortunately, the younger Pan-African structures are broadly parallel and colinear with the older Mesoproterozoic structures making them indistinguishable in the absence of time markers such as pegmatites and mafic dykes (Grantham et al. 1995). Pan-African structures trend mainly NE–SW, but locally an older generation of tight to isoclinal folds is wrapped on steeply NE-dipping axes. In central Kirwanveggen, numerous brittle–ductile shear zones transect all structures; their kinematics are top-to-NW or top-to-SE, respectively. Maximum metamorphic conditions of c. 6.5 kbar and 650°C for the Pan-African event have been deduced from different geothermobarometers, followed by a well documented H_2O-fluid dominated cooling phase (Grantham et al. 1991).

Prior to the final tectonic shortening, at least 1300 m fluvial sediments were deposited in SW Kirwanveggen (Urfjell Group). They were locally preserved in a small strike-slip basin, formed by sinistral transpressional plate boundary movements at c. 550 Ma (Croaker et al. 1999). Post-depositional transpression resulted in folding on NNE- to E-trending axes and steeply to overturned bedding within NE–SW striking fault zones. Rb–Sr dating of detrital muscovites (634 Ma) and a whole-rock errorchron (531 ± 25 Ma) bracket the age of these sedimentary sequence (Moyes et al. 1997). The detrital minerals are probably derived from juxtaposed metamorphic rocks in Kirwanveggen or eastern Heimefrontfjella.

Metamorphic basement of Gjelsvikfjella and western Mühlig-Hofmann–Gebirge

The westernmost region of central DML covers the mountain ranges of Gjelsvikfjella and western Mühlig-Hofmann–Gebirge between 2 and 6°E. The oldest rock type in this area is probably a thick sequence called the Grey Migmatic Gneisses, which is of tonalitic to granitic composition, containing abundant boudinaged remnants of mafic dykes (Bauer & Jacobs 2000; Jacobs & Bauer 2001). Recent studies on the zircon morphology confirmed a magmatic origin of these gneisses (Mitterer pers. comm.). This sequence forms large parts of the metamorphic basement and might have originally been overlain by sedimentary rocks, now forming a series of (garnet)–biotite–plagioclase gneisses, sillimanite schists, calc-silicate rocks and marbles. These rocks were intruded by voluminous granitoids. The whole complex underwent a phase of high-grade metamorphism up to 8 kbar and 750 °C (Bucher-Nurminen & Ohta 1993), migmatization and polyphase tectonic deformation. Precise geochronologic data for the first tectonothermal event and the precursor rocks of the metamorphics are lacking. The migmatitic basement was locally intruded by a younger generation of sheet-like granitoids (now exposed as augen gneisses).

At least five deformation phases have been recognized by Bauer & Jacobs (2000) for this area. The first deformation phase recognized postdates the migmatization, suggesting that possible older structures have been completely obliterated. Correlations of different deformation phases to the Late Mesoproterozoic and Pan-African events as given in Table 1 are preliminary and a geochronological program is underway.

The oldest mesoscopic structures are small, E–W oriented recumbent folds (D2). Due to later refolding by small-scale N–S-trending folds (D3), the axes plunge either to the east or west. A third generation of folds (D4) is characterized by large, open folds in a 10–100 m scale. The latest compressive tectonic event (D5) is marked by a 100–200 m wide ductile shear zone at Armlenet, Gjelsvikfjella. In contrast to the NW-vergent D4 folds, indicating a NW-directed tectonic transport, this shear zone has a gently NE-plunging stretching lineation and a top-to-SW sense of shear. The microfabric of the mylonite is completely recrystallized, i.e. the mylonitization occured at temperatures >450 °C.

In most parts of the central and eastern Mühlig-Hofmann–Gebirge, post-tectonic granitoids are exposed. Adjacent to the batholiths, migmatites with undeformed agmatitic textures are developed, indicating very high intrusion temperatures.

Metamorphic basement of Orvinfjella and Wohlthatmassiv

Further to the east, between 8 and 14°E, a well-studied metamorphic basement is exposed in the mountain chains of Orvinfjella and Wohlthatmassiv (Fig. 3). The region has been studied by several Russian, Indian and German expeditions (e.g. Ravich & Kamenev 1975; Kaul et al. 1991; Sengupta 1993; Paech et al. 1999; Bauer et al. 2003).

The metamorphic basement consists of a thick supracrustal pile which is intruded by various late- to post-kinematic plutonic and hypabyssal rocks, later deformed to augen gneisses (Bauer et al. 2003). The metamorphosed supracrustal pile is composed of paragneisses, sillimanite–cordierite–bearing metapelites and forsterite–phlogopite–spinel marbles, alternating with felsic gneisses and amphibolites which may be interpreted as bimodal metavolcanic rocks. The supracrustal sequence was intruded pre- or syntectonically by tabular granitoids of tonalitic to granitic composition. The occurrence of mineral parageneses of orthopyroxene + quartz and two-pyroxenes ± garnet in intermediate and mafic rocks indicates granulite facies peak metamorphic conditions.

At the Otto-von-Gruber-Gebirge, the basement complex was intruded by a voluminous anorthosite suite with norites and ferrodiorites (Jayaram & Bejarniya 1991; Markl et al. 2003). The Gruber Anorthosite intruded at c. 600 Ma (Jacobs et al. 1998) and was subsequently affected by intense ductile deformation. Xenoliths of gneisses prove the existence of an already metamorphosed host rock prior to the anorthosite intrusion.

The earliest deformation structures (D1) in Orvinfjella and Wohlthatmassiv (Fig. 4) are close to isoclinal intrafolial, often rootless folds in the paragneisses and metavolcanic rocks. Such folds have been locally found as trails of mafic minerals within single gneissic layers and as incorporated folded gneiss xenoliths in the Gruber Anorthosite. D1 deformation is therefore assumed to represent a Late Mesoproterozoic event. Jacobs et al. (1998) dated the associated metamorphism and the intrusion of syntectonic granitoids (now augen geisses) at c. 1085–1075 Ma. In the Schirmacher Oasis, Sengupta (1993) described rotated internal structures in mafic enclaves representing the oldest structures; all such early structures are typically rotated and overprinted by later deformation phases.

There is no geological information for the time between 1075 and 600 Ma. Assuming the existence

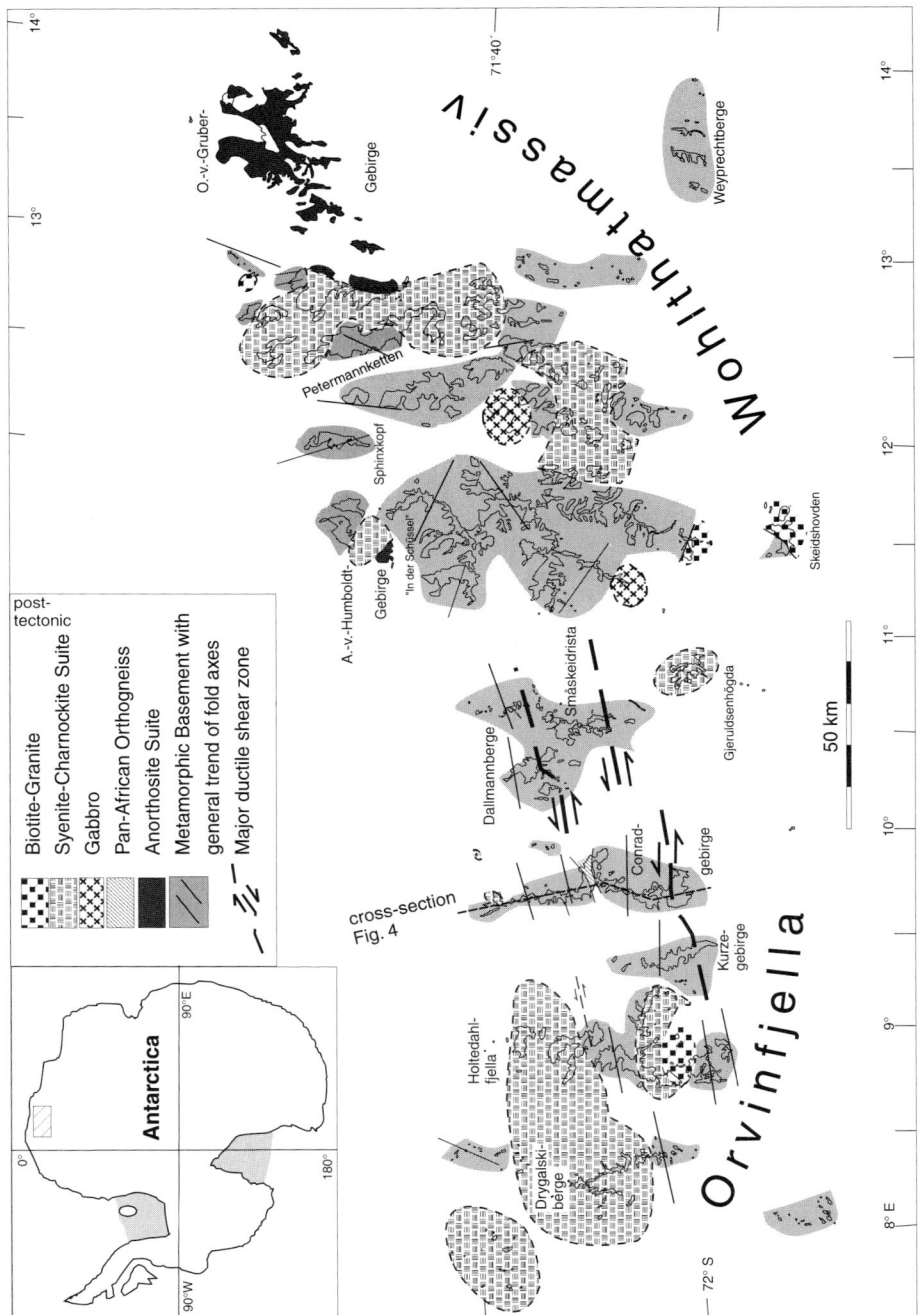

Fig. 3. Geological map of Orvinfjella and Wohlthatmassiv, according to the results of the GeoMaud expedition 1995/1996.

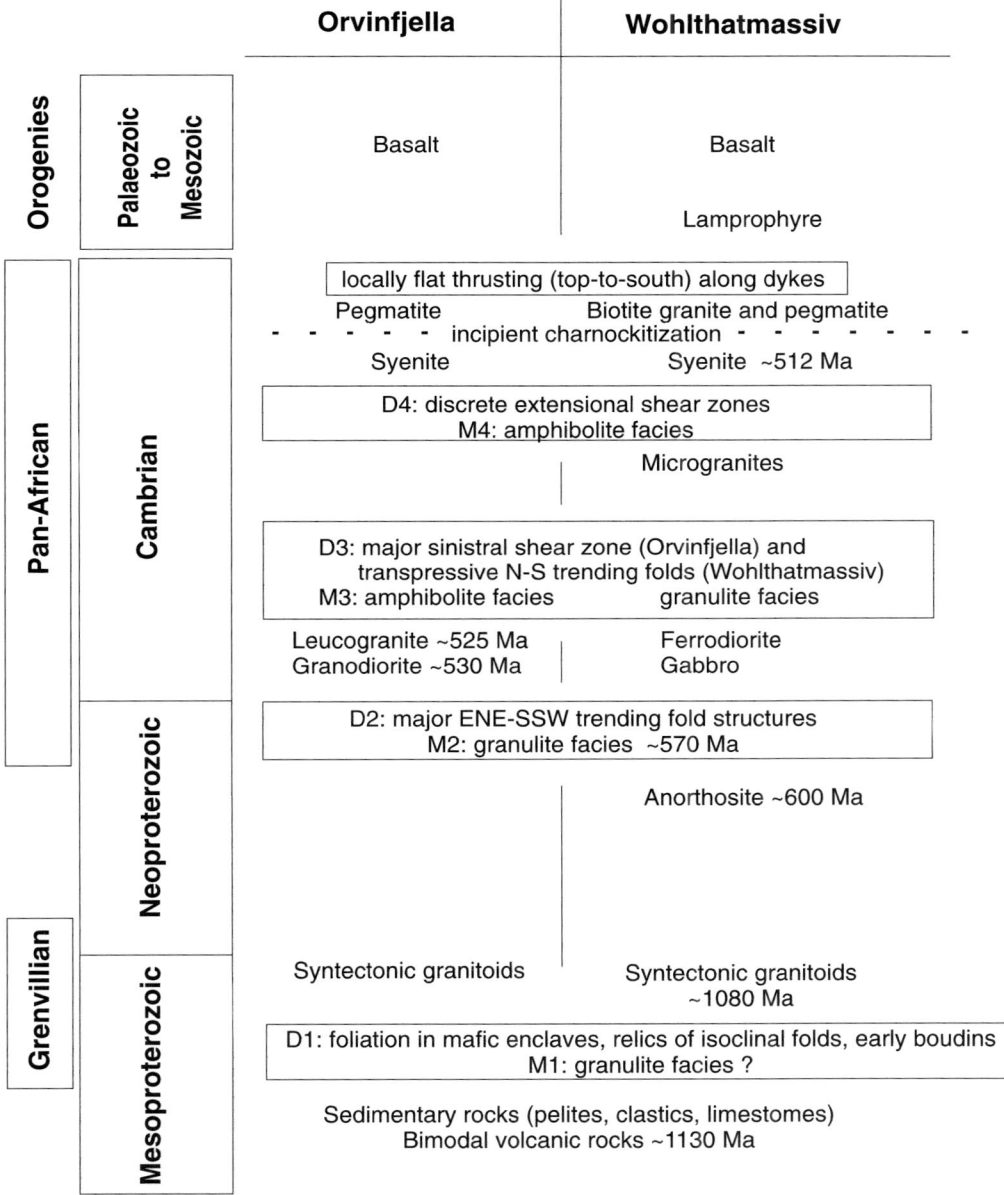

Fig. 4. Structural summary for central Dronning Maud Land. Ages are approximate and are derived from Wand *et al.* (1988), Dayal & Hussain (1997), Mikhalsky *et al.* (1997) and Jacobs *et al.* (1998).

of a Mozambique Ocean (e.g. Dalziel 1997) between c. 750 and 550 Ma in that part of Antarctica, and in formerly adjacent parts of SE Africa, sediments and relics of oceanic crust have not yet been identified, or are hidden beneath the ice.

The Pan-African orogenic cycle started with the intrusion of the Gruber Anorthosite c. 600 Ma ago.

The most prominent, main deformation phase (D2) produced major north-vergent folds with gently NE- to east-plunging axes (Fig. 6a). The wavelength of D2 folds varies between 5 km and a few centimetres. Parasitic minor folds are developed on the limbs of major folds and parallel to the axial planes of close to tight D2 folds a new S2 foliation occurs.

Syntectonic migmatization, dehydration melting of metapelites and expulsion of leucosomes, which intruded as garnet-bearing leucogranite veins in the adjacent gneisses or into boudin necks of mafic metamorphic rocks, are common features of the contemporaneous metamorphic phase. Zircon cores were overgrown by new metamorphic zircon rims during this event. The age of the first Pan-African metamorphic event is dated at c. 570 Ma (Jacobs et al. 1998). Thermobarometrical analyses (Piazolo & Markl 1999) prove granulite facies conditions (6.8 kbar and 830°C) for the ferrosilite–garnet assemblages which grew during this metamorphic phase.

A well-studied, representative example for the tectonic style of Orvinfjella is exposed in the Conradgebirge (Bauer et al. 2003). The N–S cross-section shown in Figure 5 covers the entire 35 km length of the mountain chain. The northernmost part of the cross-section is poorly exposed. North and south of the intrusive syenite, steeply northward dipping bimodal metavolcanic rocks crop out. Further to the south at Petersenegga supracrustal gneisses and amphibolites are folded with a wavelength of c. 1.5 km, forming two upright syn- and antiforms. A pronounced stretching lineation on the foliation planes is oriented parallel to the general fold axis direction. In quartz-rich lithologies on the limbs of these tight folds a subvertical stretching lineation is developed perpendicular to the fold axis direction, indicating high internal strain. The symmetrical upright shape of these major folds is not representative for minor folds at the centimetre to metre scale, which show a NNW vergence. Contrasting rheologic conditions within the amphibolite/felsic gneiss succession resulted in boudinage of the competent amphibolite layers at the southern limb of NNW-vergent folds. The necks between individual boudins are filled in part with orthopyroxene-bearing leucosomes, indicating granulite facies conditions during the deformational event. In the central part of the Conradgebirge, the form of a large anticline can be traced by the contact of a Pan-African-aged granitic orthogneiss to overlying metavolcanic gneisses. This anticline turns into a large NNW-vergent syncline with a relatively thin augen gneiss layer and metasedimentary rocks in its core. To the south, at Bjerkenuten, the folded sequence terminates at a nearly 500 m wide sinistral shear zone (Bauer & Siemes 2003) known as the South Orvinfjella Shear Zone (SOSZ). Here, metavolcanic gneisses and orthogneisses have undergone intense shearing (D3), resulting in dark ortho- to ultramylonites with light to pinkish feldspar porphyroclasts. The mylonitic foliation strikes E–W and dips at moderate angles south. A mylonitic lineation parallel to the main fold axis direction (ENE), as well as southwesterly plunging lineations and minor fold axes, have been observed (Fig. 6b). Sinistral shear kinematics were deduced from asymmetric σ-clasts and quartz c-axis measurements (Bauer & Siemes 2003).

A 530 Ma old granodiorite (Jacobs et al. 1998), anatectic melts in the necks of D2 boudins and the ferrodiorite intrusions within the Gruber Anorthosite were penetratively deformed during D3. Within the SOSZ the granulite facies protoliths are retrogressed to biotite–garnet parageneses. The sinistral, top-to-NE-directed tectonic movement in the SOSZ, led to the transposition of highly migmatitic orthogneisses of southern Kurzegebirge, Conradgebirge and Dallmannberge to anhydrous granulite facies gneisses in the central part of these ranges. The sinistral transpression (D3) induces a horizontal strain ellipse with its long axis at an angle <45° to the strike-slip zones. Such oblique fold belts, disrupting the overall

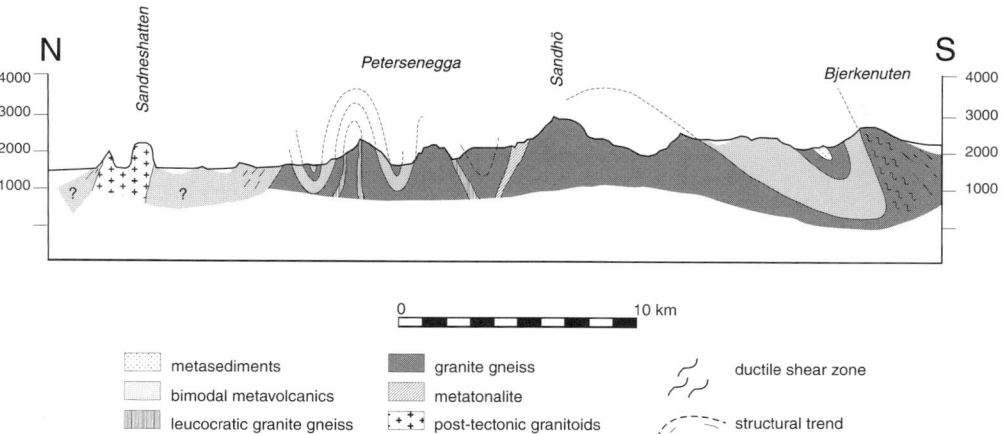

Fig. 5. Cross-section through Conradgebirge–Orvinfjella.

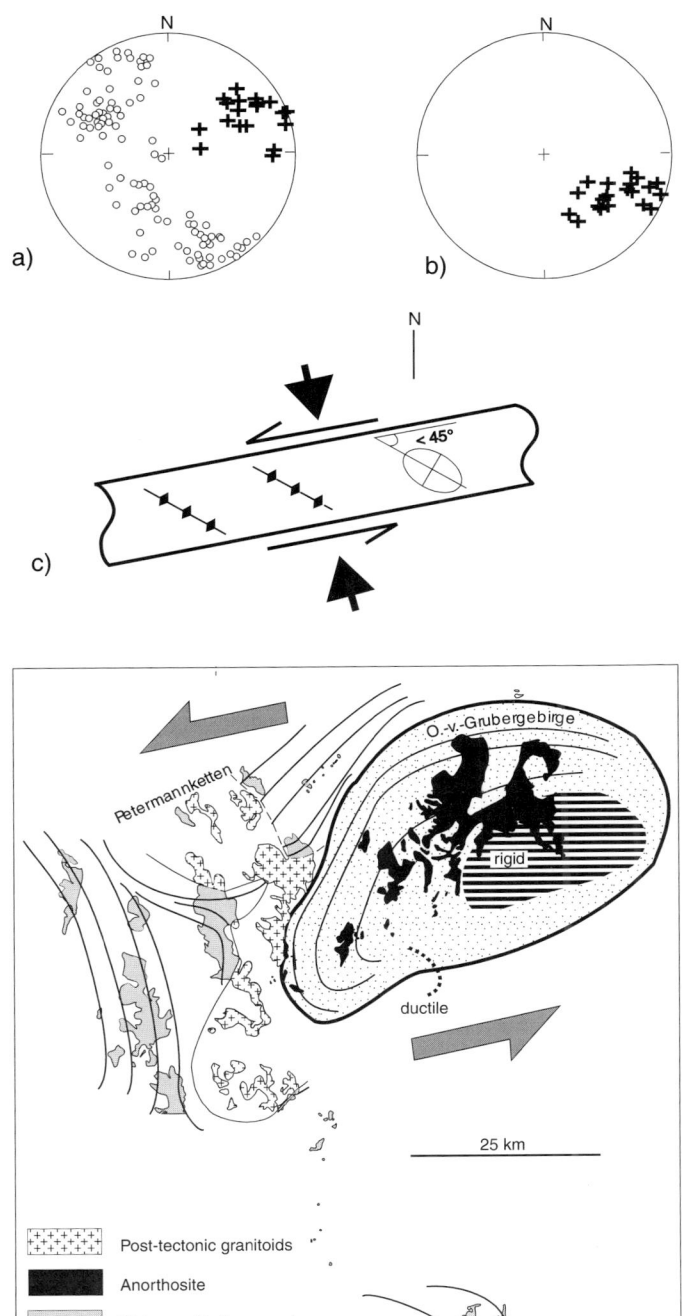

Fig. 6. (**a**) Pole figures of (B2) fold axes (crosses) and foliation planes (open circles), northern Conradgebirge. (**b**) Pole figure of transpressive, oblique D3 fold axes in the South Orvinfjella Shear Zone at southern Conradgebirge. (**c**) Orientation of folds in a horizontal section through a vertical strike-slip zone undergoing transpression; redrawn after Woodcock & Schubert (1994). (**d**) Sinistral transpression around the rigid Gruber Anorthosite, central Wohlthatmassiv, according to the model of Paech *et al.* (1999).

E-W structural trend (Fig. 6c), are developed in the Petermannketten, northern Drygalskiberge, at Sphinxkopf and in eastern parts of the A.-v.-Humboldt–Gebirge (Figs 3 & 5d), where a single major shear zone was not identified.

Due to the restricted outcrop conditions it is difficult to get an overview of the deformation patterns around the rigid Gruber Anorthosite. During the Pan-African high-grade ductile deformation, the huge rigid Gruber Anorthosite body is affected only at the margins. According to Paech et al. (1999), bending of the metamorphic foliation within the Gruber Anorthosite is parallel to the foliation in the country rocks (Fig. 6d). This can be best explained by an architecture known as 'rolling structure' (Driessche & Brun 1987), where the anorthosite rotated like a huge δ-clast during sinistral shearing. Such large-scale phenomena are also known from other anorthosite intrusions within Gondwana. For example, Martelat et al. (1997) described the Saririaky Anorthosite in southwestern Madagascar as an undeformed rigid body in a ductile matrix of the surrounding gneisses that had experienced high shear strains under high-grade conditions.

The final deformation phase (D4) in central DML is indicated by discrete, late mylonite zones. Numerous shear zones have been recorded in Orvinfjella and in the Wohlthatmassiv which are typically NW-SE trending and steeply to moderately SW dipping. In the Wohlthatmassiv, a NE-SW orientation with mainly southeasterly dipping mylonitic foliations is also recorded. The formation of shear zones parallel to the opening of tension gashes, filled with undeformed quartz and quartz/feldspar mobilisates is very common. The thickness of these mylonites ranges from 20cm to 2m. Generally, the displacements are small and southerly or southeasterly in direction. The deformation took place under amphibolite facies conditions, which is indicated by the ductile deformation and recrystallization of K-feldspar and plagioclase, as well as the mineral paragenesis of garnet + biotite + quartz (Bauer & Siemes 2003). CO_2-rich fluids led to a recrystallization of feldspars in distinct channel-like zones 1 m in diameter (maximum). Gneisses which are affected by this reaction lost their gneissic texture (incipient charnockitization sensu Yoshida et al. 1991). Many of the D4 shear zones are associated with undeformed pegmatoids and smaller syenite/alkali–granite intrusive bodies, such as at Gjeruldsenhögda, where the extensional kinematics forms pull-apart settings which faciliated the emplacement of the plutonic rocks. Referring to the U–Pb zircon date of 512 ± 2Ma (Mikhalsky et al. 1997), from a syenite of the A.-v.-Humbolt– Gebirge, a Pan-African, synmagmatic age of this D4 deformation phase in central DML is proposed. In summary, the field observations and geochronological data are shown in Table 2.

Metamorphic basement of Sør Rondane

The Sør Rondane Mountains in eastern DML were studied by Belgian and Japanese geologists (e.g. Van Autenboer & Loy 1972; Shiraishi et al. 1991). These teams reported gneisses of pelitic and psammitic origin in the northern part of the mountains, mixed with intermediate gneisses of probable volcanic origin. Subordinate marbles, calc-silicates and ultramafic rocks are also present. Four Sm–Nd and Rb–Sr dates from hornblende gneisses have proved a Late Mesoproterozoic origin for the rocks (Shiraishi & Kagami 1992).

The metamorphic rocks exhibit mineral parageneses of upper amphibolite to granulite facies (7–8.5 kbar and 750–800°C), grading partially into migmatites. Strongly deformed schists of tonalitic composition prevail in the southern part, separated from the metasedimentary rocks in the north by a wide E–W striking shear zone.

The main structural trend of the isoclinally folded gneisses is E–W (D1), but these structures are regionally deformed by N–S-trending, steeply plunging folds (D2), themselves refolded by open E–W-trending folds which dominate the present structural trend of the range. Finally, the southern part of Sør Rondane was transposed against the northern part along the major shear zone under conditions of upper greenschist facies. Shiraishi et al. (1991) reported gently SSW- to SW-plunging stretching lineations in the shear zone, indicating sinistral movement.

Table 2: Geochronologic data from late- to post-tectonic granitoid intrusions in Dronning Maud Land

Locality	Method	Ages (Ma)	Authors
H.U. Sverdrupfjella	Rb–Sr/Sm–Nd*	518±15/522±120	Grantham et al. 1991
Mühlig-Hofmann–Gebirge	Rb–Sr*	500±24	Ohta et al. 1990
Wohlthatmassiv	U–Pb[†]	512±2	Mikhalsky et al. 1997
Sør Rondane	U–Pb[†]	520±20	Pasteels & Michot 1970

*Whole-rock method.
[†]Conventional zircon method.

The metamorphic basement is intruded by post-tectonic batholiths of syenitic and granitic composition with contact aureoles in the adjacent gneisses.

General characteristics of the late- to post-tectonic magmatism in central DML

Voluminous coarse-grained to megacrystic granitoid bodies are exposed over an estimated area of at least 12000 km^2 in central and eastern DML between 2 and 28°E (Fig. 1), thus representing a major post-tectonic magmatic event. They are mainly undeformed, with small discrete shear zones locally seen at their margins (Bauer et al. 2003). The Cambrian age of these intrusives is well constrained (Table 2).

The most abundant rocks are charnockites, containing mesoperthite, plagioclase and quartz. In magnesian varieties, the mafic phases are of pigeonite and augite, whereas in iron-rich varieties fayalite + hedenbergite + quartz coexist (Frost & Bucher 1993). Granites, monzonites, quartz monzonites and syenites are also common. Thermobarometrical studies by Frost & Bucher (1993) and Bucher & Frost (1995) on charnockites, granites and gneissic enclaves suggest that the charnockites were emplaced as relatively dry melts at temperatures >900°C and pressures of c. 5 kbar. When the charnockite cooled down to <900°C, melt pockets of granitic magma remained, especially around gneiss inclusions and host rocks, which expelled fluids to flux the remaining melt. When the granitic melts solidified they expelled aqueous fluids that invaded the charnockite and retrogressed them to biotite–hornblende–granodiorite. Such parts appear as bleached zones in the otherwise brownish charnockites.

Several papers on the bulk geochemical characteristics of the charnockites and their retrogressed varieties have been published (Klimov 1964; Ravich & Soloviev 1966; Ravich & Kamenev 1975; Joshi et al. 1991; Poulsson et al. 1999; Roland 1999) but their results are ambiguous. Geochemically, the rocks are characterized as peraluminous to metaluminous, or subalkaline with a weak trend to alkaline A-type granites. In contrast to the contemporaneous, Mg-rich charnockites of India and Sri Lanka, the rocks from DML are very enriched in Fe, represented by coexisting ferrosilite and quartz. Due to low Ca, Rb and Ga values, the DML charnockites do not represent a normal A-type granite, typically associated to a continental extensional setting. A spatial and genetic link between these granitoids and an adjacent Gruber Anorthosite body, proposed by Joshi et al. (1991), is disproved, since the anorthosite is c. 90 Ma older (Jacobs et al. 1998) and was intruded prior to the main deformation. According to Roland (1999), the charnockites were derived from lower continental or underplated crust, explaining their relatively heterogeneous geochemical signature. The megacrystic, sometimes rapakivi-type, textures furthermore suggest crustal growth of a dry melt at a lower crustal level. According to Roland (1999), after the continent–continent collision between East and West Gondwana, the continental lithospheric mantle delaminated, allowing the asthenosphere to rise up and heat the adjacent crust on a large scale. The crust was composed of juvenile Mesoproterozoic rocks which produced a highly fractionated magma of charnockitic composition.

A Cambrian orogenic belt between Lützow-Holmbukta and Prydz Bay?

Numerous latest Neoproterozoic–Cambrian ages have been reported from outcrops along the Antarctic coast from DML to the Windmill Islands at 100°E (see discussion in Fitzsimons 2000), but their significance and interpretation remains unclear. The best studied areas include the outcrops at Lützow-Holmbukta, Enderby and Kemp Lands, and the Prydz Bay area, which play an important role in the Antarctica–India fit and for a possible easterly continuation of the East Antarctic Orogen (Fig. 7). Recently, two different models have been proposed for the belt, based on available geochronological, structural and thermobarometrical data. Fitzsimons (2000) suggested a wide Pan-African belt as the continuation of the East Antarctic Orogen (the Lützow-Holm Belt), with another belt between Prydz Bay and the Denman Glacier region (the Prydz Belt). The area between these belts constitutes an older microcontinent (the Rayner Province) which was thermally reworked during the Pan-African Orogeny. The western margin of the Lützow-Holm Belt is marked by the Jutulstraumen Glacier and the HSZ. Its eastern margin, c. 300–400 km east of Lützow-Holmbukta, is less well defined. Shiraishi et al. (1997) proved the presence of a high-grade Pan-African event at the Thala Hills (46°E) with U–Pb SHRIMP zircon and Sm–Nd whole-rock data. Further east in the Nye Mountains, c. 100 km away, Shiraishi et al. (1997) obtained 980–910 Ma zircon ages, but the evidence of a later Pan-African overprint is lacking.

On the other hand, Hiroi et al. (1991) interpreted the Lützow-Holm Complex and its westerly adjacent region, the Yamato–Belgica Complex, as a paired metamorphic belt. The Lützow-Holm Complex is characterized by early high-pressure/low-temperature peak metamorphic conditions, whereas widespread igneous activity and low-pressure/high-temperature conditions have been recorded in the Yamato–Belgica Complex. The metamorphism of the Lützow-Holm Complex

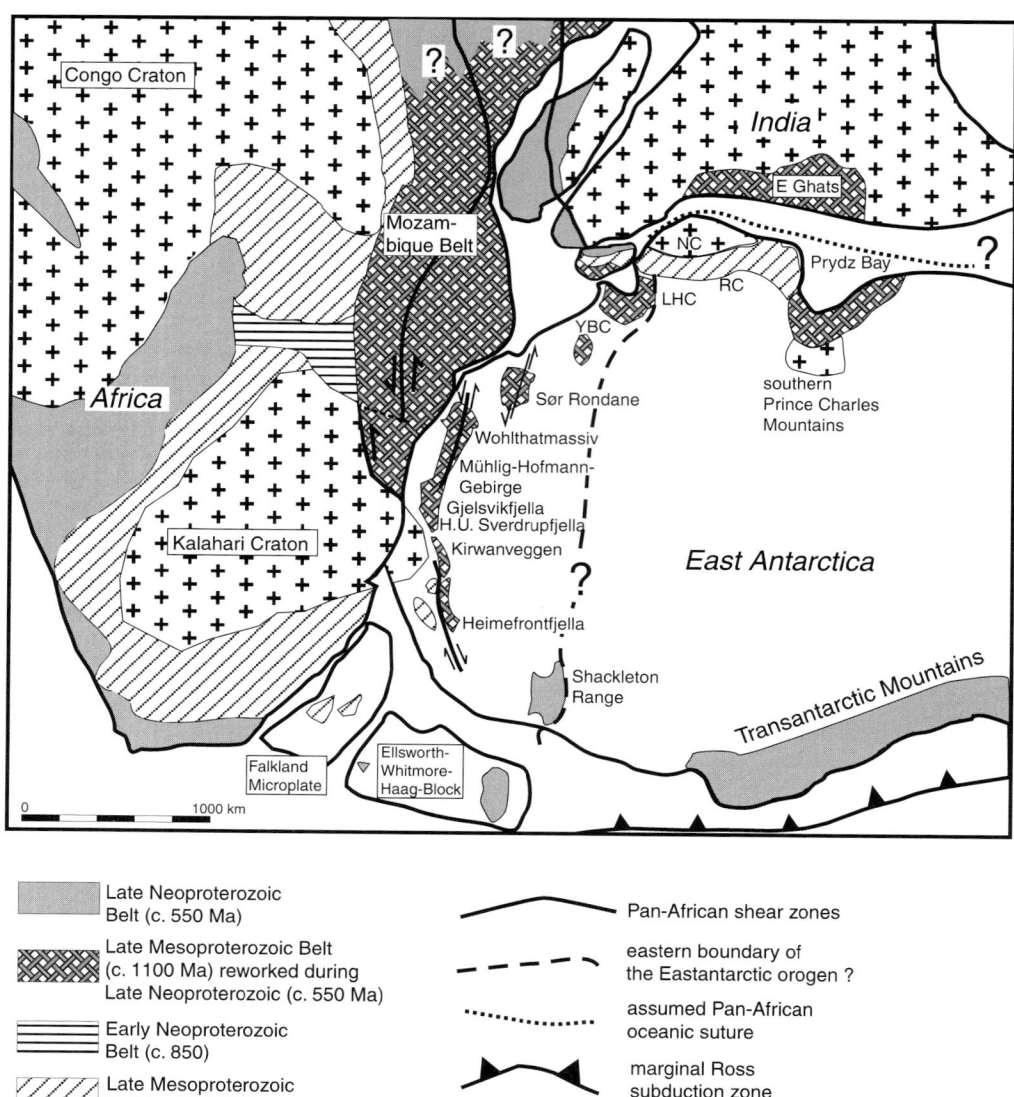

Fig. 7. Location of the Dronning Maud Land sector of Antarctica in a Gondwana reconstruction after Lawver & Scotese (1987), Shackleton (1996) and Mezger & Cosca (1999). LHC, Lützow-Holm Complex; NC, Napier Complex; RC, Rayner Complex; YBC, Yamato–Belgica Complex

decreases to the east. According to Fitzsimons (2000), the Lützow-Holm Belt runs from the coastal region through the ice-covered inner part of Antarctica towards the Shackleton Range, where Cambrian tectonism has long been known (e.g. Buggisch et al. 1993; Tessensohn et al. 1999). Furthermore, Talarico et al. (1999) reported a possible ophiolite marking a Pan-African suture in the Shackleton Range. West of the Shackleton Range, the Pan-African collision orogen passes into the marginal Ross Orogen.

The other arm of the Pan-African Orogen, named the Prydz Belt by Fitzsimons (2000), is well documented by geochronological and thermbarometric data, but neither the extent of this belt in central Antarctica and India nor its nature, i.e. continental collision, intracontinental tectonism or failed rift, has been proved. The western margin of this belt is located along the Lambert Graben, while the eastern margin may be located in the vicinity of the Mirny Station (92°E), where a Cambrian-aged syenite pluton intrudes Archaean tonalitic orthogneisses (Black *et al.* 1992). As in central DML, in this region a Late Mesoproterozoic basement was strongly overprinted during Pan-African times (Dirks & Wilson 1995; Fitzsimons 2000).

Hensen & Zhou (1997) and Jacobs & Thomas (2002) suggested another model with a subperpendicular branch of the Mozambique Ocean separating East Antarctica from the Indian Craton. Taking into account previous works of Yoshida & Vintanage (1993), Shiraishi *et al.* (1994) and Kriegsmann (1995) – the latter of whom realized that a close relation existed between the Yamato–Belgica, Lützow-Holm and Rayner Complexes in Antarctica and the Vijayan, Highland and Wanni Complexes in Sri Lanka, respectively – suggested that the Cambrian closure of this oceanic branch may have led to the observed structural and geochronological patterns. This idea remains to be dicussed, as the presence of Neoproterozoic sedimentary rocks in the Lützow-Holm Complex, stressed by Shiraishi *et al.* (1994), has been questioned (Yoshida 1994). Other evidence for an oceanic-basin-like ophiolites have not yet been recognized.

The Archaean Napier Complex in Enderby Land may represent a part of the Dharwar Craton of India, with the structures of the Rayner Complex wrapping around the craton fragment, however, this is inconsistent with the assumed Lützow-Holm Belt of Fitzsimons (2000). The basement of the Rayner Complex comprises juvenile Proterozoic rocks but in its eastern parts Pan-African high-grade events have been disproved by Shiraishi *et al.* (1997). To avoid this problem, Hensen & Zhou (1997) draw the oceanic branch between the Napier Complex and India, i.e. on the oceanward side of the present Antarctic coast, which also considers the objection of Mezger & Cosca (1999) regarding the lack of analogy between the Dharwar Craton and the Napier Complex. With this interpretation the Rayner Complex is a Late Mesoproterozoic collision orogen between the Napier microcontinent and the East Antarctic continent, which is exposed at the Archaean basement in the southern Prince Charles Mountains (Sheraton *et al.* 1996). In this case, parts of the Pan-African belt should be located beneath younger sediments of SE India.

Northerly continuation of the East Antarctic Orogen into East Africa

Until the mid-1990s, the nature and extent of a Pan-African Orogeny in DML, and also further east to the Prydz Bay area, was controversial. For example, Stüwe & Sandiford (1993) denied pervasive deformation in the East Antarctic Shield, suggesting lithospheric thinning and magmatic underplating as the cause for a major isotopic resetting at $c.\ 500\,Ma$. The closure of the Neoproterozoic Mozambique Ocean between East and West Gondwana around a rotation pole in the Weddell Sea was a widely accepted model (e.g. Hoffman 1991). First hints for Late Neoproterozoic–Cambrian tectonometamorphic activity in East Antarctica came from U–Pb SHRIMP data (Shiraishi *et al.* 1994, 1997; Jacobs *et al.* 1998) proving the existence of high-grade metamorphism, as well as pre- and syntectonic granitoids of latest Neoproterozoic–Cambrian age. At the same time, palaeomagnetic data suggested that western DML was part of West Gondwana (Gose *et al.* 1997), which supported the idea of a Pan-African suture through East Antarctica.

The continuation of the East Antarctic Orogen into Africa is represented by the East African Orogen (Stern 1994). The East African Orogen stretches from Mozambique and Madagascar through Tanzania into the Arabian peninsula (Fig. 7). The southern part of the East African Orogen shares great similarities with the basement of central DML; both are composed of Late Mesoproterozoic–Early Neoproterozoic rocks that were variably overprinted during the Pan-African event up to granulite facies (e.g. Pinna 1995). However, the northern continuation within the Arabian–Nubian Shield shows radically different rocks. The Arabian–Nubian Shield comprises Neoproterozoic juvenile crust that probably originated within the Mozambique Ocean. The lithological succession includes passive margin rift-related sedimentary sequence as evidence of early extention, volcanic rocks with island-arc affinities, and ophiolites (Stern 1994). The Arabian–Nubian Shield is thought to be composed of a number of terranes, each bound by ophiolites. These terranes were juxtaposed by oblique sinistral transpression to cause major, mainly N–S-trending, sinistral shear zones identified throughout the length of the East African Orogen. This succession was metamorphosed to greenschist and lower amphibolite facies grade. Thus, although the southern part of the East African Orogen is very similar to the East Antarctic Orogen, the northern part of the East African Orogen appears to be very different.

The position of East/West Gondwana suture in Africa is as controversial as its identification in Antarctica. Shackleton (1996) argued that the Nabitah–Bargaloi Shear Zone within the Arabian–

Nubian Shield is the most likely candidate. Two localities on the eastern side of the East African Orogen in Madagascar are potential locations for the suture further south. The Vohibori Terrane of southeastern Madagascar represents Neoproterozoic rocks that contain an interesting rock assemblage of granitic migmatites, marble, granodiorite gneiss and abundant ultramafic rocks (Besarie 1968–1971). However, it remains to be shown whether the ultramafic rocks represent a suture zone. A second potential location for a suture is situated on the east coast of Madagascar, where high-pressure metapelites are associated with ultramafic rocks (Collins 2000).

Discussion and conclusions

Several contraints have recently been placed on crustal evolution in DML. Age relationships and similarities of the cover sequences indicate that the Grunehogna Craton is a detracted part of the Kaapvaal Craton. The craton was stabilized prior to the Palaeoproterozoic, but little is known about the tectonothermal evolution of DML until the Late Mesoproterozoic (c. 1.2 Ga), when juvenile crust was formed along subduction-related magmatic arcs south of the craton (Fig. 2b). This crust-forming period was very widespread as calc-akaline metavolcanics, metaplutonic rocks and subordinate paragneisses have been recorded from the Namaqua–Natal belt of South Africa (Thomas 1989b; Thomas et al. 1994), the Haag Nunatak Block (Grantham et al. 1997), Heimefrontfjella (Jacobs et al. 1996), H.U. Sverdrupfjella (Grantham et al. 1995), central DML (Jacobs et al. 1998) and Sør Rondane (Shiraishi et al. 1991).

East of the HSZ (Fig. 1), the Pan-African overprint increases and numerous pre-, syn- and post-tectonic granitoids soak the Mesoproterozoic basement. In Natal and western Heimefrontfjella, the Pan-African overprint is weak (Jacobs & Thomas 1996), thus here it is easier to reconstruct the Late Mesoproterozoic Orogeny. In DML, Mesoproterozoic subduction was directed southward beneath the East Antarctic Craton and the arc was accreted onto the Kaapvaal–Grunehogna Craton at c. 1100 Ma. Possible ophiolithic remnants of the ocean between the Kottas–Mzumbe Arc and the Kaapvaal–Grunehogna Craton are exposed in Natal (Tugela Nappes; Matthews 1972; Thomas 1989a). Adjacent to the Kottas Arc, a terrane mainly composed of juvenile bimodal metavolcanics and minor metasedimentary sequences (Sivorg Terrane) is assumed to represent a back-arc basin (Bauer 1995; Groenewald et al. 1995; Jacobs et al. 1996). The termination of the Late Mesoproterozoic Orogeny is indicated by post-tectonic pegmatites in Heimefrontfjella at 1060 Ma (Arndt et al. 1991) and minor post-tectonic granites in Natal at 1025 Ma (Jacobs et al. 1993). At the same time, molasse-like debris from the rising mountain range were deposited on the Kaapvaal–Grunehogna Craton (Umkondo Group and Ritscherflya Supergroup).

Gose et al. (1997) have shown that western DML and the Kaapvaal Craton had similar Neoproterozoic palaeopoles but the position of the Kalahari continent (Kaapvaal–Grunehogna Craton + Zimbabwe Craton) within Rodinia is still controversial (Dalziel 1997; Hoffman 1991; Meert 2001). Wilson et al. (1997) discussed the problem of the relationship between the Kalahari and Congo Cratons during Neoproterozoic times and concluded that the Damara–Zambesi–Lufilian orogenic zone represents an array of narrow ocean branches or rift basins rather than a wide ocean. Prior to West/East Gondwana collision, southern Africa had approximately its present shape but with a prominent promontory at the southeastern edge – the Grunehogna Craton of Antarctica.

The Pan-African Orogeny affected a mainly consolidated Late Mesoproterozoic belt with a strikingly homogenous composition between Heimefrontfjella and Sør Rondane. In the whole area, no component older than 1200 Ma has so far been reported, and that might have been derived from an East Antarctic craton. Older components are described from metasediments of the southwesterly tip of Heimefrontfjella, giving U–Pb ages of 1250–2000 Ma from single detrital zircons (Arndt et al. 1991). Further east in the Yamato–Belgica Mountains (Fig. 7), Rb–Sr whole-rocks ages of c. 700 Ma, with a high initial Sr ratio, are considered to represent older components (Yoshida 1995).

In central DML the collision of West and East Gondwana resulted in an E–W-trending fold belt which shows strong similarities to the Pan-African belt in SE Africa (Jacobs & Thomas 2002). At least three phases of folding and ductile thrusting in a NNW–SSE compressional regime and polyphase high-grade metamorphism predate the intrusion of the voluminous granitoid suite (Table 1 & Fig. 4). The first Pan-African deformation event resulted in fold axes parallel to the margins of the Late Mesoproterozoic cratons. In central DML, from Orvinfjella to Sør Rondane, these (D2) folds have an E–W orientation, whereas the NE–SW fold axis orientation in Kirwanveggen (D4) and H.U. Sverdrupfjella (D3) follow the orientation of the promontory formed by the Grunehogna Craton. During a second major Pan-African phase a sinistral transpressive component led to structures which are oriented oblique to the E–W and NE–SW trends. Such structural domains are comparable with the Lurio Belt in Mozambique (Pinna et al. 1993) and around the rigid anorthosite bodies in southern Madagascar (de Wit et al. 2001). In Wohlthatmassiv and Orvinfjella, the shear strain is concentrated on major shear zones with transpressional oblique folds (D3), whereas in Gjelsvikfjella (D3 and 4) and H.U. Sverdrupfjella (D4) open NW-vergent folds prevail. Further to the west in Kirwanveggen the same deformation resulted in discrete subhorizontal shear zones with a top-to-NW-directed transport.

Final stages of the Pan-African orogenic event are documented in central DML. The formation of extensional collapse structures (D5) and the intrusion of post-tectonic charnockites and granites were followed by a cooling to <300°C between 500 and 450 Ma.

Relicts of the Mozambique Ocean (Hoffman 1991) have not been as yet identified in the region. Ophiolites and sedimentary rocks are present only in the Arabian–Nubian part of the orogen, where they may have squeezed out during an oblique collision. The position of the suture in Antarctica is still under debate (e.g. Shackleton 1996; Kleinschmidt et al. 1996). Suggested here is a southward projection of the suture between East and West Gondwana in central DML, because:

(1) the Neoproterozoic palaeopoles for western DML are different from those for East Gondwana (Gose et al. 1997);
(2) in recent Gondwana reconstructions the major structural trends of central DML and the East African Orogen are co-linear;
(3) an increase of metamorphic grade was observed from western to central DML reaching polyphase granulite facies conditions at Orvinfjella and Wohlthatmassiv;
(4) on the other side the metamorphic grade decreases along the Lützow-Holmbukta from west to east;
(5) post-tectonic, probably delamination-related, granitoids occur from H.U. Sverdrupfjella to Sør Rondane.

Financial support by the Deutsche Forschungsgemeinschaft (grants Sp 235/9, Ba1636/5 and Ja617/14) for the expeditions to Dronning Maud Land and the laboratory work is gratefully acknowledged. The expeditions were equipped and supported by the Alfred-Wegener-Institute for Marine and Polar Research. Critical and valuable comments by G. Grantham, Y. Ohta and S. Sengupta are gratefully acknowledged. This is a contribution to IGCP 368 and 440.

References

ARNDT, N. T., TODT, W., CHAUVEL, M., TAPFER, M. & WEBER, K. 1991. U–Pb zircon age and Nd isotopic composition of granitoids, charnockites and supracrustal rocks from Heimefrontfjella, Antarctica. *Geologische Rundschau*, **80**, 759–777.

BARTON, J. M. & MOYES, A. B. 1990. Cooling patterns in western Dronning Maud Land, Antarctica, and southeastern Africa and their implications to Gondwana. *Zentralblatt für Geologie und Paläontologie Teil 1*, 1/2, 33–43.

BARTON, J. M., KLEMD, R., ALLSOPP, H. L., AURET, S. H. & COPPERTHWAITE, Y. E. 1987. The geology and geochronology of the Annandagstoppane granite, Western Dronning Maud Land, Antarctica. *Contributions to Mineralogy and Petrology*, **97**, 488–496.

BAUER, W. 1995. Strukturentwicklung und Petrogenese des Grundgebirges der nördlichen Heimefrontfjella (westliches Dronning Maud Land/Antarktika). *Berichte zur Polarforschung*, **171**, 1–222.

BAUER, W. & JACOBS, J. 2000. German expedition 1999/2000 to Gjelsvikfjella and western Mühlig-Hofmann–Gebirge, central Dronning Maud Land, Antarctica. *Gondwana Research*, **3**, 557–559.

BAUER, W. & JACOBS, J. 2001. The Neoproterozoic suture between East and West Gondwana – new results from central Dronning Maud Land. *Gondwana Research*, **4**, 147–149.

BAUER, W. & SIEMES, H. 2003. Kinematics and geothermometry of mylonitic shear zones in the Orvinfjella, central Dronning Maud Land, East Antarctica. *Geologisches Jahrbuch*, **B96**, in press.

BAUER, W., JACOBS, J & PAECH, H.-J. 2003. Structural evolution of the Proterozoic basement of central Dronning Maud Land. *Geologisches Jahrbuch*, **B96**, in press.

BESARIE, H. 1968–1971. Description Geologique du Massif Ancien de Madagascar. *Document Bureau Geologique Madagascar*. Bureau Geologique Madagascar, Antananarivo.

BLACK, L. P., SHERATON, J. W., TINGEY, R. J. & MCCULLOCH, M. T. 1992. New U–Pb zircon ages from the Denman Glacier area, East Antarctica, and their significance for Gondwana reconstruction. *Antarctic Science*, **4**, 447–460.

BUCHER, K. & FROST, B. R. 1995. Charnockites and granulites of the Thor Range, Queen Maud Land, Antarctica – water recycling in high-grade metamorphism. *Terra Abstracts*, **7**, 314.

BUCHER-NURMINEN, K. & OHTA, Y. 1993. Granulites and garnet–cordierite gneisses from Dronning Maud Land, Antarctica. *Journal of Metamorphic Geology*, **11**, 691–703.

BUGGISCH, W., KLEINSCHMIDT, G., KREUZER, H. & KRUMM, S. 1993. Stratigraphy and facies of sediments and low-grade metasediments in the Shackleton Range, Antarctica. *Polarforschung*, **63**, 9–32.

COLLINS, A. S. 2000. The tectonic evolution of Madagascar: its place in the East African Orogen. *Gondwana Research*, **3**, 549–552.

CROAKER, M., JACKSON, C., ARMSTRONG, R., WHITMORE, G., LINDSAY, P. & MARÉ, L. 1999. The evolution of the Urfjell Group, Western Dronning Maud Land – sedimentology, structure, isotope and geochemical contraints on the evolution of a strike-slip basin. *In*: SKINNER, D. N. B. (ed.) *Programme & Abstracts of the 8th International Symposium on Antarctic Earth Science*, 72.

DALZIEL, I. W. D. 1991. Pazific margins of Laurentia and

East Antarctica – Australia as a conjugate rift pair. Evidence and implications for an Eocambrian supercontinent. *Geology*, **19**, 598–601.

DALZIEL, I. W. D. 1997. Neoproterozoic–Paleozoic geography and tectonics. Review, hypothesis, environmental speculation. *Geological Society of America Bulletin*, **109**, 16–42.

DAYAL, A. M. & HUSSAIN, S. M. 1997. Rb–Sr ages of lamprophyre dykes from Schirmacher Oasis, Queen Maud Land, East Antarctica. *Journal of the Geological Society of India*, **50**, 457–460.

DE WIT, M. J., BOWRING, S., ASHWAL, L. D., RANDRIANASOLO, L. G., MOREL, V. P. I. & RAMBELOSON, R. A. 2001. Age and tectonic evolution of Neoproterozoic ductile shear zones in southwestern Madagascar, with implications for Gondwana studies. *Tectonics*, **20**, 1–45.

DIRKS, P. H. G. M. & WILSON, C. J. L. 1995. Crustal evolution of the East Antarctic mobile belt in Prydz Bay: continental collision at 500 Ma? *Precambrian Research*, **75**, 189–200.

DRIESSCHE, J. VAN DEN & BRUN. J.-P 1987. Rolling structures at large shear strain. *Journal of Structural Geology*, **9**, 691–704.

FITZSIMONS, I. C. W. 2000. A review of tectonic events in the East Antarctic Shield and their implications for Gondwana and earlier supercontinents. *Journal of African Earth Sciences*, **31**, 3–23.

FROST, B. R. & BUCHER, K. 1993. Charnockites and granulites of the Thor Range, Queen Maud Land, Antarctica; the granulite uncertainty principle exemplified. *Abstracts of the Annual Meeting of the GSA*, A-448.

GLÆVER, J. 1954. *The White Desert*. Chatto & Windus, London.

GOLYNSKY, A. & JACOBS. J. 2001. Grenville-age versus Pan-African magnetic anomaly imprints in western Dronning Maud Land, East Antarctica. *Journal of Geology*, **109**, 136–142.

GOSE, W. A., HELPER, M. A., CONNELLY, J. N., HUTSON, F. E. & DALZIEL, I. W. D. 1997. Paleomagnetic data and U–Pb isotopic age determinations from Coats Land, Antarctica. Implications for late Proterozoic plate reconstructions. *Journal of Geophysical Research*, **102**, 7887–7902.

GRANTHAM, G. H. & HUNTER, D. R. 1991. The geology of the western and central H.U. Sverdrupfjella, Antarctica: a Proterozoic collision zone. *Proceedings of the 6th International Symposium on Antarctic Earth Sciences*, NIPR, Tokyo, 178–183.

GRANTHAM, G. H., JACKSON, C., MOYES, A. B., GROENEWALD, P. B., HARRIS, P. D., FERRAR, G. & KRYNAUW, J. R. 1995. The tectonothermal evolution of the Kirwanveggen–H.U. Sverdrupfjellas areas, Dronning Maud Land, Antarctica. *Precambrian Research*, **75**, 200–231.

GRANTHAM, G. H., MOYES, A. B. & HUNTER, D. R. 1991. The age, petrogenesis and emplacement of the Dalmatian Granite, H.U. Sverdrupfjella, Dronning Maud Land, Antarctica. *Antarctic Science*, **3**, 197–204.

GRANTHAM, G. H., STOREY, B. C., THOMAS, R. J. & JACOBS, J. 1997. The pre-break-up position of Haag Nunataks within Gondwana. Possible correlations in Natal and Dronning Maud Land. *In*: RICCI, C. A. (ed.) *The Antarctic Region. Geological Evolution and Processes*. Terra Antartica Publication, Siena, 13–20.

GROENEWALD, P. B., GRANTHAM, G. H. & WATKEYS, M. K. 1991. Geological evidence for a Proterozoic to Mesozoic link between southeastern Africa and Dronning Maud Land, Antarctica. *Journal of the Geological Society, London*, **148**, 1115–1123.

GROENEWALD, P. B., MOYES, A. B., GRANTHAM, G. H. & KRYNAUW, J. R. 1995. East Antarctic crustal evolution. geological constraints and modelling in western Dronning Maud Land. *Precambrian Research*, **75**, 231–251.

GRUNOW, A., HANSON, R. & WILSON, T. 1996. Were aspects of Pan-African deformation linked to Iapetus opening? *Geology*, **24**, 1063–1066.

HALPERN, M. 1970. Rb/Sr date of possible three billion years for a granite rock from Antarctica. *Science*, **169**, 977–978.

HANSON, R. E., MARTIN, M. W., BOWRING, S. A. & MUNYANYIWA, H. 1998. U–Pb zircon age for the Umkondo dolerites, eastern Zimbawe: 1.1 Ga large igneous province in southern Africa–East Antarctica and possible Rodinia correlations. *Geology*, **26**, 1143–1146.

HENSEN, B. J. & ZHOU, B. 1997. East Gondwana amalgamation by Pan-African collision? Evidence from Prydz Bay, East Antarctica. *In*: RICCI, C. A. (ed.) *The Antarctic Region. Geological Evolution and Processes*. Terra Antartica Publication, Siena, 115–120.

HIROI, Y., SHIRAISHI, K. & MOTOYOSHI, Y. 1991. Late Proterozoic paired metamorphic complexes in East Antarctica, with special reference to the tectonic signifance of ultramafic rocks. *In*: THOMPSON, M. R. A., CRAME, J. A. & THOMPSON, J. W. (eds) *Geological Evolution of Antarctica*, Cambridge University Press, Cambridge, 83–87.

HOFFMAN, P. F. 1991. Did the breakout of Laurentia turn Gondwanaland inside out? *Science*, **252**, 1409–1412.

HOLMES, A. 1951. The sequence of Pre-Cambrian orogenic belts in south and central Africa. *Proceedings of the 18th International Geological Congress*, London, 1948, 254–269.

JACOBS, J. 1991. Strukturelle Entwicklung und Abkühlungsgeschichte der Heimefrontfjella (Westliches Dronning Maud Land/Antarktika). *Berichte zur Polarforschung*, **97**, 1–141.

JACOBS, J. & BAUER, W. 2001. Gjelsvikfjella and Mühlig-Hofmann–Gebirge (E-Antarctica): another piece of the East Antarctic Orogen? *Zeitschrift der deutschen geologischen Gesellschaft*, **152**, 249–259.

JACOBS, J. & THOMAS, R. J. 1996. Pan-African rejuvenation of the c. 1.1 Ga Natal Metamorphic Province (South Africa): K–Ar muscovite and titanite fission track evidence. *Journal of the Geological Society, London*, **153**, 971–974.

JACOBS, J. & THOMAS, R. J. 2002. The Mozambique Belt from an East Antarctic perspective. *Proceedings of the 1999 International Antarctic Science Meeting*, Wellington, New Zealand.

JACOBS, J. & WEBER, K. 1993. Geologische Karte (Luftbildkarte) Scharffenbergbotnen, Heimefrontfjella, Antarktis, 1:25 000. Institut für Angewandte Geodäsie, Frankfurt.

JACOBS, J., AHRENDT, H., KREUTZER, H. & WEBER, K. 1995. K–Ar, ^{40}Ar–^{39}Ar and apatite fission track evidence for Neoproterozoic and Mesozoic basement rejuvenation events in the Heimefrontfjella and Mannefallknausane (East Antarctica). *Precambrian Research*, **75**, 251–263.

JACOBS, J., BAUER, W., SPAETH, G., THOMAS, R. J. & WEBER, K. 1996. Lithology and structure of the Grenville-aged (~1.1 Ga) basement of Heimefrontfjella (East Antarctica). *Geologische Rundschau*, **85**, 800–821.

JACOBS, J., FALTER, M., WEBER, K. & JEßBERGER, E. K. 1997. ^{40}Ar–^{39}Ar evidence for the structural evolution of the Heimefront Shear Zone (Western Dronning Maud Land), East Antarctica. *In*: RICCI, C. A. (ed.) *The Antarctic Region. Geological Evolution and Processes*. Terra Antartica Publication, Siena, 37–44.

JACOBS, J., FANNING, C. M., HENJES-KUNST, F., OLESCH, M. & PAECH, H.-J. 1998. Continuation of the Mozambique Belt into East Antarctica. Grenville-age metamorphism and polyphase Pan-African high-grade events in central Dronning Maud Land. *Journal of Geology*, **106**, 385–406.

JACOBS, J., HANSEN, B. T., HENJES-KUNST, F. *ET AL*. 1999. New age constraints on the Proterozoic/Lower Palaeozoic evolution of Heimefrontfjella, East Antarctica, and its bearing on Rodinia/Gondwana correlations. *Terra Antartica*, **6**, 377–389.

JACOBS, J., THOMAS, R. J. & WEBER, K. 1993. Accretion and indentation tectonics at the southern edge of the Kaapvaal craton during the Kibaran (Grenville) orogeny. *Geology*, **21**, 203–206.

JAYARAM, J. & BEJARNIYA, B. R. 1991. *Geology of the Schirmacher–Wohlthat Region, Central Dronning Maud Land, Antarctica, 1.250000*. Geological Survey of India, Hyderabad.

JOSHI, A., PANT, N. C. & PARIMOO, M. L. 1991. Granites of the Petermann Ranges, East Antarctica, and implications on their genesis. *Journal of the Geological Society of India*, **36**, 169–181.

KAUL, M. K., SINGH, R. K., SRIVATAVA, D., JAYARAM, S. & MUKERJI, S. 1991. Petrographic and structural characteristics of a part of the East Antarctic craton, Queen Maud Land, Antarctica. *In*: THOMSON, M. R. A., CRAME, J. A. & THOMSON, J. W. (eds) *Geological Evolution of Antarctica*. Cambridge University Press, Cambridge, 89–94.

KLEINSCHMIDT, G., HELFRICH, S., HENJES-KUNST, F., JACKSON, C. & FRIMMEL, H. E. 1996. The pre-Permo-Carboniferous rocks and structures from southern Kirwanveggen, Dronning Maud Land, Antarctica. *Polarforschung*, **66**, 7–18.

KLIMOV, L. V., RAVICH, M. G. & SOLOVIEV, D. S. 1964. Charnockites of East Antarctica. *In*: ADIE, R. J. (ed.) *Antarctic Geology*. North Holland Publishing, Amsterdam, 455–462.

KRIEGSMANN, L. M. 1995. The Pan-African event in East Antarctica. a view from Sri Lanka and the Mozambique Belt. *Precambrian Research*, **75**, 263–279.

KRYNAUW, J. R., WATTERS, B. R., HUNTER, D. R. & WISLON, A. H. 1991. A review of the field relations, petrology and geochemistry of the Borgmassivet intrusions in the Grunehogna Province, western Dronning Maud Land, Antarctica. *In*: THOMSON, M. R. A., CRAME, J. A. & THOMSON, J. W. (eds) *Geological Evolution of Antarctica*. Cambridge University Press, Cambridge, 33–39.

LAWVER, L. A. & SCOTESE, C. R. 1987. A revised reconstruction of Gondwanaland. *In*: MCKENZIE, G. D. (ed.) *Gondwana Six. Structure, Tectonics and Geophysics*. American Geophysical Union, 17–23.

MARKL, G., PIAZOLO, S., BAUER, W., KRAUß, U. & PAECH, H.-J. 2003. Panafrican massif-type anorthosites from Central Dronning Maud Land, Antarctica. *Geologisches Jahrbuch*, **B96**, in press.

MARTELAT, J.-E., NICOLLET, C., LARDEAUX, J.-M., VIDAL, G. & RAKOTONDRAZAFY, R. 1997. Lithospheric tectonic structures developed under high-grade metamorphism in the southern part of Madagascar. *Geodinamica Acta*, **10**, 94–114.

MARTIN, A. K. & HARTNADY, C. J. H. 1986. Plate tectonic development of the south west Indian Ocean. A revised reconstruction of east Antarctica and Africa. *Journal of Geophysical Research*, **91**, 4767–4778.

MATTHEWS, P. E. 1972. Possible Precambrian obduction and plate tectonics in southeastern Africa. *Nature*, **240**, 37–39.

MEERT, J. G. 2001. Growing Gondwana and rethinking Rodinia: a paleomagnetic perspective. *Gondwana Research*, **4**, 279–288.

MEZGER, K. & COSCA, M. A. 1999. The thermal history of the eastern Ghats Belt (India) as revealed by U–Pb and ^{40}Ar/^{39}Ar dating of metamorphic and magmatic minerals: implications for the SWEAT correlation. *Precambrian Research*, **94**, 251–271.

MIKHALSKY, E. V., BELIATSKY, B. V., SAVVA, E. V., WETZEL, H.-U., FEDOROV, L. V., WEISER, T. & HAHNE, K. 1997. Reconnaissance geochronologic data on polymetamorphic and igneous rocks of the Humboldt Mountains, Central Queen Maud Land, East Antarctica. *In*: RICCI, C. A. (ed.) *The Antarctic Region. Geological Evolution and Processes*. Terra Antartica Publication, Siena, 45–54.

MOYES, A. B. & BARTON, J. M. 1990. A review of isotopic data from western Dronning Maud Land, Antarctica. *Zentralblatt Geologie Paläontologie, Teil 1*, **1/2**, 19–31.

MOYES, A. B., GROENEWALD, P. B. & BROWN, R. W. 1993. Isotopic constraints on the age and origin of the Brattskarvet intrusive suite, Dronning Maud Land, Antarctica. *Chemical Geology (Isotope Geoscience Section)*, **106**, 453–466.

MOYES, A. B., KNOPER, M. W. & HARRIS, P. D. 1997. The age and signifance of the Urfjell Group, Western Dronning Maud Land. *In*: RICCI, C. A. (ed.) *The Antarctic Region. Geological Evolution and Processes*. Terra Antartica Publication, Siena, 31–36.

MOYES, A. B., KRYNAUW, J. R. & BARTON, J. M. 1995. The age of the Ritscherflya Supergroup and Borgmassivet Intrusions, Dronning Maud Land, Antarctica. *Antarctic Science*, **7**, 87–97.

OHTA, Y. 1993. *Nature environment map, Gjelsvikfjella and Western Mühlig-Hofmannfjella, Dronning Maud Land, Antarctica, 1:100000. Sheet 1 and 2*. Norsk Polarinstitutt Temakart nr. 24, Oslo.

OHTA, Y., GROENEWALD, P. B. & GRANTHAM, G. H. 1996. *Nature environment map H.U. Sverdrupfjella 1:150000*. Norsk Polarinstitutt Temakart nr. 28, Oslo.

OHTA, Y., TØRUDBAKKEN, B. O. & SHIRAISHI, K. 1990.

Geology of Gjelsvikfjella and western Mühlig-Hofmannfjella, Dronning Maud Land, east Antarctica. *Polar Research*, **8**, 99–126.

PAECH, H.-J., DAMASKE, D., DAMM, V. ET AL. 1999. Geological, geophysical and glaciological mapping results recently obtained in central Dronning Maud Land, East Antarctica. *In*: SKINNER, D. N. B. (ed.) *Programme & Abstracts of the 8th International Symposium on Antarctic Earth Science*, 234.

PASTEELS, P. & MICHOT, J. 1970. Uranium–lead radioactive dating and lead isotope study on sphene and K-feldspar in the Sør Rondane Mountains, Dronning Maud Land, Antarctica. *Eclogae Geologicae Helvetiae*, **63**, 239–254.

PETERS, M., HAVERKAMP, B., EMMERMANN, R., KOHEN, H. & WEBER, K. 1991. Palaeomagnetism, K–Ar dating and geodynamic setting of igneous rocks in western and central Neuschwabenland, Antarctica. *In*: THOMPSON, M. R. A., CRAME, J. A. & THOMPSON, J. W. (eds) *Geological Evolution of Antarctica*. Cambridge University Press, Cambridge, 549–555.

PIAZOLO, S. & MARKL, G. 1999. Humite- and scapolite-bearing assemblages in marbles and calcsilicates of Dronning Maud Land, Antarctica: new data for Gondwana reconstructions. *Journal of Metamorphic Geology*, **17**, 91–107.

PINNA, P. 1995. On the dual nature of the Mozambique Belt, Mozambique to Kenya. *Journal of African Earth Sciences*, **21**, 477–480.

PINNA, P., JOURDE, G., CALUEZ, J.-Y., MROZ, J. P. & MARQUES, J. M. 1993. The Mozambique Belt in northern Mozambique. Neoproterozoic (1100–850 Ma) crustal growth and tectogenesis and superimposed Pan-African (800–550 Ma) tectonism. *Precambrian Research*, **62**, 1–59.

POULSSON, O., AUSTRHEIM, H. & OHTA, Y. 1999. Pan-African late orogenic intrusions in Jutulsessen, western Dronning Maud Land, Antarctica. *Journal of Conference Abstracts*, **4**, 118.

RAVICH, M. G. & KAMENEV, E. N. 1975. *Crystalline Basement of the Antarctic Platform*. Wiley, New York.

RAVICH, M. G. & KRYLOV, A. Y. 1964. Absolute ages of rocks from East Antarctica. *In*: ADIE, R. J. (ed.) *Antarctic Geology*. North Holland Publ., Amsterdam, 590–596.

RAVICH, M. G. & SOLOVIEV, D. S. 1966. Geologiya i petrologiya central'noj chasti gor zemli Korolevy Mod [Geology and petrology of the central part of the mountains of central Dronning Maud]. *Trudy Naucno-Isseldovatel'skogo Instituta Geologii Arktiki*, **141**.

ROLAND, N. 1999. Pan-African granitoids in central Dronning Maud Land, East Antarctica. petrography, geochemistry and their plate tectonic setting. *In*: SKINNER, D. N. B. (ed.) *Programme & Abstracts of the 8th International Symposium on Antarctic Earth Science*, 272.

ROOTS, E. F. 1953. Preliminary note on the geology of western Dronning Maud Land. *Norsk Geologisk Tidsskrift*, **32**, 17–33.

SCHULZE, P. 1992. Petrogenese des metamorphen Grundgebirges der zentralen Heimefrontfjella (westliches Dronning Maud Land/Antarktis). *Berichte zur Polarforschung*, **117**, 1–321.

SENGUPTA, S. 1993. Tectonothermal history recorded in mafic dykes and enclaves of gneissic basement in the Schirmacher Hills, East Antarctica. *Precambrian Research*, **63**, 273–291.

SHACKLETON, R. M. 1996. The final collision zone between East and West Gondwana. Where is it? *Journal of African Earth Sciences*, **23**, 271–287.

SHERATON, J. W., TINDLE, A. G. & TINGEY, R. J. 1996. Geochemistry, origin and tectonic settings of granitic rocks of the Prince Charles Mountains, Antarctica. *AGSO Journal of Australian Geology and Geophysics*, **16**, 345–370.

SHIRAISHI, K. & KAGAMI, H. 1992. Sm–Nd and Rb–Sr ages of metamorphic rocks from the Sør Rondane Mountains, East Antarctica. *In*: YOSHIDA, Y., KAMINUMA, K. & SHIRAISHI, K. (eds) *Recent Progress in Antarctic Earth Science*. Terra Scientific Publishing Company, Tokyo, 29–35.

SHIRAISHI, K., ASAMI, M., ISHIZUKA, H. ET AL. 1991. Geology and metamorphism of the Sør Rondane Mountains, East Antarctica. *In*: THOMSON, M. R. A., CRAME, J. A. & THOMSON, J. W. (eds) *Geological Evolution of Antarctica*. Cambridge University Press, Cambridge, 77–82.

SHIRAISHI, K., ELLIS, D. J., FANNING, C. M., HIROI, Y., KAGAMI, H. & MOTOYOSHI, Y. 1997. Re-examination of the metamorphic and protolith ages of Rayner Complex, Antarctica: Evidence for the Cambrian (Pan-African) regional metamorphic event. *In*: RICCI, C. A. (ed.) *The Antarctic Region. Geological Evolution and Processes*. Terra Antartica Publication, Siena, 79–88.

SHIRAISHI, K., ELLIS, D. J., HIROI, Y., FANNING, C. M., MOTOYOSHI, Y. & NAKAI, Y. 1994. Cambrian orogenic belt in east Antarctica and Sri Lanka. Implications for Gondwana assembly. *Journal of Geology*, **102**, 47–65.

SPAETH, G. & FIELITZ, W. 1987. Structural investigations in the Precambrian of Western Neuschwabenland, Antarctica. *Polarforschung*, **57**, 71–92.

STERN, R. J. 1994. Arc assembly and continental collision in the Neoproterozoic East African Orogen. Implications for the consolidation of Gondwana. *Review Earth and Planetary Science*, **22**, 319–351.

STÜWE, K. & SANDIFORD, M. 1993. A preliminary model for the 500 Ma event in the East Antarctic Shield. *In*: FINDLAY, R. H., UNRUG, R., BANKS, M. R. & VEEVERS, J. J. (eds) *Gondwana Eight. Assembly, Evolution and Dispersion*. Balkema, Rotterdam, 125–130.

TALARICO, F., KLEINSCHMIDT, G. & HENJES-KUNST, F. 1999. An ophiolite complex in the northern Shackleton Range, Antarctica. EUG 10, Strasbourg, *Journal of Conference Abstracts*, **4**, 122.

TESSENSOHN, F., KLEINSCHMIDT, G., TALARICO, F. ET AL. 1999. Ross amalgamation of East and West Gondwana: evidence from the Shackleton Range, Antarctica. *Terra Antartica*, **6**, 317–325.

THOMAS, R. J. 1989*a*. A tale of two tectonic terranes. *South African Journal Geology*, **92**, 306–321.

THOMAS, R. J. 1989*b*. The petrogenesis of the Mzumbe Gneiss Suite, a tonalite–trondhjemite orthogneiss suite from the southern part of the Natal Structural and Metamorphic Province. *South African Journal of Geology*, **92**, 322–338.

THOMAS, R. J. & EGLINGTON, B. M. 1990. A Rb–Sr, Sm–Nd and U–Pb zircon isotopic study of the Mzumbe Suite, the oldest intrusive granitoid in southern Natal, South Africa. *South African Journal of Geology*, **93**, 761–765.

THOMAS, R. J., AGENBACHT A. L. D., CORNELL, D. H. & MOORE, J. M. 1994. The Kibaran of southern Africa: tectonic evolution and metallogeny. *Ore Geology Reviews*, **9**, 131–160.

VAN AUTENBOER, T. & LOY, W. 1972. Recent geological investigations in the Sør Rondane mountains, Belgicafjella and Sverdrupfjella, DML. *In*: ADIE, R. J. (ed.) *Antarctic Geology and Geophysics*. Universitetsforlaget, Oslo, IUGS, **B2**, 563–572.

WAND, U., BECKER, S. & KAISER, G. 1988. Zur Altersstellung und Geochemie der Basaltgänge in der Schirmacheroase, Dronning Maud Land, Ostantarktika. *Freiberger Forschungshefte*, **C421**, 41–64.

WILSON, T. J., GRUNOW, A. M. & HANSON, R. E. 1997. Gondwana Assembly. The view from southern Africa and East Gondwana. *Journal of Geodynamics*, **23**, 263–286.

WOLMARANS, L. G. & KENT, K. E. 1982. Geological investigations in western Dronning Maud Land, Antarctica – a synthesis. *South African Journal of Antarctic Research, Supplement*, **2**.

WOODCOCK, N. H. & SCHUBERT, C. 1994. Continental strike-slip tectonics. *In*: HANCOCK, P. L. (ed.) *Continental Deformation*. Pergamon Press, Oxford, 251–263.

YOSHIDA, M. 1994. Tectonothermal history and tectonics of Lützow-Holm Bay area, East Antarctica: a re-interpretation. *Journal of Geological Society of Sri Lanka*, **5**, 81–93.

YOSHIDA, M. 1995. Cambrian orogenic belt in East Antarctica and Sri Lanka: implications for Gondwana assembly: a discussion. *Journal of Geology*, **103**, 467–468.

YOSHIDA, M. 1995. Assembly of East Gondwanaland during the Mesoproterozoic and its rejuvenation during the Pan-African period. *Memoir, Geological Society of India*, **34**, 25–45.

YOSHIDA, M. & VINTANAGE, P. W. 1993. A review of the Precambrian geology of Sri Lanka and its comparison with Antarctica. *In*: FINDLEY, R. H., UNRUG, R., BANKS, M. R. & VEEVERS J. J. (eds) *Gondwana Eight. Assembly, Evolution and Dispersion*. Balkema, Rotterdam, 97–109.

YOSHIDA, M., SANTOSH, M. & SHIRAHATA, H. 1991. Geochemistry of gneiss–granulite transformation in the 'incipient charnockite' zone of southern India. *Mineralogy and Petrology*, **45**, 69–83.

Extensional collapse of the late Neoproterozoic–early Palaeozoic East African–Antarctic Orogen in central Dronning Maud Land, East Antarctica

J. JACOBS[1], R. KLEMD[2], C. M. FANNING[3], W. BAUER[4] & F. COLOMBO[5]

[1]*Fachbereich Geowissenschaften, Universität Bremen, PF 330440, 28334 Bremen, Germany, (e-mail: jojacobs@uni-bremen.de)*
[2]*Institut für Mineralogie, Universität Würzburg, Am Hubland, 97074 Würzburg, Germany*
[3]*The Australian National University, Canberra, ACT 02000, Australia*
[4]*Geologisches Institut der RWTH Aachen, Wüllnerstr. 2, 52056 Aachen*
[5]*Dipartimento di Scienze della Terra, Universitá di Siena, Via del Laterino 8, 53100 Siena, Italy*

Abstract: The East African–Antarctica Orogen resulted from the continent–continent collision of East and West Gondwana, or parts thereof, during the Pan-African event at $c.$ 650–510 Ma. The collision overprinted large areas of older, mainly Mesoproterozoic, crust up to granulite facies grade in East Antarctica. The collision history is well documented by folding and thrusting, isothermal decompression and metamorphic zircon growth at $c.$ 580–560 Ma (Pan-African I). The convergence was succeeded by an extensional phase, probably representing orogenic collapse. This Pan-African II event at $c.$ 530–510 Ma is characterized by large-scale extensional structures, finally resulting in the post-tectonic intrusion of voluminous A2–type granitoids. In central Dronning Maud Land the Pan-African II event started with the intrusion of syntectonic igneous rocks within an overall extensional setting. Two new SHRIMP data from gabbro zircons of the Zwiesel Gabbro give ages of 521 ± 5.6 and 527 ± 5.1 Ma. These ages are interpreted as crystallization ages and confirm the interpretation that the gabbro was emplaced early during the Pan-African II event. The gabbro was intruded by a network of leucogranite dykes and veins. Whereas the gabbro appears entirely undeformed, the leucogranite dykes are strongly mylonitized along extensional shear zones, indicating pronounced strain partitioning of the gabbro complex. Within the leucogranite mylonites, large tension gashes developed during mylonitization, indicating very high strain rates. Quartz c-axis orientations from quartz of the tension gashes show a distinct cross-girdle that formed during pure shear deformation. Fluid inclusion data from the leucogranite mylonites and the associated tension gashes mainly reveal recrystallization-related intracrystalline CO_2-dominant inclusions with relatively low densities of $<1\,\mathrm{g\,cm^{-3}}$. The fluid inclusion data are interpreted to represent the last stages of a retrograde P–T path that is characterized by simultaneous cooling and decompression during extensional exhumation, probably succeeding the collapse of overthickened crust. A comparable orogenic collapse of the East African–Antarctic Orogen is reported from other parts of the orogen, such as from western Madagascar and the northern Arabian–Nubian Shield.

The Late Neoproterozoic–Early Palaeozoic tectonometamorphic overprint of Archaean–Mesoproterozoic crust in eastern Africa, southern India and East Antarctica have been interpreted to represent the now dismembert fragments of an once continuous orogenic belt. This orogenic belt is thought to have formed as a consequence of the continent–continent collision between East and West Gondwana that resulted in a Himalayan-type orogen, the East African–Antarctic Orogen (Stern 1994; Shackleton 1996; Jacobs et al. 1998; Jacobs & Thomas 2002). However, the outline, exact extent and the position of the suture of the East African–Antarctic Orogen has been a matter of debate (e.g. Stern 1994; Grunow et al. 1996; Wilson et al. 1997; Tessensohn et al. 1999; Jacobs & Thomas 2002; Yoshida et al. 2002). One of the important issues is whether the main suture of the East African Orogen runs N–S along eastern Africa, continuing in Dronning Maud Land, Lützow-Holm Bay and the Shackleton Range (East Antarctica), or whether the main suture has an entirely different location, such as the Zambesi Belt (e.g. Yoshida et al. 2002).

However, palaeomagnetic data clearly indicate that the Grunehogna Craton and Coats Land in East Antarctica were part of West Gondwana and that there must be a Late Neoproterozoic suture east of these areas. Thus, the interpretation that the East Antarctic Orogen forms the southern continuation of the East African Orogen into East Antarctica, to form the combined up to 8000 km long and 1000 km wide East African–Antarctic Orogen, is favoured by the present authors. In Dronning Maud Land and Lützow-Holm Bay the East African–Antarctic Orogen has led to pervasive high-grade overprint of

From YOSHIDA, M., WINDLEY, B. F. & DASGUPTA, S. (eds) 2003. *Proterozoic East Gondwana: Supercontinent Assembly and Breakup.* Geological Society, London, Special Publications, **206**, 271–287. 0305–8719/03/$15 © The Geological Society of London.

older, mainly Grenville-age (c. 1.1 Ga) crust (Shiraishi et al. 1992, 1994; Jacobs et al. 1998). Relicts of the Mozambique Ocean have been identified in the Shackleton Range (e.g. Tessensohn et al. 1999). The collision is documented by medium-pressure (P) granulites at c. 580–550 Ma (Pan-African I) and was succeeded by a low-P granulite facies event (Pan-African II) at c. 530–515 Ma (Jacobs et al. 1998; Colombo & Talarico 2003). In this article, the extensional deformation that post-dates the main collision event is outlined. New SHRIMP U–Pb zircon data on a gabbro intrusion from central Dronning Maud Land are presented (Fig. 2), demonstrating the nature of Pan-African extensional processes. Also presented are structural and fluid inclusion data on leucogranite dykes and veins cross-cutting the gabbro to further constrain the Early Palaeozoic extension that dominates the late tectonic evolution of the East Antarctic Orogen. Finally, the late orogenic extensional processes in adjacent areas of the East African–Antarctic Orogen, are briefly reviewed.

Geological setting

First reconnaissance geological investigations of the crystalline basement in central Dronning Maud Land were conducted by Soviet geologists in the early 1960s (Ravich & Solov'ev 1966). The first major regional study of the basement geology of central Dronning Maud Land was carried out during the international GeoMaud 1995/96 expedition (Paech et al. 2003). The study area is situated in the eastern portion of central Dronning Maud Land, within the Petermannketten. The study area comprises a Mesoproterozoic (c. 1.2–1.05 Ga), dominantly juvenile, basement that is composed of an oldest metavolcanic sequence intercalated with sedimentary rocks. It has been suggested that this Mesoproterozoic crust is part of a several thousand kilometre long mobile belt that rims the Zimbabwe–Kaapvaal–Grunehogna Craton (e.g. Jacobs et al. 1998) (Fig. 1). This mobile belt was intensely overprinted during the collision of East and West Gondwana. The overprint reached granulite facies grade in central Dronning Maud Land. Time constraints on this complex tectonometamorphic evolution were given by Jacobs et al. (1998), who reported 12 U–Pb SHRIMP zircon ages, as briefly summarized in Table 1.

The oldest rocks identified are metavolcanic rocks that are intercalated with quartzites, marble, paragneisses and metapelites. The metavolcanics are mainly felsic gneisses, with intercalations of amphibolites. Zircons from these felsic gneisses gave crystallization ages of c. 1130 Ma. Some of the magmatic zircons from the supracrustal gneisses show metamorphic zircon overgrowth at c. 1080 Ma. Relict structures of this Grenville-age event are a first foliation in mafic enclaves, rootless isoclinal folds and boudinage of mafic bodies (Bauer et al. 2003). This first tectonometamorphic event is contemporaneous with the intrusion of voluminous granitoids, which were therefore interpreted to be syntectonic (D1, M1; Jacobs et al. 1998). The $P–T$ conditions of this first metamorphic event are preserved in low-strain mafic boudins (M1; Colombo & Talarico 2003).

There is no record of igneous or metamorphic rocks for the time interval between c. 1050 and 600 Ma. At 600 Ma, the metamorphic basement of central Dronning Maud Land was intruded by a major anorthosite body (Gruber Anorthosite, Fig. 2). Large magmatic zircons from the relatively undeformed core of the anorthosite body gave zircon crystallization ages of c. 600 Ma. The margins of the anorthosite and the surrounding basement was strongly deformed at c. 580–560 Ma during a first Pan-African overprint, which was associated with medium-P granulite facies metamorphism ($P > 7$ kbar and $T > 730°C$), a clockwise $P–T$ path and was interpreted to represent the collision stage (D2, M2) (Jacobs et al. 1998; Colombo & Talarico 2003). A second Pan-African overprint was recorded at c. 530–515 Ma (related to D3, M3). This Pan-African II event is associated with low-P granulite facies metamorphism and the intrusion of a number of small granitoids, such as a metagranodiorite at Conradgebirge and a number of leucogranite sheets and veins (Fig. 2). These metagranitoids are usually strongly deformed into gneisses or mylonites. They are aligned in a general N–S-trending regional foliation and often contain monophase sector-zoned zircons, indicative of the high-grade conditions of M3 (Jacobs et al. 1998; Jacobs 1999). Whilst the Pan-African I deformation is clearly associated with a contractional structural regime, this is not evident for the Pan-African II deformation. The Pan-African II event is associated with steeply to moderately inclined, generally N–S-trending (African coordinates used throughout) and up to 2 m wide, extensional shear zones and the formation of quartz- feldspar mobilizates (e.g. Bauer et al. 2003). Fine-grained kelyphitic coronas of Opx-Cpx-Plag ± spinel around garnet and symplectic aggregates of Cpx-Plag indicate near-isothermal decompression of M3 from high-grade conditions (Colombo & Talarico 2003). The latest Pan-African II event is characterized by the intrusion of voluminous post-tectonic A2-type granitoids including charnockites, biotite–granites and an anorthosite body (Roland 1999) at c. 510 Ma (Mikhalsky et al. 1997), and the hydration of high- to medium-grade mineral paragenesis. These lower crustal granitoids could be the result of large-scale asthenospheric heating of the lower crust caused by the delamina-

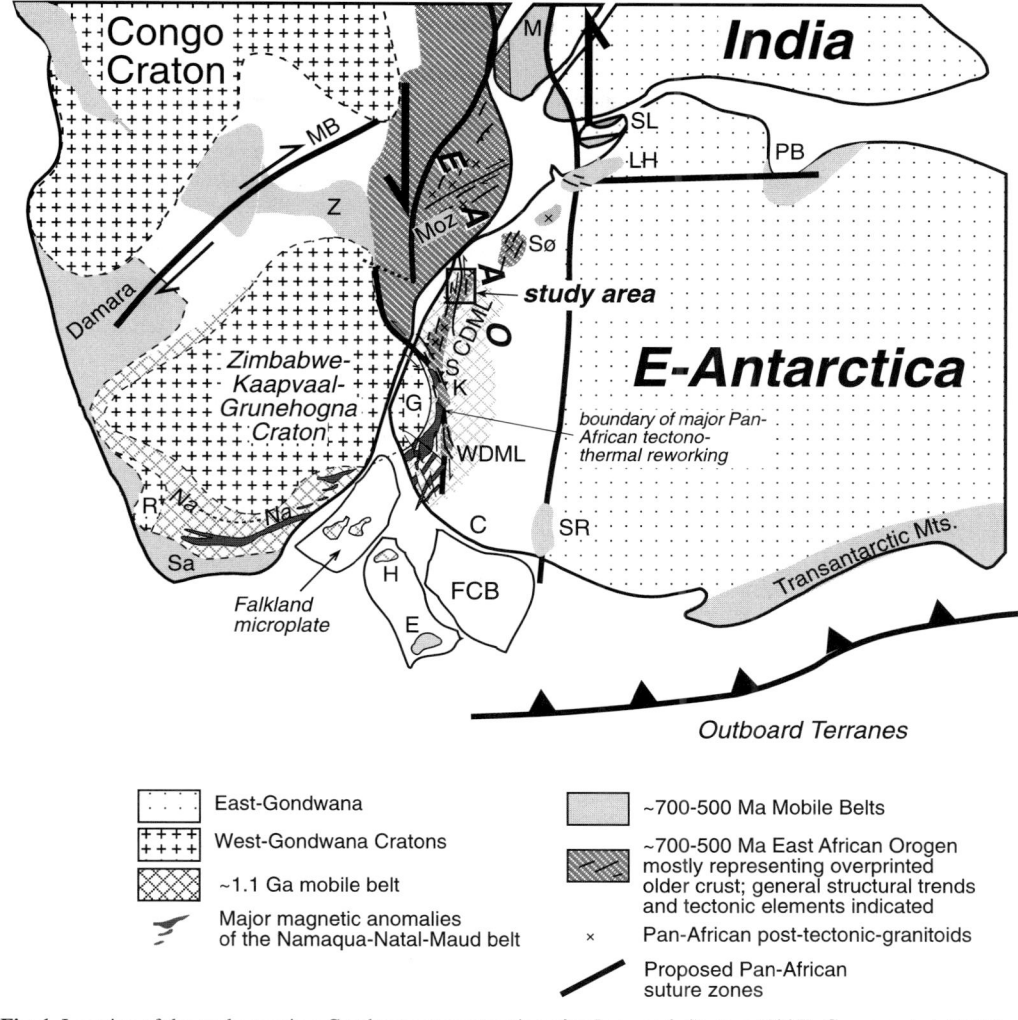

Fig. 1. Location of the study area in a Gondwana reconstruction after Lawver & Scotese (1987), Grunow et al. (1996) and Shackleton (1996), with major Pan-African belts indicated. African geology after Hartnady et al. (1985) and Porada (1989). Central Dronning Maud Land forms the immediate southern continuation of the Mozambique Belt. Possible positions of Mozambique suture after Shackleton (1996) and Grunow et al. (1996). C, Coats Land; CDML, central Dronning Maud Land; EAAO, East African–Antarctic Orogen; EM, Ellsworth Mountains; FCB, Filchner Crustal Block; G, Grunehogna Archean cratonic fragment; H, Haag Nunatak; HF, Heimefrontfjella; K, Kirwanveggen; LH, Lützow–Holm Bay; MB, Mwembeshi Shear Zone; Na–Na, Namaqua-Natal Belt; PB, Prydz Bay; R, Richtersveld Craton; M, Madagascar; Moz, Mozambique Belt; S, Sverdrupfjella; Sa, Saldania Belt; SL, Sri Lanka; Sø, Sør Rondane; SR, Shackleton Range; WDML, western Dronning Maud Land; Z, Zambezi Belt.

tion of an overthickened orogenic crust (Roland 1999).

A different interpretation of the metamorphic evolution was presented by Piazolo & Markl (1998), who interpreted their P–T data on calc-silicates and marbles to indicate a single long-lasting (100 Ma) thermal event without significant deformation. Piazolo & Markl (1998) have recorded significantly higher temperatures for M2, resulting in an overall P–T path characterized by long-lasting simultaneous cooling and decompression. Unfortunately, this study has entirely neglected the protracted Late Neoproterozoic–Early Palaeozoic structural evolution of the area, and has not taken into account that T

Table 1. *Summary of U–Pb SHRIMP zircon ages from central Dronning Maud Land (after Jacobs et al. 1998)*

	Inheritance (Ma)	Magmatic crystallization age (Ma)	Metamorphic age (Ma)	Interpretation	
Post-tectonic granitoids		$512 \pm 2^*$		Post-tectonic magmatism	Pan-African
Felsic leucosome			516 ± 5	Syntectonic magmatism; D3, M3, c. 530–515 Ma	
Leucogranite	c. 1170	527 ± 6			
Granodiorite		530 ± 8	530 ± 8		
Anorthosite (strongly deformed)			583 ± 7	D2, M2, c. 580–560 Ma	
Anorthosite (no mesoscopic deformation)		600 ± 12	555 ± 11	AMCG magmatism, c. 600 Ma	
Charnockite		608 ± 9	544 ± 15		
Felsic gneiss (stratoid granite)	c. 1190	1076 ± 14	557 ± 11a	Syntectonic granitoids; D1, M1, c. 1180 Ma	Grenvillian
Felsic gneiss (stratoid granite)		1073 ± 9			
Syntectonic granite (augen gneiss)		1086 ± 20	c. 570		
Syntectonic granite (augen gneiss)		1087 ± 28			
Felsic metavolcanic rock		1130 ± 12	c. 575	Felsic volcanism, c. 1130 Ma	
Felsic metavolcanic rock		1137 ± 21	1084 ± 8; 522 ± 10		

*Conventional U–Pb zircon age of Mikhalsky *et al.* (1997).

and the geothermal gradient could have significantly been influenced by the various syntectonic intrusions.

Zwiesel Gabbro Complex

Zwiesel Gabbro Complex is a pluton c. 10 km in diameter in the central Petermannketten (Fig. 2). It consists of a medium- to coarse-grained gabbro and a network of metaleucogranite dyke intrusions (Fig. 3). The gabbro is composed of plagioclase (varying in composition from $Ab_{51}An_{49}$ to $Ab_{67}An_{31}Or_{02}$) and clinopyroxenes consisting of augitic pyroxene lamellae ($Wo_{45}En_{20}Fs_{35}$) hosted in a pigeonite ($Wo_{13}En_{23}Fs_{64}$). Minor amounts of opaque minerals (relict magnetite and ilmenite), amphiboles (cummingtonite and ferrotschermakite), orthopyroxene, apatite, orthoclase (as antiperthitic exsolution in plagioclase), zircon and rare olivine relicts occur. In some more retrogressed parts of the gabbro, biotite is present and increasing amounts of amphiboles are found. The retrogressed parts show limited solid-state deformation structures, such as undulose extintion, tapering edges, deformation lamellae and local subgrain crystallization. Whilst the gabbro mostly appears mesoscopically undeformed, the leucogranite dykes are strongly deformed and are often transformed to leucogranite mylonites (Fig. 3).

The (meta-)leucogranites occur as a network of up to 50 m wide sheets and dykes within the gabbro. Deformation of the Zwiesel Gabbro Complex is entirely taken up by the leucogranite veins, transforming them to leucogranite mylonites. The leucogranite mylonites are LS-tectonites with a straight downdip to oblique stretching lineation and a well-developed mylonitic foliation. Unfortunately, the Zwieselberg forms steep cliffs and thus is not easily accessible, preventing the documentation of a three-dimensional shear pattern of the leucogranite dyke network. However, the leucogranite mylonites show a general N–S trend and dip both easterly and westerly. Various shear sense indicators show a uniform extensional sense of shear, which in general is characteristic for the Pan-African II event.

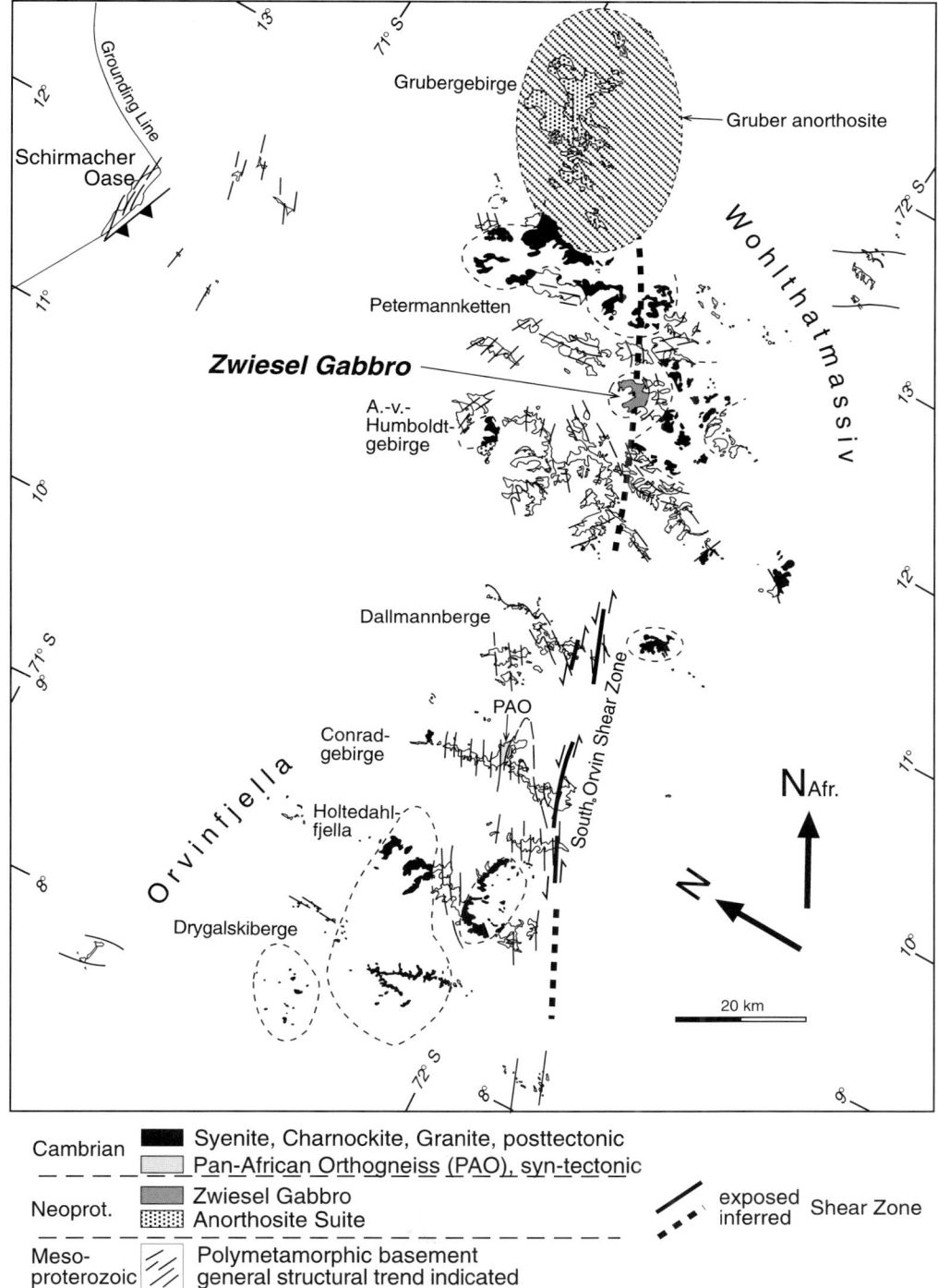

Fig. 2. Geological overview map of central Dronning Maud Land and location of the Zwiesel Gabbro Complex (after Jacobs & Thomas 2002).

Fig. 3. Zwiesel Gabbro Complex. (**a**) The Zwiesel Gabbro Complex is characterized by, in part, major leucogranite dykes. (**b**) Largely undeformed medium- and coarse-grained gabbro association. (**c**) Leucogranite mylonite in which tension gashes have developed (filled by quartz), probably at very high strain rates. (**d**) Leucogranite mylonite with quartz bands parallel to mylonitic foliation (probably representing an older generation of tension gashes that were rotated into the mylonitic foliation); also present are newly formed tension gashes, filled by quartz (shear sense is top-to-the-left).

U–Pb SHRIMP geochronology of Zwiesel Gabbro Complex

Zircons from two gabbro samples were analysed by SHRIMP in order to constrain the time of gabbro emplacement and to provide a maximum age for the extension along the leucogranite dykes and veins. Zircons were separated from two gabbro samples using standard separation techniques. Ion microprobe analyses of these zircons were carried out using SHRIMP II at the Research School of Earth Sciences, The Australian National University, Australia. SHRIMP analytical methods follow those given by Williams (1998 and refs cited therein). Prior to analysis, cathodoluminescence images (CL) of the sectioned zircon grains were taken so that specific areas within grains could be analysed. The analyses consist of six scans through the mass range using a spot size of $c.$ 20 µm in diameter. The U/Pb ratios have been calibrated relative to 1099 Ma Duluth Gabbro reference zircons (see Paces & Miller 1993) and the data reduced using the SQUID Excel Macro of Ludwig (2000). Common Pb has been corrected using the measured $^{207}Pb/^{206}Pb$ and $^{238}U/^{206}Pb$ ratios, following Tera & Wasserburg (1972) and described in detail by Compston et al. (1992). Uncertainties in the measured ratios are given at the one sigma (σ) level; however, weighted mean age uncertainties are given at the 95% confidence level (plots and calculation using ISOPLOT/EX; Ludwig 1999).

The zircons in the two samples are small clear crystals of up to $c.$ 200 µm in diameter. Grain boundaries are usually irregular. Internally, CL images indicate irregular zoning, similar to sector zoning, probably indicating rapid grain growth.

For sample J1817, 21 areas were analysed in 16 zircons. The analyses are low in common Pb and on a relative probability plot the radiogenic $^{206}Pb-^{238}U$ ages form a simple bell curve, with a slight skew towards the younger age end. A weighted mean of all analyses has an MSWD of 2.5, but if the two slightly younger analyses of grain 12 are excluded, together with the slightly older analysis of grain 16, the weighted mean $^{206}Pb-^{238}U$ age is 521.0±5.6 Ma, with a MSWD of 1.6 (Fig. 4a; Table 2). For sample J1818, 25 areas were analysed in 22 zircons. Here, the relative probability plot of the $^{206}Pb-^{238}U$ ages shows a clear skew to the younger end and several analyses are interpreted to have lost radiogenic Pb (analyses 12.1, 16.1 and 10.1); further, analysis 9.2 is significantly older at $c.$ 555 Ma. A weighted mean of the remaining 21 analyses gives 527.1±5.1 Ma (MSWD = 1.3) (Fig. 4b; Table 3). The two ages are within error and are interpreted as the crystallization age of the magmatic zircon and enclosing gabbro.

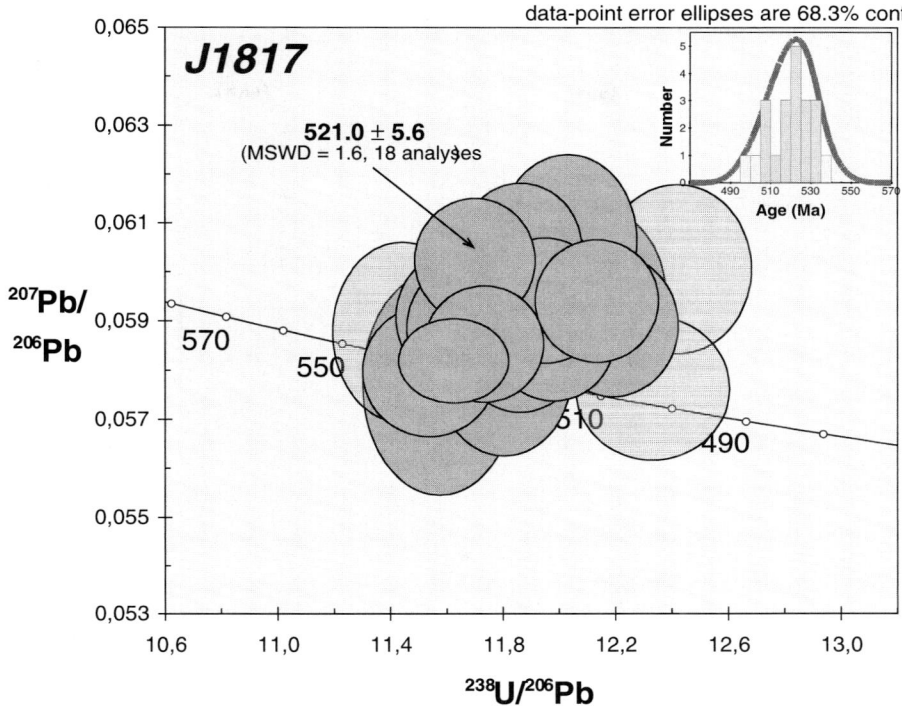

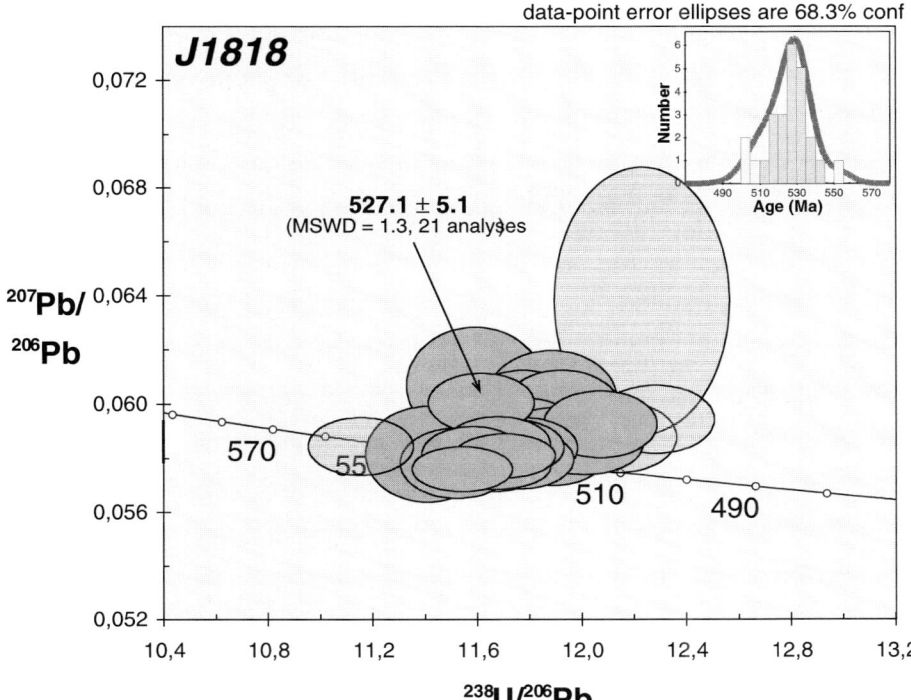

Fig. 4. U–Pb SHRIMP zircon analyses of the Zwiesel Gabbro Complex samples J1817 and J1818.

Table 2. Summary of SHRIMP U–Pb zircon results for sample J1817

Grain spot	U* (ppm)	Th (ppm)	Th/U	Pb[†] (ppm)	$^{204}Pb/^{206}Pb$	f_{206}[‡] (%)	$^{238}U/^{206}Pb$	±	$^{207}Pb/^{206}Pb$	±	$^{206}Pb/^{238}U$	±	Age (Ma) $^{206}Pb/^{238}U$	±
							Total[§]				Radiogenic			
1.1	117	68	0.60	8.5	0.000448	0.23	11.838	0.156	0.0596	0.0010	0.0843	0.0011	521.7	6.7
1.2	170	93	0.81	12.5	0.000172	0.14	11.642	0.147	0.0591	0.0009	0.0858	0.0011	530.5	6.6
2.1	91	40	0.57	6.6	0.000492	0.02	11.802	0.157	0.0580	0.0011	0.0847	0.0012	524.2	6.8
3.1	158	136	0.46	11.5	0.000038	0.35	11.862	0.143	0.0606	0.0008	0.0840	0.0010	520.0	6.1
3.2	74	34	0.90	5.5	0.000329	<0.01	11.573	0.179	0.0578	0.0015	0.0864	0.0014	534.4	8.1
4.1	161	143	0.89	11.6	0.000213	0.22	11.948	0.144	0.0595	0.0008	0.0835	0.0010	517.0	6.1
5.1	163	144	0.48	12.0	0.000110	0.29	11.701	0.141	0.0603	0.0008	0.0852	0.0010	527.2	6.2
6.1	81	40	0.91	5.8	0.000165	0.26	12.122	0.167	0.0596	0.0012	0.0823	0.0012	509.7	6.9
7.1	142	104	0.91	10.2	0.000353	0.23	11.962	0.148	0.0595	0.0008	0.0834	0.0011	516.4	6.2
8.1	186	164	0.51	13.6	—	0.09	11.732	0.139	0.0586	0.0008	0.0852	0.0010	526.9	6.1
9.1	155	127	0.75	11.1	0.000528	0.14	11.978	0.145	0.0588	0.0008	0.0834	0.0010	516.2	6.1
10.1	173	144	0.90	12.2	0.000415	0.25	12.131	0.145	0.0595	0.0008	0.0822	0.0010	509.4	6.0
11.1	87	64	0.85	6.2	0.000286	0.38	12.029	0.163	0.0607	0.0011	0.0828	0.0011	512.9	6.8
12.1	121	108	0.86	8.4	0.000011	0.06	12.327	0.179	0.0577	0.0009	0.0811	0.0012	502.5	7.2
12.2	116	102	0.82	8.0	0.000076	0.36	12.416	0.172	0.0601	0.0011	0.0802	0.0011	497.6	6.8
13.1	129	101	0.76	9.1	0.000268	0.19	12.188	0.153	0.0590	0.0009	0.0819	0.0011	507.4	6.3
14.1	139	110	0.92	10.2	0.000136	0.12	11.677	0.144	0.0589	0.0009	0.0855	0.0011	529.1	6.4
14.2	256	215	0.50	18.7	0.000165	0.16	11.539	0.155	0.0581	0.0009	0.0848	0.0010	524.6	6.0
15.1	129	91	0.90	9.3	0.000448	0.11	11.860	0.149	0.0587	0.0009	0.0842	0.0011	521.3	6.4
15.2	414	200	0.73	30.6	0.000152	0.03	11.624	0.129	0.0583	0.0005	0.0860	0.0010	531.9	5.8
16.1	90	78	0.87	6.8	—	0.07	11.443	0.163	0.0589	0.0012	0.0873	0.0013	539.7	7.5

* Uncertainties given at the one sigma level.
[†] Error in the Duluth Gabbro reference zircon calibration was 0.69 and 0.28% for the two analytical sessions (not included in above errors but required when comparing data from different mounts).
[‡] The percentage of ^{206}Pb that is common Pb.
[§] Correction for common Pb made using the measured $^{238}U/^{206}Pb$ and $^{207}Pb/^{206}Pb$ ratios following Tera & Wasserburg (1972), as outlined in Compston et al. (1992).

Table 3. Summary of SHRIMP U–Pb zircon results for sample J1818

Grain spot	U* (ppm)	Th (ppm)	Th/U	Pb† (ppm)	$^{204}Pb/^{206}Pb$	f_{206}‡ (%)	Total§ $^{238}U/^{206}Pb$	±	Total§ $^{207}Pb/^{206}Pb$	±	Radiogenic $^{206}Pb/^{238}U$	±	Age (Ma) $^{206}Pb/^{238}U$	±
1.1	136	88	0.67	10.0	0.000221	0.05	11.669	0.145	0.0584	0.0009	0.0857	0.0011	529.8	6.4
2.1	114	88	0.81	8.2	0.000207	0.27	11.921	0.154	0.0599	0.0010	0.0837	0.0011	517.9	6.5
3.1	78	60	0.79	5.7	0.000463	0.21	11.774	0.164	0.0596	0.0012	0.0848	0.0012	524.4	7.2
4.1	210	88	0.43	15.5	0.000575	0.03	11.634	0.138	0.0583	0.0007	0.0859	0.0010	531.4	6.2
5.1	257	187	0.75	18.9	0.000049	0.09	11.676	0.133	0.0588	0.0007	0.0856	0.0010	529.2	5.9
6.1	369	223	0.63	27.1	0.000115	<0.01	11.666	0.139	0.0579	0.0005	0.0857	0.0010	530.2	6.2
7.1	211	165	0.81	15.4	0.000408	0.09	11.774	0.137	0.0586	0.0007	0.0849	0.0010	525.0	6.0
8.1	246	140	0.59	18.0	0.000317	0.09	11.727	0.134	0.0586	0.0007	0.0852	0.0010	527.1	5.9
9.1	181	136	0.77	12.9	0.000194	0.24	12.068	0.143	0.0595	0.0008	0.0827	0.0010	512.0	6.0
9.2	213	162	0.78	16.4	0.000183	0.01	11.154	0.134	0.0587	0.0008	0.0896	0.0011	553.5	6.5
10.1	159	68	0.78	11.3	0.000151	0.19	12.136	0.148	0.0590	0.0008	0.0822	0.0010	509.5	6.1
10.2	146	57	0.87	10.9	—	<0.01	11.522	0.143	0.0581	0.0008	0.0868	0.0011	536.6	6.5
11.1	214	84	0.44	15.3	0.000137	0.14	12.006	0.148	0.0588	0.0007	0.0832	0.0010	515.0	6.2
12.1	174	148	0.40	12.2	0.000166	0.29	12.279	0.148	0.0597	0.0008	0.0812	0.0010	503.3	5.9
12.2	76	58	0.62	5.7	0.000148	0.01	11.408	0.155	0.0584	0.0011	0.0876	0.0012	541.6	7.2
13.1	81	68	0.41	5.8	0.000969	0.32	11.883	0.165	0.0603	0.0012	0.0839	0.0012	519.3	7.1
14.1	159	96	0.88	11.5	0.000223	0.11	11.890	0.144	0.0587	0.0009	0.0840	0.0010	520.0	6.2
15.1	318	115	0.78	23.6	0.000111	0.03	11.580	0.129	0.0584	0.0007	0.0863	0.0010	533.8	5.8
16.1	187	153	0.37	13.1	0.000022	0.82	12.227	0.221	0.0640	0.0032	0.0811	0.0015	502.8	9.1
17.1	283	145	0.85	20.7	0.000203	0.04	11.759	0.136	0.0582	0.0007	0.0850	0.0010	526.0	6.0
18.1	292	213	0.53	21.6	0.000050	0.27	11.615	0.134	0.0603	0.0007	0.0859	0.0010	531.0	6.0
19.1	509	289	0.76	37.4	0.000114	0.05	11.703	0.128	0.0583	0.0005	0.0854	0.0010	528.3	5.6
20.1	138	156	0.59	10.1	0.000496	0.13	11.782	0.150	0.0589	0.0010	0.0848	0.0011	524.5	6.5
21.1	448	264	1.17	33.3	—	<0.01	11.542	0.126	0.0579	0.0005	0.0867	0.0010	535.8	5.7
22.1	71	48	0.61	5.2	0.001258	0.35	11.590	0.172	0.0609	0.0014	0.0860	0.0013	531.7	7.8

* Uncertainties given at the one sigma level.
† Error in the Duluth Gabbro reference zircon calibration was 0.69 and 0.28% for the two analytical sessions (not included in above errors but required when comparing data from different mounts).
‡ Percentage of ^{206}Pb that is common Pb.
§ Correction for common Pb made using the measured $^{238}U/^{206}Pb$ and $^{207}Pb/^{206}Pb$ ratios following Tera and Wasserburg (1972), as outlined in Compston et al. (1992).

Granite dykes and veins and quartz-filled tension gashes

Both protomylonitic leucogranite gneisses and leucogranite ultamylonites were sampled for structural and fluid-inclusion studies. Samples J1814 and J1820 represent protomylonitic leucogranite gneiss veins, of which J1820 also contains a tension gash filled with quartz. The protomylonitic leucogranite gneiss is made up of c. 50% K-feldspar (mostly microcline), 25% plagioclase (finely twinned oligoclase), 20% quartz and accessory components consisting of dark-brown biotite, muscovite, apatite, rutile and zircon. The feldspars are equigranular and subhedral, quartz shows an anhedral morphology. The sizes of quartz and feldspars range from 0.3 to 0.5 mm. Relicts of the original magmatic zoning are preserved in some larger plagioclase grains (Fig. 5a). The latest tectonic overprint is proven by the occurrence of subgrain boundaries in quartz, brittle fracturing of K-feldspar and plagioclase, and kinking of plagioclase albite-law twins. The total amount of dynamically recrystallized feldspars is <20%.

The ultramylonitic variety of the leucogranite veins is represented by samples J1824 and J1827. Plagioclase and K-feldspar are completely recrystallized, forming an equigranular polygonal fabric with 120° triple junctions (Fig. 5b). The grain diameters range from 0.05 to 0.2 mm. Quartz grains are arranged in monocrystalline and polycrystalline ribbons [type A and B3 according to the classification of Boullier & Bouchez (1978)]. Relicts of large K-feldspars (up to 0.3 mm) show high-strain deformation patterns. Very common features are strain-induced myrmekitization (Fig. 5c) and healed microfractures. These microfractures follow the crystallographic cleavage planes and are filled with (exsolved?) plagioclase (Fig. 5d). Plagioclase grains often show undulatory extinction and deformation twins. These textures are indicative of rocks which underwent mylonitization at mid- to upper amphibolite facies conditions (e.g. Pryer 1993).

The ultramylonitic leucogranites often contain multistage tension gashes filled with coarse-grained pure quartz (Fig. 3c & d). Close to the tension gashes, the leucogranite mylonite is strongly altered. Plagioclase crystals are saussuritized and new muscovite has grown. These tension gashes occur as up to 4 m long and 30 cm wide bodies. Some tension gashes were rotated into the horizontal during successive shearing and mylonitization, whilst new tension gashes were forming at the same time (Fig. 3d). The formation of such tension gashes within a generally ductile metagranitoid rock at high T can evolve at very high strain rates. The quartz of the tension gashes is formed of individual crystals, 0.1–1 cm in diameter. The grains usually show straight or slightly curved boundaries and 120° triple junctions. Bulged boundaries (Fig. 5e), indicating grain boundary migration, are less common. The undulatory extinction is a result of the presence of prism- and rhombohedron-parallel subgrain boundaries. Individual quartz grains are crossed by numerous fluid-inclusion trails, which are mostly truncated at the grain boundaries (Fig. 5f). These intracrystalline fluid-inclusion trails indicate trapping during or after the strong recrystallisation of quartz grains (see discussion below).

Quartz c-axes orientations of a tension gash filling (sample J1820a) were measured using a conventional U-stage polarizing microscope. Due to the relatively coarse grain size, only 140 quartz c-axes were measured in the XZ section of the finite strain ellipsoid. The resulting quartz c-axis pattern is a type II crossed-girdle (Fig. 6). According to Schmid & Casey (1986), type II crossed-girdles have been observed in naturally deformed quartzites that underwent constrictional pure shear deformation.

Fluid inclusion study

In order to understand the fluid regime associated with the extensional processes, fluid inclusion studies were carried out on leucogranite mylonites and the tension gashes. These studies also allowed the construction of the retrograde P–T path, which is not covered by conventional geothermobarometry. About 140 inclusions in two mylonitic leucogranite samples (J1824 and J1827) and two samples from the tension-gash quartz (J1820 and J1820a) were microthermometrically investigated. Fluid inclusions were examined using the U.S. Geological Survey gas-flow heating–freezing stage [for description see Roedder (1984)] at the Universität Bremen. Gas densities and compositions were determined using the phase diagrams of van den Kerkhof & Thiery (2001). Isochores were calculated after the MRK equation of Holloway (1981) and modified after Brown (1989). A detailed description of the microthermometry and calibration procedure is given in Klemd (1989).

All investigated inclusions occur in quartz and are <30 μm in diameter. The inclusions are all CO_2–dominant and are mostly monophase at room temperature, and show homogenization as a final-phase transition during heating after cooling. The inclusions occur as H2–type after van den Kerkhof's (1988) classification. They may contain up to 15 vol.% H_2O, which may coat inclusions as thin films and is therefore microscopically undetectable (Roedder 1984). However clathrate melting was not observed, dismissing the possible presence of H_2O + salts. Fluid inclusions are extremely rare and many appear to be empty. This fact is probabaly due

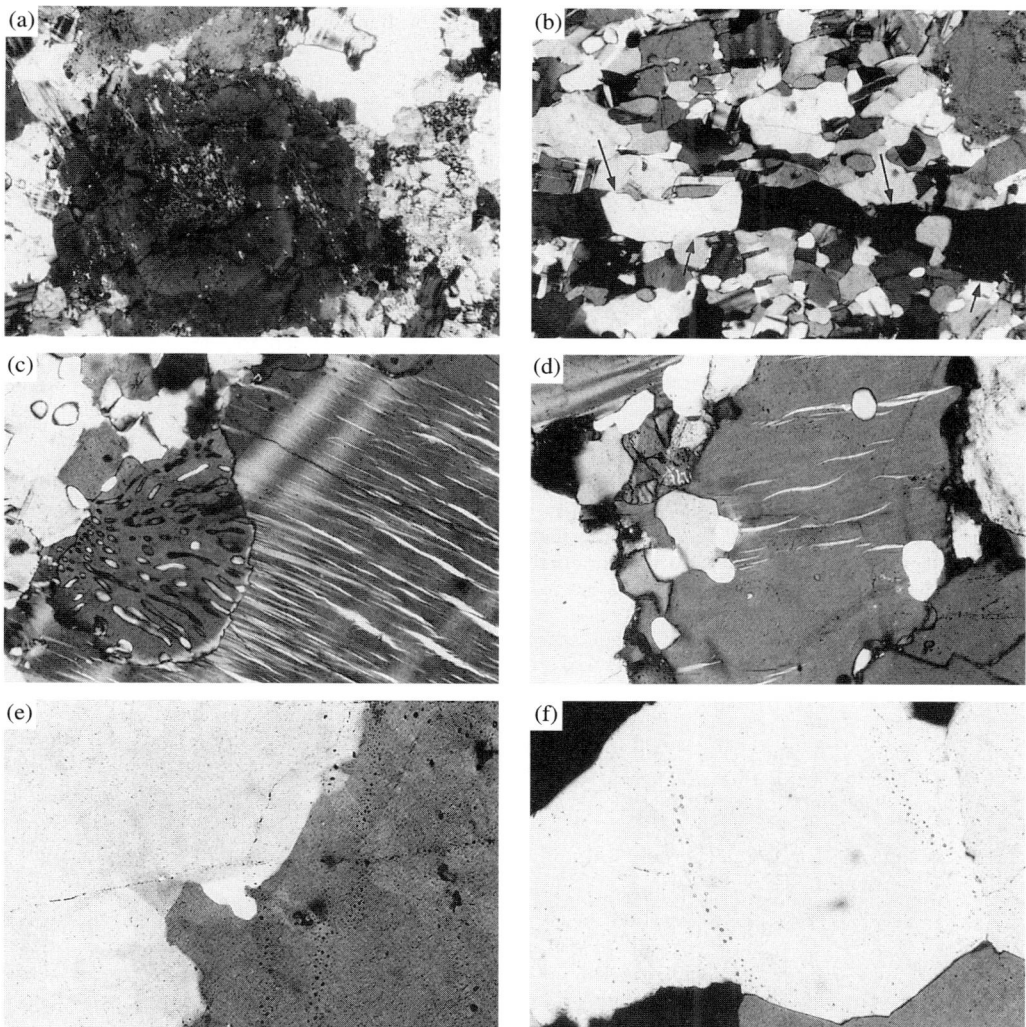

Fig. 5. Microphotographs of mylonitic leucogranites, all photographs taken at crossed polarizers, width of view 1.4 mm. (**a**) Protomylonitic leucogranite with a large magmatically zoned plagioclase surrounded by recrystallized quartz and K-feldspar (J1814). (**b**) Ultramylonitic leucogranite with quartz ribbon type B3 (arrows) in an equigranular matrix of quarz, K-feldspar and plagioclase (J1827). (**c**) Large perthitic K-feldspar rimmed by strain-induced myrmekite in ultramylonitic leucogranite (J1827). (**d**) Large K-feldspar with quartz inclusions and sygmoidal shaped microcracks (probably filled with exsolved plagioclase, J1827) in the ultramylonitic leucogranite. (**e**) Strongly bulged quartz–quartz boundary, indicating grain–boundary migration as the main fabric-forming process (J1820) in the pure quartz vein filling of tension gashes within the protomylonitic leucogranite. (**f**) Trails of pseudosecondary fluid inclusions within a large quarz crystal, note that fluid inclusion trails terminate within the quartz grain or at its boundary in the pure quartz vein filling of gashes within protomylonitic leucogranite (J1828).

to dynamic and static recrystallization processes, as indicated by the textured quartz grains and the abundant development of 120° triple junctions (e.g. Johnson & Hollister 1995). Most inclusions occur in small clusters or along healed fractures which do not cross-cut grain boundaries (Fig. 5f). The shapes vary from irregular to spherical and negative crystal shapes are rare. Due to the rare occurrence of the inclusions and the absence of intersecting healed fractures, it was impossible to establish a relative age-relationship from petrographic observations. Usually, cluster-bound inclusions are interpreted to

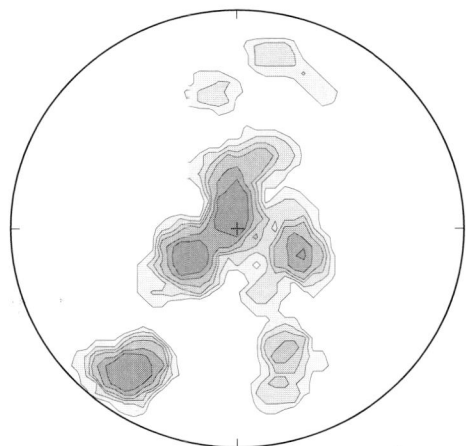

Fig. 6. Quartz c-axis diagram of quartz from gash filling (sample J1820) representing a type II crossed-girdle. Schmidt net lower hemisphere; XZ cut of the finite strain ellipsoid; $n = 140$ measurements; contours at 1, 2, 3, 4, 6 and 8% (counting method after Schmidt 1% fixed circle).

have been trapped earlier than fracture-bound inclusions (e.g. Touret 1981). Cluster-bound CO_2-rich inclusions probably formed as a result of grain-boundary migration recrystallization during ductile deformation. Therefore, fluid inclusions in the samples here investigated were trapped during and after recrystallization, as is evidenced by texturally early cluster-bound and texturally later fracture-bound CO_2-rich inclusions. However, fluid inclusions in healed fractures and clusters do not show compositional differences, nor differences in size or microthermometrical behaviour.

Homogenization temperatures (T_h) ranged between $+4.6$ and $-30.5°C$ (Fig. 7). Homogenization usually occurred into the liquid phase. However, two inclusions homogenized at $+29$ and $+20°C$ into the vapour phase (Fig. 7), and final melting (T_{mf}) occurred between -58.2 and $-56.6°C$ (Fig. 7), indicating only minor amounts of gas species such as N_2 and CH_4 in addition to CO_2. In this study it is assumed that the depression of melting temperatures (T_m) below $-56.6°C$ is mainly caused by CH_4 rather than N_2. The reason for this being that graphite is absent in the inclusions and associated wall rocks, as well as that T_h and T_m curves (i.e. $T_{mf} = -58.2°C / T_h = +20°C$) frequently do not intersect when using the CO_2–N_2 diagrams of van den Kerkhof & Thiery (2001). It should be noted that ternary mixtures of CO_2–CH_4–N_2 cannot be ruled out with certainty. Calculated densities are relatively low, ranging from 0.93 to 0.28 $g\,cm^{-3}$.

Interpretation and discussion

The two U–Pb SHRIMP zircon ages of 521 ± 5.6 and 527 ± 5.1 Ma indicate that the gabbro intruded early during the Pan-African II event. The two ages are identical, within error, with U–Pb SHRIMP zircon ages from the Conrad Granodiorite (530 ± 8 Ma) and a leucogranite dyke at Conradgebirge (52 ± 6 Ma; Table 1; Jacobs et al. 1998). This diverse synchronous magmatism characterizes the onset of large-scale extensional tectonism that is distinctive for the Pan-African II event in central Dronning Maud Land. Moreover, initial T were high enough to cause the growth of orthopyroxene and sector-zoned zircons in structural boudins within the Conrad Granodiorite at 516 ± 5 Ma (Jacobs et al. 1998; Jacobs 1999 Fig. 3). This large-scale extension finally culminated in voluminous anorogenic A2-type magmatism (Roland 1999) at c. 510 Ma (Mikhalsky et al. 1997). Many of the early and late Pan-African II granitoids are aligned along or even lay on discrete extensional dextral shear zones that reactivate sinistral transpressional shear zones. It might be speculated that these structures were operative in a transtensional pull-apart-type sense to produce the space required for the intrusion of the granitoids. This type of magmatism is not restricted to the present study area, but can be traced for at least 800 km in Dronning Maud Land (2–25°E), forming a wide zone within the East Antarctic Orogen (Jacobs et al. 1998).

The Pan-African II deformation of the Zwiesel Gabbro Complex was strongly partitioned. Strain was almost entirely taken up by leucogranite dykes and veins to produce leucogranite mylonites. Strain rates reached the ductile–brittle transition to form tension gashes that were mineralized with coarse-grained quartz. The recrystallization of K-feldspar and plagioclase are indicative for $T\,c. >450°C$. The type II crossed-girdles of quartz c-axis textures from quartz ribbons in leucogranite mylonite indicate quartz prism- and rhombohedral-$<a>$ slip during deformation. These systems are active under conditions not exceeding the upper amphibolite facies (Schmid & Casey 1986). Grain boundaries of recrystallised quartz and plagioclase are generally straight and 120° triple junctions are common. These microfabrics point to static recrystallization, postdating their mylonitization.

The present fluid inclusion study document the retrograde P–T path of the medium- to low-P granulites, with the highest densities of CO_2-rich inclusions $<1\,g\,cm^{-3}$. The calculated isochores for the highest density inclusions indicate $P = 3.2$–3.7 kbar at $T = 700$–$800\ °C$. Thus, the highest density isochores plot just below the M3-pressure estimate of 4–6 kbar of Colombo & Talarico (2002). The lower density inclusions are interpreted to be the result of

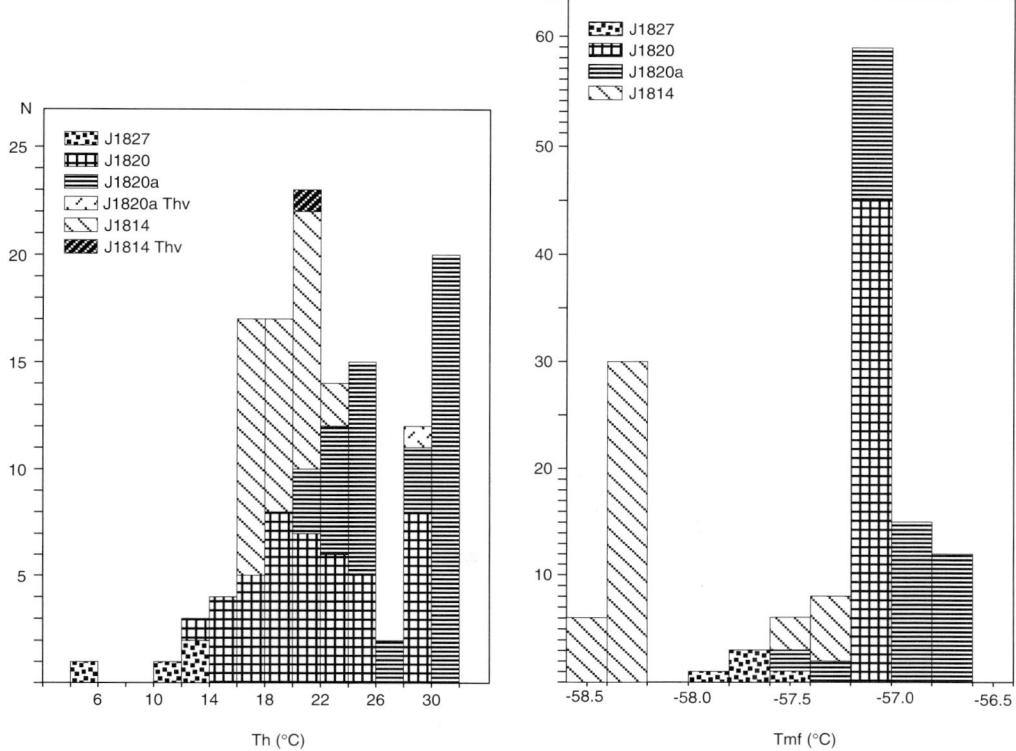

Fig. 7. Melting and homogenization temperatures of fluid inclusions in protomylonitic leucogranits (J1814), ultramylonitic leucogranite (J1827) and quartz from tension gashes (J1820, J1820a).

modifications as a consequence of dynamic and static recrystallization during and after the Pan-African II extensional deformation. Such a modification process would explain the variation of homogenization temperatures and, hence, the resulting density range. This is supported by the fact that cluster- and fracture-bound fluid inclusions do not show compositional differences, as would have been expected in non-mylonitized rocks (e.g. Touret 1981; Klemd & Bröcker 1999).

An important result of the fluid inclusion study is that the investigated rocks experienced simultaneous cooling and decompression after initial isothermal decompression (Fig. 8). The isothermal decompression between M2 and M3 is most likely the result of crustal thickening and isostatic rebound following continent–continent collision, whilst the simultaneous cooling and decompression at <530 Ma is probably the consequence of the extensional collapse of a hot lower continental crust.

Late orogenic extension within other parts of the East African–Antarctic Orogen

Extensional tectonics of the mid- to lower continental crust often produce subhorizontal structures that are difficult to identify as extensional structures (e.g. Sandiford 1989). Only small amounts of syn- to post-tectonic rotation can result in an apparent thrust setting from an original extensional setting. In contrast to contractional deformation, extension tends to be localised in shear zones. Therefore, one might not expect to see extensional deformation throughout the entire orogen. In the ultimate case, extensional structures might be confined to a single detachment zone.

In the following section the potential areas of Lower Palaeozoic deformation from west to east, from Dronning Maud Land to the Lützow-Holm Bay area, are reviewed. Adjacent parts of the East African–Antarctic Orogen are then considered. For a review of the entire Pan-African overprint of these areas see Jacobs & Thomas (2002) and Bauer et al. (2003).

In western Dronning Maud Land, the western

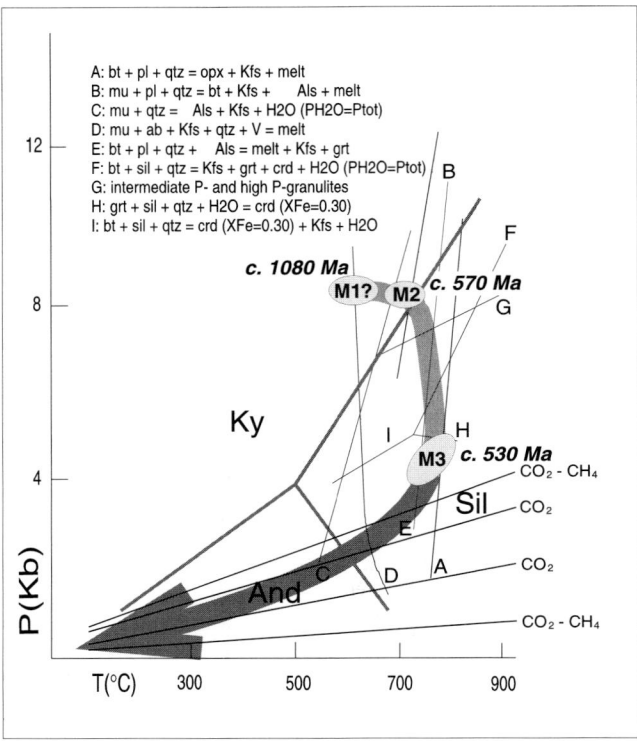

Fig. 8. *P–T* path for the Zwiesel Gabbro Complex and its country rocks. M1–M3 derived by geothermobarometry data of Colombo & Talarico (2003). Lower retrograde path from M3 derived by fluid-inclusion studies from this study.

front of the East African–Antarctic Orogen is exposed as the Heimefront Shear Zone in Heimefrontfjella (Jacobs *et al.* 1997; Golynsky & Jacobs 2001; Jacobs & Thomas 2002). East of and within the Heimefront Shear Zone, K–Ar and Ar–Ar mineral ages of the *c.* 1.1 Ga basement have been entirely reset at *c.* 560–470 Ma. Within this zone no extensional structures are apparent. However, in the immediate foreland, subhorizontal structures are present with a top-to-the-west displacement. The <2 μm fraction of an ultramylonite gave a K–Ar age of 473 ± 11 Ma (Jacobs *et al.* 1995) and was interpreted as a minimum age of the top-to-the-west displacement. Thus far, these structures were interpreted as compressional; however, as argued above, these structures could also be the result of late orogenic extension.

North of Heimefrontfjella, in Kirwanveggen and Sverdrupfjella, the Late Neoproterozoic–Early Palaeozoic overprint is difficult to delineate, since the Pan-African structures are co-linear with the older *c.* 1.1 Ga structures (Jackson pers. comm.). However, some of the late subhorizontal structures with a top-to-the-west displacement could well represent extensional structures.

East of central Dronning Maud Land, in Sør Rondane, a similar high-grade overprint is recorded as in the present study area (e.g. Shiraishi & Kagamai 1992; Osanai *et al.* 1996). The structure of the area is dominated by a number of major shear zones that are not well studied at present. However, voluminous late tectonic intrusions similar to those in central Dronning Maud Land seem to be related to theses structures.

Lützow-Holm Bay is the area where a Lower Palaeozoic high-grade overprint of older basement was first recorded in East Antarctica (Shiraishi *et al.* 1992). Although no extensional structures are reported from Lützow-Holm Bay, it is noteworthy that U–Th–Pb SHRIMP zircon data of metamorphic zircons are very similar to those in central Dronning Maud Land, ranging from *c.* 550 to 510 Ma. It would be interesting to see whether the younger ages are at least in part associated with a similar extensional setting to those in central Dronning Maud Land.

A comparison of the W–E section of Late Neoproterozoic–Early Palaeozoic tectonometamorphic overprint in Dronning Maud Land and Lützow-Holm Bay with other areas within the East African Orogen indicates that the eastern parts of the East

African Orogen have great similarities with the present study area (cf. Jacobs & Thomas 2002). For example, the Late Neoproterozoic–Early Palaeozoic structural evolution in western Madagascar seems to be grossly identical with that in central Dronning Maud Land. A period of crustal thickening between c. 580 and 550 Ma was followed by a phase of sinistral transcurrent shearing at c. 530–500 Ma (Martellat et al. 2000). Collins et al. (2000) identified an additional phase of E–W-directed extension in central Madagascar, interpreted to indicate a phase of orogenic collapse, but so far, no time constraints have been given. However, the same structure was interpreted as a refolded thrust by other workers (Schreuers pers. comm.).

Late Proterozoic collapse has also been identified in the northern part of the East African–Antarctic Orogen. Blasband et al. (2000) have identified an extensional metamorphic core complex in the northern Arabian–Nubian Shield (Wadi Kid area) that resulted from NW–SE-directed extension. It appears that the original NW–SE-directed sinistral transpression finaly reversed to a NW–SE-directed extension. This situation is very similar to central Dronning Maud Land, where a sinsitral transpressional period finally reversed into a dextral extensional setting.

Conclusions

The Pan-African collision stage of the East Antarctic Orogen at c. 580–560 Ma was succeeded by major extension between c. 530 and 515 Ma. Extension was accompanied by diverse syntectonic magmatism, including gabbros, granodiorites and leucogranites at low-P granulite facies conditions. Extensional strain strongly partitioned into shear zones, in the case of the Zwiesel Gabbro Complex into leucogranite dykes and veins, within an overall constrictional deformation regime. Fluid inclusions of the leucogranite dykes are pseudosecondary and CO_2–dominant. They verify the final part of an overall clockwise P–T path. Whilst the collision stage was characterized by isothermal decompression, the <530 Ma extensional phase was characterized by simultaneous decompression and cooling. Finally, extension resulted in the intrusion of large volumes of post-tectonic A2–type granitoids, possibly aided by major dextral pull-apart transtension. Post-tectonic high T are documented in static recrystallization of deformation fabrics. This general setting of the East African–Antarctic Orogen is comparable to other areas throughout Dronning Maud Land, but has also been described in Madagascar and the northern part of the Arabian–Nubian Shield.

This project was funded in part by Deutsche Forschungsgemeinschaft grant Ja 617/15 and 16 to J. Jacobs. We acknowledge thorough reviews by M. Santosh and M. Yoshida.

References

BAUER, W., JACOBS, J. & PAECH, H.-J. 2003. Structural evolution of the metamorphic basement of Central Dronning Maud Land. *Geologisches Jahrbuch*, **B96**, in press.

BAUER, W., THOMAS, R. J. & JACOBS, J. 2003. Proterozoic–Cambrian history of Dronning Maud Land in the context of Gondwana assembly. *In*: YOSHIDA, M., WINDLEY, B. F. & DASGUPTA, S. (eds) *Proterozoic East Gondwana: Supercontinent Assembly and Breakup*. Geological Society, London, Special Publication, **206**, 247–269.

BLASBAND, B., WHITE, S., BROOIJMANS, P., DE BOORDER, H. & VISSER, W. 2000. Late Proterozoic collapse in the Arabian–Nubian Shield. *Journal Geological Society, London*, **157**, 615–628.

BOULLIER, A. M. & BOUCHEZ, J. L. 1978. Le quartz en rubans dans les mylonites. *Bulletin de la Societe Géologique de France*, **20**, 253–262.

BROWN, P. E. 1989. FLINCOR: a microcomputer program for the reduction and investigation of fluid inclusion data. *American Mineralogist*, **74**, 1390–1393.

COLLINS, A., RAZAKAMANANA, T. & WINDLEY, B. 2000. Neoproterozoic extensional detachment in central Madagascar: implications for the collapse of the East African Orogen. *Geological Magazine*, **137**, 39–51.

COLOMBO, F. & TALARICO, F. 2003. Regional metamorphism in the high-grade basement of central Dronning Maud Land (Antarctica). *Geologisches Jahrbuch*, **B96**, in press.

COMPSTON, W., WILLIAMS, I. S., KIRSCHVINK, J. L., ZHANG, Z. & MA, G. 1992. Zircon U-Pb ages for the Early Cambrian time-scale. *Journal of the Geological Society, London*, **149**, 171–184.

GOLYNSKY, A. & JACOBS, J. 2001. Grenville-age versus Pan-African magnetic anomaly imprints in western Dronning Maud Land, East Antarctica. *Journal Geology*, **109**, 136–142.

GRUNOW, A., HANSON, R. & WILSON, T. 1996. Were aspects of Pan-African deformation linked to Iapetus opening? *Geology*, **24**, 1063–1066.

HOLLOWAY, J. R. 1981. Compositions and volumes of supercritical fluids in the earth's crust. *In*: HOLLISTER, L. S. & CRAWFORD, M. L. (eds) *Fluid Inclusions: Applications to Petrology*. Mineralogical Association of Canada, **6**, 13–38.

HARTNADY, C., JOUBERT, P. & STOWE, C. 1985. Proterozoic crustal evolution in southwestern Africa. *Episodes*, **8**, 236–244.

JACOBS, J. 1999. Neoproterozoic/Lower Paleozoic events in central Dronning Maud Land (East Antarctica). *Gondwana Research*, **2**, 473–480.

JACOBS, J. & THOMAS, R. J. 2002. The Mozambique Belt from an East Antarctic perspective. *In*: GAMBLE, J. A., SKINNER, D. N. B., HENRYS, S. & LYNCH, R. (eds) *Antarctica at the Close of a Millennium*. Proceedings Volume of the 8th International Symposium on

Antarctic Earth Sciences, Royal Society of New Zealand Bulletin, **35**.

JACOBS, J., AHRENDT, H., KREUTZER, H. & WEBER, K. 1995. K–Ar, ^{40}Ar–^{39}Ar and apatite fission track evidence for Neoproterozoic and Mesozoic basement rejuvenation events in the Heimefrontfjella and Mannfallknausane (East Antarctica). *Precambrian Research*, **75**, 251–262.

JACOBS, J., FALTER, M., WEBER, K. & JEẞBERGER, E. K. 1997. ^{40}Ar–^{39}Ar evidence for the structural evolution of the Heimefront Shear Zone (Western Dronning Maud Land), East Antarctica. *In*: RICCI, C. A. (ed.) *The Antarctic Region: Geological Evolution and Processes*. Terra Antarctica Publication, Siena, 37–44.

JACOBS, J., FANNING, C. M., HENJES-KUNST, F., OLESCH, M. & PAECH, H.-J. 1998. Continuation of the Mozambique Belt into East Antarctica: Grenville-age metamorphsim and polyphase Pan-African high-grade events in central Dronning Maud Land. *Journal Geology*, **106**, 385–406.

JOHNSON, E. L. & HOLLISTER, L. S. 1995. Syndeformational fluid trapping in quartz: determining the pressure–temperature conditions of deformation from fluid inclusions and the formation of pure CO_2 fluid inclusions during grain boundary migration. *Journal Metamorphic Geology*, **13**, 239–249.

KLEMD, R. 1989. *P–T* evolution and fluid inclusion characteristics of retrograded eclogites, Münchberg Gneiss Complex, Germany. *Contribution Mineralogy Petrology*, **102**, 221–229.

KLEMD, R. & BRÖCKER, M. 1999. Fluid influence on mineral reactions in ultrahigh-pressure granulites: a case study in the Sniezknik Mts. (West Sudetes, Poland). *Contribution Mineralogy Petrology*, **136**, 358–373.

KLEMD, R., BRÖCKER, M. & SCHRAMM, J. 1995. Characterisation of amphibolite-facies fluids of Variscan eclogites from the Orlica–Sniezknik dome Sudetes, Poland. *Chemical Geology*, **119**, 101–113.

LAWVER, L. A. & SCOTESE, C. R. 1987. A revised reconstruction of Gondwanaland. *In*: MCKENZIE, G. D. (eds) *Gondwana Six: Structure, Tectonics and Geophysics*. American Geophysical Union Geophysics Monograph, **40**, 17–23.

LUDWIG, K. R. 1999. *User's manual for Isoplot/Ex, Version 2.10, a geochronological toolkit for Microsoft Excel*. Berkeley Geochronology Center Special Publication No. 1a, 2455 Ridge Road, Berkeley CA 94709, USA.

LUDWIG K. R. 2000. *SQUID 1.00, a user's manual*. Berkeley Geochronology Center Special Publication. No. 2, 2455 Ridge Road, Berkeley, CA 94709, USA.

MARTELAT, J.-E., LARDEAUX, L.-M., NICOLLET, C. & RAKOTONDRAZAFY, R. 2000. Strain pattern and late Precambrian deformation history in southern Madagascar. *Precambrian Research*, **102**, 1–20.

MIKHALSKY, E. V., BELIATSKY, B. V., SAVVA, E. V., WETZEL, H.-U., FEDOROV, L. V., WEISER, T. & HAHNE, K. 1997. Reconnaissance geochronologic data on polymetamorphic and igneous rocks of the Humboldt Mountains, Central Queen Maud Kand, East Antarctica. *In*: RICCI, C. A. (ed.) *The Antarctic Region: Geological Evolution and Processes*. Terra Antarctica Publication, Siena, 45–54.

OSANAI, Y., SHIRAISHI, K., TAKAHASHI, Y. *ET AL.* 1996. *Antarctic geological map series, sheet 34*. Brattnipene, National Institute of Polar Research, Tokyo.

PACES, J. B. & MILLER, J. D. 1993. Precise U–Pb ages of Duluth Complex and related mafic intrusions, northeastern Minnesota: geochronological insights to physical, petrogenetic, paleomagnetic, and tectonomagmatic process associated with the 1.1 Ga Midcontinent Rift System. *Journal of Geophysical Research*, **98**, 13 997–14 013.

PEACH, H.-J. 2003. Present knowledge on the geology of central Dronning Maud Land, East Antarctica: main results of the GeoMaud expedition 1995/96. *Geologisches Jahrbuch*, **B96**, in press.

PIAZOLO, S. & MARKL, G. 1998. Humite- and scapolite-bearing assemblages in marbles and calcsilicates of Dronning Maud Land, Antarctica: new data for Gondwana reconstructions. *Journal Metamorphic Geology*, **17**, 91–107.

PORADA, H. 1989. Pan-African rifting and orogenesis in southern to equatorial Africa and eastern Brazil. *Precambrian Research*, **44**, 103–136.

PRYER, L. L. 1993. Microstructures in feldspars from a major crustal thrust zone: the Grenville Front, Ontario, Canada. *Journal Structural Geology*, **15**, 21–36.

RAVICH, M. G. & SOLOV'EV, D. S. 1966. *Geologiya i petrologiya tsentral'noy chasti gor zemli Korelevy Mod* [Geology and petrology of the central Dronning Maud Land]. Nedra, Leningrad.

ROEDDER, E. 1984. Fluid inclusions. *Mineralogical Society of America Review*, **12**, 1–644.

ROLAND, N. 1999. Pan-African granitoids in central Dronning Maud Land, East Antarctica: petrography, geochemistry and their plate tectonic setting. *In*: SKINNER, D. N. B. (ed.) *Programme & Abstracts of the 8th International Symposium on Antarctic Earth Science*, 272.

SANDIFORD, M. 1989. Horizontal structures in granulite terrains: a record of mountain building or mountain collapse? *Geology*, **17**, 449–452.

SCHMID, S. M. & CASEY, M. 1986. Complete fabric analysis of some commonly observed quartz c-axis patterns. *Geophysical Monograph*, **36**, 263–286.

SHACKLETON, R. M. 1996. The final collision zone between East and West Gondwana: where is it? *Journal of African Earth Sciences*, **23**, 271–287.

SHIRAISHI, K. & KAGAMI, H. 1992. Sm–Nd and Rb–Sr ages of metamorphic rocks from the Sør Rondane Mountains, East Antarctica. *In*: YOSHIDA, Y., KAMINUMA, K. & SHIRAISHI, K. (eds) *Recent Progress in Antarctic Earth Science*. Terra Scientific Publishing Company, Tokyo, 29–35.

SHIRAISHI, K., HIROI, Y., ELLIS, D. J., FANNING, C. M., MOTOYOSHI, Y. & NAKAI, Y. 1992. The first report of a Cambrian orogenic belt in east Antarctica – an ion microprobe study of the Lützow-Holm Complex. *In*: YOSHIDA, Y., KAMINUMA, K. & SHIRAISHI, K. (eds) *Recent Progress in Antarctic Earth Science*. Terra Scientific Publishing Company, Tokyo, 67–73.

SHIRAISHI, K., ELLIS, D. J., HIROI, Y., FANNING, C. M., MOTOYOSHI, Y. & NAKAI, Y. 1994. Cambrian orogenic belt in east Antarctica and Sri Lanka: implications for

Gondwana assembly. *Journal of Geology*, **102**, 47–65.

STERN, R. J. 1994. Arc assembly and continental collision in the Neoproterozoic East African Orogen: implications for the Consolidation of Gondwana. *Review of Earth and Planetary Science*, **22**, 319–351.

TERA, F. & WASSERBURG, G. 1972. U–Th–Pb systematics in three Apollo 14 basalts and the problem of initial Pb in lunar rocks. *Earth and Planetary Science Letters*, **14**, 281–304.

TESSENSOHN, F., KLEINSCHMIDT, G., TALARICO, F. ET AL. 1999. Ross-age amalgamation of East and West Gondwana: evidence from the Shackleton Range. *Terra Antarctica*, **6**, 317–325.

TOURET, J. L. R. 1981. Fluid inclusions in high grade metamorphic rocks. *In*: HOLLISTER, L. S. & CRAWFORD, M. L. (eds) *Fluid Inclusions: Applications to Petrology. Mineralogical Association of Canada*, **6**, 182–208.

VAN DEN KERKHOF, A. M. 1988. *The system CO_2–CH_4–N_2 in fluid inclusions: theoretical modelling and geological applications*. PhD Thesis, Free University Press, Amsterdam.

VAN DEN KERHOF, A. M. & THIERY, T. 2001. Carbonic inclusions. *Lithos*, **55**, 27–47.

WILLIAMS, I. S. 1998. U–Th–Pb geochronology by ion microprobe. *In*: MCKIBBEN, M. A., SHANKS III, W. C. AND RIDLEY, W. I. (eds) *Applications of Microanalytical Techniques to Understanding Mineralizing Processes*. Reviews in Economic Geology, **7**, 1–35.

WILSON, T. J., GRUNOW, A. M. & HANSON, R. E. 1997. Gondwana assembly: the view from southern Africa and East Gondwana. *Journal of Geodynamics*, **23**, 263–286.

YOSHIDA, M., JACOBS, J., SANTOSH, M. & RAJESH, H. M. 2003. Role of Pan-African events in the Circum-East Antarctic Orogen of East Gondwana: a critical overview. *In*: YOSHIDA, M., WINDLEY, B. F. & DASGUPTA, S. (eds) *Proterozoic East Gondwana: Supercontinent Assembly and Breakup*. Geological Society, London, Special Publication, **206**, 57–76.

Development of the Arabian–Nubian Shield: perspectives on accretion and deformation in the northern East African Orogen and the assembly of Gondwana

PETER R. JOHNSON[1] & BERAKI WOLDEHAIMANOT[2]

[1]*Saudi Geological Survey, PO Box 54141, Jiddah 21514, Saudi Arabia*
(e-mail: petergeo@dmp.net.sa)
[2]*College of Science, University of Asmara, PO Box 1220, Asmara, Eritrea*
(e-mail: beraki@asmaram.uoa.edu.er)

Abstract: The Arabian–Nubian Shield forms the suture between East and West Gondwana at the northern end of the East African Orogen (EAO). The older components of the shield include Archaean and Palaeoproterozoic continental crust, and Neoproterozoic ($c.$ 870–670 Ma) continental-marginal and juvenile intraoceanic magmatic-arc terranes that accumulated in an oceanic environment referred to as the Mozambique Ocean. Subduction, starting $c.$ 870 Ma, and initial arc–arc convergence and terrane suturing at $c.$ 780 Ma, marked the beginning of ocean-basin closure and Gondwana assembly. Terrane amalgamation continued until $c.$ 600 Ma, resulting in the juxtaposition of East and West Gondwana across the deformed rocks of the shield, and final assembly of Gondwana was achieved by $c.$ 550 Ma following overlapping periods of basin formation, rifting, compression, strike-slip faulting, and the creation of gneiss domes in association with extension and/or thrusting. Most post-amalgamation basins contain molasse deposits, but those in the eastern Arabian Shield and Oman have marine to glaciomarine deposits, which indicate that seaways penetrated the orogen soon after orogeny. The varied character of the post-amalgamation events militate against any simple tectonic model of final Gondwana convergence at the northern end of the EAO, and requires that models accommodate alternating periods of Late Neoproterozoic extension and shortening, uplift and depression, deposition and erosion.

The Arabian–Nubian Shield (ANS) consists of Precambrian rocks exposed on either side of the Red Sea in western Arabia and northeastern Africa (Fig. 1), and comprises a collage of Neoproterozoic terranes and older crust caught up as the suture between East and West Gondwana (Fig. 2; Stern 1994; Kusky *et al.* 2003). The eastern part of the shield is now preserved in the Arabian Plate; the western part is a segment of the African Plate. In the 1970s, geologists interpreted the shield in terms of Wilson Cycle plate tectonics; terrane concepts became current in the mid-1980s and since the mid-1990s broader tectonic views have become common, which place the ANS at the northern end of the East African Orogen (EAO) and interpret its development in terms of the rifting of Rodinia, closure of the ensuing Mozambique Ocean, arc accretion, orogeny, extension and orogenic collapse (Stern 1994; Blasband *et al.* 2000; Greiling *et al.* 2000). The growth of the shield lasted $c.$ 300 Ma, initiated at $c.$ 870 Ma by the breakup of Rodinia and the deposition of oceanic volcanic arcs on thin juvenile crust of the Mozambique Ocean, and terminated at $c.$ 550 Ma by the transformation of the northern EAO into a passive margin on the southern shore of palaeo-Tethys. West Gondwana is represented by Archaean and Palaeoproterozoic continental crust belonging to the ghost East Saharan Craton, which was strongly reworked by Neoproterozoic thermal and deformational events and is discontinuously exposed west of the Nile. However, representatives of East Gondwana at the northern end of the EAO are not yet identified, although general palaeogeographic considerations point to the eastern part of the Arabian Plate as a reasonable place to search. Models about supercontinent assembly and breakup at the northern end of the EAO depend, consequently, on information from the East Saharan Craton together with inferences drawn from the Neoproterozoic history of the ANS. This paper reviews the tectonic history of the ANS with the objective of adding to the body of data that constrains the history of East Gondwana. It discusses the cycle of accretion and transformation that created the shield, describes stages of terrane formation and amalgamation, and phases of post-amalgamation faulting, deposition and exhumation, and presents details of chronology and structure that constrain the timing and style of possible interactions between the shield and the converging blocks of East and West Gondwana. The emphasis is on the Arabian Shield north of Yemen, the Nubian Shield, and Precambrian rocks in the eastern part of the Arabian Plate in Oman. The paper includes brief comments about the character of the Yemen basement and its correlations with Precambrian rocks,

From: YOSHIDA, M., WINDLEY, B. F. & DASGUPTA, S. (eds) 2003. *Proterozoic East Gondwana: Supercontinent Assembly and Breakup*. Geological Society, London, Special Publications, **206**, 289–325. 0305–8719/03/$15 © The Geological Society of London.

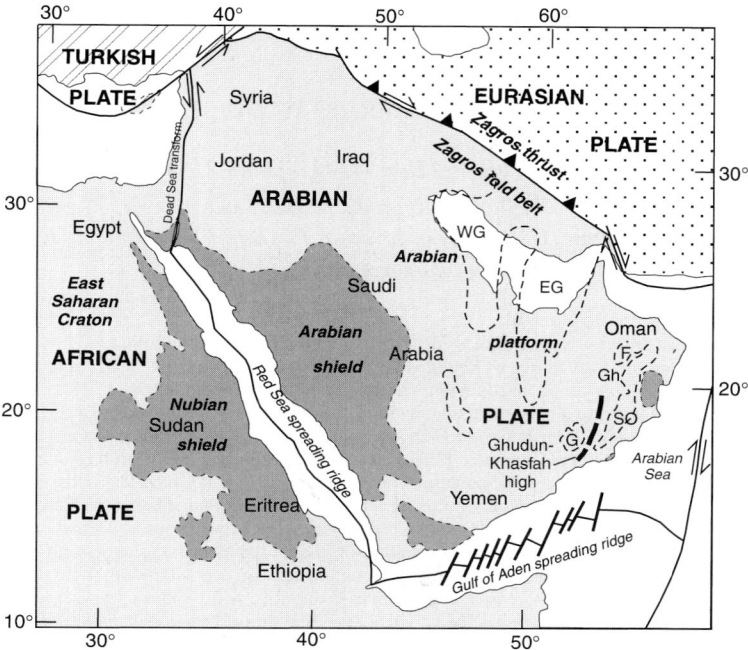

Fig 1. Present-day tectonic setting of Precambrian rocks exposed on either side of the Red Sea and in Oman, and Precambrian deposition basins in the subsurface (dashed outline). EG, Eastern Gulf Basin; F, Fahud Basin; G, Ghudun Basin; Gh, Ghaba Basin; SO, South Oman Basin; WG, Western Gulf Basin.

but the reader is referred to Windley et al. (1996, 2001) and Whitehouse et al. (1998, 2001b) for details on the Precambrian in Yemen.

Terrane analysis

Although mainly Neoproterozoic, the ANS includes Archaean and Palaeoproterozoic gneisses that have depositional and faulted contacts with Neoproterozoic rocks in the west (Schandelmeier et al. 1994b), Palaeoproterozoic plutonic rocks and gneisses that have intrusive contacts with Neoproterozoic rocks in central Arabia (Stoeser et al. 2001), and Archaean–Mesoproterozoic terranes structurally intercalated with Neoproterozoic terranes in Yemen, Somalia and southern Ethiopia (Kröner & Sassi 1996; Windley et al. 1996; Tekley et al. 1998). Robust ages for the exposed rocks therefore span a long period of geological time (Fig. 3). Together with Pb, Nd and Sr isotope information (e.g. Fleck et al. 1980; Stuckless et al. 1984; Stoeser & Stacey 1988; Sultan et al. 1992; Reischmann & Kröner 1994; Stein & Goldstein 1996; Whitehouse et al. 2001a, b), the geochronological data divide the shield into crustal blocks that variously represent Neoproterozoic juvenile oceanic environments, are parts of old continental crust or, because of reworking or the mixing of old and juvenile material, are transitions between old crust and juvenile oceanic crust (Fig. 4). Differences in stratigraphy, structure (Fig. 5), potential field signatures and seismic wave velocities (e.g. Gettings et al. 1986; Georgel et al. 1990; Johnson & Stewart 1995; Nehlig et al. 2002) add to these distinctions, and are criteria used in conjunction with the presence of ophiolite-decorated shear zones or sutures (e.g. Berhe 1990) to justify analysis of the shield in terms of tectonostratigraphic terranes (Jones et al. 1983). Despite this wealth of information, it must be noted that the varying detail and reliability of geological investigations, and the allochthonous and non-unique character of many of the ophiolite complexes that are used to identify sutures, mean that existing terrane analysis and correlations in the ANS are provisional (Church 1988; Shackleton 1988; Harris et al. 1990; Stern et al. 1990). Nonetheless, stemming from work in the 1980s (Johnson & Vranas 1984; Stoeser & Camp 1985; Vail 1985; Johnson et al. 1987) that built on earlier recognition of volcanic arcs and sutures (Al-Shanti & Mitchell 1976; Frisch & Al-Shanti 1977; Gass 1977; Camp 1984; Kröner 1985), it is widely accepted that the ANS is a collage of terranes.

The principal terranes are shown in Figure 6 and

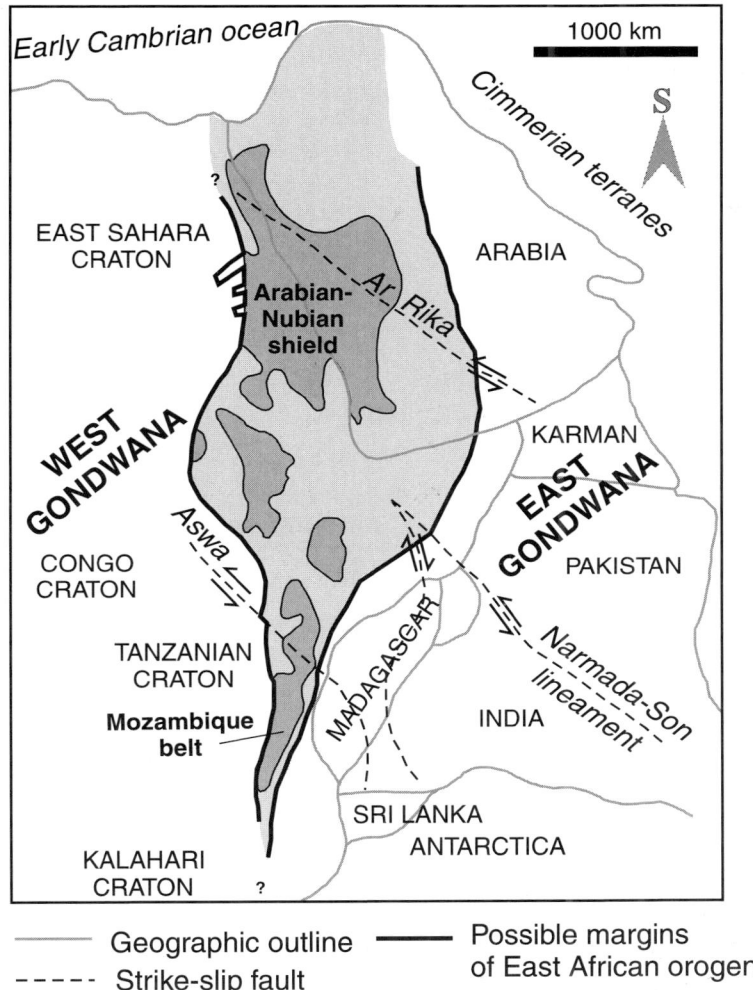

Fig. 2. Sketch map of the East African Orogen showing the location of the Arabian–Nubian Shield relative to the Mozambique Belt and adjacent cratonic margins (after Stern 1994). Note the orientation shown as at the end of the Precambrian

are described in the following paragraphs. For the purpose of this review, information is given in some detail because the evolution of the terranes, especially with regard to the accumulation of juvenile material, governed the history of the Mozambique Ocean, and the accretion of juvenile arcs in the ANS paced the growth and consumption of oceanic lithosphere at spreading centres and subduction zones. Arc evolution can therefore be viewed as a measure of the Mozambique Ocean lithospheric budget, constraining the onset of convergence to the period when the oceanic budget started to decrease by a preponderance of subduction over spreading.

The **Halfa Terrane** comprises high-grade ortho- and paragneiss belonging to the East Saharan Craton, amphibolite-grade supracrustal rocks and ophiolitic mafic–ultramafic rocks (Denkler et al. 1984; Schandelmeier et al. 1994b; Stern et al. 1994). Because of pre-Neoproterozoic plutonic precursors, the gneisses yield 2.82–1.26 Ga Nd model ages, Type III Pb ratios and strongly negative $\varepsilon_{Nd}(t)$ values (-27.0–-1.3), but also yield younger Rb–Sr ages, indicating extensive Neoproterozoic reworking (Harms et al. 1990, 1994; Stern et al. 1994). Type III Pb ratios are one of three groups of whole-rock and feldspar Pb isotopic compositions recognized in the Arabian Shield (Stoeser & Stacey 1988). Type I ratios characterize juvenile arcs (Fig. 4); Type II ratios reflect crust that is intermediate between oceanic and continental; and Type III ratios (e.g.

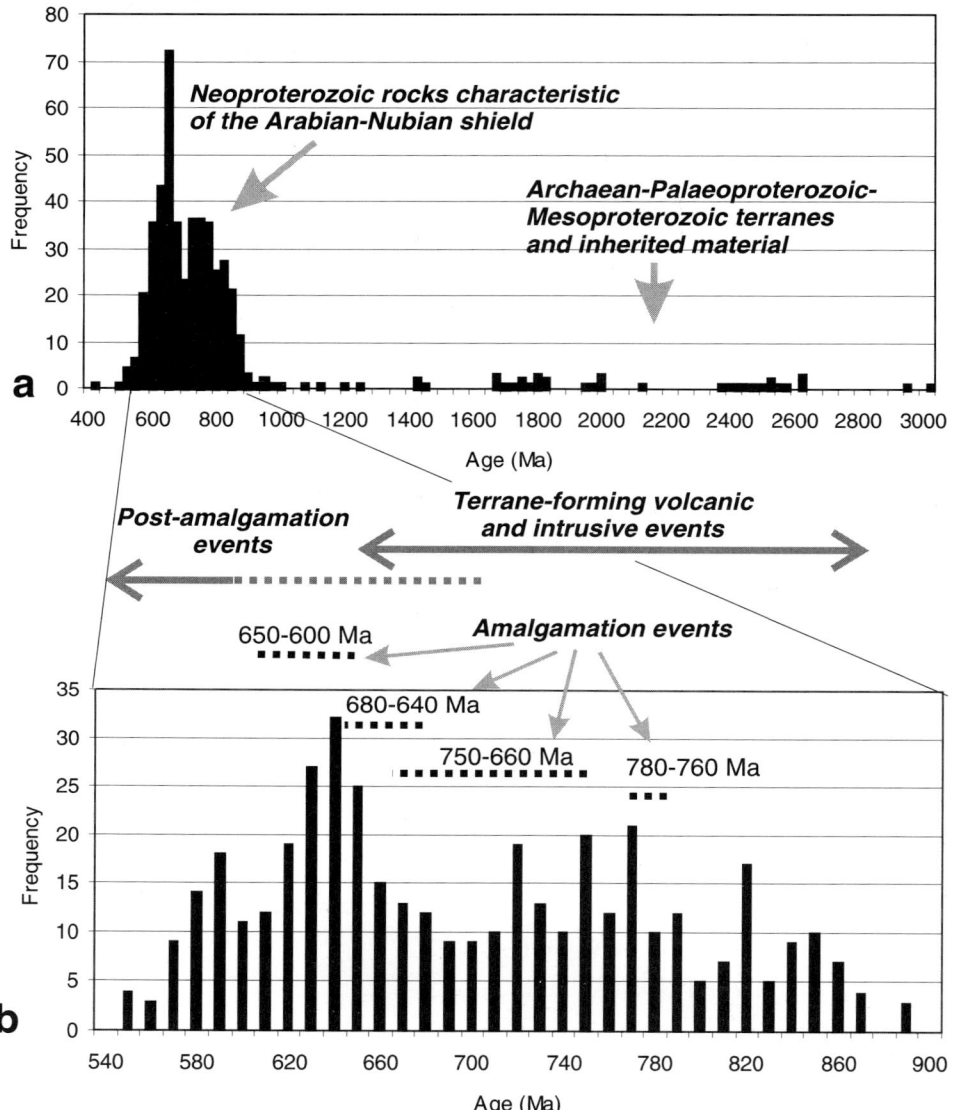

Fig. 3. Histogram of robust U–Pb, Pb–Pb, and Rb–Sr ages for rocks of the Arabian–Nubian Shield. (**a**) Entire age range. (**b**) Neoproterozoic subset. Data from references cited in text and from compilation by Johnson *et al.* (1997). Robust ages are analytically reliable and geologically meaningful; for other criteria see Johnson *et al.* (1997).

obtained from the Khida Subterrane that makes up the core of the Afif Terrane in the Arabian Shield, and from the Halfa and Aswan regions of the Nubian Shield) reflect a contribution of evolved Archaean and Palaeoproterozoic crustal material. Although now in fault contact, it is believed that the supracrustal rocks of the Halfa terrane were originally deposited against the East Saharan gneisses. They are discontinuous metasedimentary sequences, arc-related volcanic units and dismembered ophiolite complexes (Denkler *et al.* 1984; Schandelmeier *et al.* 1994*b*), and represent passive margin (Stern *et al.* 1994) and/or intraoceanic magmatic-arc rocks (Harms *et al.* 1994) deposited during the opening and closing of an oceanic basin or re-entrant above a NW-dipping subduction zone at the eastern margin of the East Saharan Craton (Schandelmeier *et al.* 1994*b*). Dating of ophiolitic gabbro, Type I grani-

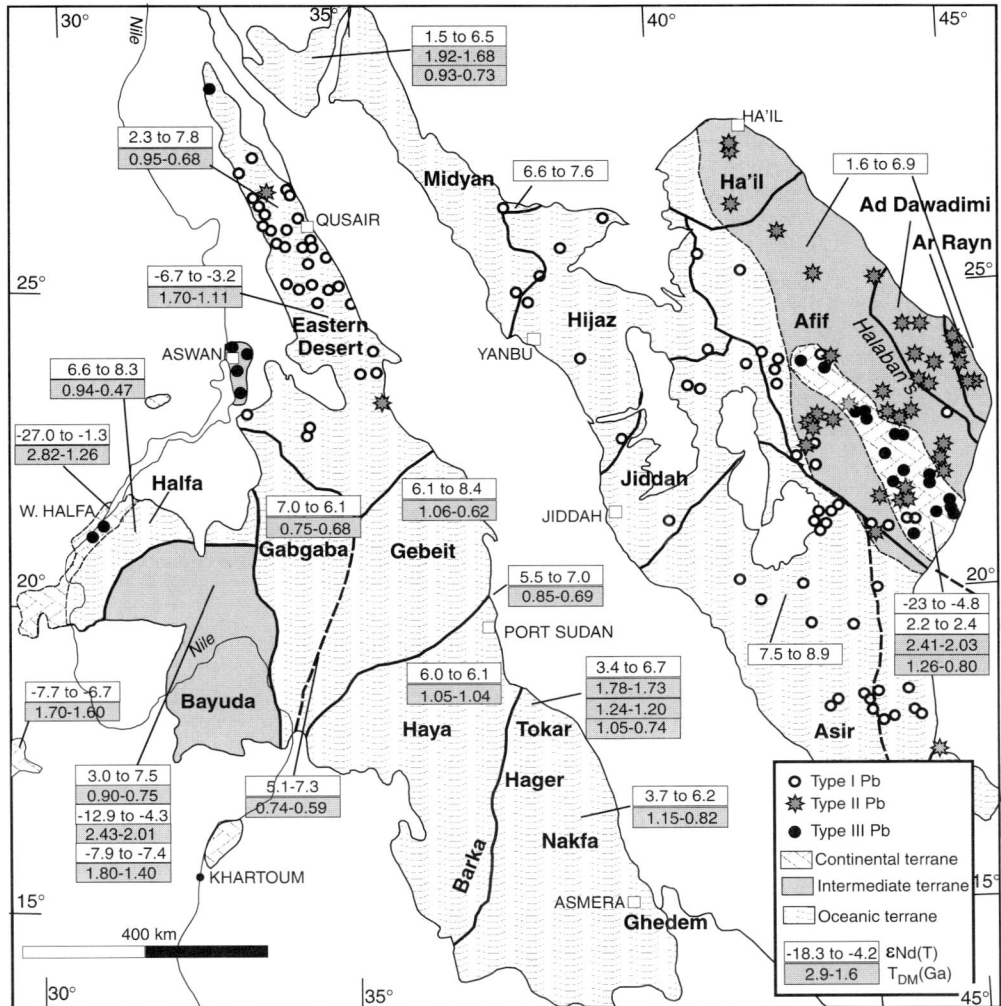

Fig. 4. Pb and Nd isotope data that divide the Arabian–Nubian Shield into regions of oceanic, continental and intermediate settings. The Pb isotope classification, after Stoeser & Stacey (1988), is based on the ratios of $^{206}Pb/^{204}Pb$, $^{207}Pb/^{204}Pb$ and $^{208}Pb/^{204}Pb$ from whole-rock, feldspar and galena samples with respect to the orogen growth curve. Type III Pb plots above the curve and reflects the presence of evolved crustal material; Type I plots below the curve and reflects the presence of juvenile (oceanic and mantle) material. Data after Bokhari & Kramers (1981), Duyverman et al. (1982), Stacey & Stoeser (1983), Harris et al. (1984), Schandelmeier et al. (1988), Stoeser & Stacey (1988 and refs cited therein), Harms et al. (1990, 1994), Sultan et al. (1990), Kröner et al. (1991), Agar et al. (1992), Stern & Kröner (1993), Reischman & Kröner (1994), Stern et al. (1994), Stein & Goldstein (1996), Teklay (1997), Moghazi et al. (1998), Stern & Abedelsalam (1998), Küster & Liégeois (2001). Tectonostratigraphic terrane boundaries from Figure 6 are shown for reference.

toids and metamorphic garnets suggests that rifting was active at c. 750 Ma, subduction occurred between c. 760 and 650 Ma, and basin closure, ophiolite obduction and metamorphism had taken place by c. 700 Ma (Harms et al. 1994; Abdelsalam et al. 2003).

The **Bayuda Terrane** is underlain by amphibolite facies granitoid gneiss and paragneiss, and subordinate amounts of amphibolite, schist, quartzite and marble. The high-grade rocks are similar to the gneisses in the Halfa Terrane and may be a rifted part of the East Saharan rocks found there. However, Ries et al. (1985) documented a Neoproterozoic arc assemblage at the eastern edge of the terrane and

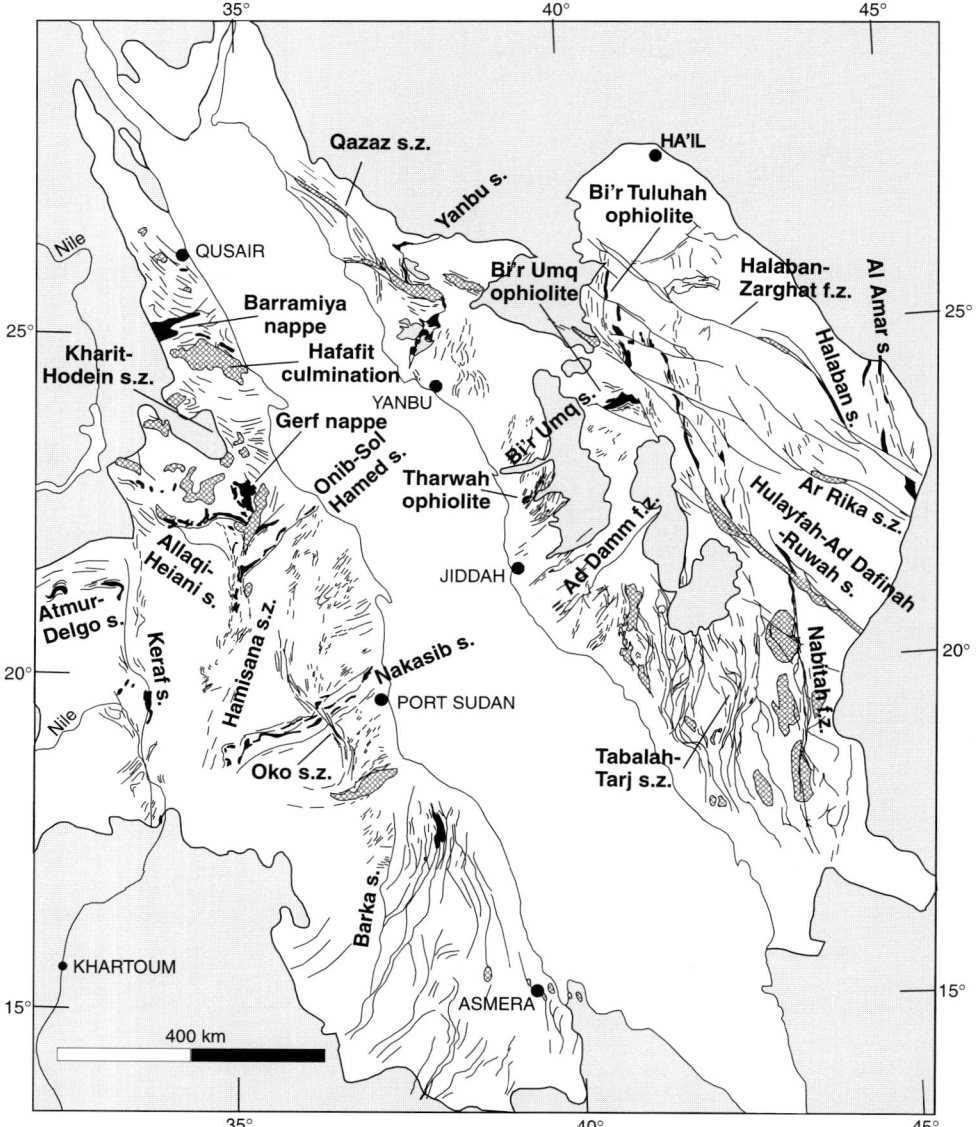

Fig. 5. Structures in the Arabian–Nubian Shield that help to define terranes, showing trends of foliations, fold axes, shear zones (s.z.), fault zones (f.z.) and sutures (s). Ultramafic bodies in black; gneiss domes and antiforms crosshatched. The terms fault zones and shear zones reflect terminology precedencies in the literature and not structural differences.

Küster & Liégeois (2001) report isotopic data suggesting that many of the high-grade gneisses in the Bayuda Desert within the bend of the Nile are derived from Neoproterozoic rocks. Amphibolite and epidote–biotite gneisses with a magmatic age of c. 806 Ma are interpreted as metamorphosed tholeiitic basalt and low- to medium-K dacite and rhyodacite extruded in a juvenile oceanic island-arc or back-arc basin environment. They yield positive $\varepsilon_{Nd}(t)$ values (+3.0–+7.5) and Nd model ages (900–750 Ma) close to the magmatic age. Leucocratic gneisses and muscovite and garnet–biotite schist that yield negative $\varepsilon_{Nd}(t)$ values (−4.3 – −12.9) and Palaeoproterozoic model ages (2.43–2.01 Ga) are believed to be metasedimentary rocks derived from local volcanic and distant older

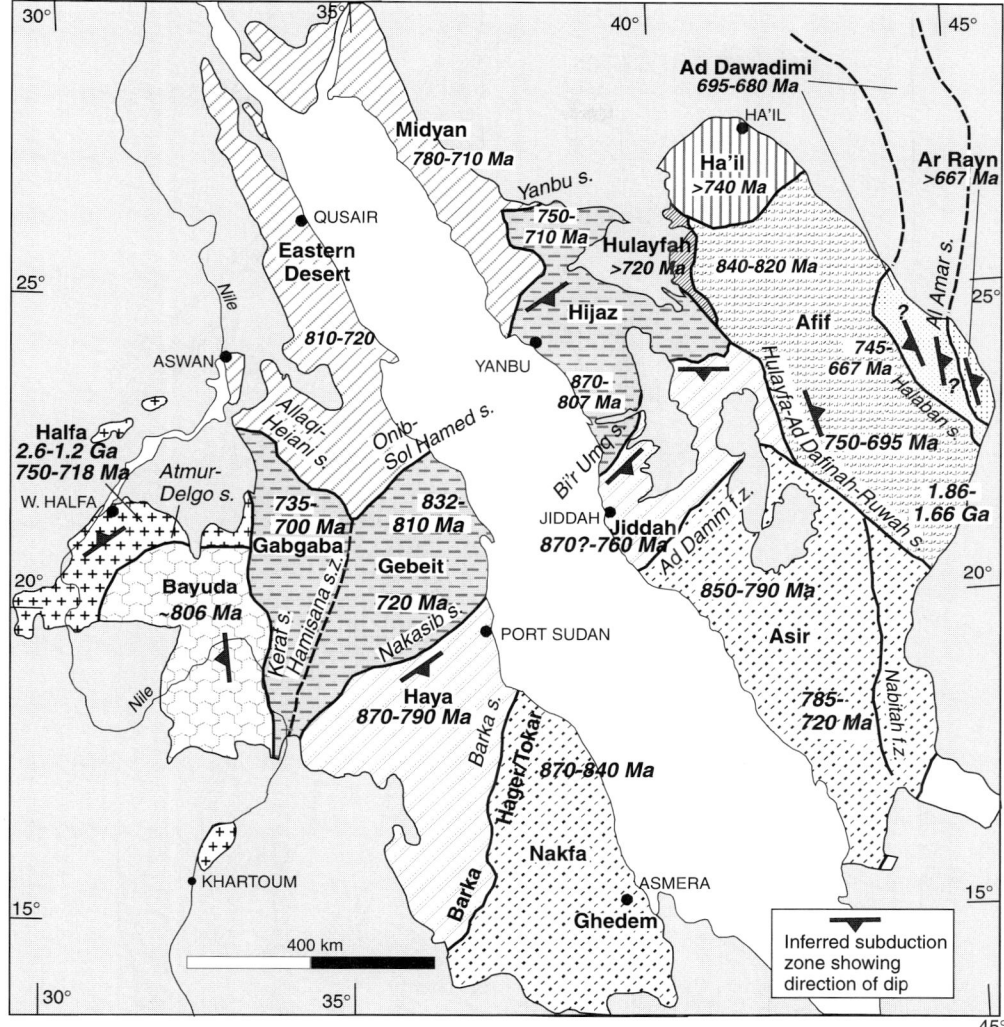

Fig. 6. Collage of suspect tectonostratigraphic terranes that make up the Arabian–Nubian Shield showing estimated terrane protolith ages and subduction orientations during terrane formation. Similarity of pattern indicates correlations accepted in this report.

continental sources (Küster & Liégeois 2001). Late Neoproterozoic granitoids (585–562 Ma) represent reworked older crust [1.8–1.4 Ga Nd model ages; $\varepsilon_{Nd}(t) = -7.4 - -7.9$] (Harms et al. 1990). Küster & Liégeois (2001) concluded that the rocks were probably metamorphosed under amphibolite facies conditions during a period of frontal convergence with an East Saharan ghost craton farther west, concurrent with the 700 Ma metamorphic-collisional event in the Halfa Terrane and coincident with the emplacement of syntectonic granitoids that are now preserved as muscovite–biotite gneiss. A low-grade passive margin assemblage, the Bailateb Group (Stern et al. 1993), composed of thinly bedded distal to proximal turbidites, and subordinate quartzites, quartzitic schist, carbonates, carbonate conglomerates and breccia, and paragneiss (Meinhold 1979; Stern et al. 1993), crops out at the northern margin of the Bayuda Terrane.

The **Barka Terrane** (Fig. 7) consists of upper amphibolite to hornblende–granulite facies orthogneiss, amphibolite, marble, pelitic schist and orthoquartzite cut by a swarm of late E–W felsic dykes (De Souza Filho & Drury 1998). Although one of the

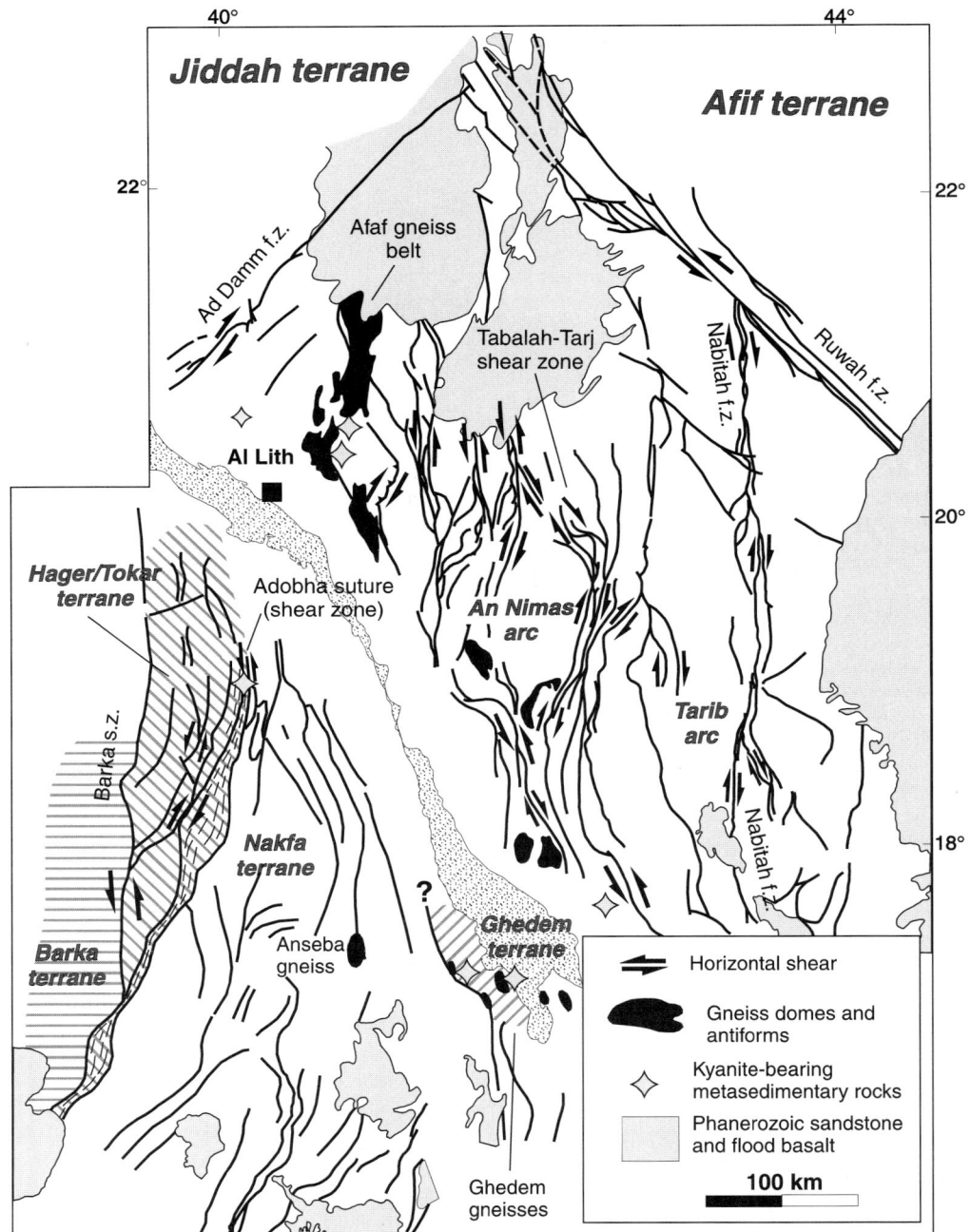

Fig. 7. Structural sketch map of the composite Asir Terrane and terranes in the southern Nubian Shield, plotted on a map showing north-trending, commonly serpentinite-decorated, shear zones and other spatial relations prior to Red Sea opening.

gneisses yields a single zircon Pb–Pb age of 700 Ma (Teklay 1997), their general protolith ages are unknown and De Souza Filho & Drury (1998) consider the orthogneisses to be an exotic, far-travelled crustal mass. The terrane is overlain on the south by the Phanerozoic and has an ill-defined transition to the Haya Terrane in the north. The structural grain of the terrane is dominantly N–S but, to the north, swings eastward and becomes broadly conformable with the structural trend of the Haya Terrane. Nonetheless, De Souza Filho & Drury (1998) treat the Barka and Haya as separate terranes because of the pronounced differences in their metamorphic grades and overall structural styles.

An assemblage of middle-greenschist facies mafic and felsic metavolcanic rocks, and mafic intrusive rocks including diorite, tonalite, granodiorite, gabbro and pyroxenite, make up the **Hager–Tokar Terrane** east of the Barka Terrane. The volcanic and mafic intrusive protoliths date from 870 to 840 Ma with partial resetting at c. 770–670 Ma; syntectonic granite dates from 827 Ma (Kröner et al. 1991; Teklay 1997). Rocks in the southern part of the terrane have a tholeiitic, mid-ocean ridge basalt (MORB)-type affinity and, in places, include chlorite schist, Fe–Mn chert, marble, pillowed metabasalt, metagabbro and serpentinite that are interpreted to be an accretionary prism deposited at or near a trench interspersed with popped-up subducted MORB-type ocean floor (De Souza Filho & Drury 1998; Woldehaimanot 2000). Nd isotopic data indicate moderate depletion in the mantle sources of the terrane protoliths [$\varepsilon_{Nd}(t) = +3.4 - +6.7$] and Mesoproterozoic Nd model ages suggest a contribution from old crust in the north (Kröner et al. 1991). The rocks are strongly deformed and, at a regional scale, appear to be an imbricated stack of east-vergent thrusts cut and displaced by strike-slip shear zones. The terrane is separated from the Nakfa Terrane on the east by a zone of sheared and thrusted, locally kyanite-bearing, metasedimentary rocks, referred to here as the Adobha Suture, and is separated from the Barka and Haya Terranes to the west by a major sinistral transpressional strike-slip system, the Barka Shear Zone (Suture). Assuming coast-to-coast closure of the Red Sea, the Hager–Tokar Terrane projects along-strike into the Al Lith area of the Asir Terrane (Fig. 7), which also contains kyanite-bearing metasedimentary units, a feature that forms a basis for correlation between the two regions (Kröner et al. 1991).

The **Nakfa Terrane** comprises variably deformed mafic and ultramafic cumulates, and diorites and granodiorites structurally overlain by greenschist facies metavolcanic and volcaniclastic metasedimentary rocks with calc-alkalic island-arc affinities (Woldehaimanot 1995, 2000; Teklay 1997; De Souza Filho & Drury, 1998). Although constrained by very limited data (a single zircon Pb–Pb age of c. 850 Ma; Teklay 1997), the rocks may be contemporary with the Hagar–Tokar Terrane, but are distinguished by a lesser degree of deformation and younger Nd model ages. Sources of the terrane protoliths were a moderately depleted mantle [$\varepsilon_{Nd}(t) = +3.7 - +6.2$] similar to those for the Hager–Tokar Terrane (Teklay 1997).

The **Ghedem Terrane** in eastern Eritrea is characterized by amphibolite facies, garnetiferous orthogneiss structurally overlain by garnet-, staurolite-, and kyanite-bearing gneisses and schists (Ghebreab 1999), which are tentatively correlated with granitoid gneisses and kyanite-bearing rocks in the Afaf Belt of the Asir Terrane (Fig. 7; cf. Beyth et al. 1997). Single zircon Pb–Pb dating of garnetiferous muscovite gneiss gives an age of 796 Ma (Teklay 1997), compatible with gneiss in the Afaf Belt. The transition to the Nakfa terrane is a subhorizontal to moderately westward dipping, up to 3 km wide, top-to-east shear zone (Ghebreab 1999). Farther north, the terrane disappears beneath the Red Sea coastal plain but in places is separated from the Nakfa Terrane by Tertiary extensional faults (De Souza Filho & Drury 1998).

The **Haya Terrane** consists of 870–850 Ma plutonic and greenschist- to amphibolite-grade volcano-sedimentary rocks in the SE (Kröner et al. 1991; Reischmann et al. 1992), a 790 Ma volcanic assemblage in the north (Abdelsalam & Stern 1993a, b; Stern & Abdelsalam 1998) and an intervening zone of intermediate-age 810 Ma plutonic rocks (Schandelmeier et al. 1994a). The rocks in the SE comprise juvenile intraoceanic island-arc rocks and syntectonic, arc-related, calc-alkaline Type I plutons, characterized by low initial Sr ratios, moderately depleted $\varepsilon_{Nd}(t)$ values (c. +6.0) and T_{DM} model ages approaching the formation ages (1.05–1.04 Ga) (Embleton et al. 1983; Klemenic 1985; Klemenic & Poole 1988; Kröner et al. 1991; Reischmann et al. 1992). The 790 Ma volcanic assemblage is a tholeiitic to calc-alkaline, within-plate and/or MORB bimodal volcanic succession, interpreted to result from crustal extension and rifting at the northern margin of the Haya Terrane (Abdelsalam & Stern 1993a). The intermediate plutonic rocks are diorite and granodiorite, geochemically resembling granitoids found in modern convergent plate margins (Schandelmeier et al. 1994a; Stern & Abdelsalam 1998). The spatial relations of these constituent rock units suggest that the Haya Terrane was formed by a northward-migrating, SE-dipping subduction zone, accompanied with some rifting at the northern margin.

In early terrane models (Kröner et al. 1987), the **Gebeit** and **Gabgaba** regions were interpreted as separate terranes, but recent work shows that the intervening Hamisana structure is a post-amalgamation zone of E–W shortening and not a suture

(Abdelsalam & Stern 1996; de Wall et al. 2001; Miller & Dixon 1992), and the Gebeit and Gabgaba rocks are treated here as a single terrane. Terrane-forming rocks are relatively sparse because younger granitoids dominate the region. Terrane protoliths in the NE comprise a c. 832 Ma assemblage of subalkaline, calc-alkaline and tholeiitic, mostly subduction-related, volcanic rocks, although some rocks with high Ti/V ratios (>20) may reflect an incipient back-arc basin in an immature, intraoceanic island-arc environment (Gaskell 1985; Reischmann & Kröner 1994). The volcanic rocks are broadly contemporary with tonalite along the Hamisana Shear Zone (Stern & Kröner 1993), and the volcanic and plutonic rocks may represent a c. 830–810 Ma arc system. Low initial Sr ratios and Nd model ages similar to crystallization ages are strong evidence of a juvenile oceanic setting, and the strongly positive $\varepsilon_{Nd}(t)$ values (+6.1–+8.4) (Fig. 4) indicate that mantle sources were more depleted than in adjacent parts of the shield (Reischmann & Kröner 1994). The northwestern part of the terrane includes immature volcanic-arc rocks (Abdeen et al. 2000), yielding Rb–Sr whole-rock isochron and zircon-evaporation ages of 735–697 Ma (Stern & Kröner 1993), and a succession of carbonate-rich, amphibolite facies marble, quartzite and amphibolite close to and along the Keraf Suture Zone, forming an extension of the passive margin sequence found in the Bayuda Terrane (Stern et al. 1989; Abdelsalam et al. 1998). Terrane protoliths in the SE include an assemblage of c. 720 Ma basalt, basaltic–andesite, andesite, dacite, rhyolite, and interbeds of agglomerate and tuff intruded by coeval granite and mafic dykes (Klemenic 1985; Klemenic & Poole 1988).

Terrane analysis in the **Eastern Desert** [alternatively named the Gerf Terrane (Kröner et al. 1987), the Southeastern Desert Terrane (Abdeen et al. 2000) or the Aswan Terrane (Greiling et al. 1994)] is problematic because of disruption of early structures by later thrusting and intrusions, uncertainties about the number of subduction zones in the region, and the locations of root zones of the ophiolite nappes and mélanges that constitute a dominant feature of the area (Greiling et al. 1994; Shackleton 1994). Stratigraphically, the region is underlain by gneiss complexes interpreted by some workers to be pre-Pan-African basement (e.g. El-Gaby & Hashad 1990). However, because they do not yield radiogenic Pb, elevated initial Sr ratios, negative $\varepsilon_{Nd}(t)$ values, or old Nd model ages (Stern et al. 1989; Sultan et al. 1992), recent compilations treat the gneisses as the metamorphic equivalents (Tier 1 infracrustal rocks) of juvenile Neoproterozoic protoliths that were probably similar to the overlying greenschist facies volcanosedimentary rocks and ophiolite nappes and mélange (Tier 2 supracrustals) (Bennet & Mosley 1987; Greiling et al. 1988, 1994). Exceptional negative $\varepsilon_{Nd}(t)$ values, old Nd model ages and inherited zircons in certain post-tectonic granites (Fig. 4); (Sultan et al. 1990; Hassanen & Harraz 1996) indicate local involvement of pre-Pan-African crustal components. In the SW, the terrane includes strongly folded ortho- and paragneisses, low-grade oceanic island-arc rocks and imbricated serpentinite, talc-carbonate schist and metagabbro ophiolite nappes, together with a shelf assemblage of marble and conglomerate (Greiling et al. 1994; Abdeen et al. 2000). Adjacent to the Onib-Sol Hamed Suture, the terrane-forming rocks are steeply dipping sheared gabbro, cumulate ultramafic rock (mainly pyroxenite), serpentinite, sheeted dykes and pillow lavas belonging to a disrupted SE-facing ophiolite succession (Fitches et al. 1983; Kröner et al. 1987). An accretionary prism to island-arc volcanic–plutonic sequence is recognized at the southeastern end of the Hafafit culmination (Greiling et al. 1994). In the central parts of the region, the terrane protoliths are rootless ophiolite nappes, the largest of which make up the Gerf and Barramiya Nappes. Zircon evaporation and Sm–Nd whole-rock dating of layered gabbro and plagiogranite from these ophiolites weakly constrains ocean-crust formation to between c. 810 and 720 Ma (Kröner et al. 1992b; Stern & Kröner 1993), compatible with an Rb–Sr isochron of 768 ± 31 Ma obtained from felsic volcanic rocks in the Abu Swayel area (Stern & Hedge 1985). Intrusions in the Allaqi–Heiani–Onib–Sol Hamed Suture Zone constrain basin closure as prior to or c. 735–720 Ma (Kröner et al. 1992b; Stern & Kröner 1993).

The **Asir Terrane** consists of juvenile oceanic rocks characterized by low initial Sr ratios and Type I Pb ratios, and $\varepsilon_{Nd}(t)$ values of +7.5 and +8.9 (Fig. 4), suggesting strong mantle depletion (Bokhari & Kramers 1981). The oldest rocks (c. 850–790 Ma) crop out in the west and are mainly assemblages of convergent-margin volcanic and plutonic rocks (Reischmann et al. 1984; Kröner et al. 1992a), such as the large An Nimas Arc (Fig. 7); (Stoeser & Stacey 1988). Younger volcanic and plutonic rocks form the Tarib Arc (785–720 Ma; Stoeser & Stacey 1988), and smaller areas of ocean spreading and convergent-margin assemblages (Bookstrom pers. comm.) underlie the eastern part of the terrane. Possible oceanic-plateau volcanic rocks (Reischmann et al. 1984) and kyanite–staurolite-bearing metasedimentary units in the NW of the terrane are candidates for correlation with the Hager–Tokar Terrane. The terrane is composite, created by the amalgamation of several subterranes along one or more of the serpentinite-decorated shear zones that dominate the region (Fig. 7). Because of intense deformation and the reconnaissance nature of existing mapping, neither subduction polarities nor convergence trajectories of these subterranes are known.

The **Jiddah Terrane** is dominated by northeasterly structural trends, similar to those in the Haya Terrane, and, although few robust ages are reported, is estimated to have a similar time span (c. 870–760 Ma). Structural and intrusive relations suggest that the oldest rocks, composed of diorite, tonalite and granite, are in the south. Resembling the 870–850 Ma plutonic rocks in the south of the Haya Terrane, they have tholeiitic to calc-alkaline affinities and mantle-type initial Sr ratios compatible with an active-arc setting (Fleck 1985; Moore & Al-Rehaili 1989). Younger, c. 815–810 Ma, calc-alkaline plutonic rocks (Calvez & Kemp 1982; Stoeser & Stacey 1988) in the north and NE correlate with plutonic rocks in the north of the Haya Terrane. A para-autochthonous fold-thrust belt, comprising volcanosedimentary rocks and syntectonic granite gneiss, crops out along the northwestern margin of the terrane (Ramsay 1986; Johnson 1998), and an autochthonous unit of immature volcanic-arc rocks (Mahd Group; <810–c. 780 Ma) intruded by 770–765 Ma granophyre, and 760 Ma tonalite and granodiorite occurs in the NE (Calvez & Kemp 1982; Afifi 1989). As in the Haya Terrane, the apparent temporal and spatial relations of the rocks suggest that the terrane was created by a northward-migrating SE-dipping subduction zone.

The **Hijaz Terrane** comprises volcanic arcs and younger volcanic sedimentary successions that may also have formed above a southerly dipping subduction system (Stoeser & Camp 1985). An older arc (< 870–807 Ma) crops out in the southern part of the terrane, containing bimodal, low-K to tholeiitic greenschist facies pillow basalt, rhyolite, tuffs and volcaniclastic sedimentary rocks deposited in an oceanic environment (Camp 1986). A younger arc (c. 750–710 Ma) is present in the northern part of the terrane as a sequence of basalt, rhyolite and large amounts of volcaniclastic sedimentary rocks that locally may have accumulated in extensional basins above older arc complexes (Kemp 1981). Mafic-ultramafic ophiolitic complexes (Bi'r Umq, Jabal Tharwah) that originated between c. 870 and 830 Ma in mixed mid-ocean-ridge and island-arc environments (Pallister et al. 1988) are thrust over the Jiddah Terrane at its southern margin (Al-Rehaili & Warden 1980; Nassief et al. 1984).

As in the Eastern Desert, the structure and distribution of protoliths in the **Midyan Terrane** are significantly disrupted by 725–696 Ma granitoid intrusions and post-terrane amalgamation deformation, but available information indicates that they include several assemblages of subduction-related volcanic–volcaniclastic and calc-alkaline intrusive rocks weakly constrained between >725 and >710 Ma (Hedge 1984), and ophiolite complexes dating from 780 to 740 Ma (Claesson et al. 1984; Pallister et al. 1988). In the north and west, the volcanic–volcanosedimentary rocks uniquely include a banded-iron formation, a rock type in the ANS that is confined to the Midyan and Eastern Desert Terranes (Sims & James 1984), and is regarded by the present authors as a strong correlation feature. The geochemical signature of some of the volcanic rocks indicates intraoceanic subduction environments. The ophiolitic units [Jabal Ess (Shanti 1983) and

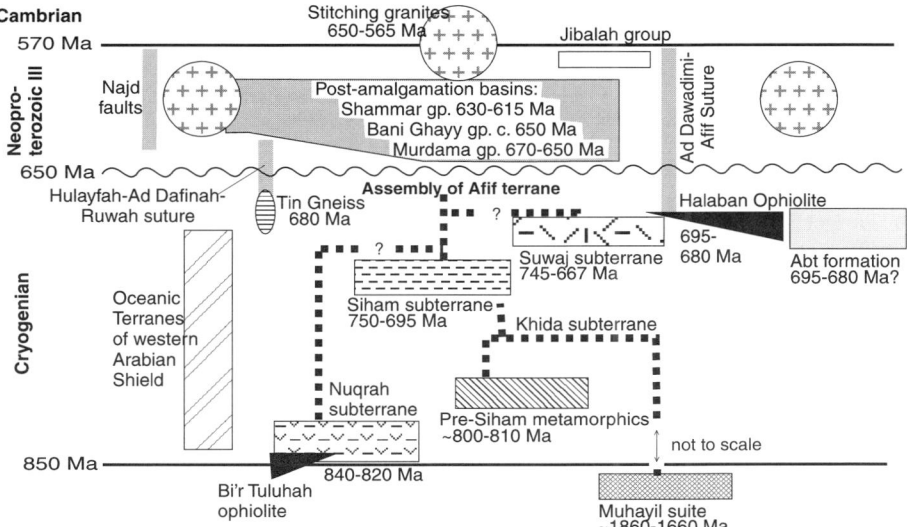

Fig. 8. Afif Terrane assembly showing estimated ages of amalgamation of subterranes, and of overlap assemblages and stitching granites that constrain the minimum age of assembly.

Jabal Wask (Bakor et al. 1976)] include discontinuous lenses and thrust sheets of serpentinite, peridotite, cumulus and non-cumulus gabbro, dyke complexes and pillow basalt. They originated in fore-arc to back-arc environments (Pallister et al. 1988) and were structurally emplaced in a subvertical shear zone at the contact between the Midyan and Hijaz Terranes, which marks the Yanbu Suture.

The **Afif Terrane** is a composite tectonostratigraphic unit in the northeastern part of the Arabian Shield assembled from four possible subterranes (Fig. 8; Johnson & Kattan 2001). The Khida Subterrane, in the SE, includes Palaeoproterozoic biotite alkali–feldspar granite, meta-anorthosite and biotite granitic orthogneiss (c. 1860–1660 Ma; Stoeser et al. 2001; Whitehouse et al. 2001a) and c. 800 Ma almandine–sillimanite amphibolite facies metavolcanic rocks, paraschist, paragneiss and orthogneiss (Stoeser pers. comm.) characterized by strongly negative $\varepsilon_{Nd}(t)$ values and old Nd model ages. The Siham Subterrane, in the SW, consists of 750–695 Ma greenschist facies volcanosedimentary rocks, gabbro, diorite, tonalite, granodiorite and monzogranite, and is inferred to be a continental-margin arc developed above a subduction zone inclined east beneath the Khida crust (Agar 1985). The Nuqrah Subterrane is a cryptic 840–820 Ma tectonostratigraphic unit in the NW composed of lower greenschist to amphibolite facies island-arc-type volcanic and sedimentary rocks (Nuqrah Formation) (Delfour 1977), mafic to intermediate plutonic rocks, and tectonically disrupted ophiolite complexes (Calvez et al. 1983; Le Metour et al. 1983; Quick 1991; Pallister et al. 1988). The Suwaj Subterrane, in the east, is a 745–667 Ma assemblage of weakly metamorphosed but strongly cataclasized diorite, quartz diorite, tonalite, sodic granodiorite, and subordinate basalt and dacite (Cole & Hedge 1986).

The **Ad Dawadimi Terrane** constitutes one of the most homogeneous crustal units in the Arabian Shield and can be traced by its distinctive, subdued, aeromagnetic signature some 300 km north beneath the Phanerozoic (Johnson & Stewart 1995). It is commonly modelled as a thin-skinned allochthon bounded and internally sliced by listric thrusts (e.g. Al-Shanti & Mitchell 1976). Terrane protoliths include thinly layered sericite–chlorite phyllite and schist derived from fine-grained sandstone and siltstone (Delfour 1979) that were deposited in a possible accretionary prism, and c. 695 Ma ocean-floor mafic–ultramafic rocks (Stacey et al. 1984) exposed as linear belts along the margins of, and within, the terrane. The terrane was thrust at c. 680 Ma over the adjacent Afif Terrane (Al-Saleh et al. 1998). The tectonic setting of the terrane is uncertain and it is debated whether the oceanic-floor material represents back-arc or fore-arc oceanic crust and whether the metasedimentary rocks are related to a west- or east-dipping subduction zone (Al-Shanti & Mitchell 1976; Al-Shanti & Gass 1983; Camp et al. 1984; Quick 1991; Johnson & Stewart 1995; Al-Saleh et al. 1998; Al-Saleh & Boyle 2001a). Westward obduction of the ocean crust marked the closure of the Ad Dawadimi Basin at c. 680 Ma and suggests that the basin had a relatively short life of c. 15 Ma.

The **Ar Rayn Terrane** crops out at the eastern margin of the shield, and extends north and east beneath the Phanerozoic (Johnson & Stewart 1995). It is bounded to the west by the Al Amar Fault and to the east by a magnetically defined contact with a possible continental block in the concealed basement. The terrane includes a bimodal volcanic–volcanosedimentary assemblage transitional between the tholeiitic and calc-alkalic series (Vaslet et al. 1983) and large mafic to intermediate calc-alkaline intrusions. The rocks may have formed in a mature island arc or more likely, as suggested by their metallogenic signature, in a continental-margin environment (Doebrich et al. 2001) above an east-dipping subduction zone. The layered rocks are not directly dated but were evidently deposited prior to the emplacement of syntectonic granitoids at 667–640 Ma (Stacey et al. 1984).

East of the Arabian Shield, Precambrian rocks dip beneath Phanerozoic strata. They descend to structural lows as much as 12 km below sea level in central Arabia (Konert et al. 2001), but rise in structural highs to within a few kilometres of the surface farther east and crop out in Oman (Fig. 9). Referred to in the literature as 'basement', crystalline Precambrian rocks in Oman consist of dolerite, granodiorite, granite, migmatite, and greenschist and amphibolite facies metasedimentary rocks and gneiss unconformably below the mainly sedimentary Late Neoproterozoic Huqf Supergroup (Al-Doukhi & Divi 2001; Allen & Hildebrand 2001; Al-Kathiri 2001; Cozzi et al. 2001; Mercolli et al. 2001). The rocks have c. 1000–730 Ma crystallization ages (Gass et al. 1990; Pallister et al. 1990; Würsten et al. 1991; Mercolli et al. 2001) and are evidently Neoproterozoic, but their exact tectonic provenance is unknown. Pallister et al. (1990) reported Pb isotope data suggesting that the region contains evolved crustal material and Johnson & Stewart (1995) speculatively inferred that much of eastern Arabia is underlain by continental crust. However, the extent of continental crust, if truly present, is not established. Nor is it established whether such crust was part of a plate east of the juvenile rocks of the ANS (perhaps as a candidate for a fragment of East Gondwana) or was a detached continental terrane intercalated with Pan-African terranes similar to the terrane assemblage in Yemen (Windley et al. 1996; Whitehouse et al. 2001b). It is likewise unknown whether the plutonic rocks

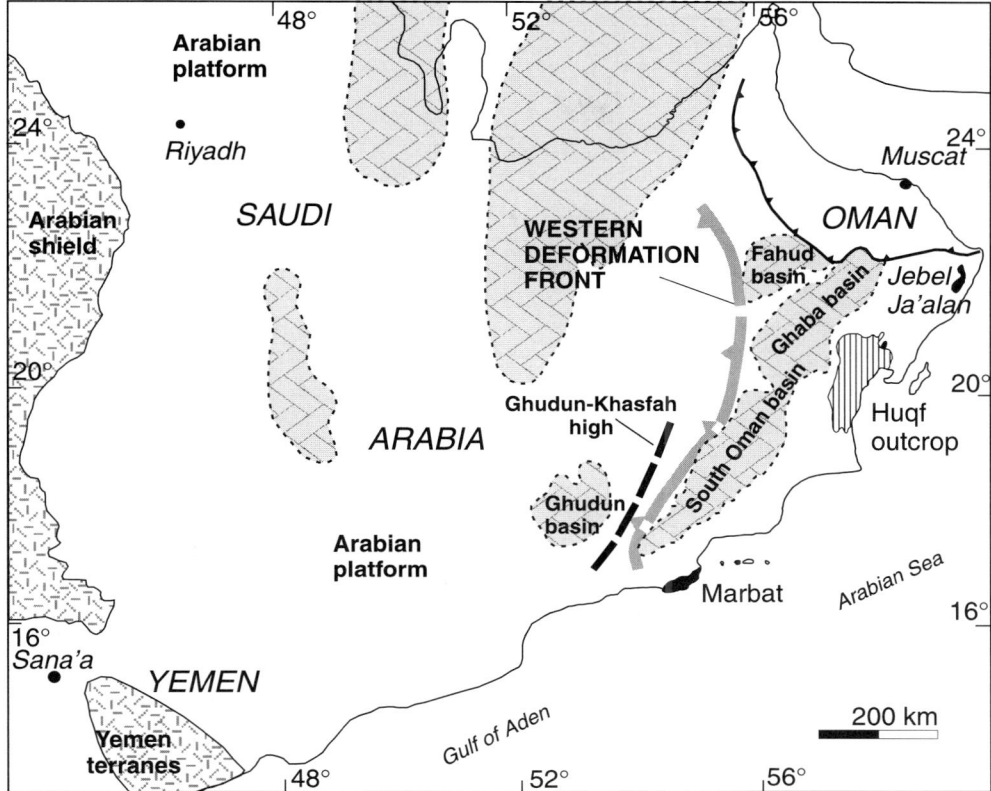

Fig. 9. Precambrian structures beneath the Arabian Platform in the eastern part of the Arabian Plate showing outcropping crystalline basement (c. 1000–750 Ma) (black), the Huqf Supergroup (c. 735–545 Ma) (vertical lines), concealed parts of the Huqf Supergroup represented by salt basins (shaded chevron pattern) and the Western Deformation Front (thick long dashes). After Loosveld et al. (1996), Blood (2001) and Konert et al. (2001).

sampled for age dating in Oman are Neoproterozoic intrusions into older continental crust or denote an entirely Neoproterozoic basement.

Terrane amalgamation and suture zones

The contacts between terranes in the ANS include (1) sutures composed of ophiolite-decorated shear zones; (2) cryptic or unimpressive shear zones that may be original sutures, faults superimposed on original sutures or post-suturing structures; and (3) post-amalgamation fault zones that may be unrelated to original suturing events. The sections below describe shear zones believed to relate to convergence and amalgamation among the ANS terranes. They are treated as sutures or modified sutures and record a c. 180 Ma period of terrane amalgamation. Structural information is sufficiently detailed so as to indicate the sense of convergence at some of the suture zones, which range from fontal (involving large amounts of thrusting) to transpressional (involving large amounts of strike-slip coupled with thrusting), and provides some constraints on the possible direction of convergence of the larger, flanking blocks of East and West Gondwana.

Amalgamation events at 786–760 Ma

The oldest structures believed to reflect convergence and amalgamation in the ANS include the Tabalah-Tarj Shear Zone and the Afaf Belt of syntectonic gneiss in the central part of the Asir Terrane, and similar rocks in Eritrea. The Tabalah-Tarj Shear Zone (Figs 5 & 10) comprises phyllonite, schist and gneiss resulting from ductile mylonitic deformation and dextral shear prior to 755 Ma (Johnson et al. 2001). Syntectonic gneiss dated at 778–763 Ma (Cooper et al. 1979) suggest a contemporary deformation and

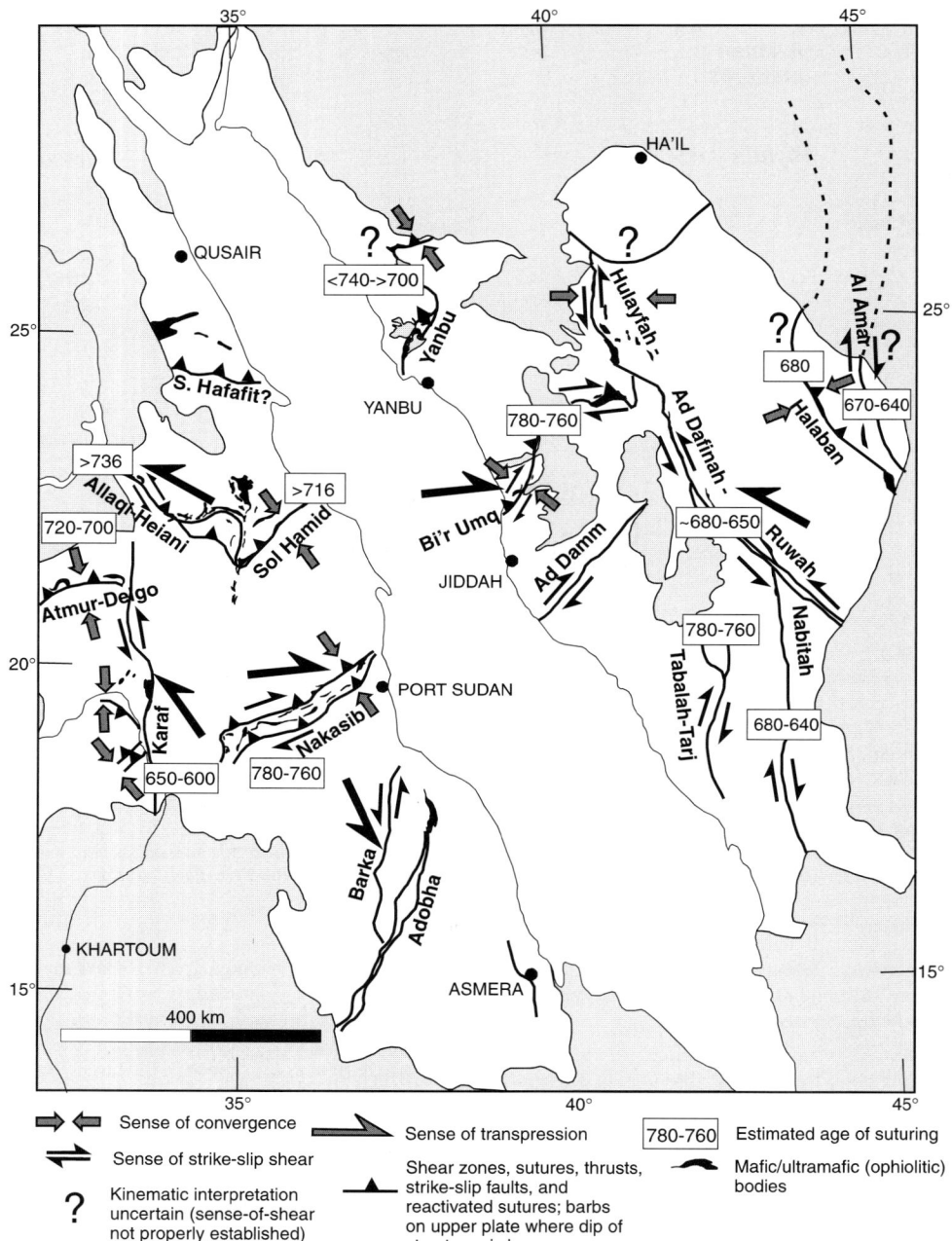

Fig. 10. Terrane assembly in the Arabian–Nubian Shield, showing inferred ages and trajectories of amalgamation.

magmatic event west and SW of the Tabalah-Tarj Shear Zone. The syntectonic gneiss is one of several gneiss antiforms that make up the Afaf Gneiss Belt in the western Asir Terrane (Fig. 7). The Afaf Belt possibly extends into the Ghedum Terrane of Eritrea, where one of the gneisses is dated at 796 Ma (Teklay 1997), and the combined structure of gneisses and shear zones is inferred to mark a period of early amalgamation in the southern Arabian and Nubian Terranes.

Contemporary accretion events to the south and north are suggested by Pb loss at c. 760 Ma from zircon grains in the Al-Mahfid Gneisses in Yemen (Whitehouse et al. 1998, 2001b), and by pre-, syn- and post-tectonic rock units along the **Bi'r Umq-Nakasib Suture** between the Jiddah-Haya and Hijaz-Gebeit Terranes. The Bi'r Umq-Nakasib Suture is an ophiolite-decorated fold-shear zone, 5–65 km wide and >600 km long, which is characterized by a commonality of structure and gold and base-metal deposits along its length, and constitutes one of the best-documented sutures in the ANS, separating regions of contrasted stratigraphy and Nd-isotope signatures (Stoeser & Camp 1985; Vail 1985; Camp 1984; Ramsay 1986; Abdelsalam & Stern 1993a; Stern & Kröner 1993; Johnson 1994; Johnson et al. 2002). The ophiolites are allochthons of mid-oceanic to suprasubduction oceanic crust (Nassief et al. 1984; Abdel Rahman 1993; Schandelmeier et al. 1994a) that formed at c. 870–830 Ma (Pallister et al. 1988) and, together with intercalated volcanosedimentary rocks, were deformed during a period of dextral transpression involving orthogonal convergence, thrusting and non-coaxial dextral shear (Abdelsalam & Stern 1993a; Wipfler 1996; Johnson 1998; Johnson et al. 2002). As constrained by the ages of terrane protoliths, suturing began some time after 810–780 Ma, was active between 780 and 760 Ma, the ages of putative syntectonic intrusions, and was complete by c. 760–750 Ma, the ages of stitching plutons and ophiolite obduction (Pallister et al. 1988; Schandelmeier et al. 1994a; Stern & Abdelsalam 1998).

Amalgamation events at 750–660 Ma

Convergence in the vicinity of the East Saharan Craton began soon after creation of the Bi'r Umq-Nakasib Structure. It was driven by N–S shortening and closure of the ocean basin in the southeastern part of the Halfa Terrane at c. 700–650 Ma (Abdelsalam et al. 2003), and caused frontal collision between the Halfa and Bayuda Terranes. The resulting **Atmur-Delgo Suture** is marked by a chain of ophiolite nappes thrust SE over the Bayuda Terrane and locally back thrust over the Halfa volcanic supracrustal rocks (Harms et al. 1994; Schandelmeier et al. 1994b). Collision and suturing, weakly constrained between 720 and 700 Ma, was associated with progressive deformation, migmatization of granitoids and upper greenschist to amphibolite facies metamorphism, which reached a peak at c. 702 Ma (Denkler et al. 1984; Harms et al. 1994; Stern et al. 1994; Abdelsalam et al. 1998). Küster & Liégeois (2001) propose concurrent accretion between the Bayuda and the East Saharan Terranes farther west. Continued downgoing, after terrane convergence, of the (detached?) oceanic crust subducted along the suture zone is suggested by the emplacement of c. 760 Ma–650 Ma Type I calc-alkaline granitoids (Harms et al. 1994), following which there was a transition to intraplate rifting and volcanism (Stern et al. 1994).

Concurrent terrane amalgamation farther north created the **Allaqi–Heiani–Sol Hamed–Onib–Yanbu Suture**, which comprises nappes and slivers of dismembered ophiolite in a sinuous but broadly east-trending shear zone between the Gebeit-Hijaz and Eastern Desert–Midyan Terranes. Continuity along this zone is generally accepted (e.g. Kröner et al. 1987; Stern et al. 1990; de Wall et al. 2001), although Shackleton (1996) argued that the Allaqi Ophiolites are obducted klippen with roots north of, rather than along, the Allaqi–Heiani Zone and Greiling et al. (1994) suggested that the Allaqi–Heiani–Sol Hamed–Onib sector of the suture may continue in a postulated South Hafafit Suture, rather than the Yanbu Suture (Fig. 10). The presence of a metamorphic-sole complex in the Wadi Haimur–Abu Swayel area is evidence, however, that the Allaqi Nappes are close to their origin, i.e. not far travelled (Abd El-Naby & Frisch 1999). The Allaqi-Heiani segment of the suture zone contains gneiss and discontinuous thrust duplexes of ophiolitic, metamorphosed island-arc rocks and mylonite linked by the Allaqi Shear Zone (Taylor et al. 1993; Greiling et al. 1994). Metamorphism is mainly in the greenschist facies, although higher grades are reported locally, and high-pressure/low-temperature blueschist assemblages occur in the footwall of the Allaqi Structure (Taylor et al. 1993). Regionally, the structure dips east to NE, but the nappes are folded and, in detail, dip south and north where the suture zone has an easterly trend, and east and west where the suture swings to the north (Greiling et al. 1994). The Allaqi Shear is a steeply dipping sinistral strike-slip shear with gently NW- and SE-plunging stretching lineations in east-trending sectors, and a reverse fault in more northerly trending sectors. These structural and kinematic changes imply a combination of orthogonal shortening and strike-slip and, although it is not clear whether the Allaqi Shear Zone itself is the original suture or a later structure that modified the suture, are compatible with terrane convergence and top-to-the-NW transport during sinistral NW-directed transpression (Taylor et al. 1993; Greiling et al. 1994). The Sol Hamed segment of the suture comprises a subvertical south-facing ophiolite. The ophiolite has a ductile flattening fabric that may reflect pre-obduction suboceanic deformation, and was tilted to the vertical and deformed by NE–SW shears prior to folding about steeply plunging axes and shearing in association with the development of steeply plunging stretching lineations (Fitches et al. 1983). The ophiolite was locally thrust SE over younger volcanic rocks and restructured as a tectonic mélange. The Yanbu segment of the suture is a subvertical to steeply dipping shear zone containing nappes and fault-bounded lenses of mafic and ultramafic rocks (Bakor et al. 1976; Shanti 1983; Pallister et al. 1988). The segment is discontinuously exposed because of late- to post-amalgamation granitoids and Cenozoic flood basalt, and is deformed by left-lateral strike-slip faults belonging to the later Najd system and by folding around the ESE-plunging nose of a gneiss antiform. The trajectory of suturing has not been established. Stoeser & Camp (1985) suggested a southerly vergence but abundant S–C fabrics indicate a significant component of dextral strike-slip. The timing of convergence along the suture is not well constrained, but amalgamation probably occurred some time between $c.$ 700 and 600 Ma, following 808–721 Ma ophiolite formation, and preceding emplacement of 730–690 Ma granodiorite and tonalite (Fitches et al. 1983; Claesson et al. 1984; Kröner 1985; Stern & Hedge 1985; Pallister et al. 1988; Kröner et al. 1992b; Stern & Kröner 1993). This time period is somewhat old, however, for K–Ar ages of $c.$ 600 and 585 Ma that represent cooling of hornblende and biotite following peak metamorphism in paragneiss 20–50 km north of the suture zone in the Allaqi area (Abd El-Naby & Frisch 2002), and robust dating of the suturing event is needed.

Amalgamation events at 680–640 Ma

Younger suturing events in the ANS reflect the convergence of oceanic terranes and mixed continental–oceanic terranes in the central and eastern parts of the Arabian Shield. The **Hulayfah–Ad Dafinah–Ruwah Suture** between the Afif Terrane and oceanic terranes to the south and west is a subvertical shear zone 5–30 km wide and 600 km long created during an episode of sinistral transpression (Johnson & Kattan 2001). Putative syntectonic orthogneiss dated at 683 ± 9 Ma (Stacey & Agar 1985) and overlap-assemblage rhyolite dated $c.$ 650 Ma (Doebrich pers. comm.) suggest that convergence was in progress by 683 Ma and completed by 650–630 Ma. Later faulting and thrusting modified the suture, particularly along its SE segment, and Najd faults offset the suture along its northern segments. The **Halaban Suture** between the Afif and Ad Dawadimi Terranes (Urd Suture of Pallister et al. 1988) is a nappe of Halaban ophiolite thrust westwards over the Suwaj Subterrane at $c.$ 680 Ma (Al-Saleh et al. 1998). Aeromagnetic data suggest that the suture originally extended north of the shield, as shown in Figure 10 (Johnson & Stewart 1995), but it has been extensively modified by post-amalgamation Najd faulting. The **Ar Amar Suture** between the Ad Dawadimi and Ar Rayn Terranes is a high-angle fault zone (the Al Amar Fault) containing narrow lenses of ophiolites and carbonate-altered ultramafic rock (listwaenite) (Al-Shanti & Mitchell 1976; Nawab 1978), which is also inferred to extend north of the shield. Syn- and post-tectonic plutons (670–640 Ma) in the Ar Rayn Terrane (Stacey et al. 1984) weakly constrain the timing of convergence; however, the trajectory of convergence is unknown, although limited observations on shear fabrics suggest that the convergence included a component of dextral horizontal shear. The **Nabitah Fault Zone** is a north-trending, serpentinite-decorated shear zone in the eastern part of the Asir Terrane (Figs 7 & 10). It is one of the classic suture zones described in the Arabian Shield literature, giving its name to the 680–640 Ma Nabitah Orogeny and to the Nabitah Orogenic (or mobile) Belt, which is mapped as a zone of deformation and magmatism extending north across the entire shield (Stoeser & Camp 1985; Stoeser & Stacey 1988). The tectonic significance of the Nabitah Fault Zone is, however, ambiguous. It may be a suture in the northern part of the Asir Terrane, south of the Ruwah Fault Zone, where it separates greenschist facies, mainly volcanosedimentary, rocks to the west from amphibolite facies paragneiss and orthogneiss to the east, but farther south it extends as a ductile shear zone through the middle of the Tarib Batholith. Bodies of syntectonic gneiss on either side of and along the fault zone are conspicuous features of the structure, and together with late tectonic granites constrain fault movement to between $c.$ 680 and <640 Ma (Stoeser et al. 1984; Stoeser & Stacey 1988; Johnson et al. 2001). Shear fabrics in gneiss, granite and volcanosedimentary rocks, as well as offsets of passive markers, indicate that shearing was dextral during both early ductile and later brittle phases of deformation.

Amalgamation events at 650–600 Ma

The youngest arc–arc amalgamation event in the ANS is represented by the **Keraf Suture**, a north-trending ophiolite-decorated belt of folded and sheared rocks at the contact between the Bayuda-Halfa and Gebeit–Gabgaba Terranes (Fig. 10;

Abdelsalam et al. 1998). Originally identified by reconnaissance ground surveys and shuttle imaging data (Almond & Ahmed 1987; Abdelsalam et al. 1995), the belt of deformed rocks was proposed to mark the arc–continent suture between Neoproterozoic rocks of the Nubian Shield and the East Saharan Craton (Stern 1994). However, if the Bayuda Terrane is largely Neoproterozoic, as inferred by Küster & Liégeois (2001) on the basis of its isotopic characteristics, the cratonic margin lies farther west. Together with the Nakasib–Bi'r Umq, Hulayfah–Ad Dafinah–Ruwah and Halaban sutures, the Keraf Zone is one of the more clearly evidenced sutures in the ANS, extending for considerable length, separating regions of contrasted crustal compositions and metamorphic grade [amphibolite–granulite facies rocks yielding $\varepsilon_{Nd}(t)$ values of $-12.9-+7.5$ to the west and greenschist facies rocks yielding $\varepsilon_{Nd}(t)$ values of $+6.1-+8.4$ to the east; Fig. 4], and coinciding, at least in the north, with slope facies sedimentary rocks that denote a strong change in depositional environments at the time of deposition of the terrane-forming rocks. The suture zone truncates the Atmur–Delgo Suture and fold-thrust belts in the Bayuda Terrane. Whether it truncates and/or converges with the Hamisana Zone in the Gebeit–Gabgaba Terrane and the Nakasib Suture is unresolved because of sand cover. Suturing was caused by c. 650–600 Ma E–W shortening and NW–SE oblique collision (Abdelsalam et al. 1998, 2003), concurrent with isotopic rehomogenization and cooling and/or uplift of high-grade basement in the Bayuda and Halfa Terranes, which was coincident with the closing stage of metamorphism along the Atmur–Delgo Suture, the emplacement of high-level granitoid plutons and a period of wrench faulting that correlates with the Najd Fault System in the Arabian Shield (Bernau et al. 1987; Harms et al. 1994). Its cessation at c. 580 Ma was marked by the rapid uplift and cooling of a deformed granite (Abdelsalam et al. 1998).

Amalgamation in Oman

Evidence from drillholes and seismic profiles suggest that a broadly contemporary convergent event may have occurred to the east of the shield. Assembly of the Oman crystalline basement by 750 Ma was followed by uplift to shallower levels concurrent with extension and the emplacement of a granite dyke swarm (Mercolli et al. 2001), the probable onset of rifting (750–590 Ma) and the unconformable deposition of the Huqf Supergroup (Allen & Hildebrand 2001). Allen & Hildebrand (2001) and Cozzi et al. (2001) proposed that the upper part of the Huqf Supergroup was deposited in a distal foreland basin created by down-flexing of the lithosphere as a result of c. 590–540 Ma thrusting involving transport of Pan-African orogenic rocks from the west. The Western Deformation Front (WDF; Fig. 9), a structural boundary inferred from subsurface drillhole and seismic data, and modelled as a west-dipping thrust, is conceivably the leading edge of such an allochthon (Allen & Hildebrand 2001). Conversely, Loosveld et al. (1996) argued that the presence of Huqf Supergroup rocks over 600 km to the west in Saudi Arabia (Fig. 9) implies that the WDF is an intra- rather than an interterrane structure, and that it probably reflects shortening and deformation during a period of compression related to movement on the Najd Fault System rather than transport of an orogenic allochthon. Confirmation of the existence of a thrust front and foreland basin in Oman would be tectonically important because, from the perspective of this paper, it would allow the hypothesis that the Oman basement is a remnant of continental foreland and a possible fragment of East Gondwana.

Amalgamation in Yemen

As extensively documented by Windley et al. (1996) and Whitehouse et al. (1998, 2001b), the Precambrian basement of Yemen is a collage of Archaean gneissic and Pan-African island-arc terranes. The boundaries between the terranes are prominent shear zones marked by structural dislocations, mylonite and, at Hajjah and NE of Sada (Fig 11), ophiolite. Correlation across the Gulf of Aden between the terrane collage in Yemen and equivalent rocks in the Horn of Africa is reasonably well established (Kröner & Sassi 1996; Whitehouse et al. 2001b; Windley et al. 2001), but correlation to the north with terranes in Saudi Arabia is less certain. Windley et al. (1996) correlate the ophiolite-decorated Shear Zone at Hajjah and Sada with the Nabitah Fault Zone, but much of the intervening ground is covered by Phanerozoic sediments and flood basalt so that the true extension of the Hajjah–Sada Shear Zone is problematic and, as shown in Figure 10, is more likely to be in the region east of Najran rather than along the Nabitah Fault Zone, unless there is a hidden intervening Najd-type fault. As noted above, the Asir Terrane is characterized by north-trending shear zones, numbers of which are decorated by slivers of mafic–ultramafic rocks (Fig. 11). In this context, the Hajjah–Sada Shear Zone is not extraordinary and is viewed by the present authors as a typical structural element of what may be a southern continuation of the Asir Terrane into Yemen. The northward extents of the Archaean Abas and Al Mahfid Terranes are concealed by Phanerozoic sedimentary rocks and eolian sand of the Rub al Khali Basin, and are likely limited

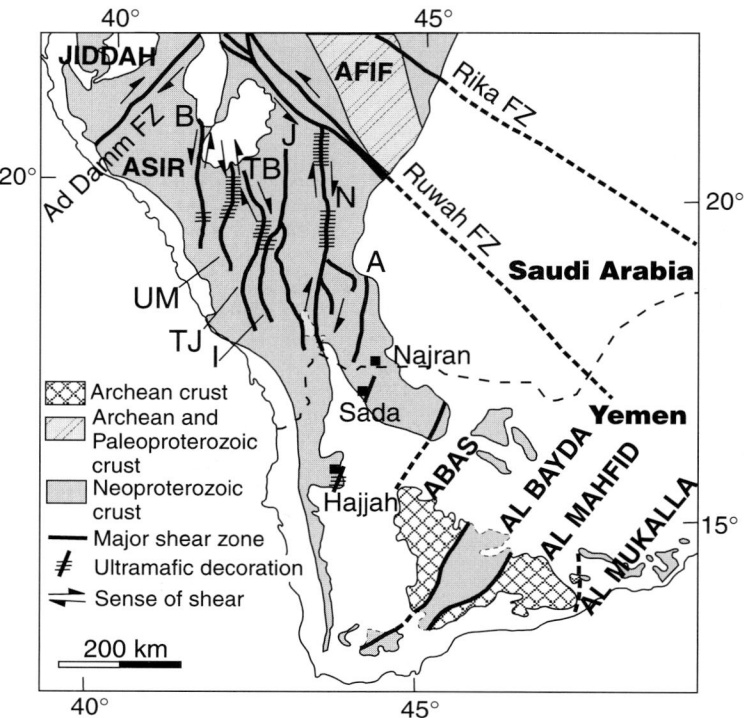

Fig. 11. Map of the southern part of the Arabian Shield showing relations between the Precambrian exposures in Saudi Arabia and Yemen. A, Asharah Fault Zone; B, Baydah Shear Zone; I, Ibran Shear Zone; J, Junaynah Fault Zone; N, Nabitah Fault Zone; UM, Umm Farwah Shear Zone; TB, Tabalah Shear Zone; TJ, Tarj Shear Zone.

by NW-trending Najd faults (Fig. 11). Proposed correlation of these terranes with the Khida Subterrane of the Afif Terrane is therefore problematic, not least because of isotopic and geochronological differences between the crustal units, as pointed out by Whitehouse et al. (2001b). Given the available data, it is clear that the Abas, Al-Mahfid and Khida Terranes are continental in character. They are conceivable fragments rifted from an older continent, but strict correlation of the terranes, in the sense that they are pieces of the same crust, is not currently supported.

Post-amalgamation overlap assemblages

The 150–100 Ma span between the 680–640 Ma amalgamation event in the eastern Arabian Shield and the transformation of the entire northern EAO into a passive margin on the southern flank of palaeo-Tethys is one of noteworthy tectonic heterogeneity involving diachronous deformational and crust-forming events that, from the Gondwana perspective, provide a large amount of information about the closing stages of Gondwana convergence and assembly. One expression of this heterogeneity was the development of post-amalgamation volcano-sedimentary basins (Fig. 12; Johnson 2003). About 40 such basins are known, ranging in size from large basins in Oman and the Arabian Gulf (Figs. 1 & 9) (underlying an area $>200 \times >600$ km) to small basins in the Midyan Terrane (5×15 km), and varying in age, based on the onset of deposition, from c. 723 to 580 Ma (Table 1). Using the criteria of more than/less than 500 m thickness of carbonate succession, the relative abundances of grey-green and red-purple rocks, and other sedimentary characteristics, the overlap assemblages are provisionally divided into marine and epicontinental basins.

Marine basins

The largest marine post-amalgamation basins occur in Oman and western Saudi Arabia. The Omani Assemblage, the Huqf Supergroup (c. 732–540 Ma: Brasier et al. 2000), comprises epiclastic, carbonate, subordinate volcanic rocks and thick successions of salt that crop out along the Arabian Sea coast and are intersected in drillholes or imaged on seismic pro-

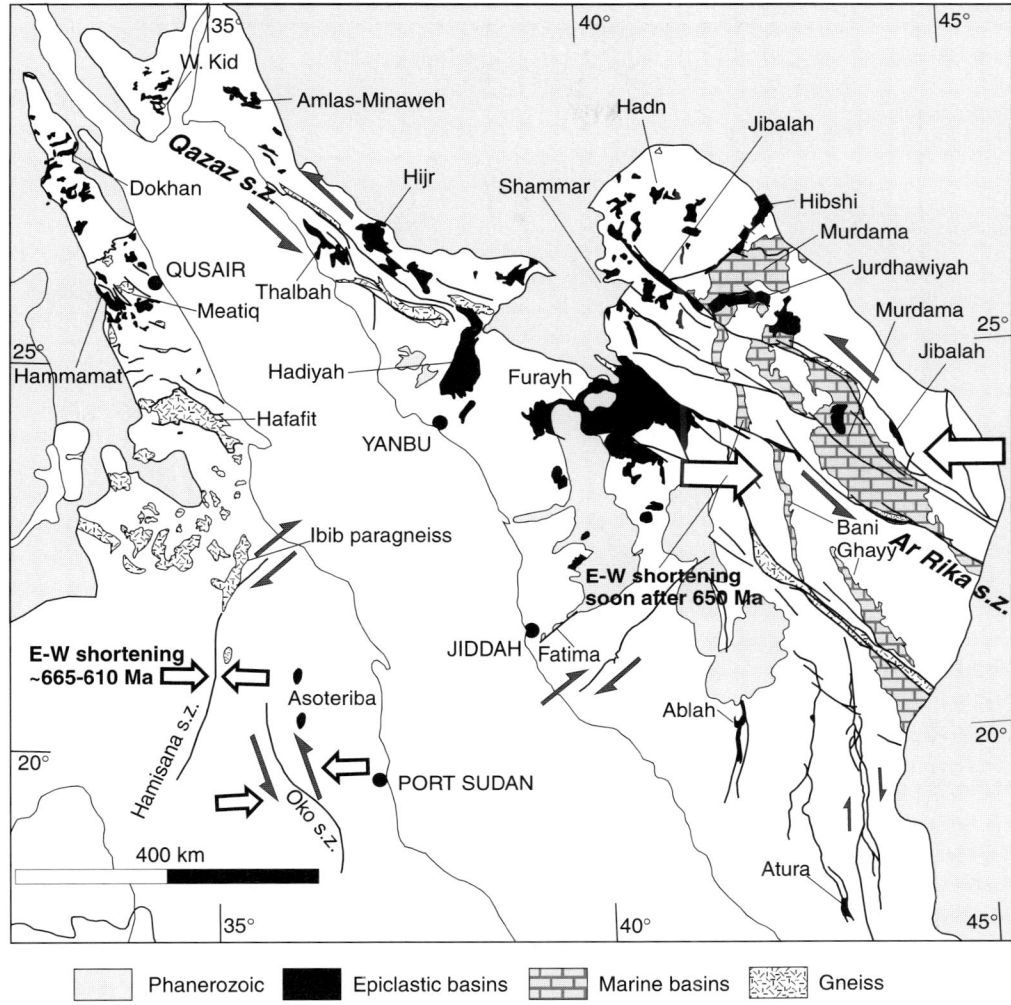

Fig. 12. Late Neoproterozoic post-amalgamation features in the Arabian–Nubian Shield showing depositional basins, shear zones, reactivated sutures, gneiss domes and sense of transcurrent shearing (half arrows) and shortening (large arrows). Ar Rika–Qazaz Shear Zone system (bold) is the main trans-Arabian Shield Najd Structure.

files beneath Phanerozoic cover inland (Loosveld et al. 1996; Blood 2001). The supergroup is unconformable on the older crystalline basement. The upper and lower parts accumulated in fault-controlled NE-trending basins, whereas the middle part was deposited in a platform-and-slope environment (Husseini & Husseini 1990; Loosveld et al. 1996; Allen & Hildebrand 2001; Leather et al. 2001). Glaciomarine deposits in the basal 1100 m, including diamictites and dropstone-bearing laminated mudstone, are associated with cap dolomites and deepwater siltstone and argillaceous sandstone, and have Sr isotopic excursions that appear to correlate with the Surtian and Marinoan glaciations (Brasier et al. 2000; Amthor et al. 2001).

The Murdama group (<670–>650 Ma; Cole & Hedge 1986) is a shallow-marine succession of deformed, but virtually unmetamorphosed, grey-green feldspathic sandstone and siltstone, polymict conglomerate, carbonate, subordinate basalt and rhyolite, and carbonate build-ups >1000 m thick that crops out in basins in the eastern Arabian Shield (Fig. 12; Johnson 2003). Conglomerate is typically massive and probably originated as fanglomerate, but the sandstone and siltstone are well bedded and represent tidal-flat to turbiditic deposits (Wallace

Table 1. *Ages of post-amalgamation depositional basins in the Arabian–Nubian Shield*

Basin name	Depositional age (Ma)	Reference
Marine Basins		
Huqf Supergroup – Oman basins	c. 723–c. 540	Brasier et al. 2000 Cozzi et al. 2001 Husseini & Husseini 1990
Murdama Group – Maslum and Maraghan Basins	<670–>655	Cole 1988 Cole & Hedge 1986
Fatimah Group	688–680?	Darbyshire et al. 1983 Duyverman et al. 1982 Grainger 2001
Bani Ghayy Group	c. 650	Agar 1986 Doebrich pers. comm.
Epicontinental basins		
Hadiyah Group	<680?	Kemp 1981
Furayh Group	c. 670	Delfour 1981
Asoteriba Group	c. 670–650	Cavanagh 1979 Stern & Kröner 1993
Ibib paragneiss protoliths	c. 663	Stern et al. 1989 Stern & Kröner 1993
Atura Formation	<664–>641	Johnson 2000
Ablah Group	641–610	Genna et al. 1999 Johnson et al. 2001
Jurdhawiyah Group	640–615	Cole 1988 Cole & Hedge 1986
Thalbah Group	640–610	Davies 1985
Shammar Group	630–625	Calvez & Kemp 1987 Kemp 1996, 1998
Hibshi Formation	c. 630	Cole & Hedge 1986 Williams et al. 1986
Dokhan Volcanics	602–593	Stern et al. 1984 Wilde & Youssef 2000
Hammamat Group	590–585	El-Kaliouby 1996 Stern & Hedge 1985 Wilde & Youssef 2001 Willis et al. 1988
Jibalah Group	c. 580–560	Delfour 1970 Matsah & Kusky 1999
Saramuj Conglomerate	c. 580–560	Jarrar et al. 1991, 1993

1986). The carbonates are locally stromatolitic. The basins may have been locally controlled by faults, but on the regional scale appear to be large down sags. Smaller marine basins are represented by the Fatima and Bani Ghayy groups, weakly constrained at 688–675 Ma (Duyverman et al. 1982; Darbyshire et al. 1983) and c. 650 (Doebrich pers. comm.), respectively. Both groups contain significant amounts of terrestrial conglomerate and sandstone, and bimodal volcanic rocks consistent with their deposition in fault-controlled basins (Agar 1986), but also contain up to 750 and 800 m, respectively, of stromatolitic carbonate (Grainger 2001). The large volume of Murdama Basin sedimentary material, testified by its surface area of c. 72 000 km^2, including a concealed 200 km extension beneath Phanerozoic rocks SE of the Arabian Shield (Johnson & Stewart 1995), a reported thickness of >8000 m, and exhumation of amphibolite and granulite facies rocks at the sub-Murdama unconformity, imply that the sub-Murdama Afif and adjacent terranes were extensively eroded soon after orogenic uplift (Cole 1988). Furthermore, the carbonate build-ups preserved in the Murdama, Fatima and Bani Ghayy Basins suggest that even if terrane assembly caused orogenic uplift, large parts of the northeastern Arabian Shield were reduced to sea level and developed connections to the ocean flanking the emerging Gondwana supercontinent within a few million years of orogeny. Basin deposition was terminated by the onset of folding, which, in the case of the Murdama and Bani Ghayy groups, denotes significant E–W shortening in the eastern shield soon after 650 Ma (Johnson 2003).

Epicontinental terrestrial basins

Other overlap assemblages in the ANS were deposited between 680 and c. 580 Ma in shallow water and terrestrial, fault-controlled down sags, pull aparts, rifts, half-grabens and calderas (e.g. Stern et al. 1984; Willis et al. 1988; Jarrar et al. 1993; El-Kaliouby 1996; Wilde & Youssef 2000). The rocks are molassic, characterized by abundant hematite stain and cement, a coarse clastic grain size, including cobbles and boulders in many conglomerates, and bimodal volcanic chemistry. Rocks in some basins were gently folded, but in others were intensely deformed and metamorphosed to greenschist and amphibolite facies depending on basin proximity to Late Neoproterozoic shear zones (Fowler & Osman 2001). Some, but not all, of the basins contain thin successions of thin-bedded stromatolitic limestone and some contain diamictites, which together with reported dropstones might be evidence of Marinoan glaciation. The stratigraphic distribution of the basins (Fig. 13) implies that several periods of extension affected the ANS after the peak of orogeny, but deformation of the basins, conversely, indicates that extension was interrupted by compression and brittle–ductile shearing, and gives evidence that Late Neoproterozoic extension was not steady state but alternated with shortening, diastrophism and, in some cases, intense metamorphism (cf. Greiling et al. 1994). Although many basins were surrounded by local high relief, as evidenced by the coarse grain size of typical deposits, the Hammamat Basin, and possibly others, may have been fed by far-flowing river systems rather than by local transport into restricted intermontane basins (Wilde & Youssef 2001).

Late Neoproterozoic gneiss belts and domes

Another expression of diachronous post-amalgamation deformation and crust-forming events is the development of Late Neoproterozoic antiforms and domes of ortho- and paragneiss (Figs. 12 & 14). Well known in the Eastern Desert of Egypt and Sinai as examples of metamorphic core complexes associated with late-orogenic extension (e.g. Fritz et al. 1996; Blasband et al. 1997) or imbricated stacks of antiformal thrust duplexes (e.g. Greiling 1997; Fowler & Osman 2001), Late Neoproterozoic gneiss domes and antiforms also occur along Najd faults in the Arabian Shield (Fig. 14) and farther south in the Nubian Shield.

The Kirsh gneiss (Fig. 15), an example of such gneiss in the Arabian Shield, is an antiform of strongly foliated biotite monzogranite orthogneiss (Al Hawriyah Anticlinorium) intruded into steeply dipping kyanite–quartz schist and subvertical zones of mylonite and ultramylonite at the southeastern end of the Ar Rika-Qazaz Shear System, the main trans-Arabian Shield Najd Structure (Fig. 14). The gentle plunge of the northwesterly trending mineral and elongate-pebble lineations indicates a large component of constriction, or unidirection stretching, and in several sectors of the shear zone the deformed rocks are L-tectonites. The monzogranite is located in a zone of extension between left-stepping faults along the sinistral Ar Rika Fault Zone. On the presumption that the Al Khushaymiyah complex to the NE, which is composed of massive monzogranite, is an undeformed equivalent of the orthogneiss, activity on this section of the Ar Rika Shear Zone is dated at c. 610 Ma.

Well-documented examples of gneiss domes in the Eastern Desert and Sinai (Fig. 12) include the structures at Meatiq (Sturchio et al. 1983; Habib et al. 1985), Gebel el-Shalul (Hamimi et al. 1994; Osman 1996), Gebel El-Sibai (Kamal El-Din et al. 1992; Bregar 1996), Um Had (Fowler & Osman 2001) and Wadi Kid (Blasband et al. 1997), as well

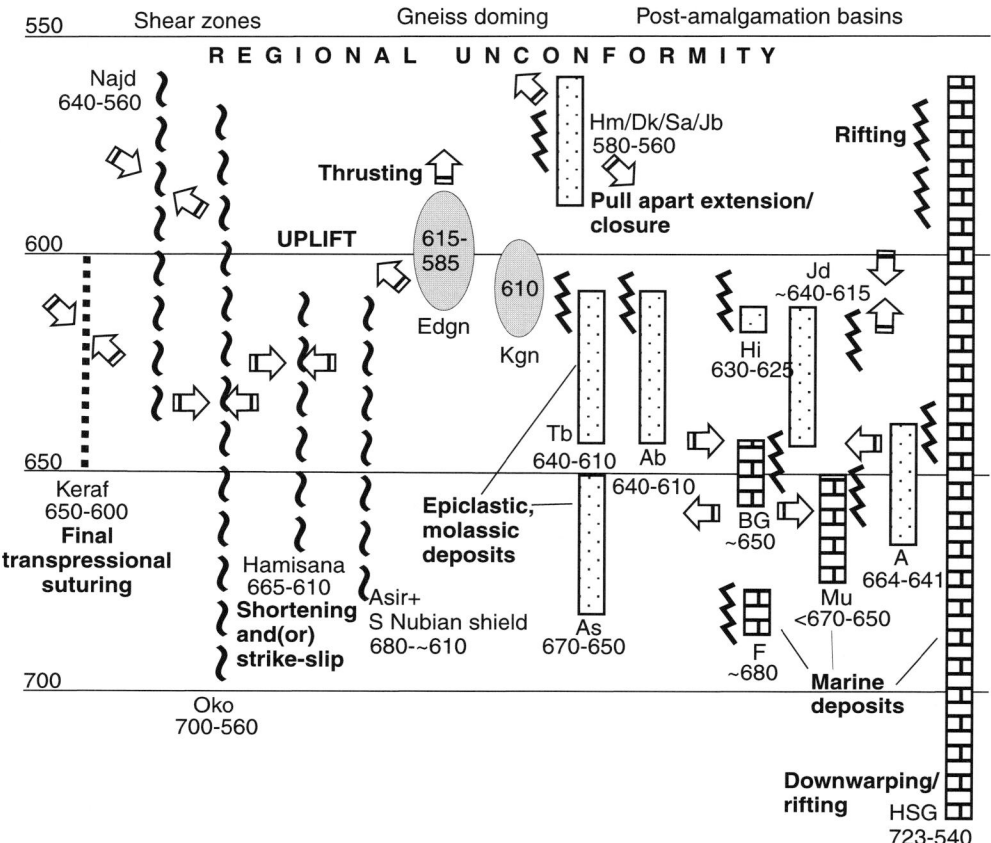

Fig. 13. Diagram illustrating the range of diachronous tectonic events that affected the Arabian–Nubian Shield following the main period of terrane amalgamation. A, Atura formation; Ab, Ablah group; As, Asoteriba and Ibib rocks; BG, Bani Ghayy group; Edgn, Eastern Desert–Sinai gneiss domes; F, Fatima group; Hi, Hibshi formation; Hm/Dk/Sa/Jb, Hammamat/Dokhan/Saramuj/ Jibalah groups; HSG, Huqf Supergroup; Jd, Jurdhawiyah group; Kgn, Kirsh gneiss; Mu, Murdama group; Tb, Thalbah group. Jagged symbol denotes basin deformation. Arrows are indicative of E–W shortening, NW–SE extension and shortening, N–S extension and shortening, and NW thrusting.

as the Hafafit Culmination (El-Ramly et al. 1984; Rashwan 1991; Greiling 1997), but though closely studied, their origins, ages and tectonic significances are debated. The chief common structural elements include: (1) an antiformal structure; (2) a prevailing NNW to NW orientation of stretching lineations and thrust transport direction; and (3) NW-trending sinistral transcurrent shears at the margins of and within the domes, which may be true subvertical shears or folded and steepened mylonitic thrust zones. Less common elements observed at some of the domes include: lineation and transport directions to the SW or NE, as well as to the NW; SW- or NE-vergent thrusts in addition to NW-vergent thrusts; NW-directed thrusts on their southern margin and NW-directed low-angle normal faults on their northern margin; or low-angle normal faults on both margins.

Geothermobarometric experiments on samples from the Meatiq Dome indicate high P–T core conditions (8 kbar and 750–600°C; Neumayr et al. 1996), and the juxtaposition of high-grade core rocks and greenschist facies envelopes implies domal uplift of as much as 10–13 km (Fowler & Osman 2001). Conditions at Wadi Kid were not as extreme, reaching P–T conditions of 3.4–4.2 kbar and 684–488°C during dome exhumation (Brooijmans et al. 2003). The presence of Hammamat and Dhokan rocks in pull-apart basins on the flanks of the domes and in thrust sheets above the domes, as well as dating on pre-, syn- and post-tectonic plutonic rocks and metamorphic minerals, constrain doming, thrusting, shearing and exhumation to sometime after 615 Ma and prior to 585 Ma (Fritz et al. 1996).

Gneiss domes that may also reflect Late

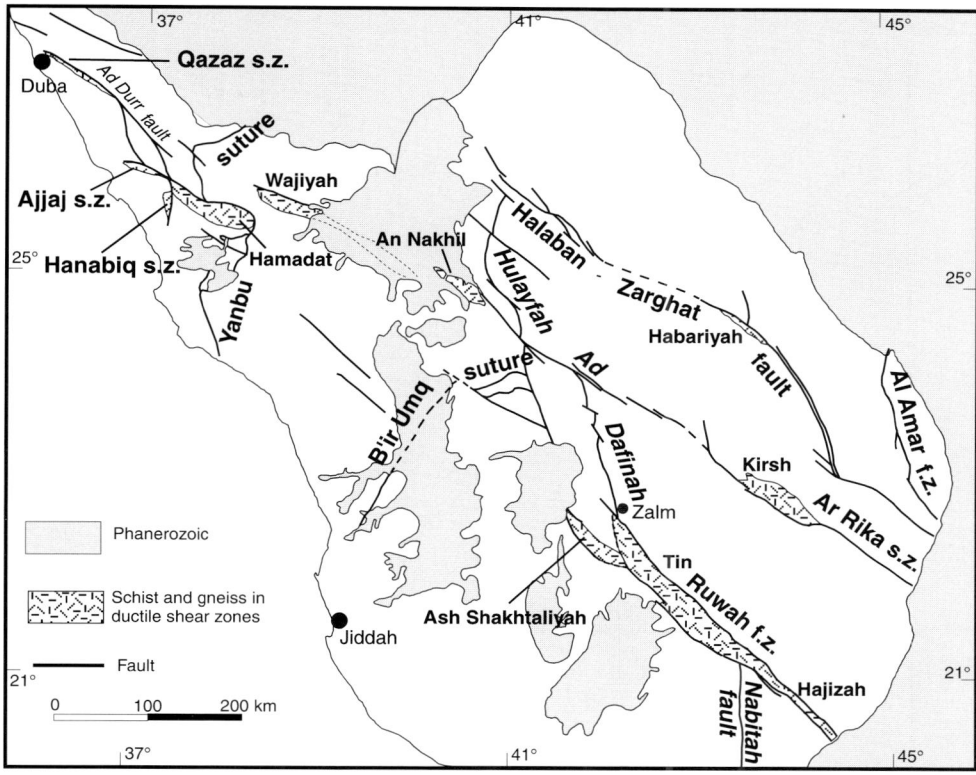

Fig. 14. Late Neoproterozoic gneiss–schist belts in the Arabian Shield showing their spatial relation with Najd faults. (s.z., shear zone; f.z., fault zone).

Neoproterozoic extension and exhumation have been described recently from Eritrea (Fig. 7). The Aneseba Gneiss (protoliths aged 700–796 Ma; Teklay 1997) comprises well-banded high-grade garnet–hornblende gneiss, two-mica gneiss and migmatite (Woldehaimanot 2001). D1 deformation affected the hornblende gneiss only; D2 deformation affected hornblende and mica gneisses; and D3 deformation affected all rocks. An early thrust event manifested by imbricate structures showing an east-directed shearing is superimposed by a widespread extensional event marked by mesoscopic ductile shear zones with normal sense-of-slip coincident with moderately NW-plunging stretching lineations, and kinematic markers (S-type asymmetrically folded quartz/pegmatite veins, rotated, pre-shearing foliation and σ-type tailed porphyroblasts of K-feldspar) indicating top-to-NW shear. This extensional event deformed earlier synmetamorphic planar fabrics and is interpreted as a late, post-metamorphic event reflecting late-stage exhumation and extension. In the Ghedem area (Ghebreab 1999) the dominant exposures are low domes of garnetiferous orthogneiss structurally overlain by garnet-bearing staurolite–kyanite paragneisses. An anticlockwise P–T path of metamorphism, determined from mineral assemblages included in cores, outer cores and rim matrices of garnets, indicate that the rocks of the Ghedem Terrane were subjected to heating with loading at intermediate crustal levels followed by near-isothermal loading at deep crustal levels, and were subsequently subjected to decompression and cooling associated with rapid exhumation from a minimum depth of 44 km (Ghebreab 1999). A different study of the high-grade metasedimentary rocks in the Ghedem area (Beyth et al. 1997) deduces a clockwise P–T path of thermobarometry with peak conditions of 8–10 kbar and c. 700°C, and proposes a collision setting for the metamorphism followed by extension on the low-angle shear zones that are observed in the area (Beyth et al. 2003).

Late Neoproterozoic shear zones

Shears are a third class of structure that reflect tectonic heterogeneity during the Late Neoproterozoic history of the ANS. They include: (1) north-trending

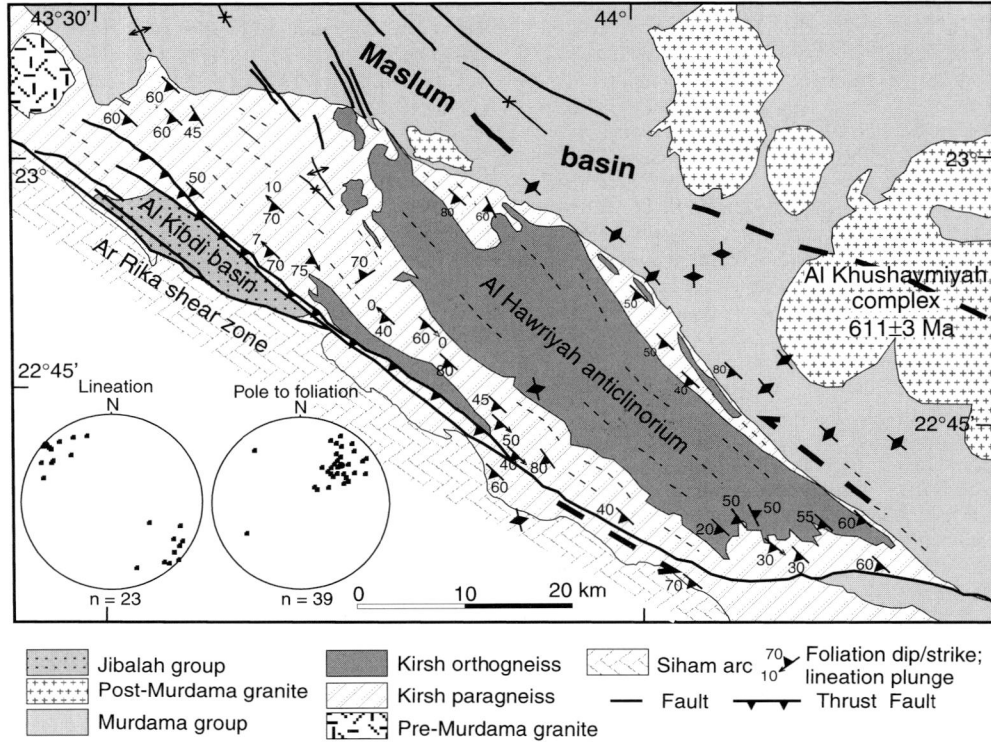

Fig. 15. Kirsh gneiss along the Ar Rika Fault Zone (after Delfour 1979, 1980). Lineation and pole-to-foliation data plotted on lower hemisphere, Schmidt nets.

subvertical brittle–ductile shear zones in the southern part of the shield; (2) the Hamisana and Oko Shear Zones in the Haya and Gebeit Terranes; and (3) NW- and NE-trending transcurrent faults of the Najd system.

North-trending shear zones

The brittle–ductile shears in the southern shield (Fig. 7) are zones of phyllonite, schist, gneiss, mylonite, and subordinate serpentinite and serpentinite schist. They crop out as belts of strongly deformed rocks, 400 km long and 1 km wide and contain an abundance of dextral and sinistral kinematic markers (S–C fabric, rotated and winged porphyroclasts). Some belts contain sheared serpentinite (Fig. 11) and certain of them, the easternmost Nabitah Fault Zone, in particular, are interpreted as sutures within the Asir Composite Terrane, but the structures are currently topics of research because recent work shows that similar shear zones in western Ethiopia and the Mozambique Belt in Madagascar formed late in the EAO orogenic cycle as a result of E–W shortening and orogen-parallel extension (Martelat et al. 2000; Braathen et al. 2001; de Wit et al. 2001). A few of the shear zones in the Asir Terrane, such as the Tabalah-Tarj Zone discussed above, appear to be old and related to an early phase of terrane amalgamation. Others underwent displacements as late as <640 Ma (Johnson et al. 2001) and may be examples of Mozambique-Belt-type shears that reworked older sutures or part of the Late Neoproterozoic deformation event that created the Najd faults (Nehlig et al. 2002).

Prominent sinuous north-trending shear zones in the Haya and Gebeit Terranes (Fig. 12) accommodated large amounts of E–W shortening. The Hamisana Shear Zone postdates and deforms the Allaqi–Heiani–Onib–Sol Hamed Suture, and was active between 665 and 610 Ma (Stern & Kröner 1993). It is a belt of strongly foliated and lineated amphibolite and amphibolite-grade paragneiss and biotite–muscovite orthogneiss (de Wall et al. 2001) that resulted from a high-T metamorphic event superimposed on the greenschist metamorphic assemblages of the adjacent rocks. Pervasive L–S fabrics and an absence of S-C fabrics indicate that deformation was mainly constrictional, consistent with inferred E–W regional compression (Miller & Dixon 1992; de Wall *et al.* 2001). Subsequent low-T retrograde metamorphism was associated with local NE-trending dextral shearing (de Wall *et al.* 2001). The longitudinal extent of the Hamisana Shear Zone is unknown: to the south it passes under eolian sand so that its relations to the Nakasib or Keraf Sutures are obscured; to the north, the zone projects into outcrops of ortho- and paragneiss extensively exposed between the Onib–Sol Hamed Suture and the Gebel Gerf Ophiolite Nappe, in which the Hamisana Shear Zone loses its character as a discrete narrow zone of high strain. de Wall *et al.* (2001) conclude that the Hamisana Shear Zone is an expression of orogenic compressional deformation unrelated to either large-scale transpression or major escape tectonics; Stern & Kröner (1993) suggest that it may be a Mozambique-type zone of shortening.

The Oko Shear Zone, superimposed on the Nakasib Suture, likewise accommodated a large amount of E–W shortening (Abdelsalam 1994; Abdelsalam & Stern 1996), but contains additional structures that denote a more complex evolution than that of the Hamisana Shear Zone. The shear zone was active between 700 and 560 Ma (Abdelsalam & Stern 1993*b*) and developed in stages of: (1) E–W shortening marked by the development of north-trending upright interference folds in pre-existing Nakasib Suture fabrics; (2) NW-trending sinistral and subordinate NE-trending dextral subhorizontal strike-slip faulting, which displaced the Nakasib Suture *c.* 10 km and rotated E–W structures into N–S trends; and (3) terminal east- and west-vergent thrusting and buckling that created a flower structure (Abdelsalam & Stern 1996).

Najd Fault System – issues of timing and origin

The Najd Fault System was originally defined for Late Neoproterozoic NW- and subordinate NE-trending transcurrent faults in the northern part of the Arabian Shield (Fig. 12; Brown & Jackson 1960; Brown 1972; Moore 1979). Faults correlated with the Najd System are known in the southern Arabian Shield, in the subsurface of Oman and in the Nubian Shield, and similar NW-trending sinistral faults are known in the EAO in southern Ethiopia, Tanzania and Madagascar (Fig. 2).

The faults are conventionally interpreted as having a common history and tectonic setting but, in detail, have significant differences in age of activity and structural style (Johnson & Kattan 1999). The Halaban–Zarghat Fault Zone was chiefly active between < 640 and 620 Ma (Cole & Hedge 1986); the Ar Rika-Qazaz Shear Zone was strongly active *c.* < 640–610 Ma; Najd faults that bound the Meatiq and other gneiss domes in the Eastern Desert were active between 615 and 585 Ma (Fritz *et al.* 1996); and the so-called Najd Rift System that operated in Oman as the putative cause of the Ara Group salt basins is inferred to have been active between 570 and 530 Ma (Al-Husseini 2000). Movement persisted on the Halaban–Zarghat Fault until the cessation of Jibalah deposition at *c.* 570 Ma. Final movements on the Ruwah Fault Zone, interpreted by Johnson & Kattan (2001) to be a Najd-reactivated segment of a *c.* 680 Ma suture zone, cataclastically deformed a 592 Ma granite (Stoeser & Stacey 1988). The Ar Rika–Qazaz Shear Zone contains gabbro and quartz syenite plugs that yield whole-rock K–Ar ages of 512 ± 17 and 487 ± 17 Ma (Brown *et al.* 1989), indicating significant Cambrian cooling. The NE-trending faults are dextral and the NW-trending faults are largely, but not entirely, sinistral: Agar (1987), Matsah & Kusky (2001) Kusky & Matsah (2003) argue that a dextral phase preceded the dominant sinistral phase, although the chronologic data of Cole & Hedge (1986) would question this argument in the case of the Halaban–Zarghat Fault. The two fault sets appear to form a conjugate pair. Complementary vertical movements of > 10 km on parts of the Najd faults are indicated by the juxtaposition of amphibolite-grade mylonitic gneiss along the fault axes, with greenschist facies rocks at the fault margins, and by ductile–brittle transitions along the fault zones that, in typical transcurrent fault systems, would have originated at depths of 10–15 km (Davis & Reynolds 1996). Vertical uplifts of as much as 5 km across the faults are implied also by the regional gentle plunge of stretching lineations (Davies 1984).

Many workers infer that the Najd Fault System formed under general conditions of compression. Soon after the publication of work by Molnar & Tapponnier (1977) on the application of slip-line theory to faulting north of the Himalayas, proposals were made that the Najd faults resulted from indentation of the ANS by a rigid continental plate – from the east in the case of Schmidt *et al.* (1979) and Davies (1984), and from the west in the case of Fleck *et al.* (1980). Agar (1987) and Abdelsalam (1994) argued that the fault system resulted from E–W shortening and collision, whereas Burke & Sengör (1986) advocated an origin during E–W collision and simultaneous orogen-parallel extension and escape. Abdelsalam & Stern (1996) modified the collisional interpretation by postulating that the direction of maximum strain during Najd faulting (σ_1) was oriented NW–SE, oblique to the axis of the EAO rather than orthogonal, as a result of regional transpression caused by the entrapment of the ANS between the East Saharan Craton on the west and an Ar Rayn microplate on the east. Stern (1994) suggested that the NW-trending shear zones were a mechanism that allowed northward orogen-parallel extension as the ANS escaped from collision between East and West Gondwana. In Oman, Loosveld et al. (1996) concluded that Late Precambrian north-trending folds and thrust faults, as well as the NE-trending Ara Group salt basins (part of the Huqf Supergroup), resulted from E–W directed compression associated with movement on presumed extensions of the Najd Fault System into the eastern part of the Arabian Peninsula shown in magnetic and gravity data. This concept was supported by Hall et al. (2001), who favoured a compressive origin of the Ghaba Basin as a push-down basin between overriding thrusts. Fowler & Osman (2001) presented a model that joins Najd faulting, extension, bidirectional thrusting and gneiss doming in the Eastern Desert in a concept of escape tectonics engendered by general E–W compression, and alternating NNW-directed extrusion and NE–SW-directed thrusting in the EAO away from a pivotal collision zone in the Mozambique Belt.

Other interpretations situate the Najd faults in an extensional stress field. Husseini & Husseini (1990) and Al-Husseini (2000), in discussing the origin of the Omani Salt Basins, envisaged that the Najd structures were part of a 570–530 Ma transform fault system on the northern flank of the Gondwana supercontinent that linked rifts in the eastern Arabian Plate (Oman), the northwestern Indian Plate (Punjab) and the Eurasian Plate (Dibbah Rift and Derik Rift), with extensional-collapse structures in the ANS. Based on the kinematic implications of NE-trending dyke swarms and volcanic grabens in the northern Eastern Desert, the Midyan Terrane and Jordan, Stern (1985) proposed that the Najd System was a set of transform faults caused by pervasive NW–SE-directed Late Neoproterozoic extension during the period of 600–575 Ma. Mercolli et al. (2001) report two periods of dyke emplacement in the basement in Oman: one event at c. 750 Ma when the crystalline basement was uplifted and intruded by a large mass of granite dykes along a conjugate set of compressional brittle faults and a second event at c. 550 Ma marked by the emplacement of a suite of calc-alkaline basalt, andesite and rhyolite dykes. Both events are consistent with extension and potential rifting. Emplacement of the Mukeiras Dyke Swarm (c. 715–615 Ma) in the Al Bayda Terrane in Yemen is believed to reflect a similar period of post-collisional rifting in the southern part of the Arabian Shield (Whitehouse et al. 1998).

Clearly, there are conflicting issues of timing and kinematics concerning the Najd Fault System that need to be resolved by additional structural mapping (e.g. Shackleton 1994) or by redefining the system or excluding some faults that conventionally are included in the system (Johnson & Kattan 1999).

The core issue is whether the Najd faults formed under conditions of shortening and compression or extension and rifting. Structurally, the two conditions are not mutually exclusive, and temporally and spatially may grade from one to the other, but conceptually the difference between 'push' and 'pull' is major and raises the question of whether the faults relate to the final convergence of East and West Gondwana or the breakup and rifting of the newly formed Gondwana supercontinent. In the present authors' opinion, Late Neoproterozoic transcurrent faults that meet broad criteria for inclusion in the Najd Fault System in the Najd region of the central Arabian Shield (the type area for the system), the Midyan Terrane and the Eastern Desert have structural features that strongly favour a compressive origin. These include: (1) a conjugate system geometry with common indications of sinistral horizontal movement on NW-trending faults and dextral movement on NE faults; (2) the presence, in the Murdama Group adjacent to the Ar Rika Fault, of north-trending en-echelon folds of the type expected from sinistral transcurrent movement (Johnson 2003); (3) an abundance of NW-trending stretching lineations along the fault zones; (4) gradations between amphibolite grade, ductile and greenschist facies, brittle conditions of metamorphism and deformation along the faults; and (5) the location of Jibalah Group depositional basins and orthogneiss domes at releasing bends between left-stepping sinistral faults. The orientations of the faults and folds are consistent with a broad E–W orientation of σ_1 and would be consistent with the general regime envisaged by many workers of Late Neoproterozoic E–W shortening, which resolved into NW- and NE-directed shearing.

Whatever their origin, most movements on the Najd faults ceased by the Early Cambrian, at which time a thick succession of post-fault continental to shallow-marine Lower Palaeozoic siliciclastic rocks began to cover the region. The siliciclastic rocks were deposited on a regional unconformity created by the erosion and depression of what had become a stable platform, and were distributed by north- and NE-flowing fluvial systems from sources in interior Gondwana (Konert et al. 2001). Subsequent movements on the Najd faults were entirely brittle and presumably caused by adjustments of the faults, as zones of crustal weakness, to Phanerozoic plate movements. Phanerozoic movements are evidenced by post-sedimentary faulting in the Wajid Group on the line of the Ruwah Fault at the SE flank of the Arabian Shield and by displacement of Cretaceous formations on the line of the Ar Rika Shear Zone in central Arabia (Andre 1989).

Discussion

A large body of substantial data indicates that the ANS evolved in the Mozambique Ocean between the converging blocks of East and West Gondwana (Stern 1994). However, representative rocks of West Gondwana only crop out in poorly exposed, commonly reworked form in the East Saharan Craton and representatives of East Gondwana in Arabia are unknown. By default, therefore, the ANS is the chief source of information about convergence and supercontinental assembly at the Tethyan margin of Gondwana. The timing of terrane accretion and amalgamation (Fig. 3) helps to constrain the history of closure of the Mozambique Ocean and the initial convergence of East and West Gondwana; the post-amalgamation events constrain models about the final contact of East and West Gondwana, and Gondwana assembly.

As indicated by the weakly constrained age of the oldest (Jabal Tharwah, Bi'r Umq, Bi'r Tuluhah) and the better constrained age of the youngest (Halaban) ophiolites in the ANS, the Mozambique Ocean existed by 870 Ma and continued to grow until c. 695 Ma, suggesting that Rodinia, in the region of the eventual ANS, had started to rift by 870 Ma. Oceanic-basin closure at subduction zones, indicated by the creation of juvenile oceanic and continental-margin arcs, commenced soon after rifting and continued until completion c. 600 Ma by creation of the Keraf Suture. Because of a paucity of palaeopole data, the pre-amalgamation spatial relationships among the terranes in the ANS are unknown. Simplistically working back from their amalgamated relations, and acknowledging that terrane analysis in the ANS is still provisional, relations such as those shown in Figure 16 are envisaged. The figure depicts rifted segments of Rodinia – a large segment in the west (present-day coordinates), which eventually became West Gondwana and a smaller segment in the east, which was entrained in the Afif Terrane as part of the Khida Subterrane – and a juvenile ocean between them. Rocks that became the Halfa and Bayuda Terranes were evolving at or close to the western Rodinia fragment, acquiring both mature and juvenile isotopic signatures, and volcanic and plutonic rocks with juvenile signatures were forming within the ocean. By 780–760 Ma (Fig. 16a), the earliest intra-oceanic arcs were in contact and amalgamation was in progress along probably intraterrane sutures within the Asir Terrane and, during a period of dextral transpression, along the Bi'r Umq–Nakasib Suture. Over the next 100 Ma (Fig. 16b), the Bayuda/Halfa Terranes converged with each other and with the East Saharan Craton, marking the onset of approach of West Gondwana, and composite intraoceanic crustal units composed of the Asir–Barka–Hager/Tokar–Nakfa–Ghedum Terranes, the Haya–Gebeit–Eastern Desert–Midyan Terranes and the Afif Subterranes assembled and converged. Approach of the continental-margin Ar Rayn Terrane from the east completed amalgamation of the Arabian Shield by 680–640 Ma (Fig. 16c) and marked closure of the eastern part of the Mozambique Ocean. By 600 Ma (Fig. 16d), the assembly of the ANS was complete and the Mozambique Ocean was fully consumed by the creation of the Keraf Suture. It is envisaged that by this stage East and West Gondwana were in contact across the amalgamated rocks of the ANS, although the absence of proven representatives of East Gondwana in the region makes its tectonic role speculative. Metallogenic and geochemical investigations in the Ar Rayn Terrane (Doebrich et al. 2001) interpretations of the Huqf Supergroup as a type of foreland basin (Cozzi et al 2001), and the magnetic character of eastern Arabia (Johnson & Stewart 1995) are permissive of continental crust east of the Arabian Shield, but its presence and provenance are yet to be established.

The amalgamated terranes exposed at the present level of erosion in the ANS constitute an orogenic belt minimally 1200 km wide in a E–W direction that at the time of orogeny would have formed part of an enormous mountain chain with a large root extending deep into the lithosphere. Since then, the root has been lost, and the present-day crust of the Arabian and northeastern African Plates displays a layered structure and a uniform Moho depth of 35–45 km (Gettings et al. 1986; Rodgers et al. 1999) that retains no vestige of crustal thickening. Modern geophysical measurements consequently provide little clue as to the Late Neoproterozoic fate of the mountain belt. The geological data, conversely, gives ample evidence of Late Neoproterozoic uplift, erosion, extension, depression and compression before transformation into a palaeo-Tethys passive margin.

The ANS lacks large areas of exhumed granulite facies rocks of the type found farther south in the Mozambique Belt (Stern 1994), implying there was no large-scale isostatic rebound following the relaxation of confining orogenic stress, but small exposures of granulite facies and more extensive amphibolite facies regional metamorphism are evidence that parts of the shield were, indeed, uplifted by 10–20 km. Additional evidence of uplift comes from the P–T conditions of gneiss domes described along the Najd Fault System and from $^{40}Ar/^{39}Ar$ evidence of rapid cooling in the eastern Arabian Shield (Al-Saleh et al. 1998; Al-Saleh & Boyle 2001b), Sinai (Cosca et al. 1999) and the western Nubian Shield (Abdelsalam et al. 1998). Judging by the onset of post-amalgamation basin deposition, uplift and erosion occurred by 725 Ma east of the shield (in

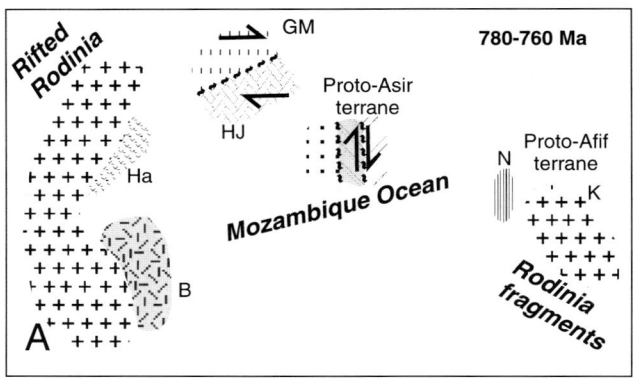
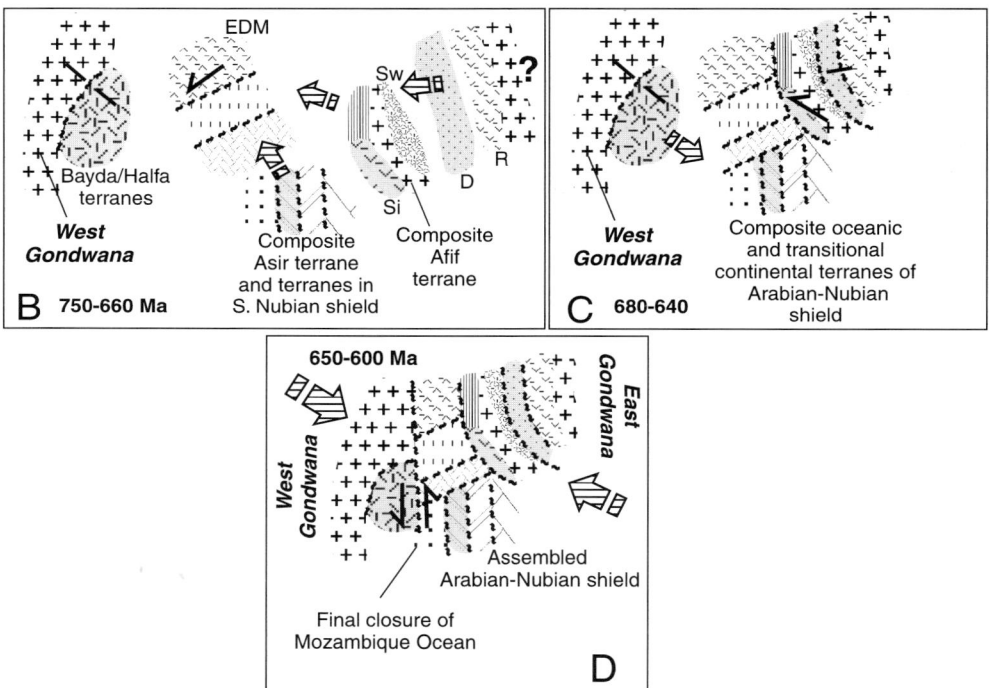

Fig. 16. Sketch of the assembly of the Arabian–Nubian Shield in four time slices showing progress of terrane amalgamation and eventual accretion to East and West Gondwana. A, 780–760 Ma; B, 750–660 Ma; C, 680–640 Ma; D, 650–600 Ma. Terranes and subterranes: B, Bayda; D, Ad Dawadimi; EDM, Eastern Desert–Midyan; GH, Gebeit–Hijaz; Ha, Halfa; HJ, Haya–Jiddah; K, Kirsh; N, Nuqrah; R, Ar Rayn; Si, Siham; Sw, Suwaj. Small arrows show sense of frontal convergence; half arrows show sense of transpression; large arrows show inferred convergence trajectories of terranes and composite terranes.

Oman), and by 680–670 Ma in the Arabian and Nubian Shields. The gneiss domes denote uplifts of as much as 44 km between 615 and 585 Ma, and the $^{40}Ar/^{39}Ar$ data indicate uplifts of between 600 and 580 Ma (Fig. 13).

Post-amalgamation erosion was intense, producing enormous volumes of debris that filled the Murdama Basin and others, and, coupled with depression, effectively reduced large parts of the eastern ANS to sea level shortly after peak orogeny. The large numbers of fault-controlled basins, starting with the Fatima Group and ending with the Hammamat, Dokhan, Saramuj and Jibalah Groups (Fig. 13), are evidence that depression was linked to extension and local rifting throughout the period of c. 680–560 Ma. On the other hand, depression of the large Murdama and Huqf Supergroup Basins may have been driven by more complex mechanisms, such as thrusting and downflexing, as well as rifting in the case of the Huqf and thermal contraction and subduction delamination in the case of the Murdama. But evidence from the same basins demonstrates that extension was periodic and alternated with compression, which folded successively and sometimes sheared, thrusted and metamorphosed all the post-amalgamation basins (with the possible exception of the Asoteriba), causing unconformities in places where younger basins are atop older basins. As late as 600 Ma and possibly younger in the case of the Oko zone, the southern part of the ANS underwent E–W shortening and north-directed shearing resulting in the Hamisana Zone, the transpressional Keraf Suture and late shear zones in Asir. Between 640 and 560 Ma the entire ANS underwent NW- and NE-directed Najd transcurrent faulting, and between 615 and 585 Ma parts of the ANS, particularly the NW, experienced thrusting and(or) low-angle normal faulting during the creation of gneiss domes.

The range of post-amalgamation events and the variations in orientations of the local strain ellipsoids in time and space implied by the varying orientations of the observed structures, militates against any simple model for the final assembly of Gondwana. Nonetheless, the far-field motion (Fig. 16d) implied by sinistral transpression on the Keraf Suture, the dominant NW–SE direction of thrusting evident at some gneiss domes and the predominant NW strike of Najd faults, are consistent with the broad E–W to NW–SE uniform or scissor-like direction of Gondwana convergence modelled by a number of authors (e.g. Stern 1994; Greiling et al. 2000; Fowler & Osman 2001). Post-amalgamation tectonic models that envisage only extension and collapse driven by gravitational instabililty (e.g. Blasband et al. 2000) do not account for the periods of shortening that interrupted extension, and models of Najd faulting that rely on passive-margin rifting do not adequately account for E–W compression. Many details of the timing and kinematics of events during the final convergence of East and West Gondwana are still problematic. Nonetheless, it is clear that adequate models of Late Neoproterozoic Gondwana convergence at the northern end of the EAO will need to accommodate both E–W and NW–SE extension and shortening soon after orogenic climax, as well as orogen-parallel extension because of slip on the Najd faults, crustal depression, the creation of depositional basins and a westerly penetration of seaways into the core of the orogen.

This paper is published, with respect to work by PRJ, with the kind permission of M. A. Tawfiq, President, Saudi Geological Survey. The authors are grateful for the support and encouragement of their many colleagues engaged in the fascinating investigation of the Arabian–Nubian Shield and acknowledge, with thanks, reviews of an early version of the manuscript by M. Whitehouse and S. Johnson.

References

ABD EL-NABY, H. H. & FRISCH, W. 1999. Metamorphic sole of Wadi Haimur–Abu Swayel ophiolite: implications on late Proterozoic accretion. *In*: DE WALL, H. & GREILING, R. O. (eds) *Aspects of Pan-African Tectonics*. Bilateral Seminars of the International Bureau, Forschungszentrum Jülich GmbH, **32**, 9–14.

ABD EL-NABY, H. H. & FRISCH, W. 2002. Origin of the Wadi Haimur–Abu Swayel gneiss belt, south Eastern Desert, Egypt: petrological and geochronologicalal constraints. *Precambrian Research*, **113**, 307–322.

ABDEEN, M. M., ABDELSALAM, M. G., DOWIDAR, H. M. & STERN, R. J. 2000. Evolution of the Neoproterozoic Allaqi-Heiana Suture zone, southern Egypt (abstract). *Journal of African Earth Sciences*, **30**, 1.

ABDEL RAHMAN, M. 1993. *Geochemical and Geotectonic Controls of the Metallogenic Evolution of Selected Ophiolite Complexes from the Sudan*. Berliner Geowissenschaftliche Abhandlungen, **145**.

ABDELSALAM, M. G. 1994. The Oko shear zone, Sudan: post-accretionary deformation in the Arabian–Nubian shield. *Journal of the Geological Society, London*, **151**, 767–776.

ABDELSALAM, M. G. & STERN, R. J. 1993a. Tectonic evolution of the Nakasib Suture, Red Sea Hills, Sudan: evidence for a late Precambrian Wilson Cycle. *Journal of the Geological Society, London*, **150**, 393–404.

ABDELSALAM, M. G. & STERN, R. J. 1993b. Timing of events along the Nakasib Suture and the Oko shear zone, Sudan. *In:* THORWEIHE, U. & SCHANDELMEIER, H. (eds) *Geoscientific Research in Northeast Africa*, Proceedings of the International Conference on Geoscientific Research in Northeast Africa, Berlin. Balkema, Rotterdam, 99–103.

ABDELSALAM, M. G. & STERN, R. J. 1996. Sutures and shear zones in the Arabian–Nubian shield. *Journal of African Earth Sciences*, **23**, 289–310.

ABDELSALAM, M. G., ABDEL-RAHMAN, E. M., EL-FAKI, E. M., AL-HUR, B., EL-BASHIER, F. M., STERN, R. J. & THURMOND, A. K. 2003. Neoproterozoic deformation in the northeastern part of the Saharan metacraton, northern Sudan. *In:* KUSKY, T. M., ABDELSALAM, M., TUCKER, R. & STERN, R. J. *Evolution of the East African and Related Orogens, and the Assembly of Gondwana*, Precambrian Research Special Issue, in press.

ABDELSALAM, M. G., STERN, R. J., COPELAND, P., ELFAKI, E. M., ELHUR, B. & IBRAHIM, F. M. 1998. The Neoproterozoic Keraf Suture in NE Sudan: sinistral transpression along the eastern margin of West Gondwana. *Journal of Geology*, **106**, 133–147.

ABDELSALAM, M. G., STERN, R. J., SCHANDELMEIER, H. & SULTAN, M. 1995. Deformation history of the Neoproterozoic Keraf Zone in NE Sudan, revealed by shuttle imaging radar. *Journal of Geology*, **103**, 475–491.

AFIFI, A. A. 1989. *Geology of the Mahd adh Dhahab district, Kingdom of Saudi Arabia*. Saudi Arabian Directorate General of Mineral Resources, Open-File Report USGS-OF-09-2.

AGAR, R. A. 1985. Stratigraphy and palaeogeography of the Siham group: direct evidence for a late Proterozoic continental microplate and active continental margin in the Saudi Arabian shield. *Journal of the Geological Society, London*, **142**, 1205-1220.

AGAR, R. A. 1986. The Bani Ghayy group; sedimentation and volcanism in pull-apart grabens of the Najd strike-slip orogen, Saudi Arabian Shield. *Precambrian Research*, **31**, 259-274.

AGAR, R. A. 1987. The Najd fault system revisited: a two-way strike-slip orogen in the Saudi Arabian Shield. *Journal of Structural Geology*, **9**, 41-48.

AGAR, R. A., STACEY, J. S. & WHITEHOUSE, M. J. 1992. *Evolution of the southern Afif Terrane - a geochronological study*. Saudi Arabian Directorate General of Mineral Resources, Open-File Report DMMR-OF-10-15.

AL-DOUKHI, H. & DIVI, R. 2001. Polyphse deformational structures in the Proterozoic rocks near Hadbin in southern Oman. *Gondwana Research*, **4**, 137-139.

AL-HUSSEINI, M. I. 2000. Origin of Arabian Plate structures: Amar collision and Najd rift. *GeoArabia*, **5**, 527-542.

AL-KATHIRI, A. 2001. Geology of the crystalline basement of the Hasik region (Dhofar, Sultanate of Oman). *GeoArabia*, **6**, 279-280.

ALLEN, Ph. & HILDEBRAND, P. 2001. Neoproterozoic basin development in Oman: an Afro-Arabian perspective. *GeoArabia*, **6**, 283.

ALMOND D. C. & AHMED, F. 1987. Ductile shear zones in the northern Red Sea Hills, Sudan and their implications for crustal collision. *Geological Journal*, **22**, 175-184.

AL-REHAILI, M. H. & WARDEN, A. J. 1980. Comparison of the Bi'r Umq and Hamdah ultrabasic complexes, Saudi Arabia. *Bulletin of the Institute of Applied Geology King Abdulaziz University, Jiddah*, **3**, 143-156.

AL-SALEH, A. M. & BOYLE, A. P. 2001a. Neoproterozoic ensialic back-arc spreading in the eastern Arabian Shield: geochemical evidence from the Halaban ophiolite. *Journal of African Earth Sciences*, **33**, 1-15.

AL-SALEH, A. M. & BOYLE, A. P. 2001b. Structural rejuvenation of the eastern Arabian shield during continental collision: $^{40}Ar/^{39}Ar$ evidence from the Ar Ridaniyah ophiolite mélange. *Journal of African Earth Sciences*, **33**, 135-141.

AL-SALEH, A. M., BOYLE, A. P. & MUSSETT, A. E. 1998. Metamorphism and $^{40}Ar/^{39}Ar$ dating of the Halaban ophiolite and associated units: evidence for two-stage orogenesis in the eastern Arabian shield. *Journal of the Geological Society*, London, **155**, 165-175.

AL-SHANTI, A. M. & GASS, I. G. 1983. The upper Proterozoic ophiolite mélange zones of the easternmost Arabian shield. *Journal of the Geological Society*, London, **140**, 867-876.

AL-SHANTI, A. M. S. & MITCHELL, A. H. G. 1976. Late Precambrian subduction and collision in the Al Amar-Idsas region, Arabian shield, Kingdom of Saudi Arabia. *Tectonophysics*, **30**, T41-T47.

AMTHOR, J. E., GROTZINGER, J. P., SCHRÖDER, S. BOWRING, S. A. & MATTER, A. 2001. Biogeochemical significance of the Precambrian-Cambrian boundary carbon isotope anomaly: constraints from Oman. *GeoArabia*, **6**, 284.

ANDRE, C. G. 1989. Evidence for Phanerozoic reactivation of the Najd fault system in AVHRR, TM, and SPOT images of central Arabia: *Photogrammetric Engineering and Remote Sensing*, **55**, 1129-1136.

BAKOR, A. R., GASS, I. G. & NEARY, C. R. 1976. Jabal al Wask, northwest Saudi Aabia: an Eocambrian back-arc ophiolite. *Earth and Planetary Science Letters*, **30**, 1-9.

BENNET, J. D. & MOSLEY, P. N. 1987. Tiered-tectonics and evolution, E Desert and Sinai, Egypt. *In*: MATHEIS, G. & SCHANDELMEIER, H. (eds) *Current Research in African Earth Sciences*. Balkema, Rotterdam, 79-82.

BERHE, S. M. 1990. Ophiolites in the northeast and east Africa: implication for Proterozoic crustal growth. *Journal of the Geological Society*, London, **147**, 41-57.

BERNAU, R., DARBYSHIRE, D. P. F., FRANZ, G. ET AL. 1987. Petrology, geochemistry and structural development of the Bir Safsaf-Aswan uplift, Southern Egypt. *Journal of African Earth Sciences*, **6**, 79-90.

BEYTH, M., AVIGAD, D., WETZEL, H., MATTHEWS, A. & BERHE, S. M. 2003. Aspects of the late orogenic evolution of the East African Orogen in northeast Ethiopia and east Eritrea with emphasis on the exhumation of kyanite-bearing rocks. *In*: KUSKY, T. M., ABDELSALAM, M., TUCKER, R. & STERN, R. J. *Evolution of the East African and Related Orogens, and the Assembly of Gondwana*. Precambrian Research Special Issue, in press.

BEYTH, M., STERN, R. J. & MATTHEWS, A. 1997. Significance of high-grade metasediments from the Neoproterozoic basement of Eritrea. *Precambrian Research*, **86**, 45-58.

BLASBAND, B., BROOIJMANS, P., DIRKS, P., VISSER, W. & WHITE, S. 1997. A Pan-African core complex in the Sinai, Egypt. *Geologie en Mijnbouw*, **73**, 247-266.

BLASBAND, B., WHITE, S., BROOIJMANS, P., DE BOORDER, H. & VISSER, W. 2000. Late Proterozoic extensional collapse in the Arabian Nubian shield. *Journal of the Geological Society*, London, **157**, 615-628.

BLOOD, M. F. 2001. Exploration for a frontier salt basin in southwest Oman. *GeoArabia*, **6**, 159-176.

BOKHARI, F. Y. & KRAMERS, J. D. 1981. Island-arc character and late Precambrian age of volcanics at Wadi Shwas, Hijaz, Saudi Arabia: geochemical and Sr and Nd isotopic evidence. *Earth and Planetary Science Letters*, **54**, 409-422.

BRAATHEN, A., GREENE, T., SELASSIE, M. G. & WORKU, T. 2001. Juxtaposition of Neoproterozoic units along the Baruda-Tulu Dimtu shear-belt in the East African Orogen of western Ethiopia. *Precambrian Research*, **107**, 215-234.

BRASIER, M., MCCARRON, G., TUCKER, R., LEATHER, J., ALLEN, P. & SHIELDS, G. 2000. New U-Pb zircon dates for the Neoproterozoic Ghubrah glaciation and for the top of the Huqf Supergroup, Oman. *Geology*, **28**, 175-178.

BREGAR, M. 1996. The core complex of the Gabal Sibai crystalline dome in the Eastern Desert of Egypt.

Abstracts of the Geological Survey of Egypt Centennial Conference 1996, Cairo, Egypt, 25–27.

BROOIJMANS, P., BLASBAND, B., WHITE, S. H., VISSER, W. J. & DIRKS, P. 2003. Geothermobarometric evidence for a metamorphic core-complex in Sinai, Egypt. *In*: KUSKY, T. M., ABDELSALAM, M., TUCKER, R. & STERN, R. J. *Evolution of the East African and Related Orogens, and the Assembly of Gondwana*. Precambrian Research Special Issue, in press.

BROWN, F. B., SCHMIDT, D. L. & HUFFMAN Jr, A. C. 1989. Geology of the Arabian Peninsula: shield area of western Saudi Arabia. U.S. Geological Survey Professional Paper 560–A.

BROWN, G. B. 1972. Tectonic map of the Arabian Peninsula Map Ap-2. Saudi Arabian Directorate General of Mineral Resources.

BROWN, G. F. & JACKSON, R. O. 1960. The Arabian shield. *International Geological Congress, XXI Session, Norden 1960*, **IX**, 69–77.

BURKE, K. & ŞENGÖR, C. 1986. Tectonic escape in the evolution of the continental crust. *In*: BARAZANGI, M. & BROWN, L. (eds) *The Continental Crust*. American Geophysical Union Geodynamics Series, **14**, 41–53.

CALVEZ, J. Y. & KEMP, J. 1982. Geochronologicalal investigations in the Mahd adh Dhahab quadrangle, central Arabian shield. Saudi Arabian Deputy Ministry for Mineral Resources, Open-File Report BRGM-OF-02-5.

CALVEZ, J. Y. & KEMP, J. 1987. Rb–Sr geochronology of the Shammar group in the Hulayfah area northern Arabian shield. Saudi Arabian Deputy Ministry for Mineral Resources, Open-File Report BRGM-OF-07-11.

CALVEZ, J. Y, ALSAC, C., DELFOUR, J., KEMP. J. & PELLATON, C. 1983. Geological evolution of western, central, and eastern parts of the northern Precambrian shield, Kingdom of Saudi Arabia. Saudi Arabian Deputy Ministry for Mineral Resources, Open-File Report BRGM-OF-02–26.

CAMP, V. E. 1984. Island arcs and their role in the evolution of the western Arabian shield. *Geological Society of America Bulletin*, **95**, 913–921.

CAMP, V. E. 1986. Geological map of the Umm al Birak quadrangle, Sheet 23D, Kingdom of Saudi Arabia. Map GM87. Saudi Arabian Directorate General of Mineral Resources.

CAMP, V. E., JACKSON, N. J., RAMSAY, C. R., ROOBOL, M. J., STOESER, D. B. & WHITE, D. L. 1984. Discussion on the upper Proterozoic ophiolite mélange zones of the easternmost Arabian Shield. *Journal of the Geological Society, London*, **141**, 1083–1085.

CAVANAGH, B. J. 1979. Rb–Sr geochronology of some pre-Nubian igneous complexes of central and northeastern Sudan. PhD Thesis, University of Leeds.

CHURCH, W. R. 1988. Ophiolites, structures, and microplates of the Arabian–Nubian shield: a critical comment. *In*: EL-GABY, S. & GREILING, R. O. (eds) *The Pan-African Belt of Northeast Africa and Adjacent Areas*. Vieweg, Braunschweig/Wiesbaden, 289–316.

CLAESSON, S., PALLISTER, J. S. & TATSUMOTO, M. 1984. Samarium–neodymium data on two late Proterozoic ophiolites of Saudi Arabia and implications for crustal and mantle evolution. *Contributions to Mineralogy and Petrology*, **85**, 244–252.

COLE, J. C. 1988. Geological map of the Aban al Ahmar quadrangle, Sheet 25F, Kingdom of Saudi Arabia. Map GM89. Saudi Arabian Deputy Ministry for Mineral Resources.

COLE, J. C., & HEDGE, C. E. 1986. Geochronological investigation of late Proterozoic rocks in the northeastern shield of Saudi Arabia. Saudi Arabian Deputy Ministry for Mineral Resources, Technical Record USGS-TR-05–5.

COOPER, J. A., STACEY, J. S., STOESER, D. B. & FLECK, R. J. 1979. An evaluation of the zircon method of isotopic dating in the southern Arabian Craton. *Contributions to Mineralogy and Petrology*, **68**, 429–439.

COSCA, M. A., SHIMRON, A. & CABY, R. 1999. Late Precambrian metamorphism and cooling in the Arabian–Nubian Shield: petrology and $^{40}Ar/^{39}Ar$ geochronology of metamorphic rocks of the Elat area (southern Israel). *Precambrian Research*, **98**, 107–127.

COZZI, A., ALLEN, P. H. & MCCARRON, G. 2001. From carbonate ramp differentiation to basin evolution: the late Precambrian Buah formation (Nafun group), north-central Oman. *GeoArabia*, **6**, 295–296.

DARBYSHIRE, D. P. F., JACKSON, N. J., RAMSAY, C. R. & ROOBOL, M. J. 1983. Rb–Sr isotope study of latest Proterozoic volcano-sedimentary belts in the Central Arabian shield. *Journal of the Geological Society, London*, **140**, 203–213.

DAVIES, F. B. 1984. Strain analysis of wrench faults and collision tectonics of the Arabian–Nubian shield. *Journal of Geology*, **92**, 37–53.

DAVIES, F. B. 1985. Geological map of the Al Wajh quadrangle, Sheet 26B, Kingdom of Saudi Arabia. Map GM83. Saudi Arabian Deputy Ministry for Mineral Resources.

DAVIS, G. H. & REYNOLDS, S. J. 1996. *Structural Geology of Rocks and Regions* (2nd edition). Wiley, New York.

DE SOUZA FILHO, C. R. & DRURY, S. A. 1998. A Neoproterozoic supra-subduction terrane in northern Eritrea, NE Africa. *Journal of the Geological Society, London*, **155**, 551–566.

DE WALL, H., GREILING, R. O. & SADEK, M. F. 2001. Post-collisional shortening in the late Pan-African Hamisana high strain zone, SE Egypt: field and magnetic fabric evidence. *Precambrian Research*, **107**, 179–194.

DE WIT, M. J., BOWRING, S. A., ASHWAL, L. D., RANDRIANASOLO, L. G., MOREL, V. P. I. & RAMBELOSON, R. A. 2001. Ages and tectonic evolution of Neoproterozoic ductile shear zones in southwestern Madagascar, with implications for Gondwana studies. *Tectonics*, **20**, 1–45.

DELFOUR, J. 1970. Le groupe de J'Balah, une nouvelle unite du bouclier arabe. *Bulletin BRGM France*, **1**, 51–67.

DELFOUR, J. 1977. Geology of the Nuqrah quadrangle, Sheet 25E, Kingdom of Saudi Arabia. Map GM28. Saudi Arabian Directorate General of Mineral Resources.

DELFOUR, J. 1979. Geological map of the Halaban quadrangle, Sheet 23G, Kingdom of Saudi Arabia. Map GM46. Saudi Arabian Directorate General of Mineral Resources.

DELFOUR, J. 1980. Geological map of the Ar Rika quadrangle, Sheet 22G, Kingdom of Saudi Arabia. Map

GM51. Saudi Arabian Directorate General of Mineral Resources.

DELFOUR, J. 1981. *Geological map of the Al Hissu quadrangle, Sheet 24E, Kingdom of Saudi Arabia. Map GM58*. Saudi Arabian Deputy Ministry for Mineral Resources.

DENKLER, T., FRANZ, G. & SCHANDELMEIER, H. 1984. Tectonometamorphic evolution of the Neoproterozoic Delgo Suture zone, northern Sudan. *Geologische Rundshau*, **83**, 578–590.

DOEBRICH, J. L., HAYES, T. S., SIDDIQUI, A. A. ET AL. 2001. The Ar Rayn Terrane: geotectonic implications of unique metallogeny in the Arabian Shield. *Gondwana Research*, **4**, 127–128.

DUYVERMAN, H. J., HARRIS, N. B. W. & HAWKESWORTH, C. J. 1982. Crustal accretion in the Pan African: Nd and Sr isotope evidence from the Arabian shield. *Earth and Planetary Science Letters*, **59**, 315–326.

EL-GABY, M. A. & HASHAD, A. H. 1990. The basement complex of the Eastern Desert and Sinai. *In*: SAID, R. (ed.) *The Geology of Egypt*. Balkema, Rotterdam.

EL-KALIOUBY, B. 1996. Provenance, tectonic setting and geochemical character of the Hammamat molasse sediments around Um Had pluton, Eastern Desert, Egypt. *Middle East Research Centre, Ain Shams University Earth Science Series*, **10**, 75–88.

EL-RAMLY, M. F., GREILING, R., KRÖNER, A. & RASHWAN, A. A. 1984. On the tectonic evolution of the Wadi Hafafit area and environs, Eastern Desert, Egypt. *Bulletin of the Faculty of Earth Sciences King Abdulaziz University, Jiddah*, **6**, 113–126.

EMBLETON, J. C. B., HUGHES, D. J., KLEMENIC, P. M., POOLE, S. & VAIL, J. R. 1983. New approach to the stratigraphy and tectonic evolution of the Red Sea Hills, Sudan. *Bulletin of the Faculty of Earth Sciences, King Abdulaziz University, Jiddah*, **6**, 101–112.

FITCHES, W. R., GRAHAM, R. H., HUSSEIN, I. M., RIES, A. C., SHACKLETON, R. M. & PRICE, R. C. 1983. The Late Proterozoic ophiolite of Sol Hamed, NE Sudan. *Precambrian Research*, **19**, 385–411.

FLECK, R. J. 1985. *Age of diorite–granodiorite gneisses of the Jiddah–Makkah region, Kingdom of Saudi Arabia*. Saudi Arabian Deputy Ministry for Mineral Resources Professional Papers PP-2.

FLECK, R. J., GREENWOOD, W. R., HADLEY, D. G., ANDERSON, R. E. & SCHMIDT, D. L. 1980. *Rubidium–strontium geochronology and plate-tectonic evolution of the southern part of the Arabian Shield*. U.S. Geological Survey Professional Paper 1131.

FOWLER, T. J. & OSMAN, A. F. 2001. Gneiss-cored interference dome associated with two phases of late Pan-African thrusting in the central Eastern Desert, Egypt. *Precambrian Research*, **108**, 17–43.

FRISCH, W. & AL-SHANTI, A. 1977. Ophiolite belts and the collision of island arcs in the Arabian Shield. *Tectonophysics*, **43**, 293–306.

FRITZ, H., WALLBRECHER, E., KHUIDEIR, A. A., EL ELA, A. & DALLMEYER, D. R. 1996. Formation of Neoproterozoic metamorphic core complexes during oblique convergence (Eastern Desert, Egypt). *Journal of African Earth Sciences*, **23**, 311–329.

GASKELL, J. L. 1985. Reappraisal of the Gebeit Gold Mine, northeast Sudan: a case history. *In*: JONES, M. (ed.) *Prospecting in Areas of Desert Terrain*. The Institution of Mining and Metallurgy. London, 49–58.

GASS, I. G. 1977. The evolution of the Pan African crystalline basement in NE Africa and Arabia. *Journal of the Geological Society, London*, **134**, 129–138.

GASS, I. G., RIES, A. C., SHACKLETON, R. M. & SMEWING, J. D. 1990. Tectonics, geochronology and geochemistry of the Precambrian rocks of Oman. *In*: ROBERTSON, A. H. F., SEALE, M. P. & RIESE, A. S. (eds) *The Geology and Tectonics of the Oman Region*. Geological Society, London, Special Publication, **49**, 585–599.

GENNA, A., GUERROT, C., DESCHAMPS, Y., NEHLIG, P. & SHANTI, M. 1999. Le formations Ablah d'Arabie Saudite (datation et implication géologique). *C.R. Acad, Sci, Paris. Earth and Planetary Sciences*, **329**, 661–667.

GEORGEL, J.-M., BOBILLIER, J., DELOM, J., BOURLIER, M. & GILOT, J.-L. 1990. *Total-intensity residual aeromagnetic maps of the Precambrian shield reduced to the pole and upward continued to 800m above ground level*. Saudi Arabian Directorate General of Mineral Resources Open-File Report BRGM-OF-09–15.

GETTINGS, M. E., BLANK, H. R. & MOONEY, W. D. 1986. Crustal structure of southwestern Saudi Arabia. *Journal of Geophysical Research*, **91**, 6491–6512.

GHEBREAB, W. 1999. Tectono-metamorphic history of Neoproterozoic rocks in eastern Eritrea. *Precambrian Research*, **98**, 83–105.

GRAINGER, D. L. 2001. The late Proterozoic Fatima group near Jiddah. *GeoArabia*, **6**, 103–114.

GREILING, R. O. 1997. Thrust tectonics in crystalline domains: the origin of a gneiss dome. *Proceedings of the Indian Academy of Science. Earth and Planetary Sciences*, **106**, 209–220.

GREILING, R. O., ABDEEN, M. M., DARDIR, A. A. ET AL. 1994. A structural synthesis of the Proterozoic Arabian–Nubian shield in Egypt. *Geologische Rundshau*, **83**, 484–501.

GREILING, R. O., DE WALL, H., KONTNY, A., KOBER, B., EL HINNAWI, M., SADEK, M. F. & ABDEEN, M. M. 2000. Pan-African orogeny in the Arabian–Nubian shield: convergence, escape, and collapse. *Journal of African Earth Sciences*, **30**, 36.

GREILING, R. O., KRÖNER, A., EL-RAMLY, M. F. & RASHWAN, A. A. 1988. Structural relationships between the southern and central parts of the Eastern Desert of Egypt: details of a fold and thrust belt. *In*: EL-GABY, S & GREILING, R. O. (eds) *The Pan-African Belt of Northeast Africa and Adjacent Areas*. Vieweg, Braunschweig/Wiesbaden, 122–145.

HABIB, M. E., AHMED, A. A. & EL-NADY, O. M. 1985. Two orogenies in the Meatiq area of the CED, Egypt. *Precambrian Research*, **30**, 83–111.

HALL, S., BRANNAN, J. & FLANAGAN, S. 2001. Tectonic development of the Gabba Salt Basin, Block 3 Oman. *GeoArabia*, **6**, 508.

HAMIMI, Z., EL-AMAWY, M. A. & WETAIT, M. 1994. Geology and structural evolution of El-Shalul dome and environs, Central Eastern Desert, Egypt. *Journal of Geology*, **38**, 575–595.

HARMS, U., DARBYSHIRE, D. P. F., DENKLER, T., HENGST, M. & SCHANDELEMEIER, H. 1994. Evolution of the Neoproterozoic Delgo Suture zone and crustal growth in Northern Sudan: geochemical and radiogenic isotope constraints. *Geologische Rundschau*, **83**, 591–603.

HARMS, U., SCHANDELMEIER, H. & DARBYSHIRE, D. P. F. 1990. Pan-African reworked early/middle Proterozoic crust in NE Africa west of the Nile: Sr and Nd isotope evidence. *Journal of the Geological Society, London*, **147**, 859–872.

HARRIS, N. B. W., GASS, I. G. & HAWKESWORTH, C. J. 1990. A geochemical approach to allochthonous terranes: a Pan-African case study. *Philosophical Transactions of the Royal Society, London*, **A331**, 533–548.

HARRIS, N. B. W., HAWKESWORTH, C. J. & RIES, A. C. 1984. Crustal evolution in northeast and east Africa from model Nd ages. *Nature*, **309**, 773–776.

HASSAN, M. A. & HASHAD, A. H. 1990. Precambrian of Egypt. *In*: SAID, R. (ed.) *The Geology of Egypt*. Balkema, Rotterdam, 201–245.

HASSANEN, M. A. & HARRAZ, H. Z. 1996. Geochemistry and Sr- and Nd-isotopic study on rare-metal-bearing granitic rocks, central Eastern Desert, Egypt. *Precambrian Research*, 80, 1–22.

HEDGE, C. E. 1984. *Precambrian geochronology of part of northwestern Saudi Arabia*. Saudi Arabian Deputy Ministry for Mineral Resources, Open-File Report USGS-OF-04-31.

HUSSEINI, M. I. & HUSSEINI, S. I. 1990. Origin of the Infracambrian salt basins of the Middle East. *In*: BROOKS, J, (ed.) *Classic Petroleum Provinces*. Geological Society, London, Special Publication, **50**, 279–292.

JARRAR, G., WACHENDORF, H. & ZACHMANN 1993. A Pan-African alkaline pluton intruding Saramuj Conglomerate, south-west Jordan. *Geologische Rundschau*, **82**, 121–135.

JARRAR, G. WACHENDORF, H. & ZELLMER, H. 1991. The Saramuj Conglomerate: evolution of a Pan-African molasse sequence from southwest Jordan. *Neues Jahrbuch für Geologie und Palaeontologie, Monatshefte*, **6**, 335–356

JOHNSON, P. R. 1994. *The Nakasib suture: a compilation of recent information about a Sudanese fold and thrust belt and implications for the age, structure, and mineralization of the Bi'r Umq suture, Kingdom of Saudi Arabia*. Saudi Arabian Deputy Ministry for Mineral Resources, Open-File Report USGS-OF-94-6.

JOHNSON, P. R. 1998. *The structural geology of the Samran–Shayban area, Kingdom of Saudi Arabia*. Saudi Arabian Deputy Ministry for Mineral Resources, Technical Report USGS-TR-98-2.

JOHNSON, P. R. 2000. *Proterozoic geology of Western Saudi Arabia – Southern sheet*. Saudi Geological Survey, Open-File Report USGS-OF-99-7.

JOHNSON, P. R. 2003. Post-amalgamation basins of the NE Arabian Shield and implications for Neoproterozoic III tectonism in the northern East African Orogen. *In*: KUSKY, T. M., ABDELSALAM, M., TUCKER, R. & STERN, R. J. *Evolution of the East African and Related Orogens, and the Assembly of Gondwana*. Precambrian Research Special Issue, in press.

JOHNSON, P. R. & KATTAN, F. 1999. The Ruwah, Ar Rika, and Halaban-Zarghat fault zones: northwest-trending Neoproterozoic brittle–ductile shear zones in west-central Saudi Arabia. *In*: DE WALL, H. & GREILING, R.O. (eds) *Aspects of Pan-African Tectonics*. Bilateral Seminars of the International Bureau, Forschungszentrum Jülich GmbH, **32**, 75–79.

JOHNSON, P. R. & KATTAN, F. 2001. Oblique sinistral transpression in the Arabian shield: the timing and kinematics of a Neoproterozoic suture zone. *Precambrian Research*, **107**, 117–138.

JOHNSON, P. R. & STEWART, I. C. F. 1995. Magnetically inferred basement structure in central Saudi Arabia. *Tectonophysics*, **245**, 37–52.

JOHNSON, P. R. & VRANAS, G. J. 1984. *The origin and development of late Proterozoic rocks in the Arabian shield – an analysis of terranes and mineral environments*. Saudi Arabian Deputy Ministry for Mineral Resources, Open-File Report RF-OF-04-32.

JOHNSON, P. R., ABDELSALAM, M. & STERN, R. J. 2002. *The Bi'r Umq–Nakasib shear zone: geology and structure of a Neoproterozoic suture in the northern East African orogen, Saudi Arabia and Sudan*. Saudi Geological Survey Technical Report SGS-TR-2002-1.

JOHNSON, P. R., CARTEN, R. B. & JASTANIAH, A. 1997. *Tabulation of previously published U–Pb, Rb–Sr, and Sm–Nd numerical age data for the Precambrian of Northeast Africa and Arabia*. Saudi Arabian Deputy Ministry for Mineral Resources, Open-File Report USGS-OF-97-1.

JOHNSON, P. R., KATTAN, F. H. & WOODEN, J. L. 2001. Implications of SHRIMP and microstructural data on the age and kinematics of shearing in the Asir terrane, southern Arabian shield, Saudi Arabia. *Gondwana Research*, **4**, 172–173.

JOHNSON, P. R., SCHEIBNER, E. & SMITH, E. A. 1987. Basement fragments, accreted tectonostratigraphic terranes, and overlap sequences: elements in the tectonic evolution of the Arabian Shield. *In*: LEITCH, E. C. & SCHEIBNER, E. (eds) *Terrane Accretion and Orogenic Belts*. American Geophysical Union, Washington, Geodynamics Series, **19**, 323–343.

JONES, D. L., HOWELL, D. G., CONEY, P. J. & MONGER, J. W. H. 1983. Recognition, character, and analysis of tectonostratigraphic terranes in western North America. *In*: HASHIMOTO, M. & UYEDE, S. (eds) Accretion Tectonics in the Circum-Pacific regions. Terra Scientific Publishing Company, Tokyo, 21–35.

KAMAL EL-DIN, G. M., KHUDAEI, A. A. & GREILING, R. O. 1992. Tectonic evolution of a Pan-African gneiss culmination, Gebal El-Sibai area, Central Eastern Desert, Egypt. Zbl. *Geologisch-Paläontologischer*, **1**, 2637–2640.

KEMP, J. 1981. *Geological map of the Wadi al Ays quadrangle, Sheet 25C, Kingdom of Saudi Arabia. Map GM53*. Saudi Arabian Deputy Ministry for Mineral Resources.

KEMP, J. 1996. The Kura Formation (northern Arabian shield); definition and interpretation: a probable fault-trough sedimentary succession. *Journal of African Earth Sciences*, **22**, 507–523.

KEMP, J. 1998. Caldera-related volcanic rocks in the Shammar group, northern Arabian shield. *Journal of African Earth Sciences*, **26**, 551–572.

KLEMENIC, P. M. 1985. New geochronologicalal data on volcanic rocks from northeast Sudan and their implication for crustal evolution. *Precambrian Research*, **30**, 263–276.

KLEMENIC, P. M. & POOLE, S. 1988. The geology and geochemistry of Upper Proterozoic granitoids from the Red Sea Hills, Sudan. *Journal of the Geological Society, London*, **145**, 635–643.

KONERT, G., AFIFI, A. M., AL-HAJRI, S. A. & DROSTE, H. J. 2001. Palaeozoic stratigraphy and hydrocarbon habitat of the Arabian Platform. *GeoArabia*, **6**, 407–442.

KRÖNER, A. 1985. Ophiolites and the evolution of tectonic boundaries in the late Proterozoic Arabian–Nubian shield of northeastern Africa and Arabia. *Precambrian Research*, **27**, 277–300.

KRÖNER, A. & SASSI, F. P. 1996. Evolution of the northern Somali basement: new constraints from zircon ages. *Journal of African Earth Sciences*, **22**, 1–15.

KRÖNER, A., GREILING, R., REISCHMANN, T. ET AL. 1987. Pan-African crustal evolution in the Nubian segment of northeast Africa. *In*: KRÖNER, A. (ed.) *Proterozoic Lithospheric Evolution*. Geodynamics Series, **17**. American Geophysical Union, Washington, 235–257.

KRÖNER, A., LINNEBACHER, P., STERN, R. J., REISCHMANN, T., MANTON, W. & HUSSEIN, I. M. 1991. Evolution of Pan-African island arc assemblages in the southern Red Sea Hills, Sudan and in southwestern Arabia as exemplified by geochemistry and geochronology. *Precambrian Research*, **53**, 99–118.

KRÖNER, A., PALLISTER, J. S. & FLECK, R. J. 1992a. Age of initial oceanic magmatism in the Late Proterozoic Arabian Shield. *Geology*, **20**, 803–806.

KRÖNER, A., TODT, W., HUSSEIN, I. M., MANSOUR, M. & RASHWAN, A. A. 1992b. Dating of late Proterozoic ophiolites in Egypt and the Sudan using the single grain zircon evaporation technique. *Precambrian Research*, **59**, 15–32.

KÜSTER, D. & LIÉGEOIS, J.-P. 2001. Sr, Nd isotopes and geochemistry of the Bayuda Desert high-grade metamorphic basement (Sudan): an early Pan-African oceanic convergent margin, not the edge of the East Saharan ghost craton? *Precambrian Research*, **109**, 1–23.

KUSKY, T. M. & MATSAH, M. I. 2003. Neoproterozoic dextral faulting on the Najd fault system, Saudi Arabia, preceded sinistral faulting and escape tectonics related to closure of the Mozambique Ocean. *In*: KUSKY, T. M., ABDELSALAM, M., TUCKER, R. & STERN, R. J. *Evolution of the East African and Related Orogens, and the Assembly of Gondwana*. Precambrian Research Special Issue, in press.

KUSKY, T. M., ABDELSALAM, M., TUCKER, R. D. & STERN, R. J. (eds) 2003. Preface *In*: KUSKY, T. M., ABDELSALAM, M., TUCKER, R. & STERN, R. J. *Evolution of the East African and Related Orogens, and the Assembly of Gondwana*. Precambrian Research Special Issue, in press.

Le Metour, J. Johan, V. & Tegyey, M. 1983. *Geology of the ultramafic–mafic complexes in the Bi'r Tuluhah and Jabal Malhijah areas*. Saudi Arabian Deputy Ministry for Mineral Resources, Open-File Report BRGM-OF-03–40.

LEATHER, J. ALLEN, PH. & BRASIER, M. 2001. Sedimentology of the glaciomarine sediments of the Neoproterozoic Ghadir Manqil formation, Jebel Akhdar, Oman. *GeoArabia*, **6**, 313–314.

LOOSVELD, R. J. H., BELL, A. & TERKEN, J. J. M. 1996. The tectonic evolution of interior Oman. *GeoArabia*, **1** 28–51.

MARTELAT, J.-E., LARDEAUX, J.-M., NICOLLET, C. & RAKOTONDRAZAFY, R. 2000. Strain pattern and late Precambrian deformation history in southern Madagascar. *Precambrian Research*, **102**, 1–120.

MATSAH, M. I. & KUSKY, T. M. 1999. Sedimentary facies of the Neoproterozoic Al-Jifn Basin, NE Arabian shield: relationship to the Halaban-Zarghat (Najd) fault system and the closure of the Mozambique Ocean. *In*: DE WALL, H. & GREILING, R. O. (eds) *Aspects of Pan-African Tectonics*. Bilateral Seminars of the International Bureau, Forschungszentrum Jülich GmbH, **32**, 17–21.

MATSAH, M. I. & KUSKY, T. M. 2001. Analysis of Landsat TM ratio imagery of the Halaban–Zarghat fault and related Jifn Basin, NE Arabian shield: implications for the kinematic history of the Najd fault system. *Gondwana Research*, **4**, 182.

MEINHOLD, K.-D. 1979. The Precambrian basement complex of the Bayuda Desert, northern Sudan. *Revue de Géologie Dynamique et de Geographie Physique*, **21**, 395–401.

MERCOLLI, I, BRINER, A., FREI, R., MARQUER, D. & PETERS, T. 2001. The geology of the late Proterozoic crystalline basement of the Salalah area, south Oman. *GeoArabia*, **6**, 317.

MILLER, M. M. & DIXON, T. H. 1992. Late Proterozoic evolution of the northern part of the Hamisana zone, northeast Sudan: constraints on Pan-African accretionary tectonics. *Journal of the Geological Society, London*, **149**, 743–750.

MOGHAZI, A. M., ANDERSEN, T., OWEISS, G. A. & EL-BOUSEILY, A. M. 1998. Geochemical and Sr-Nd-Pb isotopic data bearing on the origin of Pan-African granitoids in the Kid area, southeast Sinai, Egypt. *Journal of the Geological Society, London*, **155**, 697–710.

MOLNAR, P. & TAPPONNIER, P. 1977. Relation of the tectonics of eastern China to the India–Eurasia collision: application of slip-line field theory to large-scale continental tectonics. *Geology*, **5**, 212–216.

MOORE, J. M. 1979. Tectonics of the Najd transcurrent fault system, Saudi Arabia. *Journal of the Geological Society, London*, **136**, 441–454.

MOORE, T. A. & AL-REHAILI, M. H. 1989. *Geological map of the Makkah quadrangle, Sheet 21D, Kingdom of Saudi Arabia*. Map GM107. Saudi Arabian Directorate General of Mineral Resources.

NASSIEF, M. O., MACDONALD, R. & GASS, I. G. 1984. The Jebel Thurwah Upper Proterozoic ophiolite complex, western Saudi Arabia. *Journal of the Geological Society, London*, **141**, 537–546.

NAWAB, Z. A. H. 1978. *Evolution of the Al Amar-Idsas region of the Arabian shield, Kingdom of Saudi Aabia*. PhD Thesis, University of Western Ontario.

NEHLIG, P., GENNA, A. & ASFIRANE, F. 2002. A review of the Pan-African evolution of the Arabian Shield. *GeoArabia*, **7**, 103–124.

NEUMAYR, P., HOINKES, G., PUHL, J. & MOGESSIE, A. 1996. Polymetamorphism in the Meatiq Basement Complex (eastern Desert, Egypt): P–T variations and implications for tectonic evolution. *Abstracts of the Geological Survey of Egypt Centennial Conference 1996*, 139–141.

OSMAN, A. F. 1996. *Structural geology and geochemical*

studies on Pan-Arican basement rocks, Wadi Zeidun, Central Eastern Desert, Egypt. Scientific Series of the International Bureau, Forschungszentrum Jülich GmbH, **39**.

PALLISTER, J. S., COLE, J. C., STOESER, D. B. & QUICK, J. E. 1990. Use and abuse of crustal accretion calculations. *Geology*, **18**, 35–39.

PALLISTER, J. S., STACEY, J. S., FISCHER, L. B. & PREMO, W. R. 1988. Precambrian ophiolites of Arabia: geological settings, U–Pb geochronology, Pb-isotope characteristics, and implications for continental accretion. *Precambrian Research*, **38**, 1–54.

QUICK, J.E. 1991. Late Proterozoic transpression on the Nabitah fault system – implications for the assembly of the Arabian shield. *Precambrian Research*, **53**, 119–147.

RAMSAY, C. R. 1986. *Geological map of the Rabigh quadrangle, Sheet 22D, Kingdom of Saudi Arabia. Map GM84*. Saudi Arabian Deputy Ministry for Mineral Resources.

RASHWAN, A. A. 1991. *Petrography, geochemistry and petrogenesis of the Migif–Hafafit gneisses at Hafafit Mine area, Egypt*. Scientific Series of the International Bureau, Forschungszentrum Jülich GmbH, **5**.

REISCHMANN, T. & KRÖNER, A. 1994. Late Proterozoic island arc volcanics from Gebeit, Red Sea Hills, northeast Sudan. *Geologische Rundschau*, **83**, 547–563.

REISCHMANN, T., BACHTADSE, V., KRÖNER, A. & LAYER, P. 1992. Geochronology and palaeomagnetism of a late Proterozoic island arc terrane from the Red Sea Hills, northeast Sudan. *Earth and Planetary Science Letters*, **114**, 1–15.

REISCHMANN, T., KRÖNER, A. & BASAHEL, A. 1984. Petrography, geochemistry, and tectonic setting of metavolcanic sequences from the Al Lith area, southwestern Arabian Shield. *Bulletin of the Faculty of Earth Sciences, King Abdulaziz University, Jiddah*, **6**, 366–378.

RIES, A. C., SHACKLETON, R. M. & DAWOUD, A. S. 1985. Geochronology, geochemistry and tectonics of the NE Bayuda Desert, N. Sudan: implications for the western margin of the Late Proterozoic fold belt of NE Africa. *Precambrian Research*, **30**, 43–63.

RODGERS, A. J., WALTERS, W. R., MELLORS, R. J., AL-AMRI, A. M. S. & ZHANG, Y.-S. 1999. Lithospheric structure of the Arabian Shield and Platform from complete regional waveform modeling and surface wave group velocities. *Geophysics Journal International*, **138**, 871–878.

SCHANDELMEIER, H., ABDEL RAHMAN, E. M., WIPFLER, E., KÜSTER, D., UTKE, A. & MATHEIS, G. 1994*a*. Late Proterozoic magmatism in the Nakasib suture, Red Sea Hills, Sudan. *Journal of the Geological Society, London*, **151**, 485–497.

SCHANDELMEIER, H., DARBYSHIRE, D. P. F., HARMS, U. & RICHTER, A. 1988. The East Saharan Craton: evidence for pre-Pan-African crust in NE Africa west of the Nile. *In*: EL-GABY, S. & GREILING, R. O. (eds) *The Pan-African Belt of Northeast Africa and Adjacent Areas*. Vieweg, Braunschweig/Wiesbaden, 69–94.

SCHANDELMEIER, H., WIPFLER, E., KÜSTER, D., SULTAN, M., BECKER, R., STERN, R. J. & ABDELSALAM, M. G. 1994*b*. Atmur-Delgo suture: a Neoproterozoic oceanic basin extending into the interior of northeast Africa. *Geology*, **22**, 563–566.

SCHMIDT, D. L., HADLEY, D. G. & STOESER, D. B. 1979. Late Proterozoic crustal history of the Arabian shield, southern Najd province, Kingdom of Saudi Arabia. *Bulletin of the Institute for Applied Geology, King Abdulaziz University, Jiddah*, **3**, 41–58.

SHACKLETON, R. M. 1988. Contrasting structural relationships of Proterozoic ophiolites in northeast and eastern Africa. *In*: EL-GABY, S. & GREILING, R. O. (eds) *The Pan-African Belt of Northeast Africa and Adjacent Areas*. Vieweg, Braunschweig/Wiesbaden, 227–288.

SHACKLETON, R. M. 1994. Review of Late Proterozoic sutures, ophiolitic mélanges and tectonics of eastern Egypt and north-east Sudan. *Geologische Rundschau*, **83**, 537–546.

SHACKLETON, R. M. 1996. The final collision zone between East and West Gondwana: where is it? *Journal of African Earth Sciences*, **23**, 271–287.

SHANTI, M. 1983. The Jabal Ess ophiolite complex. *Bulletin of the Faculty of Earth Sciences King Abdulaziz University, Jiddah*, **6**, 289–317.

SIMS, P. K. & JAMES, H. L. 1984. Banded iron-formations of late Proterozoic age in the central Eastern Desert, Egypt: geology and tectonic setting. *Economic Geology*, **79**, 1777–1784.

STACEY, J. S. & AGAR, R. A. 1985. U–Pb isotopic evidence for the accretion of a continental microplate in the Zalm region of the Saudi Arabian shield. *Journal of the Geological Society, London*, **142**, 1189–1203.

STACEY, J. S. & STOESER, D. B. 1983. Distribution of oceanic and continental leads in the Arabian–Nubian Shield. *Contributions to Mineralogy and Petrology*, **84**, 91–105.

STACEY, J. S., STOESER, D. B., GREENWOOD, W. R. & FISCHER, L. B. 1984. U–Pb zircon geochronology and geological evolution of the Halaban–Al Amar region of the Eastern Arabian Shield, Kingdom of Saudi Arabia. *Journal of the Geological Society, London*, **141**, 1043–1055.

STEIN, M. & GOLDSTEIN, S. L. 1996. From plume head to continental lithosphere in the Arabian–Nubian shield. *Nature*, **382**, 773–778.

STERN, R. J. 1985. The Najd fault system, Saudi Arabia and Egypt: a late Precambrian rift-related transform system? *Tectonics*, **4**, 497–511.

STERN, R. J. 1994. Arc assembly and continental collision in the Neoproterozoic East African orogen: implications for the consolidation of Gondwanaland. *Annual Review of Earth and Planetary Science*, **22**, 319–51.

STERN, R. J. & ABDELSALAM, M. G. 1998. Formation of juvenile continental crust in the Arabian–Nubian shield: evidence from granitic rocks of the Nakasib suture, NE Sudan. *Geologische Rundschau*, **87**, 150–160.

STERN, R. J. & HEDGE, C. E. 1985. Geochronological and isotopic constraints on late Precambrian crustal evolution in the Eastern Desert of Egypt. *American Journal of Science*, **285**, 97–127.

STERN, R. J. & KRÖNER, A. 1993. Late Precambrian crustal evolution in NE Sudan: isotopic and geochronological constraints. *Journal of Geology*, **101**, 555–574.

STERN, R. J., ABDELSALAM, M. G., SCHANDELMEIER, H., SULTAN, M. & WICKHAM, S. 1993. Carbonates of the

Keaf zone, NE Sudan: a Neoproterozoic (ca. 750 Ma) passive margin on the eastern flank of West Gondwanaland? *Geological Society of America, Abstracts with Programs*, **25**, 49.

STERN, R. J., GOTTFRIED, D. & HEDGE, C. E. 1984. Late Precambrian rifting and crustal evolution in the Northeastern Desert of Egypt. *Geology*, **12**, 168–172.

STERN, R. J., KRÖNER, A., BENDER, R., REISCHMANN, T. & DAWOUD, A. S. 1994. Precambrian basement around Wadi Halfa, Sudan: a new perspective on the evolution of the East Saharan Craton. *Geologische Rundschau*, **83**, 564–577.

STERN, R. J., KRÖNER, A., MANTON, W. I., REISCHMANN, T., MANSOUR, M. & HUSSEIN, I. M. 1989. Geochronology of the late Precambrian Hamisana shear zone, Red Sea Hills, Sudan and Egypt. *Journal of the Geological Society, London*, **146**, 1017–1029.

STERN, R. J., NIELSEN, K. C., BEST, E., SULTAN, M., ARVIDSON, R. E. & KRÖNER, A. 1990. Orientation of late Precambrian sutures in the Arabian–Nubian shield. *Geology*, **18**, 1103–1106.

STOESER, D. B. & CAMP, E. 1985. Pan-African microplate accretion of the Arabian shield. *Geological Society of America Bulletin*, **96**, 817–826.

STOESER, D. B. & STACEY, J. S. 1988. Evolution, U–Pb geochronology, and isotope geology of the Pan-African Nabitah orogenic belt of the Saudi Arabian shield. *In*: EL-GABY, S. & GREILING, R. O. (eds) *The Pan-African Belt of Northeast Africa and Adjacent Areas*. Vieweg, Braunschweig/Wiesbaden, 227–288.

STOESER, D. B. & STACEY, J. S., GREENWOOD, W. R. & FISCHER, L. B. 1984. *U/Pb zircon geochronology of the southern portion of the Nabitah mobile belt and Pan-African continental collision in the Saudi Arabian shield*. Saudi Arabian Deputy Ministry for Mineral Resources Technical Record USGS-TR-04-05.

STOESER, D. B., WHITEHOUSE, M. J. & STACEY, J. S. 2001. The Khida Terrane – geology of Palaeoproterozoic rocks in the Muhayil area, eastern Arabian shield, Saudi Arabia. *Gondwana Research*, **4**, 192–194.

STUCKLESS, J. S., HEDGE, C. E., WENNER, D. B. & NKOMO, I. T. 1984. *Isotopic studies of postorogenic granites from the northeastern Arabian shield, Kingdom of Saudi Arabia*. Saudi Arabian Deputy Ministry for Mineral Resources, Open-File Report USGS-OF-04-42.

STURCHIO, N. C., SULTAN, M. & BAIZA, R. 1983. Geology and origin of the Meatiq Dome, Egypt: a Precambrian metamorphic core complex? *Geology*, **11**, 72–76.

SULTAN, M., BICKFORD, M. E., EL-KALIOUBY, B. & ARVIDSON, R. E. 1992. Common Pb systematics of Precambrian granitic rocks of the Nubian shield (Egypt) and tectonic implications. *Geological Society of America Bulletin*, **104**, 456–470.

SULTAN, M., CHAMBERLAIN, K. R., BOWRING, S. A. & ARVIDSON, R. E. 1990. Geochronological and isotopic evidence for involvement of pre-Pan-African crust in the Nubian shield. *Geology*, **18**, 761–764.

TAYLOR, W. E. G., EL KAZZAZ, Y. A. H. & RASHWAN, A. A. 1993. An outline of the tectonic framework for the Pan-African orogeny in the vicinity of Wadi Um Relan, SE Desert, Egypt. *In*: EL-GABY, S. & GREILING, R. O. (eds) *The Pan-African Belt of Northeast Africa and Adjacent Areas*. Vieweg, Braunschweig/Wiesbaden, 31–34.

TEKLAY, M. 1997. *Petrology, geochemistry, and geochronology of Neoproterozoic Magmatic arc rocks from Eritrea: implications for crustal evolution in the southern Nubian shield*. Department of Mines, Eritrea, Memoir **1**.

TEKLAY, M., KRÖNER, A., MEZGER, K. & OBERHAENSLI, R. 1998. Geochemistry, Pb–Pb single zircon ages and Nd–Sr isotope composition of Precambrian rocks from southern and eastern Ethiopia: implications for crustal evolution in East Africa. *Journal of African Earth Sciences*, **26**, 207–227.

VAIL, J. R. 1985. Pan-African (late Precambrian) tectonic terrains and the reconstruction of the Arabian–Nubian Shield. *Geology*, **13**, 839–842.

VASLET, D., DELFOUR, J., MANIVIT, J., LE NINDRE, Y.-M., BROSSE, J.-M. & FOURNIGUET, J. 1983. *Geological map of the Wadi Ar Raun quadrangle, Sheet 23H. Kingdom of Saudi Arabia. Map GM65*. Saudi Arabian Deputy Ministry for Mineral Resources.

WALLACE, C. A. 1986. *Lithofacies and depositional environment of the Maraghan formation, and speculation on the origin of gold in ancient mines, An Najadi area, Kingdom of Saudi Arabia*. Saudi Arabian Deputy Ministry for Mineral Resource, Open-File Report USGS-OF-06-6.

WHITEHOUSE, M. J., WINDLEY, B. F., BA-BTTAT, M. A. O., FANNING, C. M. & REX, D. C. 1998. Crustal evolution and terrane correlation in the eastern Arabian Shield, Yemen: geochronological constraints. *Journal of the Geological Society, London*, **155**, 281–295.

WHITEHOUSE, M. J., STOESER, D. B. & STACEY, J. S. 2001a. The Khida terrane – geochronological and isotopic evidence for Palaeoproterozoic and Archaean crust in the Eastern Arabian Shield of Saudi Arabia. *Gondwana Research*, **4**, 200–202.

WHITEHOUSE, M. J., WINDLEY, B. F., STOESER, D. B., AL-KHIRBASH, S., BA-BTTAT, M. A. O. & HAIDER, A. 2001b. Precambrian basement character of Yemen and correlations with Saudi Arabia and Somalia. *Precambrian Research*, **105**, 357–369.

WILDE, S. A. & YOUSSEF, K. 2000. Significance of SHRIMP U–Pb dating of the Imperial Porphyry and associated Dokhan Volcanics, Gebel Dokhan North Eastern Desert, Egypt. *Journal of African Earth Sciences*, **31**, 403–413.

WILDE, S. A. & YOUSSEF, K. 2001. SHRIMP U–Pb dating of detrital zircons from the Hammamat group at Gebel Umm Tawat, Northeastern Desert, Egypt. *Gondwana Research*, **4**, 202–206.

WILLIAMS, P. L., VASLET, D., JOHNSON, P. R., BERTHIAUX, A., LE STAT, P. & FOURNIGUET, J. 1986. *Geological map of the Jabal Habashi quadrangle, Sheet 26F, Kingdom of Saudi Arabia. Map GM98*. Saudi Arabian Deputy Ministry for Mineral Resources.

WILLIS, K. M., STERN, R. J. & CLAUER, N. 1988. Age and geochemistry of late Precambrian sediments of the Hammamat Series from the Northeastern Desert of Egypt. *Precambrian Research*, **42**, 173–187.

WINDLEY, F. B., WHITEHOUSE, M. J. & BA-BTTAT, M. A. O. 1996. Early Precambrian gneiss terranes and Pan-African island arcs in Yemen: crustal accretion of the eastern Arabian shield. *Geology*, **24**, 131–134.

WINDLEY, B. F., WHITEHOUSE, M. J., STOESER, D. B., AL-KHIRBASH, S., BA-BTTAT, M. A. O. & AL-GHOTBAH, A.

2001. The Precambrian terranes of Yemen and their correlation with those of Saudi Arabia and Somalia: implications for the accretion of Gondwana. *Gondwana Research*, **4**, 206–207.

WIPFLER, E. L. 1996. Transpressive structures in the Neoproterozoic Ariab–Nakasib Belt, northeast Sudan: evidence for suturing by oblique collision. *Journal of African Earth Sciences*, **23**, 347–362.

WOLDEHAIMANOT, B. 1995. *Structural geology and geochemistry of the Neoproterozoic Adobha and Adola belts (Eritrea and Ethiopia)*. Giessener geologischen Schriften, Lenz-Verlag Giessen, **54**.

WOLDEHAIMANOT, B. 2000. Tectonic setting and geochemical characterisation of Neoproterozoic volcanics and granitoids from the Adobha belt, northern Eritrea. *Journal of African Earth Sciences*, **30**, 817–831.

WOLDEHAIMANOT, B. 2001. Extensional tectonics in the high-grade metamorphic terrain of upper Anseba region, central Eritrea. *Gondwana Research*, **4**, 207–208.

WÜRSTEN, F. FLISCH, M., MICHALSKI, I., LEMETOUR, J., MERCOLLI, I, MATTHAEUS, U. & PETERS, TJ. 1991. The uplift history of the Precambrian crystalline basement of the Jabal J'alan (Sur area). *In*: PETERS, TJ., NICOLAS, A. & COLEMAN, R. G. (eds) *Ophiolite Genesis and Evolution of the Oceanic Lithosphere*. Kluwer Academic Press, Dordrecht, 613–626.

Neoproterozoic dextral faulting on the Najd Fault System, Saudi Arabia, preceded sinistral faulting and escape tectonics related to closure of the Mozambique Ocean

TIMOTHY M. KUSKY[1] & MOHAMED I. MATSAH[2]

[1]*Department of Earth and Atmospheric Sciences, St Louis University, St Louis, MO, 63103, USA*
[2]*Department of Structural Geology and Remote Sensing, King Abdul Aziz University, Jeddah, Saudi Arabia*

Abstract: The Neoproterozoic Najd Fault System extends for 2000 km across the East African Orogen, yet its history of motion and tectonic significance are widely debated. The Halaban–Zarghat Fault is the northeastern-most of the major NW-striking Najd faults in the Arabian Shield. Several sedimentary basins of the Neoproterozoic Jibalah Group are bounded by strands of the Halaban–Zarghat Fault and other Najd faults, particularly along right steps in the fault trace. Among the largest of the basins is the Jifn. The geometry of the Jifn Basin and the sedimentary facies of Jibalah Group indicate that it is a dextral pull-apart basin between strands of the Halaban–Zarghat Fault. A zone of high-grade mylonitic gneiss is located along a left step in the fault zone and may be a deeply eroded pop-up structure related to dextral transpression. Analysis of structural data from around and within the Jifn Basin, the position of other pull-apart basins and high-grade mylonite zones along the Halaban–Zarghat Fault are all consistent with early dextral movement along the Halaban–Zarghat Fault. Offsets of distinctive older rock units and transection of the Jifn Basin by sinistral faults, however, show that the latest and most significant sense of offset on the Halaban–Zarghat Fault and other Najd faults was sinistral.

A U–Pb zircon date of 624.9 ± 4.2 Ma from rhyolitic basement of the Jifn Basin gives a lower limit for the formation of the basin and initiation of dextral movement along the Halaban–Zarghat Fault. This age is interpreted as the earliest age for the collision of East and West Gondwana. A 621 ± 7 Ma pluton is offset 10 km dextrally along the Halaban–Zarghat Fault, showing that dextral motions continued for some time past 621 Ma, before switching to sinistral motions, and accreted terranes caught between the two continents were forced toward an oceanic-free face to the north. A 576.6 ± 5.3 Ma U–Pb zircon date from an undeformed felsite dyke that intrudes the Jibalah Group gives an upper time limit for movement along the Halaban–Zarghat Fault. This may mark the time that collision and escape tectonics ended, or it may reflect the time that displacements were transferred to other Najd faults in more interior parts of the East African Orogen.

The Arabian–Nubian Shield (ANS) is a complex amalgam of arcs and microcontinents assembled during Neoproterozoic closure of the Mozambique Ocean (Hoffman 1991; Dalziel 1992; Stern 1994; Abdelsalam & Stern 1996). Some of the arc terranes represent juvenile additions to the continental crust (Reymer & Schubert 1984; c.f. Stein & Goldstein 1996), whereas others may have been built on older continental basement (Vail 1983; Stoser & Camp 1985; Bowen & Jux 1987; Kroner *et al.* 1987; Stern 1994).

One of the main structural features of the Arabian Shield is a 300-km wide zone of NW-striking transcurrent faults (Fig. 1). These faults, known as the Najd Fault System (NFS), extend across the 1200–2000 km length of the shield (Moore 1979; Davies 1984; Stacey & Agar 1985; Stern 1985; Agar 1987; Stoser & Stacey 1988; Sultan *et al.* 1988; Berhe 1990; Bonovia & Chorowicz 1993). Some investigators suggest that the NFS has a length of 3000 km (Andre 1989), whereas Stern (1994) proposed that the NFS is part of a larger system which extends for 4000 km across the ANS and India (Fig. 1). Following Burke & Sengor (1986), Stern (1994) interpreted these faults to be related to escape tectonics from a collision between West and East Gondwana at the end of the Precambrian (Fig. 2).

The NFS (Fig. 3) has mapped sinistral strike-slip offsets, as revealed by displaced units along its length. In earlier studies, the NFS was considered an orogen by itself, formed during the *c.* 570–520 Ma Najd Orogeny (Brown & Coleman 1972). Later on it was considered only as a transcurrent fault system developed in the absence of contractional orogenesis (Al-Shanti 1993).

Different hypotheses have been suggested for the origin of the NFS. Moore (1979) postulated that deformation along the NFS began with the establishment of Riedel shears which later merged into continuous faults. Schmidt *et al.* (1979), Fleck *et al.*

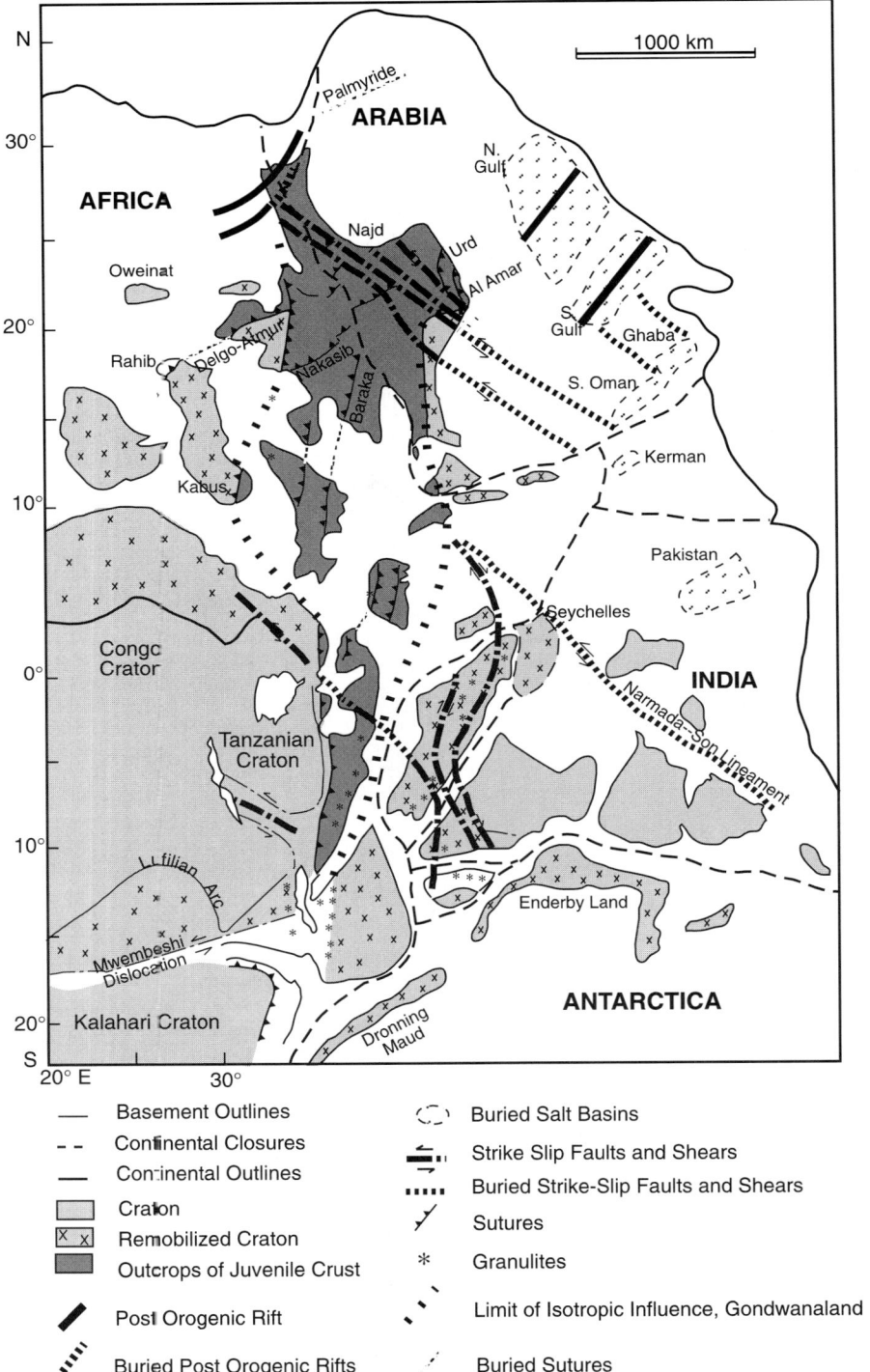

Fig. 1. Map of the East African Orogen (after Stern 1994), showing the Najd Fault System and related strike-slip faults and the location of Figure 4.

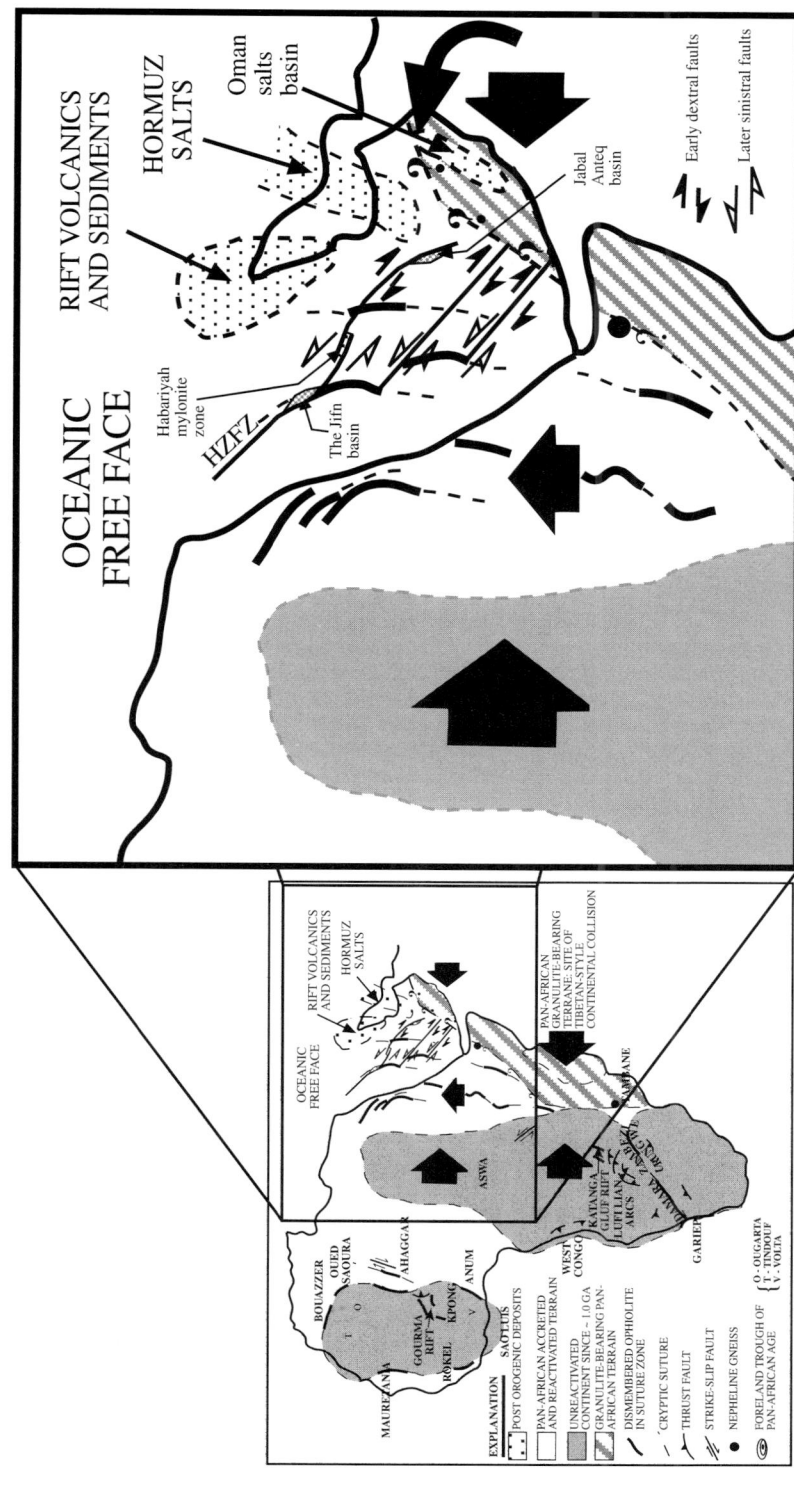

Fig. 2. Modification of the early escape tectonics model for the origin of the Najd strike-slip Fault (after Burke & Sengor 1986). The inset shows the Najd Fault System and how movement along each fault started as dextral and ended as sinistral. The Jifn and Jabel Anteq Basins and the Habariyah Mylonite Zone formed in the early dextral phase.

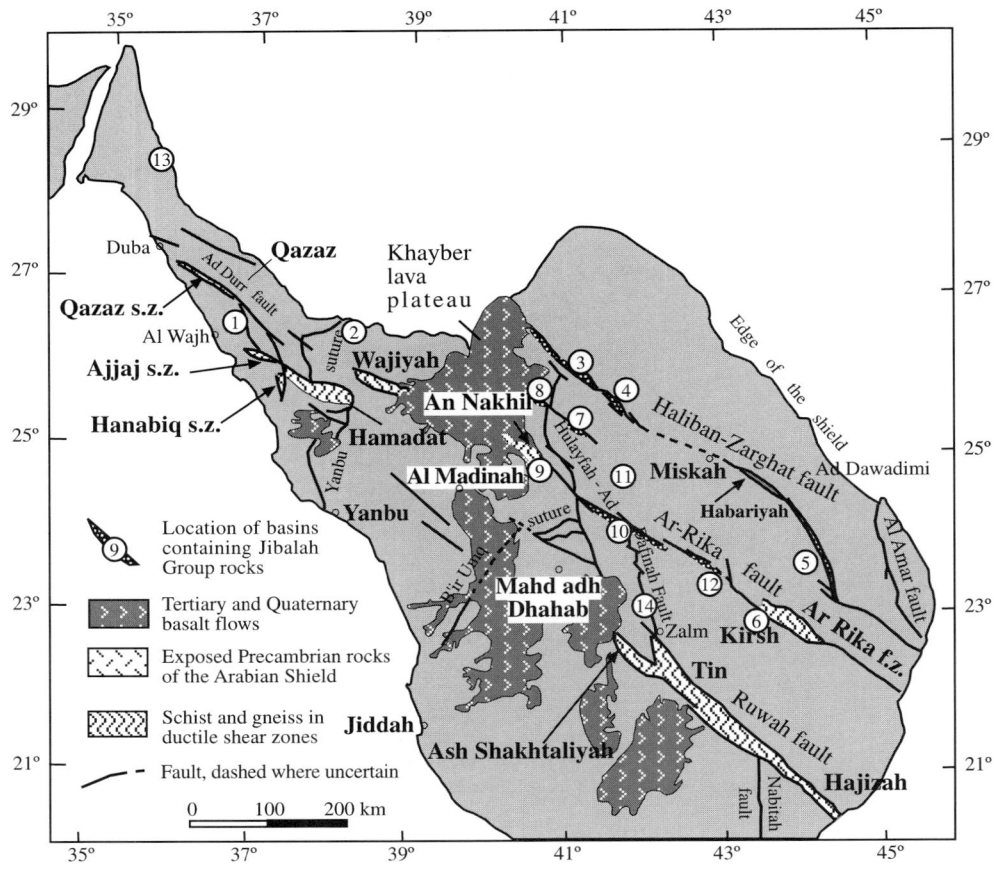

Fig. 3. NW-striking faults of the Najd Fault System in the Arabian Shield, and associated basins and ductile shear zones (after Hadley 1974; Delfour 1983; Clark 1985; Agar 1986; Johnson 1997). Locations of different basins discussed in text as follows: 1, Jabal Liban; 2, Al-Mashad; 3, Sumamiyyah–Zarghat; 4, Jifn; 5, Jabal Anteq; 6, Jabal Kibdi; 7, Bir Arja; 8, Jabal Jibalah; 9, Jabal Hawaqah–Al-Abed; 10, Sukhaibrat; 11, Al-Hissu; 12, Bir Sija; 13, Minawah Formation; 14, Bani Ghayy in the Alm area.

(1980), Davies (1984) and Agar (1987) proposed a rigid indentor hypothesis, contending that the NFS formed during collision between the ANS and a rigid indentor to the east. Stern (1985) discounted this model and argued that the predictions for this model are inconsistent with the evidence. Alternatively, Stern (1985) argued that the NFS formed as a result of a major episode of extension in the northwestern part of the ANS. Shackleton (1986) suggested that an east-dipping subduction zone beneath East Gondwana might have generated the NFS. Bonavia & Chorowicz (1992) returned to the idea of a rigid indentor, in this case the Tanzania Craton to the west. Stern (1994) agreed with this hypothesis, noting that the NW orientation of the sinistral faults in the ANS could be reconciled with the collision along the Mozambique Belt by consid-

ering that West Gondwana was the rigid indentor. Several authors have incorporated elements of escape tectonics to explain the fault system between East and West Gondwana (Burke & Sengor 1986; Agar 1987; Berhe 1990; Abdelsalam & Stern 1993; Stern 1994; Abdelsalam et al. 1995; Abdelsalam & Stern 1996). Also, the inferred conjugate geometry of these regional fault systems has been explored (Husseini 1989; Al-Shanti 1993).

Sense of movement and offset along the Najd Fault System (NFS)

The general sense of offset along the NFS is sinistral (Brown & Jackson 1960; Delfour 1970; Moore 1979; Sengor & Natal'in 1996), but some investigators (Howland 1979; Moore 1979; Agar 1986, 1987) suggest that an incipient dextral phase predated sinistral movement. Moore (1979) noticed variations in the degree of offset along the length of Najd faults and that some of these faults show dextral displacement.

Davies (1980) noted dextral displacements along some of the Najd faults in the northwestern part of the Arabian Shield. Cole & Hedge (1986) reported a 10 km right-lateral offset of a diorite along the Halaban–Zarghat Fault, one of the Najd faults in the northeastern part of the Arabian Shield. Agar (1986, 1987) concluded that the Bani Ghayy Group volcanics and sediments were deposited in pull-apart basins during an incipient dextral phase of Najd strike-slip faulting. He suggested that the NFS was a dextral fault until $c.$ 600 Ma and then the motion changed to sinistral.

Individual displacements of the Najd faults range from 2 to 25 km on average, but in some instances offset is measured as high as 40 km, with greater displacement in the middle parts of the large faults and lesser at their ends (Moore 1979). The cumulative sinistral displacement along these faults in the Arabian Shield is estimated to be $c.$ 250 km (Brown et al. 1963, 1989; Brown & Coleman 1972), as calculated from the displacement measured along offsets of the Hamdah–Nabitah Ophiolite Belt which strikes north through the central part of the Arabian Shield.

Age of the Najd Fault System (NFS)

Most investigators agree that the Najd transcurrent faults formed at the end of the Proterozoic (Hadley 1974; Delfour 1977, 1979; Schmidt et al. 1979; Schmidt & Brown 1984). Fleck et al. (1976) suggested that the period of activity of the NFS was between 580 and 530 Ma based on K–Ar results. Delfour (1977) concluded that the NFS activity continued until 507 Ma. This conclusion was based on K–Ar whole-rock dates of 493 and 502 Ma obtained for andesite flows found interbedded within the Jibalah Group. Stacey & Agar (1985) extended the initiation of fault motion to 630 Ma based on a U–Pb zircon age of 620 ± 5 Ma of a rhyolite, that lies midway through the Bani Ghayy succession, which they interpreted to be deposited in pull-apart basins along the NFS. They also report a U–Pb zircon age of 632 ± 3 Ma of Hufayrah granite which underlies an unconformity at the base of Bani Ghayy rocks. Johnson (2002) cites new unpublished SHRIMP data which suggests that the Bani Ghayy Group may be as old as 650 Ma. Cole & Hedge (1986) dated a diorite cut by one of the faults belonging to the NFS at 621 ± 7 Ma (U–Pb Zircon age). Stacey & Agar (1985) reported two separate plutons in the Zalm area which show emplacement structures that were affected and influenced by the NFS. They found that both plutons yield U–Pb zircon ages >630 Ma. Agar (1987) concluded that the NFS started soon after the Nabitah and Al-Amar Orogenies, i.e. between 640 and 630 Ma, and had a total longevity of >100 Ma.

Andre (1989) described folds and faults affecting Palaeozoic rocks, and suggests that the NFS was reactivated in Phanerozoic time, perhaps as late as Tertiary, accompanying Red Sea rifting.

Major Najd-related faults

Ar-Rika and Ruwah Shear Zones

The NFS consists of several major strands of linked faults (Fig. 3). The Ar-Rika and Ruwah Shear Zones are located to the SW of the Halaban–Zarghat Fault zone. Both fault zones contain mylonite and other metamorphic rocks, with foliations striking NNW and dipping SW, and shallow plunging stretching lineations trending NW and SE. The Ar-Rika Fault Zone is a brittle–ductile shear zone with a sharp fault boundary on the SW (Ar-Rika Fault) and a transitional boundary with a gradual diminishing of deformation intensity marked by a decrease of fault-related structures on the northeastern boundary (Johnson 1997). The Ruwah Fault Zone (Fig. 3) is a NW-striking brittle–ductile shear zone bounded by brittle faults that separate domains of high strain and ductile deformation within the fault zone (Johnson 1997). The schist and gneiss contain upright folds formed by the warping of the foliations with subhorizontal axes parallel and co-linear with the regional stretching lineations. Furthermore, these folds have subvertical axial planes and cleavage, which are coplanar with the regional foliation.

Ajjaj Shear Zone (ASZ)

The ASZ (Davies 1984; Davies & Grainger 1985; Duncan et al. 1990) is located in the northwestern part of the Arabian Shield (Fig. 3). According to Davies (1984), it consists, in the western part, of Precambrian metavolcanoclastic rocks and syntectonic granites, varying in metamorphic grade from greenschist to amphibolite facies. Duncan et al. (1990) noted that the higher metamorphic grades correspond more to the central parts of the shear zone.

Duncan et al. (1990) suggested that the boundaries of the ASZ are dominantly brittle-displacement surfaces, except along its northern side, and central and eastern parts, where it is largely ductile or transitional in nature. They inferred a minimum sinistral ductile displacement of 25–30 km in the westernmost part of the ASZ (Duncan et al. 1990). The structural style changes from a ductile shear zone in the east, to a more complex mix of brittle and ductile features in the centre and to ductile again in the west (Duncan et al. 1990). The layering has largely been rotated into parallelism to the boundary of the shear zone. They also documented that the ASZ changes in orientation from E–W to NW in the central portion due to the intersection of the ASZ with the Hanabiq Shear Zone (HSZ) and that ASZ superimposed a ductile structural fabric on the HSZ. This led Duncan et al. (1990) to the conclusion that the HSZ is an older structural weakness that existed before the Najd deformation, and that the HSZ and ASZ are not conjugate sets. Duncan et al. (1990) suggested that the HSZ belongs, in general, to the older N–S trending fabrics related to the Hejaz Orogeny (Brown & Coleman 1972) and predates the NFS by 100 Ma. This conclusion refutes the rigid-indentor model of Davies (1984).

Halaban–Zarghat Fault Zone (HZFZ)

The HZFZ is the northeasternmost strand of the Najd system (Figs 4 & 5). The fault includes areas of high-grade deformation along restraining bends and pull-apart basinal deposits along releasing bends. Among the areas of ductile deformation along the Halaban– Zarghat Fault is the Habariyah Mylonite Zone between latitudes 24°30' and 24°50'N and between longitudes 43°15' and 43°50'E (Fig. 5; Pellaton 1985; Johnson 1997). This elongate area is c. 4–8 km wide (Cole & Hedge 1986). The present authors suggest that the Habariyah Mylonite Zone formed along a transpressional-restraining bend at a left step on a fault strand of the Najd strike-slip faults. Since it is mapped on a left bend on the fault strand and the Halaban-Zarghat Fault is considered to be sinistral, it is suggested here that the mylonites formed during an earlier dextral phase of the NFS. Like other Najd-related faults in the shield, strands of the HZFZ form the boundaries of several elongate Neoproterozoic basins, the largest of which is known as the Jifn Basin (Figs 5 & 6). The Jifn Basin and Jabal Anteq Basin (NW of the town of Halaban) are situated on right bends along the HZFZ and it is suggested here that they formed in transtensional zones during a dextral phase of movement along the fault.

Pull-apart basins related to the Najd Fault System (NFS)

Pull-apart basins are features that develop in transtensional regions (Reading 1980; Aydin & Nur 1982). In transtensional settings, the principal stresses are compressional, but some areas within the region are under tension due to the obliquity of the major stress direction with respect to the plane of failure. This results in extension in the crust along releasing bends, which leads to the formation of basins. Reading (1980) provided a progressive model for the formation of pull-apart basins. A pull-apart basin starts with a fracture, progresses into a lazy Z or S fracture and then into a fault-bounded basin with aspect ratios of 2:1 to 10:1 (Mann et al. 1983). Steep sides on major fault boundaries characterize these types of basins, with normal faults developing on their shorter sides. Continuous movement along the major faults tends to offset deposits from their source inlet. These basins are characterized by rapid deposition and abrupt facies changes along or across the basin, and gradual facies changes along the longest axis of the basin (Reading 1980). Basin deposits are made mostly of coarse fanglomerate, conglomerate, sandstone, shales and shallow water limestones. Bimodal volcanics and volcaniclastic rocks may also be interbedded within the basin deposits. These bimodal volcanics are typical of those found in rift settings, but here they are in a transtensional regime. Transcurrent faults can penetrate deep into the crust, reaching the upper mantle and providing a conduit for magma.

Fault-controlled grabens have developed along several parts of the NFS and some investigators have suggested that these are pull-apart basins (Johnson et al. 1987; Sengor & Natal'in 1996; Kusky & Matsah 1999). These grabens were sites of deposition of volcanic and sedimentary rocks that are the youngest Precambrian rocks of the ANS (Agar 1986; Matsah & Kusky 1999), and are correlative with the extensional evaporitic basins of eastern Arabia (Husseini 1988). These rocks are assigned different names. The Bani Ghayy Group (U–Pb zircon age of 630–610 Ma; SHRIMP age of 650 Ma; Stacey & Agar 1985; Agar 1986; Johnson 2002) is located in the south-central part of the ANS, whereas the Minawah Formation (U–Pb zircon age of 600–575 Ma on possibly correlative plutonic rocks; Clark 1985) and the Jibalah Group (c. 600–570 Ma; Delfour 1970, 1977), are located in the northeastern part of the ANS (Figs 3 & 4; Al-Shanti 1993). The volcanosedimentary sequence associated with the NFS is thickest and best exposed in the northeastern part of the ANS and the present study is focused on this region (Fig. 3).

Agar (1986) suggested that the Bani Ghayy Group and similar groups, including the Hormuz Basin (Figs 2 & 3), are the product of sedimentation and volcanism in pull-apart grabens during the incipient dextral phase of the NFS. The Bani Ghayy basins are elongate in a N–S direction, and may represent sedimentation during small amounts of extension during

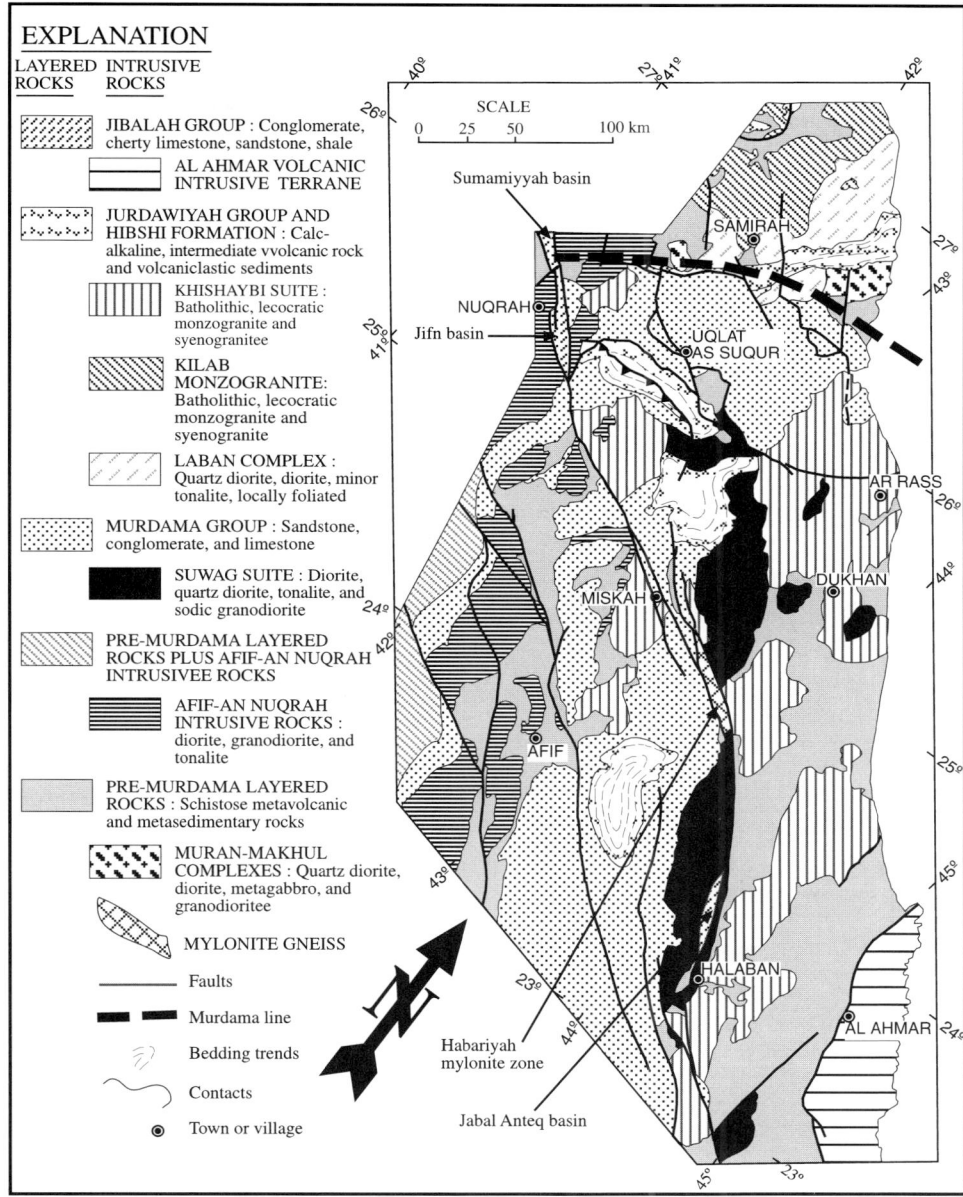

Fig. 4. Simplified map of the structural features in the northeastern part of the Arabian Shield showing the Halaban–Zarghat Fault (modified after Cole & Hedge 1986). The location of the Jifn Basin is indicated on the upper left. The Haliban–Zarghat Fault extends from near Haliban in the south to the Jifn Basin in the north. Compare with Figure 3, which shows the Haliban–Zarghat Fault as the northeasternmost Najd Fault in the Arabian Shield.

dextral faulting. The geometry of the Bani Ghayy basins is similar to the present setting of Lake Baikal, where an elongate N–S basin extends between distant strike-slip faults at either end of the lake. The basins are now displaced by sinistral faults (Agar 1986). Sengor & Natal'in (1996) proposed a basin and range model relating the NW–SE Najd–Zagros trend to post Pan-African collisional deformation (Fig. 2). Sengor & Natal'in's (1996) model suggests an early phase of dextral strike-slip

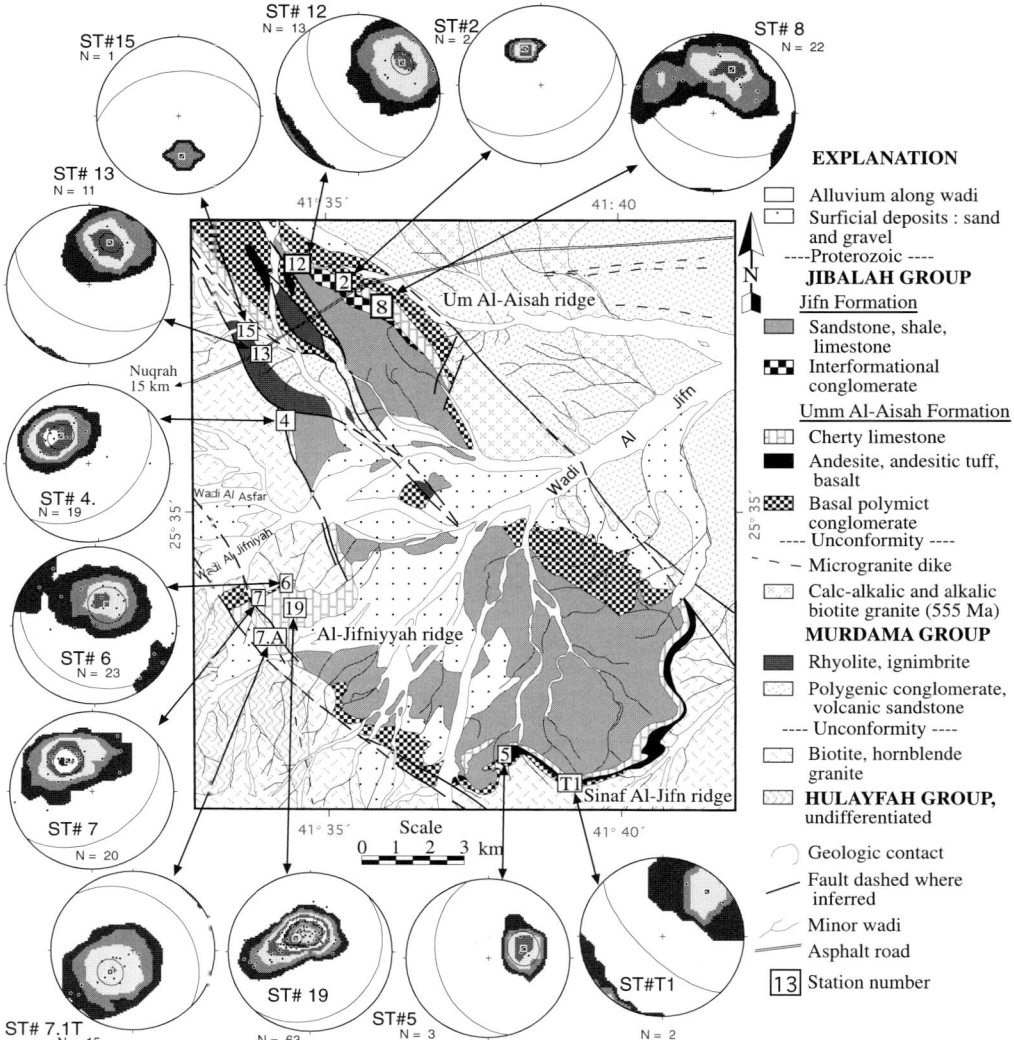

Fig. 5. Geological map of the Jifn Basin showing the distribution of the Jibalah Group and major structural features. Lower hemisphere equal area projections shows poles to bedding.

movement through which the NFS formed. They tried to link the formation of the Hormuz Basin faults and evaporites to the same model. Clark (1985) suggested that the Minawah Formation, composed of clastic sedimentary and silicic volcanic and pyroclastic rocks, was deposited in small grabens, which are considered by Johnson *et al.* (1987) to be pull-apart basins. Delfour (1970) suggested that the Jibalah Group was structurally controlled throughout its deposition by major epirogenic fractures (NFS), giving it the character of a graben deposit.

This work presents a detailed study of the Jibalah Group in the Jifn Basin in the Nuqrah area (Fig. 4) of the northeastern part of the ANS. The study includes several parts: (1) analysis of sedimentary facies of Jibalah Group; (2) the determination of the age of deposition of Jibalah Group through U–Pb zircon geochronology of volcanic/hypabyssal rocks that intrude, and lie below, the Jibalah Group rocks; and (3) detailed field-based structural observations and synthesis of data aimed at determining the kinematic history of the NFS. These ages are used to estimate the timing of basin development and to infer timing of fault motion along the NE portion of the NFS. These data also provide constraints on the final stages of amalgamation of East and West Gondwana.

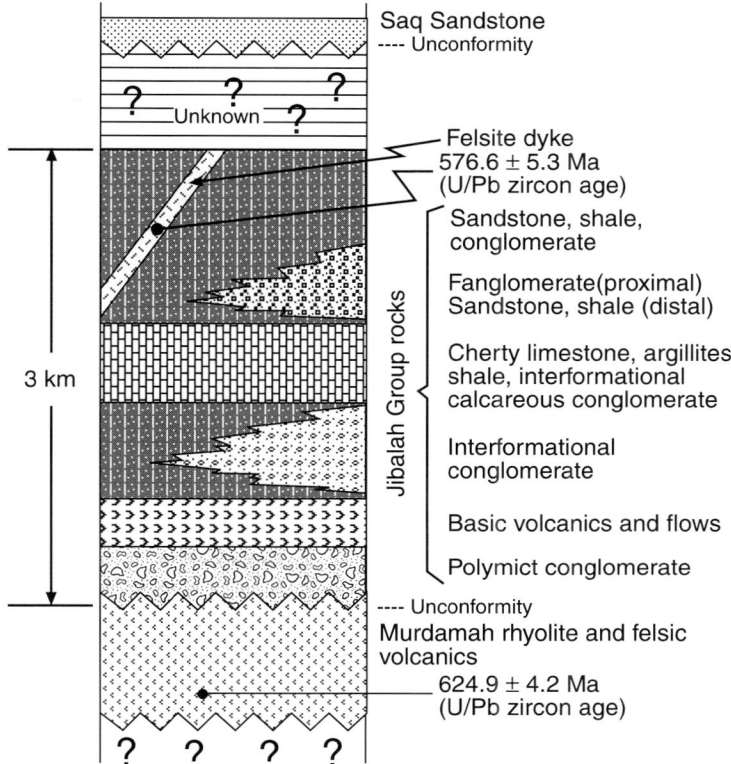

Fig. 6. Stratigraphic column of the Jibalah Group rocks in the Jifn Basin.

Geology of the Jibalah Group

The Jibalah Group is located in grabens formed along the NFS. Brown *et al.* (1963) suggested that the Jibalah Group was deposited conformably over the Shammar Group, whereas Delfour (1970) considered it to be unconformably overlying the Shammar Group. He also suggested that it represents the youngest Precambiran sedimentary and volcanic rocks of the ANS. Previous Rb–Sr and K–Ar age estimates of Jibalah Group rocks range between 600 and 570 Ma (Binda 1981). Outcrops of the Jibalah Group form three main belts along the NFS in the central and northern parts of the Arabian Shield (Fig. 3).

Most previous workers suggested that the Jibalah Group was deposited over a great part of the ANS but is now only preserved in grabens and has been eroded in other places (Al-Shanti 1993). The present authors suggest that the Jibalah Group was never regionally extensive but was deposited locally in fault-controlled pull-apart basins, influenced by shallow-marine or lacustrine conditions. Johnson *et al.* (1987) and Johnson (2002) also considered the Jibalah Group to occur only in fault-controlled ensialic pull-apart basins. The thickness of the Jibalah Group varies greatly from one place to another, consistent with deposition in fault-controlled pull-apart basins. For example, in the northeastern part of the shield it is 155 m thick, in the Al-Mashhad area it is 750–850 m thick (Hadley 1974), in Al-Sawrah Fort area, west of the Khaybar Tertiary Lava Plateau, it is *c.* 300 m thick (Hadley 1975) and in the Jibalah area, east of the Khaybar Lava Plateau, it is 3300 m thick (Delfour 1977).

The stratigraphic sequence of individual basins along the Najd faults is similar, grading from coarse clastic rocks including conglomerate, coarse greywacke and sandstone in the lower parts, basalt, andesite, dacite, rhyolite and layers of tuff in the middle part, through stromatolitic limestone and chert in the upper parts (Al-Shanti 1993).

Jifn Basin

The Jifn Basin is one of two elongate adjacent basins along the Halaban–Zarghat Fault. The other basin is called the Sumamyyah Basin (Delfour 1977) and it is located to the NW, forming an extension of the Jifn Basin (Fig. 4). The Jifn Basin has an elongate sigmoidal shape, in which an assemblage of dominantly sedimentary and mafic volcanic rocks of the Jibalah Group (Delfour 1977) was deposited (Fig. 5). The boundaries of the basin can be identified by low limestone hills that bound the basin from the east, SE, SW and north. The approximate length of the basin is 35 km and the width ranges from *c*. 1.5 km in the NW to 5 km along the Nuqrah–Buraydah highway, to *c*. 10 km maximum width along the main course of Wadi Al-Jifn across the central part of the basin. The topography is low relief within the inner parts of the basin except for the limestone ridges that surround it.

Jibalah Group in the Jifn Basin

In the Jifn Basin (Figs 4, 5 & 6), the Jibalah Group is *c*. 3000 m thick and it is composed of two formations: the Um Al-Aisah and the Jifn Formations (Delfour 1977). The Um Al-Aisah Formation includes polymict conglomerates with a calcareous matrix in the lower parts, which grade upwards into beds of variable thickness of dolomitic limestone and chert (Figs 6 & 7). In some parts, layers of basalt flows and tuff are located beneath the lower conglomerate (Al-Shanti 1993) and in some places below the limestone. The Jifn Formation includes conglomerates composed of mixed clasts of limestone and chert in the lower parts, and grades upward into shale, fine-grained and calcareous sandstones. Flows of basalt, andesite and tuff are interbedded with these sedimentary rocks at different levels, and in some places these volcanic rocks have a greater thickness than those of the shale and sandstones of the Jifn Formation (Al-Shanti 1993).

Gabbroic and dioritic intrusions cut the Jibalah Group, and geochemical analyses suggest an alkaline affinity (Al-Shanti 1993) typical of rifts. These intrusions are believed to be subvolcanic feeders of the lava flows and tuffs that are found higher in Jibalah Group (Al-Shanti 1993).

The Um Al-Aisah beds dip towards the inner parts of the basin (Fig. 5), which led Delfour (1977) to suggest that the two formations form a syncline with Um Al-Aisah Formation overlain by the Al-Jifn Formation. However, the present authors have found that the Um El-Aisah Formation is restricted to the sides of the basin while Al-Jifn Formation is found in the central parts of the basin. It is suggested that these rock formations represent a facies change from the faulted sides of the basin towards the centre of the basin, with relatively low-energy regimes controlling deposition in the central area of the basin and high-energy flow regimes on the faulted basin margins. Therefore, the following description of rock units based on lithological variations, and not the stratigraphic units of Delfour (1977), is presented. On the maps, both the lithological units and, for historical and comparative purposes, the formations proposed by Delfour (1977) are shown. Comparison of the map (Fig. 5) with the proposed stratigraphic section (Fig. 6) shows how Delfour's vertical formations are intereperted as lateral facies changes.

Conglomerates in the Jifn Basin

The Jifn Basin includes different types of conglomerate, including polymict basal conglomerate and interformational conglomerate [*sensu* Delfour (1977), for a conglomerate derived largely from units within and alongside the basin]. The interformational conglomerate of the Umm Al-Aisah Formation is divided into three units: fanglomerates, coarse- to medium-grained conglomerate beds with rounded pebbles brought from outside the basin, and calcareous reworked conglomeratic beds dominated by calcareous and cherty fragments.

Polymict basal conglomerate This facies is found on the sides of the basin (especially the faulted sides), as well as along the base of the basin (Figs 5, 6, 7a & 7b). The polymict conglomerate lies unconformably over older rocks and also rests on top of, and is locally intercalated with, mafic volcanics of the Jibalah Group. The conglomerate is characterized by different grain sizes of different rock types. The conglomerate clasts and grains range in size from sand size up to boulder size, with clasts as large as 20–35 cm in diameter. The matrix grains and clasts are rounded to subrounded in shape. Clast types include granite, rhyolite, flow-banded rhyolite, andesite, and clasts of sedimentary rocks that belong to the Murdamah Sandstone and Murdamah Conglomerate (Fig. 7). The polymict basal conglomerates are more abundant on the northeastern sides of the basin along the main northeastern strand of the Halaban–Zarghat Fault. More limited exosures of this conglomerate crop out on the southern, southwestern and western sides of the basin (Fig. 5). On the south and southeastern sides of the basin these conglomerates form a medium to thin bed below the limestone ridges and above or intercalated with the mafic volcanics.

The thickest accumulation of polymict conglomerate is found along the faulted sides of the basin, and formed in response to the abrupt change in elevation across the fault. Hence, these conglomerates represent a high-energy environment preserved on the fault's downthrown sides.

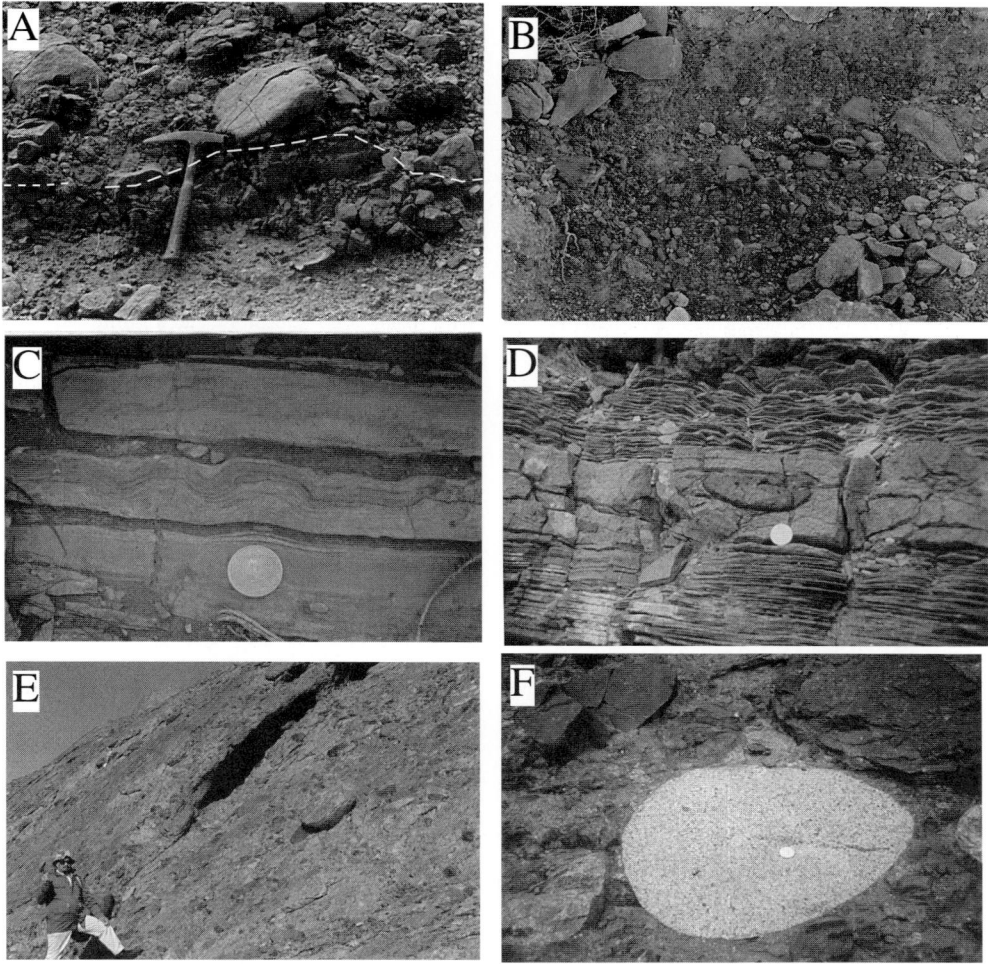

Fig. 7. (**a**) Polymict basal conglomerate, unconformably overlying Murdamah Rhyolite (dashed line is non-conformity surface); (**b**) polymict basal conglomerate; (**c**) stromatolitic limestone of the Umm Al-Aish Formation; (**d**) sandstone and shale of the Jifn Formation, showing possible glacial dropstone; (**e**) conglomerate of fanglomerate facies; (**f**) large clast of granite (possibly Rahadha Granite) in fanglomerate.

Fanglomerate facies Coarse-grained polymict fanglomerate deposits are exposed in one area of outcrop on the northeastern side of the basin; they are not recognized on the western or southwestern sides of the basin (Fig. 5). Outcrops located just north of the Nuqrah–Buraydah highway form small hummocky knobs made of coarse boulder- to cobble-conglomerate with some very large clasts of granite (c. 1 m diameter; Fig. 7) and finer-grained material, including clasts and matrix, derived from the same granite. The source of the granitic clasts is the Rahadha Batholith (part of the c. 615–621 ± 7 Ma Idah Suite; Cole & Hedge 1986), located east and NE of the basin (Delfour 1977). Clasts of Murdamah Sandstone and Conglomerate, as well as clasts of limestone, are found within these fanglomerates, but the limestone clasts are less rounded than other clasts and are interpreted to be from the nearby Jibalah Limestones. This fanglomerate includes clasts of andesite, porphyritic andesite, and clasts of rhyolite and felsic volcanics. The granitic clasts are well-rounded in the fanglomerate. The Murdamah Clasts and the mafic clasts are less rounded and the limestone clasts are angular to subangular in shape. The granitic clasts are the most abundant in the fanglomerate deposits and the matrix in many places is characterized by arkosic material. The fanglomerates were deposited along the faulted side of the basin.

The fanglomerates are coarser than the basal polymict conglomerate and they contain angular to subangular cherty limestone clasts of Jibalah Group rocks, whereas the basal conglomerate does not. The existence of angular clasts of the Jibalah Limestone in the fanglomerate denotes that these fanglomerates were deposited after deposition of some of the Jibalah Limestones. The fanglomerates also crop out only on the northeastern side of the basin, whereas the basal polymict conglomerate is found on several sides of the basin (Fig. 5). Both the basal polymict conglomerate and fanglomerate are coarse along the NE sides of the basin, indicating a proximal depositional environment as well as an abrupt change in elevation along the faulted basin margins.

Coarse-grained interformational conglomerate facies The coarse-grained interformational conglomerate facies includes beds of interformational conglomerate interbedded with lesser amounts of the limestone and argillites. They are similar to the fanglomerate deposits except that they are finer in grain size, with the maximum clast sizes up to 30 cm in diameter. These conglomerates form rounded grains and clasts (granite and rhyolites) to angular clasts of the Jibalah Limestone. These deposits are interpreted to be the lateral extension of the fanglomerates. These deposits are found below and within the limestones. The matrix is cemented by calcite.

Limestone and argillite facies The basal conglomerate deposits are overlain by a succession of alternating beds of cherty limestone, shale, interformational conglomerate, calcareous sandstone and calcareous reworked conglomerates (Figs 6, 7c & 7d). The limestone and argillaceous layers contain alternating beds of cherty and siliceous limestone and calcareous argillite. The silicic limestone layers are more resistant to erosion than the argillaceous beds, which produces a ribbbed appearance to outcrops. Some large clasts in the sandstone and argillite may be glacial dropstones (Fig. 7d).

The thickness of the limestone beds ranges from millimetres to decimetres, and their alternation with the soft shale layers and the calcareous reworked conglomerate may represent calcareous debris flows. The limestone and argillite layers form homoclinal ridges along the sides of the basin and their general dip is towards the inner parts of the basin. These deposits are folded in many places with different styles of folding.

A large limestone ridge bounds the basin on the northeastern side and is known as Jabal Um Al-Aisah. This ridge is the type locality for the Um Al-Aisah Formation (Delfour 1977). There are no outcrops of limestone in the central parts of the basin (Fig. 5). Limestone forms outcrops adjacent to the boundaries of the basin with dips toward the inside of the basin. Interformational conglomerates that grade into finer grained sandstone, siltstone and shale toward the central parts of the basin overlie the limestone. Beds and nodules of chert are mixed within the limestone layers and the cherty rocks range in colour from white, grey, red and black (Delfour 1977).

Calcareous reworked conglomerate facies Lenses of reworked calcareous conglomerate with angular broken chert and limestone clasts are intercalated with rocks of the limestone and argillite facies. It is suggested that these deposits resulted from slumping of previously deposited units. These rocks are interpreted as debris flows, which may have been triggered by shocks from movements along the faults that resulted in the collapse of parts of the deposits found on the basin sides and floor. This material is highly calcareous and contains broken cherty and clasts, as well as calcareous clasts. Medium- to fine-grained material derived from the underlying conglomerates is also found in these calcareous reworked conglomerates, but with very low percentages and usually at the base of individual beds. These calcareous reworked sediments show clear graded bedding, indicating that they are in an upright position.

Sandstone and shale-bed facies In the central parts of the basin the deposits become finer grained, including sandstone, shale and fine-grained conglomerates. These deposits were described by Delfour (1977) as the Al-Jifn Formation – the upper part of Jibalah Group. In the present study it is suggested that these rocks are a distal facies variation of the marginal conglomerates.

The sandstone and shale are brown to purplish brown in colour and exhibit different primary structures, including ripple marks, graded bedding and cross-beds (Fig. 7d). Ripple marks and cross-laminations from sandstones of the Jifn Formation show palaeocurrent directions from the margins toward the centre of the basin. These sandstones and shales indicate a lower energy environment within the central parts of the basin. The facies denotes a shallow- to deeper water environment of low energy, which may be explained as a result of basin subsidence. These beds are intercalated in some places with beds of limestone and fine conglomerates.

Volcanic rocks Volcanic rocks in the Jifn Basin include mafic volcanics below and intercalated with the basal conglomerate beds, as well as two rhyolite horizons found in and along the margin of the basin. In the southeastern part of the basin, mafic volcanic rocks include vesicular and amygdaloidal basaltic to andesitic flows; these are intercalated with the basal conglomerate. Locally these flows are found below the limestones. No sign of contact metamorphism was found in the sedimentary deposits and it is suggested that the basic volcanics formed in the early stages of the basin formation in close association with the basal conglomerate, but before the limestones were deposited.

Felsic volcanic rocks below the Jibalah Group include flow-banded rhyolite of the Murdamah Group NW of the basin. These rhyolites form a series of small ridges (Fig. 8) extending in a NW direction, and are bounded and cut by the major fault strands of the Halaban–Zarghat System (Fig. 5). Previous investigators (Delfour 1977) suggested that the Murdamah Rhyolites were formed just before the Najd faulting event. The present authors agree with this and assume that the age of these rhyolites represents a maximum age limit to the Najd faults, or at least the HZFZ, and a sample of this rhyolite was collected for U–Pb dating (a maximum age constraint on the formation of Al-Jifn Basin).

Another fine-grained felsic dyke was mapped intruding the Jibalah Group in the southern part of the basin near Jabal Sinaf Al-Jifn (Figs 5 & 8). This was sampled for U–Pb dating because its age will give a minimum constraint on the time of movement on the Halaban–Zarghat Fault that led to formation of the Al-Jifn Basin. The two dating samples yield upper and lower age limits for the formation of the Al-Jifn Basin and the deposition of the Jibalah Group rocks. It is assumed that these dates also constrain the phase of movement on the Halaban–Zarghat Fault.

Geochronology

Sample location and local structural relationship

Two samples of felsic rocks were collected from each of two outcrops to decipher the age of deposition of the Jibalah Group in the Jifn Basin. One of these samples was collected from the Murdamah Rhyolites on the west-central side of the Jifn Basin and the other is a felsite dyke that intrudes the Jibalah Group rocks in the southern part of the basin. The Murdamah Rhyolites are older than the Jibalah Group (Delfour 1977) and are cut and affected by the Halaban–Zarghat Fault. Dating of such rocks is assumed to give a maximum age limit for the Halaban–Zarghat Fault as well as for the deposition of Jibalah Group rocks in the basin. The other felsite dyke intrudes the Jibalah Group and is apparently not cut by fractures or faults related to the Halaban–Zarghat Fault. Its age will give us a minimum age limit for the Halaban–Zarghat Fault as well as the formation of the basin and the deposition of Jibalah Group rocks in the Jifn Basin. The relationship between the felsic volcanics of Murdamah Rhyolite, felsite dyke and the Jibalah Group rocks in the Jifn Basin is shown in Figure 6.

Samples were crushed, powdered, separated and abraded in the geochronology laboratories at Massachusetts Institute of Technology (MIT). The techniques used are described in Matsah (2000). After abrasion, all zircon fractions were washed with dilute nitric acid prior to dissolution. Sample weights were estimated using a video monitor with a gridded screen and were measured to within 10%. After air abrasion, all samples were spiked using a mixed ^{205}Pb–^{233}U–^{235}U spike and dissolved following the methods of Krogh (1973) and Parrish (1987). Separation of Pb and U was accomplished using standard HCl chemistry chromatographic techniques using a 15 μl column. Purified U and Pb fractions were loaded on single Re filaments with silica gel and graphite, respectively. All analyses were performed on the VG Sector 54 multicollector mass spectrometer housed at MIT.

Analysis of Pb fractions was accomplished in single-collector mode using the Daly knob and ion counting: analysis of U fractions was accomplished in static multicollector mode using Faraday detectors. Common Pb corrections for all analyses were considered zero for the two samples because the two rocks are igneous and are not metamorphosed, so it is considered that the initial radiogenic Pb at the time of crystalization of the magma was zero.

Blank Pb is estimated as 2.75 pg ± 12.7% and 2.7 pg ± 16.6% for zircon analyses of the felsite dyke and the Murdamah Rhyolite, respectively. Data reduction and error analysis was accomplished using PbMacDat-2 by D. Coleman and C. Isachsen using the algorithms of Ludwig (1989). Replicate reduction of data sets using PbMacDat-2σ and Isoplt (Ludwig 1989) yields identical results. All errors are reported at the 2σ confidence interval. Figure 9 shows zircon U–Pb concordia diagrams for the two analysed samples (Mat-2 and Ryo-1); Tables 1 and 2 show the resulting data of the analyses of the two samples.

U–Pb Results and discussion

The sample from Murdamah felsic volcanics (Mat-2) is represented by three discordant points on the concordia diagram (Fig. 9). The first trial for linear regression resulted in a best-fit line with an upper intercept with a concordia time line of age 623.5 ± 2474.6 Ma, which is considered to be unreliable due to the large marginal error. In a final regression, point (Z2) was omitted because it shows Pb loss (probably due to inheritance). Accordingly, the

Fig. 8. (**a**) Photograph showing Murdamah Rhyolites that form the basement to the Jifn Basin. Photograph taken looking north from the west rim of the basin (Jifn Basin to the right). These rhyolites yielded a U–Pb zircon age of 624.9 ± 4.2 Ma, interpreted as a maximum age for the basin. (**b**) Photograph of cross-cutting felsite dyke intruding the Jibalah Group in the Jifn Basin. This dyke yielded an age of 576.7 ± 6.5 Ma, interpreted as being a minimum age for the basin.

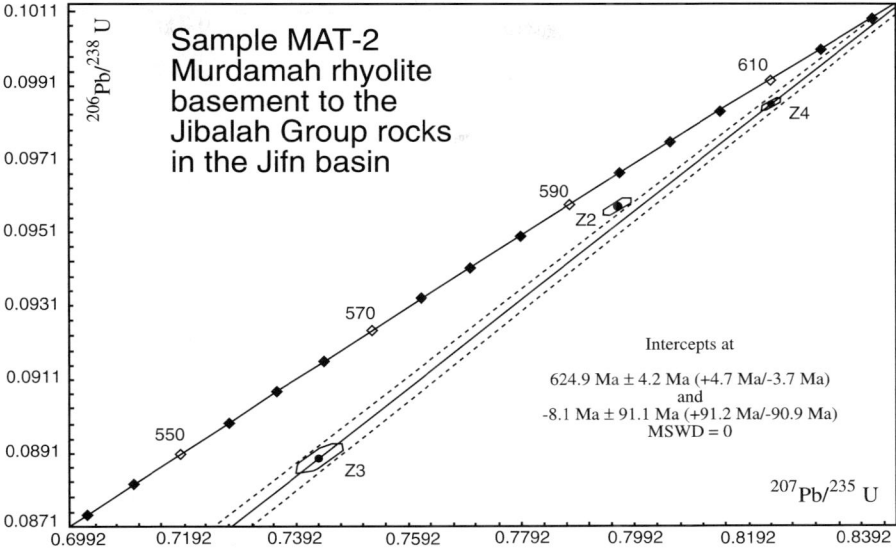
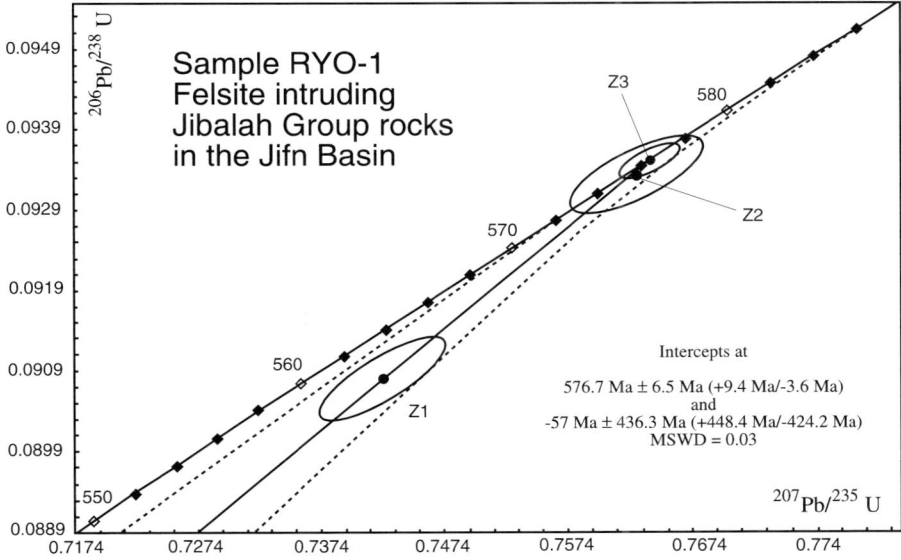

Fig. 9. Concordia diagrams for the two samples that were analysed for age controls on the Jifn Basin.

Table 1. *U–Pb isotopic data for zircons from sample Ryo-1 of Murdama rhyolites*

Fractions and properties	Weight (mg)	Concentrations					Atomic ratios						Ages (Ma)			Correction coefficient	Total common Pb (pg)
		U (ppm)	Pb* (ppm)	$^{206}Pb^\dagger/^{204}Pb$	$^{208}Pb*/^{206}Pb$	$^{206}Pb^\ddagger/^{238}U$	(% error)	$^{207}Pb^\ddagger/^{235}U$	(% error)	$^{207}Pb^\ddagger/^{206}Pb$	(% error)	$^{206}Pb-^{238}U$	$^{207}Pb-^{235}U$	$^{207}Pb-^{206}Pb$		
Z1	0.0014	331.24	31.65	959.19	0.130	0.00343	(0.23)	0.76333	(0.30)	0.05926	(0.20)	575.8	575.9	576.6	0.764	3.0
Z2	0.0003	728.20	68.93	412.62	0.122	0.09312	(0.49)	0.76122	(0.70)	0.05929	(0.46)	573.9	574.7	577.8	0.749	3.1
Z3	0.0004	356.66	33.51	391.36	0.147	0.09052	(0.54)	0.74009	(0.68)	0.05930	(0.40)	558.6	562.5	578.0	0.813	2.4

* Radiogenic Pb.
† Measured ratio corrected for fractionation only. Mass fractionation correction of $0.15 \pm 0.08\%$/amu was applied to all Pb analyses.
‡ Corrected for fractionation, spike, blank and initial common Pb

All samples were single zircon grains analysed at Massachusetts Institute of Technology using mixed $^{205}Pb-^{235}U-^{235}U$ spike. Zircon grains are designated as diamagnetic (d) and magnetic (m) in terms of degrees tilt on a Frantz LB-1 magnetic separator. Sample weights were estimated using a video monitor with a gridded screen and were known to within 10%. Zircon dissolution followed the methods of Krogh (1973) and Parrish (1987). Separation of Pb and U was accomplished using HCl chemistry. Decay constants used are $^{238}U = 0.15513 \times 10^{-9}$, and $^{235}U = 0.98485 \times 10^{-9}$ (Steiger & Jäger 1977). Common Pb corrections were made using the model of Stacey & Kramers (1975) for the interpreted crystallization age. U blank = 1 pg $\pm 50\%$; Pb blank = 3.5 pg $\pm 50\%$. Data reduction and error analysis was accomplished using the algorithms of Ludwig (1989, 1990) and all errors are reported in percentages at the 2σ confidence interval.

Table 2. U–Pb isotopic data for zircons from sample MAT-2 of felsite dyke

Fractions and properties	Concentrations			Atomic ratios								Ages (Ma)			Correction coefficient	Total common Pb (pg)
	Weight (mg)	U (ppm)	Pb* (ppm)	$^{206}Pb^{\dagger}/^{204}Pb$	$^{208}Pb^*/^{206}Pb$	$^{206}Pb^{\ddagger}/^{238}U$	(% error)	$^{207}Pb^{\ddagger}/^{235}U$	(% error)	$^{207}Pb^{\ddagger}/^{206}Pb$	(% error)	$^{206}Pb/^{238}U$	$^{207}Pb/^{235}U$	$^{207}Pb/^{206}Pb$		
Z2	0.01	21.33	2.21	1143.13	0.195	0.09580	(0.25)	0.79639	(0.33)	0.06029	(0.2)	589.8	594.8	614.0	0.782	1.2
Z3	0.01	25.14	2.36	469.36	0.166	0.08895	(0.45)	0.74348	(0.58)	0.06062	(0.35)	549.3	564.4	625.8	0.795	3.2
Z4	0.01	47.19	4.99	2174.02	0.167	0.09857	(0.15)	0.82365	(0.21)	0.06060	(0.13)	606.0	610.1	625.1	0.763	1.4

* Radiogenic Pb.
† Measured ratio corrected for fractionation only. Mass fractionation correction of $0.15 \pm 0.08\%$/amu was applied to all Pb analyses.
‡ Corrected for fractionation, spike, blank and initial common Pb

All samples were single zircon grains analysed at Massachusetts Institute of Technology using mixed $^{205}Pb-^{233}U-^{235}U$ spike. Zircon grains are designated as diamagnetic (d) and magnetic (m) in terms of degrees tilt on a Frantz LB-1 magnetic separator. Sample weights were estimated using a video monitor with a gridded screen and were known to within 10%. Zircon dissolution followed the methods of Krogh (1973) and Parrish (1987). Separation of Pb and U was accomplished using HCl chemistry. Decay constants used are $^{238}U = 0.15513 \times 10^{-9}$ and $^{235}U = 0.98485 \times 10^{-9}$ (Steiger & Jäger 1977). Common Pb corrections were made using the model of Stacey & Kramers (1975) for the interpreted crystallization age. U blank = 1 pg $\pm$ 50%; Pb blank = 3.5 pg $\pm$ 50%. Data reduction and error analysis was accomplished using the algorithms of Ludwig (1989, 1990) and all errors are reported in percentages at the 2σ confidence interval.

sample from Murdamah Felsic Volcanics gave a discordia U–Pb zircon age of 624.9 ± 4.2 Ma, representing the time of cooling of the magma of the Murdamah Rhyolites. This age tells us that the Jifn Basin formed after this date and that the Halaban–Zarghat strike-slip Fault Zone can be inferred to be younger than this age. This age is suggested to be the lower limit for the formation of the HZFZ as well as the Jifn Basin.

The felsite dyke sample (Ryo-1) yielded three points on the concordia diagram (Fig. 9). A regression line with 85% probability of fit based on a MSWD (Mean Squares of Weighted Deviates) of 0.03 passed through the points and intercepted the concordia time line, leading to a concordant age of 576.7 ± 5.3 Ma. This age is suggested to be the upper time limit for the formation of the Jifn Basin and its deposits because this felsite dyke intrudes the upper parts of the Jibalah Group rocks in the Jifn Basin. This age also gives an indication that the major movements along the HZFZ had ended around or before that time. From the two obtained ages, it is suggested that movement on the HZFZ took place between 624.9 ± 4.2 and 576.6 ± 5.3 Ma, and that the Jifn Basin formed within these two age limits.

Agar (1987) suggested earlier dates for the start of the NFS in the Zalm area, where he suggested that the Bani Ghayy Group rocks are sediments characteristic of a pull-apart basin. Following the suggestion of Johnson (1957), the present authors concur that the NFS should be considered as a group of different fault zones that were forming in different or successive times. From the U–Pb dates obtained in this study it is likely that the Zalm Fault Zone is older than the HZFZ, and that the time of initiation of movement on fault zones across the NFS become younger towards the east and NE.

Structural features in Al-Jifn Basin

Several cross-sections were drawn across the width and the length of the Jifn Basin (Fig. 10). These cross-sections show the change of the deposit's characteristics from the basin boundaries of coarse-grained proximal deposits to the central fine-grained distal environment. Cherty limestones are found along the sides of the basin and dip toward the central or inner parts of the basin. It is suggested here that the limestones are buried below the fine-grained sediments in the central parts of the basin, consistent with the synclinal structure suggested by Delfour (1977). The limestone indicates that the basin was submerged sometime during its formation. It could be that it was submerged under local small seas or even lakes, since correlative limestone deposits are only found in these structurally controlled basins of the NFS and not elsewhere in the shield. In contrast, the conglomerates are interpreted to be basin margin facies that grade basinward into sandstone and shale.

Faults

The major HZFZ (Najd) forms pronounced lineaments on enhanced TM (Thematic Mapper) images (Matsah 2000) and in many places sinistral offsets of rock units mark the fault strands. The Jifn Basin is marked by two bounding lineaments that correspond to two major faults giving the rhombic shape to the basin (Fig. 5).

The major faults are obscured in most places by wadi fill and debris. Minor faults were mapped, measured and examined for kinematic indicators such as slickensides and displaced grains (Fig. 11). Many faults were mapped cutting the Jifn Basin deposits (Fig. 12), and most of these have strike-slip offset with both sinistral and dextral senses. These faults are most recognizable in the limestone and conglomerate ridges along the boundaries of the basin. The fanglomerate deposits found north of the Nuqrah–Buraydah highway are also cut by numerous minor faults and fracture zones. Figure 13 shows stereographic projections of faults and fractures in the fanglomerate; the sense of offset in these faults indicates both sinistral and dextral movement. Faults in the limestone are observed but, due to the solubility of the limestone, most of these fault surfaces are jagged and fault striations are not clear, except where silicic or cherty material is present in the limestone and along the fault plane. Fault breccias are made up of a high percentage of calcareous fragments and calcite cement. Faults also cut the limestone ridges in the basin. The major fault-strike in the limestone ridge of Jabal Umm Al-Aisah, south of the road, is NW with steep dips. Among the faults that cut the limestone ridges are those that bound the basin from the west and are separated from the western side of Jabal Al-Jifniyah by one of the major faults (Fig. 12). Other faults cutting the basal conglomerate were mapped along a road cut along the Nuqrah–Buraydah highway (Fig. 12). Figure 14 shows stereographic projections of the faults in this station documenting different senses of movement; Figure 15 shows complex dextral faults displacing parts of Jabal Al-Jifniyyah ridge.

Some of the faults and shear zones cut the Murdamah Rhyolites bounding the basin on the west (Fig. 12). The fracture zones in these rhyolites coincide in orientation with the fractures and faults in the fanglomerate on the other side of the basin. The majority of sinistral faults are found on the northeastern side of the basin, while predominantly dextral faults are preserved on the southwestern side of the basin.

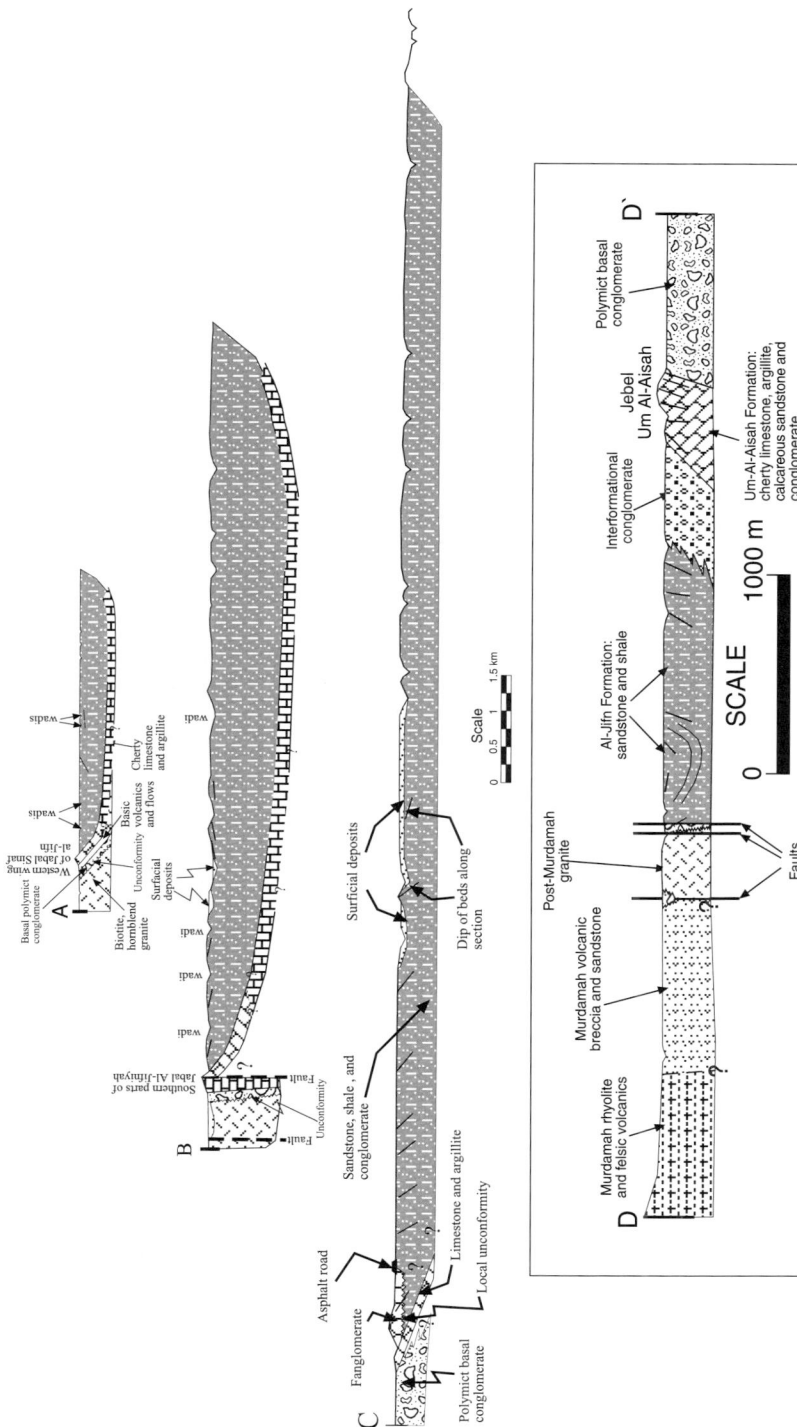

Fig. 10. Structural cross-sections drawn across the Jifn Basin showing the relationships of faults and Jibalah Group rocks. Coarse-grained proximal deposits are found along the basin margins, whereas fine-grained distal deposits are found in the centre of the basin.

Fig. 11. Photographs showing slickensides and faulted cobbles in the Jifn Basin.

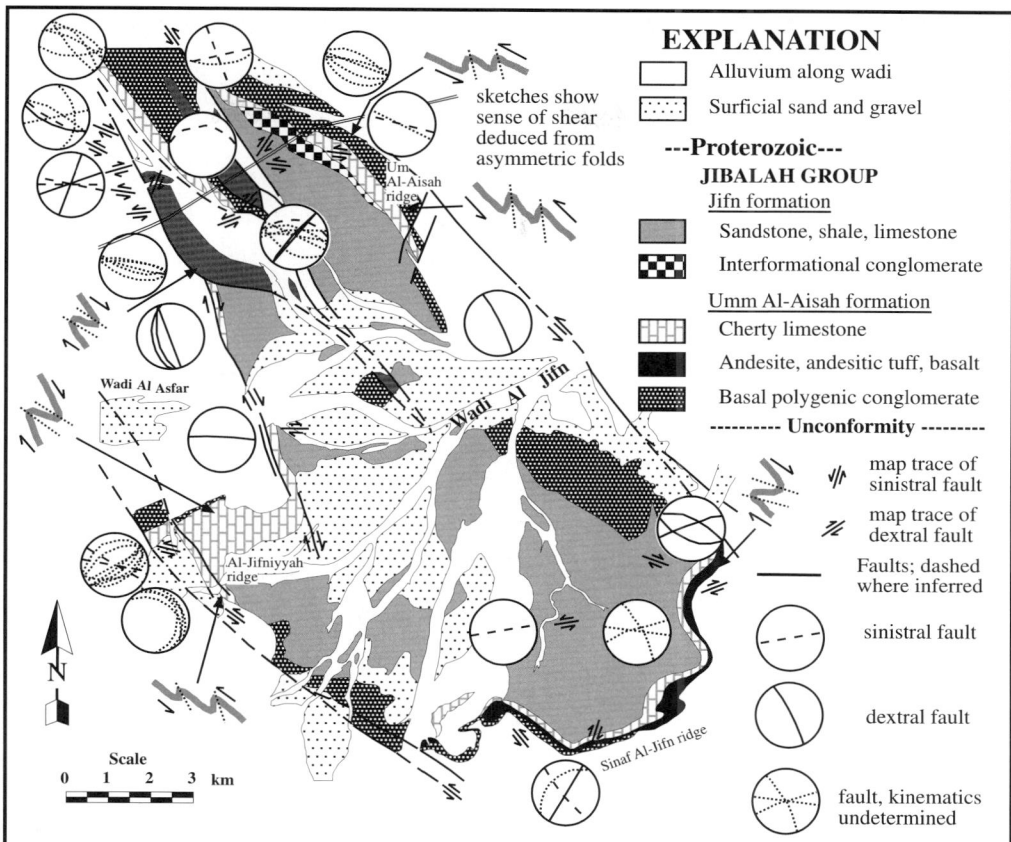

Fig. 12. Structural map showing the faults mapped in the Jifn Basin, along with their kinematics (where known).

Folding

Folding of the Al-Jifn Basin deposits is very pronounced, especially on the limestone ridges and the basal conglomerates. Different styles of folding are found, including disharmonic folds, en-echelon folds, and S and Z asymmetric folds interpreted to have formed in sinistral and dextral simple shear, respectively.

Limestone beds are folded disharmonically along the northwestern end of the Jabal Um Al-Aisah Ridge and in an asymmetric geometry along-strike. Another limestone ridge bounds the basin at the southeastern end and it is called Jabal Sinaf Al-Jifn (Figs 5 & 12). Most limestone beds in Jabal Sinaf Al-Jifn dip towards the basin; however, they are gently folded with a Z (dextral) asymmetry. A third ridge of limestone is found in the western parts of the basin and it is called as Jabal Al-Jifniyah (Fig. 12). This limestone shows a Z-fold shape (Fig. 12). The general dip of layers in Jabal Al-Jifniyah is towards the south, towards the basin. On the western side of Jabal Al-Jifniyah, an elongate, nearly vertically dipping limestone ridge strikes NW–SE and is separated from the Jabal Al-Jifniyah limestone by a major fault (Fig. 12). The southeastern part of this ridge is cut by the main course of Wadi Al-Jifn and the small continuation of the limestone ridge towards the south is folded with en-echelon folds with S (sinistral) asymmetry. Small hills of dissected inclined cherty grey limestone also outcrop on the western side of the basin as a strip of east- to NE-dipping homoclinal limestone beds. Folds within these cherty limestones, near the contact with Murdamah Rhyolites in the northwestern parts of the basin, show Z (dextral) asymmetry (Fig. 12). In the northern most parts of the basin the limestone also strikes NW and dips towards the west and SW. In these areas the limestones are folded in many places and they do not form ridges but only small low-lying outcrops.

Disharmonic folding Figures 16 and 17 show disharmonic folds mapped on the northwestern end of

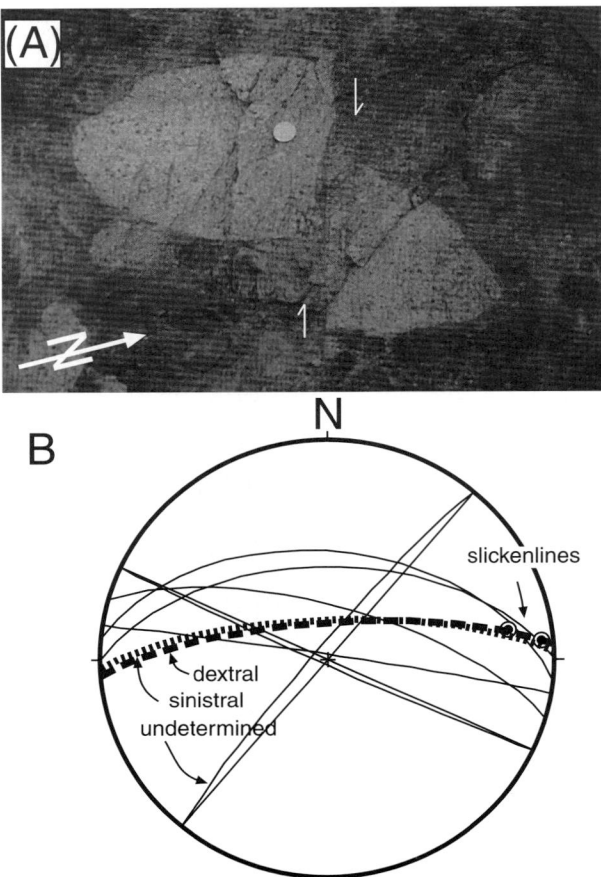

Fig. 13. (a) Photograph of clast that is offset dextrally on the northeast side of the Jifn Basin. (b) Lower hemisphere projection of faults from same outcrop as offset clast, showing both dextral and sinistral faults.

the limestone ridge of Jabal Um Al-Aisah, SE of the fanglomerate deposit; the rocks are less strongly folded NW of the fanglomerate ridge. The folds are cut by several faults in different orientations. The major fault direction is parallel to the major faults elsewhere in the basin (NW–SE). Detailed mapping of parts of these folds shows that the fanglomerate seems to 'push aside' the folded limestone (Fig. 16). Therefore, it is suggested that these folds formed by synsedimentary extrusion and sliding of the semiconsolidated limestone and argillite sequence during rapid deposition of the fanglomerate (Figs 16 & 17). The limestone is folded disharmonically along the two sides of the fanglomerate and the folding intensifies at the boundaries of the fanglomerate.

En-echelon folds in Al-Jifn Basin En-echelon folds in Al-Jifn Basin include sets of folds that are geometrically related to the strike-slip motion along the major faults, such that the fold traces are parallel and make an acute angle with the strike of the major fault strands. These folds occur in different parts of the basin but most are located at the boundaries of the basin. For example, en-echelon folds are prominent along the eastern and southeastern sides of the main ridge of Jabal Um Al-Aisah, and their axial traces trend north (Fig. 12). These folds are found in the basal conglomerate as well as the limestone argillite sequence.

West of Jabal Um Al-Aisah, a small elongate hill trending NW parallel to the main Um Al-Aisah Ridge is made up of a set of en-echelon folds. The folds on this side of the basin are characterized by S asymmetry, with long and steep limbs striking NW and shallow-dipping limbs striking north to NE. Their position with respect to the major faults and the general shape of the basin suggests a sinistral sense of movement.

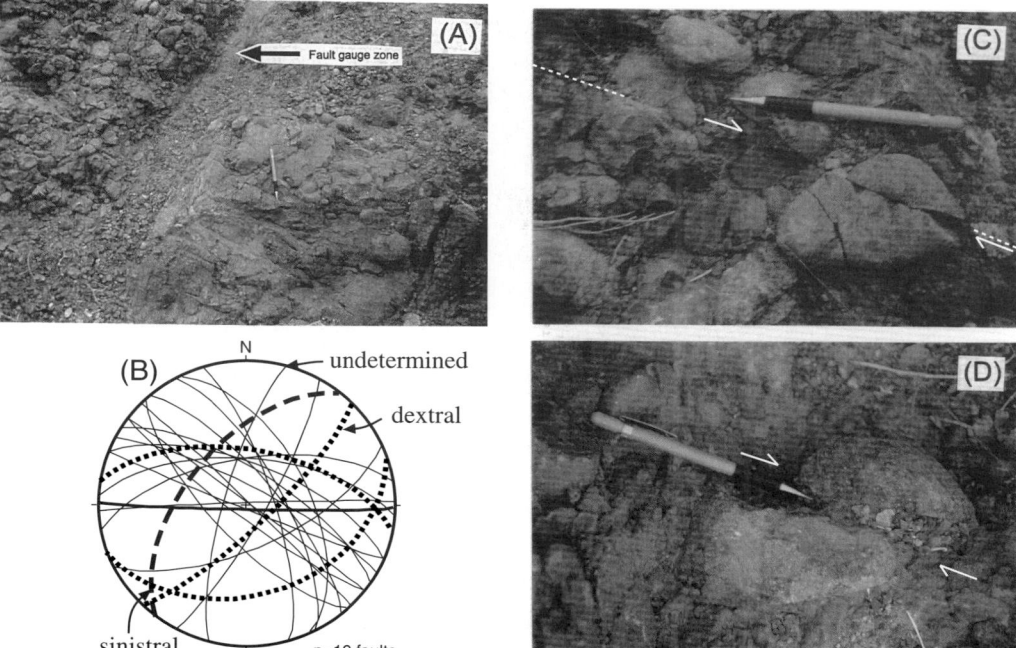

Fig. 14. (**a**) Photograph showing gouge in fault cutting basal conglomerate of the Jifn Basin. (**b**) Lower hemisphere equal-angle plot of faults mapped from this location. Note sinistral, dextral and indeterminate faults. (**c**) and (**d**) Clasts offset by faults from this outcrop.

Jifn Basin as a pull-apart basin

The Jifn and Sumamyyah Basins are arranged in an en-echelon geometry with their longest axes oriented NW. These two basins were formed by the enlargement and extension of large mega-releasing bends (e.g. Koide & Bhattacharji 1977; Mann et al. 1983) along the HZFZ. The geometry of these basins form right steps along the HZFZ, indicating a dextral sense of movement along the fault zone. The HZFZ (Fig. 4) is made up of anastomozing fault strands that form a lazy S-shaped (Mann et al. 1983) sigmoidal fault zone (Fig. 5). There are three major pull-apart basins along the HZFZ, located at two right bends along the fault zone. Furthermore, the Habariyah Mylonite Zone (Pellaton 1985; Cole & Hedge 1986), containing higher grade rocks than elsewhere along the faults, is located on a left bend on the same fault zone, suggesting that this may be a pop-up structure found along a transpressional part of the fault (Fig. 4). The setting of such basins and mylonite zones along the HZFZ is consistent with a dextral strike-slip setting. If the movement along the HZFZ were sinistral all the time then the Jifn and Sumamyyah Basins would have been areas of transpression, developing contractional structures such as thrust belts instead of being pull-apart basins, and they would have been areas of erosion rather than areas of sedimentation. The geometry of the Jifn Basin is also remarkably similar to experimental pull-apart basins formed in analogous dextral simple-shear experiments (e.g. McClay & Dooley 1995; Dooley et al. 1999)

The Jifn Basin has an elongate rhombic geometry with a length:width ratio of c. 1:4. The deposits along the long sides of the basin are coarse grained and include polymict conglomerates and fanglomerates interpreted to represent deposits at the base of fault escarpments. These deposits become finer in grain size both vertically and horizontally towards the centre of the basin. Fine-grained conglomerate, sandstone, limestone and shale, denoting a lower energy depositional environment, occupy the central part of the basin. Limestone, sandstone, conglomerate and shale form ridges and homoclines near the basin sides. The greater abundance of coarse-grained deposits on the northeastern side of the basin suggests that the basin's northeastern side was eroded more than the opposite southwestern side. The southwestern side of the basin may have had a relatively small elevation difference from inside to outside the basin, or it may have been made up of several steep fault steps.

The conglomerate and sandstone in the basin were

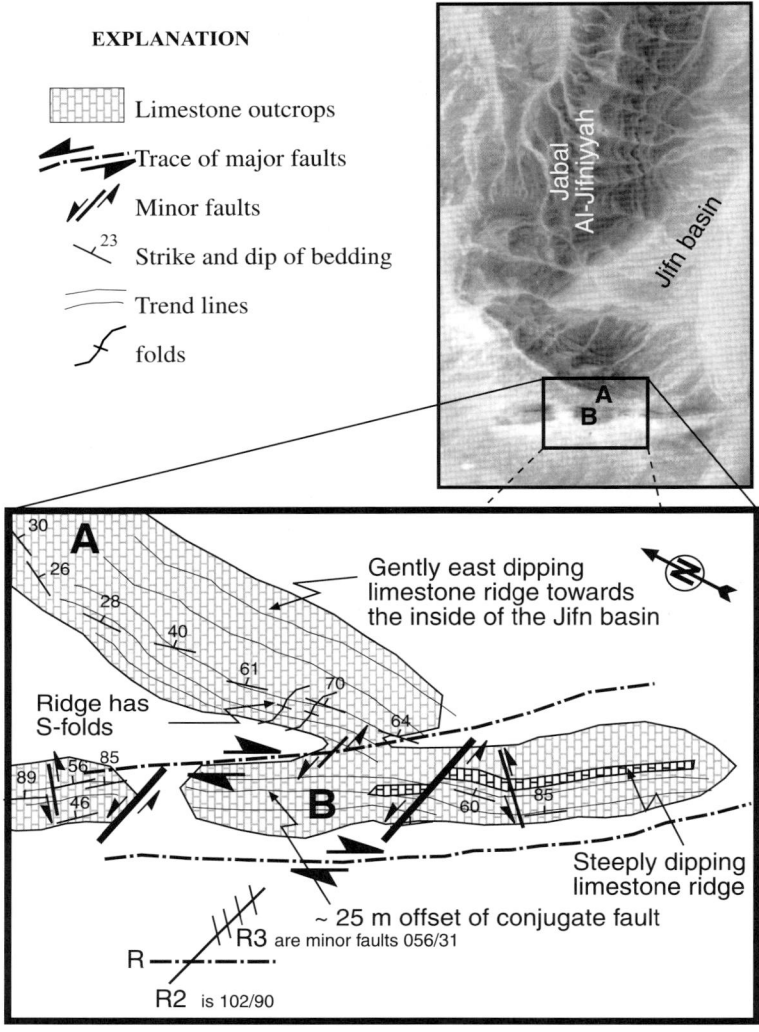

Fig. 15. Sketch map and aerial photograph of the Al-Jifniyyah Ridge in the Jifn Basin showing limestone offset by major faults.

derived from the surrounding rocks, including clasts of granite (including the post-Murdamah or Rahadha Granite, tentatively part of the Idah Suite), rhyolite, andesite and clasts of the adjacent Murdamah Group, as well as clasts of limestones of Jibalah rocks from the basin. The limestone clasts occur in the upper fanglomerate deposits.

Mafic and felsic volcanics also occur within the basin deposits. Basaltic and andesitic flows and volcanics are intercalated with the basal conglomerate of the basin and outcrop in the central part of it. These mafic volcanics are suggested to have formed in the early stages of basin formation. Undeformed felsic dykes intrude the Jibalah Group rocks denoting that volcanism took place during later stages of sedimentation: their age (576.6 ± 5.3 Ma) represents the end of significant movement and fault activity in the basin. It also indicates the evolution of the magma from mafic to silicic, perhaps reflecting greater crustal involvement.

The folding of the Jibalah Group rocks in Al-Jifn Basin shows that the sinistral movement developed later, denoting a reversal in the movement along the faults. This is clear in Jabal Sinaf Al-Jifn in the southeastern parts of the basin and in Jabal Al-Jifniyyah on the western side of the basin.

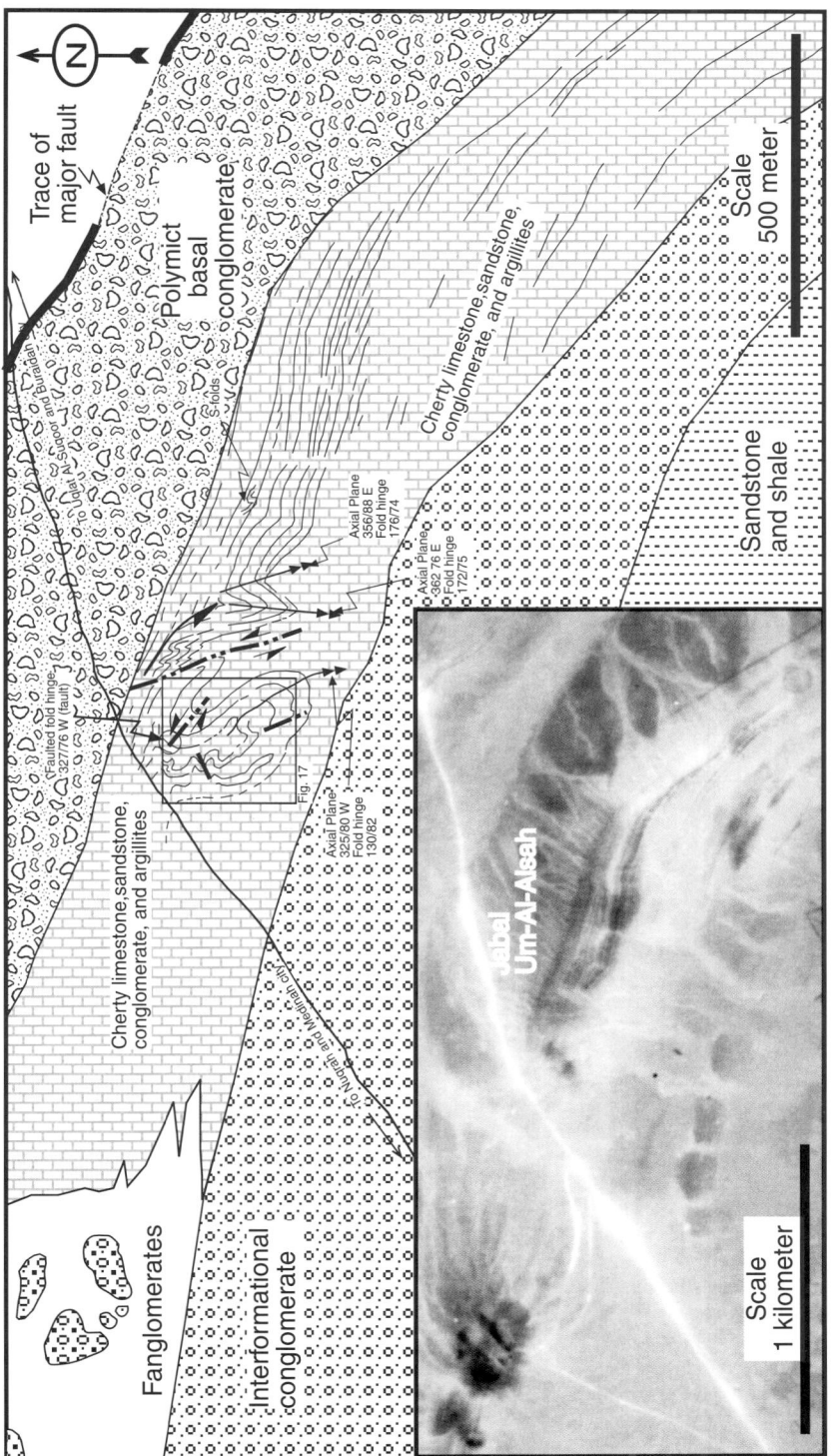

Fig. 16. Map of the Jabal Um–Al-Aisah area in the Jifn Basin, showing disharmonic folds in thinly bedded limestone and argillite. The folds are most strongly developed in places adjacent to thick wedges of fanglomerate that entered the basin from its margins, suggesting a causal link.

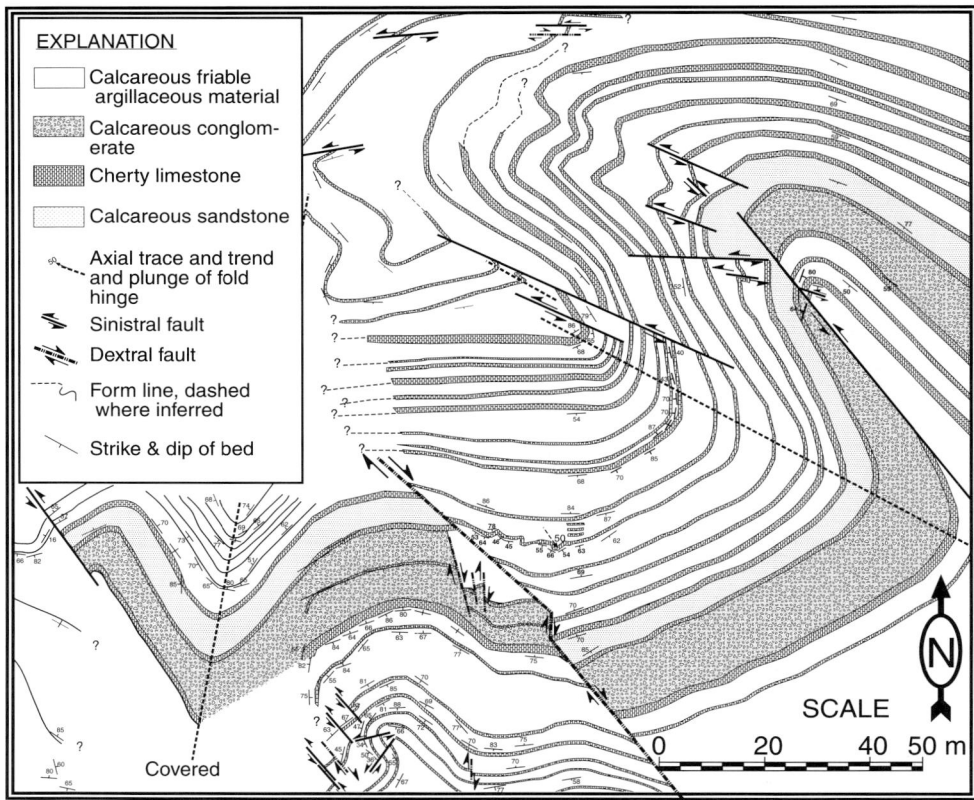

Fig. 17. Detailed map of disharmonically folded limestone and argillite beds in the northeastern parts of Umm Al-Aisah Ridge (location shown on Fig. 16).

It is suggested that the three pull-apart depositional basins along the HZFZ were formed in an early dextral phase of movement along the fault zone between 624.9 ± 4.2 and 576.6 ± 5.3 Ma. This idea might be applicable for other basins along other fault zones in the NFS. Agar (1987) reached similar conclusions in his study of the Bani Ghayy pull-apart Basin in the Zalm area of the Arabian Shield. He postulated an early phase of dextral movement along parts of the NFS in the southeastern parts of the shield. Davies (1980) reported dextral displacements along parts of the NFS in the northwestern part of the Najd fault zone in the Duba area. Howland (1979) and Moore (1979) observed that the termination and the end of the master faults in the Najd Fault System change strike in the opposite sense to that expected for propagation during sinistral shear (Agar 1987).

In the Jifn Basin, Z-asymmetry folds and dextral faults, as well as the overall shape of the basin (Mann et al. 1983), are strong evidence that the Jifn Basin formed in a dextral shear regime and that sinistral movements occurred in later stages. Figure 18 shows a model for the development of the Jifn Basin and how it was first formed in a dextral shear regime, with the basin forming along a releasing bend in the Halaban–Zarghat Fault. Deposition of the Jibalah Group rocks in the Jifn Basin occurred in the early dextral phase. Then, as the movement along the fault strands changed direction, folds with S asymmetry and sinistral minor faults overprinted older dextral structures. The fault on which the reversal of motion occurred is suggested to cut through the basin diagonally from NW to SE (Fig. 18).

Other likely pull-apart basins related to the Najd Fault System (NFS) in the Arabian Shield

There are many small Neoproterozoic basins located along steps in the NFS in the Arabian Shield (Fig. 3).

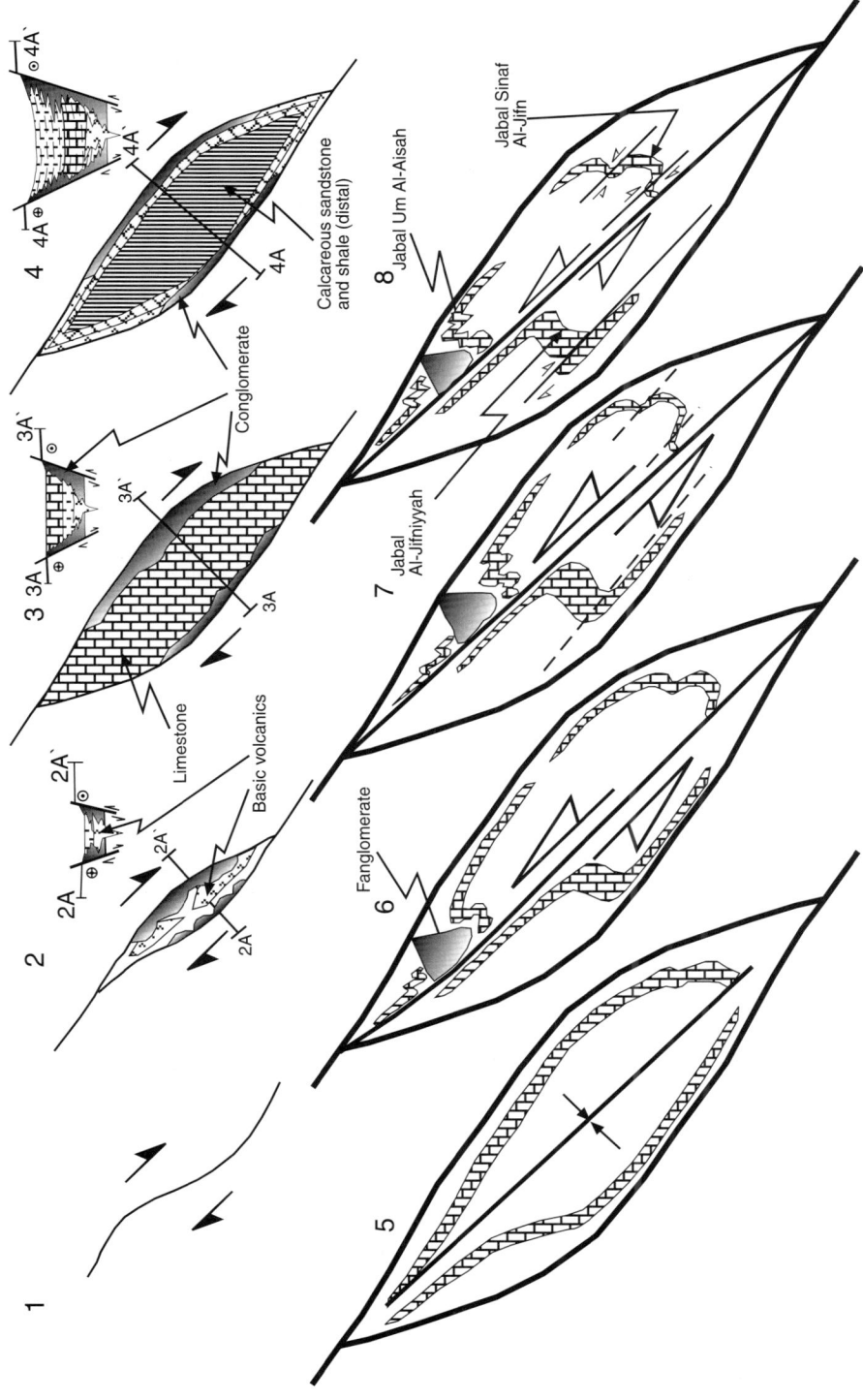

Fig. 18. Cartoon showing evolution of the Jifn Basin.

In this section the general characteristics of the more prominent of these basins are described in order to assess whether or not conclusions drawn for the Jifn Basin may be generally applicable to the NFS as a whole.

Agar (1986) mapped the Bani Ghayy Group (14 on Fig. 3) in the Zalm area in the south central Arabian shield and considered these rocks to be the youngest Proterozoic succession in this area. The rocks crop out in three N–S-trending elongate belts and are in fault contact with the older rocks; these faults are part of the NFS. The rocks are similar in all belts, comprising conglomerates, greywackes, limestone, subordinate siltstones, and bimodal volcanics and volcanosedimentary rocks. However, particular rock types occupy different positions in each succession, indicating that each belt evolved separately. This supports the idea that these deposits were formed as pull-apart basins and not as blanket deposits covering the whole shield, as suggested by Al-Shanti (1993). Agar (1986) suggested that there is an E–W (across the width of the basin) facies change in the lower part of the belts.

The conglomerates occur along the strike-slip fault margins of the belts and grade inwards to greywackes (Agar 1986), similar to the trends observed in the Jifn Basin. Agar (1986) suggested that the conglomerates are subaqueous fanglomerate deposits typical of steep-sided sedimentary basins (van Houton 1974; Mann et al. 1983). The conglomerates are polymict and composed of very coarse (>1 m) subangular clasts in a matrix of smaller cobbles and grit (Agar 1986). Red porphyritic rhyolite dykes, sills and flows are intercalated with the conglomerate, and clasts of these dykes are also found within the conglomerate. The dykes, conglomerate, and development of the pull-apart basins are therefore contemporaneous. The greywackes form upward-fining units with clasts of 10–20 cm at the base that fine to <1 cm in diameter (Agar 1986). The greywackes increase in abundance but decrease in grain size towards the centre of the belts. The graywacke, conglomerates, grits and coarser siltstones all show 'truncated' a, ab or abc cycles of Bouma (1962), and are typical proximal types (Agar 1986). The limestones are pure, dark grey and well bedded. Clastic textures in some places in the limestones are associated with interbedded calcareous shales and siltstones (Agar 1986). These limestones pass into massive stromatolitic limestone (Agar 1986).

The volcanogenic rocks comprise intercalated lavas, pyroclastic, volcanoclastic and tuffaceous units. These volcanics are dominated by basaltic andesites but rhyodacitic rocks are also found, and both volcanic and volcanoclastic rocks are compositionally bimodal. The southern part of the basins are older than the northern parts (Agar 1986), implying growth of the basins from south to north. Agar (1986) suggested that the basins had a contrast from high-energy conditions in the west relative to low-energy conditions in the east. He suggested that the western fault margin was steep whereas the eastern one was more gentle with multiple-stepped faults, indicating asymmetrical basins. Agar (1986) proposed that there is more than one succession of conglomerate and greywacke in some of the basins, which he attributed to a renewal of faulting and movements along the faults. The Bani Ghayy Group has a U–Pb zircon age of c. 630–610 Ma (Stacey & Agar 1985), although Johnson (2002) reports that they may be as old as 650 Ma. Agar (1986) suggested that the Bani Ghayy Group is the product of sedimentation and volcanism in pull-apart grabens of an incipient dextral phase of the Najd strike-slip Orogen.

The Al-Mashhad area is located in the northwestern part of the Arabian Shield between latitudes 26°30' N and longitudes 38°15' E (2 on Fig. 3). The area is dissected by north- to NW-striking Najd faults and by east to NE faults – the latter are offset by the Najd faults. According to Hadley (1974), the thickness of the Jibalah Group is between 750 and 850 m and is divided into three formations – terrigenous clastic rocks, volcanic rocks and limestone – which he interpreted as grabens.

The Sumamiyah–Zarghat Basin (3 and 4 on Fig. 3) is among the longest Jibalah Group basin in the Arabian Shield. Its length reaches c. 100 km with a width of c. 8.5 km, and is located between latitudes 25°44' and 26°02'N and longitudes 41°10' and 41°27'E. It is located next to the Jifn Basin in an en-echelon geometry, with the basin's longest axes trending NW, parallel to the HZFZ. The deposits found in this basin resemble, but are thinner than, those found in the Jifn Basin (Brosset 1970; Delfour 1970, 1983).

The Jabal Antaq basin (5 on Fig. 3) is an asymmetrical structural basin, or half-graben. It is c. 45 km long and c. 8 km wide, and is located between latitudes 22°32' and 23°41'N and longitudes 42°35' and 42°52'E. It is located along the HZFZ, with the Jifn and the Sumamiyyah Basins c. 300 km away to the NW. It is suggested that these three basins formed in a dextral strike-slip regime along two major releasing bends along the HZFZ. According to Bois (1971) and Delfour (1983), the Jabal Antaq Basin is made up of c. 3000 m of arkose, coarse- and fine-grained sandstone, argillite and shale, besides the basal polymict conglomerate, and c. 100 m of volcanic andesite flows occur within the sequence.

The Jabal Jibalah Basin (8 on Fig. 3) hosts the type locality for Jibalah Group rocks (Delfour 1977). It is c. 35 km long and c. 5 km wide, and is located between latitudes 25°25' and 25°35'N and between longitudes 40°40' and 40°59'E. It comprises, with Bir Arja Basin, a belt of Jibalah Group basins along the Al-Rika–An Nakhil–Wajiyah Fault Zone parallel to the HZFZ (Johnson 1997). The Jibalah Basin is characterized by the abundant andesite flows intercalated with polymict basal conglomerate which locally overlie limestone beds (Delfour 1983). The andesitic–basaltic flows and volcanics in this belt are more abundant and better exposed than those on the Halaban-Zarghat Fault.

Bounding faults to the Jabal Jibalah Basin sinistrally offset the Nabitah–Hulayfah ophiolite belt by $c.$ 23 km. The Jibalah Basin is an elongate basin, narrow at its southeastern end and wider at its northwestern end. Andesitic volcanics are exposed in great abundance in the Jibalah-Arja Fault Belt in comparison to the Jifn, Sumamayyah and Jabal Antaq Basins along the Halaban-Zarghat Fault Belt.

The Bir Arja Basin (7 on Fig. 3) is a southeastern extension of the Jibalah Basin. It is located on the Jibalah-Arja Fault Zone between latitudes 25°13′ and 25°26′N and longitudes 41°27′ and 41°42′E. Its approximate length is 35 km and its maximum width is 6 km. Its long axis trends NW and it contains abundant andesitic–basaltic volcanics in the central areas of the basin bounded by limestones and conglomerates along the sides of the basin. Faulting and folding are pronounced within the Bir-Arja and Jibalah Basins, similar to the Sumamiyah, Al-Jifn and Jabal Antaq Basins on the HZFZ. It appears that finer grained materials are preserved more in the basins along the HZFZ than in their counterparts on the Jibalah–Bir Arja Fault Zone.

The Bir Al-Hissu Basin (11 on Fig. 3) is located along the southeastern part of the Jibalah–Bir Arja Fault Zone between latitudes 24°20′ and 24°26′N and between longitudes 41°27′ and 41°42′E, with a length of $c.$ 25 km and a width of $c.$ 6 km (Delfour 1983). It is composed of strongly folded Jibalah Group rocks and it is almost entirely bounded by faults. Coarse conglomerates, cherty limestone dominate the deposits, and argillites with flows of andesite intercalated in the sequence (Duhamel & Petot 1972; Delfour 1983).

The Jabal Hawaqah and Jabal Al-Abd Basins (9 on Fig. 3) forms a triangular area (Delfour 1983) along the Wajiyah–An-Nakhil–Ar-Rika Fault Zone (Johnson 1997). It is located between latitudes 24°38′ and 24°52N and between longitudes 40°35′ and 40°57′E, with a maximum length of 41 km and a maximum width of $c.$ 12 km. According to Delfour (1983), this basin is filled with deposits of conglomerate, arkosic sandstone followed by andesite flows and overlain by cherty limestone with argillite intercalations. Fine-grained sandstone and argillite beds are found high in the section. A belt of andesite 10 km long and 2 km wide on the eastern side of the basin is succeeded by red arkosic sandstone and conglomerate (Delfour 1983).

The Sukhaybarat basin (10 on Fig. 3) is situated on the Wajiyah–An–Nakhil–Ar-Rika Fault Zone SE of the Jabal Hawaqah and Jabal Al-Abd Basins. Delfour (1983) suggested that it is the continuation of the Jabal Hawaqah and Jabal Al-Abd Basin. It is located between latitudes 23°45′ and 24°25′N and between longitudes 41°05′ and 42°35′E. The basin is 70 km long and its maximum width is 10 km (Delfour 1983). The Jibalah Group rocks found in the Sukhaybarat Basin include basal conglomerate that overlays unconformably parts of Hamdah–Nabitah ophiolites, which passes into arkosic and conglomeratic sandstone and grades up into cherty limestone and argillites in the southern parts of the basin (Delfour 1983). Andesitic flows are poorly represented but some gabbro stocks intrude the sandstone and argillites, forming hornfels zones 2 m thick around them (Delfour 1983).

The Bir Sija Basin (12 on Fig. 3) is a small 26 × 4 km basin on the Wajiyah–An–Nakhil–Ar-Rika Fault Zone SE of the Sukhaybarat Basin. It is located between latitudes 23°35′ and 23°41′N and between longitudes 42°35′ and 42°52′E. The basin's long axis trends NW (Delfour 1983). The Jibalah Group rocks in this basin are made up of polymict basal conglomerate, andesite flows, and a thick series of fine-grained sandstone and argillites with some cherty limestone (Conraux et al. 1969; Letalenet 1975; Delfour 1983).

The Jabal Liban Basin (1 on Fig. 3) is located on the northwestern part of the shield along the Ajjaj Shear Zone (Duncan et al. 1990). It is located between latitudes 26°22′ and 26°50′N and betweens longitudes 36°21′ and 36°45′E. It has a length of 60 km and a width of 20 km. The deposits found in this basin include conglomerate, siltstone and sandstone (Delfour 1970; Frets 1977; Davis 1980). Felsic pyroclastics are found intercalated with the Liban Basin deposits (Delfour 1970).

The Minawah Formation is found in a basin (13 on Fig. 3) located in the northwestern part of the Arabian Shield. Clark (1985) suggested that it is a pull-apart basin bounded by parts of the NFS. It is located between latitudes 27°55′ and 28°30′N and between longitudes 35°30′ and 36°00′E. The general trend of the maximum axis of the basin is in a west–NW direction, with a maximum length of 26 km and a maximum width of 10 km (Clark 1985). It is located on the Ad-Durr Fault Zone and its deposits include conglomerate, sandstone, and shale intercalated with silicic lava, tuffs, breccia and ignimbrite (Clark 1985).

Discussion

The shape and geometry of the Jifn Basin does not support formation along a sinistral strike-slip fault system. Instead an origin as a pull-apart basin on a right bend in a dextral strike-slip fault setting is suggested. The Jifn Basin resembles other basins (e.g. the Sumamayyah Basin) along two right bends along the same HZFZ. Furthermore a mylonite zone (Habariyah) occurs in a left step along the same fault zone (Pellaton 1985). The setting of these basins and the mylonite zone coincide with settings in a dextral strike-slip fault. The present-day setting of the faults and the apparent offset of rock units along the fault, especially the Rahadha Granites (Delfour 1977; Matsah & Kusky 2001), show a sinistral sense of offset. The contradiction in the evidence suggests that movement along the HZFZ changed, and is therefore divided into two stages. It is suggested that there was an early dextral phase of movement along the Halaban–Zarghat Fault, and a later sinistral phase had the most obvious offsets. It is also suggested that the basin formed and most of the Jibalah Group rocks were deposited in the dextral early stage, followed by more significant sinistral displacements.

Moore (1979) and Howland (1979) noticed that the ends of some Najd master faults change strike in sense, opposite to what is expected for propagation during sinistral shear. Moore (1979) noticed variations in the degree of offset along the length of Najd faults and that some of these faults show dextral displacement. Agar (1986, 1987) also found that deposits of Bani Ghayy Group rocks formed in a pull-apart basin that formed at an incipient dextral phase of motion along a Najd fault zone. Later these deposits were affected by sinistral movement along the same fault, resulting in deformation of these deposits. Davies (1980) noted dextral displacements along some of the Najd faults in the northern part of the Arabian Shield. Pellaton (1985) and Cole & Hedge (1986) mapped and reported a 10 km right-lateral offset of a quartz–diorite along the Halaban–Zarghat Fault north of the town of Miskah. A U–Pb zircon age of that diorite of 621 ± 7 Ma was reported by Cole & Hedge (1986). Johnson & Kattan (1999) also recognized a complex evolution for the Halaban–Zarghat Fault, allowing for some early dextral shear.

The style and deformation of deposits in the Jifn Basin was found in the present study to be unusual. Dextral and sinistral faults and folds with S and Z asymmetry cannot be related to one phase of movement along the Halaban–Zarghat Fault. Faults and folds in the Jifn Basin are consistent with formation during dextral shearing and deformation during later sinistral shear.

Movement along the Halaban–Zarghat Fault was initiated at or after $c.$ 624.9 ± 4.2 Ma along a dextral strike-slip fault zone. The date represents a U–Pb zircon age of a rhyolite basement to the Jibalah Group rocks in the Jifn Basin, where dextral motion continued past 621 ± 7 Ma, as indicated by the age of the Badai–Mushrifah quartz–diorite Pluton (possibly early Idah Suite) which is offset by 10 km dextrally. Deposits of conglomerate, carbonates, sandstone and shales were accumulating in the basin as it was lengthening. Mafic volcanics were erupted during the early stages, as they are intercalated with the basal polymict conglomerate.

The later history of movement is not clear. However, according to Stacey & Agar (1985) and Agar (1987), movement along the NFS switched to a sinistral phase at $c.$ 620 Ma. It is suggested here that the sinistral movement along the Halaban–Zarghat Fault took place between 621 and 576.6 ± 5.3 Ma. If it is assumed that the 10 km dextral offset of the Badai–Mushrifah Pluton took $c.$ 10 Ma to accumulate, then it is estimated that dextral slip occurred between 625 and 610 Ma. Sinistral movement in the Jifn Basin is younger than the Rahadha Granites based on the following observations: (1) the Rahadha Granites are sinistrally offset; (2) the Rahadha granite clasts are only found in the upper fanglomerate deposits of the Jifn Basin, which are thought to have been deposited in the sinistral stage; (3) Cole & Hedge (1986) considered the Rahadha Granites to be part of the Idah Suite, is dated (with Rb–Sr whole-rock and U–Pb zircon ages) as 615–620 Ma. If the Badai–Mushrifah quartz–diorite Pluton is part of the Idah Suite, then this suite apparently intruded during the change from dextral to sinistral motions.

The fanglomerates at the top of the Jibalah Sequence in the Al-Jifn Basin are not folded and are less deformed than older deposits in the basin, e.g. the folding of the basal polymict conglomerate is not observed in the top fanglomerate deposits. Hence, it is suggested that the top fanglomerates were deposited during the late sinistral stage of movement along the faults. The other noticeable feature is that the limestone and carbonates are folded on both sides of the fanglomerate deposit, particularly north of the Nuqrah–Buraydah highway. The style of folding differs on both sides of the fanglomerate. For example, the carbonates on the southern side of the fanglomerate are highly folded and faulted with soft-sediment disharmonic folding and en-echelon folds, and a decollment is also apparent (Figs 16 & 17). The folding of the carbonates on the northern side of the fanglomerate is less pronounced.

The sinistral faults and the S-asymmetry folds are more pronounced on the northeastern side of the basin while the dextral faults and Z-asymmetry folds are more common on the southwestern side of the basin, especially the Z folding of Jabal Al-Jifniyah in the west of the basin (Fig. 12). A major sinistral fault runs through the central part of the basin which offsets the older Jibalah rocks in the basin, including the basal conglomerate and the carbonates as well as the sandstone and shale (Fig. 12). A major dextral

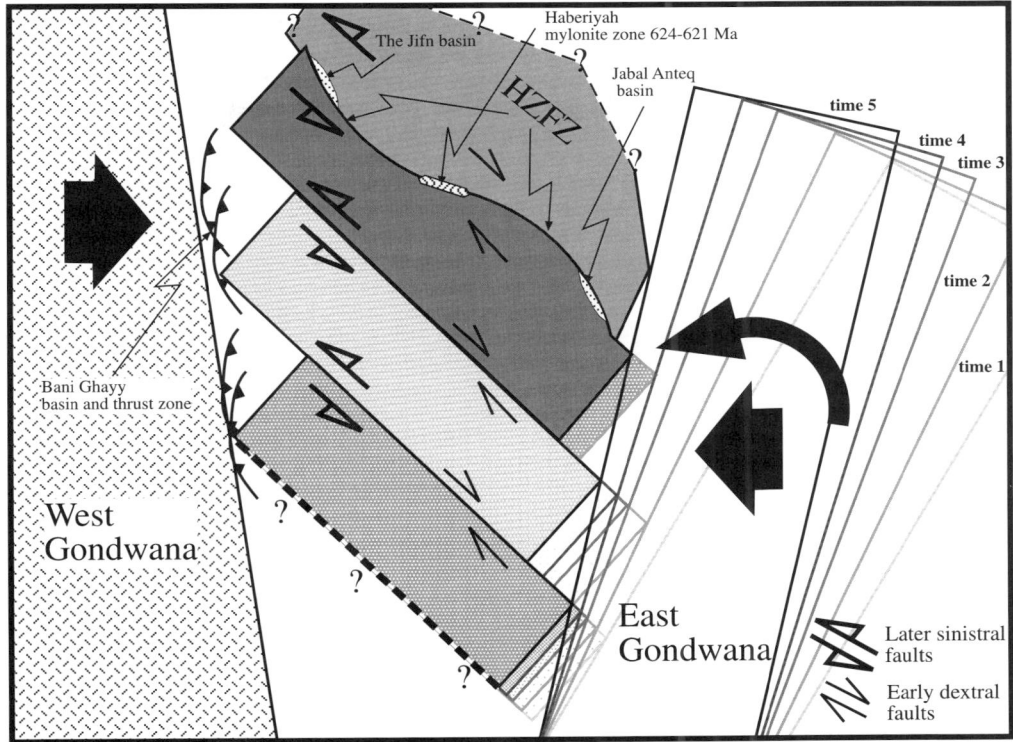

Fig. 19. Cartoon showing movement of the Arabian–Nubian Shield blocks along the Najd fault zones in response to contraction and amalgamation between East and West Gondwana. Dextral movement started on each zone, which later switches to sinistral motion.

fault runs along the western side of the basin (Fig. 12).

The evidence therefore suggests that movement along the Halaban–Zarghat Fault included two stages of movement, an early dextral stage and a late sinistral stage. The movement on the fault is suggested to have ended by 576 Ma, when a felsite dyke intruded the Jibalah Group rocks in the Jifn Basin.

Implications for the evolution of the East African Orogen

The continental collision of India and Siberia (Eurasia) can be compared with the Neoproterozoic collision of East and West Gondwana. Like SE Asia, caught between and escaping laterally from India and Asia, the ANS was trapped between East and West Gondwana causing it to tectonically escape toward an oceanic-free face to the north (Fig. 2). The collision between East and West Gondwana along the East African Orogen started first and was more intense in the south, as inferred from the change in degree of metamorphism from granulite grade in the south (Mozambique Belt) to greenschist facies in the north (ANS).

Early stages of the collision saw the amalgamation of arc terranes and collision between East and West Gondwana. E–W shortening (Abdelsalam & Stern 1996) was accommodated by escape tectonics where the trapped terranes of the ANS between East and West Gondwana started to escape to the north (Burke & Sengor 1986; Hoffman 1991; Storey 1993). According to the evidence along the NFS, the initial motion of the Najd faults was dextral (Fig. 19). The fault sequence began in the southwestern parts of the NFS in the Zalm and Bani Ghayy areas, perhaps as early as 650 Ma and certainly by 630 Ma. The southwestern blocks moved first towards the north–NW, and the other blocks moved similarly later. This sequence was related to the oblique accretion of crustal blocks against West Gondwana, as the Mozambique Ocean was closing in a scissor-like manner from south to north (Fig. 19). East Gondwana was progressively sutured to these newly accreted blocks. As the first block could no longer move against the rigid parts of West Gondwana (Congo Craton) the next block started to move

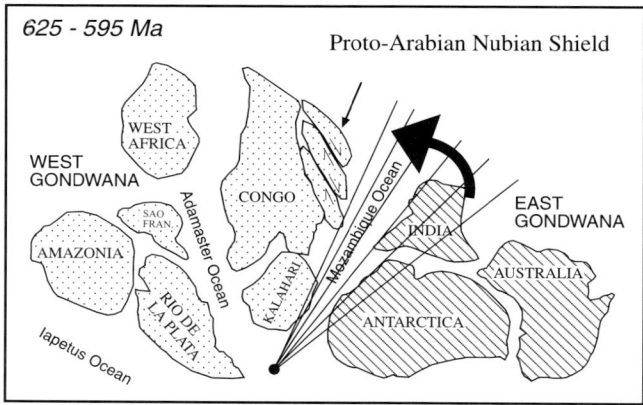

Fig. 20. Model for the evolution of the Jifn Basin and the Najd Fault System. Early dextral faulting (625–c. 610 Ma) is related to oblique accretion, and later sinistral motion is related to escape tectonics of arc and microcontinental terranes of the Arabian–Nubian Shield. Map compiled form Unrug (1996, 1997) and Grunow et al. (1996).

sinistrally with respect to the previous southwestern block and dextrally with respect to the next northeastern block (Figs 19 & 20). Hence, the sense of shear on the fault between the first and second blocks switched from dextral to sinistral.

The later sinistral stage might alternatively be a result of a change of direction of compression between East and West Gondwana. The reversal of movement along the same faults indicates that the direction of compression might also have changed from NE to SE, utilizing the same faults for the dextral and sinistral phases. Abdelsalam (1994) suggested that the Ar-Rayn continental terrain came from the SE and collided with the other parts of the Arabian Shield, causing the movement direction to change along the Halaban-Zarghat Fault in particular and along the NFS in general. This stage ended just before 576 Ma, representing the ending of the amalgamation of East and West Gondwanaland.

This work was supported by grants from the Royal Embassy of the Kingdom of Saudi Arabia and the Saudi Arabian Cultural Ministry to the U.S.A. We thank S. Bowring and D. Coleman for help with the U-Pb geochronology. The Prince of Qasim, Saudi Arabia, is thanked for facilitating field work in the northern shield. The Amir of Al-Nuqrah, Hadee bin Sultan is also thanked for his hospitality. Reviews by P. Johnson and K. Nielson greatly improved this manuscript.

References

ABDELSALAM, M. G. 1994. The Oko shear zone: post-accretionary deformations in the Arabian–Nubian Shield. *Journal of the Geological Society, London,* **151**, 767–776.

ABDELSALAM, M. G. & STERN, R. J. 1993. Structure of the late Proterozoic Nakasib suture, Sudan. *Journal of the Geological Society, London,* **150**, 1065–1074.

ABDELSALAM, M. G. & STERN, R. J. 1996. Sutures and shear zones in the Arabian–Nubian Shield. *Journal of African Earth Sciences,* **23**, 289–310.

ABDELSALAM, M. G., STERN, R. J. SCHANDELMEIER, H. & SULTAN, M. 1995. Deformational history of the Keraf Zone in northeast Sudan, revealed by Shuttle Imaging Radar. *Journal of Geology,* **103**, 475–491.

AGAR, R. A. 1986. The Bani Ghayy Group; sedimentation and volcanism in pull-apart grabens of the Najd strike-slip orogen, Saudi Arabian Shield. *Precambrian Research,* **31**, 259–274.

AGAR, R. A. 1987. The Najd fault system revisited: a two way strike-slip orogen in the Saudi Arabia Shield, *Journal of Structural Geology,* **9**, 41–48.

AL-SHANTI, A. M. S. 1993. T*he Geology of The Arabian Shield* [in Arabic]. Center for Scientific Publishing, King Abdlaziz University, Saudi Arabia.

ANDRE, C. G. 1989. Evidence for Phanerozoic reactivation of the Najd Fault System in AVHRR, TM, and SPOT Images of Central Arabia. *Photogrammetry Engineering and Remote Sensing,* **55**, 1129–1136.

AYDIN, A. & NUR, A. 1982. Evolution of pull-apart basins and their scale independence. *Tectonics,* **1**, 91–105.

BERHE, S. M. 1990. Ophiolites in northeast and east Africa: implications for Proterozoic crustal growth. *Journal of the Geological Society, London,* **147**, 41–57.

BINDA, P. L. 1981. *The Precambrian Boundary in the Arabian Shield. A Review.* King Abdulaziz University, Jeddah, Faculty of Earth Sciences Bulletin, **4**, 107–120.

BOIS, J. 1971. *Geology and mineral exploration of the northwestern part of the Jabal Damkh quadrangle, Photomozaic 115 W.* French Bureau de Recherches Geologiques et Minieres Technical Record, 71–JED-2.

BONAVIA F. F. & CHOROWICZ J. 1992. Northward explusion of the Pan-African of northeast Africa guided by re-entrant zone of the Tanzania craton. *Geology,* **20**, 1023–1026.

BONAVIA F. F. & CHOROWICZ J. 1993. Neoproterozoic structures in the Mozambique orogenic belt of south Ethiopia *Precambrian Research*, **62**, 307–322.

BOUMA, A. H. 1962. *Sedimentology of some Flysch Deposits. A Graphic Approach to Facies Interpretation*. Elsevier, Amsterdam.

BOWEN, R. & JUX, U. 1987. *Afro-Arabian Geology: A Kinematic View*. Chapman & Hall, London.

BROSSET, R. 1970. *Zarghat magnesite deposit*. French Bureau de Recherches Geologiques et Minieres Technical Record, 70–JED-2.

BROWN, G. F. & COLEMAN, R. G. 1972. The tectonic framework of the Arabian Peninsula. *XXIV International Geological Congress*, Montreal, **Part III**, 300–305.

BROWN, G. F. & JACKSON, R. O. 1960. The Arabian Shield. *XXI International Geological Congress*, Copenhagen (Norden), **Part IX**, 69–77.

BROWN, G. F., JACKSON, R. O., BOGUE, R. G. & BERGY JR, E. L. 1963. *Geological map of the northwestern Hijaz quadrangle, Kingdom of Saudi Arabia, scale 1:500000*. Miscellaneous Geologic Investigation Map, 1–204 A, United States Geological Survey.

BROWN, G. F., SCMIDT, D. L. & HUFFMAN, JR, C. 1989. *Geology of the Arabian Peninsula, Shield area of Western Saudi Arabia*. United States Geological Survey Professional Paper, 560–A.

BURKE K. & SENGOR C. 1986. Tectonic escape in the evolution of the continental crust. *In*: BARAZANGI, M. & BROWN, L. (eds) *Reflection Seismology: The Continental Crust*. American Geophysical Union, Geodynamics Series 14. American Geophysical Union, Washigton, DC, 41–53.

CLARK, M. D. 1985. Late Proterozoic crustal evolution of the Midyan region northwestern Saudi Arabia. *Geology*, **13**, 611–615.

COLE, J. C. & HEDGE, C. E. 1986. *Geochronologic investigation of late Proterozoic rocks in the northeastern shield of Saudi Arabia. Deputy Ministry for Mineral Resources, Jeddah, Saudi Arabia*. United States Geological Survey, Technical Record, TR-05-5.

CONRAUX, J., DELFOUR, J. & EIJKELBOOM, G. 1969. *The Jabal Sayid prospect*. French Bureau de Recherches Geologiques et Minieres Technical Record, 69–JED-43.

DALZIEL, I. W. D. 1992. Antarctica: a tale of two supercontinents? *Annual Reviews of Earth and Planetary Sciences*, **20**, 501–26.

DAVIES, F. B. 1980. *Reconnaissance geology of the Duba quadrangle, sheet 27/35 D, Kingdom of Saudi Arabia, scale 1:100000. Map GM57*. Saudi Arabian Deputy Ministry for Mineral Resources.

DAVIES, F. B. 1984. Strain analysis of wrench faults and collision tectonics of the Arabian Shield, *Geology*, **82**, 37–53.

DAVIES, F. B. & GRAINGER, D. J. 1985. Geologic map of the Al-Muwaylih quadrangle, sheet 27A, Kingdom of Saudi Arabia (with text), *Scale 1:250000. Map GM46A*. Saudi Arabia Deputy Ministry of Mineral Resources.

DELFOUR, J. 1970. Le group de j'balah une nouvelle unite du bouclier Arabe, *[France] Bureau des Recherches Geologiques et Miniers Bulletin*, **ser. 2, sec. 4**, 19–32, 5 figures, 2 tables; see also The J'Balah Group, A new unit of the Arabian Shield, *[France] Bureau des Recherches Geologiques et Miniers Open-File Report* **70–JED-4**: 31pp, 5 figures.

DELFOUR, J. 1977. *Geology of the Nuqrah quadrangle, sheet 25E, Kingdom of Saudi Arabia, Scale 1:250000. Map GM28*. Saudi Arabian Deputy Ministry for Mineral Resources.

DELFOUR, J. 1979. *Geology of the Halaban quadrangle, sheet 23G, Kingdom of Saudi Arabia, Scale 1:250000. Map GM46A*. Saudi Arabian Deputy Ministry for Mineral Resources Geologic.

DELFOUR, J. 1983. *Geology and mineral resources of the northern Arabian Shield, A Synopsis of BRGM investigations, 1965–1975*. BRGM Technical Record. TR-03–1.

DOOLEY, T., MCCLAY, K. & BONORA, M. 1999. 4D evolution of segmented strike-slip fault systems: applications in NW Europe. *In*: FLEET, A. J. & BOLDY, S. A. R. (eds) *Petroleum Geology of Northwest Europe: Proceedings of the 5th Conference*, Geological Society of London, 215–225.

DUHAMEL, M. & PETOT, J. 1972. *Geology and mineral exploration of the Al Hissu quadrangle*. French Bureau de Recherches Geologiques et Minieres Technical Record 72–JED-7.

DUNCAN I. J., RIVARD B., ARVIDSON R. E. & SULTAN M. 1990. Structural interpretation and tectonic evolution of a part of the Najd Shear Zone (Saudi Arabia) using Landsat thematic mapper data. *Tectonophysics*, **178**, 309–335.

FLECK, R. J., COLEMAN, R. G., CORNWALL, H. R. ET AL. 1976. Geochronology of the Arabian Shield, Western Saudi Arabia: K–Ar results. *Geological Society of America Bulletin*, **87**, 9–21.

FLECK, R. J., GREENWOOD, W. R., HADLEY, D. G., ANDERSON, R. E. AND SCMIDT, D. L. 1980. Rubidium–strontium geochronology and plate-tectonic evoluton of the southern part of the Arabian Shield. *United States Geological Survey, Professional Paper*, **1131**.

FRETS, B. C. 1977. *Stratigraphic, structural, and economic aspects of Aba al Qazaz quadrangle, northwestern Hijaz, Kingdom of Saudi Arabia*. Saudi Arabian Directorate General of Mineral Resources, Open-File Report DGMR-597.

GRUNOW, A., HANSON, R. & WILSON, T. 1996. Were aspects of Pan-African deformation linked to Iapetus opening? *Geology*, **24**, 1063–1066.

HADLEY, D. G. 1974. The taphrogeosynclinal Jubaylah group in the Mashhad area, Northwestern Hijaz, Kingdom of Saudi Arabia. *Saudi Arabian Directorate General of Mineral Resources Bulletin* **10**, 18pp., 19 figures, 2 tables.

HADLEY, D. G. 1975. *Geology of Qal'at As-Sawrah quadrangle, sheet 26/38 D, Kingdom of Saudi Arabia, scale 1:100000. Map GM24*. Saudi Arabian Deputy Ministry for Mineral Resources.

HOFFMAN, P. F. 1991. Did the Breakout of Laurentia turn Gondwanaland inside-out? *Science*, **252**, 1409–1412.

HOWLAND, A. F. 1979. Discussion of Tectonics of the Najd Transcurrent Fault System, Saudi Arabia. *Journal of the Geological Society, London*, **136**, 453–454.

HUSSEINI, M. I. 1988. The Arabian Infracambrian extensional system. *Tectonophysics*, **148**, 93–103.

HUSSEINI, M. I. 1989. A seismic–seismogenic coupling for

long strike-slip fault. *Geophysical Journal of the Royal Astronomical Society*, **97**, 391–407.

JOHNSON, P. R. 1997. *Late NeoProterozoic shears in west-central Arabia: brittle–ductile deformation during the Najd event*. Saudi Arabian Ministry of Petroleum and Mineral Resources, Technical Report 97–1. Proceedings of the Third Annual Meeting for the Saudi Society for Earth Sciences, King Saud University, Riyadh, Oct. 15–17, 1996, pp. 99–105.

JOHNSON, P. R. 2002. Post-amalgamation basins of the NE Arabian Shield and implications for Neoproterozoic III tectonism in the northern East African orogen. *In*: KUSKY, T. M., ABDELSALAM, M., TUCKER, R. & STERN, R. (eds). *The Evolution of the East African and Related Orogens*. Precambrian Research, Special Issue, in press.

JOHNSON, P. R. & KATTAN, F. 1999. The Ruwah, Ar Rika & Halaban-Zarghat fault zones: northwest-trending Neoproterozoic brittle–ductile shear zones in west-central Saudi Arabia. *In*: DE WALL, H. & GREILING, R. (eds) *International Cooperation, Bilateral Seminars of the International Bureau 32*. Forschungszentrum Julich, Germany.

JOHNSON, P. R., SCHEIBNER, E. & SMITH, A. 1987. Basement fragments, accreted tectonostrarigraphic terranes, and overlap sequences: elements in the tectonic evolution of the Arabian Shield. *Geodynamics Series, American Geophysical Union*, **17**, 324–343.

KOIDE, H. & BATTACHARJI, S. 1977. Geometric patterns of active strike-slip faults and their significance as indicators for areas of energy release, *In*: SAXENA, S. K. (ed.) *Energetics of Geological Processes*, New York, Springer Verlag, 46–66.

KROGH, T. E. 1973. A low-contamination method for hydrothermal decomposition of zircon and extraction of U and Pb for isotopic age determinations. *Geochemica et Cosmochemica Acta*, **37**, 485–494.

KRONER, A., STERN, R. J., DAWOUD, A. S., COMPSTON, W. & REISCHMANN, T. 1987. The Pan-African continental margin in northeastern Africa: evidence from a geochronological study of granulites at Sabaloka, Sudan. *Earth and Planetary Sciences Letters*, **85**, 91–104.

KUSKY, T. M. & MATSAH, M. 1999, NeoProterozoic dextral faulting on the Najd Fault System, Saudi Arabia, preceded sinistral faulting and escape tectonics related to closure of the Mozambique Ocean. *Geological Society of America, Abstracts with Programs*, **31**, A118.

LETALENET, J. 1975. *Geology and mineral exploration of the 'Afif quadrangle, 23/42 B*. French Bureau de Recherches Geologiques et Minieres Technical Record 75–JED-2.

LUDWIG, K. 1989. *A plotting and a regression program for radiogenic – isotopic data, for IBM-PC compatible computers, Version 1.05*. United States Geological Survey Open File Report, 88–557.

LUDWIG, K. R. 1990. *ISOPLOT a plotting and regression program for radiogenic – isotope data, for IBM-PC compatible computers version 2.02*. U.S. Geological Survey, Open File Report **88–557**.

MANN, P., HEMPTON, M. R., BRADLEY, D. C. & BURKE, K. 1983. Development of pull-apart basins. *Journal of Geology*, **91**, 529–554.

MATSAH, M. I. 2000, *The deposition of the Jibalah Group in pull-apart basins of the Najd Fault System as a final stage of the consolidation of Gondwanaland*. PhD Thesis, Boston University.

MATSAH, M. & KUSKY, T. M. 1999. Sedimentary facies of the NeoProterozoic Al-Jifn Basin, NE Arabian Shield; relationships to the Halaban–Zarghat (Najd) faults system and the closure of the Mozambique Ocean. *In*: GREILING, R. (ed.) *Pan-African of Northern Africa–Arabia*. Forschungszentrum Jülich, Geologisch-Palaeontologisches Institut, Ruprecht-Karls-Universitaet Heidelberg, Germany, Proceedings of a Workshop, October 22–23 1998.

MATSAH, M. & KUSKY, T. M. 2001. *Analysis of Landsat TM ratio imagery of the Halaban–Zarghat fault and related Jifn Basin, NE Arabian Shield: Implications for the kinematic histroy of the Najd fault system*. Field Workshop on the Geology and Tectonics of the Arabian Shield, King Abdul Aziz University, Jeddah, Saudi Arabia.

McCLAY, K. & DOOLEY, T. 1995. Analogue models of pull-apart basins. *Geology*, **23**, 711–714.

MOORE, J. M. 1979. Tectonics of the Najd Transcurent Fault System, Saudi Arabia. *Journal of the Geological Society, London*, **136**, 441–454.

PARRISH, R. R. 1987. An improved micro-capsule for zircon dissolution in U-Pb geochronology, *Isotope Geoscience*, **66**, 99–102.

PELLATON, C. 1985. *Geology of the Miskah quadrangle, sheet 24F, Kingdom of Saudi Arabia, scale 1:250000*. Map GM99A. Saudi Arabian Deputy Ministry for Mineral Resources.

READING, H. G. 1980. Characteristics and recognition of strike-slip faults systems. *In*: BALLANCE, P. F. & READING, H. G. (eds) *Sedimentation in Oblique-slip Mobile Zones*. International Association of Sedimentologists, Special Publication, **4**, 7–26.

REYMER, A. & SCHUBERT, G. 1984. Phanerozoic addition rates to the continental crust and crustal growth. *Tectonics*, **3**, 63–77.

SCHMIDT, D. L. & BROWN, G. F. 1984. Major-element chemical evolution of the late Proterozoic Shield of Saudi Arabia, in Pan-African crustal evolution in the Arabian–Nubian Shield: Jiddah. *King Abdulaziz University, Institute of Applied Geology Bulletin*, **6**, 1–22.

SCHMIDT, D. L., HADLEY, D. G. & STOESER, D. B. 1979. Late Proterozoic crustal history of the Arabian Shield, southern Najd Province, Kingdom of Saudi Arabia. *In*: TAHOUN, A. (ed.) *Evolution and Mineralization of the Arabian Nubian Shield*. Pergamon Press/King Abdulaziz University, Institute of Applied Geology, Bulletin **3**.

SENGOR, A. M. C. & NATAL'IN, B. A. 1996. Paleotectonics of Asia: fragments of a synthesis. *In*: YIN, A. & HARRISON, M. (eds) *The Tectonic Evolution of Asia*. Cambridge University Press, Cambridge, 486–640.

SHACKELTON, R. M. 1986. Precambrian collision tectonics in Africa. *In*: COWARD, M. P. & RIES, A. C. (eds) *Collision Tectonics*. Geological Society, London, Special publication, 329–349.

STACEY, J. S. & AGAR, R. A. 1985. U–Pb isotopic evidence for the accretion of a continental microplate in the Zalm region of the Saudi Arabian Shield. *Journal of the Geological Society, London*, **142**, 1189–204.

STACEY, J. S. & KRAMERS, J. D. 1975. Approximation of terrestrial lead isotope evolution by a two-stage model, *Earth and Planetary Science Letters*, **26**, 207–221.

STEIGER, R. H. & JAGER, E. 1977, Subcommission on geochronology: Convention of the use of decay constants in geo- and cosmochronology, *Earth and Planetary Science Letters*, **36**, 359–362.

STEIN, M. & GOLDSTEIN, S. L. 1996. From plume head to continental lithosphere in the Arabian–Nubian shield. *Nature*, **382**, 773–778.

STERN, R. J. 1985. The Najd Fault System, Saudi Arabia and Egypt: a Late Precambrian rift-related transform system. *Tectonics*, **4**, 497–511.

STERN, R. J. 1994. Arc assembly and continental collision in the Neoproterozoic East African Orogen: implications for consolidation of Gondwanaland. *Annual Review of Earth and Planetary Sciences*, **22**, 319–351.

STOESER, D. B. & CAMP, V. E. 1985. Pan-African microplate accretion of the Arabian Shield. *Geological Society of America Bulletin*, **96**, 817–826.

STOESER, D. B. & STACEY, J. S. 1988. Evolution, U–Pb geochronology, and isotope geology of the Pan-African Nabitah orogenic belt of the Saudi Arabian Shield. *In:* EL GABY, S. AND GREILING, R. O. (eds) *The Pan-African Belts of Northeast Africa and Adjacent Areas*, Friedr Vieweg and Sohn, Heidelberg, 227–288.

STOREY, B.C. 1993. The changing face of late Precambrian and early Palaeozoic reconstructions. *Journal of the Geological Society, London*, **150**, 665–668.

SULTAN, M., ARVIDSON, R. E., DUNCAN, I. J., STERN, R. & EL KALIOUBY, B. 1988. Extension of the Najd Fault System from Saudi Arabia to the central Eastern Desert of Egypt based on integrated field and Landsat observations. *Tectonics*, **7**, 1291–1306.

UNRUG, R. (ed.) 1996. *The geodynamic map of Gondwana Supercontinent assembly*, scale 1:10 000 000. Bureau de Recherches Geologiques et Minieres, Orleans, France.

UNRUG, R. 1997. Rodinia to Gondwana: the geodynamic map of Gondwana supercontinent assembly. *GSA Today*, **7**, 1–6.

VAIL, J. R. 1983. Pan-African crustal accretion in northeast Africa. *Journal of African Earth Sciences*, **1**, 285–294.

VAN HOUTEN, F. B. 1974. Northern Alpine molasse and similar Cenozoic sequences of Southern Europe. *In:* DOTT, Jr. R. H. & SHAVER R. H. (eds) *Modern and Ancient Geosynclinal Sedimentation*, Special Publication of the Society of Economic, Paleontology & Mineralealogy, **19**, 260–273.

Neoproterozoic deformation in central Madagascar: a structural section through part of the East African Orogen

ALAN S. COLLINS[1], SIMON JOHNSON[2,4], IAN C. W. FITZSIMONS[1], CHRIS McA. POWELL[2,†], BREGJE HULSCHER[2], JENNY ABELLO[2] & THÉODORE RAZAKAMANANA[3]

[1] *Tectonics SRC, Department of Applied Geology, Curtin University of Technology, GPO Box U1987, Perth, WA 6845, Australia (e-mail: alanc@lithos.curtin.edu.au)*
[2] *Tectonics SRC, Department of Geology and Geophysics, The University of Western Australia, Nedlands, WA 6907, Australia*
[3] *Département des Sciences de la Terre, Université de Toliara, Toliara, Madagascar*
[4] *Present address: Institute for Frontier Research on Earth Evolution, Japan Marine Science Technology Centre, 2–15 Natsushima-cho, Yohosuka, Kanagaua, 237–0061, Japan*
† *Deceased*

Abstract: The Itremo region of central Madagascar has an importance in the evolution of the East African Orogen (EAO) that belies its size. Unusually for the southern EAO (Mozambique Belt), it is made up of low-grade metasedimentary rocks and therefore preserves an almost unique window into upper crustal deformation during this key period of the Gondwana supercontinent cycle.
In this paper new field mapping of three linked regions in the Itremo Sheet and in the upper part of the underlying mid-crustal Antananarivo Block are presented. From these a complete structural section through the eastern Itremo Sheet is produced and the complex deformation record preserved there is then discussed.
An early deformation (D1) consists of 10 km scale recumbent isoclinal folds that predate intrusion of a *c.* 780–800 Ma igneous suite. Metamorphic aureoles around these plutonic bodies overprint D1-related fabrics. Local deformation accompanies intrusion of the *c.* 780–800 Ma, suite (D2). Extensive E–W contractional deformation occurs between 780 and *c.* 570 Ma, that is here amalgamated as a composite D3 event, which includes thrusts and at least two phases of upright folds. Post-551 Ma, normal shearing (D4) marks the boundary between the Itremo Sheet and the underlying Antananarivo Block (the Betsileo Shear Zone), and may have also been responsible for formation of the Saronara Shear Zone. Finally, E–W open folding and dextral shear zone development marks a late N–S contractional event that is interpreted as a far-field response to collision between the northern Bemarivo Belt and central Madagascar.

Over the last decade knowledge of Pre-Cambrian and Palaeozoic palaeogeography has increased dramatically. With this knowledge has come a growing realization of the existence of pre-Pangaean supercontinents, such as Rodinia, and the apparent episodic pulse of supercontinent assembly and dispersal. In understanding the details of these supercontinent transformations, the Neoproterozoic–Cambrian East African Orogen (EAO) (Stern 1994) has a pivotal role to play, as it is a major collisional zone that formed during breakup of Rodinia and amalgamation of Gondwana (Powell *et al.* 1993; Shackleton 1996; Dalziel 1997; Hoffman 1999; Collins & Windley 2002).

In reconstructions of Gondwana (Lawver *et al.* 1998; Reeves & de Wit 2000), Madagascar lies within the EAO (Fig. 1), close to the NE (azimuth relative to present-day Africa) limit of Neoproterozoic thermal reworking (Bartlett *et al.* 1998; Collins & Windley 2002). Collins and co-workers (Collins *et al.* 2000a; Kröner *et al.* 2000; Collins & Windley 2002) recently suggested that part of the Dharwar Craton was left behind in Madagascar as Gondwana broke up. An eastern colliding continent is not seen further north in the EAO, either in mainland Africa or in Arabia. Consequently, Madagascar forms a rare direct link between the eastern and western parts of Gondwana and also lies in a key region for understanding the along-strike change in character of the EAO from the southern high-grade Mozambique Belt to the low-grade, ophiolite-rich Arabian–Nubian Shield.

Fundamental to unravelling the evolution of the Malagasy EAO is the need to understand the structural framework of the orogen. Satellite-image-based structural studies have been undertaken in specific regions of Madagascar (Martelat *et al.* 1995; Martelat *et al.* 1997; 2000; Goncalves *et al.* 2003; Tucker *et al.* 2003) and these have produced a broad understanding of the gross structure of parts of the island. Detailed field-based structural studies of southern Madagascar (de Wit *et al.* 2001), eastern Madagascar (Collins *et al.* 2003) and central Madagascar (Daso 1986;

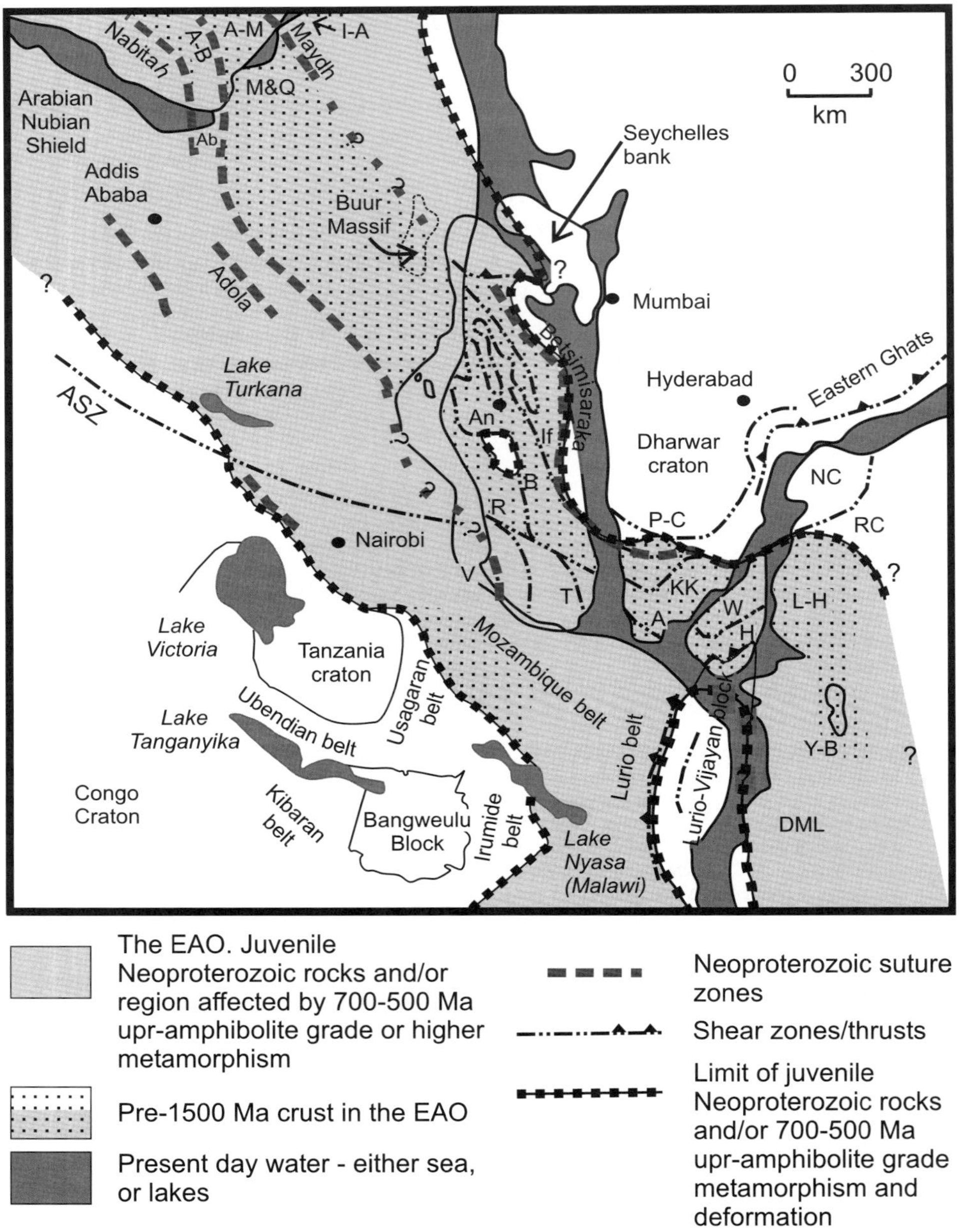

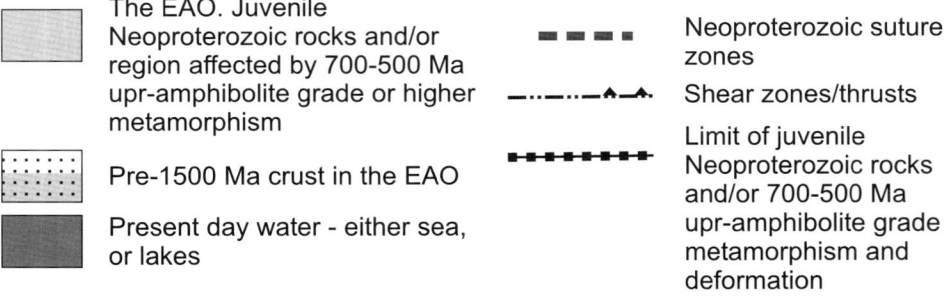

Fig. 1. The East African Orogen (EAO) superimposed on the tight-fit reconstruction of Gondwana by Lawver et al. (1998). Ab, Abdulkadir Terrane; A, Achankovil Shear Zone; A-B, Al-Bayda Terrane; A-M, Al-Mahfid Terrane; an, Antananarivo; ASZ, Aswa Shear Zone; DML, Dronning Maud Land; H, Highland Complex; I-A, Inda Ad Complex; If, Ifanadiana Shear Zone; KK, Karur–Kambam–Painavu–Trichur Shear Zone; L-H, Lützow-Holm Complex; M&Q, Mora and Qabri Bahar Terrane; NC, Napier Complex; P–C, Palghat–Cauvery Shear Zone system; R, Ranotsara Shear Zone; RC, Rayner Complex; T, Tranomaro Shear Zone; V, Vohibory Belt; W, Wanni Complex; Y–B, Yamato–Belgica complex; Betsimisaraka, Betsimisaraka Suture. Data from Kröner & Williams (1993), Shiraishi et al. (1994), Kröner & Sassi (1996), Shackleton (1996), Kröner et al. (1997), Bartlett et al. (1998), Jacobs et al. (1998), Teklay et al. (1998), Tucker et al. (1999a,b), de Wit et al. (2001), Muhongo et al. (2001) and Whitehouse et al. (2001). The blank ornamentation defines regions of pre-Neoproterozoic protoliths that were unaffected by 700–500 Ma amphibolite-grade or higher metamorphism, note that the low-grade eastern Itremo Sheet falls into this category.

Nédélec et al. 1994; Raoelison 1997; Collins et al. 2000b; Nédélec et al. 2000) reveal a complex story of multiple deformations that can, in some cases, be geochronologically separated. In this contribution, many of these studies are complemented by presenting the results of detailed mapping in a number of linked areas through the eastern Itremo Sheet of central Madagascar and down into the underlying Antananarivo Block. By doing this, a cross-strike section is produced through much of central Madagascar that, when combined with previous work (Collins et al. 2003), completes a cross-section across 90% of the width of the Malagasy Basement.

Central Madagascar was originally mapped by geologists from the Service Géologique de Madagascar and was summarized in the 1:500 000 map and memoir of Besairie (1969–1971, 1970). Moine (1966, 1974) concentrated on the metasedimentary sequence [the Itremo Group of Emberger (1955) and Cox et al. (1998); the 'groupe schisto–quartzo–dolomitique' of Moine (1968), and the 'série schisto–quartzo–calcaire' of Besairie (1970)] that is exposed over a significant part of central Madagascar. His impressive map (Moine 1968) still forms the basic framework for geological investigation in the region. Recently, U–Pb zircon and monazite studies of the Itremo Group (Cox et al. 1998, 2000; Huber 2000) and cross-cutting intrusions (Handke et al. 1999) have bracketed deposition of the Itremo Group between c. 1680 and 800 Ma and have identified a younger, highly deformed, sedimentary sequence to the west of the Itremo Sheet (Cox et al. 2001).

Much controversy has surrounded the recent attempts at documenting and interpreting the structural history of the Itremo region (Collins et al. 2000b; Fernandez et al. 2000; Cox et al. 2001; Hulscher et al. 2001; Tucker et al. 2001b; Fernandez & Schreurs 2003; Tucker et al. 2003.) In this paper a new, field-based, structural section through the whole of the eastern Itremo Group is presented and how these new data relate to the main controversies is discussed.

Tectonic architecture of central Madagascar

Collins et al. (2000a,b) and Collins & Windley (2002) divided central and north Madagascar into five tectonic units (Fig. 2). All rocks in each of these units share a similar tectonic history and each unit is separated from the others by shear zones. The five tectonic units are summarized below.

(1) The Antongil Block, consisting of gneiss with c. 3.2 Ga protoliths, intruded by c. 2.5 Ga granites that have only experienced low-grade metamorphism since c. 2.5 Ga.

(2) The Antananarivo Block that consists of orthogneiss with c. 2.5 Ga protoliths interlayered with 820–740 Ma granitoids and gabbros, pervasively deformed and metamorphosed to granulite facies conditions between 700 and 550 Ma.

(3) The Itremo Sheet containing metasedimentary rocks of the Mesoproterozoic–Early Neoproterozoic Itremo Group (Cox et al. 1998, 2000) and the Neoproterozoic Molo Group (Cox et al. 2001) thrust over, and imbricated with, rocks of the Antananarivo Block. Metasediments of the Itremo Sheet non-conformably overlie paragneiss (Cox et al. 1998) and orthogneiss that, in places, have a similar protolith age to orthogneisses in the Antananarivo Block (Tucker et al. 1999b).

(4) The Tsaratanana Sheet, which is formed of mafic gneiss with protolith ages of 2.7–2.5 Ga that contain Mid–Late Archaean Sm–Nd ages and zircon xenocrysts (Tucker et al. 1999b; Collins et al. 2001). These rocks were deformed and metamorphosed at c. 2.5 Ga (Goncalves et al. 2000) and cut by 800–760 Ma gabbro intrusions (Guerrot et al. 1993). Later contractional deformation continued until after 630 Ma (Collins et al. 2003; Goncalves et al. 2003).

(5) The Bemarivo Belt, which consists of E–W-striking metasediments, granites and gneisses, overlain by contractionally deformed metavolcanics. Young granulite facies metamorphism in the Bemarivo Belt is dated at 510–520 Ma (Tucker et al. 1999a; Buchwaldt & Tucker 2001).

Central and northern Madagascar are separated from southern Madagascar by the Ranotsara Shear Zone. Sinistral displacement of >100 km is implied by the regional drag of foliation to either side of the shear zone (Windley et al. 1994).

Southern Madagascar is split into a number of tectonic units. From west to east these are: the Vohibory, Ampanihy, Bekily, Betroka, Tranomaro and Fort Dauphin–Anosyan Belts (Fig. 2; Windley et al. 1994). The belts consist of rock packages that preserve different structural styles; the Ampanihy and Betroka Belts are subvertical shear zones [the latter is called the Vorokafotra Shear Zone by Rolin (1991) and de Wit et al. (2001), and the Beraketa Shear Zone by Martelat et al. (2000)]. Subsequently, Martelat et al. (2000) identified the Tranomaro Shear Zone that separates the Tranomaro Belt from the Fort Dauphin–Anosyan Belt (Fig. 2). With the exception of the Vohibory Belt [Vohibory Terrain of de Wit et al. (2001)], these belts are derived from mainly sedimentary protoliths and consist of granulite- and upper amphibolite-grade gneisses.

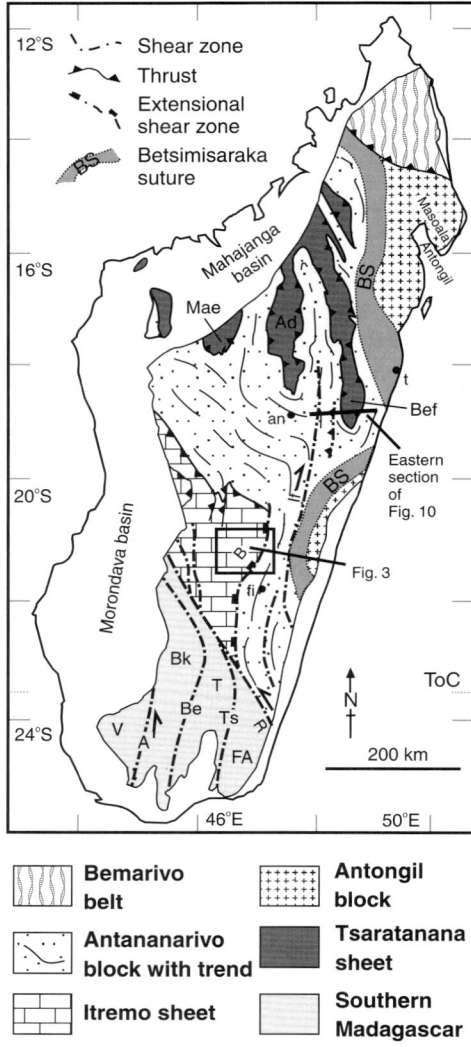

Fig. 2. Map of the tectonic units of Madagascar [after Collins *et al.* (2000*b*)], based on original mapping summarized by Besairie (1973) and the interpretations of Hottin (1976). an, Antananarivo; t, Toamasina; fi, Fianarantsoa; B, Betsileo Shear Zone; If, Ifanadiana Shear Zone; R, Ranotsara Shear Zone; Masoala, Masoala Peninsula; Antongil, Bay of Antongil; Mae, Maevatanana Belt; Bef, Beforona Belt; Ad, Andriamena Belt; BS, Betsimisaraka Suture Zone. Shear zones south of the Ranotsara Shear Zone [after Windley *et al.* (1994), Martelat *et al.* (2000) and de Wit *et al.* (2001)]: V, Vohibory Belt; A, Ampanihy Shear Zone; Bk, Bekily Belt; Be, Betroka Shear Zone; T, Tranomaro Belt; Ts, Tranomaro Shear Zone; FA, Fort Dauphin–Anosyan Belt; ToC, Tropic of Capricorn.

Study areas

Three areas were mapped (Fig. 3) that form along-strike continuations of each other, separated only by the Itsindro Gabbro. When combined, the three areas give a near-continuous cross-section through the Itremo Sheet and the top of the structurally underlying Antananarivo Block.

Upper Mania River

The Upper Mania River section (Figs 3 & 4) includes part of the upper Antananarivo Block that contains within it strips of quartzite and pelite that may represent high-grade equivalents of the Itremo Group metasediments. The section passes west and structurally upsection to the 790 Ma Ilaka Granite

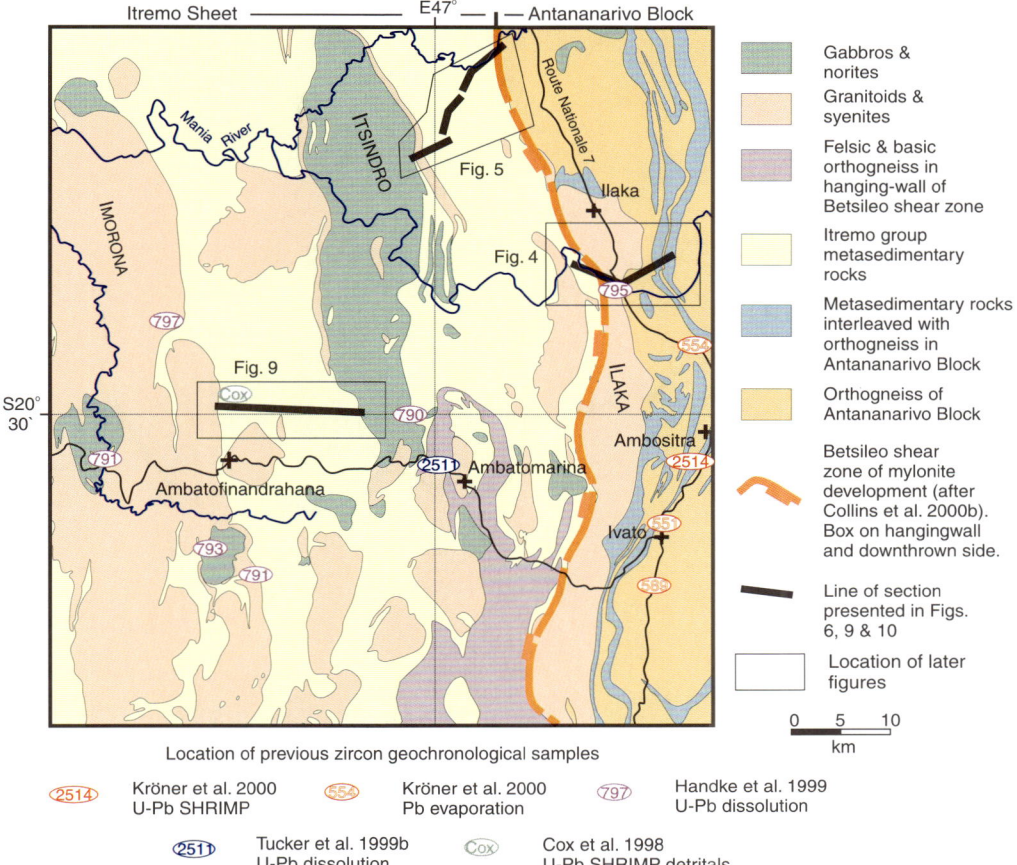

Fig. 3. Geological map of part of central Madagascar with location of previous geochronolgogical samples marked (see Fig. 2 for location). Lithology distribution after Moine (1968) with simplified rock-type nomenclature. Tectonostratigraphic division after Collins & Windley (2002) and the location of the highest strain part of the Betsileo Shear Zone after Collins *et al.* (2000*b*).

and the top-to-the-west Betsileo Shear Zone (Collins *et al.* 2000*b*).

The Betsileo Shear Zone stretches N–S, along-strike, for over 200 km and marks the eastern margin of the Itremo Sheet (Figs 2 & 3). Collins *et al.* (2000*b*) described many top-down-foliation kinematic indicators from this shear zone and, as it coincides with a large jump in metamorphic grade, suggested that the shear zone formed a crustal-scale extensional detachment. To the east, sillimanite + muscovite quartzites and highly weathered garnet + biotite schists occur as sheets interleaved with biotite granite gneisses and amphibolites (Fig. 4). Gneissic foliation and schistosity are approximately parallel to bedding in the quartzite and plot around a broad great circle striking NNW–SSE (Fig. 4, stereonet b). Mineral lineations cluster around the pole to this great circle (Fig. 4, stereonet b), suggest-

ing that the finite extension axis was parallel to late fold axes that deformed the composite bedding/foliation fabric. Whether this open folding was coeval with foliation formation and represents contraction along the intermediate strain axis or whether it represents a late superimposed phase of NNW–SSE contraction is unknown. Top-to-the-east directed σ-type structures [terminology after Passchier & Trouw (1996)] around feldspar porphyroblasts occur along the contact between a sheet of quartzite and granitic gneiss on the south bank of the Mania River. This contrasts with the top-to-the-west, extensional, shear sense described by Collins *et al.* (2000*b*) from near the Betsileo Shear Zone (Fig. 4). This kinematic separation correlates with a westward swing in the plunge of mineral lineations as the Betsileo Shear Zone is approached, suggesting that early pervasive top-to-the-west thrusting may

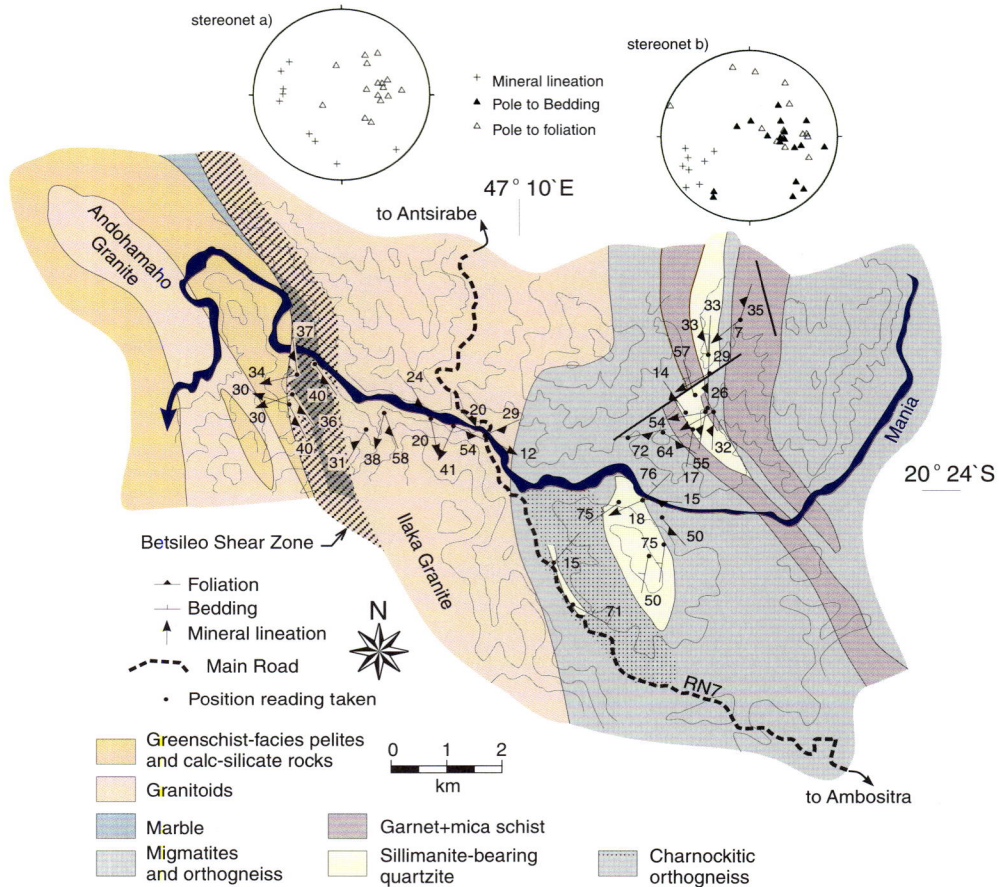

Fig. 4. Map of the Upper Mania River region. Data and mapping west of the Route Nationale 7 bridge from Collins *et al.* (2000*b*) (see Fig. 3 for location). Stereonets are plotted as equal-area lower hemisphere projections: (**a**) data from within the footwall of the Betsileo Shear Zone; (**b**) data east and outside the influence of the Betsileo Shear Zone.

here have had a different shear vector than later extensional deformation.

Iheninkenina region

The Iheninkenina map (Fig. 5) and section (Fig. 6) stretch from the Itsindro Gabbro in the west of the region through to the Betsileo Shear Zone, which marks the eastern margin of the Itremo Sheet. The region consists of greenschist facies metasediments of the Itremo Group (quartzites, marbles and pelites), the Itsindro Gabbro and a strip of syenite that lies close to the Itsindro Gabbro margin. The western half of the area is deformed primarily by dextral strike-slip faults and bivergent thrusts, whereas in the eastern half-complex three-phase interference folds are preserved (Figs 5, 6 & 7).

Directly above the trace of the Betsileo Shear Zone, which is here unexposed, lie west-dipping mylonitised quartzites with S/C structures that indicate top-to-the-east thrusting. At a number of places along the Betsileo Shear Zone similar top-to-the-east kinematic indicators are preserved in the hanging wall, interpreted by Collins *et al.* (2000*b*) as pre-existing thrust-related deformation selectively preserved in the upper plate of the extensional Betsileo Detachment. Westwards, away from the Betsileo Shear Zone, the quartzites that make up Analabe–Ampandrianombilapa Ridge are downward facing, implying that they form the inverted limb of a large-scale (>10km amplitude) recumbent fold. Within the ridge, these quartzites are folded into an asymmetric, kilometre-scale, east-verging antiform, that passes west into a region of multiple metre-scale recumbent folds that have pronounced mineral elon-

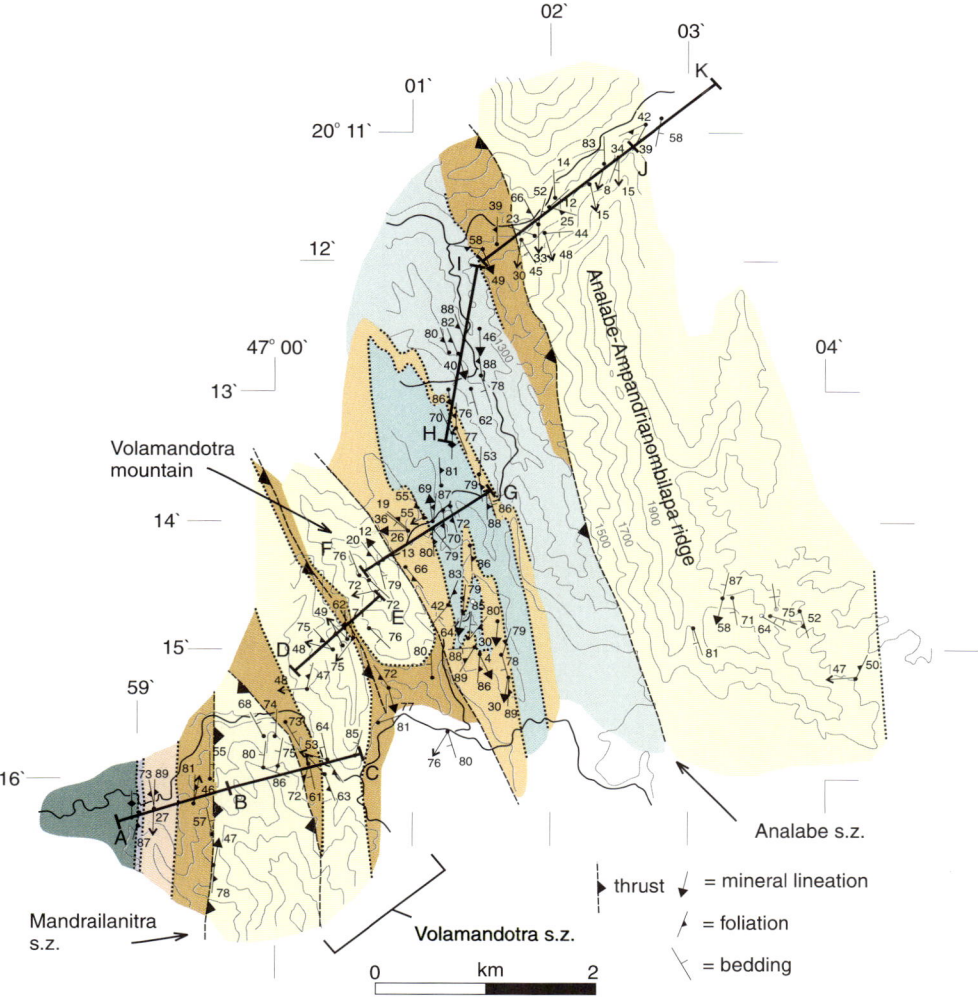

Fig. 5. Map of the Iheninkenina region (see Fig. 3 for location). Solid lines, line of sections in Fig. 6 (see Fig. 6 for key). s.z., Shear zone.

gation lineations parallel to the fold hinges and mylonitized fold limbs. This localized deformation and the map-scale pinching out of the overlying pelite suggests that significant shearing has occurred along this boundary. This shear zone separates downward-facing rocks in the footwall from upward facing hanging-wall rocks. West of the quartzites, centimetre-scale bedded pelites and metasiltstones pass upward into massive marbles locally rich in calc-silicate layers. These, in turn, pass west through a prominent ridge of centimetre–decimetre-scale bedded psammites and pelites, with graded beds and truncation surfaces that imply westward younging. The younging indicators imply that there are two marble units and two pelite units in this region, which is consistent with the recent findings of Fernandez & Scheurs (2002) but differs from the stratigraphic work of Cox *et al.* (1998). The marbles to the west of this ridge map out as a complex fold interference pattern that needs three broadly co-axial fold phases to develop. Overall the refolded fold forms an antiformal syncline with the youngest rocks preserved in the fold core. Continuing west, Volamandotra Mountain is separated from the refolded fold closure by a prominent fault. The crossbedded quartzites that form the mountain are the right way up and are deformed into asymmetric kink folds. A thrust plane cuts through the western flank of Volamandotra

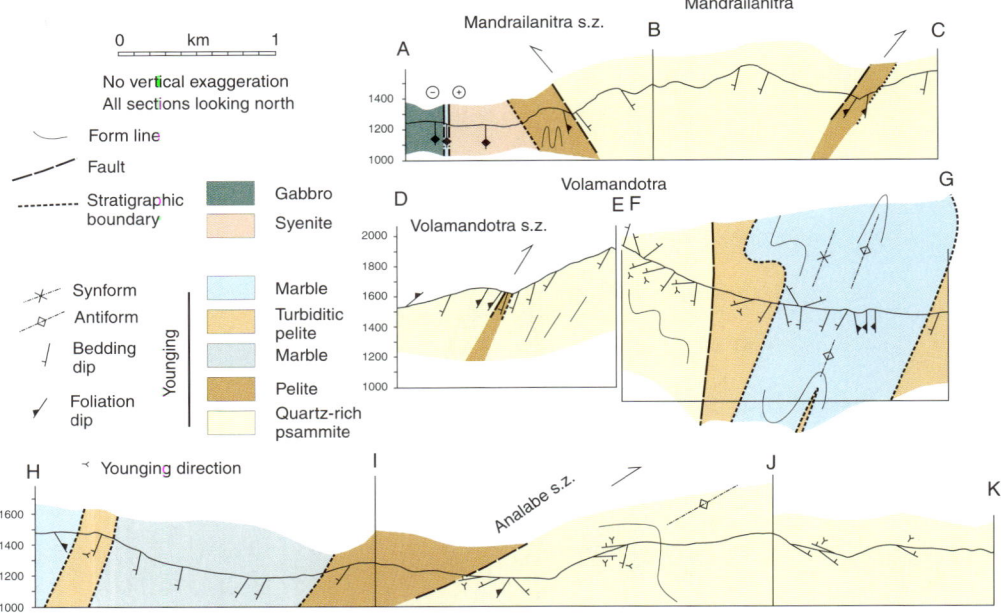

Fig. 6. Cross-sections through the Iheninkenina region (see Fig. 5 for location). For structural key see Figure 5. s.z., Shear zone.

Mountain, imbricating the quartzite and sandwiching a band of pelite along the thrust plane (Figs 5 & 8a). Along the trace of the thrust the quartzite is intensely mylonitized with a west-dipping shear fabric and pronounced down-dip stretching lineations. Quartzites makes up the next ridge west, the Mandrailanitra Ridge. On the western flank of this ridge an east-dipping fault occurs that preserves a fault-propagation fold implying top-to-the-west thrusting. Mandrailanitra is therefore bound to both the east and the west by two thrusts that verge in opposite directions. Quartzites pass west into a band of strongly foliated syenite that is separated from the Itsindro Gabbro by a <100m thick band of marble. Pronounced mineral lineations in the syenite plunge 27–30° to 157–182°. S/C fabrics and σ-type fabrics imply a dextral shear sense.

Saronara region

The Saronara region (Figs 3 & 9) lies to the west of both the Itsindro Gabbro and the Iheninkenina region. It passes from the margin of the Itsindro Gabbro through to the Amabalafampana Granite, a component of the larger, composite, Imorona Granite Massif. This region consists of predominantly greenschist facies Itremo Group metasediments and is the area where much of the stratigraphic work on the group has taken place (Cox et al. 1998). Local, high temperature–low pressure, contact metamorphism is evident at the contact between the metasediments and the Itsindro Gabbro. Within pelitic assemblages, andalusite porphyroblasts are common (Fig. 8b) and, at the contact of the gabbro, cordierite, garnet and sillimanite occur. Within carbonate units, tremolite and diopside are common while the highest grade assemblage, located at the contact with the gabbro, is represented by olivine and periclase pseudomorphed by brucite. This contact aureole in the Itremo Group metasediments indicates that the Itremo Group predates the 790 Ma Itsindro Gabbro (Handke et al. 1999). Bedding and structural facing are easily identified in low-strain zones from cross-bedding, ripple marks and graded units within the quartzites, and from variably deformed stromatolites within the marbles. Planar structures, including bedding and foliation, consistently dip moderately toward the east, while the structural facing of these units changes, indicating the presence of regional-scale folding. Parasitic, decimetre-scale folds illustrate the west-verging, isoclinal attitude of these regional structures (Fig. 9). A penetrative foliation is developed axial planar to these isoclines and is subparallel to the bedding.

In the west of the region, bedding and folding are difficult to identify due to intense mylonitization (Fig. 8c) associated with the imbrication and repetition of quartzite and pelite units in the Saronara Shear Zone [named by Cox in Ashwal (1997); Fig.

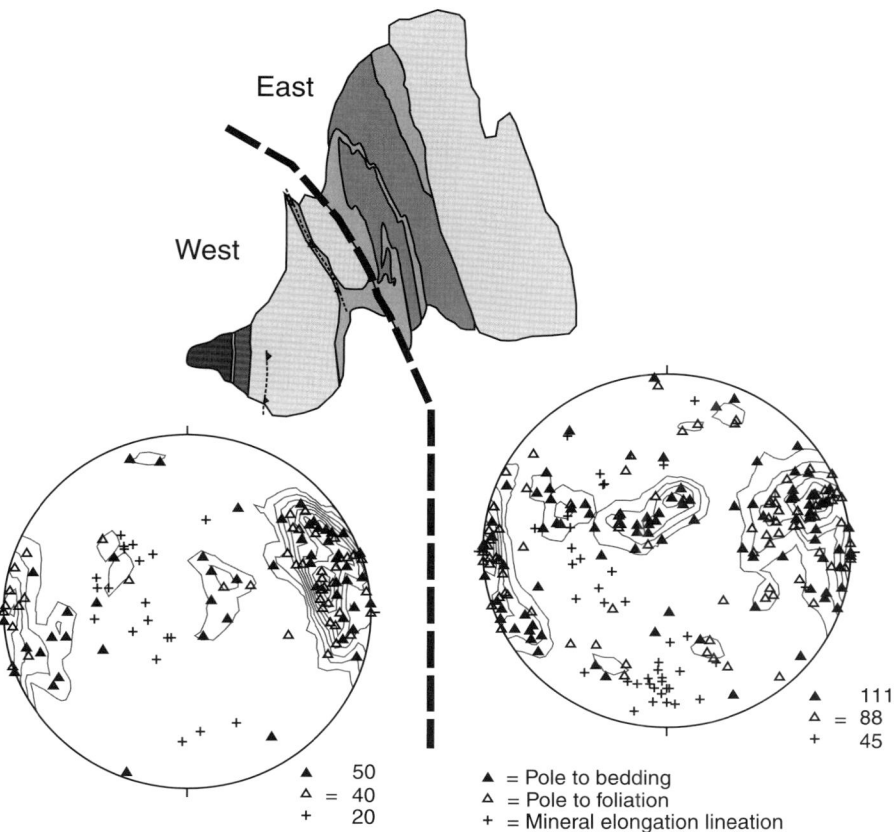

Fig. 7. Equal-area lower hemisphere projections of structural data from the Iheninkenina region.

9]. S/C fabrics within discrete high-strain zones between the thrust-bound packages reveal a normal sense of motion (i.e. top-to-the-NE; Fig. 8d). This direction is opposite to the vergence of the early folding in the region (Fig. 9), suggesting that either the Saronara Shear Zone is a post D1 feature or that it has been reactivated during subsequent deformation. The western contact with the Amabalafampana Granite is intensely mylonitized with the granite fabric focused within 20 m of the contact, whereas the quartzite and pelite mylonite fabric is >200 m thick with multiple generations of crenulation fabrics preserved in the pelites. The lack of contact metamorphism above the regional, low-greenschist facies assemblages and associated mylonitic fabrics suggest that the Amabalafampana Granite is in tectonic contact with the Itremo Group metasediments.

Towards the centre of the region, strain reaches a minimum and primary sedimentary structures are well preserved. The bedding and subparallel foliation trend c. 160° with a moderate 40° NE dip. The central ridge of quartzite (Fig. 9) preserves a regional, westward-verging isoclinal fold with well-developed decimetre-scale parasitic folds. Further east, these structures are complexly refolded and the pelites are overprinted by high-temperature contact metamorphism related to the intrusion of the Itsindro Gabbro. Bedding and foliation are difficult to establish but where evident they define a great circle with a pole plunging moderately toward the NE (Fig. 9), parallel with the measured hinges of second-phase folds of minor refolded folds. A sillimanite-bearing high-temperature metamorphic fabric is developed axial planar to these second folds. The proximity to the Itsindro Gabbro of this high-temperature deformation suggests that the gabbro intrusion was responsible for both this second deformation and the overprinting metamorphism. The regional isoclinal folding must have occurred before intrusion of the Itsindro Gabbro and associated deformation. However, as the absolute age of the early isoclinal folding is unknown, the time gap between the two deformations is also unknown.

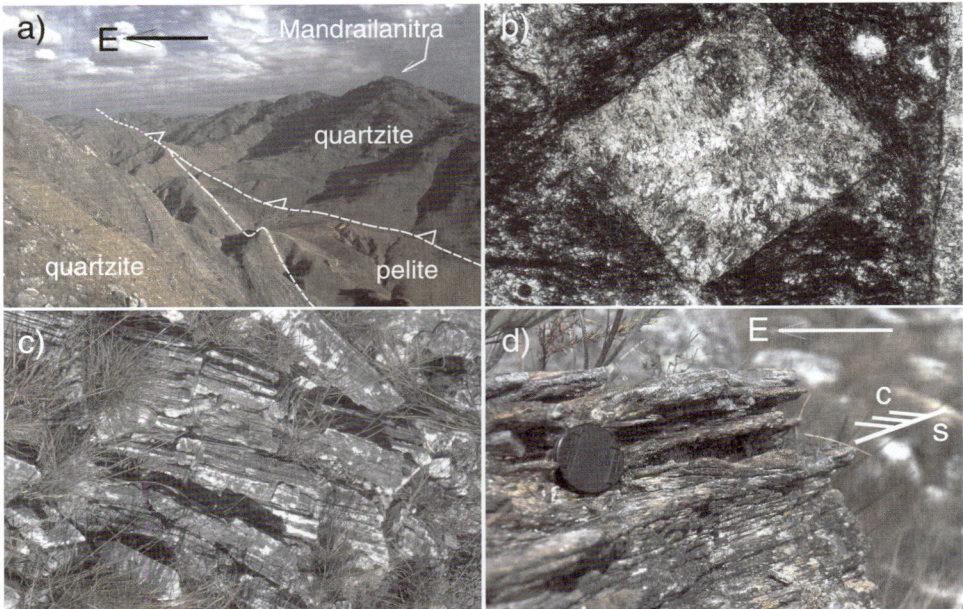

Fig. 8. (**a**) Photograph looking south from Volamandotra Mountain at part of the Volamandotra Shear Zone and Mandrailanitra Mountain. A band of pelite is thrust imbricated within the shear zone. Quartzites on Mandrailanitra Mountain are thrust east during the composite D3 deformation. (**b**) Photomicrograph of graphitic, chiastolite hornfels in the contact aureole of the Itsindro Gabbro (for location see Fig. 9). XPL (cross polarized light) view: field of view, 2 mm. The andalusite porphyroblast is set within a fine-grained matrix of biotite, quartz and graphite, and is itself partially pseudomorphed by sericite. (**c**) Rodded quartzite in the Saronara Shear Zone. The photograph is looking at a foliation surface showing large amounts of extension parallel to the top-to-the-east transport direction. (**d**) Well-developed S/C fabric developed in the Saronara Shear Zone indicating top-to-the-east non-coaxial shear. Fine S surfaces are preserved between the dominant C surfaces that have a pronounced mineral lineation on them, plunging 25° towards 084° (for location see Fig. 9).

Discussion: structural evolution of the eastern Itremo Sheet and the directly underlying Antananarivo Block

The three sections detailed above are laterally offset from one another but can be linked to provide a structural transect (Fig. 10a), which illustrates the structural evolution of the eastern Itremo Sheet and the upper part of the Antananarivo Block (Table 1).

The region preserves an approximately bedding-parallel foliation formed during an early deformation (D1). In the Saronara and Iheninkenina areas this foliation is axial planar to parasitic isoclinal folds that are related to kilometre-scale fold closures that verge west (see also Fernandez & Schreurs 2003). D1-related foliation (S1) is overprinted by low-pressure porphyroblasts in the 790 Ma Itsindro Gabbro aureole and in similar aged aureoles to the south of Ambatofinandrahana (Fig. 10; Handke *et al.* 1999; Hulscher *et al.* 2001). These relationships indicate that this deformation occurred before 790 Ma.

Local post-D1 folding and foliation development occurs near the west margin of the Itsindro Gabbro, where aligned porphyroblasts in a second axial planar foliation indicate that deformation and contact metamorphism occurred together. These observations indicate that this localized D2 deformation was coeval with intrusion of the gabbro.

After intrusion of the Itsindro Gabbro, a phase of E–W contraction occurred that produced upright folding and thrusting. Because of the difficulty of linking the different areas and the complexity of the structures developed, this phase of E–W shortening is here amalgamated as a composite D3 event. It is sealed by the intrusion of a suite of *c.* 570–520 Ma granitoids (Handke *et al.* 1997).

Two D3 upright *c.* N–S fold phases are seen in the Iheninkenina area where a three-phase interference fold has been produced (Fig. 10). This impressive structure is developed by the overprinting of two approximately coaxial and almost coplanar fold phases on an original D1 synclinal closure to produce a complex antiformal syncline. The axis of

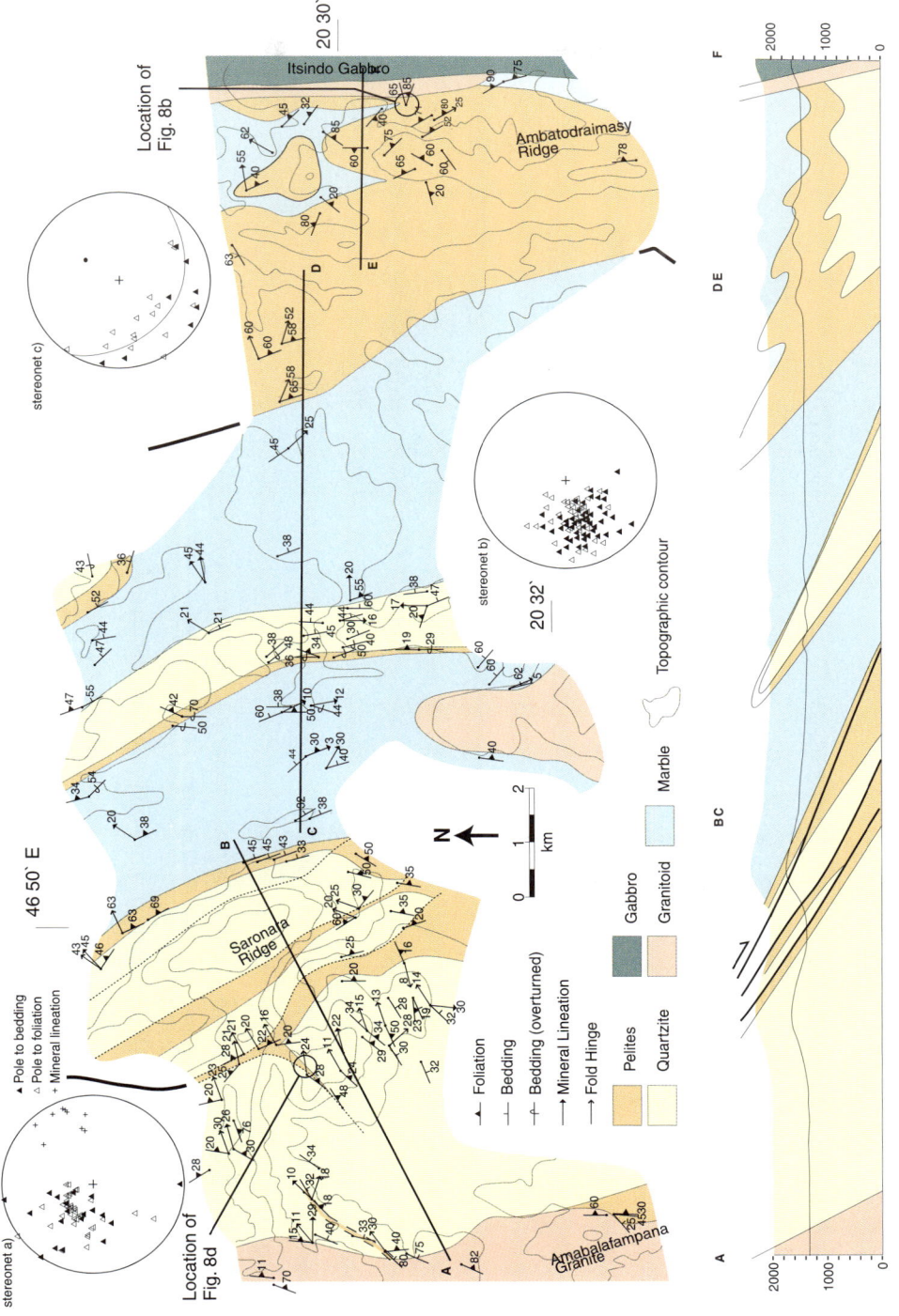

Fig. 9. Map of the Saronara Mountain region (see Fig. 3 for location). Stereonets are plotted as equal-area lower hemisphere projections. s.z., Shear zone.

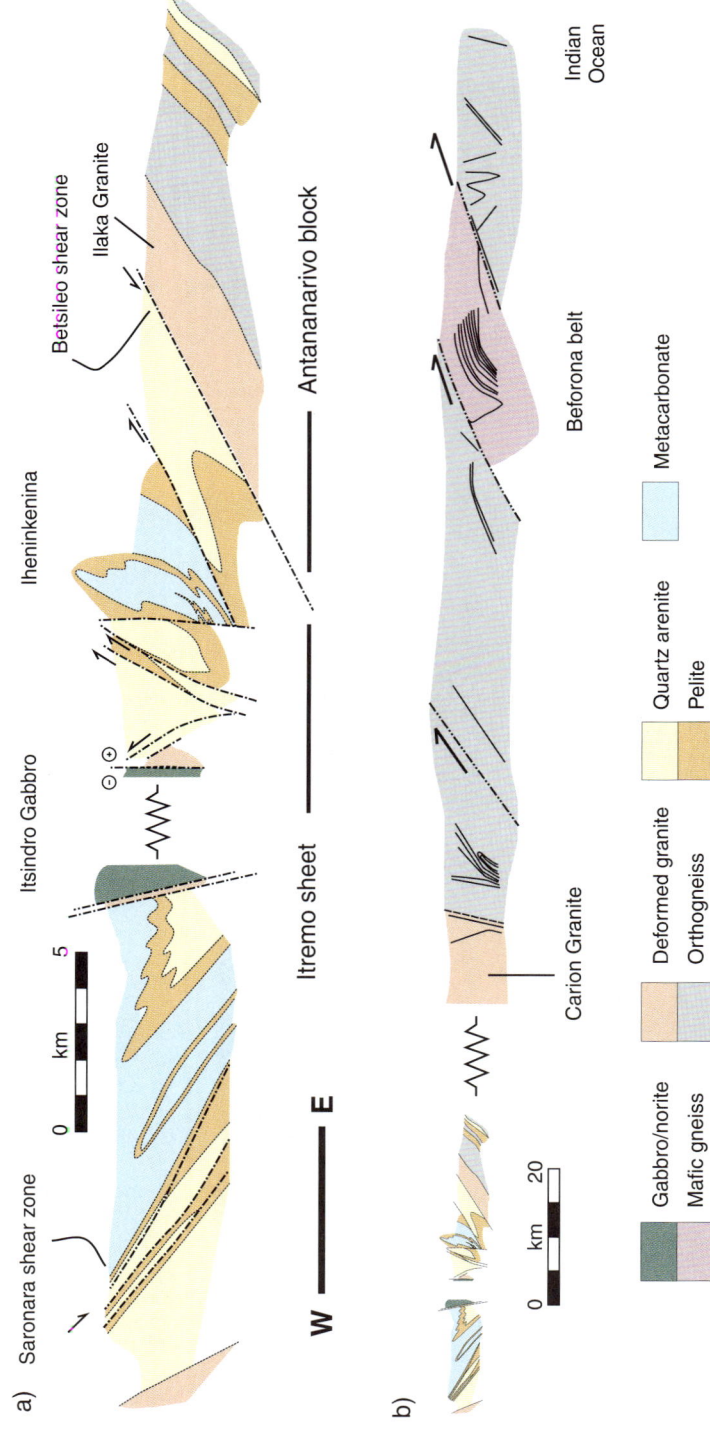

Fig. 10. Cross-sections through central Madagascar: (a) a composite section combining the mapping presented in this paper; (b) to scale combination of the section constructed in (a) with that through eastern Madagascar presented by Collins *et al.* (2003).

Table 1. *Summary of the deformation record preserved in the eastern Itremo Sheet and upper Antananarivo Block*

Deformation phase (timing constraints)	Characteristic features	
	Itremo Sheet	Upper Antananarivo Block
D5 (<551 Ma)	Broad E–W open folds Dextral shearing along east margin of Itsindro Gabbro	Broad open folding
D4 (<551 Ma)	Normal shearing along the Betsileo Shear Zone and possibly the Saronara Shear Zone	Broad zone of normal shearing along the Betsileo Shear Zone and in thin mylonite zones
D3 (c. 780–570 Ma)	Bivergent thrusting producing the: Analabe Shear Zone (top-to-the-east); Volamandotra Shear Zone. (top-to-the east); Mandrailanitra Shear Zone (top-to-the-west). Two phases of upright coaxial folding producing, amongst others, the refolded fold at Iheninkenina	Top-to-the-east shearing
D2 (c. 800–780 Ma)	Localized deformation during intrusion of the Itsindro Gabbro	
D1 (c. 1680–800 Ma)	10 km scale recumbent isoclinal west-verging folds	

maximum finite contractional strain for both the later upright fold phases is approximately horizontal and oriented E–W.

A number of discrete shear zones are interpreted to have formed during D3. Highly strained mylonites along the western flank of the Analabe–Ampandrianombilapa Ridge (Analabe Shear Zone), Volamandotra Mountain (Volamandotra Shear Zone) and Mandrailanitra Mountain (Mandrailanitra Shear Zone) all have thrust geometries, although they dip and verge in opposing directions. In the regions mapped, these three shear zones have not been deformed after formation. However, the Analabe Shear Zone may be folded by D3 folding around the west side of Ibity Mountain, where Moine (1968) mapped bedding truncations in quartzite.

The Saronara Shear Zone shows evidence for top-to-the-east non-coaxial shear. Fernandez and co-workers (Fernandez et al. 2000; Fernandez & Schreurs 2003) suggested that this shear zone formed as an early top-to-the-west thrust during the first deformation seen in the area. The present work contradicts this, as top-to-the-east kinematics have been shown in the shear zone that are opposite to the vergence of D1 folds, and unlike D1 the Saronara Shear Zone is younger than 790 Ma as it deforms the margin of the Imorona Composite Batholith. For it to have been active as a D1 top-to-the-west thrust it must have subsequently been reactivated as a normal shear zone and all trace of its earlier kinematic history destroyed. A second possibility is that it formed as an east-verging thrust during D3 contraction and has subsequently rotated into its present normal shear zone geometry. Alternatively, and by far the simplest explanation, is that it formed after D3 contraction as an extensional shear zone, possibly in the same extensional regime that created the Betsileo Shear Zone (see below, Collins et al. 2000b).

The upper Antananarivo Block preserves rocks pervasively deformed under amphibolite facies conditions. Top-to-the-east, non-coaxial shearing along west-dipping shear planes is here interpreted to be related to D3 deformation in the Itremo Sheet, as the 790 Ma Ilaka Granite (Handke et al. 1999) is locally deformed by it. Collins et al. (2000b) also reported locally preserved top-to-the-east shear indicators in the west-dipping Betsileo Shear Zone, which they interpreted as preserved relicts of a previous thrust history for the shear zone. It is therefore, possible that the Betsileo Shear Zone began its life as a D3 top-to-the-east thrust.

D4 deformation is manifest as the discrete post–551 Ma west-dipping Betsileo Shear Zone that shows excellent top-to-the-west kinematic indicators and can be traced for >200 km along-strike. For much of its length it juxtaposes relatively low-grade metasediments (down to lower greenschist facies) in the hanging wall against upper amphibolite facies orthogneisses (Collins et al. 2000b). Collins et al. (2000b) interpreted this structure as a crustal-scale extensional shear zone that was active during collapse of the EAO.

Late N–S to NE–SW contraction is seen throughout the region by the presence of open E–W-plunging folds forming broad deflections in lithological boundaries, and N–S spreads of poles to foliations and bedding. Dextral shearing along the eastern margin of the Itsindro Gabbro (Figs 5 & 6) may also be related to this D5 event, as the shear zone shows no sign of post-formation deformation and requires NE–SW contraction to form. This late contraction postdates the Betsileo Shear Zone, which it locally folds, and may be a far-field effect of the Cambrian collision of the northern Bemarivo Belt with central Madagascar (Buchwaldt & Tucker 2001; Collins & Windley 2002).

When scaled and placed next to a recently published cross-section through the eastern Antananarivo Block and part of the Tsaratanana Sheet (for location see Fig. 2; Collins et al. 2003), deformation in the Itremo Sheet appears considerably more complex than that preserved in the underlying rocks (Fig. 10b). This is presumably a factor of the difference in metamorphic grade between the upper crustal Itremo Sheet and the upper amphibolite- to granulite-grade Antananarivo Block. Collins et al. (2003) identified four deformation events in their traverse through the east of the island. These four deformations all postdate 818 Ma, with D2, D3 and D4 all formed during E–W contraction. Because of this similarity in age and structural geometry, these deformations are correlated with the composite D3 of the Itremo Sheet.

Conclusions

The main conclusion of the present work is that central Madagascar was deformed by complex polyphase deformation that lasted throughout much of the Neoproterozoic. Within this broad timescale a number of specific deformation phases are isolated. The earliest recognizable deformation predated intrusion of a gabbro and granitoid suite at c. 780–800 Ma (Handke et al. 1999; Hulscher et al. 2001) and involved the formation of 10 km scale recumbent isoclinal folds that, in the Saronara region, face west (see also Fernandez & Schreurs 2003). The precise age of this deformation is poorly constrained to between 1680 ± 34 Ma (youngest detrital mineral, in this case monazite, in the Itremo Group; Huber 2000) and 797 ± 5 Ma (Handke et al. 1999). However, palaeomagnetic evidence shows that India (and the Seychelles) were widely separated from the Congo Craton until after c. 750 Ma (Torsvik et al. 2001; Tucker et al. 2001a) and were not fully amalgamated in Gondwana until the Cambrian (Powell et al. 1993). D1 deformation cannot then be a result of this continental collision. Instead, it may have been related to the 1.1–0.9 Ga amalgamation of Rodinia, or possibly a result of accretion-related deformation along the middle Neoproterozoic active margin postulated by a number of workers (Handke et al. 1999; Kröner et al. 2000; Collins & Windley 2002).

This interpretation of D1 differs from a number of previous works in that it is here presented as conclusive evidence that D1 occurred before 790 Ma (cf. Fernandez et al. 2000; Fernandez & Schreurs 2003). The present authors agree with Cox et al. (1998) in so far as D1 predates intrusion of the Itsindro Gabbro, but point out that the fabric in the margin of the gabbro may not be of D1 origin and therefore should not be used as evidence for syntectonic intrusion. The present authors also broadly agree with Tucker et al. (2001b, 2003) in that the deformation here resulted from accretion of Gondwana along the EAO between c. 720 and c. 570 Ma, but emphasize that much earlier (at least 80 Ma) deformation exists that, at present, is only bracketed between 1680 and 790 Ma.

The major deformation associated with amalgamation of Gondwana produced much contractional deformation in the Itremo Sheet and underlying Antananarivo Block (Nédélec et al. 2000; Collins et al. 2003), including upright folding and thrusting (top-to-the-east in the Antananarivo Block and bivergent in the Itremo Sheet). After this major contractional phase the orogen collapsed, producing the Betsileo Shear Zone (Collins et al. 2000a,b) and possibly the Saronara Shear Zone. Finally, a late phase of N–S to NE–SW contraction deformed the region, producing open folds and a dextral shear zone along the eastern margin of the Itsindro Gabbro.

Chris Powell died tragically whilst this manuscript was in preparation. Chris's enthusiasm and drive were the main driving forces behind the project from which this paper derives. The other authors wish to acknowledge a great debt they owe to Chris, and will sorely miss his tempestuous field debates and keen support.

This project was funded by the Australian Research Council through the Tectonics Special Research Centre (TSRC) and a postdoctoral fellowship to ASC. ASC and ICWF would like to thank the villagers of Antanifotsy for feeding them. Tsu and Gabi of 'Les Lezard des Tana' are thanked for their long-suffering logistic support. D. Bishop helped keep camp and collect data. R. Rambeloson and M. Rakotondrazafy of the University of Antananarivo provided facilities for rock cutting, interesting discussions and help with rock export. Reviews by T. Kusky and R. Cox considerably improved the original manuscript. This publication forms TSRC publication number 178.

References

ASHWAL, L. D. 1997. *Proterozoic Geology of Madagascar. Guidebook to Field Excursions*. Miscellaneous Publications, 6. Gondwana Research Group.

BARTLETT, J. M., DOUGHERTY-PAGE, J. S., HARRIS, N. B. W.,

HAWKESWORTH, C. J. & SANTOSH, M. 1998. The application of single zircon evaporation and model Nd ages to the interpretation of polymetamorphic terrains: an example from the Proterozoic mobile belt of south India. *Contributions to Mineralogy and Petrology*, **131**, 181–195.

BESAIRIE, H. 1969–1971. *Carte géologique à 1/500000, de Madagascar, in 8 sheets: 1: Diego Suarez; 2: Antalaha; 3: Majunga; 4: Tamatave; 5: Tananarive; 6: Morondava; 7: Fianarantsoa; 8: Ampanihy.* Bureau Géologique Madagascar, Antananarivo.

BESAIRIE, H. 1970. *Déscription géologique du massif ancien de Madagascar. Quatrième volume: la région centrale. 2. Le Système du Vohibory, Série schisto-quartzo-calcaire.* Groupe d'Amborompotsy, Documentation du Bureau Géologique. Service Géologique de Madagascar, Tananarive.

BESAIRIE, H. 1973. *Carte géologique à 1\2000000 de Madagascar.* Service Géologique de Madagaskara., Antananarivo.

BUCHWALDT, R. & TUCKER, R. D. 2001. P–T–Time constraints on the metamorphic rocks of north Madagascar and their relevance on the assembly of Gondwanaland. *Geological Society of America Abstracts with Programs*, **33**, A436.

COLLINS, A. S. & WINDLEY, B. F. 2002. The Tectonic evolution of central and northern Madagascar and its place in the Final Assembly of Gondwana. *Journal of Geology*, **110**, 325–339.

COLLINS, A. S., FITZSIMONS, I., HULSCHER, B. & RAZAKAMANANA, T. 2002. Structure of the eastern East African Orogen in central Madagascar. *Precambrian Research*, in press.

COLLINS, A. S., FITZSIMONS, I., KINNY, P. D., BREWER, T. S., WINDLEY, B. F., KRÖNER, A. & RAZAKAMANANA, T. 2001. The Archaean rocks of Central Madagascar: their Place in Gondwana. *In*: CASSIDY, K. F., DUNPHY, J. M. & VAN KRANENDONK, M. J. (eds) *Proceedings of the 4th International Archaean Symposium 2001*, Extended Abstracts. AGSO–Geoscience Australia, Record 2001/37.

COLLINS, A. S., KRÖNER, A., RAZAKAMANANA, T. & WINDLEY, B. F. 2000a. The tectonic architecture of the East African Orogen in Central Madagascar – a structural and geochronological perspective. *Journal of African Earth Sciences*, **30**, 21.

COLLINS, A. S., RAZAKAMANANA, T. & WINDLEY, B. F. 2000b. Neoproterozoic extensional detachment in central Madagascar: implications for the collapse of the East African Orogen. *Geological Magazine*, **137**, 39–51.

COX, R., ARMSTRONG, R. A. & ASHWAL, L. D. 1998. Sedimentology, geochronology and provenance of the Proterozoic Itremo Group, central Madagascar, and implications for pre-Gondwana palaeogeography. *Journal of the Geological Society, London*, **155**, 1009–1024.

COX, R., COLEMAN, D. S., WOODEN, J. L. & CHOKEL, C. B. 2000. SHRIMP data from detrital zircons with metamorphic overgrowths reveal tectonic history of the Proterozoic Itremo Group, central Madagascar. *Geological Society of America Abstracts with Programs*, **37**, A248.

COX, R., COLEMAN, D. S., WOODEN, J. L. & DEOREO, S. B. 2001. A newly recognised Late Neoproterozoic metasedimentary sequence in Central Madagascar suggests terrane juxtaposition at 560 ± 7 Ma during Gondwana assembly. *Geological Society of America Abstracts with Programs*, **33**, A436.

DALZIEL, I. W. D. 1997. Neoproterozoic–Paleozoic geography and tectonics: review, hypothesis, environmental speculation. *Geological Society of America Bulletin*, **109**, 16–42.

DASO, A. A. H. 1986. *Géologie d'une plateforme carbonatée metamorphique: Valée de la Sahatany, centre de Madagascar: etude structurale pétrographique et géochimique.* PhD Thesis, Université Paul Sabatier.

DE WIT, M. J., BOWRING, S. A., ASHWAL, L. D., RANDRIANASOLO, L. G., MOREL, V. P. I. & RAMBELOSON, R. A. 2001. Age and tectonic evolution of Neoproterozoic ductile shear zones in southwestern Madagascar, with implications for Gondwana studies. *Tectonics*, **20**, 1–45.

EMBERGER, A. 1955. *Les terrains cristallins du pays Betsileo et de ses confins orientaux.* Thèse d'Etat, Université Clermont II, Clermont-Ferrand, France.

FERNANDEZ, A. & SCHREURS, G. 2003. Tectonic evolution of the Proterozoic Itremo Group metasediments in central Madagascar. *In*: YOSHIDA, M. & WINDLEY, B., DASGUPTA, S. (eds) *Proterozoic East Gondwana: Supercontinent Assembly and Breakup.* Geological Society, London, Special Publication, **206**, 381–399.

FERNANDEZ, A., HUBER, S. & SCHREURS, G. 2000. Evidence for Late Cambrian–Ordovician final assembly of Gondwana in central Madagascar. *Geological Society of America Abstracts with Programs*, **37**, A175.

GONCALVES, P., NICOLLET, C. & LARDEAUX, J.-M. 2000. In-situ electron microprobe monazite dating of the complex retrograde evolution of UHT granulites from Andriamena (Madagascar): apparent petrographical path vs PTt path. *Geological Society of America abstracts with Programs*, **32**, A174–A175.

GONCALVES, P., NICOLLET, C. & LARDEAUX, J.-M. 2003. Late Neoproterozoic strain pattern in the Andriamena unit (North-Central Madagascar): evidence for thrust tectonics and cratonic convergence. *Precambrian Research*, in press.

GUERROT, C., COCHERIE, A. & OHNENSTETTER. 1993. Origin and evolution of the west Andriamena Pan African mafic-ultramafic complexes in Madagascar as shown by U-Pb, Nd isotopes and trace element constraints. *Terra Abstracts*, **5**, 387.

HANDKE, M., TUCKER, R. D. & ASHWAL, L. D. 1999. Neoproterozoic continental arc magmatism in west-central Madagascar. *Geology*, **27**, 351–354.

HANDKE, M., TUCKER, R. D. & HAMILTON, M. A. 1997. Early Neoproterozoic (800–790 Ma) intrusive igneous rocks in central Madagascar; geochemistry and petrogenesis. *Abstracts with Programs*, **29**, 468.

HOFFMAN, P. F. 1999. The break-up of Rodinia, birth of Gondwana, true polar wander and the snowball Earth. *Journal of African Earth Sciences*, **28**, 17–33.

HOTTIN, G. 1976. Présentation et essai d'interprétation du Précambrien de Madagascar. *Bulletin du Bureau de Recherches Géologiques et Minières, 2nd series*, **IV**, 117–153.

HUBER, S. 2000. *Geologie und Geochronologie der Itremo Sedimente im Gebiet des Mont Ibity (Zentral*

Madagaskar). MSc Thesis, Universität Bern, Bern, Switzerland.

HULSCHER, B., COLLINS, A. S., DAHL, K. L. ET AL. 2001. Evidence for 800 Ma and possibly older Deformation and Plutonism in Madagascar. *Abstracts of the Structural Geology and Tectonics Studies Group Meeting*, Ulverstone, 91–92.

JACOBS, J., FANNING, C. M., HENJES-KUNST, F., OLESCH, M. & PAECH, H. J. 1998. Continuation of the Mozambique Belt into East Antarctica: Grenville-age metamorphism and polyphase Pan-African high-grade events in central Dronning Maud Land. *Journal of Geology*, **106**, 385–406.

KRÖNER, A. & SASSI, F. P. 1996. Evolution of the northern Somali basement: new constraints from zircon ages. *Journal of African Earth Sciences*, **22**, 1–15.

KRÖNER, A. & WILLIAMS, I. S. 1993. Age of metamorphism in the high-grade rocks of Sri Lanka. *Journal of Geology*, **101**, 513–521.

KRÖNER, A., HEGNER, E., COLLINS, A. S., WINDLEY, B. F., BREWER, T. S., RAZAKAMANANA, T. & PIDGEON, R. T. 2000. Age and Magmatic History of the Antananarivo Block, Central Madagascar, as Derived from Zircon Geochronology and Nd Isotopic Systematics. *American Journal of Science*, **300**, 251–288.

KRÖNER, A., SACCHI, R., JAECKEL, P. & COSTA, M. 1997. Kibaran magmatism and Pan-African granulite metamorphism in northern Mozambique: single zircon ages and regional implications. *Journal of African Earth Sciences*, **25**, 467–484.

LAWVER, L. A., GAHAGAN, L. M. & DALZIEL, I. W. D. 1998. A tight-fit early Mesozoic Gondwana, a plate reconstruction perspective. *Memoir of the National Institute for Polar Research, Tokyo*, **53**, 214–229.

MARTELAT, J.-E., LARDEAUX, J.-M., NICOLLET, C. & RAKOTONDRAZAFY, R. 2000. Strain pattern and late Precambrian deformation history in southern Madagascar. *Precambrian Research*, **102**, 1–20.

MARTELAT, J.-E., NICOLLET, C., LARDEAUX, J.-M., VIDAL, G. & RAKOTONDRAZAFY, R. 1997. Lithospheric tectonic structures developed under high-grade metamorphism in the Southern part of Madagascar. *Geodinamica Acta (Paris)*, **10**, 94–114.

MARTELAT, J.-E., VIDAL, G., LARDEAUX, J.-M., NICOLLET, C. & RAKOTONDRAZAFY, R. 1995. Images spatiales et tectonique profonde des continents: l'exemple du Sud-Ouest de Madagascar. *Comptes Rendus de l'Académie des Sciences de Paris*, **321**, 325–332.

MOINE, B. 1966. Grands traits structuraux du massif schisto-quartzo-calcaire (centre ouest de Madagascar). *Compte Rendus Semaine Géologique Madagascar*, 93–97.

MOINE, B. 1968. *Carte du Massif Schisto-Quartzo-Dolomitique. 1\200000*. Service Géologique de Madagasikara, Antananarivo.

MOINE, B. 1974. Caractères de sédimentation et de métamorphisme des séries précambriennes épizonales à catazonales du centre de Madagascar (Région d'Ambatofinandrahana). Sciences de la Terre, Mémoire, Nancy.

MUHONGO, S., KRÖNER, A. & NEMCHIN, A. A. 2001. Single zircon evaporation and SHRIMP ages for granulite-facies rocks in the Mozambique belt of Tanzania. *Journal of Geology*, **109**, 171–189.

NÉDÉLEC, A., PAQUETTE, J. L., BOUCHEZ, J. L., OLIVIER, P. & RALISON, B. 1994. Stratoid granites of Madagascar: structure and position in the Panafrican orogeny. *Geodinamica Acta (Paris)*, **7**, 48–56.

NÉDÉLEC, A. RALISON, B., BOUCHEZ, J.-L. & GRÉGOIRE, V. 2000. Structure and metamorphism of the granitic basement around Antananarivo: a key to the Pan-African history of central Madagascar and its Gondwana connections. *Tectonics*, **19**, 997–1020.

PASSCHIER, C. W. & TROUW, R. A. J. 1996. *Micro-tectonics*. Springer-Verlag, Berlin.

POWELL, C. M., LI, Z. X., MCELHINNY, M. W., MEERT, J. G. & PARK, J. K. 1993. Paleomagnetic constraints on timing of the Neoproterozoic breakup of Rodinia and the Cambrian formation of Gondwana. *Geology*, **21**, 889–892.

RAOELISON, I. L. 1997. *Structure and metamorphism of the Itremo Group, central Madagascar*. MSc Thesis, Rand Afrikaans University, Johannesburg.

ROLIN, P. 1991. Présence de décrochements précambriens dans le bouclier méridional de Madagascar: implications structurales et géodynamiques. *Compte Rendu Academie des Sciences, Paris*, **312**, 625–629.

REEVES, C. & DE WIT, M. J. 2000. Making ends meet in Gondwana: retracing the transforms of the Indian Ocean and reconnecting continental shear zones. *Terra Nova*, **12**, 272–280.

SHACKLETON, R. M. 1996. The final collision between East and West Gondwana; where is it? *Journal of African Earth Sciences*, **23**, 271–287.

SHIRAISHI, K., ELLIS, D. J., HIROI, Y., FANNING, C. M., MOTOYOSHI, Y. & NAKAI, Y. 1994. Cambrian orogenic belt in east Antarctica and Sri Lanka: implications for Gondwana assembly. *Journal of Geology*, **102**, 47–65.

STERN, R. J. 1994. Arc assembly and continental collision in the Neoproterozoic East African orogeny – implications for the consolidation of Gondwana. *Annual Review of Earth and Planetary Sciences*, **22**, 319–351.

TEKLAY, M., KRÖNER, A., MEZGER, K. & OBERHÄNSLI, R. 1998. Geochemistry, Pb–Pb single zircon ages and Nd–Sr isotope composition of Precambrian rocks from southern and eastern Ethiopia: implications for crustal evolution in East Africa. *Journal of African Earth Sciences*, **26**, 207–227.

TORSVIK, T. H., CARTER, L. M., ASHWAL, L. D., BHUSHAN, S. K., PANDIT, M. K. & JAMTVEIT, B. 2001. Rodinia refined or obscured: palaeomagnetism of the Malani igneous suite (NW India). *Precambrian Research*, **108**, 319–333.

TUCKER, R. D., ASHWAL, L. D., HAMILTON, M. A., TORSVIK, T. H. & CARTER, L. M. 1999a. Neoproterozoic silicic magmatism of northern Madagascar, Seychelles, and NW India: clues to Rodinia's assembly and dispersal. *Geological Society of America Abstracts with Programs*, **31**, 317.

TUCKER, R. D., ASHWAL, L. D., HANDKE, M. J., HAMILTON, M. A., LE GRANGE, M. & RAMBELOSON, R. A. 1999b. U–Pb geochronology and isotope geochemistry of the Archean and Proterozoic rocks of north-central Madagascar. *Journal of Geology*, **107**, 135–153.

TUCKER, R. D., ASHWAL, L. D. & TORSVIK, T. H. 2001a. U–Pb geochronology of Seychelles granitoids: a

Neoproterozoic continental arc fragment. *Earth and Planetary Science Letters*, **187**, 27–38.

TUCKER, R. D., KUSKY, T. M., BUCHWALDT, R. & HANDKE, M. J. 2001*b*. Neoproterozoic nappes and superimposed folding of the Itremo Group, west-central Madagascar. *Geological Society of America Abstracts with Programs*, **33**, A448.

TUCKER, R. D., KUSKY, T. M., BUCHWALDT, R. & HANDKE, M. J. 2003. Neoproterozoic nappes and superimposed folding of the Itremo Group, west-central Madagascar. *Precambrian Research*, in press.

WHITEHOUSE, M. J., WINDLEY, B. F., STOESER, D. B., AL-KHIRBASH, S., BA-BTTAT, M. A. O. & HAIDER, A. 2001. Precambrian basement character of Yemen and correlations with Saudi Arabia and Somalia. *Precambrian Research*, **105**, 357–369.

WINDLEY, B. F., RAZAFINIPARANY, A., RAZAKAMANANA, T. & ACKERMAND, D. 1994. Tectonic framework of the Precambrian of Madagascar and its Gondwana connections: a review and reappraisal. *Geologische Rundschau*, **83**, 642–659.

Tectonic evolution of the Proterozoic Itremo Group metasediments in central Madagascar

ALAIN FERNANDEZ & GUIDO SCHREURS

Institute of Geological Sciences, University of Bern, Baltzerstrasse 1, CH-3012 Bern, Switzerland (e-mail: fernandez@geo.unibe.ch)

Abstract: The major geologic units of the Itremo region in central Madagascar include: (1) upper amphibolite to granulite facies (higher grade) Precambrian rocks, mainly para- and orthogneisses, and migmatites; (2) the newly defined Itremo Nappes, a fold-and-thrust belt containing the Proterozoic Itremo Group sediments, metamorphosed at greenschist to lower amphibolite facies (lower grade) conditions; (3) Middle Neoproterozoic and Late Neoproterozoic–Cambrian intrusives. The stratigraphic succession of the Itremo Group in the eastern part of the Itremo region is, from bottom to top: quartzites, metapelites, metacarbonates and metapelites overlain by metacarbonates. During D1 the Itremo Group sediments were detached from their continental substratum, deformed into a fold-and-thrust nappe (Itremo Nappes), and transported on top of higher grade rocks that are intruded by Middle Neoproterozoic (c. 797–780 Ma) granites and gabbros. A second phase of deformation shortening (D2) affected both the Itremo Sedimentary Nappes and structurally underlying higher-grade rocksunits, and formed large-scale N–S-trending F2 folds. S1 axial plane foliations in Itremo Group sediments are truncated by Late Neoproterozoic–Cambrian granites (c. 570–540 Ma). The age of the formation of the Itremo Nappes is not well constrained: they formed in Neoproterozoic times between 780 and 570 Ma.

The assembly of the supercontinent Gondwana started in the Neoproterozoic and ultimately resulted in a large orogenic belt stretching over a distance of c. 6000 km [i.e. the East African Orogen of Stern (1994)]. Due to the subsequent breakup of Gondwana, remnants of this belt are nowadays exposed in the Arabian–Nubian Shield, eastern Africa (Pan-African Mozambique Belt), Madagascar, India, Sri Lanka and probably also Antarctica. The East African Orogen is generally considered to have formed either by closure of an oceanic basin and subsequent collision of two major crustal blocks (East and West Gondwana) or by accretion of a series of microcontinental blocks. However, in many parts of the East African Orogen, detailed tectonometamorphic studies and reliable geochronological data are lacking, and the exact nature and timing of the East African Orogen remains poorly understood.

Madagascar occupied a central position within Gondwana, and a better understanding of its Precambrian evolution could provide constraints on the timing and nature of events that led to Gondwana assembly. Although the geochronological database of Madagascar has improved significantly in recent years (e.g. Nédélec *et al.* 1994, 2000; Paquette *et al.* 1994; Ashwal & Tucker 1997; Handke *et al.* 1997, 1999; Martelat 1998; Paquette & Nédélec 1998; Tucker *et al.* 1999; Kröner *et al.* 2000; Martelat *et al.* 2000; de Wit *et al.* 2001; Handke 2001; Meert *et al.* 2001a, b; Fernandez *et al.* 2003), uncertainties and controversies regarding the tectonic interpretation of the complexly deformed and metamorphosed rocks of central Madagascar persist. This paper describes the tectonic evolution of the East African Orogen in the Itremo region of central Madagascar, south of Antananarivo and north of the Bongolava–Ranotsara Shear Zone (Fig. 1). The present authors' structural, sedimentological and remote sensing studies allow the recognition of a low-grade fold-and-thrust belt consisting of metasediments, which is structurally underlain by high-grade gneisses, migmatites and middle Neoproterozoic granites and gabbros.

Geological setting of the Itremo region

The Itremo region is situated between c. 46° and 47°15′E and 19°45′ and 21°00′S, and covers an area of almost 20 000 km² in the highlands of central Madagascar (Fig. 2). The Precambrian rocks in this region include: (1) a lower grade supracrustal sequence comprising greenschist to lower amphibolite facies Proterozoic metasediments, dominated by quartzites, metapelites and dolomitic carbonates; (2) higher grade rocks including upper amphibolite para- and orthogneisses, migmatites, and sillimanite- and K-feldspar-bearing quarzites; (3) middle and Late Neoproterozoic–Cambrian intrusives. The lower grade metasediments of central Madagascar form part of the Itremo Group (Emberger 1955; Hottin 1976; Cox *et al.* 1995, 1998), which has also been referred to as groupe schisto–quartzo–dolomitique (SQD) (Moine 1974) and séries schisto–quartzo–calcaire (SQC) (Moine 1966,

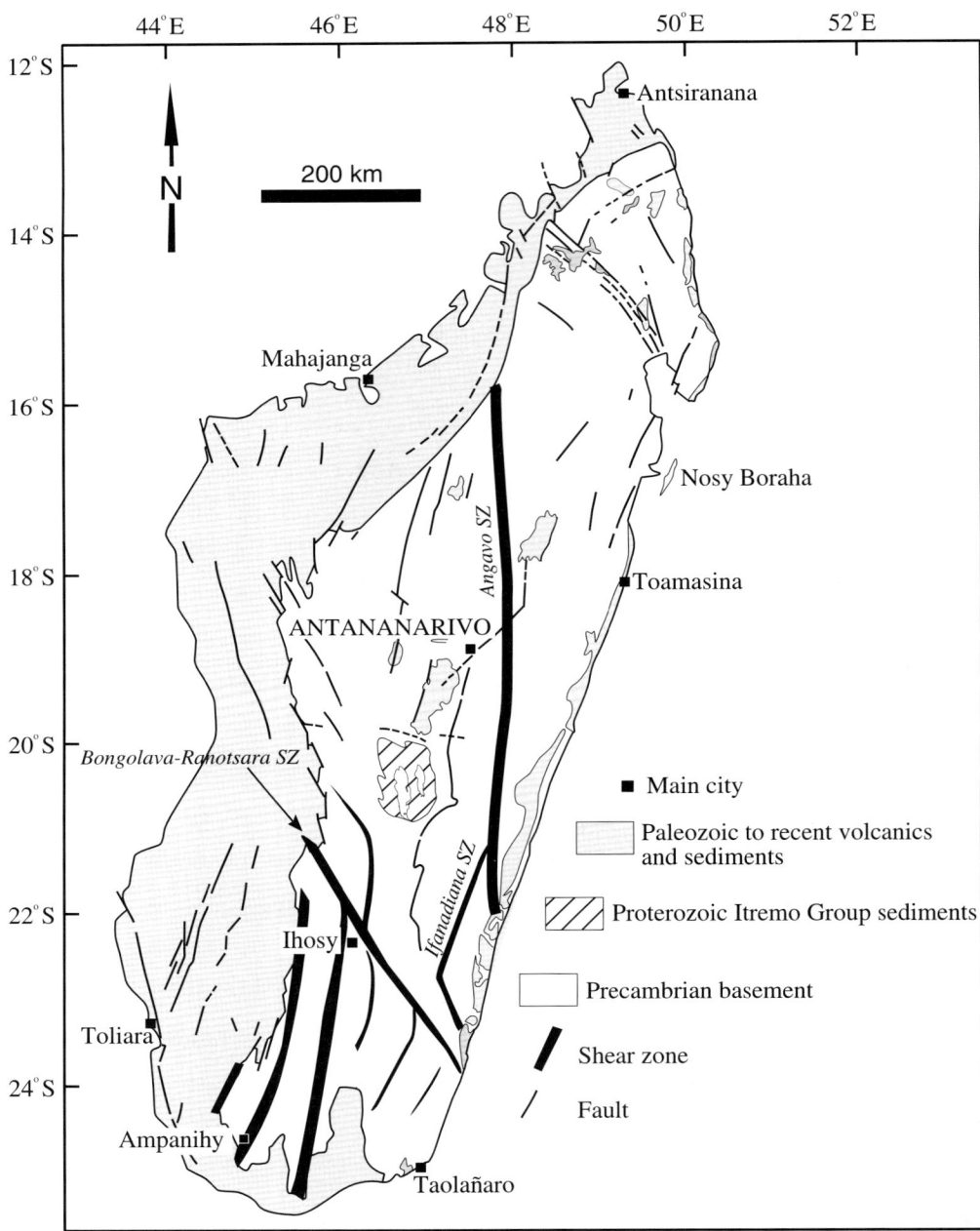

Fig. 1. Simplified geological map of Madagascar showing major shear zones and faults, and the location of the Itremo Group metasediments in central Madagascar (modified after Hottin 1976; Ralison & Nédélec 1997; Martelat 1998; Martelat et al. 2000; Nédélec et al. 2000).

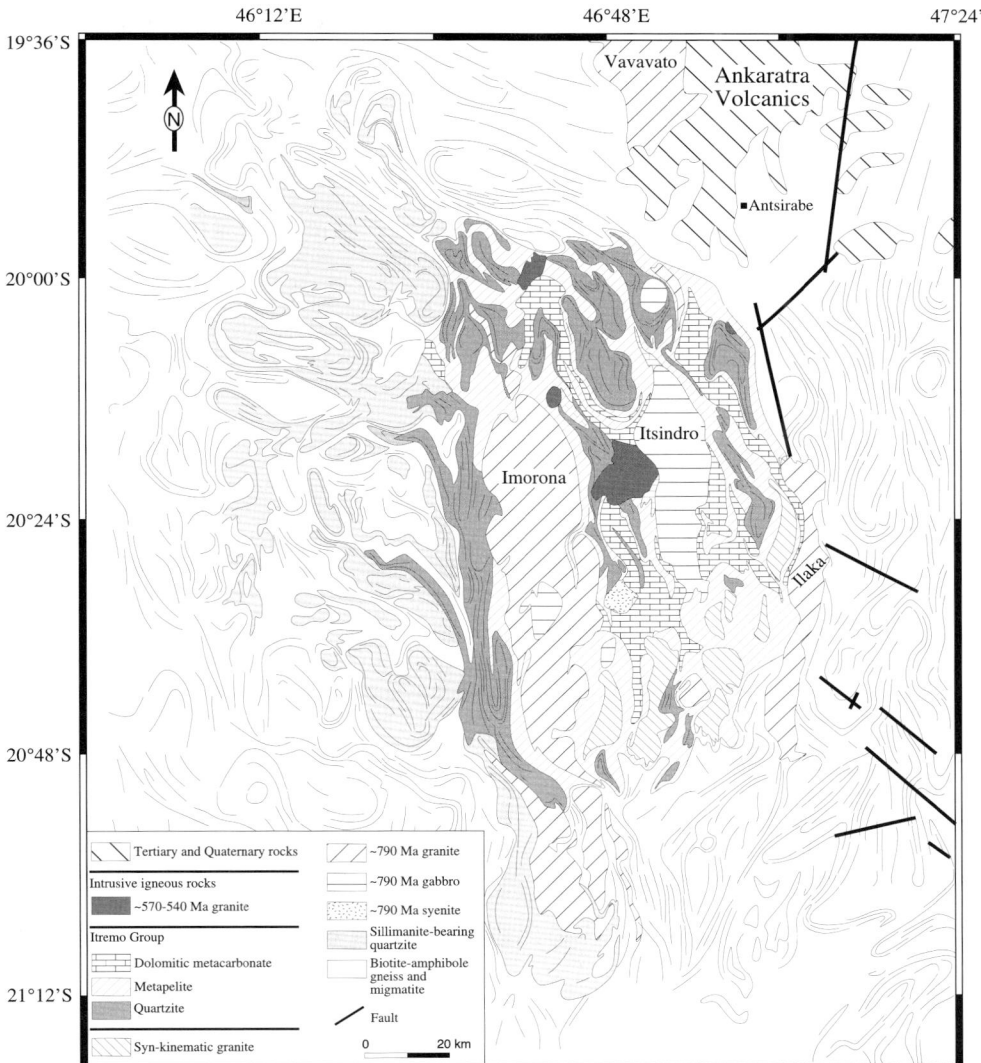

Fig. 2. Geological map of the Itremo region showing main structural features and lithological units [modified after Moine (1974, p. 185), 1/100 000 geological maps, and the present authors' remote sensing and field data]. Geochronological data on intrusive rocks are from Handke *et al.* (1999) and Fernandez *et al.* (2003).

1967). Moine (1974) correlated lower grade metasediments in the central and eastern part of the Itremo region (his groupe schisto–quartzo–dolomitique) with higher grade metasediments of the Amborompotsy and Ikalamavony Groups further west, assuming both a lateral facies change and a westward increase in metamorphic grade. Other authors have included part or most of the higher grade metasediments in the Itremo Region and to the west of it into the Itremo Group (e.g. Emberger 1955; Cox *et al.* 1998; Ashwal & Tucker 1999; Collins *et al.* 2000), and the geographic extent of the Itremo Group varies accordingly. As will be seen later on, the present structural, metamorphic and remote sensing studies indicate that the lower grade Itremo Group metasediments underwent a common tectonometamorphic evolution, which, in part, was clearly different from the evolution of the surrounding higher grade rocks. In this paper, the term Itremo Group has been restricted to the lower grade (greenschist to lower amphibolite facies) metasediments. The contact between lower and higher grade rocks is a major tectonic contact. The lower grade Itremo Group metasedimentary rocks were deformed into a fold-and-thrust belt, which was subsequently overprinted by large-scale folding. The fold-and-thrust belt will be referred to in this paper as the newly defined Itremo Nappes.

Lithologies of the Itremo Group

The Itremo Group consist of quartzites, metapelites and metacarbonates, with minor intercalations of conglomerates in the quartzites and calc-silicates within the metacarbonates. The metasediments have been intensely folded and internally thrusted, obscuring mostly the initial stratigraphic relations between different lithologies. The presence of numerous well-preserved sedimentary structures permitted the stratigraphic polarity in many outcrops to be established and, coupled with a detailed structural analysis, also allowed determination of the following stratigraphic succession, from bottom to top: quartzites, metapelites, metacarbonates and metapelites overlain by metacarbonates (Fig. 3). Estimations of the thicknesses of individual lithological units are difficult because of the complex polyphase deformation and possible lateral facies changes. Moreover, lithological boundaries are rarely exposed. In the eastern part of the Itremo region, thickness estimates are: 750 m for the basal quartzite, 450 m for the overlying lower metapelite, 600 m for the lower metacarbonates, 250 m for the upper metapelites and a minimum of 200 m for the upper metacarbonates.

Quartzites

The quartz arenites are generally well sorted and have a fine- to medium-sized average grain size. Despite the intense deformation and the greenschist facies metamorphism, primary sedimentary structures are often well preserved (Fig. 4), and include parallel lamination, cross-bedding (Fig. 4a & b) and ripple marks (Fig. 4c). Set thickness of crossbedding varies from several centimetres to several metres. Herringbone cross-bedding sets (Fig. 4a) suggest tide-generated current transport. Both wave and current ripples are observed: the ripple index varies between c. 3 and 6. Conglomerates occur occasionally in the upper part of the quartzite sequence. They are mostly monomictic, consisting of quartzite cobbles in a matrix of fine- to medium-grained quartz sand. Conglomerates near Mount Kiboy (NE part of the Itremo region), however, are polymictic and contain both gneiss and quartzite cobbles. The lateral extension of the conglomerates is limited and their thickness is generally only a few metres. Thin pelitic interbeds appear in the upper part of the quartzite sequence. They become thicker and more common upsection, and mark the transition towards the overlying metapelites.

The sedimentary structures in the quartzite unit indicate deposition under submarine conditions. The lenticular shape of conglomerate beds and the quartz matrix suggest an affinity with beach deposits. The good sorting of the quartzites, their regular grain size, the presence of wave ripples, the evidence for reversing currents, the abundant medium- to large-scale cross-bedding and the paucity of interbedded fine material are consistent with a shallow-marine environment.

Metapelites

The metapelites of the Itremo Group have been referred to as micaschistes or gneiss pélitiques by previous workers (e.g. Moine 1974). In the eastern part of the Itremo region, the metapelites can be divided into a lower metapelite (overlying the quartzite sequence) and an upper metapelite unit (overlying the lower metacarbonate unit). Fine-scale sedimentological features in the metapelites have mostly been obliterated by metamorphism and intense deformation. The metapelites show a well-developed schistosity and bedding is only in some cases preserved, and consists of thin alternating light, quartz-rich layers and dark, biotite-rich layers, which are interpreted as initially finely laminated siltstones and mudstones. Fine-grained quartzites (0.5–25 cm) occur in the lower metapelite unit as interbeds. Normal grading and wavy asymmetric cross-lamination (hummocky cross-stratification) occur and could be an indicator for storm-controlled deposits. The upper metapelite unit is dominated by micaschists with rare quartzitic interbeds. Graphite-containing schists occur in the upper metapelite unit. In the central Itremo region (north of Ambatofinandrahana) rare, isolated mafic igneous layers, 1–5 m in thickness, intercalated in metapelites, have been described by Cox et al. (1998) and were interpreted by them as flows.

Metacarbonates

The metacarbonates of the Itremo Group consist of calc-silicates and dolomitic marbles. The calc-silicates represent initially sandy to shaly carbonates and may record periods of increased terrigeneous input. The calc-silicates generally mark the transition from metapelites to the overlying marbles, but also occur interlayered with marbles. The calc-sili-

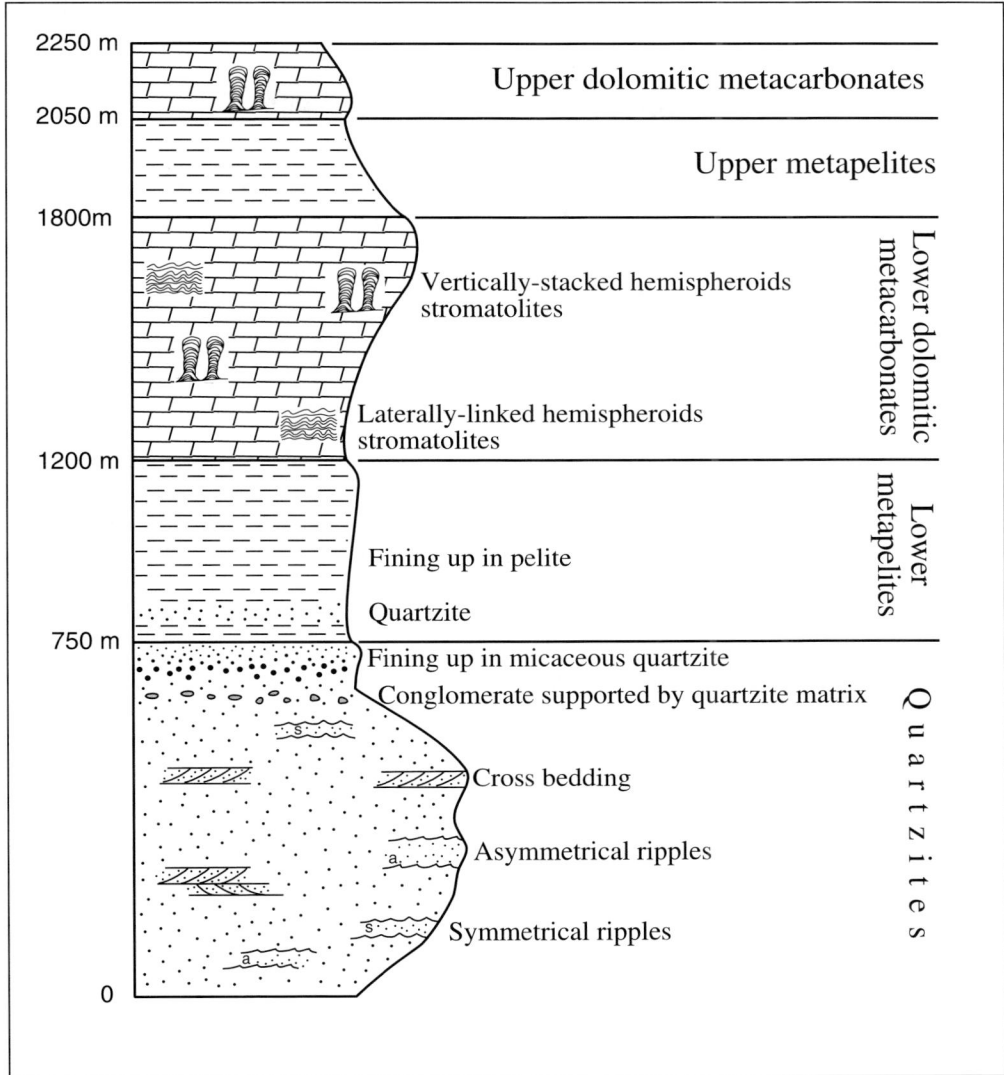

Fig. 3. Generalized stratigraphy and thickness estimates of the Itremo Group metasediments in the eastern Itremo region.

cates contain subordinate carbonate and varying proportions of quartz and mica. Tremolite and diopside occur as metamorphic minerals and the calc-silicates have been referred to by previous workers as pyroxeno–amphibolites (e.g. Moine 1974; Daso 1986). The metacarbonates occur in two different stratigraphic levels: a lower metacarbonate unit overlying the lower metapelite unit and an upper metacarbonate unit overlying the upper metapelite unit. The marbles consist predominantly of dolomite, with varying proportions of diopside, tremolite, calcite, quartz and phlogopite. The presence of stromatolites in the metacarbonates was first described by Trottereau (1969): they can be recognized in many places and are mainly domal or pseudocolumnar (Fig. 4d), which suggests that they formed in a lower intertidal to subtidal setting.

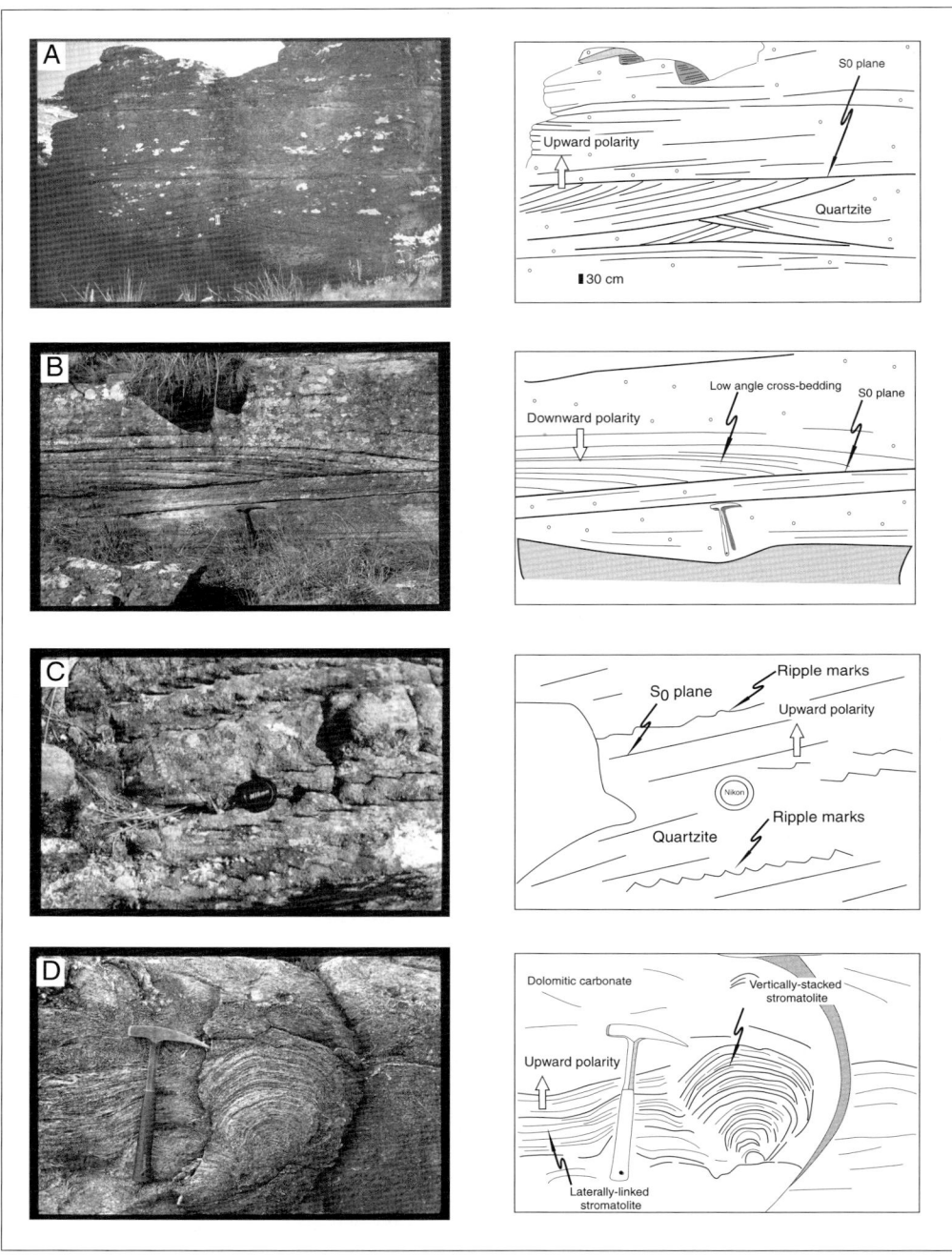

Fig. 4. Sedimentary features in Itremo Group metasediments: (a) herringbone cross-bedding in quartzites; (b) low-angle cross-bedding in quartzites (hammer for scale); (c) ripple marks in quartzites (lens cap for scale); (d) stromatolites in dolomitic carbonates (hammer for scale).

Depositional setting and age of the Itremo Group

The lithological association of quartzites, metapelites and metacarbonates is characteristic of a continental-shelf sequence. The transition from quartzites to lower metapelites, and the repetition of metapelites and metacarbonates in the upper part of the sequence, is possibly the result of relative sea-level changes. The sedimentary structures in the quartzites and metapelites, the well-sorted nature of the quartzites and the association with stromatolites suggest that the Itremo Group sediments were derived from a cratonic source area and deposited on a shallow continental shelf.

The depositional age of the Itremo Group is not well constrained. Based on SHRIMP ages on detrital zircons, separated from low-grade quartzites, Cox et al. (1998) concluded that the maximum depositional age must be c. 1850 Ma. U–Th–Pb dating of detrital monazite grains in low-grade metapelites constrained the maximum depositional age at 1637 ± 24 Ma (Huber 2000). Handke et al. (1999) considered gabbroic and granitic plutons (e.g. the Itsindro Gabbro and the Imorona Composite Batholith), dated between 797 and 780 Ma (U–Pb zircon and baddeleyite ages), as intrusive into Itremo Group metasediments, providing a minimum depositional age of 797 Ma. The present studies, however, indicate that the contacts between c. 800 Ma intrusives and Itremo Group metasediments are tectonic, and that their age does not, therefore, provide a minimum depositional age. In fact, the Itsindro Gabbro and the Imorona Composite Batholith occupy second-phase (with respect to deformation in the Itremo Nappes) antiformal cores and are structurally overlain by the Itremo Nappes. Late Neoproterozoic–Early Cambrian intrusives granites, however, clearly intrude the Itremo Group. They vary in age between c. 570 and 540 Ma (Handke et al. 1999; Handke 2001; Fernandez et al. 2003) and provide a minimum depositional age of c. 570 Ma for the Itremo Group.

Metamorphic grade of the Itremo Group

Moine (1965, 1968, 1974) assumed that low-grade quartzites in the east of the Itremo region were the lateral equivalents of high-grade quartzites in the west of the region (the Amborompotsy and Ikalamavony Groups), and proposed a westward increase in metamorphism from greenschist to granulite facies. Most present-day tectonic models of the Itremo region are based on this assumption (e.g. Windley & Razakamanana 1996; Raoelison 1997; Collins et al. 2000; Collins & Windley 2002). Mineral assemblages in the various lithologies of the Itremo Nappes, however, indicate that the rocks experienced only greenschist to lower amphibolite facies metamorphism. Evidence was not found for a major westward increase in metamorphic grade within the metasediments of the Itremo Nappes. There is, however, an abrupt increase in metamorphic grade across the tectonic contact between the Itremo Nappes and the structurally underlying rocks.

Mineral assemblages in the metapelites of the Itremo Nappes are commonly characterized by quartz + muscovite + tourmaline or quartz + muscovite + biotite + tourmaline. Theses parageneses are typical of the greenschist to lower amphibolite facies and suggest minimum temperatures of c. 500°C. Locally the assemblage quartz + biotite + muscovite + garnet is also observed in the metapelites. Biotite, mica and tourmaline define the S1 axial plane schistosity in the metapelites of the Itremo Nappes. A later generation of muscovite and/or biotite is sometimes observed to overgrow S1. In some cases they define a weakly developed S2 axial plane planar foliation, in other cases they seem randomly oriented. The presence of the late-stage mica indicates that greenschist facies conditions persisted after D1. The quartzites of the Itremo Nappes show the mineral assemblage quartz + muscovite + biotite. Typical assemblages in the metacarbonates consist of quartz + dolomite + tremolite + calcite and quartz + dolomite + diopside + calcite + phlogopite. These parageneses indicate upper greenschist and amphibolite facies, respectively, with minimum temperatures of c. 450°C.

Higher grade basement rocks of the Itremo region

The higher grade basement rocks of the Itremo region are dominated by upper amphibolite facies para- and orthogneisses, and migmatites, and locally granulite facies gneisses and quartzites. Sedimentary-derived rock types include sillimanite- and K-feldspar-bearing quartzites and pyroxene–amphibole gneisses. These rocks lack any clear primary sedimentary structures, except lithological alternations that possibly represent initial bedding surfaces. Quartzites containing sillimanite and muscovite are widespread throughout the Itremo region, especially west of the Itremo Group metasediments. This paragenesis indicates at least upper amphibolite facies metamorphic conditions. Mineral assemblages in calc-silicates include quartz + scapolite + grossular + wollastonite + clinopyroxene and are typical of granulite facies (Raoelison 1997).

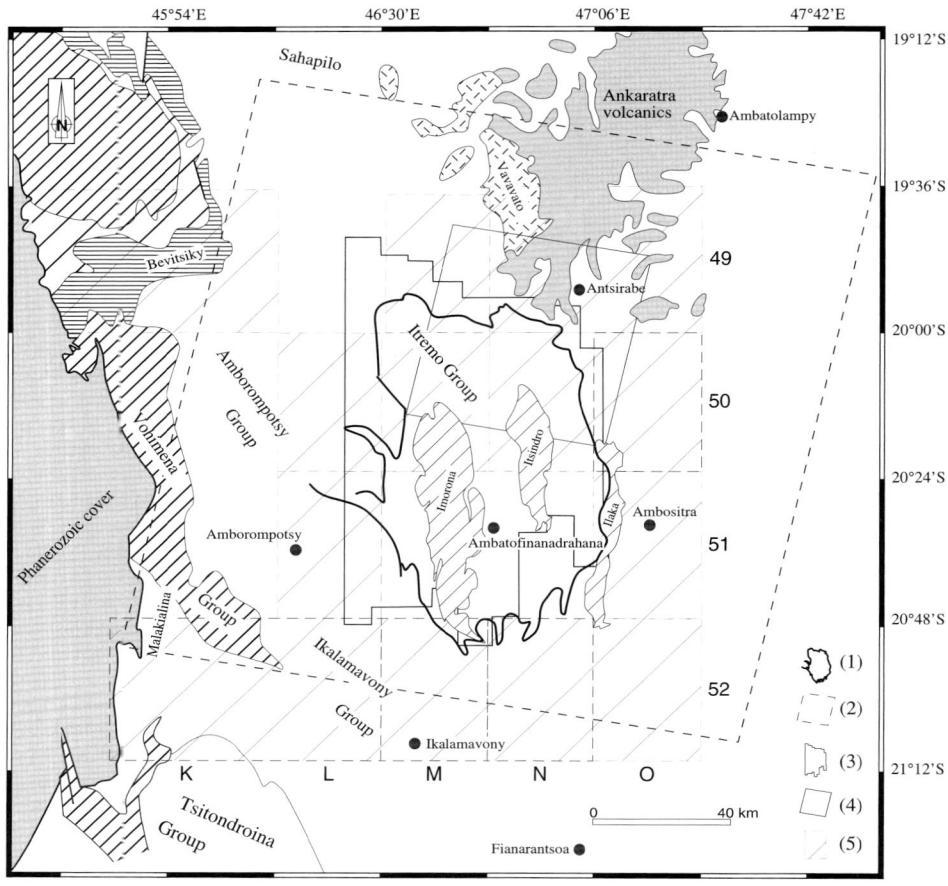

Fig. 5. Location of remote sensing data and 1/100 000 scale geological maps used in this study. IJK49, Ankotrofotsy–Dabolava–Anjoma; NO49, Antsirabe–Antanifotsy; L50, Lazarivo; M50, Tsangandrano; N50, Manandona; L51, Amborompotsy; MN51, Itremo–Ambatofinandrahana; O51, Ambositra; K52, Salajea; L52, Fitampito; M52, Ikalamavony; N52, Fanjakana; O52, Ambohimasoa. (1) Boundary of Itremo Group; (2) Landsat TM and ETM + image; (3) mosaic of 220 aerial photographs (mission IGN 1991; scale 1:50 000); (4) SPOT image; (5) 1:100 000 geological maps. In addition, a 1:200 000 geological map (Moine 1974) that covers the central part of the Itremo region was used.

Structures of the Itremo region

Introduction

The structures of the Itremo region have been determined by extensive field work, remote sensing studies and interpretations of existing 1:100 000 and 1:200 000 geological maps (Fig. 5). Remote sensing studies involved interpretation of a georeferenced mosaic consisting of 220 aerial photographs (scale 1:50 000), and Landsat TM 5, Landsat ETM + (Fig. 6) and SPOT digital satellite data. Main foliation trends and deformation features of the Itremo region are shown in Figure 7. Longhi *et al.* (2000) used reflectance spectra from siliceous muscovite-bearing rock samples and Landsat Thematic Mapper spectral properties to distinguish high-grade rocks from low-grade rocks in the western part of the Itremo region.

The present authors' investigations indicate that the greenschist/lower amphibolite facies metasediments of the Itremo region underwent a tectonometamorphic evolution which, in part, was different from the one experienced by the high-grade basement rocks. On a regional scale, the difference in the tectonometamorphic evolution is manifested by the trends of large-scale folds. In the area covered by lower grade Itremo Group metasediments (central-eastern part of the Itremo region), N–S-trending

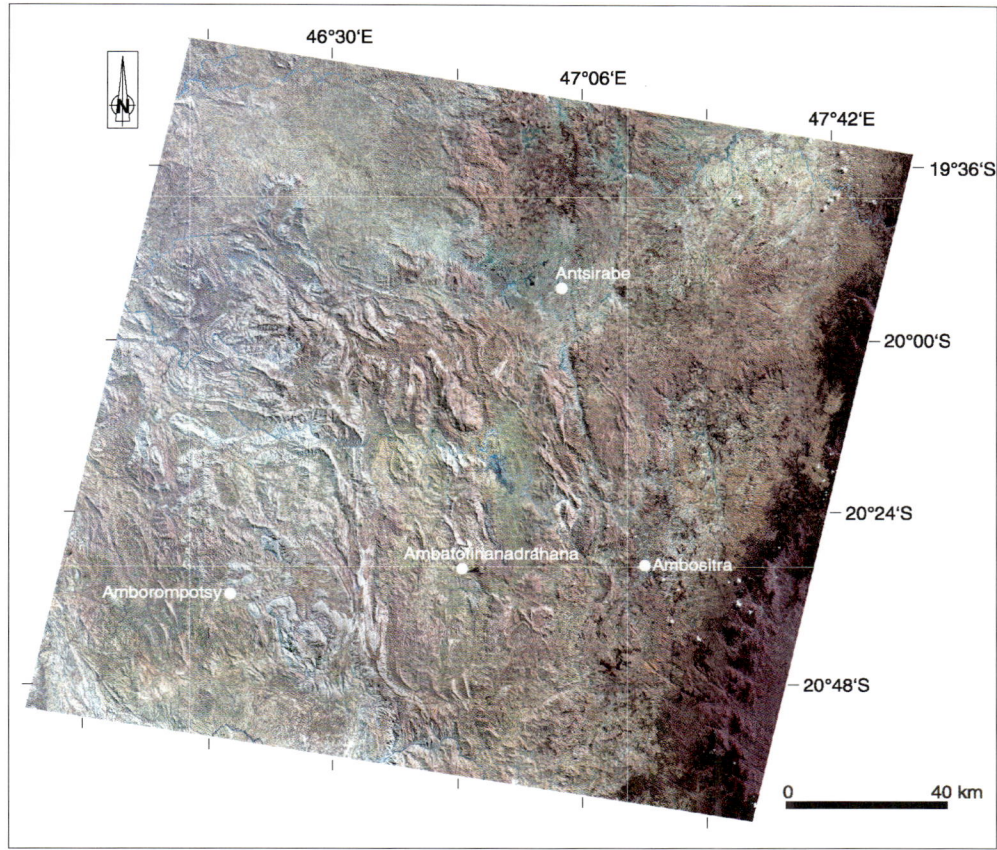

Fig. 6. Landsat ETM + false-colour composite image 532 (red, green, blue) of the Itremo region. Landsat TM data can be used to distinguish high-grade quartzites from low-grade quartzites (cf. Longhi *et al.* 2000). Predominant bluish colours in the western part of the image characterize high-grade quartzites, whereas whitish-reddish colours in the central part of the image characterize low-grade quartzites.

major folds dominate, whereas the remainder of the region shows both E–W- and N–S-trending folds.

The contact between lower grade Itremo Group metasediments and higher grade basement rocks is tectonic, and is marked by ductile shear zones. The lower grade metasediments form part of a coherent tectonic unit (the Itremo Nappes), whose structural evolution will now be discussed.

Structures of the Itremo Nappes

The lower grade Itremo Group metasediments of the Itremo Nappes contain numerous well-preserved primary sedimentary structures that allow the determination of structural facing directions. Overprinting relationships and the pattern of structural facing directions indicate clearly that the lowItremo Group was affected by two main deformation phases (D1 and D2). The major N–S-trending folds are second-phase folds that refold first-phase folds and ductile shear zones.

An example of a well-exposed large-scale F1–F2 fold interference pattern occurs in the Ampozana area located in the central-northern Itremo region (Fig. 8). The lower grade quartzites in this area have preserved abundant primary sedimentary features such as ripple marks and cross-bedding. The major fold in the southwestern part of the Ampozana area is an F1 fold: a steeply north-plunging synformal anticline (facing south). Its axial plane is folded by a large-scale, north-plunging F2 antiform with a subvertical S2 axial plane. The overprinting is approximately intermediate between a Type II and III interference pattern (Ramsay 1967). Unfolding of the F2 fold results in an approximately WSW-facing

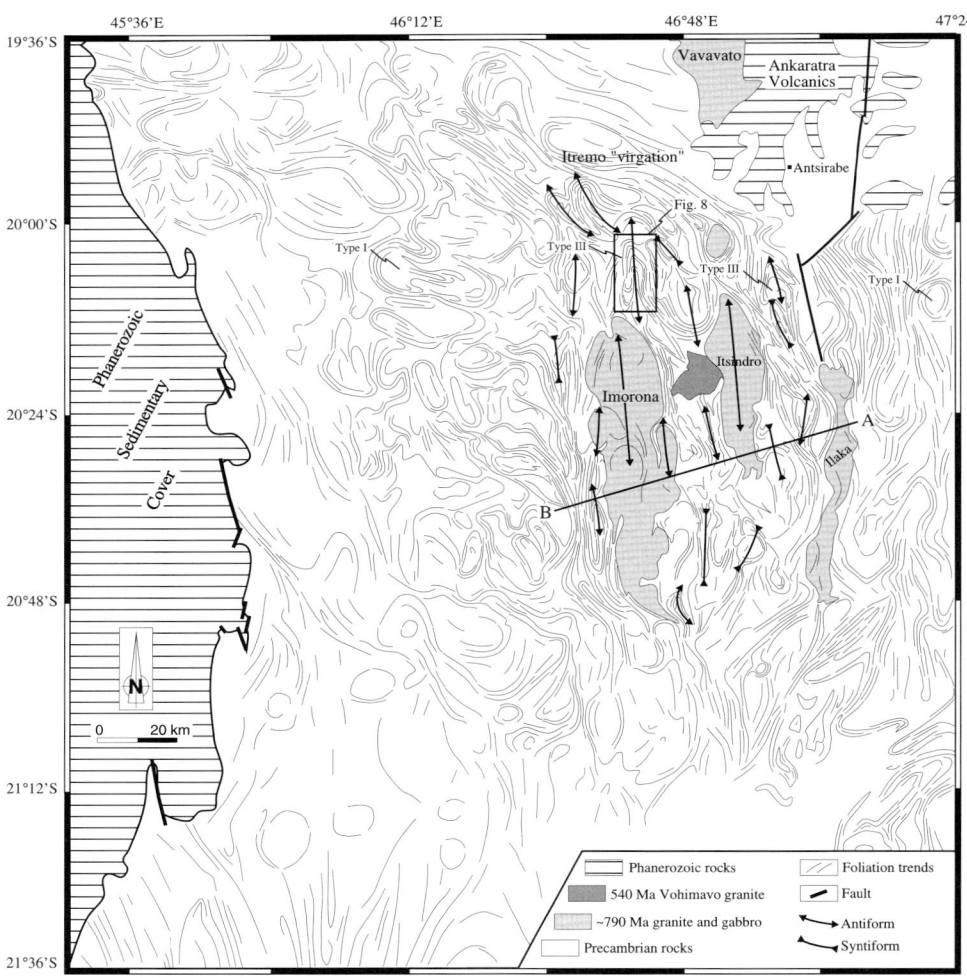

Fig.7. Simplified geological map of the Itremo region with foliation traces obtained from field data, remote sensing studies and interpretations of 1:100000 and 1:200000 geological maps. Major antiforms and synforms indicated are D2 structures. A–B denotes the position of the schematic cross-section shown in Figure 13.

F1 fold, suggesting a west-directed component of tectonic transport during D1. An S1 axial plane schistosity was formed in suitable lithologies and consists of aligned muscovite and/or biotite in metapelites and quartzites.

Based on such overprinting relationships in lower grade metasediments metasedimentary rocks throughout the Itremo region, it is concluded that WSW-facing F1 folds at greenschist to lower amphibolite facies conditions affected the Itremo Group sedimentary sequence. In metapelites the S1 axial plane foliation is usually penetratively developed and the original sedimentary bedding is often difficult to distinguish. A second phase of shortening subsequently deformed the Itremo Group and resulted in large-scale, roughly N–S trending, F2 folds with steeply dipping S2 axial planes. An S2 axial plane foliation is rarely developed, except in metapelites where biotite and muscovite sometimes define an S2 crenulation cleavage, indicating that greenschist to lower amphibolite facies conditions persisted. The plunge of second-phase fold axes and intersection lineations varies considerably, depending on the orientation of sedimentary bedding and S1 axial plane schistosity before F2 folding. The azimuth of F2 folds, however, trend roughly N–S.

Large-scale F2 folds overprint D1 mylonitic shear zones that are typically c. 50–100m wide. These mylonitic zones mark the basal contact with the

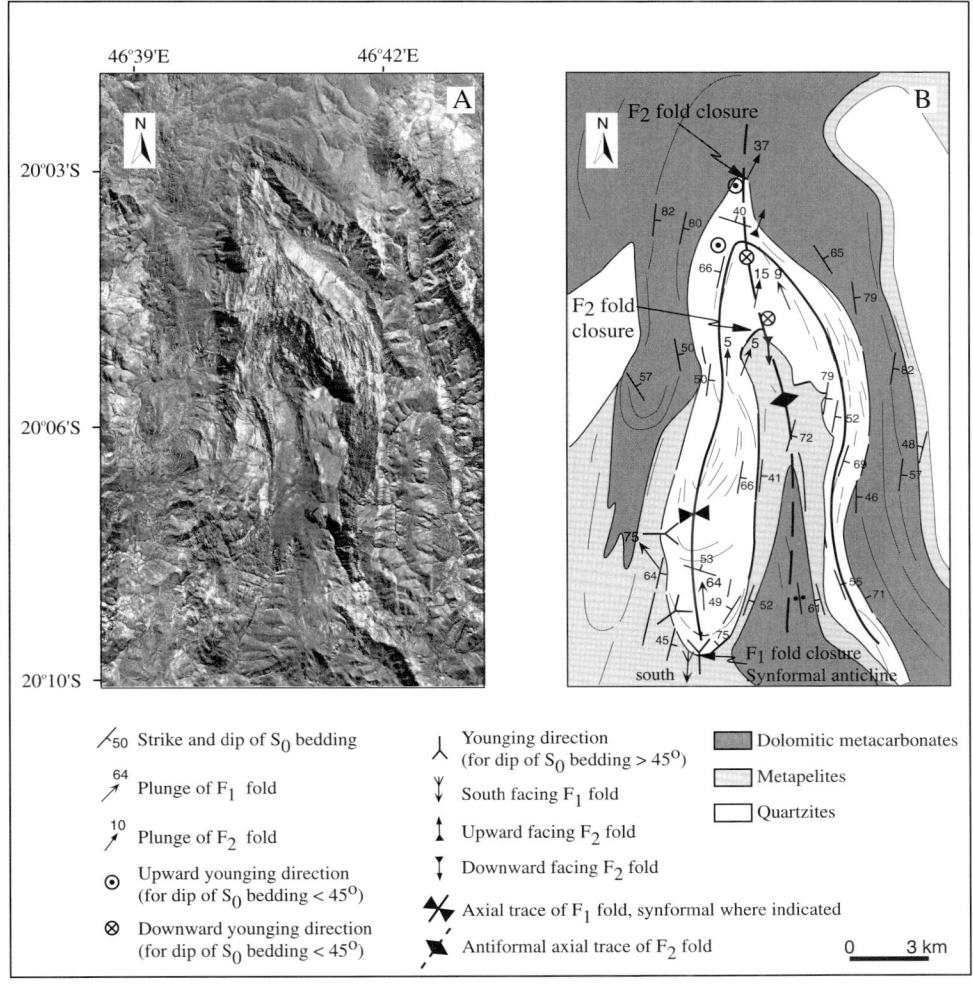

Fig. 8. D1–D2 fold interference pattern in Itremo Group metasediments. (**a**) Aerial photograph mosaic of the Ampozana area (central-northern Itremo region); location indicated in Fig. 7). (**b**) Structural interpretation of the Ampozana area; retrodeformation of the N–S trending F2 antiform results in a *c.* WSW-facing F1 fold.

structurally underlying higher grade rocks, but also occur internally within the Itremo Nappes (Fig. 9). Quartz–mylonites display a prominent L1 stretching lineation (Fig. 10), defined by elongated quartz crystals or stretched conglomerate pebbles. L1 stretching lineations within ductile shear zones are oriented parallel to centimetre-scale isoclinal F1 folds (Fig. 10a), but are perpendicular to the trend of large-scale (>decametre) F1 folds outside the mylonite zones axes. The orientation of D1 mylonitic foliations and associated L1 stretching lineations depend on their position with respect to F2 folds. In those areas where the S1 mylonitic foliation dips westward (on the western limbs of major N–S-trending F2 anti-

forms), the L1 stretching lineation plunges approximately WSW-ward (Figs 9 & 11). Where the mylonitic foliation dips eastward (on the eastern limbs of N–S-trending F2 antiforms), the L1 stretching lineation plunges approximately ENE-ward (Figs 9 & 11). The present authors consider the L1 stretching lineations in these mylonite zones to indicate the tectonic transport direction during D1. The sense of shear was determined for two quartz-mylonites. The crystallographic-preferred orientations of quartz grains yielded a top-to-the-WSW shear sense (Fig. 12). In areas of relatively low non-coaxial strain, L1 stretching lineations are locally observed that are parallel to large-scale F1 fold axes. These lineations

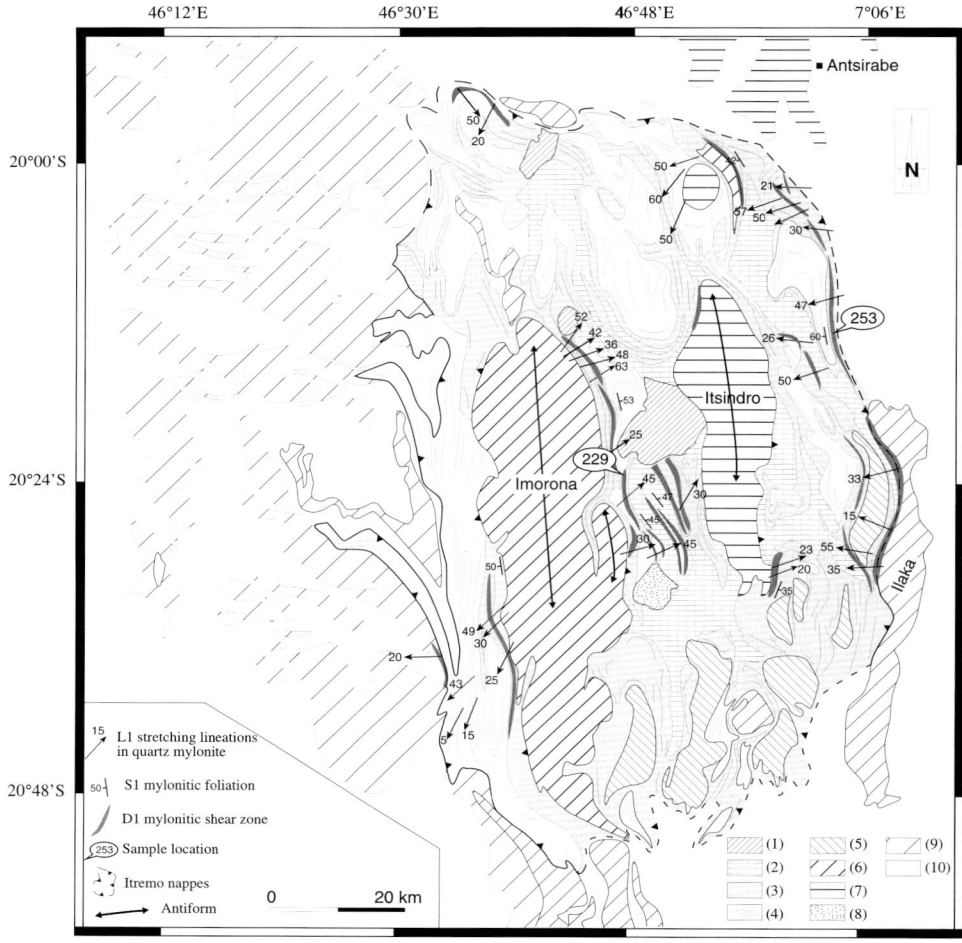

Fig. 9. Map of the Itremo region showing the location and orientation of D1 mylonite zones, and the orientation of L1 stretching lineations contained within these shear zones. Note that shear zones occur both at the basal contact between higher and lower grade rocks, and within lower grade Itremo Group metasediments.

are not considered to indicate a tectonic transport direction but probably reflect fold-axis parallel stretching. They are parallel to intersection lineations formed by the S1 and S0 foliations. In the hinge zones of large-scale F2 folds, aligned tourmaline and biotite minerals in quartzites define subhorizontal, N–S oriented L2 lineations. These lineations are considered to be the result of E–W shortening of the units in the Itremo region during D2.

In contrast to areas in southern Madagascar, the Itremo region lacks clear N–S-trending shear zones. However, satellite images show deflections of foliations from N–S to NW–SE in the northern part of the Itremo region (Figs 6 & 7). This zone is referred to here as the Itremo Virgation (Fig. 7): S2 axial traces are deflected by it and the virgation is interpreted as being either syn- or post-D2.

Structures of the higher grade rocks

The higher grade basement rocks of the Itremo region high-grade basement include Archaean tonalitic gneisses (2522–2494 Ma; Tucker *et al.* 1999) that were intruded by Middle Neoproterozoic igneous rocks dated between 797 and 780 Ma (Handke *et al.* 1999). Structures are dominated by both E–W and N–S fold axes, and can be interpreted as Type I (Ramsay 1967) dome-and-basin structures resulting from two phases of deformation. The N–S-

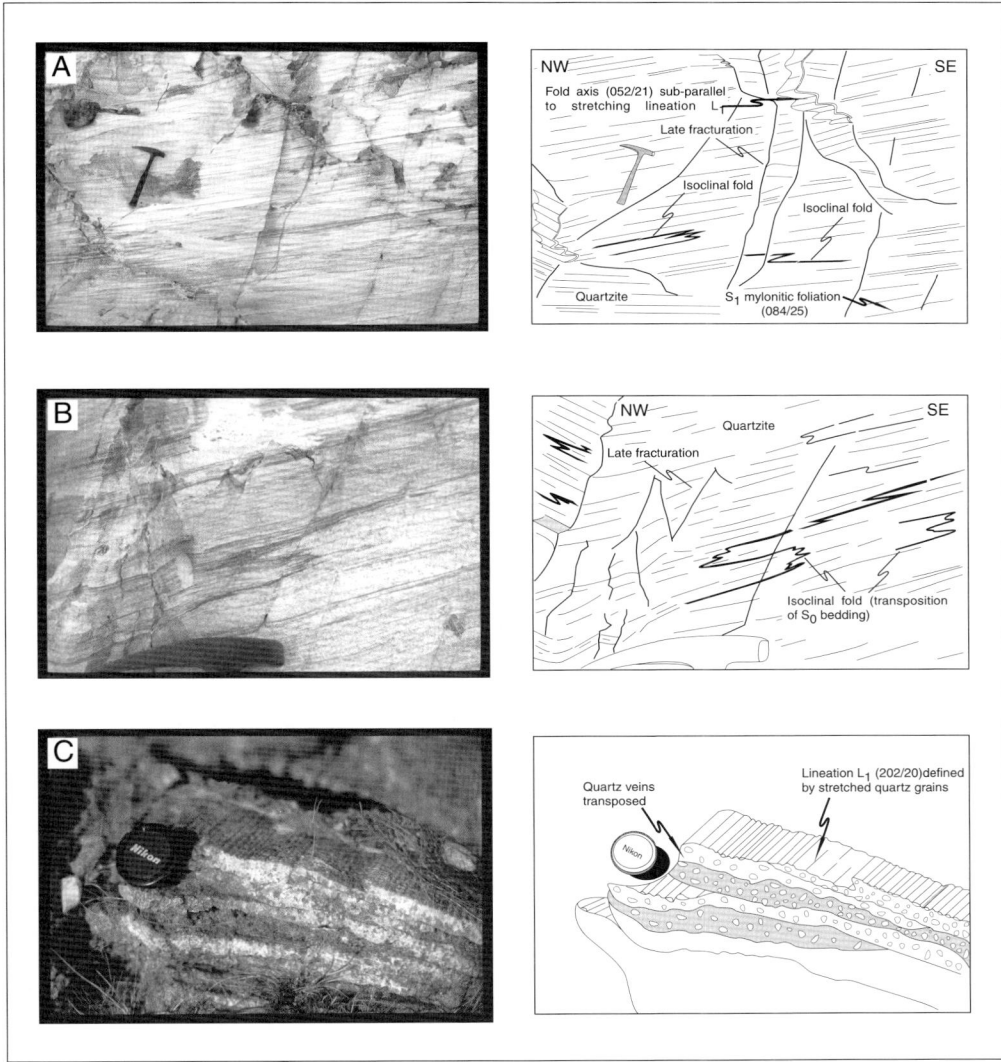

Fig. 10. Photographs and line drawings of D1 quartz mylonites. (**a**) Fold axes of centimetre-scale folds are subparallel to a stretching lineation defined by stretched quartz grains (46°49′33.7″E, 20°30′24.5″S) (hammer for scale). (**b**) Transposition of S0 sedimentary bedding parallel to S1 mylonitic foliation (46°49′33.7″E, 20°30′24.5″S) (hammer for scale). (**c**) L-tectonite quartzite with L1 lineation (46°34′38.4″E, 20°41′31.7″S) (lens cap for scale).

directed folding corresponds to the D2 deformation event in the Itremo Nappes and E–W-trending folds probably belong to an earlier phase. Prominent lineations are observed on a penetratively developed foliation plane in many parts of the higher grade basement. In quartzites it is defined by preferentially oriented sillimanite crystals or aggregates that are parallel to a stretching lineation defined by elongated quartz crystals. In granitoids the lineation is defined by elongated feldspar and quartz crystals. It is not clear whether this lineation, which is mostly E–W oriented, represents a tectonic transport direction or if it is the result of stretching parallel to large-scale E–W-trending folds.

The age of pre-D2 deformation (with respect to deformation in the Itremo Nappes) in the structurally underlying higher grade rocks is not well known. Foliations in higher grade rocks in the immediate vicinity of the tectonic contact with the overlying Itremo Nappes are usually concordant with the

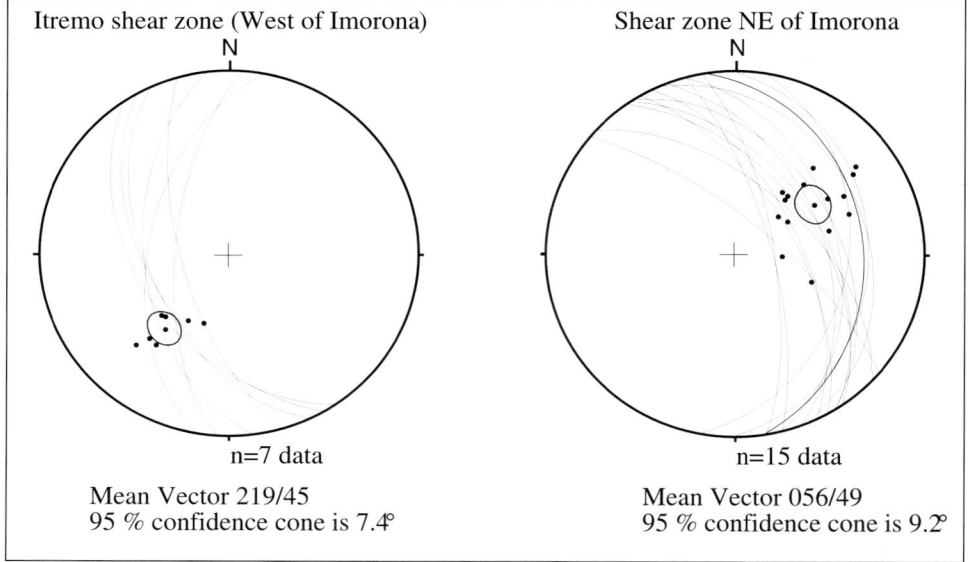

Fig. 11. Stereographic projections of S1 mylonitic foliations (great circles) and associated L1 stretching lineations (black dots) west and NE of the Imorona Composite Batholith (Schmid, equal-area lower hemisphere). Mean orientation of stretching lineation is given for each stereographic projection.

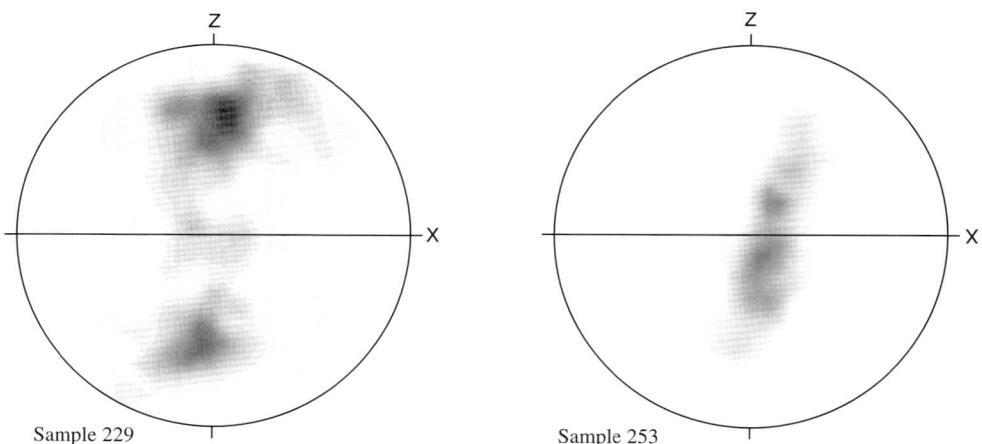

Fig. 12. Quartz c-axis textures from D1 mylonite zones suggesting a top-to-the-WSW sense of shear. Location of samples is indicated in Figure 9.

mylonitic foliation, suggesting that they are related to the overthrusting. Away from the tectonic contact, older foliations may have been preserved in the higher grade rocks and it is quite likely that the E–W-trending folds in these rocks reflect an earlier, higher grade event that is not present in the Itremo Nappes.

Relationship between Neoproterozoic intrusive rocks and lower grade Itremo Group metasediments

The Itremo region underwent two separate periods of magma emplacement – during the middle Neoproterozoic (c. 797–780 Ma; Handke et al. 1999) and during the Late Neoproterozoic and/or

Early Cambrian (570–540 Ma; Tucker et al. 1997; Ashwal & Tucker 1999; Fernandez et al. 2003). The Itsindro, Ranomandry, and Ifasina Gabbros, and igneous granitoid rocks that form part of the Imorona Batholith Complex (Fig. 2), have been dated as middle Neoproterozoic with U–Pb zircon and baddeleyite ages of between 797 and 780 Ma (Handke et al. 1999). These igneous rocks were considered by Handke et al. (1999) to be intrusive into both higher grade migmatitic gneisses and into lower grade Itremo Group metasediments [the quartzo–schisto–calcaire series of Handke et al. (1999)], thus providing a minimum depositional age for the Itremo Group sediments. The present investigations indicate that the c. 800 Ma Neoproterozoic igneous rocks are intrusive into higher grade migmatitic gneisses but not into lower grade Itremo Group metasediments. This is based on the following arguments:

(1) Mylonitic zones occur at the contact between middle Neoproterozoic igneous intrusives and Itremo Group metasediments. Such mylonites have been observed along the western margin of the Ilaka Granitic Intrusion (west-dipping foliation with WSW- to west-plunging stretching lineation), along the eastern margin of the Itsindro Gabbro (east-dipping foliation with ENE-plunging lineation), and along the eastern and western margins of the Imorona Complex (east-dipping foliation with ENE-plunging lineation and west-dipping foliation with WSW-plunging lineation, respectively). These mylonite zones are associated with the emplacement of the Itremo Nappes. Variations in orientation of mylonites and associated stretching lineations are due to F2 folding.

(2) Most of the igneous rocks of the Imorona Complex are orthogneisses displaying a tectonic foliation. Near the margins of the Imorona Intrusive Complex this foliation is concordant with the first-phase foliation trend in adjacent lower grade Itremo Group metasedimentsmetasedimentary rocks. Both foliations are folded by large-scale N–S-trending folds (corresponding to second-phase folding in the Itremo Nappes). If middle Neoproterozoic rocks would have intruded into lower grade Itremo Group metasediments, and both units were subsequently deformed, one would expect discordant relationships to remain preserved. This does not seem to be the case and the present authors consider that the tectonic foliation in the igneous intrusive rocks of the Imorona Complex either developed in a high-strain zone related to overthrusting of lower grade Itremo Group metasediments (first-phase deformation in Itremo Nappes) or formed during an earlier deformation event that is lacking in the overlying metasediments.

(3) The absence of large contact metamorphic aureoles in lower grade Itremo Group metasediments bordering the huge middle Neoproterozoic Imorona and Itsindro Intrusives. Although Handke et al. (1999) noted the presence of relict andalusite in schists adjacent to a presumed Neoproterozoic granitoid on the eastern margin of the Imorona Complex, the locality is c. 9 km south of a dated middle Neoproterozoic intrusive. The intrusive rocks of the complicated Imorona Batholith vary greatly in composition and it cannot be excluded that they also vary in age. Therefore, it is likely that that the relict andalusite is related to the Late Neoproterozoic–Early Cambrian (570–540 Ma) magmatic event, which is widespread throughout central Madagascar. Hulscher et al. (2001) observed non-oriented large mica overgrowing a main tectonic foliation in Itremo Group metasediments. The mica overgrowth, however, appears to be a regional event and is not just restricted to the area close to the contact between the Imorona Composite Batholith and adjacent metasediments. On the basis of c. 470 Ma Ar–Ar data on such late-stage mica overgrowth (Fernandez et al. 2003), it is believed that mica growth is related to a Palaeozoic thermal event and not to a c. 800 Ma contact metamorphic event.

Discussion and conclusions

Structural, metamorphic and remote sensing studies indicate that the Itremo region consists of higher grade continental basement, lower grade Itremo Nappes and middle–Late Neoproterozoic–Early Cambrian intrusives. The Itremo Nappes contain a thick sequence of metasediments (the Itremo Group), whose stratigraphic sequence is, from bottom to top: quartzites, lower metapelites, lower metacarbonates, upper metapelites and upper metacarbonates. The sedimentary structures in the quartzites and metapelites, the well-sorted nature of the quartzites and the association with stromatolites suggest that the Itremo Group sediments were derived from a cratonic source area and deposited on a shallow continental shelf. Maximum and minimum depositional ages for the Itremo sediments are c. 1640 and c. 570 Ma, respectively. The upper age bracket is provided by U–Th–Pb ages on detrital monazite cores (Huber 2000; Fernandez et al. 2001, 2003), while the lower age bracket is provided by granites that intrude the Itremo sediments (Handke et al. 1999; Fernandez et al. 2003). During a first deformation phase, the Mesoproterozoic and/or

Neoproterozoic metasediments of the Itremo Nappes were metamorphosed at greenschist to lower amphibolite facies conditions and thrusted and internally folded on top of a higher grade continental basement. A second phase of deformation affected both the Itremo Nappes and the structurally underlying higher grade continental basement, resulting in large-scale N–S-trending folding. This overprinting gave rise to a complex outcrop distribution of lower and higher grade rocks. The Itremo Nappes are nowadays mainly preserved as klippen in the cores of large-scale N–S-trending synforms, while higher grade rocks and middle Neoproterozoic orthogneisses and metagabbros occupy the antiformal cores (Figs 9 & 13).

The present authors' structural and published isotopic age data suggest that D1 Itremo Nappe formation occurred between c. 780 Ma and c. 570 Ma. The Itremo Nappes are thrusted on top of c. 797–780 Ma (Handke et al. 1999) orthogneisses and metagabbros, and D1 structures in the Itremo Nappes are truncated by 570–540 Ma (Handke 2001; Fernandez et al. 2003) granites. The D2 deformation resulting from E–W shortening seems not to be restricted just to the Itremo region, but most likely affected the whole of the Precambrian units of Madagascar. On the basis of U–Th–Pb chemical dating of monazite, Goncalves et al. (2003) suggested an age of c. 500 Ma for N–S trending folds in central-northern Madagascar (the Andriamena series). Using the same age-dating method, Martelat (1998) and Martelat et al. (2000) proposed an age of c. 530–500 Ma for N–S striking shear zones in southern Madagascar, that were interpreted as coeval with N–S trending folds. However, more recently, de Wit et al. (2001) dated the shear zones in southern Madagascar at between 609 and 607 Ma. In the Itremo region the c. 540 Ma Vohimavo Granite (Handke 2001) seems to cut D2 structures in the Itremo Group. This suggests that D2 is older than c. 540 Ma and could therefore be coeval with the age of D3 in southern Madagascar (de Wit et al. 2001), and with the D3 N–S-striking Angavo Shear Zone (Ralison & Nédélec 1997), dated by a syntectonic granite at c. 556 Ma (Kröner et al. 1999).

Taking into account the present-day outcrop area of the allochthonous Itremo Group metasediments in a WSW–ENE direction (parallel to the D1 tectonic transport direction) and the fact that the Itremo Nappes were additionally shortened during D2, a minimum thrust distance of at least 100 km is suggested. Although the analysis of facing directions suggests a WSW-directed thrusting during D1, further reliable shear-sense criteria from high-strain zones are needed to confirm this. If thrusting was towards the WSW, the continental crust on which the sediments of the Itremo Group were deposited has to be sought in eastern Madagascar, or even India.

Lateral correlatives of the Itremo Group metasediments are not exposed east of the Itremo region: this area is dominated by high-grade gneisses that form part of the Antananarivo Block (Collins et al. 2000; Collins & Windley 2002). Late N–S-trending brittle faults (e.g. between Antsirabe and Ambositra in Fig. 7) occur east of the lower grade Itremo Group metasediments. Although their relative sense of movement is unknown, normal offset along this fault and erosion of the eastern, hanging wall could explain the absence of Itremo Group metasediments further east.

Collins et al. (2000) proposed a laterally extensive west-dipping Neoproterozoic shear zone (the Betsileo Shear Zone) in central Madagascar (>200 km in a N–S distance). On the basis of contrasting metamorphic grades on either side of this shear zone in the Itremo region and based on top-to-the-SW-directed shear sense criteria, a major extensional detachment horizon associated with crustal-scale collapse was inferred. However, the present authors interpret the contact between lower grade Itremo metasediments and higher grade basement rocks (incl. sillimanite quartzites and granitoids) as a D1 basal thrust nappe contact. Subsequently, this thrust was deformed by large-scale N–S-trending steeply inclined to upright, N–S-trending F2 folds (Fig. 13). Therefore, the orientation of the thrust contact varies in an E–W section through the Itremo region, and can dip to the west or the east depending on its orientation with respect to F2 folds. According to this study, there is no evidence for a major post-orogenic extensional detachment horizon, at least not in the Itremo region.

Research was funded by a grant from the Swiss National Science Foundation (Nr. 2100–054112.98/1). Fieldwork in Madagascar was largely facilitated through the help of M. Rakotondrazafy and R. Rambeloson. We thank A. Collins and A. Nédélec for their constructive reviews, and I. Blechschmidt, M. Herwegh, J. Kramers and I. Villa of the Institute of Geological Sciences (University of Bern) for stimulating discussions. This paper is a contribution to UNESCO–IUGS–IGCP 368 and is dedicated to the memory of Chris Powell.

References

ASHWAL, L. D. & TUCKER, R. D. 1997. Archean to Neoproterozoic events in Madagascar: implications for the assembly of Gondwana. *Terra Nova*, **9**, 163–164.

ASHWAL, L. D. & TUCKER, R. D. 1999. Geology of Madagascar: a brief outline. *Gondwana Research*, **2**, 335–339.

COLLINS, A. S. & WINDLEY, B. F. 2002. The tectonic evolution of central and northern Madagascar and its place in the final assembly of Gondwana. *Journal of Geology*, **110**, 325–339.

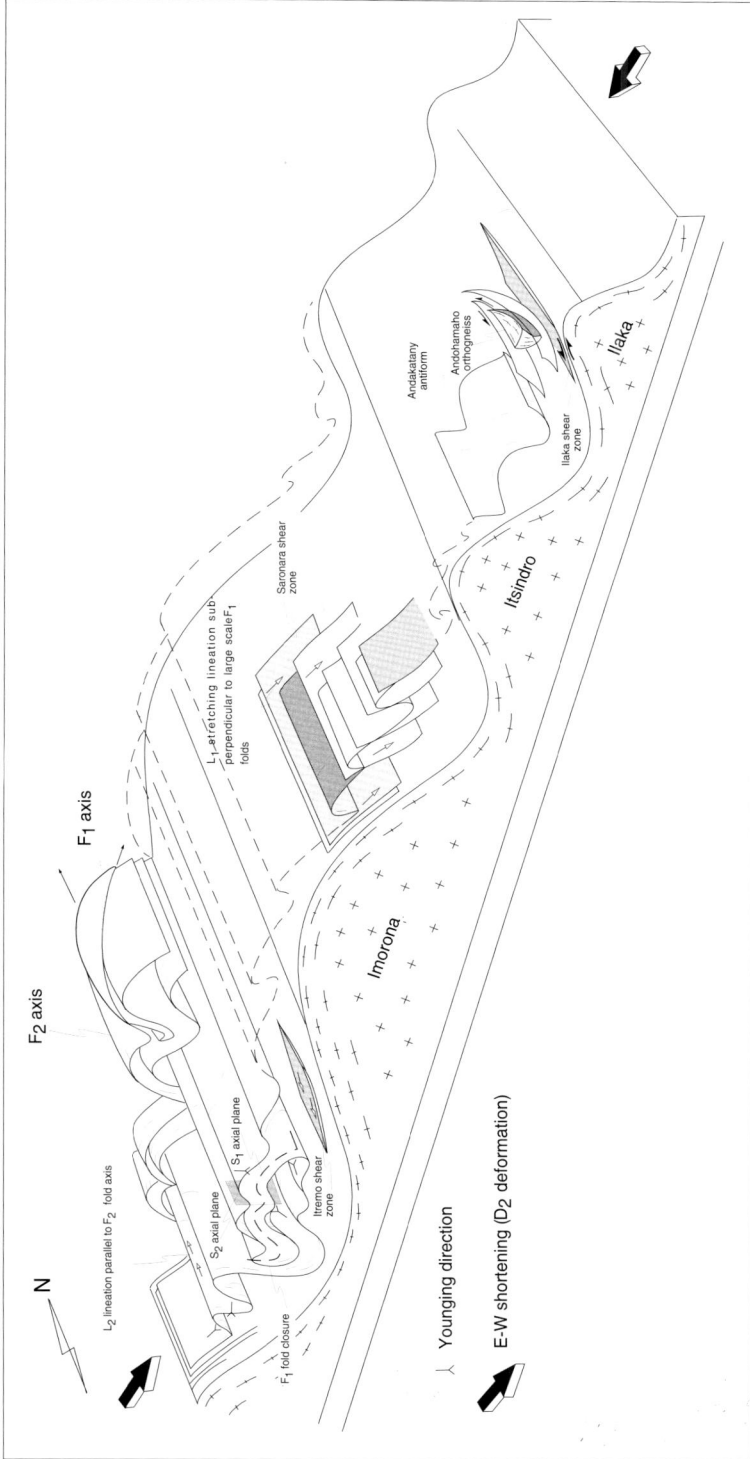

Fig. 13. Schematic block diagram through the central part of the Itremo region showing large-scale structural relations. First-phase Itremo Nappes are refolded by second-phase, large-scale N–S-trending folds. Location of frontal part of block diagram is indicated in Figure 7.

COLLINS, A. S., RAZAKAMANANA, T. & WINDLEY, B. F. 2000. Neoproterozoic crustal-scale extensional detachment in central Madagascar: implications for extensional collapse of the East African orogen. *Geological Magazine*, **137**, 39–51.

COX, R., AMSTRONG, R. A. & ASHWAL, L. D. 1998. Sedimentology, geochronology and provenance of the Proterzoic Itremo Group, central Madagascar, and implications for pre-Gondwana palaeogeography. *Journal of the Geological Society, London*, **155**, 1009–1024.

COX, R., ARMSTRONG, R. A., ASHWAL, L. D. & DE WIT, M. J. 1995. Proterozoic shelf sediments of the Itremo Group, Central Madagascar. Implication for understanding the evolution of the Mozambique belt and the assembly of East and West Gondwana. *Geogical Society of South Africa, Centennial Geocongress 1995, Extended Abstracts*, **1**, 214–217.

DASO, A. 1986. *Géologie d'une plateforme carbonatée métamorphique: valleé de la Sahatany – centre de Madagascar*. Thèse, Université Paul Sabatier, Toulouse.

DE WIT, M. J., BOWRING, S. A., ASHWAL, L. D., RANDRIANASOLO, L. G., MOREL, V. P. & RAMBELOSON, R. A. 2001. Age and tectonic evolution of Neoproterozoic ductile shear zones in southwestern Madagascar, with implications for Gondwana studies. *Tectonics*, **20**, 1–45.

EMBERGER, A. 1955. *Les terrains cristallins du pays Betsileo et de ses confins orientaux*. Thèse d'Etat, Clermont-Ferrand.

FERNANDEZ, A., HUBER, S., SCHREURS, G., VILLA, I. M. & RAKOTONDRAZAFY, M. 2001. Tectonic evolution of the Itremo Region (Central Madagascar) and implications for Gondwana assembly. *Gondwana Research*, **4**, 165–168.

FERNANDEZ, A., SCHREURS, G., VILLA, I. M., HUBER, S., KÖCHLI, M., LENGACHER, M. & RAKOTONDRAZAFY, M. 2003. Age constraints on the tectonic evolution of the Itremo region in Central Madagascar. *Precambrian Research*, in press.

GONCALVES, P., NICOLLET, C. & LARDEAUX, J.-M. 2003. Finite strain pattern in Andriamena unit (North-Central Madagascar): evidence for Late Neoproterozoic–Cambrian thrusting during continental convergence. *Precambrian Research*, in press.

HANDKE, M. J. 2001. *Neoproterozoic magmatism in the Itremo region, Central Madagascar: geochronology, geochemistry, and petrogenesis*. PhD Thesis, Washington University, St. Louis.

HANDKE, M. J., TUCKER, R. D. & ASHWAL, L. D. 1999. Neoproterozoic continental arc magmatism in west-central Madagascar. *Geology*, **27**, 351–354.

HANDKE, M. J., TUCKER, R. D. & HAMILTON, M. A. 1997. Age, geochemistry, and petrogenesis of the Early Neoproteorzoic (800–790 Ma) intrusive igneous rocks of the Itremo Region, Central Madagascar. In: COX, R. & ASHWAL, L. D. (eds) *Proceedings of UNESCO–IUGS–IGCP-348/368 International Field Workshop on Proterozoic Geology of Madagascar*. Gondwana Research Group Miscellaneous Publications, **5**, 28–29.

HOTTIN, G. 1976. Présentation et essai d'interprétation du Précambrien de Madagascar. Bull. etin du Burureau de recherches géologiques et minières (deuxième série), Section IV, **2**, 117–153.

HUBER, S. 2000. *Geologie und Geochronologie der Itremo Sedimente im Gebiet des Mont Ibity (Zentral Madagaskar)*. Diplomarbeit, Universität Bern.

HULSCHER, B., COLLINS, A. S., DAHL, K. L. ET AL. 2001. Evidence for 800 Ma and possible older deformation and plutonism in Madagascar. *Geological Society of Australia Abstracts*, **64**, 91–92.

KRÖNER, A., BRAUN, I. & JAECKER, P. 1996. Zircon geochronology of anatectic melts and residues from a high-grade pelitic assemblate at Ihosy, southern Madagascar: evidence for Pan-African granulite metamorphism. *Geological Magazine*, **133**, 311–323.

KRÖNER, A., HEGNER, E., COLLINS, A. S., WINDLEY, B. F., BREWER, T. S., RAZAKAMANANA, T. & PIDGEON, R. T. 2000. Age and magmatic history of the Antananarivo block, central Madagascar, as derived from zircon gechronology and Nd isotopic systematics. *American Journal of Science*, **300**, 251–288.

KRÖNER, A., JAECKEL, P., WINDLEY, B. F., BREWER, T. & RAZAKAMANANA, T. 1999. New zircon ages and regional significance for the evolution of the Pan-African orogen in Madagascar. *Journal of the Geological Society, London*, **156**, 1125–1135.

LONGHI, I., MAZZOLI, C. & SGAVETTI, M. 2000. Determination of metamorphic grade in siliceous muscovite-bearing rocks in Madagascar using reflectance spectroscopy. *Terra Nova*, **12**, 21–27.

MARTELAT, J.-E. 1998. *Evolution thermomécanique de la croûte inférieure du sud de Madagascar*. Thèse, Université Blaise Pascal, Clermont Ferrand.

MARTELAT, J.-E., LARDEAUX, J.-M., NICOLLET, C. & RAKOTONDRAZAFY, R. 2000. Strain pattern and late Precambrian deformation history in southern Madagascar. *Precambrian Research*, **102**, 1–20.

MEERT, J. G., HALL, C., NÉDÉLEC, A. & MADISON RAZANATSEHENO, M. O. 2001a. Cooling of a late synorogenic pluton: evidence from laser K-feldspar modelling of the Carion granite, Madagascar. *Gondwana Research*, **4**, 541–550.

MEERT, J. G., NÉDÉLEC, A., HALL, C., WINGATE, M. T. D. & RAKOTONDRAZAFY, M. 2001b. Paleomagnetism, geochronology and tectonic implications of the Cambrian-age Carion granite, Central Madagascar. *Tectonophysics*, **340**, 1–21.

MOINE, B. 1966. Grands traits structuraux du massif schisto-quartzo-calcaire (Centre-Ouest de Madagascar). *Comptes Rendus de la semaine géologique de Madagascar*, 93–97.

MOINE, B. 1967. Relations stratigraphiques entre la série Schisto–Quartzo–Calcaire et les gneiss environnants (centre-ouest de Madagascar). *Comptes Rendus de la semaine géologique de Madagascar*, 49–53.

MOINE, B. 1974. Caractères de sédimentation et de métamorphism des séries précambriennes épizonales à catazonales du centre de Madagascar (région d'Ambatofinandrahana). *Sciences de la Terre Mémoires*, **31**, 1–293.

NÉDÉLEC, A., PAQUETTE, J. L., BOUCHEZ, J. L., OLIVIER, P. & RALISON, B. 1994. Stratoid granites of Madagascar: structure and position in the Panafrican orogeny. *Geodinamica Acta*, **7**, 48–56.

NÉDÉLEC, A., RALISON, B., BOUCHEZ, J. L. & GRÉGOIRE, V.

2000. Structure and metamorphism of the granitic basement around Antananarivo: a key to the Pan-African history of central Madagascar and its Gondwana connections. *Tectonics*, **19**, 997–1020.

NÉDÉLEC, A., STEPHENS, W. E. & FALLICK, A. E. 1995. The Panafrican stratoid granites of Madagascar: alkaline magmatism in a post collisional extensional setting. *Journal of Petrology*, **36**, 1367–1391.

PAQUETTE, J. L. & NÉDÉLEC, A. 1998. A new insight into Pan-African tectonics in the East–West Gondwana collision zone by U–Pb zircon dating of granites from central Madagascar. *Earth and Planetary Science Letters*, **155**, 45–56.

PAQUETTE, J. L., NÉDÉLEC, A., MOINE, B. & RAKOTONDRAZAFY, M. 1994. U–Pb, single zircon Pb-evaporation, and Sm–Nd isotopic study of a granulite domain in SE Madagascar. *Journal of Geology*, **102**, 523–538.

RALISON, B & NÉDÉLEC, A. 1997. Contrasted Pan-African structures near Antananarivo (Madagascar). *In*: COX, R. & ASHWAL, L. D. (eds) *Proceedings of UNESCO–IUGS–IGCP-348/368 International Field Workshop on Proterozoic Geology of Madagascar*. Gondwana Research Group Miscellaneous Publications, **5**, 83–84.

RAMSAY, J. 1967. *Folding and Fracturing of Rocks*. McGraw Hill, New York.

RAOELISON, I. L. 1997. *Structure and metamorphism of the Itremo Group, Central Madagascar*. MSc Thesis, Rand Afrikaans University, Johannesburg.

STERN, R. J. 1994. Arc assembly and continental collision in the Neoproterozoic East African Orogen: implications for the consolidation of Gondwanaland. *Annual Review of Earth and Planetary Sciences*, **22**, 319–351.

TROTTEREAU, G. 1969. Note préliminaire relative à la présence de stromatolithes dans les cipolins de la série supérieure du socle, à l'Est et au Nord d'Ambatofinandrahana. *Comptes Rendus de la semaine géologique de Madagascar*, 131–132.

TUCKER, R. D., ASHWAL, L. D., HANDKE, M. J. & HAMILTON, M. A. 1997. A geochronologic overview of the Precambrian rocks of Madagascar: a record from the middle Archean to the late Neoproterozic. *In*: COX, R. & ASHWAL, L. D. (eds) *Proceedings of UNESCO–IUGS–IGCP-348/368 International Field Workshop on Proterozoic Geology of Madagascar*. Gondwana Research Group Miscellaneous Publications, **5**, 99–100.

TUCKER, R. D., ASHWAL, L. D., HANDKE, M. J., HAMILTON, M. A., LEGRANGE, M. & RAMBELOSON, R. A. 1999. U–Pb geochronology and isotope geochemistry of the Archean granite–greenstone belts of Madagascar. *Journal of Geology*, **107**, 135–153.

WINDLEY, B. F. & RAZAKAMANANA, T. 1996. The Madagascar–India connection in a Gondwana framework. *In*: SANTOSH, M. & YOSHIDA, M. (eds) *The Archaean and Proterozoic Terrains of Southern India within East Gondwana*. Gondwana Research Group Memoir, **3**, 25–37.

A review of the evolution of the Mozambique Belt and implications for the amalgamation and dispersal of Rodinia and Gondwana

G. H. GRANTHAM[1], M. MABOKO[2] & B. M. EGLINGTON[1]

[1]*Council for Geosciences, P/Bag X112, Pretoria 0001, South Africa*
(e-mail: grantham@geoscience.org.za)
[2]*Department of Geology, University of Dar-es-Salaam, Dar-es-Salaam, Tanzania*

Abstract: Geochronological, isotopic, lithological and structural data from the Mozambique Belt, and its extensions in Antarctica, Sri Lanka, India and Madagascar, are summarized and reviewed within a Gondwana framework. Much of the southern Mozambique Belt is dominated by Rodinian–Grenvillian-age juvenile magmatism and crustal genesis, with a strong metamorphic overprint during the Pan-African. Magmatism at $c.$ 800 Ma, possibly related to Rodinian fragmentation, is restricted to the Zambesi Belt, a few examples in northern Mozambique, southern Malawi, Tanzania, Madagascar and the Rayner Complex in Antarctica. Significantly, no crust of this age is recognized in western Dronning Maud Land. Amalgamation of Gondwana initially involved the closure of the Mozambique Ocean between the Tanzanian–Congo Cratons and the Madagascar–India–Enderby Cratons forming the East African Orogeny between $c.$ 580 and $c.$ 800 Ma. This was followed by the $c.$ 500–580 Ma transpressional sinistral collision of the combined Kalahari–East Antarctic Cratons and associated Grenville-age belts in the south, with the northern block comprising the combined Tanzanian–Congo–Madagascar–India–Enderby Block.

The Mozambique Belt was first named by Holmes (1951), who recognized a structural discontinuity between the Tanzanian Craton and gneisses to the east, which were found to have an age of $c.$ 1300 Ma. Subsequent work by many authors over wide areas using Rb–Sr and K–Ar methods yielded ages of $c.$ 500–600 Ma, which resulted in Kennedy (1964) coining the term 'Pan-African thermotectonic episode' and resulted in the perception that the Mozambique Belt was largely formed at that time. Subsequent work (e.g. Pinna *et al.* 1993; Moller *et al.* 1998, 2000; Manhica *et al.* 2001) has shown most of these $c.$ 550 Ma ages are metamorphic blocking temperatures, or cooling ages, and that in many areas the crystallization ages of the metamorphosed and deformed rocks are commonly $c.$ 1000–1200 Ma and older.

At the same time that geochronological methods improved, resulting in the recognition of a more complex tectonothermal history for the Mozambique Belt, so plate reconstruction models evolved. Early studies, which concentrated on reconstruction of Gondwana fragments that dispersed at $c.$ 180–190 Ma, were followed by current studies that concentrate on the amalgamation of Gondwana during the Pan-African, as well as concentrating on the amalgamation of Rodinia at $c.$ 1000–1200 Ma and its subsequent dispersal into the fragments which later coalesced to form Gondwana. The Mozambique Belt and its extensions southwards into Antarctica are viewed by some researchers as being the site of amalgamation of East and West Gondwana at $c.$ 550–600 Ma (Shackleton 1996; Wilson *et al.* 1997; Jacobs *et al.* 1998), whereas others see the locus of amalgamation being partially located in the Mozambique Belt in the north, then turning westward along the Zambezi Belt and extending into the Damara Belt in Namibia (e.g. Trompette 1994). The first option would require the Kalahari and Tanzanian–Congo Cratons as having behaved as a unitary contiguous block since Rodinian amalgamation at $c.$ 1000–1200 Ma, with both blocks being viewed as part of West Gondwana. In this configuration, the Mozambique Ocean, which was formed by Rodinian fragmentation at $c.$ 800 Ma, would have been positioned along the eastern margin of the combined Kalahari–Congo–Tanzanian Cratons, with amalgamation of Gondwana having occurred along the Mozambique Belt, including the East African Orogen (EAO) of Stern (1994) and its extensions into Dronning Maud Land, Antarctica (Jacobs *et al.* 1998).

The second option would have the Kalahari and East Antarctic Blocks behaving as a single entity, at least along the western Dronning Maud Land interface, having been joined during Rodinian amalgamation and forming part of East Gondwana. In Hoffman's (1991) option, the Mozambique Ocean would have separated the Kalahari Craton in the south from the Congo–Tanzanian Craton in the north, with Gondwana amalgamation having occurred along the Mozambique, Zambesi and Damara Belts.

This paper summarizes current literature with the aim of understanding the evolution of the Mozambique Belt, concentrating on those aspects

From YOSHIDA, M., WINDLEY, B. F. & DASGUPTA, S. (eds) 2003. *Proterozoic East Gondwana: Supercontinent Assembly and Breakup*. Geological Society, London, Special Publications, **206**, 401–425. 0305-8719/03/$15 © The Geological Society of London.

which shed light on lithological and structural characteristics which may define palaeoplate boundaries. The paper concentrates on the areas approximately south of the equator in an African framework and does not review possible Mozambique Belt extensions in North Africa and Saudi Arabia, i.e. the northern portion of the EAO of Stern (1994). The lithologies considered to be important for palaeoplate boundary identification include those which one might associate with plate margins, i.e. carbonates which commonly form on continental shelves, volcanic-arc lithologies typical of convergent subduction zone settings [trondhjemite–tonalite–granite (TTG) suites] eclogite occurrences typical of the exhumation of the root zones of orogenic belts containing deeply buried ocean-floor sequences and, similarly, ophiolite sequences which are interpreted as ocean-floor lithologies which have been obducted onto continental margins at convergent margin orogenic belts. The structural aspects include large-scale fault and/or shear structures including strike-slip, oblique-slip (transpressional or transtensional) and dip-slip (normal or thrust fault) zones, as well as fold or orogenic belts. For both of the structural and lithological aspects, the ages of the lithologies, their protoliths and the structures and associated metamorphic assemblages are crucial characteristics in determining where original plate boundaries were located. From the comparisons of data it is hoped to be able to recognize lithologies and structures formed during the amalgamation of Rodinia at c. 1000–1200 Ma, those formed during the fragmentation of Rodinia at c. 800 Ma and those generated during the amalgamation of Gondwana at c. 470–600 Ma.

Consequently, this paper concentrates on these aspects with the lithologies and their crystallization/deposition/source ages being considered concurrently. Similarly, the structural and metamorphic aspects will be considered in tandem with ages related to deformation, metamorphism and cooling.

The paper looks at the Mozambique Belt in a broad sense, combining data from juxtaposed areas in Antarctica, Madagascar, India and Sri Lanka. Besides drawing conclusions on the evolution of the Mozambique Belt, problems and areas which require further study to refine understanding of the Mozambique Belt, Gondwana dispersal and amalgamation, and Rodinia dispersal and amalgamation will be identified. The data collected and reviewed are collated on a reconstruction of Gondwana from Lawver et al. (1998) (Fig. 1), which places the various continental fragments as close as possible, allowing overlap where varying degrees of post-Gondwana sedimentation have altered, obscured and extended continental margins subsequent to Gondwana fragmentation. Recent publications which support the use of reconstructions based on the tightest possible fit of crustal blocks include Tikku et al. (2002), Reeves et al. (2002). Figure 1 shows the necessary location of various localities referred to in the text.

LITHOLOGIES

Magmatic arc sequences

Subduction zones are commonly characterized by rocks generated in magmatic arcs which typically have calc-alkaline signatures. The rocks originated as volcanic piles dominated by andesites and dacites with subordinate basalts and rhyolites or as their plutonic equivalents in the form of TTG suites. Depending on whether the magmatic arcs were located on juvenile crust (intra-oceanic island arcs) or on significantly older continental crust (continental arcs), the ages of the protoliths of the rocks can vary significantly. The high heat flows normally associated with magmatic arcs cause the high-temperature/low-pressure metamorphism typical of such areas, producing mineral assemblages characteristic of upper amphibolite to granulite facies metamorphism, generally accompanied by extensive migmatization. The distribution of such calc-alkaline TTG suites of Grenvillian and younger ages are shown in Figure 2.

In the Kenya–Tanzania sector, the Eastern Granulites exposed in the Pare Mountains, the Usambara Mountains, and the Wami and Furua areas include calc-alkaline orthogneisses (Maboko 2001a; Maboko & Nakamura 2002) which have been metamorphosed to granulite facies. The rocks have juvenile Grenvillian Sm–Nd model ages and contain zircons which yield crystallization ages of c. 725–850 Ma (Muhongo et al. 2001) with marginally younger (c. 650 Ma) metamorphic overprints. These rocks are therefore comparable in age to those of the EAO described by Stern (2002) from northern Kenya, Ethiopia, Sudan and Egypt.

Progressing southwards towards northern Mozambique, there is an area apparently devoid of TTG suite rocks. In northern Mozambique, rocks having calc-alkaline chemistry and consisting of tonalitic and trondjhemitic gneisses form a significant component of the Nampula Supergroup (Pinna et al. 1993). The rocks yield Rb–Sr whole-rock dates of between 1100 and 800 Ma (Pinna et al. 1993; Cadoppi et al. 1987) and mostly have low $^{87}Sr/^{86}Sr$ ratios typical of juvenile crustal additions. The rocks are commonly migmatized and locally have orthopyroxene developed in zones of patchy metamorphic charnockite. Kröner et al. (2002) report calc-alkaline rocks of both c. 900 and c. 550 Ma in southern Malawi, although no detailed description of the

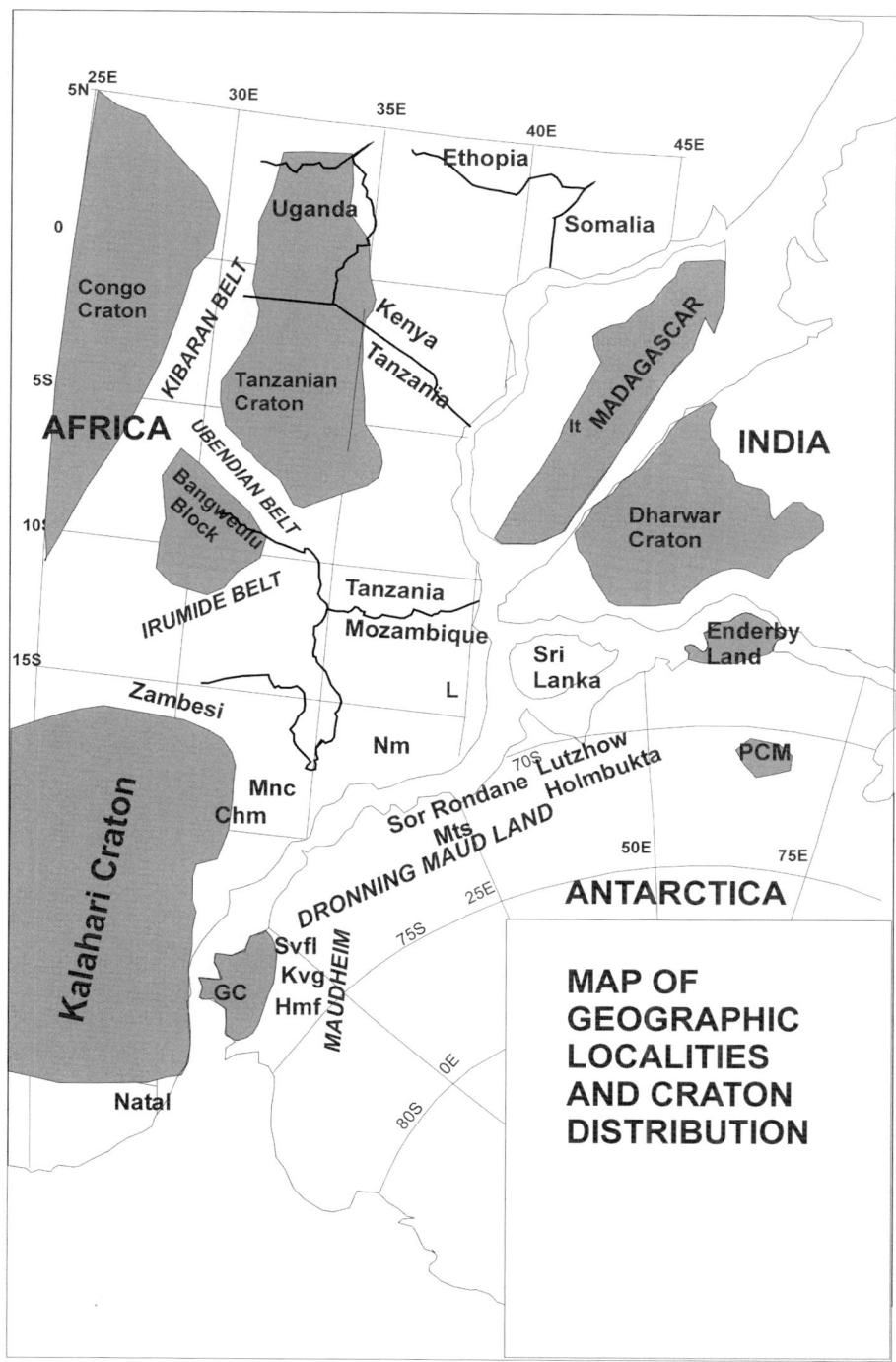

Fig. 1. Locality map of various localities mentioned in the text. Chm, Chimanimani, Zimbabwe; Mnc, Manica, Mozambique; Nm, Nampula; It, Itremo; Svfl, Sverdrupfjella; Kvg, Kirwanveggan; Hmf, Heimefrontfjella; PCM, Prince Charles Mountains; L, Lurio. GC, Grunehogna Craton. The heavy lines are international boundaries between countries; the shaded areas represent Palaeoproterozoic–Archaean Cratons.

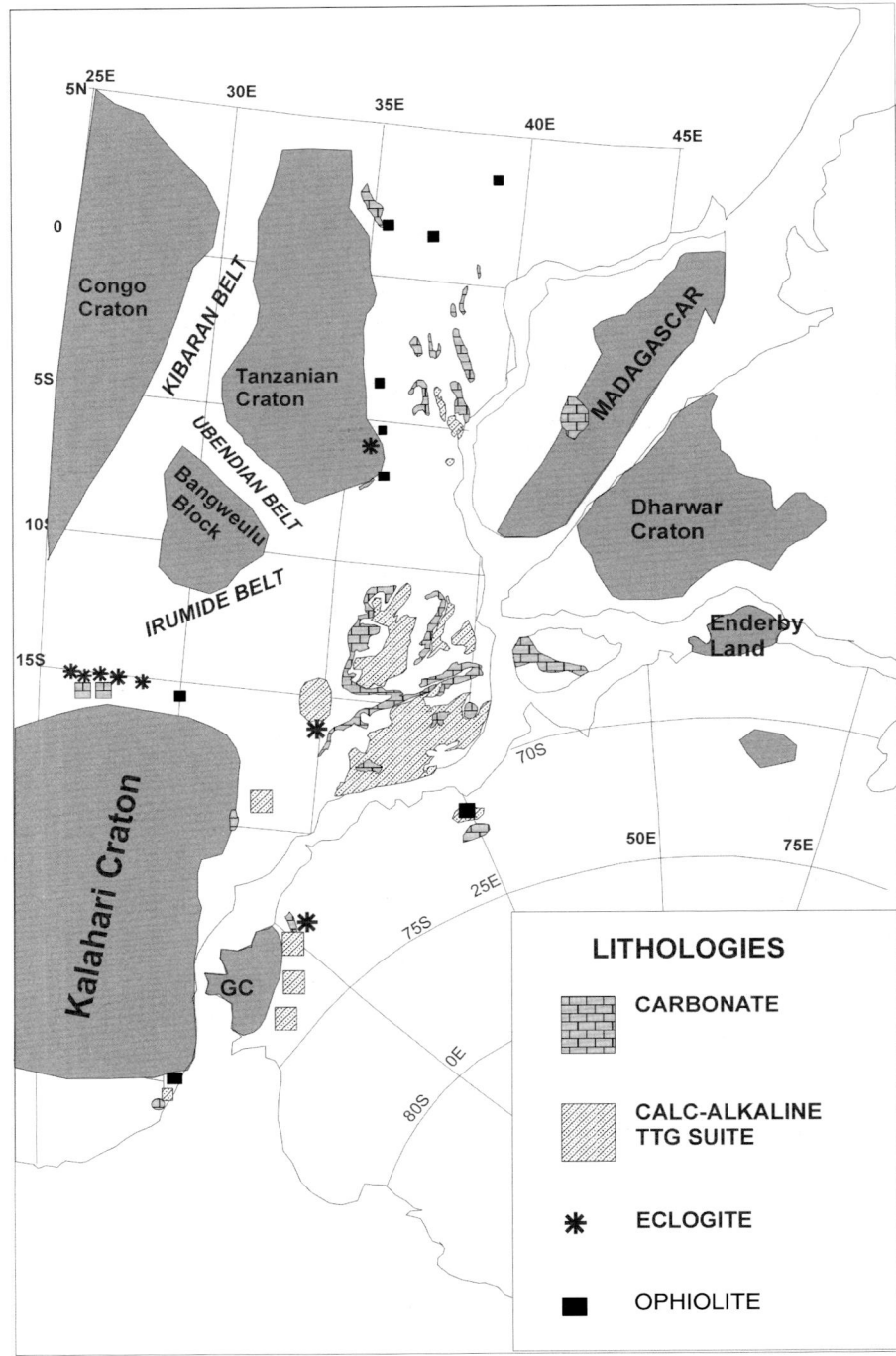

Fig. 2. Gondwana reconstruction based on Lawver *et al.* (1998) showing the distribution of formations containing carbonates, calc-alkaline trondhjemite–tonalite–granite (TTG) suites, eclogites and ophiolites.

chemistry is provided and consequently the tectonic setting of these rocks is uncertain.

In central Mozambique, rocks with calc-alkaline signatures, mineralogy and chemistry typical of TTG suites with relatively low $^{87}Sr/^{86}Sr$ ratios are exposed c. 60 km east of Manica, near Chimoio (Manhica 1998; Manhica et al. 2001). These rocks have yielded a U–Pb SHRIMP crystallization age of c. 1110 Ma (Manhica et al. 2001).

In western Dronning Maud Land (WDML) rocks that have these characteristics include the Grey Gneiss Complex in Sverdrupfjella (Grantham et al. 1988) and the Kvervelknatten Orthogneiss (Grantham et al. 1995). Rocks from the former area are interpreted as forming part of a layered volcano-sedimentary supracrustal sequence (Grantham 1992), whereas those from the latter area are clearly plutonic and intrusive in origin (Grantham et al. 1995). Unpublished U–Pb SHRIMP ages on zircon from these rocks yield crystallization ages of c. 1140 Ma. In both Sverdrupfjella and Kirwanveggen, the rocks are migmatitic and preserve mineral assemblages typical of upper amphibolite facies metamorphism. In Heimefrontfjella, rocks of similar composition and age are interpreted as representing metamorphosed volcanic rocks (Arndt et al. 1991; Jacobs et al. 1996).

In Natal, significant portions of the c. 1200 Ma Mzumbe Terrane (Thomas 1989) are underlain by the Mzumbe Suite, which is interpreted to be a plutonic TTG suite with associated metavolcanic rocks (Thomas & Eglington 1990). The distribution of these TTG calc-alkaline rocks in Natal, WDML and the Haag Nunatacks provided the basis of comparison and correlation described by Grantham et al. (1997) and Wareham et al. (1998). Almost all the gneisses described by these authors (except the Grey Gneiss Complex in Sverdrupfjella) are characterized by low $^{87}Sr/^{86}Sr$ ratios typical of juvenile crustal additions. The ages of the rocks in Antarctica are c. 1140 Ma (Grantham unpublished data) whereas those from Natal are c. 1200 Ma (Thomas & Eglington 1990).

In central Sør Rondane, central Dronning Maud Land (CDML), Osanai et al. (1992) reported metamorphosed volcanosedimentary gneisses in which the orthogneisses are largely basic to intermediate in composition. The basic gneisses are described as having mixed characteristics, suggesting mid-ocean ridge basalt (MORB) and island-arc basalt origins, whereas the intermediate gneisses suggest island-arc andesitic and continental-margin andesitic origins.

The distribution of such rock types in other sectors of East Antarctica, Madagascar, India and Sri Lanka is uncertain. Handke et al. (1999) suggest that the c. 770 Ma granitoids in central Madagascar are related to a continental-arc setting; however, they recognize that the absence of typical TTG suite rocks may require an alternative explanation, particularly since the rocks show a bimodal basic–acid distribution more typical of extensional settings.

The above data indicate therefore at least two generations of juvenile TTG suite rocks, i.e. those extending down from the EAO to Tanzania of c. 750 Ma age and those further south along the Namaqua–Natal–Maud–Mozambique–CDML Belt of c. 1200–1100 Ma age. The former contributed to the amalgamation of Gondwana whereas the latter represent lithologies are related to Rodinian/Grenvillian amalgamation. This is consistent with the conclusion of Kröner (2001) that no reliable crystallization ages c. >900 Ma have been recognized from juvenile rocks in the Mozambique Belt north of northern Mozambique.

Carbonates

Carbonates constitute only c. 2% of rocks in the Earth's crust (Wyllie 1971) and their genesis, particularly in the Phanerozoic, is largely restricted to shallow-marine environments (Sellwood 1986). Recognizing their limited contribution to the crustal lithology budget and their environment of genesis, their distribution can therefore be useful in defining palaeocontinental margins. The distribution of supracrustal sequences with significant contents of carbonate-bearing rocks is shown in Figure 2.

In Tanzania, extensive metacarbonates exist in the Uluguru Mountains (associated with metapelites and juvenile orthogranulites), in the Wami River area (flanking juvenile calc-alkaline orthogranulites), Mahenge (south of the Uluguru Mountains) and in Merelani (NW of the Pare Mountains). The Mahenge occurrence corresponds to the inferred boundary between juvenile Neoproterozoic crust and reworked Archaean crust, whereas the Merelani occurrence appears to be associated with reworked Archaean basement. Extensive carbonates are also present in the Turoka Series of southeastern and south-central Kenya (Mosley 1993).

In northern Mozambique and southern Tanzania, Pinna et al. (1993) report significant metacarbonates in the Chiure Supergroup, which is additionally characterized by quartzites, graphite schists and conglomerates. These rocks occupy extensive areas north of the Lurio Shear Zone with many of the exposures south of the Lurio Belt being interpreted as klippe thrust southwards from the NW. The distribution of the Chiure Supergroup is shown in Figure 2 after Pinna et al. (1993).

In central Mozambique and southeastern Zimbabwe, strongly deformed quartzites, carbonates and phyllosilicate schists of the Frontier Formation (Watson 1969), which are correlated as a lower, underlying facies of the Umkondo Group

(Watson 1969), are intruded by dolerites that have yielded an age of 1105 Ma (Hanson et al. 1998).

Carbonates form significant components of the c. 1400 Ma oceanic crust recognized in the Zambesi Belt (Oliver et al. 1998). Significantly younger (c. 775 Ma; Trompette 1994) carbonate sequences flank the northern and southern (Nama) margins of the Damara Fold Belt in Namibia.

In Dronning Maud Land (DML), the Fuglefjellet Formation comprises extensive metacarbonates with subordinate metaconglomerate, quartzites, calc-silicates and quartzofeldspathic gneisses (Hjelle 1972; Grantham 1992). These carbonates are interlayered with the c. 1140 Ma old calc-alkaline TTG suit gneisses. No exposures of similar rocks are recorded in Kirwanveggen and Heimefrontfjella, the absence of which can partly be ascribed to poor exposure, as well as their location in the footwall of the thrust-fault stack in Sverdrupfjella. In Kirwanveggen and Heimefrontfjella it is possible that only the upper portions of the thrust stack are exposed.

To the east in CDML, relatively extensive carbonate-bearing lithologies have been reported in Sør-Rondane (Van Autenboer & Loy 1966; Van Autenboer 1972; Shiraishi et al. 1991). Van Autenboer & Loy (1966) suggested that the carbonate sequences in Sør-Rondane could be younger than gneisses they appeared to overlie discordantly. Similar relationships have been observed in Sverdrupfjella; however, in that area, low-angle thrust faults are commonly layer-parallel and appear to be located along carbonate layer boundaries (Grantham unpublished data.)

In Natal, the Marble Delta and the Muckelbraes Formation represent similar carbonate-dominated rocks in the Natal Metamorphic Province (Thomas 1989). Due to limited distribution and poor exposures, their relationship to the calc-alkaline TTG suite Mzumbe Granodiorite Gneisses to the north is uncertain; however, isotope studies by Eglington & Harmer (1987) suggest that they are of similar age.

In central Madagascar, carbonates with stromatolitic structures form a significant component of the Itremo Group, which also includes metapelitic phyllosilicate schists and quartzites (Cox et al. 1998). The age of the Itremo Group has been constrained between c. <1855 Ma, the age of the youngest detrital zircons recognized, and >c. 833 Ma, the age of metamorphism recognized in the rims of the zircons (Cox et al. 1998). The Itremo Group (Cox et al. 1998) is virtually identical in composition and appearance to the Frontier Formation in eastern Zimbabwe and central Mozambique.

In Sri Lanka, the centrally located Highland Group contains paragneissic sequences that include metapelites, metapsammites, marbles and quartzites (Yoshida & Vitanage 1993). These authors recognized that these metasediments might correlate with rocks of the Ongul and Skallen Groups in the Lützhow-Holmbukta area of East Antarctica.

The distribution of the carbonate-bearing lithologies presents an enigma in that they appear to be distributed: (1) along the eastern margins of the Kalahari and Tanzanian Cratons in a N–S orientation; (2) E–W through the central highlands of Sri Lanka into East Antarctica and through the Zambesi Belt; and (3) NE–SW along the margins of the Damara Orogen in Namibia. It is possible/probable that the two orientations are of different ages. Those in the Zambesi and Natal Belts of southern Africa and Maud Province, and the extensions through Sri Lanka into CDML in Antarctica, are probably of Grenvillian age (Eglington & Harmer 1987). Those in the north in Tanzania are possibly of younger age, being similarly distributed to the TTG suites in that area. The ages of the carbonates in the Frontier Formation (Watson 1969) of eastern Zimbabwe/central Mozambique and the Itremo Group in Madagascar (Cox et al. 1998) is uncertain. The latter are clearly older than c. 800 Ma (Cox et al. 1998), whereas the former are c. >1100 Ma, because of their lower stratigraphic position in the Mkondo Group which is intruded by c. 1100 Ma gabbros and diorites (Hanson et al. 1998).

Eclogites and ophiolites

Eclogites in metamorphic belts are commonly perceived as originating from mafic rocks, possibly of ocean-floor basalt origin which have been buried to depths c. >10kb (Newton 1986) and exhumed during plate margin collisional orogenies. The preservation of eclogitic assemblages in orogenic belts has been ascribed to deep burial in subduction settings followed by rapid exhumation to prevent re-equilibration during exhumation (Behrmann & Ratschbacher 1989), with the exhumation commonly being facilitated by thrust faulting at convergent margins. Ophiolites are commonly perceived as originating from ultramafic rocks which have either been exhumed or obducted during plate margin collisional orogenies. Consequently, the distribution and age of these lithologies may give an indication to possible palaeoplate margins; however, the scarcity of such rocks and the difficulty related to determining their age of crystallization or formation is a limiting factor in their usefulness. In addition, the occurrence and distribution of eclogites is limited in the Precambrian, in contrast to their greater abundance in the Phanerozoic. The limited occurrence in the Precambrian has commonly been ascribed to deeper levels of erosion in the case of ophiolites and to high heat flows contributing to the re-equilibration to lower pressure assemblages in the case of eclogites.

The distribution of eclogites in the Mozambique and neighbouring belts is shown in Figure 2. Data sources include Vrana et al. (1975), Andreoli & Hart (1985), Mosley (1993), Groenewald (1995), Moller et al. (1995) and Dirks (1997). The ages of the eclogite are not always clear in the context of this paper; e.g. those reported by Andreoli & Hart (1985) are described as being introduced into the crust between 1.1 and 0.5 Ga ago. Similarly, those reported by Vrana et al. (1975) and Dirks (1997) are hosted in the Zambezi Belt, the age of which is uncertain with ages varying between c. 500 and 1300 Ma (Oliver et al. 1998; Johnson & Oliver 2000). Those reported by Groenewald (1995) are hosted in gneisses clearly of c. 1100 Ma age. The eclogites reported by Moller et al. (1995) are c. 2000 Ma old and therefore are clearly not related to the time period under review here.

Two unequivocal ophiolites (Moyale and Baragoi) have been identified in northwestern Kenya (Shackleton 1986; Stern 1994 and refs cited therein). The Baragoi Ophiolite occurs as a dismembered body along the interface between the Mozambique Belt and the Tanzania Craton in western Kenya (Vearncombe 1983), whereas the Moyale Ophiolite occurs as a disrupted body along the Kenya–Ethiopia border (Kazmin 1976). The position of the Baragoi Ophiolite is particularly significant as it occurs along a clear boundary separating reworked Archaean crust in the west and Neoproterozoic crust, dated by the c. 1020–1190 Ma depleted mantle mean crustal ages of the Sekerr Pelites and Amphibolite. Shackleton (1986) shows three smaller possible examples in southern Tanzania. These rocks are therefore probably of similar age to the TTG suite gneisses

In Natal, the mafic to ultramafic rocks of the Tugela Terrane (Thomas 1989) have been interpreted as being a possible ophiolite (Matthews 1972) obducted onto the southern margin of the Kaapvaal Craton (Barkhuizen & Matthews 1990). However, the chemistry of these rocks reflects limited MORB-like characteristics, being more comparable to island-arc volcanic sequences (Arima et al. 2001). The gneisses are intruded by the c. 1200 Ma old Tugela Rand layered intrusion (Wilson 1990) and yield Ar–Ar (hornblende) metamorphic ages of >1100 Ma (Jacobs et al. 1997).

Ultramafic blocks contained within metasedimentary gneisses in the Lützow-Holm Complex in CDML, Antarctica, are interpreted as being of oceanic crustal origin and therefore possibly ophiolitic (Shiraishi & Kojima 1987).

The distribution of rocks of eclogitic and ophiolitic affinity in Madagascar, Sri Lanka, southern India and the Rayner Complex of Antarctica is uncertain.

Isotopic data

Crystallization and metamorphic ages

Figures 3 & 5 show the distribution of metamorphic ages in the Mozambique Belt and surrounding areas. These ages include data from isotopic systems which generally have blocking temperatures in the field of metamorphism, i.e. <650°C, and include most mineral-based data using Ar–Ar, K–Ar, Rb–Sr and Sm–Nd methods. Data from single zircon U–Pb samples from migmatitic veins are also included. Data sources are from Andriessen et al. (1985), Maboko et al. (1985, 1989), Eglington et al. (1989), Moyes and Barton (1990), Grantham et al. (1991), Santosh et al. (1992), Shiraishi & Kagami (1992), Shiraishi et al. (1992, 1994), Moyes (1993), Moyes et al. (1993, 1997), Shackleton (1993 and refs cited therein), Muhongo & Lenoir (1994), Maboko & Nakamura (1995), Moyes & Groenewald (1996), Jacobs et al. (1997), Kröner et al. (1997), Jacobs et al. (1998, 1999), Dirks et al. (1999), Jamal et al. (1999), Vinyu et al. (1999), Elworthy et al. (2000), Manhica et al. (2001), Markl et al. (2000 and refs cited therein), Moller et al. (2000), Maboko (2001a,b) and Muhongo et al. (2001).

The data are not complicated, with almost all the ages falling in the Pan-African period of 450–700 Ma. It is this simple pattern of cooling ages which led to the description of the Pan-African thermotectonic episode by Kennedy (1964), which many workers now interpret as having formed during the amalgamation of Gondwana. Limited metamorphic data from the 900–1100 Ma period are recognized from Natal (Jacobs et al. 1997; Elworthy et al. 2000) and along the eastern margin of the Kalahari Craton in central Mozambique, where a Grenville-age overprint is recognized in biotite from Archaean gneisses (Manhica et al. 2001).

Figures 4 & 5 show the distribution of igneous crystallization ages based largely on single zircon analyses, but also include whole-rock Rb–Sr data from northern Mozambique and Sm–Nd data from central Mozambique and Madagascar where single zircon data are sparse. Data sources include Nicolaysen & Burger (1965), Coolen (1980), Coolen et al. (1982), Hart & Barton (1984), Sacchi et al. (1984), Maboko et al. (1985), Black (1986, 1988), Black et al. (1987), Muhongo et al. (1987, 2001), Hanson et al. (1988, 1993, 1994), Eglington & Kerr (1989), Eglington et al. (1989, 1993, 2000a, b), Naidoo et al. (1989), Bigioggero et al. (1990), Moyes & Barton (1990), Thomas & Eglington (1990), Arndt et al. (1991), Costa et al. (1992, 1994), Grantham & Eglington (1992), Santosh et al. (1992), Shiraishi & Kagami (1992), Shiraishi et al. (1992, 1994), Thomas et al. (1992, 1993a, b, 1999),

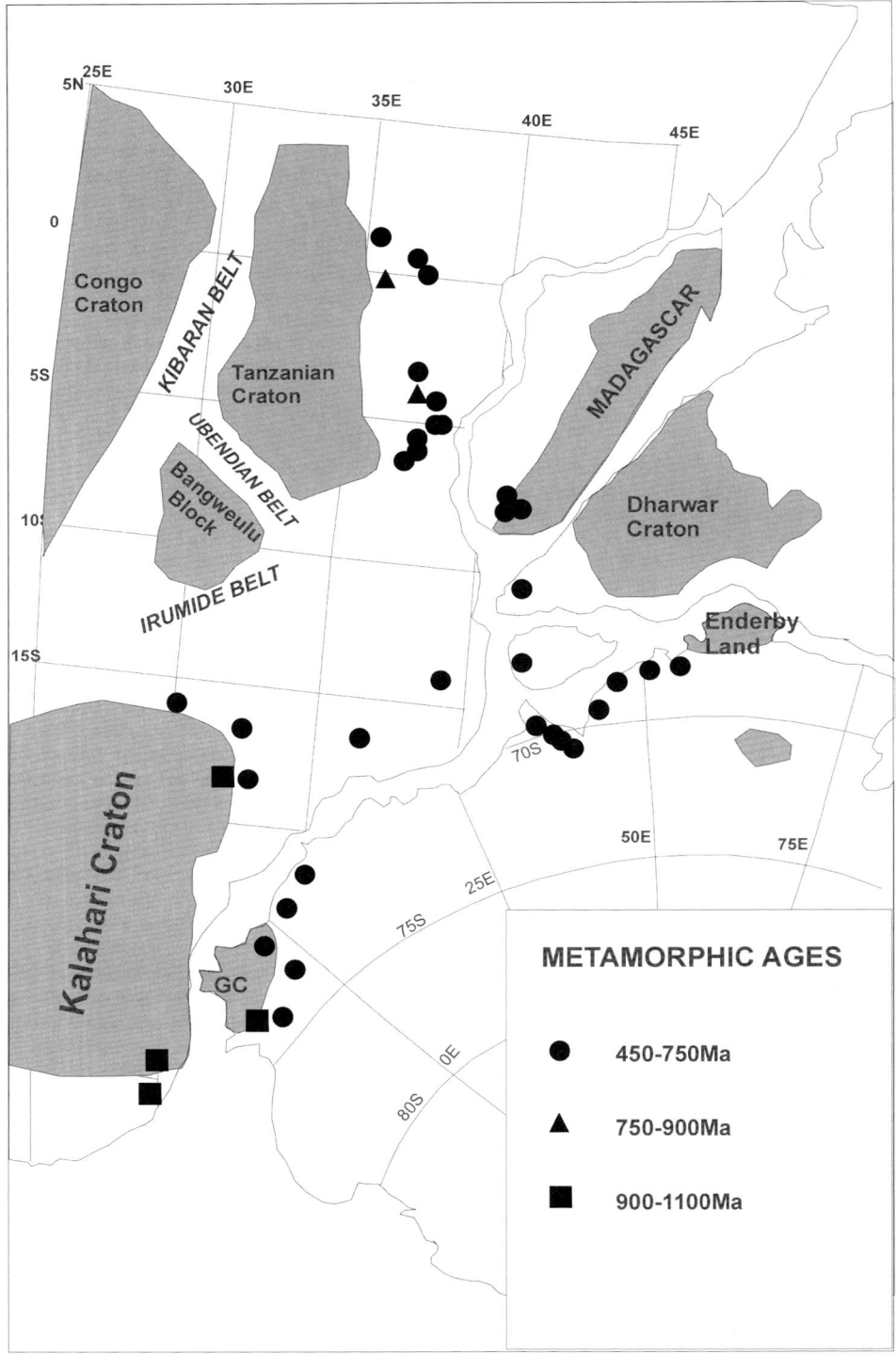

Fig. 3. Gondwana reconstruction based on Lawver *et al.* (1998) showing the locations of geochronological sample points which yielded metamorphic ages.

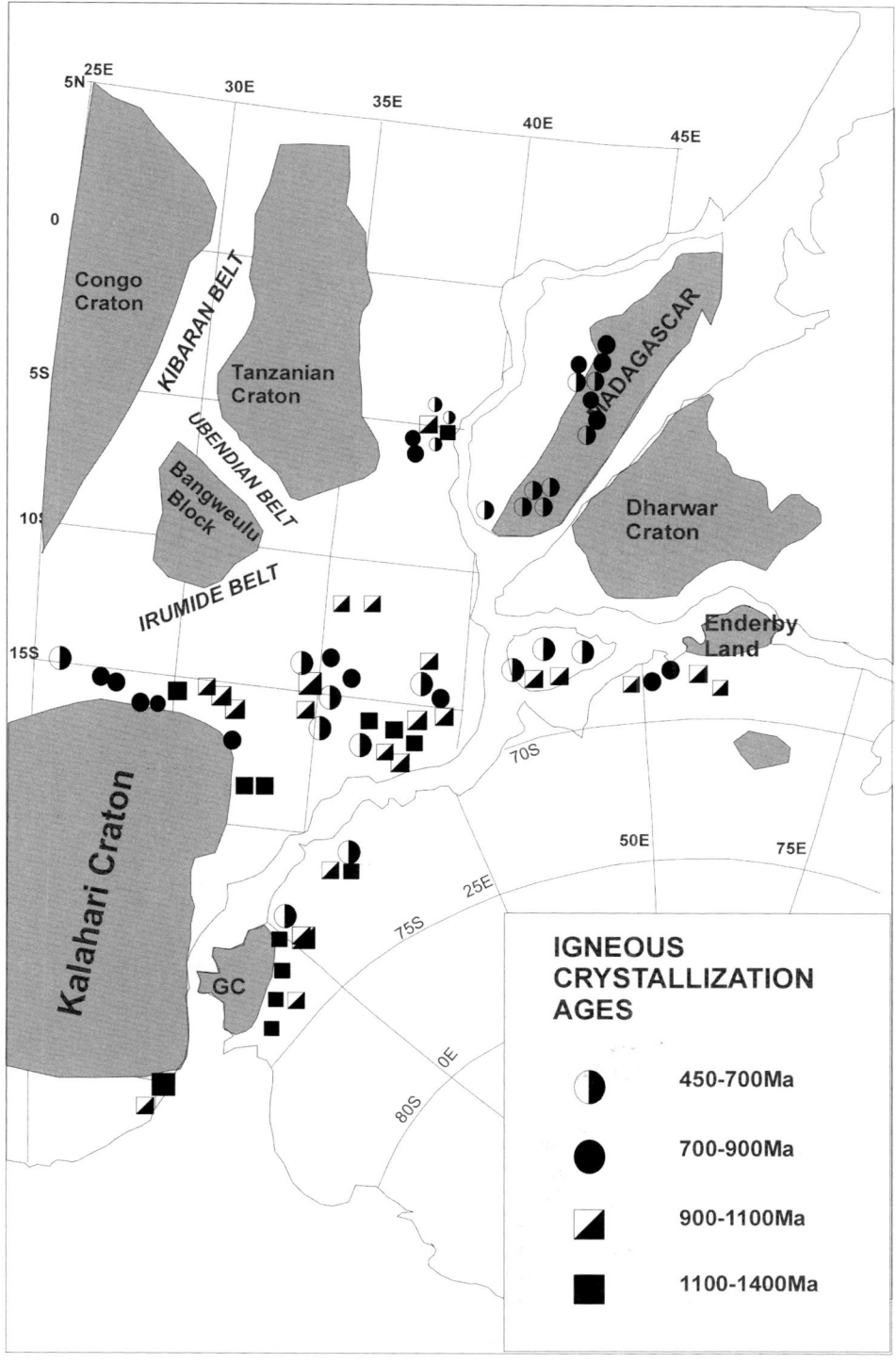

Fig. 4. Gondwana reconstruction based on Lawver *et al.* (1998) showing the locations of geochronological sample points of igneous crystallization ages.

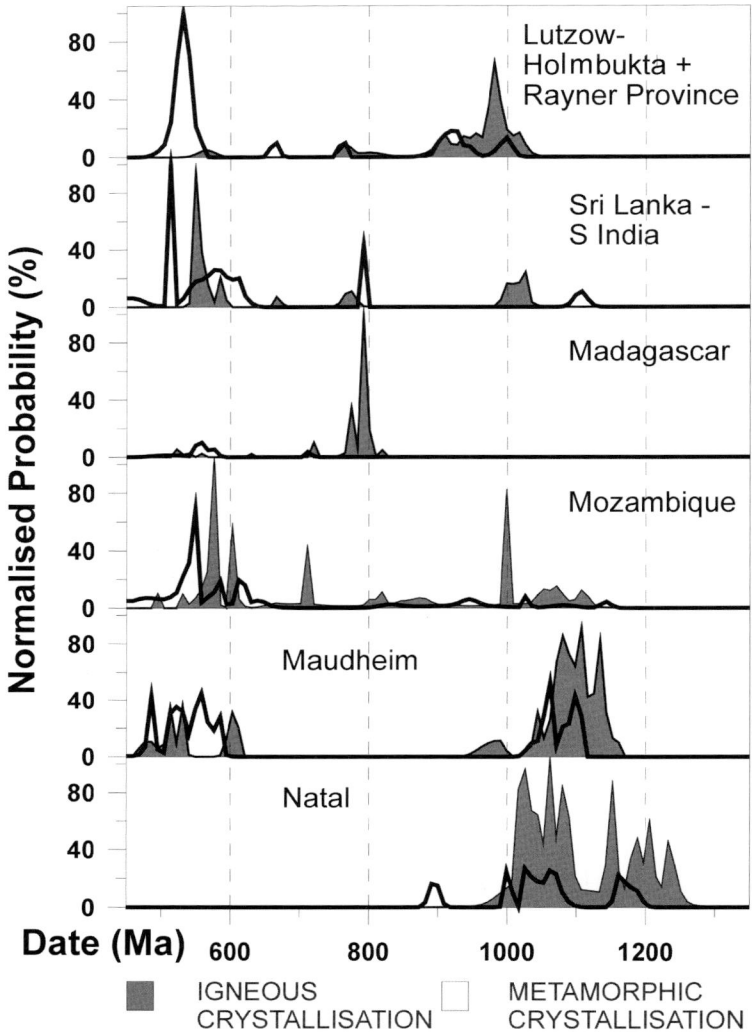

Fig. 5. Probability graphs of the various age ranges recorded from the various crustal fragments reflected in the Gondwana reconstructions from Figures 1–4. The shaded areas represent crystallization ages, the unshaded areas represent metamorphic ages.

Moyes et al. (1993, 1995), Pinna et al. (1993), Muhongo & Lenoir (1994), Paquette et al. (1994), Harris et al. (1995), Jacobs & Thomas (1996), Kröner et al. (1996, 1997, 2000, 2001), Krynauw & Jackson (1996), Jackson & Armstrong (1997), Jacobs et al. (1997, 1998), Braun et al. (1998), Paquette & Nédélec (1998), Dirks et al. (1999), Evans et al. (1999), Handke et al. (1999), Jamal et al. (1999), Vinyu et al. (1999), Eglington & Armstrong (2000), Johnson & Oliver (2000), Manhica et al. (2001) and Tucker et al. (2001).

The data can be divided into two areas: (1) south of c. 12°S in an African frame of reference; and (2) north of c. 8°S in an African frame of reference. There is an extensive gap in information in the Mozambique Belt between 8 and 12°S.

In the area south of c. 12°S, the crystallization ages cover the range from <1400 to >450 Ma. The most widespread ages in this area are those in the 900–1100 Ma range followed by those in the 1100–1400 Ma range. The spatial distribution of ages in the 700–900 Ma range forms a linear zone stretching along the northern margin of the Kalahari Craton in the Zambesi Valley and across northern Mozambique, with some representatives in the Rayner Complex, East Antarctica. The distribution of 450–700 Ma ages stretches from northern Sverdrupfjella, WDML, northwards and eastwards to Sri Lanka and westwards into the Zambesi Belt.

In the area north of 12°S crystallization ages > 900 Ma are few, with the ages falling approximately equally into the 450–700 and 700–900 Ma groups. It is uncertain whether the paucity of measurements >900 Ma is an artefact of sampling or whether it is real. This difference is also reported by Kröner (2001).

Graphs comparing recent crystallization and metamorphic ages based on U–Pb single zircon (SHRIMP evaporation) studies from the various areas, which were juxtaposed prior to Gondwana breakup, show some similarities as would be expected but also some anomalous features (Fig. 5 – data sources as above for Figs 3 & 4). The Grenvillian period data for Mozambique is broadly similar to that for the Maudheim Province in Antarctica, showing a long-lived intrusion history between c. 1150 and 980 Ma. The Pan-African in Mozambique appears to start earlier than in Maudheim, Antarctica with small peaks at c. 800 and 720 Ma and a main event at 630–500 Ma, whereas the dates in Maudheim stretch from c. 620 Ma and last longer to c. 480 Ma. Progressing southwards, the Grenvillian event is longer lived in Natal–Namaqua, lasting from c. 1230 to c. 980 Ma, with no crystallization event during the Pan-African with minor metamorphism.

There is little comparison between the ages from Mozambique and Madagascar, with significant c. 800 Ma ages being recorded in Madagascar in contrast to a small peak at 800 Ma in Mozambique. The c. 700–800 Ma ages recognized from Mozambique are from NE Zimbabwe near the border with the Zambesi Belt, with others being derived further west in the Zambezi Belt and some in southern Malawi and central Tanzania. The structural style for the c. 800 Ma event in the Zambezi Belt areas is interpreted as being extensional (Dirks et al. 1998), whereas the data from Tanzania (Maboko 2001a) and Madagascar (Handke et al. 1999) are interpreted as arc-accretion related. These authors reported U–Pb zircon and baddeleyite ages of between 804 and 776 Ma for a 450 km long belt of plutonic rocks in west-central Madagascar and interpreted the rocks as constituting the root of a c. 800 Ma continental magmatic arc at the time of, or slightly preceding, the breakup of Rodinia. A few other c. 800 Ma ages have been recorded from orthogneisses from Mount Vechernyaya area of the Rayner Complex in East Antarctica by Shiraishi et al. (1997), although the tectonochemical characteristics of these rocks is uncertain

Comparison between Madagascar and India and Sri Lanka shows similarities at c. 800, 1000 Ma and 600–500 Ma. Sri Lanka/India show similar age patterns to Lützow-Holmbukta (Antarctica) at c. 1000 and c. 600–500 Ma, however, c. 800 Ma ages are not recognized in Lützow-Holmbukta.

A recent study of geochronology dominantly from the Mozambique Belt and EAO (Meert 2002) recognizes three broad age groups related to tectonic phases. These phases include the initial oceanic arc/ophiolite genesis in the EAO between c. 710 and 800 Ma. This was followed by the EAO between c. 690 and 580 Ma and the Kuunga Orogeny from c. 580 to 460 Ma. Although there is some overlap in the age ranges of EAO and Kuunga Orogenies, the geographic distribution of the data points shows that the former, and marginally older, episode occurred along a c. N–S-oriented zone stretching from the Arabian–Nubian Shield to northern Mozambique. In contrast, the latter defines both c. E–W- and c. N–S-oriented zones along the northern and eastern margins of the Kalahari Craton, respectively.

Sm–Nd (T_{DM}) model ages

A substantial body of data are available from areas surrounding the Mozambique Belt, however, data from within the belt in Mozambique are extremely limited. Data from WDML and CDML, Antarctica, Madagascar and southern Mozambique have been recalculated after Michard et al. (1985). Data from Lützow-Holmbukta, the Rayner and Napier Complexes, Antarctica, and Sri Lanka and southern India are from Möller et al. (1998) and Rickers et al. (2001). The data are shown in Figure 6 with data sources in Antarctica, Madagascar and southern Mozambique from the various sectors being as follows: Antarctica (Arndt et al. 1991; Shiraishi & Kagami 1992; Moyes 1993; Moyes et al. 1993, 1995; Moyes & Groenewald, 1996; Wareham et al. 1998; Harris 1999; Grantham et al. 2002 and Rickers et al. 2001 and refs cited therein), Mozambique, Malawi and Tanzania (Barr & Brown 1987; Maboko, 1995, 2000, 2001a, b; Maboko & Nakamura 1995, 1996, 2002; Moller et al. 1998; Kröner et al. 2002), India and Sri Lanka (Moller et al. 1998 and refs cited therein; Rickers et al. 2001 and refs cited therein) and Madagascar (Paquette et. al. 1994; Ashwal et al. 1998; Kröner et al. 2000).

The data can broadly be subdivided into the following four categories: (1) areas where the T_{DM} ages are >2.0 Ga; (2) areas that have T_{DM} <1.7 Ga; (3) areas which have T_{DM} values intermediate between >1.7 and <2.0 Ga; and (4) areas that have wide ranges of ages possibly implying complex mixed-source relationships at depth. The latter two categories have limited distributions and probably reflect relatively high degrees of mixing between old protoliths and juvenile magmas.

The areas where T_{DM} ages are >2.0 Ga are

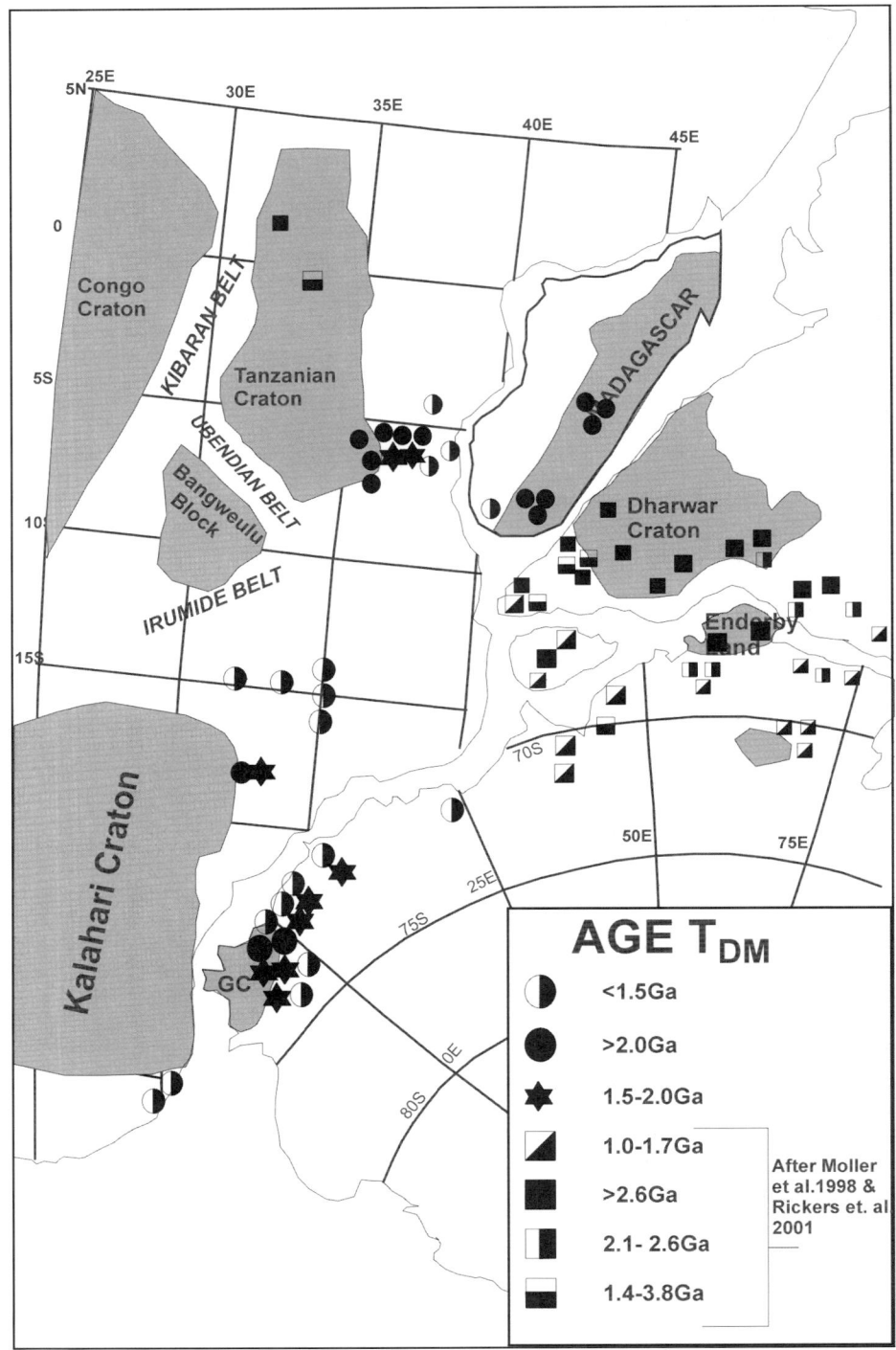

Fig. 6. Gondwana reconstruction based on Lawver *et al.* (1998) showing the Nd T_{DM} model ages from various areas. The figure also incorporates data from Moller *et al.* (1998) and Rickers *et al.* (2001), which were calculated by those authors using different Nd T_{DM} model-age calculation methods.]

dominantly located north of c. 12°S in an African frame of reference and include much of Tanzania, Madagascar, India and Enderby Land, Antarctica. The areas with T_{DM} ages >2.0 Ga are dominantly north of the locus of the Kuunga Orogeny defined by Meert (2002). This area can potentially be divided into two – a western Tanzania–Congo Craton and an eastern Madagascar–India–Enderby Craton separated by the c. 700–800 Ma EAO zone areas. Kröner (2001) and Stern (2002) both recognized that in these areas extensive reworking of Palaeoproterozoic and Archaean crust has occurred in the Pan-African–EAO and that it is only in the limited areas that juvenile additions occurred at c. 700–800 Ma. In central Tanzania, Hepworth & Kennerly (1970) and Hepworth (1972), using structural arguments, concluded that the western part of the Mozambique Belt was underlain by rocks of the Archaean Tanzania Craton, which were progressively reworked eastwards during the Neoproterozoic. This interpretation was latter confirmed by the Nd isotopic data reported by Maboko (1995, 2000) and Moller et al. (1998). These authors reported Late Archaean depleted-mantle mean crustal residence (T_{DM}) ages similar to those obtained in the craton for most of the amphibolite facies granitic gneisses in the central and western parts of the Mozambique Belt in Tanzania (Fig. 6). More recently, Muhongo et al. (2001) reported single crystal zircon U–Pb ages of c. 2600–2700 Ma from granitic gneisses in the central-eastern part of the belt, further corroborating a Late Archaean protolith age for most of these rocks. Some of these gneisses share identical geochemical and isotopic signatures to granitoids of the same age in the Tanzania craton, providing further evidence for a reworked Archaean basement origin (Maboko 2000).

In the southern highlands of Tanzania, Wendt et al. (1972) reported a gradual decrease in mica K–Ar ages from c. 1850 Ma near the cratonic margin in the west to c. 500 Ma well within the Mozambique Belt in the east. Implicit in the results of Wendt et al. (1972) is that the Mozambique Belt in southeastern Tanzania is composed of progressively reworked Palaeoproterozoic rocks of the Usagaran Belt. The latter have been shown by Maboko & Nakamura (1995) to be composed of material that was extracted from the mantle at c. 2.0 Ga but had assimilated as much as 40% of pre-existing Archaean crust. The available evidence therefore shows that most of the central and western part of the Mozambique Belt in Tanzania is composed of either Palaeoproterozoic or Late Archaean rocks that were reactivated during Neoproterozoic time.

In Tanzania, the only evidence for Neoproterozoic protoliths has been obtained from the Eastern Granulites, a c. 900 km long, c. N–S-trending discontinuous belt of high-pressure granulites that extend from near the Mozambique border northwards to the Kenyan border (Fig. 6). These rocks yield T_{DM} ages of between 970 and 1100 Ma (Maboko 1995, 2001a, b; Moller et al. 1998; Maboko & Nakamura, 2002). Maboko & Nakamura (2002) also reported a Sm–Nd whole-rock isochron age of 815 ± 58 Ma and an initial εNd value of 4.1 for calc-alkaline granulites of basaltic to dacitic composition from the Usambara Mountains granulites close to the Tanzania–Kenya border. Maboko & Nakamura (2002) interpreted this age as dating crystallization of the granulite protolith during an event of regional calc-alkaline magmatism in the area. Muhongo et al. (2001) reported similar zircon U–Pb emplacement ages for the protoliths of orthogranulites in the Wami River, and the Uluguru and Usambara Mountains. Thus, the available data suggest that the rocks yielding Neoproterozoic T_{DM} ages in the Eastern Granulites were emplaced at c. 800–850 Ma. This age, however, is within range of the 650–850 Ma phase of convergent margin calc-alkaline magmatic activity which led to widespread crust formation in the Arabian–Nubian Shield (Bentor 1985; Stein & Goldstein 1996). This similarity has been used by Maboko & Nakamura (2002) as evidence that Neoproterozoic crust formation in the Arabian–Nubian Shield and northern parts of the Mozambique Belt was broadly contemporaneous.

Despite the unambiguous evidence for Neoproterozoic juvenile crust in eastern Tanzania, the extent of the area underlain by such material remains unclear. Maboko & Nakamura (2002), however, surmise that most of the Eastern Granulites represent Neoproterozoic juvenile material. Elsewhere in the northern part of the Mozambique Belt, evidence for juvenile Neoproterozoic protoliths have been reported by Harris et al. (1984) from the high-grade metamorphic rocks of northwestern Kenya. These workers reported T_{DM} ages of between 1022 to 1193 Ma (after Michard et al. 1985) from metapelites and amphibolites from the Sekerr area, similar to those obtained from the Tanzanian Granulites. The existence of metasedimentary rocks with a juvenile isotopic signature is particularly significant as it indicates the existence, in western Kenya, of a considerable area of exposed juvenile crust during Neoproterozoic time, the erosion of which supplied the sediments. Similar T_{DM} ages (1160 and 1190 Ma; after Michard et al. 1985) have been reported by Key et al. (1989) from the Il Poloi Metagranitoids of southwestern Kenya. These granitoids, which yield a Rb–Sr whole-rock errorchron age of c. 830 Ma, form small plutons that intrude banded gneisses at higher tectonic levels but form conformable sheets enveloped by migmatites at lower levels (Key et al. 1989). This mode of occurrence suggests formation of the granites by anatexis of juvenile Neoproterozoic protolith rocks at c. 860 Ma. Other evidence for juvenile protoliths come from central Kenya where Shibata & Suwa (1979) reported a four-point Rb–Sr isochron age of 766±40Ma from granitic gneisses in the Mbooni Hills near Machakos. These workers interpreted the c. 770Ma age as dating the intrusion of a granitic mass that was later transformed into orthogneiss. The low $^{86}Sr/^{87}Sr$ initial ratio of c. 0.7041 obtained from the isochron precludes the possibility of the protolith being reworked Archaean material. Rocks with similarly low $^{86}Sr/^{87}Sr$ initial ratios and which yield Neoproterozoic Rb–Sr whole-rock errorchrons occur over a large area of Kenya (Key et al. 1989). Thus, combined together, the available data suggest that apart from the extreme western part of the country, where reworked Archaean basement occurs, most rocks of the Mozambique Belt in Kenya cannot have been derived from protoliths that are much older than c. 1200 Ma.

Coeval granitoids with similar geochemical and isotopic affinities have also been reported in the Arabian–Nubian Shield of northern Ethiopia and Eritrea (Tedesse et al. 2000) and in the high-grade metamorphic basement of the of the Bayuda Desert in Sudan. These occurrences have been used by Maboko & Nakamura (2002) as evidence that Neoproterozoic crust formation in the EAO extended from Tanzania in the south, through the southern part of the Arabian–Nubian Shield and into the better documented northern part of the shield.

Those areas where T_{DM} ages are <1.5Ga are generally related to the juvenile accretion zones which either have Grenvillian (c. 1000–1200Ma) crystallization ages (e.g. Namaqua–Natal–Maud–CDML) or c. 700–800Ma crystallization ages (e.g. Tanzania and northwards; Stern 2002) (the latter group have been described above). The former group represents areas where juvenile crustal addition occurred during the amalgamation of Rodinia and include an area stretching from the Natal Metamorphic Province in South Africa, through the Maud Province in WDML and into southern Mozambique, with restricted examples from Sri Lanka, southern India, continuing eastwards into CDML and the Rayner Complex and Prince Charles Mountains, Antarctica. Significantly, the distribution of this group shows a similar distribution to the c. ≤580Ma Kuunga Orogeny area defined by Meert (2002).

No data are available from the crucial area of northern Mozambique and southern Tanzania. From the reconstruction (Fig. 6) it could be expected that the rocks of Sri Lanka which possibly extend into northern Mozambique are >2.0Ga with the Highland Complex of Sri Lanka having older Palaeoproterozoic ages (Moller et al. 1998 and refs cited therein). Extensions of this zone into Mozambique could possibly suggest that northern Mozambique and southern Tanzania may be underlain by Palaeoproterozoic–possibly Archaean crust which has been reworked by the Pan-African Orogeny.

In Malawi and Mozambique there is no isotopic database comparable to that existing in Tanzania. As a result, the protolith ages of rocks involved in the Neoproterozoic tectonism cannot be ascertained. However, Pinna et al. (1993) and Kröner et al. (1997) reported Rb–Sr and zircon U–Pb ages of between c. 850 and 1100Ma for the extensive granitic gneisses of northern Mozambique, which they interpreted as dating the emplacement age of the protolith. As these ages are not constrained by Nd isotopic studies, it is unclear whether the granitoids represent older reworked material or Neoproterozoic additions to the continental crust. More clear-cut evidence for Neoprotererozoic crustal additions are found in central and southern Malawi, where Kröner et al. (2002) reported T_{DM} ages of between 1.0 and 1.5Ga for calc-alkaline granitic orthogneisses with emplacement ages of 1040–929Ma and younger calc-alkaline intrusions aged between c. 555 and 710Ma.

In conclusion, the picture that emerges is one of two provinces. The northern province is dominantly underlain by Palaeoproterozoic–Archaean crust, represented by the Tanzanian–Congo and the Madagascar–India–Enderby Cratons. These two cratonic areas are separated by the c. N–S-oriented c. 710–800Ma EAO, which appears to die out in central Tanzania or possibly continues into southern Madagascar or northern Mozambique where data are lacking.

The southern province comprises the Palaeoproterozoic–Archaean-age Kalahari Craton and its metamorphic belts along its margins. The northern margin includes the Zambesi Belt in Zimbabwe, Zambia, Malawi and possibly northern Mozambique, continuing eastwards through Sri Lanka into CDML and beyond. Its eastern margin extends southwards through southern Mozambique into the Maud Province of WDML, Antarctica, and into Natal in southern Africa. Besides the Kalahari Craton all of these areas are characterized by crystallization ages dominantly of c. Rodinian–Grenvillian age and T_{DM} ages of <1.7Ga. Subordinate crystallization ages of 750–800 and c. 550–600Ma are present, the former being interpreted as due to extension in the Zambesi Belt and the latter being related to anatexis during the Kuunga Orogeny. In this southern province, samples with T_{DM} ages of between 1.5 and 2.0Ga are recognized but are mostly close to the craton–mobile belt interface, suggesting that they have originated from mixing between a juvenile (1100–1200Ma?) and an older Palaeoproterozoic–Archaean component.

Structural aspects

The definition of major crustal structures (strike-slip, oblique-slip and dip-slip) and their continuations across crustal fragments have long been recognized as being particularly useful in constraining possible reconstructions. The age and nature of the structural data described below are summarized in Figure 7.

The Mozambique Belt in Mozambique has many major crustal structures including the Lurio Belt (Pinna et al. 1993), a large shear east of Manica (Manhica et al. 2001) and the Namama Belt (Cadoppi et al. 1987). All three of these belts have yielded geochronological evidence for significant deformation and metamorphism of c. 1100 Ma rocks during the Pan-African Orogeny between c. 460 and 650 Ma (Kröner et al. 1997; Jamal et al. 1999; Manhica et al. 2001). In addition, the sense of movement of these zones is typically sinistral (Cadoppi et al. 1987; Manhica et al. 2001) or transpressional with both strike-slip and dip-slip geometries (Thomas et al. 2001). It is significant that the c. 550 Ma deformation in northern Mozambique has a top-to-the-SE geometry (Thomas et al. 2001) which is directly opposite to the top-to-the-NW geometries of the Rodinian-age structures in Sverdrupfjella, WDML. Localized top-to-the-SE c. 480–600 Ma structures are recognized in NE Sverdrupfjella in Antarctica (Grantham 1992; Grantham et al. 1995, 2003). Manhica et al. (2001) described a strong N–S-oriented shear zone along the boundary between the Kalahari Craton and the Mozambique Belt in central Mozambique. Manhica et al. (2001) suggest that the sense of shear is sinistral with the shear zone possibly being active as late as c. 470 Ma, as indicated by Ar–Ar studies on mica. In northern Mozambique (Thomas et al. 2001) stretching lineation orientations define a quadrant which plunges shallowly to the west through to the north, suggesting that there is a strong degree of strike-slip transpression along this structure with both strike-slip and dip-slip movement. Measurements of lineation directions within the Lurio Shear Belt plunge shallowly towards the WSW, near-parallel to strike and possibly implying a strike-slip sense of movement (Thomas et al. 2001). Figure 7 suggests that a possible strike extension of the Lurio Belt is present in Sri Lanka and separates the Highland Complex from the Vijyan Complex. If the correlation of the shear belt in Sri Lanka with that in the Lurio is correct, then the age of the shear must similarly be c. 600 Ma, the age of the deformation and migmatization in the Lurio Belt (Jamal et al. 1999).

In Sri Lanka, the Highlands Complex in the centre of the island is separated from the Vijayan Complex to the south by a top-to-the-SE thrust zone in which the dominant lineation direction is almost parallel to strike (Kriegsman 1995) possibly suggesting a significant strike-slip component.

In the Zambesi Belt in Zambia, Hanson et al. (1988) have constrained the age of the Mwembeshi shearing to Pan-African age. Similarly, in the Zambesi Belt in NE Zimbabwe, Dirks et al. (1998) have described NE extensional shear zones coeval with the emplacement of c. 800–830 Ma granite sheets and mafic dykes. The chemistry of the sheeted granitoids is reportedly consistent with extensional emplacement (Dirks et al. 1998). Similarly, the Makuti Group exposed in the Zambezi Belt, NW Zimbabwe, contains c. 800 Ma quartzofeldspathic gneisses which were emplaced into Archaean crust in a NW–SE-directed extensional setting. Extensional tectonism in the Zambesi Belt at c. 800–980 Ma was also recognized by Wilson et al. (1993), who indicated that the extensional phase was followed by sinistral transcurrent shearing, which they suggested occurred at c. 820 Ma. The Mwembeshi Shear Zone is also a large-scale Pan-African (Hanson et al. 1993) sinistral shear zone (de Swardt et al. 1965; Coward & Daly 1984; Daly 1986) separating the Kalahari Craton from the Congo Craton.

In WDML large-scale shear belts in Heimefrontfjella (strike-slip and dip-slip – Jacobs et al. 1993, 1995, 1996, 1997, 1998), Sverdrupfjella (dip-slip – Groenewald et al. 1991; Grantham et al. 1995) and Kirwanveggen (strike-slip – Croaker 1999; Croaker et al. 1999; dip-slip – Grantham et al, 1995) are recognized. The dip-slip shear belts in Heimefrontfjella, Kirwanveggen and Sverdrupfjella are interpreted to be the result of top-to-the-NE thrust faulting of Rodinian age, i.e. c. 1000–1150 Ma. The dextral strike-slip deformation in Heimefrontfjella is related to the same deformation in a transpressional setting. This contrasts with the sinistral 530 Ma (Moyes et al. 1997) brittle strike-slip deformation recognized in the c. 500 Ma Urfjell sediments in Kirwanveggen (Croaker 1999). Ductile to brittle ductile deformation of similar age to that in Urfjell, i.e. 470–500 Ma, but involving top-to-the-SE reverse faulting and asymmetric folding at mid-crustal levels, has been described in Sverdrupfjella by Grantham et al. (1991, 2003). This deformation comprises steep reverse faults with top-to-the-SE geometry developed synchronously with the emplacement of the c. 490 Ma Dalmatian Granite (Grantham et al. 1991; Krynauw & Jackson 1996). In WDML, dextral transpressional strike-slip shears in Heimefrontfjella are reported (Jacobs et al. 1995, 1996, 1997, 1999).

In CDML, Bauer et al. (2002) describe a sinistral strike-slip shear zone in Orvinfjella in the Muhlig Hoffman Mountains. The orientation, location, age and sense of shear of this shear zone suggests that it could be continuous in southern Mozambique with the Namama Shear Zone described by Cadoppi et al.

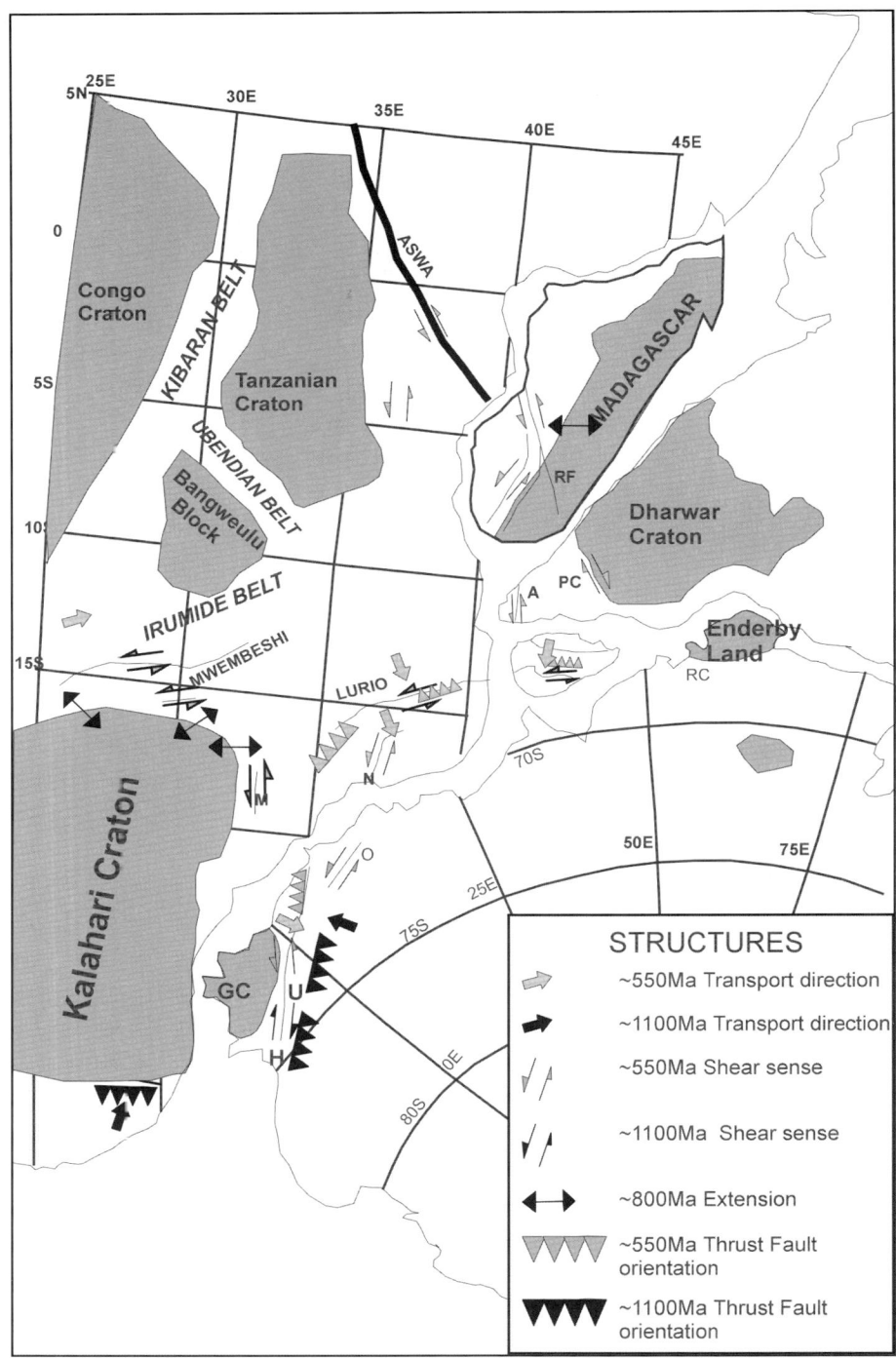

Fig. 7. Gondwana reconstruction based on Lawver *et al.* (1998) showing the ages and locations of various large-scale shear belts in the various continental fragments. PC, Palgat–Cauvery Shear Zone; RC, Rayner Complex; A, Achankovil Shear Zone; GC, Grunehogna cratonic fragment; H, Heimefrontfjella; RF, Ranotsara Shear Zone; U, Urfjell; N, Namama Shear Belt; O, Orvinfjella Shear Zone; M, Manica.

(1987). Both structures are strike-slip, sinistral and are clearly Pan-African in age.

In southern Madagascar, Martelat et al. (2000) recognized two phases of deformation, both of Pan-African age. D1 at 590–530 Ma comprised E–W-directed deformation to produce a flat-lying planar foliation with related stratoid granites and isoclinal folding, whereas D2 was effective from 500–530 Ma and comprised sinistral strike-slip deformation in a transpressional setting in the Ranotsara Shear Zone. Collins et al. (2000) recognized c. 550 Ma dip-slip extensional structures at the base of the >800 Ma old Itremo Group, termed the Betsileo Shear Zone. The shear zone becomes asymptotic to the Ranotsara Shear Zone, suggesting that it marginally predated that structure. The Ranotsara Shear Zone has possible continuations to the north in Africa as the Aswa Shear Zone in Kenya and to the south as the Palgat–Cauvery Shear Zone.

In India, shear zones which transect southern Peninsular India include the Palgat–Cauvery Shear Zone and the Achankovil Shear Zone. The age of the former shear zone, which has a dominantly dextral sense of shear, is c. 500–630 Ma (Bhaskar Rao et al. 1996; Narayana et al. 1996; Meisner et al. 2002). Bhaskar Rao et al. (1996) suggested that the shear zone continues into Antarctica and separates the Rayner Complex from the Napier Complex of Enderby Land. The sense of movement on the Achankovil Shear Zone is not clear but it is recognized as being of a strike-slip nature with an age of 480–550 Ma (Chetty 1996).

In Natal, various authors (Matthews 1972; Jacobs & Thomas 1994) have recognized and described the thrust–nappe sequences of the Tugela Terrane of the Natal Metamorphic Province. The tectonic transport direction related to the emplacement of the thrust slices is interpreted to be from the SW. The age of this deformation is undoubtedly c. >1100 Ma, as reflected by the Ar–Ar age of hornblende from the mafic gneisses (Jacobs et al. 1997).

Numerous authors (e.g. Jacobs et al. 1998; Bauer et al. 2002) have recognized the large-scale sinistral nature of c. Pan-African-age shear zones in Antarctica, Mozambique, Madagascar, Kenya and Sri Lanka, with most suggesting that this shearing is consistent with the amalgamation of East and West Gondwana along the eastern margin of Africa, although none of these authors have identified a specific suture of Pan-Africa age or suture-related rocks in these areas. Another feature which requires consideration is that the Pan-African shear zones in India, Madagascar, Kenya and Tanzania are oriented c. N–S and at right angles to the Zambesi–Lurio–Highlands Complex orientation (Fig. 7).

Another problem arises from shear zone correlation in that disagreement is evident in the literature as to whether some of the strike-slip zones are sinistral or dextral, e.g. the Heimefrontfjella Shear Zone is reportedly dextral (Jacobs et al. 1996), however, tectonic interpretations of both Rodinian and Gondwanan amalgamation models would suggest that the shear zone could be expected to be sinistral. Similarly, Stern (1994), Kriegsman (1995), Shackleton (1996), Möller et al. (1998), Kröner (2001) and Muhongo et al. (2001) all show the Palghat–Cauvery Shear Zone in southern India as being dextral, which contrasts with the statement by Chetty (1996) that it is dominantly dextral with local sinistral orientations. In the broader scheme of the shear zone patterns shown in Figure 7 and the reported Pan-African age of the shearing, one would expect the shear zone to be sinistral.

Discussion and conclusions

The available data suggest that the Kalahari and the East Antarctica Cratons were amalgamated during the formation of Rodinia at c. 1200–1000 Ma. This is supported by the distribution of mostly juvenile TTG-type lithologies, carbonates, eclogites and ophiolites in the vicinity of the contact zones of these cratons. Similarly, c. 1400 Ma old rocks of juvenile origin (ophiolites) separate the Congo Craton from the Kalahari Craton in the Zambesi Valley, also suggesting possible Rodinian amalgamation. Similarly, East Antarctica, Australia and India were thought to have been amalgamated before c. 1050 Ma (Powell et al. 1993); the position of Madagascar at this time is uncertain. Various authors have suggested the fragmentation of Rodinia at c. 700–800 Ma to form the Mozambique Ocean. Magmatism of this age, which conceivably could be related to fragmentation, is recorded in the Zambesi Belt, Malawi, northen Mozambique and CDML, and in Madagascar, Tanzania and Kenya northwards. The tectonic setting of this c. 700–800 Ma magmatism is reportedly extensional in the Zambesi Belt (Dirks et al. 1998), and arc related in Tanzania (Maboko 2001a, b) and Madagascar (Handke et al. 1999). These data would suggest that while the Kalahari Craton was possibly separating from an unknown entity at 700–800 Ma (Congo Craton?), the Tanzanian Craton was joining with the Madagascar–India–Enderby Craton to form the EAO. Southwards, c. 800 Ma granitic magmatism is not recognized in the southern Mozambique Belt, DML, Antarctica, or in Natal: nor is evidence for extensional deformation recognized in these areas, the only possible extensional rocks being mafic dykes of uncertain age. The absence of extensional intrusions and structures from southern Mozambique and WDML, Antarctica, would suggest that the Kalahari and East Antarctic Cratons remained as a unitary contiguous block when Rodinia fragmented.

These events were then followed by the amalgamation of 'North' and 'South' Gondwana along the c. E–W oriented Kuunga Orogeny Zone at c. 580–480 Ma. This amalgamation involved sinistral transpression along the continental interface between the Kalahari and Congo–Tanzanian Cratons, which is recognized from the west along the Mwembeshi Shear Zone through the Zambesi Belt. The N–S sinistral shearing along the eastern margin of the Kalahari Craton and further to the east can be interpreted as 'escape' tectonic structures, with East Antarctica–Madagascar–India 'escaping' around the eastern margin of the combined Congo–Tanzanian Craton, resulting in the transpressional structures of the Lurio Belt in northern Mozambique, the shearing in Sri Lanka and the extension of the Palghat–Cauvery Shear Zone into the Rayner Complex in Antarctica. It is significant that many of the younger metamorphic and crystallization ages of c. 460–480 Ma are recorded along the SE sector (southern Mozambique – Manhica et al. 2001; WDML, Antarctica – Grantham et al. 1991; Groenewald 1995), reflecting the terminal stages of the Kuunga Orogeny. The distribution of Pan-African shear zones suggests the presence of cratonic material underlying southern Tanzania/northern Mozambique, around which many of the shear zones appear to have been deflected. The existence of older crust in this area is possibly supported by the recognition of old source rocks in the Highlands Complex of Sri Lanka, directly along-strike to the east of northern Mozambique and southern Tanzania.

Consequently, the Mozambique/Adamastor Ocean is likely to have separated the Kalahari Craton from the Congo–Tanzanian Craton and separated the Congo–Tanzanian Craton from the Madagascar-India Block. The closure of this ocean occurred at c. 700–580 Ma along the latter cratonic interface and closed at c. 600–500 Ma along the former cratonic interface.

From the above study, the following aspects are anomalous and require consideration and further investigation. The crystallization ages of rocks in Madagascar do not correlate closely with its neighbours in a Gondwana setting, a feature recognized by Kröner et al. (2000) and discussed by Torsvik et al. (2001). This aspect may suggest that Madagascar is 'out of place' with regard to Rodinian plate reconstructions: Hoffman (1991) placed Madagascar next to the Kibaran–Grenville Belt in East Antarctica; Dalziel (1991) placed Madagascar next to western India; Piper (1976, 2000) placed Madagascar farther south, adjacent to southern Africa prior to the Pan-African amalgamation of Gondwana with Madagascar, being positioned against Kenya/Tanzania during the Pan-African amalgamation of Gondwana. The position suggested by Piper (1976, 2000) would certainly facilitate closer proximity to, and therefore possible correlation between, the Itremo Group in Madagascar and the Frontier Formation of the Mkondo Group. In addition, the c. 800 Ma magmatism recognized in Madagascar could then conceivably be related to that which is recognized in the Zambesi and Damara Belts. Palaeomagnetic testing of the c. 800 Ma magmatic event rocks between Madagascar and elsewhere could confirm or reject this possibility.

The strong shearing recognized in the Lurio Belt appears to have a strike continuation into and through Sri Lanka. In addition to this, the paragneissic Chiure Group of northern Mozambique appears to be a strike extension of the Highland Complex in Sri Lanka.

The paucity of structural, geochronological and isotopic data from northern Mozambique and southern Tanzania needs to be addressed by new research initiatives. The limited data available suggest the possibility of cratonic material at depth. This has significant economic implications with regard to the possibility of diamond and gold deposits.

We would like to acknowledge discussions with numerous colleagues during the compilation of this paper. We would also like to record our sense of loss at the passing of Chris Powell: he was on an IGCP fieldworkshop to Natal with one of us (GHG) a few days before his premature death, during which his interest and enthusiasm was an inspiration to us all. We would also like to acknowledge the constructive criticisms of K. Shiraishi and R. J. Stern, who also alerted us to additional data sources.

References

ANDREOLI, M. A. G. & HART, R. J. 1985. Metasomatized granulites of the Mozambique belt: implications for mantle devolatilisation. *In*: HERBERT H. K. & HO S. E. *Stable Isotopes and Fluid Processes in Mineralisation*. University of Western Australia, Publication 23, 121–140.

ANDRIESSEN, P. A. M., COOLEN, J. J. M. M. M. & HEBEDA, E. H. 1985. K–Ar hornblende dating of late Pan African metamorphism in the Furua Granulite Complex of southern Tanzania. *Precambrian Research*, **30**, 351–360.

ARIMA, M., TANI, K., KAWATE, S. & JOHNSTON, S. 2001. Geochemical characteristics and tectonic setting of metamorphosed rocks in the Tugela Terrane, Natal Belt, South Africa. *Memoir of the National Institute of Polar Research*, **55**, 1–39.

ARNDT, N. T., TODT, W., CHAUVEL, C., TAPFER, M. & WEBER, K. 1991. U–Pb zircon age and Nd isotopic composition of granitoids, charnockites and supracrustal rocks from Heimefrontfjella, Antarctica. *Geologische Rundschau*, **80/3**, 759–777.

ASHWALL, L. D., HAMILTON, M. A., MOREL, V. P. I. & RAMBELOSON, R. A. 1998. Geology, petrology and isotope geochemistry of massif-type anorthosite from southwest Madagascar. *Contributions to Mineralogy and Petrology*, **133**, 389–401.

BARKHUIZEN, J. G. & MATTHEWS, P. E. 1990. Gravity modelling of the Natal Thrust-front: a Mid-Proterozoic crustal suture in southeastern Africa. *Geocongress '90 Abstracts University of Cape Town*, 32–35.

BARR, M. W. C. & BROWN, M. A. 1987. Precambrian gabbro–anorthosite complexes, Tete Province, Mozambique. *Geological Journal*, **22**, 139–159.

BAUER, W., THOMAS, R. J. & JACOBS, J. 2003. Proterozoic–Cambrian History of Dronning Maud Land in the Context of Gondwana Assembly. *In*: YOSHIDA, M., WINDLEY, B. F. & DASGUPTA, S. (eds) *Proterozoic East Gondwana: Supercontinent Assembly and Breakup*. Geological Society, London, Special Publication, **206**, 247–269.

BEHRMANN, J. H. & RATSCHBACHER, L. 1989. Archimedes revisited: a structural test of eclogite emplacement models in the Austrian Alps. *Terra Nova*, **1**, 242–252.

BENTOR, Y. K. 1985. The crustal evolution of the Arabian–Nubian Massif with special reference to the Sinai Peninsula. *Precambrian Research*, **28**, 1–74.

BHASKAR RAO, Y. J., CHETTY, T. R. K., JANARDHAN, A. S. & GOPALAN, K. 1996. The Cauvery shear zone, south India: evidence for Late Archaean rocks and their Proterozoic reworking. *Proceedings of the IGCP 368th International workshop on The Proterozoic Continental Crust of southern India*. Gondwana Research Group miscellanous publication 4. Trivandrum India, 7–9.

BIGIOGGERO, B., CADOPPI, P., COSTA, M., OMENETTO, P. & SACCHI, R. 1990. Granites of Zambesia (Mozambique). *GeologickyZbornik – Geologica Carpathica*, **41**, 605–618.

BLACK, L. P. 1986. Isotopic resetting of the systems Rb–Sr and Sm–Nd total rock and U–Pb zircon in Antarctica – the cold facts. *Terra Cognita*, **6**.

BLACK, L. P. 1988. Isotopic resetting of U–Pb zircon and Rb–Sr and Sm–Nd whole-rock systems in Enderby Land, Antarctica: implications for the interpretation of isotopic data from polymetamorphic and multiply deformed terrains. *Precambrian Research*, **38**, 355–365.

BLACK, L. P., HARLEY, S. L., SUN, S. S. & MCCULLOCH, M. T. 1987. The Rayner Complex of East Antarctica: complex isotopic systematics within a Proterozoic mobile belt. *Journal of Metamorphic Geology*, **5**, 1–26.

BRAUN, I., MONTWEL, J. M. & NICOLLET, C. 1998. Electron microprobe dating of monazites from high grade gneisses and pegmatites of the Kerala khondalite belt, southern India. *Chemical Geology*, **146**, 65–85.

CADOPPI, P., COSTA, M. & SACCHI, R. 1987. A cross section of the Namama Thrust belt (Mozambique). *Journal of African Earth Sciences*, **6**, 493–504.

CHETTY, T. R. K. 1996. Proterozoic shear zones in southern granulite Terrain, India. *In*: SANTOSH, M. & YOSHIDA, M. (eds) *The Archaean and Proterozoic Terrains of Southern India within East Gondwana*. Gondwana Research Group, Memoir, **3**, 77–89.

COLLINS, A. S., RAZAKAMANANA, T. & WINDLEY, B. F. 2000. Neoproterozoic crustal-scale extensional detachment in central Madagascar: implications for extensional collapse of the East African Orogen. *Geological Magazine*, 137, 39–51.

COOLEN, J. J. M. M. M. 1980. Chemical petrology of the Furua granulite complex, southern Tanzania. GUA papers in Geology, Series 1, 13.

COOLEN, J. J. M. M. M., PRIEM, H. N. A., VERDURMEN, E. A. TH. & VERSCHURE, R. H. 1982. Possible zircon U–Pb evidence for Pan African granulite facies metamorphism in the Mozambique Belt of southern Tanzania. *Precambrian Research*, **17**, 31–40.

COSTA, M., CADOPPI, P., SACCHI, R. & FANNING, C. M. 1994. U–Pb SHRIMP dating of zircons from Mozambique Gneiss. *Bollettino della Societa Geologica Italiana*, **113**, 173–178.

COSTA, M., FERRARA, G., SACCHI, R. & TONARINI, S. 1992. Rb/Sr dating of the Upper Proterozoic basement of Zambesia, Mozambique. *Geologische Rundschau*, **81**, 487–500.

COWARD, M. P. & DALY, M. C. 1984. Crustal lineaments and shear zones in Africa: their relationship to plate movements. *Precambrian Research*, **24**, 27–45.

COX, R., ARMSTRONG, R. A. & ASHWAL, L. D. 1998. Sedimentology, geochronology and provenance of the Proterozoic Itremo Group, central Madagascar, and implications for pre-Gondwana palaeogeography. *Journal of the Geological Society, London*, **155**, 1009–1024.

CROAKER, M. 1999. *Geological constraints on the evolution of the Urfjell Group, southern Kirwanveggen, western Dronning Maud Land, Antarctica*. M.Sc. Thesis, University of Natal.

CROAKER, M., JACKSON, C., ARMSTRONG, R., WHITMORE, G., LINDSAY, P. & MARE, L. 1999. The evolution of the Urfjell Group, western Dronning Maud Land – sedimentology, structure, isotope and geochemical constraints on the evolution of a strike-slip basin. *Proceedings of the 8th International symposium on Antarctic Earth Sciences*, 5–9th July, Victoria University, Wellington, New Zealand, 72 (abstracts).

DALY, M. C. 1986. Crustal shear zones and thrust belts: their geometry and continuity in central Africa. *Philosophical Transactions of the Royal Society of London A*, **317**, 111–128.

DALZIEL, I. W. D. 1991. Pacific margins of Laurentia and East Antarctica–Australia as a conjugate rift pair: evidence and implications for an Eocambrian supercontinent. *Geology*, **19**, 598–601.

DE SWARDT, A. M. J., GARRARD, P. & SIMPSON, J. G. 1965. Major zones of transcurrent dislocation and Superposition of orogenic Belts in part of Central Africa. *Geological Society of America Bulletin*, **76**, 89–102.

DIRKS, P. H. G. M. 1997. The Zambezi Belt in NW Zimbabwe: the Makuti Group. *In*: DIRKS, P. H. G. M., JELSMA, H. A. (eds) *The Zambezi Belt of Zimbabwe. Excursion Guide*. Geological Society Zimbabwe, Harare, 14–26.

DIRKS, P. H. G. M., JELSMA, H. A., VINYU, M. L. & MUNYANYIWA, H. 1998. The structural history of the Zambezi Belt in northeast Zimbabwe; evidence for crustal extension during the early Pan-African. *South African Journal of Geology*, **101**, 1–16.

DIRKS, P. H. G. M., KRÖNER, A., JELSMA, H. A., SITHOLE, T. A. & VINYU, M. L. 1999. Structural relations and Pb-Pb zircon ages for the Makuti Gneisses: evidence for a crustal-scale Pan African shear zone in the Zambezi

Belt, northwest Zimbabwe. *Journal of African Earth Sciences*, **28**, 427–442.

EGLINGTON, B. M. & ARMSTRONG, R. A. 2000. *U–Pb SHRIMP and Rb–Sr, Sm–Nd isotope investigation of Proterozoic basement intersections from boreholes penetrating the Karoo Supergroup in South Africa.* Council for Geoscience Open File Report, 2000–0073-O, 1–20.

EGLINGTON, B. M. & HARMER, R. E. 1987. Sr isotopes in Proterozoic carbonate metasediments from Natal: constraints on the formation of the Natal Structural and Metamorphic Province. *South African Journal of Geology*, **90**, 514–519.

EGLINGTON, B. M. & KERR, A. 1989. Rb–Sr and Pb–Pb geochronology of Proterozoic intrusions from the Scottburgh area of southern Natal. *South African Journal of Geology*, **92**, 400–409.

EGLINGTON, B. M., ARMSTRONG, R. A. & THOMAS, R. J. 2000a. *Geochronology of the Oribi Gorge Suite, South Africa.* Council for Geoscience Open File Report, 2000-0071-O, 1–28.

EGLINGTON, B. M., DE BEER, J. H., PITTS, B. E., MEYER, R., GEERTHSEN, K. & MAHER, M. J. 1993. Geological, geophysical and isotopic constraints on the nature of the Mesoproterozoic Namaqua–Natal Belt of southern Africa, *In*: PETERS, J. W., KESSE, G. O. & ACQUAH, P. C. (eds) *Proceedings of the 9th International Geological Conference of the Geological Society of Africa*, Geological Society of Africa, Accra, 114–135.

EGLINGTON, B. M., HARMER, R. E. & KERR, A.1989. Rb–Sr isotopic constraints on the ages of the Mgeni and Nqwadolo granites, Valley of a Thousand Hills, Natal. *South African Journal of Geology*, **92**, 393–399.

EGLINGTON, B. M., HARMER, R. E. & KERR, A. 2000b. Isotopic evidence for crustal accretion during the Kibaran Namaqua–Natal orogeny in eastern South Africa. *Terra Cognita*, **7**, 150–150.

ELWORTHY, T., EGLINGTON, B. M., ARMSTRONG, R. A. & MOYES, A. B. 2000. Rb–Sr isotope constraints on the timing of late- to post-Archaean tectono-metamorphism affecting the south eastern Kaapvaal Craton. *Journal of African Earth Sciences*, **30**, 641–650.

EVANS, R. J., ASHWAL, L. D. & HAMILTON, M. A. 1999. Mafic, ultramafic and anorthositic rocks of the Tete Complex, Mozambqiue: petrology, age and significance. *South African Journal Geology*, **102**, 153–166

GRANTHAM, G. H. 1992. *Geological Evolution of western H.U. Sverdrupfjella, Dronning Maud Land, Antarctica.* PhD Thesis, University of Natal.

GRANTHAM, G. H., ARMSTRONG, R. A. & MOYES, A. B 2003. The age, chemistry and structure of mafic dykes at Roerkulten, H.U. Sverdrupfjella, western Dronning Maud Land, Antarctica. International Dyke Conference Volume, in press.

GRANTHAM, G. H. & EGLINGTON, B. M. 1992. Mineralogy, chemistry, and age of granitic veins at Nicholson's Point, South Coast Natal. *South African Journal of Geology*, **95**, 88–93.

GRANTHAM, G. H., EGLINGTON, B. M., THOMAS, R. J. & MENDONIDIS, P. 2001. The nature of the Grenville-age charnockitic A-type magmatism from the Natal, Namaqua and Maud Belts of southern Africa and western Dronning Maud Land, Antarctica. *Memoir of the National Institute of Polar Research*, **55**, 59–86.

GRANTHAM, G. H., GROENEWALD, P. B. & HUNTER, D. R. 1988. Geology of the northern H.U. Sverdrupfjella, western Dronning Maud land and implications for Gondwana reconstructions. *South African Journal Antarctic Research*, **18**, 2–10.

GRANTHAM, G. H., JACKSON, C., MOYES, A. B., GROENEWALD, P. B., HARRIS, P. D., FERRAR, G. & KRYNAUW, J. R. 1995. The tectonothermal evolution of the Kirwanveggan–H.U. Sverdrupfjella areas, Dronning Maud Land, Antarctica. *Precambrian Research*, **75**, 209–230.

GRANTHAM, G. H., MOYES, A. B. AND HUNTER, D. R. 1991. The age, petrogenesis and emplacement of the Dalmatian Granite, H.U. Sverdrupfjella, Dronning Maud Land, Antarctica. *Antarctic Science*, **3**, 197–204.

GRANTHAM, G. H., STOREY, B.C., THOMAS, R. J. & JACOBS, J. 1997. The pre-breakup position of Haag Nunataks within Gondwana: possible correlatives in Natal and Dronning Maud Land. *In*: RICCI, C. A. (ed.) *The Antarctic Region: Geological Evolution and processes.* Proceedings of the VII International Symposium on Antarctic Earth Sciences, Siena, 13–20.

GROENEWALD, P. B. 1995. *The geology of northern H.U. Sverdrupfjella and its bearing on crustal evolution in Dronning Maud Land, Antarctica.* PhD Thesis, University of Natal.

GROENEWALD, P. B., GRANTHAM, G. H. & WATKEYS, M. K. 1991. Geological evidence for a Proterozoic to Mesozoic link between southeastern Africa and Dronning Maud land, Antarctica. *Journal of the Geological Society, London*, **148**, 1115–1123.

HANDKE, M. J., TUCKER, R. D. & ASHWAL, L. D. 1999. Neoproterozoic continental arc magmatism in west central Madagascar. *Geology*, **27**, 351–354.

HANSON, R. E., MARTIN, M. W., BOWRING S. A. & MUNYANYIWA, H. 1998. U–Pb zircon age for the Umkondo dolerites, eastern zimbabwe: 1.1 Ga large igneous province in southern Africa–East Antarctica and possible Rodinia correlations. *Geology*, **26**, 1143–1146.

HANSON, R. E., WARDLAW M. S., WILSON, T. J. & MWALE, G. 1993. U–Pb zircon ages from the Hook granite massif and Mwenbeshi dislocation: constraints on Pan African deformation, plutonism and transcurrent shearing in central Zambia. *Precambrian Research*, **63**, 189–209.

HANSON, R. E., WILSON, T. J. & MUNYANIWA, W. 1994. Geologic evolution of the Neoproterozoic Zambezi orogenic belt in Zambia. *Journal of African Earth Sciences*, **18**, 135–150.

HANSON, R. E., WILSON, T. J. & WARDLAW, M. S. 1988. Deformed batholiths in the Pan-African Zambezi belt, Zambia: age and implications for regional Proterozoic tectonics. *Geology*, **16**, 1134–1137.

HARRIS, N. B. W., HAWKESWORTH, C. J. & RIES, A. C. 1984. Crustal evolution in north-east and east Africa from model Nd ages. *Nature*, **309**, 773–776.

HARRIS, P. D. 1999. *The geological evolution of Neumayerskarvet in the northern Kirwanveggen, western Dronning Maud Land, Antarctica.* PhD Thesis, Rand Afrikaans University.

HARRIS, P. D., MOYES, A. B., FANNING, C. M. &

ARMSTRONG, R. A. 1995. Zircon ion microprobe results from the Maudheim High Grade Gneiss Terrane, Western Dronning Maud Land, Antarctica. *Geocongress '95 Abstracts*, Rand Afrikaans University, Johannesburg, 240–243.

HART, R. J. & BARTON, E. S. 1984. The application of U–Th–Pb isotope systematics in the investigation of potential uranium source rocks in the Natal Precambrian basement. *Transactions of the Geological Society of South Africa*, **87**, 73–78.

HEPWORTH, J. V. 1972. The Mozambique orogenic belt and its foreland in northeast Tanzania: a photogeologically-based study. *Journal of the Geological Society, London*, **128**, 461–500.

HEPWORTH, J. V. & KENNERLY, J. B. 1970. Photogeology and structure of the Mozambique orogenic front in northeast Tanzania: a photogeologically-based study. *Journal of the Geological Society, London*, **125**, 447–479.

HJELLE, A. 1972. Some observations on the geology of H. U. Sverdrupfjella, Dronning Maud Land. *Norsk Polarinstitutt Arbok*, 1972, 7–22.

HOFFMAN, P. 1991. Did the breakout of Laurentia turn Gondwanaland Inside-out. *Science*, **252**, 1409–1412.

HOLMES, A. 1951. The sequence of Precambrian orogenic belts in south and central Africa. *Proceedings of the 18th International Geological Congress*, London. Association des Services geologiques africans, **XIV**, 254–269.

JACKSON, C. & ARMSTRONG, R. A. 1997. The tectonic evolution of the central Kirwanveggen, Dronning Maud Land, Antarctica: temporal resolution of deformation episodes using SHRIMP U–Pb zircon geochronology. *Tectonics Division of the Geological Society of South Africa 13th Anniversary Conference*, 21–22 (abstracts).

JACOBS, J. & THOMAS, R. J. 1994. Oblique collision at about 1.1 Ga along the southern margin of the Kaapvaal Craton continent, south east Africa. *Geologische Rundschau*, **83**, 322–333.

JACOBS, J. & THOMAS, R. J. 1996. Pan-African rejuvenation of the c.1.1 Ga Natal Metamorphic Province (South Africa): K–Ar muscovite and titanite fission track evidence. *Journal of the Geological Society, London*, **153**, 971–978.

JACOBS, J., AHRENDT, H., KREUTZER, H. & WEBER, K. 1995. K–Ar, $^{40}Ar/^{39}Ar$ and apatite fission-track evidence for Neoproterozoic and Mesozoic basement rejuvenation events in the Heimefrontfjella and Mannefallknausane (East Antarctica). *Precambrian Research*, **75**, 251–262.

JACOBS, J., BAUER, W., SPAETH, G., THOMAS, R. J. & WEBER, K. 1996. Lithology and structure of the Grenville-aged (c. 1.1Ga) basement of Heimefrontfjella. *Geologische Rundschau*, **85**, 800–821.

JACOBS, J., FALTER, M., WEBER, K. & JESSBERGER, E. K. 1997. $^{40}Ar/^{39}Ar$ Evidence for the structural evolution of the Heimefront Shear Zone (western Dronning Maud Land) East Antarctica. *In*: RICCI, C. A. (ed.) *The Antarctic Region: Geological Evolution and Processes*. Proceedings of the VII International Symposium on Antarctic Earth Sciences, Siena, 13–20.

JACOBS, J., FANNING, C. M., HENJES-KUNST, F., OLESCH, M. & PAECH, H. J. 1998. Continuation of the Mozambique Belt into East Antarctica: Grenville age metamorphism and Polyphase Pan-African high grade events in Central Dronning Maud Land. *Journal of Geology*, **106**, 385–406.

JACOBS, J., HANSEN, B. T., HENJES-KUNST, F. ET AL. 1999. New age constraints on the Proterozoic/Lower Paleozoic Evolution of Heimefrontfjella, East Antarctica, and its bearing on Rodinia/Gondwana Correlations. *Terra Antarctica*, **6**, 377–389.

JACOBS, J., THOMAS, R. J. & WEBER, K. 1993. Accretion and indentation tectonics at the southern edge of the Kaapvaal craton during the Kibaran (Grenville) orogeny. *Geology*, **21**, 203–206.

JAMAL, D. L., ZARTMAN, R. E. & DE WIT, M. J. 1999. U–Pb single zircon dates from the Lurio Belt, northern Mozambique: Kibaran and Pan African orogenic events highlighted. GSA 11: Earth Resources for Africa (abstracts). *Journal of African Earth Sciences*, **32**.

JOHNSON, S. P. & OLIVER, G. J. H. 2000. Mesoproterozoic oceanic subduction, island arc formation and the initiation of back-arc spreading in the Kibaran belt of central, southern Africa: evidence from the opiolite Terrane, Chewore Inliers, northern Zimbabwe. *Precambrian Research*, **103**, 125–146.

KAZMIN, V. 1976. *Ophiolites in the Ethiopian Basement*. Ethiopian Institute of Geological Survey Note, 35.

KENNEDY, W. Q. 1964. *The structural differentiation of Africa in the Pan African (500Ma) tectonic episode*. Leeds University Research Institute for African Geology Annual Report, **8**, 48–49.

KEY, R. M., CHARSLEY, T. J., HACKMAN, B. D., WILKINSON, A. F. & RUNDLE, C. C. 1989. Superimposed upper Proterozoic collision-controlled orogenies in the Mozambique Orogenic Belt of Kenya. *Precambrian Research*, **44**, 197–225.

KRIEGSMAN, L. M. 1995. The Pan-African event in East Antarctica: a view from sri Lanka and the Mozambique Belt. *Precambrian Research*, **75**, 263–277.

KRÖNER, A. 2001. The Mozambique belt of East Africa and Madagascar significance of zircon and Nd model ages for Rodinia and Gondwana supercontinent formation and dispersal. *South African Journal of Geology*, **104**, 151–166.

KROENER, A., HEGNER, E., COLLINS, A., WINDLEY, B. F., BREWER, T. S., RAZAKAMANANA, T. & PIDGEON, R. T. 2000. Age and magmatic history of the Antananarivo Block, Central Madagascar, as derived from zircon chronology and Nd isotopic systematics. *American Journal of Science*, **300**, 251–288.

KRÖNER, A., BRAUN, I. & JAEKEL, P. Z. 1996. Zircon geochronology of anatectic melts and residues from a high grade pelitic assembly at Ihosy, southern Madagascar: evidence for Pan African granulite metamorphism. *Geological Magazine*, **133**, 311–323.

KRÖNER, A., SACCHI, R., JAECKEL, P. & COSTA, M. 1997. Kibaran magmatism and Pan African granulite metamorphism in northern Mozambqiue: single zircon ages and regional implications. *Journal of African Earth Sciences*, **25**, 467–484.

KRÖNER, A., WILLNER, A. P., HEGNER, E., JAECKEL, P. & NEMCHIN, A. 2002. Single zircon ages, *PT* evolution and Nd isotopic systematics of high grade gneisses in southern Malawi and their bearing on the evolution of the Mozambique belt in southeastern Africa. *Precambrian Research*, in press.

KRYNAUW, J. R. & JACKSON, C. 1996. Geological evolution of western Dronning Maud land within a Gondwana framework. South African National Antarctic Programme Final Report: 1991–1996: Geology Subsection, 1–48.

LAWVER, L. A., GAHAGEN, L. M. & DALZIEL, I. W. D. 1998. A tight fit early Mesozoic gondwana, a plate reconstruction perspective. *Special Issue, Memoir of the National Institute of Polar Research*, **53**, 214–229.

MABOKO, M. A.H. 1995. Neodymium isotopic constraints on the protolith ages of rocks involved in Pan-African tectonism in the Mozambique Belt of Tanzania. *Journal of the Geological Society, London*, **152**, 911–916.

MABOKO, M. A. H. 2000. Nd–Sr isotopic investigation of the Archaean–Proterozoic boundary in north eastern Tanzania constraints on the nature of Neoproterozoic tectonism in the Mozambique Belt. *Precambrian Research*, **102**, 87–98.

MABOKO, M. A.H. 2001a. Isotopic and geochemical constraints on Neoproterozoic crust formation in the Wami River area, eastern Tanzania. *Journal of African Earth Sciences*, **33**, 91–101.

MABOKO, M. A. H. 2001b. Dating post-metamorphic cooling of the Eastern Granulites in the Mozambique Belt of northern Tanzania using the garnet Sm–Nd method. *Gondwana Research*, **4**, 229–336.

MABOKO, M. A. H. & NAKAMURA, E. 1995. Sm–Nd garnet ages from the Uluguru granulite complex of Eastern Tanzania: further evidence for post-metamorphic slow cooling in the Mozambique belt. *Precambrian Research*, **74**, 195–202.

MABOKO, M. A. H. & NAKAMURA, E. 1996. Nd and Sr isotopic mapping of the Archean–Proterozoic boundary in southeastern Tanzania using granites as probes for crustal growth. *Precambrian Research*, **77**, 105–115.

MABOKO, M. A. H. & NAKAMURA, E. 2002. Isotopic dating of Neoproterozoic crustal growth in the Usambara mountains of north eastern Tanzania: evidence for coeval crust formation in the Mozambique Belt and the Arabian–Nubian shield. *Precambrian Research*, **113**, 227–242.

MABOKO, M. A. H., BOELRIJK, N. A. I. M., PRIEM, H. N. A. & VERDURMEN, E. A. T. H. 1985. Zircon U–Pb and biotite Rb–Sr dating of the Wami River granulites, eastern Granulites, Tanzania: evidence for approximately 715 Ma old granulite facies metamorphism and final Pan African cooling approximately 475 Ma ago. *Precambrian Research*, **30**, 361–378.

MABOKO, M. A. H., MCDOUGALL, I. & ZEITLER, P. K. 1989. Dating late Pan African cooling in the Uluguru granulite complex of Eastern Tanzania using the $^{40}Ar-^{39}Ar$ technique. *Journal of African Earth Sciences*, 159–167.

MANHICA, A. S. T. D. 1998. *The geology of the Mozambique Belt and the Zimbabwe Craton around Manica, western Mozambique*. M.Sc. Thesis, University of Pretoria.

MANHICA, A. S. T. D., GRANTHAM, G. H., ARMSTRONG, R. A., GUISE, P. G. & KRUGER, F. J. 2001. Polyphase deformation and metamorphism at the Kalahari Craton–Mozambique Belt Boundary. *In*: MILLER, J. A., HOLDSWORTH, R. E., BUICK, I. S. & HAND, M. (eds) *Continental Reactivation and Reworking*. Geological Society, London, Special Publication, **184**, 303–321.

MARKL, G., BAUERLE, J. & GRUJIC, D. 2000. Metamorphic evolution of Pan-African granulite facies metapelites from southern Madagascar. *Precambrian Research*, **102**, 47–68.

MARTELAT, J, LARDEAUX, J., NICOLLET, C. & RAKOTONDRAZAFY, R. 2000. Strain pattern and late Precambrian deformation history in southern Madagascar. *Precambrian Research*, **102**, 1–20.

MATTHEWS, P. E. 1972. Possible pre-Cambrian obduction and plate tectonics in southeastern Africa. *Nature*, **240**, 37–39.

MEERT, J. 2002. A synopsis of events related to the assembly of Eastern Gondwana. *Tectonophysics*, in press.

MEISSNER, B., DETERS, P., SRIKANTAPPA, C. & KOHLER, H. 2002. Geochronological evolution of the Moyar, Bhavani and Phalgat shear zones of southern India: implications for east Gondwana correlations. *Precambrian Research*, **114**, 149–175.

MICHARD, A, GURRIET, P., SOUDANT, M. & ALBAREDE, F. 1985. Nd isotopes in French Phanerozoic shales: external vs. internal aspects of crustal evolution. *Geochimica et Cosmochimica Acta*, **49**, 601–610.

MOLLER, A., MEZGER, K. & SCHENK, V. 1998. Crustal age domains and the evolution of the continental crust in the Mozambique Belt of Tanzania: combined Sm–Nd, Rb–Sr, and Pb–Pb isotopic evidence. *Journal of Petrology*, **39**, 749–783.

MOLLER, A., APPEL, P., MEZGER, K. & SCHENK, V. 1995. Evidence for a 2 Ga subduction zone: eclogites in the Usagaran belt of Tanzania. *Geology*, **23**, 1067–1070.

MOLLER, A., MEZGER, K. & SCHENK, V. 2000. U–Pb dating of metamorphic minerals: Pan-African metamorphism and prolonged slow cooling of high pressure granulites in Tanzania, East Africa. *Precambrian Research*, **104**, 123–147.

MOSELY, P. N. 1993. Geological evolution of the late Proterozoic 'Mozambique Belt' of Kenya. *Tectonophysics*, **221**, 223–250.

MOYES, A. B. 1993. The age and origin of the Jutulsessen granitic gneiss, Gjelsvikfjella, Dronning Maud Land South African. *Journal of Antarctic Research*, **23**, 25–32.

MOYES, A. B. & BARTON, J. M. 1990. A review of isotopic data from western Dronning Maud Land, Antarctica. *Zentralblatt fur Geologie und Palaeontologie, Teil I*, 19–31.

MOYES, A. B. & GROENEWALD, P. B. 1996. Isotopic constraints on Pan African Metamorphism in Dronning Maud Land, Antarctica. *Chemical Geology*, **129**, 247–256.

MOYES, A. B., BARTON, J. M. & GROENEWALD, P. B. 1993. Late Proterozoic to early Palaeozoic tectonism in Dronning Maud Land, Antarctica: supercontinental fragmentation and amalgamation. *Journal of the Geological Society, London*, **150**, 833–842.

MOYES, A. B., KNOPER, M. W. & HARRIS, P. D. 1997. The age and significance of the Urfjell Group, western Dronning Maud Land. *In*: RICCI, C. A. (ed.) *The Antarctic Region: Geological Evolution and Pro-*

cesses. Proceedings of the VII International Symposium on Antarctic Earth Sciences, Siena, 13–20.

MOYES, A. B., KRYNAUW, J. R. & BARTON, J. M. 1995. The age of the Ritscherflya Supergroup and Borgmassivet Intrusions, Dronning Maud Land, Antarctica. *Antarctic Science*, **7**, 87–97.

MUHONGO, S. & LENOIR, J.-L. 1994. Pan-African granulite facies metamorphism in the Mozambique Belt of Tanzania: U–Pb zircon geochronology. *Journal of the Geological Society, London*, **151**, 343–347.

MUHONGO, S., KRÖNER, A. & NEMCHIN, A. A., 2001. Zircon ages from granulite facies rocks in the Mozambique belt of Tanzania and implications for Gondwana assembly. *Journal of Geology*, **109**, 171–190.

MUHONGO, S., SCHANDELMEIER, H. R. & KRAMERS, J. D. 1987. Rb/Sr whole rock dating of the granulite complex of the Uluguru Mountains, Tanzania and some speculations on the age of the Mozambique orogeny. *In*: MATHEIS, G. & SCHANDELMEIER, H. (eds) *Current research in African sciences*, Balkema, Rotterdam.

NAIDOO, D. D., EGLINTON, B. M. & HARMER, R. E. 1989. A Rb–Sr isotope study of a young granite sheet at Marble Delta, southern Natal. *South African Journal Geology*, **92**, 389–392.

NICOLAYSEN, L. O. & BURGER, A. J. 1965. Note on the extensive zone of 1000 million year old metamorphic igneous rocks in southern Africa. *Science de la Terre (Nancy)*, **10**, 497–516.

NARAYANA, B. L., SIVARAMAN, T. V., REDDY, G. L. N., GOVIL, P. K. & BALARAM, V. 1996. Sankari granite Rb–Sr age, tectonic setting, origin and implictions for the upper age of Cauvery Suture zone. *Proceedings of the 29th Annual Convention*, Indian Geophysical Union, Hyderabad (abstracts).

NEWTON, R. C. 1986. Metamorphic temperatures and pressures of group B and C eclogites. *Geological Society of America Memoir*, **164**, 17–30.

OLIVER, G. J. H., JOHNSON, S. P., WILLIAMS, I. S. & HERD, D. A. 1998. Relict 1.4 Ga oceanic crust in the Zambezi Valley, northern Zimbabwe. Evidence for mesoproterozoic supercontinental fragmentation. *Geology*, **26**, 571–573.

OSANAI, Y., SHIRAISHI, K., TAKAHASHI, H. *ET AL.* 1992. Geochemical characteristics of Metamorphic rocks from the Central Sor Rondane Mountains, East Antarctica. *In*: YOSHIDA, Y. (eds) *Recent Progress in Antarctic Earth Science*. Terra Scientific Publishing Company, Tokyo, 17–27.

PAQUETTE, J.-L. & NÉDÉLEC, A. 1998. A new insight into Pan-African tectonics in the East–West Gondwana collision zone by U–Pb zircon dating of granites from central Madagascar. *Earth and Planetary Science Letters*, **155**, 45–56.

PAQUETTE, J. L., NEDELEC, A., MOINE, B. & RAKOTONDRAZAFY, M. 1994. U–Pb, single zircon Pb-evaporation and Sm–Nd isotopic study of a granulite domain in SE Madagascar. *Journal of Geology*, **102**, 523–538.

PINNA, P., JOURDE, G., CALVEZ, J. Y., MROZ, J. P. & MARQUES, J. M. 1993. The Mozambique Belt in northern Mozambique: Neoproterozoic (1100–850 Ma) crustal growth and tectogenesis, and superimposed Pan-African (800–550 Ma) tectonism. *Precambrian Research*, **62**, 1–59.

PIPER, J. D. A. 1976. Paleomagnetic evidence for a Proterozoic supercontinent. *Philosophical Transactions of the Royal Society of London, Series A*, **280**, 469.

PIPER, J. D. A. 2000. The Neoproterozoic supercontinent: Rodinia or Palaeopangea? *Earth and Planetary Science Letters*, **176**, 131–146.

POWELL, C. MCA, LI, Z. X., MCELHINNEY, M. W., MEERT, J. G. & PARK, J. K. 1993. Paleomagnetic constraints on timing of the Neoprterozoic breakup of Rodinia and the Cambrian formation of Gondwana. *Geology*, **21**, 889–892.

REEVES, C. V., SAHU, B. K. & DE WIT, M. 2002. A re-examination of the paleoposition of Africa's eastern neighbors in Gondwana. *Journal of African Earth Sciences*, in press.

RICKERS, K., MEZGER, K. & RAITH, M. 2001. Evolution of the continental crust in the Proterozoic Eastern Ghats Belt, India and new constraints for rodinia reconstruction: implications from Sm–Nd, Rb–Sr and Pb–Pb isotopes. *Precambrian Research*, **112**, 183–210.

SACCHI, R., MARQUES, J., COSTA, M. & CASATI, C. 1984. Kibaran events in the southern Mozambique Belt. *Precambrian Research*, **25**, 141–159.

SANTOSH, M., KAGAMI, H., YOSHIDA, M. & NANDA-KUMAR, V. 1992. Pan African charnockite formation in East Gondwana: geochronologic (Sm–Nd and Rb–Sr). and petrological constraints. *Bulletin of the Indian Geological Association*, **25**, 1–10.

SELLWOOD, B. W. 1986. Shallow-marine carbonate environments. *In*: READING, H. G. (ed.) *Sedimentary Environments and Facies*. Blackwell Science, London, 283–342.

SHACKLETON, R. M. 1996. The final collision between East and West Gondwana: where is it? *Journal of African Earth Sciences*, **23**, 271–287.

SHACKLETON, R. M. 1986. Precambrian collision tectonics in Africa. *In*: COWARD, M. P. & RIES, A. C. (eds) *Collision Tectonics*. Geological Society, London, Special Publication, **19**, 329–353.

SHACKLETON, R. M. 1993. Tectonics of the Mozambique Belt in East Africa. *In*: PRICHARD, H. M., ALABASTER, T., HARRIS, N. B. W. & NEARY, C. R. (eds) *Magmatic Processes and Plate Tectonics*. Geological Society, London, Special Publication, **76**, 345–362.

SHIBATA, K. & SUWA, K. 1979. *A geochronological study on granitoid gneiss from the Mbooni Hills, Machakos area, Kenya*. Fourth Preliminary Report on African Studies, Department of Earth Sciences, Nagoya University, Japan, 163–167.

SHIRAISHI, K. & KAGAMI, H. 1992. Sm–Nd and Rb–Sr ages of metamorphic rocks from the Sør Rondane Mountains, east Antarctica. *In*: YOSHIDA, Y. (eds) *Recent Progress in Antarctic Earth Science*. Terra Scientific Publishing Company, Tokyo, 29–35.

SHIRAISHI, K. & KOJIMA, S. 1987. Basic and intermediate gneisses from the western part of the Sor Rondane Mountains, East Antarctica. *Proceedings of the National Institute of Polar Research Symposium on Antarctica Geosciences*, **1**, 129–149.

SHIRAISHI, K., ASAMI, M., ISHIZUKA, H. *ET AL.* 1991. Geology and metamorphism of the Sor Rondane mountains, east Antarctica. *In*: THOMSON, M. R. A., CRAME, J. A. & THOMSON, J. W. (eds) *Geological*

Evolution of Antarctica. Cambridge University Press, Cambridge, 77–82.

SHIRAISHI, K., ELLIS, D. J., FANNING, C. M., HIROI, Y., KAGAMI, H. & MOTOYOSHI, Y. 1997. Re-examination of the metamorphic and protolith ages of the Rayner Complex, Antarctica: evidence for the Cambrian (Pan-African) regional metamorphic event. *In*: RICCI, C. A. (ed.) *The Antarctic Region: Geological Evolution and Processes.* Proceedings of the VII International Symposium on Antarctic Earth Sciences, Siena, Terra Antarctica.

SHIRAISHI, K., ELLIS, D. J., HIROI, Y., FANNING, C. M., MOTOYOSJI, Y. & NAKAI, Y. 1994. Cambrian orogenic belt in East Antarctica and Sri Lanka: implications for Gondwana assembly. *Journal of Geology*, **102**, 47–65.

SHIRAISHI, K., HIROI, Y., ELLIS, D. J., FANNING, C. M., MOTOYOSHI, Y. & NAKAI, Y. 1992. The first report of a Cambrian orogenic belt in east Antarctica – an ion microprobe study of the Lützow-Holm Complex. *In*: YOSHIDA, Y. (eds), *Recent Progress in Antarctic Earth Science.* Terra Scientific Publishing Company, Tokyo, 67–73.

STEIN, M. & GOLDSTEIN, S. L. 1996. From plume head to continental lithosphere in the Arabian–Nubian shield. *Nature*, **382**, 773–778.

STERN, R. J. 1994. Arc assembly and continental collison in the Neoproterozoic East African Orogen: Implications for the consoildation of Gondwanaland. *Annual Reviews Earth Science*, **22**, 319–351.

STERN, R. J. 2002. Crustal evolution in the East African Orogen: a neodymium isotopic perspective. *Journal of African Earth Sciences*, in press.

TADESSE, T., HOSHINO, M., SUZUKI, K. & IIZUMI, S. 2000. Sm–Nd, Rb–Sr and Th–U–Pb zircon ages of syn- and post-tectonic granitoids from the Axum area of northern Ethiopia. *Journal of African Earth Sciences*, **30**, 313–327.

THOMAS, R. J. 1989. A tale of two tectonic terranes. *South African Journal of Geology*, **92**, 306–321.

THOMAS, R. J. & EGLINGTON, B. M. 1990. A Rb–Sr, Sm–Nd and U–Pb zircon isotopic study of the Mzumbe suite, the oldest intrusive granitoid in southern Natal, South Africa. *South African Journal of Geology*, **93**, 761–765.

THOMAS, R. J., CORNELL, D. H. & ARMSTRONG, R. A. 1999. Provenance age and metamorphic history of the Quha Formation, Natal Metamorphic Province: a U–Th–Pb zircon SHRIMP study. *South African Journal of Geology*, **102**, 83–88.

THOMAS, R. J., EGLINGTON, B. M. & BOWRING, S. A. 1992. U–Pb zircon data from the Mbizana microgranite: dating the cessation of Kibaran magmatism in Natal, South Africa, *Proceedings of the 9th International Geology Conf*erence, Geological Society of Africa, 30 (abstracts).

THOMAS, R. J., EGLINGTON, B. M. & BOWRING, S. A. 1993a. Dating the cessation of Kibaran magmatism in Natal, South Africa. *Journal of African Earth Sciences*, **16**, 247–252.

THOMAS, R. J., EGLINGTON, B. M., BOWRING, S. A., RETIEF, E. A. & WALRAVEN, F. 1993b. New isotopic data from a neoproterozoic porphyritic granitoid–charnockite suite from Natal, South Africa. *Precambrian Research*, **62**, 83–101.

THOMAS, R. J., GRANTHAM, G. H, INGRAM, B., MACEY, P.,

CRONWRIGHT, M. & DU TOIT, M. 2001. The Geology of NE Mozambique – new structural inferences from recent fieldwork. *International Symposium and Fieldworkshop on the Assembly and Breakup of Rodinia and Gondwana and Growth of Asia*, Osaka, October 2001 (abstracts).

TIKKU, A. A., MARKS, K. M. & KOVACS, L. C. 2002. An early Cretaceous extinct spreading center in the northern Natal, Valley. *Tectonophysics*, in press.

TROMPETTE, R. 1994. *Geology of Western Gondwana (2000–500 Ma): Pan-Africa–Brasiliano Aggregation of South America and Africa.* Balkema, Rotterdam.

TORSVIK, T. H., CARTER, L. M., ASHWAL, L. D., BHUSHAN, S. K., PANDIT, M. K. & BAMTVEIT, B. 2001. Rodinia refined or obscured: paleomagnetism of the Malani igneous suite (NW India). *Precambrian Research*, **108**, 319–333.

TUCKER, R. D., ASHWAL, L. D. & TORSVIK, T. H. 2001. U–Pb geochronology of Seychelles granitoids: a Neoproterozoic continental arc fragment. *Earth and Planetary Science Letters*, **187**, 27–38.

VAN AUTENBOER, T. 1972. Recent geological investigations in the Sor-Rondane Mountains, Belgicafjella, Dronning Maud Land. *In*: ADIE, R. J. (ed.) *Antarctic Geology and Geophysics.* Universiteits Forlaget, Oslo, 563–572.

VAN AUTENBOER, T. & LOY, W. 1966. *The geology of Sor-Rondane, Antarctica. Data report: central part of the range.* Centre National de Recherches Palaires de Belgique, 3, Avenue Circulaire, Uccle.

VEARNCOMBE, J. R. 1983. A dismembered ophiolite from the Mozambique Belt, West Pokot, Kenya. *Journal of African Earth Sciences*, **1**, 133–143.

VINYU, M. L., HANSEN, R. E., MARTIN, M. W., BOWRING, S. A., JELSMA, H. A., KROL, M. A. & DIRKS, P. H. G. M. 1999. U–Pb and $^{40}Ar/^{39}Ar$ geochronological constraints on the tectonic evolution of the eastern part of the Zambezi orogenic belt, northeast Zimbabwe. *Precambrian Research*, **98**, 67–82.

VRANA, S., PRASAD, R. & FEDIUKOVA, E. 1975. Metamorphic kyanite eclogites in the Lufilian arc of Zambia. *Contributions to Mineralogy and Petrology*, **51**, 139–160.

WAREHAM, C. D., PANKHURST, R. J., THOMAS, R. J., STOREY, B. C., GRANTHAM, G. H., JACOBS, J. & EGLINGTON, B. M. 1998. Pb, Nd, and Sr isotope mapping of Grenville-age crustal provinces in Rodinia. *Journal of Geology*, **106**, 647–659.

WATSON, R. L. A. 1969. The geology of the Cashel Melsetter and Chipinga areas. *Rhodesia Geological Survey, Bulletin*, **60**.

WENDT, I., BESANG, C., HARRE, W., KREUZER, H., LENZ, H. & MULLER, P. 1972. Age determinations of granitic intrusions and metamorphic events in the early Precambrian of Tanzania. *Proceedings of the 24th International Geological Congress*, Montreal, Section 1, 295–314.

WILSON, A. H. 1990. Tugela Rand Layered Suite. *In*: *Catalogue of South African Lithostratigraphic Units* 2, 47–48.

WILSON, T. J., GRUNOW, A. M. & HANSON, R. E. 1997. Gondwana assembly: the view from southern Africa and East Gondwana. *Journal of Geodynamics*, **23**, 263–268.

WILSON, T. J., HANSON, R. E. & WARDLAW, M. S. 1993. Late Proterozoic evolution of the Zambesi belt, Zambia: implications for Regional Pan-African Tectonics and shear displacements in Gondwana. *In*: FINDLAY, R. H., UNRUG, R., BANKS, M. R. & VEEVERS, J. J. (eds) *Gondwana Eight*. Balkema Press, Rotterdam, 69–82.

WYLLIE, P. J. 1971. *The Dynamic Earth*. John Wiley & Sons, New York.

YOSHIDA, M. & VITANAGE, K. 1993. A review of the Precambrian geology of Sri Lanka and its comparison with Antarctica. *In*: FINDLAY, R. H., UNRUG, R., BANKS, M. R. & VEEVERS, J. J. (eds) *Gondwana Eight*. Balkema Press, Rotterdam, 97–109.

Proterozoic geochronology and tectonic evolution of southern Africa

R. E. HANSON

Department of Geology, Texas Christian University, Fort Worth, Texas 76129, USA
(e-mail: r.hanson@tcu.edu)

Abstract: The Precambrian foundation of southern Africa consists of Archaean cratonic nuclei surrounded by belts formed during three separate Proterozoic orogenic episodes. Crust that formed or was reworked at 2.05–1.8 Ga is represented by orogenic belts (e.g. Kheis–Magondi, Ubendian–Usagaran) that partly wrap the ancient craton margins or form extensive basement in younger belts. Orogenic belts formed at 1.35–1.0 Ga record arc magmatism and collisional events during assembly of the Rodinia supercontinent. The Namaqua–Natal Belt defines a major convergent plate boundary active at this time along the southern margin of the Archaean Kaapvaal Craton, and the western part of this belt is inferred to link with a largely buried, NE-trending orogen present in the Kalahari region. Orogenesis in the same time frame farther north is recorded by the Kibaran and Irumide Belts, which are separated by the Palaeoproterozoic Bangweulu Block but are inferred to have undergone a linked tectonic evolution. East of the Irumide Belt, extensive Mesoproterozoic arc crust occurs in Malawi and Mozambique, although strong Neoproterozoic overprinting of the arc rocks makes their original relations unclear. Breakup of Rodinia is signalled by widespread Neoproterozoic alkaline and bimodal magmatism associated with rift zones which, in some cases, evolved into major ocean basins. Subsequent amalgamation of Gondwana led to collisional orogenesis culminating at 575–505 Ma in the Mozambique and West Congo–Kaoko–Gariep–Saldania Belts along the present eastern and western margins of southern Africa. Coeval deformation in the interconnecting, transcontinental Damara–Lufilian–Zambezi Orogen may reflect destruction of linked rifts and narrow ocean basins driven by farfield stresses from the collisional plate margins.

This paper provides an overview of the Proterozoic geochronology and tectonic evolution of Africa south of the Equator (Fig. 1), emphasizing broad patterns of orogenesis and crustal growth, and highlighting outstanding problems. The discussion focuses on the major Proterozoic belts of the region, although some events affecting the adjacent cratons are also examined.

Recent years have seen a substantial increase in the understanding of the Proterozoic tectonic framework of southern Africa, but there are still many gaps in our knowledge. One problem is that large areas of Precambrian basement are blanketed by Phanerozoic deposits in the Congo Basin in Central Africa and in the Kalahari Desert and surrounding regions in southern Africa (Fig. 1). Subsurface data provide some insight into the basement geology beneath the Congo Basin (e.g. Daly *et al.* 1992), but the Precambrian structural framework throughout much of that vast region remains poorly known. More complete geophysical data, research boreholes and scattered basement exposures provide better constraints on the subsurface distribution of Proterozoic belts in the Kalahari region (Fig. 1; Hutchins & Reeves 1980; Meixner & Peart 1984; Key & Ayres 2000; Singletary *et al.* 2002).

Isotopic ages for Proterozoic belts in Africa cluster at 2.05–1.8 Ga, 1.35–1.0 Ga and 650–450 Ma (e.g. Clifford 1970; Cahen *et al.* 1984). These three orogenic episodes are generally referred to as Eburnian, Kibaran, and Pan-African in the literature. In order to avoid confusion, the first two of these terms are not used in this general sense here because they were originally coined for specific belts or domains (the Eburnian Domain in West Africa and the Kibaran Belt in Central Africa). The term Pan-African, in contrast, was coined for a continentwide thermotectonic episode (Kennedy 1964) and provides a useful general designation for Neoproterozoic–Early Palaeozoic orogenesis that can be traced over large parts of Africa and occurred during assembly of the Gondwana supercontinent (Fig. 2). This episode in Africa equates, for example, with widespread Brasiliano orogenesis in South America (e.g. Trompette 1994, 2000; Brito Neves *et al.* 1999). It should be noted, however, that usage of Pan-African in this general sense is not followed by some workers (e.g. de Wit 1998; de Wit *et al.* 2001).

Archaean cratonic nuclei in southern Africa include the Tanzania Craton, and the Zimbabwe and Kaapvaal Cratons welded along the Archaean Limpopo Belt (Fig. 1). Archaean crust is also exposed on the flanks of the Congo Basin, including the Angola–Kasai Craton, parts of the Angola basement farther west and the Chaillu Massif in Gabon. More extensive cratonic entities that became stabilized following Mesoproterozoic orogenesis in southern Africa are termed the Congo and Kalahari Cratons (Fig. 2; Clifford 1970), which represent two of the main components of the Rodinian and Gondwana supercontinents (e.g. Hoffman 1991;

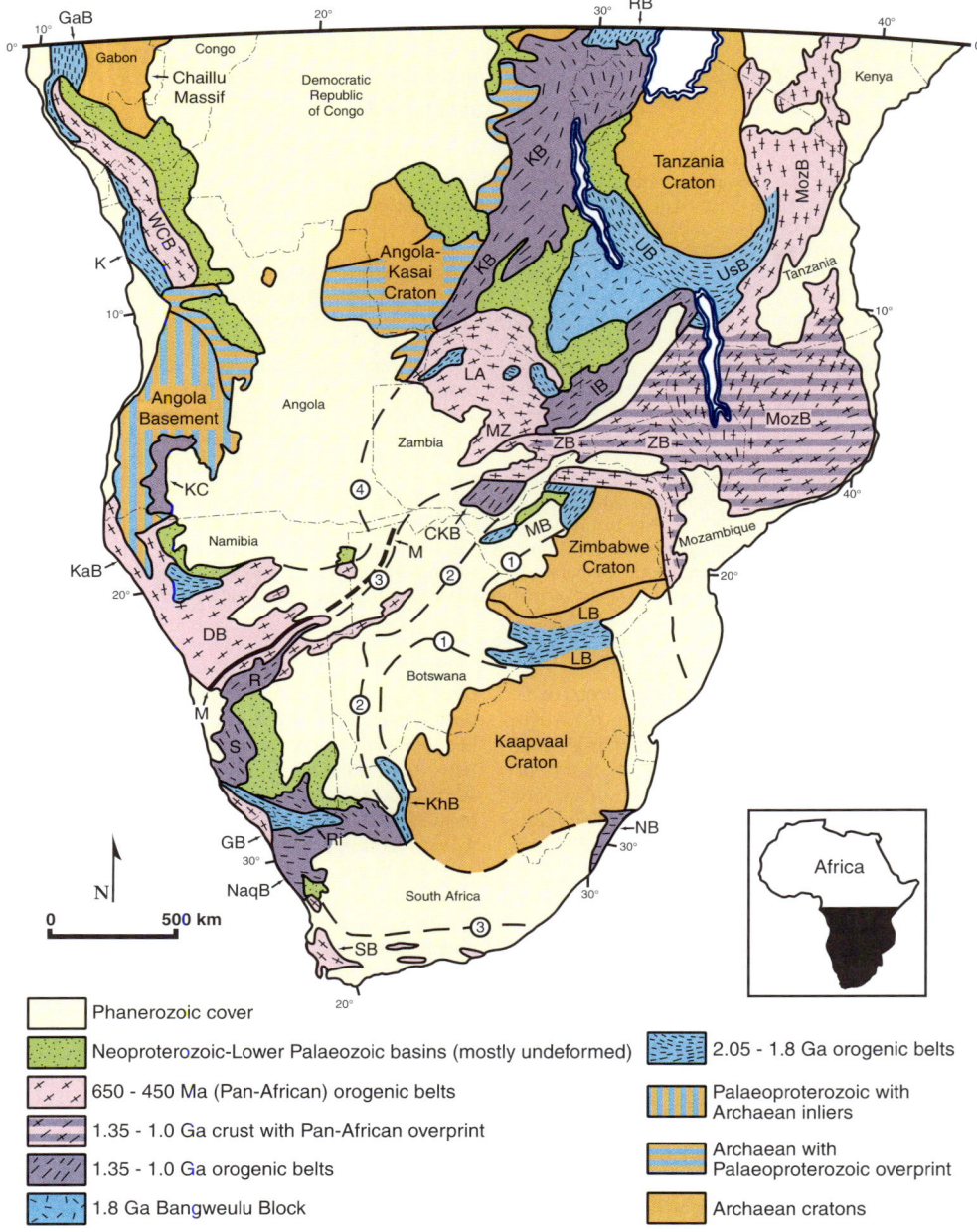

Fig. 1. Precambrian tectonic framework of southern Africa, modified from various sources including Unrug (1996). Structural trends are shown schematically. Dashed lines with circled numbers in the Kalahari region indicate inferred positions of major subsurface tectonic boundaries and are after Majaule *et al.* (2001) and Singletary *et al.* (2002). 1, western edge of Archaean cratons; 2, boundary between 2.05–1.8 Ga orogenic belts and 1.35–1.0 Ga belts; 3, boundary between 1.35–1.0 Ga belts and Pan-African belts; 4, western edge of Pan-African belts. CKB – Choma–Kalomo Block; DB, Damara Belt; GaB, Gabon Belt; GB, Gariep Belt; IB, Irumide Belt; K, Kimezian; KaB, Kaoko Belt; KB, Kibaran Belt; KC, Kunene Complex and associated granites; KhB, Kheis Belt; LA, Lufilian Arc; LB, Limpopo Belt; M, Matchless Amphibolite; MB, Magondi Belt; MozB, Mozambique Belt; MZ, Mwembeshi Shear Zone; NaqB, Namaqua Belt; NB, Natal Belt; R, Rehoboth Inlier; RB, Ruwenzori Belt; Ri, Richtersveld Terrane; S, Sinclair Sequence; SB, Saldania Belt; UB, Ubendian Belt; UsB, Usagaran Belt; WCB, West Congo Belt; ZB, Zambezi Belt.

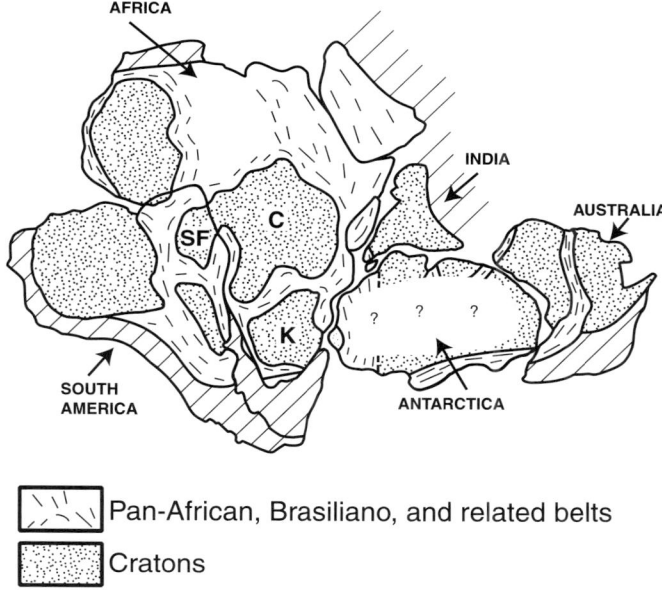

Fig. 2. Gondwana reconstruction at end of Neoproterozoic, from Hoffman (1991) and Fitzsimons (2000). C, Congo Craton; K, Kalahari Craton; SF, São Francisco Craton.

Dalziel 1997). Prior to Gondwana breakup, the Congo Craton originally linked with the São Francisco Craton to form a single large mass of continental crust wrapped by Pan-African–Brasiliano orogenic belts (Fig. 2; Trompette 1994, 2000). In southern Africa, the Congo and Kalahari Cratons are separated by a transcontinental Pan-African orogen comprising the Damara Belt, the Lufilian Arc and the Zambezi Belt (Figs 1 & 2). The nature and magnitude of Neoproterozoic plate displacements recorded by this orogen are controversial, and the older terranes to each side are therefore discussed separately below.

The best available geochronological constraints known to the author on the evolution of Proterozoic orogenic belts in southern Africa are presented in Figures 3–8. U–Pb zircon or monazite ages are emphasized whenever possible, but in some regions reliance still must be placed on less robust geochronometers, such as the Rb–Sr whole-rock isochron technique. Many of the U–Pb zircon ages discussed here, particularly those taken from the older literature, are based on multigrain zircon fractions that in some cases are highly discordant, with large errors in concordia intercepts. The use of multigrain fractions also means that zircons of different ages may have been mixed, leading to possibly meaningless intercepts. Hence, some of the U–Pb zircon data should be viewed with a certain degree of caution until confirmed by single-crystal or SHRIMP U–Pb analyses. Single-crystal $^{207}Pb/^{206}Pb$ zircon evaporation ages are distinguished here from conventional U–Pb zircon ages because the former technique does not allow testing for concordance, although it has been proven to yield reliable results in many cases (e.g. Kröner et al. 2001 and refs cited therein).

All cited ages have been recalculated, where necessary, using currently recommended constants (Steiger & Jäger 1977). Many of the Rb–Sr and U–Pb results published prior to 1984 were recalculated by Cahen et al. (1984) using more modern regression techniques, and these recalculated ages are used here. Cited errors come from the original works, unless recalculated by Cahen et al. (1984).

Greatest attention is given in this paper to the Pan-African orogenic episode because of its importance in Gondwana assembly, and because much of the present large-scale structural framework of Africa is inherited from this episode.

Orogenic belts formed at 2.05 to 1.8 Ga

Ubendian–Usagaran Orogen and the Bangweulu Block

An outstanding example of Palaeoproterozoic accretionary tectonics in southern Africa is represented by the linked Usagaran and Ubendian Belts along the eastern and southern margins of the Tanzania Craton (Figs 1 & 3). The Usagaran Belt contains a thick supracrustal assemblage that preserves a complex record of sedimentation, volcanism and tectonism along the eastern craton margin. The lower part of the assemblage consists of semipelitic to pelitic paragneisses that were deformed and metamorphosed in the upper amphibolite to granulite facies prior to deposition of unconformably overlying volcanosedimentary sequences (Priem et al. 1979; Gabert 1984; Shackleton 1986; Mruma 1989). The entire supracrustal assemblage was subsequently isoclinally folded and imbricated under greenschist to amphibolite facies conditions and thrust against the craton margin (Mruma 1989). Extensive granitoid rocks intruded during and after the deformation have Nd isotopic signatures suggesting derivation from mixed juvenile and Archaean crustal sources (Maboko & Nakamura 1996). Rb–Sr whole-rock isochron (or errorchron) dates of 1920 ± 62 and 1894 ± 33 Ma have been obtained for the metamorphosed supracrustal rocks, and the granitoid plutons have yielded Rb–Sr isochron dates ranging from 1910 ± 89 to 1747 ± 47 Ma (Fig. 3; Wendt et al. 1972; Gabert & Wendt 1974; Priem et al. 1979; Maboko & Nakamura 1996). Although these results are difficult to interpret in detail, they document a Palaeoproterozoic orogenic history for the belt. Metapelitic whiteschists and eclogites with mid-ocean ridge basalt (MORB)-like trace element contents are locally structurally interleaved with the paragneisses and record metamorphic pressures $\geq$ 18 kbar (Möller et al. 1995). U–Pb monazite ages of 2000 ± 2 and 1999 ± 2 Ma for the whiteschists are interpreted to record the high-pressure metamorphism, and a U–Pb titanite age of 1996 ± 2 Ma for the eclogites may record rapid cooling during decompression (Möller et al. 1995).

The eastern and northern limits of the Usagaran Belt are uncertain because of strong overprinting during Pan-African orogenesis in the Mozambique Belt (Fig. 1). To the SW, the Usagaran structural trends can be traced around the southern margin of the Tanzania Craton to merge with the Ubendian Belt (e.g. Priem et al. 1979; Gabert 1984). The latter belt is characterized by a series of orogen–scale, terrane-bounding, NW-trending transcurrent or transpressional dextral ductile shear zones that were initiated during Palaeoproterozoic orogenesis but show evidence of significant later reactivation. Individual terranes consist of migmatitic felsic orthogneiss and paragneiss, anorthosite, metamorphosed mafic and ultramafic rock, and relatively minor amounts of marble, quartzite and mica schist. Granulites are common in some areas and show variable degrees of amphibolite facies retrogression that accompanied dextral shearing (Fitches 1970; Ray 1974; Daly 1988; Lenoir et al. 1994; Theunissen et al. 1996). Mafic rocks with MORB geochemical signatures contain relict high-pressure granulite and eclogite facies assemblages recording pressures $\geq$ 17 kbar (Boven et al. 1999).

Geochronological constraints on early stages in the evolution of the Ubendian Belt are restricted to its southeastern part (Fig. 3), which has been partly overprinted by the younger Irumide Belt. Ring et al. (1997) obtained a series of ^{207}Pb/^{206}Pb zircon evaporation ages from this region in northern Malawi that are interpreted to record arc-type plutonism at 2093 ± 0.6 to 2048 ± 0.7 Ma, followed by high-pressure (9–11 kbar) granulite facies metamorphism at 2002 ± 0.3 Ma, and lower pressure (5.0–5.5 kbar) granulite facies metamorphism associated with syntectonic granite intrusion and migmatization at 1995 ± 0.4 to 1988 ± 0.6 Ma. Lenoir et al. (1994) reported a protolith U–Pb zircon crystallization age of 2084 ± 86 Ma for tonalite gneiss in the adjacent part of Tanzania, near the junction with the Usagaran Belt. The c. 2093–2048 Ma granitoid rocks most likely represent parts of an Andean-type arc incorporated within the Ubendian Belt. Structural juxtaposition of high-pressure rocks against lower pressure cordierite-bearing granulites reflects imbrication of tectonic slices derived from different crustal levels following peak metamorphism (Ring et al. 1997). Two-mica granites emplaced during, or shortly after, the end stages of granulite facies metamorphism have yielded U–Pb zircon and ^{207}Pb/^{206}Pb zircon evapora-

Fig. 3. Geochronological data for 2.05–1.8 Ga orogenic belts in southern Africa. Known or inferred 2.05–1.8 Ga crust in younger belts is also shown; isolated ages in younger belts represent basement inliers too small to show at the scale of the figure. G, Gweta borehole; KI, Kubu Island; OI, Okwa Inlier; Ru, Rusizian. Other abbreviations as in Figure 1. References for isotopic ages: 1, Vachette (1964), Bonhomme et al. (1982); 2, Caen-Vachette et al. (1988); 3, Maurin et al. (1990), Djama et al. (1992); 4, Delhal & Ledent (1976); 5, Cahen et al. (1984); 6, Tegtmeyer & Kröner (1985); 7, Seth et al. (1998); 8, Franz et al. (1999), Luft et al. (2001); 9, Burger et al. (1976); 10, Burger & Jacob, unpublished data cited in Kröner et al. (1991); 11, Becker et al. (1996); 12, Welke et al. (1979); 13, Reid (1982, 1997); 14, Barton (1983); 15, Robb et al. (1999); 16, Kröner et al. (1983); 17, Kabengele et al. (1990); 18, Brewer et al. (1979a), Schandelmeier (1983); 19, Pasteels (1971); 20, Key et al. (2001); 21, Ngoyi et al. (1991); 22, Rainaud et al. (1999); 23, Mapeo et al. (2001); 24, Singletary et al. (2002); 25, Majaule et al. (2001); 26, Ramokate et al. (2000); 27, Gérards & Ledent (1976); 28, Lenoir et al. (1994); 29, Möller et al. (1995); 30, Wendt et al. (1972), Gabert & Wendt (1974), Priem et al. (1979), Maboko & Nakamura (1996); 31, Dodson et al. (1975), Ring et al. (1997); 32, Dirks et al. (1999); 33, Munyanyiwa et al. (1995); 34, McCourt et al. (2001); 35, Barton & Sergeev (1997), Jaeckel et al. (1997), Holzer et al. (1998); 36, Kröner et al. (1999); 37, McCourt & Armstrong (1998), Chavagnac et al. (2001).

tion ages ranging from 2026 ± 8 to 1970 ± 30 Ma (Dodson *et al.* 1975; Lenoir *et al.* 1994; Ring *et al.* 1997). Subsequent transpressional shearing and amphibolite facies retrogression in the Ubendian Belt is inferred to have occurred at *c.* 1900–1850 Ma, based on U–Pb zircon and Rb–Sr geochronological results from sheared orthogneisses and late tectonic granites (Fig. 3; Lenoir *et al.* 1994). An ^{40}Ar/^{39}Ar barroisite cooling age of 1848 ± 6 Ma from mafic tectonite associated with one of the main shear zones is also interpreted to record this event (Boven *et al.* 1999).

Eclogites and high-pressure mafic granulites with MORB affinities in the Ubendian–Usagaran Orogen

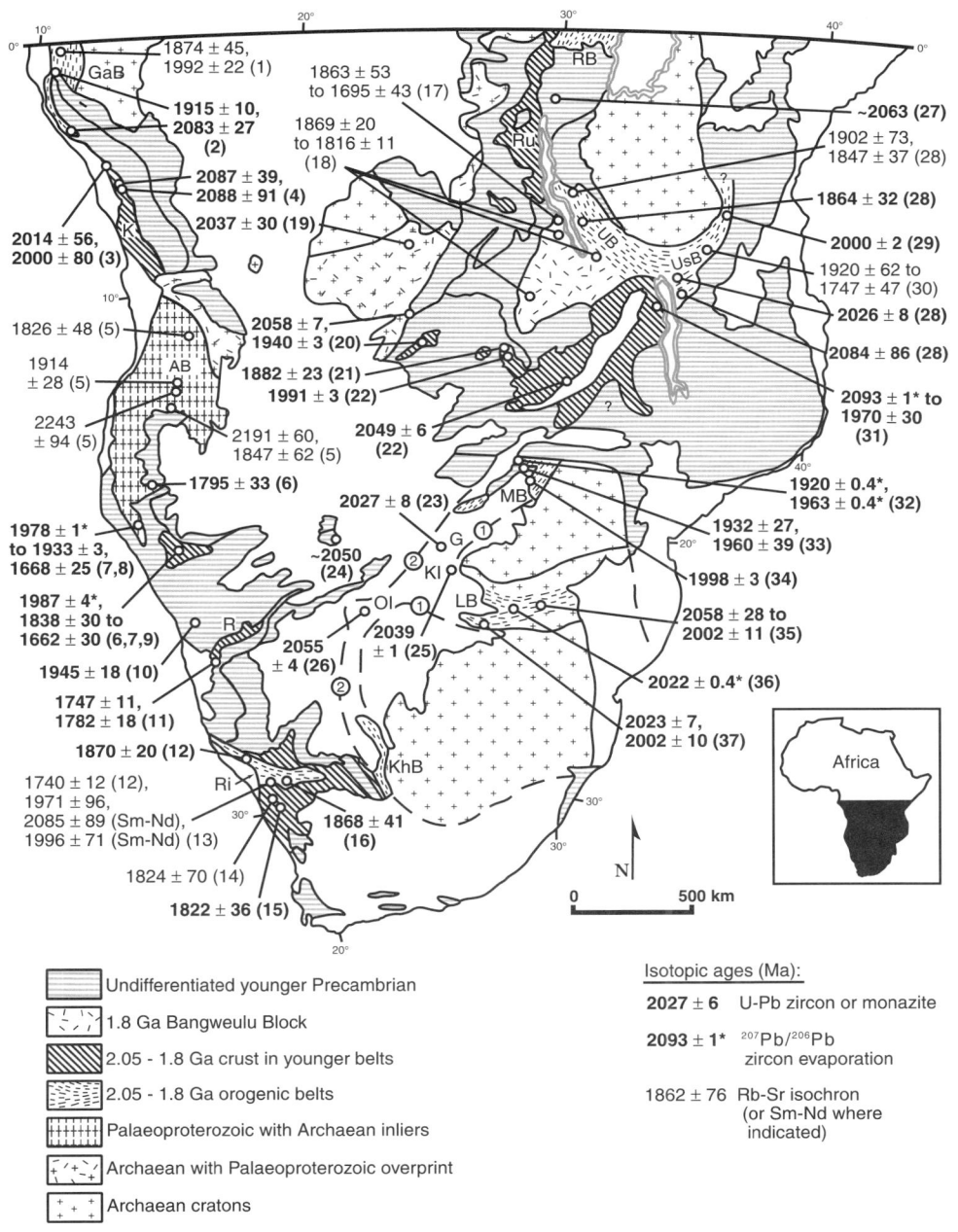

provide evidence for subduction of oceanic crust prior to collisional orogenesis. The kinematic evolution of the orogen is explained by a model involving oblique convergence and collision between the Tanzania Craton and another, unidentified, craton to the SE. In this scenario, west-vergent frontal thrusting and crustal thickening in the Usagaran Belt are linked to lateral accretion of terranes in a dextral shear regime in the Ubendian Belt along the southwestern margin of the Tanzania Craton (Daly 1988; Theunissen et al. 1996; Boven et al. 1999).

Voluminous, late to post-tectonic calc-alkaline granites and granodiorites were emplaced along both the northeastern margin of the Ubendian Belt against the Tanzania Craton and along the southwestern margin in contact with the Bangweulu Block (Fig. 3). They are widespread in the Bangweulu Block, where they intrude cogenetic dacitic to rhyolitic ignimbrites that locally rest nonconformably on sheared Ubendian gneisses (Schandelmeier 1981; Andersen & Unrug 1984; Lenoir et al. 1994). The ignimbrites and granitoid rocks have yielded Rb–Sr whole-rock isochron dates of 1869 ± 20 to 1816 ± 11 Ma (Brewer et al. 1979a; Schandelmeier 1983; Kabengele et al. 1990), and one post-tectonic intrusive complex has a Rb–Sr whole-rock isochron date of 1695 ± 43 Ma (Kabengele et al. 1990). Generation of these voluminous felsic magmas may reflect late to postcollisional anatexis of thickened crust. The basement beneath the Bangweulu Block is generally hidden, but low-grade schists that predate the main felsic magmatism are exposed in the eastern part of the block and presumably represent a southern extension of the Ubendian Belt. Initial Sr isotope ratios for some of the granites are as high as 0.707, suggesting that older sialic crust may underlie the block (Andersen & Unrug 1984).

The Rusizian gneisses (Fig. 3) are commonly considered to represent the northwestern continuation of the Ubendian Belt where it has been overprinted by the Kibaran Belt (e.g. Cahen & Snelling 1966; Ray 1974), and they provide a link between the Ubendian Belt and the broadly coeval Ruwenzori Belt farther north (Gabert 1984). Partly reworked Palaeoproterozoic basement also occurs in younger belts south of the Bangweulu Block and in places has yielded U–Pb zircon ages that overlap with ages for Ubendian orogenesis (Fig. 3). This basement is widespread in the Irumide Belt (Daly 1986a; Rainaud et al. 1999; De Waele & Mapani 2002) and is also exposed in structural domes in the Lufilian Arc (Ngoyi et al. 1991; Rainaud et al. 1999; Key et al. 2001). Extensive Palaeoproterozoic basement is inferred to occur still farther SE, in southeastern Zambia and adjacent parts of Malawi and Mozambique, based on structural evidence and imprecise Rb–Sr errorchron results (e.g. Haslam et al. 1986; Johns et al. 1989). The Champira Dome within this area in Malawi (Haslam et al. 1980) may contain even older, Archaean crust (Cahen et al. 1984). Whether all of the Palaeoproterozoic basement in this complex, poorly understood region is directly related to the Ubendian–Usagaran–Bangweulu province is unclear because of strong overprinting during younger orogenic events.

Kheis–Magondi Orogen

Palaeoproterozoic orogenesis in the Magondi Belt affected a sequence ≥ 5 km thick of sedimentary and volcanic rocks (Magondi Supergroup) deposited along the western margin of the Zimbabwe Craton (Fig. 3). Rift-related tholeiitic basalts and clastic rocks at the base of the sequence are overlain by a passive-margin assemblage that contains abundant carbonates and, prior to tectonic disruption of the sequence, is inferred to have interfingered westward with deeper marine pelites and turbiditic sandstones (Leyshon & Tennick 1988; Treloar 1988; Master 1991). The Magondi Supergroup was deformed at low metamorphic grade in an east-vergent fold-thrust belt along the craton margin, but metamorphic grade increases north and west within the orogen, culminating in granulite facies assemblages in places. In higher grade regions, the supracrustal rocks are tectonically interleaved with migmatitic basement gneisses (Treloar 1988).

The Magondi supracrustal rocks initially appear to have followed a typical prograde Barrovian-type pressure–temperature (P–T) path into the amphibolite facies during contractional deformation (Treloar 1988). However, available geothermobarometric data show that the amphibolite–granulite facies transition occurred isobarically at c. 6 kbar (Treloar & Kramers 1989; Munyanyiwa & Maaskant 1998). These data imply that the granulites do not record crustal thickening during collisional orogenesis and instead may reflect high heat flow at mid-crustal levels caused by magmatic intraplating or underplating. How this event relates to the overall tectonic evolution of the belt is yet to be resolved.

Master (1991) has argued that the Magondi Supergroup was deposited behind a west-facing volcanic arc in a back-arc basin that was destroyed when the arc collided with the continental margin. This model is consistent with the presence of andesitic and felsic volcanic and volcaniclastic rocks intercalated with deep-marine strata deposited off the continental margin (Leyshon & Tennick 1988). Remnants of the arc may be represented by extensive granitoid rocks in the western part of the belt. Archaean basement is at least locally present in that region (Dirks et al. 1999), implying that the arc may have been built on a fragment of older crust.

The timing of Magondi orogenesis is partly constrained by U–Pb zircon ages of 1960 ± 39 to 1932 ± 27 Ma for syntectonic charno-enderbite plutons intruded during granulite facies metamorphism in the northern part of the belt (Munyanyiwa et al. 1995); orthogneisses in the same general area have yielded comparable $^{207}Pb/^{206}Pb$ zircon evaporation ages (Fig. 3; Dirks et al. 1999). Farther south, the late tectonic Urungwe Granite, which forms a major batholith intruded into the supracrustal rocks, has a U–Pb zircon age of 1998 ± 3 Ma (McCourt et al. 2001).

Geophysical evidence shows the Magondi Belt to continue SW into northern Botswana beneath extensive cover in the Kalahari region (Carney et al. 1994). U–Pb geochronological data from high-grade paragneisses penetrated by the Gweta borehole in this buried part of the belt (Fig. 3) record metamorphic zircon growth at c. 2027 Ma (Mapeo et al. 2001), and small outcrops of c. 2039 Ma granite occur at Kubu Island (Majaule et al. 2001). Still farther SW along-strike, limited exposures of granite, metarhyolite and orthogneiss in the Okwa Inlier have yielded U–Pb zircon ages of 2055 ± 4 Ma (Ramokate et al. 2000).

South of the extensive Kalahari cover, the Kheis Belt occupies an analogous tectonic position to the Magondi Belt (Figs 1 & 3). The eastern part of the Kheis Belt is a thin-skinned fold-thrust belt affecting the Olifantshoek Supergroup, a low-grade sequence ≥ 5 km thick that is dominated by quartzarenites and was deposited on the Kaapvaal Craton margin in fluvial and marginal marine environments (Stowe 1986; Altermann & Hälbich 1991; Moen 1999). The timing of Kheis deformation is not well constrained. Rift-related tholeiitic basalts near the base of the Olifantshoek Supergroup have yielded a $^{207}Pb/^{206}Pb$ zircon evaporation age of 1928 ± 4 Ma (Cornell et al. 1998), providing an older age limit for Kheis orogenesis. A younger limit for the timing of deformation comes from a Rb–Sr date of 1750 ± 60 Ma from igneous biotite within a dolerite dyke that cuts the east-vergent structures (Cornell et al. 1998).

Metamorphic grade and ductile strain increase westward in a domain lying between the low-grade Kheis fold-thrust belt and the core of the younger Namaqua Belt farther west. It has generally been thought that the main ductile fabric for as much as 50 km west of the low-grade fold-thrust belt formed during Kheis orogenesis, with the Namaqua imprint becoming dominant farther west (Hartnady et al. 1985; Stowe 1986; Thomas et al. 1994). Metamorphic pressures reached 8–11 kbar in this supposed western extension of the Kheis Belt, pointing to significant crustal thickening during the deformation (Humphreys et al. 1991).

Recently, Moen (1999) has shown that there is no unequivocal evidence for a Kheis age for the ductile deformation west of the low-grade fold-thrust belt and has suggested instead that the penetrative fabric farther west records Mesoproterozoic Namaqua deformation. Detailed geochronology is needed to resolve the extent to which Namaqua structures are superimposed on older Kheis basement near the boundary between the two belts.

Subsurface data show the Kheis fold-thrust belt to extend north along the Kaapvaal Craton margin into Botswana toward the Okwa Inlier (Hutchins & Reeves 1980; Meixner & Peart 1984). The simplest interpretation of the available data is that the Kheis and Magondi Belts and the Okwa Inlier represent parts of a single, largely buried, Palaeoproterozoic orogenic province that records accretionary crustal growth outward from the Archaean Kaapvaal and Zimbabwe cratonic nuclei (e.g. Hartnady et al. 1985; Carney et al. 1994). This Palaeoproterozoic orogen evolved diachronously along-strike, with c. 1930 Ma rifting affecting the Kheis Belt at roughly the same time as orogenesis in the Magondi Belt to the north (McCourt et al. 2001). There is also evidence for an earlier episode of pre-2.2 Ga east-vergent thrusting that affected Palaeoproterozoic strata (Griqualand West Supergroup) along the Kaapvaal Craton margin prior to deposition of the Olifantshoek Supergroup (Altermann & Hälbich 1991). The regional significance of this older contractional deformation is unclear.

Limpopo Belt and Bushveld Igneous Province

In recent years it has become apparent that an important Palaeoproterozoic tectonothermal event affected the Archaean Limpopo Belt between the Kaapvaal and Zimbabwe Cratons (Fig. 1). U–Pb zircon and monazite and $^{207}Pb/^{206}Pb$ zircon evaporation ages record widespread deformation, granulite facies metamorphism and anatexis in the central zone of the belt at c. 2058–2002 Ma (Fig. 3; Barton & Sergeev 1997; Jaeckel et al. 1997; Holzer et al. 1998; Kröner et al. 1999; Chavagnac et al. 2001), together with emplacement of the large Mahalapye Granite at c. 2023 Ma in the western part of the belt (McCourt & Armstrong 1998). U–Pb titanite, apatite and rutile ages, Pb stepwise leaching dates for metamorphic silicate phases, $^{40}Ar/^{39}Ar$ and Rb–Sr hornblende and mica dates, and Sm–Nd mineral isochron dates from different parts of the belt also record this major event, as well as subsequent lower grade ductile shearing and post-tectonic cooling extending to ≤ 1800 Ma (e.g. Barton & van Reenen 1992; Barton et al. 1994; Kamber et al. 1995a, b, 1996; Holzer et al. 1998, 1999; Schaller et al. 1999; Chavagnac et al. 2001). Holzer et al. (1998, 1999) suggested that the Kaapvaal and Zimbabwe Cratons were first juxtaposed during this event,

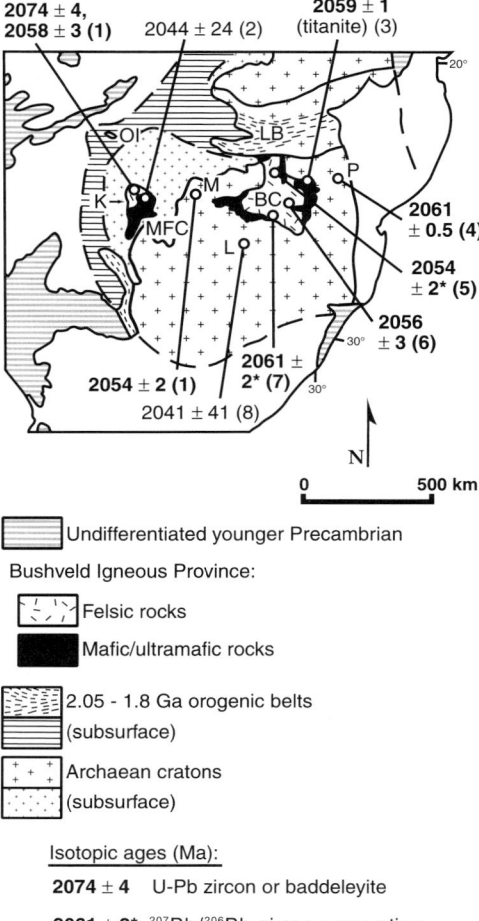

Fig. 4. Geochronological data for the Bushveld Igneous Province. BC, Bushveld Complex; K, Kukong Felsites; L, Losberg Complex; LB, Limpopo Belt; OI, Okwa Inlier; M, Moshaneng Complex; MFC, Molopo Farms Complex; P, Phalaborwa Complex. References for isotopic ages: 1, Key & Ayres (2000); 2, Reichhardt (1994); 3, Buick et al. (2001); 4, Reischmann (1995); 5, Walraven & Hattingh (1993); 6, de Waal & Armstrong (2000); 7, Walraven (1997); 8, Coetzee & Kruger (1989).

along a collisional plate boundary undergoing dextral transpression. However, the Limpopo Belt clearly records a major Archaean orogenic imprint, and the weight of evidence suggests that the Palaeoproterozoic tectonism represents reactivation of Archaean structural trends (e.g. McCourt & Armstrong 1998; Kröner et al. 1999). Transpressional reactivation was presumably kinematically linked to deformation in the Kheis–Magondi Orogen along the convergent margin to the west. Such a model is supported by the observation that aeromagnetic anomalies defining the subsurface continuation of the Magondi Belt wrap continuously around the Zimbabwe Craton margin to connect with the western part of the Limpopo Belt (Majaule et al. 2001).

The enormous Bushveld Complex was emplaced south of the Limpopo Belt at c. 2061–2054 Ma (Fig. 4; Walraven & Hattingh 1993; Walraven 1997; de Waal & Armstrong 2000; Buick et al. 2001) and was associated with widespread magmatism in adjacent parts of the Kaapvaal Craton. Other members of the Bushveld Igneous Province include the Moshaneng plutonic complex and the subsurface Molopo Farms layered mafic intrusion and Kukong felsites farther west (Fig. 4; Reichhardt 1994; Key & Ayres 2000).

The timing of this extensive igneous event overlaps with tectonothermal activity in parts of the Kheis–Magondi Orogen (e.g. Okwa Inlier), and possible relations between the Bushveld event and plate convergence along the craton margin to the west require further investigation (e.g. Uken & Watkeys 1997). Thermal weakening of the crust during the Bushveld event may help explain the extensive reactivation that affected the Limpopo Belt during regional Palaeoproterozoic orogenesis (e.g. McCourt & Armstrong 1998).

Richtersveld Terrane and Rehoboth Inlier

Extensive, variably reworked Palaeoproterozoic basement appears to be widespread in the Namaqua Belt (Thomas et al. 1994). It is best preserved in low-grade parts of the Richtersveld Terrane (Fig. 3), which contains a calc-alkaline volcano–plutonic complex that developed in a mature island-arc environment (Reid et al. 1987b; Reid 1997). Along the southern margin of the Damara Belt, the Rehoboth Inlier (Fig. 3) also contains calc-alkaline magmatic arc rocks, which have U–Pb zircon crystallization ages of 1782 ± 18 to 1747 ± 11 Ma (Becker et al. 1996). This Palaeoproterozoic basement outboard of the Kheis–Magondi Orogen may comprise one or more exotic terranes accreted during Mesoproterozoic collisional orogenesis in the Namaqua Belt.

Palaeoproterozoic basement north of the Damara Belt

U–Pb zircon and $^{207}Pb/^{206}Pb$ zircon evaporation ages from gneisses, granitoid rocks and volcanic sequences within basement inliers along the northern margin of the Damara Belt and in the Kaoko Belt record magmatism at c. 2050–1660 Ma (Fig. 3; Burger et al. 1976; Tegtmeyer & Kröner 1985; Seth et al. 1998; Franz et al. 1999; Luft et al. 2001; Singletary et al. 2002). These inliers appear to represent the southern part of an extensive, but poorly understood, Palaeoproterozoic crustal province that can be traced far to the north. Variably reworked Archaean crust occurs within this province in northern Namibia, Angola and the Democratic Republic of Congo (Cahen et al. 1984; Seth et al. 1998; Franz et al. 1999; Carvalho et al. 2000). In Angola, granites with Rb–Sr whole-rock isochron dates ranging from 2243 ± 94 to 1847 ± 62 Ma are developed on a regional scale within the province and intrude less abundant supracrustal rocks showing variable metamorphic grades (Cahen et al. 1984; Carvalho & Alves 1993; Carvalho et al. 2000). The ages of the supracrustal rocks are not well defined, but Carvalho et al. (2000) consider some of them to be Archaean. The main Palaeoproterozoic tectonothermal activity in Angola was followed by intrusion of late tectonic to anorogenic granites that have Rb–Sr whole-rock isochron dates of 1763 ± 21 to 1596 ± 86 Ma (Cahen et al. 1984; Carvalho & Alves 1993; Carvalho et al. 2000) and also occur in the Damaran Basement farther south (Burger et al. 1976). A relatively large number of Rb–Sr whole-rock isochron dates are available for the Palaeoproterozoic province in Angola, and only a few representative examples are shown in Figure 3. The geological significance of the Rb–Sr dates is unclear, and detailed U–Pb zircon geochronological studies are needed in Angola to determine the extent of Archaean basement and elucidate the timing of Palaeoproterozoic events.

Farther north, c. 2090–2000 Ma Kimezian basement in the West Congo Belt (Delhal & Ledent 1976; Maurin et al. 1990; Djama et al. 1992) provides a link between the Angola basement and the Gabon Belt (Fig. 3). To the east, this Palaeoproterozoic province is separated from the Ubendian–Usagaran–Bangweulu Province by the younger Kibaran Belt (Fig. 1). As discussed below, some workers have argued that the latter feature represents a deformed intracontinental basin. If so, this would imply that the Ubendian–Usagaran–Bangweulu Province is the eastern continuation of an extensive swath of Palaeoproterozoic (and reworked Archaean) crust that can be traced from Gabon, through Angola and the Democratic Republic of Congo, all the way to the eastern margin of the Tanzania Craton.

Orogenic belts formed at 1.35 to 1.0 Ga

Namaqua–Natal Orogen

In South Africa and adjacent parts of Namibia, the Namaqua and Natal Belts form exposed segments of a single, continuous orogen that can be traced geophysically along the southern margin of the Kaapvaal Craton beneath younger cover (Figs 1 & 5; Thomas et al. 1994). Except for a zone of deformation along the craton margin, the Natal Belt consists entirely of accreted juvenile crust (Eglington et al. 1989; Thomas & Eglington 1990). In the northern part of the belt, the craton margin is structurally overlain by an imbricated slab that contains both island-arc and ophiolitic components, and has a complex pre-obduction magmatic and deformational history (Matthews 1972; McCourt et al. 2000). Pre-obduction magmatism in the slab is recorded by a U–Pb zircon crystallization age of 1209 ± 5 Ma from arc-related tonalite gneisses (McCourt et al. 2000), and by a Sm–Nd mineral and whole-rock isochron date of 1189 ± 13 Ma from a layered mafic–ultramafic intrusion (Wilson 1990). South of the obducted slab, amphibolite to granulite facies supracrustal gneisses and deformed I-type plutons record magmatism and volcaniclastic sedimentation in at least two separate island arcs. Volcanic and plutonic components of the arc assemblages have U–Pb zircon crystallization ages ranging from 1235 ± 9 to 1163 ± 12 Ma (Thomas & Eglington 1990; Cornell et al. 1996; Thomas et al. 1999).

The timing of arc accretion and obduction of the ophiolitic slab is constrained by ^{40}Ar/^{39}Ar hornblende dates of c. 1135 Ma from sheared amphibolites (Jacobs et al. 1997). Continued convergence led to development of steep, transcurrent ductile shear zones that controlled emplacement of voluminous, late tectonic A-type granites within a regime of high heat flow (Jacobs & Thomas 1994). Associated high-temperature/low–moderate-pressure metamorphism resulted in formation of granulites with near-isobaric cooling paths (Grantham et al. 1994). U–Pb zircon and ^{207}Pb/^{206}Pb zircon evaporation ages for the granites range from 1068 ± 2 to 1029 ± 11 Ma (Thomas et al. 1993). ^{40}Ar/^{39}Ar hornblende dates indicate that shearing continued to c. 980 Ma (Jacobs et al. 1997).

In the Namaqua Belt, a relatively narrow terrane near the eastern Namaqua front contains tholeiitic, calc-alkaline and shoshonitic metavolcanic rocks with juvenile isotopic signatures. These rocks are inferred to represent an island arc accreted to the craton margin during Namaqua orogenesis (Barton & Burger 1983; Cornell et al. 1986; Geringer et al. 1994). The timing of arc magmatism is constrained by a ^{207}Pb/^{206}Pb zircon evaporation age of 1285 ± 14 Ma from a metadacite (Fig. 5; Cornell et al. 1990a). Much of the Namaqua Belt farther west apparently incorporates extensive amounts of Palaeoproterozoic basement, as noted above. Supracrustal sequences containing terrigenous, chemical sedimentary, and less abundant bimodal volcanic rocks (e.g. Bushmanland and Okiep Groups) were deposited in extensional basins developed on this basement (Reid et al. 1987a; Moore et al. 1990; Thomas et al. 1994), but the intense Namaqua deformation makes it difficult to unravel the original stratigraphy (e.g. Colliston et al. 1989). The terrigenous rocks contain detrital zircons with U–Pb ages ranging from c. 2020 to 1180 Ma (Robb et al. 1999; Raith & Cornell 2000). Amphibolite facies metabasaltic rocks in the Bushmanland Group have yielded a Sm–Nd whole-rock isochron date of 1649 ± 90 Ma, which is tentatively interpreted as the protolith crystallization age (Reid et al. 1987a).

Mesoproterozoic orogenesis in the main part of the Namaqua Belt involved polyphase ductile deformation that occurred in the amphibolite and granulite facies, and was accompanied and followed by intrusion of voluminous granitoid plutons (Thomas et al. 1994). Thrusting was to the NE near the Kaapvaal Craton margin (Stowe 1986), whereas the main Namaqua structures farther west record SW to south-vergent ductile thrusting and nappe emplacement (e.g. Blignault et al. 1983; Colliston et al. 1991; Colliston & Schoch 1998).

The available geochronological constraints for the Namaqua Belt indicate a prolonged tectonothermal history. Low-pressure mafic granulites in reworked Archaean basement near the Kaapvaal Craton margin have yielded a Rb–Sr whole-rock isochron date of 1353 ± 33 Ma, which may reflect an early extensional event affecting the continental margin (Humphreys & Cornell 1989). A Sm-Nd garnet–whole-rock isochron date of 1215 ± 50 Ma constrains the timing of prograde amphibolite facies metamorphism in the juvenile arc terrane to the west and is inferred to record crustal thickening during arc accretion (Cornell et al. 1990b, 1992). The main regional metamorphism in this part of the Namaqua Belt occurred prior to eruption of unconformably overlying, bimodal volcanic rocks of the Koras Group, which have a U–Pb zircon age of 1171 ± 7 Ma (Gutzmer et al. 2000). Farther west, sheet-like granites derived by partial melting of older crust (e.g. Barton 1983) are inferred to have been emplaced prior to or during the main phase of thrusting. These granites have yielded U–Pb zircon crystallization ages ranging from 1212 ± 11 to 1135 ± 30 Ma (Fig. 5; Armstrong et al. 1988; Robb et al. 1999).

A subsequent magmatic and thermal episode that involved low-pressure granulite facies metamorphism, and occurred in a compressional regime (e.g. Kisters et al. 1996), is well documented by a number of U–Pb zircon and monazite ages from western parts of the Namaqua Belt in South Africa (Fig. 5). Igneous activity during this episode included emplacement of granite plutons and the mafic to anorthositic Koperberg Intrusive Suite at c. 1064–1029 Ma (Clifford et al. 1995; Robb et al. 1999). Metamorphic zircon and monazite growth under high-grade conditions occurred at c. 1047–1022 Ma (Raith et al. 1999; Robb et al. 1999; Knoper et al. 2000; Raith & Cornell 2000). Tungsten deposits inferred to be related to late-stage granitic intrusions emplaced during the retrograde stages of this episode have yielded a Re–Os isochron date of 1019 ± 6 Ma (Raith & Stein 2000). This second major magmatic and metamorphic episode significantly postdates c. 1212–1135 Ma arc accretion and south-vergent thrusting associated with emplacement of pre- to syntectonic granitoid sheets. Possibly it reflects renewed subduction or collision along the new continental margin outboard of previously accreted terranes (e.g. Robb et al. 1999).

An important segment of the Namaqua Belt that

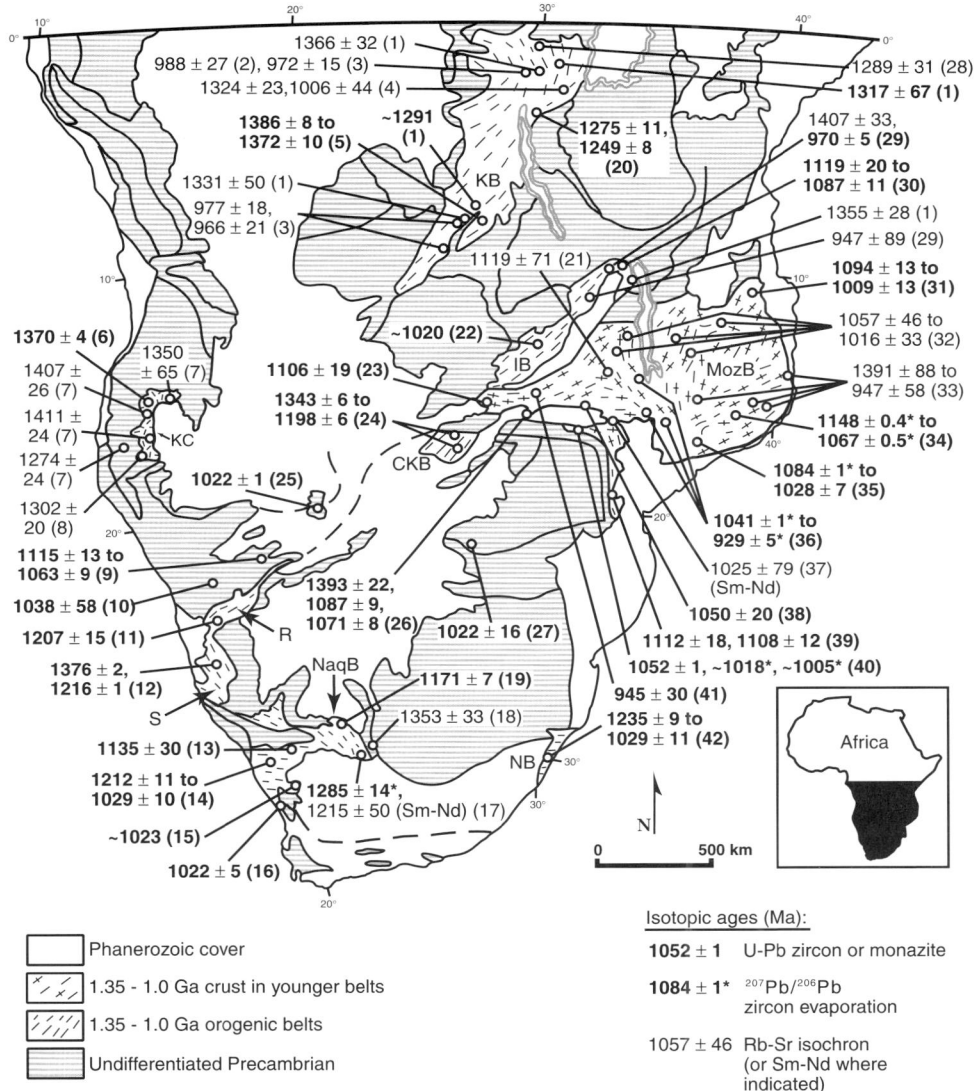

Fig. 5. Geochronological data for 1.35–1.0 Ga orogenic belts in southern Africa. Extent of 1.35–1.0 Ga crust in the Pan-African Mozambique and Zambezi Belts is also shown. Abbreviations as in Figure 1. References for isotopic ages: 1, Cahen *et al.* (1984); 2, Lavreau & Liégeois (1982); 3, Cahen & Ledent (1979); 4, Ikingura *et al.* (1990); 5, Kokonyangi *et al.* (2001); 6, Mayer *et al.* (2000); 7, Carvalho *et al.* (1979); 8, Carvalho *et al.* (1987); 9, Steven *et al.* (2000); 10, Kröner *et al.* (1991); 11, Pfurr *et al.* (1991); 12, Hoal & Heaman (1995); 13, Armstrong *et al.* (1988); 14, Clifford *et al.* (1995), Raith *et al.* (1999), Robb *et al.* (1999); 15, Knoper *et al.* (2000); 16, Raith & Cornell (2000); 17, Cornell *et al.* (1990*a, b*, 1992); 18, Humphreys & Cornell (1989); 19, Gutzmer *et al.* (2000); 20, Tack *et al.* (1994); 21, Brewer *et al.* (1979*b*); 22, De Waele & Tembo (2000); 23, Hanson *et al.* (1988*b*); 24, Hanson *et al.* (1988*a*); 25, Singletary *et al.* (2002); 26, Oliver *et al.* (1998), Goscombe *et al.* (2000); 27, van der Wel *et al.* (1998); 28, Vernon-Chamberlain & Snelling (1972); 29, Daly (1986*a*); 30, Ring *et al.* (1999); 31, Kröner (2001); 32, Haslam *et al.* (1983), Pinna *et al.* (1993), Piper (1996); 33, Cahen *et al.* (1984), Pinna *et al.* (1993); 34, Kröner *et al.* (1997), Jamal *et al.* (1999); 35, Costa *et al.* (1994), Kröner *et al.* (1997); 36, Kröner *et al.* (2001); 37, Evans *et al.* (1999); 38, Barr *et al.* unpublished data cited in Evans *et al.* (1999); 39, Manhica *et al.* (2001); 40, Dirks *et al.* (2000), Hargrove *et al.* (2002); 41, Barr *et al.* (1978); 42, Thomas & Eglington (1990), Thomas *et al.* (1993, 1999), Cornell *et al.* (1996), McCourt *et al.* (2000).

escaped this second high-grade episode is exposed farther to the NW in Namibia, where the belt comprises amphibolite facies granitoid orthogneisses, and metavolcanic and metasedimentary rocks, together with unconformably overlying, lower grade volcanic and volcaniclastic rocks of the Sinclair Sequence (Fig. 5). Geochemical data show that magmatic rocks in both assemblages formed in arc-related environments (Hoal 1993a, b). The older gneisses have a U–Pb zircon crystallization age of 1376 ± 2 Ma, whereas high-level granitoid intrusions related to the Sinclair Sequence have yielded several U–Pb zircon ages indicating emplacement at 1216 ± 1 Ma (Hoal & Heaman 1995). The sequence appears to have been preserved in arc-related extensional basins that developed along the same convergent plate boundary as coeval, higher grade parts of the Namaqua Belt to the SE (Hoal 1993a).

To the NE, the Sinclair volcano–plutonic complex continues into the Rehoboth Inlier along the southern margin of the Pan-African Damara Belt, where Sinclair-type granitoid rocks have yielded a U–Pb zircon age of 1207 ± 15 Ma (Pfurr et al. 1991). Farther north in the Damara Belt, basement inliers have yielded U–Pb zircon ages of 1115 ± 13 to 1038 ± 58 Ma (Kröner et al. 1991; Steven et al. 2000). The overall extent of this Mesoproterozoic basement within the Damara Belt is unclear.

The Namaqua–Natal Orogen contains significant amounts of juvenile crust and clearly represents one of the main convergent margins active during assembly of the Rodinia supercontinent. The presence of extensive Palaeoproterozoic basement in much of the Namaqua segment of the orogen is consistent with models involving collision of one or more exotic terranes or microcontinents in this part of the orogen (e.g. Hartnady et al. 1985; Thomas et al. 1994). There is a general similarity in the timing of events from Namaqualand to Natal, involving arc magmatism, accretion of outboard terranes and initial stages of collision over a prolonged interval from c. 1376 to 1135 Ma. This was followed by a distinct thermal pulse involving high-grade metamorphism and voluminous plutonism, still within an overall compressional regime, at c. 1068–1020 Ma. This latter event marks the culminating orogenic phase and coincides with final stages in assembly of Rodinia. In both the Namaqua and Natal Belts, the main contractional structures are overprinted by steep transcurrent shear zones. The regional kinematic patterns of these shear zones are consistent with a model in which the southwestern margin of the Kaapvaal Craton acted as an indentor within an overall NE–SW-directed compressional regime during Rodinia assembly (Jacobs et al. 1993; Dalziel et al. 2000).

Choma-Kalomo Block and subsurface continuation

North of extensive cover in the Kalahari region, part of a Mesoproterozoic orogenic belt is exposed in the Choma–Kalomo Block of southern Zambia (Fig. 1). The block contains amphibolite facies paragneisses, pelitic schists and metabasic rocks deformed along NE trends and intruded by syn- to post-tectonic granites that have yielded U–Pb zircon ages of 1343 ± 6 to 1198 ± 6 Ma (Fig. 5; Hanson et al. 1988a). Subsurface data indicate that the structural trends in the Choma-Kalomo Block continue beneath Kalahari cover into northern Botswana (e.g. Hutchins & Reeves 1980; Meixner & Peart 1984; Singletary et al. 2002). Although thick cover prevents this Mesoproterozoic Belt from being traced farther SW geophysically, it most probably connects in the subsurface with the Namaqua Belt (Fig. 1; Singletary et al. 2002).

Umkondo Igneous Province

Isolated outcrops in the Kalahari Desert in Botswana expose volcanic and sedimentary rocks (Kgwebe Formation) that were deposited in a younger rift basin superimposed on buried parts of the Mesoproterozoic orogen represented by the Choma–Kalomo Block (Modie 1996; Kampunzu et al. 1998a; Key & Mapeo 1999). Parts of the rift fill can be traced farther SW into Namibia, along the southern margin of the Damara Belt. Magmatic rocks within the rift in Botswana and Namibia have yielded U–Pb zircon ages of 1108 ± 1 to 1094 ± 20 Ma (Fig. 6 and refs cited in the caption). Bimodal metavolcanic rocks within reworked basement in the main part of the Damara Belt that have U–Pb zircon ages of 1115 ± 13 to 1081 ± 10 Ma (Steven et al. 2000) may also belong to this rift assemblage.

The rift-related magmatic rocks in Botswana and Namibia are considered by Hanson et al. (1998b) to represent the western extension of the Umkondo Igneous Province, which farther east comprises widespread tholeiitic dolerite and gabbro intrusions in parts of eastern Botswana, Zimbabwe and South Africa. Available palaeomagnetic and geochronological evidence indicates that the Umkondo igneous rocks were emplaced in a narrow time frame at c. 1100 Ma throughout this region (Fig. 6; Jones & McElhinny 1966; Allsopp et al. 1989; Hanson et al. 1998b; Reimold et al. 2000; Pancake et al. 2001; Wingate 2001). The province developed behind coeval parts of the Namaqua–Natal convergent plate boundary and, in western Botswana and Namibia, was superimposed on the slightly older Mesoproterozoic orogen now present in the subsurface in that region. Formation of this major within-plate

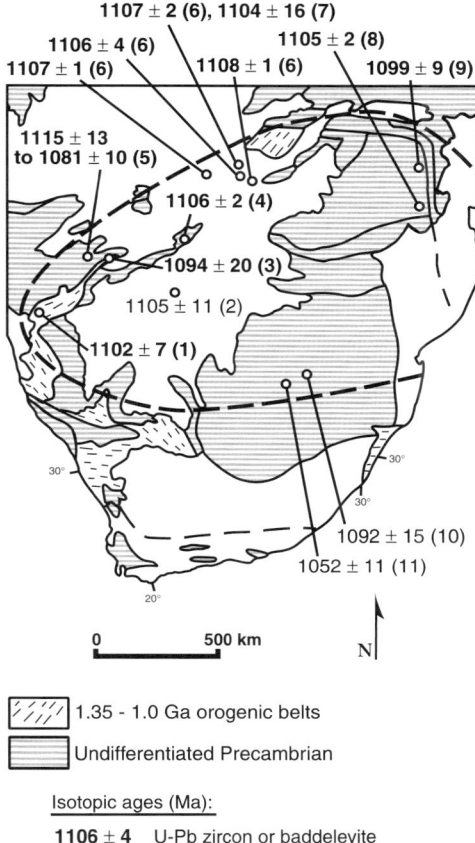

Fig. 6. Geochronological data for the Umkondo Igneous Province. Inferred extent of province indicated by heavy dashed line. References for isotopic ages: 1, Pfurr *et al.* (1991); 2, Key & Mapeo (1999); 3, Hegenberger & Burger (1985); 4, Schwartz *et al.* (1996); 5, Steven *et al.* (2000); 6, Singletary *et al.* (2002); 7, Kampunzu *et al.* (2000a); 8, Hanson *et al.* (1998b); 9, Wingate (2001); 10, Allsopp *et al.* (1967); 11, Reimold *et al.* (2000).

igneous province during assembly of Rodinia may reflect an interplay between an uprising mantle plume and stresses transmitted inboard from the Namaqua–Natal plate boundary (Hanson *et al.* 1998b).

An intriguing recent result is the report by van de Wel *et al.* (1998) of a U–Pb zircon crystallization age of 1022 ± 16 Ma from an areally restricted array of leucogranite dykes in eastern Botswana (Fig. 5). Xenocrystic cores of the zircons have ages that range from c. 1780 to 1036 Ma, and the source for this Proterozoic xenocrystic material within what has always been considered an integral part of the Archaean Zimbabwe Craton remains a mystery. It is possible that the leucogranite was emplaced during final stages of the Umkondo Igneous event, although it is significantly younger than the main Umkondo magmatism.

Kibaran and Irumide Belts and adjacent terranes

Mesoproterozoic magmatism and orogenesis are recorded over large regions west and south of the Tanzania Craton, extending down to the younger Zambezi Belt to the south (Fig. 5). One of the main Mesoproterozoic belts in this region is the Kibaran Belt, which separates the Tanzania Craton from Archaean and Palaeoproterozoic crust farther west (Fig. 1). The Kibaran Belt contains a metamorphosed supracrustal assemblage that has a structural thickness > 10 km and is dominated by pelites and psammites (typically quartzites). Carbonate rocks generally occur only in minor amounts, and basaltic to rhyolitic volcanic rocks are locally present. Parts of the assemblage were deposited in fluvial to shallow-marine environments, but some of the psammite–pelite units show typical features of turbidites, suggesting deposition in deeper marine settings (Klerkx et al. 1987; Rumvegeri 1991; Pohl 1994). Along the northern margin of the Kibaran Belt, the sedimentary rocks rest unconformably on the southern portion of the Palaeoproterozoic Ruwenzori Belt (Fig. 1; Tanner 1973; Pohl 1994). The supracrustal assemblage has commonly been termed the Kibaran Supergroup (e.g. Cahen et al. 1984), but original stratigraphic relations between different parts of the assemblage are unclear because of the strong deformational overprint.

Reworked Palaeoproterozoic (Rusizian) basement occurs in significant amounts in places within the Kibaran Belt (Fernandez-Alonso & Theunissen 1998). Ductile strain and metamorphic grade increase inward from the margins of the belt, and granulite facies assemblages have been documented in the core of the orogen in the Democratic Republic of Congo (Rumvegeri 1991). Structural vergence is to the west along the western margin of the belt exposed in the Democratic Republic of Congo, but polyphase folds and thrusts verge to the east in the eastern part of the belt in Burundi and Rwanda (Rumvegeri 1991; Tack et al. 1994). The role of extension versus contraction in forming the main ductile fabric in parts of the belt is, however, contentious (Klerkx et al. 1987; Tack et al. 1994; Fernandez-Alonso & Theunissen 1998).

Voluminous, pre- to syntectonic granitoid plutons occur in higher grade parts of the belt and show various degrees of solid-state deformation (Cahen et al. 1984; Fernandez-Alonso et al. 1986; Theunissen 1988). Peraluminous granites produced by crustal anatexis predominate (Fernandez-Alonso et al. 1986), although I-type granitoid rocks are also present (Kampunzu 1989). Published U–Pb zircon ages and Rb–Sr whole-rock isochron dates from the pre- to syntectonic plutons range from 1386 ± 8 to 1289 ± 31 Ma (Fig. 5), but the Rb–Sr dates must be treated with caution as there is evidence for significant post-crystallization resetting of whole-rock Rb–Sr systems in some Kibaran plutonic rocks (Tack et al. 1994). The best published constraints on the timing of early deformation in the belt are U–Pb zircon ages ranging from 1386 ± 8 to 1372 ± 10 Ma for granitoid rocks inferred to have been intruded during ductile deformation that preceded development of the main NE structural trends in the belt (Kokonyangi et al. 2001).

Layered mafic–ultramafic intrusions and A-type granites were emplaced along transcurrent shear zones that cut the main thrusts and folds in the northeastern part of the Kibaran Belt in Burundi. These intrusions have yielded U–Pb zircon ages of 1275 ± 11 and 1249 ± 8 Ma, indicating that the main contractional deformation had ended by that time. This magmatic event may have been triggered by late orogenic extensional collapse of the thickened crust. It was followed by another episode of east-vergent thrusting that locally sheared the intrusions under greenschist facies conditions. A Rb–Sr whole-rock isochron date of 1137 ± 39 Ma for the intrusion that yielded the c. 1275 Ma zircon crystallization age may constrain the timing of retrograde thrusting (Tack et al. 1994).

Younger magmatism is recorded by a distinctive suite of strongly peraluminous, Sn-bearing leucogranites that are present throughout the length of the Kibaran Belt and have yielded Rb–Sr whole-rock isochron dates of 1006 ± 44 to 966 ± 21 Ma (Fig. 5; Cahen & Ledent 1979; Ikingura et al. 1990). These leucogranites were emplaced during a late phase of contractional deformation (Pohl 1994).

The Kibaran Belt appears to terminate to the north against the Palaeoproterozoic Ruwenzori Belt and adjacent Archaean basement (Fig. 1), and the manner in which crustal shortening in the Kibaran Belt is accommodated along this boundary is unclear. To the SW, in the Democratic Republic of Congo, the belt is partly covered by Neoproterozoic strata. Regional maps generally show the belt to continue farther SW into Angola as a narrow strip exposed between the Angola–Kasai Craton and the Pan-African Lufilian Arc (e.g. Carvalho et al. 2000). However, recent mapping by Key et al. (2001) in northwestern Zambia has shown that the main part of the belt exposed in the Democratic Republic of Congo terminates against Archaean and Palaeoproterozoic granitic and migmatitic basement before reaching the Zambia border (Fig. 1). NE-trending shear zones cutting the basement may reflect limited Kibaran overprinting. Similar Palaeoproterozoic basement is exposed in a large structural dome to the SE within the Lufilian Arc (Fig. 3; Key et al. 2001). Whether parts of the Kibaran Belt continue beneath Neoproterozoic cover in the Lufilian Arc to connect with Mesoproterozoic terranes farther SW in Namibia and Angola remains unclear, although Kampunzu et al. (1999) have argued for such a link.

The Kibaran Belt is separated from the Irumide Belt to the east by the Bangweulu Block (Fig. 1), where a flat-lying Palaeo- to Mesoproterozoic terrigenous continental succession stratigraphically overlies the 1.8 Ga Bangweulu felsic volcano–plutonic complex. The succession thickens markedly to the SE into the Irumide Belt, where it has a structural thickness >13 km (De Waele & Mapani 2002), and there is evidence for increasing marine influence in the same direction (Daly & Unrug 1982). This succession, which is termed the Muva Supergroup, contains pelites intercalated with abundant, laterally continuous quartzites and shows marked similarities to the Kibaran supracrustal assemblage. There is a comparable paucity of carbonate rocks, and mafic or bimodal volcanic assemblages are present in places

(De Waele et al. 2000; De Waele & Mapani 2002). The Muva quartzites can be traced discontinuously around the southern edge of the Bangweulu Block, where they overlie Palaeoproterozoic basement exposed in structural domes in the eastern part of the Lufilian Arc (e.g. Daly et al. 1984). It is possible that the Kibaran and Muva sedimentary basins were originally continuous around the southwestern margin of the Bangweulu Block.

Irumide structures define a typical low-grade foreland fold-thrust belt along the southeastern margin of the Bangweulu Block, and related NE-verging thrusts cut the central part of the block (Daly 1986b, 1988). As in the Kibaran Belt, the intensity of deformation and grade of metamorphism increase toward the core of the orogen, where upper amphibolite facies assemblages developed in a regime of thick-skinned thrusting that affected both the Muva Supergroup and underlying Palaeoproterozoic basement (De Waele et al. 2000; De Waele & Mapani 2002). Vergence changes from NW to SE across a central zone of upright structures (Daly 1986b). Extensive orthogneisses represent pre- to syntectonic granitoid plutons, and late to post-tectonic granites are also present (Daly 1986a; De Waele & Tembo 2000; De Waele et al. 2000; De Waele & Mapani 2002). Although little has been published on the geochemistry of these rocks, some of them show S-type characteristics and have high initial Sr isotope ratios (Daly 1986a), making them similar to the granites in the Kibaran Belt.

Metamorphic grade decreases along-strike in the northern part of the Irumide Belt, which terminates against the older Ubendian Belt (Daly 1986a, b). The Muva quartzites nonconformably overlie Ubendian gneisses and granites in that region. Folding of these northernmost Muva rocks occurred along typical NE Irumide trends and was associated with shearing of the underlying basement (Fitches 1971). Daly (1986b) has argued convincingly that the large amount of crustal shortening across the Irumide Belt must be accommodated by transcurrent shearing in the Ubendian Belt. This model is supported by evidence for Mesoproterozoic transcurrent reactivation of major Ubendian shear zones (e.g. Klerkx et al. 1998).

There are few robust published age constraints on the timing of events in the Irumide Belt. Daly (1986a) obtained a Rb–Sr whole-rock isochron date of 1407 ± 33 Ma from pre- to syntectonic gneissic granite in the northern part of the belt. At the northern edge of the belt, Ubendian granite sheared along NE trends has yielded a Rb–Sr whole-rock isochron date of 1355 ± 28 Ma, which Cahen et al. (1984) interpreted to record resetting of the Rb–Sr system during Irumide deformation. Slightly farther north, several small granite plutons with A-type affinities have yielded U–Pb zircon ages ranging from 1119 ± 20 to 1087 ± 11 Ma (Ring et al. 1999), but the tectonic significance of these granites is unclear. Daly (1986a) obtained a U–Pb zircon age of 970 ± 5 Ma and a Rb–Sr whole-rock isochron date of 947 ± 89 Ma for two of the late to post-tectonic granites in the belt (Fig. 5), and other such granites have yielded preliminary U–Pb zircon ages of c. 1020 Ma (De Waele & Tembo 2000). The available data thus suggest a broadly contemporaneous tectonothermal evolution for both the Kibaran and Irumide Belts, including late granite plutonism at c. 1000–950 Ma. Both belts developed from basins that underwent a similar depositional history, and it seems likely that Mesoproterozoic deformation in the two belts was kinematically linked on a large scale.

The distinctive Muva quartzites can be traced into southeastern Zambia and adjacent parts of Mozambique (Johns et al. 1989), suggesting that the original Irumide basin extended at least that far. Quartzites within reworked Mesoproterozoic basement in the Zambezi Belt in northern Zimbabwe (Goscombe et al. 2000) may also correlate with the Muva Supergroup. Farther east, Mesoproterozoic calc-alkaline meta-igneous rocks with isotopic signatures recording input of significant volumes of juvenile magma are present in Mozambique and Malawi (Andreoli 1984; Sacchi et al. 1984, 2000; Costa et al. 1992; Pinna et al. 1993; Piper 1996; Kröner et al. 2001), although in the latter area they appear to be less abundant than younger, Pan-African granitoid rocks (Kröner et al. 2001). Many of the Mesoproterozoic meta-igneous rocks are dated only by Rb–Sr whole-rock isochrons (Fig. 5), but a significant number of U–Pb zircon and ^{207}Pb/^{206}Pb zircon evaporation ages are also now available, and document magmatism between c. 1148 and 929 Ma (Fig. 5; Costa et al. 1994; Kröner et al. 1997, 2001). Although other scenarios are possible (Kröner et al. 2001), these rocks are most reasonably interpreted to have developed in arc-related settings. Basement orthogneisses with U–Pb zircon crystallization ages that fall in the same time frame extend south into the younger Zambezi Belt (Fig. 5). Sheared plagiogranite within a dismembered ophiolitic/volcanic-arc assemblage in that area has yielded a U–Pb zircon age of 1393 ± 22 Ma (Oliver et al. 1998; Johnson & Oliver 2000), suggesting that arc magmatism occurred over a prolonged time interval. The arc-type rocks in Malawi and Mozambique show a strong Pan-African overprint. It is unclear whether they represent a continuous swath of Mesoproterozoic crust that was present in this region prior to Pan-African orogenesis, as implied by Figures 1 and 5, or whether they consist at least partly of smaller crustal slivers or exotic terranes tectonically incorporated within a zone of Pan-African accretion and collision (cf. Kröner 2001).

In this paper, the Kibaran and Irumide Belts are

considered to represent parts of a single, long-lived Mesoproterozoic orogenic province that formed during assembly of the Rodinia supercontinent, with widespread, broadly coeval calc-alkaline igneous rocks farther east defining the position of a major convergent margin active at this time. Several workers have suggested that the Kibaran and Irumide Belts developed from essentially intracontinental basins that were deformed in response to far-field stresses transmitted inward from the convergent margin (e.g. Daly 1986b; Klerkx et al. 1987; Pohl 1994), which may have experienced repeated accretionary events.

Southwestern Angola and northwestern Botswana

Older basement in southwestern Angola was affected by a major Mesoproterozoic magmatic event that involved emplacement of one of the world's largest anorthositic intrusions, the Kunene Complex, together with a series of anorogenic granites. The Kunene Complex has generally been considered as Palaeoproterozoic in age, but a mangerite dyke inferred to be cogenetic with the complex has recently yielded a U–Pb zircon crystallization age of 1370 ± 4 Ma (Mayer et al. 2000). The spatially associated anorogenic granites have yielded Rb–Sr whole-rock isochron dates ranging from 1411 ± 24 to 1302 ± 20 Ma (Fig. 5; Carvalho et al. 1979, 1987). Mayer et al. (2000) interpreted the Kunene Complex to have formed in an extensional environment by upward injection of plagioclase-rich magma batches derived from differentiation of underplated basaltic magma; the associated granites are inferred to reflect concomitant lower crustal anatexis. The cause of this magmatic event is unknown.

Carvalho et al. (1979) reported a Rb–Sr whole-rock isochron date of 1274 ± 24 Ma from mylonites along the Namibia-Angola border (Fig. 5), but it is uncertain whether this result records a deformational event of regional significance. Also of uncertain significance are Mesoproterozoic granites present in northwestern Botswana, one of which has yielded a U–Pb zircon age of 1022 ± 1 Ma (Singletary et al. 2002). The granites intrude Palaeoproterozoic gneiss exposed in small outcrops in a remote part of the Kalahari Desert. They are cut by Pan-African ductile shear zones but show no evidence of Mesoproterozoic deformation. The extensive Kalahari cover makes it difficult to place these granites into a regional context.

Pan-African orogenic belts

A variety of different names are used for Pan-African orogenic belts in southern Africa (Fig. 1), but their overall tectonic evolution is best approached by considering them as segments of three major orogens that surround and separate the Congo and Kalahari Cratons (Fig. 2). The East African Orogen extends along the present-day eastern margin of Africa (Stern 1994) and, in the area of concern here, is commonly termed the Mozambique Belt (Fig. 1; see Grantham et al. 2002). A second major orogen is discontinuously exposed along the western margin of southern Africa and comprises the West Congo, Kaoko, Gariep and Saldania Belts, which were originally connected to coeval Brasiliano belts in South America prior to Gondwana breakup (Trompette 1994). These two N–S-trending orogens are linked by a third, transcontinental orogen comprising the Damara Belt, the Lufilian Arc and the Zambezi Belt.

The Lufilian Arc and Zambezi Belt are separated by the transcurrent Pan-African Mwembeshi Shear Zone (Fig. 1), which is generally considered to accommodate a change in structural vergence between the two belts (Unrug 1983; Daly 1986c). The junction between the Lufilian Arc and Pan-African belts to the west is hidden beneath Phanerozoic cover. Daly (1986c) suggested that the Lufilian Arc connects with the West Congo Belt in the subsurface and took this inferred boundary to define the northern edge of a separate 'Angolan microplate', delimited to the south by the Damara Belt and lying between the Congo and Kalahari Cratons. This model was adopted by Porada (1989) and Porada & Berhorst (2000). As far as the present author is aware, no direct evidence exists for this proposed connection between the West Congo Belt and the Lufilian Arc. SW structural trends in the western part of the Lufilian Arc in Zambia can be traced in the subsurface to the Angola border by geophysical means (e.g. Mazac 1974). Farther SW, geophysical data and limited exposures in the Kalahari Desert show the NE-trending Damara Belt in Namibia to continue beneath Phanerozoic cover into northwestern Botswana (Carney et al. 1994; Eberle et al. 1996; Key & Ayres 2000), although its width decreases significantly around a promontory in the buried Congo Craton margin (Fig. 1; Singletary et al. 2002). The Damara Belt almost certainly connects directly with the Lufilian Arc north of this promontory (Fig. 1), although lack of published geophysical data from the intervening part of Angola precludes direct tracing of structures between the two belts.

Within-plate magmatism and basin initiation

All of the Pan-African Belts in southern Africa deform thick Neoproterozoic supracrustal successions (Fig. 7) that are dominated by sedimentary rocks but also contain locally important volcanic units. Where the successions are relatively well preserved, typical continental rift assemblages are present in the lower parts and rest unconformably on older basement. Within-plate magmatism preceded and/or accompanied the rifting.

The oldest known rift-related magmatism within these Neoproterozoic basins is represented by bimodal igneous rocks in the West Congo Belt (Franssen & André 1988), which have U–Pb zircon ages of 999 ± 7 to 912 ± 7 Ma (Fig. 7; Tack et al. 2001). Farther south, a variety of within-plate igneous rocks were emplaced prior to and during initiation of the Gariep depositional basin. The alkaline Richtersveld Igneous Complex, which is unconformably overlain by the Gariep succession, has U–Pb zircon ages ranging from 833 ± 2 to 771 ± 6 Ma (Frimmel et al. 2001). Bimodal volcanic rocks intercalated with rift deposits near the base of the Gariep succession have a $^{207}Pb/^{206}Pb$ zircon evaporation age of 741 ± 6 Ma (Frimmel et al. 1996b). The tholeiitic to alkaline Gannakouriep mafic dyke swarm in the same region has yielded a Rb–Sr isochron date of 717 ± 11 Ma (Fig. 7; Reid et al. 1991; Frimmel et al. 1996a).

Continental rift deposits are extensively developed in the basal parts of the supracrustal succession within the Kaoko and Damara Belts (Miller 1983a; Porada 1985; Hoffman 1989). Fault-controlled accumulations of alkaline rhyolitic ignimbrites ≥ 6.6 km thick occur within the rift sequence along the northern margin of the Damara Belt, near its junction with the Kaoko Belt, and are associated with less voluminous basalts. Syenite, nepheline syenite and carbonatite intrusions are also present (Miller 1983a). The rhyolite and syenite have U–Pb zircon ages ranging from 756 ± 2 to 746 ± 2 Ma (Hoffman et al. 1996). Carbonatite in the central part of the Damara Belt has yielded a U–Pb titanite igneous crystallization age of $837 + 60/-49$ Ma (Bühn et al. 2001). Renewed extension-related magmatism is locally recorded by basalts at higher stratigraphic levels in the Kaoko Belt and in the central part of the Damara Belt, and by extensive, fissure-fed continental tholeiitic and alkaline basalts along the southern margin of the Damara Belt (Miller 1983b; Breitkopf 1989; Kampunzu et al. 2000b).

To the NE, along-strike in the Lufilian Arc, initial anorogenic magmatism is represented by the large Nchanga Granite, which intrudes older basement and has A-type geochemical affinities (Tembo et al. 2000). The granite has a U–Pb zircon crystallization age of 877 ± 11 Ma (Armstrong et al. 1999) and is nonconformably overlain by Neoproterozoic strata of the Katangan succession. Bimodal volcanic rocks occur within continental rift deposits at the base of the Katangan, and basaltic volcanic rocks and associated metagabbros are locally abundant at higher stratigraphic levels (Tembo et al. 1999; Kampunzu et al. 2000b; Porada & Berhorst 2000). In northwestern Zambia, Katangan basalts and strongly altered intermediate volcanic rocks have yielded U–Pb zircon ages of 765 ± 5 and 735 ± 5 Ma (Key et al. 2001).

The Katangan basaltic rocks record a progressive change in composition with decreasing age, ranging from typical continental tholeiites near the base of the succession, to tholeiitic and alkaline rocks at higher stratigraphic levels, and then to tholeiites showing geochemical similarities to E-type MORB (Tembo et al. 1999; Kampunzu et al. 2000b). Based on the local presence of basalts with E-MORB affinities, Kampunzu et al. (2000b) inferred that rifting may have reached the stage where a narrow ocean basin formed. There is no direct evidence, however, that any of the presently exposed Katangan succession was deposited on oceanic crust.

Deformed and metamorphosed bimodal and felsic magmatic assemblages are abundant in the Zambezi Belt to the SE and have U–Pb zircon ages ranging from 881 ± 7 to 795 ± 2 Ma; inclusion of $^{207}Pb/^{206}Pb$ zircon evaporation ages extends this range to 737 ± 1 Ma (Fig. 7 and refs cited in the caption). In some cases, original textures are well enough preserved to permit recognition of both volcanic and plutonic protoliths (e.g. Mallick 1966; Barton et al. 1991). Geochemical data indicate typical within-plate affinities, and the felsic units have A-type characteristics and in some cases are peralkaline (Barton et al. 1991; Munyanyiwa et al. 1997; Hargrove et al. 2002). Deep-crustal equivalents of these units are represented by A-type felsic orthogneisses that have U–Pb zircon ages of c. 870 Ma and occur within thrust slices of high-pressure granulites that structurally overlie Neoproterozoic supracrustal rocks in the eastern part of the belt (Barton et al. 1991; Carney et al. 1991; Vinyu et al. 1999; Hargrove et al. 2002). U–Pb zircon geochronology shows that the granulite facies metamorphism occurred between c. 870 and 850 Ma, and it presumably records thermal effects in the lower crust associated with the A-type magmatism (Mariga et al. 1998; Mariga 1999; Vinyu et al. 1999; Hargrove et al. 2002).

Hanson et al. (1988b) interpreted a mylonitized, sheet-like granite batholith within the Zambezi Belt in Zambia to have been intruded during contractional deformation, and argued that a U–Pb zircon crystallization age of 820 ± 7 Ma for the granite dates the main Zambezi orogenesis. However, this age is comparable to ages for widespread igneous rocks elsewhere in the belt that have geochemical

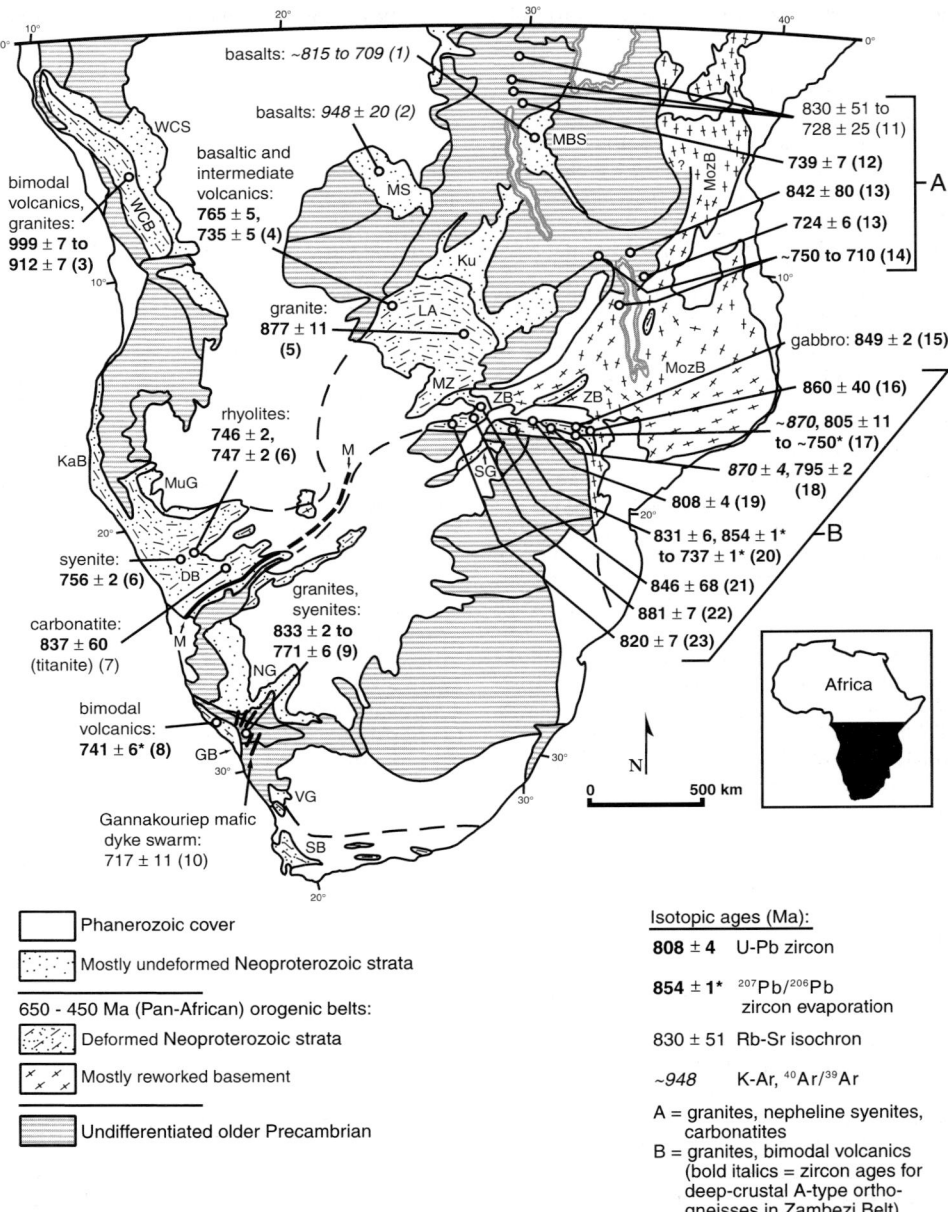

Fig. 7. Geochronological data for Neoproterozoic within-plate and rift-related igneous rocks in southern Africa. Extent of Neoproterozoic strata is shown (except in parts of the Mozambique Belt). Ku, Kundelungu rift basin; MBS, Malagarazi/Bukoba Supergroups; MS, Mbuji Mayi Supergroup; MuG, Mulden Group; NG, Nama Group; SG, Sijarira Group; VG, Vanrhynsdorp Group; WCS, West Congo Supergroup. Other abbreviations as in Figure 1. References for isotopic ages: 1, Cahen & Snelling (1974), Deblond *et al.* (2001); 2, Cahen *et al.* (1975); 3, Tack *et al.* (2001); 4, Key *et al.* (2001); 5, Armstrong *et al.* (1999); 6, Hoffman *et al.* (1996); 7, Bühn *et al.* (2001); 8, Frimmel *et al.* (1996b); 9, Frimmel *et al.* (2001); 10, Reid *et al.* (1991); 11, Kampunzu *et al.* (1998b); 12, Tack *et al.* (1984); 13, Lenoir *et al.* (1994); 14, Eby *et al.* (1998); 15, Mariga *et al.* (1998); 16, Barr *et al.* unpublished data cited in Pinna *et al.* (1993); 17, Vinyu *et al.* (1999), Dirks *et al.* (2000); 18, Hargrove *et al.* (2002); 19, Hanson *et al.* (unpublished data); 20, Hanson *et al.* (1998a), Dirks *et al.* (1999); 21, Barr *et al.* (1978), recalculated in Hanson *et al.* (1988b); 22, Hanson *et al.* (1994); 23, Hanson *et al.* (1988b).

characteristics more consistent with an anorogenic, within-plate setting, as discussed above. The crystallization ages for these A-type rocks fall within the range of ages for igneous assemblages in adjacent parts of the Pan-African network that clearly formed prior to or during continental rifting (Fig. 7). These regional relations suggest that the interpretation of Hanson et al. (1988b) is in error, and that the Zambezi Belt was probably undergoing extension in the same time frame as other parts of the Pan-African network. Dirks et al. (1998, 1999) have argued that much of the amphibolite facies ductile structure in the belt in Zimbabwe formed during extension associated with the A-type magmatism, but their interpretation conflicts with other structural work (see below).

Farther north, a variety of intrusive complexes containing peralkaline granite, nepheline syenite, carbonatite and ijolite intrude parts of the older Ubendian and Kibaran Belts, and have yielded U–Pb zircon ages and Rb–Sr isochron dates ranging from c. 842 to 710 Ma (Fig. 7 and refs cited in the caption). Basalts in the Malagarazi and Bukoba Supergroups along this trend are considered by Deblond et al. (2001) to have been emplaced between c. 815 and 709 Ma, based on K–Ar, and $^{40}Ar/^{40}Ar$ whole-rock and mineral dates. Basalt lavas also occur at the top of the Mbuji Mayi Supergroup on the west side of the Kibaran Belt and are inferred to have been extruded at c. 948 Ma, based on K–Ar whole-rock dates (Cahen et al. 1975), but these results from ancient, altered basalts should be treated with great caution.

In summary, Neoproterozoic successions in southern Africa preserve evidence of widespread within-plate magmatism and rifting at c. 1000–710 Ma. This records initiation of major Neoproterozoic basins and, on a global scale, broadly overlaps in time with breakup of the Rodinia supercontinent. Alkaline plutonic centres in the Ubendian and Kibaran Belts reflect uprise of mantle-derived magmas along older lines of structural weakness in the same time frame. Comparable A-type magmatic assemblages related to early stages of rifting have not been documented from the main part of the Mozambique Belt in the area of Figure 1. However, the Neoproterozoic succession in the Zambezi Belt wraps the northeastern corner of the Zimbabwe Craton to continue directly into adjacent parts of the Mozambique Belt (Fig. 7; Johnson & Vail 1965; Whittington et al. 1999). Hence, the southern part of the Mozambique Belt presumably had a similar early history of basin formation to that recorded in the Zambezi Belt.

The alkaline complexes in the Ubendian and Kibaran Belts, in some cases, show evidence of emplacement during contractional deformation that reactivated older structures in these belts (e.g. Tack et al. 1984). How this contractional deformation ties in with the regional framework of extension-related magmatism discussed above is unclear.

Neoproterozoic basin evolution and seafloor spreading

Continental rifting in the Kaoko and Gariep Belts was followed by thermal subsidence and the formation of carbonate-rich passive-margin sequences (Miller 1983a; Stanistreet et al. 1991; Frimmel et al. 1996b). A well-developed passive-margin sequence is lacking in the Saldania Belt, possibly because continental breakup occurred there in a transtensional regime involving formation of pull-apart basins oblique to the continental margin (Rozendaal et al. 1999). Both the Gariep and Saldania Belts contain fragments of accreted oceanic crust (Frimmel et al. 1996a; Rozendaal et al. 1999). In a Gondwana reconstruction, these belts are continuous with Brasiliano belts in South America that contain Neoproterozoic volcanic-arc and ophiolite assemblages, and represent the trace of a major ocean (the Adamastor Ocean) that was consumed during Gondwana assembly (e.g. Trompette 1994, 2000; Brito Neves et al. 1999). The West Congo Belt to the north shows a broadly similar depositional history to the Kaoko and Gariep Belts (Schermerhorn 1981), but developed within an embayment extending into the Congo–São Francisco Craton (Fig. 2). Continental rifting in the West Congo Belt led to formation of a narrow ocean basin that linked southward (in present coordinates) with the larger Adamastor Ocean (Pedrosa-Soares et al. 2001).

Remnants of Neoproterozoic oceanic crust generally have not been documented from that part of the Mozambique Belt shown in Figure 1, although rocks with ophiolitic affinities locally occur in a major thrust zone in the centre of the belt in Mozambique south of latitude 15°S (Sacchi et al. 2000). Farther north within the East African Orogen, Neoproterozoic ophiolites, deformed passive-margin sequences and volcanic-arc assemblages testify to the existence of a long-lived ocean (the Mozambique Ocean) that was destroyed during Pan-African collisional orogenesis (e.g. Vearncombe 1983; Mosley 1993; Shackleton 1993; Stern 1994).

The nature of the basin (or basins) originally located along the trend of the Damara–Lufilian–Zambezi Orogen is controversial. Some workers have argued that this transcontinental orogen represents the site of another major ocean destroyed during Gondwana assembly (e.g. Burke et al. 1977; Barnes & Sawyer 1980). In contrast, Shackleton (1973), Kröner (1983) and Hanson et al. (1988a) suggested that the Zambezi Belt in the eastern part of the transcontinental orogen cuts directly across an older, Mesoproterozoic orogen defined by the Irumide Belt and Choma–Kalomo Block (Fig. 1). Such a relation would support an essentially intracontinental mode of development for the Zambezi Belt, perhaps as an aulacogen connected to the Mozambique Ocean (Hanson et al. 1994; Stern 1994). It should be noted, however, that a direct correlation of older terranes on each side of the Zambezi Belt is discounted by some workers (e.g. Johnson & Oliver 2000).

The supracrustal assemblage in the Zambezi Belt is perhaps best preserved in the western part of the belt in Zambia, where bimodal metavolcanic rocks at the structural base of the assemblage are overlain by metaconglomerates, metasandstones and pelitic schists, which are in turn overlain by extensive carbonate and calc-silicate rocks; widespread scapolite is inferred to reflect the original presence of evaporites (e.g. Hanson et al. 1994). Although the original stratigraphic relations of these units have been obscured by intense Pan-African deformation, the lithological assemblage is compatible with deposition in an evolving intracontinental rift subjected to marine incursion during thermal subsidence. Porada & Berhorst (2000) have assigned parts of the Zambezi supracrustal assemblage to the underlying basement, but there is currently no direct geochronological evidence to support this interpretation, which is not followed here.

The Katangan succession in the Lufilian Arc to the NW is generally better preserved, and records sedimentation in continental, marginal marine and marine environments (e.g. Binda 1994; Cailteux 1994; Porada & Berhorst 2000). Evidence for deep-marine sedimentary rocks deposited on oceanic crust is lacking, but such rocks could be covered beneath younger parts of the succession. The Kundelungu rift basin extends NE from the main Katangan depository (Fig. 7). Its geometrical relations with the Lufilian Arc suggest that a rift–rift–rift triple junction may have existed in this area (e.g. Kampunzu et al. 2000b), although, based on stratigraphic evidence, Porada & Berhorst (2000) have argued that the Kundelungu basin formed later than the main Katangan rifts to the south.

In the Damara Belt, continental rift sedimentation gave way to deposition on marine platforms on the margins of a deep-marine basin that received large volumes of terrigenous turbidites (Miller 1983a; Porada 1985; Hoffman 1989; Stanistreet et al. 1991; Kukla 1992). This basin may have been floored at least partly by oceanic crust and presumably linked with the larger Adamastor Ocean to the west (in present coordinates) (Kukla & Stanistreet 1991; Stanistreet et al. 1991). A distinctive unit of metabasaltic rocks within the turbidite sequence, the Matchless Amphibolite, crops out in a narrow zone extending c. 400 km parallel to the length of the basin (Miller 1983b); it can be traced an additional c. 500 km in the subsurface into northwestern Botswana (Figs 1 & 7; Lüdtke et al. 1986). The amphibolites are metamorphosed tholeiitic gabbros and basalts, the latter exhibiting relict pillow structures in places. Both N-MORB and E-MORB compositions are present (Miller 1983b; Breitkopf 1989; Kukla 1992; Kampunzu et al. 2000b). Parts of the turbidite sequence are intimately interleaved with the amphibolites, and in places the metagabbros show relict chilled margins against the metasedimentary rocks (Miller 1983b). These relations can be explained by a model in which the Matchless Amphibolite represents a nascent mid-ocean ridge that was buried by terrigenous sediments within a narrow, Red-Sea-type ocean basin (e.g. Miller 1983a; Breitkopf & Maiden 1988).

The preserved supracrustal assemblages in the Damara–Lufilian–Zambezi Orogen are thus compatible with deposition in a series of linked intracontinental rifts that, in the Damara Belt, probably evolved into a narrow ocean basin. In this interpretation, the Congo and Kalahari Cratons were already juxtaposed prior to Gondwana assembly, and their present configuration is inherited from the earlier Rodinia supercontinent.

Pan-African orogenesis

Pan-African orogenesis in southern Africa involved destruction of the Neoproterozoic basins described above, together with extensive structural reworking of older basement. Associated synorogenic foreland-basin deposits on the craton margins are only partly preserved but are present in the Nama, Vanrhynsdorp and Mulden Groups in Namibia and South Africa, the West Congo Supergroup, parts of the Katangan succession in the Lufilian Arc, and the Sijarira Group in Zimbabwe (Figs 7 & 8).

The regional extent of Pan-African tectonothermal activity in southern Africa was originally recognized largely on the basis of K–Ar and Rb–Sr whole-rock and mineral dates, which typically fall in the 650–420 Ma range throughout the Pan-African orogenic network (e.g. Cahen & Snelling 1966; Miller 1983a; Cahen et al. 1984; Goscombe et al. 2000). Space does not permit detailed discussion of these data, the tectonic significance of which is not always clear. The younger dates are generally inferred to reflect final uplift and cooling of the orogens.

Pan-African structures in the West Congo Belt record east-vergent thrusting onto the Congo Craton margin, and associated regional metamorphism reached the amphibolite facies in the core of the orogen (Franssen & André 1988; Maurin 1993; Tack et al. 2001). Dextral and sinistral transcurrent shear

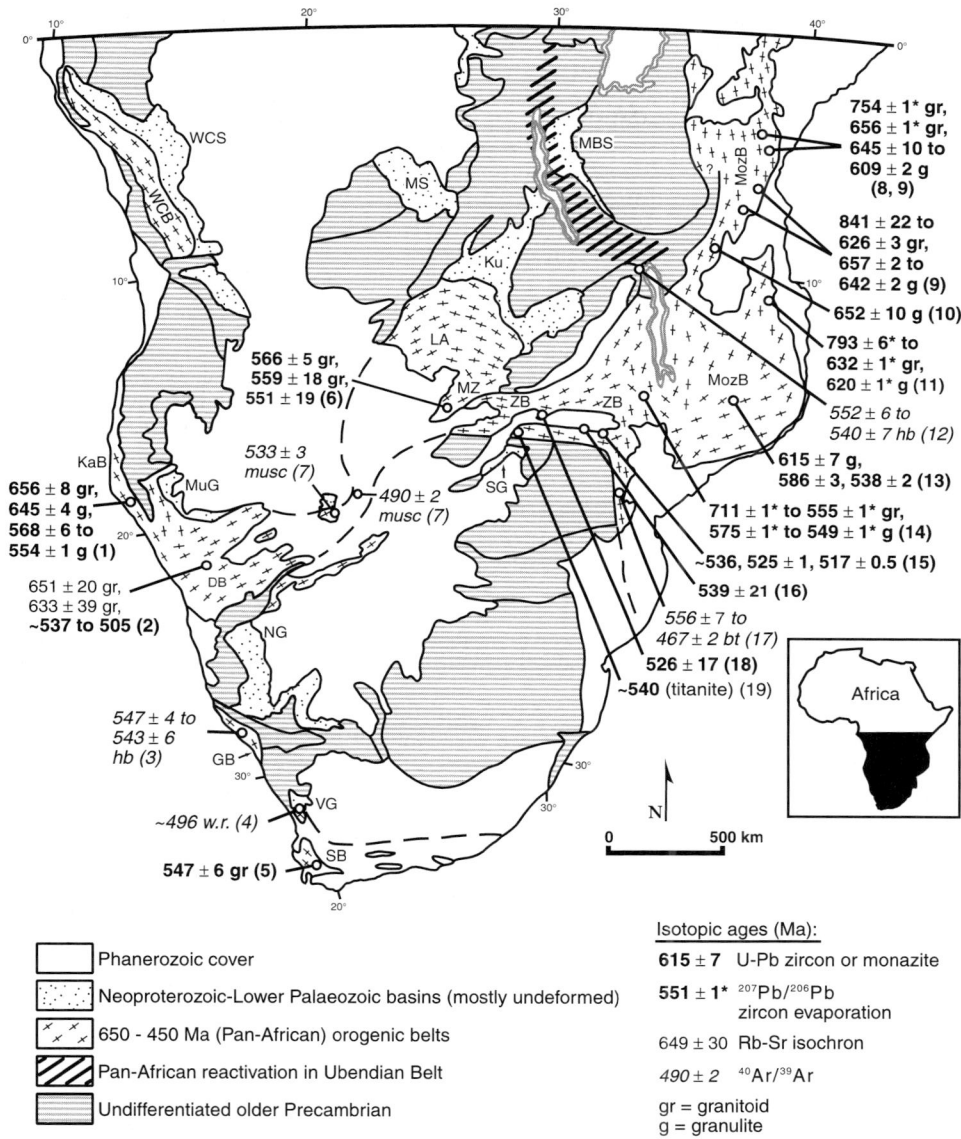

Fig. 8. Geochronological data for Pan-African orogenic belts in southern Africa. Abbreviations as in Figures 1 & 7. Ages for post-tectonic plutons and hydrothermal deposits are not shown. ^{40}Ar/^{39}Ar dates are shown only in areas where other geochronological data are not available (bt, biotite; hb, hornblende; musc, muscovite; w.r., whole rock). The range in U–Pb zircon and monazite ages shown for the Damara Belt represents only those ages considered to record the main orogenic event and peak metamorphism (see text for further discussion of the abundant geochronological data from this belt). References for isotopic ages: 1, Seth et al. (1998), Franz et al. (1999), Luft et al. (2001); 2, Downing & Coward (1981), Kröner (1982), Miller (1983a), Kukla et al. (1991), Bühn et al. (1994), Jung (2000), Jung et al. (2000a, b, 2001), Jung & Mezger (2001); 3, Singletary et al. (2002); 4, Gresse et al. (1988); 5, Rozendaal et al. (1999), da Silva et al. (2000); 6, Hanson et al. (1993); 7, Singletary et al. (2002); 8, Muhongo & Lenoir (1994); 9, Möller et al. (2000), Muhongo et al. (2001); 10, Coolen et al. (1982); 11, Kröner (2001); 12, Ring et al. (1999); 13, Kröner et al. (1997), Jamal et al. (1999); 14, Kröner et al. (2001); 15, Vinyu et al. (1999), Mariga (1999); 16, Müller et al. (2000); 17, Manhica et al. (2001); 18, Goscombe et al. (2000); 19, Hanson et al. (1998a).

zones within the belt are inferred by Maurin (1993) to reflect lateral escape of blocks during contractional deformation. Compressional stresses transmitted cratonward caused inversion of a separate Neoproterozoic rift basin now buried beneath younger strata in the Congo Basin (Daly et al. 1992).

To the south, sinistral transpressional deformation in the Kaoko, Gariep and Saldania Belts is interpreted to record oblique closure of the Adamastor Ocean (Dürr & Dingeldey 1996; Frimmel & Frank 1998; Rozendaal et al. 1999). Deep-marine fan deposits and accretionary prism, oceanic seamount and ophiolitic assemblages were thrust onto the adjacent craton margins during this event (Frimmel et al. 1996a; Rozendaal et al. 1999). The main orogenesis in these belts occurred at c. 568–543 Ma (Fig. 8), based on U–Pb zircon and monazite ages from granulites and syntectonic granitoid rocks in the Kaoko Belt (Seth et al. 1998; Franz et al. 1999; Luft et al. 2001), $^{40}Ar/^{39}Ar$ hornblende cooling ages from lower grade rocks in the Gariep Belt (Frimmel & Frank 1998), and U–Pb zircon ages from late syntectonic S-type granites emplaced in transpressional shear zones in the Saldania Belt (Rozendaal et al. 1999; da Silva et al. 2000). Post-tectonic I- and A-type granitoid rocks in the Saldania and Gariep Belts have U–Pb zircon ages ranging from 536 ± 5 to 507 ± 6 Ma (Rozendaal et al. 1999; da Silva et al. 2000; Frimmel 2000). U–Pb zircon geochronology also documents a separate granulite facies metamorphic event in the Kaoko Belt at c. 656–645 Ma (Seth et al. 1998; Franz et al. 1999). Franz et al. (1999) suggested that the older granulites may have developed in an extensional environment related to Neoproterozoic basin formation, or may be a crustal fragment derived from South America during subsequent collisional orogenesis.

The Damara Belt underwent north-vergent thrusting along its northern margin, polyphase ductile deformation involving sinistral transpression in a central zone, and major south-vergent thrusting and nappe emplacement in the southern part of the belt (Coward 1983; Miller 1983a). Some of the ductile deformation in the central zone is inferred to reflect extensional shearing related to tectonic escape of upper crust during transpression (Oliver 1994). Metamorphic grade increases from subgreenschist facies along the margins to upper amphibolite and granulite facies in the interior of the belt (e.g. Kasch 1983; Jung et al. 2000a).

Voluminous plutons intruded the central part of the Damara Belt at various times relative to the deformation. Some workers have interpreted the plutonic rocks to record the presence of a volcanic arc generated by northward-directed subduction of oceanic lithosphere prior to collisional orogenesis (e.g. De Kock 1991; Stanistreet et al. 1991). However, 96% of the exposed intrusive rock consists of granite, with only minor amounts of gabbro, diorite, tonalite and granodiorite (Miller 1983a), making the plutonic suite unlike batholiths formed in typical continental margin arcs. Sr, Nd and O isotopic data indicate that much of the magma was derived from, or extensively contaminated by, older continental crust, and initial Sr isotope ratios are significantly higher on average than those shown by volcanic rocks from the present-day Andean continental margin arc (Haack et al. 1982; Hawkesworth et al. 1986; McDermott et al. 1996; Jung et al. 2001). These data suggest derivation of the plutonic suite predominantly from anatexis of crust thickened during closure of the Damara Basin.

A few of the Damaran plutons have yielded Rb–Sr whole-rock isochron dates as old as c. 651–633 Ma (Fig. 8; Downing & Coward 1981; Kröner 1982) and are inferred to postdate an early phase of deformation (Miller 1983a). Kröner (1982) obtained an Rb–Sr whole-rock isochron date of 750 ± 35 Ma from one diorite pluton, but Miller (1983a) has argued that this result is inconsistent with other geological and geochronological evidence. That the other Rb–Sr isochron results record an event of some significance is supported by the fact that detrital muscovite grains in Nama foreland strata derived from the Damara Belt have K–Ar dates as old as c. 670 Ma (Horstmann et al. 1990). Younger stages in the evolution of the Damara Belt are constrained by a variety of isotopic ages too numerous to show individually on Figure 8. The Damaran granites have yielded U–Pb zircon and monazite ages ranging from c. 590 to 488 Ma (Allsopp et al. 1983; Miller & Burger 1983; Kukla et al. 1991; Jacob et al. 2000; Jung et al. 2000b, 2001), although the older ages are based on concordia intercepts with large error bars. The main orogenic event occurred at c. 534–516 Ma, based on U–Pb zircon and monazite ages from syntectonic granites (Miller 1983a; Kukla et al. 1991; Jung 2000; Jung et al. 2000b). U–Pb zircon and monazite ages and Sm–Nd garnet–whole-rock isochron dates from metapelites and migmatites in the central part of the belt indicate that high-grade metamorphism had begun by c. 539 Ma and peaked at c. 520–500 Ma (Kukla et al. 1991; Bühn et al. 1994; Jung 2000; Jung et al. 2000a; Jung & Mezger 2001; Nex et al. 2001). The large, late tectonic Donkerhuk Granite has a U–Pb monazite age of 505 ± 5 Ma (Kukla et al. 1991), and other A- and S-type granites intruded after peak metamorphism have U–Pb monazite ages and Sm–Nd garnet–whole-rock isochron dates of c. 497–484 Ma (Jung et al. 2000b, 2001). Some migmatites have yielded U–Pb monazite ages as young as c. 470 Ma (Jung 2000), and late, highly differentiated granites (alaskites) have Rb–Sr whole-rock isochron dates as young as c. 460 Ma (Miller 1983a).

To the NE, the outer part of the Lufilian Arc is a

north-verging, thin-skinned, fold-thrust belt that links to the south with basement-involved thrusts developed at higher metamorphic grades (e.g. Daly 1986c; Cosi et al. 1992; Kampunzu & Cailteux 1999; Wendorff 2000). Still farther south, transpressional deformation in the inner part of the arc occurred during transcurrent displacement (of uncertain magnitude) along the Mwembeshi Shear Zone and was associated with emplacement of the composite Hook granitic batholith (Hanson et al. 1993).

Cahen et al. (1984) inferred that early phases of contractional deformation affected the Lufilian Arc at c. 950 and 850 Ma, but these age assignments are now known to be in error (Richards et al. 1988b; Kampunzu & Cailteux 1999; Porada & Berhorst 2000). Cosi et al. (1992) argued that the main Lufilian thrusting is constrained by a Rb–Sr muscovite date of 744 ± 8 Ma from sheared basement and a K–Ar biotite date of 708 ± 7 Ma from metamorphosed Katangan mafic rocks in the northern part of the arc. Kampunzu et al. (1998c) suggested that the thrusting occurred at c. 790–740 Ma, based on model Pb ages from Pb–Zn deposits inferred to be synorogenic. These proposed ages for Lufilian thrusting are contradicted by recently obtained U–Pb zircon ages of 765 ± 5 to 735 ± 5 Ma for volcanic rocks in the Katangan succession that were emplaced prior to contractional deformation (Key et al. 2001).

In the outer part of the Lufilian Arc, U–Pb ages are available for different generations of uraninite inferred by Kampunzu & Cailteux (1999) to postdate the main contractional structures. Cahen et al. (1984) reviewed a large number of these uraninite ages and considered the most reliable to range from c. 656 to 503 Ma; more recent uraninite age determinations fall in the same range (e.g. Richards et al. 1988a; Loris et al. 1997). Kampunzu & Cailteux (1999) used these data to argue that contractional deformation affected the outer part of the Lufilian Arc prior to 656 Ma. However, the interpretation that uraninite as old as 656 Ma formed after the thrusting has been disputed by Porada & Berhorst (2000), based on local structural relations of the uraninite mineralization. The timing of transpressional deformation in the inner part of the arc is constrained by U–Pb zircon ages of 566 ± 5 and 559 ± 18 Ma for syntectonic granites in the Hook batholith (Fig. 8), and by a U–Pb zircon age of 551 ± 19 Ma for associated hypabyssal rhyolite intruded during displacement along the Mwembeshi Shear Zone; post-tectonic felsic magmatism in the same area occurred at c. 538–533 Ma (Hanson et al. 1993). Late to post-tectonic hydrothermal mineralization occurred in the outer part of the Lufilian Arc at c. 514–502 Ma, as shown by U–Pb rutile, monazite and uraninite ages and Re–Os molybdenite ages from vein minerals (Richards et al. 1988a, b; Torrealday et al. 2000).

Based on the available evidence, Porada & Berhorst (2000) concluded that the Lufilian Arc was first deformed at c. 560–550 Ma, followed by post-tectonic thermal effects and vein mineralization.

In the Zambezi Belt to the SE, most workers agree that Pan-African deformation involved south-vergent thrusting along the southern margin of the belt, associated in places with transcurrent or transpressional shearing (e.g. Daly 1986c; Barton et al. 1991, 1993; Wilson et al. 1993; Hanson et al. 1994; Munyanyiwa & Blenkinsop 1994; Goscombe et al. 2000). Metamorphic grade is typically in the amphibolite facies, but whiteschists and eclogites are locally present along a discontinuous trend that can be traced into higher grade parts of the Lufilian Arc (Vrána et al. 1975; Andreoli & Hart 1990; Cosi et al. 1992; Johnson & Oliver 1998; Dirks & Sithole 1999; John et al. 2000). Some of the eclogites record metamorphic pressures of c. 23 kbar (John et al. 2000). There are currently no robust age constraints for these rocks, and Dirks & Sithole (1999) have suggested that the eclogite facies metamorphism may predate Pan-African orogenesis.

As noted previously, high-pressure, 870–850 Ma granulites structurally overlie Neoproterozoic supracrustal rocks in the eastern part of the Zambezi Belt. The granulites show a strong amphibolite facies overprint inferred to have developed at the same time that the underlying supracrustal rocks underwent prograde metamorphism into the amphibolite facies (e.g. Barton et al. 1991, 1993; Carney et al. 1991). Based partly on studies of shear-sense indicators in this structurally complex area, Dirks et al. (1998) argued that the regional amphibolite facies ductile fabrics, including those in the Neoproterozoic supracrustal rocks, formed during an early extensional episode. However, other mapping in the same area has produced convincing evidence that the fabrics record contractional deformation during Pan-African orogenesis, with penetrative deformation of the supracrustal rocks occurring during south-vergent thrusting of the older granulites, followed by north-vergent back-thrusting (Barton et al. 1991, 1993; Carney et al. 1991; Mariga et al. 1998; Mariga 1999; Hargrove et al. 2002).

The timing of the main phase of Pan-African orogenesis in the Zambezi Belt is constrained by U–Pb geochronology from several areas in Zimbabwe (Fig. 8). Metamorphic zircon growth occurred within older, partly reworked basement in the central part of the belt at 526 ± 17 Ma (Goscombe et al. 2000). In the eastern part of the belt, the timing of thrust emplacement of high-pressure granulites is constrained by a U–Pb monazite age of 525 ± 1 Ma from syntectonic pegmatite beneath the thrust sheets (Mariga 1999), and by a U–Pb zircon age of 539 ± 21 Ma from a late syntectonic pegmatite within one of the thrust sheets (Müller et al. 2000). Amphibolite facies retrogression of the granulites and prograde metamorphism of the structurally underlying supracrustal rocks are recorded by U–Pb zircon and monazite ages of c. 536–517 Ma (Mariga 1999; Vinyu et al. 1999). In southeastern Zambia, the granitic to syenitic Sinda batholith, which is late to post-tectonic with respect to Pan-African deformation (Johns et al. 1989), has yielded a Rb–Sr whole-rock isochron date of 489 ± 42 Ma (Haslam et al. 1986).

Pan-African deformation in the Mozambique Belt involved west-vergent thrusting along the margins of the Congo and Kalahari Cratons, and may have included an important component of strike-slip shearing along the Kalahari Craton margin (Grantham 2001). The core of the orogen contains highly deformed supracrustal rocks, migmatitic gneisses and granulites characterized by recumbent to east-dipping ductile structures (Shackleton 1993; Muhongo 1994). There is isotopic evidence for involvement of Archaean crust in much of the belt in Tanzania, together with addition of juvenile crust during the Pan-African event (Maboko 1995, 2000, 2001; Möller et al. 1998; Kröner 2001; Muhongo et al. 2001; Maboko & Nakamura 2002). It is unknown whether the Archaean crustal components were derived from the Tanzania Craton to the east or represent one or more terranes accreted during collisional orogenesis. Farther south in Malawi and Mozambique, Pan-African structures overprint Mesoproterozoic arc crust discussed previously (Figs 1 & 5).

Möller et al. (2000), Kröner (2001) and Muhongo et al. (2001) reported U–Pb zircon and $^{207}Pb/^{206}Pb$ zircon evaporation ages from predominantly calc-alkaline, granulite facies mafic to felsic orthogneisses in the Mozambique Belt in Tanzania that range from 841 ± 22 to 626 ± 3 Ma (Fig. 8) and are inferred to record the crystallization ages of the igneous protoliths. U–Pb geochronology also records granulite facies metamorphism in the same region at c. 657–609 Ma (Coolen et al. 1982; Muhongo & Lenoir 1994; Möller et al. 2000; Kröner 2001; Muhongo et al. 2001). U–Pb zircon ages of c. 715 and 695 Ma that were originally interpreted to date granulite facies metamorphism (Maboko et al. 1985; Muhongo & Lenoir 1994) are now believed to reflect igneous emplacement ages. Geothermobarometry on the granulites documents deep-crustal metamorphism at pressures ≥ 9–12 kbar, with counterclockwise, isobaric cooling paths (Muhongo & Tuisku 1996; Maboko 1997; Appel et al. 1998; Muhongo et al. 1999).

Farther south in Mozambique, U–Pb zircon geochronology shows that amphibolite and granulite facies metamorphism and thrusting occurred at c. 615–586 Ma (Kröner et al. 1997; Jamal et al. 1999), with emplacement of late tectonic pegmatites at c. 538 Ma (Jamal et al. 1999). In Malawi, U–Pb zircon and $^{207}Pb/^{206}Pb$ zircon evaporation ages are interpreted to record emplacement of a widespread calc-alkaline plutonic assemblage at c. 711–555 Ma. High-pressure (c. 9.5 kbar) granulite facies metamorphism occurred at c. 575–549 Ma, followed by isobaric cooling, and affected both the Pan-African plutonic rocks and Mesoproterozoic basement in this region (Kröner et al. 2001). Final stages of the Pan-African episode in Malawi involved intrusion of a series of post-tectonic granites and syenites that are exposed near Lake Malawi. These rocks have yielded Rb–Sr whole-rock isochron dates of 489 ± 14 to 443 ± 30 Ma (Haslam et al. 1983).

The chemical compositions and style of metamorphism shown by the c. 841–555 Ma plutonic rocks in Tanzania and Malawi are consistent with emplacement in volcanic-arc settings. Ages for the oldest members of these predominantly calc-alkaline plutonic suites overlap with ages for within-plate magmatism farther west (Fig. 7), and possible relations between the magmatism in these two different geotectonic settings require further study.

The older Ubendian Belt and parts of the Kibaran Belt to the north underwent an important episode of Pan-African thermal and structural overprinting involving transpressional deformation (e.g. Lenoir et al. 1994; Theunissen et al. 1996), which is recorded by K–Ar, $^{40}Ar/^{39}Ar$ and Rb–Sr mineral dates that fall between c. 650 and 540 Ma (Cahen et al. 1984; Ring et al. 1999). According to Ring (1999) and Ring et al. (1999), Pan-African metamorphic overprinting in the southeastern part of the Ubendian Belt in Malawi occurred under amphibolite, granulite and eclogite facies conditions. The intensity of the deformational and metamorphic overprint in that area suggests a direct link with high-grade Pan-African terranes to the south and east in Malawi and Mozambique. Farther to the NW in the Ubendian and Kibaran Belts, Pan-African deformation occurred at lower metamorphic grades (typically greenschist facies) and appears to be more restricted in extent, suggesting that it records limited reactivation of older lines of structural weakness in response to stresses transmitted inward from the plate boundary to the east.

In summary, robust age constraints for most parts of the Pan-African orogenic network in southern Africa indicate that culminating phases of orogenesis occurred at c. 575–505 Ma (Fig. 8). Granulite facies metamorphism affected the Kaoko Belt at c. 650 Ma, but the tectonic setting of this metamorphism is uncertain. The Damara Belt is inferred to have undergone an early phase of deformation prior to emplacement of granitoid plutons that have Rb–Sr dates of c. 651–633 Ma, but the regional significance of this event is unclear. Appel et al. (1998) argued that c. 657–609 Ma granulites in Tanzania formed prior to collisional orogenesis in the Mozambique

Belt, possibly in the root zones of one or more volcanic arcs (see also Kröner 2001; Kröner et al. 2001). Other workers have inferred that the Tanzania granulites record continent–continent collision during Gondwana assembly, which accords with timing constraints for this event in northeastern Africa and Arabia (e.g. Stern 1994; Meert & Van der Voo 1997; Meert 2002). In this model, post-600 Ma granulites in the southern part of the Mozambique Belt (e.g. Malawi) are interpreted to record subsequent collision of another continent during progressive amalgamation of Gondwana. The second collision is also recorded by widespread isotopic ages of c. 565–515 Ma for high-grade terranes in parts of Madagascar, Sri Lanka, southern India and East Antarctica (e.g. Jacobs et al. 1998; Fitzsimons 2000; Kröner 2001; Kröner et al. 2001; Meert 2002 and refs cited therein).

Pan-African collisional belts along both the present-day western and eastern margins of southern Africa clearly represent sites where major oceans (Adamastor and Mozambique Oceans) were consumed during Gondwana assembly. In contrast, the Damara–Lufilian–Zambezi Orogen may have developed by closure of essentially intracontinental rift basins (probably involving oceanic crust in the Damara Belt), in response to stresses transmitted inward from the convergent margins. It remains to be seen whether such a model can accommodate the structural and metamorphic evidence for significant Pan-African shortening and crustal thickening within this transcontinental orogen.

My research in Africa has been funded over the past several years by National Science Foundation grants EAR 95-08013 and EAR 99-09269, by a Fulbright Research Scholarship, and by Texas Christian University Research Foundation grants, which I gratefully acknowledge. I thank the many colleagues I have had the opportunity to collaborate with on this research, who are too numerous to mention here. U. Hargrove, Z. Mussleman and M. Williams drafted the figures, and C. Philips typed the references. The idea for this paper stems from an invitation by A. Kröner to present a review paper on this topic at the 31st International Geological Congress in Rio de Janeiro. Reviews by P. Dirks and P. Treloar helped to improve the manuscript, and A. Kröner provided constructive comments on Figure 1.

References

ALLSOPP, H. L., BARTON, E. S., KRÖNER, A., WELKE, H. J. & BURGER, A. J. 1983. Emplacement versus inherited isotopic age patterns: a Rb–Sr and U–Pb study of Salem-type granites in the central Damara belt. In: MILLER, R. McG. (ed.) *Evolution of the Damara Orogen of South West Africa/Namibia*. Geological Society of South Africa Special Publications, **11**, 281–287.

ALLSOPP, H. L., BURGER, A. J. & VAN ZYL, C. 1967. A minimum age for the Premier kimberlite pipe yielded by biotite Rb–Sr measurements, with related galena isotopic data. *Earth and Planetary Science Letters*, **3**, 161–166.

ALLSOPP, H. L., KRAMERS, J. D., JONES, D. L. & ERLANK, A. J. 1989. The age of the Umkondo Group, eastern Zimbabwe, and implications for palaeomagnetic correlations. *South African Journal of Geology*, **92**, 11–19.

ALTERMANN, W. & HÄLBICH, I. W. 1991. Structural history of the southwestern corner of the Kaapvaal Craton and the adjacent Namaqua realm: new observations and a reappraisal. *Precambrian Research*, **52**, 133–166.

ANDERSEN, L. S. & UNRUG, R. 1984. Geodynamic evolution of the Bangweulu Block, northern Zambia. *Precambrian Research*, **25**, 187–212.

ANDREOLI, M. A. G. 1984. Petrochemistry, tectonic evolution and metasomatic mineralisations of Mozambique belt granulites from S. Malawi and Tete (Mozambique). *Precambrian Research*, **25**, 161–186.

ANDREOLI, M. A. G. & HART, R. J. 1990. Metasomatized granulites and eclogites of the Mozambique belt: implications for mantle devolatilization. In: HERBERT, H. K. & HO, S. E. (eds) *Stable Isotopes and Fluid Processes in Mineralization*. Geology Department and University Extension, University of Western Australia, Special Publications, **23**, 121–140.

APPEL, P., MÖLLER, A. & SCHENK, V. 1998. High-pressure granulite facies metamorphism in the Pan-African belt of eastern Tanzania: P–T evidence against granulite formation by continent collision. *Journal of Metamorphic Geology*, **16**, 491–509.

ARMSTRONG, R. A., REID, D. L., WATKEYS, M. K., WELKE, H. J., LIPSON, R. D. & COMPSTON, W. 1988. Zircon U–Pb ages from the Aggeneys area, central Bushmanland. *Extended Abstracts, Geocongress '88, Geological Society of South Africa*, Durban, 493–496.

ARMSTRONG, R. A., ROBB, L. J., MASTER, S., KRUGER, F. J. & MUMBA, P. A. C. C. 1999. New U–Pb age constraints on the Katangan Sequence, Central African Copperbelt. *Journal of African Earth Sciences, Special Abstracts Issue, GSA 11: Earth Resources for Africa*, **28**, 6–7.

BARNES, S.-J. & SAWYER, E. W. 1980. An alternative model for the Damara mobile belt: ocean crust subduction and continental convergence. *Precambrian Research*, **13**, 297–336.

BARR, M. W. C., CAHEN, L. & LEDENT, D. 1978. Geochronology of syntectonic granites from central Zambia: Lusaka granite and granite NE of Rufunsa. *Annales de la Société Géologique de Belgique*, **100**, 47–54.

BARTON, C. M., CARNEY, J. N., CROW, M. J., DUNKLEY, P. N. & SIMANGO, S. 1991. *The Geology of the Country around Rushinga and Nyamapanda*. Zimbabwe Geological Survey Bulletin **92**.

BARTON, C. M., CARNEY, J. N., CROW, M. J., EVANS, J. A. & Simango, S. 1993. Geological and structural framework of the Zambezi Belt, northeastern Zimbabwe. In: FINDLAY, P. R. H., UNRUG, R., BANKS, M. R. & VEEVERS, J. J. (eds) *Gondwana Eight: Assembly, Evolution and Dispersal*. Balkema, Rotterdam, 55–68.

BARTON, E. S. 1983. Reconnaissance isotopic investigations in the Namaqua mobile belt and implications for Proterozoic crustal evolution – Namaqualand geotraverse. In: BOTHA, B. J. V. (ed.) *Namaqualand Metamorphic Complex*. Geological Society of South Africa Special Publications, **10**, 45–66.

BARTON, E. S. & BURGER, A. J. 1983. Reconnaissance isotopic investigations in the Namaqua mobile belt and implications for Proterozoic crustal evolution – Upington geotraverse. In: BOTHA, B. J. V. (ed.) *Namaqualand Metamorphic Complex*. Geological Society of South Africa Special Publications, **10**, 173–191.

BARTON, JR, J. M. & SERGEEV, S. 1997. High precision, U–Pb analyses of single grains of zircon from quartzite in the Beit Bridge Group yield a discordia. *South African Journal of Geology*, **100**, 37–41.

BARTON, JR, J. M. & VAN REENEN, D. D. 1992. The significance of Rb–Sr ages of biotite and phlogopite for the thermal history of the Central and Southern Marginal Zones of the Limpopo Belt of southern Africa and the adjacent portions of the Kaapvaal Craton. *Precambrian Research*, **55**, 17–31.

BARTON, JR, J. M., HOLZER, L., KAMBER, B., DOIG, R., KRAMERS, J. D. & NYFELER, D. 1994. Discrete metamorphic events in the Limpopo belt, southern Africa: implications for the application of P–T paths in complex metamorphic terrains. *Geology*, **22**, 1035–1038.

BECKER, T., HANSEN, B. T., WEBER, K. & WIEGAND, B. 1996. U–Pb and Rb–Sr isotopic data for the Mooirivier Complex, Weener Igneous Suite and Gaub Valley Formation (Rehoboth Sequence) in the Nauchas area and their significance for Palaeoproterozoic crustal evolution in Namibia. *Namibia Geological Survey Communications*, **11**, 31–46.

BINDA, P. L. 1994. Stratigraphy of Zambian Copperbelt orebodies. *Journal of African Earth Sciences*, **19**, 251–264.

BLIGNAULT, H. J., VAN ASWEGEN, G., VAN DER MERWE, S. W. & COLLISTON, W. P. 1983. The Namaqua geotraverse and environs: part of the Proterozoic Namaqua mobile belt. In: BOTHA, B. J. V. (ed.) *Namaqualand Metamorphic Complex*. Geological Society of South Africa Special Publications, **10**, 1–29.

BONHOMME, M. G., GAUTHIER-LAFAYE, F. & WEBER, F. 1982. An example of Lower Proterozoic sediments: the Francevillian in Gabon. *Precambrian Research*, **18**, 87–102.

BOVEN, A., THEUNISSEN K., SKLYAROVA, E., KLERKX, J., MELNIKOV, A., MRUMA, A. & PUNZALAN, L. 1999. Timing of exhumation of high-pressure mafic granulite terrane of the Palaeoproterozoic Ubende belt (West Tanzania). *Precambrian Research*, **93**, 119–137.

BREITKOPF, J. H. 1989. Geochemical evidence for magma source heterogeneity and activity of a mantle plume during advanced rifting in the southern Damara Orogen, Namibia. *Lithos*, **23**, 115–122.

BREITKOPF, J. H. & MAIDEN, K. J. 1988. Tectonic setting of the Matchless belt pyritic copper deposits, Namibia. *Economic Geology*, **83**, 710–723.

BREWER, M. S., HASLAM, H. W., DARBYSHIRE, D. P. F. & DAVIS, A. E. 1979a. *Rb–Sr Age Determinations in the Bangweulu Block, Luapula Province, Zambia*. Institute of Geological Sciences Report **79/5**.

BREWER, M. S., HASLAM, H. W., DARBYSHIRE, D. P. F. & DAVIS, A. E. 1979b. *The Petrology and Geochronology of Hypersthene Granites in the Mchinji Area, Malawi*. Institute of Geological Sciences Report **79/1**.

BRITO NEVES, B. B., CAMPOS NETO, M. C. & FUCK, R. A. 1999. From Rodinia to Western Gondwana: an approach to the Brasiliano–Pan African cycle and orogenic collage. *Episodes*, **22**, 155–166.

BÜHN, B., DÖRR, W. & BRAUNS, C. M. 2001. Petrology and age of the Otjisazu Carbonatite Complex, Namibia: implications for the pre- and syn-orogenic Damaran evolution. *Journal of African Earth Sciences*, **32**, 1–17.

BÜHN, B., HÄUSSINGER, H., KRAMM, U., KUKLA, C., KUKLA, P. A. & STANISTREET, I. G. 1994. Tectonometamorphic patterns developed during Pan-African continental collision in the Damara inland belt, Namibia. *Chemie der Erde*, **54**, 329–354.

BUICK, I. S., MAAS, R. & GIBSON, R. 2001. Precise U–Pb titanite age constraints on the emplacement of the Bushveld Complex, South Africa. *Journal of the Geological Society, London*, **158**, 3–6.

BURGER, A. J., CLIFFORD, T. N. & MILLER, R. McG. 1976. Zircon U–Pb ages of the Franzfontein granitic suite, northern South West Africa. *Precambrian Research*, **3**, 415–431.

BURKE, K., DEWEY, J. F. & KIDD, W. S. F. 1977. World distribution of sutures – the sites of former oceans. *Tectonophysics*, **40**, 69–99.

CAEN-VACHETTE, M., VIALETTE, Y., BASSOT, J.-P. & VIDAL, P. 1988. Apport de la géochronologie isotopique à la connaissance de la géologie gabonaise. *Chronique de la Recherche Minière*, **491**, 35–54.

CAHEN, L. & LEDENT, D. 1979. Précisions sur l'age, la pétrogenèse et la position stratigraphique des 'granites à étain' de l'est de l'Afrique Centrale. *Bulletin Société Belge Géologique*, **88**, 33–49.

CAHEN, L. & SNELLING, N. J. 1966. *The Geochronology of Equatorial Africa*. North-Holland, Amsterdam.

CAHEN, L. & SNELLING, N. J. 1974. Potassium–argon ages & additions to the stratigraphy of the Malagarasian (Bukoban System of Tanzania) of SE Burundi. *Journal of the Geological Society, London*, **130**, 461–470.

CAHEN, L., LEDENT, D. & SNELLING, N. J. 1975. Données géochronologiques dans le Katangien inférieur du Kasai oriental et du Shaba nord-oriental (République du Zaïre). Musée Royal de l'Afrique Centrale, Tervuren, Belgique, Département Géologie et Minéralogie, Rapport Annuel (1974), 59–69.

CAHEN, L., SNELLING, N. J., DELHAL, J. & VAIL, J. R. 1984. *The Geochronology and Evolution of Africa*. Clarendon Press, Oxford.

CAILTEUX, J. 1994. Lithostratigraphy of the Neoproterozoic Shaba-type (Zaire) Roan Supergroup and metallogenesis of associated stratiform mineralization. *Journal of African Earth Sciences*, **19**, 279–301.

CARNEY, J. N., ALDISS, D. T. & LOCK, N. P. 1994. *The Geology of Botswana*. Botswana Geological Survey Bulletin **37**.

CARNEY, J. N., TRELOAR, P. J., BARTON, C. M., CROW, M. J., EVANS, J. A. & SIMANGO, S. 1991. Deep-crustal granulites with migmatitic and mylonitic fabrics from the Zambezi Belt, northeastern Zimbabwe. *Journal of Metamorphic Geology*, **9**, 461–479.

CARVALHO, H. & ALVES, P. 1993. The Precambrian of SW Angola and NW Namibia. *Comunicações do Instituto de Investigação Científica Tropical, Lisboa, Portugal, Série de Ciencias da Terra*, **4**.

CARVALHO, H., CRASTO, J. P., SILVA, Z. C. G. & VIALETTE, Y. 1987. The Kibaran cycle in Angola – a discussion. *Geological Journal*, **22**, 85–102.

CARVALHO, H., FERNANDEZ, A. & VIALETTE, Y. 1979. Chronologie absolue du Précambrien du sud-ouest de l'Angola. *Comptes Rendus de l'Académie des Sciences*, **288**, 1647–1650.

CARVALHO, H., TASSINARI, C., ALVES, P. H., GUIMARÃES, F. & SIMÕES, M. C. 2000. Geochronological review of the Precambrian in western Angola: links with Brazil. *Journal of African Earth Sciences*, **31**, 383–402.

CHAVAGNAC, V., KRAMERS, J. D., NÄGLER, T. F. & HOLZER, L. 2001. The behaviour of Nd and Pb isotopes during 2.0 Ga migmatization in paragneisses of the Central Zone of the Limpopo Belt (South Africa and Botswana). *Precambrian Research*, **112**, 51–86.

CLIFFORD, T. N. 1970. The structural framework of Africa. *In*: CLIFFORD, T. N. & GASS, I. G. (eds) *African Magmatism and Tectonics*. Oliver and Boyd, Edinburgh, 1–26.

CLIFFORD, T. N., BARTON, E. S., RETIEF, E. A., REX, D. C. & FANNING, C. M. 1995. A crustal progenitor for the intrusive anorthosite–charnockite kindred of the cupriferous Koperberg Suite, O'okiep District, Namaqualand, South Africa; new isotope data for the country rocks and the intrusives. *Journal of Petrology*, **36**, 231–258.

COETZEE, H. & KRUGER, F. J. 1989. The geochronology, Sr- and Pb-isotope geochemistry of the Losberg Complex, and the southern limit of Bushveld Complex magmatism. *South African Journal of Geology*, **92**, 37–41.

COLLISTON, W. P. & SCHOCH, A. E. 1998. Tectonostratigraphic features along the Orange River in the western part of the Mesoproterozoic Namaqua mobile belt. *South African Journal of Geology*, **101**, 91–100.

COLLISTON, W. P., PRAEKELT, H. E. & SCHOCH, A. E. 1989. A broad perspective (Haramoep) of geological relations established by sequence mapping in the Proterozoic Aggeneys Terrane, Bushmanland, South Africa. *South African Journal of Geology*, **92**, 42–48.

COLLISTON, W. P., PRAEKELT, H. E. & SCHOCH, A. E. 1991. A progressive ductile shear model for the Proterozoic Aggeneys Terrane, Namaqua mobile belt, South Africa. *Precambrian Research*, **49**, 205–215.

COOLEN, J. J. M. M. M., PRIEM, H. N. A., VERDURMEN, E. A. TH. & VERSCHURE, R. H. 1982. Possible zircon U–Pb evidence for Pan-African granulite facies metamorphism in the Mozambique belt of southern Tanzania. *Precambrian Research*, **17**, 31–40.

CORNELL, D. H., ARMSTRONG, R. A. & WALRAVEN, F. 1998. Geochronology of the Proterozoic Hartley Basalt Formation, South Africa: constraints on the Kheis tectogenesis and the Kaapvaal Craton's earliest Wilson Cycle. *Journal of African Earth Sciences*, **26**, 5–27.

CORNELL, D. H., HAWKESWORTH, C. J., VAN CALSTEREN, P. & SCOTT, W. D. 1986. Sm–Nd study of Precambrian crustal development in the Prieska–Copperton region, Cape Province. *Geological Society of South Africa Transactions*, **89**, 17–28.

CORNELL, D. H., HUMPHREYS, H., THEART, H. F. J. & SCHEEPERS, D. J. 1992. A collision-related pressure–temperature–time path for Prieska Copper Mine, Namaqua–Natal tectonic province, South Africa. *Precambrian Research*, **59**, 43–71.

CORNELL, D. H., KRÖNER, A., HUMPHREYS, H. & GRIFFIN, G. 1990a. Age of origin of the polymetamorphosed Copperton Formation, Namaqua–Natal Province, determined by single grain zircon Pb–Pb dating. *South African Journal of Geology*, **93**, 709–716.

CORNELL, D. H., THEART, H. F. J. & HUMPHREYS, H. C. 1990b. Dating a collision-related metamorphic cycle at Prieska Copper Mines, South Africa. *In*: SPRY, P. G. & BRYNDZIA, L. T. (eds) *Regional Metamorphism of Ore Deposits and Genetic Implications*. VSP, Utrecht, 97–116.

CORNELL, D. H., THOMAS, R. J., BOWRING, S. A., ARMSTRONG, R. A. & GRANTHAM, G. H. 1996. Protolith interpretation in metamorphic terranes: a back-arc environment with Besshi-type base metal potential for the Quha Formation, Natal Province, South Africa. *Precambrian Research*, **77**, 243–271.

COSI, M., DE BONIS, A., GOSSO, G. *ET AL*. 1992. Late Proterozoic thrust tectonics, high-pressure metamorphism and uranium mineralization in the Domes area, Lufilian Arc, northwestern Zambia. *Precambrian Research*, **58**, 215–240.

COSTA, M., CADOPPI, P., SACCHI, R. & FANNING, C. M. 1994. U–Pb SHRIMP dating of zircons from Mozambique gneiss. *Bolletino Societá Geologica Italiana*, **113**, 173–178.

COSTA, M., FERRARA, G., SACCHI, R. & TONARINI, S. 1992. Rb/Sr dating of the Upper Proterozoic basement of Zambesia, Mozambique. *Geologische Rundschau*, **81**, 487–500.

COWARD, M. P. 1983. The tectonic history of the Damara belt. *In*: MILLER, R. MCG. (ed.) *Evolution of the Damara Orogen of South West Africa/Namibia*. Geological Society of South Africa Special Publications, **11**, 409–421.

DALY, M. C. 1986a. *The tectonic and thermal evolution of the Irumide belt, Zambia*. Ph.D. Thesis, University of Leeds.

DALY, M. C. 1986b. The intracratonic Irumide belt of Zambia and its bearing on collision orogeny during the Proterozoic of Africa. *In*: COWARD, M. P. & RIES, A. C. (eds) *Collision Tectonics*. Geological Society, London, Special Publications, **19**, 321–328.

DALY, M. C. 1986c. Crustal shear zones and thrust belts: their geometry and continuity in Central Africa. *Philosophical Transactions of the Royal Society of London Series A*, **317**, 111–128.

DALY, M. C. 1988. Crustal shear zones in Central Africa: a kinematic approach to Proterozoic tectonics. *Episodes*, **11**, 5–11.

DALY, M. C. & UNRUG, R. 1982. The Muva Supergroup of northern Zambia: a craton to mobile belt sedimentary sequence. *Geological Society of South Africa Transactions*, **85**, 155–165.

DALY, M. C., CHAKRABORTY, S. K., KASOLO, P. *ET AL*. 1984. The Lufilian arc and Irumide belt of Zambia: results of a geotraverse across their intersection. *Journal of African Earth Sciences*, **2**, 311–318.

DALY, M. C., LAWRENCE, S. R., DIEMU-TSHIBAND, K. & MATOUANA, B. 1992. Tectonic evolution of the Cuvette Centrale, Zaire. *Journal of the Geological Society, London*, **149**, 539–546.

DALZIEL, I. W. D. 1997. Neoproterozoic–Paleozoic geography and tectonics: review, hypothesis, environmental speculation. *Geological Society of America Bulletin*, **109**, 16–42.

DALZIEL, I. W. D., MOSHER, S. & GAHAGAN, L. M. 2000. Laurentia–Kalahari collision and the assembly of Rodinia. *Journal of Geology*, **108**, 499–513.

DA SILVA, L. C., GRESSE, P. G., SCHEEPERS, R., MCNAUGHTON, N. J., HARTMANN, L. A. & FLETCHER, I. 2000. U–Pb SHRIMP and Sm–Nd age constraints on the timing and sources of the Pan-African Cape Granite Suite, South Africa. *Journal of African Earth Sciences*, **30**, 795–815.

DEBLOND, A., PUNZALAN, L. E., BOVEN, A. & TACK, L. 2001. The Malagarazi Supergroup of southeast Burundi and its correlative Bukoba Supergroup of northwest Tanzania: Neo- and Mesoproterozoic chronostratigraphic constraints from Ar–Ar ages on mafic intrusive rocks. *Journal of African Earth Sciences*, **32**, 435–449.

DE KOCK, G. S. 1991. The mafic Audawib Suite in the Central Damara orogen of Namibia; geochemical evidence for volcanic arc volcanism. *Journal of African Earth Sciences*, **12**, 593–599.

DELHAL, J. & LEDENT, D. 1976. Age et évolution comparée des gneiss migmatitiques pré-Zadiniens des régions de Boma et de Mpozo-Tombagadio (Bas-Zaire). *Annales de la Société Géologique de Belgique*, **99**, 165–187.

DE WAAL, S. A. & ARMSTRONG, R. A. 2000. The age of the Marble Hall diorite, its relationship to the Uitkomst Complex, and evidence for a new magma type associated with the Bushveld igneous event. *South African Journal of Geology*, **103**, 128–140.

DE WAELE, B. & MAPANI, B. 2002. Geology, stratigraphy and correlation of the central part of the Irumide belt. *Journal of African Earth Sciences*, in press.

DE WAELE, B. & TEMBO, F. 2000. Geochemical and petrological characteristics of granitic rocks in the Serenje area, central Irumide Belt, Zambia. *Journal of African Earth Sciences, Special Abstracts Issue, 18th Colloquium of African Geology*, **30**, 16.

DE WAELE, B., TEMBO, F. & KEY, R. 2000. Towards a better understanding of the Mesoproterozoic Irumide Belt of Zambia: report on a geotraverse across the belt. *Episodes*, **23**, 126–130.

DE WIT, M. J. 1998. Clues to Kennedy's Pan-African thermo-tectonism. *Journal of African Earth Sciences, Special Abstracts Issue, Gondwana 10: Event Stratigraphy of Gondwana*, **27**, 55–58.

DE WIT, M. J., BOWRING, S. A., ASHWAL, L. D., RANDRIANASOLO, L. G., MOREL, V. P. I. & RAMBELOSON, R. A. 2001. Age and tectonic evolution of Neoproterozoic ductile shear zones in southwestern Madagascar, with implications for Gondwana studies. *Tectonics*, **20**, 1–45.

DIRKS, P. H. G. M. & SITHOLE, T. A. 1999. Eclogites in the Makuti gneisses of Zimbabwe: implications for the tectonic evolution of the Zambezi belt in southern Africa. *Journal of Metamorphic Geology*, **17**, 593–612.

DIRKS, P. H. G. M., JELSMA, H. A., VINYU, M. & MUNYANYIWA, H. 1998. The structural history of the Zambezi Belt in northeast Zimbabwe: evidence for crustal extension during the early Pan-African. *South African Journal of Geology*, **101**, 1–16.

DIRKS, P. H. G. M., KRÖNER, A., JELSMA, H. A., MANEYA, C. & JAMAL, D. L. 2000. Pb–Pb zircon ages and a tectonic framework for the Zambezi Belt, Zimbabwe. *Journal of African Earth Sciences, Special Abstracts Issue, 18th Colloquium of African Geology*, **30**, 23–24.

DIRKS, P. H. G. M., KRÖNER, A., JELSMA, H. A., SITHOLE, T. A. & VINYU, M. L. 1999. Structural relations and Pb–Pb zircon ages for the Makuti gneisses: evidence for a crustal-scale Pan-African shear zone in the Zambezi Belt, northwest Zimbabwe. *Journal of African Earth Sciences*, **28**, 427–442.

DJAMA, L. M., LETERRIER, J. & MICHARD, A. 1992. Pb, Sr and Nd isotope study of the basement of the Mayumbian belt (Guena gneisses and Mfoubou granite, Congo): implications for crustal evolution in Central Africa. *Journal of African Earth Sciences*, **14**, 227–237.

DODSON, M. H., CAVANAGH, B. J., THATCHER, E. C. & AFTALION, M. 1975. Age limits for the Ubendian metamorphic episode in Northern Malawi. *Geological Magazine*, **112**, 403–410.

DOWNING, K. N. & COWARD, M. P. 1981. The Okahandja lineament and its significance for Damaran tectonics in Namibia. *Geologische Rundschau*, **70**, 972–1000.

DÜRR, S. B. & DINGELDEY, D. P. 1996. The Kaoko belt (Namibia): part of a late Neoproterozoic continental-scale strike-slip system. *Geology*, **24**, 503–506.

EBERLE, D., HUTCHINS, D. G., REBBECK, R. J. & SOMERTON, I. 1996. Compilation of the Namibian airborne magnetic surveys: procedures, problems and results. *Journal of African Earth Sciences*, **22**, 191–205.

EBY, G. N., WOOLLEY, A. R., DIN, V. & PLATT, G. 1998. Geochemistry and petrogenesis of nepheline syenites: Kasungu–Chipala, Ilomba, and Ulindi nepheline syenite intrusions, North Nyasa Alkaline Province, Malawi. *Journal of Petrology*, **39**, 1405–1424.

EGLINGTON, B. M., HARMER, R. E. & KERR, A. 1989. Isotope and geochemical constraints on Proterozoic crustal evolution in south-eastern Africa. *Precambrian Research*, **45**, 159–174.

EVANS, R. J., ASHWAL, L. D. & HAMILTON, M. A. 1999. Mafic, ultramafic, and anorthositic rocks of the Tete Complex, Mozambique: petrology, age, and significance. *South African Journal of Geology*, **102**, 153–166.

FERNANDEZ-ALONSO, M. & THEUNISSEN, K. 1998. Airborne geophysics and geochemistry provide new insights in the intracontinental evolution of the Mesoproterozoic Kibaran belt (Central Africa). *Geological Magazine*, **135**, 203–216.

FERNANDEZ-ALONSO, M., LAVREAU, J. & KLERKX, J. 1986. Geochemistry and geochronology of the Kibaran granites in Burundi, Central Africa: implications for the Kibaran orogeny. *Chemical Geology*, **57**, 217–234.

FITCHES, W. R. 1970. A part of the Ubendian Orogenic Belt in northern Malawi and Zambia. *Geologische Rundschau*, **59**, 444–458.

FITCHES, W. R. 1971. Sedimentation and tectonics at the northeast end of the Irumide Orogenic Belt, N. Malawi and Zambia. *Geologische Rundschau*, **60**, 589–619.

FITZSIMONS, I. C. W. 2000. A review of tectonic events in the East Antarctic Shield and their implications for Gondwana and earlier supercontinents. *Journal of African Earth Sciences*, **31**, 3–23.

FRANSSEN, L. & ANDRÉ, L. 1988. The Zadinian Group (Late Proterozoic, Zaire) and its bearing on the origin of the West-Congo orogenic belt. *Precambrian Research*, **38**, 215–234.

FRANZ, L., ROMER, R. L. & DINGELDEY, D. P. 1999. Diachronous Pan-African granulite facies metamorphism (650Ma and 550Ma) in the Kaoko belt, NW Namibia. *European Journal of Mineralogy*, **11**, 167–180.

FRIMMEL, H. E. 2000. New U–Pb zircon ages for the Kuboos pluton in the Pan-African Gariep Belt, South Africa: Cambrian mantle plume or far field collision effect. *South African Journal of Geology*, **103**, 207–214.

FRIMMEL, H. E. & FRANK, W. 1998. Neoproterozoic tectono-thermal evolution of the Gariep Belt and its basement, Namibia and South Africa. *Precambrian Research*, **90**, 1–28.

FRIMMEL, H. E., HARTNADY, C. J. H. & KOLLER, F. 1996a. Geochemistry and tectonic setting of magmatic units in the Pan-African Gariep Belt, Namibia. *Chemical Geology*, **130**, 101–121.

FRIMMEL, H. E., KLÖTZLI, U. S. & SIEGFRIED, P. R. 1996b. New Pb–Pb single zircon age constraints on the timing of Neoproterozoic glaciation and continental break-up in Namibia. *Journal of Geology*, **104**, 459–469.

FRIMMEL, H. E., ZARTMAN, R. E. & SPÄTH, A. 2001. The Richtersveld Igneous Complex, South Africa: U–Pb zircon and geochemical evidence for the beginning of Neoproterozoic continental breakup. *Journal of Geology*, **109**, 493–508.

GABERT, G. 1984. Structural-lithologic units of Proterozoic rocks in East Africa, their base, cover, and mineralization. *In*: KLERKX, J. & MICHOT, J. (eds) *African Geology*. Musée Royal de l'Afrique Centrale, Tervuren, 11–22.

GABERT, G. & WENDT, I. 1974. Datierung von granitischen Gesteinen in Dodoman- und Usagaran-System und in der Ndembera-Serie (Tanzania). *Geologisches Jahrbuch*, **B11**, 3–55.

GÉRARDS, J. & LEDENT, D. 1976. Les réhomogénéisations isotopiques d'âge lufilien dans les granites du Rwanda. Musée Royal de l'Afrique Centrale, Tervuren, Belgique, Département Géologie et Minéralogie, Rapport Annuel (1975), 91–103.

GERINGER, G. J., HUMPHREYS, H. C. & SCHEEPERS, D. J. 1994. Lithostratigraphy, protolithology, and tectonic setting of the Areachap Group along the eastern margin of the Namaqua Mobile Belt, South Africa. *South African Journal of Geology*, **97**, 78–100.

GOSCOMBE, B., ARMSTRONG, R. & BARTON, J. M. 2000. Geology of the Chewore Inliers, Zimbabwe: constraining the Mesoproterozoic to Palaeozoic evolution of the Zambezi belt. *Journal of African Earth Sciences*, **30**, 589–627.

GRANTHAM, G. H. 2001. Cryptic Pan-African transpression along the eastern margin of the Kalahari craton and beyond – a possible suture between East and West Gondwana. *Gondwana Research*, **4**, 623.

GRANTHAM, G. H., MABOKO, M. & EGLINGTON, B. M. 2003. A review of the evolution or the Mozambique Belt and implications for the amalgamation and dispersal of Rodinia and Gondwana. *In:* YOSHIDA, M., WINDLEY, B. F. & DASGUPTA, S. (eds) *Proterozoic East Gondwana: Supercontinent Assembly and Breakup*. Geological Society, London, Special Publications, **206**, 401–425.

GRANTHAM, G. H., THOMAS, R. J. & MENDONIDIS, P. 1994. Contrasting P–T–t loops from southern East Africa, Natal and East Antarctica. *Journal of African Earth Sciences*, **19**, 225–235.

GRESSE, P. G., FITCH, F. J. & MILLER, J. A. 1988. $^{40}Ar/^{39}Ar$ dating of the Cambro–Ordovician Vanrhynsdorp tectonite in southern Namaqualand. *South African Journal of Geology*, **91**, 257–263.

GUTZMER, J., BEUKES, N. J., PICKARD, A. & BARLEY, M. E. 2000. 1170Ma SHRIMP age for Koras Group bimodal volcanism, Northern Cape Province. *South African Journal of Geology*, **103**, 32–37.

HAACK, U., HOEFS, J. & GOHN, E. 1982. Constraints on the origin of Damaran granites by Rb/Sr and $\delta^{18}O$ data. *Contributions to Mineralogy and Petrology*, **79**, 279–289.

HANSON, R. E., HARGROVE, U. S., MARTIN, M. W. ET AL. 1998a. New geochronological constraints on the tectonic evolution of the Pan-African Zambezi belt, south-central Africa. *Journal of African Earth Sciences, Special Abstracts Issue, Gondwana 10: Event Stratigraphy of Gondwana*, **27**, 104–105.

HANSON, R. E., MARTIN, M. W., BOWRING, S. A. & MUNYANYIWA, H. 1998b. U–Pb zircon age for the Umkondo dolerites, eastern Zimbabwe: 1.1 Ga large igneous province in southern Africa/East Antarctica and possible Rodinia correlations. *Geology*, **26**, 1143–1146.

HANSON, R. E., WARDLAW, M. S., WILSON, T. J. & MWALE, G. 1993. U–Pb zircon ages from the Hook granite massif and Mwembeshi dislocation: constraints on Pan-African deformation, plutonism, and transcurrent shearing in central Zambia. *Precambrian Research*, **63**, 189–209.

HANSON, R. E., WILSON, T. J. & MUNYANYIWA, H. 1994. Geologic evolution of the Neoproterozoic Zambezi orogenic belt in Zambia. *Journal of African Earth Sciences*, **18**, 135–150.

HANSON, R. E., WILSON, T. J., BRUECKNER, H. K., ONSTOTT, T. C., WARDLAW, M. S., JOHNS, C. C. & HARDCASTLE, K. C. 1988a. Reconnaissance geochronology, tectonothermal evolution, and regional significance of the Middle Proterozoic Choma–Kalomo block, southern Zambia. *Precambrian Research*, **42**, 39–61.

HANSON, R. E., WILSON, T. J. & WARDLAW, M. S. 1988b. Deformed batholiths in the Pan-African Zambezi belt, Zambia: age and implications for regional Proterozoic tectonics. *Geology*, **16**, 1134–1137.

HARGROVE, U. S., HANSON, R. E., MARTIN, M. W., BLENKINSOP, T. G., BOWRING, S. A. & MUNYANYIWA, H.

2002. Tectonic evolution of the Zambezi orogenic belt: geochronological, structural, and petrological constraints from northern Zimbabwe. *Precambrian Research*, in press.

HARTNADY, C., JOUBERT, P. & STOWE, C. 1985. Proterozoic crustal evolution in southwestern Africa. *Episodes*, **8**, 236–244.

HASLAM, H. W., BREWER, M. S., DARBYSHIRE, D. P. F. & DAVIS, A. E. 1983. Irumide and post-Mozambiquian plutonism in Malawi. *Geological Magazine*, **120**, 21–35.

HASLAM, H. W., BREWER, M. S., DAVIS, A. E. & DARBYSHIRE, D. P. F. 1980. Anatexis and high-grade metamorphism in the Champira Dome, Malawi: petrological and Rb–Sr studies. *Mineralogical Magazine*, **43**, 701–714.

HASLAM, H. W., RUNDLE, C. C. & BREWER, M. S. 1986. Rb–Sr studies of metamorphic and igneous events in eastern Zambia. *Journal of African Earth Sciences*, **5**, 447–453.

HAWKESWORTH, C. J., MENZIES, M. A. & VAN CALSTEREN, P. 1986. Geochemical and tectonic evolution of the Damara Belt, Namibia. *In*: COWARD, M. P. & RIES, A. C. (eds) *Collision Tectonics*. Geological Society, London, Special Publications, **19**, 305–319.

HEGENBERGER, W. & BURGER, A. J. 1985. The Oorlogsende Porphyry Member, South West Africa/Namibia: its age and regional setting. *South West Africa/Namibia Geological Survey Communications*, **1**, 23–29.

HOAL, B. G. 1993a. The Proterozoic Sinclair Sequence in southern Namibia: intracratonic rift or active continental margin setting? *Precambrian Research*, **63**, 143–162.

HOAL, B. G. 1993b. The Proterozoic Awasib Mountain terrain – northwestward extension of a pre-Sinclair active continental margin across southern Namibia. *Extended Abstracts, 16th International Colloquium On Africa Geology*, Swaziland Geological Survey and Mines Department, 155–158.

HOAL, B. G. & HEAMAN, L. M. 1995. The Sinclair Sequence: U–Pb age constraints from the Awasib Mountain area. *Namibia Geological Survey Communications*, **10**, 83–91.

HOFFMAN, K. H. 1989. New aspects of lithostratigraphic subdivision and correlation of Late Proterozoic to Early Cambrian rocks of the southern Damara Belt and their correlation with the central and northern Damara Belt and the Gariep Belt. *Namibia Geological Survey Communications*, **5**, 59–67.

HOFFMAN, P. F. 1991. Did the breakout of Laurentia turn Gondwanaland inside-out? *Science*, **252**, 1409–1412.

HOFFMAN, P. F., HAWKINS, D. P., ISACHSEN, C. E. & BOWRING, S. A. 1996. Precise U–Pb zircon ages for early Damaran magmatism in the Summas Mountains and Welwitschia Inlier, northern Damara belt, Namibia. *Namibia Geological Survey Communications*, **11**, 47–52.

HOLZER, L., BARTON, J. M., PAYA, B. K. & KRAMERS, J. D. 1999. Tectonothermal history of the western part of the Limpopo Belt: tectonic models and new perspectives. *Journal of African Earth Sciences*, **28**, 383–402.

HOLZER, L., FREI, R., BARTON, JR, J. M. & KRAMERS, J. D. 1998. Unraveling the record of successive high grade events in the Central Zone of the Limpopo Belt using Pb single phase dating of metamorphic minerals. *Precambrian Research*, **87**, 87–115.

HORSTMANN, U. E., AHRENDT, H., CLAUER, N. & PORADA, H. 1990. The metamorphic history of the Damara Orogen based on K/Ar data of detrital white micas from the Nama Group, Namibia. *Precambrian Research*, **48**, 41–61.

HUMPHREYS, H. C. & CORNELL, D. H. 1989. Petrology and geochronology of low-pressure mafic granulites in the Marydale Group, South Africa. *Lithos*, **22**, 287–303.

HUMPHREYS, H. C., SCHLEGEL, G. C.-J. & STOWE, C. W. 1991. High-pressure metamorphism in garnet–hornblende–muscovite–plagioclase–quartz schists from the Kheis Belt. *South African Journal of Geology*, **94**, 170–173.

HUTCHINS, D. G. & REEVES, C. V. 1980. Regional geophysical exploration of the Kalahari in Botswana. *Tectonophysics*, **69**, 201–220.

IKINGURA, J. R., BELL, K., WATKINSON, D. H. & VAN STRAATEN, P. 1990. Geochronology and chemical evolution of granitic rocks, NE Kibaran (Karagwe–Ankolean) belt, NW Tanzania. *In*: ROCCI, G. & DESCHAMPS, M. (eds) *Recent Data in African Earth Sciences*. International Center for Training and Exchanges in the Geosciences Occasional Publication, **22**, 97–99.

JACOB, R. E., MOORE, J. M. & ARMSTRONG, R. A. 2000. SHRIMP dating and implications for the timing of Au mineralisation, Navachab, Namibia. *Journal of African Earth Sciences, Special Abstracts Issue, GSSA 27: Geocongress 2000*, **31**, 32–33.

JACOBS, J. & THOMAS, R. J. 1994. Oblique collision at about 1.1 Ga along the southern margin of the Kaapvaal continent, southeast Africa. *Geologische Rundschau*, **83**, 322–333.

JACOBS, J., FALTER, M., THOMAS, R. J., KUNZ, J. & JEßBERGER, E. K. 1997. ^{40}Ar/^{39}Ar thermochronological constraints on the structural evolution of the Mesoproterozoic Natal Metamorphic Province, SE Africa. *Precambrian Research*, **86**, 71–92.

JACOBS, J., FANNING, C. M., HENJES-KUNST, F., OLESCH, M. & PAECH, H.-J. 1998. Continuation of the Mozambique Belt into East Antarctica: Grenville-age metamorphism and polyphase Pan-African high-grade events in central Dronning Maud Land. *Journal of Geology*, **106**, 385–406.

JACOBS, J., THOMAS, R. J. & WEBER, K. 1993. Accretion and indentation tectonics at the southern edge of the Kaapvaal craton during the Kibaran (Grenville) orogeny. *Geology*, **21**, 203–206.

JAECKEL, P., KRÖNER, A., KAMO, S. L., BRANDL, G. & WENDT, J. I. 1997. Late Archaean to Early Proterozoic granitoid magmatism and high-grade metamorphism in the central Limpopo Belt, South Africa. *Journal of the Geological Society, London*, **154**, 25–44.

JAMAL, D. L., ZARTMAN, R. E. & DE WIT, M. J. 1999. U–Pb single zircon dates from the Lurio belt, northern Mozambique: Kibaran and Pan-African orogenic events highlighted. *Journal of African Earth Sciences, Special Abstracts Issue, GSA 11: Earth Resources for Africa*, **28**, 32.

JOHN, T., SCHENK, V. & TEMBO, F. 2000. MORB-type geochemical signatures of eclogites from central Zambia: evidence for a Precambrian suture-zone. *Journal of*

African Earth Sciences, Special Abstracts Issue, 18th Colloquium of African Geology, **30**, 42–43.

JOHNS, C. C., LIYUNGU, K., MABUKU, S. ET AL. 1989. The stratigraphic and structural framework of Eastern Zambia: results of a geotraverse. *Journal of African Earth Sciences*, **9**, 123–136.

JOHNSON, R. L. & VAIL, J. R. 1965. The junction between the Mozambique and Zambezi orogenic belts, northeast Southern Rhodesia. *Geological Magazine*, **102**, 489–495.

JOHNSON, S. P. & OLIVER, G. J. H. 1998. A second natural occurrence of yoderite. *Journal of Metamorphic Geology*, **16**, 809–818.

JOHNSON, S. P. & OLIVER, G. J. H. 2000. Mesoproterozoic oceanic subduction, island-arc formation and the initiation of back-arc spreading in the Kibaran Belt of central, southern Africa: evidence from the Ophiolite Terrane, Chewore Inliers, northern Zimbabwe. *Precambrian Research*, **103**, 125–146.

JONES, D. L. & MCELHINNY, M. W. 1966. Paleomagnetic correlation of basic intrusions in the Precambrian of southern Africa. *Journal of Geophysical Research*, **71**, 543–552.

JUNG, S. 2000. High-temperature, low/medium-pressure clockwise *P–T* paths and melting in the development of regional migmatites: the role of crustal thickening and repeated plutonism. *Geological Journal*, **35**, 345–359.

JUNG, S. & MEZGER, K. 2001. Geochronology in migmatites – a Sm–Nd, U–Pb and Rb–Sr study from the Proterozoic Damara belt (Namibia): implications for polyphase development of migmatites in high-grade terranes. *Journal of Metamorphic Geology*, **19**, 77–97.

JUNG, S., HOERNES, S. & MEZGER, K. 2000a. Geochronology and petrology of migmatites from the Proterozoic Damara Belt – importance of episodic fluid-present disequilibrium melting and consequences for granite petrology. *Lithos*, **51**, 153–179.

JUNG, S., HOERNES, S. & MEZGER, K. 2000b. Geochronology and petrogenesis of Pan-African, syntectonic, S-type and post-tectonic A-type granite (Namibia): products of melting of crustal sources, fractional crystallization and wall rock entrainment. *Lithos*, **50**, 259–287.

JUNG, S., MEZGER, K. & HOERNES, S. 2001. Trace element and isotopic (Sr, Nd, Pb, O) arguments for a midcrustal origin of Pan-African garnet-bearing S-type granites from the Damara orogen (Namibia). *Precambrian Research*, **110**, 325–355.

KABENGELE, M., TSHIMANGA, K., LUBALA, R. T., KAPENDA, D. & WALRAVEN, F. 1990. Geochronology of the calcalkaline granitoids of the Marungu plateau (eastern Zaire – central Africa). *In*: ROCCI, G. & DESCHAMPS, M. (eds) *Recent Data in African Earth Sciences*. International Center for Training and Exchanges in the Geosciences Occasional Publication, **22**, 51–55.

KAMBER B. S., BIINO, G. G., WIJBRANS, J. R., DAVIES, G. R. & VILLA, I. M. 1996. Archaean granulites of the Limpopo Belt, Zimbabwe: one slow exhumation or two rapid events? *Tectonics*, **15**, 1414–1430.

KAMBER B. S., BLENKINSOP, T. G., VILLA, I. M. & DAHL, P. S. 1995a. Proterozoic transpressive deformation in the Northern Marginal Zone, Limpopo Belt, Zimbabwe. *Journal of Geology*, **103**, 493–508.

KAMBER B. S., KRAMERS, J. D., NAPIER, R., CLIFF, R. A. & ROLLINSON, H. R. 1995b. The Triangle Shear Zone, Zimbabwe, revisited: new data document an important event at 2.0 Ga in the Limpopo Belt. *Precambrian Research*, **70**, 191–213.

KAMPUNZU, A. B. 1989. Geochemistry of granitoids from northern Kibaran Belt in eastern Zaire (Central Africa). *In*: POHL, W. & DELHAL, J. (eds) *Metallogeny of the Kibara Belt, Central Africa, Bulletin – Newsletter*, Musée Royal de l'Afrique Centrale, Tervuren, **2**, 35–36.

KAMPUNZU, A. B. & CAILTEUX, J. 1999. Tectonic evolution of the Lufilian Arc (Central Africa Copper Belt) during Neoproterozoic Pan African orogenesis. *Gondwana Research*, **2**, 401–421.

KAMPUNZU, A. B., AKANYANG, P., MAPEO, R. B. M., MODIE, B. N. & WENDORFF, M. 1998a. Geochemistry and tectonic significance of the Mesoproterozoic Kgwebe metavolcanic rocks in northwest Botswana: implications for the evolution of the Kibaran Namaqua–Natal Belt. *Geological Magazine*, **135**, 669–683.

KAMPUNZU, A. B., ARMSTRONG, R., MODISI, M. P. & MAPEO, R. B. 1999. The Kibaran belt in southwest Africa: ion microprobe U–Pb zircon data and definition of the Kibaran Ngamiland belt in Botswana, Namibia and Angola. *Journal of African Earth Sciences, Special Abstracts Issue, GSA 11: Earth Resources for Africa*, **28**, 34.

KAMPUNZU, A. B., ARMSTRONG, R., MODISI, M. P. & MAPEO, R. B. 2000a. Ion microprobe U–Pb ages on detrital zircon grains from the Ghanzi Group: implications for the identification of a Kibaran-age crust in northwest Botswana. *Journal of African Earth Sciences*, **30**, 579–587.

KAMPUNZU, A. B., KRAMERS, J. D. & MAKUTU, M. N. 1998b. Rb–Sr whole rock ages of the Lueshe, Kirumba and Numbi igneous complexes (Kivu, Democratic Republic of Congo) and the break-up of the Rodinia supercontinent. *Journal of African Earth Sciences*, **26**, 29–36.

KAMPUNZU, A. B., TEMBO, F., MATHEIS, G., KAPENDA, D. & HUNTSMAN-MAPILA, P. 2000b. Geochemistry and tectonic setting of mafic igneous units in the Neoproterozoic Katangan basin, Central Africa: implications for Rodinia break-up. *Gondwana Research*, **3**, 125–153.

KAMPUNZU, A. B., WENDORFF, M., KRUGER, F. J. & INTIOMALE, M. M. 1998c. Pb isotopic ages of sediment-hosted Pb–Zn mineralisation in the Neoproterozoic Copperbelt of Zambia and Democratic Republic of Congo (ex-Zaire): reevaluation and implications. *Chronique de la Recherche Minière*, **530**, 55–61.

KASCH, K. W. 1983. Regional *P–T* variations in the Damara orogen with particular reference to early high-pressure metamorphism along the southern margin. *In*: MILLER, R. McG. (ed.) *Evolution of the Damara Orogen of South West Africa/Namibia*. Geological Society of South Africa Special Publications, **11**, 243–253.

KENNEDY, W. Q. 1964. The structural differentiation of Africa in the Pan-African (±500 m.y.) tectonic episode. *University of Leeds, Research Institute of African Geology, Annual Report*, **8**, 48–49.

KEY, R. M. & AYRES, N. 2000. The 1998 edition of the National Geological Map of Botswana. *Journal of African Earth Sciences*, **30**, 427–451.

KEY, R. M. & MAPEO, R. 1999. The Mesoproterozoic history of Botswana and the relationship of the NW Botswana Rift to Rodinia. *Episodes*, **22**, 118–122.

KEY, R. M., LIYUNGU, A. K., NJAMU, F. J., SOMWE, V., BANDA, J., MOSLEY, P. N. & ARMSTRONG, R. A. 2001. The Western arm of the Lufilian Arc in NW Zambia and its potential for copper mineralization. *Journal of African Earth Sciences*, **33**, 503–528..

KISTERS, A. F. M., CHARLESWORTH, E. G., GIBSON, R. L. & ANHAEUSSER, C. R. 1996. Steep structure formation in the Okiep Copper District, South Africa: bulk inhomogeneous shortening of a high-grade metamorphic granite-gneiss sequence. *Journal of Structural Geology*, **18**, 735–751.

KLERKX, J., LIÉGEOIS, J.-P., LAVREAU, J. & CLAESSENS, W. 1987. Crustal evolution of the northern Kibaran belt, eastern and central Africa. *In*: KRÖNER, A. (ed.) *Proterozoic Lithospheric Evolution.* American Geophysical Union, Geodynamics Series, **17**, 217–233.

KLERKX, J., THEUNISSEN, K. & DELVAUX, D. 1998. Persistent fault controlled basin formation since the Proterozoic along the Western Branch of the East African Rift. *Journal of African Earth Sciences*, **26**, 347–361.

KNOPER, M., ARMSTRONG, R. A., ANDREOLI, M. A. G. & ASHWAL, L. D. 2000. The Steenkampskraal monazite vein: a subhorizontal stretching shear zone indicating extensional collapse of Namaqualand at 1033 Ma? *Journal of African Earth Sciences, Special Abstracts Issue, GSSA 27: Geocongress 2000*, **31**, 38–39.

KOKONYANGI, J., ARMSTRONG, R., KAMPUNZU, A. B. & YOSHIDA, M. 2001. SHRIMP U–Pb zircon geochronology of granitoids in the Kibaran type-area, Mitwaba – central Katanga (Congo). *Gondwana Research*, **4**, 661–663.

KRÖNER, A. 1982. Rb–Sr geochronology and tectonic evolution of the Pan-African Damara belt of Namibia, southwestern Africa. *American Journal of Science*, **82**, 1471–1507.

KRÖNER, A. 1983. Proterozoic mobile belts compatible with the plate tectonic concept. *In:* MEDARIS, JR, L. G., BYERS, C. W., MICKELSON, D. M. & SHANKS, W. C. (eds) *Proterozoic Geology: Selected Papers from an International Proterozoic Symposium.* Geological Society of America Memoir, **161**, 59–74.

KRÖNER, A. 2001. The Mozambique belt of East Africa and Madagascar: significance of zircon and Nd model ages for Rodinia and Gondwana supercontinent formation and dispersal. *South African Journal of Geology*, **105**, 151–168.

KRÖNER, A., BARTON, E. S., BURGER, A. J., ALLSOPP, H. L. & BERTRAND, J. M. 1983. The ages of the Goodhouse Granite and grey gneisses from the marginal zone of the Richtersveld Province and their bearing on the timing of tectonic events in the Namaqua mobile belt. *In*: BOTHA, B. J. V. (ed.) *Namaqualand Metamorphic Complex.* Geological Society of South Africa Special Publications, **10**, 123–129.

KRÖNER, A., JAECKEL, P., BRANDL, G., NEMCHIN, A. A. & PIDGEON, R. T. 1999. Single zircon ages for granitoid gneisses in the Central Zone of the Limpopo Belt, Southern Africa and geodynamic significance. *Precambrian Research*, **93**, 299–337.

KRÖNER, A., RETIEF, E. A., COMPSTON, W., JACOB, R. E. & BURGER, A. J. 1991. Single-grain and conventional zircon dating of remobilized basement gneisses in the central Damara belt of Namibia. *South African Journal of Geology*, **94**, 379–387.

KRÖNER, A., SACCHI, R., JAECKEL, P. & COSTA, M. 1997. Kibaran magmatism and Pan-African granulite metamorphism in northern Mozambique: single zircon ages and regional implications. *Journal of African Earth Sciences*, **25**, 467–484.

KRÖNER, A., WILLNER, A. P., HEGNER, E., JAECKEL, P. & NEMCHIN, A. 2001. Single zircon ages, PT evolution and Nd isotopic systematics of high-grade gneisses in southern Malawi and their bearing on the evolution of the Mozambique belt in southeastern Africa. *Precambrian Research*, **109**, 257–291.

KUKLA, C., KRAMM, U., KUKLA, P. A. & OKRUSCH, M. 1991. U–Pb monazite data relating to metamorphism and granite intrusion in the northwestern Khomas Trough, Damara Orogen, central Namibia. *Namibia Geological Survey Communications*, **7**, 49–54.

KUKLA, P. A. 1992. *Tectonics and Sedimentation of a Late Proterozoic Damaran Convergent Continental Margin, Khomas Hochland, Central Namibia.* Geological Survey of Namibia Memoir **12**.

KUKLA, P. A. & STANISTREET, I. G. 1991. Record of the Damaran Khomas Hochland accretionary prism in central Namibia: refutation of an 'ensialic' origin of a Late Proterozoic orogenic belt. *Geology,* **19**, 473–476.

LAVREAU, J. & LIÉGEOIS, J.-P. 1982. Granites à étain et granito-gneiss Burundiens au Rwanda (région de Kibuye): âge et signification. *Annales de la Société Géologique de Belgique*, **105**, 289–294.

LENOIR, J. L., LIÉGEOIS, J.-P., THEUNISSEN, K. & KLERKX, J. 1994. The Palaeoproterozoic Ubendian shear belt in Tanzania: geochronology and structure. *Journal of African Earth Sciences*, **19**, 169–184.

LEYSHON, P. R. & TENNICK, F. P. 1988. The Proterozoic Magondi Mobile Belt in Zimbabwe – a review. *South African Journal of Geology*, **91**, 114–131.

LORIS, N.-B.-T., CHARLET, J.-M., PECHMANN, E., CLARE, C., CHABU, M. & QUINIF, Y. 1997. Caractéristiques minéralogiques, cristallographiques, physico-chimiques et âges des minéralisations uranifères de Lwiswishi (Shaba, Zaïre). *In*: CHARLET, J.-M. (ed.) *International Cornet Symposium Strata-bound Copper Deposits and Associated Mineralizations*, 285–306.

LÜDTKE, G., EBERLE, D. & VAN DER BOOM, G. 1986. *Geophysical, Geochemical and Geological Investigations in the Ngami and Kheis Areas of Botswana, 1980–1983. Final Report.* Botswana Geological Survey Bulletin **32**.

LUFT, JR, J. L., CHEMALE, F. JR & BITENCOURT, M. F. 2001. Significance of Palaeoproterozoic to Neoproterozoic ages in the Kaoko belt, NW Namibia. *Gondwana Research*, **4**, 693.

MABOKO, M. A. H. 1995. Neodymium isotopic constraints on the protolith ages of rocks involved in Pan-African tectonism in the Mozambique Belt of Tanzania.

Journal of the Geological Society, London, **152**, 911–916.

MABOKO, M. A. H. 1997. P–T conditions of metamorphism in the Wami River granulite complex, central coastal Tanzania: implications for Pan-African geotectonics in the Mozambique Belt of eastern Africa. *Journal of African Earth Sciences*, **24**, 51–64.

MABOKO, M. A. H. 2000. Nd and Sr isotopic investigation of the Archaean–Proterozoic boundary in north eastern Tanzania: constraints on the nature of Neoproterozoic tectonism in the Mozambique Belt. *Precambrian Research*, **102**, 87–98.

MABOKO, M. A. H. 2001. Isotopic and geochemical constraints on Neoproterozoic crust formation in the Wami River area, eastern Tanzania. *Journal of African Earth Sciences*, **33**, 91–101.

MABOKO, M. A. H. & NAKAMURA, E. 1996. Nd and Sr isotopic mapping of the Archaean–Proterozoic boundary in southeastern Tanzania using granites as probes for crustal growth. *Precambrian Research*, **77**, 105–115.

MABOKO, M. A. H. & NAKAMURA, E. 2002. Isotopic dating of Neoproterozoic crustal growth in the Usambara Mountains of northeastern Tanzania: evidence for coeval crust formation in the Mozambique Belt and the Arabian–Nubian Shield. *Precambrian Research*, **113**, 227–242.

MABOKO, M. A. H., BOELRIJK, N. A. I. M., PRIEM, H. N. A. & VERDURMEN, E. A. TH. 1985. Zircon U–Pb and biotite Rb–Sr dating of the Wami River granulites, Eastern Granulites, Tanzania: evidence for approximately 715 Ma old granulite-facies metamorphism and final Pan-African cooling approximately 475 Ma ago. *Precambrian Research*, **30**, 361–378.

MAJAULE, T., HANSON, R. E., KEY, R. M., SINGLETARY, S. J., MARTIN, M. W. & BOWRING, S. A. 2001. The Magondi Belt in northeast Botswana: regional relations and new geochronological data from the Sua Pan area. *Journal of African Earth Sciences*, **32**, 257–267.

MALLICK, D. I. J. 1966. The stratigraphy and the structural development of the Mpande Dome, southern Zambia. *Geological Society of South Africa Transactions*, **69**, 211–230.

MANHICA, A. D. S. T., GRANTHAM, G. H., ARMSTRONG, R. A., GUISE, P. G. & KRUGER, F. J. 2001. Polyphase deformation and metamorphism at the Kalahari Craton–Mozambique Belt boundary. *In*: MILLER, J. A., HOLDSWORTH, R. E., BUICK, I. S. & HAND, M. (eds) *Continental Reactivation and Reworking*. Geological Society, London, Special Publications, **184**, 303–321.

MAPEO, R. B. M., ARMSTRONG, R. A. & KAMPUNZU, A. B. 2001. SHRIMP U–Pb zircon geochronology of gneisses from the Gweta borehole, northeast Botswana: implications for the Palaeoproterozoic Magondi Belt in southern Africa. *Geological Magazine*, **138**, 299–308.

MARIGA, J. 1999. *Structural and geochronological evolution of deep-crustal granulites, supracrustal rocks, and deformed plutons in the Zambezi orogenic belt, Rusambo Mission area, northeastern Zimbabwe*. MS thesis, Texas Christian University.

MARIGA, J., HANSON, R. E., MARTIN, M. W., SINGLETARY, S. J. & BOWRING, S. A. 1998. Timing of polyphase ductile deformation at deep to mid-crustal levels in the Neoproterozoic Zambezi belt, NE Zimbabwe. *Geological Society of America Abstracts with Programs*, **30**, A292.

MASTER, S. 1991. *Stratigraphy, Tectonic Setting, and Mineralization of the Early Proterozoic Magondi Supergroup, Zimbabwe: a Review*. University of the Witwatersrand, Economic Geology Research Unit, Information Circular **238**.

MATTHEWS, P. E. 1972. Possible Precambrian obduction and plate tectonics in southeastern Africa. *Nature: Physical Science*, **240**, 37–39.

MAURIN, J. C. 1993. The Pan-African West-Congo belt: links with eastern Brazil and geodynamical reconstruction. *International Geology Review*, **35**, 436–452.

MAURIN, J. C., MPEMBA-BONI, J., PIN, C. & VICAT, J.-P. 1990. La granodiorite de Les Saras, un témoin de magmatisme éburnéen (2 G.a.) au sein de la chaîne ouest-congolienne: conséquences géodynamiques. *Comptes Rendus de l'Académie des Sciences*, **310**, 571–575.

MAYER, A., SINIGOI, S. & MORAIS, E. 2000. The Kunene gabbro-anorthosite complex: coalescence of crystal mush intrusions during the early Kibaran. *American Geophysical Union Transactions*, **81**, 1248–1249.

MAZAC, O. 1974. *Reconnaissance Gravity Survey of Zambia*. Geological Survey of Zambia Technical Report **76**.

MCCOURT, S. & ARMSTRONG, R. A. 1998. SHRIMP U–Pb zircon geochronology of granites from the Central Zone, Limpopo Belt, southern Africa: implications for the age of the Limpopo Orogeny. *South African Journal of Geology*, **101**, 329–338.

MCCOURT, S., BISNATH, A., PATHER, S. *ET AL.* 2000. Geology and tectonic setting of the Tugela Terrane, Natal Belt, South Africa. *Journal of African Earth Sciences, Special Abstracts Issue, GSSA 27: Geocongress 2000*, **31**, 48–49.

MCCOURT, S., HILLIARD, P., ARMSTRONG, R. A. & MUNYANYIWA, H. 2001. SHRIMP U–Pb zircon geochronology of the Hurungwe granite, north-west Zimbabwe: age constraints on the timing of the Magondi orogeny and implications for the correlation between the Kheis and Magondi Belts. *South African Journal of Geology*, **104**, 39–46.

MCDERMOTT, F., HARRIS, N. B. W. & HAWKESWORTH, C. J. 1996. Geochemical constraints on crustal anatexis: a case study from the Pan-African Damara granitoids of Namibia. *Contributions to Mineralogy and Petrology*, **123**, 406–423.

MEERT, J. 2002. A synopsis of events related to the assembly of eastern Gondwana. *Tectonophysics*, in press.

MEERT, J. & Van der Voo, R. 1997. The assembly of Gondwana 800–550 Ma. *Journal of Geodynamics*, **23**, 223–235.

MEIXNER, H. M. & PEART, R. J. 1984. *The Kalahari Drilling Project: a Report on the Geophysical and Geological Results of Follow-up Drilling to the Aeromagnetic Survey of Botswana*. Botswana Geological Survey Bulletin **27**.

MILLER, R. MCG. 1983a. The Pan-African Damara orogen of South West Africa/Namibia. *In*: MILLER, R. MCG. (ed.) *Evolution of the Damara Orogen of South West Africa/Namibia*. Geological Society of South Africa Special Publications, **11**, 431–515.

MILLER, R. McG. 1983b. Tectonic implications of the contrasting geochemistry of Damaran mafic volcanic rocks, Southwest Africa/Namibia. In: MILLER, R. McG. (ed.) *Evolution of the Damara Orogen of South West Africa/Namibia*. Geological Society of South Africa Special Publications, **11**, 115–138.

MILLER, R. McG. & BURGER, A. J. 1983. U–Pb zircon ages of members of the Salem granitic suite along the northern edge of the central Damara granite belt. In: MILLER, R. McG. (ed.) *Evolution of the Damara Orogen of South West Africa/Namibia*. Geological Society of South Africa Special Publications, **11**, 273–280.

MODIE, B. N. 1996. *The Geology of the Ghanzi Ridge*. Botswana Geological Survey District Memoir **7**.

MOEN, H. F. G. 1999. The Kheis Tectonic Subprovince, southern Africa: a lithostratigraphic perspective. *South African Journal of Geology*, **102**, 27–42.

MÖLLER, A., APPEL, P., MEZGER, K. & SCHENK, V. 1995. Evidence for a 2 Ga subduction zone: eclogites in the Usagaran belt of Tanzania. *Geology*, **23**, 1067–1070.

MÖLLER, A., MEZGER, K. & SCHENK, V. 1998. Crustal age domains and the evolution of the continental crust in the Mozambique Belt of Tanzania: combined Sm–Nd, Rb–Sr, and Pb–Pb isotopic evidence. *Journal of Petrology*, **39**, 749–783.

MÖLLER, A., MEZGER, K. & SCHENK, V. 2000. U–Pb dating of metamorphic minerals: Pan-African metamorphism and prolonged slow cooling of high pressure granulites in Tanzania, East Africa. *Precambrian Research*, **104**, 123–146.

MOORE, J. M., WATKEYS, M. K. & REID, D. L. 1990. The regional setting of the Aggeneys/Gamsberg base metal deposits, Namaqualand, South Africa. In: SPRY, P. G. & Bryndzia, L. T. (eds) *Regional Metamorphism of Ore Deposits and Genetic Implications*. VSP, Utrecht, 77–95.

MOSLEY, P. N. 1993. Geological evolution of the Late Proterozoic 'Mozambique Belt' of Kenya. *Tectonophysics*, **221**, 223–250.

MRUMA, A. H. 1989. Stratigraphy, metamorphism and tectonic evolution of the Early Proterozoic Usagaran Belt, Tanzania. *Res Terrae, Series A*, **2**.

MUHONGO, S. 1994. Neoproterozoic collision tectonics in the Mozambique Belt of East Africa: evidence from the Uluguru mountains (Tanzania). *Journal of African Earth Sciences*, **19**, 153–168.

MUHONGO, S. & LENOIR, J.-L. 1994. Pan-African granulite-facies metamorphism in the Mozambique Belt of Tanzania: U–Pb zircon geochronology. *Journal of the Geological Society, London*, **151**, 343–347.

MUHONGO, S. & TUISKU, P. 1996. Pan-African high pressure isobaric cooling: evidence from the mineralogy and thermobarometry of the granulite-facies rocks from the Uluguru Mountains, eastern Tanzania. *Journal of African Earth Sciences*, **23**, 443–463.

MUHONGO, S., KRÖNER, A. & NEMCHIN, A. A. 2001. Single zircon evaporation and SHRIMP ages for granulite-facies rocks in the Mozambique Belt of Tanzania. *Journal of Geology*, **109**, 171–189.

MUHONGO, S., TUISKU, P. & MTONI, Y. 1999. Pan-African pressure-temperature evolution of the Merelani area in the Mozambique Belt in northeast Tanzania. *Journal of African Earth Sciences*, **29**, 353–365.

MÜLLER, M. A., KRÖNER, A., BAUMGARTNER, L. P., DIRKS, P. H. G. M. & JELSMA, H. A. 2000. Evolution of Neoproterozoic high-grade rocks in the Mavuradonha Mountains, Zambezi Belt, northeast Zimbabwe. *Journal of African Earth Sciences, Special Abstracts Issue, 18th Colloquium of African Geology*, **30**, 64–65.

MUNYANYIWA, H. & BLENKINSOP, T. G. 1994. Pan-African structures and metamorphism in the Makuti Group, north-west Zimbabwe. *Journal of African Earth Sciences*, **19**, 185–198.

MUNYANYIWA, H. & MAASKANT, P. 1998. Metamorphism of the Palaeoproterozoic Magondi mobile belt north of Karoi, Zimbabwe. *Journal of African Earth Sciences*, **27**, 223–240.

MUNYANYIWA, H., HANSON, R. E., BLENKINSOP, T. G. & TRELOAR, P. J. 1997. Geochemistry of amphibolites and quartzofeldspathic gneisses in the Pan-African Zambezi belt, northwest Zimbabwe: evidence for bimodal magmatism in a continental rift setting. *Precambrian Research*, **81**, 179–196.

MUNYANYIWA, H., KRÖNER, A. & JAECKEL, P. 1995. U–Pb and Pb–Pb single zircon ages for charno–enderbites from the Magondi mobile belt, northwest Zimbabwe. *South African Journal of Geology*, **98**, 52–57.

NEX, P. A. M., OLIVER, G. J. H. & KINNAIRD, J. A. 2001. Spinel-bearing assemblages and P–T–t evolution of the Central Zone of the Damara Orogen, Namibia. *Journal of African Earth Sciences*, **32**, 471–489.

NGOYI, K., LIÉGEOIS, J.-P., DEMAIFFE, D. & DUMONT, P. 1991. Late-Ubendian age (Lower Proterozoic) for the granitic domes of the Zairian–Zambian Copperbelt. *Comptes Rendus de l'Académie des Sciences*, **313**, 83–89.

OLIVER, G. J. H. 1994. Mid-crustal detachment and domes in the central zone of the Damara orogen, Namibia. *Journal of African Earth Sciences*, **19**, 331–344.

OLIVER, G. J. H., JOHNSON, S. P., WILLIAMS, I. S. & HERD, D. A. 1998. Relict 1.4 Ga oceanic crust in the Zambezi Valley, northern Zimbabwe: evidence for Mesoproterozoic supercontinental fragmentation. *Geology*, **26**, 571–573.

PANCAKE, J., HANSON, R., GOSE, W., RAMEZANI, J. & BOWRING, S. 2001. Paleomagnetism and geochronology of widespread dolerite sills in Botswana related to the 1.1 Ga Umkondo large igneous province. *Geological Society of America Abstracts with Programs*, **33**, A436.

PASTEELS, P. 1971. Age du granite de la Lunge (près de Kamina, Katanga). *Musée Royal de l'Afrique Centrale, Département Géologie et Minéralogie, Tervuren, Belgique, Rapport Annuel (1970)*, 41.

PEDROSA-SOARES, A. C., NOCE, C. M., WIEDEMANN, C. M. & PINTO, C. P. 2001. The Araçuaí–West Congo Orogen in Brazil: an overview of a confined orogen formed during Gondwanaland assembly. *Precambrian Research*, **110**, 307–323.

PFURR, N., AHRENDT, H., HANSEN, B. T. & WEBER, K. 1991. U–Pb and Rb–Sr isotopic study of granitic gneisses and associated metavolcanic rocks from the Rostock massifs, southern margin of the Damara orogen: implications for lithostratigraphy of this crustal segment. *Namibia Geological Survey Communications*, **7**, 35–48.

PINNA, P., JOURDE, G., CALVEZ, J. Y., MROZ, J. P. & MARQUES, J. M. 1993. The Mozambique Belt in northern Mozambique: Neoproterozoic (1100–850 Ma) crustal growth and tectogenesis, and superimposed Pan-African (800–550 Ma) tectonism. *Precambrian Research*, **62**, 1–59.

PIPER, D. P. 1996. *The geotectonic evolution of the crystalline rocks of central Malawi*. PhD thesis, University of Wales.

POHL, W. 1994. Metallogeny of the northeastern Kibara belt, Central Africa – recent perspectives. *Ore Geology Reviews*, **9**, 105–130.

PORADA, H. 1985. Stratigraphy and facies in the Upper Proterozoic Damara Orogen, Namibia, based on a geodynamic model. *Precambrian Research*, **29**, 235–264.

PORADA, H. 1989. Pan-African rifting and orogenesis in southern to equatorial Africa and eastern Brazil. *Precambrian Research*, **44**, 103–136.

PORADA, H. & BERHORST, V. 2000. Towards a new understanding of the Neoproterozoic–Early Palaeozoic Lufilian and northern Zambezi Belts in Zambia and the Democratic Republic of Congo. *Journal of African Earth Sciences*, **30**, 727–771.

PRIEM, H. N. A., BOELRIJK, N. A. I. M., HEBEDA, E. H., VERDURMEN, E. A. TH. & VERSCHURE, R. H. 1979. Isotopic age determinations on granitic and gneissic rocks from the Ubendian–Usagaran System in southern Tanzania. *Precambrian Research*, **9**, 227–239.

RAINAUD, C., MASTER, S., ROBB, L. J. & ARMSTRONG, R. A. 1999. A fertile Palaeoproterozoic magmatic arc beneath the Central African Copperbelt. *In*: Stanley, C. J. ET AL. (eds) *Mineral Deposits: Processes to Processing*. Balkema, Rotterdam, **1**, 1427–1430.

RAITH, J. G. & CORNELL, D. H. 2000. The first U–Pb ion probe zircon ages from a migmatitic metapelite, central zone, western Namaqualand, South Africa. *Journal of African Earth Sciences, Special Abstracts Issue, 18th Colloquium of African Geology*, **30**, 73.

RAITH, J. G. & STEIN, H. J. 2000. Re-Os dating and sulfur isotope composition of molybdenite from tungsten deposits in western Namaqualand, South Africa: implications for ore genesis and the timing of metamorphism. *Mineralium Deposita*, **35**, 741–753.

RAITH, J. G., STEIN, H. J. & FINGER, F. 1999. Timing and duration of high-grade metamorphism in western Namaqualand, South Africa. *Journal of Conference Abstracts*, world wide web address: http://www.campublic.co.uk/publications.html.

RAMOKATE, L. V., MAPEO, R. B. M., CORFU, R. & KAMPUNZU, A. B. 2000. Proterozoic geology and regional correlation of the Ghanzi–Makunda area, western Botswana. *Journal of African Earth Sciences*, **30**, 443–466.

RAY, G. E. 1974. The structural and metamorphic geology of northern Malawi. *Journal of the Geological Society, London*, **130**, 427–440.

REICHHARDT, F. J. 1994. The Molopo Farms Complex, Botswana: history, stratigraphy, petrography, petrochemistry and Ni–Cu–PGE mineralization. *Exploration and Mining Geology*, **3**, 263–284.

REID, D. L. 1982. Age relationships within the Vioolsdrif batholith, Lower Orange River region II. A two stage emplacement history and the extent of Kibaran overprinting. *Geological Society of South Africa Transactions*, **85**, 105–110.

REID, D. L. 1997. Sm–Nd age and REE geochemistry of Proterozoic arc-related igneous rocks in the Richtersveld Subprovince, Namaqua Mobile Belt, Southern Africa. *Journal of African Earth Sciences*, **24**, 621–633.

REID, D. L., RANSOME, I. G. D., ONSTOTT, T. C. & ADAMS, C. J. 1991. Time of emplacement and metamorphism of Late Precambrian mafic dykes associated with the Pan-African Gariep orogeny, Southern Africa: implications for the age of the Nama Group. *Journal of African Earth Sciences*, **13**, 531–541.

REID, D. L., WELKE, H. J., ERLANK, A. J. & BETTON, P. J. 1987a. Composition, age and tectonic setting of amphibolites in the central Bushmanland Group, Western Namaqua Province, southern Africa. *Precambrian Research*, **36**, 99–126.

REID, D. L., WELKE, H. J., ERLANK, A. J. & MOYES, A. 1987b. The Orange River Group: a major Proterozoic calcalkaline volcanic belt in the western Namaqua Province, southern Africa. *In*: PHARAOH, T. C., BECKINSALE, R. D. & RICKARD, D. (eds) *Geochemistry and Mineralization of Proterozoic Volcanic Suites*. Geological Society, London, Special Publications, **33**, 327–346.

REIMOLD, W. U., PYBUS, G. Q. J., KRUGER, F. J., LAYER, P. W. & KOEBERL, C. 2000. The Anna's Rust Sheet and related gabbroic intrusions in the Vredefort Dome–Kibaran magmatic event on the Kaapvaal Craton and beyond? *Journal of African Earth Sciences*, **31**, 499–521.

REISCHMANN, T. 1995. Precise U/Pb age determination with baddeleyite (ZrO_2), a case study from the Phalaborwa Igneous Complex, South Africa. *South African Journal of Geology*, **98**, 1–4.

RICHARDS, J. P., CUMMING, G. L., KRISTIC, D., WAGNER, P. A. & SPOONER, E. T. C. 1988a. Pb isotope constraints on the age of sulfide ore deposition and U–Pb age of late uraninite veining at the Musoshi stratiform copper deposit, Central African Copper Belt, Zaire. *Economic Geology*, **83**, 724–741.

RICHARDS, J. P., KROGH, T. E. & SPOONER, E. T. C. 1988b. Fluid inclusion characteristics and U–Pb rutile age of late hydrothermal alteration and veining at the Musoshi stratiform copper deposit, Central African Copper Belt, Zaire. *Economic Geology*, **83**, 118–139.

RING, U. 1999. Volume loss, fluid flow, and coaxial versus noncoaxial deformation in retrograde, amphibolite facies shear zones, northern Malawi, east-central Africa. *Geological Society of America Bulletin*, **111**, 123–142.

RING, U., KRÖNER, A., LAYER, P., BUCHWALDT, R. & TOULKERIDIS, T. 1999. Deformed A-type granites in northern Malawi, east-central Africa: pre- or syntectonic? *Journal of the Geological Society, London*, **156**, 695–714.

RING, U., KRÖNER, A. & TOULKERIDIS, T. 1997. Palaeoproterozoic granulite-facies metamorphism and granitoid intrusions in the Ubendian-Usagaran Orogen of northern Malawi, east-central Africa. *Precambrian Research*, **85**, 27–51.

ROBB, L. J., ARMSTRONG, R. A. & WATERS, D. J. 1999. The history of granulite-facies metamorphism and crustal growth from single zircon U–Pb geochronology: Namaqualand, South Africa. *Journal of Petrology*, **40**, 1747–1770.

ROZENDAAL, A., GRESSE, P. G., SCHEEPERS, R. & LE ROUX, J. P. 1999. Neoproterozoic to Early Cambrian crustal evolution of the Pan-African Saldania Belt, South Africa. *Precambrian Research*, **97**, 303–323.

RUMVEGERI, B. T. 1991. Tectonic significance of Kibaran structures in Central and Eastern Africa. *Journal of African Earth Sciences*, **13**, 267–276.

SACCHI, R., CADOPPI, P. & COSTA, M. 2000. Pan-African reactivation of the Lurio segment of the Kibaran Belt system: a reappraisal from recent age determinations in northern Mozambique. *Journal of African Earth Sciences*, **30**, 629–639.

SACCHI, R., MARQUES, J., COSTA, M. & CASATI, C. 1984. Kibaran events in the southernmost Mozambique belt. *Precambrian Research*, **25**, 141–159.

SCHALLER, M., STEINER, O., STUDER, I., HOLZER, L., HERWEGH, M. & KRAMERS, J. D. 1999. Exhumation of Limpopo Central Zone granulites and dextral continent-scale transcurrent movement at 2.0 Ga along the Palala Shear Zone, Northern Province, South Africa. *Precambrian Research*, **96**, 263–288.

SCHANDELMEIER, H. 1981. The Precambrian of NE Zambia in relation to the dated Kate, Mambwe and Luchewe intrusives. *Geologische Rundschau*, **70**, 956–971.

SCHANDELMEIER, H. 1983. *The Geochronology of Post-Ubendian Granitoids and Dolerites from the Mambwe Area, Northern Province, Zambia*. Institute of Geological Sciences Report **83/1**.

SCHERMERHORN, L. J. G. 1981. The West Congo orogen: a key to Pan-African thermotectonism. *Geologische Rundschau*, **70**, 850–867.

SCHWARTZ, M. O., KWOK, Y. Y., DAVIS, D. W. & AKANYANG, P. 1996. Geology, geochronology and regional correlation of the Ghanzi Ridge, Botswana. *South African Journal of Geology*, **99**, 245–250.

SETH, B., KRÖNER, A., MEZGER, K., NEMCHIN, A. A., PIDGEON, R. T. & OKRUSCH, M. 1998. Archaean to Neoproterozoic magmatic events in the Kaoko belt of NW Namibia and their geodynamic significance. *Precambrian Research*, **92**, 341–363.

SHACKLETON, R. M. 1973. Correlation of structures across Precambrian orogenic belts in Africa. *In*: TARLING, D. H. & RUNCORN, S. K. (eds) *Implications of Continental Drift to the Earth Sciences*. Academic Press, London, **2**, 1091–1095.

SHACKLETON, R. M. 1986. Precambrian collision tectonics in Africa. *In*: COWARD, M. P. & RIES, A. C. (eds) *Collision Tectonics*. Geological Society, London, Special Publications, **19**, 329–349.

SHACKLETON, R. M. 1993. Tectonics of the Mozambique Belt in East Africa. *In*: PRICHARD, H. M., ALABASTER, T., HARRIS, N. B. W. & NEARY, C. R. (eds) *Magmatic Processes and Plate Tectonics*. Geological Society, London, Special Publications, **76**, 345–362.

SINGLETARY, S. J., HANSON, R. E., MARTIN, M. W. ET AL. 2002. Geochronology of basement rocks in the Kalahari Desert, Botswana, and implications for regional Proterozoic tectonics. *Precambrian Research*, in press.

STANISTREET, I. G., KUKLA, P. A. & HENRY, G. 1991. Sedimentary basinal responses to a Late Proterozoic Wilson Cycle: the Damara Orogen and Nama Foreland, Namibia. *Journal of African Earth Sciences*, **13**, 141–156.

STEIGER, R. H. & JÄGER, E. 1977. Subcommission on geochronology; convention on the use of decay constants in geo- and cosmochronology. *Earth and Planetary Science Letters*, **36**, 359–362.

STERN, R. J. 1994. Arc assembly and continental collision in the Neoproterozoic East African orogen: implications for the consolidation of Gondwanaland. *Annual Reviews of Earth and Planetary Sciences*, **22**, 319–351.

STEVEN, N., ARMSTRONG, R., SMALLEY, T. & MOORE, J. 2000. First geological description of a Late Proterozoic (Kibaran) metabasaltic andesite-hosted chalcocite deposit at Omitiomire, Namibia. *In*: CLUER, J. K., PRICE, J. G., STRUHSACKER, E. M., HARDYMAN, R. F. & MORRIS, C. L. (eds) *Geology and Ore Deposits 2000: The Great Basin and Beyond: Symposium Proceedings*, Reno, Nevada, Geological Society of Nevada, **2**, 711–734.

STOWE, C. W. 1986. Synthesis and interpretation of structures along the north-eastern boundary of the Namaqua Tectonic Province, South Africa. *Geological Society of South Africa Transactions*, **89**, 185–198.

TACK, L., DE PAEPE, P., DEUTSCH, S. & LIÉGEOIS, J.-P. 1984. The alkaline plutonic complex of the Upper Ruvubu (Burundi): geology, age, isotopic geochemistry and implications for the regional geology of the western Rift. *In*: KLERKX, J. & MICHOT, J. (eds) *African Geology*. Musée Royal de l'Afrique Centrale, Tervuren, 91–114.

TACK, L., LIÉGEOIS, J.-P., DEBLOND, A. & DUCHESNE, J. C. 1994. Kibaran A-type granitoids and mafic rocks generated by two mantle sources in a late orogenic setting (Burundi). *Precambrian Research*, **68**, 323–356.

TACK, L., WINGATE, M. T. D., LIÉGEOIS, J.-P., FERNANDEZ-ALONSO, M., & DEBLOND, A. 2001. Early Neoproterozoic magmatism (1000–910 Ma) of the Zadinian and Mayumbian Groups (Bas-Congo): onset of Rodinia rifting at the western edge of the Congo craton. *Precambrian Research*, **110**, 277–306.

TANNER, P. W. G. 1973. Orogenic cycles in East Africa. *Geological Society of America*, **84**, 2839–2850.

TEGTMEYER, A. & KRÖNER, A. 1985. U–Pb zircon ages for granitoid gneisses in northern Namibia and their significance for Proterozoic crustal evolution of southwestern Africa. *Precambrian Research*, **28**, 311–326.

TEMBO, F., KAMPUNZU, A. B. & MUSONDA, B. M. 2000. Geochemical characteristics of Neoproterozoic A-type granites in the Lufilian belt, Zambia. *Journal of African Earth Sciences, Special Abstracts Issue, 18th Colloquium of African Geology*, **30**, 85.

TEMBO, F., KAMPUNZU, A. B. & PORADA, H. 1999. Tholeiitic magmatism associated with continental rifting in the Lufilian Fold Belt of Zambia. *Journal of African Earth Sciences*, **28**, 403–425.

THEUNISSEN, K. 1988. Kibaran thrust fold belt (D1–2) and shear belt (D2). *In*: POHL, W. & KLERKX, J. (eds) *Metallogeny of the Kibara Belt, Central Africa, Bulletin – Newsletter*. Musée Royal de l'Afrique Centrale, Tervuren, **1**, 55–64.

THEUNISSEN, K., KLERKX, J., MELNIKOV, A. & MRUMA, A.

1996. Mechanisms of inheritance of rift faulting in the western branch of the East African Rift, Tanzania. *Tectonics*, **15**, 776–790.

THOMAS, R. J. & EGLINGTON, B. M. 1990. A Rb–Sr, Sm–Nd and U–Pb zircon isotopic study of the Mzumbe Suite, the oldest intrusive granitoid in southern Natal, South Africa. *South African Journal of Geology*, **93**, 761–765.

THOMAS, R. J., AGENBACHT, A. L. D., CORNELL, D. H. & MOORE, J. M. 1994. The Kibaran of southern Africa: tectonic evolution and metallogeny. *Ore Geology Reviews*, **9**, 131–160.

THOMAS, R. J., CORNELL, D. H. & ARMSTRONG, R. A. 1999. Provenance age and metamorphic history of the Quha Formation, Natal Metamorphic Province: a U-Th-Pb zircon SHRIMP study. *South African Journal of Geology*, **102**, 83–88.

THOMAS, R. J., EGLINGTON, B. M., BOWRING, S. A., RETIEF, E. A. & WALRAVEN, F. 1993. New isotope data from a Neoproterozoic porphyritic granitoid– charnockite suite from Natal, South Africa. *Precambrian Research*, **62**, 83–101.

TORREALDAY, H. I., HITZMAN, M. W., STEIN, H. J., MARKLEY, R. J., ARMSTRONG, R. & BROUGHTON, D. 2000. Re–Os and U–Pb dating of the vein-hosted mineralization at the Kansanshi copper deposit, northern Zambia. *Economic Geology*, **95**, 1165–1170.

TRELOAR, P. J. 1988. The geological evolution of the Magondi mobile belt, Zimbabwe. *Precambrian Research*, **38**, 55–73.

TRELOAR, P. J. & KRAMERS, J. D. 1989. Metamorphism and geochronology of granulites and migmatitic granites from the Magondi Mobile Belt, Zimbabwe. *Precambrian Research*, **45**, 277–289.

TROMPETTE, R. 1994. *Geology of Western Gondwana (2000–500 Ma)*. Rotterdam, Balkema.

TROMPETTE, R. 2000. Gondwana evolution: its assembly at around 600 Ma. *Comptes Rendus de L'Académie des Sciences, Série II. Earth and Planetary Sciences*, **330**, 305–315.

UKEN, R. & WATKEYS, M. K. 1997. An interpretation of mafic dyke swarms and their relationship with major mafic magmatic events on the Kaapvaal Craton and Limpopo Belt. *South African Journal of Geology*, **100**, 341–348.

UNRUG, R. 1983. The Lufilian Arc: a microplate in the Pan-African collision zone of the Congo and Kalahari cratons. *Precambrian Research*, **21**, 181–196.

UNRUG, R. (ed.) 1996. *Geodynamic Map of Gondwana Supercontinent Assembly*. Bureau de Recherches Géologiques et Minières, Orléans, France.

VACHETTE, M. 1964. Ages radiométriques des formations cristallines D'Afrique Équatoriale (Gabon, République Centrafricaine, Tchad, Moyen Congo). *Annales Scientifiques de l'Université de Clermont, No. 25, Géologie, Minéralogie, Part 8, Étude Géochronologiques*, **1**, 31–38.

VAN DE WEL, L., BARTON, JR, J. M. & KINNY, P. D. 1998. 1.02 Ga granite magmatism in the Tati Granite–Greenstone Terrane of Botswana: implications for mineralization and terrane evolution. *South African Journal of Geology*, **101**, 67–72.

VEARNCOMBE, J. R. 1983. A proposed continental margin in the Precambrian of western Kenya. *Geologische Rundschau*, **72**, 663–670.

VERNON-CHAMBERLAIN, V. E. & SNELLING, N. J. 1972. Age and isotope studies on the arena granites of S.W. Uganda. *Musée Royal de l'Afrique Centrale, Tervuren, Belgique, Sciences Géologiques*, **73**, 1–44.

VINYU, M. L., HANSON, R. E., MARTIN, M. W., BOWRING, S. A., JELSMA, H. A., KROL, M. A. & DIRKS, P. H. G. M. 1999. U–Pb and $^{40}Ar/^{39}Ar$ geochronological constraints on the tectonic evolution of the easternmost part of the Zambezi orogenic belt, northeast Zimbabwe. *Precambrian Research*, **98**, 67–82.

VRÁNA, S., PRASAD, R. & FEDIUKOVÁ, E. 1975. Metamorphic kyanite eclogites in the Lufilian arc of Zambia. *Contributions to Mineralogy and Petrology*, **51**, 1–22.

WALRAVEN, F. 1997. *Geochronology of the Rooiberg Group, Transvaal Supergroup, South Africa*. University of the Witwatersrand, Economic Geology Research Unit, Information Circular **316**.

WALRAVEN, F. & HATTINGH, E. 1993. Geochronology of the Nebo Granite, Bushveld Complex. *South African Journal of Geology*, **96**, 31–41.

WELKE, H. J., BURGER, A. J., CORNER, B., KRÖNER, A. & BLIGNAULT, H. J. 1979. U–Pb and Rb–Sr age determinations on Middle Proterozoic rocks from the lower Orange River area, south-western Africa. *Geological Society of South Africa Transactions*, **82**, 205–214.

WENDORFF, M. 2000. Genetic aspects of the Katangan megabreccias: Neoproterozoic of Central Africa. *Journal of African Earth Sciences*, **30**, 703–715.

WENDT, I., BESANG, C., HARRE, W., KREUZER, H., LENZ, H. & MÜLLER, P. 1972. Age determinations of granitic intrusions and metamorphic events in the early Precambrian of Tanzania. *Proceedings of the 24th International Geological Congress*, Montreal, Section 1, 295–314.

WHITTINGTON, A., BUSBEY, A., HANSON, R. & MORGAN, K. 1999. Remote-sensing study of structural relations between the Pan-African Zambezi and Mozambique orogenic belts in NE Zimbabwe and adjacent parts of Mozambique. *Geological Society of America Abstracts with Programs*, **31**, A37.

WILSON, A. H. 1990. Tugela Rand Layered Suite. *In*: JOHNSON, M. R. (ed.) *South African Committee for Stratigraphy, Catalogue of South African Lithostratigraphic Units, Report*, **2**, 47–48.

WILSON, T. J., HANSON, R. E. & WARDLAW, M. S. 1993. Late Proterozoic evolution of the Zambezi belt, Zambia: implications for regional Pan-African tectonics and shear displacements in Gondwana. *In*: FINDLAY, R. H., UNRUG, R., BANKS, M. R. & VEEVERS, J. J. (eds) *Gondwana Eight: Assembly, Evolution and Dispersal*. Balkema, Rotterdam, 69–82.

WINGATE, M. T. D. 2001. SHRIMP baddeleyite age for an Umkondo dolerite sill, Nyanga Mountains, eastern Zimbabwe. *South African Journal of Geology*, **104**, 13–22.

Index

Page numbers in *italic* refer to tables, and those in **bold** to figures.

accretionary orogens, defined 23
Africa, East
  Congo–Sao Francisco Craton 4, 5, 35, 45–6, 49, **64**
    palaeomagnetic poles at 1100–700 Ma *37*
  East African(–Antarctic) Orogen
    accretion and deformation, Arabian–Nubian Shield (ANS) 327–61
    continuation of East Antarctic Orogen 263
    E/W Gondwana suture 263–5
    evolution 357–8
    extensional collapse in DML 271–87
      deformations 283–5
    Heimefront Shear Zone, DML 208, **251**, *252–3*, 284, 415, 417
    structural section, Neoproterozoic deformation, Madagascar 365–72
    see also Arabian–Nubian Shield (ANS); Mozambique Belt
  Mozambique Belt evolution 60–1, **291**, 401–25
    carbonates 405–6
    Dronning Maud Land 62–3
    eclogites and ophiolites 406–7
    isotopic data
      crystallization and metamorphic ages 407–11
      Sm–Nd (T^{DM}) 411–14
    lithologies 402–7
    magmatic arc sequences 402–5
    and relationship with Rodinia and Gondwana 417–18
    structural aspects 414–17
      crustal structures 414
    see also East Africa(–Antarctic) Orogen
Africa, South
  Angola, and NW Botswana, orogenic belts (1.35–1.0 Ga) 442
  Angola–Kasai Craton 427
  Archaean cratonic nuclei 417–9
  Barberton Greenstone Belt 28
    geochronology and tectonic evolution 427–63
  Kalahari Craton **2–5**, 35, 42, **64**, 208, 417–18
    amalgamation with EAC, Rodinia formation 417–18
    and Antarctica 208
    and Australia 42
    Namaqua Orogen 4, 42
  Kalahari Plate 4
    palaeomagnetic poles at 1100–700 Ma *37*, **43**
  orogenic belts (2.05–1.8 Ga) 429–35
    Kheis–Magondi Orogen 432–3
    Limpopo Belt and Bushveld Igneous Province 433–5
    Palaeoproterozoic basement N of Damara Belt 435
    Richtersveld Terrane and Rehoboth Inlier 435
    Ubendian–Usagaran Orogen and Bangweulu Block **404**, **408–9**, 429–32
  orogenic belts (1.35–1.0 Ga) 435–42
    Choma–Kalomo Block and subsurface continuation 438
    Kibaran and Irumide Belts and adjacent terranes 25, 439–42
  Namaqua–Natal Orogen 435–8
  SW Angola and NW Botswana 442
  Umkondo Igneous Province 438–9
  Pan-African orogenic belts (650–450 Ma) 442–50
    Damara–Lufilian Arc–Zambezi Belt **3**, 435, 442–50
    Katangan basaltic rocks 443, 446
    Mwembeshi Shear Zone 442
    Neoproterozoic basin evolution and seafloor spreading 445–6
    orogenesis 446–51
    Ubendian and Kibaran Belts 445
    within-plate magmatism and basin initiation 443–5
  Zambezi Belt 27, 415
  Zambezi Orogen **3**, 5
  Zambezi–Damara Belt 65, 67, 442–50
  Zimbabwe Belt, ophiolites 27
  Zimbabwe Craton 427, 433
  Zimbabwe–Kapvaal–Grunehogna Craton 42, 208, **250**, 272–3
  see also Pan-African
Africa, West 40–1
  Amazonia–Rio de la Plata megacraton **2–3**, 40–1
  Birimian Orogen 24
Al-Jifn Basin *see* Najd Fault System
Albany–Fraser–Wilkes Orogen, Australia 67, 83, 103–10
Aldan Shield, Siberia 44
  Thelon Magnetic Belt 45
Altaid Orogen, central Asia 28
Amazonia–Laurentia collision 4, 40
Amazonia–Rio de la Plata megacraton, W Africa **2–3**, 40–1
anatexis
  Grenvillean events, Prydz Bay 237–8
  Kuunga Orogeny Zone 414
Antananarivo Block, Madagascar 61, 365, 372–6, **382**
Antarctic Orogen *see* Africa, East, East African(–Antarctic) Orogen; East Antarctica
Antarctica, East 57–75, 251–61
  Archaean–Cambrian crustal development 203–30
    Cambrian juxtaposition of Proterozoic crustal provinces 220–5
    correlations, Rayner Province and Eastern Ghats 219–20
    East Antarctic Shield 204–6
  Circum-East Antarctic Orogen vi 57–9, 66–9
  East Antarctic Orogen 57–75, 251–61
    Conradbjerge–Orvinfjella 255–60, **275**, 415–16
    Gjelsvikfjella 255
    pathways for Pinjarra Orogen **119**
    UHT rocks, reaction textures **222**
  East Antarctic Shield 204–6
  East Antarctica–Australia, Gondwana fit 117
  Gondwana setting 117, **205**
  magmatism, late to post-tectonic, central DML 261
  metamorphic basements
    Gjelsvikfjella and Western Muhlig–Hofmann–Gebirge 255
    Orvinfjella and Wohlthatmassiv 255–60, **275**, 415–16

466 INDEX

Antarctica, East (*cont.*)
　Pan-African Suture
　　definition 60–3, 263
　　Shackleton Range to Lützow–Holm Bay 60–2, 102–3
　　Southern Prydz Bay area 62–3
　Progress Granite 62–3, 66, 233–6, 238
　Proterozoic–Cambrian history of Dronning Maud Land 251–61
　　metamorphic basements, Sør Rondane 209, *252–3*, 260–1
　Prydz Bay, Pan-African events, inference in East Gondwana Tectonics 231–45
　terranes of Antarctic–Africa–Madagascar sector 208–9
　　Grunehogna Craton, Dronning Maud Land 42, 208, **250**, 272–3, **403–4**
　　Mesoproterozoic–Neoproterozoic Maud Province 208
　　Pan-African high-grade tectonism in Dronning Maud Land Belt 208–9
　terranes of Antarctic–Australia sector 207–8
　　Denman Glacier/Obruchev Hills area 207
　　Pan-African high-grade events in Denman Glacier Belt 208–9
　　Wilkes Province 207
　terranes of Antarctic–India–Sri Lanka sector 209–19
　　Archaean Napier Complex, Enderby Land 61, 67, 133, **136**, 209–12
　　Archaean–Mesoproterozoic, Rauer Terrane, Prydz Bay 213–14, 239
　　Pan-African high-grade tectonism in Lützow–Holm Bay Belt 216–18
　　Pan-African high-grade tectonism in Prydz Bay Belt 218–19
　　Rayner Province 61, 67, 133, **136**, 209–12, 214–16
　　Ruker Terrane, Southern Prince Charles Mountains 103, 213
　　Vestfold Block, Prydz Bay 212–13
　Terre Adelie, zircon U–Pb SHRIMP ages 100–2
　Windmill Islands 109–10
　see also Dronning Maud Land; Lützow–Holm Bay; Prydz Bay
APWP paths 35–8
　see also palaeomagnetism/palaeopoles
Ar–Ar zircon ages 66, 135
Arabian–Nubian Shield (ANS) 4, 24, **289**
　age range, Pb and Nd data **293**
　assembly, 4 time slices **316**
　and assembly of Gondwana 263–4, 289–325
　crust formation rates 28
　diachronous tectonic events **310**
　Late Neoproterozoic
　　gneiss belts and domes 309–11
　　north-trending shear zones 312–13
　　shear zones 311–15
　movement along Najd fault zones **357**
　N continuation 263
　Nabitah–Bargaloi Shear Zone 263–4
　Neoproterozoic origin 263
　ophiolites 27
　post-amalgamation overlap assemblages 306–9
　　ages of basins *308*
　　epicontinental terrestrial basins 309
　　marine basins 306–9
　present-day tectonic setting **290**

terrane amalgamation and suture zones 301–6
　Oman 305
　Yemen 305–6
terrane amalgamation and sutures (650–600 Ma) 304–5
　Keraf Suture 304–5
terrane amalgamation and sutures (680–640 Ma) 304
　Ar Amar Suture 304
　Halaban Suture 304
　Hulayfah–Ad Dafinah–Ruwah Suture 304
　Nabitah Fault Zone 304
terrane amalgamation and sutures (750–660 Ma) 303–4
　Allaqi–Heiani–Sol Hamed–Onib–Yanbu Suture 303–4
　Atmur–Delgo Suture 303
terrane amalgamation and sutures (786–760 Ma) 301–3
　Bi'r Umq–Nakasib Suture 303
　Tabalah–Tarj Shear Zone 301–2
terrane analysis 290–301
　Ad Dawadimi Terrane 300
　Afif Terrane 300
　Ar Rayn Terrane 300
　Asir Terrane **296**, 298
　Barka Terrane 295–7
　Bayuda Terrane 293–5, 414
　Eastern Desert/Gerf Terrane 298
　Gebeit and Gabgaba regions 297–8
　Ghedem Terrane 297
　Hager–Tokar Terrane 297
　Halfa Terrane 290–3
　Haya Terrane 297
　Hijaz Terrane 299
　Jiddah Terrane 299
　Midyan Terrane 299–300
　Nafka Terrane 297
terrane assembly **302**
　U–Pb, Pb–Pb, Rb–Sr ages 292
　see also Najd Fault System (NFS)
Archaean cratonic nuclei
　Proterozoic cover, Gondwana assembly context, DML 249
　S Africa 417–9
Archaean cratonic terranes, Eastern Ghats Granulite Belt (EGB) 146–8
Archaean domains, India 170–3
Archaean Napier Complex, Enderby Land 209–12
Archaean–Cambrian crustal development 203–30
Archaean–Mesoproterozoic, Antarctic–India–Sri Lanka sector 213–14, 239
Arctida, Precambrian craton 43
Arizona, Payson Ophiolite 27
AUSMEX Rodinia reconstruction 41, 46
Australia
　1.8–0.7 Ma deformation **80**
　Albany–Fraser–Wilkes Orogen 103–10
　　Biranup Complex **105**, 107
　　Bunger Hills 110
　　Bunger Hills–Denman Glacier region 63, 102, 115–16
　　Corumup Gneiss **105**, 107
　　Fraser Complex **105**, 107
　　Nornalup Complex **105**, 108
　　Northern Foreland **105**, 106–7

time–space diagram **105**
Windmill Islands 109–10
Barramundi Orogeny 78
Centralian Superbasin **83**
cratonic assemblages **79**
  North Australian (NAC) 78–81
  South Australian (SAC) 81–2
  West Australian (WAC) 78
Curnamona Province 82
Eyre Peninsula **97, 98**
Gairdner Dyke Swarm 27, 46, 83
Gawler Craton 82, 95–100
Gawler Range Volcanics **96**, 100
  pole, APWP **87**
Gondwana fit 117
Kararan Orogeny 100
Kimberley Block 78
Leeuwin–Darling Belt–Denman Glacier Belt 62, 67, **112**, 113–14
Lincoln Batholith 81–2, 98–9
Mawson Craton **41, 64**, 95–103, 117–18, 207
  Australia–Antarctica **41**
  correlation with Terre Adelie and King George V Land **96**, 100–2, 207
  Gawler Craton 82, 95–100
  geographical extent 117–18
  Miller Range 102
  other potential fragments 102–3
  time–space diagram **96**
models, single vs multiple block assembly 77
Musgrave Block 83
orogenic provinces **94**
palaeomagnetic constraints on tectonic evolution 77–91
  Mesoproterozoic amalgamation 82–3
  Neoproterozoic events 83
  tectonic summary 78–83
palaeomagnetic poles at 1100–700 Ma *37*
palaeomagnetic segments 83–6
Petermann Ranges 83
Pinjarra Orogen 62–3, 67, 110–17
  Darling Fault Zone 110, **112**, 114–15
  Darling Mobile Belt 110
  Denman Glacier region 63, 102, **111, 112**, 115–16
  low-grade schist 114
  Mullingarra Complex 113–14
  Northampton, Mullingarra and Leeuwin Complexes **112**, 113–14
  pathways, East Antarctica **119**
  Prydz Bay region 62–3, **112**, 116
  Southern extension 118–21
  Southern Prince Charles Mountains, E Antarctica 62, 102–3, 116–17
  tectonic setting and its Antarctic counterparts 121–2
  time–space diagram **112**
precambrian rocks 81
Proterozoic tectonic evolution 77–87
S and SW basement provinces, correlation with Antarctica 93–130
Stirling Range Formation 107, 114
Strangeways Orogeny 78
Yilgarn Craton 82, **106**, 109
Australia–Antarctica, Mawson Craton **41, 64**, 95–103, 117–18, 207

AUSWUS Rodinia configuration 2, 41, 46
Avalonia, accretionary orogen 4, 25

Baltic Shield 24
  flood basalts 26
Baltica
  and Amazonia 3
  APWP paths **39**
  breakup from Laurentia 27
  crust production rates 28
  palaeomagnetic poles at 1100–700 Ma *36–7*
Bangweulu Block, Africa, South **404, 408–9**, 429–32
Barberton Greenstone Belt, South Africa 28
Barentsia plate 39
Bayuda Terrane, Arabian–Nubian Shield (ANS) 293–5, 414
Belcher Islands, Canada 26
Bhandara Craton, Eastern Ghats Granulite Belt 146, 148, **150**
Bhavani Shear Zone, India 176–7
Birimian Orogen, West Africa 24
Botswana, and Angola, orogenic belts, S Africa 442
Bunger Hills–Denman Glacier, Australia 63, 102, 110, 115–16
Bushveld Igneous Province, S Africa 433–5

Cadomian–Avalonian accretionary orogen 25
Canada
  Belcher Islands 26
  Coppermine River basalts 26
  Flin Flon Belt 24, 25, 27
  Grenville Dyke Swarm 27
  Kapuskasing Uplift 28
  Portuniq Ophiolite 27
  Trans-Hudson Orogen 24, 28
  Ungava Orogen 24
carbonates 405–6
Chewore Ophiolite, Zimbabwe Belt, S Africa 27
China, S
  and Laurentia 42–3
  palaeomagnetic poles at 1100–700 Ma *37*
  Sibao Orogen 42
Choma–Kalomo Block, and subsurface continuation 438
Cimmerian terranes 291
conglomerates, Al-Jifn Basin, Najd Fault System (NFS) 336–9
Congo Craton, E Africa 4, 5, 35, 45–6, 49, **64**
Conradbjerge–Orvinfjella, East Antarctic Orogen 255–60, **275**, 415–16
continental flood basalts 26
continental growth 1–21
  see also Proterozoic continental growth
Coppermine River basalts, Canada 26
Cuddapah Basin, Eastern Ghats Granulite Belt 146, **151**

Damara–Lufilian Arc–Zambezi belt, S Africa 442, 445–50
Darling Fault Zone, Pinjarra Orogen, Australia 110, **112**, 114–15
Darling Mobile Belt, Pinjarra Orogen, Australia 83, 110
Denman Glacier region, Australia, S and SW 63, 102, **111**, 115–16
  Obruchev Hills 207–9

Dharwar Craton, Eastern Ghats Granulite Belt 146, 170–2, 177
Dronning Maud Land 65, 247–69, 271–87
  Antarctic–Africa–Madagascar sector 208–9
  extensional collapse of Neoproterozoic–Palaeozoic E African–Antarctic Orogen 271–87
    fluid inclusion study 280–2
    geological setting 272–4
    late orogenic extension 283–5
    Zwiesel Gabbro Complex 274–80
  geological map **248**
  Gondwana reconstruction **262**
  Grunehogna Craton 42, 208, **250**, 272–3, **403–4**
  Heimefront Shear Zone 208, **251**, *252–3*, 284, 415, 417
  late Mesoproterozoic orogenic belt, W continuation in S Africa 249–51
  magmatism, late to post-tectonic, central DML 261
  Mozambique Belt 62–3
  Orvinfjella 255–60, **275**, 415–16
  Pan-African high-grade tectonism 208–9
  Pan-African lithostratigraphy *252–3*
  Proterozoic–Cambrian history, Gondwana assembly context 247–69
    Annandagstoppane 249
    Archaean craton and its Proterozoic cover 249
    Cambrian orogenic belt, Lützow-Holmbukta and Prydz Bay 261–3
    East Antarctic Orogen 251–61
    N continuation of E Antarctic Orogen into E Africa 263–4
    Ritscherflya Supergroup 249
    Western DML 254–5
  structural summary **257**
  summary 264–5
  zircon U–Pb SHRIMP ages *274*, 276–9
dykes and dyke swarms 26–7

Eastern Ghats, India 131–68
  Archaean cratonic terranes 146–8
    Bhandara Craton 146, 148, **150**
    Cuddapah Basin 146, **151**
    Dharwar Craton 146, 170–2, 177
    geological outline 148, **150**
    Proterozoic intracratonic basins 146–8
    Singhbhum Craton 146, **150**
  boundaries of crustal provinces 159–60
  Eastern Ghats Granulite Belt (EGB) 131–68
    geological framework of EGB 131–5, **146**
      lithology 131–3
      subdivisions 133–5, *149*
  Eastern Ghats Province 156–9
    Angul Domain 158
    Chilka Lake Domain 159
    Khariar Domain 158
  geological evolution
    high-grade crustal terranes, Rodinia/Gondwana context 162–3
    pre-Rodinia 160–1
    Rodinia assembly 161–2
  Godaveri Rift 131, **134**, **147**
  Indo–Antarctica correlation 138–9
  isotopic domains **134**
  Jeypore Province 149–51, 159
  Krishna Province 153–6
    Nellore–Khammam schist belt 153–5
    Ongole Domain 155–6
    Udayagiri Domain 155
    Vinjamuru Domain 153
  location of Grenvillean orogenic front 139
  Mahanadi Rift 131, **134**, **145**
  metamorphic history updated 135–8
  Nagavalli–Vamasdhara Shear Zone 133
  Napier–Rayner Block 61, 67, 133, **136**, 209–12, 214–16
  new subdivision 149
  pre-Rodinia evolution 160–1
  Rengali Province 151–2
  Sileru Shear Zone 133, 158–9
  spatial and temporal variations in P–T trajectories **136**, 138–9
  suture zones 162
  zircon Ar–Ar ages 135
  zircon U–Pb SHRIMP ages 135
eclogites 29, **404**, 406–7
Ediacaran trace fossils 191
Egersund Dyke Swarm, Norway 27
Enderby Land, Napier–Rayner Block, E Antarctica 61, 67, 133, **136**, 209–12, 214–16
Euler rotation parameters *47*

fanglomerates, Al Jifn Basin, Najd Fault System (NFS) 337–8
Fennoscandia, dolerite sill complexes 27
Finland, Jormua Ophiolite 27
Flin Flon Belt, Canada 24, 25, 27
flood basalts 26
FMAS assemblages 210, 223
French Guiana 25

Gairdner Dyke Swarm, Australia 27, 46, 83
Gawler Craton, Australia 82, 95–100
Gjelsvikfjella, East Antarctic Orogen 255
Gondwana
  assembly *vi*, 49, 123
  assembly
    Arabian–Nubian Shield (ANS) 263–4, 289–325
    East Antarctic Orogen 117, **205**
    EGB geological evolution 162
    Mozambique Belt evolution 401–25, **403**
    Proterozoic–Cambrian history of Dronning Maud Land 247–69
    see also Arabian–Nubian Shield
  reconstruction
    500 Ma **242**
    750–500 Ma **364**
    Dronning Maud Land **262**
    E/W suture 263–4
    igneous crystallization ages **409**, **410**
    metamorphic crystallization ages **408**, **410**
    N and S, Kuunga Orogeny Zone 413–14, 418
    Nd T^{DM} model ages 177, 411–14, **412**
    probability graphs **410**
    shear belts **416**
Gondwana, East
  defined 93
  Pan-African amalgamation 63
  reevaluation of Pan-African events 57–75

INDEX

tectonics, Prydz Bay 240–3
see also Pan-African
Gondwana, West, defined 93
greenstone belts 26, 28
  summary 28
Grenville Dyke Swarm, Canada 27
Grenvillean Orogen 1–4, 12, 57–9
  front, location in Eastern Ghats Granulite Belt (EGB) 139
  inference in East Gondwana Tectonics, Prydz Bay 237–40
Grove Mountains, E Antarctica 233, **235**, 239
Grunehogna Craton, Dronning Maud Land 42, 208, **250**, 272–3, **403–4**

Harohalli Dykes, India 49
Heimefront Shear Zone, Dronning Maud Land 208, **251**, *252–3*, 284, 415, 417
HFSE patterns 24

Iapetus Ocean opening 9–10, 27
India
  Achankovil Shear Zone 61, 175, 180, 417
  Archaean domains 170–3
  and Australia juxtaposition 103
  crustal architecture 145–68
  crustal evolution 169–80
  Dharwar Craton 146, 170–2, 177
    Biligirirangan (BR) Hills 171
    granulites west of Chennai 171–2
    Niligiri Hills 171, 176
  Eastern Ghats Granulite *see* Eastern Ghats
  Harohalli Dykes 49
  palaeomagnetic poles at 1100–700 Ma *37*
  peninsular, units **145**
  Proterozoic domains 173–80
  Southern Granulite Terrain **145, 170**, 172–80
    Achankovil Shear Zone 61, 175, 180, 417
    Archaean domains 172–3
    Bhavani Shear Zone 176–7
    boundaries 176–7
    Cardamom Hills Massif 173
    charnockite–enderbite massifs 173–4
    crust-forming processes and products 177–8
    Dharwal craton to Anamalai Hills **174**
    geochronology 178–9
    geological evolution and assembly 180
    gneiss–migmatic terrains 174–5
    Kerala Khondalite Belt (KKB) 173, **176**
    Krishnagiri–Salem Zone 172–3
    lithologies 173–6
    Madurai block 61, 173, 178–80
    metamorphic evolution 179–80
    Moyar Shear Zone 179
    Neoproterozoic igneous rocks 176
    Palghat–Cauvery Shear Zone 61, 172–4, 177, 417–18
    Ponmudi Unit 174, 180
    Proterozoic domains 173–80
    and Sri Lanka
      Gondwana context 190–3
      Rodinia context 190–1
    Western Ghats 173
    zircon U–Pb SHRIMP ages 61–3
    summary 189–93

Southern Granulite Terrain (SGT) **145**, 172–80
terranes of Antarctic–India–Sri Lanka sector 209–19
see also Eastern Ghats, India
Indo–Antarctic correlation 138–9
Iratsu Body, Japan 29
Irumide Belt, and adjacent terranes, S Africa 439–42
island arcs 24–5

Japan
  Iratsu Body 29
  subduction of Sørachi Plateau 29
Jormua Ophiolite, Finland 27
juvenile crust material 12–14, 23

Kaapval Craton, S Africa 427, **428**, 433
Kadugannawa Complex (KC), Sri Lanka 183–5
Kalahari Craton, S Africa **2–5**, 35, 42, **64**, 208, 417–18
Kalahari Plate, S Africa 4, *37*, **43**
Kalahari–Grunehogna Craton, Dronning Maud Land 42, 208, **250**, 272–3, **403–4**
Kapuskasing Uplift, Canada 28
Kararan Orogeny, Australia 100
Katangan basaltic rocks, S Africa 443, 446
Kheis–Magondi Orogen, S Africa 432–3
Kibaran Belt, and adjacent terranes, S Africa 25, 439–42
Kimban Orogen, Australia 99–100
Kimberley Block, Australia 78
King George V Land, correlation with Mawson Craton, Australia 100–2
Kola Peninsula
  lamprophyres 26
  Pechenga Greenstone Belt 26
  Umba Complex 25
Krishna Province (EGB), India 153–5
Kuunga Orogeny Zone
  anatexis 414
  Gondwana reconstruction 413
  N and S Gondwana amalgamation 418

Lambert Glacier, drainage region, E Antarctica 63, 103
Larsemann Hills, Progress Granite, E Antarctica 62–3, 66, 233–6, 238
Laurentia
  APWP **39**
  and Australia 41
  crust production rates 28
  Laurentia–Amazonia collision 4, 40
  Laurentia–Baltica 38–40
    breakup from Baltica 27
  Laurentia–Siberia
    connection 2–4, 43–5
    palaeopositions **44**
  palaeomagnetic poles at 1100–700 Ma **36, 43**
  Western, and South China 42–3
Leeuwin–Darling Belt–Denman Glacier Belt, Australia 62, 67, **112**, 113–14
Lewisian Orogen, Scotland 24, 25, 28
Limpopo Belt, S Africa 433–5
Lincoln Batholith, Australia 81–2, 98–9
Loch Maree Group, Scotland 25–6
Lurio Belt, Mozambique 418
Lützow–Holm Bay
  E Antarctica, zircon U–Pb data 284
  and Prydz Bay, Cambrian orogenic belt? 261–3

Lützow–Holm Bay (*cont.*)
  suture 60–2, 65–7
  terranes of Antarctic–India–Sri Lanka sector 216–18

Madagascar 49, 60–1, **64**, 68
  Itremo Group metasediments 387–95
    Antananarivo Block 365, 372–6, **382**
    Antongil Block 365
    Bemarivo Belt 365
    deformation 417
      summary *375*
    depositional setting and age 387
    geological setting 381–4
    higher grade basement rocks 387, 392–4
    Itheninkenina region 368–70
    Itremo Nappes 389–96, **397**
    Itremo Sheet 365
    lithologies 384–6
    metacarbonates 384–6
    metamorphic grade 387
    metapelites 384–5
    quartzites 384–6
    Saronara region 370–2
    structural evolution 372–6
    structures 388–95
    study areas 366
    Tsaratanana Sheet 365
    Upper Mania River 366–8
  Neoproterozoic deformation, East African Orogen structural section 365–72
  Neoproterozoic intrusive rocks 394–5
Madurai Block, Southern Granulite Terrain 61, 173, 178–80
mafic dykes/rocks 26–7, 406
magmatic arc sequences, Mozambique Belt evolution 402–5
magmatism
  East Antarctica/Orogen 261
  within-plate, and basin initian Pan-African orogen 443–5
Magondi Supergroup Orogen, S Africa 432–3
mantle plume model, crust formation 29
Mawson Craton, Australia **41**, **64**, 95–103, 117–18, 207
Mesoproterozoic (Late) juvenile crust **15**
Miller Range 102
Miltalie Gneiss 98
Mongolia, Bayankhongur Ophiolite 28
Mozambique Belt, E Africa 60–1, **291**, 401–25
Mozambique Ocean
  closure 263, 327–61, 418
    4 time slices **316**
  Najd Fault System, ANS 327–61
Mt Harding, E Antarctica 233, **235**, 239
Muhlig–Hofmann–Gebirge, East Antarctic Orogen 255
Mullingarra Complex, Pinjarra Orogen 113–14
Mwembeshi Shear Zone, S Africa 442
mylonitic leucogranites, DML **276**, **281**
  fluid inclusion study 280–2

Nabitah–Bargaloi Shear Zone, Arabian–Nubian Shield 263–4
Nagavalli–Vamasdhara Shear Zone, Eastern Ghats, India 133–4

Najd Fault System (NFS), Arabian–Nubian Shield (ANS) 327–61
  age of NFS 331
  Al-Jifn Basin 335–9, 344–52
    calcareous reworked facies 338
    coarse-grained interformational facies 338
    conglomerates 336–9
    fanglomerate facies 337–8
    faults 344–7
    folding 347–8
    limestone and argillite facies 338
    polymict basal facies 336
    pull-apart basin 332–5, 349–52
    sandstone and shale-bed facies 338
    structural features 344–52
    volcanic rocks 338–9
  dextral/sinistral faulting 327–61
  East African Orogen, implications for evolution 357–8
  geochronology 339–44
    sample location and local structural relationship 339
    U–Pb results and discussion 339–44
  Habariyah Mylonite Zone **329**
  Jibalah Group 335–9
  major Najd-related faults 331–2
    Ajjaj Shear Zone (ASZ) 331–2
    Ar-Rika and Ruwah Shear Zones 331
    Halaban–Zarghat Fault Zone (HZFZ) 332, **333**
  movement and offset along NFS 330–1
  timing and origin 313–14
  see also Arabian–Nubian Shield (ANS)
Namaqua Orogen, Kalahari Craton 4, 42
Namaqua–Natal Orogen 435–8
Napier–Rayner Block, Enderby Land, E Antarctica 61, 67, 133, **136**, 209–12, 214–16
  as part of Dharwar Craton 263
Nd isotopic mapping 24
  data summary 29
  Pb and Nd data **293**
  Sm–Nd (T^{DM}) 66, 68, 177, **239**, 411–14
Nellore–Khammam schist belt, India 153–5
Neoproterozoic 1–21
  basin evolution and seafloor spreading 445–6
  and Early Palaeozoic East African–Antarctic Orogen, extensional collapse in DML 271–87
  gneiss belts and domes 309–11
  island arcs 25
  juvenile crust, Rodinia *7–8*, **15**
  Madagascar 394–5
    deformation 363–79
  palaeomagnetic constraints on tectonic evolution, Australia 83
  shear zones 311–15
  Southern Granulite Terrain, India 176
Nile Craton 4
Nimrod Orogeny, S Antarctica 102
Northampton Complex, Pinjarra Orogen **112**, 113–14
Norway, Egersund Dyke Swarm 27

Oaxaquia 3
  palaeomagnetic poles at 1100–700 Ma *37*
obducted oceanic plateaus 29–30
oceanic plateaus 25–6
  subductability 29–30
Oman, terrane amalgamation and suture zones, ANS 305

Ongole Domain, Krishna Province 155–6
Ontong Java Plateau, Solomon Islands 29
ophiolites 29, **404**, 406–7
  island arcs 25
  and ocean floor 27–8
Orvinfjella, East Antarctic Orogen 255–60, **275**, 415–16
Oubanguide Belt 45

P–T (pressure–temperature) path(s)
  Antarctica–Indian sector **23**, 138–9, **206**, 220–5
  vs extensional collapse **224**
  metamorphic evolution, Sri Lanka 186–8
  Zwiesel Gabbro Complex, DML **284**
palaeomagnetism segments (Australia) 83–6
  I:1.8–1.7 Ga 85
  II:1.6–1.5 Ga 85
  III:1.36–1.32 Ga 85
  IV:1.07 Ga 85–6
palaeomagnetism/palaeopoles 83–6
  1100–700 Ma *36–7*
  APWP construction 83–4
  APWP paths 35, 35–8, **39**
    Australia 83–6, **84**
Palghat–Cauvery Shear Zone (PCSZ), India 61, 172–4, 177, 417–18
Pampean–Rio de la Plata Craton **2–3**, 40–1
Pan-African events 57–75, 59–60
  650–450 Ma 442–51
  750–550 Ma 9–11
  DML, terranes of Antarctic–Africa–Madagascar sector 208–9
  lithostratigraphy in DML *252–3*
  Lützow–Holm Bay Belt **61–2**, 65, 216–18
  Pan-African amalgamation 63
  Pan-African orogeny 264–5
  possible constraining approach 67–9
  Prydz Bay 62–3, 240–3
  recyclicity of orogens 66–7
  sinistral nature 417
  suture identification 63
  sutures within East Gondwana **59**, 63–6
Pan-African Suture 60–3, 261–3
  definition 60–3, 263
  Lützow–Holm Bay to Prydz Bay 60, 261–3
  S Prydz Bay to S Prince Charles Mountains 62–3, **104**, **262**
  Shackleton Range to Lützow–Holm Bay 60–2, 102–3
Pangaea, fragmentation, extent 26
Pannotia–Gondwana 580 Ma 13
Payson Ophiolite, Arizona 27
Pb diffusion method 68
Pechenga Greenstone Belt, Kola Peninsula 26
Pinjarra Orogen, Australia 62–3, 67, 110–17
Portuniq Ophiolite, Canada 27
Prince Charles Mountains, Pinjarra Orogen, E Antarctica 62, 102–3, **104**, 116–17, 213
  Pan-African Suture 62–3, **104**, **262**
probability graphs, Gondwana reconstruction **410**
Progress Granite
  E Antarctica 62–3, 66, 233–6, 238
    zircon Ar–Ar and Pb–Pb ages 66
Proterozoic continental growth 23–33
  continental flood basalts 26
  crustal growth rates 28–9

island arcs 24–5
mafic dykes 26–7
Neodymium isotopic mapping 24
oceanic plateaus 25–6
ophiolites and the ocean floor 27–8
seismic reflection studies 24
Proterozoic intracratonic basins, India 146–8
Prydz Bay, E Antarctica 62–3, 65–6, 116, 231–45, **262**
  accretionary model **241**
  Archaean Vestfold Block 212–13
  Grenvillean events 237–40
    anatexis and high-grade metamorphism 237–8
    basement and cover 238–9
    multiple metamorphism or terranes amalgamation 239–40
  location map **232**
  and Lützow–Holm Bay, Cambrian orogenic belt? 261–3
  Pan-African events 240–3
  Pan-African Suture, Prince Charles Mountains 60, 62–3, **104**, **262**
  Rauer Terrane 213–14, 239
  regional geological background 231–3
  syn-and post-tectonic granites 233–7
  temperature–time *(T–t)* cooling paths **218**
  Vestfold Hills 212–13, 218
  zircon Pb–Pb SHRIMP ages 62, *240*
  zircon U–Pb SHRIMP ages 65–6, 237–8
pull-apart basins, Najd Fault System (NFS) 332–5, 349–52

Rauer Terrane, East Antarctica 213–14, 239
Rayner Complex, Eastern Ghats, India 67, 214–16
  correlations 219–20, 263
  Napier–Rayner Block 61, 133, **136**, 209–12
REE abundances 24
Rehoboth Inlier, S Africa 435
Rengali Province, Eastern Ghats, India 151–2
Richtersveld Terrane, S Africa 435
Rio de la Plata Craton, Amazonia–Rio de la Plata megacraton **2–3**, 40–1
Rockall Plateau 38
Rodinia 1–21, 35–55
  700–550 Ma **10**
  800–700 Ma **6**
  900–800 Ma **3**
  990 Ma **40**, **41**, **46**
  1100 Ma **2**, 5
  assembly 161–2
    AUSMEX 41, 46
    AUSWUS 2, 41, 46
    Eastern Ghats Granulite Belt (EGB) 161–2
    Kalahari Craton amalgamation with EAC 417–18
  assembly and fragmentation model 35–55
  breakup 46–9
  configuration 1–4, 38–46
    Amazonia, West Africa and Rio de la Plata **2–3**, 40–1
    Australia, Antarctica and India 41–2
    Congo–Sao Francisco Craton 45–6
    Kalahari and Australia 42
    Laurentia and Australia 41
    Laurentia and Baltica 38–40
    Laurentia and Siberia 2–4, 43–5

Rodinia (*cont.*)
  Laurentia (Western) and South China 42–3
  Laurentia–Baltica and Amazonia 40
  SWEAT orogen/hypothesis 1, 26, 41, 131, **132**
  connection, Sri Lanka 190–1
  crustal growth rate 14–16
  juvenile continental crust (1350–500 Ma) 7–8, 12–14, **15**
  Mozambique Belt evolution 401–25
  Neoproterozoic 7–8, 12–14, **15**
  Pan-African orogenic system (750–550 Ma) 9–11
  rifting and collisional events, 900–500 Ma **5**
  Sr isotopic record 11–12
  superplume events 16–17
  tectonic history 4–9
    breakup (850–600 Ma) 5–9
    early rifting (1100–900 Ma) 4–5
Ross Orogeny, Australia 102
Ruker Terrane, S Prince Charles Mountains, E Antarctica 103, 213

Sao Francisco Craton, Africa, East 45–6
Saudi Arabia *see* Arabian–Nubian Shield (ANS)
Scotland
  Lewisian Orogen 24, 25, 28
  Loch Maree Group 25–6
seismic reflection studies 24
Shackleton Range, E Antarctica 102–3
  and Dronning Maud Land 62–3, 65, 67
  to Lützow–Holm Bay, Pan-African Suture 60–2, 102–3
Sibao Orogen, China 42
Siberia
  Aldan Shield 44
  palaeomagnetic poles at 1100–700 Ma *37*
  Siberian Craton 43–4
  Thelon Magnetic Belt 45
Siberia–Laurentia
  connection 2–4
  palaeopositions **44**
Sileru Shear Zone 133, 158–9
  Eastern Ghats, India 133, 158–9
Singhbhum Craton, Eastern Ghats Granulite Belt (EGB) 146
Sm–Nd isochron ages 66, 68, **239**
Sm–Nd (T^{DM}), Mozambique Belt evolution 411–14
Solomon Islands, obduction of Ontong Java Plateau 29
Sør Rondane, metamorphic basement, Dronning Maud Land 209, *252–3*, 260–1, 284
Sr isotopic record, Rodinia 11–12
Sri Lanka 49, 61, 181–200
  age of metamorphism 188
  correlation between India, Madagascar and Antarctica 190–3
  correlation of nomenclature *181*
  Highland Complex (HC) 181–3
  Kadugannawa Complex (KC) 183–5
  large-scale structures and scheme of structural phenomena 185–6
  metamorphic evolution and P–T path(s) 186–8
  Rodinia connection 190–1
  SGT and India, Gondwana context 189–90, 191–3
  structural evolution 415
  tectonic evolution 188–9

  terranes of Antarctic–India–Sri Lanka sector 209–19
  units and lithologies 181–3
  Vijayan Complex (VC) 185
  Wanni Complex (WC) 183–5
Stanovoy Province 45
Strangeways Orogeny, Australia 78
subduction model 29–30
supercontinent cycle, superplume events **17**
Superior Province
  Canada
    basalts 26
    greenstone belts 28
superplume events, Rodinia 16–17
sutures
  defined 63
  within East Gondwana 63–6
Svecofennian Orogen 24, 28
  crustal thickness 28
Sveconorwegian Orogen 38
  APWP Loops 38–9
SWEAT orogen/hypothesis, Rodinia configuration 1, 26, 41, 131, **132**

Tanzania Craton, S Africa 427, **428**, 430
temperature–time (T–t) cooling paths, Prydz Bay **218**
Terre Adelie, East Antarctica **96**, **98**, 100–2, 207
Thelon Magnetic Belt, Siberia 45
Tournefort Dyke Swarm, Australia 99
Trans-Hudson Orogen, Canada 24, 28
Trivandrum Block, India 61
trondjhemite–tonalite–granite (TTG) suites
  Kalahari Craton amalgamation with EAC, Rodinia formation 417–18
  magmatic arc sequences 402–5

U–Pb isotopic mapping, data summary 29
Ubendian–Usagaran Orogen, S Africa **404**, **408–9**, 429–32
Umba Complex, Kola Peninsula 25
Umkondo Igneous Province, S Africa 438–9
Ungava Orogen, Canada 24

Vestfold Block, Prydz Bay 212–13, **232**

Wilkes Land–Albany–Fraser Orogen, Australia 67, 83, 103–10
Wilkes Province, East Antarctica 207
Windmill Islands, Australia 109–10
Wohlthatmassiv, East Antarctic Orogen 255–60, **275**

Yamato–Belgica Block, India 61
Yemen, terrane amalgamation and suture zones, ANS 305–6

Zambezi *see* Africa, South
Zimbabwe *see* Africa, South
Zimbabwe–Kapvaal–Grunehogna Craton, S Africa 42, 208, **250**, 272–3
zircon grains, CL images **236–7**
Zwiesel Gabbro Complex, Dronning Maud Land 274–80
  granite dykes, veins and quartz-filled tension gashes 280
  P–T path **284**
  U-Pb SHRIMP geochronology 276–9